BIOLOGY OF PLANTS

FOURTH EDITION

The Plain at Auvers was painted by Vincent van Gogh in 1890, shortly after leaving the mental institution at Saint-Rémy-de-Provence. From Auvers, he wrote his brother Théo, "I have painted huge wheat fields under troubled skies, and I did not exert myself to express sadness and extreme solitude." Van Gogh committed suicide a few months later.

BIOLOGY OF PLANTS

FOURTH EDITION

PETER H. RAVEN

Missouri Botanical Garden and
Washington University, St. Louis

RAY F. EVERT

University of Wisconsin, Madison

SUSAN E. EICHHORN

University of Wisconsin, Madison

WORTH PUBLISHERS, INC.

BIOLOGY OF PLANTS, FOURTH EDITION

COPYRIGHT © 1971, 1976, 1981, 1986 BY WORTH PUBLISHERS, INC.

PRINTED IN THE UNITED STATES OF AMERICA

LIBRARY OF CONGRESS CATALOG CARD NO. 85-51292

ISBN: 0-87901-315-X   *1511026*

FIRST PRINTING, MARCH 1986

EDITORS: Sally Anderson, Gunder Hefta

PRODUCTION: George Touloumes, Patricia Lawson

DESIGN: Malcolm Grear Designers

TYPOGRAPHER: New England Typographic Service, Inc.

PRINTING AND BINDING: Von Hoffmann Press, Inc.

COVER: *The Plain at Auvers* BY VINCENT VAN GOGH, 1890,
PRIVATE COLLECTION, ZURICH, SWITZERLAND

WORTH PUBLISHERS, INC.

33 IRVING PLACE

NEW YORK, NEW YORK 10003

Preface

In the late 1980s, botany remains an exciting field, one that holds greater promise for human welfare than ever before. A rapidly growing human population is placing enormous pressure on the productive capacity of the global ecosystem. Human beings must respond to that pressure by building sustainable and productive systems in agriculture and forestry by which they may continue to support themselves. And so the need for botanical knowledge becomes ever more critical.

In this fourth edition, BIOLOGY OF PLANTS remains true to its original goals: to present the basic information about plants in a solid, well-illustrated, and interesting way. As in previous editions, we pay special attention to what we think of as the book's interlocking themes: (1) the plant body as the dynamic result of the processes of growth and development mediated by chemical interactions; (2) evolutionary relationships as the guide to understanding form and function in organisms; and (3) ecology as a way of emphasizing our dependence on plants to sustain our lives and those of all other living organisms.

Despite its title, the book treats all the groups of organisms traditionally considered by university departments of botany: viruses, bacteria, a selection of protists, fungi, and the plants themselves. We continue to define plants as consisting only of bryophytes and vascular plants, a coherent evolutionary line derived from ancestors which would be classified as green algae if we knew what they were. We do not include red and brown algae with the plants because they were independently derived from unicellular ancestors. We have also left the green algae in the kingdom Protista, an arbitrary decision but one that makes the kingdom Plantae easier to visualize. Fungi, of course, are treated as a wholly independent group.

While adhering to many of the features of previous editions, we have made a number of significant changes. We have, for example, added information on the multicellular algae, particularly the green algae, devoting an entire chapter to them. We have also added a new chapter (Chapter 30) on plants and human beings, in which we trace the history of the domestication of the major crops, and examine the potential for finding new plant sources for food, energy, and chemicals and

the prospects for improving plants through genetic engineering. This chapter has been grouped with the chapters on general evolution and the evolution of flowering plants in order to explain not only evolution as a process but also the role of human beings in modifying that process. Throughout the book, the treatment of physiological ecology has been strengthened and updated.

In addition, we have tried to make this edition of BIOLOGY OF PLANTS more succinct, more vivid and more accurate than ever before. The book has grown somewhat in length, not because its coverage is more comprehensive or detailed, but because (1) new diagrams, paintings, and photographs have been added—many of them in color; (2) more information on the relevance of plants to practical human affairs has been included (see, for example, the new section on wood in Chapter 23, *Secondary Growth*); and (3) more pedagogical aids—tables, summaries, overviews—have been provided throughout. In particular, we would like to call your attention to the new life cycles (pages 260–261, 328–329, 346–347, 366–367, etc.), much larger and easier to follow than before, all drawn in a uniform style.

The present edition of BIOLOGY OF PLANTS is organized in such a way that topics can be assigned in any preferred sequence. Great care has been taken to enhance the book's usefulness for the many courses in which less detail is stressed and only parts of the book are assigned. We have tried to provide students with a clear and interesting introduction to botany and with an up-to-date reference to topics that may catch their imagination and into which they may wish to delve more deeply.

By introducing young people not only to what is known about botany but also to the many unsolved problems, we hope to enlist new talents and new enthusiasms in working toward the solutions on which all of our futures depend.

PETER H. RAVEN
RAY F. EVERT
SUSAN E. EICHHORN
March 1986

Acknowledgments

We greatly appreciate the suggestions and comments made by teachers who used the previous edition of this book and who took the time to advise us regarding the new edition. Without their help and the help of other experts in various areas of botanical research and teaching, this book could not have been written. We had substantial assistance from the following: George B. Johnson, Washington University, who helped with the genetics and evolution chapters; Kenneth G. Keegstra and Michael R. Sussman, both of the University of Wisconsin, who reviewed the material on membranes; Paul W. Ludden, University of Wisconsin, who reviewed the chapters on energy; John W. Einset, Harvard University, who made suggestions for the chapters on growth regulation and growth responses; Peter H. Quail, University of Wisconsin, who reviewed the material on photobiology; Gerald C. Gerloff, University of Wisconsin, who helped with the chapter on plant nutrition and soils; and Jennifer L. Parke, University of Wisconsin, who made suggestions regarding the material on mycorrhizae. Gary L. Floyd, Ohio State University, Karl R. Mattox, Miami University, Linda E. Graham, University of Wisconsin, Karl J. Niklas, Cornell University, Max Hommersand, University of North Carolina, and Rudolf Schmidt, University of California, Berkeley, provided special assistance with the material on algae. We also wish to thank Sue A. Tolin and George H. Lacy, both of Virginia Polytechnic Institute and State University, and Paul G. Ahlquist, University of Wisconsin, for help with the viruses; Peter Crane of the Field Museum of Natural History, Chicago, and Charles B. Beck, University of Michigan, who provided expert guidance for our treatment of the ancestors of seed plants; Charles B. Heiser, Jr., Indiana University, William L. Brown, Pioneer Hi-bred International, Inc., Jack R. Harlan, University of Illinois, and Ernest Jaworski, Monsanto Company, for help with the chapter on plants and people; and Dwight Billings, Duke University, Peter Vitousek, Stanford University, and Lawrence Bliss, University of Washington, for special help with the ecology section.

Others to whom we wish to express our thanks for reviewing parts of the manuscript or supplying information for this fourth edition are:

Vernon Ahmadjian, *Clark University*
Linda Berg, *University of Maryland*
John T. Bonner, *Princeton University*
Chris H. Bornman, *Hilleshög Aktiebolag, Landskrona, Sweden*
Robert Bowman, *Colorado State University*
Winslow Briggs, *Carnegie Institution, Washington*
Raymond Buck, *Montgomery College*
Jack H. Burk, *California State University, Fullerton*
John Burris, *Pennsylvania State University*
Allan Campbell, *Stanford University*
Leeds Carluccio, *Central Connecticut State University*
Theodore S. Cochrane, *University of Wisconsin*
Grant Cottam, *University of Wisconsin*
Marshall Crosby, *Missouri Botanical Garden*
Billy J. Cumbie, *University of Missouri*
William Darden, *University of Alabama*
Marshall Darley, *University of Georgia*
Jerry Davis, *University of Wisconsin, LaCrosse*
T. O. Diener, *U.S. Department of Agriculture*
David Dilcher, *Indiana University*
John Dwyer, *St. Louis University*
Walter Eschrich, *University of Göttingen, Federal Republic of Germany*
Sharon Eversman, *Montana State University*
Richard Eyde, *U.S. Museum of Natural History, Smithsonian Institution*
Thomas J. Givnish, *University of Wisconsin*
Ursula Goodenough, *Washington University*
Winston Hackbarth, *Louisiana Tech University*
John Heslop-Harrison, *University College of Wales*

Yolande Heslop-Harrison, *University College of Wales*
Arthur Kelman, *University of Wisconsin*
David L. Kirk, *Washington University*
A. H. Knoll, *Harvard University*
Ross E. Koning, *Rutgers University*
Mark Littler, *U.S. Museum of Natural History, Smithsonian Institution*
Robert Magill, *Missouri Botanical Garden*
Michael Mesler, *Humboldt State University*
Eldon H. Newcomb, *University of Wisconsin*
Robert Ornduff, *University of California, Berkeley*
Stanley J. Peloquin, *University of Wisconsin*
James Phillips, *University of Wisconsin*
Tom L. Phillips, *University of Illinois*
K. A. Pirozynski, *National Museums of Canada, Ottawa*
Jeff Pommerville, *Texas A & M University*
David Porter, *University of Georgia*
Don Prusso, *University of Nevada, Reno*
Kenneth G. Raper, *University of Wisconsin*
D.B.O. Saville, *Canada Department of Agriculture*
W. B. Schofield, *University of British Columbia*
William Schopf, *University of California, Los Angeles*
K. D. Stewart, *Miami University*
Ruth Stockey, *University of Alberta*
John Thomas, *Stanford University*
John Thomson, *University of Wisconsin*
James M. Trappe, *U.S.D.A., Corvallis, Oregon*
Dwain Vance, *North Texas State University*
Joseph E. Varner, *Washington University*
Thomas Volk, *University of Wisconsin*
Jerry Weis, *Kansas State University*
Mark A. Wetter, *University of Wisconsin*
George M. Woodwell, *Marine Biology Laboratory, Woods Hole, Massachusetts*
Michael Wynne, *University of Michigan*
Clarice Yentsch, *Bigelow Laboratory for Ocean Services, West Boothbay Harbor, Maine*

We would like, once again, to thank L.A.S. Johnson, Royal Botanic Gardens, Sydney, Australia, for his detailed help with the previous edition of this book, which has also appreciably improved this edition.

Once again, we are grateful to Rhonda Nass for her many superb drawings and to Damian S. Neuberger for his outstanding photographic work. Ellen Marie Dudley also supplied a number of fine drawings for this edition.

We also wish to thank the following people at the University of Wisconsin: Claudia Lipke, for color photographs; Jean Arnold, for typing; and Diep Hoang, Janice Varner, Dianne Wentland, Louise Wang, Garland Williams, Olga Zizich French, Mary Bauschelt, and Von Mansfield, for general assistance.

The first author would like to express his appreciation to his wife, Tamra Engelhorn Raven, for her assistance and encouragement. The second author would like to express his gratitude to his wife, Mary, for her patience and encouragement, and the third author would like to thank her husband, Henry, for being supportive and especially for never complaining during the preparation of the fourth edition.

We would like to offer our special thanks to Helena Curtis for her outstanding contributions to earlier editions of this book, which have helped greatly to make it so successful.

Finally, we wish to express our sincere thanks to Sally Anderson, Gunder Hefta, George Touloumes, Patricia Lawson, Geraldine McGowan, and the many other people at Worth Publishers who have made outstanding contributions to the quality and usefulness of this book.

PETER H. RAVEN
RAY F. EVERT
SUSAN E. EICHHORN
March 1986

x

Contents in Brief

Contents

BIOLOGY OF PLANTS

FOURTH EDITION

C H A P T E R   1

An Introduction to Botany

When a particle of light strikes a molecule of chlorophyll, an electron is raised to a higher energy level and then transferred to an acceptor molecule to initiate a flow of electrons. Within a fraction of a second, the electron returns to its previous energy state. With very few exceptions, all life on this planet is dependent upon the energy momentarily gained by the electron. The process by which some of the energy given up by the electron in returning to its original energy level is converted into chemical energy—energy in a form usable by living systems—is known as photosynthesis. Photosynthesis is the vital link between the physical and the biological worlds, or, as Nobel laureate Albert Szent-Györgyi said, more poetically: "What drives life is . . . a little current, kept up by the sunshine."

Only a few types of organisms—plants, algae, and some bacteria—possess chlorophyll, which can, when embedded in the membranes of a living cell, carry out photosynthesis. Once light energy is trapped in chemical form, it becomes available as an energy source to all other organisms, including human beings. We are totally dependent upon photosynthesis, a process for which plants are exquisitely adapted.

THE EVOLUTION OF PLANTS

Like all other living organisms, plants have had a long evolutionary history. The planet earth itself—an accretion of dust and gases swirling in orbit around the star that is our sun—is some 4.5 billion years old. The earliest known fossils are about 3.5 billion years old and consist of several kinds of small, relatively simple cells (Figure 1–2). These fossils have been found in some of the oldest rocks on earth.

As events are reconstructed, these first cells were formed by a series of chance events. The early atmosphere of the earth is thought to have consisted primarily of gases that were released by the numerous

1–1

All life on earth depends on the ability of plants to capture the sun's energy.

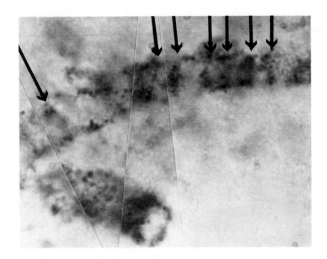

1–2

The earliest known fossils, chainlike bacteria from ancient rocks in Australia, dated at 3.5 billion years of age. These fossils, shown here in ultrathin sections as viewed with a transmission electron microscope, are of very simple structure. The arrows indicate the cross walls between the individual cells. They are about a billion years younger than the earth itself, but there are few suitable older rocks in which to look for earlier evidence of life. More complex organisms— those with eukaryotic cellular organization —did not evolve until about 1.5 billion years ago. For at least 2 billion years, therefore, bacteria were the only forms of life on earth.

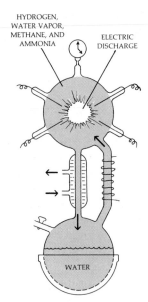

HYDROGEN, WATER VAPOR, METHANE, AND AMMONIA

ELECTRIC DISCHARGE

WATER

1–3

Stanley Miller, while a graduate student at the University of Chicago in the 1950s, used apparatus such as that diagrammed here, to simulate conditions he believed existed on the primitive earth. Methane and ammonia were circulated continuously between a lower "ocean," which was heated, and an upper "atmosphere," through which an electric discharge was transmitted. At the end of 24 hours, about half of the carbon originally present in the methane gas was converted to amino acids and other organic molecules.

volcanoes of the time. This atmosphere seems to have consisted primarily of nitrogen gas, with relatively large amounts of carbon dioxide and water vapor in it. These three molecules contain the chemical elements carbon, oxygen, hydrogen, and nitrogen, which make up about 98 percent of the material found in living organisms today.

Through the thin atmosphere, the rays of the sun beat down on the harsh, bare surface of the young earth, bombarding it with light, heat, and ultraviolet radiation. Molecules of gases such as hydrogen sulfide, ammonia, and methane also seem to have been present in the early atmosphere. In contrast, oxygen gas, which now makes up about 21 percent of our atmosphere, was not formed until living organisms evolved and began to carry out photosynthesis. Thus the first steps in the evolution of life took place in an *anaerobic* (without oxygen) atmosphere.

As the crust of the earth was cooling and stabilizing, violent storms raged, accompanied by lightning and the release of electrical energy. Radioactive substances in the earth emitted energy, and molten rock and boiling water erupted from beneath the earth's surface. The energy in this vast crucible broke apart the simple gases

of the atmosphere and re-formed them into larger and more complicated molecules. Ultraviolet light bathed the surface of the earth, breaking these molecules and the gases, and causing the formation of still more new molecules.

According to present hypotheses, the compounds that were formed in the early atmosphere tended to be washed out of it by the driving rains and to collect in the oceans, which grew larger as the earth cooled. Experiments have been performed with the same gases hypothesized to be present in the primitive atmosphere and under conditions presumably simulating those of the time (Figure 1–3). Under such experimental conditions, complex organic molecules similar to those of the fundamental building blocks of life are formed. On the early earth, the oceans became an increasingly rich mixture of such organic molecules.

Some organic molecules have a tendency to aggregate in groups. In the primitive oceans these groups probably took the form of droplets, similar to the droplets formed by oil in water. Such droplets of organic molecules appear to have been the forerunners of primitive cells, the first forms of life.

According to present theories, these organic molecules also served as the source of energy for the earliest forms of life. The primitive cells or cell-like structures were able to use these abundant compounds to satisfy their energy requirements. As they evolved and became more complex, they became increasingly able to control their own destinies. With increasing complexity, they acquired the ability to grow, to reproduce, and to pass on their characteristics to subsequent generations.

Cells that satisfy their energy requirements by consuming the organic compounds produced by external sources are known as *heterotrophs* (Gk. *heteros*, other + *trophos*, feeder). A heterotrophic organism is one that is dependent on an outside source of organic molecules for its energy. This category of organisms today includes all living things classified as animals or fungi (Figure 1–4) and many of the one-celled organisms—most bacteria and some protists.

As the primitive heterotrophs increased in number, they began to use up the complex molecules on which their existence depended—and which had taken millions of years to accumulate. Organic molecules in free solution (not inside a cell) became more and more scarce, and competition began. Under the pressure of this competition, cells that could make efficient use of the limited energy sources now available were more likely to survive than cells that could not. In the course of time, by the long, slow process of elimination of the less fit, cells evolved that were able to make their own energy-rich molecules out of simple nonorganic materials. Such organisms are called *autotrophs*, "self-feeders." Without the evolution of these early autotrophs, life on earth would soon have come to an end.

The most successful of the autotrophs were those that evolved a system for making direct use of the sun's energy—that is, the process of photosynthesis (Figure 1–4). The earliest photosynthetic organisms, although simple in comparison to plants, were much more complex than the primitive heterotrophs. To capture and use the sun's energy required a complex pigment system that could catch and hold the energy of light and, linked to this system, a way of fixing the energy in an organic molecule.

Evidence of the activities of photosynthetic organisms has been found in rocks 3.4 billion years old, about 100 million years after the first fossil evidence of life on earth. We can be almost certain, however, that both life and photosynthetic organisms evolved considerably earlier than the evidence would suggest. In addition, there seems to be no doubt that heterotrophs evolved before autotrophs. With the arrival of autotrophs, the flow of energy in the biosphere came to assume its modern form: radiant energy channeled through the photosynthetic autotrophs to all other forms of life.

(a)

(b)

1–4

A modern heterotroph and a photosynthetic autotroph. (a) A fungus, Coprinus atramentaris, *growing on a forest floor in California.* Coprinus, *which, like other fungi, absorbs its food (often from other organisms), is heterotrophic.* (b) *Large-flowered trillium (*Trillium grandiflorum*), one of the first plants to flower in spring in the deciduous woods of eastern and midwestern North America. Like most vascular plants, trilliums are rooted in the soil; photosynthesis occurs chiefly in the leaves of this autotrophic organism. The flowers are produced in well-lighted conditions before the leaves appear on the sur-rounding trees. The underground portions (rhizomes) of the plant live for many years and spread to produce new plants vegetatively under the thick cover of decaying leaves and other organic material on the forest floor. Trilliums also reproduce by producing seeds, which are dispersed by ants.*

Photosynthesis and the Evolution of Atmospheric Oxygen

As photosynthetic organisms increased in number, they changed the face of the planet. This biological revolution came about because one of the most efficient strategies of photosynthesis—the one employed by nearly all living autotrophs—involves splitting the water molecule (H_2O) and releasing its oxygen. Thus, as a result of photosynthesis, the amount of oxygen gas (O_2) in the atmosphere increased. This had two important consequences.

First, some of the oxygen molecules in the outer layer of the atmosphere were converted to ozone (O_3) molecules. When there is a sufficient quantity of ozone molecules in the atmosphere, they filter the ultraviolet rays—rays highly destructive to living organisms—from the sunlight that reaches the earth. By about 450 million years ago, organisms, protected by the ozone layer, could survive in the surface layers of water and on the land.

Second, the increase in free oxygen opened the way to a much more efficient utilization of the energy-rich carbon-containing molecules formed by photosynthesis—it enabled organisms to break down those molecules by respiration. As is discussed in Chapter 6, respiration yields far more energy than can be extracted by any anaerobic process.

Before the atmosphere became *aerobic,* the only cells that existed were *prokaryotic*—simple cells that lacked nuclear envelopes and did not have their genetic material organized into complex chromosomes. Another name for prokaryotes is "bacteria," and all of the kinds of organisms that existed on earth prior to about 1.5 billion years ago were bacteria, some heterotrophic and others autotrophic. According to the fossil record, the increase of relatively abundant free oxygen was accompanied by the first appearance of *eukaryotic* cells—cells that have nuclear envelopes, complex chromosomes, and membrane-bound organelles. Eukaryotic organisms, in which the individual cells are usually much larger than those of the bacteria, appeared about 1.5 billion years ago and were well established and diverse by 1 billion years ago. All living systems, except for the bacteria, are composed of one or more eukaryotic cells.

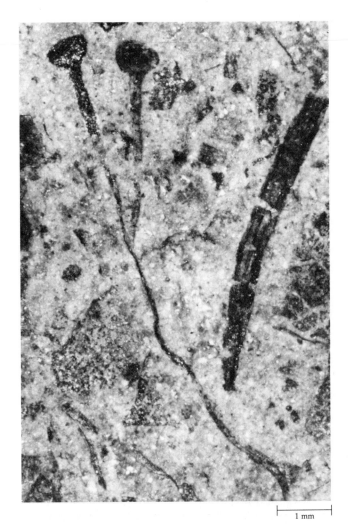

1–5

A fossil of Cooksonia, *one of the earliest and simplest plants known, from the late Silurian period (414–408 million years ago).* Cooksonia *consisted of little more than a branched axis with terminal sporangia, or spore-producing structures.*

The short, straight line at the bottom of this micrograph and those that follow provides a reference for size; a millimeter, abbreviated mm, is 1/10 centimeter. The same system is used to indicate distances on a road map.

The Sea and the Shore

Early in evolutionary history, the principal photosynthetic organisms were microscopic cells floating below the surface of the sunlit waters. Energy abounded, as did carbon, hydrogen, and oxygen, but as the cellular colonies multiplied, they quickly depleted the mineral resources of the open ocean. (It is this shortage of essential minerals that is the limiting factor in any modern plans to harvest the seas.) As a consequence, life began to develop more abundantly toward the shores, where the waters were rich in nitrates and minerals carried down from the mountains by rivers and streams and scraped from the coasts by the ceaseless waves.

The rocky coast presented a much more complicated environment than the open sea, and, in response to these evolutionary pressures, living organisms became increasingly complex in structure and more diversified. Not less than 650 million years ago, organisms evolved in which many cells were linked together to form an integrated, multicellular body (Figure 1–5). In these primitive organisms we see the early stages in the evolution of plants, fungi, and animals. Fossils of multicellular organisms are much easier to detect than those of simpler ones; thus, the history of life on earth is much better documented from the time of their first appearance.

On the turbulent shore, multicellular photosynthetic organisms were better able to maintain their position against the action of the waves, and, in meeting the challenge of the rocky coast, new forms developed. Typically, they evolved relatively strong cell walls for support and specialized structures to anchor their bodies to the rocky surfaces (Figure 1–6). As these organisms increased in size, they were confronted with the problem of how to supply food to the dimly lit, more deeply submerged portions of their bodies where photosynthesis was not taking place. As a result of these new pressures, specialized food-conducting tissues evolved that extended down the center of their bodies and connected the upper photosynthesizing parts with the lower, nonphotosynthesizing structures.

The Transition to Land

The body of the familiar plant can best be understood in terms of its long history and, in particular, in terms of the evolutionary pressures involved in the transition to land. The requirements of a photosynthetic organism are relatively simple: light, water, carbon dioxide for photosynthesis, oxygen for respiration, and a few minerals, or inorganic ions. On land, light is abundant, as are oxygen and carbon dioxide—both of which circulate more freely in air than in water—and the soil is generally rich in inorganic ions. The critical factor, then, for the transition to land is water.

Land animals, generally speaking, are mobile and able to seek out water just as they seek out food. Fungi, though immobile, remain largely below the surface of the soil or within whatever damp organic material they feed upon. Plants utilize an alternative evolutionary strategy. _Roots_ anchor the plant in the ground and collect the water required for maintenance of the plant body and for photosynthesis, while the _stems_ provide support for the principal photosynthetic organs, the _leaves_. A continuous stream of water moves into the root hairs, up through the roots and stems, and then out through the leaves. All of the aboveground portions of the plant that are ultimately involved in photosynthesis are covered with a waxy _cuticle_ that retards water loss. However, the cuticle also tends to prevent the necessary exchange of gases between the plant and the surrounding air. The solution to this dilemma is found in specialized openings called _stomata_ (singular: stoma), which open and close in response to environmental and physiological signals, thus helping the plant maintain a balance between its water losses and its oxygen and carbon dioxide requirements (Figure 1–7).

1–6
Multicellular photosynthetic organisms anchored themselves to rocky shores early in the course of their evolution. These kelp, seen at low tide on the rocks at Bodega Head, north of San Francisco, California, are brown algae (Phaeophyta), a group in which multicellularity evolved independently of other groups of organisms.

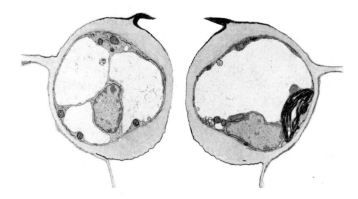

1–7
Transverse section of a mature stoma of a sugar beet (Beta vulgaris) leaf. The stomata, or small openings, in the aerial parts of the plant are regulated by the guard cells, which border the stomata. Each stoma is bordered by two guard cells.

1–8

Diagram of a young broad bean (Vicia faba) plant, showing the principal organs and tissues of the modern vascular plant body. The organs—root, stem, and leaf—are composed of tissues, which are groups of cells with distinct structures and functions. Collectively, the roots make up the root system, and the stems and leaves together make up the shoot system of the plant. In the great majority of vascular plants, the shoot system is above the soil surface and the root system is below. Unlike roots, stems are divided into nodes and internodes. The node is the part of the stem at which one or more leaves are attached, and the internode is the part of the stem between two successive nodes. (In the broad bean, the first few foliage leaves are divided into two leaflets each.) Buds (embryonic shoots) commonly arise in the axils (upper angle between leaf and stem) of the leaves. Lateral, or branch, roots arise from the inner tissues of the roots. The vascular tissues—xylem and phloem—occur together and form a continuous vascular system throughout the plant body. (It is the mesophyll tissue of leaves that is specialized for photosynthesis.)

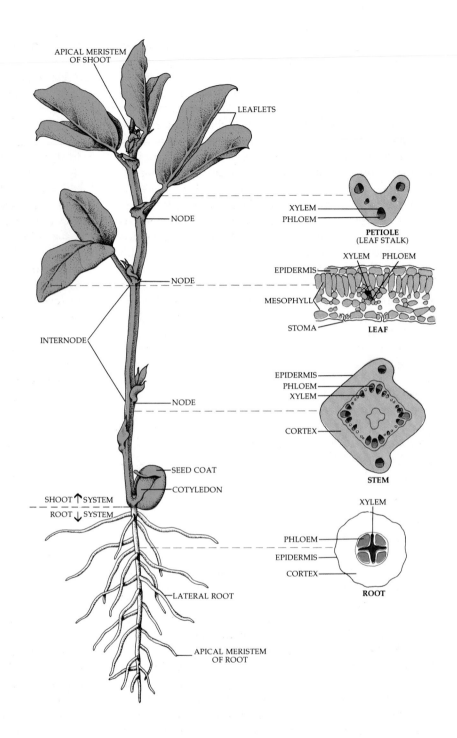

In younger plants and those with a life span of one year (annuals), the stem is also a photosynthetic organ. In longer-lived plants (perennials), the stem may become thickened and woody and covered with cork, which like the cuticle retards water loss. In both cases, the stem serves to conduct, via the *vascular system* (conducting system), a variety of substances between the photosynthetic and nonphotosynthetic parts of the plant body. The vascular system has two major components: the *xylem*, through which water passes upward through the plant body; and the *phloem*, through which food manufactured in the leaves and other photosyn-

thetic parts of the plant is transported throughout the plant body. It is this efficient conducting system that has given the main group of plants—the vascular plants—their name.

Plants, unlike animals, continue to grow throughout their lives. All plant growth originates in *meristems*—localized regions of perpetually embryonic tissues. Meristems located at the tips of all roots and shoots—*apical meristems*—are involved with the extension of the plant body. Thus the roots are continually reaching new sources of water and minerals, and the photosynthetic regions are continually extending toward the light. The type of growth that originates from apical meristems is known as *primary growth*. *Secondary growth,* which results in a thickening of stems, branches, and roots, originates from two lateral meristems—the *vascular cambium* and the *cork cambium.*

Plants also had to overcome new challenges in order to reproduce on land. These were met at first mainly by drought-resistant spores, and later by the evolution of complex, multicellular structures in which the gametes, or reproductive cells, were held. In the *seed plants,* which include almost all familiar plants except the ferns, mosses, and liverworts, the young plant, or embryo, is enclosed within specialized coverings provided by the parent. There the embryo is protected from both drought and predators.

Thus, in summary, the vascular plant (Figure 1–8) is characterized by a root system that serves to anchor the plant in the ground and to collect water and inorganic ions from the soil; a stem or trunk that raises the photosynthetic parts of the plant body toward its energy source, the sun; and leaves, highly specialized photosynthetic organs. Roots, stems, and leaves are interconnected by a complicated and efficient system for the transport of food and water. The reproductive cells of plants are enclosed within multicellular structures, and in seed plants the embryos are protected by resistant coverings. All of these characteristics are adaptations to a photosynthetic existence on land.

THE EVOLUTION OF COMMUNITIES

The invasion of the land by plants changed the face of the continents. Looking down from an airplane on one of the earth's great expanses of desert or on one of its mountain ranges, one can begin to imagine what the world looked like before the appearance of plants. Yet even in these regions, the traveler who goes by land will find an astonishing variety of plants punctuating the expanses of rock and sand. In those parts of the world where the climate is more temperate and the rains are more frequent, communities of plants dominate the land and determine its character. In fact, to a large extent, they *are* the land. Rainforest, meadow, woods, prairie, tundra—each of these words brings to mind a portrait of a landscape (see Figure 1–9 on the next page). The main features of each landscape are its plants, enclosing us in a dark green cathedral in our imaginary rainforest, carpeting the ground beneath our feet with wild flowers in a meadow, moving in great golden waves as far as the eye can see across our imaginary prairie. Only when we sketch these *biomes*—natural communities of wide extent, characterized by distinctive, climatically controlled groups of plants and animals—in terms of trees and shrubs and grasses can

we fill in other features, such as deer, antelope, rabbits, or wolves.

How do vast plant communities, such as those seen on a continental scale, come into being? To some extent we can trace the evolution of the different kinds of plants and animals that populate them. Even with accumulating knowledge, however, we have only begun to glimpse the far more complex pattern of development, through time, of the whole system of organisms that make up these various communities. Such communities, along with the nonliving environment of which they are a part, are known as ecological systems, or *ecosystems.* Ecosystems are discussed later in greater detail. For now, it is sufficient to regard an ecosystem as a kind of corporate entity made up of transient individuals. Some of these individuals, the larger trees, live as long as several thousand years; others, the microorganisms, live only a few hours or even minutes. Yet the ecosystem as a whole tends to be remarkably stable (although not static); once in balance, it does not change for centuries. Our grandchildren will someday walk along a woodland path once followed by our great-grandparents, and where they saw a pine tree, a mulberry bush, a meadow mouse, wild blueberries, or a towhee, these children, if this woodland still exists, will see roughly the same kinds of plants and animals in the same numbers.

An ecosystem functions as an integrated unit, although many of the organisms in the system compete for resources. Virtually every living thing, even the smallest bacterial cell or fungal spore, provides a food source for some other living organism. In this way, the energy captured by green plants is transferred in a highly regulated way through a number of different types of organisms before it is dissipated. Moreover, interactions among the organisms themselves, and between the organisms and the nonliving environment, produce an orderly cycling of elements such as nitrogen and phosphorus. Energy must be added to the ecosystem constantly, but the elements are cycled through the organisms, returned to the soil, decomposed by soil bacteria and fungi, and recycled. These transfers of energy and the cycling of elements involve complicated sequences of events, and in these sequences each group of organisms has a highly specific role. As a consequence, it is impossible to change a single element in an ecosystem without the risk of destroying the balance upon which its stability depends.

At the base of productivity in virtually all ecosystems are the plants, algae, and photosynthetic bacteria that alone have the ability to capture energy from the sun and to manufacture organic molecules that they and all other kinds of organisms may require for life. There are roughly half a million kinds of organisms capable of photosynthesis, and at least eight or ten times that many heterotrophic organisms, which are completely

(a)

(b)

(c)

(d)

1–9

Some of the enormous diversity of biological communities on earth is suggested by these scenes. (a) The tropical rainforest, shown here in Trinidad, is the richest and most diverse biome on earth. At least two-thirds of all species of organisms occur in the tropics, where a single hectare (2.4 acres) may contain as many species of trees as all of the United States and Canada—approximately 700. (b) The temperate deciduous forest,
which covers most of the eastern United States and southeastern Canada, is dominated by trees that lose their leaves in the cold winters. Here, in early spring, the trees are a bright, fresh green.
(c) Savannas are tropical biomes that feature a strongly marked dry season. In Africa, as seen here in the famous Ngorongoro Crater of Tanzania, they are inhabited by huge herds of grazing mammals, such as these wildebeest. The
tree is a euphorbia. (d) Mediterranean climates are rare on a world scale. They are characterized by cool, moist winters, during which the plants grow, and hot, dry summers, during which they become dormant. At the western edge of the Mojave Desert in California, near Lancaster, the desert is covered with California poppies and other herbs that flourish in climates of this sort.

dependent upon the photosynthesizers. For animals, including human beings, there are many kinds of molecules—including essential amino acids, vitamins, and minerals—that can be obtained only through plants or other photosynthetic organisms. Furthermore, the oxygen that is released into the atmosphere by such organisms makes it possible for life to exist on the land and in the surface layers of the ocean. Oxygen is necessary for the energy-producing metabolic activities of the great majority of organisms, including photosynthetic organisms.

THE APPEARANCE OF HUMAN BEINGS

Human beings are relative newcomers to the world of living organisms (Figure 1–10). If the entire history of the earth were measured on a 24-hour time scale starting at midnight, cells would appear in the warm seas before dawn. The first multicellular organisms would not be present until well after sundown, and the earliest appearance of humans (about 2 million years ago) would be half a minute before the day's end. Yet humans more than any other animal—and almost as much as the plants that invaded the land—have changed the surface of the planet, shaping the biosphere according to their own needs, ambitions, or follies.

The development of agriculture, starting at least 11,000 years ago, in time made it possible to maintain large populations of people in towns and cities. This development (reviewed in detail in Chapter 30) allowed specialization and the diversification of human culture. One of the characteristics of this culture is that it looks at itself and the nature of other living things, including plants. Eventually, the science of biology developed within the human communities that had been made possible with the domestication of plants. That part of biology that deals with plants and, by tradition, with bacteria, fungi, and photosynthetic protists—algae—is called *botany*.

The Science of Botany

The study of plants has been pursued for thousands of years, but like all branches of science, it has become diverse and specialized only during the past three centuries. Until little more than a century ago, botany was a branch of medicine, pursued chiefly by physicians as an avocation or principal specialization. Today, however, it is an important scientific discipline that has many subdivisions: *plant physiology*, which is the study of how plants function, that is, how they capture and transform energy and how they grow and develop; *plant morphology*, the study of the form of plants; *plant anatomy*, the study of their internal structure; *plant classification*, also called taxonomy or systematics, the naming and classifying of plants; and many other specialized fields, including *cytology* (the study of cells), *genetics* (the study of heredity), and *ecology* (the study of the relationships of organisms to their environment).

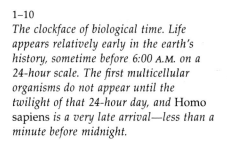

1–10
The clockface of biological time. Life appears relatively early in the earth's history, sometime before 6:00 A.M. on a 24-hour scale. The first multicellular organisms do not appear until the twilight of that 24-hour day, and Homo sapiens *is a very late arrival—less than a minute before midnight.*

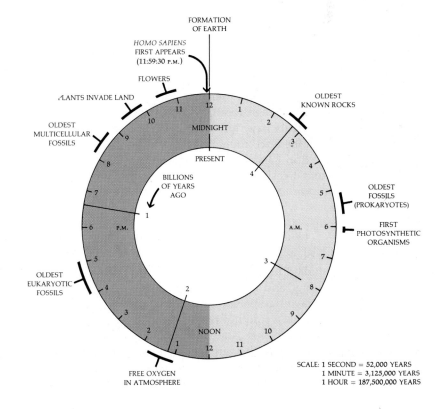

FORMATION OF EARTH

HOMO SAPIENS FIRST APPEARS (11:59:30 P.M.)

FLOWERS

PLANTS INVADE LAND

OLDEST KNOWN ROCKS

OLDEST MULTICELLULAR FOSSILS

MIDNIGHT

PRESENT

BILLIONS OF YEARS AGO

OLDEST FOSSILS (PROKARYOTES)

FIRST PHOTOSYNTHETIC ORGANISMS

P.M.

A.M.

OLDEST EUKARYOTIC FOSSILS

NOON

FREE OXYGEN IN ATMOSPHERE

SCALE: 1 SECOND = 52,000 YEARS
1 MINUTE = 3,125,000 YEARS
1 HOUR = 187,500,000 YEARS

In times past, all organisms were considered to be either plants or animals, and microscopic organisms were assigned to either the plant or animal kingdom as they were discovered. Fungi were considered as plants, presumably because most of them do not move around and their growth form is vaguely more like that of the familiar green plants than it is like that of the animals. The differences between bacteria and other living organisms are much more fundamental than those that separate other groups of organisms, as is explained in the following chapter and in greater detail in Chapter 11. Viruses (Chapter 12) are not really living organisms but are pieces of the genetic apparatus of other organisms; viruses replicate themselves by directing the metabolic processes of cells. Some viruses were originally derived from bacteria, others probably from eukaryotes.

Among the eukaryotes, there are many kinds of unicellular organisms that differ greatly among themselves. The heterotrophic eukaryotes, which have traditionally been called protozoa, are grouped with the animals, whereas the autotrophic eukaryotes, which have traditionally been called algae, are grouped with the plants. The interrelations between the heterotrophic and autotrophic groups, however, are obvious to those who have studied them in detail; they do not really represent different evolutionary lines. Instead, all basically unicellular (eukaryotic) organisms are now grouped in the kingdom Protista, which is considered in detail in Chapters 14 and 15. Among the algae, several evolutionary lines have become multicellular: the green, brown, and red algae. All plants are multicellular, but they are not directly related to the multicellular algae, except for the green algae from which they have evolved during their invasion of the land. Because of the unique features of plants—multicellular, terrestrial, nonmobile, photosynthetic organisms—they are recognized as a distinct kingdom, but with a narrower definition than has been traditional in the past.

Whereas plants obtain their food by photosynthesis (a few exceptional plants have lost this ability but are clearly derived from the rest), animals ingest their food, and fungi (discussed in Chapter 13) absorb it after secreting enzymes and digesting it externally. These three multicellular lines are each treated as a distinct kingdom of eukaryotes, whereas all other eukaryotes—a very diverse group—are assigned to the kingdom Protista.

Included in this book are all organisms that have traditionally been studied by botanists: plants as well as bacteria, viruses, fungi, and autotrophic protists (algae). Only the animals have traditionally been the province of zoologists. Although we do not regard algae, fungi, bacteria, or viruses as plants, and shall not refer to them as plants in this book, they are included here because of tradition, and because they are normally considered as part of the botanical portion of the curriculum, just as botany itself used to be considered a part of medicine. *Virology, bacteriology, phycology* (the study of algae), and *mycology* (the study of fungi) are well-established fields in their own right, but they still fall loosely under the umbrella of botany.

Botany and the Future

In this chapter, we have ranged from the beginnings of life on this planet to the evolution of plants and ecosystems to the development of agriculture and civilization. These broad topics are of interest to many people other than botanists. The urgent efforts of botanists and agricultural scientists will be needed to feed the rapidly growing world population of human beings, as is discussed in Chapter 30. Extant plants offer the best hope of providing a renewable source of energy for human activities, just as extinct plants have been responsible for the massive accumulations of gas, oil, and coal on which our modern industrial civilization depends. In an even more fundamental sense, the role of plants, along with that of algae and photosynthetic bacteria, commands our attention. As the producers of the global ecosystem, these organisms are the route by which all other living things, including ourselves, obtain energy, oxygen, and many other kinds of material that are necessary for our continued existence. As students of botany, you should be in a better position to assess the important ecological and environmental issues of the day and, by understanding, help to build a healthier world.

Marvelous new possibilities have been developed during the past few years for the better utilization of plants by people—these are discussed throughout this book. It is now possible to stimulate the growth of plants, to deter their pests, to control weeds in crops, and to form hybrids between plants with much more precision and far wider possibilities than ever before. The potential of these new discoveries is growing with every passing year, as new discoveries are made and new applications are developed. For example, the methods of genetic engineering (discussed further in Chapter 30) now make it possible, in principle, to transfer either natural or synthetic genes between plants and animals in order to produce certain characteristics. These methods, first applied in 1973, have already been the basis of billions of dollars in investments and increased hope for the future. Discoveries still to be made undoubtedly will far exceed our wildest dreams and go far beyond the facts that are available to us now.

As we turn to Chapter 2, in which our attention narrows to a cell so small it cannot be seen by the unaided eye, it is well to keep these broader concerns in mind. A basic knowledge of plant biology is useful in its own right and is essential in many fields of endeavor, but it is also increasingly relevant to some of society's most crucial problems and to the difficult decisions that will face us in choosing among the proposals for diminishing them. Our own future, the future of the world, and the future of all kinds of plants—as individual species and as components of the life-support systems into which we all have evolved—depends upon our knowledge. Thus this book is dedicated not only to the botanists of the future, whether teachers or researchers, but also to the informed citizens, scientists and laypeople alike, in whose hands such decisions lie.

SUMMARY

Only a few kinds of organisms—plants, algae, and some bacteria—have the capacity to capture energy from the sun and to fix it in organic molecules by the process of photosynthesis. Virtually all life on earth depends, directly or indirectly, on the products of this process.

The planet earth is about 4.5 billion years old. Initially, its atmosphere is thought to have consisted primarily of nitrogen gas, with relatively large amounts of water vapor and carbon dioxide. The four elements present in these gases—carbon, hydrogen, nitrogen, and oxygen—make up about 98 percent of the material found in all living organisms. In the turbulent early atmosphere, the gases were spontaneously recombined into new, larger molecules. Oxygen (which now makes up nearly 21 percent of the earth's atmosphere) was essentially absent until photosynthetic organisms began to produce it in quantity. As a result, ultraviolet rays [now largely blocked from the surface of the earth by ozone (O_3), a form of oxygen] bombarded the surface of the earth and assisted in the synthesis of molecules.

Heterotrophs, organisms that feed on organic molecules or other organisms, were the first to evolve. The oldest known fossils date back 3.5 billion years. Autotrophic organisms, those that could produce their own food by photosynthesis, evolved no less than 3.4 billion years ago. Until about 1.5 billion years ago, bacteria—the prokaryotes—were the only organisms that existed. Eukaryotes, with larger, much more complex cells, evolved at that time. Multicellular eukaryotes began to evolve at least 650 million years ago, and they began to invade the land about 450 million years ago.

Plants are basically a terrestrial group, one of several evolutionary lines that consist entirely or mostly of multicellular organisms. The other major groups are fungi, which absorb their food, and animals, which ingest it. The unicellular eukaryotic organisms are placed in the kingdom Protista, along with three smaller groups of multicellular eukaryotes—the red, brown, and green algae. All six of these multicellular lines evolved independently from unicellular protists.

Plants, which evolved from green algae, have achieved a number of specialized characteristics that suit them for life on land. These are best developed among the members of the dominant group known as the vascular plants. Among them are a waxy cuticle, penetrated by specialized openings known as stomata through which gas exchange takes place; an efficient conducting system, consisting of xylem, in which water and absorbed nutrients pass from the roots to the stems and leaves, and phloem, which transports the products of photosynthesis to all parts of the plant. Plants increase in length by primary growth and in girth by secondary growth, both of which take place in zones of rapid cell division known as meristems.

As plants have evolved, they have come to constitute biomes, great terrestrial assemblages of plants and animals. The interacting systems made up of biomes and their nonliving environments are called ecosystems. Human beings, which appeared about 2 million years ago, developed agriculture at least 11,000 years ago and have subsequently become the dominant ecological force on earth. Humans have used their knowledge of plants to foster their own development and will continue to do so with increasingly greater importance in the future.

SUGGESTIONS FOR FURTHER READING

DICKERSON, R.: "Chemical Evolution and the Origin of Life," *Scientific American* 239(3): 62–78, September 1978.

An excellent brief account of the chemical changes thought to have occurred during the evolution of life on earth.

GROVES, DAVID I., JOHN S. R. DUNLOP, and ROGER BUICK: "An early Habitat of Life," *Scientific American* 245(4): 64–73, October 1981.

A vivid description of a mud flat in Australia that may have sheltered microorganisms some 3.5 billion years ago.

MARGULIS, LYNN: *Early Life,* Science Books International, Inc., Boston, 1982.

An outstanding essay concerning the evolution of life on earth and its history from about 3.5 billion to about 700 million years ago; semipopular in style and easily read.

MILNE, DAVID, DAVID RAUP, JOHN BILLINGHAM, KARL NIKLAS, and KEVIN PADIAN (Eds.): *The Evolution of Complex and Higher Organisms,* National Aeronautics and Space Administration Special Publication SP-478, U.S. Government Printing Office, 1985.

Traces of history of life on earth and the course of its evolution; up-to-date and informative about directions for future research.

SCHOPF, J. WILLIAM: *Earth's Earliest Biosphere: Its Origin and Evolution,* Princeton University Press, Princeton, N.J., 1983.*

A splendid collection of papers documenting the earliest history of earth prior to the evolution of eukaryotic cells.

* Available in paperback.

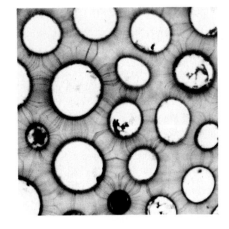

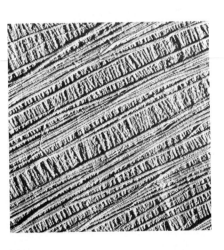

SECTION ONE

The Plant Cell

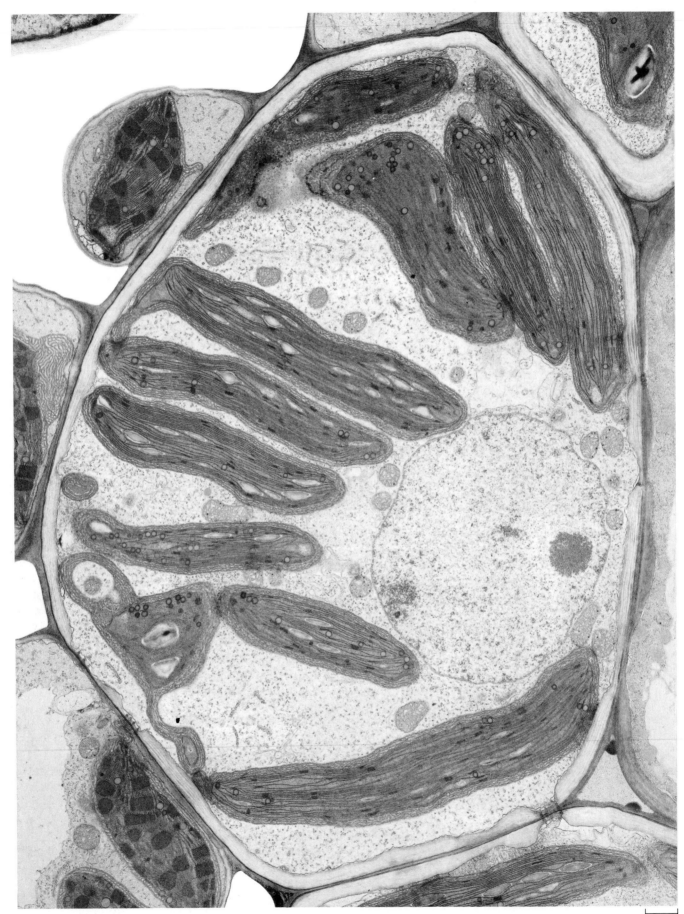

CHAPTER 2

Introduction to the Eukaryotic Cell

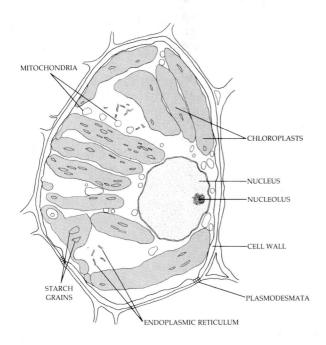

MITOCHONDRIA

CHLOROPLASTS

NUCLEUS

NUCLEOLUS

CELL WALL

STARCH GRAINS

PLASMODESMATA

ENDOPLASMIC RETICULUM

2–1

At left is a cross section of a plant cell as seen with an electron microscope. In the labeled drawing (above), some of the cell's components are identified.

This particular cell is a bundle-sheath cell from the leaf of a corn plant (Zea mays). The nucleus—site of the cell's genetic information—can be seen at the lower right. A single, well-defined nucleolus is visible within the nucleus. Several chloroplasts (organelles involved in photosynthesis) and mitochondria (organelles involved in the conversion of food to usable energy) can also be seen in this cell. Some of the chloroplasts contain oval starch grains that are lighter in appearance.

Cells are the structural and functional units of life (Figure 2–1). The smallest living organisms are composed of single cells. The largest are made up of billions of cells, each of which still lives a partly independent existence. The realization that all organisms are composed of cells was one of the most important conceptual advances in the history of biology because it provided a unifying theme for the study of all living things. When studied at the cellular level, even the most diverse organisms are remarkably similar to one another, both in their physical organization and in their biochemical properties. The *cell theory* was formulated early in the nineteenth century, well before the presentation of Darwin's *theory of evolution,* but these two great unifying concepts are, in fact, closely related. In the similarities among cells, we catch a glimpse of a long evolutionary history that links modern organisms, including plants and ourselves, with the first cellular units that took shape on earth billions of years ago.

There are many different kinds of cells. Within our own bodies there are more than 100 distinct cell types. In a teaspoon of pond water, one can find several different one-celled organisms, and in an entire pond, there are probably several hundred clearly different kinds. Plants are composed of cells that are superficially quite different from those of our own body, and insects have many kinds of cells not found in plants or in vertebrates. Thus, one remarkable fact about cells is their diversity.

A second, even more remarkable fact about cells is their similarity. Every living cell is a self-contained and at least partially self-sufficient unit, and each is bounded by an outer membrane—the plasma membrane, or plasmalemma (often simply called the cell membrane)—that controls the passage of materials into and out of the cell and so makes it possible for the cell to differ biochemically and structurally from its surroundings. Within this membrane is the cytoplasm which, in most cells, includes a variety of discrete bodies and various dissolved or suspended molecules. In addition, every cell contains DNA (deoxyribonucleic acid), which encodes the genetic information (see Chapter 8), and this code is the same for every organism, whether bacterium, oak tree, or human.

PROKARYOTES AND EUKARYOTES

Two fundamentally distinct groups of organisms can be recognized: *prokaryotes* and *eukaryotes*. These terms are derived from the Greek word *karyon,* meaning "kernel" (nucleus). The name *prokaryote* means "before a nucleus," and *eukaryote,* "with a good, or true, nucleus."

The prokaryotes are also known as bacteria; included are the cyanobacteria, or "blue-green algal algae" (see Chapter 11). Prokaryotic cells differ most notably from eukaryotic cells in that their DNA is not organized into chromosomes—complex, protein-containing, thread-like bodies—nor is it surrounded by a membranous envelope (Figure 2–2). Also, bacteria have no specialized membrane-bound structures to perform specific functions.

Eukaryotic cells are divided into distinct compartments that perform different functions (Figure 2–1). The DNA, combined with protein, is located in chromosomes, which are in a nucleus bounded by a double membrane called the nuclear envelope. Eukaryotic cells are usually larger than prokaryotic cells.

Compartmentation in eukaryotic cells is accomplished by means of membranes, which, when seen with the aid of an electron microscope, look remarkably similar in various organisms. When suitably preserved and stained, these membranes have a three-layered appearance (Figure 2–3). They consist of two dark layers (each about 25 Å thick) separated by a lighter layer about 35 Å thick. The term *unit membrane* is commonly used to designate such a visually definable, three-layered membrane.

THE PLANT CELL

The plant cell typically consists of a more or less rigid *cell wall* and a *protoplast*. The term *protoplast* is derived from the word *protoplasm*, which has long been used to refer to the living contents of cells. A protoplast is the protoplasm of an individual cell—in the case of a plant cell, the unit of protoplasm inside the cell wall.

A protoplast consists of *cytoplasm* and a *nucleus* (see Table 2–1). The cytoplasm includes certain distinct entities, or organelles (such as ribosomes, microtubules, plastids, and mitochondria), and systems of membranes (the endoplasmic reticulum and dictyosomes), all of which can be seen in detail only with the aid of

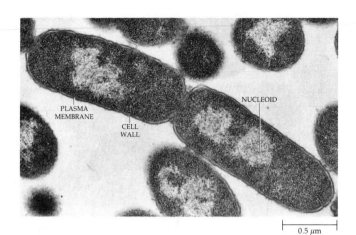

PLASMA MEMBRANE   CELL WALL   NUCLEOID

|—————| 0.5 μm

2–2

Escherichia coli, *a bacterium that is a common, usually harmless inhabitant of the human digestive tract, is an example of a prokaryotic cell. Each of these rod-shaped organisms has a cell wall, a plasma membrane, and cytoplasm. The two cells in the center have just undergone cell division and have not yet separated completely. The DNA is found in the less granular area—the nucleoid— in the center of each cell. The densely granular appearance of the cytoplasm is largely due to the presence of numerous ribosomes (organelles involved in protein synthesis). These small bodies are only about 20 nanometers in diameter.*

an electron microscope. In addition, the cytoplasm includes the cytoplasmic matrix, or *ground substance*, in which these entities and membrane systems are suspended. The cytoplasm is delimited from the cell wall by the *plasma membrane*, which is a unit membrane. In contrast to most animal cells, plant cells develop one or more liquid-filled cavities, or *vacuoles*, within their cytoplasm. The vacuole is bounded by a unit membrane called the *tonoplast.*

In a living plant cell, such as that of the pondweed *Elodea* (Figure 2–4), the ground substance is frequently in motion; the organelles, as well as various substances suspended in the ground substance, can be observed being swept along in an orderly fashion in the moving currents. This movement is known as cytoplasmic streaming, or *cyclosis*, and it continues as long as the cell is alive. Cyclosis undoubtedly facilitates the exchange of materials within the cell and between the cell and its environment; however, it is not known if this is a primary function.

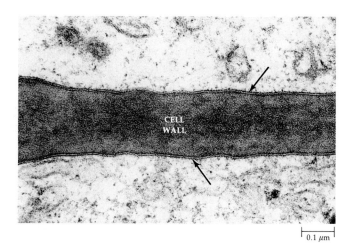

0.1 μm

2–3
Under high magnification, cellular membranes have a three-layered (dark-light-dark) appearance, as seen in the plasma membranes (see arrows) on either side of the common wall between two endodermal cells from a leaf of the fern Vittaria guineensis. *Such membranes are called unit membranes.*

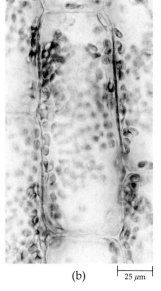

(a)   (b)   25 μm

2–4
An Elodea *cell. (a) Upper surface of the cell. (b) Middle of the cell. The numerous disk-shaped structures are chloroplasts located in the cytoplasm along the wall. In surface view (a), the disk-shaped chloroplasts appear to be circular in outline. In (b) the chloroplasts have their broad surfaces facing the surface of the wall and appear to be elongated. Notice in (b) the absence of chloroplasts in the center of the cell—that is, within the vacuole.*

ORIGIN OF THE CELL THEORY

In the seventeenth century, the English physicist Robert Hooke, using a microscope of his own construction, noted that cork and other plant tissues are made up of small cavities separated by walls. He called these cavities "cells," meaning "little rooms." The word did not take on its present meaning, however, until 150 years later.

In 1838, Matthias Schleiden, a German botanist, came to the conclusion that all plant tissues are organized in the form of cells. In the following year, zoologist Theodor Schwann extended Schleiden's observation to animal tissues and proposed a cellular basis for all life. The cell theory is of tremendous and central importance to biology because it emphasizes the basic similarity of all living systems and so brings an underlying unity to widely varied studies involving many different kinds of organisms.

In 1858, the cell theory took on broader significance when the great pathologist Rudolf Virchow generalized that cells can arise only from preexisting cells: "Where a cell exists, there must have been a preexisting cell, just as the animal arises only from an animal and the plant only from a plant. . . . Throughout the whole series of living forms, whether entire animal or plant organisms or their component parts, there rules an eternal law of continuous development."

In the broad perspective of evolution, Virchow's concept takes on an even larger significance. There is an unbroken continuity between modern cells—and the organisms that they compose—and the primitive cells that first appeared on earth at least 3.5 billion years ago.

PLASMA MEMBRANE

Among the various membranes of the cell, the plasma membrane typically has the clearest dark-light-dark, or unit membrane, appearance. Invaginations (infoldings) of the plasma membrane are common in plant cells, and they have been given a variety of names.

The plasma membrane has several important functions: (1) it mediates the transport of substances into and out of the protoplast (see Chapter 4); (2) it coordinates the synthesis and assembly of cell wall microfibrils (cellulose); and (3) it translates hormonal and environmental signals involved in the control of cell growth and differentiation.

NUCLEUS

The nucleus is often the most prominent structure within the cytoplasm of eukaryotic cells. The nucleus performs two important functions: (1) it controls the ongoing activities of the cell by determining which protein molecules are produced by the cell and when they are produced, as we shall see in Chapter 8; and (2) it stores the genetic information, passing it on to the daughter cells in the course of cell division.

In eukaryotic cells, the nucleus is bounded by a pair of unit membranes called the *nuclear envelope*. The nuclear envelope, as seen with the aid of an electron microscope, contains a large number of circular pores that are 30 to 100 nanometers in diameter (see Figure 2–5); the inner and outer membranes are joined around each pore to form the margin of its opening. The pores are not merely holes in the envelope; each has a complicated structure. In various places the outer membrane of the envelope may be continuous with the endoplasmic reticulum. The nuclear envelope may be considered a specialized, locally differentiated portion of the endoplasmic reticulum.

If the cell is treated by special staining techniques, thin threads and grains of *chromatin* can be distinguished from the *nucleoplasm*, or nuclear ground substance. Chromatin is made up of DNA combined with large amounts of proteins called histones. During the process of cell division the chromatin becomes progressively more condensed until it takes the form of *chromosomes*. Recently it has been suggested that the chromosomes (chromatin) of nondividing (interphase nuclei) are attached at one or more sites to the nuclear envelope. As in bacteria, the hereditary information is carried in molecules of DNA. The content of DNA per cell is much higher in eukaryotic organisms than in bacteria. In bacteria, the DNA molecules are essentially free in the cytoplasm. In eukaryotes, they are organized into much larger units—the chromosomes—contained within the nucleus.

Different organisms vary in the number of chromosomes present in their somatic (body) cells. *Haplopappus gracilis*, a desert annual, has 4 chromosomes per cell; cabbage, 20; the common sunflower, 34; bread wheat, 42; humans, 46; and one species of the fern *Ophioglossum*, about 1250. The reproductive cells, or gametes, however, have only half the number of chromosomes that is characteristic of the somatic cells of the organism. The number of chromosomes in the gametes is referred to as the *haploid* ("single") number, and that in the somatic cells is called the *diploid* ("double") number. Cells that have more than two sets of chromosomes are said to be *polyploid*.

Often the only structures within a nucleus that are discernible with the light microscope are the spherical structures known as *nucleoli* (singular: nucleolus), one or more of which are present in each nucleus. Nucleoli are present in nondividing nuclei (Figure 2–1) and are the sites of the formation of ribosomal RNA. Their very presence is due to the accumulation into large aggre-

(a) |—— 0.5 μm ——|

POLYSOME

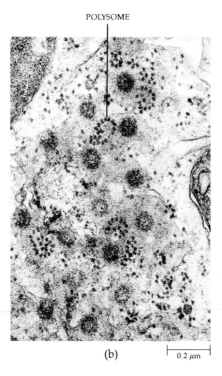

(b) |—— 0.2 μm ——|

ENDOPLASMIC RETICULUM

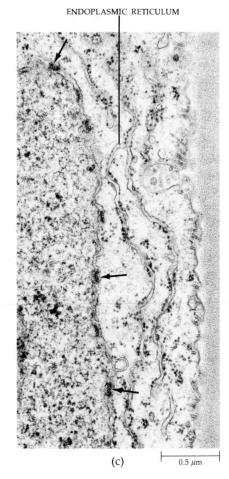

(c) |—— 0.5 μm ——|

2–5

Nuclear pores as revealed (a) in a freeze-etch preparation of the surface of the nuclear envelope of an onion (Allium cepa) root tip cell, and (b) and (c) in electron micrographs of nuclei in parenchyma cells of the seedless vascular plant Selaginella kraussiana. In (b) the *pores are shown in surface view, whereas in (c) they are shown in sectional view (see arrows). Note the polysomes (coils of granules) on the surface of the nuclear envelope in (b) and the rough endoplasmic reticulum paralleling the nuclear envelope in (c).*

VIEWING THE MICROSCOPIC WORLD

Most cells can be seen only with the aid of a microscope. The units of measurement generally used for describing cells are micrometers and nanometers (see table). Unaided, the human eye has a resolving power of about 1/10 millimeter, or 100 micrometers. This means that if you look at two lines that are less than 100 micrometers apart, they merge into a single line. Similarly, two dots less than 100 micrometers apart look like a single blurry dot. In order to separate structures closer than this, optical instruments such as microscopes must be used. The best light microscope has a resolving power of 0.2 micrometer, or about 200 nanometers, and so is about a 500-fold improvement on the naked eye. It is theoretically impossible to build a light microscope that will do better than this.

Notice that resolving power and magnification are different; using the best light microscope, if you take a photograph of two lines that are less than 0.2 micrometer, or 200 nanometers, apart, you can enlarge the image indefinitely, but the two lines will still blur together. By using more powerful lenses, you can increase the magnification, but this will not improve the resolution.

The Transmission Electron Microscope

With the electron microscope, resolving power has been increased almost 400 times over that provided by the light microscope. This is achieved by using "illumination" consisting of electron beams instead of light rays. Under the very best conditions, electron microscopy currently affords a resolving power of about 0.5 nanometer, roughly 200,000 times greater than that of the human eye. (A hydrogen atom is about 0.1 nanometer in diameter.)

In a transmission electron microscope, a beam of electrons passing through the specimen leaves its imprint on a screen. Areas in the specimen that permit the transmission of more electrons, that is, "electron-transparent" regions, show up light in color, and "electron-dense" areas are dark.

The transmission electron microscope has one great disadvantage. Electrons have a very small mass and must travel in a vacuum; electron beams can only pass through specimens that are exceedingly thin. To prepare them for the electron microscope, specimens must therefore be killed and embedded in hard materials so that they can be sliced by special cutting instruments. This means, of course, that the high resolving powers of the electron microscope can be applied only to tissues that are no longer alive. Also, it is sometimes difficult to determine what changes may be produced in the material in the course of its preparation.

A few kinds of cells, almost all viruses, and many of the structures within cells can be seen only with the transmission electron microscope. Figure 2–1 is one of the many transmission electron micrographs found in this book.

The Scanning Electron Microscope

In a scanning electron microscope, the electrons whose imprints are recorded come from the surface of the specimen. The electron beam is focused into a fine probe which is used to scan the specimen. As a result of the electron bombardment from the probe, the specimen emits low-energy secondary electrons. Variations in the surface of the specimen alter the number of secondary electrons emitted. Holes and fissures give off fewer secondary electrons and thus appear dark, whereas the larger number of secondary electrons given off by high points and ridges cause these surfaces to appear light, imparting a three-dimensional image. Electrons scattered from the surface, plus any secondary electrons, are collected, amplified, and transmitted to a screen that is scanned in synchrony with the electron probe. Figures 2–10 and 2–16 were prepared with a scanning electron microscope, as were others in this textbook.

Measurements Used in Microscopy

1 centimeter (cm) = 1/100 meter = 0.4 inch

1 millimeter (mm) = 1/1,000 meter = 1/10 cm

1 micrometer (μm)* = 1/1,000,000 meter = 1/10,000 cm

1 nanometer (nm) = 1/1,000,000,000 meter
             = 1/10,000,000 cm

1 angstrom (Å) = 1/10,000,000,000 meter
           = 1/100,000,000 cm
                or

1 meter = 10^2 cm = 10^3 mm = 10^6 μm
     = 10^9 nm = 10^{10} Å

* Micrometers were formerly known as microns, indicated by the Greek letter μ, which corresponds to the letter m in our alphabet and is pronounced "mew."

gates of the molecules being fashioned into ribosomal subunits. Commonly, diploid organisms have two nucleoli, one for each haploid set of chromosomes. The nucleoli often fuse and then appear as one large structure. Consisting mostly of protein, the nucleoli also contain about 5 percent RNA (ribonucleic acid).

PLASTIDS

Together with vacuoles and cell walls, *plastids* are characteristic components of plant cells. Each plastid is bounded by an envelope consisting of two unit membranes. Internally, the plastid is differentiated into a system of membranes and a more or less homogeneous ground substance, the *stroma*. Mature plastids are commonly classified on the basis of the kinds of pigments they contain.

Chloroplasts, the sites of photosynthesis (see Chapter 7), contain chlorophylls and carotenoid pigments. In plants, chloroplasts are usually disk-shaped and measure between 4 and 6 micrometers in diameter. A single mesophyll ("middle of the leaf") cell may contain 40 to 50 chloroplasts; a square millimeter of leaf contains some 500,000. In the cytoplasm, the chloroplasts are usually found with their broad surfaces parallel to the cell wall, as shown in Figure 2–6.

The internal structure of the chloroplast is complex (Figure 2–7). The stroma is traversed by an elaborate system of membranes in the form of flattened sacs called *thylakoids*. Each thylakoid is similar to the plastid envelope in that it consists of two membranes. The thylakoids are believed to constitute a single, interconnected system. Chloroplasts are generally characterized by the presence of *grana* (singular: granum)—stacks of disklike thylakoids that resemble a stack of coins. The thylakoids of the various grana are connected with each other by thylakoids (the stroma thylakoids, or intergrana thylakoids) that traverse the stroma. Chlorophylls and carotenoid pigments are found embedded in thylakoid membranes.

The chloroplasts of green algae and plants often contain starch grains and small lipid (oil) droplets. The starch grains are temporary storage products and accumulate only when the alga or plant is actively photosynthesizing (Figure 2–1). They may be lacking in the

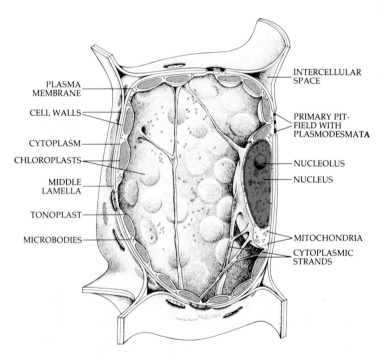

2–6

A three-dimensional diagram of a chloroplast-containing plant cell. The numerous disk-shaped chloroplasts are located in the cytoplasm along the wall and have their broad surfaces facing the surface of the wall. Most of the volume of this cell is occupied by the vacuole, which is traversed here by a few strands of cytoplasm. In this cell, the nucleus lies in the cytoplasm along the wall, although in some cells it might appear suspended by strands of cytoplasm in the center of the vacuole, surrounded by a thin layer of cytoplasm.

Table 2–1   *An Inventory of Plant Cell Components*

Cell wall	Middle lamella	
	Primary wall	
	Secondary wall	
	Plasmodesmata	
Protoplast	Nucleus	Nuclear envelope
		Nucleoplasm
		Chromatin
		Nucleolus
	Cytoplasm	Plasma membrane
		Ground substance (micro-trabecular lattice)
		Organelles bounded by two membranes
		Plastids
		Mitochondria
		Organelles bounded by one membrane
		Microbodies
		Vacuoles (tonoplast)
		Endomembrane system
		Endoplasmic reticulum
		Dictyosomes
		Vesicles
		Cytoskeleton
		Microtubules
		Microfilaments
		Ribosomes
Ergastic substances*	Crystals	
	Anthocyanins	
	Starch grains	
	Tannins	
	Fats, oils, and waxes	
	Protein bodies	

* May occur in cell wall, ground substance, and/or organelles.

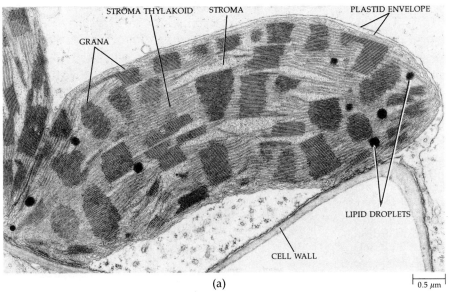

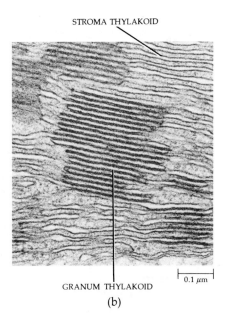

2-7

(a) *A chloroplast of a corn (Zea mays) leaf.* (b) *Detail showing grana composed* *of stacks of disklike thylakoids. The thylakoids of the various grana are inter-* *connected by other thylakoids, commonly called stroma thylakoids.*

chloroplasts of plants that have been kept in the dark for as little as 24 hours but often reappear after the plant has been in the light for only 3 or 4 hours.

Chloroplasts are semiautonomous organelles that resemble bacteria in several ways. For example, the ribosomes of both bacteria and chloroplasts are about two-thirds as large as the ribosomes of eukaryotes. Also, protein synthesis in bacterial and chloroplast ribosomes is inhibited by chloramphenicol, an antibiotic substance that has no effect on eukaryotic ribosomes. In addition, both bacteria and chloroplasts contain one or more nucleoids—clear, grana-free regions containing strands of DNA. The DNA of the plastid is organized in the same manner as that of a bacterium: it is not bounded by a membrane, it is not associated with histone proteins, and usually exists in circular form.

The genetic code of plastid DNA is currently being investigated in several laboratories. Isolated chloroplasts have the capacity to synthesize RNA, which, as discussed in Chapter 8, is usually carried out only under the direction of chromosomal DNA. The ability to form chloroplasts and the pigments associated with them is largely controlled by chromosomal DNA interacting in some poorly understood fashion with chloroplast DNA. Chloroplasts, therefore, cannot be formed in the absence of chloroplast DNA.

Chloroplasts may be considered the most important of cell organelles, for they are the ultimate sources of most of our food supplies and of our fuel. Chloroplasts are not only sites of photosynthesis; they are also involved in amino acid synthesis and fatty acid synthesis, as well as providing space for the temporary storage of starch.

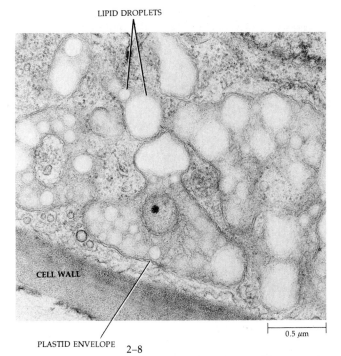

2-8

Chromoplasts from a marigold (Tagetes) petal. Each chromoplast contains numerous lipid droplets in which the pigment responsible for the color of the petal is stored.

Chromoplasts (Gk. *chroma*, color) are pigmented plastids (Figure 2-8). Of variable shape, chromoplasts lack chlorophyll but synthesize and retain carotenoid pigments, which are often responsible for the yellow, orange, or red colors of many flowers, old leaves (see

Chapter 7), some fruits, and some roots. Chromoplasts may develop from previously existing green chloroplasts by a transformation in which the chlorophyll and internal membrane structure of the chloroplast disappear and masses of carotenoids accumulate, as occurs during the ripening of many fruits. The precise functions of chromoplasts are not well understood, although at times they act as attractants to insects and other animals with which they have coevolved (see Chapter 29).

Leucoplasts (Figure 2–9) are nonpigmented plastids. Some synthesize starch (*amyloplasts;* Figures 2–10 and 2–11), but others are thought to be capable of forming a variety of substances, including oils and proteins. Upon exposure to light, leucoplasts may develop into chloroplasts.

Proplastids are small, colorless or pale green, undifferentiated plastids that occur in meristematic (dividing) cells of roots and shoots. They are the precursors of the other, more highly differentiated plastids such as chloroplasts, chromoplasts, or amyloplasts (Figure 2–12). If the development of a proplastid into a more highly differentiated form is arrested by the absence of light, it may form one or more *prolamellar bodies,* which are semicrystalline bodies composed of tubular membranes (Figure 2–13). Plastids containing prolamellar bodies are called *etioplasts*. During subsequent development of etioplasts into chloroplasts in the light, the membranes of the prolamellar bodies develop into thylakoids. Etioplasts form in leaf cells of plants grown in the dark. In nature the proplastids in the embryos of seeds first develop into etioplasts; then, upon exposure to light, the etioplasts develop into chloroplasts. The various kinds of plastids are remarkable for the relative ease with which they can change from one type to another.

Plastids reproduce by fission—by dividing into equal halves—and in this regard also they resemble bacteria. In meristematic cells, the division of proplastids roughly keeps pace with cell division. In mature cells, however, the greater proportion of the final plastid population may be derived from the division of mature plastids.

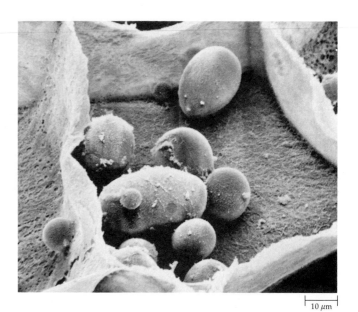

2–10
Scanning electron micrograph showing spherical and egg-shaped starch grains in a cell of a potato (Solanum tuberosum). *The starch grains form in amyloplasts, one grain per plastid.*

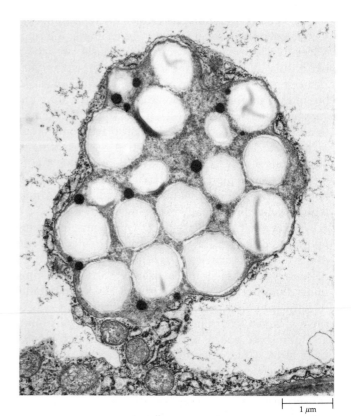

2–11
Amyloplast from the embryo sac of soybean (Glycine max). *The round, clear bodies are starch grains. The smaller, dense bodies are lipid droplets.*

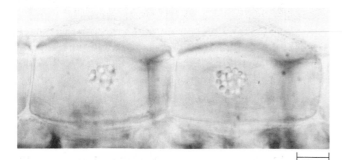

2–9
Leucoplasts clustered around the nucleus in epidermal cells of a Zebrina *leaf.*

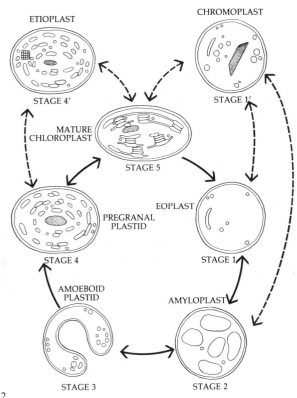

STAGE 4' ETIOPLAST
CHROMOPLAST STAGE 1'
MATURE CHLOROPLAST STAGE 5
EOPLAST
PREGRANAL PLASTID STAGE 4
STAGE 1
AMOEBOID PLASTID
AMYLOPLAST
STAGE 3
STAGE 2

2–12
Plastid developmental cycle as proposed by J. M. Whatley. The successive stages in plastid development. Stages 1–3: Eoplast (eo-, "early"), amyloplast, and amoeboid can be grouped together as nongreen proplastid stages. Stage 4: The pregranal plastid may be green or nongreen. In the absence of light, the pregranal stage may be represented by an etioplast (stage 4'). As indicated, chromoplasts (1') may be formed from several types of plastid.

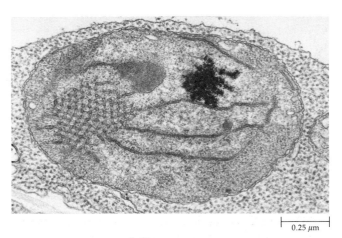

0.25 μm

2–13
Etioplast, with a semicrystalline prolamellar body composed of tubular membranes, in a leaf cell of a plant grown in the dark. When exposed to light, the membranes of the prolamellar body develop into thylakoids.

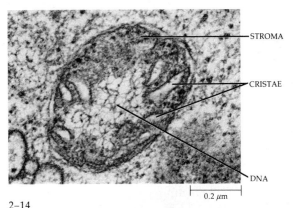

STROMA

CRISTAE

DNA

0.2 μm

2–14
Mitochondrion in a leaf cell of spinach (Spinacia oleracea), in a plane of section revealing some strands of DNA. The envelope of the mitochondrion consists of two separate membranes. The inner membrane folds inward to form cristae, which are embedded in a dense stroma. The small granules in the stroma are ribosomes.

MITOCHONDRIA

Like chloroplasts, mitochondria are bounded by two unit membranes (Figures 2–14 and 2–15). The inner membrane is extensively folded into pleats or shelf-like projections known as *cristae* (singular: crista), which greatly increase the surface area available to enzymes and the reactions associated with them. Mitochondria are generally smaller than plastids, measuring about half a micrometer in diameter, and there is great variability in length and shape. As a rule, mitochondria are barely visible with a light microscope, but they can readily be seen with the aid of an electron microscope.

Mitochondria are the sites of respiration, a process involving the release of energy from organic molecules and its conversion to molecules of ATP (adenosine triphosphate), the chief chemical energy source for all eukaryotic cells. (These processes are discussed further in Chapter 6.) Most plant cells contain hundreds or thousands of mitochondria, the number of mitochondria per cell being related to the cell's demand for ATP.

With the aid of time-lapse photography, mitochondria can be seen to be in constant motion, turning and twisting and moving from one part of the cell to another; they also fuse and divide by fission. Mitochondria tend to congregate where energy is required. In cells in which the plasma membrane is very active in transporting materials into or out of the cell, they often can be found arrayed along the membrane surface. In motile, single-celled algae, mitochondria are typically clustered at the bases of the flagella, presumably providing energy for the movement of the flagella.

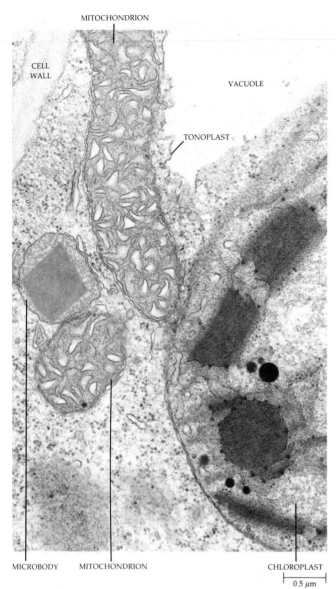

MITOCHONDRION

CELL
WALL

VACUOLE

TONOPLAST

MICROBODY   MITOCHONDRION   CHLOROPLAST

0.5 μm

2–15
*Organelles in a leaf cell of tobacco
(Nicotiana tabacum). A microbody with
a large crystalline inclusion and a single
bounding membrane may be contrasted
with the two mitochondria and the
plastid bounded by a double membrane.
The double nature of the plastid envelope
is apparent in only the lower portion of
the micrograph. A portion of a vacuole
(the clear area at the upper right) is
separated from the rest of the cytoplasm
by a single membrane, the tonoplast.*

Mitochondria, like plastids, are semiautonomous organelles, containing the components necessary for the synthesis of some of their own proteins. The inner membrane of the mitochondrion encloses a liquid matrix that contains proteins, RNA, strands of DNA, ribosomes similar to those of bacteria, and various solutes (dissolved substances). The DNA occurs as circular molecules in one or more clear areas, the nucleoids (Figure 2–14).

Evolutionary Origin of Mitochondria and Chloroplasts

On the basis of the close similarity among bacteria and the mitochondria and chloroplasts of eukaryotic cells, it seems probable that mitochondria and chloroplasts originated as bacteria that found shelter within larger heterotrophic cells. These larger cells were the forerunners of the eukaryotes. The smaller cells, which contained (and still contain) all the mechanisms necessary to trap and/or convert energy from their surroundings, donated these useful capacities to the larger cells. Cells with these respiratory "assistants" had a clear advantage over their contemporaries and undoubtedly soon multiplied at their expense. All but a very few modern eukaryotes contain mitochondria, and all autotrophic eukaryotes also contain chloroplasts; both seem to have been acquired by independent symbiotic events. (Symbiosis is the close association between two or more different organisms that may be, but is not necessarily, beneficial to each.) The larger and more complex cells of eukaryotes seem to protect their symbiotic organelles from environmental extremes. As a consequence, eukaryotes were able to invade the land and the acidic waters, where the prokaryotic cyanobacteria, or "blue-green algae," are absent but the eukaryotic green algae abound.

MICROBODIES

Unlike plastids and mitochondria, which are bounded by two membranes, *microbodies* are spherical organelles bounded by a single membrane. They range in diameter from 0.5 to 1.5 micrometers. Microbodies have a granular interior and sometimes contain a crystalline body composed of protein (Figure 2–15). Microbodies are generally associated with one or two segments of endoplasmic reticulum, the three-dimensional system of internal membranes within cells.

Some microbodies, called *peroxisomes*, play an important role in glycolic acid metabolism associated with photorespiration (see page 108); in green leaves they are closely associated spatially with mitochondria and chloroplasts (Figure 2–15). Other microbodies, called *glyoxysomes*, contain enzymes necessary for the conversion of fats into carbohydrates during germination in many seeds (see page 92).

VACUOLES

Vacuoles are membrane-bounded regions within the cell that are filled with a liquid called *cell sap*. They are surrounded by the tonoplast, or vacuolar membrane (Figures 2–6 and 2–15).

The immature plant cell typically contains numerous small vacuoles, which increase in size and fuse into a single vacuole as the cell enlarges. In the mature cell, as much as 90 percent of the volume may be taken up by the vacuole, with the cytoplasm consisting of a thin peripheral layer closely pressed against the cell wall (Figure 2–6). By filling such a large proportion of the cell with "inexpensive" vacuolar contents, plants not only save "expensive" nitrogen-consuming cytoplasm but also acquire a large surface between the thin layer of cytoplasm and the environment. Most of the increase in size of the cell results from enlargement of the vacuoles. A direct consequence of this strategy is the development of turgor pressure and the maintenance of tissue rigidity, one of the principal roles of the vacuole and tonoplast (see Chapter 4).

The principal component of the cell sap is water, with other components varying according to the type of plant and its physiological state. Vacuoles typically contain salts and sugars, and some contain dissolved proteins. The tonoplast plays an important role in the active transport of certain ions into the vacuole and their retention there. Thus, ions may accumulate in the cell sap in concentrations far in excess of those in surrounding cytoplasm. Sometimes the concentration of a particular material in the vacuole is sufficiently great for it to form crystals. Calcium oxalate crystals, which can assume several different forms, are especially common (Figures 2–16 and 2–17). Vacuoles are usually slightly acidic. Some of them, like the vacuoles in citrus fruits, are very acidic—hence the tart, sour taste of the fruit.

Vacuoles are important storage compartments for various metabolites (products of metabolism), such as the reserve proteins in seeds and the malic acid in CAM plants (see page 111). They also remove toxic secondary products of metabolism, such as nicotine, an alkaloid, from the rest of the cytoplasm (see also Chapter 29, page 617).

The vacuole is often a site of pigment deposition. The blue, violet, purple, dark red, and scarlet colors of plant cells are usually caused by a group of pigments known as the anthocyanins (see Chapter 29). Unlike most other plant pigments, the anthocyanins are readily soluble in water, and they are dissolved in the cell sap. They are responsible for the red and blue colors of many vegetables (radishes, turnips, cabbages), fruits (grapes, plums, cherries), and a host of flowers (cornflowers, geraniums, delphiniums, roses, and peonies). Sometimes the pigments are so brilliant they mask the chlorophyll in the leaves, as in the ornamental red

2–16

Druses, compound crystals composed of calcium oxalate, in epidermal cells of a cotyledon of redbud (Cercis canadensis), as seen with a scanning electron microscope.

maple. Anthocyanins are also responsible for the brilliant red colors of some leaves in autumn. These pigments form in response to cold, sunny weather, at the same time the leaves stop producing chlorophyll. As the chlorophyll that is present disintegrates, the newly formed anthocyanins are unmasked. In leaves that do not form anthocyanin pigments, the breakdown of chlorophyll in autumn may unmask more stable yellow-to-orange carotenoid pigments already present in the chloroplasts. The most spectacular autumnal coloration develops in years when cool, clear weather prevails in autumn.

Vacuoles are also involved with the breakdown of macromolecules and the recycling of their components within the cell. Entire cell organelles, such as ribosomes, mitochondria, and plastids, may be deposited and degraded in vacuoles. Because of this digestive activity, vacuoles are comparable in function with organelles known as *lysosomes* that occur in animal cells.

Vacuoles have long been considered to arise from the endoplasmic reticulum. When it was realized that vacuoles are analogous to the lysosomes of animal cells, attempts were initiated to determine whether the

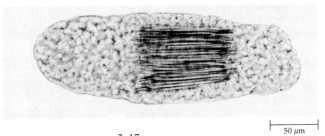

2–17

A bundle of raphides—needlelike crystals of calcium oxalate—in a vacuole of a leaf cell of the snake plant (Sansevieria). The tonoplast is not discernible in this photomicrograph. The granular substance surrounding the crystals is cytoplasm.

2–18

Stages in the formation of vacuoles as represented by F. Marty. Initially appearing as pear-shaped blebs, tubular provacuoles, which arise from the GERL complex, branch and ramify throughout the cell (I). The provacuoles then wrap themselves around portions of cytoplasm, like bars of a cage (II). The tubes of each *cage then merge together to completely isolate the engulfed cytoplasm (III). When this is accomplished, the inner membrane of the cage and the engulfed cytoplasm are digested by lysosomal enzymes (IV). Up to this point, the digestive enzymes had been kept rigorously in check within the fusing* *tubes. The other membrane of the cage remains intact, and thus digestive activities are confined within the developing vacuole. Once formed, the numerous small vacuoles fuse (IV) and finally give rise to the one or more large vacuoles characteristic of mature plant cells (V).*

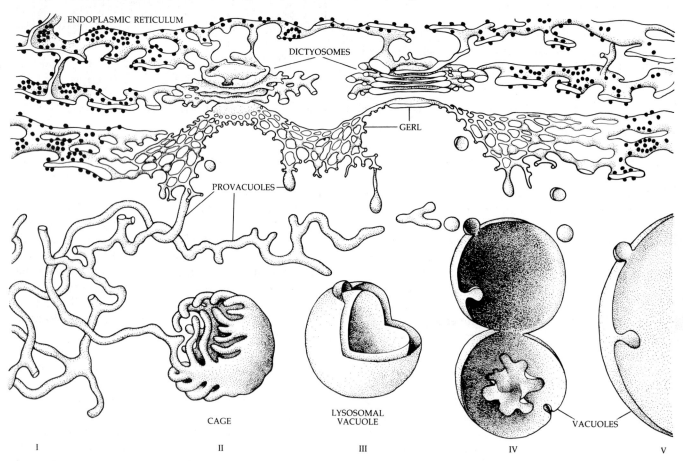

ENDOPLASMIC RETICULUM

DICTYOSOMES

GERL

PROVACUOLES

CAGE

LYSOSOMAL VACUOLE

VACUOLES

I   II   III   IV   V

vacuoles of at least some plant cells had origins similar to those of animal-cell lysosomes. In animal cells, lysosome formation is associated with a specialized region of cytoplasm called GERL, for Golgi associated–endoplasmic reticulum–lysosomal complex. GERLs have now been found in root tip cells of radish and a euphorbia and in the cotyledons of mung bean embryos. Figure 2–18 shows the stages in vacuole formation associated with a GERL. The Golgi apparatus is discussed on page 28.

R I B O S O M E S

Ribosomes are small particles, only about 17 to 23 nanometers in diameter, consisting of about equal amounts of protein and RNA. They are the sites at which amino acids are linked together to form proteins and are abundant in the cytoplasm of metabolically active cells. Ribosomes occur freely in the cytoplasm or are attached to the endoplasmic reticulum; they commonly occur at both places in the same cell. Ribosomes are also found in nuclei. As mentioned previously, plastids and mitochondria contain ribosomes similar to those in prokaryotes.

Ribosomes actively involved in protein synthesis occur in clusters or aggregates called polyribosomes, or *polysomes* (Figure 2–19). Cells that are synthesizing proteins in large quantities often contain extensive systems of polysome-bearing endoplasmic reticulum. Polysomes are often attached to the outer surface of the nuclear envelope (Figure 2–5b).

2–19

Groups of ribosomes (polysomes) on the surface of the endoplasmic reticulum. The endoplasmic reticulum is a network of membranes that is found throughout the cytoplasm of the eukaryotic cell, dividing it into compartments and providing surfaces on which chemical reactions can take place. Ribosomes are the sites at which amino acids are assembled into proteins. This electron micrograph shows a portion of a parenchyma cell from a leaf of the fern Regnellidium diphyllum.

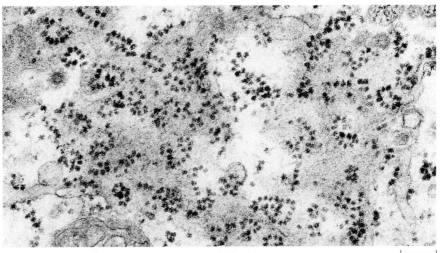

0.15 μm

ENDOPLASMIC RETICULUM

The *endoplasmic reticulum* is a complex three-dimensional membrane system of indefinite extent. In sectional view, the endoplasmic reticulum appears as two unit membranes with a narrow, transparent space between them (Figures 2–5c and 2–20). The form and abundance of the endoplasmic reticulum within a cell vary greatly depending on the cell type, its metabolic activity, and its stage of development. For example, in cells that secrete or store proteins, the endoplasmic reticulum has the form of flattened sacs, or *cisternae* (singular: cisterna), with numerous ribosomes attached to its outer surface. Endoplasmic reticulum bearing ribosomes is called *rough endoplasmic reticulum* (Figures 2–5c, 2–19, and 2–20). The ribosomes are assembled into polysomes, and the rough endoplasmic reticulum is a major site of protein synthesis. In contrast, cells that secrete lipids have an extensive system of tubules that lack ribosomes. Endoplasmic reticulum lacking ribosomes is called *smooth endoplasmic reticulum*. Smooth endoplasmic reticulum is usually tubular in form. Both rough and smooth forms of endoplasmic reticulum may occur within the same cell; usually numerous connections exist between them.

The endoplasmic reticulum appears to function as a communications system within the cell. In some electron micrographs, it can be seen to be continuous with the outer membrane of the nuclear envelope. In fact, these two structures together seem to form a single membrane system. When the nuclear envelope breaks down at cell division, its fragments are similar to portions of endoplasmic reticulum. It is easy to visualize the endoplasmic reticulum as a system for channeling materials—such as proteins and lipids—to different parts of the cell. In addition, the endoplasmic reticulum of adjacent cells is interconnected by way of cytoplasmic threads, called *plasmodesmata*, which traverse their common walls (see page 35).

The endoplasmic reticulum is the principal site of membrane synthesis within the cell. It appears to give rise to vacuolar and microbody membranes, as well as to the cisternae of dictyosomes in at least some plant cells.

2–20

Parallel arrays of rough endoplasmic reticulum—endoplasmic reticulum studded with ribosomes—seen in sectional view in a leaf cell of the fern Vittaria guineensis. *The relatively clear areas are vacuoles.*

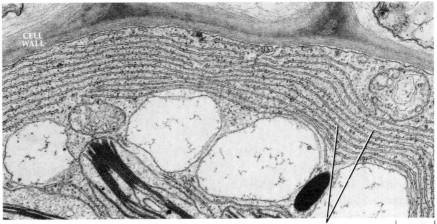

CELL WALL

ROUGH ENDOPLASMIC RETICULUM 0.5 μm

GOLGI APPARATUS

The term *Golgi apparatus* is used to refer collectively to all of the *dictyosomes,* or *Golgi bodies,* of a cell. Dictyosomes are groups of flat, disk-shaped sacs, or cisternae, which are often branched into a complex series of tubules at their margins (Figure 2–21). Those in higher plant cells usually have four to eight cisternae stacked together.

Customarily the two poles or surfaces of a stack are referred to as forming and maturing faces. The membranes of the forming cisternae are structurally similar to membranes of the endoplasmic reticulum, while those toward the opposite pole, or maturing face, become progressively more like plasma membrane (Figure 2–22).

Dictyosomes are involved in secretion, and most higher plant dictyosomes are involved in cell wall synthesis. Cell wall polysaccharides synthesized by the dictyosomes collect in vesicles that are pinched off cisternae on the maturing faces. These secretory vesicles migrate to and fuse with the plasma membrane (Figure 2–22). The vesicles then discharge their contents to the cell exterior, and their polysaccharide contents become part of the cell wall. The products secreted by the dictyosomes are not always synthesized entirely by the dictyosomes. Some products are formed in part by other cellular components, such as the endoplasmic reticulum, and then are transferred to the dictyosomes where they are modified prior to secretion. A good example of such products are glycoproteins—carbohydrate-protein compounds—that are important cell wall constituents. The protein portion is synthesized on polysomes of rough endoplasmic reticulum, and the carbohydrate portion is synthesized in the dictyosome where the glycoprotein is then assembled.

The Endomembrane Concept

Membranes are dynamic, mobile structures that continually change their shape and surface area. An example of the mobility of cellular membranes is provided by the concept of the *endomembrane system.* According to this concept, the internal cytoplasmic membranes (with the exception of mitochondrial and plastid membranes) constitute a continuum, with the endoplasmic reticulum being the initial source of membranes. New cisternae are added to the dictyosomes from the endoplasmic reticulum via transition vesicles, while secretory vesicles derived from the dictyosomes eventually contribute to growth of the plasma membrane (Figure 2–22). Hence, the endoplasmic reticulum and dictyosomes are integrated into a functional unit, in which the dictyosomes act as the main vehicles for the transformation of endoplasmic-reticulum-like membranes into plasma-membrane-like membranes.

It is important to note that even in tissues undergoing very little growth or cell division, a turnover of membrane components is continually taking place. Although the membranes persist, their constituent molecules are continuously replaced.

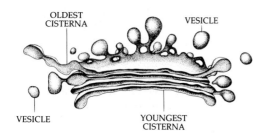

OLDEST CISTERNA   VESICLE

VESICLE   YOUNGEST CISTERNA

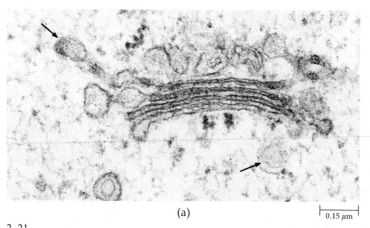

(a)   0.15 μm

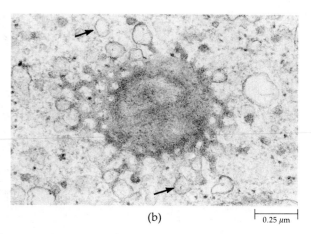

(b)   0.25 μm

2–21

The dictyosome consists of a group of flat, membranous sacs associated with vesicles that apparently bud off from the sacs. It serves as a "packaging center" for the cell and is concerned with secretory activities in eukaryotic cells. In (a) the cisternae of a dictyosome from a stem parenchyma cell of the horsetail Equisetum hyemale are shown in sectional view, whereas (b) shows a single cisterna in a surface view. The arrows indicate numerous secretory vesicles along the margins of the cisternae.

2-22

A diagrammatic representation of the endomembrane concept. This drawing illustrates the idea that new membranes are synthesized at the rough endoplasmic reticulum. Small transition vesicles pinch off the smooth surface of the endoplasmic reticulum and carry membranes and enclosed materials to the forming face of the dictyosome. Secretory vesicles derived from cisternae at the maturing face of the dictyosome migrate to the plasma membrane and fuse with it, contributing new membrane to the plasma membrane and discharging their contents in the region of the wall.

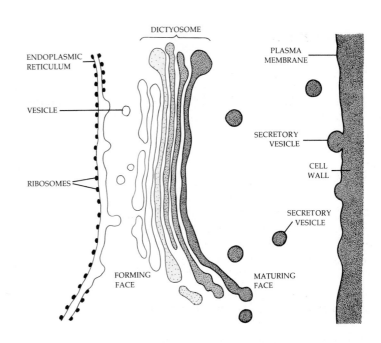

MICROTUBULES

Microtubules, which are found in virtually all eukaryotic cells, are long, thin cylindrical structures about 24 nanometers in diameter and of varying lengths. Each microtubule is built up of subunits of the protein called tubulin. The subunits are arranged in a helix to form 13 vertical filaments around a hollow core. Microtubules are dynamic structures that undergo regular sequences of breakdown and re-formation at specific points in the cell cycle (see Figure 2-45, page 43). Their assembly takes place at specific sites called microtubule organizing centers, which, in plant cells, have a poorly defined amorphous structure.

Microtubules have many functions. In enlarging and differentiating cells, microtubules just inside the plasma membrane are believed to be involved in the orderly growth of the cell wall, especially through their control of the alignment of the cellulose microfibrils that are deposited by the cytoplasm in adding to the cell wall (Figure 2-23). The direction of expansion of the cell is governed, in turn, by the alignment of cellulose microfibrils in the cell wall. Microtubules also serve to direct dictyosome vesicles toward the developing wall, comprise the spindle fibers, which form in dividing cells, and apparently play a role in the formation of the cell plate (the initial partition between dividing cells). In addition, microtubules are important components of flagella and cilia and apparently are involved in the movement of these structures.

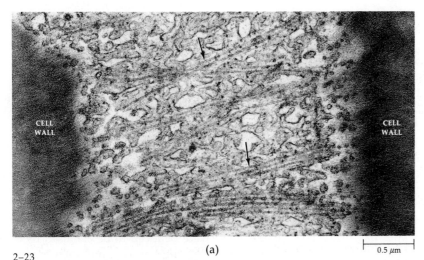

(a)

0.5 µm

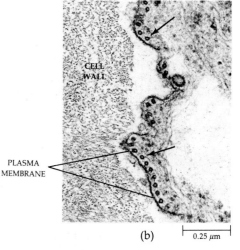

(b)

0.25 µm

2-23

Microtubules (indicated by arrows) seen in (a) longitudinal and (b) transverse views in leaf cells of Botrychium virginianum, a fern. A glancing section more or less parallel to and just inside the wall and plasma membrane is shown in (a). In (b) the microtubules can be seen to be separated from the wall by the plasma membrane.

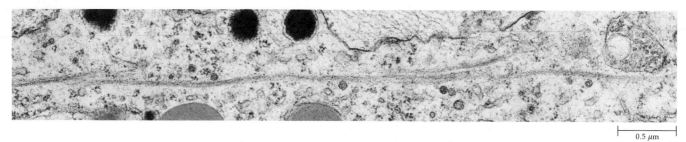

|———| 0.5 μm

MICROFILAMENTS

Microfilaments, like microtubules, are found in virtually all eukaryotic cells. They are contractile proteins composed of actin similar to that of muscle, and they occur as long filaments 5 to 7 nanometers thick. Bundles of microfilaments have been found in many cells of higher plants (Figure 2–24). It is believed that these bundles play a causative role in cytoplasmic streaming. Together the microfilaments and microtubules form a flexible framework called the *cytoskeleton.*

GROUND SUBSTANCE

Until recently, the ground substance of the cell was regarded as a homogeneous, protein-rich solution with little or no structure. However, recent studies of animal cells indicate that the ground substance possesses an intricate structure and behavior of its own. When viewed with the high-voltage (million-volt) electron microscope, the ground substance can be seen to consist of a three-dimensional lattice of slender strands, 3 to 6 nanometers in diameter, pervading the entire cell. Other cytoplasmic components, including microtubules and microfilaments, are suspended by this *microtrabecular lattice.*

The microtrabecular lattice divides the cell into two phases, a protein-rich phase (the strands of the lattice) and a fluid, water-rich phase filling the spaces of the lattice. With its high water content the lattice has a gel-like consistency; indeed, it is a living gel.

The microtrabecular lattice is believed to provide attachment points for organelles, make communication between parts of the cell possible, and give direction to intracellular transport. There are also indications that plant cells possess a microtrabecular lattice.

LIPID DROPLETS

Lipid droplets are more or less spherical structures that impart a granular appearance to the cytoplasm of a plant cell when viewed with the light microscope. In electron micrographs, the oil droplets have an amorphous appearance (Figure 2–25). Apparently similar but usually smaller droplets are common in plastids (Figure 2–7a).

2–24
A bundle of microfilaments in a leaf cell of the staghorn fern (Platycerium bifurcatum).

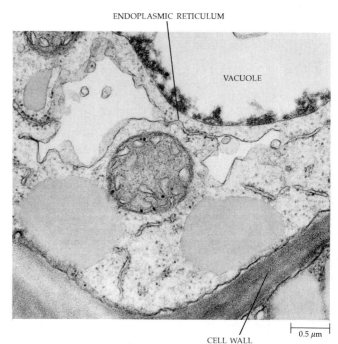

ENDOPLASMIC RETICULUM

VACUOLE

CELL WALL |———| 0.5 μm

2–25
Cytoplasmic components present in a parenchyma cell from the thickened stem, or corm, of the quillwort Isoetes muricata. *Two lipid droplets can be seen below, one on either side of the mitochondrion near the center of this electron micrograph. Above, and on either side of the mitochondrion, a cisterna of the endoplasmic reticulum appears distended. Some vacuoles are thought to arise from the endoplasmic reticulum in this way. Part of a vacuole can be seen at the top of the micrograph. The dense material lining this vacuole is tannin.*

Lipid droplets, originally described as organelles called spherosomes, were believed to be enclosed by a single unit membrane or one-half of a unit membrane. Recent data indicate that the oil droplets are not surrounded by membrane, although they may be coated with protein.

ERGASTIC SUBSTANCES

Ergastic substances are passive products of the protoplast; some are storage products, others are waste products. These substances may appear and disappear at different times in the life of a cell. We have already discussed several examples of ergastic substances: starch grains, crystals, anthocyanin pigments, and oil droplets. Other examples are resins, gums, tannins, and protein bodies. Ergastic substances are found in the cell wall, in the cytoplasmic ground substance, and in organelles, including vacuoles.

FLAGELLA AND CILIA

Flagella and *cilia* (singular: flagellum and cilium) are hairlike structures that extend from the surface of many different types of eukaryotic cells. They are relatively thin and constant in diameter (about 0.2 micrometer), but they vary in length from about 2 to 150 micrometers. By convention, the ones that are longer or are present alone or in small numbers are usually referred to as flagella, whereas the shorter ones or those occurring in greater numbers are called cilia. There is no definite distinction between these two types of organelles, however, and we will use the term *flagellum* to refer to both.

In some algae and fungi, flagella are locomotor organs, propelling the organisms through the water. In plants, these organelles are found only in the sex cells (gametes) and then only in plants that have motile sperm, such as mosses, liverworts, ferns, and certain gymnosperms. Some flagella, called tinsel flagella, bear one or two rows of minute, lateral appendages, whereas others, called whiplash flagella, lack such processes (Figure 2–26).

One of the most intriguing of the discoveries made possible by the electron microscope is the internal structure of flagella. Each flagellum has a precise inter-

nal organization (Figure 2–27). An outer ring of nine pairs of microtubules surrounds two additional microtubules in the center of the flagellum proper. Enzyme-containing "arms" extend from one of the microtubules of each outer pair. This basic 9-plus-2 pattern of organization is found in all flagella of eukaryotic organisms.

TINSEL FLAGELLUM

WHIPLASH FLAGELLUM

2.5 μm

2–26
The two types of flagella—tinsel and whiplash—are represented in this single cell of the colonial organism Synura petersenii, *a golden alga. This organism has a long tinsel flagellum (left) and a somewhat shorter whiplash flagellum (right).*

2–27
(a) *Longitudinal section of a flagellum of a gamete of the green alga* Ulvaria. *Notice that the membrane surrounding the flagellum proper is continuous with the plasma membrane.* (b) *Transverse section through the flagellum proper of the* Ulvaria *flagellum, showing the typical 9-plus-2 structure.* (c) *Transverse section through the basal body of the* Ulvaria *flagellum. Notice that the basal body has a circle of nine triplet microtubules and no central microtubules.*

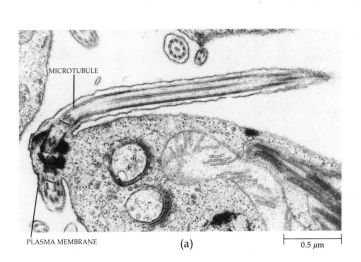

MICROTUBULE

PLASMA MEMBRANE   (a)   0.5 μm

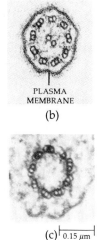

PLASMA MEMBRANE

(b)

(c)   0.15 μm

The movement of flagella and cilia originates from within the structures themselves; flagella and cilia are capable of movement after they are detached from cells. Some investigators believe that the flagellar movement is based on a sliding microtubule mechanism, in which the outer pairs of microtubules move past one another without contracting. As the pairs of microtubules slide past one another, their movement causes localized bending of the flagellum. Sliding of the pairs of microtubules is thought to result from cycles of attachment and detachment of the "arms" on one of the microtubules to the nearer microtubule of the adjacent outer pair.

Flagella grow out of cylinder-shaped structures in the cytoplasm known as *basal bodies,* which then form the basal portion of the flagellum (Figure 2–27). Basal bodies have an internal structure that resembles the structure of the flagellum proper, except that the outer tubules in the basal body occur in triplets rather than in pairs, and the two central tubules are absent.

THE CELL WALL

The presence of a cell wall, above all other characteristics, distinguishes plant cells from animal cells. Its presence is the basis of many of the characteristics of plants as organisms. The cell wall limits the size of the protoplast and prevents it from being ruptured due to enlargement from the intake of water by the vacuole.

Once regarded as merely an outer, inactive product of the protoplast, the cell wall is now recognized as having specific functions that are essential not only to the existence of the cell and the tissue in which it is found but also to the entire plant. Cell walls play important roles in the absorption, transport, and secretion of substances in plants, and they may also serve as sites of lysosomal, or digestive, activity.

Components of the Cell Wall

The most characteristic component of plant cell walls is *cellulose,* which largely determines their architecture. Cellulose is made of repeating molecules of glucose attached end to end (see Chapter 3). These long, thin cellulose molecules are united into *microfibrils* about 10 to 25 nanometers wide (Figure 2–28). Cellulose has crystalline properties (Figure 2–29) because of the orderly arrangement of cellulose molecules in certain parts, the *micelles,* of the microfibrils (Figure 2–30). The microfibrils wind together to form fine threads that may coil around one another like strands in a cable. Each "cable," or *macrofibril,* measures about 0.5 micrometer in width and may reach 4 micrometers in length. Cellulose molecules wound in this fashion are as strong as an equivalent thickness of steel.

The cellulose framework of the wall is interpenetrated by a cross-linked *matrix* of noncellulosic molecules. Some of these are polysaccharides called *hemicelluloses* and others are *pectic substances,* or *pec-*

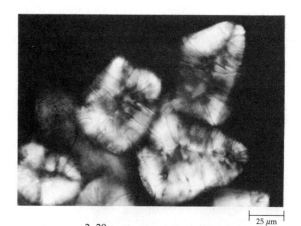

| | 25 μm |

2–29

Stone cells (sclereids) from the flesh of a pear (Pyrus communis) *seen in polarized light. Clusters of such stone cells are responsible for the gritty texture of this fruit. The stone cells have very thick secondary walls traversed by numerous simple pits, which appear as lines in the walls. The walls appear bright in polarized light because of the crystalline properties of their principal component, cellulose.*

| | 0.5 μm |

2–28

Surface of the cell wall of the algal cell Chaetomorpha, *showing cellulose microfibrils, each of which is made up of hundreds of molecules of cellulose.*

2–30

The detailed structure of a cell wall. (a) Portion of wall showing the middle lamella, primary wall, and three layers of secondary wall. Cellulose, the principal component of the cell wall, exists as a system of fibrils of different sizes. (b) The *largest fibrils, macrofibrils, can be seen with the light microscope. (c) With the aid of an electron microscope, the macrofibrils can be resolved into microfibrils about 100 Å wide. (d) Parts of the microfibrils, the micelles, are arranged in* *an orderly fashion and impart crystalline properties to the wall. (e) A fragment of a micelle shows parts of the chainlike cellulose molecules in a lattice arrangement.*

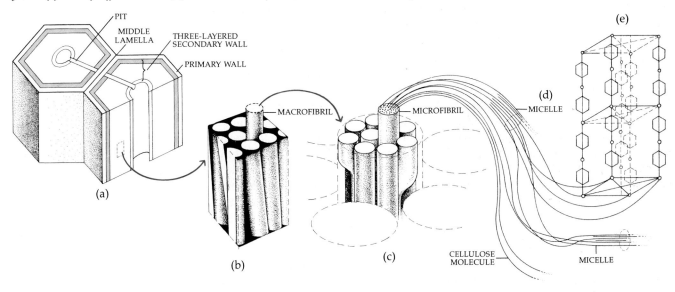

tins, which are closely related chemically to hemicelluloses (Figure 2–31).

Another important constituent of the walls of many kinds of cells is *lignin,* which, aside from cellulose, is the most abundant polymer found in plants. Physically, lignin is rigid, and it serves to add rigidity to the wall; it is commonly found in the walls of plant cells that have a supporting or mechanical function.

Cutin, *suberin,* and *waxes* are fatty substances commonly found in the walls of the outer, protective tissues of the plant body. Cutin, for example, is found in the walls of the epidermis, and suberin is found in those of the secondary protective tissue, cork. Both substances occur in combination with waxes and function largely to reduce water loss from the plant (see page 50).

Cell Wall Layers

Plant cell walls vary greatly in thickness, depending partly on the role the cells play in the structure of the plant and partly on the age of the individual cell. Developmental studies, coupled with the use of the electron microscope, polarized light, and X-rays, indicate that there are two layers in all plant cell walls: the *middle lamella* (also called intercellular substance) and the *primary wall*. In addition, many cells deposit another wall layer, the *secondary wall*. The middle lamella occurs between the primary walls of adjacent cells; the secondary wall, if present, is laid down by the protoplast of the cell on the inner surface of the primary wall (Figure 2–32).

2–31

Schematic diagram showing how the cellulose microfibrils and matrix components (pectins, hemicelluloses, and glycoproteins) of the primary wall may be interconnected. Hemicellulose molecules are linked to the surface of the cellulose microfibrils by hydrogen bonds. Some of these hemicellulose molecules are cross-linked in turn to acidic pectin molecules by neutral pectin molecules. The glycoproteins are probably attached to the pectin molecules.

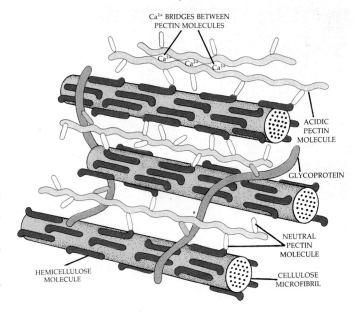

The Middle Lamella

The middle lamella is composed mainly of pectic substances. Frequently, it is difficult to distinguish the middle lamella from the primary wall, especially in cells that develop thick secondary walls. When a cell wall becomes lignified, this process characteristically begins in the middle lamella and then spreads to the primary wall and finally to the secondary wall.

The Primary Wall

The cellulosic wall layer deposited before and during the growth of the cell is called the primary wall. In addition to cellulose, hemicelluloses, and pectin, primary walls contain glycoprotein (Figure 2–31). Primary walls may also become lignified. The pectic component imparts plastic properties to the wall, which make it possible for the primary wall to be stretched permanently during elongation of a root, stem, or leaf.

Actively dividing cells commonly contain only primary walls, as do most mature cells involved with such metabolic processes as photosynthesis, respiration, and secretion. These cells—that is, cells with primary walls and living protoplasts—are able to lose their specialized cellular form, divide, and differentiate into new types of cells. For this reason, it is principally cells with only primary walls that are involved in wound healing and regeneration in the plant.

Usually, primary cell walls are not of uniform thickness throughout but have thin areas that are called *primary pit-fields* (Figure 2–32). Cytoplasmic threads, or plasmodesmata, which connect the living protoplasts of adjacent cells, are commonly aggregated in the primary pit-fields.

The Secondary Wall

Although many plant cells have only a primary wall, in others a secondary wall is deposited by the protoplast inside the primary wall. This occurs mostly after the cell has stopped growing and the primary wall is no longer enlarging in area. It is partly because of this that the secondary wall is structurally distinct from the primary wall. Secondary walls are particularly important in specialized cells that have functions such as strengthening and conduction of water; in these cells, the protoplast often dies after the secondary wall has been laid down. Cellulose is more abundant in secondary walls than in primary walls and pectic substances are lacking; the secondary wall is therefore rigid and not readily stretched. Glycoproteins, which are relatively abundant in primary cell walls, are absent in secondary cell walls. The matrix of the secondary wall is composed of hemicellulose.

Frequently, three distinct layers—designated S_1, S_2, and S_3, for the outer, middle, and inner layer, respectively—can be distinguished in the secondary wall (Figure 2–33). The layers differ from one another in the orientation of their cellulose microfibrils. Such multiple-wall layers are found in certain cells of the secondary xylem, or wood. The laminated structure of such secondary walls—like that seen in plywood—greatly increases the strength of the wall. The cellulose microfibrils are laid down in a denser pattern than in the primary wall. Lignin is common in the secondary walls of cells found in wood.

When the secondary cell wall is deposited, it is not laid down over the primary pit-fields of the primary wall. Consequently, characteristic depressions, or *pits*, are formed in the secondary wall (Figure 2–32). In

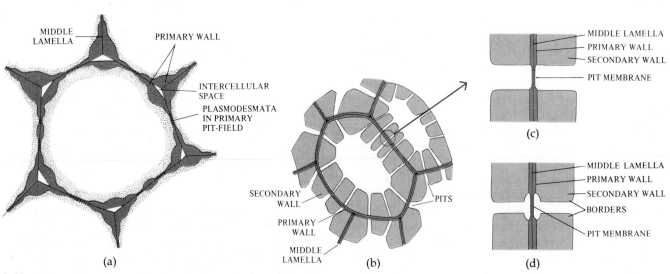

2–32

Primary pit-fields, pits, and plasmodesmata. (a) Parenchyma cell with primary walls and primary pit-fields, thin areas in *the walls. As shown here, plasmodesmata commonly traverse the wall at the primary pit-fields. (b) Cells with* *secondary walls and numerous simple pits. (c) A simple pit-pair. (d) A bordered pit-pair.*

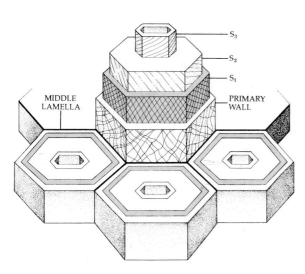

2–33
Diagram showing microfibril organization in primary walls and in three layers (S_1, S_2, S_3) of secondary wall.

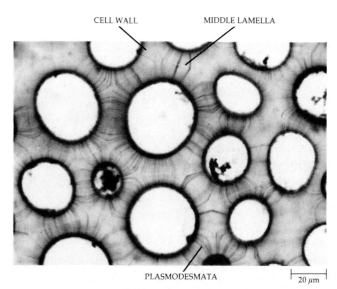

2–34
Light micrograph of plasmodesmata in the thick primary walls of persimmon (Diospyros) endosperm, the nutritive tissue within the seed. The plasmodesmata appear as fine lines extending from cell to cell across the walls.

some instances, pits are also formed in areas where there are no primary pit-fields.

A pit in a cell wall usually occurs opposite a pit in the wall of the cell with which it is in contact. The middle lamella and two primary walls between the two pits are called the *pit membrane*. The two opposite pits plus the membrane constitute a *pit-pair*. Two principal types of pits are found in cells that have secondary walls: *simple* and *bordered*. In bordered pits, the secondary wall arches over the *pit cavity*. In simple pits, there is no overarching.

Growth of the Cell Wall

Cell walls grow both in thickness and in surface area. Extension of the wall is a complex process under the close biochemical control of the protoplast. It requires loosening of wall structure—a phenomenon regulated by the hormone auxin (see Chapter 24)—as well as an increase in protein synthesis, respiration, and uptake of water by the cell. Most of the new microfibrils are placed on top of those previously formed (layer upon layer), although some may be inserted into the existing wall structure.

In cells that enlarge more or less uniformly in all directions, the microfibrils are laid down in a random array, forming an irregular network. Such cells are found in the pith of stems, in storage tissues, and in tissue cultures. In contrast, in elongating cells, the microfibrils of the side walls are deposited in a plane at right angles (transverse) to the axis of elongation. As the wall increases in surface area, the orientation cel of the outer microfibrils becomes more nearly longitudinal, or parallel to the long axis of the cell.

The matrix substances—pectic substances and hemicelluloses—are carried to the wall in dictyosome vesicles, as are the glycoproteins. The type of matrix substance synthesized and secreted by a cell at any given time depends on the stage of development. For example, pectins are more characteristic of enlarging cells, while hemicelluloses predominate in cells that are no longer enlarging.

Cellulose synthesis is poorly understood, but it is now quite clear that (1) cellulose microfibrils are synthesized at the surface of the cell by an enzyme complex embedded in the plasma membrane, and that (2) the orientation of the microfibrils is controlled by microtubules lying just inside the plasma membrane.

PLASMODESMATA

The protoplasts of adjacent plant cells are connected with one another by fine threads of cytoplasm known as *plasmodesmata* (singular: plasmodesma). Although such structures have long been seen with the light microscope (Figure 2–34), they were difficult to interpret. It was not until they could be observed with an electron microscope that their nature was confirmed (Figure 2–35).

Plasmodesmata may occur throughout the cell wall, or they may be aggregated in primary pit-fields or the membranes between pit-pairs. With the electron microscope, plasmodesmata appear as narrow canals (about 30 to 60 nanometers in diameter) lined by a plasma membrane and traversed by a tubule of endoplasmic reticulum known as the *desmotubule*. Many plasmodesmata are formed during cell division as strands of tubu-

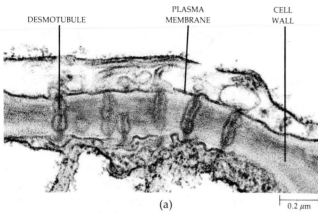

DESMOTUBULE   PLASMA
                 MEMBRANE   CELL
                                   WALL

(a)

0.2 μm

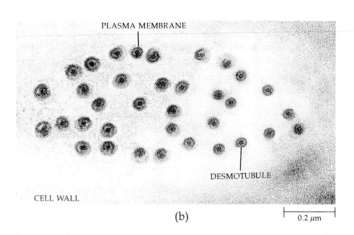

PLASMA MEMBRANE

DESMOTUBULE

CELL WALL

(b)

0.2 μm

2–35
Electron micrograph of (a) plasmodesmata, in longitudinal view, connecting the protoplasts of two corn (Zea mays) leaf cells. (b) Transverse view of plasmodesmata in primary pit-field of corn leaf cell.

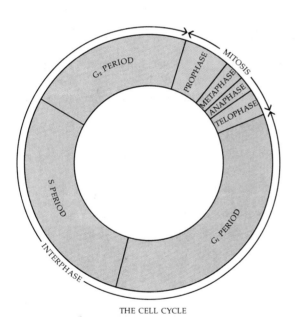

G₂ PERIOD

PROPHASE

MITOSIS

METAPHASE

ANAPHASE

TELOPHASE

S PERIOD

G₁ PERIOD

INTERPHASE

THE CELL CYCLE

2–36
The cell cycle can be divided into mitosis and interphase, the phase between successive mitotic divisions. Cells usually spend most of their time in interphase, which can be divided into three periods. The first of these (G₁) is a period of general growth and replication of cytoplasmic organelles. During the second period (S), DNA synthesis takes place. The third (G₂) is a period during which structures directly associated with mitosis, such as the spindle fibers, are formed. After the G₂ period, mitosis begins. Mitosis is divided into four phases.

lar endoplasmic reticulum become trapped within the developing cell plate (see Figure 2–44). Plasmodesmata are also formed in walls of nondividing cells. These structures are thought to provide an effective pathway for the transport of certain substances between cells (see page 66).

CELL DIVISION

In one-celled organisms, cells grow by absorbing substances from the environment and using these materials to synthesize new structural and functional molecules. When such a cell reaches a certain size, it divides. The two daughter cells, each about half the size of the original mother cell, then begin growing again. In a one-celled organism, cell division may occur every day or even every few hours, producing a succession of identical organisms. In many-celled organisms, cell division, together with cell enlargement, is the means by which the organism grows. In all of these instances, the new cells that are thus produced are structurally and functionally similar to both the parent cell and one another.

Cell division in eukaryotes consists of two overlapping stages: *mitosis* and *cytokinesis*. Mitosis is the process by which a nucleus gives rise to two daughter nuclei, each morphologically and genetically equivalent to the other. Cytokinesis involves the division of the cytoplasmic portion of a cell and the separation of daughter nuclei into separate cells.

THE CELL CYCLE

Dividing cells pass through a regular sequence of events known as the cell cycle (Figure 2–36). Completion of the cycle requires varying periods of time, depending on both the type of cell and external factors, such as temperature or availability of nutrients. Commonly, the cycle is divided into *interphase* and the four phases of *mitosis*.

Interphase

Interphase, the phase between successive mitotic divisions, was once regarded as the "resting phase" in the cell cycle. Nothing could be further from the truth, however, for interphase is a period of intense cellular activity.

Interphase can be divided into three periods, which are designated G_1, S, and G_2 (Figure 2–36). The G_1 period (G stands for gap) occurs after mitosis and is primarily a time of growth of the cytoplasmic material, including the various organelles. Also during the G_1 period, according to a current hypothesis, substances are synthesized that either inhibit or stimulate the S period and the rest of the cycle, thus determining whether cell division will actually occur.

The S (synthesis) period follows the G_1 period. During the S period the genetic material (DNA) is duplicated. During the G_2 period that follows, structures directly involved with mitosis, such as the components of the spindle fibers, are formed.

Some cells pass through successive cell cycles repeatedly. This group includes the one-celled organisms and certain cells in regions of active growth (meristems). Some specialized cells lose their capacity to replicate once they are mature. A third group of cells, such as those that form wound tissue (callus), retains the capacity to divide but does so only under special circumstances.

Mitosis

Mitosis, or nuclear division, is a continuous process that is conventionally divided into four major phases: *prophase, metaphase, anaphase,* and *telophase* (Figures 2–37 through 2–39). These four phases make up the process by which the genetic material duplicated during interphase is divided equally between two daughter nuclei.

Preprophase Band

One of the earliest signs that the cell is going to divide is the appearance of a narrow, ringlike band of microtubules just beneath the plasma membrane. (See the boxed essay on page 41.) This relatively dense band of microtubules encircles the nucleus in a plane corresponding to the equatorial plane of the future mitotic spindle (Figure 2–45). Inasmuch as it appears just before prophase, it is called the *preprophase band*. The preprophase band disappears after initiation of the mitotic spindle, long before initiation of the cell plate in late telophase. Yet, as the cell plate forms, it grows outward to fuse with the mother cell wall precisely at the zone previously occupied by the preprophase band.

Prophase

At the beginning of *prophase*, the chromosomes gradually become visible as elongated threads scattered throughout the nucleus. As prophase advances, the

2–37

Mitosis, a diagrammatic representation with four chromosomes. During early prophase, the chromosomes become visible as long threads scattered throughout the nucleus; the chromosomes shorten and thicken until each can be seen to consist of two threads (chromatids); finally, the nucleolus and nuclear membrane disappear.

The appearance of the spindle marks the beginning of metaphase, during which the chromosomes migrate to the equatorial plane of the spindle. At full metaphase (shown here) the centromeres of the chromosomes lie in that plane.

Anaphase begins as the centromeres divide and separate, providing each of the sister chromatids, which now are called daughter chromosomes, with a centromere. As shown here, the daughter chromosomes then move to opposite poles of the spindle.

Telophase—more or less the reverse of prophase—begins when the daughter chromosomes have completed their migration.

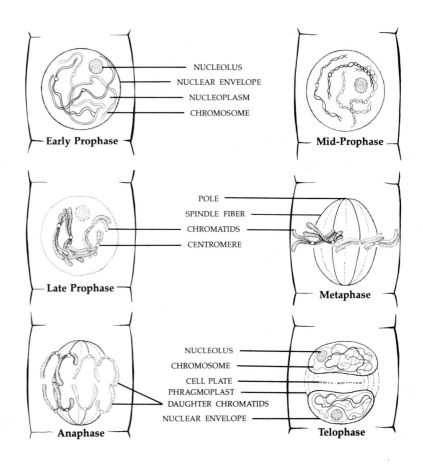

NUCLEOLUS
NUCLEAR ENVELOPE
NUCLEOPLASM
CHROMOSOME

Early Prophase

Mid-Prophase

Late Prophase

POLE
SPINDLE FIBER
CHROMATIDS
CENTROMERE

Metaphase

Anaphase

NUCLEOLUS
CHROMOSOME
CELL PLATE
PHRAGMOPLAST
DAUGHTER CHROMATIDS
NUCLEAR ENVELOPE

Telophase

threads shorten and thicken, and as the chromosomes become more distinct, each can be seen to be composed of not one but two threads, called _chromatids_, coiled about one another. By late prophase, after further shortening, the two duplicate chromatids of each chromosome lie side by side and nearly parallel, joined to one another in a narrowed area known as the _centromere_ (Figure 2–40). The centromere, which has a characteristic location on any one chromosome, divides the chromosome into two arms of various lengths.

During prophase, a _clear zone_ appears around the nuclear envelope. Microtubules arise in this zone. The microtubules are initially randomly oriented, but by late prophase they are aligned parallel to the nuclear surface along the spindle axis. This is the earliest manifestation of mitotic spindle assembly.

Toward the end of prophase, the nucleolus gradually becomes indistinct and finally disappears. Shortly afterward, the nuclear envelope breaks down, marking the end of prophase (Figure 2–41).

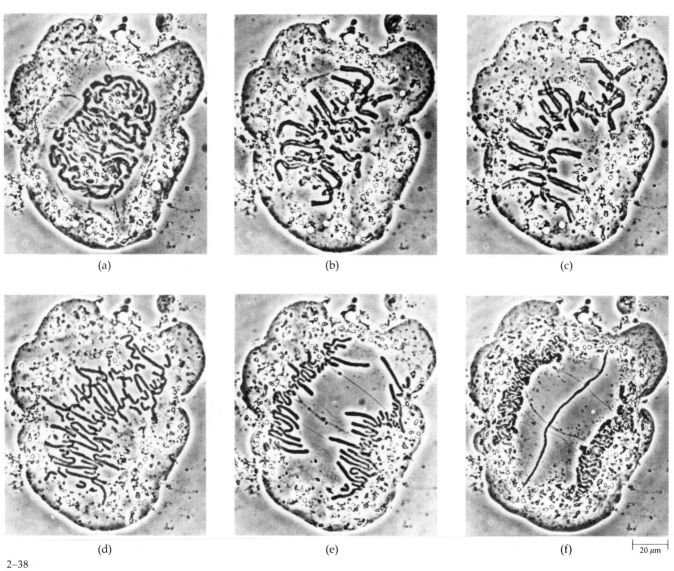

(a)　　　　　(b)　　　　　(c)

(d)　　　　　(e)　　　　　(f)　　　　　⊢———⊣ _20 μm_

2–38

Mitosis as seen with phase-contrast optics in a living cell of the African blood lily (Haemanthus katherinae). _The spindle is barely discernible in these cells, which have been flattened in order to show all of the chromosomes more clearly._ (a) _Late prophase: The chromosomes have condensed. A clear zone has developed around the nucleus._ (b) _Late_ prophase–early metaphase: _The nuclear envelope has disappeared and the ends of some of the chromosomes are protruding into the cytoplasm._ (c) _Metaphase: The chromosomes are arranged with their centromeres in the equatorial plane._ (d) _Mid-anaphase: The sister chromosomes (now called daughter chromosomes) have_ separated and are moving to opposite poles of the spindle. (e) _Late anaphase._ (f) _Telophase-cytokinesis: The daughter chromosomes have reached the opposite poles and the two chromosome masses have begun the formation of two daughter nuclei. Cell plate formation is nearly completed._

Metaphase

Metaphase begins as the *spindle*, a three-dimensional structure that is widest in the middle and tapers to a point at each pole, occupies the area formerly occupied by the nucleus. The spindle consists of *spindle fibers,* which are bundles of microtubules (Figure 2–42). During metaphase, the chromosomes—each consisting of two chromatids—become arranged so that each centromere lies on the equatorial plane of the spindle. Each chromosome appears to be attached to the spindle fibers by its centromere (Figure 2–42). Some of the spindle fibers pass from one pole to the other and have no chromosome attached to them. Whether the chromosomes are randomly arranged on the equatorial plane or have their centromeres in a fixed order is a matter of considerable current interest.

When the chromosomes have all moved to the equatorial plane, the cell has reached full metaphase. The chromatids are now in position to separate.

Anaphase

During *anaphase* (Figure 2–38d, e), the chromatids of each chromosome separate from one another; they are now called *daughter chromosomes*. First, the centromere divides and the two daughter chromosomes move away from one another toward opposite poles. Their centromeres, which still appear to be attached to the spindle fibers, move first, and the arms drag behind. The two daughter chromosomes pull apart, with the tips of the longer arms separating last.

The spindle fibers attached to the chromosomes shorten, apparently causing the chromatids to separate and the daughter chromosomes to move apart. It has been suggested that the microtubules are continuously formed at one end of the spindle fiber and disassembled at the other. In the process, the spindle fibers tug the daughter chromosomes toward the poles by their centromeres.

By the end of anaphase, the two identical sets of daughter chromosomes have been separated and moved to opposite poles.

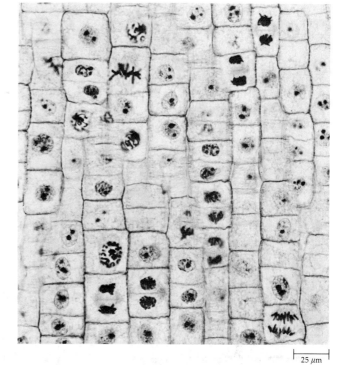

2–39

Photomicrograph of dividing cells in an onion (Allium) root tip. By comparing the phases of mitosis illustrated in Figures 2–37 and 2–38 to these cells, you should be able to identify the mitotic phases shown here.

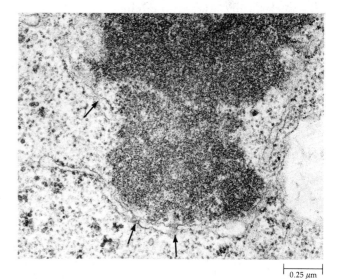

2–41

A portion of a late prophase nucleus of a parenchyma cell in a horsetail (Equisetum hyemale) stem. The arrows point to nuclear pores in fragments of the nuclear envelope, which is in the process of breaking down.

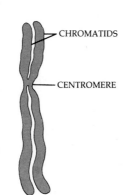

CHROMATIDS

CENTROMERE

2–40

A diagram of a prophase chromosome. The chromosomal DNA was replicated during interphase, so that each chromosome now consists of two identical parts, called chromatids, which lie parallel to each other. The chromatids are joined together at their centromere.

2–42

Portion of a chromosome at metaphase in a stem parenchyma cell of a horsetail (Equisetum hyemale). *Some microtubules (components of the spindle fibers) lead directly into the dense chromosomal material at a centromere. Other microtubules (not shown) extend from pole to pole.*

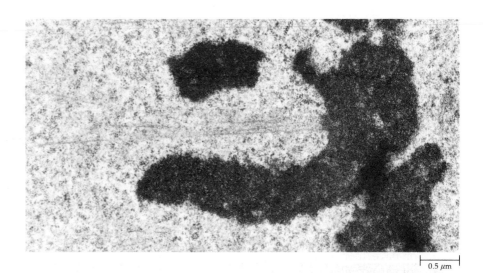

0.5 μm

Telophase

During *telophase* (Figure 2–38f) the separation of the two identical sets of chromosomes is made final as nuclear envelopes are organized around each of them (Figure 2–43). The membranes of these nuclear envelopes are derived from rough endoplasmic reticulum. The spindle apparatus disappears. In the course of telophase, the chromosomes become increasingly indistinct, elongating to become slender threads again. Nucleoli also re-form at this time. When these processes are completed and the chromosomes have once more disappeared from view, mitosis is completed and the two daughter nuclei have entered interphase.

During mitosis, the two daughter nuclei that are produced are genetically equivalent to one another and to the nucleus that divided to produce them. This is important, for the nucleus is the control center of the cell,

as is described in more detail in Chapter 8. It contains coded instructions that specify the production of proteins, many of which mediate cellular processes by acting as enzymes and some of which serve directly as structural elements in the cell. This hereditary blueprint must be exactly passed on to the daughter cells, and its exact duplication is ensured in eukaryotic organisms by the organization of chromosomes and their division in the process of mitosis.

The duration of mitosis varies with the tissue and the organism involved. However, prophase is the longest phase and anaphase is the shortest. In a root tip, the relative lengths of time for each of the four phases may be as follows: prophase, 1 to 2 hours; metaphase, 5 to 15 minutes; anaphase, 2 to 10 minutes; and telophase, 10 to 30 minutes. By contrast, the duration of interphase may range from 12 to 30 hours.

2–43

In plant cells, separation of the daughter chromosomes is followed by formation of a cell plate, which completes the separation of the dividing cells. Here numerous Golgi vesicles can be seen fusing in an early stage of cell-plate formation. The two groups of chromosomes on either side of the developing cell plate are at telophase. Arrows point to portions of the nuclear envelope reorganizing around the chromosomes.

CHROMOSOMES   CELL PLATE   CHROMOSOMES

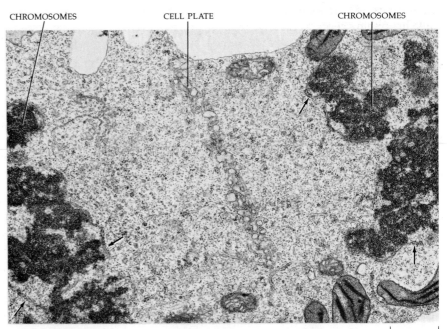

1 μm

IMMUNOFLUORESCENCE MICROSCOPY OF TUBULIN AND MICROTUBULES

Immunofluorescence microscopy has been used to reveal the location of microtubules at different stages of cell division. Unlike other procedures, it has the capacity to provide three-dimensional views. Here root-tip cells of onion (*Allium cepa*) were stained with specially prepared fluorescence antibodies to tubulin and photographed by fluorescence microscopy. The fluorescent antibodies, which became attached to the protein tubulin, revealed the location of the microtubules. (a) A preprophase band of microtubules (horizontal arrows) encircles the cell just before prophase.

Other microtubules (arrowheads) outline the nuclear envelope (not visible). (b) The preprophase band has disappeared and the area it previously occupied (arrows) appears as a nonfluorescing band. Microtubules of the spindle extend from this region to the spindle poles. (c) In early telophase, after the chromosomes have separated, new microtubules form a phragmoplast, in which cell plate formation takes place. (d) As cell plate formation continues, the microtubules become concentrated where the cell plate is extending outward toward the walls of the dividing cell.

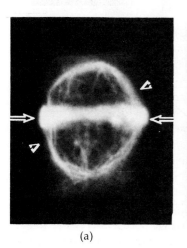

(a)

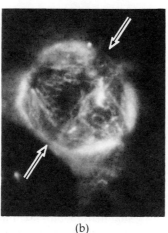

(b)

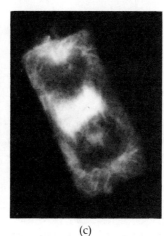

(c)

(d)

Organization of the Mitotic Spindle

As mentioned previously, the assembly of microtubules takes place at specific sites known as microtubule organizing centers. In many eukaryotic cells, the microtubule organizing centers responsible for the formation of the mitotic spindle are associated with structures called *centrioles*. Until recently, it was believed that the centrioles themselves were the microtubule organizing centers. It is now clear, however, that the organizing centers are poorly defined amorphous structures at the poles of the spindles.

Cytokinesis

As noted previously, *cytokinesis* is the process by which the cytoplasm is divided. In most organisms, cells divide by ingrowth of the cell wall, if present, and constriction of the plasma membrane, a process that cuts through the spindle fibers. In all plants (bryophytes and vascular plants) and in a few algae, cell division occurs by the formation of a *cell plate* (Figures 2–43 through 2–45).

In early telophase, an initially barrel-shaped system of fibrils called the *phragmoplast* forms between the two daughter nuclei. The fibrils of the phragmoplast, like those of the mitotic spindle, are composed of microtubules. As seen with the light microscope, small droplets appear across the equatorial plane of the phragmoplast and gradually fuse, forming the cell plate. The cell plate grows outward until it reaches the wall of the dividing cell, completing the separation of the two daughter cells. With the electron microscope, the fusing droplets can be seen to be vesicles derived from the Golgi apparatus. The vesicles presumably contain pectic substances that form the middle lamella and contribute their membranes to the formation of the plasma membrane on either side of the plate. Plasmodesmata are formed at this time, as segments of tubular endoplasmic reticulum are "caught" between fusing vesicular contents (Figure 2–44).

Following the formation of the middle lamella, each protoplast deposits a primary wall next to the middle lamella. In addition, each daughter cell deposits a new wall layer around the entire protoplast; this new wall is continuous with the wall at the cell plate (Figure 2–45). The original wall of the parent cell stretches and ruptures as the daughter cells enlarge.

2–44

Progressive stages in cell plate formation in root cells of lettuce (Lactuca sativa), *showing association of the endoplasmic reticulum with the developing cell plate and the origin of plasmodesmata. (a) A relatively early stage, with numerous*

small, fusing vesicles of dictyosome origin and loosely arranged elements of tubular endoplasmic reticulum. (b) An advanced stage, revealing a persistent close relationship between the endoplasmic reticulum and fusing vesicles.

Strands of tubular endoplasmic reticulum become trapped during cell plate consolidation. (c) Mature plasmodesmata, which consist of a plasma-membrane-lined canal and a tubule, the desmotubule, of endoplasmic reticulum.

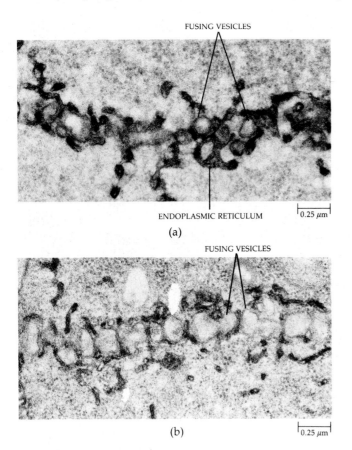

FUSING VESICLES

ENDOPLASMIC RETICULUM   0.25 μm

(a)

FUSING VESICLES

(b)   0.25 μm

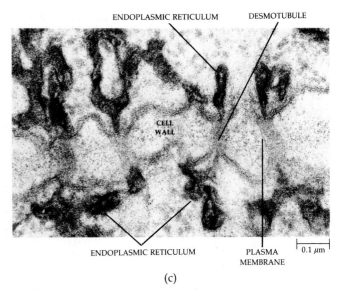

ENDOPLASMIC RETICULUM   DESMOTUBULE

CELL
WALL

ENDOPLASMIC RETICULUM   PLASMA
MEMBRANE   0.1 μm

(c)

SUMMARY

All living matter is composed of cells. Cells are extremely varied, ranging in structure and function from independent, single-celled organisms to the highly specialized, interdependent cell types found in complex multicellular plants and animals. However, cells are remarkably similar in their basic structure. All cells are bounded by an outer membrane (known as the plasma membrane, or plasmalemma). Within this membrane is the cytoplasm and the hereditary information in the form of DNA. Collectively, the living contents of the cell are called protoplasm.

Cells are of two fundamentally different types: prokaryotic and eukaryotic. Prokaryotic (bacterial) cells lack nuclei and membrane-bound organelles, and are represented today by bacteria, including cyanobacteria. Eukaryotic cells have true nuclei and are compartmented, containing distinct structures that perform different functions.

Eukaryotic cells are divided into compartments by membranes that generally have a three-layered appearance. Such membranes are called unit membranes. Membranes not only control the passage of materials into and out of the cell but also into and out of organelles.

Plant cells typically have a semirigid cell wall and a protoplast, the unit of protoplasm inside the cell wall. The protoplast includes the cytoplasm and the nucleus. The cytoplasmic matrix, or ground substance, of plant cells is frequently in motion, a phenomenon known as cytoplasmic streaming, or cyclosis. The plasma membrane, a unit membrane, separates the cytoplasm from the cell wall.

In addition to cell walls, plant cells are characterized by the presence of vacuoles within their cytoplasm. Vacuoles are cavities that are filled with cell sap, an aqueous solution of materials including a variety of salts, sugars, and other substances. Vacuoles play an important role in cell enlargement and the maintenance of tissue rigidity. In addition, many vacuoles are involved in the breakdown of macromolecules and the recycling of their components within the cell. The vacuole is bounded by a unit membrane called the tonoplast. Young cells commonly contain numerous small vacuoles, which increase in size and fuse into a single large vacuole as the cell matures. The enlargement of the vacuoles is the major cause of increase in the size of the cell.

2–45

Changes in the distribution of microtubules during the cell cycle and cell wall formation during cytokinesis. (a) During interphase, and in enlarging and differentiating cells, the microtubules lie just inside the plasma membrane. (b) Just before prophase, a ringlike band of microtubules, the preprophase band, encircles the nucleus in a plane corresponding to the equatorial plane of the future mitotic spindle. (c) During metaphase, the microtubules occur in the form of a spindle. (d) As the spindle microtubules disappear, new microtubules give rise to the phragmoplast between the two daughter nuclei. The cell plate forms at the equator of the phragmoplast between the two nuclei and grows outward until it reaches the wall of the dividing cell. (e) Each sister cell forms its own primary wall. (f) With enlargement of the daughter cells (only the upper one is shown here), the mother cell wall is torn. In (e) and (f) the microtubules once more lie just inside the plasma membrane.

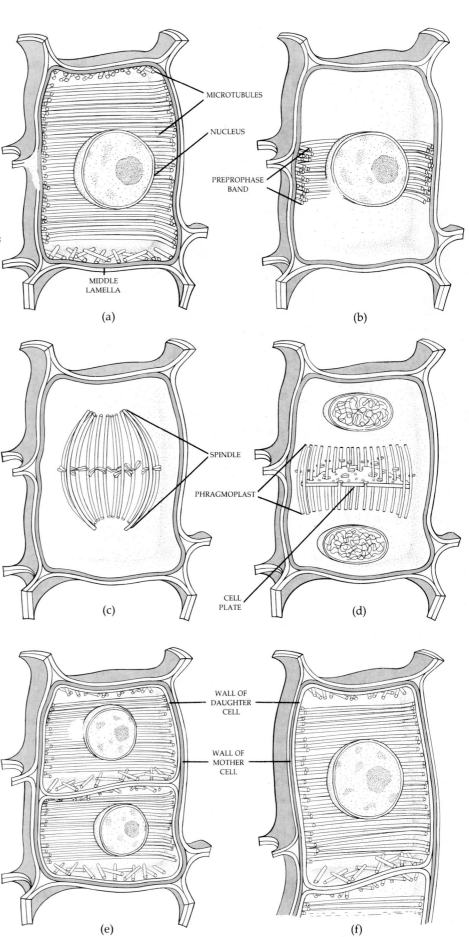

The nucleus, which is the control center of the cell, is often the most prominent structure within the protoplast. It is surrounded by a nuclear envelope composed of two unit membranes. Within the envelope is the chromatin which becomes visible as distinct chromosomes during nuclear division. Chromatin and chromosomes consist of DNA and proteins.

Together with vacuoles and cell walls, plastids are characteristic components of plant cells. Each plastid is bounded by an envelope consisting of two unit membranes. Mature plastids are classified on the basis of the kinds of pigments they contain: chloroplasts contain chlorophylls and carotenoid pigments; chromoplasts contain carotenoid pigments; and leucoplasts are nonpigmented. Chloroplasts contain grana—stacks of flattened sacs of membranes called thylakoids—in which the pigments are located. The various kinds of plastids may develop from small, colorless bodies known as proplastids.

Like plastids, mitochondria are bounded by two unit membranes; the inner one is folded to form an extensive inner membrane system, increasing the surface area available to enzymes and the reactions associated with them. Mitochondria are the principal sites of respiration in eukaryotic cells.

In addition to organelles, the cytoplasm of eukaryotic cells contains two membrane systems: the endoplasmic reticulum and the Golgi apparatus. The endoplasmic reticulum is an extensive three-dimensional system of membranes. It may have numerous connections with the nuclear envelope and often has numerous ribosomes attached to it. Ribosomes are also found free (not attached to endoplasmic reticulum) and, in eukaryotes, are present in both the cytoplasm and nucleus. Ribosomes are the sites at which amino acids are linked together to form proteins. During protein synthesis, the ribosomes occur in clusters called polyribosomes, or polysomes.

The term Golgi apparatus refers collectively to all of the dictyosomes, or Golgi bodies, of a cell. Dictyosomes are groups of flat, disk-shaped membranes from which numerous vesicles bud off. The dictyosomes apparently serve as collection and packaging centers for complex carbohydrates and other substances, which are transported to the surface of the cell in the vesicles. The vesicles also serve as a source of plasma membrane material. The endoplasmic reticulum and dictyosomes are integrated into a functional unit called the endomembrane system.

Microtubules are thin structures of variable length and are composed of subunits of the protein tubulin. They play a role in mitosis, cell-plate formation, the growth of the cell wall, and the movement of flagella.

Microfilaments are contractile proteins composed of actin. These long filaments occur in bundles believed to play a causative role in cyclosis. Together the microfilaments and microtubules form the cytoskeleton of the cell.

Flagella are hairlike structures that project from the surface of many different types of eukaryotic cells, serving as locomotor organelles. All flagella of eukaryotic cells have the same highly characteristic 9-plus-2 internal structure; that is, an outer ring of nine pairs of microtubules surrounding an inner pair of microtubules in the center of the flagellum.

The cell wall is the major distinguishing feature of the plant cell. It determines the structure of the cell, the texture of plant tissues, and many important characteristics that distinguish plants as organisms. The cell wall may be composed of three layers: the middle lamella, the primary wall, and the secondary wall. Cellulose is present in the cell walls of plants as rigid fibrils composed of a large number of cellulose molecules. These cellulose fibrils exist in a cross-linked matrix of noncellulosic molecules, such as hemicelluloses and pectin. Lignin may also be present in all three wall layers but is especially characteristic of secondary cell walls.

Cells with secondary walls often die after the wall is laid down, leaving the cell to serve as a conduit for water or as a rigid supporting structure.

Integration of the plant body is accomplished by means of the plasmodesmata, which cross the walls and connect the protoplasts of adjacent cells.

Dividing cells pass through a cell cycle consisting of interphase and mitosis. Interphase can be divided into three periods: a G_1 period of general growth and replication of organelles; an S period, during which the genetic material, DNA, is duplicated; and a G_2 period of synthesis of the structures involved in mitosis.

When the cell is in interphase, the chromosomes are in an uncoiled state and are difficult to distinguish from the nucleoplasm. During mitosis, the chromosomal material, or chromatin, condenses, and each chromosome can be seen to consist of two parallel threads—the chromatids—held together at the centromere. The centromere divides, and the duplicate chromatids—now called daughter chromosomes—move to opposite poles. The separation of the two identical sets of chromosomes is completed when new nuclear envelopes are formed around them. In this way, the genetic material is equally distributed between the two new nuclei.

Mitosis is generally followed by cytokinesis, the division of the cytoplasm. In plants and certain algae, the cytoplasm is divided by a cell plate that begins to form during mitotic telophase. Once the cytoplasm is divided, the protoplasts lay down new cell walls.

CHAPTER 3

The Molecular Composition of Cells

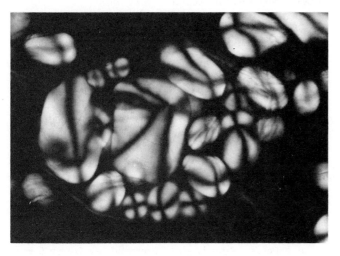

3–1
Starch grains in amyloplasts of the potato tuber, photographed with polarized light. Starch, which is composed of carbon, hydrogen, and oxygen, is the chief storage polysaccharide in plants.

As discussed in the previous chapter, cells are remarkably similar in their basic structures, despite their variety. These similarities become even more striking when we examine cells at the molecular level.

The earth and its atmosphere contain 92 different types of naturally occurring atoms, or elements. (The basic chemical principles that describe the behavior of these elements are reviewed in Appendix A.) Of all these available types of atoms, only a relatively few were selected in the course of evolution to form the complex, highly organized material of the cells of living organisms. In fact, as shown in Table 3–1, about 99 percent (by weight) of living matter (protoplasm) is composed of only six elements. In cells, these elements are found chiefly in organic compounds and, in particular, in water.

Water makes up more than half of all living tissue and more than 90 percent of most plant tissues. (See Appendix A for a review of some important properties of water.) By contrast, ions such as potassium (K^+), sodium (Na^+), and calcium (Ca^{2+}) account for no more than 1 percent. Almost all the rest of the tissue, chemically speaking, is composed of organic molecules.

ORGANIC COMPOUNDS

Organic compounds, by definition, are compounds that contain carbon. In addition to carbon, nearly all organic compounds contain hydrogen, and most of them contain oxygen as well (Figure 3–1). Nitrogen and sulfur occur less frequently than oxygen, and only a small percentage of organic compounds contain other elements. Just as protoplasm is composed of relatively few types of substances, the atoms of these few elements are arranged to form a relatively few organic compounds that, in various combinations, make up most of the dry weight of living organisms. The four principal types of organic compounds are carbohydrates, lipids, proteins, and nucleic acids (see Table 3–2, page 53).

Table 3–1   *Atomic Composition of Three Representative Organisms (Percentage by Fresh Weight)*

ELEMENT	HUMANS	ALFALFA	BACTERIA
C	19.37	11.34	12.14
H	9.31	8.72	9.94
N	5.14	0.83	3.04
O	62.81	77.90	73.68
P	0.63	0.71	0.60
S	0.64	0.10	0.32
C H N O P S total:	97.90	99.60	99.72

SOURCE: H. J. Morowitz, *Energy Flow in Biology* (New York: Academic Press, 1968).

CARBOHYDRATES

Carbohydrates (sugars, starches, and various related substances) are compounds that contain carbon combined with hydrogen and oxygen. Carbohydrates are the most abundant organic compounds in nature. Large molecules that are made up of similar or identical subunits are known as *polymers;* the subunits are called *monomers.* In the case of carbohydrates, the monomers are monosaccharides, and the polymers are polysaccharides.

Monosaccharides and Disaccharides

Monosaccharides are the simplest carbohydrates; they are made up of a chain of carbon atoms to which H atoms and O atoms are attached in the proportion of one carbon atom to two hydrogen atoms to one oxygen atom (CH_2O). Examples of several common monosaccharides are shown in Figure 3–2. As the figure indicates, the five-carbon and six-carbon sugars can also exist in ring form; in fact, they are normally found in this form when dissolved in water.

Disaccharides consist of two monosaccharides. As shown in Figure 3–3, the union is accomplished by the removal of a molecule of water from the pair of monosaccharide molecules. Such joined molecules can be broken apart by *hydrolysis*—the addition of a molecule of water at each linkage—to form the monosaccharide units again. Hydrolysis is an exergonic, or "downhill,"

reaction; the chemical bonding energy of its products is less than that of the original molecule. Thus, energy is released in such a reaction. Conversely, linking two monosaccharides together to form a disaccharide requires the input of energy.

Glucose, a monosaccharide, is the form in which sugar is most often transported through animal systems. A combination of glucose and fructose forms sucrose, the form in which most sugar is transported in plants. Sucrose is common table sugar (cane or beet sugar). Lactose (milk sugar) is a disaccharide composed of glucose and galactose.

The ultimate source of sugar in all plant cells is photosynthesis. In the process of photosynthesis, energy from the sun is converted to the chemical bond energy needed to form a sugar molecule. When the sugar molecule is broken down again, the energy of the chemical bonds is released.

Polysaccharides

Polysaccharides are made up of monosaccharides linked together in long chains. Some polysaccharides are storage forms of sugar. *Starch,* which is built of many glucose molecules (glucose residues), is the chief storage polysaccharide in plants (Figures 3–1 and 3–4); and glycogen is the common storage form of sugar in fungi, bacteria, and animals. These carbohydrates must be hydrolyzed before they can be used as energy sources or transported through living systems.

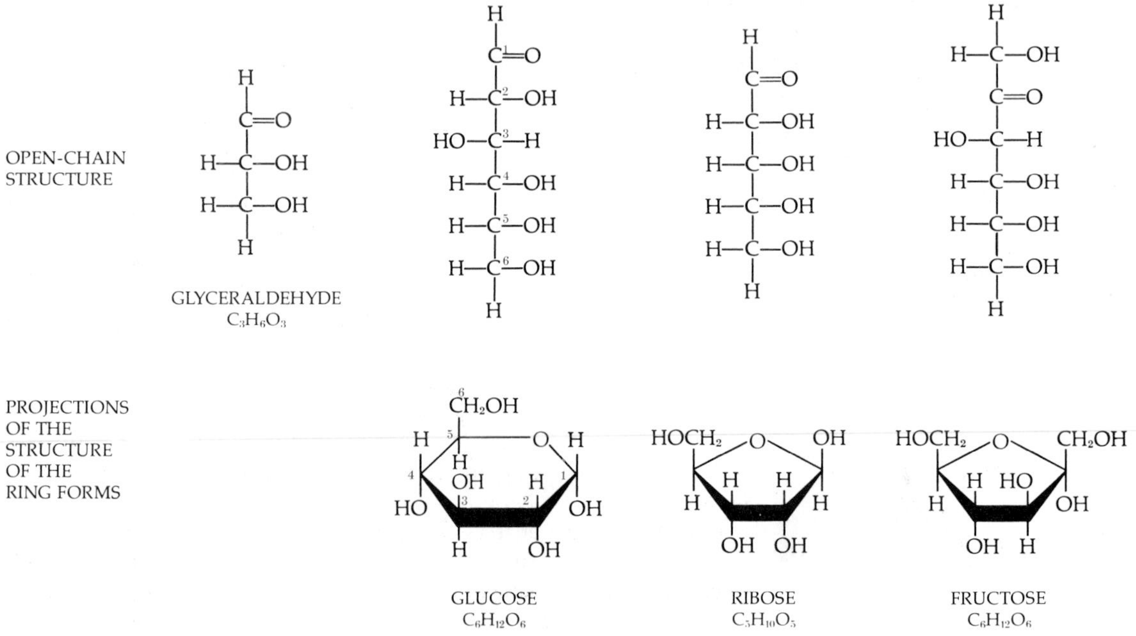

3–2

Examples of biologically important monosaccharides. Five-carbon sugars (pentoses) and six-carbon sugars (hexoses) can exist in both chain and ring forms, as shown here. The position occupied by each carbon atom has a number. The numbering of the carbon atoms is shown for the two forms of the glucose molecule. By convention, carbon atoms are not shown in the ring forms; their presence is assumed at each empty corner of the ring.

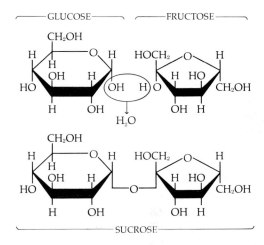

GLUCOSE   FRUCTOSE

SUCROSE

3–3

The monosaccharides glucose and fructose can combine to form the disaccharide sucrose. Sucrose is the form in which sugar is generally transported in plants. Bonds between monosaccharides are formed by the removal of a molecule of water. Formation of sucrose requires an energy input of 5.5 kcal per mole by the cell.

RADIOCARBON DATING

All organic materials contain carbon. All of this carbon previously existed in the form of carbon dioxide and made its way into the living organisms by means of photosynthesis. Most of the carbon atoms present in carbon dioxide have an atomic weight of 12 (^{12}C), but a fixed proportion of the atoms are carbon 14 (^{14}C), a radioactive isotope of carbon. Carbon 14 is produced as a result of bombardment by high-energy particles from outer space and occurs in small amounts as heavy carbon dioxide. Plant cells use carbon dioxide to make glucose and other organic molecules, accepting ^{14}CO$_2$ almost as readily as ^{12}CO$_2$. All animals are directly or indirectly dependent upon the products of plant metabolism for food; thus a fixed proportion of carbon atoms in the tissues of all living things is radioactive carbon 14. After death, carbon is no longer ingested and the proportions shift, with the radioactive carbon 14 decaying slowly and the carbon 12 remaining the same. Carbon 14 has a half-life of 5730 years, so a fossil of this age should contain just half the carbon 14 of a living plant. Objects, such as a fossil or even a manmade structure of wood or some other once-living material, can be dated quite accurately by measuring the ratio of ^{14}C to ^{12}C. Radiocarbon dating is particularly useful for studying archeological remains. This method of dating depends on the assumption that the proportion of ^{14}C to ^{12}C has remained constant in the atmosphere within the time span under study. Atmospheric nuclear testing has made it impossible for future archeologists to date current history in this way.

(a)

AMYLOSE

(c)

(b)

AMYLOPECTIN

3–4

In plants, accumulated sugars are stored in the form of starch. Starch is composed of two different types of polysaccharides: (a) amylose and (b) amylopectin. A single molecule of amylose contains 1000 or more glucose units in a long, unbranched chain that winds to form a helix (c). A molecule of amylopectin has a molecular weight of 1 to 6 million and branches approximately every 20 to 25 glucose units. The glucose units in starch are alpha-glucose monomers (see Figure 3–5). Perhaps because of their helical nature, starch molecules tend to cluster into grains (see Figures 2–10 and 3–1).

The common storage form for sugar in animals, fungi, and bacteria is glycogen, which resembles amylopectin in its general structure with the exception that branching occurs every 6 to 12 glucose units. The molecular weight of glycogen ranges up to 100 million.

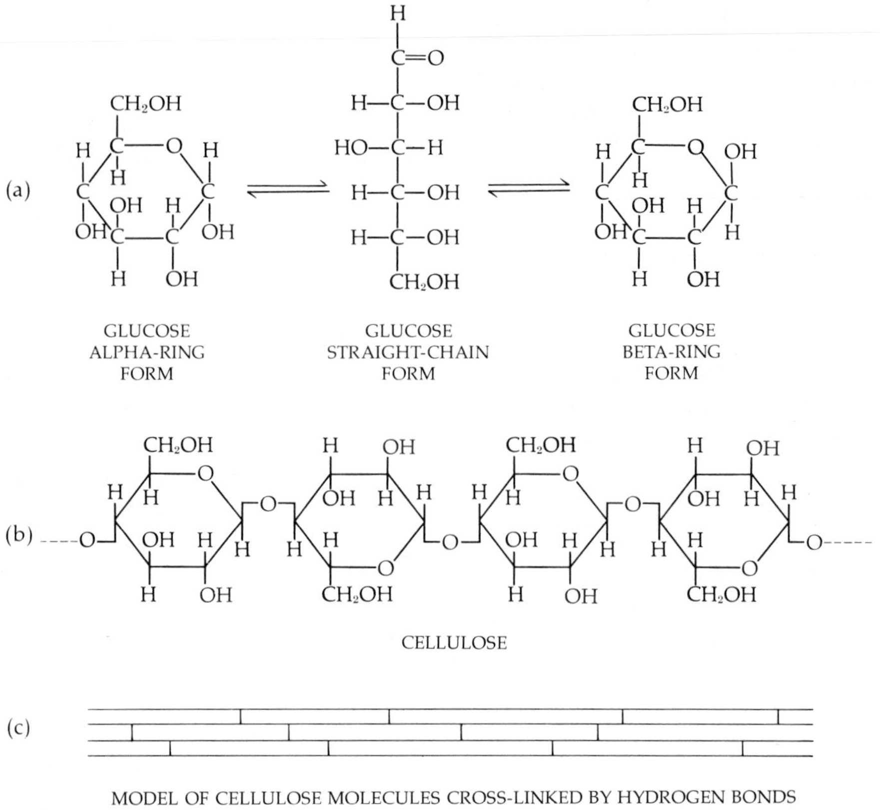

3–5

(a) In living systems, alpha-ring and beta-ring forms of the glucose molecule occur in equilibrium. The molecules pass through the straight-chain form to get from one ring structure to another. Whereas starch consists of alpha-glucose monomers, cellulose (b) consists entirely of beta-glucose monomers. The —OH groups (in color), which project from both sides of the chain, form hydrogen bonds with neighboring —OH groups, resulting in the formation of bundles of cross-linked parallel chains (c). In the starch molecule, most of the —OH groups capable of forming hydrogen bonds face toward the exterior of the helix.

(a)

GLUCOSE
ALPHA-RING
FORM

GLUCOSE
STRAIGHT-CHAIN
FORM

GLUCOSE
BETA-RING
FORM

(b)

CELLULOSE

(c)

MODEL OF CELLULOSE MOLECULES CROSS-LINKED BY HYDROGEN BONDS

Polysaccharides are also important structural compounds. In plants, the principal structural polysaccharide is cellulose (Figure 3–5). Moreover, cellulose is the most abundant polysaccharide in nature. Although cellulose is made of the same building materials as starch, the arrangement of its long-chain molecules makes it rigid, and so its biological role is very different. Also, because the bonds linking the glucose units in cellulose are different from those found in starch, cellulose is not readily hydrolyzed. In addition to cellulose, plant cell walls may contain a variety of other polysaccharides (Figure 3–6), pectic substances (pectins) (Figure 3–7), and lignin.

Once monosaccharides are incorporated into the plant cell wall in the form of cellulose, they are no longer available to the plant as an energy source. In fact, only some bacteria, fungi, and protozoa, and a very few animals (silverfish, for example), possess enzyme systems capable of breaking down cellulose. Other organisms, such as cattle (and other ruminants),

termites, and cockroaches, are able to utilize cellulose as a source of energy only because of the microorganisms (which possess the necessary enzyme systems) that inhabit their digestive tracts.

Chitin is another important structural polysaccharide; it is the principal structural component of fungal cell walls and also of the relatively hard outer coverings, or exoskeletons, of insects and crustaceans. The monomer of chitin is a six-carbon sugar to which a nitrogen-containing group has been added (see Figure 13–5, page 198).

LIPIDS

Lipids are fats and fatlike substances. They have two principal distinguishing characteristics: (1) they are generally hydrophobic ("water-fearing") and thus are insoluble in water (although they are soluble in other lipids), and (2) they contain a large number of carbon-

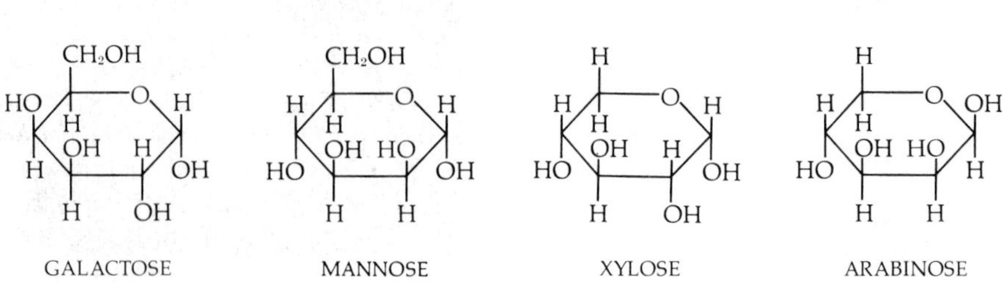

3–6

Building blocks of some polysaccharides other than cellulose that are commonly found in the cell walls of plants. The monosaccharides shown here are the building blocks of the polysaccharides called, respectively, galactans, mannans, xylans, and arabinans.

GALACTOSE

MANNOSE

XYLOSE

ARABINOSE

3–7

*Pectic compounds are built up of
residues of alpha-galacturonic acid (a),
which is a derivative of glucose. Pectic
acid is the pectic compound shown in
(b). Calcium and magnesium salts of
pectic acid make up most of the middle
lamella that binds adjacent plant cells.
In pectin and protopectin, different
proportions of the hydrogen ions (shown
in color) are replaced with methyl (—CH₃)
groups. Protopectin is a common
constituent of the cell wall and is less
soluble than pectin, which is commonly
found dissolved in plant juices.*

α-GALACTURONIC
ACID

(a)

PECTIC ACID

(b)

hydrogen bonds and, as a consequence, release a larger amount of energy in oxidation than other organic compounds. Fats yield, on the average, about 9.3 kcal per gram, as compared to a yield of about 3.8 kcal per gram for carbohydrates. These two characteristics of lipids determine their roles as structural materials and as energy reserves.

Fats

Fat is the principal form in which lipids are used for storage (Figure 3–8). Some plants store food energy as fats (oils), especially in seeds and fruits. Cells synthesize fats from sugars. A fat consists of three fatty acids joined to a glycerol molecule (Figure 3–9). Fatty acids are long hydrocarbon chains that carry a terminal carboxyl group, giving them the characteristics of a weak acid. The glycerol forms a link with the carboxyl group, releasing a molecule of water. The glycerol thus serves as a binder or carrier for the fatty acids. (Like polysaccharides and proteins, fats are broken down by hydrolysis.)

The physical nature of the fat is determined by the chain lengths of the fatty acids and by whether the acids are *saturated* or *unsaturated.* In saturated fatty acids, all the carbon atoms hold as many hydrogen atoms as possible. Unsaturated fatty acids contain carbon atoms joined by double bonds; such carbon atoms are able to form additional bonds with other atoms (hence the term "unsaturated"). Unsaturated fats, which tend to be oily liquids, are more common in plants than in animals; examples are safflower oil, peanut oil, and corn oil. Animal fats, such as lard and butter, contain saturated fatty acids and usually have higher melting temperatures than unsaturated fats.

3–8

*Two cambial cells from the fleshy,
underground stem (corm) of the quillwort
Isoetes muricata. During winter these
cells contain a large quantity of stored fat
in the form of lipid droplets. In addition,
carbohydrate in the form of starch grains
is stored within amyloplasts. Several
vacuoles can be seen in each of these
cells. The dense material lining the
vacuoles is tannin.*

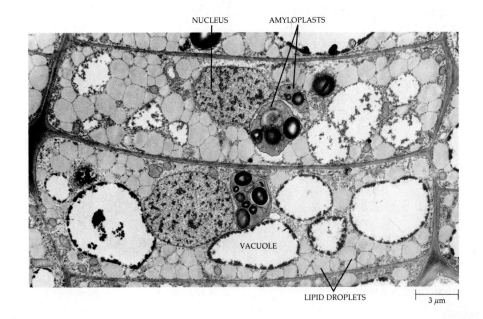

NUCLEUS   AMYLOPLASTS

VACUOLE

LIPID DROPLETS   3 μm

$$H-\overset{\displaystyle H}{\underset{\displaystyle H}{C}}-OH \qquad HO-\overset{\displaystyle O}{C}-CH_2-CH_2-CH_2-CH_2-CH_2-CH_2-CH_2-CH_2-CH_2-CH_2-CH_2-CH_2-CH_2-CH_2-CH_2-CH_2-CH_3$$

STEARIC ACID

$$H-\overset{\displaystyle }{C}-OH \qquad HO-\overset{\displaystyle O}{C}-CH_2-CH_2-CH_2-CH_2-CH_2-CH_2-CH_2-CH=CH-CH_2-CH_2-CH_2-CH_2-CH_2-CH_2-CH_2-CH_3$$

OLEIC ACID

$$H-\overset{\displaystyle }{\underset{\displaystyle H}{C}}-OH \qquad HO-\overset{\displaystyle O}{C}-CH_2-CH_2-CH_2-CH_2-CH_2-CH_2-CH_2-CH_2-CH_2-CH_2-CH_2-CH_2-CH_2-CH_3$$

CARBOXYL   PALMITIC ACID
GROUP

GLYCEROL   FATTY ACID PORTION

3–9

A fat molecule consists of three fatty acids joined to a glycerol molecule. As with the polysaccharides, these bonds are formed by the removal of water molecules (the atoms involved in this hydrolysis are shown in color). A fatty acid such as stearic acid, in which each carbon atom holds as many hydrogen atoms as possible, is known as saturated. A fatty acid such as oleic acid, in which any carbon atom holds only one instead of two hydrogens, is unsaturated. Unsaturated fatty acids are more common in plants than in animals.

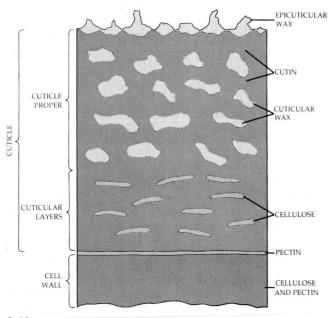

3–10

Diagram showing various layers of the cuticle. The first layer, the epicuticular layer, is composed entirely of wax. The second layer, the cuticle proper, consists of wax and cutin, while the following layer, the cuticular layer, is made up of cellulose and cutin. Wax may also be present in the cuticular layer. A layer of pectin may occur between the cuticle and cell wall, which itself is composed of cellulose and pectin.

Cutin, Suberin, and Waxes

Cutin and *suberin* are unique, insoluble lipid polymers that are important structural components of many plant cell walls. The major function of these polymers is that they form the matrices (singular: matrix) in which *waxes*—long-chain lipid compounds—are embedded. Together the cutin and waxes or suberin and waxes form barrier layers that help prevent the loss of moisture and other molecules from the aerial or above-ground parts of the plant. In particular, it is the waxes that constitute the major barrier.

Cutin, together with its embedded waxes, forms the *cuticle*, which covers the outer walls of epidermal cells. As seen in Figure 3–10, the cuticle has a complicated structure and consists of several layers. The first layer is composed of waxes deposited on the surface as epicuticular wax (Figure 3–11). Beneath the epicuticular wax is the cuticle proper consisting of cuticular wax and cutin. This may be followed by one or more so-called cuticular layers consisting of cellulose, cutin, and wax. Finally, a layer of pectin may occur between the cuticle and the cell wall.

Suberin, a major component of all cork cell walls, is also found in the Casparian strip of endodermal cells (see page 406) and the bundle-sheath cell walls in the leaves of grasses. As seen with the electron microscope, suberized walls have a lamellar, or layered, appearance, with light bands alternating with dark bands (see Figure 21–13). The light bands are believed to be composed of waxes and the dark bands of suberin. Some cuticles also have a lamellar appearance.

Phospholipids

Closely related to the fats are the phospholipids—various compounds in which glycerol is attached to only two fatty acids, with the third space occupied by a molecule containing phosphorus (Figure 3–12). Phospholipids are very important in cellular structure, particularly in the cellular membranes.

The phosphate end of the phospholipid molecule is hydrophilic ("water-loving") and thus is soluble in water, in contrast to the hydrophobic fatty acids. If phospholipids are added to water, they tend to form a

3–11
The upper leaf surface of Eucalyptus cloeziana *showing deposits of epicuticular wax. Beneath these deposits is the cuticle, a wax-containing layer covering the outer walls of the epidermal cells. Waxes help protect exposed plant surfaces from water loss.*

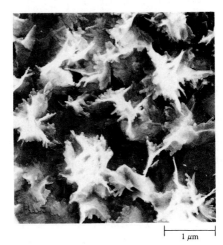

1 μm

3–12
A phospholipid molecule consists of two fatty acids linked to a glycerol molecule, as in a fat, and a phosphate group (indicated by color) linked to the third carbon of the glycerol. The phosphate group also usually contains an additional chemical group, indicated by the letter R.

The fatty acid tails are nonpolar (lacking positive and negative charges) and are therefore hydrophobic (insoluble in water); the polar head containing the phosphate and R groups is hydrophilic (soluble).

POLAR HEAD   NONPOLAR TAILS

$$R-O-\overset{\overset{O}{\|}}{\underset{\underset{O}{\|}}{P}}-O-CH_2$$

$$H-C-O-\overset{O}{\overset{\|}{C}}-(CH_2)_7-CH=CH-(CH_2)_7-CH_3$$

$$H-C-O-\overset{O}{\overset{\|}{C}}-(CH_2)_{16}-CH_3$$

H

GLYCEROL

film along its surface, with their polar heads under the water and the insoluble fatty acid chains (tails) protruding above the surface (Figure 3–13a). In the watery interior of the cell, phospholipids tend to align themselves in rows, with the insoluble fatty acids oriented toward one another and the phosphate ends directed outward (Figure 3–13b). Such configurations are important in the structure of cellular membranes (see page 62).

3–13
The glycerol-phosphate combination is hydrophilic (soluble in water), whereas the fatty acids are hydrophobic. Consequently, when placed in water, the molecules tend to form a film (a) on the water surface, with the polar, or hydrophilic, heads beneath the surface and the nonpolar, or hydrophobic, tails protruding above the surface. In the cell, they tend to align themselves in layers (b) with their soluble heads pointing outward.

SYMBOL OF PHOSPHOLIPID MOLECULE

TAIL
HEAD

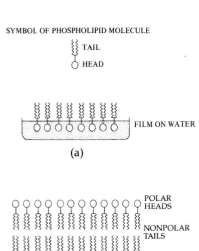

FILM ON WATER

(a)

POLAR HEADS
NONPOLAR TAILS
POLAR HEADS

(b)

PROTEINS

Proteins, like polysaccharides, are polymers made up of a number of similar units. In proteins, the molecular monomers are nitrogen-containing molecules known as *amino acids*. The same basic set of 20 standard amino acids are found in all proteins, but because protein molecules are large and complex—often containing several hundred amino acids—the number of different amino acid sequences and, hence, the possible variety of protein molecules are enormous. As shown below, the amino acid sequence in a particular protein is very important because it determines the structure and therefore the function of that protein. A single cell of the bacterium *Escherichia coli* can contain 600 to 800 different kinds of proteins at any one time, and the cell of a plant or animal probably has several times that number.

In plants, the largest concentration of proteins is found in seeds, in which as much as 40 percent of the dry weight may be protein. These proteins appear to function as storage forms of amino acids that are used by the embryo when it resumes growth upon germination of the seed.

NONPOLAR

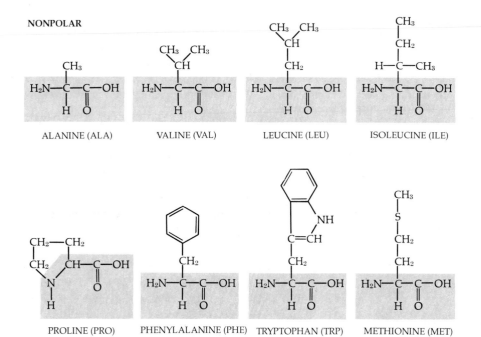

ALANINE (ALA)   VALINE (VAL)   LEUCINE (LEU)   ISOLEUCINE (ILE)

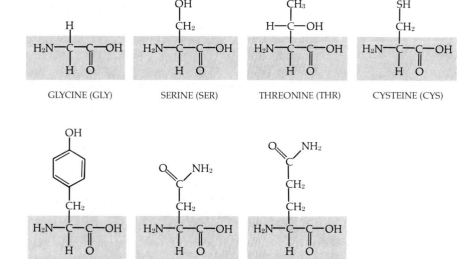

PROLINE (PRO)   PHENYLALANINE (PHE)   TRYPTOPHAN (TRP)   METHIONINE (MET)

POLAR BUT UNCHARGED

GLYCINE (GLY)   SERINE (SER)   THREONINE (THR)   CYSTEINE (CYS)

TYROSINE (TYR)   ASPARAGINE (ASN)   GLUTAMINE (GLN)

ACIDIC (NEGATIVELY CHARGED)

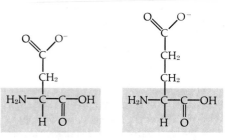

ASPARTIC ACID (ASP)   GLUTAMIC ACID (GLU)

BASIC (POSITIVELY CHARGED)

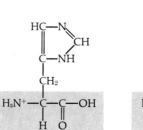
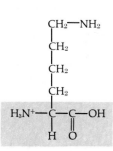
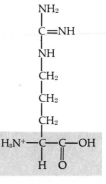

HISTIDINE (HIS)   LYSINE (LYS)   ARGININE (ARG)

3–14

Every amino acid contains an amino group ($-NH_2$) and a carboxyl group ($-COOH$) bonded to a central carbon atom. A hydrogen atom and a side group R are also bonded to the same carbon atom. The R represents the side group, which is different in each kind of amino acid. This basic structure is the same in all amino acids:

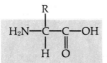

The 20 kinds of amino acids found in proteins are shown to the left and below. As you can see, their basic structures are the same; they differ only in their side groups. Because of differences in their side groups, amino acids may be nonpolar (with no difference in charge between one part of the molecule and another), polar (but with the two charges balancing one another so that the amino acid as a whole is uncharged), negatively charged (acidic), or positively charged (basic). The nonpolar molecules are insoluble in water, whereas the charged and polar molecules are soluble.

Table 3–2 *Some Important Classes of Organic Compounds*

ORGANIC COMPOUNDS	FUNCTIONS	COMPONENTS	COMPOSITION
Carbohydrates	Energy source, structural material, building blocks for other molecules	Simple sugars	Carbon, hydrogen, and oxygen
Lipids	Energy storage, structural material, barrier to moisture loss	Fatty acids, glycerol	Carbon, hydrogen, and oxygen
Proteins	Structural materials, enzymes	Amino acids	Carbon, hydrogen, oxygen, nitrogen, and sulfur
Nucleic acids	Protein synthesis	Nucleotides (nitrogenous bases, sugars, and phosphates)	Carbon, hydrogen, oxygen, nitrogen, and phosphorus

Amino Acids

The basic structure of an amino acid is shown within the caption of Figure 3–14. It consists of an amino group and a carboxyl group bonded to a carbon atom, the so-called alpha-carbon. The "R" represents the rest of the molecule, which varies in structure from one amino acid to another. It is this R group that determines the identity of any particular amino acid. Figure 3–14 shows the complete structure of the 20 amino acids found in proteins.

The amino acids are grouped according to their electrical charges. These charges are important in determining the properties of the various amino acids and, in particular, of the proteins formed from their combination.

Polypeptides

A chain of amino acids is known as a *polypeptide*. As in the case of polysaccharides, a molecule of water is removed to create each link in the chain. The amino group of one amino acid always links to the carboxyl group of the next amino acid; this bond is known as the *peptide bond* (Figure 3–15). The amino acid subunits of a polypeptide are often referred to as amino acid residues. Because of the way in which the amino acids are bonded together, there is always a free amino group at one end of the chain and a free carboxyl group at the other. The end with the amino group is called the N terminal; the one with the carboxyl group is called the C terminal.

Amino acids can also be cross-linked by disulfide bonds, which are covalent bonds formed between two sulfur atoms of two cysteine residues. A disulfide bond may link two cysteine residues in the same polypeptide chain or in two different chains (Figure 3–16).

Proteins are large polypeptides; they have molecular weights ranging from 10^4 (10,000) to more than 10^6 (1,000,000). In comparison, water has a molecular weight of 18, and glucose has a molecular weight of 180. (The average molecular weight of an amino acid residue is about 120; if you remember this figure, you can rapidly calculate the approximate number of amino acids in a protein on the basis of its total molecular weight. The way in which molecular weights are determined is described in Appendix A.)

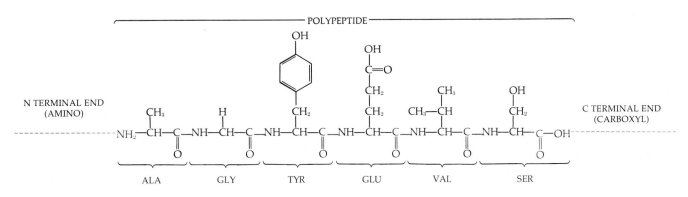

3–15
The links between amino acid residues are known as peptide bonds. Peptide bonds are formed by the removal of a molecule of water, as are the bonds between sugar residues. The bonds, shown here in color, always form between the carboxyl (—COOH) group of one amino acid and the amino (—NH₂) group of the next. Consequently, the basic structure of a protein is a long, unbranched molecule. This linear arrangement of the amino acids is known as the primary structure of the protein. The end with the amino group is called the N terminal; the one with the carboxyl group is called the C terminal.

Levels of Protein Organization

The sequence of amino acids in the polypeptide chain is referred to as the *primary structure* of the protein. The polypeptide chains often coil in a helix *(secondary structure)*, and the helix may be folded to produce a so-called globular protein molecule *(tertiary structure)*. Finally, the polypeptide chains of a protein having two or more chains may interact to form a *quaternary structure*.

Primary Structure

The primary structure of a protein is simply the linear sequence of amino acids in the polypeptide chain. Each kind of protein has a different primary structure—a unique protein "word" consisting of a unique sequence of amino acid "letters." The primary structure of one protein, the enzyme lysozyme, is shown in Figure 3–16.

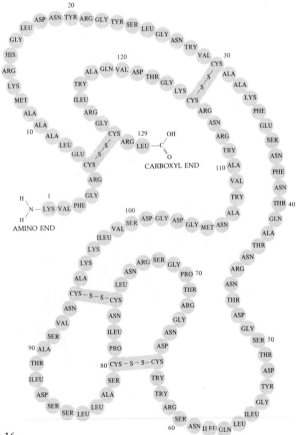

3–16

The primary structure of the enzyme lysozyme is shown in this two-dimensional model. Lysozyme is a protein containing 129 amino acid residues. These residues form a polypeptide chain that is cross-linked at four places by disulfide (–S–S–) bonds. (From The Three-Dimensional Structure of an Enzyme Molecule, *by D. C. Phillips. Copyright © 1966 by Scientific American, Inc.*

Secondary Structure

In the cell, polypeptide chains do not lie flat, as they usually appear in most diagrams; they spontaneously assume regular coiled structures. This three-dimensional coiled arrangement is the secondary structure of a protein. The most common secondary structure is the alpha helix, which resembles a spiral staircase (Figure 3–17). The alpha helix is very uniform in its geometry, with a turn of the helix occurring every 3.6 amino acids. The helical structure of polypeptide chains is maintained by hydrogen bonds across successive turns of the spiral, with the hydrogen atom in the amino group of one amino acid bonded to an oxygen atom of the carboxyl group of an amino acid in the next coil.

Tertiary and Quaternary Structure

Polypeptide chains may also fold up to form globular structures—the tertiary structure of protein. Figure 3–18 shows the folding of the main chain in the enzyme lysozyme. Most biologically active proteins, such as enzymes, are globular. The specialized proteins of cellular membranes are also globular in structure. In addition, microtubules are composed of a large number of spherical subunits, each of which is a globular protein.

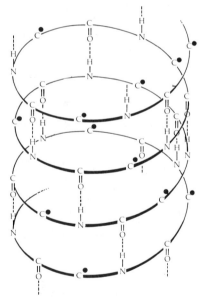

3–17

The alpha helix. The helix is held in shape by hydrogen bonds, indicated by the dashed lines. The bonds form between the oxygen atom in the carboxyl group in one amino acid and the amino group in another amino acid that occurs four amino acids farther along the chain. The R groups, not shown in this diagram, are attached to the carbons indicated by the colored dots. The groups extend out from the helix. One turn of the helix occurs every 3.6 residues.

The tertiary structure of protein is determined by the primary structure. It is maintained largely by hydrogen bonds and by interactions among the various amino acid residues, based on their various charges (see Figure 3–14), and between the amino acid residues and water molecules. These bonds are relatively weak and can be broken quite easily by physical or chemical changes in the environment, such as heat or increased acidity. This breakdown is called denaturation. The coagulation of egg white upon boiling is a common example of protein denaturation. When proteins are denatured, the poly-peptide chains are unfolded, causing a loss of the biological activity of the protein. Organisms cannot live at extremely high temperatures because their enzymes and other proteins become unstable and nonfunctional due to denaturation.

The quaternary structure refers to how the polypeptide chains of a protein consisting of two or more chains are arranged in relation to each other. Most larger proteins contain two or more polypeptide chains, which may or may not be identical. Many enzymes consist of two or more polypeptide chains.

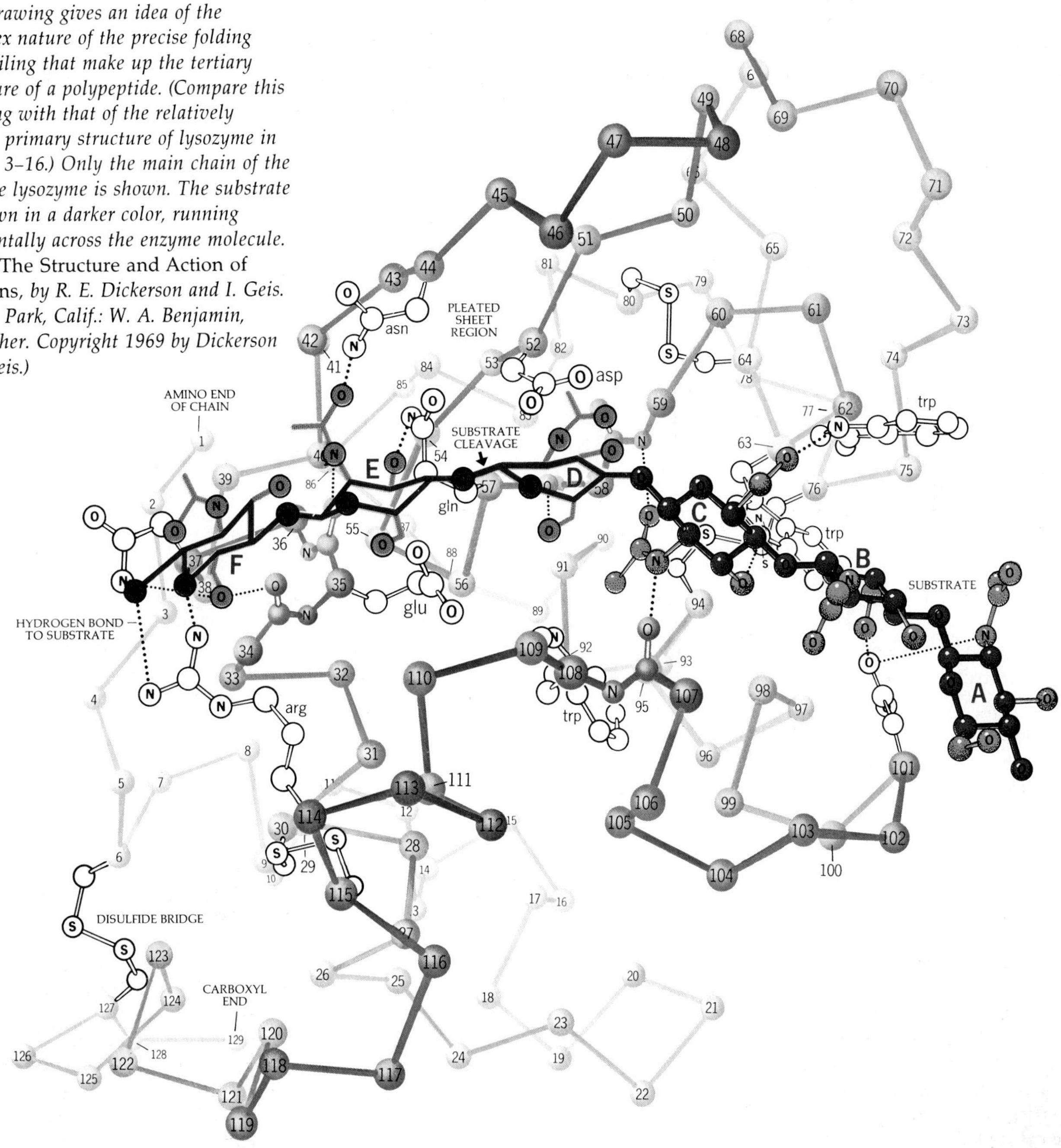

3–18

This drawing gives an idea of the complex nature of the precise folding and coiling that make up the tertiary structure of a polypeptide. (Compare this drawing with that of the relatively simple primary structure of lysozyme in Figure 3–16.) Only the main chain of the enzyme lysozyme is shown. The substrate is shown in a darker color, running horizontally across the enzyme molecule. (From The Structure and Action of Proteins, *by R. E. Dickerson and I. Geis. Menlo Park, Calif.: W. A. Benjamin, Publisher. Copyright 1969 by Dickerson and Geis.)*

Enzymes

Enzymes are large globular proteins that act as catalysts. (Different enzymes vary in their molecular weight from about 12,000 to more than 1 million.) Catalysts are substances that accelerate the rate of a chemical reaction by lowering the energy of activation (see Appendix A), but which remain unchanged in the process. Because they remain unaltered, catalysts can be used over and over again and so are typically effective in very small amounts.

In the laboratory, the rates of chemical reactions are usually accelerated by the application of heat, which increases the force and frequency of collisions among molecules. In a cell in nature, however, hundreds of different reactions are going on at the same time, and heat would speed up all these reactions indiscriminately. Moreover, heat would melt the lipids, denature the proteins, and have other generally destructive effects on the cell. Because of catalytic enzymes, cells are able to carry out chemical reactions at great speeds and at relatively low temperatures. If enzymes were not present, the reactions would occur anyway, but at a rate so slow that their effects would be negligible.

Enzymes are generally named by adding the ending *-ase* to the root of the name of the substrate. Amylase catalyzes the hydrolysis of amylose (starch), and sucrase catalyzes the hydrolysis of sucrose into glucose and fructose. Nearly 2,000 different enzymes are now known, and each of them is capable of catalyzing some specific chemical reaction. The behavior of enzymes in biological reactions is further explained in Chapter 5.

NUCLEIC ACIDS

Nucleic acids are polymers of nucleotides—complex molecules made up of three subunits: (1) a phosphate group, (2) a five-carbon sugar (pentose), and (3) a nitrogenous base, so named because its ring structure contains nitrogen as well as carbon.

There are two types of nucleic acids—DNA (deoxyribonucleic acid) and RNA (ribonucleic acid). DNA is the molecule in which the genetic information is stored. RNA serves as a translator and transmitter of this genetic information. It is through RNA that the DNA dictates the structure of proteins and thus the structure and function of the cell.

Nucleotides contain two types of nitrogenous bases: *pyrimidines*, which have a single ring, and *purines*, which have two fused rings (Figure 3–19). The three pyrimidines found in nucleotides are thymine, cytosine, and uracil. DNA contains thymine and cytosine; RNA contains cytosine and uracil. The two purines in nucleotides are adenine and guanine; RNA and DNA both contain these purines. DNA and RNA also contain slightly different types of sugar; DNA contains deoxyribose, and RNA contains ribose. "Deoxy" means

"minus one oxygen," and as is shown in Figure 3–19, this is the only difference between these two pentose sugars.

The way in which these molecules are put together, the significance of their structure, their role in the cell, and how this information was discovered are presented in Chapter 8.

OTHER NUCLEOTIDE DERIVATIVES

Nucleotides and compounds derived from nucleotides serve a variety of functions within the cell. Two nucleotide derivatives of prime importance are ATP (adenosine triphosphate; see Figure 5–10) and ADP (adenosine diphosphate). Almost all energy exchanges within the cell involve the transfer of phosphate groups; ATP and ADP are the most important molecules in phosphate transfer. The role of these energy-exchange molecules is discussed in Chapter 5.

Nucleotides play important roles in a variety of other energy-exchange molecules. One such compound is NAD (nicotinamide adenine dinucleotide), which contains a nucleotide of adenine plus a phosphate (see Figure 5–8). Widespread use of the purine-sugar-phosphate combination in energy transfers may have to do with the fact that these comparatively large, charged molecules do not pass through membranes and thus cannot "escape" into and out of cells or membrane-bound organelles.

SUMMARY

Living matter is composed of only a few of the naturally occurring elements. The bulk of living matter is water. Most of the rest of living material is composed of organic compounds—carbohydrates, lipids, proteins, and nucleic acids.

Carbohydrates serve as a primary source of chemical energy for living systems and as important structural elements in cells. The simplest carbohydrates are the monosaccharides, such as glucose and fructose. Monosaccharides can be combined to form disaccharides, such as sucrose, and polysaccharides (chains of many submolecules of sugar), such as starch and cellulose. These molecules can usually be broken apart by the addition of a water molecule at each linkage, a chemical reaction known as hydrolysis.

Lipids are another source of energy and structural material for cells. Compounds in this group—fats, cutin, suberin, waxes, and phospholipids—are generally insoluble in water.

Proteins are very large molecules composed of long chains of amino acids known as polypeptides. Twenty different amino acids are found in proteins, and from these amino acids, enormous numbers of different protein molecules are built. The principal levels of protein organization are: (1) primary structure, the linear amino acid sequence; (2) secondary structure, the coiling or spiraling of the polypeptide chain; (3)

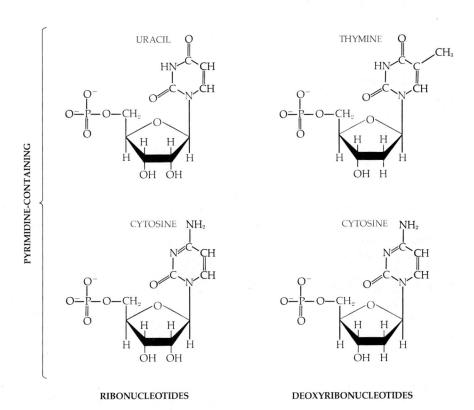

3–19

The building blocks of RNA and DNA. Each nucleotide building block contains a phosphate group, a sugar, and a nitrogenous base, which can be either a purine or a pyrimidine.

The purines in both groups are the same, but one type of DNA nucleotide contains thymine, whereas its RNA counterpart contains uracil. The only other difference between the two is the presence of one more oxygen atoms on the sugar (ribose) component of the RNA. As described in Chapter 8, however, the biological roles of the two are profoundly different.

tertiary structure, the folding of the coiled chain into globular shapes; and (4) quaternary structure, which results from specific interactions between two or more polypeptide chains.

Enzymes are globular proteins that act as catalysts in chemical reactions. Because of enzymes, cells are able to accelerate the rate of chemical reactions at relatively low temperatures.

Nucleic acids consist of nucleotides linked together in long chains. Nucleotides are composed of three subunits: a nitrogenous base, a five-carbon sugar, and a phosphate group. Nucleotides containing deoxyribose sugar form DNA; those containing ribose sugar form RNA.

Two nucleotide derivatives—ATP and ADP—are involved in most of the energy exchanges within the cell.

C H A P T E R   4

The Movement of Substances
into and out of Cells

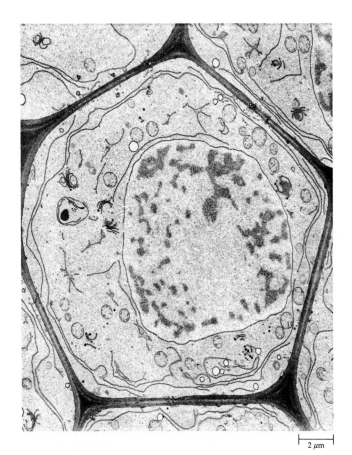

|⎯| 2 μm

4–1

*Electron micrograph of a cell from the
root tip of a corn plant (Zea mays).
This cell has been fixed in potassium
permanganate, a chemical that combines
selectively with biological membranes;
hence, the endoplasmic reticulum,
dictyosomes, and bounding membranes of
various organelles are clearly defined.
These membranes regulate the movement
of substances into and out of the cell and
control their passage from one part of the
cell to another.*

All cells are separated from their surroundings by a
surface membrane—the plasma membrane; eukaryotic
cells are further divided internally by a variety of membranes, including those of the endoplasmic reticulum,
dictyosomes, and the bounding membranes of organelles (Figure 4–1). These cellular membranes are not
impenetrable barriers, for cells are able to regulate the
amount, kind, and often the direction of movement of
substances that pass across membranes. This is an essential capacity of living cells because few metabolic
processes could occur at reasonable rates if they depended upon the concentration of necessary substances
found in the cell's surroundings. In fact, one criterion
by which we identify living systems is the difference in
concentration of a variety of substances in living matter
and in the surrounding, nonliving environment.

Control of the exchange of substances across membranes depends on the physical and chemical properties of the membranes and of the ions or molecules that
move through them. Water is the most important of the
molecules moving into and out of cells.

PRINCIPLES OF WATER
MOVEMENT

The movement of water, whether in living systems or
in the nonliving world, is governed by three basic principles: bulk flow, diffusion, and osmosis.

Bulk Flow

Bulk flow is the overall movement of water (or some
other liquid). It occurs in response to differences in the
potential energy of water, usually referred to as *water
potential.*

A simple example of water that has potential energy
is water behind a dam or at the top of a waterfall. As

this water runs downhill, its potential energy can be converted to mechanical energy by a waterwheel or to mechanical and then electrical energy by a hydroelectric turbine (Figure 4–2).

Pressure is another source of water potential. If we put water into a rubber bulb and squeeze the bulb, this water, like the water at the top of a waterfall, has water potential, and it will move to an area of less water potential. Can we make the water that is running downhill run uphill by means of pressure? Yes, we can, but only so long as the water potential produced by the pressure exceeds the water potential produced by gravity. *Water moves from an area where water potential is greater to an area where water potential is less, regardless of the reason for the difference.*

The concept of water potential is a useful one because it enables physiologists to predict the way in which water will move under various conditions. Measurements of water potential are usually made in terms of the pressure required to stop the movement of water —that is, the hydrostatic pressure—under the particular circumstances involved. The unit used to express this pressure is usually the bar. (A bar is a metric unit of pressure equal to the average pressure of the air at sea level.)

Diffusion

Diffusion is a familiar phenomenon. If a few drops of perfume are sprinkled in one corner of a room, the scent will eventually permeate the entire room even if the air is still. If a few drops of dye are put in one end of a glass tank full of water, the dye molecules will slowly become evenly distributed throughout the tank. This process may take a day or more, depending on the size of the tank, the temperature, and the size of the dye molecules.

Why do the dye molecules move apart? If you could observe the individual dye molecules in the tank (Figure 4–3), you would see that each one of them moves individually and at random. Looking at any single molecule—at either its rate of motion or its direction of motion—gives you no clue at all as to where the molecule is located with respect to the others. So how do molecules get from one side of the tank to the other? Imagine a thin section through the tank, running from top to bottom. Dye molecules will move into and out of the section, some moving in one direction, some moving in the other. But you would see more dye molecules moving from the side of greater dye concentration. Why? Simply because there are more dye molecules at that end of the tank. Since there are more dye molecules on the left, more dye molecules will move randomly to the right, even though there is an equal probability that any one molecule of dye will move from right to left. Consequently, the overall (net) movement of dye molecules will be from left to right. Simi-

4–2
Water at the top of a waterfall is said to possess potential energy. As the water falls, this potential energy becomes energy of motion (kinetic energy), which can be converted into mechanical energy to do work.

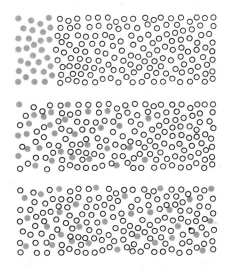

4–3
A diagram of the diffusion process. Diffusion is the result of the random movement of individual molecules, which produces a net movement from a more concentrated area to a less concentrated area with respect to that particular type of molecule. Notice that as one type of molecule (indicated by color) diffuses to the right, the other diffuses in the opposite direction. The result will be an even distribution of both types of molecules. Can you see why the net movement of molecules will slow down as equilibrium (even distribution) is reached?

IMBIBITION

Water molecules exhibit a tremendous cohesiveness because of the difference in charge between one end of a water molecule and the other (see Appendix A). Similarly, because of this difference in charge, water molecules can cling either to positively charged or to negatively charged surfaces. Many large biological molecules, such as cellulose, tend to develop charges when they are wet and so attract water molecules. The adherence of water molecules is responsible for another biologically important phenomenon called imbibition or, sometimes, hydration.

Imbibition (from the Latin *imbibere*, "to drink in") is the movement of water molecules into substances such as wood or gelatin, which swell or increase in volume as a result of adhesion to the water molecules. The pressures developed by imbibition can be astonishingly large. It is said that stone for the ancient Egyptian pyramids was quarried by driving wooden pegs into holes drilled in the rock face and then soaking the pegs with water. The swelling wood created a force that split the slab of stone. In living plants, imbibition occurs particularly in seeds, which may increase to many times their original size as a result.

state of equal distribution, that is, when there are no more gradients, they continue to move, but there is no net movement in either direction. In other words, the net transfer of the molecules is zero; the system may be said to be in a state of *dynamic equilibrium.*

The concept of water potential is also useful in understanding diffusion. A high concentration of solute (dissolved substance) in one region, such as one corner of the tank, means low water concentration there and, thus, low water potential. If the pressure is equal everywhere, water molecules, as they move along the gradient, are moving from a region of higher water potential to a region of lower water potential. The region of the tank in which there is pure water has a greater water potential than the region containing water plus some dissolved substance. When dynamic equilibrium is reached, water potential is equal in all parts of the tank.

The essential characteristics of diffusion are: (1) each molecule moves independently of the others, and (2) these movements are random. The net result of diffusion is that the diffusing substance eventually becomes evenly distributed. Briefly, *diffusion* may be defined as *the dispersion of substances by a movement of their ions or molecules, which tends to equalize their concentrations throughout the system.*

larly, if you could see the movement of the individual water molecules in the tank, you would see that net movement is from right to left.

What happens when all the molecules are evenly distributed throughout the tank? The even distribution does not affect the behavior of the molecules as individuals; they still move at random. But there are now as many molecules of dye and as many molecules of water on one side of the tank as on the other, so there is no net direction of motion. There is, however, just as much individual motion as before, provided the temperature has not changed.

Substances that are moving from a region of higher concentration to a region of lower concentration are said to be moving *along a gradient.* Diffusion only occurs along a gradient. A substance moving in the opposite direction, toward a higher concentration of its own molecules, would be moving *against a gradient,* which is analogous to being pushed uphill. The steeper the downhill gradient—that is, the larger the difference in concentration—the more rapid the net flow. Also, diffusion is more rapid in gases than in liquids and is more rapid at higher than at lower temperatures. Can you explain why?

Notice that there are two gradients in our imaginary tank; the dye molecules are moving along one of them, and the water molecules are moving in the opposite direction along the other. In both cases, the movement is along a gradient. When the molecules have reached a

Cells and Diffusion

Diffusion is essentially a slow process, except over very short distances. It is efficient only if the concentration gradient is steep and the volume is relatively small. For instance, the rapid spread of a perfume through the air is not due primarily to diffusion but, rather, to the circulation of air currents. Similarly, in many cells, the transport of materials is speeded by active streaming of the cytoplasm. Cells also hasten diffusion by their own metabolic activities. For example, in a nonphotosynthesizing cell, oxygen is used up within the cell almost as rapidly as it enters, thereby maintaining a steep oxygen gradient from outside to inside. Carbon dioxide is produced by the cell, and so a gradient from inside to outside is maintained for carbon dioxide.

Similarly, within a cell, materials are often produced in one place and used in another. Thus, a concentration gradient can be established between two areas, and materials can diffuse along the gradient from the site of production to the area of use.

Most organic molecules are polar (hydrophilic) and so cannot freely diffuse through the lipid barrier of cellular membranes. However, carbon dioxide and oxygen, which are soluble in lipids, move freely through these same membranes. Water also moves in and out freely. Since water is not soluble in lipids, the fact that water moves so freely has led biologists to postulate the presence of pores in the membrane that permit the passage of water molecules (and some small ions).

Osmosis

While permitting the passage of water, cellular membranes block the passage of most dissolved substances. (These substances are the _solutes_, and the water is the _solvent_ of the solution.) Such a membrane is known as a _differentially permeable membrane_, and the diffusion of water through it is known as _osmosis_. Osmosis involves a net flow of water from a solution that has a high water potential to a solution that has a low water potential. In the absence of other factors that influence water potential (such as pressure), the movement or diffusion of the water by osmosis will be from a region of lower solute concentration (and therefore of higher water concentration) into a region of higher solute concentration (and lower water concentration). The presence of solute decreases the water potential, creating a gradient of water potential along which water moves.

Osmosis can result in a buildup of pressure as water molecules continue to diffuse across the membrane into regions of lower water concentration. As shown in Figure 4–4, if water is separated from a solution by a semipermeable membrane (a membrane that allows the ready passage of water but no passage of solute), water will move across the membrane and cause the solution to rise in the tube until equilibrium is reached—that is, until the water potential on both sides of the membrane is equal. If enough pressure is applied from the upper part of the tube, it is possible to prevent the net movement of water into it. The pressure that must be applied to the solution to stop the movement of water is called the _osmotic pressure_. (Increasingly, botanists working with plant-water relationships use the terms "osmotic potential" or "solute potential" rather than "osmotic pressure." The terms are numerically equal, but osmotic potential is negative.) The term "osmotic pressure" is also used to express the reduction of water potential caused by the solute. An increase in solute concentration increases the osmotic pressure and decreases the water potential of a solution.

The movement of water is affected not by what is dissolved in the water but only by how much solute it contains—the number of particles of solute (molecules or ions). The word _isotonic_ was coined to describe solutions that have equal numbers of dissolved particles and therefore exert equal osmotic pressures. There is no net movement of water across a membrane separating two solutions that are isotonic to one another, unless, of course, physical pressure is exerted on one side. In comparing solutions with different concentrations, the solution that contains less solute and therefore exerts a lower osmotic pressure is known as _hypotonic_, and the one that has more solute and exerts a higher osmotic pressure is known as _hypertonic_. (Note that _iso-_ means "the same"; _hyper-_ means "more"—in this case, more molecules of solute; and _hypo-_ means "less"—in this case, fewer molecules of solute.)

Since solutes decrease water potential, a hypotonic solution has a higher water potential than a hypertonic solution. _In osmosis, water molecules move through a differentially permeable membrane into a hypertonic solution until the water potential is equal on both sides of the membrane._

Osmosis and Living Organisms

The movement of water across the plasma membrane from a hypotonic to a hypertonic solution causes some crucial problems for living systems, particularly those in an aqueous environment. These problems vary according to whether the cell or organism is hypotonic, isotonic, or hypertonic in relation to its environment. For example, one-celled organisms that live in salt water are usually isotonic with the medium they inhabit, which is one way of solving the problem. Similarly, the cells of higher animals are isotonic with the blood and lymph that constitute the watery medium in which they exist.

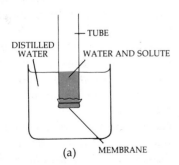

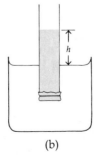

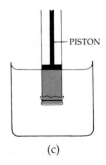

4–4

Osmosis and osmotic pressure. (a) The tube contains a solution; the beaker contains distilled water. (b) The semipermeable membrane permits the passage of water but not solute. The movement of water into the solution _causes the solution to rise in the tube until the osmotic pressure, resulting from the tendency of water to move into a region of lower water concentration, is counterbalanced by the height,_ h, _and density of the column of solution. (c) The_ _force that must be applied to the piston to oppose the rise of the solution in the tube is a measure of the osmotic pressure. It is proportional to the height and density of the solution in the tube._

Many types of cells live in a hypotonic environment. In many freshwater single-celled organisms, such as *Euglena,* the interior of the cell is hypertonic relative to the surrounding water; consequently, water tends to move into the cell by osmosis. If too much water moves into the cell, it can dilute the cell contents to the point of interfering with function and can eventually even rupture the plasma membrane. This is prevented by a specialized organelle known as a contractile vacuole, which collects water from various parts of the cell body and pumps it out with a rhythmic contraction.

Turgor

If a plant cell is placed in a hypotonic solution, the protoplast expands and the plasma membrane stretches and exerts pressure against the cell wall. However, the plant cell will not rupture because it is restrained by the relatively tough cell wall.

Plant cells tend to concentrate relatively strong solutions of salts within their vacuoles, and they can also accumulate sugars, organic acids, and amino acids. As a result, plant cells constantly absorb water by osmosis and build up their inner hydrostatic pressure. This pressure against the cell wall keeps the cell stiff, or *turgid.* Consequently, the hydrostatic pressure in plant cells is commonly referred to as turgor pressure. *Turgor pressure* may be defined as *the pressure that develops in a plant cell as a result of osmosis and/or imbibition* (see "Imbibition," page 60). Equal to and opposing the turgor pressure is the inwardly directed mechanical pressure of the cell wall, called the *wall pressure.*

Turgor in the plant is especially important in the support of nonwoody plant parts. As discussed in Chapter

2, most of the growth of a plant cell is the direct result of water uptake, with most of the increase in size of the cell resulting from enlargement of the vacuoles. The hormone auxin presumably contributes to this water uptake by relaxing the cell wall and thus decreasing the resistance the wall exerts to turgor pressure.

Turgor is maintained by most plant cells because they generally exist in a hypotonic medium. However, if a turgid plant cell is placed in a hypertonic solution, water will leave the cell by osmosis, and the vacuole and protoplast will shrink, thus causing the plasma membrane to pull away from the cell wall (Figure 4–5). This phenomenon is known as *plasmolysis.* The process can be reversed if the cell is then transferred to pure water. Figure 4–6 shows *Elodea* leaf cells before and after plasmolysis. Although the plasma membrane and the tonoplast—the membrane surrounding the vacuole—are, with few exceptions, permeable only to water, the cell walls allow both solutes and water to pass freely through them. The loss of turgor by plant cells may result in *wilting,* or drooping of the leaves and stems.

STRUCTURE OF CELLULAR MEMBRANES

The model of membrane structure most widely accepted today is the *fluid-mosaic model.* As shown in Figure 4–7, the membrane is composed of a *lipid bilayer* in which globular proteins are embedded. These proteins, known as *integral proteins,* often extend across the bilayer and protrude on either side. The portion of the protein embedded in the bilayer is hydrophobic, while

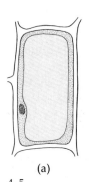 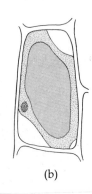 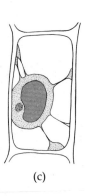

(a)   (b)   (c)

4–5

Plasmolysis in a leaf epidermal cell. (a) Under normal conditions, the protoplasm fills the space within the cell walls. (b) When the cell is placed in a relatively strong sucrose solution, water passes out of the cell into the hypertonic medium and the plasma membrane contracts slightly. (c) When immersed in a stronger (more concentrated) sucrose solution, the cell loses even larger amounts of water and contracts still further.

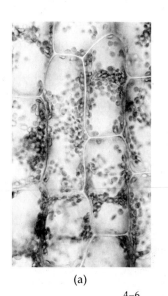 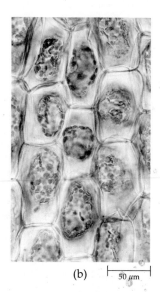

(a)   (b)   50 μm

4–6

Elodea leaf cells. (a) Turgid cells and (b) cells after being placed in a relatively strong sucrose solution. The cells in (b) are plasmolyzed.

that exposed on the outside is hydrophilic. Although the evidence remains inconclusive, there may be hydrophilic channels traversing some of the integral proteins. All membranes of a cell, including the plasma membrane and the internal membranes, have this same basic structure.

The two surfaces of a membrane differ considerably in chemical composition. For example, there are two major types of lipids in the plasma membrane of plant cells—phospholipids (the more abundant) and sterols—and the two layers of the bilayer have differing concentrations of them. Moreover, the integral proteins have definite orientations within the bilayer, and the portions protruding on either side have differing amino acid compositions and tertiary structures. On the inner surface of the membrane, additional protein molecules, known as *peripheral proteins*, are attached to some of the integral proteins protruding from the bilayer. On the outer surface, short-chain carbohydrates are attached to the protruding proteins. The carbohydrates, which form a coat on the outer surface of the membranes of some eukaryotic cells, are believed to play important roles in cell-to-cell adhesion processes and cell-surface transformations.

Whereas the lipid bilayer provides the basic structure of cellular membranes, the proteins are responsible for most membrane functions. Most membranes are composed of 40 to 50 percent lipid (by weight) and 50 to 60 percent protein, but the amounts and types of proteins in a membrane reflect its function. Some of the proteins are enzymes that regulate membrane-associated reactions, while others are carriers involved in the transport of specific molecules into and out of the cell. Still others act as receptors for receiving and transducing chemical signals from the cell's environment. Although some of the integral proteins appear to be anchored in place, the lipid bilayer is generally quite fluid. Some of the proteins float more or less freely in the bilayer, and they and the lipid molecules can move laterally within it, forming different patterns, or mosaics, that vary from time to time and place to place.

TRANSPORT ACROSS MEMBRANES

Molecules move across membranes by three different processes: simple *diffusion, facilitated diffusion,* and *active transport.* Nonpolar (hydrophobic) substances, including O_2, that are soluble in lipids usually cross a membrane by simple diffusion. (The observation that hydrophobic molecules diffused readily across plasma membranes provided the first evidence of the lipid nature of the membrane.)

Water and other polar (hydrophilic) molecules and ions might be expected to be excluded by a membrane's lipid bilayer, and yet hydrophilic molecules and ions do cross the membrane. How is this accomplished? In the case of water and certain other polar molecules, such as CO_2, diffusion across a membrane is possible in part because these molecules are small and uncharged.

The diffusion of a nonpolar or small uncharged molecule across a membrane is an example of *passive transport,* and the direction of that transport is determined only by the difference in the concentration of the molecule on the two sides of the membrane (the *concentration gradient*). If a solute carries a net charge, however, both the concentration gradient and the total electrical gradient across the membrane (the membrane poten-

4–7

Fluid-mosaic model of membrane structure. The membrane is composed of a bilayer (double layer) of lipid molecules—with their hydrophobic "tails" facing inward—and large protein molecules. The proteins embedded in the bilayer are known as integral proteins. On the inside of the bilayer other proteins, called peripheral proteins, are attached to some of the integral proteins. The portion of a protein's molecule embedded in the lipid bilayer is hydrophobic; the portion exposed on the outside is hydrophilic. Short carbohydrate chains are attached to the outside of the plasma membrane. The whole structure is quite fluid and, hence, the proteins can be thought of as floating in a lipid "sea."

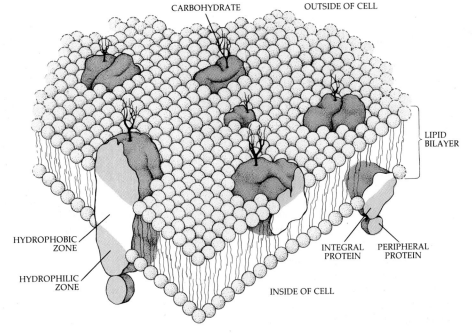

CARBOHYDRATE   OUTSIDE OF CELL

LIPID BILAYER

HYDROPHOBIC ZONE

HYDROPHILIC ZONE

INTEGRAL PROTEIN   PERIPHERAL PROTEIN

INSIDE OF CELL

tial) influence its transport. Together, both gradients constitute the *electrochemical gradient*. Plant cells typically maintain electrical gradients across the plasma membrane and the tonoplast. The ground substance is negative to both the aqueous medium outside the cell and the solution (cell sap) inside the vacuole.

Most substances required by cells are polar in nature and require *transport proteins* (carrier proteins) that are embedded within the membrane to transfer them across. Some transport proteins, called *uniports*, simply transport one solute from one side of the membrane to the other. Others function as *cotransport systems*, in which the transfer of one solute depends on the simultaneous or sequential transfer of a second solute. The second solute may be transported in the same direction (*symport*) or in the opposite direction (*antiport*) (Figure 4–8). All transport proteins apparently form a continuous protein pathway across the membrane. Hence, the solutes they transport do not come into direct contact with the hydrophobic interior of the lipid bilayer.

Two basically different categories of carrier-assisted transport have been proposed: facilitated diffusion and active transport. *Facilitated diffusion* is driven by the concentration gradient and so moves molecules down the concentration gradient. Neither simple diffusion nor facilitated diffusion (both examples of passive transport) is capable of moving solutes *against* a concentration gradient (Figure 4–9) or an electrochemical gradient. The ability to move solutes against a concentration gradient or an electrochemical gradient requires energy, and this process is called *active transport* (Figure 4–10). In plant and fungal cells, active transport is driven by a proton pump energized by ATP and mediated by an H^+-ATPase located in the membrane. The enzyme generates a large electrical potential and a pH gradient that provide the driving force for the uptake of all the H^+-coupled cotransport systems.

Carrier-assisted transport, whether by facilitated diffusion or active transport, is highly selective; the transport protein may accept one molecule and exclude a

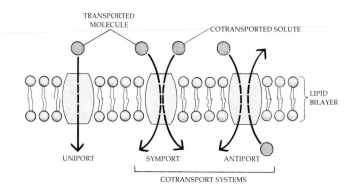

4–8

Diagram illustrating how the three types of transport proteins function. A uniport transport protein simply transfers a solute across the membrane. In the cotransport systems, the transport of one solute depends on the simultaneous or sequential transfer of a second solute, in either the same direction (symport) or opposite direction (antiport).

nearly identical one. The transport protein is not permanently altered in the process. In this respect, transport proteins are like enzymes; to emphasize this aspect of their function, transport proteins have been named *permeases*. However, they are unlike enzymes in the fact that they do not necessarily produce chemical change in the molecules with which they are temporarily bound.

The Sodium-Potassium Pump

One of the most important and best understood active transport systems in animal cells is the sodium-potassium pump. (Mechanisms that perform active transport are commonly called pumps.) Most animal cells maintain a differential concentration gradient of sodium ions

4–9

Diagram illustrating passive transport down an electrochemical gradient and active transport against such a gradient. Both simple diffusion and facilitated diffusion are passive processes, whereas active transport requires the expenditure of metabolic energy.

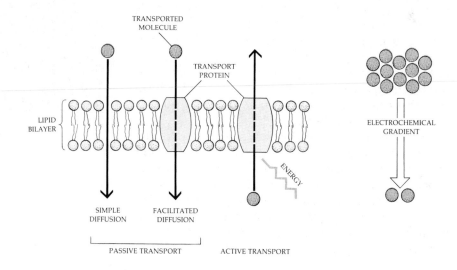

4–10

A model of the sodium-potassium pump. (a) An Na⁺ ion in the cytoplasm fits precisely into the transport protein. (b) A chemical reaction involving ATP then attaches a phosphate group (P) to the protein, releasing ADP (adenosine diphosphate). This process results in (c) a change of shape that causes the Na⁺ to be released outside the cell. (d) A K⁺ ion in the extracellular space is bound to the transport protein (e), which in this form provides a better fit for K⁺ than for Na⁺. (f) The phosphate group is then released from the protein, inducing conversion back to the other shape, and the K⁺ ion is released into the cytoplasm. The transport protein is now ready to transport another Na⁺ ion out of the cell.

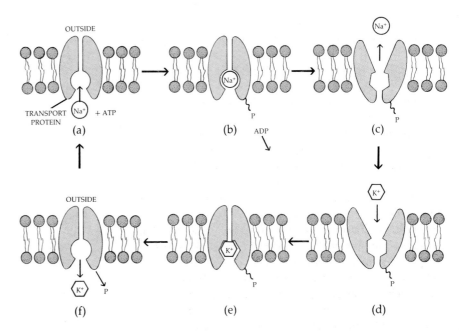

(Na^+) and potassium ions (K^+) across the plasma membrane: Na^+ is maintained at a low concentration inside the cell, and K^+ is kept at a high concentration. The Na^+ gradient is used in animal cells the same way the H^+ gradient is used in plants and fungi—to provide the driving force for the uptake of all the Na^+-coupled cotransport systems. The energy needed to drive the sodium-potassium pump is provided by ATP molecules manufactured in respiration. A measure of the importance of this mechanism to an organism is that more than a third of the ATP used by a resting animal is consumed by the sodium-potassium pump.

The transport of Na^+ and K^+ ions is accomplished by a transport protein that is believed by some investigators to exist in two alternative forms. One has a cavity opening to the inside of the cell, into which a Na^+ ion can fit; the other has a cavity opening to the outside, into which a K^+ ion fits. As shown in Figure 4–10, Na^+ binds to the transport protein. ATP is broken down to ADP, and the released phosphate is attached to the protein (the protein is phosphorylated). This causes a change in the conformation (form) of the protein such that the bound Na^+ is brought to the outer surface of the membrane and released. The transport protein then picks up a K^+ ion, which results in a dephosphorylation of the protein, thus causing it to return to the first conformation and release the K^+ ion to the inside of the cell. This process generates a gradient of Na^+ and K^+ ions across the membrane.

The proton pump in the plasma membrane of plants and fungi is chemically very similar to the sodium pump of animal plasma membranes. The plant protein is phosphorylated during the reaction cycle, in a manner similar to the animal pump described above. Despite the similarity in chemical structure and enzyme mechanism, the two pumps utilize different cations, H^+ in plants and fungi and Na^+ in animals.

ENDOCYTOSIS AND EXOCYTOSIS

In *endocytosis*, materials are taken into cells by means of invagination of the plasma membrane. The invaginations form small, saclike structures that are pinched off the plasma membrane and carried, with their enclosed material, into the cytoplasm.

When the substance to be taken in is a solid, such as a bacterial cell, the process is usually called *phagocytosis*, from the Greek word *phagein*, "to eat." Many one-celled organisms, such as amoebas, feed in this way. Among the organisms discussed in this book, the true slime molds and cellular slime molds exhibit phagocytosis (Figure 4–11).

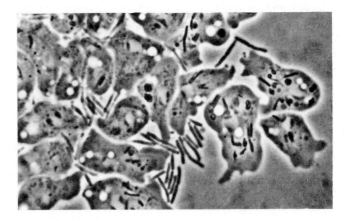

4–11

The amoeboid feeding stage of the cellular slime mold Dictyostelium aureum. *These amoebas have been feeding upon the rod-shaped bacterium* Escherichia coli *by phagocytosis. Notice the whole bacterial cells both inside and outside the amoebas.*

The taking in of dissolved molecules, as distinct from solid particles, is sometimes given the special name of *pinocytosis* (from the Greek *pinein*, "to drink"), although it is, in principle, the same process as phagocytosis. Pinocytosis occurs not only in single-celled organisms but also in multicellular plants and animals.

Phagocytosis and pinocytosis can also work in reverse. Many substances are exported from cells in vesicles or special vacuoles. One example cited in Chapter 2 is the role of dictyosome vesicles in cell wall formation. The vesicles, with their enclosed cell wall precursors, move to the surface of the cell. When they reach the plasma membrane, their bounding membranes fuse with it, and their contents are expelled into the region of the developing cell wall. This reverse endocytosis is referred to as *exocytosis*.

Although phagocytosis and pinocytosis appear to be superficially different from membrane transport systems involving carrier molecules, they are fundamentally the same; all three depend on the capacity of the membrane to "recognize" particular molecules.

TRANSPORT VIA PLASMODESMATA

As described in Chapter 2, neighboring cells of the plant body are interconnected by narrow strands of cytoplasm called plasmodesmata, which provide potential pathways for the passage of substances from cell to cell. The term *symplast* is used to refer to the interconnected protoplasts and their plasmodesmata. The movement of substances from cell to cell by means of the plasmodesmata is called *symplastic transport*. In contrast, the movement of substances in the cell wall continuum, or *apoplast*, surrounding the symplast is called *apoplastic transport*.

Plasmodesmata may provide a more efficient pathway between neighboring cells than the less direct, alternative route of plasma membrane, cell wall, and plasma membrane. It is believed that cells and tissues that are far removed from direct sources of nutrients can be supplied with nutrients either by simple diffusion or bulk flow through plasmodesmata. In addition, some substances are believed to move through plasmodesmata to and from the xylem and phloem—the tissues concerned with long-distance transport in the higher plant body. The substances may be transported within the desmotubules, which are continuous with the endoplasmic reticulum of adjacent cells and/or by way of the canals surrounding the desmotubules, provided that the desmotubules and/or the canals are not constricted or blocked (Figure 4–12).

Evidence for transport between cells via plasmodesmata comes from studies involving the use of fluorescent dyes and electrical currents that are injected into cells. The dyes, which do not easily cross the plasma membrane, can be observed moving from the injected cells into neighboring cells and beyond through adjacent cells (Figure 4–13). The passage of the electrical current pulses is monitored by means of receiver electrodes placed in neighboring cells. The extent of the electrical force detected is found to vary with the frequency of plasmodesmata and number of cells and

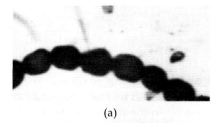

(a)

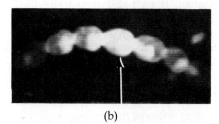

(b)

4–13

Staminal hairs of Setcreasea purpurea *before* (a) *and after* (b) *injection of one of the hair cells with the fluorescent dye disodium fluorescein. When the dye was injected into the cytoplasm of the cell indicated by the arrow, it passed in both directions into the cytoplasm of its neighboring cells.* (b) *Two minutes after injection. Inasmuch as the plasma membrane is impermeable to the dye, the movement of dye from cell to cell must have occurred via plasmodesmata, which occur in the walls between cells.*

ENDOPLASMIC RETICULUM

DESMOTUBULE

PLASMA MEMBRANE

CYTOPLASMIC ANNULUS

PRIMARY WALL

MIDDLE LAMELLA

PRIMARY WALL

NECK CONSTRICTION

(a)   (b)   (c)

4–12

Diagrams illustrating possible variation in the structure of plasmodesmata. (a) *Both the desmotubule and the "cytoplasmic annulus" between the desmotubule and plasma membrane lining the plasmodesmatal canal are open pathways.*

(b) *The cytoplasmic annulus is open but the desmotubule is closed.* (c) *The desmotubule is open but the cytoplasmic annulus is closed at either end by "neck constrictions."*

length of cells between the injection and receiver electrodes. Whether the plasmodesmata have any control over the movement of substances from cell to cell has not yet been determined, although some investigators have found what might be "valves" in some plasmodesmata.

SUMMARY

The plasma membrane regulates the passage of materials into and out of the cell, a function that makes it possible for the cell to maintain its structural and functional integrity. This regulation depends on interaction between the membrane and the materials that pass through it.

Water is one of the principal materials passing into and out of cells. Water moves by bulk flow, diffusion, and osmosis. Water potential determines the direction the water moves; that is, the movement of water is from where the water potential is higher to where it is lower. Bulk flow is the overall movement of water molecules, as when water flows downhill or moves in response to pressure.

Diffusion involves the random movement of molecules and results in net movement along a concentration gradient. Carbon dioxide and oxygen are two important molecules that move into and out of cells by diffusion across the membrane. Diffusion is most efficient when the distance involved is small and when the concentration gradient is steep. The rate of movement of substances within cells is increased by cytoplasmic streaming.

Osmosis is the diffusion of water through a membrane that permits the passage of water but inhibits the movement of solutes; such a membrane is called a differentially permeable membrane. In the absence of other forces, the movement of water by osmosis is from a region of lower solute concentration (a hypotonic medium, one of higher water potential) to a region of higher solute concentration (a hypertonic medium, one of lower water potential). The turgor (rigidity) of plant cells is a result of osmosis.

According to the fluid-mosaic model of membrane structure, cellular membranes are composed of lipid bilayers in which globular proteins are suspended. Some of these proteins act as carriers, transferring molecules across the membrane. If the carrier-assisted transport is driven by a concentration gradient, the process is known as facilitated diffusion. If it requires energy, it is known as active transport. Active transport can move substances against a concentration gradient. One of the most important active-transport systems in animal cells is the sodium-potassium pump, which maintains sodium ions at a low concentration and potassium ions at a high concentration in the cytoplasm.

Controlled movement into and out of a cell may also occur by endocytosis or by exocytosis, processes in which substances are transported in vesicles. Endocytosis of solids is called phagocytosis, whereas endocytosis of dissolved molecules is called pinocytosis.

Movement of substances between plant cells may also occur by way of narrow strands of cytoplasm, the plasmodesmata, which interconnect the protoplasts of neighboring cells. Such movement is known as symplastic transport.

SUGGESTIONS FOR FURTHER READING

ALBERTS, BRUCE, DENNIS BRAY, JULIAN LEWIS, MARTIN RAFF, KEITH ROBERTS, and JAMES D. WATSON: *Molecular Biology of the Cell*, Garland Publishing, Inc., New York, 1983.*

A large book covering both the molecular biology of the cell and the behavior of cells in multicellular animals and plants. A truly modern approach to the cell, this book is beautifully illustrated. Written for introductory courses in cell biology.

CRAM, JANE M., and DONALD J. CRAM: *The Essence of Organic Chemistry*, Addison-Wesley Publishing Co., Reading, Mass., 1978.

An elementary textbook emphasizing the features of organic chemistry most relevant to life processes.

GIESE, ARTHUR C.: *Cell Physiology*, 5th ed., W. B. Saunders Co., Philadelphia, 1979.

A description of the major problems in cell physiology; intended for the more advanced student.

GUNNING, B. E. S., and A. W. ROBARDS (Eds.): *Intercellular Communications in Plants: Studies on Plasmodesmata*, Springer-Verlag, New York, 1976.

An outstanding series of reviews of plasmodesmatal structure, distribution, and probable function by leading specialists in the field.

GUNNING, B. E. S., and M. W. STEER: *Ultrastructure and the Biology of Plant Cells*, Crane, Russak & Co., Inc., New York, 1975.

A collection of plant electron micrographs and accompanying explanations.

LEHNINGER, ALBERT L.: *Principles of Biochemistry*, Worth Publishers, Inc., New York, 1982.

This introductory text is outstanding both for its clarity and for its consistent focus on the living cell. There are numerous medical and practical applications throughout.

O'BRIEN, T. P., and MARGARET E. McCULLY: *Plant Structure and Development: A Pictorial and Physiological Approach*, The Macmillan Company, London, 1969.*

A beautifully illustrated atlas of cellular structure and development in the higher plants.

ROBINSON, DAVID G.: *Plant Membranes*, John Wiley & Sons, Inc., New York, 1985.

An up-to-date monograph on the structure and function of plant membranes.

STUMPF, P. K., and E. E. CONN (Eds.): *The Biochemistry of Plants: A Comprehensive Treatise*, Vol. 1, The Plant Cell, N. E. Tolbert (Ed.), Academic Press, New York, 1980.

A multiauthored book considering the structure and function of plant cells. The first chapter is introductory in nature and discusses the cell in general. Each of the remaining chapters is devoted to a different subcellular component.

* Available in paperback.

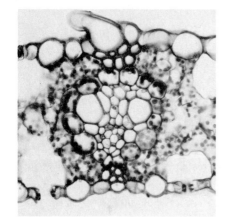

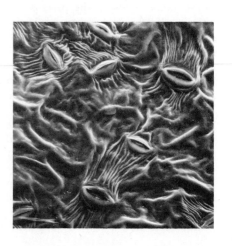

SECTION TWO

Energy and the Living Cell

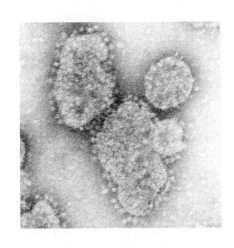

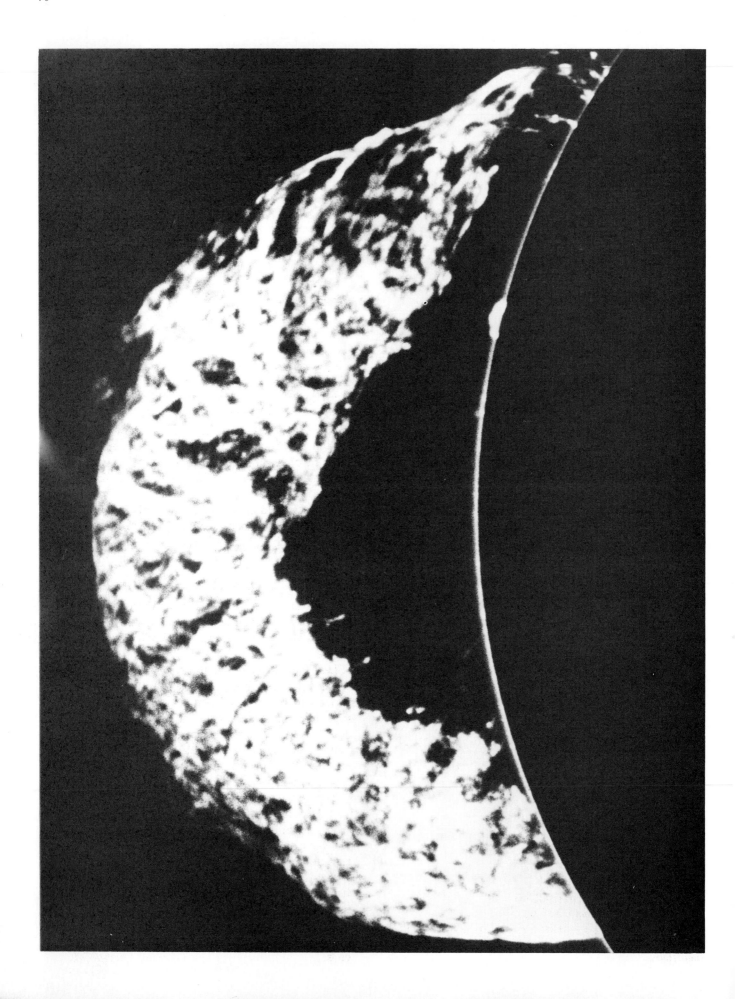

C H A P T E R   5

The Flow of Energy

5-1

The sun, our nearest star. During photosynthesis, the radiant energy from the sun is changed to chemical energy that is then available to drive the metabolic processes associated with life on earth. Shown here is an immense arching plume of hot gas—a solar prominence—produced during a violent solar storm.

Life here on earth depends on the flow of energy from thermonuclear reactions taking place at the heart of the sun (Figure 5–1). The amount of radiant energy delivered by the sun to the earth is 13×10^{23} (the number 13 followed by 23 zeros) calories per year. This is certainly a difficult quantity to imagine. For example, the amount of energy striking the earth every day is the equivalent of about 1 million Hiroshima-sized atomic bombs.

About a third of this solar energy is immediately reflected back into space as light (just as light is from the moon). Much of the remaining two-thirds is absorbed by the earth and converted to heat. Some of this absorbed heat energy serves to evaporate the waters of the oceans, producing the clouds that, in turn, produce rain and snow. Solar energy, in combination with other factors, is also responsible for the movements of air and water that help set patterns of climate over the surface of the earth.

Less than 1 percent of the solar energy reaching the earth becomes, through a series of operations performed by the cells of plants and other photosynthetic organisms, the energy that drives nearly all the processes of life. Living systems change energy from one form to another, transforming the radiant energy of the sun into the chemical and mechanical energy used by living organisms (see Figure 5–2).

These concepts of a vital relationship between plants and animals, and between energy and life, are relatively recent ones. They form part of the study of *thermodynamics*—the science of energy exchanges. Energy is such a common word in our energy-conscious world that it is surprising to discover that the word itself did not come into existence until about 100 years ago. Let us first consider a few key thermodynamic principles and then explore the manner in which enzymes mediate many of the reactions and processes carried out by cells.

5–2

An example of the flow of biological energy. The radiant energy of sunlight is produced by the fusion reactions taking place in the sun. Chloroplasts, present in all photosynthetic eukaryotic cells, capture this energy and use it to convert water and carbon dioxide into carbohydrates, such as glucose, starch, and other foodstuff molecules. Oxygen is released into the air as a product of these photosynthetic reactions. Mitochondria—organelles present in all eukaryotic cells—break down these carbohydrates and capture their stored energy in ATP molecules. This process—cellular respiration—consumes oxygen and produces carbon dioxide and water, completing the cycle.

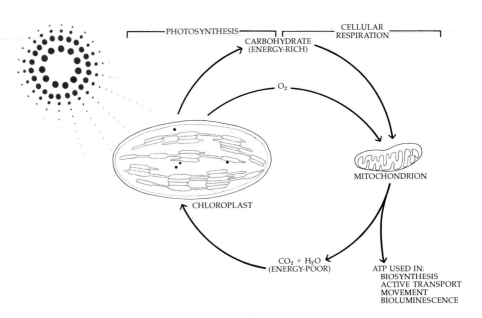

THE LAWS OF THERMODYNAMICS

Energy is an elusive concept. It is usually defined operationally, that is, by what it does rather than what it is. Until less than 200 years ago, heat—the form of energy most readily studied—was even considered to be a separate, though weightless, substance called "caloric." An object was hot or cold depending upon how much caloric it contained; when a cold object was placed next to a hot one, the caloric flowed from the hot object into the cold one; when metal was pounded with a hammer, it became warm because the caloric was forced to the surface. Even though the idea of a caloric substance proved incorrect, the concept is surprisingly useful.

The First Law

The rapid development of the steam engine in the latter part of the eighteenth century, more than any other single chain of events, changed scientific thinking about the nature of energy. Energy came to be associated with work, and heat and motion came to be seen as forms of energy. The way was opened for the formulation of the laws of thermodynamics. The *first law of thermodynamics* states, quite simply: *Energy can be changed from one form to another but cannot be created or destroyed.*

In engines, for instance, chemical energy (such as in coal or gasoline) is converted to heat energy, which is then partially converted to mechanical movements (kinetic energy). Some of the energy is converted back to heat by the friction of these movements, and some leaves the engine in the form of exhaust products. Unlike the heat in the engine or the boiler, the heat produced by friction and lost in the exhaust cannot produce "work"—that is, it cannot turn the gears—be-

cause it is dissipated into the environment. But it is nevertheless part of the total equation. In fact, engineers calculate that most of the energy consumed by an engine is dissipated randomly as heat; most engines work with less than 25 percent efficiency.

The notion of potential energy developed in the course of such engine-efficiency studies. A barrel of oil or a ton of coal could be assigned a certain amount of potential energy, expressed in terms of the amount of heat it would liberate when burned. The efficiency of the conversion of the potential energy to usable energy depended on the design of the energy conversion system.

Although these concepts were formulated in terms of engines running on heat energy, they apply to other systems as well. For example, a boulder pushed to the top of a hill contains energy—potential energy. Given a little push (the energy of activation), it rolls down the hill again, converting the potential energy to energy of motion and to heat produced by friction. As mentioned previously, water can also possess potential energy (page 59). As it moves by bulk flow from the top of a waterfall or over a dam, it can turn waterwheels that turn gears, for example, to grind corn. Thus the potential energy of water, in this system, is converted to the kinetic energy of the wheels and gears and to heat, which is produced by the movement of the water itself as well as the turning wheels and gears.

Light is another form of energy, as is electricity. Light can be changed to electrical energy, and electrical energy can be changed to light (say, by letting it flow through the tungsten wire in a light bulb).

The first law of thermodynamics states that in energy exchanges and conversions, wherever they take place and whatever they involve, the energy of the products of the reaction plus the energy released in the reaction is equal to the energy possessed by the initial reactants.

The Second Law

In a biological sense, the *second law of thermodynamics* is the more interesting one. It predicts the direction of all events involving energy exchanges; thus it has been called "time's arrow." One way of stating the second law is: *In all energy exchanges and conversions, if no energy leaves or enters the system under study, the potential energy of the final state will always be less than the potential energy of the initial state.* The second law is in keeping with everyday experience (Figure 5–3). Of its own accord, a boulder will roll downhill but never uphill. Heat will flow from a hot object to a cold one and never the other way. Our cells can process glucose enzymatically to yield carbon dioxide and water but—since we cannot capture the energy of the sun as plants do—our enzymatic processes cannot produce glucose from carbon dioxide and water.

A process in which the potential energy of the final state is less than that of the initial state is one that yields energy (otherwise it would be in violation of the first law). An energy-yielding process is called an *exergonic reaction.* As the second law predicts, only exergonic reactions can take place spontaneously. (The word "spontaneously" says nothing about the rate of reaction but only whether or not it can occur.) In contrast, *endergonic reactions* are energy-requiring reactions, and in order for them to proceed, an input of energy is required that is greater than the difference in energy between product and reactants.

One important factor in determining whether or not a reaction is exergonic is ΔH, the change in heat content of the system (Δ stands for change, H for heat content). In general, the change in heat content is approximately equal to the change in potential energy. As noted in Appendix A, the energy change that occurs when glucose, for instance, is oxidized can be measured in a calorimeter and expressed in terms of ΔH. The oxidation of a mole of glucose yields 673 kilocalories. (A mole is the amount of a substance, in grams, that equals its molecular weight. For example, the atomic weight of carbon is 12 and the atomic weight of oxygen is 16, so the molecular weight of CO_2 is 44. One mole of CO_2 is therefore 44 grams of CO_2.)

$$C_6H_{12}O_6 + 6O_2 \longrightarrow 6CO_2 + 6H_2O$$

$$\Delta H = -673 \text{ kcal/mole}$$

The minus sign indicates that energy has been released. Generally, a chemical reaction with a negative ΔH is an exergonic reaction.

Another factor besides the gain or loss of heat can determine the direction of a process. This factor, called *entropy,* is a measurement of the disorder, or randomness, of a system. Let us return to water as an example. The change from ice to liquid water and the change from liquid water to water vapor are both endothermic processes—a considerable amount of heat is removed

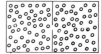

5–3
Some illustrations of the second law of thermodynamics. In each case, a concentration of energy—in the hot copper block, in the gas molecules under pressure, and in the neatly organized books—is dissipated. In nature, processes tend toward randomness, or disorder. Only an input of energy can reverse this tendency and reconstruct the initial state from the final state. Ultimately, however, disorder will prevail, since the total amount of energy in the universe is finite.

from the surroundings as they occur. Yet, under the appropriate conditions, they proceed spontaneously. The key factor in these processes is the increase in entropy. In the case of the change of ice into liquid water, a solid is being turned into a liquid, and some of the bonds that hold the water molecules together in a crystal (ice) are being broken. As the liquid water turns to vapor, the rest of the hydrogen bonds are ruptured as the individual water molecules separate, one by one. In each case, the disorder of the system has increased.

The notion that there is more disorder associated with more numerous and smaller objects than with fewer, larger ones is in keeping with our everyday experience. If there are 20 papers on a desk, the possibilities for disorder are greater than if there are 2 or even 10. If each of the 20 papers is cut in half, the entropy of the system—the capacity for randomness—increases. The relationship between entropy and energy is also a commonplace idea. If you were to find your room tidied up and your books in alphabetical order on the shelf, you would recognize that someone had been at work—that energy had been expended. To organize the papers on a desk similarly requires the expenditure of energy.

$E = mc^2$

Protons and neutrons are arbitrarily assigned an atomic weight of 1 (see Appendix A). One would therefore expect that an element with, for example, twice as many protons and neutrons as another element would weigh twice as much. This assumption is true—almost. If the weights of nuclei are measured with great accuracy, as they can be by instruments developed by modern physics, small nuclei always have, proportionately, slightly greater weights than larger nuclei. For example, the most common isotope of carbon has a combined total of 12 protons and neutrons; and so carbon, by convention, is assigned an atomic weight of 12. The hydrogen atom, however, has an atomic weight not of 1, as would be expected, but of 1.008. Helium has two protons and two neutrons. However, it does not have a weight of 4, or of 4.032 (four times the weight of hydrogen); it weighs, in relation to carbon, 4.0026. Similarly, oxygen, with a combined total of 16 protons and neutrons, has an atomic weight, in relation to that of carbon, of 15.995. In short, when protons and neutrons are assembled into an atomic nucleus, there are slight changes in weight, which reflect changes in mass.

One of the oldest and most fundamental concepts of chemistry is the law of conservation of mass—that mass is never created or destroyed. Yet under conditions of extremely high temperature, atomic nuclei fuse to make new elements. What happens to the mass "lost" in the course of this fusion? This is the question answered by Einstein's fateful equation $E = mc^2$, where E stands for energy, m stands for mass, and c is a constant equal to the speed of light. Einstein's equation simply means that under certain extreme and unusual conditions, mass is turned into energy.

Albert Einstein in 1905, the year he published his paper on the theory of relativity. He was 26 years old and working at the Swiss Patent Office in Bern as a technical expert third class.

The sun consists largely of hydrogen nuclei. At the extremely high temperatures at the core of the sun, hydrogen nuclei strike each other with enough velocity to fuse. In a series of steps, four hydrogen nuclei fuse to form one helium nucleus. In the course of these reactions, enough energy is released to keep the fusion reaction going and to emit tremendous amounts of radiant energy into space. Life on this planet depends on energy emitted by the sun in the course of this reaction. This same reaction provides—as Einstein foresaw—the energy of the hydrogen bomb.

Now let us return to the question of the energy changes that determine the course of chemical reactions. As discussed, both the change in the heat content of the system (ΔH) and the change in entropy (which is symbolized as ΔS) contribute to the overall change in energy. This total change—which takes into account both heat and entropy—is called the *free-energy change* and is symbolized as ΔG, after the American physicist Josiah Willard Gibbs (1839–1903), who was one of the first to integrate all of these ideas.

Keeping ΔG in mind, let us examine once again the oxidation of glucose. The ΔH of this reaction is −673 kcal/mole. The ΔG is −686 kcal/mole. Thus, the entropy factor has contributed 13 kcal/mole to the free-energy change of the process. The change in heat content and in entropy both contribute to the lower energy state of the products of the reaction.

The free-energy change, ΔG, can also enable one to predict processes that occur when ΔH is zero or even positive. For instance, it confirms our earlier observa-

tions that heat will flow from a warm object to a cold one or that dye molecules will diffuse in a beaker of water. In each of these processes, the final state has more entropy—and therefore less potential energy—than the initial state.

The relationship between ΔG, ΔH, and entropy is given in the following equation:

$$\Delta G = \Delta H - T\,\Delta S$$

This equation states that the free-energy change is equal to the change in heat content (a minus figure in exothermic reactions, which give off heat) minus the change in entropy, which is affected by the absolute temperature T. In exergonic reactions, ΔG is always negative, but ΔH may be zero or even positive. Since T is always positive, the greater the change in entropy, the more negative ΔG will be; that is, the more exergonic the reaction will be. Therefore it is possible to state the second law in another, simpler way: *In general, all natural processes are exergonic.*

Biology and the Second Law

The laws of thermodynamics are of crucial importance to biology, as they are to physics and chemistry. They are the organizing principles under which a number of different kinds of processes can be unified. Also, as we shall see in the following chapter, they permit a kind of biochemical bookkeeping.

The most interesting implication of the second law, as far as biology is concerned, is the relationship between entropy, on the one hand, and order and organization on the other. Living systems are continually expending large amounts of energy to maintain order. Stated in terms of chemical reactions, living systems continually expend energy to maintain a position far from equilibrium. If equilibrium were to occur, the chemical reactions in a cell would, for all practical purposes, stop and no further work could be done. At equilibrium, a cell would soon die.

OXIDATION-REDUCTION

Chemical reactions are essentially energy transformations in which energy stored in chemical bonds is transferred to other, newly formed chemical bonds. In such transfers, electrons shift from one energy level to another (see Appendix A). In many reactions, electrons pass from one atom or molecule to another. These reactions, known as oxidation-reduction (or redox) reactions, are of great importance in living systems. The *loss* of an electron is known as *oxidation*, and the atom or molecule that loses the electron is said to be oxidized. The reason electron loss is called oxidation is that oxygen, which attracts electrons very strongly, is most often the electron acceptor.

Reduction involves the *gain* of an electron. Oxidation and reduction occur simultaneously; the electron lost by the oxidized atom is accepted by another atom, which is thus reduced.

Redox reactions may involve only a solitary electron, as when sodium loses an electron and becomes oxidized to Na^+, and chlorine gains an electron and is reduced to Cl^-. Often, however, the oxidation of organic molecules involves the removal of both electrons and hydrogen ions (protons), and the reduction of such molecules involves the gain of electrons and protons. For example, when glucose is oxidized, electrons and hydrogen ions are lost by the glucose molecule and are gained by oxygen to form water:

$$C_6H_{12}O_6 + 6O_2 \longrightarrow 6CO_2 + 6H_2O + energy$$

The electrons are moved to a lower energy level, and energy is released.

Conversely, during photosynthesis, electrons and hydrogen ions are transferred from water to carbon dioxide, thereby reducing the carbon dioxide to form glucose:

$$6CO_2 + 6H_2O + energy \longrightarrow C_6H_{12}O_6 + 6O_2$$

In this case, the electrons are moved to a higher energy level, and an input of energy is required for the reaction to take place.

In living systems, the energy-capturing reactions (photosynthesis) and the energy-releasing reactions (glycolysis and respiration) are oxidation-reduction reactions. As we have seen, the complete oxidation of a mole of glucose releases 686 kilocalories of free energy (conversely, the reduction of carbon dioxide to form a mole of glucose stores 686 kilocalories of free energy in the chemical bonds of glucose). If the energy released during the oxidation of glucose were to be released all at once, most of it would be dissipated as heat. Not only would it be of no use to the cell, but the resulting high temperature would destroy the cell. Mechanisms have evolved in living systems, however, that regulate these chemical reactions—and a multitude of others—in such a way that energy is stored in particular chemical bonds from which it can be released in small amounts as required by the cell.

ENZYMES AND LIVING SYSTEMS

In any living system, thousands of different chemical reactions occur, many of them simultaneously. The sum of all these reactions is referred to as *metabolism* (from the Greek *metabolē*, meaning "change"). If one merely listed the individual chemical reactions, it would be difficult to understand a cell's metabolic activities. Fortunately, there are some guiding principles that lead one through the maze of cell metabolism. First, virtually all the chemical reactions that take place in a cell involve *enzymes*—the catalysts and regulators for the metabolic processes of living systems. Second, biochemists group these reactions in an ordered series of steps, called a pathway, often containing a dozen or more sequential reactions. Each pathway serves a function in the overall life of the cell or organism. Furthermore, certain pathways have many steps in common, such as those that are concerned with the synthesis of the different amino acids or the various nitrogenous bases. Some pathways converge; for example, the pathway by which fats are broken down to yield energy leads to the same pathway by which glucose is broken down to yield energy (Figure 5–4).

Many living systems have unique pathways. Plant cells expend energy building cellulose walls, an activity not engaged in by animal cells. Red blood cells specialize in the synthesis of hemoglobin molecules, which are made nowhere else in the animal body. It is not sur-

5–4
*A modern highway and the meandering
old road to its right are alternative paths
through these mountains. Identical
chemical reactions also can occur via
alternative pathways that involve vastly
different rates.*

prising that the distinctive differences among cells and
organisms are reflected not only in their forms and
functions but also in their biochemistry. What is surprising, however, is that much of the metabolism of
even the most diverse of organisms is exceedingly similar; the differences in many of the metabolic pathways
of humans, oak trees, mushrooms, and jellyfish are
very slight. Some pathways are found in virtually all
living systems.

The magnitude of the chemical work carried out by a
cell is illustrated by the fact that, for the most part, the
thousands of different molecules found within a cell
are synthesized there. The total of chemical reactions
involved in this synthesis is called *anabolism.* Anabolic
reactions usually involve an increase in atomic order (a
decrease in entropy) and are almost always energy-requiring (endergonic). Cells are also constantly involved
in the breakdown of larger molecules; these activities,
known collectively as *catabolism,* involve a decrease in
atomic order (an increase in entropy) and are usually
energy-liberating (exergonic). Catabolic reactions serve
two purposes: (1) they release the energy for anabolism
and other work of the cell, and (2) they serve as a
source of raw materials for anabolic processes. Hence,
both aspects of metabolism—anabolism and catabolism—are essential for normal cellular activities.

Living systems carry out this multitude of chemical
activities under extraordinarily difficult conditions.
Most of their chemical reactions are carried out within
living cells, and thousands of different kinds of molecules are intermingled. Temperatures cannot be high;
otherwise, many of the fragile structures on which life
depends would be destroyed. How is all this complex
chemical work accomplished? The question can be answered in a single word: enzymes. Without enzymes,
biochemical reactions would take place so slowly that,
for all practical purposes, they would not occur at all,
and the activities we associate with life would cease to
exist.

ENZYMES AS CATALYSTS

Enzymes are the catalysts of biological reactions. They
differ from other catalysts in that they are very selective
in their action. Some enzymes catalyze a reaction with
only a single set of reactants. In enzyme-catalyzed reactions, the reactant (or reactants) on which an enzyme
operates is called its *substrate.* The selectivity an enzyme exhibits in choosing a substrate is known as its
specificity. Enzymes otherwise resemble other catalysts
in that they are not used up in the course of the reaction and so can be used over and over again.

Enzymes enormously accelerate the rate at which reactions occur. For instance, the reaction of carbon dioxide with water,

$$CO_2 + H_2O \rightleftharpoons H_2CO_3$$

can occur spontaneously, as it does in the oceans. In the
human body, however, this reaction is catalyzed by an
enzyme, carbonic anhydrase (names of enzymes
usually end with the suffix "ase"). Carbonic anhydrase
is one of the most effective enzymes known—each enzyme molecule leads to the production of 10^5 (100,000)
molecules of the product, carbonic acid, per second.
The catalyzed reaction is 10^7 times faster than the uncatalyzed one. In animals, this reaction is essential in
the transfer of carbon dioxide from the cells, where it is
produced, to the bloodstream, which transports it to the
lungs.

THE ACTIVE SITE

Enzymes are complex globular proteins consisting of
one or more polypeptide chains (see Chapter 3). They
are folded so as to form a groove or pocket into which
the reacting molecule or molecules—the substrate—fit
and where the reactions occur. This portion of the enzyme is known as the *active site.* The active site is
formed by the very exact folding of the polypeptide
chain. The relationship between the active site and the
substrate is very precise and is often compared to that
of a lock and key (Figure 5–5).

5–5

A model of the lock-and-key hypothesis of enzyme action. A molecule of sucrose is hydrolyzed to yield one molecule of glucose and one molecule of fructose. The enzyme involved in the reaction is specific for this process; the active site of the enzyme exactly fits the opposing surface of the sucrose molecule.

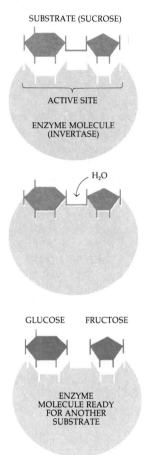

5–6

A model of an enzyme. This enzyme (the digestive enzyme chymotrypsin) is composed of three polypeptide chains. The amino (—NH₂) and the carboxyl (—COOH) groups of each are labeled. The numbers represent the positions of particular amino acids in the chains, and disulfide bridges connect certain amino acids. The three-dimensional shape of the molecule is the result of a combination of disulfide bonds and of interactions among the chains and between the chains and the surrounding water molecules, based on the positive or negative charges (the polarity) of the various amino acids. As a result of this bending and twisting of the polypeptide chains, particular amino acids come together in a highly specific configuration to form the active site of the enzyme. Two amino acids known to be part of the active site are shown in color.*

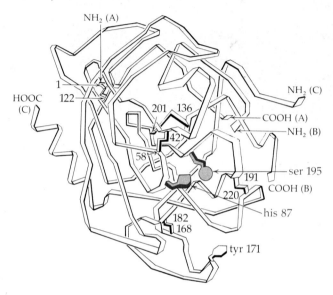

5–7

A model of the induced-fit hypothesis of enzyme action. The active site is believed to be flexible and thus adjustable in shape, or conformation, to that of the substrate molecule. This induces a particularly close fit between the active site and the substrate.*

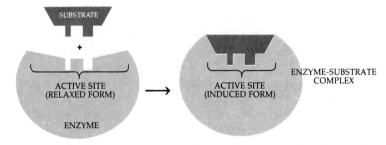

The active site not only has a precise three-dimensional shape but also has exactly the correct array of charged and uncharged, or hydrophilic and hydrophobic, areas on its binding surface. If a particular portion of the substrate has a negative charge, the corresponding feature on the active site has a positive charge, and so on. Thus, the active site not only confines the substrate molecule but also orients it in the correct direction.

The amino acids involved in the active site need not be adjacent to one another on the polypeptide chains. In fact, in an enzyme with quaternary structure, they may even be on different polypeptide chains, as shown in Figure 5–6. The amino acids are brought together at the active site by the precise folding of the polypeptide chains in the molecule.

The Induced-Fit Hypothesis

During the last several years, studies of enzyme structure have suggested that the binding that takes place between enzyme and substrate alters the conformation of the enzyme, thus inducing an even closer fit between the active site and the reactants. It is believed that this induced fit may place some strain on the reacting molecules and so further facilitate the reaction (Figure 5–7).

COFACTORS IN ENZYME ACTION

The catalytic activity of some enzymes appears to depend only on their structure as proteins. Many enzymes, however, require one or more nonprotein components, known as *cofactors*, in order to function.

Ions as Cofactors

Certain ions are cofactors for particular enzymes. For example, the magnesium ion (Mg^{2+}) is required in most enzymatic reactions involving the transfer of a phosphate group from one molecule to another. Its two positive charges hold the negatively charged phosphate group in position. Other ions, such as Na^+ and K^+, play similar roles in other reactions. In some cases, ions serve to hold the enzyme protein together.

Coenzymes and Vitamins

Nonprotein organic cofactors can also play a crucial role in enzyme-catalyzed reactions. Such cofactors are called *coenzymes*. For example, in some oxidation-reduction reactions, electrons are passed along to a molecule that serves as an electron acceptor. There are several different electron acceptors in any given cell, and each is tailor-made to hold the electron at a slightly different energy level. As an example, let us look at just one, nicotinamide adenine dinucleotide (NAD), which is shown in Figure 5–8. At first glance, NAD looks complex and unfamiliar, but if you look at it more closely, you will find that you recognize most of its component parts. The two units labeled ribose are five-carbon sugars; they are linked by two phosphate groups. One of the sugars is attached to the nitrogenous base adenine. The other is attached to another nitrog-

enous base, nicotinamide. (A nitrogenous base plus a sugar plus a phosphate is called a *nucleotide*; a molecule that contains two of these combinations is called a dinucleotide.)

The nicotinamide ring is the active end of NAD, that is, the part that accepts electrons. Nicotinamide is a vitamin—niacin. Vitamins are compounds that are required in small quantities by many living organisms; humans and other animals cannot synthesize vitamins and so must obtain them in their diets. When nicotinamide is present, our cells can use it to make NAD. Many vitamins are coenzymes or parts of coenzymes.

Nicotinamide adenine dinucleotide, like many other coenzymes, is recycled. That is, NAD^+ is regenerated when $NADH + H^+$ passes its electrons to another electron acceptor. Thus, although this coenzyme is involved in many cellular reactions, the actual number of NAD molecules required is relatively small.

Some enzymes use cofactors that remain attached to the protein. Such cofactors are referred to as prosthetic groups. Examples of prosthetic groups are the iron-sulfur clusters of ferredoxins (page 104), or the pyridoxal phosphate (vitamin B_6) of some transaminases.

ENZYMATIC PATHWAYS

Enzymes work in an ordered series of steps—the pathways we referred to earlier. Consequently, a living organism carries out its chemical activities with remarkable efficiency. First, there is little accumulation of waste products, since each product tends to be used up in the next reaction along the pathway. A second advantage of such sequential reactions is understandable when one considers that chemical reactions can go in either direction; that is, they are reversible (see Ap-

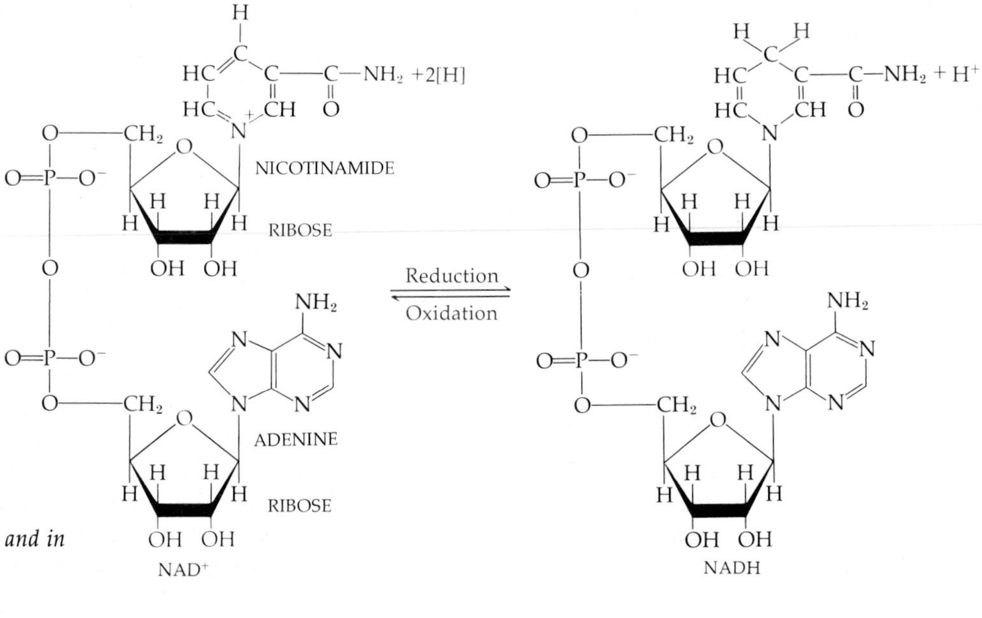

5–8

Nicotinamide adenine dinucleotide (NAD) in its oxidized form, NAD⁺, and in its reduced form, NADH.

pendix A). If each product along a series of reactions is used up by the next reaction almost as rapidly as it is formed, the tendency for reversal of the reactions will be minimized. Moreover, if the eventual end product is also used up rapidly, the whole series of reactions will move toward completion. Another advantage is that the groups of enzymes making up a common pathway can be segregated within the cell. Some are found in small vesicles, or membrane-surrounded sacs, in the cytoplasm. Others are embedded in the membranes of specialized organelles, such as mitochondria and chloroplasts.

REGULATION OF ENZYME ACTIVITY

Another remarkable feature of the metabolic activity of cells is the extent to which each cell regulates the synthesis of the products necessary for its well-being, making them in the appropriate amounts and at the rates required. At the same time, cells avoid overproduction, which would waste both energy and raw materials. The availability of reactant molecules or of cofactors is a principal factor in limiting enzyme action, and for this reason most enzymes probably work at a rate well below their maximum.

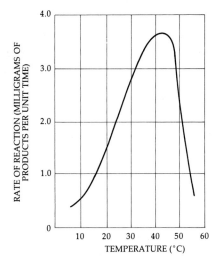

5–9
The effect of temperature on the rate of an enzyme-controlled reaction. The concentrations of enzyme and reacting molecules (substrate) were kept constant. The rate of the reaction, as in most chemical reactions, approximately doubles for every 10°C rise in temperature up to about 40°C. Above this temperature, the rate decreases, and at about 60°C, the reaction stops altogether, presumably because the enzyme is denatured.

Temperature affects enzymatic reactions. An increase in temperature increases the rate of enzyme-catalyzed reactions, but only to a point. As can be seen in Figure 5–9, the rate of most enzymatic reactions approximately doubles for each 10°C rise in temperature and then drops off very quickly after about 40°C. The increase in reaction rate occurs because of the increased energy of the reactants; the decrease in the reaction rate occurs as the enzyme molecule itself begins to move and vibrate, disrupting the hydrogen bonds and other relatively fragile forces that hold it together.

The pH of the surrounding solution also affects enzyme activity. Among other factors, the conformation of an enzyme depends on the attraction and repulsion between negatively charged (acidic) and positively charged (basic) amino acids. As the pH changes, these charges change, and so the shape of the enzyme changes until it is so drastically altered that it is no longer functional. More important, probably, the charges of the active site and substrate are changed so that the binding capacity is affected. Some enzymes are frequently found at a pH that is not their optimum, suggesting that this discrepancy may not be an evolutionary oversight but a way of controlling enzyme activity.

Living systems also have more precise ways of turning enzyme activity on and off. Some enzymes are produced only in an inactive form and are activated only when they are needed, usually by another enzyme. Specific mechanisms by which enzymes may be activated or inactivated are discussed in Chapter 8.

THE ENERGY FACTOR: ATP

All of the biosynthetic (anabolic) activities of the cell (and many other activities as well) require energy. A large proportion of this energy is supplied by a single molecule, *adenosine triphosphate* (ATP), a nucleotide derivative that is the cell's chief energy currency. Glucose, glycogen, and starch are like money in the bank. ATP is like change in your pocket.

At first glance, ATP (shown in Figure 5–10 on page 80) also appears to be a complex molecule. However, as with NAD, you will find that its component parts are familiar. ATP is made up of adenine, the five-carbon sugar ribose, and three phosphate groups. These three phosphate groups, with strong negative charges, are covalently bonded to one another; this is an important feature in ATP function.

To understand the role of ATP, let us briefly review the concept of the chemical bond and bond energies. A chemical bond is the total of the forces that hold the constituent atoms together in a molecule. Because a bond is a stable configuration, energy is needed to break one bond and form a new one. This energy is the energy of activation (see Appendix A). Because of en-

5–10

Adenosine triphosphate (ATP) is the cell's chief energy currency. The bonds between the three phosphate groups in the molecule are important in ATP function. The symbol ~ designates a high-energy bond.

zymes, which greatly reduce the energy of activation required, the reactions essential to life can proceed at an appropriate rate. However, there is an important additional limitation on chemical reactions in living systems: the bond energies of the products must always be less than the bond energies of the reactants. Following this statement to its logical conclusion, one might think that *biosynthetic reactions* could never occur. This, of course, is not true. Cells circumvent this requirement by *coupled reactions* in which energy-requiring chemical reactions are linked to energy-releasing reactions. The molecule that participates most frequently in such coupled reactions is ATP.

Because of its internal structure, the ATP molecule is well suited to this role in living systems. Energy is released from the ATP molecule when one phosphate group is removed by hydrolysis, producing a molecule of ADP (adenosine diphosphate):

$$\text{ATP} + \text{H}_2\text{O} \longrightarrow \text{ADP} + \text{phosphate}$$

In the course of this reaction, some 7.3 kcal/mole of ATP are released—a relatively large amount of chemical energy. Removal of a second phosphate group produces AMP (adenosine monophosphate) and releases an equivalent amount of chemical energy:

$$\text{ADP} + \text{H}_2\text{O} \longrightarrow \text{AMP} + \text{phosphate}$$

To indicate that relatively large amounts of chemical energy are involved, the covalent bonds linking these two phosphates to the rest of the molecule are often called high-energy bonds (Figure 5–10). However, this name is somewhat misleading because the energy released during the reaction does not arise entirely from the bond. The difference in energy between reactants and products is due only in part to the energy in the bonds. It is also partly the result of the rearrangement of the electron orbitals of the ATP or ADP molecules. The phosphate groups each carry a negative charge and so tend to repel each other. When a phosphate group is removed, the molecule undergoes a change in electron configuration that results in a structure with less energy.

In most reactions that take place within a cell, the terminal phosphate group of ATP is not simply removed but is transferred to another molecule. This addition of a phosphate group to a molecule is known as *phosphorylation;* the enzymes that catalyze such transfers are known as kinases. This reaction transfers some of the energy in the high-energy bond to the phosphorylated compound, which, thus energized, can participate in the reaction.

Let us look at a simple example of energy exchange involving ATP in the formation of sucrose in sugarcane. Sucrose is formed from the monosaccharides glucose and fructose; under standard thermodynamic conditions, its synthesis is strongly endergonic, requiring an input of 5.5 kcal for each mole of sucrose formed:

$$\text{glucose} + \text{fructose} \rightleftharpoons \text{sucrose} + \text{H}_2\text{O}$$

However, the synthesis of sucrose in sugarcane, when coupled with the breakdown of ATP, is actually exergonic. During the series of reactions involved in the formation of sucrose, a phosphate group is transferred to a glucose molecule and to a fructose molecule, thus energizing each of them; hence we have the overall equation:

$$\text{glucose} + \text{fructose} + 2\text{ATP} \longrightarrow$$
$$\text{sucrose} + 2\text{ADP} + 2\text{ phosphates} + \text{H}_2\text{O}$$

In this reaction, 5.5 kcal are utilized in the formation of sucrose, and the overall energy difference in products and reactants is about 8.5 kcal. Thus, the sugarcane plant is able to form sucrose by coupling the breakdown of two molecules of ATP to the synthesis of a covalent bond between glucose and fructose.

Where does the ATP originate? The energy released in the cell's catabolic reactions, such as in the breakdown of glucose, is used to "recharge" the ADP molecule. Of course, the energy released in these catabolic reactions is originally derived from the sun as radiant energy which is converted, during photosynthesis, into chemical energy. Some of this chemical energy appears as the high-energy bonds of ATP before being converted to chemical bond energies of other organic molecules. Thus the ATP/ADP system serves as a universal energy-exchange system, shuttling between energy-releasing and energy-requiring reactions.

SUMMARY

Life on this planet is dependent on the flow of energy from the sun. A small fraction of this energy, captured in the process of photosynthesis, is converted to the energy that drives the many other metabolic reactions associated with living systems and from which living systems derive their order and organization.

The thermodynamic relationship between photosynthesizers and nonphotosynthesizers is exceedingly complicated. Stated in its simplest form: In the course of photosynthesis, the energy of the sun is used to forge high-energy carbon-carbon and carbon-hydrogen bonds; then, in the course of respiration, these bonds are once more broken down to carbon dioxide and water, and energy is released. As in machines, some useful energy is lost in each step of these energy conversions.

Living systems thus operate in accord with the laws of thermodynamics. The first law of thermodynamics states that energy can never be created or destroyed, although it can be changed from one form to another. The potential energy of the initial state (or reactants) is equal to the potential energy of the final state (or products) plus the energy released in the process or reaction. The second law of thermodynamics states that, in the course of energy conversions, the potential energy of the final state will always be less than the potential energy of the initial state. The difference in energy between the initial and final state is known as the free-energy change and is symbolized as ΔG. Exergonic (energy-yielding) reactions have a negative ΔG. Factors that determine the ΔG include ΔH, the change in heat content, and ΔS, the change in entropy, which, multiplied by the absolute temperature T, is a measure of randomness, or disorder:

$$\Delta G = \Delta H - T\,\Delta S$$

Energy transformations in living cells involve the transfer of electrons from one energy level to another and, often, from one atom or molecule to another. Reactions involving the transfer of electrons from one atom or molecule to another are known as oxidation-reduction reactions. An atom or molecule that loses electrons is oxidized; one that gains electrons is reduced.

Metabolism is the total of all the chemical reactions that take place in cells. Reactions resulting in the breakdown or degradation of molecules are known, collectively, as catabolism. Biosynthetic reactions—the building of new molecules—are called anabolism. Metabolic reactions take place in an ordered series of steps, called pathways, each of which serves a particular function in a cell. Each step in the pathway is controlled by a particular enzyme.

Enzymes serve as catalysts; they enormously increase the rate at which reactions take place but remain unchanged in the process. Enzymes are large protein molecules folded in such a way that particular groups of amino acids form an active site. The reacting molecules, known as the substrate, fit precisely into this active site. Many enzymes require cofactors, which may be simple ions, such as Mg^{2+} or Na^+, or nonprotein organic molecules, such as NAD. The latter are known as coenzymes.

Enzyme-catalyzed reactions are under tight cellular control. The rate of enzymatic reactions is affected by temperature and pH.

ATP supplies the energy for most of the activities of the cell. The ATP molecule consists of a nitrogenous base, adenine; the five-carbon sugar ribose; and three phosphate groups. The phosphate groups are linked by two high-energy bonds—bonds that release a relatively large amount of energy when they are broken. ATP participates as an energy carrier in most series of reactions that occur in living systems.

C H A P T E R   6

Respiration

Respiration is the means by which the energy of carbohydrates is transferred to ATP—the universal energy-carrier molecule—and is thus made available for the immediate energy requirements of the cell (Figure 6–1). In the following pages, we will describe in some detail how a cell breaks down carbohydrates and captures and stores much of the released energy in the high-energy phosphate bonds of ATP. The process is described in detail because it provides an excellent illustration both of chemical principles described in previous chapters and of the way in which cells perform biochemical work.

As mentioned in Chapter 3, energy-yielding carbohydrate molecules are generally found stored in plants as sucrose or starch. A necessary preliminary step to the respiratory sequence is the hydrolysis of these storage molecules to monosaccharides. Respiration itself is generally considered to begin with glucose, the building block of sucrose and starch.

6–1
Mitochondria from a leaf cell of the fern Regnellidium diphyllum. *Mitochondria are the sites of cellular respiration by which chemical energy is transferred from carbon-containing compounds to ATP. Most of the ATP is produced on the surfaces of the cristae by enzymes that form a part of the structure of these membranes.*

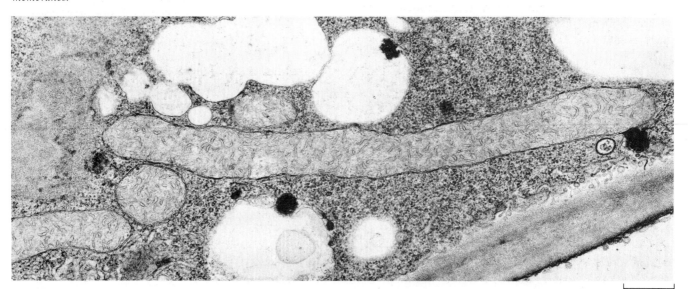

0.5 μm

Glucose can be used as a source of energy under both aerobic (with O_2) and anaerobic (without O_2) conditions. However, maximum energy yields for oxidizable organic compounds are generally achieved only under aerobic conditions. Consider, for example, the overall reaction for the complete oxidation of glucose:

$$C_6H_{12}O_6 + 6O_2 \longrightarrow 6CO_2 + 6H_2O + energy$$

With oxygen as the ultimate electron acceptor, this reaction is highly exergonic (energy-yielding), with a ΔG of -686 kcal/mole. This reaction represents the process called _respiration_. (When energy is extracted from organic compounds without the involvement of oxygen, the process is called _fermentation_, which is discussed later on in this chapter.)

Respiration involves three distinct stages: glycolysis, the Krebs cycle, and the electron transport chain. In _glycolysis_, the six-carbon glucose molecule is broken down to a pair of three-carbon molecules of pyruvic acid or pyruvate. (Pyruvic acid dissociates, producing pyruvate and a hydrogen ion. Pyruvic acid and pyruvate exist in dynamic equilibrium, and the two terms can be used interchangeably.) In the _Krebs cycle_ and the _electron transport chain_, the pyruvate molecules are further broken down to carbon dioxide and water.

As the glucose molecule is oxidized, some of its energy is extracted in a series of small, discrete steps and is stored in the high-energy bonds of ATP.

In accord with the second law of thermodynamics, some of this chemical energy is dissipated as heat energy. In birds, mammals, and to some extent, other vertebrates, the heat generated by cellular respiration is conserved by various mechanisms so that the body temperature of the organism generally remains above that of the environment. In plants, respiration is generally so slow that any heat produced has virtually no effect on body temperature. However, during the rapid growth associated with flowering, parts of some plants, such as _Philodendron_ and eastern skunk cabbage (_Symplocarpus foetidus_), have been found to have temperatures as much as 20°C above the air temperature.

GLYCOLYSIS

Glycolysis (from _glyco_, meaning "sugar," and _lysis_, meaning "splitting") occurs in a series of nine steps, each catalyzed by a specific enzyme (Figure 6–2). This series of reactions is carried out by virtually all living cells, from bacteria to the eukaryotic cells of plants and animals. Glycolysis is an anaerobic process that occurs in the ground substance of the cytoplasm (sometimes referred to as the cytosol). Biologically, glycolysis may be considered a primitive process, in the sense that it most likely arose before the appearance of atmospheric oxygen and before the origin of cellular organelles.

6–2

The first of the nine steps of glycolysis described on the following pages involves the transfer of a "high-energy" phosphate group from ATP to glucose. By this step, energy is put into the reaction. Like all the steps in glycolysis, this one is catalyzed by a specific enzyme.

The glycolytic pathway is shown in detail in Figure 6–3. As the series of reactions is discussed, notice how the carbon skeleton of the molecule is disassembled as its atoms are rearranged step by step. Remember, it is not intended that you memorize these steps: simply follow them closely. Note especially the formation of ATP from ADP and the formation of $NADH_2$ from NAD. (To be exact, NAD and $NADH_2$ should be written as NAD^+ and $NADH + H^+$, as in Figure 5–8, but we will follow the simpler convention.) ATP and $NADH_2$ represent the cell's net energy harvest from this process.

Step 1 The first step in glycolysis requires an input of energy. This activation energy is supplied by the hydrolysis of ATP to ADP. The terminal phosphate group is transferred from an ATP molecule to the glucose molecule, producing glucose 6-phosphate. The combining of ATP with glucose to produce glucose 6-phosphate and ADP is an energy-yielding reaction. Some of the energy released from the ATP is conserved in the chemical bond linking the phosphate to the sugar molecule. This reaction is catalyzed by a specific enzyme (hexokinase), and each of the reactions that follows is similarly catalyzed by a different specific enzyme.

Step 2 In this step, the molecule is rearranged, again with the help of a particular enzyme. The six-sided ring characteristic of glucose becomes a five-sided fructose ring. As shown in Figure 3–2, glucose and fructose have the same number of atoms ($C_6H_{12}O_6$) and differ only in the arrangement of their atoms. This reaction can proceed in either direction; it is pushed forward by the accumulation of glucose 6-phosphate from step 1 and the disappearance of fructose 6-phosphate as it enters step 3.

Step 3 This step, which is similar to step 1, results in the attachment of a phosphate to the first carbon of the fructose molecule, which produces fructose, 1,6-bisphosphate, that is, fructose with phosphates in the 1 and 6 positions. The conversion of the glucose molecule to the higher-energy fructose 1,6-bisphosphate compound is accomplished at the expense of two molecules of ATP. Thus far, no energy has been recovered, but the overall yield will more than compensate for this initial investment.

Step 4 This is the cleavage step of glycolysis. The molecule is split into two interconvertible three-carbon molecules—glyceraldehyde 3-phosphate and dihydroxyacetone phosphate. However, because the glyceraldehyde 3-phosphate is used up in subsequent reactions, all of the dihydroxyacetone phosphate is eventually converted to glyceraldehyde 3-phosphate. Thus, all subsequent steps must be considered twice to account for the fate of each glucose molecule. With the completion of step 4, the preparatory reactions that require an input of ATP energy are complete.

Step 5 Next, glyceraldehyde 3-phosphate molecules are oxidized—that is, hydrogen atoms with their electrons are removed—and NAD is converted to $NADH_2$. This is the first reaction from which the cell gains energy. Energy from this oxidation reaction is also used to attach an additional

6–3

The steps of glycolysis.

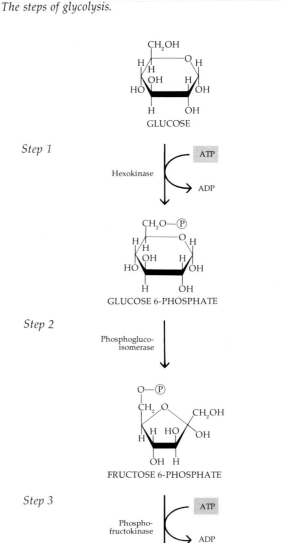

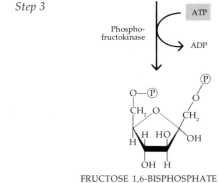

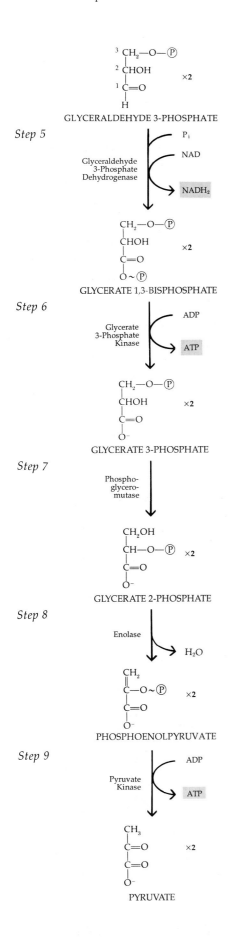

phosphate group to what is now the 1 position of each of the glyceraldehyde molecules. (The designation P_i indicates inorganic phosphate that is available as a phosphate ion in solution in the cytoplasm.) Note that a high-energy bond ($\sim$) is formed.

Step 6 The high-energy phosphate is released from the glycerate 1,3-bisphosphate molecule and is used to recharge a molecule of ADP (a total of two molecules of ATP are formed per molecule of glucose). This is a highly exergonic reaction and pulls all the previous reactions forward.

Step 7 The remaining phosphate group is transferred from the number 3 to the number 2 carbon.

Step 8 In this step, a molecule of water is removed from the three-carbon compound, and as a consequence of this internal rearrangement of the molecule, a high-energy phosphate bond is formed.

Step 9 The high-energy phosphate is transferred to a molecule of ADP, forming another molecule of ATP (again, a total of two molecules of ATP are formed per molecule of glucose). This is also a highly exergonic reaction, and so the sequence runs to completion with an accelerating force.

Summary of Glycolysis

The complete sequence of glycolysis begins with one molecule of glucose (Figure 6–4). Energy enters the sequence at steps 1 and 3 by the transfer of a phosphate group from an ATP molecule—one at each step—to the sugar molecule. The six-carbon molecule splits at step 4, and after this point the sequence yields energy. At step 5, two molecules of NAD, in being reduced to two molecules of $NADH_2$, store some of the energy from the oxidation of glyceraldehyde 3-phosphate. At both steps 6 and 9, two molecules of ADP take energy from the system, form additional phosphate bonds, and are phosphorylated to become two molecules of ATP. (Phosphorylation that occurs during glycolysis is known as *substrate phosphorylation*.)

Glycolysis (from glucose to pyruvate) can be summarized by the overall equation:

$$C_6H_{12}O_6 + 2NAD + 2ADP + 2P_i \longrightarrow$$
GLUCOSE

$$2C_3H_4O_3 + 2NADH_2 + 2ATP$$
PYRUVATE

Thus, one glucose molecule is converted to two molecules of pyruvate. The net harvest—the energy recovered—is two molecules of ATP and two molecules of $NADH_2$. The two molecules of pyruvate have a total energy content of about 546 kcal, which is a large portion of the 686 kcal stored in the original glucose molecule.

6–4

A summary of glycolysis. Two molecules
of ATP and two molecules of NADH₂
represent the energy yield. Most of the
energy stored in the original glucose
molecule is found in the two molecules of
pyruvate.

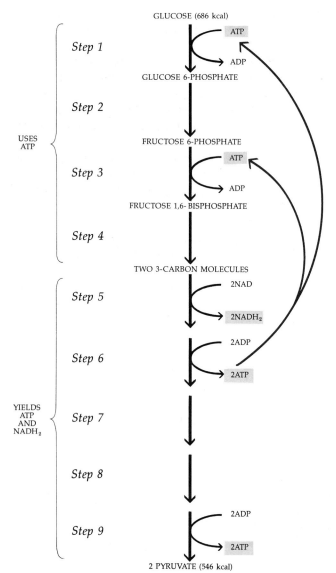

AEROBIC PATHWAY

Pyruvate is a key intermediate compound in cellular energy metabolism, and it can be utilized in one of several pathways. Which pathway it follows depends in part upon the conditions under which metabolism takes place and in part upon the specific organism involved. The principal environmental factor that operates to affect which pathway is followed is the availability of oxygen.

In the presence of oxygen, pyruvate is oxidized completely to carbon dioxide, and glycolysis is but the initial phase of respiration. This aerobic pathway results in the complete oxidation of glucose and a much greater ATP yield than can be achieved by glycolysis alone. These reactions take place in two stages—the Krebs cycle and the electron transport chain—both occurring within mitochondria in eukaryotic cells (see Figure 6–1).

Recall that mitochondria are surrounded by two membranes, the inner one convoluted inwardly into folds called cristae. Within the innermost compartment, surrounding the cristae, is a dense solution containing enzymes, coenzymes, water, phosphates, and other molecules involved in respiration. Thus, a mitochondrion is a self-contained chemical factory. The outer membrane allows most small molecules to move in or out freely, but the inner one only permits the passage of certain molecules, such as pyruvate and ATP, while it prevents the passage of others. The enzymes of the Krebs cycle are found in solution within the inner compartment. The enzymes and other components of the electron transport chain are built into the surfaces of the cristae.

The Krebs Cycle

The Krebs cycle is named in honor of Sir Hans Krebs, who was largely responsible for its elucidation. Krebs postulated this metabolic pathway in 1937 and later received a Nobel Prize in recognition of his brilliant work. The Krebs cycle is also called the tricarboxylic acid (TCA) cycle because it is initiated with the formation of an organic acid (citrate) that has three carboxylic acid groups.

Before entering the Krebs cycle, pyruvate is both oxidized and decarboxylated. In the course of this exergonic reaction, a molecule of NADH₂ is produced from NAD. The original glucose molecule has now been oxidized to two acetyl (CH_3CO) groups; two molecules of CO_2 have been liberated; and two molecules of NADH₂ have been formed from NAD (Figure 6–5).

Each acetyl group is then temporarily attached to coenzyme A (CoA)—a large molecule, a portion of which is a nucleotide and a portion of which is pantothenic acid, one of the B-complex vitamins. The com-

Note that glycolysis is also an internal oxidation-reduction reaction. Compare the original substrate for glycolysis (glucose) to the product molecules (pyruvate). The —CH₃ (methyl group) of pyruvate arises from the first and last carbons of the glucose molecule, and they are more reduced in pyruvate than in glucose. In contrast, the —COOH (carboxyl group) of pyruvate arises from the two middle carbons of glucose, and these carbon molecules are more oxidized than the carbons of the glucose molecule from which they were derived.

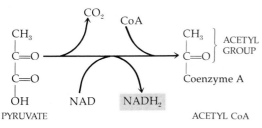

6-5
The three-carbon pyruvate molecule is oxidized and decarboxylated to form the two-carbon acetyl group, which then combines with coenzyme A to form acetyl CoA. The oxidation of the pyruvate molecule is coupled to the production of $NADH_2$ from NAD. Acetyl CoA is the molecule needed to enter the Krebs cycle.

bination of the acetyl group and CoA is known as acetyl CoA (Figure 6–5).

Fats and amino acids can also be converted to acetyl CoA and enter the respiratory sequence at this point. A fat molecule is first hydrolyzed to glycerol and three fatty acids. Then, beginning at the carboxyl end, two-carbon groups are successively removed from the fatty acids. A molecule such as palmitic acid (see Figure 3–9), which contains 16 carbon atoms, yields eight molecules of acetyl CoA.

Upon entering the Krebs cycle (Figure 6–6), the two-carbon acetyl group is combined with a four-carbon compound (oxaloacetate) to produce a six-carbon compound (citrate). In the course of the cycle, two of the six carbons are oxidized to CO_2, and oxaloacetate is regenerated—thus literally making this series a cycle. Each turn around the cycle uses up one acetyl group and regenerates one molecule of oxaloacetate, which is then ready to begin the Krebs cycle again. In the course of these steps, some of the energy released by the oxidation of the carbon atoms is used to convert ADP to ATP (one molecule per cycle), and some is used to convert NAD to $NADH_2$ (three molecules per cycle). In addition, some of the energy is used to reduce a second electron carrier—the coenzyme flavin adenine dinu-

6-6
In the course of the Krebs cycle, the carbons donated by the acetyl group are oxidized to carbon dioxide, and the hydrogen atoms are passed to electron carriers. As in glycolysis, a specific enzyme is involved at each step.

6–7

Flavin adenine dinucleotide, an electron acceptor, in its oxidized form (FAD) and its reduced form (FADH$_2$). Riboflavin is a vitamin (vitamin B$_2$) made by all plants and many microorganisms. It is a pigment and, in its oxidized form, is bright yellow.

A related electron acceptor, flavin mononucleotide (FMN), consists of riboflavin and the first phosphate group shown here. It accepts electrons from NADH in the electron transport chain.

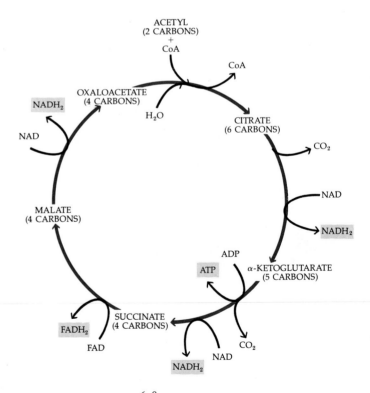

cleotide (FAD) (Figure 6–7). One molecule of FADH$_2$ is formed from FAD in each turn of the cycle. Oxygen is not directly involved in the Krebs cycle; the electrons and protons removed in the oxidation of carbon are all accepted by NAD and FAD:

oxaloacetate + acetyl CoA + ADP + 3NAD + FAD $\longrightarrow$
    oxaloacetate + 2CO$_2$ + CoA + ATP + 3NADH$_2$ + FADH$_2$

The Krebs cycle is summarized in Figure 6–8.

The Electron Transport Chain

The glucose molecule is now completely oxidized. Some of its energy has been used to produce ATP from ADP. Most of it, however, still remains in the electrons removed from the carbon atoms as they were oxidized. These electrons were passed to the electron carriers NAD and FAD and are at a high energy level. In the course of the electron transport chain, they are passed "downhill" to oxygen, and the energy released is used to form ATP from ADP. This process is known as *oxidative phosphorylation.*

The electron carriers of the electron transport chain of the mitochondria differ from NAD and FAD in their chemical structures. Some of them belong to a class of compounds known as cytochromes—protein molecules with an iron-containing porphyrin ring, or heme group, attached (Figure 6–9). Each cytochrome differs in its protein chain and in the energy level at which it holds the electrons. Cytochromes carry a single electron without a proton.

6–8

A summary of the Krebs cycle. One molecule of ATP, three molecules of NADH$_2$, and one molecule of FADH$_2$ represent the energy yield of each acetyl group passing through the cycle.

6–9

Cytochromes are molecules in which an atom of iron is held in a nitrogen-containing (porphyrin) ring. The porphyrin ring with its atom of iron is known as heme. Cytochromes are involved in electron transfer. The iron combines with the electrons, each iron atom accepting an electron as it is reduced from Fe^{3+} to Fe^{2+}.

CYTOCHROME

6–10

Iron-sulfur proteins contain bound iron and sulfur. The plant iron-sulfur protein shown here contains a two-iron, two-sulfur center.

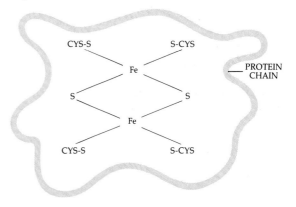

6–11

Coenzyme Q, a fat-soluble quinone in oxidized and reduced forms. The shaded groups function in carrying H atoms.

Oxidized form

$-2e^-$ $+2e^-$
$-2H^+$ $+2H^+$

Reduced form

Non-heme iron proteins—the iron-sulfur proteins—are another component of the electron transport chain. The iron of these proteins is not attached to a porphyrin ring; instead, the iron molecules are attached to sulfides and to the sulfurs of cysteines of the protein chain (Figure 6–10). Like cytochromes, the iron-sulfur proteins carry an electron but not a proton.

A third important component of the electron transport chain are quinone molecules (Figure 6–11). Quinones are the most abundant component of the electron transport chain. Unlike cytochromes and iron-sulfur proteins, quinones carry the equivalent of a hydrogen atom. By alternating electron transfer between components that carry or do not carry a proton with the electron, protons can be shuttled across the membrane. For example, every time a quinone molecule accepts an electron from a cytochrome, it also picks up a proton from the surrounding medium. When the quinone gives up its electron to the next carrier, such as a cytochrome, the proton is again released into the medium. If the electron carriers are arranged in the membrane so that protons are picked up on one side of the membrane and dumped on the other, a proton gradient is generated. Most of these quinone molecules are not attached to proteins and are thought to be able to move across the membrane.

At the top of the electron transport chain are the electrons held by $NADH_2$ and $FADH_2$. The yield of the Krebs cycle was two molecules of $FADH_2$ and six molecules of $NADH_2$. The oxidation of pyruvate to acetyl CoA yielded two molecules of $NADH_2$. Recall that an additional two molecules of $NADH_2$ were produced in glycolysis; in the presence of oxygen, these $NADH_2$ molecules are transported into the mitochondrion where they are transferred to the electron acceptor flavin mononucleotide (abbreviated FMN; see Figure 6–7) and then are fed into the electron transport chain.

As electrons flow along the electron transport chain from a higher to a lower energy level, the electron transport chain harnesses the released energy and uses it to convert ADP to ATP (Figure 6–12). At the end of the chain, the electrons are accepted by oxygen and combine with protons (hydrogen ions) to produce water. Each time one pair of electrons passes from $NADH_2$ to oxygen, three molecules of ATP are formed. Each time a pair of electrons passes from $FADH_2$, which holds them at a slightly lower energy level than $NADH_2$, two molecules of ATP are formed.

The Mechanism of Oxidative Phosphorylation: Chemiosmotic Coupling

The chemiosmotic coupling hypothesis was first proposed in the 1960s by the British biochemist Peter Mitchell (Nobel laureate in chemistry, 1978). It proposes that oxidative phosphorylation is driven by a proton gradient (a differential H^+ concentration across a membrane) produced between the two sides of the inner mitochondrial membrane during electron transport. According to this ingenious concept, protons are pumped out of the mitochondrial matrix to the outer mitochondrial compartment as electrons from $NADH_2$ are passed along the electron transport chain, which

forms a part of the mitochondrial membrane. Each pair of electrons crosses the membrane three times as it moves from one electron carrier to the next (and finally to oxygen). This generates an electrochemical gradient of H^+ across the membrane that drives the protons back into the matrix through diffusion channels in knoblike structures that protrude into the matrix.

It is now known that the channel through which the protons flow back into the matrix is provided by a large enzyme complex known as *ATP synthetase*. This enzyme complex consists of two major components, or factors, F_O and F_1 (Figure 6–13a). The F_O is embedded in the inner mitochondrial membrane, traversing it from outside to inside. F_1, a large globular structure consisting of nine polypeptide subunits, is attached to the F_O on the matrix side of the membrane. The F_1 components appear as protruding knobs in electron micrographs (Figure 6–13b). F_1 alone cannot make ATP and ADP and phosphate in its isolated form, but it can hydrolyze ATP to ADP and phosphate, thus functioning as an ATPase. Its usual function when attached to the F_O component in the intact mitochondrion is, however, the reverse. As protons flow down the electrochemical gradient from the outside into the matrix, passing through the F_O component and then the F_1 component, the free energy that is released powers the synthesis of ATP from ADP and phosphate.

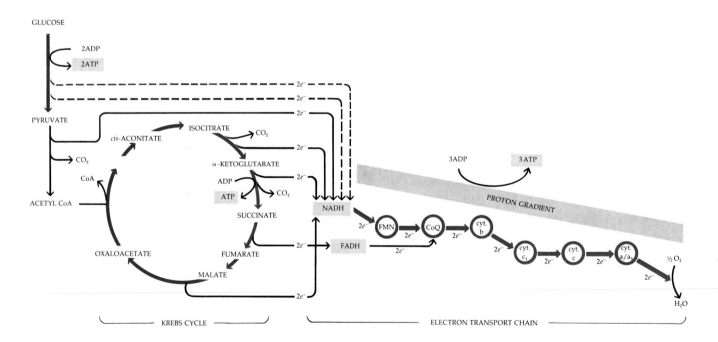

6–12

A summary of respiration. Glucose is first broken down to pyruvate, with a yield of two ATP molecules and the reduction (dashed arrows) of two NAD molecules. Pyruvate is oxidized to acetyl CoA, and one molecule of NAD is reduced (note that this and subsequent

reactions occur twice for each glucose molecule; this electron passage is indicated by solid arrows). In the Krebs cycle, the acetyl group is oxidized and the electron acceptors, NAD and FAD, are reduced. $NADH_2$ and $FADH_2$ then transfer their electrons to the electron

transport chain, which consists mostly of a series of cytochromes. As these cytochromes pass the electrons downhill, the energy released is used to form ATP from ADP, as shown in Figure 6–14. The electron transport chain is shown here in abbreviated form.

6–13

(a) *Diagram of the ATP synthetase complex. The F_O portion is embedded in the inner membrane and extends across it, and the F_1 component, which consists of nine subunits, extends into the mitochondrial matrix. (b) In this electron micrograph the knobs protruding from the membrane of these vesicles are the F_1 portions of ATP synthetase complexes.*

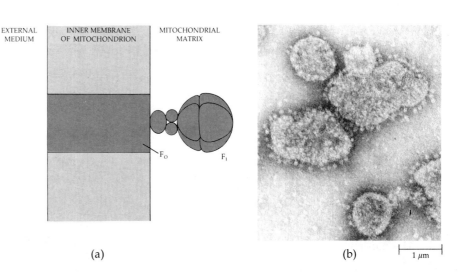

(a)

(b)   ⊢ 1 μm ⊣

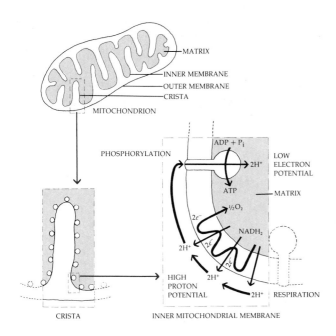

6–14

According to the chemiosmotic coupling hypothesis, protons are pumped out of the mitochondrial matrix as electrons are passed down the electron transport chain, which forms part of the inner mitochondrial membrane. The movement of protons down the electrochemical gradient as they pass through the ATP synthetase complex provides the energy by which ATP is synthesized from ADP and phosphate. The exact number of protons pumped out of the matrix as each electron pair moves down the chain remains to be determined, as does the number that must flow through ATP synthetase for each molecule of ATP formed.

Figure 6–14 summarizes the chemiosmotic coupling mechanism as it occurs in oxidative phosphorylation.

Control of Oxidative Phosphorylation

Electrons continue to flow down the electron transport chain only if ADP is available for conversion to ATP. Thus, oxidative phosphorylation is regulated by the "law of supply and demand." When the energy requirements of the cell decrease, fewer molecules of ATP are used, fewer molecules of ADP become available, and the electron flow is decreased.

Overall Energy Harvest

We are now in a position to see how much of the energy originally present in the glucose molecule has been recovered in the form of ATP. The "balance sheet" for ATP yield given in Table 6–1 may help you keep track of the discussion that follows.

Glycolysis in the presence of oxygen, yields 2 molecules of ATP directly, plus 2 molecules of $NADH_2$ (from which an additional 6 molecules of ATP are formed). The total gain, however, is not 8 ATPs, as you might calculate, but rather 6 ATPs. There is a "cost" of 2 ATPs to transport the electrons held by the 2 molecules of $NADH_2$ across the mitochondrial membranes.

Table 6–1   *Energy Yield from One Molecule of Glucose*

Glycolysis:	2 ATP	
	$2\ NADH_2 \longrightarrow 4$ ATP	$\longrightarrow$ 6 ATP
Pyruvate $\longrightarrow$		
Acetyl CoA:	$1\ NADH_2 \longrightarrow 3$ ATP	$(\times 2) \longrightarrow$ 6 ATP
	1 ATP	
Krebs cycle:	$3\ NADH_2 \longrightarrow 9$ ATP	$(\times 2) \longrightarrow 24$ ATP
	$1\ FADH_2 \longrightarrow 2$ ATP	

The conversion of pyruvate to acetyl CoA yields 2 molecules of NADH$_2$ (inside the mitochondrion) for each molecule of glucose and so produces 6 molecules of ATP.

The Krebs cycle yields, for each molecule of glucose, 2 molecules of ATP, 6 molecules of NADH$_2$, and 2 molecules of FADH$_2$, for a total of 24 molecules of ATP.

As the balance sheet shows, the complete yield from a single molecule of glucose is 36 molecules of ATP. All but 2 of the 36 molecules of ATP have come from reactions in the mitochondrion, and all but 4 involve the oxidation of NADH$_2$ or FADH$_2$ along the electron transport chain.

The total difference in free energy between the reactants (glucose and oxygen) and the products (carbon dioxide and water) is 686 kilocalories. Approximately 39 percent of this, or 263 kilocalories (7.3×36), has been captured in the high-energy bonds of the 36 ATP molecules (Figure 6–15).

OTHER IMPORTANT ENERGY-YIELDING PATHWAYS

The Glyoxylate Cycle

Many seeds store fats as an energy reserve for use during germination and the early stages of growth. When fats are broken down to two-carbon units as acetyl CoA, the acetyl CoA is available as an energy source. In the glyoxylate cycle (Figure 6–16), which is a modification of the Krebs cycle (Figure 6–6), isocitrate is split to yield one molecule of succinate and one molecule of glyoxylate. The molecule of succinate is then available for the Krebs cycle, and the molecule of glyoxylate condenses with another molecule of acetyl CoA to make a molecule of malate. The molecule of malate can also enter the Krebs cycle, or it may be converted to oxaloacetate to complete the glyoxylate cycle. The glyoxylate cycle occurs in plants, yeasts, and some bacteria, but not in animals. In plants this pathway occurs in microbodies called glyoxysomes (see page 24). The glyoxysome is also the site of fatty acid breakdown. The glyoxylate cycle is referred to as an anapleurotic pathway; anapleurotic pathways serve to replenish the intermediates of other pathways. In this case, the glyoxylate cycle serves to provide four-carbon acids to replenish the Krebs cycle.

The Pentose Phosphate Pathway

Plant tissues also have the capability to bypass glycolysis (the glycolytic pathway) and oxidize glucose by a pathway referred to as the pentose phosphate pathway. The features of this pathway are: (1) no ATP is synthesized directly, and (2) five-carbon sugars, such as ribose, are synthesized for use in the production of nu-

BIOLUMINESCENCE

In most organisms, the energy transferred to ATP is used for cellular work. In some, however, the chemical energy is sometimes reconverted to light energy. Bioluminescence is probably an accidental by-product of energy exchanges in most luminescent organisms, such as the fungus *Mycena lux-coeli* shown here, photographed by their own light. In some luminescent organisms, such as fireflies, in which the flashes of light serve as mating signals, bioluminescence serves a useful function.

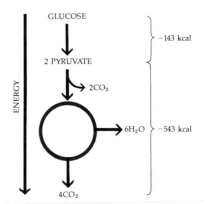

6–15
The energy changes in the oxidation of glucose. The complete respiratory sequence proceeds with an energy drop of 686 kilocalories. Of this, about 39 percent (263 kilocalories) is conserved in 36 ATP molecules. By contrast, in anaerobic respiration only two ATP molecules are produced, representing only about 2 percent of the available energy of glucose.

6–16

The glyoxylate cycle. In the glyoxylate cycle, two molecules of acetyl CoA are used in each cycle to produce four-carbon acids. The glyoxylate cycle is a modification of the Krebs cycle. It uses two

unique enzymes: isocitrate lyase, which splits isocitrate into the four-carbon acid succinate and glyoxylate, and malate synthase, which condenses the glyoxylate with a molecule of acetyl CoA to

generate another four-carbon acid, malate. The series of reactions catalyzed by these two enzymes is in color.

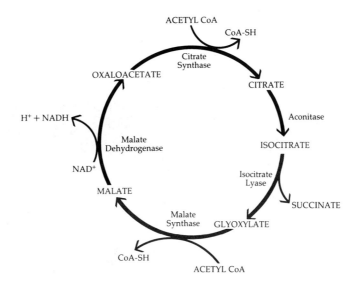

cleotide bases for DNA (see page 57) and cofactor synthesis. Instead of NAD, the enzymes of the pentose phosphate pathway use NADP as a cofactor. NADP has an extra phosphate attached to one of the riboses, and the NADPH$_2$ produced is used in biosynthesis and is oxidized by mitochondria to yield ATP.

NADP is similar to NAD (see Figure 5–8, page 78) but has an additional phosphate group. Their biological roles, however, are distinctly different. NADH$_2$ transfers its electrons to other electron carriers, which continue to pass them down in discrete steps to successively lower electron levels. In the course of this electron transfer, ATP molecules are formed. NADPH$_2$ provides energy directly to biosynthetic processes of the cell that require large energy inputs.

The pentose phosphate pathway bypasses the regulatory steps of the glycolytic pathway. Diseased plant tissue often oxidizes glucose by the pentose phosphate pathway in preference to the glycolytic pathway.

ANAEROBIC PATHWAYS

In eukaryotic cells (as well as most bacteria), pyruvate usually follows the aerobic pathway and is completely oxidized to carbon dioxide and water. However, in the absence of oxygen, pyruvate is not the end product of glycolysis because the NADH$_2$ produced during the oxidation of glyceraldehyde 3-phosphate must be reoxidized to NAD. Without this reoxidation, glycolysis would soon stop, because the cell would run out of

NAD as an electron acceptor. This oxygenless, or anaerobic, process may result in the formation of lactate in many bacteria, fungi, protists, and animal cells; it is therefore called *lactate fermentation*. For example, lactate is produced in muscle cells during bursts of unusually vigorous work, as when an athlete runs a sprint. The muscles accumulate what is known as an "oxygen debt" by producing lactate from glucose. The lactate lowers the pH of the muscles and reduces the capacity of the muscle fibers to contract, producing the sensation of muscle fatigue. The lactate diffuses into the blood and is carried to the liver where it is resynthesized to pyruvate and then to glucose or glycogen.

In yeasts and most plant cells, pyruvate is broken down to ethanol and carbon dioxide under anaerobic conditions. Under anaerobic conditions, NADH$_2$ is reoxidized by the transfer of electrons (and protons) to pyruvate. This process is preceded by the removal of carbon dioxide (decarboxylation) and results in the production of ethanol and carbon dioxide instead of lactate (Figure 6–17). Since the principal end product of glycolysis is ethanol, this process is called *alcoholic fermentation*.

Yeast cells are present as a "bloom" on the skin of grapes. When the glucose-filled juices of grapes and other fruits are extracted and stored in airtight kegs, these yeast cells turn the fruit juice to wine by converting glucose to ethanol. Yeasts, like all living things, have a limited tolerance for alcohol, and when the critical concentration (about 12 percent) is reached, the yeast cells cease to function.

$$\text{PYRUVATE} \quad\quad\quad \text{ACETALDEHYDE} \quad\quad\quad \text{ETHANOL}$$

(a)

(b)

6–17

(a) *The steps by which pyruvate is converted anaerobically to ethanol. In the first step, carbon dioxide is released. In the second, $NADH_2$ is oxidized and acetaldehyde is reduced. Most of the energy of the glucose remains in the alcohol, which is the principal end product of the sequence. However, by regenerating NAD, these steps allow glycolysis to continue, with its small but sometimes vital yield of ATP. (b) The consequences of anaerobic glycolysis. Yeast cells, which are visible on grapes as a dustlike "bloom," mix with the juice when the grapes are crushed. Storing the mixture under anaerobic conditions causes the yeast to break down the glucose in the grape juice to alcohol. In modern wine making, pure yeast cultures are added to relatively sterile grape juice for fermentation, rather than relying on the yeasts carried on the grapes.*

Thermodynamically, lactate fermentation and alcoholic fermentation are similar. In both, the $NADH_2$ is reoxidized, and the energy yield is only the 2 ATPs harvested during glycolysis. The complete, balanced equation for the fermentation of glucose can be written as follows:

$$\text{glucose} + 2\text{ADP} + 2\text{P}_i \longrightarrow \begin{Bmatrix} 2 \text{ lactate } or \\ 2 \text{ ethanol} + 2\text{CO}_2 \end{Bmatrix} + 2\text{ATP}$$

During alcoholic fermentation, approximately 7 percent of the total available energy of the glucose molecule—about 52 kilocalories—is released, with about 93 percent remaining in the two alcohol molecules. Only 14.6 kilocalories of the 52 kilocalories released during fermentation are trapped and stored in the two molecules of ATP. So, in terms of energy yield, anaerobic fermentation is relatively inefficient.

SUMMARY

Respiration, or the complete oxidation of glucose, is the chief source of energy in most cells. As the glucose is broken down in a series of small enzymatic steps, some of the energy in the molecule is packaged in the form of high-energy bonds in molecules of ATP, and the rest is lost as heat.

The first phase in the breakdown of glucose is glycolysis, in which the six-carbon glucose molecule is split into two three-carbon molecules of pyruvate; two new molecules of ATP and two of $NADH_2$ are formed. This reaction occurs in the cytoplasmic ground substance.

In the course of respiration, the three-carbon pyruvate molecules are broken down within the mitochondrion to two-carbon acetyl groups, which then enter the Krebs cycle. In the Krebs cycle, the acetyl group is broken apart in a series of reactions to yield carbon dioxide. In the course of the oxidation of each acetyl group, four electron acceptors (three NAD and one FAD) are reduced, and another molecule of ATP is formed.

The final stage of respiration is the electron transport chain, which involves a series of electron carriers and enzymes embedded in the inner membranes of the mitochondrion. Along this series of electron carriers, the high-energy electrons accepted by NAD during glycolysis and by NAD and FAD during the Krebs cycle pass "downhill" to oxygen. The large quantity of free energy released during the passage of electrons down the electron transport chain powers the pumping of protons (H^+ ions) out of the mitochondrial matrix. This creates an electrochemical gradient of H^+ across the inner membrane of the mitochondrion. When the protons pass back into the matrix through the ATP synthetase complex, the energy that is released is used to form ATP from ADP and phosphate. In the course of the breakdown of the glucose molecule, 36 molecules of ATP are gained, most of them in the mitochondrion.

In the absence of oxygen, pyruvate may be converted either to lactate (in many bacteria, fungi, and animal cells) or to ethanol and carbon dioxide (in yeasts and most plant cells). These anaerobic processes—called fermentation—yield 2 ATPs for each glucose molecule (two pyruvate molecules).

C H A P T E R   7

Photosynthesis

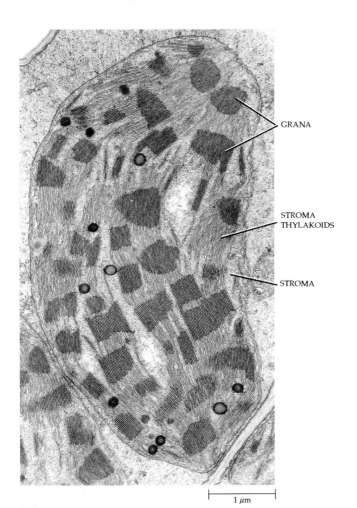

1 μm

7–1

The chloroplast is the site of photosynthesis in eukaryotic organisms. The light-capturing reactions occur in thylakoids, where the chlorophylls and other pigments are found. In chloroplasts, many of the thylakoids characteristically occur in disklike stacks called grana. The series of reactions by which this energy is transferred to carbon-containing compounds occurs in the stroma, the material surrounding the thylakoids.

GRANA

STROMA
THYLAKOIDS

STROMA

In the previous chapter, we described the breakdown of carbohydrates to yield the energy required for the many kinds of activities carried out by living systems. In the pages that follow, we will complete the circle by describing the way in which energy coming from the sun in the form of light is captured and converted to chemical energy.

This process—photosynthesis—is the route by which virtually all energy enters our biosphere. Each year more than 150 billion metric tons of sugar are produced worldwide by the photosynthetic process. The importance of photosynthesis, however, extends far beyond the sheer weight of this product. Without this flow of energy from the sun, channeled largely through the chloroplasts of eukaryotic cells (Figure 7–1), the pace of life on this planet would swiftly diminish and then, following the inexorable second law of thermodynamics, would virtually cease altogether.

OVERVIEW OF PHOTOSYNTHESIS

The importance of photosynthesis in the economy of nature was not recognized until comparatively recent times. Aristotle and other Greeks, observing that the life processes of animals were dependent upon the food they ate, thought that plants derived their food from the soil.

A little over 300 years ago, in one of the first carefully designed biological experiments ever reported, the Belgian physician Jan Baptista van Helmont (ca. 1577–1644) offered the first experimental evidence that soil alone does not nourish the plant. Van Helmont grew a small willow tree in an earthenware pot, adding only water to the pot. At the end of 5 years, the willow had increased in weight by 74.4 kilograms, whereas the earth had decreased in weight by only 57 grams. On the basis of these results, van Helmont incorrectly concluded that all the substance of the plant was produced from the water and none from the soil or the air!

Toward the end of the eighteenth century, the English scientist Joseph Priestley (1733–1804) reported that he had "accidently hit upon a method of restoring air that had been injured by the burning of candles." On 17 August 1771, Priestley "put a [living] sprig of mint into air in which a wax candle had burned out and found that, on the 27th of the same month, another candle could be burned in this same air." The "restorative which nature employs for this purpose," he stated, was "vegetation." Priestley extended his observations and soon showed that air "restored" by vegetation was not "at all inconvenient to a mouse." Priestley's experiments offered the first logical explanation of how air remained "pure" and able to support life despite the burning of countless fires and the breathing of many animals. When he was presented with a medal for his discovery, the citation read in part: "For these discoveries we are assured that no vegetable grows in vain . . . but cleanses and purifies our atmosphere." Today we would explain Priestley's experiments simply by saying that plants take up the CO_2 produced by combustion or exhaled by animals, and that animals inhale the O_2 released by plants.

Later, the Dutch physician Jan Ingenhousz (1730–1799) confirmed Priestley's work and showed that the air was restored only in the presence of sunlight and only by the green parts of plants. In 1796, Ingenhousz suggested that carbon dioxide is split in photosynthesis to yield carbon and oxygen, with the oxygen then released as gas. Subsequently, the proportion of carbon, hydrogen, and oxygen atoms in sugars and starches was found to be about one atom of carbon per molecule of water (CH_2O), as the word "carbohydrate" indicates. Thus, in the overall reaction for photosynthesis,

$$CO_2 + H_2O + \text{light energy} \longrightarrow (CH_2O) + O_2$$

it was generally assumed that the carbohydrate came from a combination of the carbon and water molecules and that the oxygen was released from the carbon dioxide molecule. This entirely reasonable hypothesis was widely accepted; but, as it turned out, it was quite wrong.

The investigator who upset this long-held theory was C. B. van Niel of Stanford University. Van Niel, then a graduate student, was investigating the activities of different types of photosynthetic bacteria. In their photosynthetic reactions, one particular group of such bacteria—known as the purple sulfur bacteria—reduces carbon to carbohydrates but does not release oxygen. The purple sulfur bacteria require hydrogen sulfide for their photosynthetic activity. In the course of photosynthesis, globules of sulfur accumulate inside the bacterial cells (Figure 7–2). In these bacteria, van Niel found that the following reaction takes place during their photosynthesis:

$$CO_2 + 2H_2S \xrightarrow{\text{light}} (CH_2O) + H_2O + 2S$$

This finding was simple and did not attract much attention until van Niel made a bold extrapolation. Van Niel proposed the following generalized equation for photosynthesis:

$$CO_2 + 2H_2A \xrightarrow{\text{light}} (CH_2O) + H_2O + 2A$$

In this equation, H_2A represents either water or some oxidizable substance, such as hydrogen sulfide or free hydrogen. In algae and green plants, H_2A is water (Figure 7–3). In short, van Niel proposed that water, *not* carbon dioxide, was split in photosynthesis.

This brilliant speculation, proposed in the early 1930s, was proven years later, when investigators using a heavy isotope of oxygen ($^{18}O_2$) traced the oxygen from water to oxygen gas:

$$CO_2 + 2H_2^{18}O \xrightarrow{\text{light}} (CH_2O) + H_2O + {}^{18}O_2$$

Thus, in the case of algae and green plants, in which water serves as the electron donor, the complete, balanced equation for photosynthesis becomes:

$$6CO_2 + 12H_2O \xrightarrow{\text{light}} C_6H_{12}O_6 + 6O_2 + 6H_2O$$

About two hundred years ago, as noted above, light was discovered to be required for the process we now call photosynthesis. It is now known that photosynthesis occurs in two stages, only one of which actually requires light. Evidence for this two-stage process was first presented in 1905 by the English plant physiologist F. F. Blackman, as the result of experiments in which he measured the individual and combined effects of changes in light intensity and temperature on the rate of photosynthesis.

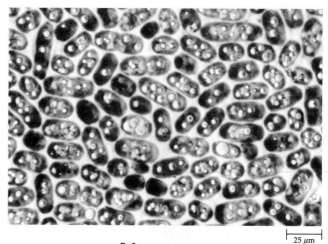

7–2

Purple sulfur bacteria. In these cells, hydrogen sulfide plays the same role as water does in the photosynthetic process of plants. The hydrogen sulfide is split, and the released sulfur accumulates in globules, visible within these cells.

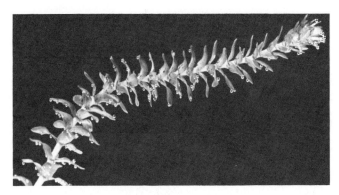

7–3

The bubbles on the leaves of this submersed pondweed, Elodea, *are bubbles of oxygen, one of the products of photosynthesis. Van Niel was the first to propose that the oxygen produced in photosynthesis came from the splitting of water rather than the breakdown of carbon dioxide.*

Blackman made the following conclusions on the basis of his experiments: (1) There was one set of light-dependent reactions that were temperature-independent. The rate of these reactions in the dim-to-moderate light range could be accelerated by increasing the amount of light (Figure 7–4a), but it was not accelerated by increases in temperature (Figure 7–4b). (2) There was a second set of reactions that were dependent, not on light, but on temperature. When both light and temperature were increased, the rate of photosynthesis was greatly accelerated (Figure 7–4b). Both sets of reactions seemed to be required for the process of photosynthesis. Increasing the rate of only one set of reactions increased the rate of the entire process, but only to the point at which the second set of reactions

began to hold back the first (that is, the second set became rate-limiting). Thereafter, it was necessary to increase the rate of the second set of reactions in order for the first to proceed unimpeded.

Photosynthesis was thus shown to have both a light-dependent stage, the "light reactions," as well as a light-independent stage, the "dark reactions." It is important to remember that the dark reactions normally occur in the light; they require the products of the light reactions. The expression "dark reactions" merely indicates that light is not directly involved.

The dark reactions increased in rate as the temperature was increased, but only up to about 30°C, after which the rate began to decrease. From this evidence it was concluded that these reactions were controlled by enzymes, since this is the way enzymes are expected to respond to temperature (see Figure 5–9). This conclusion has since been proven to be correct.

In the first stage of photosynthesis—the light reactions—light energy is used to form ATP from ADP and to reduce electron carrier molecules. In the second stage of photosynthesis—the dark reactions—the energy products of the first stage are used to reduce carbon from carbon dioxide to a simple sugar, thus converting the chemical energy of the carrier molecules to forms suitable for transport and storage and, at the same time, forming a carbon skeleton on which other organic molecules can be built. This conversion of CO_2 into organic compounds is known as *carbon fixation*.

THE LIGHT REACTIONS

The Role of Pigments

The first step in the conversion of light energy to chemical energy is the absorption of light. A *pigment* is any substance that absorbs visible light. Some pigments ab-

7–4

(a) *In dim-to-moderate light, increasing the light intensity increases the rate of photosynthesis, but at higher intensities, a further increase in light intensity has no effect. A curve such as the one shown here indicates that some other factor— known as a rate-limiting factor—is involved in the process under study. CO_2 concentration is commonly the rate-limiting factor in photosynthesis. (b) At a low intensity of light, an increase in temperature does not increase the rate of photosynthesis (lower curve). At a high intensity, however, an increase in temperature has a very marked effect (upper curve). Based on these data, Blackman concluded that photosynthesis included both light-dependent and light-independent reactions.*

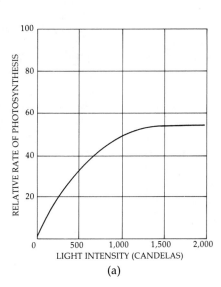

(a)

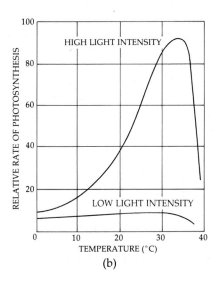

(b)

sorb all wavelengths of light and so appear black. Some absorb only certain wavelengths and transmit or reflect the wavelengths they do not absorb. Chlorophyll, the pigment that makes leaves green, absorbs light principally in the violet and blue wavelengths and also in the red; because it reflects green light, it appears green. The absorption pattern of a pigment is known as the *absorption spectrum* of that substance (Figure 7–5).

One line of evidence that chlorophyll is the principal pigment involved in photosynthesis is the similarity between its absorption spectrum and the action spectrum for photosynthesis (Figure 7–6). An *action spectrum* defines the relative effectiveness of different wavelengths of light for light-requiring processes, such as photosynthesis, flowering, and phototropism (the curving of an organism toward the light). Similarity between the absorption spectrum of a pigment and the action spectrum of a process is considered as evidence that that particular pigment is responsible for that particular process (Figure 7–7).

When pigments absorb light, electrons are boosted to a higher energy level, with three possible consequences: (1) the energy may be dissipated as heat; (2) it may be reemitted almost instantaneously as light energy of longer wavelength, a phenomenon known as fluorescence (when it is reemitted as light at a later time, it is known as phosphorescence); or (3) the energy may be captured in a chemical bond, as it is in photosynthesis.

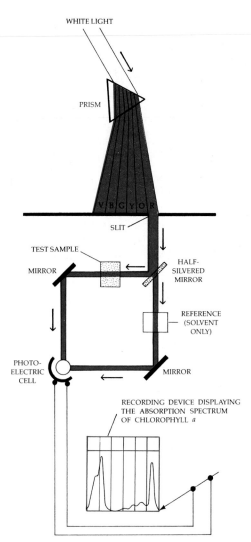

7–5

The absorption spectrum of chlorophyll a *as measured by a spectrophotometer. This device directs a beam of light of each wavelength at the object to be analyzed and records the percentage of light of each wavelength absorbed by the pigment sample as compared with a reference sample. Because the mirror is lightly (half-) silvered, half of the light is reflected and half is transmitted. The photoelectric cell is connected to an electronic device that records the percentage absorption at each wavelength.*

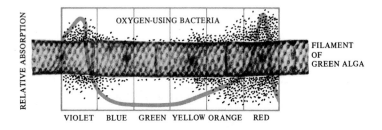

7–6

Results of an experiment performed in 1882 by T. W. Englemann revealed the action spectrum of photosynthesis in a filamentous alga. Like investigators working more recently, Englemann used the rate of oxygen production to measure the rate of photosynthesis. Unlike his successors, however, he lacked sensitive devices for detecting oxygen. As his oxygen indicator, he chose bacteria that are attracted by oxygen. In place of the mirror and diaphragm usually used to illuminate objects under view in his microscope, he substituted a "microspectral apparatus" which, as its name implies, produced a tiny spectrum of colors that it projected upon the slide under the microscope. Then he arranged a filament of algal cells parallel to the spread of the spectrum. The oxygen-seeking bacteria congregated mostly in the areas where the violet and red wavelengths fell upon the algal filament. As you can see, the action spectrum for photosynthesis that Englemann revealed in his experiment paralleled the absorption spectrum of chlorophyll. He concluded that photosynthesis depends on the light absorbed by chlorophyll. This is an example of the sort of experiment that scientists refer to as "elegant," for it was not only brilliant but also was simple in design and conclusive in its results.*

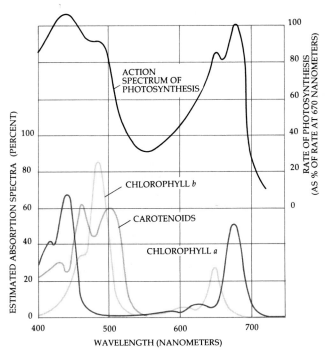

7–7

The upper curve shows the action spectrum for photosynthesis, and the lower curves show the absorption spectra for chlorophyll a, chlorophyll b, and the carotenoids in the chloroplast. The action spectrum of photosynthesis indicates that chlorophyll a, chlorophyll b, and the carotenoids absorb the light used in photosynthesis.

If chlorophyll molecules are isolated in a test tube and light is permitted to strike them, they fluoresce. In other words, the pigment molecules absorb light energy, so the electrons are momentarily raised to a higher energy level and then fall back again to a lower one. As they fall to a lower energy level, they release much of this energy as light. None of the light absorbed by isolated chlorophyll molecules is converted to any form of energy useful to living systems. Chlorophyll can convert light energy to chemical energy only when it is associated with certain proteins and embedded in a specialized membrane (a thylakoid).

Photosynthetic Pigments

Pigments that participate in photosynthesis include the chlorophylls, the carotenoids, and the phycobilins.

There are several different kinds of chlorophyll, which differ from one another only in the details of their molecular structure. Chlorophyll *a* (Figure 7–8) occurs in all photosynthetic eukaryotes and in the cyanobacteria, and it is considered to be essential for the type of photosynthesis carried out by organisms of these groups.

The vascular plants, bryophytes, green algae, and euglenoid algae also contain chlorophyll *b*. Chlorophyll *b* is an accessory pigment—a pigment that serves to broaden the spectrum of light absorption in photosynthesis. When a molecule of chlorophyll *b* absorbs light, the excited molecule transfers its energy to a molecule of chlorophyll *a*, which then transforms it into chemical energy during the course of photosynthesis. Because chlorophyll *b* absorbs light of different wavelengths than does chlorophyll *a* (see Figure 7–7), it extends the range of light that can be used for photosynthesis. In the leaves of green plants, chlorophyll *b* generally constitutes about one-fourth of the total chlorophyll content.

Chlorophyll *c* takes the place of chlorophyll *b* in some groups of algae, most notably the brown algae and the diatoms. The photosynthetic bacteria (other than cyanobacteria) cannot extract electrons from water and thus do not evolve oxygen. They contain either bacteriochlorophyll (in purple bacteria) or chlorobium

7–8

Chlorophyll a is a large molecule with a central core of magnesium held in a porphyrin ring. Attached to the ring is a long, insoluble carbon-hydrogen chain, which serves to anchor the molecule in the internal membranes of the chloroplast. Chlorophyll b differs from chlorophyll a in having a —CHO group in place of the —CH₃ group indicated in color here. Alternating single and double bonds, known as conjugated bonds, such as those in the porphyrin ring of chlorophylls, are common among pigments. Note the similarity between the chlorophyll a molecule shown here and the cytochrome molecule of Figure 6–9.

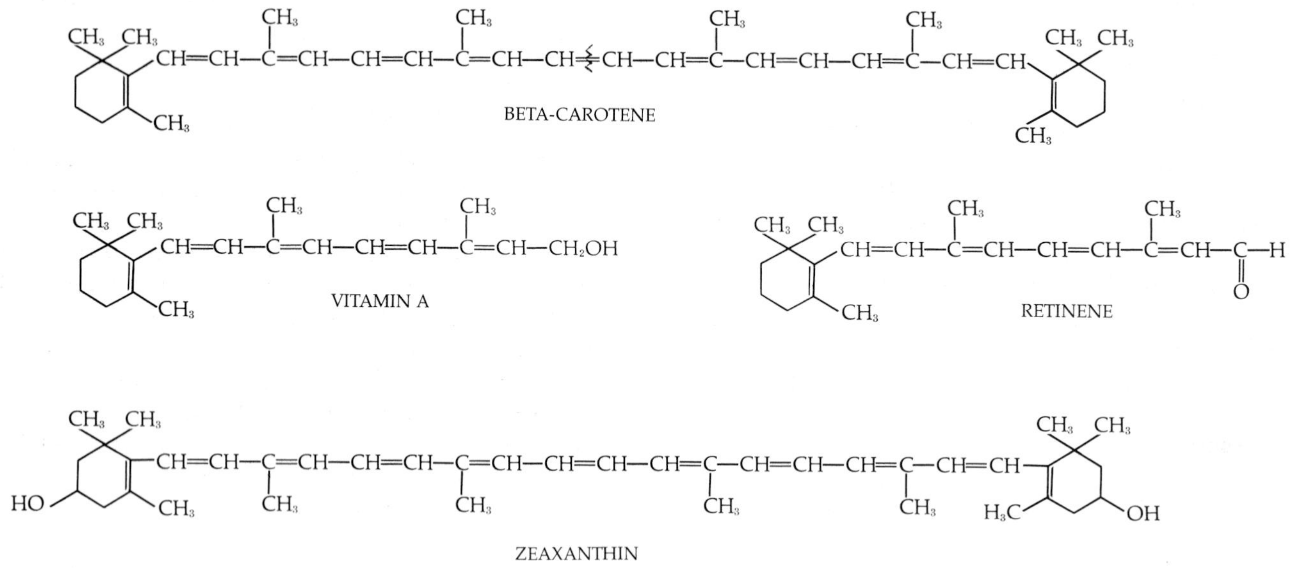

7–9

Related carotenoids. Cleavage of the beta-carotene molecule at the point shown yields two molecules of vitamin A. Oxidation of vitamin A yields retinene, the pigment involved in vision. In the carotenoids, the conjugated bonds are located in the carbon chains. Zeaxanthin is the pigment responsible for the yellow color of corn kernels.

BETA-CAROTENE

VITAMIN A

RETINENE

ZEAXANTHIN

chlorophyll (in green sulfur bacteria). Chlorophylls *b* and *c* and the photosynthetic pigments of purple bacteria and green sulfur bacteria are simply chemical variations of the basic structure shown in Figure 7–8.

Two other classes of pigments involved in the capture of light energy are the *carotenoids* and the *phycobilins*. The energy absorbed by these accessory pigments must be transferred to chlorophyll *a*; they cannot substitute for chlorophyll *a* in photosynthesis.

Carotenoids are red, orange, or yellow fat-soluble pigments found in all chloroplasts and in cyanobacteria. Like the chlorophylls, the carotenoid pigments of chloroplasts are embedded in the thylakoid membranes. Two groups of carotenoids—*carotenes* and *xanthophylls*—are normally present in chloroplasts (xanthophylls contain oxygen and carotenes do not). The betacarotene found in plants is the principal source of the vitamin A required by humans and other animals (Figure 7–9). In green leaves, the color of the carotenoids is masked by the much more abundant chlorophylls.

The third major class of accessory pigments, the phycobilins, are found in the cyanobacteria and in the chloroplasts of the red algae. Unlike the carotenoids, the phycobilins are water soluble.

The Photosystems

In the chloroplast (Figure 7–1), the chlorophyll and other pigment molecules are embedded in the thylakoids in discrete units of organization called *photosystems* (Figure 7–10). Each photosystem is an assembly of about 250 to 400 pigment molecules.

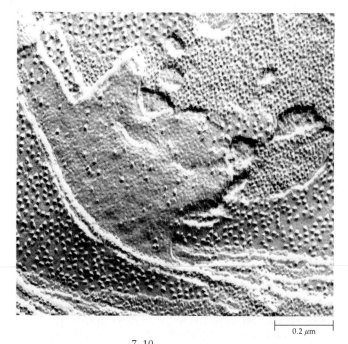

0.2 μm

7–10

Internal surface of a thylakoid, as viewed by the freeze-fracture technique. The particles embedded within the membrane are thought to be the structural units of the photosystems involved in the light reactions.

All of the pigments within a photosystem are capable of absorbing _photons_ (particles of light energy; see "Light and Life," page 102), but only one chlorophyll molecule per photosystem can actually use the energy in the photochemical reaction. This special chlorophyll molecule is called the _reaction center_ of the photosystem, and the other pigment molecules are termed _antenna pigments_, so named because they act as an antennalike network for the gathering of light.

Light energy absorbed by a pigment molecule anywhere in the network is transferred from one pigment molecule to the next until it reaches the reaction center, which is a special form of chlorophyll _a_. When this particular chlorophyll molecule absorbs the energy, one of its electrons is boosted to a higher energy level and transferred to an acceptor molecule to initiate electron flow. The chlorophyll molecule is thus oxidized and positively charged.

According to present evidence, there are two different kinds of photosystems. In Photosystem I, the reaction center is a form of chlorophyll _a_ known as P_{700}. The "P" stands for pigment and the subscript "700" designates the optimal absorption peak in nanometers (nm). The reaction center of Photosystem II also contains a special form of chlorophyll _a_. Its optimal absorption peak is at 680 nm, and accordingly, it is called P_{680}.

In general, the two photosystems work together simultaneously and continuously. As we shall see, however, Photosystem I can operate independently.

A Model of the Light Reactions

Figure 7–11 shows the present model of how the two photosystems work together. According to this model, light energy enters Photosystem II where it is trapped by the reaction center, P_{680}, either directly or indirectly via one or more of the pigment molecules. When P_{680} is excited, its energized electron is transferred to an acceptor molecule designated as "Q," because of its ability to _Q_uench the loss of energy by fluorescence of excited P_{680}. It is also probably a quinone. By an as yet poorly understood reaction, the electron-deficient P_{680} molecule is able to replace its electrons (the electrons actually move in pairs) by extracting them from a water molecule. When the electrons pass from the water molecule to the P_{680} molecule, the water molecule falls apart and is dissociated into protons and oxygen gas.

7–11

Noncyclic electron flow and photophosphorylation. This zigzag scheme (the Z scheme) shows the pathway of electron flow from water (lower left) to NADP (upper right), as well as the energy relationships. To raise the energy of electrons derived from water to the energy level required to reduce NADP to NADPH$_2$, the electrons must be boosted twice (heavy brown lines) by photons absorbed in Photosystems I and II. After each boosting step, the high-energy electrons flow downhill via the routes shown (heavy black lines). Photophosphorylation of ADP to yield ATP is coupled to electron flow in the electron transport chain connecting Photosystem II to Photosystem I (see Figure 7–12). Here ferredoxin is shown donating its electrons for the reduction of NADP to NADPH$_2$. Ferredoxin has a number of roles in the chloroplast, including the donation of electrons to enzymes involved in amino acid biosynthesis and fatty acid biosynthesis.

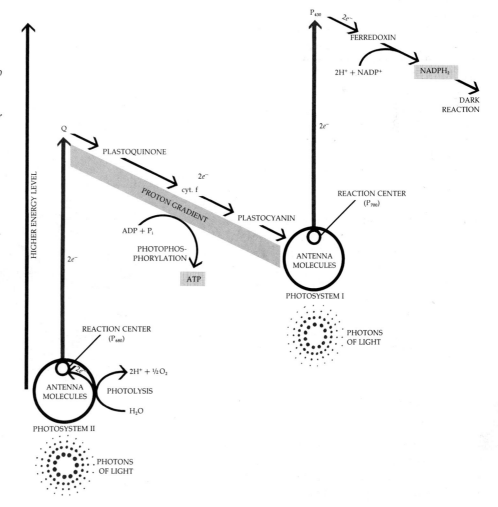

LIGHT AND LIFE

Almost 300 years ago, the English physicist Sir Isaac Newton (1642–1727) separated light into a spectrum of visible colors by letting it pass through a prism. By this experiment, Newton showed that white light is actually made up of a number of different colors, ranging from violet at one end of the spectrum to red at the other. Their separation is possible because light of different colors is bent at different angles in passing through a prism.

In the nineteenth century, through the genius of the British physicist James Clerk Maxwell (1831–1879), it came to be known that what we experience as light is but a very small part of a vast continuous spectrum of radiation, the electromagnetic spectrum. As Maxwell showed, all the radiations included in this spectrum travel in waves. The wavelengths—that is, the distances from one peak to the next—vary from those of X-rays, which are measured in nanometers, to those of low-frequency radio waves, which are measured in kilometers. The shorter the wavelength, the greater the energy. Within the spectrum of visible light, red light has the longest wavelength and violet has the shortest. Another feature that these radiations have in common is that, in a vacuum, they all travel at the same speed—300,000 kilometers per second.

By 1900 it had become clear, however, that the wave theory of light was not adequate. The key observation, a very simple one, was made in 1888: when a zinc plate is exposed to ultraviolet light, it acquires a positive charge. The metal, it was soon deduced, becomes positively charged because the radiation energy dislodges electrons, forcing them out of the metal atoms. Subsequently it was discovered that this photoelectric effect, as it is known, can be produced in all metals. Each metal has a critical wavelength for the effect; the radiation (visible or invisible) must be of that particular wavelength or shorter (must be more energetic) for the effect to occur. (The hypothesis that electrons orbit atoms at particular energy levels, as formulated by Bohr and others, is based on these observations.)

With some metals, such as sodium, potassium, and selenium, the critical wavelength is within the spectrum of visible light, and as a consequence, visible light striking the metal can set up a moving stream of electrons (an electric current). The electric eyes that open doors at supermarkets or airline terminals, exposure meters, and television cameras all operate on this principle of turning light energy into electrical energy.

Wave or Particle?

Now here is the problem. The wave theory of light would lead one to predict that the brighter the light, the greater the force with which the electrons would be dislodged. Whether or not light can eject the electrons of a particular metal, however, depends on its wavelength, not on its brightness.

A very weak beam of the critical wavelength is effective, whereas a stronger (brighter) beam of a longer wavelength is not. Furthermore, increasing the brightness of light of a critical intensity increases the number of electrons dislodged but not the velocity at which they are ejected from the metal. To increase the velocity, one must use a shorter wavelength of light. Nor is it necessary for energy to accumulate in the metal. With even a dim beam of a critical wavelength, an electron may be emitted the instant the light hits the metal.

To explain such phenomena, the particle theory of light was proposed by Albert Einstein in 1905. According to this theory, light is composed of particles of energy called photons, or quanta of light. The energy of a photon (a quantum of light) is not the same for all kinds of light but is, in fact, inversely proportional to the wavelength—the longer the wavelength, the lower the energy. Photons of violet light, for example, have almost twice the energy of photons of red light, the longest visible wavelength.

The wave theory of light permits physicists to describe certain aspects of its behavior mathematically, whereas the photon theory permits another set of mathematical calculations and predictions. These two models are no longer regarded as opposing one another; rather, they are complementary, in the sense that both are required for a complete description of the phenomenon we know as light.

The coexistence of these two theories further illustrates a facet of the scientific method. If a scientist defines light as a wave and measures it as such, it behaves like a wave. Similarly, if one defines it as a particle, one obtains information about light as a particle. As Einstein once said, "It is the theory which determines what we can observe."

The Fitness of Light

Light, as Maxwell showed, is only a tiny band in a continuous spectrum. From the physicist's point of view, the difference between light and darkness—so dramatic to the human eye—is only a few nanometers of wavelength. There is no qualitative difference marking the borders of the light

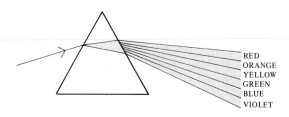

RED
ORANGE
YELLOW
GREEN
BLUE
VIOLET

When white light passes through a prism, it is sorted into a spectrum of different colors. This separation occurs because each color has slightly different wavelengths.

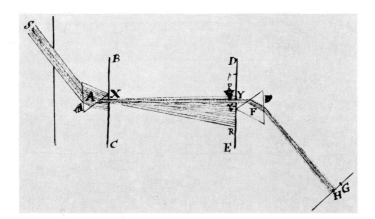

Manuscript drawing from one of Newton's papers, showing what he considered to be a "crucial experiment" in his study of the mixture of colors in "white" light. Sunlight enters a darkened room from the left, producing a spectrum of colors on passing through the first prism A. The hole at Y in the opaque board at the right permits light of a single color to pass through the second prism F, with the result that the beam of light is bent, but there is no further alteration in color. Newton thus demonstrated that a prism does not alter the light in producing a spectrum; rather, it merely "bends" the various colors to different degrees.

spectrum. Why is it that this tiny portion of the electromagnetic spectrum is responsible for vision, for phototropism (the curving of an organism toward light), for photoperiodism (the seasonal changes that take place in an organism with the changing length of day and night), and also for photosynthesis, on which all life depends? Is it an amazing coincidence that all these biological activities depend on these same wavelengths?

George Wald of Harvard University, one of the greatest living experts on the subject of light and life, says no. He thinks that if life exists elsewhere in the universe, it is probably dependent upon this same fragment of the vast spectrum of radiation. Wald bases this conjecture on two points. First, living things are composed of large, complicated molecules held in special configurations and relationships to one another by hydrogen bonds and other weak bonds. Radiation of even slightly higher energies than the energy of violet light can break these bonds and so disrupt the structure and function of the molecules. Radiation with a wavelength below 200 nanometers drives electrons out of atoms to create ions; hence, it is called ionizing radiation.

The energy of light of wavelengths longer than those of the visible band is largely absorbed by water, which makes up the bulk of all living things. When such light does reach organic molecules, its lower energy causes them to increase their motion (increasing heat) but does not trigger changes in their structure. Only that radiation within the range of visible light has the property of exciting molecules—that is, of raising electrons from one energy level to another—and so of producing biological changes.

The second reason that the visible band, above all others of the electromagnetic spectrum, was "chosen" by living things is that it, above all, is what is available. Most of the radiation reaching the earth from the sun is within this range. Higher-energy wavelengths are screened out by the oxygen and ozone high in the atmosphere. Much infrared radiation is screened out by water vapor and carbon dioxide before it reaches the earth's surface.

This is an example of what has been termed "the fitness of the environment"; the suitability of the environment for life and the suitability of life for the physical world are exquisitely interrelated. If they were not, life could not exist.

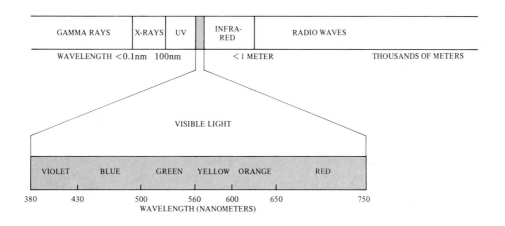

7–12

The chemiosmotic coupling mechanism of photophosphorylation. In this process protons are "pumped" across the thylakoid membrane, from the outside (stroma) to the inside (thylakoid space), as a result of the arrangement of the electron carriers in the membrane. The proton concentration in the thylakoid space builds up in part from the splitting of water there and partly because of the oxidation of plastoquinone (PQ) at the inner face of the membrane. As the protons flow down the gradient from the thylakoid space back into the stroma, ADP is phosphorylated to ATP by ATP synthetase. For every three protons that pass through the ATP synthetase, one ATP is synthesized.

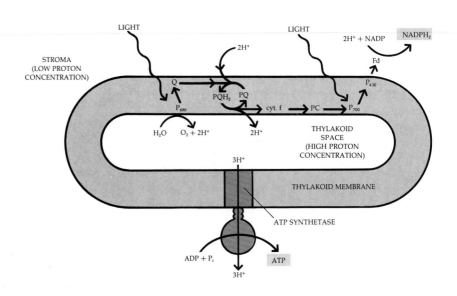

This light-dependent oxidative splitting of water molecules is called *photolysis*. The enzyme that carries out the photolysis of water is located on the inside of the thylakoid membrane. Thus, the photolysis of water molecules contributes to the generation of a proton gradient across the membrane. Manganese is an essential cofactor for the oxygen-evolving mechanism.

The electrons then pass downhill along an electron transport chain to Photosystem I. The components of the electron transport chain of the light reactions resemble those of the electron transport chain of respiration; cytochromes, iron-sulfur proteins, and quinones are involved. In addition, photosynthetic electron transport involves chlorophyll and the copper-containing protein plastocyanin. The electron transport chain between the two photosystems is arranged so that ATP can be formed from ADP and phosphate by a mechanism similar in many ways to the oxidative phosphorylation that takes place in mitochondria (see page 23). In chloroplasts, the process is called *photophosphorylation* (Figure 7–12).

In Photosystem I, light energy boosts the electrons from P_{700} to an electron acceptor designated "P_{430}," which is believed to be an iron-sulfur protein. The next electron carrier is ferredoxin ("Fe" for iron, and "redoxin" to indicate that it carries out reduction-oxidation reactions), which is also an iron-sulfur protein but is different from P_{430}. Ferredoxin then passes its electrons downhill to the coenzyme NADP, resulting in the reduction of the NADP to $NADPH_2$ and oxidation of the P_{700} molecule. The electrons removed from the P_{700} molecule are replaced by the electrons from Photosystem II.

Thus, in the light there is a continuous flow of electrons from water to Photosystem II to Photosystem I to NADP. This unidirectional flow of electrons from water to NADP is called *noncyclic electron flow*, and the production of ATP that occurs during noncyclic electron flow is called *noncyclic photophosphorylation*.

The free-energy change (ΔG) for the reaction

$$H_2O + NADP \longrightarrow NADPH_2 + 1/2\ O_2$$

is 51 kilocalories per mole. The energy of the equivalent mole of photons of 700-nm light (the equivalent of a mole of photons is called an einstein) is 40 kilocalories per einstein. Because four photons are required to boost two electrons to the level of $NADPH_2$, 160 kilocalories are available. Approximately one-third of the available energy is captured as $NADPH_2$. The total energy harvest from noncyclic electron flow (based on the passage of 12 pairs of electrons from H_2O to NADP) is 12ATP and $12NADPH_2$.

Cyclic Photophosphorylation

As mentioned previously, Photosystem I can work independently of Photosystem II (Figure 7–13). In this process, called *cyclic electron flow*, electrons are boosted from P_{700} to the electron acceptor P_{430} by illumination of Photosystem I. Instead of the electrons being passed downhill to NADP, they are shunted to the electron transport chain that connects Photosystems I and II, and then they pass downhill through that chain back into the reaction center of Photosystem I. ATP is produced in the course of this passage, and inasmuch as this process involves a cyclic flow of electrons, it is called *cyclic photophosphorylation*. It is believed that the most primitive photosynthetic mechanisms worked in this way, and this is apparently the way in which some bacteria carry out photosynthesis. Eukaryotic cells are also able to synthesize ATP by cyclic electron flow. However, no water is split, no O_2 is evolved, and no $NADPH_2$ is formed.

Cyclic electron flow and photophosphorylation are believed to occur when the cell is amply supplied with reducing power in the form of $NADPH_2$ but requires additional ATP for other metabolic needs.

7–13

Cyclic photophosphorylation involves only Photosystem 1. ATP is produced from ADP by the same process shown in Figure 7–12, but oxygen is not released and NADP is not reduced.

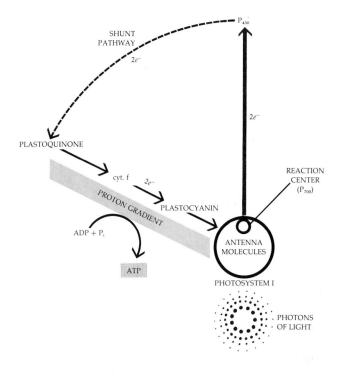

7–14

Scanning electron micrograph showing stomata on the lower surface of a cottonwood (Populus) leaf. Carbon dioxide reaches the photosynthetic cells through the stomata.

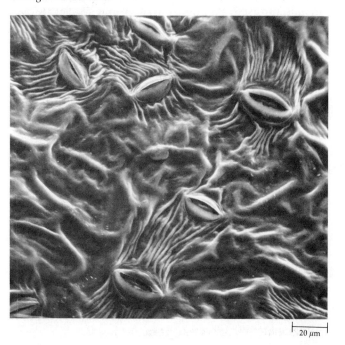

THE DARK REACTIONS

In the second stage of photosynthesis, the chemical energy harvested by the light reactions is used to reduce carbon. Carbon is available to photosynthetic cells in the form of carbon dioxide. For algae and cyanobacteria, this carbon dioxide is found dissolved in the surrounding water. In most plants, carbon dioxide reaches the photosynthetic cells through special openings, called stomata, which are found in their leaves and green stems (Figure 7–14).

The Calvin Cycle: The Three-Carbon Pathway

The reduction of carbon occurs in the stroma of the chloroplast by means of a series of reactions known as the Calvin cycle (named after its discoverer, Melvin Calvin, who received a Nobel Prize for his work on the elucidation of this pathway). The Calvin cycle is analogous to the Krebs cycle (see page 86) in that, by the end of each turn of the cycle, the starting compound is regenerated. The starting (and ending) compound in the Calvin cycle is a five-carbon sugar with two attached phosphate groups—ribulose 1,5-bisphosphate (RuBP). The process begins when carbon dioxide enters the cycle and is fixed to RuBP. The resultant compound then splits to form two molecules of 3-phosphoglycerate, or PGA (Figure 7–15). (Each PGA molecule contains three carbon atoms; hence the designation of the Calvin cycle as the C_3, or three-carbon, pathway.)

7–15

Calvin and his collaborators briefly exposed photosynthesizing algae to radioactive carbon dioxide ($^{14}CO_2$). They found that the radioactive carbon is first incorporated into ribulose 1,5–bisphosphate (RuBP). The RuBP then immediately splits to form two molecules of 3-phosphoglycerate (PGA). The radioactive carbon atom, indicated here in color, next appears in one of the two molecules of PGA. This is the first step in the Calvin cycle.

7–16

A summary of the Calvin cycle. At each full turn of the cycle, one molecule of carbon dioxide enters the cycle. Six turns are summarized here—the number required to make two molecules of glyceraldehyde 3-phosphate. Six molecules of ribulose 1,5-bisphosphate (RuBP), a five-carbon compound, are combined with six molecules of carbon dioxide, yielding twelve molecules of 3-phosphoglycerate, a three-carbon compound. These are converted to twelve molecules of glyceraldehyde 3-phosphate. Ten of these three-carbon molecules are combined and rearranged to form six five-carbon molecules of RuBP. The "extra" molecules of glyceraldehyde 3-phosphate represent the net gain from the Calvin cycle. The energy that "drives" the Calvin cycle is the ATP and $NADPH_2$ produced by the light reactions.

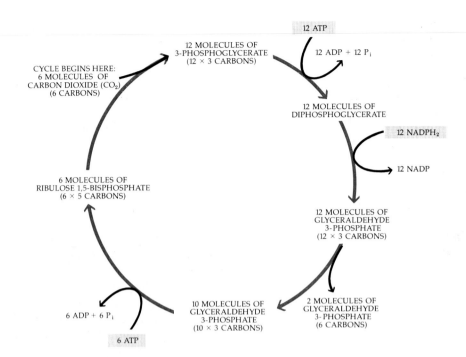

RuBP carboxylase, the enzyme catalyzing these crucial initial reactions, is quite abundant in chloroplasts, making up more than 15 percent of the total chloroplast protein. (It is said to be the most abundant protein in the world. Can you say why?) This enzyme is located on the surface of the thylakoid membranes.

The complete cycle is diagrammed in Figure 7–16. As in the Krebs cycle, each step is regulated by a specific enzyme. At each full turn of the cycle, a molecule of carbon dioxide enters the cycle and is reduced, and a molecule of RuBP is regenerated. Six revolutions of the cycle, with the introduction of six atoms of carbon, are necessary to produce a six-carbon sugar, such as glucose. On a per-glucose basis, the net overall equation is:

$$6CO_2 + 12NADPH_2 + 18ATP \longrightarrow$$
$$1 \text{ glucose} + 12NADP + 18ADP + 18P_i + 6H_2O$$

The immediate product of the cycle is glyceraldehyde 3-phosphate. This same sugar-phosphate is formed when the fructose 1,6-bisphosphate molecule is split at the fourth step in glycolysis (see page 84). These same steps, using the energy of the phosphate bond, can be reversed to form glucose from glyceraldehyde 3-phosphate.

The Four-Carbon Photosynthetic Pathway

The Calvin cycle is not the only carbon-fixation pathway used in the dark reactions. In some plants, the first product of CO_2 fixation to be detected is not the three-carbon molecule PGA, as it is in the Calvin cycle; it is the four-carbon compound oxaloacetate (which is also an intermediate in the Krebs cycle). Plants that employ this pathway are commonly called C_4 (for four-carbon plants), as distinct from the C_3 plants that use *only* the Calvin cycle. (The C_4 pathway is also referred to as the Hatch-Slack pathway after two Australian plant physiologists who played key roles in its elucidation.)

The oxaloacetate is formed when carbon dioxide is fixed to a compound known as phosphoenolpyruvate (PEP). This reaction is catalyzed by the enzyme PEP carboxylase (Figure 7–17). The oxaloacetate is then reduced to malate or converted with the addition of an amino group to the amino acid aspartate. These steps occur in mesophyll cells. The next step is a surprise: the malate (or aspartate, depending on the species) moves from the mesophyll cells to bundle-sheath cells surrounding the vascular bundles of the leaf, where it is decarboxylated to yield CO_2 and pyruvate. The CO_2 then enters the Calvin cycle and reacts with RuBP to form PGA and other intermediates of the cycle, whereas the pyruvate returns to the mesophyll cells where it reacts with ATP to form more molecules of PEP (Figure 7–18). Hence, the anatomy of the plant imparts a spatial separation between the C_4 pathway and the Calvin cycle in the leaves of C_4 plants.

The two primary carboxylating enzymes use different forms of the carbon dioxide molecule as a substrate. RuBP carboxylase uses CO_2, which is present at a concentration of about 15–20 μM in the cell in equilibrium with air. In contrast, PEP carboxylase uses the hydrated form of carbon dioxide, HCO_3^-, as its substrate. At pH 8.0, HCO_3^- is present at about 15–20 μM in the cell in equilibrium with air. RuBP carboxylase is found in the chloroplast, while PEP carboxylase occurs in the cytoplasmic ground substance.

7–17

Carbon dioxide fixation by the C$_4$ pathway. Carbon dioxide is incorporated into phosphoenolpyruvate (PEP) by the enzyme PEP carboxylase. The resulting oxaloacetate is either reduced to malate or converted to aspartate through the addition of an amino group ($-NH_2$). These steps will be reversed later, releasing carbon dioxide for use in the Calvin cycle.

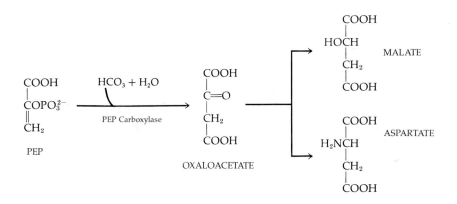

7–18

A pathway for carbon fixation in C$_4$ plants. CO$_2$ is first fixed in mesophyll cells as oxaloacetate, which is rapidly converted to malate. The malate is then transported to bundle-sheath cells, where the carbon dioxide is released. The CO$_2$ thus released enters the Calvin cycle, ultimately yielding starch and sucrose. Pyruvate returns to the mesophyll cell for regeneration of phosphoenolpyruvate (PEP). The C$_4$ leaf shown here is from a corn plant (Zea mays).

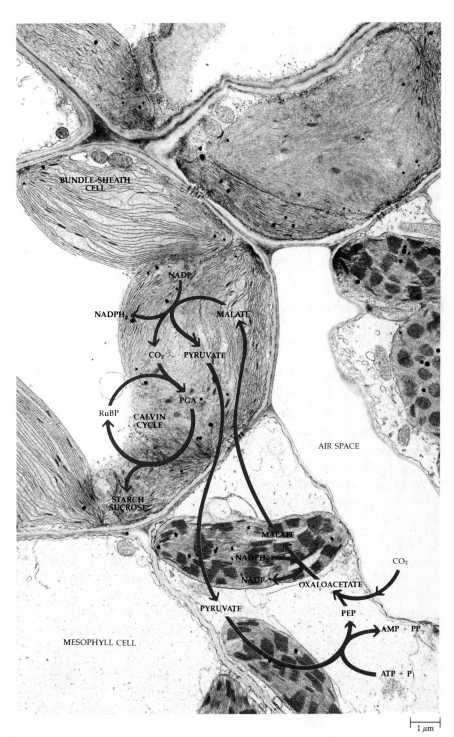

7–19

Transverse section of a portion of a corn (Zea mays) leaf. As is typical of C_4 plants, the vascular bundles are surrounded by large, chloroplast-containing, bundle-sheath cells that are, in turn, surrounded by a layer of mesophyll cells. The C_4 pathway takes place in the mesophyll cells; the Calvin cycle occurs in the bundle-sheath cells.

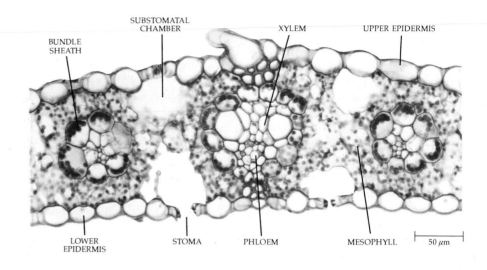

SUBSTOMATAL CHAMBER

BUNDLE SHEATH

XYLEM

UPPER EPIDERMIS

LOWER EPIDERMIS

STOMA

PHLOEM

MESOPHYLL

50 μm

Typically, the leaves of C_4 plants are characterized by an orderly arrangement of the mesophyll cells around a layer of large bundle-sheath cells, so that together the two form concentric layers around the vascular bundle (Figure 7–19). This wreathlike arrangement has been termed Kranz anatomy. (*Kranz* is the German word for "wreath.") In some C_4 plants the chloroplasts of the mesophyll cells have well-developed grana, while those of the bundle-sheath cells have poorly developed grana or none at all (Figure 7–18). In addition, when photosynthesis is occurring, the bundle-sheath chloroplasts commonly form larger and more numerous starch grains than the mesophyll chloroplasts.

Efficiency of C_4 Plants

CO_2 fixation in C_4 plants has a larger energy cost than in C_3 plants. For each molecule of CO_2 fixed in the C_4 pathway, a molecule of PEP must be regenerated at the cost of two high-energy phosphate groups of ATP. C_4 plants need five ATPs altogether to fix one molecule of CO_2; C_3 plants need only three.

One might well ask why C_4 plants should have evolved such a seemingly clumsy and energetically expensive method of providing carbon dioxide to the Calvin cycle. The C_4 pathway can be better understood when one learns that photosynthesis in C_3 plants is always accompanied by *photorespiration*, a process that *consumes* oxygen and *releases* CO_2 in the presence of light (Figure 7–20). Photorespiration is a wasteful process. Unlike mitochondrial respiration, photorespiration is not accompanied by oxidative phosphorylation; hence, it yields no ATP. In addition, photorespiration diverts some of the reducing power generated in the light reactions from the biosynthesis of glucose into the reduction of oxygen. Under normal atmospheric conditions, as much as 50 percent of the carbon fixed in pho-

tosynthesis by a C_3 plant may be reoxidized to CO_2 during photorespiration. Although photorespiration is very active in C_3 plants, seriously limiting their efficiency, it is nearly absent in C_4 plants.

The major substrate oxidized by photorespiration in C_3 plants is glycolic acid. Glycolic acid is oxidized in the peroxisomes (see page 24) of photosynthetic cells and is formed by the oxidative breakdown of RuBP by RuBP carboxylase, the identical enzyme that fixes CO_2 into PGA. How is this possible?

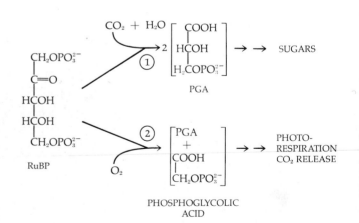

7–20

Reactions catalyzed by RuBP carboxylase. Reaction 1 is favored by high CO_2 and low oxygen concentrations. Reaction 2 has been proposed to occur in the presence of low CO_2 and high oxygen concentrations (normal atmospheric conditions).

THE CARBON CYCLE

By photosynthesis, living systems incorporate carbon dioxide from the atmosphere into organic, carbon-containing compounds. In respiration, these compounds are broken down again into carbon dioxide and water. These processes, viewed on a worldwide scale, result in the carbon cycle. The principal photosynthesizers in this cycle are plants and the phytoplankton, the marine algae, and cyanobacteria. They synthesize carbohydrates from carbon dioxide and water and release oxygen into the atmosphere. About 75 billion metric tons of carbon per year are bound into carbon compounds by photosynthesis.

Some of the carbohydrates are used by the photosynthesizers themselves. Plants release carbon dioxide from their roots and leaves, and marine algae and cyanobacteria release it into the water where it maintains an equilibrium with the carbon dioxide of the air. Some 500 billion metric tons of carbon are "stored" as dissolved carbon dioxide in the seas, and some 700 billion metric tons in the atmosphere. Some of the carbohydrates are used by animals that feed on live plants, on algae, and on one another, releasing carbon dioxide. An enormous amount of carbon is contained in the dead bodies of plants and other organisms plus discarded leaves and shells, feces, and other waste materials that settle into the soil or sink to the ocean floors where they are consumed by decomposers—small invertebrates, bacteria,

and fungi. Carbon dioxide is also released by these processes into the reservoir of the air and oceans. Another, even larger store of carbon lies below the surface of the earth in the form of coal and oil, deposited there some millions of years ago.

The natural processes of photosynthesis and respiration balance one another out. For many millions of years, the carbon dioxide of the atmosphere, as far as we can tell, has remained constant. By volume, it is a very small proportion of the atmosphere, only about 0.03 percent. It is important, however, because carbon dioxide, unlike most other components of the atmosphere, absorbs heat from the sun's rays. Since 1850, carbon dioxide concentrations in the atmosphere have been increasing, owing in part to our use of fossil fuels, to our plowing of the soil, and to our destruction of forest land, particularly in the tropics. Some environmentalists predict that this increase in the carbon dioxide "blanket" will increase the temperature here on earth, with a consequential increase in the great deserts of the world. Others, looking on the sunnier side, foresee an increase in photosynthetic activity because of the increased carbon dioxide that will be available to plants and algae. Most, however, feel alarmed by the fact that although we do not know the consequences of what we are doing, we keep right on doing it.

The carbon cycle. The arrows indicate the movement of carbon atoms. The numbers are all estimates of the amount of carbon stored, expressed in billions of metric tons. The amount of carbon released by respiration and combustion has, it is believed, begun to exceed the amount fixed by photosynthesis.

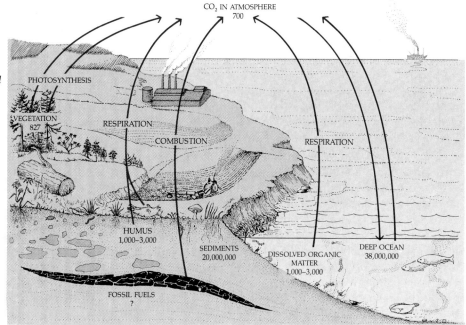

RuBP carboxylase can promote the reaction of RuBP with *either* CO_2 or O_2. When the CO_2 concentration is high and that of O_2 is relatively low, RuBP carboxylase fixes CO_2 to RuBP to yield PGA. When the CO_2 concentration is low and that of O_2 is relatively high, the enzyme acts as an oxygenase, combining RuBP and oxygen to yield phosphoglycolic acid and PGA, instead of the two PGA molecules normally formed in the carboxylation. The phosphoglycolic acid is then converted to glycolic acid—the substrate oxidized during photorespiration.

High CO_2 and low O_2 concentrations limit photorespiration. Consequently, C_4 plants have a distinct advantage over C_3 plants because CO_2 fixed by the C_4 pathway is essentially "pumped" from the mesophyll cells into the bundle-sheath cells, thus maintaining a high $CO_2:O_2$ ratio at the site of the action of RuBP carboxylase. This favors the carboxylation of RuBP. In addition, since the Calvin cycle and its associated photorespiration is localized in the inner, bundle-sheath layer, any CO_2 liberated by photorespiration into the outer, mesophyll layer can be refixed by the C_4 pathway that operates there. The CO_2 liberated by photorespiration can thus be prevented from escaping from the leaf. Moreover, compared to C_3 plants, C_4 plants are superior utilizers of available CO_2; this is in part due to the fact that the enzyme PEP carboxylase is not inhibited by O_2. As a result, the net photosynthetic rates of C_4 grasses, such as corn (*Zea mays*), sugarcane (*Saccharum officinale*), and sorghum (*Sorghum vulgare*), can be two to three times the rates of C_3 grasses, such as wheat (*Triticum aestivum*), rye (*Secale cereale*), oats (*Avena sativa*), and rice (*Oryza sativa*).

C_4 plants evolved primarily in the tropics and are especially well adapted to high light intensities, high temperatures, and dryness. The optimal temperature range for C_4 photosynthesis is much higher than that for C_3 photosynthesis; and C_4 plants flourish even at temperatures that would eventually be lethal to many C_3 species. Because of their more efficient use of carbon dioxide, C_4 plants can attain the same photosynthetic rate as C_3 plants but with smaller stomatal openings and, hence, with considerably less water loss. Analyses of the geographic distribution of C_4 species in North America have shown that they are generally most abundant in climates with high temperatures. Differences appear to exist, however, between monocots and dicots in the particular type of high temperature favored. For example, C_4 grasses are most abundant in regions having the highest temperatures during the growing season (Figure 7–21); in contrast, C_4 dicots are most abundant in regions having the greatest aridity during the growing season.

A striking illustration of differing growth patterns of C_4 plants is found in our lawns, which, in the cooler parts of the country at least, consist mainly of C_3

7–21
The numbers shown here indicate the percentages of the total number of grass species having the C_4 pathway in each of 32 local floras in North America. The highest percentages are found in those regions with the highest temperatures during the growing season.

grasses such as Kentucky bluegrass (*Poa pratensis*) and creeping bent (*Agrostis tenuis*). Crabgrass (*Digitaria sanguinalis*), which all too often overwhelms these dark green, fine-leaved grasses with patches of its yellowish green, broader leaves, is a C_4 grass that grows much more rapidly in the heat of summer than the temperate C_3 grasses mentioned above.

All of the plants now known to utilize C_4 photosynthesis are flowering plants, including at least 19 families, 3 of monocots and 16 of dicots; however, no family has been found that contains only C_4 species. This pathway has undoubtedly arisen independently many times in the course of evolution.

CARBON ISOTOPE COMPOSITION OF C₃ AND C₄ PLANTS

In the free atmosphere, carbon dioxide exists as $^{12}CO_2$, $^{13}CO_2$, and $^{14}CO_2$. Unlike ^{14}C, ^{12}C and ^{13}C are stable isotopic forms of carbon. Plants do not assimilate the isotopes of carbon equally but discriminate against the heavier ones. It is now known that C_3 plants discriminate to a greater extent against ^{13}C than do C_4 plants. The discrimination by C_3 plants against ^{13}C occurs in large part during carboxylation associated with RuBP carboxylase. The RuBP carboxylase of C_3 plants discriminates to a greater extent against ^{13}C than does the PEP carboxylase of C_4 plants. As a consequence, the compounds of C_4 plants contain greater amounts of ^{13}C than do those of C_3 plants.

The stable isotope concentrations in plants are expressed as $\delta^{13}C$ values relative to ^{12}C and ^{13}C values from a standard substance. (The values are determined with use of a mass spectrometer.) The $\delta^{13}C$ values for C_3 plants are about -27 parts per thousand (compared with the standard) and those for C_4 plants about -11 parts per thousand. The $\delta^{13}C$ value is important in determining whether or not a plant is C_3 or C_4.

$\delta^{13}C$ values can also be of importance to the physiological ecologist. Inasmuch as the $\delta^{13}C$ values are transmitted into the ecosystem from plants to herbivores, analyses of $\delta^{13}C$ values of the rumen and fecal contents of herbivores (plant-eating animals) can be used to determine which herbivores in an ecosystem prefer C_3 or C_4 plants.

Crassulacean Acid Metabolism

Crassulacean acid metabolism, commonly designated CAM, has evolved independently in many succulent plants, including the cacti (Cactaceae) and stonecrops (Crassulaceae). Plants are considered to have CAM if their photosynthetic cells have the ability to fix CO_2 in the dark via PEP carboxylase, forming malic acid, which is stored in the vacuole. During the following light period, the malic acid is decarboxylated, and the CO_2 is transferred to RuBP of the Calvin cycle *within the same cell*. Thus, CAM plants, like C_4 plants, utilize both C_4 and C_3 pathways; they differ from the C_4 plants, however, in that there is a temporal separation of the two pathways in the CAM plants, rather than a spatial one as in the C_4 plants.

CAM plants are largely dependent upon nighttime accumulation of carbon for their photosynthesis because their stomata are closed during the day, retarding water loss. This is obviously advantageous in the conditions of high light intensity and water stress under which most CAM plants live. If all atmospheric CO_2 uptake in a CAM plant occurs at night, the water-use efficiency of that plant can be many times greater than that of a C_3 or C_4 plant. During periods of prolonged drought, some CAM plants can keep their stomata closed both night and day, maintaining low metabolic rates by refixing respiratory CO_2 via the dark accumulation of malic acid followed by CO_2 fixation via the Calvin cycle the following day.

CAM is more widespread among the vascular plants than C_4 photosynthesis. It has been reported in at least 23 families of flowering plants, mostly dicots, including such familiar houseplants as the maternity plant (*Kalanchoë daigremontiana*), wax plant (*Hoya carnosa*), and snake plant (*Sansevieria zeylanica*). Not all CAM plants are highly succulent; two examples of less succulent ones are the pineapple (*Ananas comosus*) and Spanish "moss" (*Tillandsia usneoides*), both members of the family Bromeliaceae (monocots). Some nonflowering plants have also been reported to show CAM activity, including a most bizarre gymnosperm, *Welwitschia mirabilis* (see Figure 18–35), the quillwort (*Isoetes*; see Figures 17–19 and 17–20), and some ferns.

Adaptive Significance of Photosynthetic Mechanisms

From our discussions of C_3, C_4, and CAM photosynthesis, it may have become apparent that the photosynthetic mechanism is not the only factor that determines where a plant lives. Although it is an important factor, all three mechanisms have advantages and disadvantages, and a plant can compete successfully only when the benefits of its type of photosynthesis outweigh other factors. For example, although C_4 plants generally tolerate higher temperatures and dryer conditions than C_3 species, at temperatures below $25°C$ they may not compete successfully, in part because they are more sensitive to cold than C_3 species. In addition, as discussed, CAM plants, which are best adapted to very arid conditions, conserve water by closing their stomata during the day. This practice, however, severely reduces their ability to take in and fix CO_2. Hence, CAM plants grow slowly and compete poorly with C_3 and C_4 species under less extreme conditions. Thus, to some extent each photosynthetic type of plant is the victim of its own mechanism.

SUMMARY

In photosynthesis, light energy is converted to chemical energy, and carbon is "fixed" into organic compounds. The generalized equation for this reaction is

$$CO_2 + 2H_2A + \text{light energy} \longrightarrow (CH_2O) + H_2O + 2A$$

in which H_2A represents water or some other substance that can be oxidized, that is, from which electrons can be removed.

The first step in photosynthesis is the absorption of light energy by pigment molecules. The pigments involved in photosynthesis in eukaryotes include the chlorophylls and the carotenoids, which are packed in the thylakoids of chloroplasts as photosynthetic units called photosystems. Light absorbed by pigment molecules boosts their electrons to a higher energy level. Because of the way the pigment molecules are arranged in the photosystems, they are able to transfer this energy to a special pigment molecule called the reaction center. There are two photosystems, Photosystem I and Photosystem II.

Not all of the photosynthetic reactions require light. The series of reactions requiring light are referred to as the "light reactions," and those not requiring light are called the "dark reactions."

In the currently accepted model of the light reactions in photosynthesis, light energy enters Photosystem II, where it is trapped by the reaction center P_{680}. Electrons are boosted uphill from P_{680} to an electron acceptor. As the electrons are removed, they are replaced by electrons from water molecules, and oxygen is produced. Pairs of electrons then pass downhill to Photosystem I along an electron transport chain; this passage generates a proton gradient that drives the synthesis of ATP from ADP and phosphate (photophosphorylation). Meanwhile, light energy absorbed in Photosystem I is passed to its reaction center, P_{700}. The energized electrons are ultimately accepted by the coenzyme molecule NADP, and the electrons removed from P_{700} are replaced by the electrons from Photosystem II. The energy yield from the light reactions is stored in the molecules of $NADPH_2$ and in the ATP formed by photophosphorylation. Photophosphorylation also occurs in cyclic electron flow, a process that bypasses Photosystem II.

Like oxidative phosphorylation in mitochondria, photophosphorylation in chloroplasts is a chemiosmotic process. As electrons flow down the electron transport chain from Photosystem II to Photosystem I, protons are pumped from the stroma into the thylakoid space, creating a gradient of potential energy. As protons flow down this gradient from the thylakoid space back into the stroma, passing through an ATP synthetase, ATP is formed.

In the dark reactions, which take place in the stroma of the chloroplast, the $NADPH_2$ and ATP produced in the light reactions are used to reduce carbon dioxide to organic carbon. This is accomplished by means of the Calvin cycle. In the Calvin cycle, a molecule of carbon dioxide is combined with the starting material, a five-carbon sugar called ribulose 1,5-bisphosphate (RuBP), to form a three-carbon compound, 3-phosphoglycerate (PGA). At each turn of the cycle, one carbon atom enters the cycle. Three turns of the cycle produce a three-carbon molecule, glyceraldehyde 3-phosphate, two molecules of which (in six turns of the cycle) can combine to form a glucose molecule. At each turn of the cycle, RuBP is regenerated.

Plants in which the Calvin cycle is the only carbon fixation pathway, and in which the first detectable product of CO_2 fixation is PGA, are called C_3 plants. In the so-called C_4 plants, carbon dioxide is initially fixed to a compound known as PEP (phosphoenolpyruvate) to yield oxaloacetate, a four-carbon compound. The oxaloacetate is then rapidly converted to either malate or aspartate, which transfers the CO_2 to RuBP of the Calvin cycle. The Calvin cycle in C_4 plants occurs in bundle-sheath cells, whereas the C_4 pathway takes place in the mesophyll cells. C_4 plants are more efficient utilizers of CO_2 than C_3 plants, in part because PEP carboxylase is not inhibited by O_2, thus enabling C_4 plants to capture CO_2 with minimal water loss. Also, photorespiration, a process that consumes oxygen and releases CO_2 in the light, is nearly absent in C_4 plants.

Crassulacean acid metabolism (CAM) is found in many succulent plants. In CAM plants the fixation of CO_2 by PEP carboxylase into C_4 compounds occurs at night, when the stomata are open. The C_4 compounds are stored overnight, and then, during the daytime, when the stomata are closed, the fixed CO_2 is transferred to RuBP of the Calvin cycle. The Calvin cycle and the C_4 pathway occur within the same cells in CAM plants; hence, these two pathways, which are spatially separated in C_4 plants, are temporally separated in CAM plants.

SUGGESTIONS FOR FURTHER READING

ALBERTS, BRUCE, DENNIS BRAY, JULIAN LEWIS, MARTIN RAFF, KEITH ROBERTS, and JAMES D. WATSON: *Molecular Biology of the Cell*, Garland Publishing, Inc., New York, 1983.*

A large book covering both the molecular biology of the cell and the behavior of cells in multicellular animals and plants. A truly modern approach to the cell, this book is beautifully illustrated. Written for introductory courses in cell biology.

ASIMOV, I.: *Life and Energy*, Avon Books, New York, 1962.

An elementary but elegant description of the energetic basis of life by one of the greatest science writers of our time.

BECKER, WAYNE M.: *Energy and the Living Cell: An Introduction to Bioenergetics*, Harper & Row, Publishers, Inc., New York, 1977.*

An outstanding short introduction to topics in bioenergetics and cellular energy metabolism; intended for undergraduates.

CONANT, JAMES B. (Ed.): *Harvard Case Histories in Experimental Science*, Vol. 2, Harvard University Press, Cambridge, Mass., 1964.

Case No. 5, Plants and the Atmosphere, *edited by Leonard K. Nash, describes early work on photosynthesis. The narrative, often presented in the words of the investigators themselves, illuminates the historical context in which these initial discoveries were made.*

GABRIEL, MORDECAI L., and SEYMOUR FOGEL (Eds.): *Great Experiments in Biology*, Prentice-Hall, Inc., Englewood Cliffs, N.J., 1955.*

Presents many of the fundamental discoveries of biology as seen firsthand through the eyes of their discoverers. The examples are well chosen, and their value is greatly enhanced by accompanying explanatory notes and chronological tables of key developments in various areas of biology.

HINKLE, P. C., and R. E. McCARTY: "How Cells Make ATP," *Scientific American* 238(3): 104–23, March 1978.

A clearly written article that provides evidence in support of the chemiosmotic theory of ATP synthesis.

KLUGE, M., and I. P. TING: *Crassulacean Acid Metabolism: Analysis of an Ecological Adaptation,* Ecological Studies: Vol. 30, Springer-Verlag, New York, 1979.

A comprehensive treatment of all aspects of Crassulacean acid metabolism, with special emphasis on CAM as an ecological adaptation. The significance of CAM in agriculture is also considered.

LEHNINGER, ALBERT L.: *Principles of Biochemistry,* Worth Publishers, Inc., New York, 1982.

This introductory text is outstanding both for its clarity and for its consistent focus on the living cell. There are numerous medical and practical applications throughout.

SALISBURY, FRANK B., and CLEON W. ROSS: *Plant Physiology,* 3rd ed., Wadsworth Publishing Co., Inc. Belmont, Calif., 1985.

A detailed and useful review of the entire subject.

STRYER, LUBERT: *Biochemistry,* 2nd ed., W. H. Freeman and Company, San Francisco, 1981.

A good introduction to cellular energetics.

STUMPF, P. K., and E. E. CONN (Eds.): *The Biochemistry of Plants: A Comprehensive Treatise,* Vol. 1, The Plant Cell, N. E. Tolbert (Ed.), Academic Press, New York, 1980.

A multiauthored book considering the structure and function of plant cells. The first chapter is introductory in nature and discusses the cell in general. Each of the remaining chapters is devoted to a different subcellular component.

WILKINS, MALCOLM B. (Ed.): *Advanced Plant Physiology,* Pitman Press, Bath, Great Britain, 1984.

An up-to-date advanced plant physiology textbook with multiple authors.

ZELITCH, ISRAEL: "Photosynthesis and Plant Productivity," *Chemical and Engineering News* 57(6), 1979.

A concise but thorough description of the major areas of research in photosynthesis today. Discusses a number of research areas with direct relevance to agriculture.

* Available in paperback.

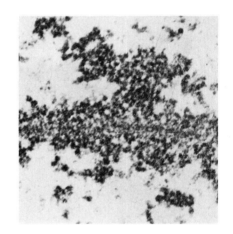

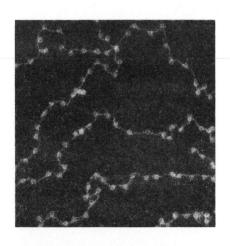

S E C T I O N T H R E E

Genetics

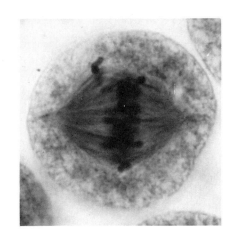

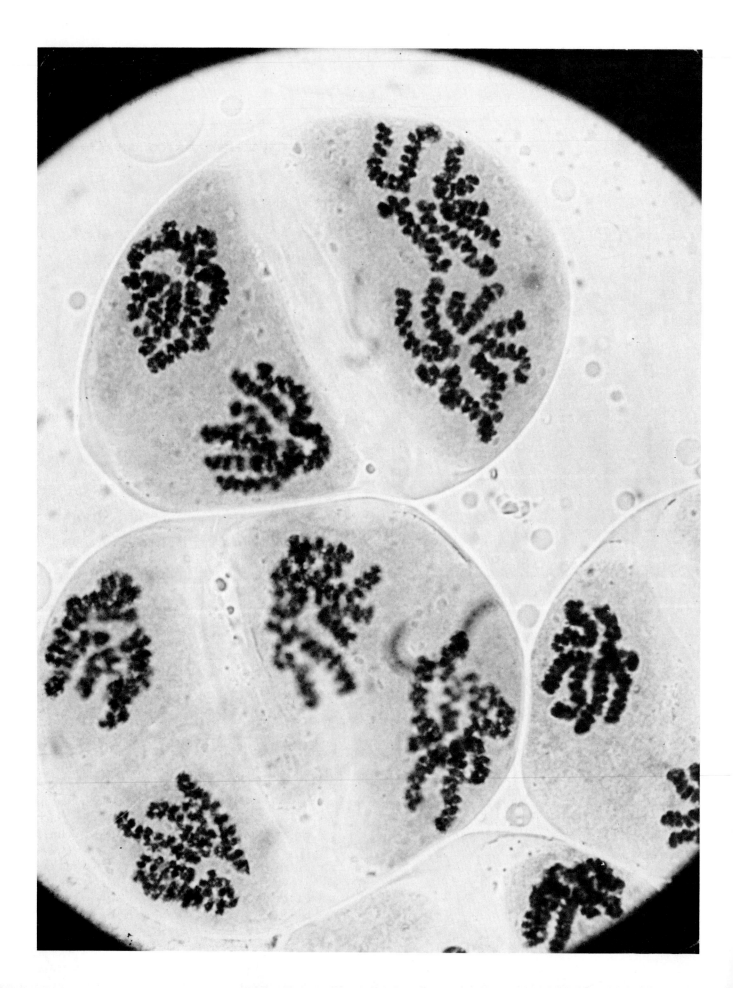

CHAPTER 8

The Chemistry of Heredity

Ever since people first started to look at the world around them, they have puzzled and wondered about heredity. Why is it that the offspring of all living things —whether dandelions, dogs, aardvarks, or oak trees— always resemble their parents and never some other species? Why does a child have her mother's eyes or her father's chin or, even more puzzling, her grandfather's nose?

These questions were recorded as far back as the writings of the Greeks, and they were probably old even then. Such questions have always been important. Throughout history, biological inheritance has been a major factor in determining the distribution of wealth, power, land, and privilege. From the point of view of biology, heredity is at the heart of any definition we may have of life.

Some say that the twentieth century will be remembered as the time in history when human beings first reached the moon, but others predict that it will be best known as the age in which we discovered the nature of DNA and thus began to unravel the secrets of heredity.

THE CHEMISTRY OF THE GENE: DNA VERSUS PROTEIN

Biologists have known for a long time that heredity is associated with the nucleus of cells and, in particular, with the chromosomes (Figure 8–1). Chromosomes in eukaryotes are complexes of DNA and protein, which often appear as slender threads when they are stained and viewed with the light microscope. Once it became clear to researchers that chromosomes carried the genetic information, the problem became one of deciding whether the protein or the DNA plays this essential role.

By the early 1950s, a great deal of evidence for the role of DNA as the genetic material had accumulated:

8–1

The second anaphase of meiosis during microspore formation in the wake-robin (Trillium erectum). Separation of the clearly visible chromosomes, which contain the genetic material, DNA, is nearly complete. Each of the newly forming nuclei will possess only half the number of chromosomes that were present in the original nucleus at the beginning of meiosis.

8–2

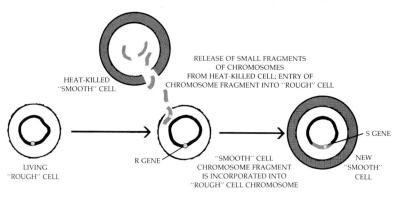

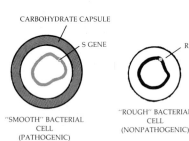

(1) As revealed by specific staining techniques, DNA is present in the chromosomes of all cells, and it is mainly found in chromosomes. (2) In general, the body cells of plants and animals contain twice as much DNA as is found in their gametes. (3) As shown in Table 8–1, the proportions of purines and pyrimidines vary from one species to another. Such variation is essential in a molecule that spells out the "language of life." Even more significantly, the proportion of guanine in the DNA of a particular species is always equal to the proportion of cytosine, and the proportion of adenine is equal to the proportion of thymine. This relationship, which has come to be known as Chargaff's rule, provided a fundamental key in unraveling the process of heredity. (4) DNA isolated from bacterial cells can provide other bacterial cells with new genetic characteristics (Figure 8–2). (5) In the course of the infection of bacterial cells by certain bacterial viruses (bacteriophages), DNA alone enters the cell and directs the formation of new viruses (Figure 8–3).

Despite all this evidence, it was not until the discovery of the structure of DNA that its genetic role came to be generally understood.

THE NATURE OF DNA

In 1951, the American geneticist James D. Watson went to England, where he arranged to work with Francis Crick of the Cavendish Laboratory in Cambridge. Watson and Crick were among the scientists who were convinced that DNA, rather than protein, is the genetic material. In Watson's words, DNA is "the most golden of all molecules."

Watson and Crick based their investigations of how the DNA molecule is put together partly on the understanding of *genes*, units of heredity in the chromosomes, that had been assembled up to that time. They knew that if DNA were the genetic material, it would have to meet at least four requirements:

1. It must carry genetic information from cell to cell and from generation to generation; furthermore, it must carry a great deal of information. (Consider how many instructions must be contained in the set of genes that directs, for example, the development of an elephant, a tree, or even a bacterium.)
2. It must copy itself, for the chromosome does this before every cell division; moreover, it must replicate itself with great precision. (For instance, we know from accumulated data on human mutation rates that any one human gene must have been copied, on average, for millions of years without a mistake.)
3. On the other hand, a gene must sometimes change, or mutate. (When a gene changes, that is, when a mistake is made, the "mistake"—rather than what was originally there—must be copied. This is a most important property, perhaps the unique attribute of living things; without the capacity to replicate "errors," there could be no evolution by natural selection.)
4. It must have some mechanism for "reading out" the stored information and translating it into action in the living individual.

Watson and Crick were well aware that the DNA molecule could be the genetic material only if it could be shown to have the size, configuration, and complex-

8-3

A summary of the Hershey-Chase experiments, a series of important experiments that demonstrate that DNA is the hereditary material of a virus. Bacteriophages (or phages, from the Greek phagein, "to eat") are viruses that infect bacteria. By growing certain virus-infected bacteria on different radioactive media, two separate samples of viruses were obtained—one in which the DNA was labeled with the radioactive isotope ^{32}P and the other in which the coat protein was labeled with the radioactive isotope ^{35}S. (DNA contains no sulfur, and proteins have no phosphorus.) The labeled phages were then permitted to infect bacteria growing in nonradioactive media—one culture of bacteria was infected with the ^{32}P-labeled phage and another with ^{35}S-labeled phage. After the infectious cycle had begun, the cells were agitated to remove any viral material remaining outside the cells and were then centrifuged to separate the cells from the viral fragments. Testing revealed that the ^{35}S had remained outside the cells and the ^{32}P had entered the cells, infected them, and caused the production of new viruses. It was therefore concluded that it must be the DNA that carries the hereditary information necessary for the production of new phage particles.

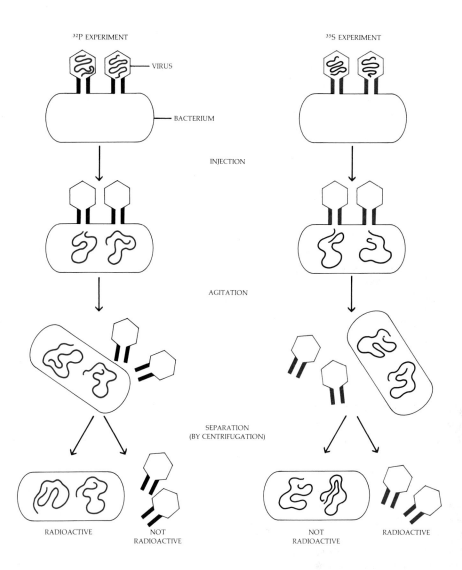

ity required to code the tremendous store of information needed by living things and to make exact copies of this code.

The information that Watson and Crick used to demonstrate that DNA was indeed the genetic material came from previous biochemical studies of the DNA molecule. Among these data were the following:

1. The molecule is very large and also long and thin.
2. The three components (a nitrogenous base, a sugar, and a phosphate) are arranged in nucleotides, as shown in Figure 3-19 on page 57.
3. According to X-ray diffraction studies conducted by Rosalind Franklin and Maurice Wilkins at King's College in London, the long molecule of DNA is composed of regularly repeating units that appear to be arranged in a spiral or, more accurately, a helix.
4. The data shown in Table 8-1 indicate that the ratio of nucleotides containing adenine to those containing thymine is 1 to 1, as is the ratio of nucleotides containing guanine to those containing cytosine.

Table 8-1   *Composition of DNA in Several Species**

SOURCE	PURINES		PYRIMIDINES	
	ADENINE	GUANINE	CYTOSINE	THYMINE
Human being	30.4	19.6	19.9	30.1
Ox	29.0	21.2	21.2	28.7
Salmon sperm	29.7	20.8	20.4	29.1
Wheat germ	28.1	21.8	22.7	27.4
Escherichia coli	24.7	26.0	25.7	23.6
Sheep liver	29.3	20.7	20.8	29.2

* In moles per 100 gram-atoms, percent; after Chargaff, Erwin, *Essays on Nucleic Acids*, 1963.

Watson and Crick did not perform experiments in the usual sense but, rather, chose to assemble all the facts then known about DNA into a meaningful whole. They worked with the accumulated data, with measurements provided by the X-ray diffraction photographs of

(a)

8–4

(a) Watson (left) and Crick with their tin model of DNA. (b) A representation of the Watson-Crick concept of the DNA molecule. On the left, the molecule is drawn in side view with the axis indicated by the vertical rod. The molecule consists of two polynucleotide chains, which form helices coiling to the right. These chains are coiled together, forming a double helix with an overall diameter of 2 nanometers (20 angstroms).

The two chains that make up the helices are composed of nucleotides containing deoxyribose sugar residues (S) in which the sugar of each nucleotide is linked by a phosphate group (P) to the sugar of the adjacent nucleotide. The regular sequence of sugars and phosphates forms the backbone of the molecule. The sugars of each helix project into the cylinder.

The nitrogenous base pairs (indicated by heavy horizontal lines) are flat molecules occupying the central area in the cylinder (the dashed rectangles in the cross sections on the right). The base pairs, represented by thymine (T) and adenine (A) at level A and by cytosine (C) and guanine (G) at level B, are linked by hydrogen bonds. The broken circular line in the cross sections indicates the outer edge of the double helix when the molecule is viewed end on.

The bases are stacked above one another at intervals of 3.4 angstroms and are rotated 36° with each step. Thus, there are 10 base pairs per complete turn of the helix. As a result of this rotation, the successive side views of the base pairs appear as lines of varying lengths, depending on the viewing angle.

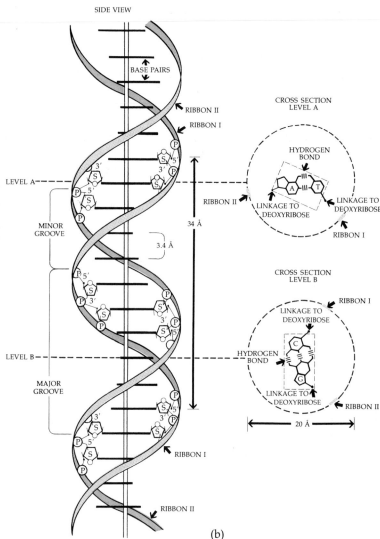

(b)

Franklin and Wilkins, and with a tin model that they attempted to construct in a way consistent with the physical and chemical data about DNA (Figure 8–4a).

The most important question for them was whether or not the chemical structure of DNA would reflect its biological function. "In pessimistic moods," Watson recalled in a review of these investigations, "we often worried that the correct structure might be dull—that is, that it would suggest absolutely nothing." The structure of DNA turned out, in fact, to be unbelievably exciting.

The Double Helix

By piecing together the various data, Watson and Crick were able to deduce that DNA is not a single-stranded helix, as are many proteins, but is instead a huge, entwined double helix. Imagine the banister of a spiral staircase, which forms a single helix. If a ladder were twisted into the shape of a helix, keeping the rungs perpendicular to the sides, a crude model of a double helix would be formed.

In a DNA molecule, the two sides are made up of alternating deoxyribose sugar molecules and phosphate groups (see Figures 8–4b and 8–5). The rungs of the ladder are formed by the nitrogenous bases—adenine (A), thymine (T), guanine (G), and cytosine (C)—one base for each sugar-phosphate, with two bases forming each rung. The paired bases meet across the double helix and are joined together by hydrogen bonds, the same kind of relatively weak chemical bonds that are such an important part of the secondary and tertiary structure of proteins. In the spiral stairway analogy, the stairs are formed of paired bases (Figure 8–5).

The distance between the two sides, or railings, according to Wilkin's measurements, is 2 nanometers. Two purines in combination would extend a distance of more than 2 nanometers, and two pyrimidines would not reach all the way across this space. But a perfect fit is obtained if each purine is paired with a pyrimidine. The paired bases—the rungs of the ladder—are therefore always purine-pyrimidine combinations (see Figure 8–5). For this reason, the ratio of purines to pyrimidines in a molecule of DNA is always 1 to 1.

Watson and Crick noticed that the nucleotides along any one chain of the double helix could be assembled in any order, such as ATGCGTACATT, and so on. Because a DNA molecule may be several thousand nu-

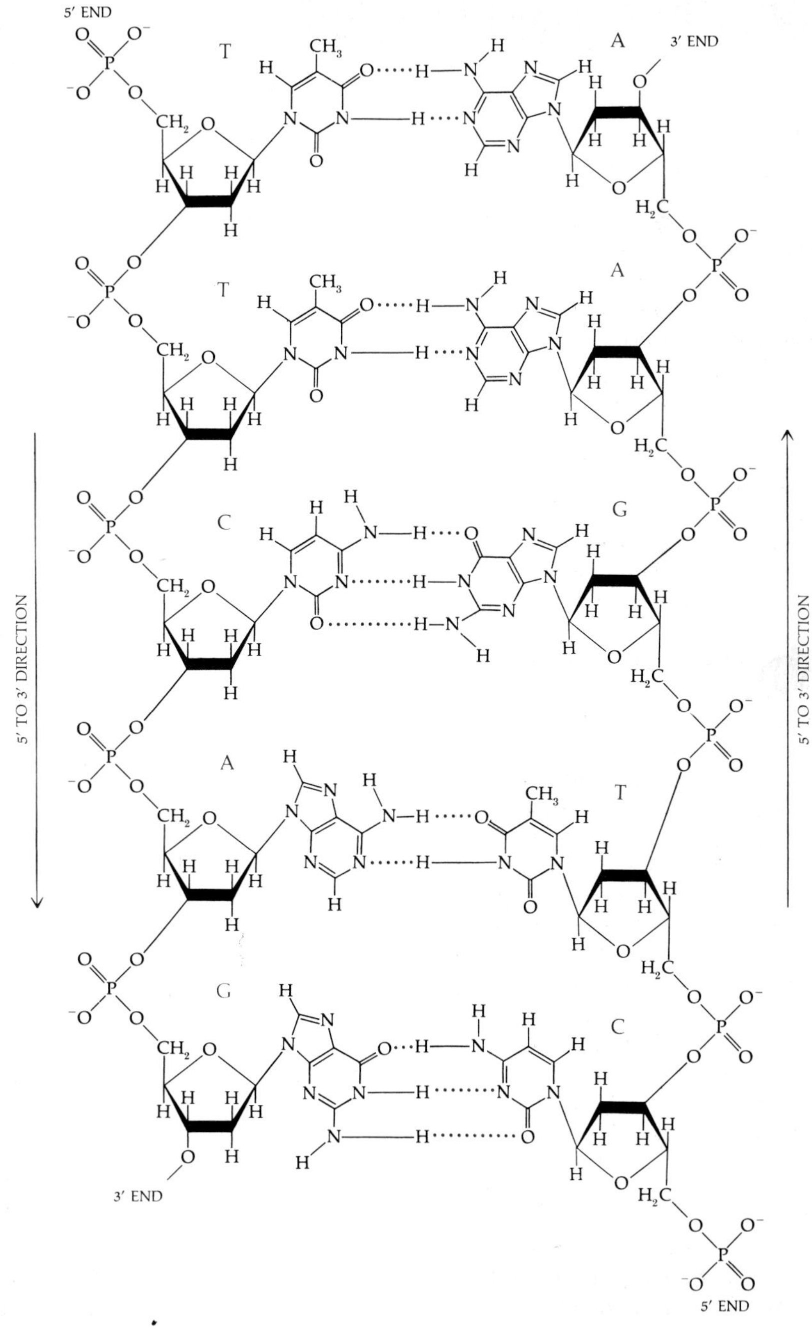

8–5

The double-stranded structure of a portion of the DNA molecule. Each nucleotide consists of a sugar (deoxyribose), a phosphate group, and a nitrogenous base (a purine or a pyrimidine). Note the repetitive sugar-phosphate-sugar-phosphate sequence that forms the backbone of the molecule. Each phosphate group is attached to the 5′ carbon of one sugar and to the 3′ carbon of the sugar in the adjacent nucleotide, so that the strand has a 5′ end and a 3′ end. The bridges formed by the phosphate groups between the nucleotides in the two strands run in opposite directions; that is, the strands are antiparallel. (Compare this figure with Figure 8–4b on the facing page.)

cleotides long, a wide variety of arrangements is possible. The number of paired bases ranges from about 5,000 for the simplest virus up to an estimated 5,000,000,000 in the 46 chromosomes of human beings. The DNA from a single human cell—which, if extended in a single thread, would be about 1.5 meters long—contains an amount of information equal to some 600,000 printed pages averaging 500 words each, the equivalent of a library of about 1,000 books. In short, the DNA molecule does indeed have the capacity to store the necessary genetic information.

The Molecule That Copies Itself

The most exciting discovery came when Watson and Crick set out to construct the matching DNA strand. In doing this, they encountered an interesting and important restriction: not only could purines not pair with purines, and pyrimidines not pair with pyrimidines, but adenine could pair only with thymine, and guanine could pair only with cytosine. Only these two combinations of nitrogenous bases form proper hydrogen bonds; adenine cannot form hydrogen bonds with cytosine, and guanine cannot form hydrogen bonds with thymine.

Now look again at Table 8-1. The Watson-Crick model explains the base composition of DNA in a simple, logical way. Perhaps the most important property of the model is that the two chains are *complementary,* that is, each chain contains a sequence of bases that is complementary to the other. When the DNA molecule reproduces itself, it simply "unzips" down the middle; the nitrogenous bases break apart at the hydrogen bonds (Figure 8-6). The two strands separate, and new strands form along each old one. If a T (thymine) is present on the old strand, only an A (adenine) can pair with the new strand; a G (guanine) will pair only with a C (cytosine), and so on. In this way, each strand forms a complementary copy of itself, and two exact replicas of the original molecule are produced. The age-old question of how hereditary information is duplicated and passed on for generation after generation was, in principle, answered by the discovery of the structure of DNA.

In what could be considered one of the great understatements of all time, Watson and Crick wrote in their original brief publication: "It has not escaped our notice that the specific pairing we have postulated immediately suggests a possible copying mechanism for the genetic material." In 1962, nine years after their original hypothesis was published, Watson, Crick, and Wilkins shared a Nobel Prize in recognition of their landmark studies.

Subsequently, it has been shown that because the DNA molecule is very long, double-stranded, and helical, the replication process is complex, requiring many enzyme-catalyzed steps. Specific enzymes are required

to separate and unwind the strands of DNA, as well as to place the newly formed nucleotides in the correct position. The way in which these enzymes function provides the key to understanding how DNA is replicated.

As the chemistry of DNA replication came to be better understood, an added complication became apparent—the ends of the two strands of the DNA double helix were different. In each strand, the phosphate group that joins two deoxyribose molecules is attached to one sugar at the 5' position (carbon 5 of deoxyribose) and to the adjacent sugar at the 3' position (the third carbon in the ring of deoxyribose). This configuration of sugar-phosphate-sugar attachments gives each strand a 5' end and a 3' end (Figure 8-5).

Moreover, while the direction of DNA synthesis runs from the 3' end to the 5' end in one strand, it runs in the opposite direction in the other strand. Thus, the two strands are antiparallel (Figure 8-5). Synthesis of the strand running from 3' to 5' is continuous, adding one

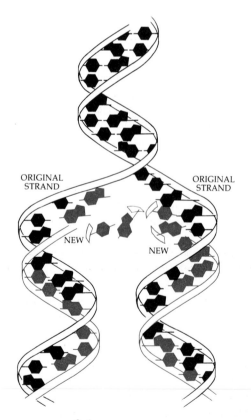

8-6
The DNA molecule shown here is in the process of replication, separating down the middle as the paired bases separate at the hydrogen bonds. (For clarity, the bases are shown out of plane.) Each of the original strands then serves as a template along which a new, complementary strand forms from nucleotides available to the cell.

nucleotide at a time, whereas that of the other strand is discontinuous, being built in the opposite direction from a series of short fragments that are eventually joined, forming a continuous strand. More than 14 different enzymes are now known to be involved in the steps that occur during the process of DNA replication in the bacterium *Escherichia coli,* and the process is even more complicated in eukaryotes.

HOW DO GENES WORK?

Watson and Crick disclosed the chemical nature of the gene and suggested the way in which it duplicated itself. However, one question remained unanswered: How does the information that is stored in the DNA molecule influence structure or function through the production of specific proteins? For example, how does DNA change a harmless pneumococcus to a virulent one, determine the shape of a leaf or the odor of a flower, or explain why your eyes are the same color as those of your mother?

The Molecules of Heredity

The structure and function of cells depend almost entirely on the proteins they make, particularly their enzymes. Therefore, the steps by which the genetic information of the DNA molecule is utilized in the formation of specific proteins are of special interest. In it, a specific linear sequence of bases is translated into a corresponding linear sequence of amino acids (the primary structure of the protein); the particular sequence of amino acids is very important in determining the shape (structure) of the protein and its function. A number of steps in protein synthesis involve the sister molecule of DNA, *ribonucleic acid* (RNA). RNA's involvement was long suspected because cells that are synthesizing large amounts of protein invariably contain large amounts of RNA.

RNA differs from DNA in several ways (discussed in Chapter 3): the sugar component of the RNA molecule is ribose rather than deoxyribose, as it is in DNA; and RNA contains the pyrimidine uracil (U) in place of the thymine of DNA. The RNA molecule occurs only rarely in a fully double-stranded form; its properties and activities are therefore quite different from those of DNA. The three principal classes of RNA are messenger RNA, transfer RNA, and ribosomal RNA.

Messenger RNA (mRNA) is a large molecule, ranging in size from a few hundred to about 10,000 nucleotides. It is formed along one strand of the DNA helix by the same base-pair copying principle that applies to the formation of a new strand of DNA. The presence of adenine in the parent DNA strand leads to the insertion of a uracil unit in the mRNA strand. Each sequence of three bases in the mRNA molecule specifies a single

RIGHT-HANDED AND LEFT-HANDED DNA

The continuous double helix of DNA that was described by James Watson and Francis Crick coils in a right-handed direction (right). However, in the late 1970s, Alex Rich and his colleagues at the Massachusetts Institute of Technology discovered a second stable configuration of DNA, one that was left-handed in its coiling direction (left). The discovery of these forms of DNA became possible because of the development of improved analytical capabilities in chemistry.

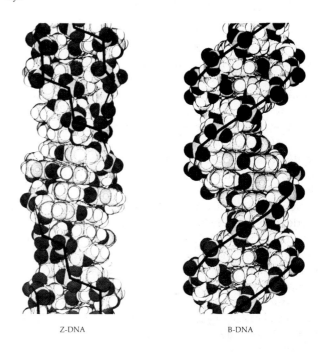

Z-DNA   B-DNA

In the models shown above, the polynucleotide backbone of the DNA is shown as a superimposed line. The backbone of the left-handed form of DNA zig-zags, and it is therefore called Z-DNA. Z-DNA is interconvertible with the right-handed form of DNA (known as B-DNA) under certain conditions. The presence of Z-DNA has been shown to be correlated with the regulated transcription of certain genes.

amino acid during translation; such a sequence is known as a *codon.*

Transfer RNA (tRNA) is sometimes called "the dictionary of the language of life." There are many different types of tRNA, apparently one specific type for each of the codons of the *genetic code,* the code by which the sequence of bases in DNA specifies the sequence of amino acids in the protein that will be produced (see Figure 8–7). One of the most remarkable discoveries of molecular biology is that the genetic code is identical in all organisms, with a very few exceptions that concern only small details. Genes from bacteria

8–7

The genetic code, consisting of 64 codons (triplet combinations of bases) and their corresponding amino acids (see Figure 3–14, page 52). Of the 64 codons, only 61 specify particular amino acids. The other three codons are "stop signals," which cause the chain to terminate. Since 61 triplets code 20 amino acids, there obviously must be "synonyms"; for example, leucine has six. Most of the synonyms differ only in the third nucleotide. The code is shown here as it would appear in the mRNA molecule.

SECOND LETTER

		U	C	A	G	
FIRST LETTER	U	UUU } phe UUC UUA } leu UUG	UCU UCC } ser UCA UCG	UAU } tyr UAC UAA stop UAG stop	UGU } cys UGC UGA stop UGG trp	U C A G
	C	CUU CUC } leu CUA CUG	CCU CCC } pro CCA CCG	CAU } his CAC CAA } gln CAG	CGU CGC } arg CGA CGG	U C A G
	A	AUU AUC } ile AUA AUG met	ACU ACC } thr ACA ACG	AAU } asn AAC AAA } lys AAG	AGU } ser AGC AGA } arg AGG	U C A G
	G	GUU GUC } val GUA GUG	GCU GCC } ala GCA GCG	GAU } asp GAC GAA } glu GAG	GGU GGC } gly GGA GGG	U C A G

THIRD LETTER

8–8

(a) The two-dimensional structure of a tRNA molecule. Such molecules consist of about 80 nucleotides linked together in a single chain. The chain always terminates in a CCA sequence. An amino acid can link to its specific tRNA at this end. Some nucleotides are the same in all tRNAs; these are shown in gray. The other nucleotides vary according to the particular tRNA. The symbols D, γ, ψ, and T represent unusual modified nucleotides characteristic of tRNA molecules. The significance of these unusual nucleotides is that they prevent certain hydrogen bonds from forming while promoting other cross-chain associations, thus determining the final folding of the molecule.

Each portion of the tRNA molecule appears to have a unique function. The acceptor end is where the amino acid is bound. The left-hand loop, known as the TψC loop, is the same in every tRNA molecule; it probably governs the binding of tRNA to ribosomes. The right-hand loop, the DHU loop, in contrast, differs from one tRNA molecule to another; it may be involved in determining which amino acid is enzymatically added to the acceptor end. Some of the nucleotides are hydrogen-bonded to one another, as indicated by the colored lines. The unpaired nucleotides at the bottom of the diagram (indicated in color) are known as the anticodon. They serve to "plug in" the tRNA molecule to an mRNA codon.

(b) The molecule folds over on itself, producing this three-dimensional structure. This is a photograph of a model.

can function perfectly, under the proper circumstances, in mammalian cells. Also, plant genes, at the proper stage of their expression, can be introduced into bacteria where they synthesize their own products. This observation not only provides a striking demonstration of the fact that all life on earth has had a common origin, it also provides the basis for the techniques of genetic engineering, which hold such promise for human progress in the future (see Chapter 30).

Each tRNA molecule consists of about 80 nucleotides, which form a long, single strand that folds back on itself (Figure 8–8). Part of the specificity of a particular tRNA molecule resides in the nature of its *anticodon*, or antiparallel triplet sequence, which determines that it will recognize only one particular triple codon on the mRNA.

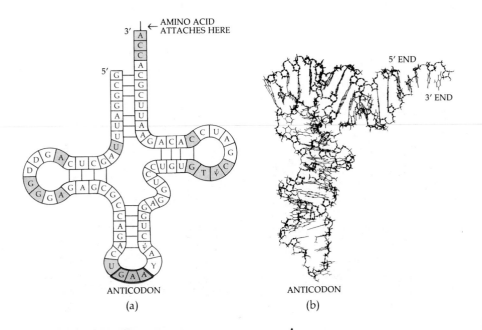

ANTICODON

(a)   (b)

The base sequences of several different tRNA molecules have been discovered. Each type of tRNA differs in its nucleotide sequence, but all have a similar number of bases and a similar shape.

Equally important to the specificity of the tRNA molecule is its ability to bind a particular amino acid—the one coded by the anticodon. This specificity is achieved by *activating enzymes*, which are specific enzymes that recognize both a particular amino acid and a particular tRNA. These activating enzymes are key elements in the translation of the genetic message; they determine which amino acid will be associated with which tRNA (and thus with which triplet anticodon).

Ribosomal RNA (rRNA) is found, along with characteristic proteins, in the ribosomes. A ribosome is composed of two subunits, each with its own characteristic kinds of RNA and proteins. For example, in *Escherichia coli* the smaller subunit of the ribosome has one type of rRNA, and the larger subunit has two other types. mRNA and tRNA come together at the ribosome. Presumably, the function of the ribosome is to position the mRNA, tRNA, amino acids, and the protein being formed in a precise relationship to one another as protein synthesis occurs. Ribosomal proteins are also active; for example, one of them is an enzyme that forms peptide bonds.

In eukaryotic cells, rRNA is formed on the DNA of the nucleolus. Although somewhat larger than those of *E. coli*, the ribosomes of eukaryotic cells have a similar structure and the same function as those of bacteria.

Transcribing RNA from DNA

RNA is transcribed, or copied, from DNA in a complementary fashion similar to that in which DNA is replicated (Figures 8–9 and 8–10). In eukaryotic cells, there are different enzymes, called RNA polymerases, for the transcription of tRNA, rRNA, and mRNA. The mRNA precursor molecules synthesized from the DNA template in the eukaryote nucleus are 10 to 20 times larger than the mRNAs found in bacteria. If newly synthesized RNA in eukaryotes is labeled by making radioactive uracil available for a short time, the long molecules of RNA precursor within the nucleus display radioactivity. However, only about 10 percent of this radioactivity occurs outside the nucleus, demonstrating that most of the RNA precursor formed within the nuclei of eukaryotes does not play a role in protein synthesis. The noncoding regions of eukaryotic genes—*introns*—transcribe the portion of the RNA that never leaves the nucleus. The portions of eukaryotic genes that contain the nucleotide sequences that encode the amino acid sequence of the protein are called *exons*.

The long mRNA precursor molecules of eukaryotes are "processed" within the nucleus after they are synthesized. Large segments of mRNA precursor transcribed from the introns are excised from the middle portions of the molecule by special enzymes. The remaining segments—those transcribed from the exons—are spliced together to form the *processed* form of eukaryotic mRNA, which passes from the nucleus into the cytoplasm. The processed mRNA in eukaryotes is often similar in size to the mRNA found in bacteria.

8–9

A bacterial gene in action. The micrograph (left) shows several different mRNA strands being formed simultaneously. (The strands are shown in color in the diagram at the right.) The longest strand was the first to begin to be synthesized. As each mRNA strand peels off the active chromosome segment (the DNA molecule), ribosomes attach to the mRNA strand, translating it into protein. The protein molecules are not visible in this micrograph.

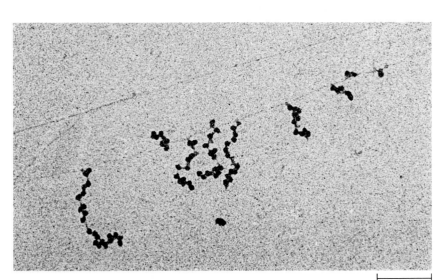

0.2 μm

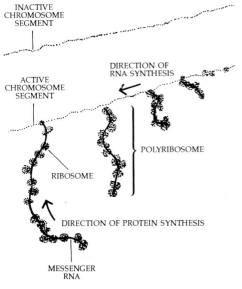

INACTIVE CHROMOSOME SEGMENT

ACTIVE CHROMOSOME SEGMENT

DIRECTION OF RNA SYNTHESIS

POLYRIBOSOME

RIBOSOME

DIRECTION OF PROTEIN SYNTHESIS

MESSENGER RNA

8–10

A schematic view of the processes of transcription and translation. (a) During transcription, a strand of the DNA helix serves as a template for the synthesis of a complementary molecule of messenger RNA (mRNA). (b) During translation, three classes of RNA interact with a specific group of enzymes and proteins to generate the formation of a new polypeptide chain. Ribosomal RNA (rRNA) is a component of the ribosomes, on which polypeptide synthesis takes place. Bacterial ribosomes contain a large (50S) and a small (30S) subunit. Transfer RNA (tRNA) causes the insertion of the correct amino acids into the growing polypeptide chain. Messenger RNA (mRNA) carries the information contained in a gene to the ribosome. The information is encoded as groups of three nucleotides, with each specifying a particular amino acid. Each codon is recognized by a complementary anticodon on a transfer RNA molecule that became associated with that particular amino acid earlier. In this figure most of the amino acids are represented by numbered circles; the amino acid glycine (gly) has just been bound to its site on the ribosome by the corresponding transfer RNA. It will form a peptide bond with leucine (leu), thus extending the growing polypeptide chain by one amino acid. The ribosome then moves the length of the codon along the messenger RNA and so comes into position to bind the transfer RNA carrying serine (ser).

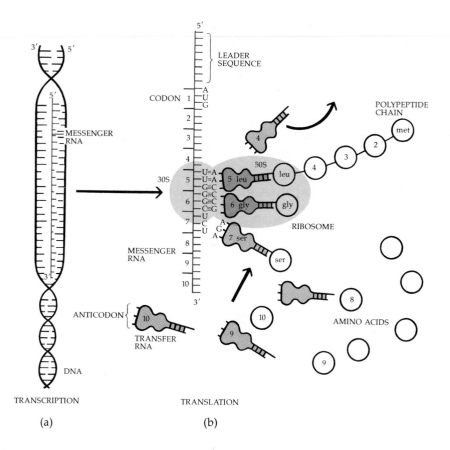

(a)   (b)

It appears that not only mRNAs but also rRNAs and tRNAs are initially transcribed from eukaryotic chromosomes as large precursor molecules, which then undergo extensive and highly specific enzymatic processing. This suggests that there are important sequences in eukaryotic genes that do not relate to the functional RNA or to the protein ultimately formed but serve another function in gene transcription.

In eukaryotes, therefore, the mRNA that serves as a template for protein synthesis is actually spliced together from segments of the mRNA precursor that is transcribed directly from chromosomal DNA. The chain of information in both bacteria and eukaryotes flows from gene to mRNA molecule to polypeptide chain—the "central dogma" of genetics. In eukaryotes, however, the processing of transcripts adds an extra level of complication: gene ⟶ primary transcript ⟶ mRNA ⟶ polypeptide chain.

Translating mRNA into Protein

In translating an mRNA molecule, the sequence of codons is read from the 5′ end to the 3′ end of the molecule, as it is in the duplication of a DNA strand (Figure 8–10). The first portion of the mRNA molecule is involved in guiding the binding of ribosomes to it. In bacteria, the smaller ribosome subunit carries a special tRNA (the one for *N*-formylmethionine) and passes down the mRNA molecule until it encounters the codon (AUG) that specifies *N*-formylmethionine; the smaller ribosome unit then binds the molecule of *N*-formylmethionine to a special site on the ribosome, forming an initiation complex. Eukaryotes, in contrast, do not employ AUG codons to signal the start of polypeptide synthesis; instead, they employ a chemically modified form of guanine, called 7-methylguanine.

In both bacteria and eukaryotes, the growing end of the protein chain is attached to the ribosome-mRNA complex by a molecule of tRNA (Figure 8–10). The next amino acid, bound to its special tRNA molecule, enters the complex and binds (by means of the tRNA anticodon) to the next three bases (the codon) of the message. The amino group of the incoming amino acid loses a proton to the ribosome and forms a peptide bond with the growing peptide chain. As a result, this chain is now bound to the newly entered tRNA–amino acid complex, while the tRNA from which the amino acid has just been removed is free to dissociate. The ribosome continues to move along the mRNA in one-codon increments, moving the polypeptide chain along with it. Each time it moves, another amino acid is added and the process is repeated.

The synthesis of polypeptide chains is terminated when a ribosome encounters one of three termination signals, or "stop" codons: UAA, UAG, or UGA. These so-called *nonsense codons* do not specify amino acids; instead, their presence initiates a process that leads to the cleavage of the bond that joins the last amino acid in the peptide to its tRNA. When this happens, protein synthesis stops.

REGULATING GENE TRANSCRIPTION

The transcription of genes into mRNA is under elaborate cellular control. Even in bacteria, several interlocking systems operate to regulate which genes are transcribed and to what extent. Eukaryotes appear to share many of the properties of bacterial gene regulation, although less is known about the details of control mechanisms in eukaryotes.

Because there has been intensive research on some bacterial genes, we know a great deal about how their transcription is regulated. Perhaps the best understood case is the lactose *(lac)* system of the bacterium *Escherichia coli.* Normally, these bacteria do not encounter the disaccharide lactose in their environment, so they do not synthesize the enzymes necessary to metabolize it. If lactose is added to the bacterial growth medium, the cells begin to produce large amounts of the enzyme β-galactosidase, which splits the disaccharide into glucose and galactose, thus permitting the lactose to be metabolized (Figure 8–11).

In the operation of this system, it is as if the lactose substrate calls forth, or induces, the formation of the enzyme necessary to metabolize it. Many cases of such *induction* of enzymes by energy-rich substrates are known. In other cases, the enzymes involved in the synthesis of particular amino acids or other metabolites are not produced by the cell when that amino acid is present; this *repression* of enzyme synthesis by key metabolites involves the same control system as induction.

The Operon

Francois Jacob and Jacques Monod, working in France on studies for which they received a Nobel Prize in 1965, developed the concept of the *operon* to explain how bacterial cells regulate enzyme biosynthesis. The operon is a group of related genes that are aligned along a single segment of DNA. The operon is a unit of translation; all its structural genes are expressed in a co-ordinated fashion as they are transcribed onto the same mRNA molecule. The *lac* operon region of *E. coli,* for example, contains the gene for β-galactosidase and two other genes involved in lactose metabolism.

The enzymes encoded by all the genes in the *lac* operon are needed at the same time a bacterium utilizes lactose for food, and they are all produced together because they are on the same mRNA molecule. In cells growing on a medium that contains lactose, approximately 3,000 molecules of β-galactosidase are present in each cell. This represents about 3 percent of all the protein in the cell. A cell does not express these genes unless lactose is available, because of the presence of a protein called a *repressor*, which acts as a roadblock. The repressor binds to a site on the DNA between the genes themselves and the site where the RNA-transcribing polymerase binds to the DNA. When lactose is present, it binds to this repressor protein, thereby changing its shape so that it no longer binds to the DNA. This leaves the polymerase free to move unimpeded along the DNA, translating the enzyme-encoding genes in the operon.

In eukaryotes, groups of related genes (operons) are not transcribed in clusters onto composite mRNA molecules, as they are in bacteria. Each cytoplasmic mRNA of a eukaryotic cell carries the information for only one peptide. Control systems do occur in plants, however. Mutations that result in the unrestrained production of groups of enzymes have been detected in both corn *(Zea mays)* and evening primrose *(Oenothera).*

8–11
In the bacterium Escherichia coli, the splitting of a molecule of lactose into galactose and glucose requires the enzyme β-galactosidase. β-galactosidase is an inducible enzyme; that is, its production is regulated by an inducer—in this case, its substrate, the disaccharide lactose.

Regulation by Feedback Inhibition

In addition to the genetic mechanisms that change the functions of cells by controlling the *production* of particular enzymes in them, there are a number of physiological control systems that work by direct feedback inhibition of enzymes—in other words, by controlling the *activity* of enzymes.

Fine adjustment control systems, such as end-product feedback inhibition, are examples of *allosteric interactions*. The binding of a particular molecule (an allosteric effector) at one site on the protein may so affect the weak interactions determining a protein's tertiary structure (see page 54) that the conformation of the protein is changed. When a protein molecule undergoes a change in shape, a second site, perhaps a site that allows the protein to function as an enzyme, may also be altered. Interactions of this sort can control the levels of particular kinds of enzymatic activity in cells, thus enhancing their effective function. The binding of lactose to the repressor protein, discussed above, brings about an allosteric change in that protein. The allosteric effector is, in some instances, the end product of the pathway that begins with the reaction catalyzed by the enzyme under consideration.

THE CONTROL OF DEVELOPMENT IN PLANTS

The genetic processes we have discussed are ultimately responsible for the characteristics of an individual organism. In achieving all of the particular characteristics of a mature organism, however, differentiation and development are necessary. *Differentiation* is the process by which a relatively unspecialized cell or tissue undergoes progressive changes, thus becoming more specialized in function and structure. Development involves the organization of the complex sets of tissues that constitute the mature organism. A mature organism consists of an intricate array of many different tissues in precise physiological and morphological relation to one another. Ultimately, all originate from the same cell, which in diploid organisms is the fertilized egg, or zygote. As more and more complex tissues are organized, increasingly complex developmental control becomes necessary.

The process of development in animals is radically different from that in plants. Animal development proceeds according to a specific blueprint and in a precise sequence; the order and timing are critical to the outcome. At each stage, highly specific cell adjustments and changes in gene expression occur. The course of development in animals is usually not affected much by outside environmental influences; once a mature animal has been produced, its development is at an end.

THE CONTROL OF MULTICELLULAR DIFFERENTIATION

Differentiation is the developmental process by which a relatively unspecialized cell or tissue undergoes progressive changes to become more specialized in function and structure. This process is the means by which genes ultimately express themselves in the formation of the mature organism. One well-studied system that illustrates this process involves the cellular slime mold *Dictyostelium discoideum.* Its life cycle shows how influences originating outside developing cells or tissues can also affect their mature characteristics.

The cellular slime molds usually exist as free-living amoebalike cells, or myxamoebas, which feed on bacteria by surrounding and engulfing them (see Figure 4–11, page 65). They reproduce by cell division and show little morphological differentiation until they exhaust the available supply of bacteria. In response to starvation, the cellular slime molds produce spores, but not in a simple way. The individual cells first aggregate to form a sluglike motile mass called a *pseudoplasmodium*, which migrates to a new place before developing and releasing spores. This is a marvelous developmental advance, ensuring that new spores will not be released into old habitats in which the bacteria have been depleted.

The aggregation of myxamoebas is initiated when one or more starved cells begin to secrete cyclic adenosine monophosphate (cAMP) into the medium. The cAMP diffuses out and establishes a concentration gradient, which causes the surrounding cells to move toward those that are secreting the cAMP. The secreting cells, in turn, are stimulated to emit a new "pulse" of cAMP after a period of about five minutes. At least three waves of cells are recruited in this fashion. As the cells accumulate at the aggregation center, their plasma membranes become sticky, which causes them to adhere to each other, and results in formation of a pseudoplasmodium surrounded by a cellulose sheath.

The eventual developmental fate of an individual cell is determined by its position in the aggregation. The first cells to aggregate usually give rise to the anterior portion of the pseudoplasmodium, whereas those cells entering the

In plants, on the other hand, the process of development is continuous: plants develop throughout their life span. Their development is directly affected both by external influences (which are minimal in animals), and by internal ones. Tissue-specific differentiation in plants is under the direction of hormones, whose production is sensitive to changes in the environment (see Chapter 24). The ability of plants to respond to external stimuli contributes to the development of individuals that are well suited to their particular habitats. This is

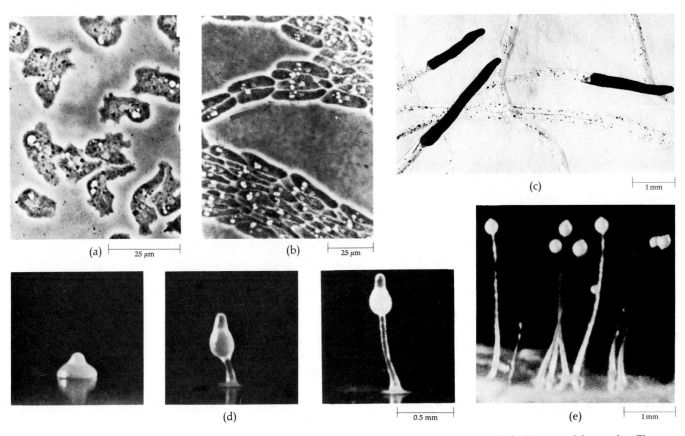

(a)   25 μm   (b)   25 μm   (c)   1 mm

(d)   0.5 mm   (e)   1 mm

aggregation last form the base. When migration ceases, the cells of the apical region become the stalk cells of the developing fruiting body. Posterior cells of the pseudoplasmodium then move to the top of the stalk and become spores. Eventually, the basal disk and stalk cells die, and the spores are dispersed. If they fall on warm damp ground, they germinate. Each spore releases a single myxamoeba, and the cycle is repeated. See also Chapter 14.

Thus, even this relatively simple eukaryotic system involves tissue recognition, differential cell migration, and localized cell death. Much more complex systems, with more precise control, characterize cellular differentiation in plants.

(a) *The feeding stage of the amoebas. The light gray area in the center of each cell is the nucleus, and the white areas are contractile vacuoles. (b) Amoebas aggregating. The direction in which the stream is moving is indicated by an arrow. (c) Migrating pseudoplasmodia, each formed of many amoebas. Each sluglike mass deposits a thick slime sheath, which collapses behind it. (d) At the end of the migration, the pseudoplasmodium begins to rise vertically, differentiating into a stalk and mass of spores suspended in a droplet (e).*

an important factor in view of the inability of plants to seek more favorable conditions by moving from place to place.

In plants, almost all differentiation is reversible, provided that the differentiated cells retain living, nucleated protoplasts and have not developed a secondary wall. This is correlated with the lack of a fixed developmental blueprint in plants. Such reversibility is rarely possible in animals.

The reversibility of development in plants can be il-

lustrated by experimental data. The German botanist Gottlieb Haberlandt suggested as early as 1902 that all living plant cells were *totipotent:* that is, that each of them possessed the potential to develop into any other kind of plant cell. Haberlandt believed that it ought to be possible to develop whole mature plants from single pieces of individual plant tissue, or even from single cells, but he was never able to accomplish this. In fact, more than half a century passed before his hypothesis was proved essentially correct.

It turned out that the problem lay in not knowing what substances to add to the media in which the plants were grown. In the 1950s, Cornell University plant physiologist F. C. Steward isolated small bits of phloem tissue from carrot (*Daucus carota*) root and placed them in liquid growth medium in a rotating flask. The medium contained sucrose and the minerals necessary for plant growth (Chapter 26), as well as certain vitamins, which are organic substances that plants cannot manufacture. Still other substances, however, are required for the process of growth and differentiation. Steward supplied these in coconut milk—known to be rich in plant growth compounds—although the nature of these compounds was not understood.

In the rotating flask, individual cells continually broke away from the growing cell mass and floated free in the medium. These individual cells were able to grow and divide. Before long, Steward observed that many of these new cell clumps had developed roots. If left in the swirling medium, the cell clumps did not continue to differentiate, but if they were transferred to a solid medium—which was agar in these experiments—some of the clumps developed shoots (Figure 8–12). If the clumps were then transplanted to soil, the little plants leafed out, flowered, and produced seed.

These results indicated that at least some of the cells of the differentiated phloem tissue still contained all the genetic potential for full plant development, although they were not expressed. These experiments also illustrated the fact that such differentiated cells are capable of expressing portions of their previously unexpressed genetic potential when suitable environmental signals are provided to trigger those particular developmental patterns. (The hormonal control of such patterns of differentiation is discussed in Chapter 24.) By achieving these results, Steward finally confirmed Haberlandt's hypothesis.

The form and structure of a mature, functioning organism is determined by the blueprint contained in its genes. However, the characteristics of such an organism reach their full expression only via the complex processes of differentiation and development. The processes are directed by the interaction of the products of genes in the course of their evolution.

Cytoplasmic Influences on Differentiation

In many organisms, cytoplasmic components play a direct role in cellular differentiation; these include organelles such as plastids and mitochondria, which contain their own DNA complements (see pages 21 and 24). When these organelles are unequally partitioned between the daughter cells during cell division, the resulting cell lines may have very different fates. Such effects account for many of the "maternal effects"—whereby identical hybrids differ depending on which sort of organism was their mother—that have been reported in the genetic literature.

In a similar way, chemical gradients within cells— that is, changes in the concentration of a substance in different parts of a cell—play a major role in differentiation. For example, in the brown alga *Fucus* (see Figure 15–9c, page 258), a gradient of stored insoluble food particles is apparently set up in the zygote by the action of gravity, or possibly by electrochemical forces. This gradient determines the position of the spindle apparatus in the first division of the zygote, establishing two distinctive cell lines from the very first cell division in the life of the individual. Unequal cell division may also be of fundamental importance in parceling out different kinds of elements present in the cytoplasm and in determining the fate of the resulting cell lines.

In plant and animal cells, many substances diffuse continually at different rates and in different directions,

8–12

Two buds forming on undifferentiated tissue (callus) from a geranium following treatment with two hormones, auxin and a cytokinin. Callus from some types of plants will continue to grow as undifferentiated tissue, or roots, or buds, depending on the relative proportions of hormones.

and tissues of various kinds are often packed together in close proximity. Because of these relationships, effects similar to the ones analyzed in *Fucus* and other simpler organisms must be common. Not only do gradients within individual cells and tissues lead to an extremely subtle control of developmental processes, but also the differentiation of any one cell may be determined largely by its position in the body of the developing plant or animal. Some of the ways in which hormones and other factors interact in plant development are discussed in Section 6.

SUMMARY

The molecular key to understanding the process of heredity was provided by the early 1950s with the discovery of the role of DNA. A variety of evidence strongly indicated that the genetic information resided in DNA, and the description by James Watson and Francis Crick in 1953 of the double-helical structure of DNA, with its complementary base pairing, provided a molecular model. This led to rapid advances in our knowledge, and the intimate workings of the cell's genetic machinery can now be described in surprising detail.

DNA is replicated by the synthesis of a complementary copy of each of the two strands of the double helix. A variety of enzymes act in concert to unravel DNA coils, unwind the double helix locally, and add new bases to each of the two separated strands.

The genetic information in the DNA is not expressed directly but is transferred via messenger RNA (mRNA). The long molecules of mRNA are assembled by complementary base-pairing along one strand of the DNA helix and then are carried to the cytoplasmic ribosomes. This process, called transcription, is under rigid genetic control. Each sequence of three nucleotides in the mRNA molecule is the codon for a specific amino acid.

At the ribosomes, mRNA is combined with a series of small molecules called transfer RNAs (tRNAs), which are bound to specific amino acids. Each tRNA has a three-base sequence (anticodon), which is complementary to a codon of the mRNA. The mRNA-designated tRNA molecule fits its complementary three-base anticodon onto the next codon of the mRNA base sequence and, guided by the ribosome, transfers its specific amino acid to the end of the growing peptide chain, which is transferred to it. The amino acid is then bound to the chain by a peptide bond formed by the enzymes in the ribosome. This process is called translation.

Each of the 20 amino acids is coded by one or more three-base sequences (codons) found in mRNA. The sequence of amino acids in a protein is determined by the sequence of codons along the mRNA molecule that directs the synthesis of that particular protein. The sequence of codons in each type of mRNA molecule depends ultimately on the sequence of bases in the DNA from which the mRNA is transcribed. Most amino acids are specified by three or four alternative codons, each with a different specific tRNA.

Not all gene information represents information about protein sequence. Much of the genetic information in eukaryotic nuclear mRNA is transcribed from DNA segments called introns; these mRNA segments are excised from the mRNA before it reaches the cytoplasm. The remaining mRNA segments, transcribed from portions of the DNA known as exons, are spliced together in the nucleus before moving into the cytoplasm.

The regulation of expression of simple bacterial gene systems, such as the *lac* system of *Escherichia coli*, is straightforward: one system activates transcription in the presence of the potential substrate—the "inducer" lactose—while another shuts down transcription in the presence of excess product—glucose. In eukaryotes, more complex developmental sequences occur.

In plants, gene transcription is organized in blocks of developmental expression, each corresponding to a different pattern of cellular differentiation. The choice of a particular pattern is influenced by the environment and may be reversible. In principle, any differentiated plant cell that retains a protoplast with a nucleus can be dedifferentiated and induced to produce a complete plant.

C H A P T E R 9

The Genetics of Eukaryotic Organisms

Much of the research on the molecular mechanisms of heredity has been done on bacteria; in this chapter, we focus on the genetics of eukaryotic organisms, primarily plants. The branch of genetics discussed here—which deals with relatively obvious traits and their genetic control—is generally referred to as Mendelian genetics, in recognition of the work of Gregor Mendel (Figure 9–1). Mendel outlined the fundamental laws of genetics in 1865, but his findings were not fully appreciated for more than 30 years.

We shall also examine the relationship and relevance of Mendelian genetics to evolutionary theory. Charles Darwin's monumental work on evolution, *On the Origin of Species,** was written in ignorance of Mendel's studies, despite the fact that these two preeminent scientists were active at the same time. Yet the study of evolution in the late twentieth century is almost as dependent upon the principles developed by Mendel and his followers as it is on Darwin.

EUKARYOTES VERSUS PROKARYOTES

One of the major differences between eukaryotes and prokaryotes (bacteria) is that most eukaryotic organisms undergo sexual reproduction, a process that does not occur in bacteria. Although some eukaryotes do not reproduce sexually, it is evident that most such organisms have lost the capacity to do so during the course of their evolutionary history.

Sexual reproduction involves a regular alternation between meiosis and syngamy. *Meiosis* is the process of nuclear division in which the chromosome number is

9–1
Gregor Mendel (1822–84), shown here standing at the right, holding a fuchsia, carried out his studies in the garden of an Austrian monastery. His work in the field of genetics, which was not understood and was therefore largely ignored during his lifetime, was rediscovered in 1900.

* The complete title of Darwin's treatise is *On the Origin of Species by Means of Natural Selection, or the Preservation of Favoured Races in the Struggle for Life.*

reduced from the diploid (2*n*) to the haploid *(n)* number. During meiosis, the nucleus of a diploid cell undergoes two divisions, one of which is a reduction division. These divisions result in the production of four daughter nuclei, each containing one-half the number of chromosomes of the original nucleus. In plants, meiosis occurs in the production of spores in flowers, cones, or similar structures; most plants we see, with the exception of mosses and liverworts, are diploid. *Syngamy,* or fertilization, is the process by which two haploid cells (gametes) fuse to form a diploid zygote. Syngamy thus reestablishes the diploid chromosome number.

All the organisms discussed in this chapter are diploid for a major portion of their life cycle. Diploid organisms have two sets of chromosomes, one derived from each parent. Corresponding chromosomes, which form pairs during the course of meiosis, are called homologous chromosomes, or *homologs.* The interaction of products indirectly derived from genes on each of these sets of chromosomes determines the genetic characteristics of the diploid plant or animal at maturity.

STRUCTURE OF EUKARYOTIC CHROMOSOMES

The chromosomes of eukaryotes are made up of DNA and protein, which together form chromatin. Most chromosomes consist of about 75 percent DNA and 25 percent protein. The absolute amount of DNA in individual chromosomes varies widely both within and between species, and this variation is reflected in great differences in chromosome size. During periods of protein synthesis (interphase), RNA is also associated with the chromosomes. Each eukaryotic chromosome consists of a single threadlike molecule of double-stranded DNA, which is continuous throughout its entire length.

This is certainly an impressive feat of packaging, for the DNA molecule may be several centimeters long.

Chromosomal DNA forms complexes with a variety of proteins, the majority of which are histones—proteins that are positively charged because of their high concentration of the amino acids arginine and lysine (see Figure 3–14, page 52). Most histones consist of 100 to 130 amino acid residues, with 20 to 40 lysine and arginine residues clustered at one end of the molecule. In contrast to histones, double-stranded DNA is negatively charged, because of its exposed phosphate groups. Beadlike structures called *nucleosomes* (Figure 9–2a) form because of the interaction of these oppositely charged molecules. Recent studies have shown that the "beads" are ellipsoidal, about 110 angstroms long and 65 to 70 angstroms thick. Each nucleosome consists of an octomer (eight-molecule complex) of four different kinds of histones, with the DNA helix wound around the octamer in a path resembling a spring. Within such nucleosomes, the DNA is efficiently packaged and protected from enzymatic degradation.

Between the nucleosomes are strings of DNA with nonhistone proteins attached to them. The nonhistone proteins that occur in proteins are very diverse. They include regulators of the transcription of specific genes, enzymes involved in the modification of nucleic acids and proteins, and proteins that dictate the structure of chromosomes and nuclei.

Some portions of chromosomes are packaged in a highly condensed state called *heterochromatin.* The rest of the chromatin, which is not condensed during interphase, is called *euchromatin.* The condensed organization of heterochromatin prevents transcription of mRNA; indeed, the expression of euchromatic genes is often inhibited if they are located near a heterochromatic region.

During early meiosis or mitosis the euchromatin condenses. DNA initially aggregates into a series of tight

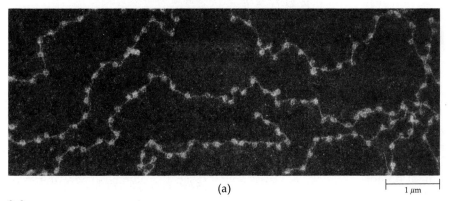

(a)

1 μm

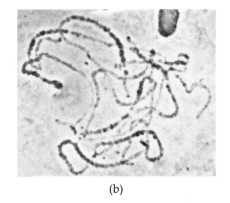

(b)

9–2

(a) *Electron micrograph of nucleosomes and connecting threads of DNA from a chicken erythrocyte (red blood cell). The nucleosomes—the structures that appear* *like beads on a string—are each approximately 1 micrometer in diameter. (b) Photomicrograph of meiotic chromosomes in a tarweed* (Hemizonia pungens). *The* *beadlike structures visible along the paired chromosomes are chromomeres, or regions of aggregated chromatin.*

"coils" separated by uncoiled regions. These coiled regions, or *chromomeres*, are much larger than nucleosomes and can be seen with a light microscope (Figure 9–2b). Even larger coils are formed at certain stages of nuclear division (see Figure 8–1, page 116).

MEIOSIS

Meiosis occurs only in specialized diploid cells and only at particular times in the life cycle of a given organism. Through meiosis and cytokinesis, a single diploid cell gives rise to four haploid cells—either gametes or spores. A *gamete* is a cell that unites with another gamete to produce a diploid zygote. The zygote may then divide, either meiotically, producing four unicellular haploid cells, or mitotically, eventually forming a multicellular diploid organism. If haploid cells are formed, they may function as (1) independent, unicellular, haploid organisms, or (2) as gametes, in which case they may eventually unite with one another by syngamy. If a multicellular diploid organism is formed it will, in most cases, eventually produce haploid spores or gametes by meiosis. A *spore* is a cell that can develop into an organism without uniting with another cell. Spores often divide mitotically, producing multicellular organisms that are entirely haploid and that eventually give rise to gametes by mitosis (see Figure 10–11, page 163).

First Meiotic Division

Meiosis consists of two successive nuclear divisions. Refer to Figure 9–6 on page 136 to help keep track of the processes described in the following paragraphs.

In *prophase I* (prophase of the first meiotic division), the chromosomes—present in the diploid number—first become visible as long, slender threads. As in mitosis (see Chapter 2), the chromosomes have already duplicated during the preceding interphase. Consequently each chromosome, at the beginning of prophase I, consists of two identical chromatids attached at the centromere. At this early stage of meiosis, however, each chromosome appears to be single rather than double.

Before the chromatids become apparent, the homologous chromosomes pair with each another. The pairing is very precise, beginning at one or more sites along the length of the chromosomes and proceeding in a zipper-like fashion, such that the same portions of the homologous chromosomes lie next to one another. Each homolog is derived from a different parent and is made up of two identical chromatids. Thus, a homologous pair consists of four chromatids. The pairing of homologous chromosomes is a necessary part of meiosis; the process cannot occur in haploid cells, because such homologs are not present. The pairing process itself is called *synapsis*, and the associated pairs of homologous chromosomes are called *bivalents*.

During the course of prophase I, the paired threads become more and more tightly contracted, and, consequently, the chromosomes shorten and thicken. With the aid of an electron microscope, it is possible to identify a densely staining axial core of protein in each chromosome (Figure 9–3a). During mid-prophase, the axial cores of a pair of homologous chromosomes approach each other to within 0.1 micrometer, forming a *synaptonemal complex* (Figure 9–3b).

In favorable material, it can now be seen that each axial core is double; that is, each bivalent is made up of four chromatids, two per chromosome. During the time when the synaptonemal complex exists, portions of the chromatids break apart and rejoin with corresponding segments from their homologous chromatids. This *crossing-over* results in chromatids that are complete but which have a different representation of genes than originally. Figure 9–4 shows the visible evidence that crossing-over has taken place—the X-like configuration called a *chiasma* (plural: chiasmata).

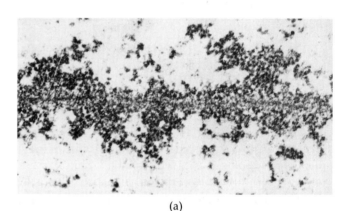

(a)

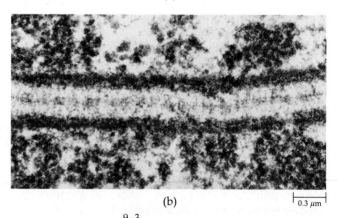

(b) |——| 0.3 μm

9–3
(a) *Portion of a chromosome of* Lilium *early in prophase I, prior to pairing. Note the dense axial core. This core, which consists mainly of proteins, may arrange the genetic material of the chromosome in preparation for pairing and genetic exchange.* (b) *Synaptonemal complex in a bivalent of* Lilium. *Only two of the four chromatids are visible.*

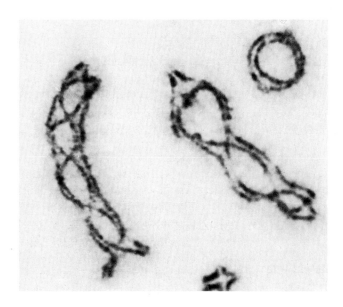

9–4

*Variations in the number of chiasmata
can be seen in the paired chromosomes of
a grasshopper,* Chorthippus parallelus.

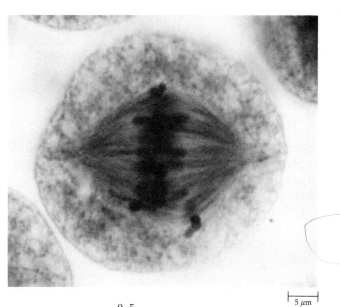

9–5

*The spindle in a pollen mother cell of
wheat* (Triticum aestivum) *during
metaphase I of meiosis.*

As prophase I proceeds, the synaptonemal complex ceases to exist. Eventually, the nuclear envelope breaks down. The nucleolus usually disappears as RNA synthesis is temporarily suspended. Finally, the homologous chromosomes appear to repulse one another. Their chromatids are held together at the chiasmata, however. These chromatids separate very slowly. As they separate, some of the chiasmata slip toward the end of the chromosome arm. One or more chiasmata may occur in each arm of the chromosome, or only one in the entire bivalent; the appearance of a particular bivalent can vary widely, depending on the number of chiasmata present (Figure 9–4).

In *metaphase I*, the spindle apparatus—an axis of microtubules similar to that which functions in mitosis—becomes conspicuous (Figure 9–5). As meiosis proceeds, individual microtubules become attached to the centromeres of the individual chromosomes of each bivalent. These paired chromosomes then move to the equatorial plane of the cell, where they line up randomly in a configuration characteristic of metaphase I. The centromeres of the paired chromosomes line up on opposite sides of the equatorial plate; in contrast, in mitotic metaphase, as we have seen, the centromeres of the individual chromosomes line up directly on the equatorial plate.

Anaphase I begins when the homologous chromosomes separate and begin to move toward the poles. (Notice again the contrast with mitosis. In mitotic anaphase, the centromeres divide and the identical chromatids separate.) In meiotic anaphase I, the centromeres do not divide and the chromatids remain together; it is the homologs that separate. Because of the exchanges of chromatid segments that result from crossing-over, however, the chromatids are not identical, as they were at the onset of meiosis.

In *telophase I*, the coiling of the chromosomes relaxes and the chromosomes become elongated and once again indistinct. The nuclear envelope is reorganized out of the endoplasmic reticulum, as telophase gradually changes to interphase. Finally, the nucleolus is reformed and protein synthesis commences again. In many organisms, however, no interphase intervenes between meiotic divisions I and II: their chromosomes pass more or less directly from telophase I to prophase II of the second meiotic division.

Second Meiotic Division

At the beginning of the second meiotic division, the chromatids are still attached by their centromeres. This division resembles a mitotic division: the nuclear envelope (if one re-formed during telophase I) becomes disorganized, and the nucleolus disappears by the end of *prophase II*. At *metaphase II*, a spindle apparatus again becomes obvious, and the chromosomes—each consisting of two chromatids—line up with their centromeres *on* the equatorial plane. At *anaphase II*, the centromeres divide and are pulled apart, and the newly separated chromatids, now called chromosomes, move to opposite poles (see Figure 8–1 on page XXX). At *telophase II*, new nuclear envelopes and nucleoli are organized, and the contracted chromosomes relax as they fade into an interphase nucleus.

The Consequences of Meiosis

The end result of meiosis is that the genetic material present in the diploid nucleus has been *replicated* only once but has *divided* twice. Therefore, each cell has only half as many chromosomes as the original diploid nucleus. But more important are the genetic consequences of the process. At metaphase I, the orientation of the bivalents is random; that is, the chromosomes derived from one parent are randomly divided between the two new nuclei. In addition, because of crossing-over, each chromosome often contains segments that have been derived from both parents. If the original diploid cell had two pairs of homologous chromosomes, $n = 2$, there are four possible ways in which the chromosome pairs could line up at metaphase with respect to one another, and consequently four possible

ways in which they can be distributed among the haploid cells. If $n = 3$, there are 8 possibilities; if $n = 4$, there are 16. The general formula is 2^n. In human beings, $n = 23$, so the number of possible combinations is 2^{23}, which is equal to 8,388,608. Many organisms have much higher numbers of chromosomes than human beings.

As the number of chromosomes increases, the chance of reconstituting the same set that was present in the original diploid nucleus becomes smaller and smaller. Quite apart from this, the existence of at least one chiasma in each bivalent makes it almost impossible that any cell produced by meiosis could be genetically the same as any one of those that fused to produce the diploid line of cells undergoing meiosis.

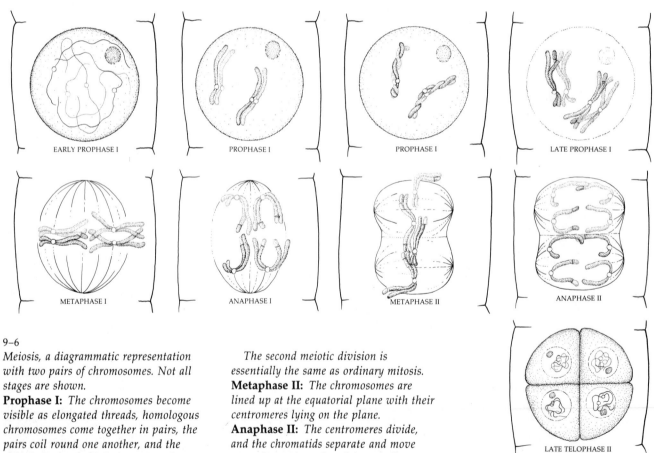

EARLY PROPHASE I   PROPHASE I   PROPHASE I   LATE PROPHASE I

METAPHASE I   ANAPHASE I   METAPHASE II   ANAPHASE II

LATE TELOPHASE II

9–6

Meiosis, a diagrammatic representation with two pairs of chromosomes. Not all stages are shown.
Prophase I: *The chromosomes become visible as elongated threads, homologous chromosomes come together in pairs, the pairs coil round one another, and the paired chromosomes become very short.*
Metaphase I: *The paired chromosomes move into position on the metaphase plate with their centromeres evenly distributed on either side of the equatorial plane of the spindle.*
Anaphase I: *The paired chromosomes separate and move to opposite poles.*

The second meiotic division is essentially the same as ordinary mitosis.
Metaphase II: *The chromosomes are lined up at the equatorial plane with their centromeres lying on the plane.*
Anaphase II: *The centromeres divide, and the chromatids separate and move toward opposite poles of the spindle.*
Telophase II: *The chromosomes have completed their migration. Four new nuclei, each with the haploid number of chromosomes, are formed.*
  Meiosis in crested wheat grass (Agropyron cristatum), n = 7, is shown opposite.

Early Prophase I. Chromosomes appear as threads. Each thread is actually double-stranded, composed of two identical chromatids.

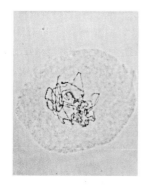

Prophase I. Homologous chromosomes pair. This is a crucial point of difference between meiosis and mitosis. Chiasmata are visible.

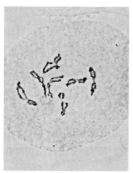

Metaphase I. Bivalents are now lined up randomly at the equatorial plane of the cell, with their centromeres evenly distributed on either side of the plane.

Anaphase I. The homologous chromosomes have separated from each other and are moving toward opposite poles of the cell.

Late Telophase I. The chromosomes are regrouped at each pole, and the cell is dividing to form two cells.

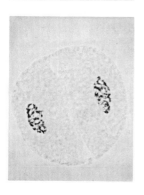

Prophase II. The chromosomes are reappearing. Each still consists of two chromatids. Because of crossing-over, the chromatids are no longer identical to each other.

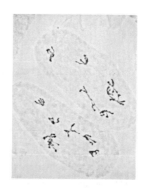

Metaphase II. The chromosomes are lined up at the equatorial plane of the cell, with their centromeres on the plane.

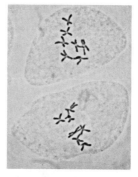

Anaphase II. The centromere of each chromosome has divided and the chromatids—now chromosomes—are moving toward opposite poles.

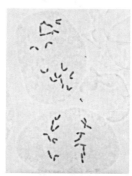

Telophase II. The chromosomes have now completely separated and new cell walls are forming.

Tetrad. New plasma membranes and cell walls form as the process of cytokinesis is completed. These four haploid cells will become pollen grains.

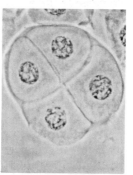

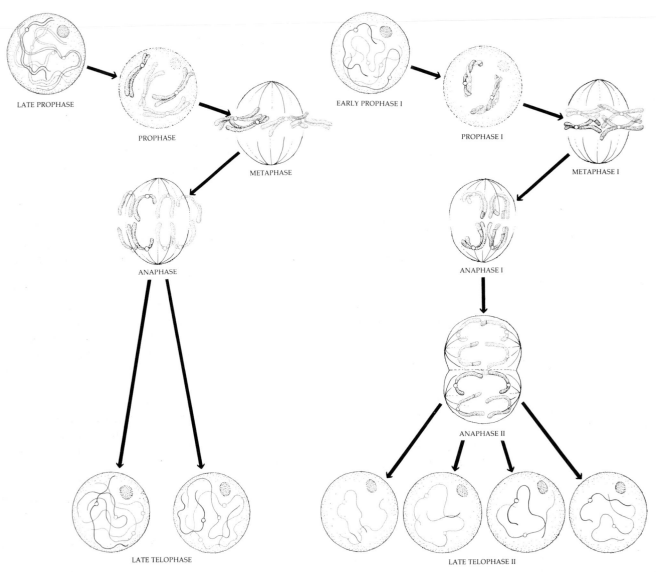

9–7

A comparison of mitosis and meiosis.

Meiosis differs from mitosis in three fundamental ways (see Figure 9–7):

1. Although the genetic material is replicated only one time, there are two nuclear divisions involved, producing a total of four nuclei.
2. Each of the four nuclei is haploid, containing only one-half of the number of chromosomes present in the original diploid nucleus from which it was produced.
3. The nuclei produced by meiosis contain entirely new combinations of chromosomes.

In meiosis, nuclei *different* from the original nucleus are produced, in contrast with mitosis, which produces nuclei with chromosome complements *identical* to those of the original nucleus. The genetic and evolutionary consequences of the behavior of chromosomes in meiosis are profound. Because of meiosis and syngamy, the populations of diploid organisms that occur in nature are far from uniform; instead, they consist of individuals that differ from one another in many characteristics.

HOW ARE CHARACTERISTICS DETERMINED?

Figure 9–8 is a plot of the distribution of ear length in a particular variety of corn, but a curve of this shape could just as well represent the distribution of weight among a random assortment of pinto beans or height among oak trees. Environmental factors, such as rain in a cornfield, might affect the actual measurements involved in preparing such a graph, but they rarely affect the shape of the curve.

9–8

Distribution of ear length of the Black Mexican variety of corn (Zea mays). This is an example of a phenotypic characteristic that is determined by the interaction of a number of genes. Such characteristics show continuous variations; if these variations are plotted as a curve, the curve is bell-shaped, with the mean, or average, falling in the center of the curve.

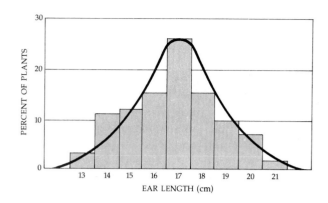

A bell-shaped curve of this sort is known as a "normal" curve. The pattern of variation is said to be *continuous*. The smooth, symmetrical shape of the curve indicates that populations cannot be divided into a series of sharply contrasting forms. Human height follows a similar pattern of inheritance. Much of the variation that is seen in nature is of exactly this kind, and the early scholars who tried to understand how the characteristics of plants and animals were inherited were baffled by them. In plants, the most important components of agricultural yield—dimensions, weight, and height—all exhibit continuous variation. We know now that patterns of this sort are produced by the combined interaction of many genes, and that they can be analyzed, at least in principle, by determining the individual roles of these genes.

Part of Gregor Mendel's genius was that he studied not continuous variation, which is frequent in nature, but rather discrete, qualitative characteristics mediated by single genes. Such characteristics are frequent in plants, which provided excellent material for Mendel's investigations; in plants, true-breeding varieties are common, crosses between them are simple to make, and large hybrid progenies can be grown easily. These advantages do not apply to many groups of animals. By looking at simple cases, he was able to unravel the complexities that so baffled his contemporaries, includ-

ing Charles Darwin. Mendel worked in the garden of the Augustinian monastery of St. Thomas at Brünn, Austria (the capital of Moravia—now called Brno), and in Czechoslovakia from 1856 to 1863. He carefully selected his principal experimental subject, the garden pea *(Pisum sativum)*. Since his youth, Mendel had collected the seeds of peas with different characteristics, and by 1856 he had already worked out which kinds of peas would cross with one another, and which would not.

For his experiments, Mendel chose well-defined, contrasting traits, such as differences in flower color or seed shape. He then made large numbers of experimental crosses and analyzed the numerical relationships among the progenies. Perhaps most important of all, he studied the offspring not only of the first generation but also of subsequent generations and their crosses. Moreover, before he made a cross between two different kinds of peas, he first obtained pure-breeding lines for each of the traits in which he was interested; this was a simple matter, since self-pollination occurs automatically in the garden pea (Figure 9–9). Earlier students of heredity had not used such well-defined characteristics, nor had they handled their experimental material so carefully; their results, consequently, were ambiguous and difficult to analyze.

The interpretation of Mendel's results was surpris-

9–9

In a flower, the pollen develops in the anthers, and the egg cells develop in the ovules. Pollination occurs when pollen grains, trapped on the stigma, germinate and grow down to the ovules, where fertilization occurs. In most species, this process involves pollen from one plant reaching the stigma of another individual, a phenomenon known as cross-pollination. The fertilized egg, or zygote, develops into an embryo within the ovule, and the mature ovule and embryo mature to form the seed.

In the flower of the garden pea, the

stigma and anthers are enclosed completely by petals. Because the pea flower, unlike those of many other species of plants, does not open until after pollination and fertilization have taken place, the plant normally self-pollinates. In his crossbreeding experiments, Mendel pried open the bud before the pollen matured and removed the anthers with forceps. Then he pollinated the flower artificially by dusting the stigma with pollen collected from another plant that had the characteristic(s) in which he was interested.

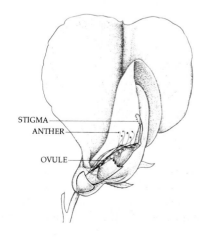

ingly clear. Table 9–1 lists seven traits of the pea plants that he used in his experiments. Crosses between individuals that differ in one single trait, like those that Mendel carried out, are called *monohybrid* crosses; those that involve two traits are called *dihybrid* crosses. When Mendel crossed plants that differed by a pair of contrasting traits of this sort, he observed that, in every case, one of the alternate traits could not be seen in the first generation (F_1). For example, the seeds of all the progeny of the cross between yellow-seeded plants and green-seeded plants were as yellow as those of the yellow-seeded parent. Mendel called the trait for yellow seeds, as well as the other traits that were seen in the F_1 generation, dominant traits. He called the traits that did not appear in the first generation recessive traits. When plants of the F_1 generation were allowed to self-pollinate, the recessive trait reappeared in the F_2 generation in ratios of approximately 3 dominant to 1 recessive (Table 9–1).

These results can be easily understood in terms of meiosis, a phenomenon that was unknown at the time Mendel was carrying out his studies. The characteristics of diploid organisms are determined by interactions between alleles. An *allele* is one of two or more alternative forms of the same gene. Alleles occupy the same site, or *locus,* on homologous chromosomes.

Consider a cross between a white-flowered plant and a red-flowered plant. The allele for white flower color, which is a recessive trait, is indicated by the lowercase letter *w*. The contrasting allele for red flower color, which is a dominant trait, is indicated by the capital let-

ter *W*. In the pure-breeding strains of garden pea with which Mendel worked, white-flowered individuals had the genetic constitution, or *genotype, ww.* Red-flowered individuals had the genotype *WW*. Individuals such as these, which have two identical alleles at a particular site, or locus, on their homologous chromosomes, are said to be *homozygous.* When plants with these contrasting traits are crossed, every individual in the F_1 generation receives a *W* allele from the red-flowered parent and a *w* allele from the white-flowered parent, and thus has the genotype *Ww.* Such an individual is said to be *heterozygous* for the gene for flower color.

In the course of meiosis, a heterozygous individual will form two kinds of gametes, *W* and *w*, which will be present in equal proportions. As indicated in Figure 9–10a, the *W* and *w* gametes derived, respectively, from the two parents, will recombine to form one *WW* individual, one *ww* individual, and two *Ww* individuals, on average, for every four offspring produced. In terms of their appearance, or *phenotype,* the heterozygous *Ww* individuals will be red-flowered; they are indistinguishable in this respect from the homozygous *WW* individuals. The products of the allele from the red-flowered parent are sufficient to mask those of the allele from the white-flowered parent. This, then, is the basis for the 3 to 1 phenotypic ratios that Mendel observed (Table 9–1).

How can you tell whether the genotype of a plant with red flowers is *WW* or *Ww?* As shown in Figure 9–10b, you can tell by crossing such a plant with a white-flowered plant and counting the progeny of the

9–10

(a) *A cross between a pea plant with two dominant alleles for red flowers* (WW) *and one with two recessive alleles for white flowers* (ww). *The phenotype of the offspring in the* F_1 *generation is red, but note that its genotype is* Ww. *The* F_2 *generation is shown in the diagram below. The W allele, being dominant, determines the appearance, or phenotype, of the flower. Only when the offspring receives a w allele from each parent, producing a ww genotype, does the recessive trait (white flowers) appear. The ratio of dominant to recessive phenotypes in the* F_2 *generation is thus 3 to 1, as expected. The genotypic ratio is 1 to 2 to 1.*
(b) *A testcross between a pea plant with red flowers and one with white flowers. Although the red-flowered plant is phenotypically identical to the* WW *plant shown in* (a), *results of the testcross reveal that it is heterozygous* (Ww) *for this gene; only a heterozygous parental genotype could produce the genotypes observed in the progeny. In this case, the genotypic ratio is 1* WW *to 1* Ww.

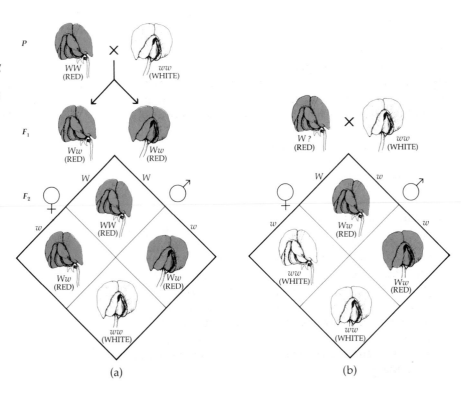

Table 9–1 *Mendel's Pea-Plant Experiment*

TRAIT	DOMINANT	RECESSIVE	F_2 GENERATION DOMINANT	RECESSIVE
Seed form (7)	Round	Wrinkled	5,474	1,850
Seed color (1)	Yellow	Green	6,022	2,001
Flower site (4)	Axial	Terminal	621	207
Flower color (1)	Red	White	705	224
Pod form (4)	Inflated	Tight	882	299
Pod color (5)	Green	Yellow	428	152
Stem length (4)	Tall	Dwarf	787	277

The numbers following the names of the traits indicate the chromosome on which the particular trait is located (see page 142 for discussion).

cross. Mendel performed just this kind of experiment, which is known as a *testcross*—the crossing of an individual showing a dominant trait with a second individual that is homozygous recessive for that trait.

The Principle of Segregation

The principle that was established by these experiments is the *principle of segregation,* which is sometimes known as Mendel's first law. According to this principle, hereditary characteristics are determined by discrete factors (now called genes) that appear in pairs, one of each pair being inherited from each parent. During meiosis, the pairs of factors are separated, or *segregated.* Hence, each gamete that is produced by an offspring at maturity contains only one member of the pair that the offspring possesses. This concept of discrete factors explained how a characteristic could persist from generation to generation without blending with other characteristics, as well as how it could seemingly disappear and then reappear in a later generation. The development of the principle of segregation, therefore, was of the utmost importance for understanding both genetics and evolution.

Incomplete Dominance

In the previous examples, the action of the dominant allele, when it was present, masked the existence of the second, recessive allele. Dominant and recessive characteristics are not always so clear-cut, however. In cases of *incomplete dominance,* for example, a given trait appears in the heterozygote in a form intermediate between that seen in the two homozygotes, since the action of one allele does not completely mask the action of the other.

For example, in snapdragons, a cross between a red-flowered plant and a white-flowered plant produces one that has pink flowers. When the F_1 generation is intercrossed, the traits segregate again, the result being one red-flowered (homozygous) plant to two pink-flowered (heterozygous) plants to one white-flowered (homozygous) plant (Figure 9–11). Thus, cases of incomplete dominance also conform to Mendel's principle of segregation.

Independent Assortment

In inheritance patterns involving more than one gene, certain differences in the patterns depend on whether the genes are located on the same chromosome or on different chromosomes. We shall consider first the relatively simple situation in which the genes are located on different chromosomes and then the more complicated one in which they are located on the same chromosome.

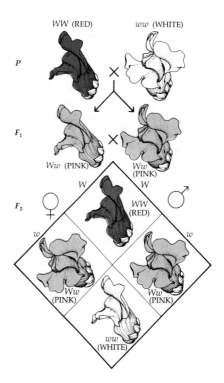

9–11

When a snapdragon with red flowers and one with white flowers are crossed, the resulting progeny have pink flowers (Ww). When the pink-flowered plants are crossed, the results show that although the gene products blend in the heterozygote, the genes themselves remain discrete and segregate according to the Mendelian ratio: 1 to 2 to 1 in both genotype and phenotype.

Mendel studied hybrids that involved two pairs of contrasting characteristics, called dihybrid crosses. For example, he crossed strains of garden peas in which one parent had seeds that were round and yellow and the other had seeds that were wrinkled and green. As Table 9–1 shows, the traits for round seeds and yellow seeds are both dominant, and the wrinkled and green traits are recessive. All the plants of the F_1 generation had seeds that were round and yellow. When the F_1 seeds were planted and the flowers were allowed to self-pollinate, 556 F_2 plants were produced. Of these, 315 plants showed the two dominant characteristics—round and yellow—in their seeds, and 32 combined the recessive traits, wrinkled and green. All the rest of the plants produced seeds that were unlike those of either parent: 101 were wrinkled and yellow, and 108 were round and green. Totally new combinations of characteristics had appeared.

Figure 9–12 shows the basis for such results. In a cross involving two pairs of dominant and recessive alleles, with each pair on a different chromosome, the ratio of distribution of phenotypes is 9:3:3:1. The 9 represents the proportion of the F_2 progeny that shows both dominant traits; 3 represents the proportions of the two possible combinations of dominant and recessive traits; and 1 is the proportion that shows both recessive traits. In the example shown, one parent carries both dominant traits and the other carries both recessive traits. Suppose that each of the parents carried one recessive and one dominant trait. Would the results be the same? If you are not sure of the answer, try making a diagram of the possibilities, as was done in Figure 9–12. Diagrams of the sort presented in that figure are called Punnett squares; they are named after the man who developed them, the English geneticist Reginald Crundall Punnett. Punnett was one of the scientists who rediscovered Mendel's work early in the present century.

From these experiments, Mendel formulated his second law *the principle of independent assortment.* This law states that the inheritance of a pair of factors for one trait is independent of the simultaneous inheritance of factors for other traits, such factors assort independently, as though no other factors were present.

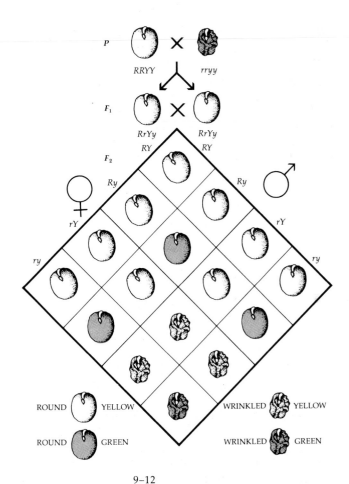

9–12

Independent assortment. In one of Mendel's experiments, he crossed a plant with round (RR) and yellow (YY) seeds with a plant that had wrinkled (rr) and green (yy) seeds. The seeds of the F_1 generation were round and yellow. In the F_2 generation, however, as shown in the diagram, the recessive traits reappeared. Furthermore, they appeared in new combinations. In a cross such as this, involving two pairs of alleles on different chromosomes, the expected phenotypic ratio in the F_2 generation is 9:3:3:1. The genotypic ratio is 1 rryy to 2 rrYy to 1 rrYY to 2 Rryy to 1 RRyy to 4 RrYy to 2 RrYY to 2 RRYy to 1 RRYY.

Linkage

When Mendel was performing his experiments, the existence of chromosomes was unknown. Knowing that genes are located on chromosomes, one can readily see that if two different genes controlling certain characteristics are located on the same chromosome pair, they generally will not segregate independently. In this case, a 9:3:3:1 ratio will not be obtained in the F_2 generation.

Either through good fortune or by careful selection of the traits he studied, Mendel avoided the phenomenon of linkage, which he probably would not have been able to explain. The garden pea has seven pairs of chromosomes, and the seven major traits studied by Mendel are located on four different chromosomes, as shown in Table 9–1 (page 141). In addition, the traits located on the same chromosome are so far apart that, with a single exception, they segregate independently of one another. The genes for pod form and stem length, which are on chromosome 4, are so close to one another that Mendel should have observed the effects of linkage between these two features. He made no report of such a dihybrid cross, however, so we do not know whether he made one or not.

The linkage of genes was first discovered by the English geneticist William Bateson and his co-workers in 1905, while they were working with the sweet pea (*Lathyrus odoratus*). These scientists crossed a doubly homozygous recessive strain of sweet peas that had red petals and round pollen with a second strain (resembling the wild form of the species) that had purple petals and elongate pollen grains. From this cross, they obtained the following characteristics in the F_2 generation:

284	purple	long
21	purple	round
21	red	long
55	red	round

If the two genes were on the same chromosome, there should have been only two types of progeny. If they were on two different chromosomes, then the ratio of the four classes should have been 216:72:72:24, or 9:3:3:1. Clearly, the genes of the "parental" types were being held together, as if they were *linked* to one another.

Linkage and the related phenomenon of crossing-over were clarified as a result of work with the fruit fly *Drosophila* by T. H. Morgan and his research group at Columbia University in the early 1900s. *Drosophila* is a particularly useful organism for genetic studies: colonies are easy to breed and maintain; a new generation of flies can be produced every two weeks; each female lays hundreds of eggs at a time; and the most common species has only four pairs of chromosomes. Female fruit flies have two homologous sex chromosomes, *XX*, whereas the males have only one *X* chromosome and a much smaller *Y* chromosome on which many of the alleles of the *X* chromosome are not present.

In the course of Morgan's experiments, an unusual type of fruit fly with white eyes was discovered. Because males always expressed this recessive trait, and male *Drosophila* are *XY*, Morgan concluded that the gene determining eye color was located on the *X* chromosome. Females, which were *XX*, expressed this trait only when they were homozygous for the allele that determined white eye color. When other recessive "sex–linked" traits, such as a yellow body, were also observed, Morgan reasoned that they must also be located on the *X* chromosome. A gene on the *X* chromosome would be the only allele affecting its particular trait in a male *Drosophila*. For this reason, the gene would always be expressed, just as if it were located in a haploid organism. In effect, the *X* chromosome in a male *Drosophila* is present in a haploid condition.

Morgan crossed a male fly *(XY)* showing two recessive sex-linked traits with a "wild-type" female fly *(XX)* —that is, one that resembled the natural wild form of the species. As expected, all the F_1 offspring—both male and female—were phenotypically "wild." A test-cross was then carried out between an F_1 female fly and its double-recessive male parent, with the expectation that the progeny would be divided between the wild type and the white-eyed, yellow-bodied type. Some of the flies that resulted from this cross, however, had white eyes and wild-type bodies, while others had wild-type eyes and yellow bodies.

The explanation for these results is that the two genes are "linked" on one chromosome but are sometimes exchanged between homologous chromosomes during crossing-over. It is now known that crossing-over—the breakage and rejoining of chromosomes that results in the appearance of chiasmata—occurs in prophase I of meiosis (see Figure 9–4). The greater the distance between two genes on a chromosome, the greater the chance that a crossover will occur between them. The closer together two genes are, the greater will be their tendency to assort together in meiosis and, thus, the greater is their "linkage." Maps can be constructed, based on the amount of crossing-over that occurs between the genes; and such *genetic maps* provide an approximation of the actual positions of the genes on the chromosomes.

Polygenic Inheritance

The first experiment that illustrated how many genes can interact to produce a continuous pattern of variation in plants (in this case, wheat) was carried out by the Swedish scientist H. Nilson-Ehle. Table 9–2 shows the phenotypic effects of various combinations of two genes, each with two alleles, which act together to control the intensity of color in wheat kernels. By this and similar studies, we are beginning to learn how many genes may interact to produce the complex characteristics of plants and animals. The principles first developed by Mendel well over a century ago can thus be applied to an analysis of the sorts of characteristics that had puzzled earlier students of heredity.

MUTATIONS

The studies on independent assortment described above depend on the existence of differences between the alleles of a gene. How do such differences arise? The first answer to this question was provided by the Dutch scientist Hugo de Vries (see Figure 9–13, page 145).

The Origin of the Concept of Mutation

In 1901, de Vries studied the inheritance of characteristics in a kind of evening primrose (*Oenothera glazioviana*), which was abundantly naturalized on the coastal dunes of Holland. In this plant, he found that, although the patterns of heredity were generally well

Table 9–2   *The Genetic Control of Color in Wheat Kernels*

Parents	$R_1R_1R_2R_2 \times r_1r_1r_2r_2$
	(dark red) (white)

F_1	$R_1r_1R_2r_2$ (medium red)

F_2		Genotype		Phenotype	
1		$R_1R_1R_2R_2$		Dark Red	
2 2	4	$R_1R_1R_2r_2$ $R_1r_1R_2R_2$		Medium-dark red Medium-dark red	
4 1 1	6	$R_1r_1R_2r_2$ $R_1R_1r_2r_2$ $r_1r_1R_2R_2$		Medium red Medium red Medium red	15 red to 1 white
2 2	4	$R_1r_1r_2r_2$ $r_1r_1R_2r_2$		Light red Light red	
1		$r_1r_1r_2r_2$		White	

ordered and predictable, a characteristic occasionally appeared that had not been observed previously in either parental line. De Vries hypothesized that this new characteristic was the phenotypic expression of a change in a gene. Moreover, according to his hypothesis, the changed gene would then be passed along just as the other genes were. De Vries spoke of this hereditary change in one of the alleles of a gene as a *mutation*, and of the organism carrying it as a *mutant*.

Ironically, only two of roughly 2,000 changes in the evening primrose that were observed by de Vries were what we would term mutations today. All of the rest were due to new genetic combinations or to the presence of extra chromosomes rather than actual abrupt changes in any particular gene. Even though most of de Vries's examples do not correspond to our current definitions, his formulation of the concept of a mutant and his recognition of the role that mutations play in producing variation were accurate and are still of fundamental importance.

The Kinds of Mutations

Today, any change in the genetic message of an organism is called a mutation. Such changes may involve either alterations in the coding sequence itself, or changes in the way in which the genetic message is organized (Chapter 8). The principal kinds of mutations are:

(1) *Point Mutations*. Point mutations involve only one or a few nucleotides. They may arise from chemical or physical damage to the DNA. *Mutagens*, such as ionizing radiation, ultraviolet radiation, or various kinds of chemicals, generally cause point mutations; in human beings and other animals, this is often how cancer starts. Point mutations may also arise by very low frequency mispairing during the replication of DNA.

(2) *Deletions*. Small sections of the chromosome may be deleted, as by X-rays. This usually results in a change in the characteristics of the organism. Many deletions apparently occur in nature as the result of crossing-over between different parts of the same chromosome, at sites where duplications occur.

(3) *Position Effects*. Genes are not necessarily in a fixed position on the chromosomes but may move around. In bacteria, *plasmids*—small circular molecules of DNA separate from the main chromosome—may enter that chromosome at places where they share a common base sequence. (The consequences of such a

process for bacterial and viral evolution will be discussed in Chapters 11 and 12, and that for genetic engineering in Chapter 30.) Both in bacteria and in eukaryotes, genes may move as small, mobile portions of the chromosome called _transposons_, which migrate from one chromosomal position to another at random. In either case, the changed position of the genes involved may disrupt the action of their new neighbors, or vice versa, and lead to effects that we recognize as mutations. The proximity of genes to areas of heterochromatin or other regions that control gene expression may likewise contribute to these effects.

(4) _Inversions and Translocations_. During the course of evolution of particular groups of organisms, a chromosomal segment may break free and then re-enter its chromosome with its sequence of genes oriented in the direction opposite to the original one. Such a change in chromosomal sequence is called an inversion. Another sort of change that is frequent in certain groups of plants involves the breaking off of a segment of one chromosome and the reattachment of that segment on another chromosome. Such a change in position is called a translocation; translocations are often reciprocal. In the case of either an inversion or a translocation, the genes on the chromosomal segment involved may express themselves differently in their new environment, thus resulting in a mutation. The potential patterns of recombination are also altered by these changes, with important consequences for plants.

(5) _Changes in Chromosome Number_. Mutationlike effects may also be associated with changes in chromosome number, which occur spontaneously and rather frequently but are usually eliminated promptly. Whole chromosomes may, under certain circumstances, be added to or subtracted from the basic set; or whole sets of chromosomes may be added, in the phenomenon of polyploidy, the duplication of whole sets of chromosomes. We shall discuss the evolutionary importance of polyploidy in plants in Chapter 28. In any of these cases, associated alterations of the phenotype may occur.

The Evolutionary Effects of Mutations

When a mutation occurs in a predominantly haploid organism, such as the fungus _Neurospora_ or a bacterium, the phenotype associated with this mutation is immediately exposed to the environment. If favorable, it tends to increase in the population as a result of natural selection; if unfavorable, as is usually the case, it is quickly eliminated from the population. Other mutations may be more nearly neutral and then may persist by chance alone; but most are either negative or positive in their effect on the organism in which they occur. In a diploid organism, the situation is very different. Each chromosome and every gene is present in duplicate, and a mutation on one of the homologs, even if it would be unfavorable in a double dose, may have much less effect or even be advantageous when present in a single dose. For this reason, such a mutation may persist in the population. The mutant gene may eventually alter its function, or the selective forces on the population may change in such a way that its effects become advantageous.

9–13

(a) _Hugo de Vries, shown standing next to_ Amorphophallus titanum, _a member of the same family as the calla lily. The plant, a native of the Sumatran jungles, has one of the most massive inflorescences, or flower clusters, of any of the angiosperms. This picture was taken in the arboretum of the Agricultural College at Wageningen, Holland, in 1932._ (b) Oenothera glazioviana, _the organism in which de Vries first reported mutations._

(a)   (b)

Although most mutations are harmful, the capacity to mutate is extremely important, for it allows the individuals in a species to vary and to adapt to changing conditions. Mutations thus provide the basis for evolutionary change. Mutations in eukaryotes occur spontaneously at a rate of about 5×10^{-6} per locus per cell division (that is, 1 mutant gene at a given locus per 200,000 cell divisions); this, together with recombination, provides abundant variation of the sort that is necessary for evolutionary change through natural selection.

GENE ORGANIZATION

Contrary to what was believed for many years, and to the prevailing situation in bacteria, eukaryotic genes are normally not present in one unique copy only. Some genes do exist as single copies, but many exist in hundreds, or even thousands, of copies. Sometimes these copies are grouped together, and sometimes they are widely scattered on one or more chromosomes. The principal kinds of duplicated genes are called satellite sequences, tandem clusters, transposons, and multigene families.

Satellite DNA consists of short sequences such as ATAAT or ATATAAT repeated millions of times, and constituting up to nearly a third of the total nucleic acids in some organisms. Almost all of the satellite DNA is clustered near the centromere and at the ends of the chromosomes, regions that remain densely coiled and thus densely staining throughout the chromosomal cycle. Apparently, satellite DNA plays a structural role and is not transcribed.

Tandem clusters are repeated sequences of genes that code for proteins that the cell requires in large amounts. Included are the genes producing ribosomal RNA (rRNA), which are present in several hundred copies in most eukaryotic organisms. The genes that are responsible for the synthesis of histones are normally present in from ten to a few hundred copies. In both cases, the highly replicated genes occur together in one or a few clusters.

Transposons are gene sequences that are repeated thousands of times. They have a normal representation of both A-T and G-C base pairs and therefore a normal density. Transposons have the ability to change their positions spontaneously, taking other genetic material with them; such shifts may have important effects on gene expression. The kinds of genetic changes that Barbara McClintock (Figure 9–14) reported in corn, starting in the late 1940s, led her to postulate the existence of transposons for the first time. On the basis of these studies, Dr. McClintock was awarded a Nobel Prize in 1984. Transposons have been discovered in recent years both in bacteria and in eukaryotes.

Multigene families are clusters of genes that occur in most eukaryotes. They seem to have arisen as a result of the duplication, relocation, and specialization of pre-existing genes, and to constitute a major evolutionary factor in eukaryotic evolution.

THE DETERMINATION OF THE PHENOTYPE

The appearance of any organism—its phenotype—is the result of many complex, interacting biochemical processes. Some of these processes occur during the course of development of the organism, and some continue at any stage in its life. In plants especially, these interactions never cease. Simple changes at the level of DNA can have results upon the final appearance of an organism that are both complex and largely unpredictable. When the science of genetics was relatively new, genes were described in terms of their more obvious phenotypic effects, that is, as purple-petaled, hairy, and so forth. It was eventually found, however, that single genes could control whole complexes of characteristics, a phenomenon known as *pleiotropy*, and that most characteristics are controlled by the combined effects of several or many genes, a phenomenon known as *epistasis*. In epistatic interactions, one gene modifies the phenotypic expression of another, nonallelic gene. In

9–14

Barbara McClintock, who was awarded a Nobel Prize in 1984 for a lifetime of creative work in genetics. She is shown here holding an ear of corn of the type she used in her studies on transposons.

fact, no gene ever acts in isolation; its effects are always modified by the internal environment, which is, of course, the end product of the interaction of thousands of genes.

We have come a long way in our understanding of the mechanism of heredity and of the ways in which the characteristics of organisms are produced. Much more remains to be learned, however, and the field will remain one of central importance to the science of botany for many years to come.

SUMMARY

Sexual reproduction involves two events: the reduction of the diploid chromosome number through meiosis and its reestablishment through syngamy.

Meiosis results in the production of either gametes or spores. A gamete is a haploid cell that fuses with another gamete, producing a diploid zygote. A spore is a cell that, without fusing with another cell, develops into a mature haploid organism.

Meiosis involves two sequential nuclear divisions and results in a total of four nuclei (or cells), each of which has the haploid number of chromosomes.

In the first meiotic division, the homologous chromosomes pair lengthwise. The chromosomes are double, each consisting of two chromatids; chiasmata form between the chromatids of the homologs. These chiasmata are the visible evidence of crossing-over—the exchange of chromatid segments between homologous chromosomes. The bivalents line up at the equatorial plane in a random manner (but with the centromeres of the paired chromosomes on either side of the plane), so that the chromosomes from the female parent and those from the male parent are completely reassorted during anaphase I. This reassortment, together with crossing-over, ensures that each of the products of meiosis differs from the parental set of chromosomes and from each other. In this way, meiosis permits the expression of the variability that is stored in the diploid genotype.

In the second meiotic division, the chromosomes divide as in mitosis.

The genetic makeup of an organism is known as its genotype; its outward characteristics constitute its phenotype. In diploid organisms—a category that includes most familiar plants and animals—every gene is present at least twice. The members of these gene pairs are known as alleles. The phenotype of diploid organisms with respect to a simple characteristic like those studied by Mendel is determined by the interaction of the two alleles that are present at the same locus on homologous chromosomes. Both alleles may be alike (homozygous), or they may be different (heterozygous). Consequently, mutations in diploid organisms are more difficult to detect than those that occur in haploid organisms, but they may be more important in an evolutionary sense. There are a number of different kinds of mutations: point mutations, deletions, position effects, inversions, translocations, and changes in chromosome number.

Although both alleles are present in the genotype, only one may be detected in the phenotype. The allele that is expressed in the phenotype is the dominant allele. The one that is concealed in the phenotype is the recessive allele. When two organisms that are each heterozygous for a given pair of alleles are crossed, the ratio of dominant to recessive in the phenotype of the offspring is 3:1. If the action of one allele is insufficient to mask the action of its alternative form (incomplete dominance), the heterozygotes are distinguishable, and the phenotypic ratio is 1:2:1.

Eukaryotic genes are normally not present in one unique copy only; many exist in hundreds, or even thousands, of copies, which may be scattered or grouped together. The principal kinds of duplicated genes are called satellite sequences, tandem clusters, transposons, and multigene families. Tandem clusters of genes are responsible for the synthesis of substances that the cell requires in large amounts, such as histones. Transposons, which are gene sequences that are repeated thousands of times, have the ability to change their positions spontaneously, taking other genetic material with them; such shifts may have important effects on gene expression.

A given characteristic is usually controlled by the interaction of the products of more than one gene, which is the major reason that pre-twentieth-century scholars found it so difficult to analyze the principles of heredity. The pattern of segregation for most features is continuous and is controlled by many genes. Some of these genes affect other genes, a phenomenon known as epistasis. Other genes affect more than one feature of the organism, a phenomenon known as pleiotropy.

SUGGESTIONS FOR FURTHER READING

GOODENOUGH, URSULA: *Genetics*, 3rd ed., Holt, Rinehart & Winston, New York, 1983.
A thoroughly up-to-date account of genetics from a molecular point of view, emphasizing bacterial genetics.

JUDSON, HORACE: *The Eighth Day of Creation*, Simon and Schuster, New York, 1979.
The story of the unraveling of modern genetics, told by the participants. A superb account of science in action.

PETERS, JAMES A. (Ed.): *Classic Papers in Genetics*, Prentice-Hall, Inc., Englewood Cliffs, N.J., 1960.*
Includes papers by most of the scientists responsible for the important developments in genetics—Mendel, Sutton, Morgan, Beadle and Tatum, Watson and Crick, Benzer, and so on. This book is very interesting and surprisingly readable, and the original papers give a feeling of immediacy that no other account can achieve.

STRICKBERGER, MONROE W.: *Genetics*, 3rd ed., Macmillan Publishing Co., Inc., New York, 1985.
A traditional and cohesive account of the science, this book is much broader in coverage than most texts at this level.

*Available in paperback.

SECTION FOUR

Diversity

C H A P T E R   1 0

The Classification of Living Things

10–1

Coral hairstreak butterfly (Harkenclenus titus) *visiting an inflorescence of butterflyweed* (Asclepias tuberosa).

At least 5 million different kinds of living organisms share our biosphere. We humans differ from these other organisms both in the degree of our curiosity and in our power of speech. As a consequence of these two characteristics, we have long sought to inquire about other creatures and to exchange information about them. As knowledge about organisms grew, it became necessary to know the names that others had given them in order to learn what was already known and to report new information.

Most familiar organisms have been given common names, but even for the simplest of purposes, such names may be inadequate. Sometimes they are misleading, particularly when we are exchanging information with people from different parts of the world. A sycamore or a cowslip in Great Britain may or may not be the same as the plants that are called sycamores and cowslips in North America. A pine in Europe or the United States is not the same as a pine in Australia. A yam in the southeastern United States is a totally different vegetable from one a few hundred kilometers away in the West Indies. When different languages are involved, the problems become hopelessly complex. For these reasons, biologists designate organisms with Latin names that are officially recognized by international organizations of botanists, bacteriologists, and zoologists.

These formal Latin names originated from informal systems of naming plants of just the sort we have been discussing. Different kinds of organisms have long been given names, corresponding to categories such as "oaks," "roses," or "dandelions." In medieval times, when an interest in communicating information about organisms was growing, Latin was the language of scholarship. Hence, the Latin names for these "kinds" of organisms were adopted as standard and were widely disseminated in books printed with the newly discovered movable type. The names were often those that the Romans had used; in other cases, they were newly invented ones or names that were put into a Latin form. These "kinds" eventually came to be called *genera* (singular: *genus*), and the individual members of these genera, such as red oaks or willow oaks, were called "species."

At first, the species were identified by descriptive Latin phrases consisting of one to many words; these phrase names are called "polynomials." The first word in a polynomial was the name of the genus to which the plant belonged. Thus, all oaks were identified by polynomials beginning with the word *Quercus,* and all roses with polynomials that started with the word *Rosa.* The ancient Latin names for these plants have continued to be used to designate genera, as we shall see, down to the present.

THE BINOMIAL SYSTEM

A simplification in the system of naming living things was made by the eighteenth-century Swedish professor and naturalist, Carolus Linnaeus (Figure 10–2), whose ambition was to name and describe all of the known kinds of plants, animals, and minerals. In 1753, Linnaeus published a two-volume work, *Species Plantarum*

10–2

Carolus Linnaeus (1707–1778), the naturalist who devised the binomial system for naming organisms. Linnaeus believed that each living thing corresponded more or less closely to some ideal model and that by classifying them, he was revealing the grand pattern of creation.

("The Kinds of Plants"). In this work, Linnaeus used polynomial designations for all species of plants as a means of describing and naming them. In many cases he changed the polynomial designations that had been used by earlier authors in order to make the polynomials in his book directly comparable with those that referred to other species in the same genus. Linnaeus regarded these polynomials as the proper names for the species he included, but, in adding an important innovation to his book, he laid the foundation for the binomial system of nomenclature—the one that we still use today. In the margin of the *Species Plantarum,* next to the "proper" polynomial name of each species, Linnaeus wrote a single word, which, combined with the generic name, formed a convenient "shorthand" designation for the species. For example, for catnip, which was formally named *Nepeta floribus interrupte spicatus pedunculatis* (meaning "*Nepeta* with flowers in an interrupted pedunculate spike"), Linnaeus placed the word "cataria" (meaning "cat-associated") in the margin of the text, thus calling attention to a familiar attribute of the plant. He and his contemporaries soon began calling this species *Nepeta cataria,* which is still its official designation today.

The convenience of this new system was obvious, and the cumbersome polynomial names were soon replaced by binomial ("two-term") names. The earliest binomial name applied to a particular species has priority over other names applied to the same species later. The rules governing the application of botanical names to plants are embodied in the *International Code of Botanical Nomenclature,* which is revised at successive International Botanical Congresses held every six years.

A species name consists of two parts—the generic name and the specific name. However, a generic name may be written alone when one is referring to the entire group of species comprising that genus. For instance, Figure 10–3 shows three species of the violet genus, *Viola.* If a species is discovered to have been placed in the wrong genus initially and must then be transferred to another genus, the second part of its name—the *specific epithet*—moves with it to the new genus. If there is already a species in that genus that has that particular specific epithet, however, an alternative name must be found. Each species has a type specimen, usually a dried plant specimen housed in a museum, which is designated by the person who originally names that species or by a subsequent author, if the original one failed to do so. The type specimen serves as a basis for comparison with other specimens in determining whether they are members of the same species or not.

A specific name is meaningless when written alone; for example, *biennis* could refer to any of the scores of species in different genera that have this word as part of their name. For example, *Artemisia biennis,* a kind of wormwood, and *Lactuca biennis,* a species of wild lettuce, are two very different members of the sunflower

10–3

Three members of the violet genus. (a) *The long-spurred violet,* Viola rostrata, *which grows in temperate regions of eastern North America as far west as the Great Lakes.* (b) Viola quercetorum, *a yellow-flowered violet of California and southern Oregon.* (c) *Pansy,* Viola

tricolor *var.* hortensis, *an annual, cultivated strain of a mostly perennial species that is native to western Europe. These photographs indicate the kinds of differences in flower color and size, leaf shape and margin, and other features*

that distinguish the species of this genus, even though there is an overall similarity between all of them. There are about 500 species of the genus Viola; *most of them are found in temperate regions of the Northern Hemisphere.*

(a)   (b)   (c)

family, and *Oenothera biennis,* an evening primrose, belongs to a different family. Because of the danger of confusing names, a specific name is always preceded by the name or the initial letter of the genus that includes it: for example, *Oenothera biennis* or *O. biennis.* Names of genera and species are printed in italics or are underlined when written or typed.

Some species consist of two or more races, called subspecies or varieties. Subspecies (varieties) of one species bear an overall resemblance to one another but exhibit one or more important differences. As a result of these subdivisions, although the binomial name is still the basis of classification, the names of some plants and animals may consist of three parts. Thus the peach tree is *Prunus persica* var. *persica,* whereas the nectarine is *Prunus persica* var. *nectarina.* Names of subspecies and varieties are also written in italics or underlined, and the first-named subspecies or variety (chronologically speaking) repeats the name of the species.

What Is a Species?

Groups of populations that resemble one another relatively closely and other groups of populations less closely are called *species,* but the application of this term differs widely from one group of organisms to another. The word "species" itself has no special connotation; it simply means "kind" in Latin. The patterns of variation that occur in different groups of organisms as a result of the evolutionary processes (discussed in Chapter 28) differ greatly from one another, so that the term "species" cannot be applied in a uniform way. For example, genetic recombination is unknown in some

groups (e.g., algae related to *Euglena,* and most bacteria), while in others outcrossing is prevalent and interspecific hybridization is frequent. Despite such differences, the term "species" provides a convenient way to talk about and catalog organisms.

Other Taxonomic Groups

Linnaeus (and earlier scientists) recognized the plant, animal, and mineral kingdoms, and the kingdom is still the most inclusive unit used in biological classification. Scientists employ several additional taxonomic categories between the levels of genus and kingdom. Thus, genera are grouped into *families,* families into *orders,* orders into *classes,* classes into *divisions,* and divisions into *kingdoms.* The kinds of groups that botanists call divisions would be called phyla (singular: phylum) by zoologists, an unfortunate difference that has historical roots.

Regularities in the form of the names for the different categories make it possible to recognize them as names at that level. For example, names of plant families end in -aceae, with a very few exceptions. Older names are allowed as alternatives for a few families, such as Fabaceae, the pea family, which may also be called Leguminosae; Apiaceae, the parsley family, Umbelliferae; and Asteraceae, the sunflower family, Compositae. Names of plant orders end in -ales. No scientific names except those of genera and species are written in italics or underlined.

Sample classifications of corn *(Zea mays)* and the commonly cultivated edible mushroom *(Agaricus campestris)* are given in Table 10–1 on page 157.

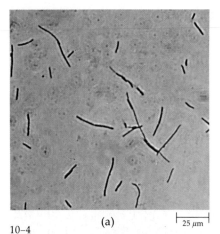

(a)   25 μm

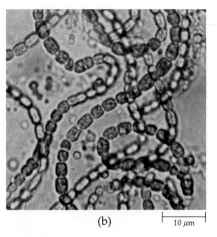

(b)   10 μm

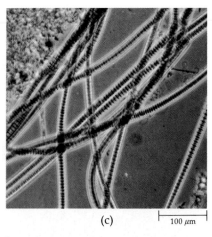

(c)   100 μm

10–4

Monera. This kingdom is made up of the bacteria, which are the only prokaryotic organisms. (a) Lactobacillus acidophilus, *bacteria that sour milk. (b) A gelatinous*

colony of the cyanobacterium Nostoc. *The cyanobacteria are one of the most abundant groups of photosynthetic*

bacteria. (c) Oscillatoria, *another cyano-bacterium, in which the growth form is filamentous.*

10–5

Protista. (a) Plasmodium of a plasmodial slime mold, Physarum, *growing on an agar culture medium. (b)* Postelsia palmiformis, *the "sea palm" (division Phaeophyta), growing on exposed*

intertidal rocks off Vancouver Island, British Columbia. (c) Volvox, *a motile colonial green alga (division Chlorophyta). (d)* Sebdenia polydactyla, *a red alga*

(division Rhodophyta). (e) A pennate dia-tom (division Chrysophyta) showing the intricately marked shell characteristic of this group.

10–6
(a) *A lichen* (Pseudocephelaria anthraspis) *growing on a tree trunk.* (b) *A flower fly* (Syrphus) *that has been killed by the fungus* Empusa muscae, *a zygomycete.* (c) *A white coral fungus (family Clavariaceae).* (d) *Mushrooms (genus Marasmius), with dew droplets, growing in a rainforest in Peru.* (e) *A puffball,* Scleroderma aurantium.

10–7
Plants. (a) Sphagnum, *the peat moss (division Bryophyta), forms extensive bogs in cold and temperate regions of the world.* (b) Marchantia *is by far the most familiar of the thallose liverworts (division Bryophyta). It is a widespread, terrestrial genus that grows on moist soil and rocks. (Continued on next page.)*

10–7 *(Continued)*
Plants. (c) Lycopodium digitatum, *a "club moss" (division Lycophyta)*. (d) *Wood horsetail,* Equisetum sylvaticum *(division Sphenophyta)*. (e) *Bulblet fern,* Cystopteris bulbifera *(division Pterophyta)*. (f) *The dandelion,* Taraxacum officinale, *and* (g) *fishhook cactus,* Mammillaria microcarpa, *are dicots (class Dicotyledones, division Anthophyta)*. (h) *Foxtail barley,* Hordeum jubatum, *and* (i) Cymbidium *orchids are monocots (class Monocotyledones, division Anthophyta)*. (j) *Sugar pine,* Pinus lambertiana, *and incense cedar,* Calocedrus decurrens, *are both conifers (division Coniferophyta)*.

Table 10–1   *Biological Classification. Notice how much you can tell about an organism when you know its place in the system. The descriptions here do not define the various categories but tell you something about their characteristics.*

Corn

CATEGORY	NAME	DESCRIPTION
Kingdom	Plantae	Organisms that are terrestrial, have chlorophyll *a* and *b* contained in chloroplasts, and show structural differentiation.
Division	Anthophyta	Vascular plants with seeds and flowers; ovules enclosed in an ovary, pollination indirect; the angiosperms.
Class	Monocotyledones	Embryo with one cotyledon; flower parts usually in threes; many scattered vascular bundles in the stem.
Order	Commelinales	Monocots with fibrous leaves; reduction and fusion in flower parts.
Family	Poaceae	Hollow-stemmed monocots with reduced greenish flowers; fruit a specialized achene (caryopsis); the grasses.
Genus	*Zea*	Robust grasses with separate staminate and carpellate flower clusters; caryopsis fleshy.
Species	*Zea mays*	Corn.

Edible Mushroom

CATEGORY	NAME	DESCRIPTION
Kingdom	Fungi	Nonmotile, multinucleate, heterotrophic, absorptive organisms in which chitin predominates in the cell walls.
Division	Basidiomycota	Dikaryotic fungi that form a basidium bearing four spores (basidiospores); the basidiomycetes.
Class	Hymenomycetes	Basidiomycetes that produce basidiocarps, or "fruiting bodies," and club-shaped, aseptate basidia that line gills or pores.
Order	Agaricales	Fleshy fungi with radiating gills or pores.
Family	Agaricaceae	Agaricales with gills.
Genus	*Agaricus*	Dark-spored soft fungi with a central stalk and gills free from the stalk.
Species	*Agaricus campestris*	The common edible mushroom.

THE MAJOR GROUPS OF ORGANISMS

In Linnaeus's time, all organisms were considered to be either plants or animals. Animals moved, ate things, and breathed, and grew until they were adults. Plants did not move, eat, or breathe; they were not observed to feed on other organisms and seemed to be able to grow indefinitely.

As new groups of organisms were discovered, they were classified either as plants or as animals. Thus fungi and bacteria were grouped with the plants, and protozoa were grouped with animals. Eventually, however, other organisms were discovered, such as *Chlamydomonas*, a swimming green alga that moves *and* manufactures its own food. Organisms of this sort could not be classified easily either as plants or animals, and by the 1930s the traditional division of living organisms into two kingdoms had clearly become little more than a historical curiosity.

Unfortunately, no completely acceptable alternative has been proposed, and scientists disagree about how many kingdoms of organisms should be recognized and what kinds of organisms should be included in each. Moreover, the old division into plants and animals is still widely reflected in the organization of college and university science departments, research projects, and textbooks (including this one). All groups traditionally considered to be plants, or those studied in departments of botany, are discussed in this book.

Prokaryotes

As discussed in Chapter 2, it is now evident that the most fundamental division in the living world is that between the prokaryotes and the eukaryotes (Figure 10–8). The prokaryotes, or bacteria, have many subgroups, including the cyanobacteria (formerly known as the "blue-green algae"). Bacteria are classified in the kingdom Monera, a group that differs far more from the eukaryotes than do plants and animals from one another.

Bacteria (discussed in detail in Chapter 11), do not have membrane-bound cellular organelles, microtubules, or the complex 9-plus-2 flagella that are characteristic of eukaryotes. Their genetic material is borne on a single circular molecule of DNA, which is not associated with proteins. Although several mechanisms leading to genetic recombination are known in bacteria, such recombination occurs infrequently; when it does, it is accomplished by means other than sexual reproduction. Bacteria also differ biochemically from eukaryotes in many ways.

Viruses

As discussed in Chapter 12, the viruses are segments of DNA or RNA that have achieved the ability to use the machinery of other cells to reproduce themselves. Viruses have the ability to manufacture a protein coat around themselves, which protects them as they move from one host to another.

Eukaryotes

All eukaryotes have a definite nucleus that is bounded by a double membrane, the nuclear envelope. Within the nuclear envelope are complex chromosomes in which the DNA is associated with proteins. These chromosomes divide regularly by mitosis. The flagella and cilia of eukaryotes have a characteristic 9-plus-2 pattern of microtubules (described in Chapter 2). Mi-

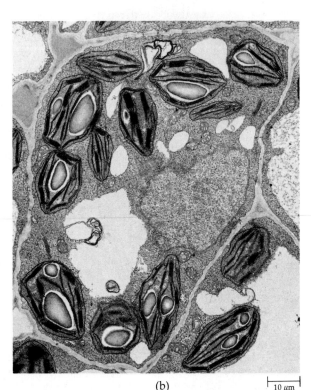

10–8

Electron micrographs of (a) a prokaryote, the cyanobacterium Anabaena, *and (b) a eukaryotic cell of a sugar beet (Beta vulgaris) leaf. Note the greater complexity of the eukaryotic cell, with its conspicuous nucleus and chloroplasts.*

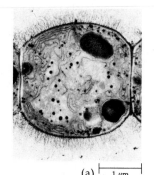

(a) |— 1 µm

(b) |— 10 µm

crotubules also occur within the cytoplasm of eukaryotes, which is subdivided in a complex way by membranes. Organelles, including mitochondria, occur in the cells of all eukaryotes; and vacuoles, which are bounded by a single membrane, or tonoplast, also occur widely in eukaryotic organisms, especially plants.

In addition, many eukaryotes exhibit two important features not found in bacteria: integrated multicellularity and sexual reproduction. Bacterial cells sometimes remain together in filamentous or even three-dimensional masses following their division, but there are usually no protoplasmic connections between the individual cells and hence no overall integration of the entire filament or mass. In plants, the protoplasts of contiguous cells are connected by plasmodesmata, which traverse their cell walls. In animals, there are no cell walls, and the cells are separated primarily by their plasma membranes.

RELATIONSHIPS WITHIN THE EUKARYOTES

One frequently used system of classification divides organisms into five kingdoms: the prokaryotic Monera (the bacteria) and four eukaryotic kingdoms (Figure 10–9). The bacteria are sharply distinct from all eukaryotic organisms, so that their classification as a separate kingdom is undoubtedly justified. Among the eukaryotes, on the other hand, the relationships are far less clear-cut. Most of the phyla and divisions (equivalent groups, given different names in animals and plants, a situation that confuses the classification of protists) consist entirely of unicellular organisms. These organisms are remarkably diverse. These unicellular phyla and divisions, and, as we shall see, some of the multicellular lines associated with them, are grouped in the kingdom Protista.

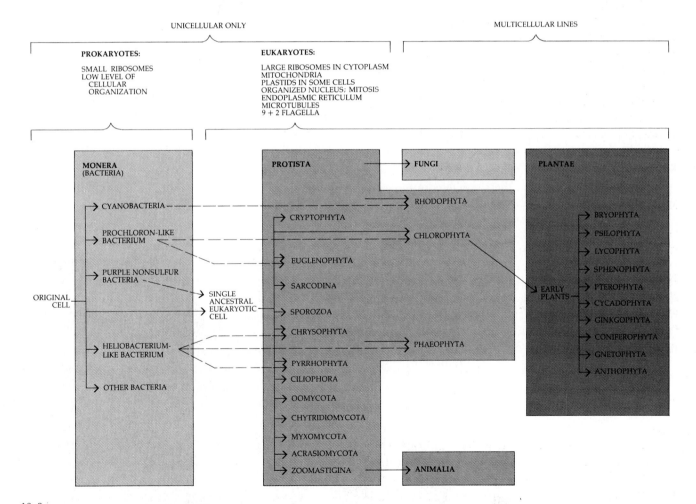

10–9

One possible scheme of the evolutionary relationships among organisms. The solid lines indicate phylogenetic relationships, and the dashed lines indicate the establishment of symbiotic relationships. All eukaryotic organisms were derived

from a single line of cells containing mitochondria, which in turn originated as symbiotic bacteria. From this single line of cells the diverse assemblage of one-celled organisms known as the Protista developed. Those protists that

developed symbiotic relationships with different groups of photosynthetic bacteria gave rise to the several different lines of modern algae. From particular single-celled protists, the fungi, plants, and animals evolved.

The Origins of Multicellularity

Multicellular organisms originated from unicellular ancestors many times in the history of the eukaryotes. In each case, the ancestors, if they are known, are classified as members of the kingdom Protista. Three of these multicellular groups are particularly large and significant—plants, animals, and fungi—and we treat them as kingdoms. These groups consist almost entirely of multicellular organisms, although there are a few unicellular fungi (the yeasts) that were probably originally derived from multicellular ancestors. Plants, animals, and fungi were certainly derived from different groups of unicellular protists. They differ fundamentally in their mode of nutrition: plants manufacture their food, animals ingest food, and fungi secrete digestive enzymes, digesting their food externally and then absorbing it.

In addition to the three major multicellular groups of organisms that we recognize as kingdoms, there are three other divisions—the red algae, green algae, and brown algae—in which multicellular organisms are well represented. The brown algae are exclusively multicellular, the red algae are almost entirely multicellular, and the green algae consist of large numbers of both unicellular and multicellular organisms. Since the members of these divisions are photosynthetic, and at least some of the organisms included are multicellular, it has been argued by some scientists that they should be included in the plant kingdom. Since the red algae and brown algae share no common ancestor with the green algae and plants other than unrelated groups of unicellular heterotrophic protists, classifying all of these groups as plants is clearly incorrect from an evolutionary point of view. The green algae include the ancestors of the plants, but they are sharply distinct from any living plants (for one thing, they are aquatic). For these reasons, we prefer to recognize the plants as a distinct kingdom and to retain the red algae, brown algae, and green algae among the Protista. Although these divisions are unusual in including so many multicellular organisms, most divisions and phyla of protists have at least some members that are multicellular.

Mitochondria and Chloroplasts

One of the most significant features in the evolution of eukaryotes has been their acquisition of mitochondria and chloroplasts. As explained in Chapter 2, it is virtually certain that these complex organelles had a symbiotic origin, which is evident in their structure and the nature of their hereditary apparatus. Among the bacteria, mitochondria most closely resemble the group called the purple nonsulfur bacteria (see Chapter 11). In these bacteria, the plasma membrane is elaborately folded throughout the cytoplasm in a way that is similar to the cristae that occur in the interior of mitochon-

dria. There are also biochemical similarities between mitochondria and the purple nonsulfur bacteria.

Chloroplasts seem to represent the descendants of at least three different kinds of symbiotic, photosynthetic, aerobic bacteria. The different groups of photosynthetic protists (called algae) are apparently not directly related to one another; they seem to have acquired their chloroplasts independently at various times in the distant past. Some of the groups of bacteria likely to have been involved in the ancestry of their chloroplasts will be discussed in Chapters 11, 14, and 15.

FORMAL CLASSIFICATION OF ORGANISMS

The following is a synopsis of the system of classification used in this book, which recognizes five kingdoms of organisms (see Table 10–2).

Kingdom Monera

The members of the kingdom Monera are the bacteria. These prokaryotic organisms do not possess nuclear envelopes, plastids, mitochondria, or 9-plus-2 flagella. They exhibit solitary unicellular or colonial unicellular organization (see Figure 10–4) and lack protoplasmic connections between the cells. Absorption is the mode of nutrition in most groups, but some are photosynthetic or chemosynthetic. Reproduction is predominantly by cell division, although genetic recombination occurs in several groups. They are either motile by simple flagella or by gliding, or are nonmotile. Monera are discussed in Chapter 11.

Kingdom Protista

Protista (see Figure 10–5) are here considered to comprise all organisms traditionally regarded as protozoa (one-celled "animals"), as well as all algae except for the group of bacteria sometimes misleadingly called "blue-green algae." The term "algae" is an informal one used for photosynthetic eukaryotes other than plants and for the cyanobacteria; nearly all algae are aquatic. Also included in the kingdom Protista are some heterotrophic groups of organisms, including the water molds and their relatives (division Oomycota), the chytrids (division Chytridiomycota), the cellular slime molds (division Acrasiomycota), and the plasmodial slime molds (division Myxomycota)—four groups of organisms that have traditionally been placed with the fungi.

The reproductive cycles of protists are varied but typically involve both cell division and sexual reproduction. Protists may be motile by 9-plus-2 flagella or cilia or by amoeboid movement, or they may be nonmotile. The predominantly unicellular groups of protists in-

Table 10-2   *Classification of Living Organisms Traditionally Regarded as Plants.*
(See Appendix C for summary descriptions of these groups.)

Prokaryotes

Kingdom Monera	Bacteria	
Eukaryotes		
Kingdom Protista	Heterotrophic protists	Division Oomycota (water molds)
		Division Chytridiomycota (chytrids)
		Division Acrasiomycota (cellular slime molds)
		Division Myxomycota (plasmodial slime molds)
	Photosynthetic protists ("algae")	Division Chrysophyta (diatoms and chrysophytes)
		Division Pyrrophyta (dinoflagellates)
		Division Euglenophyta (euglenoids)
		Division Rhodophyta (red algae)
		Division Phaeophyta (brown algae)
		Division Chlorophyta (green algae)
Kingdom Fungi	Fungi	Division Zygomycota (zygomycetes)
		Division Ascomycota (ascomycetes)
		Division Basidiomycota (basidiomycetes)
Kingdom Plantae	Bryophytes	Division Bryophyta (bryophytes)
		Class Hepaticae (liverworts)
		Class Anthocerotae (hornworts)
		Class Musci (mosses)
	Vascular plants	Division Psilophyta (psilopsids)
	Seedless vascular plants	Division Lycophyta (lycopods)
		Division Sphenophyta (horsetails)
		Division Pterophyta (ferns)
	Seed plants	Division Cycadophyta (cycads)
		Division Ginkgophyta (ginkgo)
		Division Coniferophyta (conifers)
		Division Gnetophyta (gnetophytes)
		Division Anthophyta (angiosperms)
		Class Dicotyledones (dicots)
		Class Monocotyledones (monocots)

cluded in this book are discussed in Chapter 14, and the three major groups of algae—green algae, brown algae, and red algae—in Chapter 15. Discussions of the heterotrophic protists known as protozoa—in other words, those that have historically been treated as animals instead of as fungi—are not included in this text.

In summary, the kingdom Protista includes a very heterogeneous assemblage of unicellular, colonial, and multicellular eukaryotes that do not have the distinctive characteristics of the animals, plants, or fungi.

Kingdom Animalia

The animals are multicellular organisms with wall-less eukaryotic cells lacking plastids and photosynthetic pigments. Nutrition is primarily ingestive with digestion in an internal cavity, but some forms are absorptive, and a number of groups lack an internal digestive cavity. The level of organization and tissue differentiation in complex animals far exceeds that of the other kingdoms, particularly with the evolution of complex sensory and neuromotor systems. The motility of the organism (or, in sessile forms, of its parts) is based on contractile fibrils. Reproduction is predominantly sexual. Animals are not discussed in this book, except in relation to some of their interactions with plants and the other organisms that are treated here.

Kingdom Fungi

Fungi (see Figure 10-6) are nonmotile filamentous eukaryotes that lack plastids and photosynthetic pigments and absorb their nutrients from either dead or living organisms. The fungi have traditionally been grouped with plants, but there is no longer any doubt that the fungi are in fact an independent evolutionary line.

Aside from their filamentous growth habit, fungi have virtually nothing in common with any of the groups that have been classified as algae. The cell walls of fungi include a matrix of chitin. The structures in which fungi form their spores are often complex. Fungal reproductive cycles, which also can be quite complex, typically involve both sexual and asexual processes. Fungi are discussed in Chapter 13.

Kingdom Plantae

Plants—the bryophytes (mosses, liverworts, and hornworts) and the nine divisions of vascular plants—comprise a kingdom of photosynthetic organisms adapted for a life on the land. Their ancestors were specialized green algae. All plants are multicellular and are composed of vacuolate eukaryotic cells with cellulosic cell walls. Their principal mode of nutrition is photosynthesis, although a few plants have become heterotrophic. Structural differentiation occurred during the evolution of plants on land, with trends toward the evolution of organs specialized for photosynthesis, anchorage, and support (see Figure 10–7). In more complex plants, such organization has produced specialized photosynthetic, vascular, and covering tissues. Reproduction in plants is primarily sexual, with cycles of alternating haploid and diploid generations; the haploid generation (gametophyte) has been reduced during the course of evolution of the more advanced members of the kingdom. The bryophytes are discussed in Chapter 16 and the vascular plants in Chapters 17, 18, and 29.

S E X U A L   R E P R O D U C T I O N

The DNA of bacteria is not duplicated, as it is in diploid eukaryotes; despite this, bacteria do have several means of achieving genetic recombination. As described in more detail in Chapter 11, a portion of the bacterial genetic material may be transferred from one cell to another, but there is no mechanism comparable to meiosis by which it can then be regularly transmitted along with the genetic material of the cell that it joins. The only method for regular transmission of the new genetic information in prokaryotes is the incorporation of the new fragment into the bacterial DNA molecule, a system that is neither precise nor readily repeatable.

Sexual reproduction, which involves a regular alternation between meiosis and syngamy, has a high selective advantage because it is the primary mechanism that produces and maintains variability in natural populations of eukaryotes. Variability, as we shall see in Chapter 28, is the basic trait that makes possible the adjustment of living organisms to their environment through the process of evolution.

The Evolution of Diploidy

The first eukaryotic organisms were probably haploid and asexual, but once sexual reproduction was established among them, the stage was set for the evolution of diploidy. It seems likely that this condition first arose when two haploid cells combined to form a diploid zygote; such an event probably took place repeatedly. Presumably the zygote then divided immediately by meiosis, thus restoring the haploid condition (Figures 10–10 and 10–11a). In organisms with this simple kind of life cycle, the zygote is the only diploid cell.

By "accident"—an accident that occurred in a number of separate evolutionary lines—some of these zygotes divided mitotically instead of meiotically and, as a consequence, produced an organism that was composed of diploid cells, with meiosis occurring later. In animals, this delayed meiosis results in the production of gametes—eggs and sperm. These gametes then fuse, an event that immediately restores the diploid state

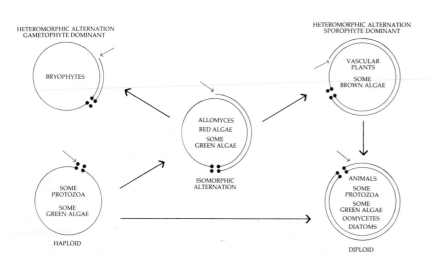

10–10

The evolution of genetic systems. Each of these circles represents a different type of life cycle. The most primitive eukaryotes were undoubtedly haploid for most of their life cycle, as indicated by the single circle (lower left). In this type of life cycle, meiosis (indicated by four black spheres) occurs right after fertilization (designated by the small arrow in color). The other life cycles differ from the haploid one in the point at which fertilization takes place, as well as in the proportion of the life cycle spent in the diploid state (indicated by the extent of the outer circle). The groups named within the circles refer to some of the modern organisms with these particular types of life cycles; not all groups are included here.

(Figures 10–10 and 10–11b). Therefore, in animals, gametes are the only haploid cells.

In plants, meiosis results in the production of spores, not gametes. Spores are cells that can divide directly by mitosis to produce a multicellular haploid organism; this is in contrast to gametes, which can develop only following fusion with another gamete. Multicellular haploid organisms that appear in alternation with diploid forms are found in plants; in some brown, red, and green algae; and in two closely related genera of chytrids and one or more other groups of protists not treated in this book. Such organisms exhibit the phenomenon known as *alternation of generations* (Figures 10–10 and 10–11c). Among the plants, the haploid, gamete-producing generation is called the *gametophyte,* and the diploid, spore-producing generation is called the *sporophyte*. This same terminology is used for the algae and sometimes other groups as well.

In some algae—most of the red algae, many of the green algae, a few of the brown algae—the haploid and diploid forms are the same in external appearance. Such types of life cycles are said to exhibit an *isomorphic* alternation of generations (Figure 10–10).

In contrast to the life cycles discussed above, there are some in which the haploid and diploid forms are not identical. During the history of these groups, mutations occurred that were expressed in only one generation, although the alleles were, obviously, present in both the diploid and the haploid generations. In life cycles of this kind, the gametophyte and sporophyte became notably different from one another, and a *heteromorphic* alternation of generations originated. Such life cycles are characteristic of plants and some brown algae (Figure 10–10).

In the bryophytes (mosses, liverworts, and hornworts), the gametophyte is dominant; it is nutritionally independent from and usually larger than the sporophyte, which may be more complex structurally. In the vascular plants, on the other hand, the sporophyte dominates; it is much larger and more complex than the gametophyte, which is nutritionally dependent on the sporophyte in nearly all groups.

(a)

(b)

(c)

10–11

Diagrams of the principal types of life cycles. In these diagrams, the diploid phase of the cycle takes place below the broad bar, and the haploid phase occurs above it. The four white arrows signify the products of meiosis; the single white arrow represents the fertilized egg.

(a) In zygotic meiosis, the zygote divides by meiosis to form four haploid cells, which divide by mitosis to produce more haploid cells or a multicellular individual that eventually gives rise to gametes by differentiation. This type of

life cycle is found in Chlamydomonas *and a number of other algae.*

(b) In gametic meiosis, the haploid gametes are formed by meiosis in a diploid individual and fuse to form a diploid zygote that divides to produce another diploid individual. This type of life cycle is characteristic of most animals and some protists (Oomycota), as well as the brown alga Fucus.

(c) In sporic meiosis, the sporophyte, or diploid individual, produces haploid spores as a result of meiosis. These spores

do not function as gametes but undergo mitotic division. This gives rise to multicellular haploid individuals (gametophytes), which eventually produce gametes that fuse to form diploid zygotes. These zygotes, in turn, differentiate into diploid individuals. This kind of life cycle, known as alternation of generations, is characteristic of the plants and many algae. A similar sort of life cycle is found in the chytrid Allomyces *and one other closely related genus and in a few other groups of protists not included in this book.*

As mentioned previously, diploidy permits the storage of more genetic information and so perhaps allows a more subtle expression of the organism's genetic background in the course of development. This may be the reason that the sporophyte is the large, complex, and nutritionally independent generation in vascular plants. One of the clearest evolutionary trends in this group, which predominates in most terrestrial habitats, is the increasing dominance of the sporophyte and suppression of the gametophyte. Among the flowering plants, the female gametophyte is a microscopic body that consists of only seven cells, and the male gametophyte consists of only three cells. Both of these gametophytes are completely dependent on the sporophyte in terms of their nutrition.

SUMMARY

Biologists have developed methods of naming and classifying organisms that permit them to designate the organism very precisely, an essential factor in scientific communication. The classification of an organism also reveals its relationship to other living things.

Organisms are designated scientifically by a name that consists of two words—a binomial. The first word in the binomial is the name of the genus (plural: genera), and the second word, the specific epithet, combined with the name of the genus completes the name of the species. Species are sometimes subdivided into subspecies or varieties. Genera are grouped into families, families into orders, orders into classes, and classes into divisions. Divisions are grouped into kingdoms, the kingdom being the largest unit used in classification of the living world. In this text, living organisms are grouped into five kingdoms, the first of which consists of prokaryotes, the others of eukaryotes: (1) Monera, the bacteria, or prokaryotes; (2) Protista, the protozoa, eukaryotic algae, slime molds, and water molds; (3) Animalia, multicellular organisms that are not photosynthetic; (4) Fungi, the fungi; and (5) Plantae, bryophytes and vascular plants, photosynthesizing organisms that are terrestrial and more complex than the algae.

Mitochondria, which are characteristic of all eukaryotes, probably originated from aerobic (oxygen-requiring) bacteria similar to the purple nonsulfur bacteria, whereas chloroplasts appear to have originated from at least three different groups of photosynthetic, aerobic bacteria. The possession of chloroplasts in different groups of eukaryotes, therefore, does not signify a direct relationship between them. Multicellularity originated independently a number of times in different groups of protists. Some of the multicellular lines, including the red algae, brown algae, and green algae, are retained among the protists in the five-kingdom system of classification used in this book, whereas three major multicellular lines —animals, plants, and fungi—are recognized as separate kingdoms.

In the evolution of organisms, diploidy evolved subsequent to the process of sexual reproduction. In primitive eukaryotes and all fungi, the zygote formed by syngamy divides immediately by meiosis. From early cycles of this sort, more complex life cycles involving diploid phases were derived on a number of occasions, when the zygote divided by mitosis. If the haploid cells produced by meiosis function immediately as gametes, the result is the type of life cycle found in animals and in some groups of protists. If they divide by mitosis, as in many algae, all plants, and two genera of chytrids (Protista), they are considered spores; the diploid generation that gives rise to such spores is called the sporophyte. The haploid generation, which develops from the spores by mitosis, is called the gametophyte; it eventually produces gametes by mitosis. If the gametophyte and sporophyte in a particular life cycle are approximately equal in size and complexity, the alternation of generations is said to be isomorphic; if they differ widely in size and complexity, it is said to be heteromorphic.

SUGGESTIONS FOR FURTHER READING

ALTSCHUL, SIRI VON REIS: "Exploring the Herbarium," *Scientific American,* May 1977, pages 96–104.
Valuable information about the uses of little-known plants that might provide food, drugs, or other useful products can be gleaned from the labels on specimens in herbaria.

BOLD, HAROLD G., C. S. ALEXOPOULOS, and T. DELEVORYAS: *Morphology of Plants and Fungi,* 4th ed., Harper & Row, Publishers, Inc., New York, 1980.
A well-illustrated and ample treatment of the diversity of plants, algae, and fungi.

CORNER, E.J.H.: *The Life of Plants,* Mentor Books, New American Library, Inc., New York, 1968.*
A renowned botanist with a flair for poetic prose describes the evolution of plant life, telling how plants modified their structures and functions to meet the challenge of a new environment as they invaded the shore and spread across the land.

JONES, S. B., and ARLENE E. LUCHSINGER: *Plant Systematics,* McGraw-Hill Book Company, New York, 1979.
A good, practical overall treatment of the field.

MARGULIS, LYNN: *Symbiosis in Cell Evolution,* W. H. Freeman and Company, New York, 1981.
A fascinating discourse on the origin of eukaryotic cells by serial symbiotic events.

MARGULIS, LYNN, and KARLENE V. SCHWARTZ: *Five Kingdoms. An Illustrated Guide to the Phyla of Life on Earth,* W. H. Freeman and Company, New York, 1982.
An excellent and concise account of the diversity of life on earth.

PARKER, SYBIL P. (Ed.): *Synopsis and Classification of Living Organisms,* 2 vols., McGraw-Hill Book Company, Inc., New York, 1982.
An encyclopedic and authoritative outline of the major groups of plants, animals, and microorganisms.

SCAGEL, ROBERT F., et al.: *Nonvascular Plants: An Evolutionary Approach,* Wadsworth Publishing Co., Inc., Belmont, Calif., 1982.
An exhaustive, thoroughly illustrated review of the diversity of bryophytes, fungi, bacteria, and those groups of protists treated in this book.

* Available in paperback.

C H A P T E R   1 1

Bacteria

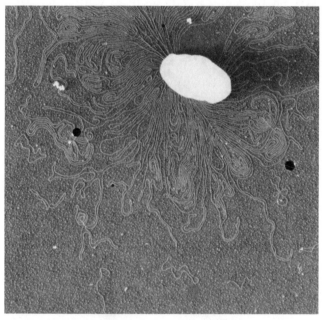

11–1
Because of the way that it has been prepared, this cell of the common colon bacterium Escherichia coli *is spilling out its DNA. The bacterial chromosome consists of a single giant molecule of DNA containing some 3 billion base pairs. The smaller circles of DNA visible in the left side of the micrograph are plasmids with approximately 5000 base pairs. (The black dots on the photograph are holes in the plastic film on which the preparation was made; the scattered white dots may be dust particles or pieces of the cell.)*

Bacteria—the prokaryotes—are the simplest, smallest, and most abundant organisms (Figure 11–1). For at least 2 billion years, they were the only form of life on earth. Because bacteria are so distinct from all other living things, they are classified as a separate kingdom, the kingdom Monera. Bacteria are metabolically diverse, and some of them are photosynthetic. One group of photosynthetic bacteria, the so-called "blue-green algae," or cyanobacteria, were previously (and incorrectly) associated with the photosynthetic Protista, or "algae," but they are simply one of the distinctive groups of bacteria. Their properties can be understood best in this context.

Bacteria lack an organized nucleus surrounded by a nuclear envelope. They do not have complex chromosomes like those of the eukaryotes, and they do not reproduce sexually, although genetic recombination occurs occasionally. Unlike eukaryotes, bacteria are never truly multicellular. Although some bacteria form filaments or masses of cells, these are connected because their cell walls fail to separate completely following cell division or because they are held together within a common mucilaginous capsule, or sheath. Plasmodesmata between such linked cells are extremely rare, occurring only in a few species of cyanobacteria.

Bacteria lack membrane-bound organelles but have other structures that play similar roles. Their plasma membranes often have folds and convolutions that extend into the interior of the cells (Figure 11–2). Such membranes increase the surface area to which enzymes are bound, and they facilitate the separation of different enzymatic functions. In some photosynthetic bacteria, the photosynthetic pigments are located in internal membranes. In others, these pigments are located in discrete spherical bodies called chromatophores. The cell walls of most bacteria contain muramic acid, which is absent in the cell walls of eukaryotes.

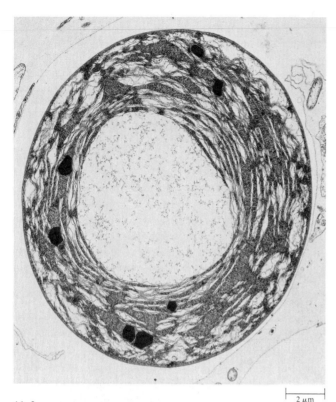

├── 2 μm ──┤

11-2

A single cell of the bacterium Prochloron, *showing the extensive internal
membrane structure.* Prochloron *is a
photosynthetic bacterium with chlorophylls a and b and carotenoids like those
found in the green algae and plants.*

GENERAL CHARACTERISTICS OF BACTERIA

Most bacteria are only about 1 micrometer in diameter, with some being only one-tenth of that size and others ranging up to 10 micrometers (or rarely, even 30 micrometers or more) in length. Even though individual bacteria are so small, the total weight of all the bacteria in the world is estimated to exceed that of all the other living organisms combined. About 2500 species of bacteria are currently recognized.

The oldest known sedimentary rocks, from Greenland, are about 3.9 billion years old, but no fossils have been found in them. Fossil bacteria as much as 3.5 billion years old are known from Western Australia (see Figure 1–2, page 2), with records almost as old also known from South Africa. Organic spheroidal bodies similar to cyanobacteria have been found in South African strata about 3.2 billion years old. Fossil and chemical evidence from the same series of deposits indicates that photosynthesis occurred as early as 3.3 billion years ago. The oldest eukaryotes, on the other hand, are *only* about 1.5 billion years old. The fossil bacteria

are studied in ultrathin sections of rocks with the aid of the electron microscope and have been known only since the 1950s.

Bacteria occur in virtually all habitats and, because of the unusual metabolic characteristics of some members of the group, can survive in many environments that support no other form of life. Some bacteria are *obligate anaerobes*; they live only in the absence of oxygen. Others are *facultative anaerobes*; that is, they can survive without oxygen but grow more vigorously in its presence. Respiration yields much more energy than fermentation, as discussed in Chapter 6.

Bacteria have been found in an amazingly diverse set of habitats: they truly occur all over the globe. For example, the bacterium *Thermoanaerobacter ethanolicus*, living in hot springs at Yellowstone National Park, can thrive at temperatures as high as 78°C. Even this extreme has been surpassed by the bacteria that live at very high pressure around deep-sea vents at temperatures as high as 360°C. When bacteria collected from around these vents were cultured in the laboratory at high pressure, several kinds grew actively and doubled in number at 250°C within 40 minutes! Such high temperatures can, of course, exist only under very high pressure and therefore are possible on a continuing basis only on the ocean floor. Other bacteria occurring in different deep-sea habitats, such as the intestines of marine animals, also require high pressures for their proper growth and development.

At another extreme, living bacteria have been found in Antarctica in samples of rock and ice recovered by drilling to depths of up to 430 meters. These bacteria, which are at least 10,000 and possibly a million years old, were lying dormant at temperatures ranging from about −7° to −14°C but immediately became active when their temperature was elevated. These findings illustrate the fact that some bacteria are capable of surviving in a state of "suspended animation" for extremely long periods of time.

Recently, the limits of tolerance of certain bacteria have been investigated in relation to the possible existence of extraterrestrial life on various planets with known atmospheres. For example, the extreme alkalinity of Jupiter's atmosphere has been considered one of the arguments against the chance of life occurring there; however, some bacteria from the Livermore Valley of California are able to grow and reproduce in alkaline water with a pH of 11.5. Other bacteria can tolerate not only extreme alkalinity but also the presence of high concentrations of ammonia, conditions that resemble those on the planet Venus. In view of these findings, the survival of certain bacteria on other planets seems possible. Most bacteria, however, do not survive well in acid conditions; for this reason, vinegar —dilute acetic acid—is an effective food preservative.

Bacteria play a vital role in the functioning of the world ecosystem. Some groups of bacteria are autotrophic and so make major contributions to the global carbon balance; some of these bacteria are being grown experimentally as possible commercial sources of protein. The role of certain bacteria in fixing atmospheric nitrogen is of crucial ecological importance. Heterotrophic bacteria, like fungi, are decomposers. Through the action of decomposers, materials incorporated into the bodies of once-living organisms are released and made available for successive generations of living organisms. In a single gram of fertile agricultural soil, there may be 2.5 billion bacteria, 400,000 fungi, 50,000 algae, and 30,000 protozoa. At all stages of the recycling of elements in the global ecosystem, certain bacteria play important roles.

Bacteria are also enormously abundant in the sea; for example, it has been estimated that 90 percent of all the biomass in the ocean is made up of organisms less than 10 micrometers long—that is, mainly bacteria. Depending on the area of the sea concerned, between 20 percent and 60 percent of the primary production in the sea passes through free-living heterotrophic bacteria.

The metabolic diversity of bacteria enables them to degrade many different kinds of organic substances. For this reason, bacteria are being investigated as a possible means of decomposing unwanted synthetic substances, such as pesticides, dyes, and petroleum, both before and after these substances are released into the environment. Nylon, first manufactured in about 1939, is broken down by a bacterium of the genus *Flavobacterium,* which has evolved two apparently novel enzymes in this relatively short period of time. Certain bacteria break down pesticides so rapidly that they have only a limited effect on the pests they are meant to control. Other bacteria are being studied for their possible use in extracting oil from rocks.

In addition to their ecological roles, bacteria are important in a number of other ways. They are responsible for many of the most serious diseases of humans and other animals—including tuberculosis, cholera, anthrax, gonorrhea, diphtheria, and tetanus—and cause a wide range of economically significant diseases in plants. More than 200 species of bacteria are currently recognized as plant pathogens in the United States alone. The disease known as fire blight (Figure 11–3), for example, destroyed thousands of pear trees as it spread across the United States following its accidental introduction on the eastern seaboard in about 1880. By the 1930s, this disease virtually wiped out the pear industry. Currently, the commercial production of pears in the United States is restricted to a few limited areas.

Bacteria are used as the commercial source of a number of important antibiotics, such as tyrothricin, bacitracin, subtilin, and polymyxin B. Many bacteria are used experimentally and, in some instances, commercially in the production of drugs and other molecules (see Chapter 30). For example, specific bacteria are used commercially in the production of acetic acid and vinegar, various amino acids, and enzymes. In addition, the production of almost all cheeses involves bacterial fermentation of lactose into lactic acid, which coagulates milk proteins. The same kinds of bacteria that are responsible for the production of cheese are also responsible for the production of yogurt, and for the lactic acid that preserves sauerkraut and pickles. *Spirulina,* one of the cyanobacteria, is being grown in many countries as a source of protein, which can be mixed with human foods or fed to animals (Figure 11–4).

One group of bacteria, the actinomycetes, is especially important in the production of antibiotics such as streptomycin, aureomycin, neomycin, and tetracycline. About two-thirds of the several thousand kinds of antibiotics that have been registered to date are produced by actinomycetes. Other actinomycetes fix nitrogen in nodules they form on the roots of certain plants, thus playing a major ecological role.

11–3
Fire blight, caused by the bacterium Erwinia amylovora *(Figure 11–10), is a common and destructive disease of pears, apples, and related members of the rose family (Rosaceae). It kills the branches of these plants and then eventually causes cankers on the stems (shown here on a pear tree). Soon after, the plants usually die. In the late nineteenth century, fire blight was the first plant disease shown to be caused by a bacterium.*

11–4

(a) *The characteristic corkscrew-shaped filaments of cells of* Spirulina, *a cyanobacterium that is being cultivated widely as a source of protein. (b) Ponds for the cultivation of* Spirulina *at Lake Texcoco, near Mexico City. (c) Collection* of wet cell paste of Spirulina. *The use of* Spirulina *for human food was recorded by the Spanish conquistadors at Mexico City as early as 1521; the paste produced from this cyanobacterium contains more protein than soybeans, and the protein is* balanced in its amino acid content. *Yields of protein per day are about 10 times that of wheat or soybeans, which accounts for a great deal of current interest in the cultivation of this cyanobacterium.*

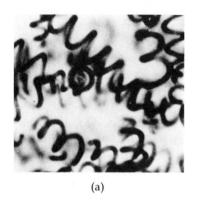

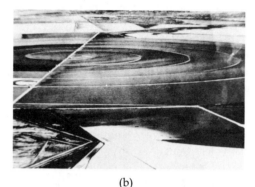

(a)   (b)   (c)

Bacterial Form

Bacteria vary greatly in form and cellular organization (see Figure 11–5). Straight, rod-shaped bacteria are known as *bacilli*, spherical ones are called *cocci*, and long, coiled ones are called *spirilla* (Figure 11–6). Spherical bacteria may stick together in pairs after division (diplococci), they may occur in clusters (staphylococci), or they may form chains (streptococci). The organism that causes pneumonia is a diplococcus, whereas staphylococci are responsible for many serious infections that are characterized by boils or abscesses.

Bacilli more commonly occur as isolated individuals than do cocci. When bacilli do remain together, they tend to form filaments because they divide transversely. Since these filaments are funguslike in appearance, the names of bacilli often have the prefix *myco-* (from the Greek word for fungus). The actinomycete *Mycobacterium tuberculosis*, for example, is a rod-shaped bacterium that forms a filamentous funguslike growth in culture, although not in its host; other actinomycetes (see Figure 11–5d) have a similar growth form. The cells of cyanobacteria, too, often adhere in filaments after they divide (Figures 10–4b, c, and 11–7).

Cell Wall

In the early 1880s, a Danish microbiologist, Hans Christian Gram, working in a morgue in Berlin, discovered that he could determine the presence of the bacteria that cause pneumonia in infected tissues by staining the tissues with a dye similar to crystal violet, the one used now. When treated with such stains, most bacteria turn purple. Gram was disappointed that not all bacteria retained his stain, but the discovery has turned out to be a valuable tool in understanding the structure of bacterial cell walls by distinguishing different groups of bacteria. As the test is performed now, the cells are first stained, then treated with a dilute solution of iodine (which decreases the solubility of the purple dye by forming a dye-iodine complex), and finally rinsed with ethanol, which readily removes the stain from the cell walls of some bacteria, but not others. The bacteria that retain the purple dye appear purple and are called gram-positive; those in which it is lost are gram-negative.

Much later it was discovered that bacterial cell walls consist of a matrix of disaccharides cross-linked by short chains of amino acids (peptides). The cell wall of a gram-positive bacterium—which ranges from about 15 to 80 nanometers in thickness depending on the species—consists of a single macromolecule of this kind, known as a *peptidoglycan*. In gram-negative bacteria, an additional layer, consisting of large molecules of lipopolysaccharide (polysaccharide chains with lipids attached to them) encases the peptidoglycan layer. The cell walls of gram-negative bacteria are only about 10 nanometers thick.

Gram-positive and gram-negative bacteria differ in their reactions to antibiotics. Penicillin, for example, blocks the formation of the peptide cross-links in gram-positive bacteria; without these cross-links, the cell ruptures when it increases in size during growth. Penicillin has no effect on gram-negative bacteria because of the lipopolysaccharide outer layer in their cell walls. Another commonly used antibiotic, actinomycin, is a large molecule that disrupts protein synthesis by binding to the DNA double helix. The actinomycin molecule passes easily through the cell wall of gram-positive bacteria, but not that of gram-negative bacteria. Thus, for gram-negative bacteria, antibiotics other than penicillin and actinomycin must be employed; one that is commonly prescribed for them is erythromycin.

11–5

Some of the diverse types of bacteria. (a) Myxobacteria, or gliding bacteria, have a pattern of organization similar to that of the slime molds (see Chapter 14). This is a scanning electron micrograph of the "fruiting bodies" of Chondromyces crocatus, *each consisting of as many as 1 million cells. Normally the myxobacteria are rods that glide together and eventually differentiate into the kinds of bodies shown here. (b) Spirochaetes are spiral bacteria that are up to 500 micrometers long, which is an enormous size for bacteria. Their motility is by means of a helical wave moving along massed filaments that line the body of the cell.* Treponema pallidum, *shown here,*

is the causative agent of syphilis. (c) Actinomycetes are abundant in soil, where they are largely responsible for the "moldy" odor of damp soil and decaying organic material. Streptomyces fradiae, *shown here, is the commercial source of the antibiotic neomycin. (d) Gliding bacteria are filamentous and have relatively large cells in which waves of contraction, responsible for their "gliding" movements, cause periodic alterations in form. In* Beggiatoa, *shown here, the conspicuous granules in the cells are sulfur, produced by the oxidation of hydrogen sulfide—an energy-yielding process.*

11–6

*The three major forms of bacteria: (a) bacilli, (b) cocci, and (c) spirilla. Cell shape is a relatively constant feature in most species of bacteria. Bacilli include those microorganisms that cause lockjaw (*Clostridium tetani*), as well as the familiar* Escherichia coli. *They also are responsible for many plant diseases, including fire blight of pears and apples (*Erwinia amylovora*) and bacterial wilt of tomatoes, potatoes, and bananas (*Pseudomonas solanacearum*). Among the cocci are* Diplococcus pneumoniae, *the cause of bacterial pneumonia;* Streptococcus lactis, *a common milk-souring agent; and* Nitrosococcus nitrosus, *soil bacteria that oxidize ammonia to nitrites. The spirilla, which are less common, are helically coiled.*

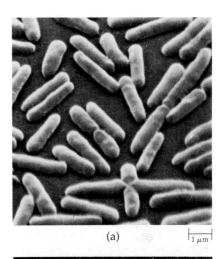

(a)

1 µm

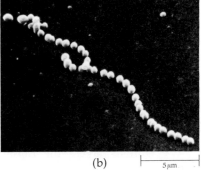

(b)

5 µm

(c)

20 µm

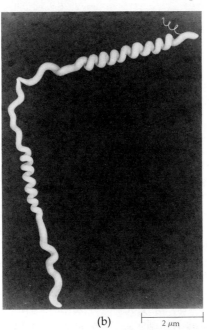

(b)

2 µm

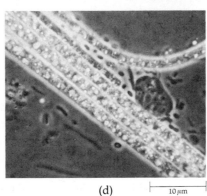

(d)

10 µm

(a)

(c)

200 µm

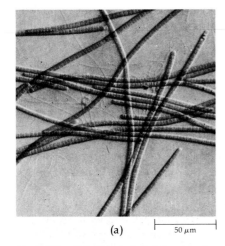

(a)   50 μm

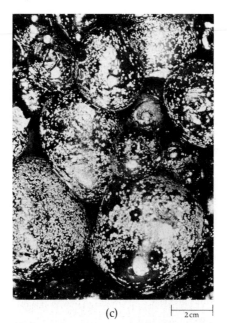

(c)   2cm

(d)   50 μm

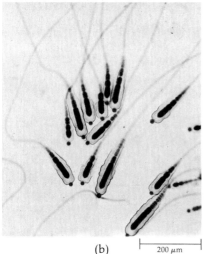

(b)   200 μm

11–7

Three common genera of cyanobacteria and a related form: (a) Oscillatoria, *in which the only form of reproduction is by means of fragmentation of the filament. (b)* Calothrix (Gloeotrichia), *a filamentous form with a basal heterocyst (see page 184).* Gloeotrichia *is capable of forming akinetes—enlarged cells that develop a resistant outer envelope—just above the heterocysts. (c) Gelatinous "balls" of* Nostoc commune, *each containing hundreds of filaments, occur frequently in freshwater habitats. (d)* Thiothrix, *a genus that lacks chlorophyll but is apparently closely allied to the cyanobacteria.* Thiothrix *obtains energy by the oxidation of H_2S. The chains of cells, each filled with particles of sulfur, are attached to the substrate at the base (center) and so form a characteristic rosette.*

The adhesive properties of individual bacterial cells are determined by a layer of tangled fibers of polysaccharides called the _glycocalyx_. Although bacteria often do not form a glycocalyx when they are grown in laboratory cultures, in nature this structure may play a key role in the initiation and progression of bacterial infection by mediating the adhesion of the bacteria to different substrates. By means of enzymatic reactions that occur in the glycocalyx, for example, certain bacteria are able to colonize the intact enamel surfaces of teeth, as well as many other seemingly resistant substrates.

A gelatinous layer—the _capsule_—is often formed outside a bacterial cell wall. The capsule is apparently secreted through the wall by the bacterial protoplast (Figure 11–8). Like the wall, the capsule is composed of polysaccharides, but it is not as firmly bound to the bacterium and can be removed by vigorous washing.

Cytoplasm

As in all cells, the bacterial cytoplasm is bounded by a plasma membrane that has many enzymes localized on its inner surface. Included in the cytoplasm are a number of ribosomes and granular inclusions, as well

as one or two areas where DNA is concentrated. Bacterial cells usually contain at least two such areas of DNA concentration, because cell division lags behind the division of the genetic material. Each of the bodies consists of a closed loop of double-stranded DNA about 700 to 1000 times as long as the bacterial cell in which it occurs.

Flagella and Pili

Certain kinds of bacteria have very slender, rigid, helical flagella, each several times longer than the cell to which it is attached. The flagella, which beat very rapidly with a rotary movement, enable the bacteria to swim constantly from place to place. The bacteria move rapidly for their size—up to 20 diameters of their own body or even more per second. Rotary motion is virtually unknown among eukaryotes, and the constant rotation of bacterial flagella is unique.

Bacterial flagella resemble the 9-plus-2 flagella found in eukaryotes only in their general shape. They are long (3 to 12 micrometers), slender, and wavy. Because they are only 10 to 20 nanometers in diameter, they are usually too fine to be seen by ordinary microscopic

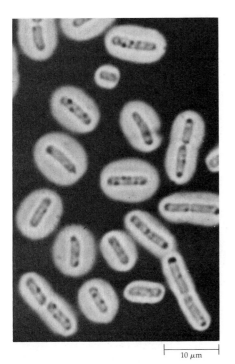

11–8
A photomicrograph of individuals of
Bacillus megaterium *dispersed in India*
ink. The capsules stand out as translucent
halos.

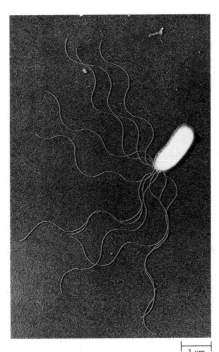

11–9
Flagella on a cell of Pseudomonas
marginalis, *a common bacterium that is*
widespread in soils. It causes the soft-rot
disease found in many fleshy and leafy
vegetables.

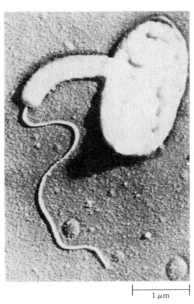

11–10
An individual of the bacterium
Bdellovibrio bacteriovorus, *which*
parasitizes other bacteria. These bacteria,
which are abundant in soil and sewage,
move by means of a single posterior
flagellum. The parasitic cell shown here
is attacking a single rod-shaped cell of
Erwinia amylovora (right), *the bacterium*
that causes fire blight of pears and
apples.

techniques. In some bacteria they are well distributed over the cell, whereas in others they are restricted to one or both ends of the cell (Figures 11–9 and 11–10). Each bacterial flagellum consists of a single rigid molecule of the protein flagellin, which extends out through a sleeve in the cell wall from the complex rotating mechanism within. In this mechanism there is a rotating ring, turned by electric forces, that is held within another, larger, stationary ring.

Pili are shorter (up to several micrometers) and straighter than flagella and are only about 7.5 to 10 nanometers in diameter (Figure 11–11). Pili occur mainly in gram-negative bacteria. Like flagella, pili consist entirely of protein, although their proteins are different from those of the flagella and their structure is distinct. Special hollow pili are formed on bacterial cells during conjugation; their exact function is not known. They may serve as a bridge for the transfer of DNA; they may pull the conjugating cells together; or they may have some other, unknown function. Ordinary pili evidently help bacteria find and attach themselves to appropriate surface membranes. A better understanding of the function of bacterial pili may aid in combating bacterial diseases.

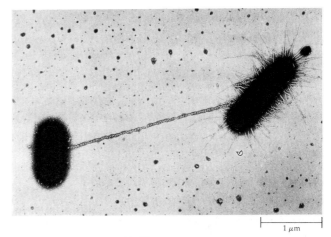

11–11
Pili in Escherichia coli. *A hollow*
conjugation pilus connects the two cells,
and many ordinary pili can be seen on
the bacterial cell at the right.

Special Features of Cyanobacteria

Members of the photosynthetic group of bacteria known as cyanobacteria often form filaments and may grow in large masses up to 1 meter or more in length. Some are unicellular, a few form branched filaments, and a very few form plates or irregular colonies (Figures 10–4b, c, 11–7, 11–12, and 11–13). Any cell of a cyanobacterium—except for the heterocysts—may divide, and the resulting subunits separate to form new colonies. As in other filamentous or colonial bacteria, the cells are usually attached to one another only by their walls or by gelatinous sheaths, so that each cell actually leads an independent life; small plasmodesmata have, however, been found in some cyanobacteria.

The cells of cyanobacteria that occur in freshwater or marine habitats—especially those that inhabit the surface layers of the water where the microscopic community of organisms known as the _plankton_ lives —commonly contain bright, irregularly shaped structures called gas vacuoles. These vacuoles provide and regulate the buoyancy of the organisms, thus allowing them to float at certain levels in the water. When numerous cyanobacteria become unable to regulate their gas vacuoles properly, for example because of extreme fluctuations of temperature or oxygen supply, they may float to the surface of the water and form visible masses called "blooms." Some of the cyanobacteria that form blooms secrete chemical substances that are toxic to other organisms; these may die in large numbers at such times.

Movement in Bacteria

Despite their simple structure, bacterial cells are able to respond to stimuli, moving in the direction of increasing concentrations of oxygen or food. Motile bacteria such as _Escherichia coli_ are able to move toward or away from chemical stimuli by rotating their flagella; individuals of _E. coli_ have about six to eight flagella. Bacteria have specific chemosensors for a variety of nutritious substances, such as sugars. After sensing the presence of attractive or repellent molecules, they are able to move toward or away from areas where the molecules are concentrated.

Julius Adler and his colleagues at the University of Wisconsin have calculated that _E. coli_ has at least 20 different chemosensors—which are proteins—12 for attractants and 8 for repellents. The proteins responsible for binding the sugars galactose, maltose, and ribose are located in the periplasmic space between the cell wall and the plasma membrane. Other chemosensors —also proteins—are associated with the plasma membrane. In some unknown way, the information that reaches these chemosensors affects the movement of the flagella.

Individuals of _E. coli_ move rapidly in gently curved lines, or "runs," each lasting approximately a second. A

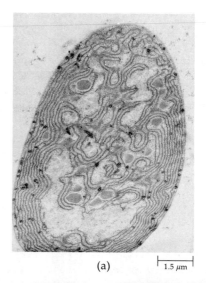

(a)   1.5 μm

(b)

11–12

(a) _A cell of the cyanobacterium_ Anabaena azollae, _showing the major features visible with the electron microscope. The gelatinous sheath of this cell has been destroyed in preparing the specimen for electron microscopy. This organism is associated with the common floating water fern_ Azolla _and is responsible for nitrogen fixation in it (see Figure 10–8a and Chapter 26). (b) A worker collecting_ Azolla _from a pond in China for cultivation and reintroduction into the rice fields._

series of runs is interrupted frequently by periods of tumbling lasting about a tenth of a second. After each tumble, the bacterium starts off in a new direction on its next run. The frequency of the tumbling controls the length of the runs and hence the overall direction of movement. If the bacterium is swimming in the "right" direction, its tumbling is inhibited; if it is swimming in the "wrong" direction, the frequency of tumbling is increased.

The end result of these movements is that the bacteria migrate toward the source of an attractant and are dispersed away from the source of a repellent. The presence of an attractant causes the flagella to rotate in a counterclockwise direction; the addition of a repellent causes a clockwise rotation. Thus sustained swimming is brought about by counterclockwise rotation of the flagella, and tumbling is brought about by clockwise rotation. In sustained swimming, the flagella work together as an organized bundle at the rear of the organism, even though they originate at sites all over the cell's surface. When the flagella rotate in a clockwise direction, the bundle falls apart and the cell tumbles. Under such circumstances, the flagella rotate independently. The responses of such bacteria to external stimuli are complex for a prokaryotic level of cellular organization and foreshadow the much more complex reactions found among the eukaryotes.

Some groups of bacteria use other means of measuring the external environment. For example, a few species of aquatic bacteria synthesize and then carry within themselves crystals of magnetite (Fe_3O_2), a mineral that is also known as lodestone. These bacteria are able to orient themselves in the earth's magnetic field with the aid of these crystals, swimming steadily in one direction. Those found in the Northern Hemisphere swim steadily north and also downward, following the magnetic field, which declines from the horizontal. In the Southern Hemisphere, similar bacteria swim in the opposite direction and also downward. Presumably this property of the bacterium assists it in seeking the food-rich bottoms of the ponds in which it lives.

A completely different form of bacterial locomotion is found in some of the filamentous cyanobacteria and other groups of bacteria that lack flagella (see Figure 11–5e). The motion of these bacteria consists of gliding and may include rotation around a longitudinal axis. Short segments that break off from a cyanobacterial colony may glide away at rates as rapid as 10 micrometers per second. This movement may be connected with the extrusion of mucilage through small pores in the cell wall, together with the production of contractile waves in one of the surface layers of the wall. Some cyanobacteria have intermittent jerky movements.

Bacterial Genetics

Cell Division

The chief mode of reproduction in the bacteria is asexual; each cell simply increases in size and splits into two cells. In this process, called *fission*, the plasma membrane and cell wall grow inward, eventually dividing the cell in two (Figure 11–13). The new wall is thicker than ordinary cell walls, and it soon splits down the middle, providing each daughter cell with a new cell wall. Chains of bacteria are formed when the new cell wall splits only partially; such chains may break into

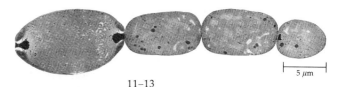

11–13

Cell division in Anabaena. *This electron micrograph shows a chain of cells held together along incompletely separated walls, as well as a single cell (at the left) undergoing cell division. The first cell on the left end of the chain is a heterocyst (see page 184). The sort of cell division shown here, in which the margins of the cell grow inward, is characteristic of all organisms except plants and a few genera of algae, in which a cell plate is formed.*

multicellular fragments, which, in cyanobacteria, are called *hormogonia*.

The single, circular, double-stranded DNA molecule that contains the hereditary information of bacteria is believed to be attached at one point to the interior surface of the plasma membrane. After this molecule replicates, there are two identical circular molecules attached side by side to the plasma membrane. New plasma membrane and cell wall material are laid down between these two points of attachment as the cell divides, and the membrane eventually pinches in between them. As a result of this process, each daughter cell is provided with an identical DNA molecule (Figure 11–14).

Some bacteria have the ability to form thick-walled *endospores*, which are resistant to heat and dehydration.

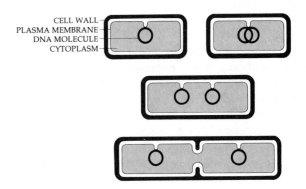

11–14

A schematic diagram of the mode of attachment of bacterial DNA to the plasma membrane, which leads to the distribution of one DNA molecule to each daughter cell. As you can readily see by comparison with Figure 11–1, the DNA is actually much longer than shown here.

Endospores are formed by the division of the protoplast of a bacterial cell into two or more portions. A very thick spore coat is formed around each portion, including the DNA. Endospores may germinate to produce new individuals after decades or even centuries. The resistant spores that are formed in cyanobacteria are called *akinetes*. Akinetes differ fundamentally from endospores in that they are enlarged vegetative cells around which a thickened envelope develops (Figure 11–15). Actinomycetes produce chains of endospores in the cells at the ends of their branches.

Genetic Recombination

Genetic recombination in bacteria results from the transfer of a portion of a DNA molecule from one bacterial cell to another. This DNA fragment may simply act in concert with the DNA molecule of the cell it enters to produce messenger RNA, or it may be incorporated into the DNA molecule (a closed circle) of the recipient cell. If such incorporation occurs, the altered DNA molecule is passed on to the daughter cells with the rest of the hereditary material. The process of recombination is the same whether the fragments are passed from cell to cell by direct contact (conjugation), are carried into a cell by a virus (transduction), or enter cells in solution as "naked" DNA (transformation).

In nearly all bacteria there are small, circular fragments of DNA—*plasmids* (see Figure 11–1)—in addition to the large circular molecule of DNA—the *bacterial chromosome*. Some plasmids can be integrated into and then replicated with the bacterial chromosome. In some bacterial strains, conjugation and the transmission of such fragments occur frequently (see Figure 11–11). This recombination is important in the spread of such characteristics as resistance to antibiotics from one kind of bacterium to another. Genetic engineering is mainly based on this process (see Chapter 30).

Genetic material can also be passed from one strain of bacteria to another strain by a process known as *transduction*. In this process, bacterial viruses, or *bacteriophages* (see Figure 12–1, page 186), may incorporate small portions of bacterial chromosomes into their own genetic material and then carry them to other bacteria. Both the viral DNA and the bacterial DNA may be incorporated into the chromosomes of the new host bacterial strain.

Another kind of genetic recombination in bacteria is known as *transformation*. The first experiments that suggested the existence of bacterial transformation involved the causative agent of one kind of pneumonia, *Diplococcus pneumoniae*. In this bacterium, there are two types of colonies: rough (R) and smooth (S). The strains that produce S colonies readily produce capsules and are virulent (disease-causing). As early as 1928, Frederick Griffith showed that it was possible to trans-

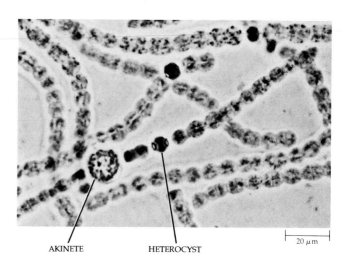

AKINETE   HETEROCYST
|———| 20 μm

11–15
The filaments of Anabaena, *which is a nitrogen-fixing cyanobacterium, are composed of barrel-shaped cells held in a gelatinous matrix. Nitrogen fixation takes place within specialized cells called heterocysts. Like* Calothrix *(see Figure 11–7b),* Anabaena *forms akinetes. Electron micrographs of* Anabaena *also can be seen in Figures 11–12 and 11–13.*

form a harmless R strain into a virulent S strain by exposing the R strain to heat-killed S-strain cells (see Figure 8–2, page 118). When bacterial cells are broken down by chemicals or by heat, fragments of DNA are released and may pass into other cells. The demonstration that DNA was the genetically active material involved in such transformations was the first direct evidence for the genetic role of DNA. Transformation is now known to occur in many different groups of bacteria.

Mutation is a far more important source of variability in bacteria than is genetic recombination (Figure 11–16). For a given gene, it has been calculated that there will be about 1 mutant cell per 10^7 (10 million) individuals. An individual of *Escherichia coli* has approximately 5000 genes. Thus, in a culture of this bacterium, there is about 1 cell that is a mutant for any gene per 2000 individuals; 0.05 percent of the individuals in the culture will have a mutant phenotype in each cell division. In a culture that approaches 10^9 cells—one that has divided about 30 times—the frequency of mutants will be about 30×0.05 percent, or as high as 1.5 percent. Bacteria multiply very rapidly. An *E. coli* population may, under optimal conditions, double every 12.5 minutes; thus the number of mutant individuals produced is very high. The rapid generation time of bacteria combined with mutation is responsible for their extraordinary adaptability.

Bacterial Metabolism

Heterotrophs

Most bacteria are heterotrophs—organisms that cannot make organic compounds from simple inorganic substances but must obtain them from organic compounds found in other organisms. The largest group of heterotrophic bacteria are the *saprobes*. Saprobes are organisms that obtain their nourishment from dead organic matter. Saprobic bacteria and fungi are responsible for the decay and recycling of organic material in the soil; many of the characteristic odors associated with soil come from substances produced by heterotrophic bacteria.

Photosynthetic Bacteria

There are at least five groups of photosynthetic bacteria: (1) cyanobacteria, (2) green sulfur bacteria, (3) purple sulfur bacteria (see Figure 7–2, page 96), (4) purple nonsulfur bacteria, and (5) *Prochloron*. Like plants, photosynthetic bacteria contain chlorophyll. The cyanobacteria and *Prochloron* contain chlorophyll *a*, as do all photosynthetic eukaryotes. The chlorophylls that are present in the other groups of photosynthetic bacteria differ in several ways from chlorophyll *a*, but they all have the same basic structure (see Figure 7–8, page 99).

The colors characteristic of these groups of bacteria are associated with the presence of several different accessory pigments that function in photosynthesis. In the two groups of purple bacteria, these pigments are yellow and red carotenoids; in the cyanobacteria, in addition to carotenoids, there is always a blue pigment, phycocyanin, and generally a red one, phycoerythrin.

Photosynthesis in cyanobacteria is identical to that found in photosynthetic eukaryotes. The cells of cyanobacteria contain numerous layers of membranes, often parallel to one another, and masses of ribosomes. The membranes are photosynthetic thylakoids that resemble those found in chloroplasts. In cyanobacteria, however, they are not organized into a chloroplast. The main carbohydrate storage product of the cyanobacteria, like that of all other bacteria, is glycogen.

In the green sulfur and purple sulfur bacteria, sulfur compounds play the same role in photosynthesis that water does in organisms that contain chlorophyll *a*. That is,

$$CO_2 + H_2S \xrightarrow{Light} (CH_2O) + H_2O + 2S$$

As discussed in Chapter 7, an understanding of the course of photosynthesis in the purple sulfur bacteria was the key that led C. B. van Niel to propose the generalized equation for photosynthesis,

$$CO_2 + 2H_2A \xrightarrow{Light} (CH_2O) + H_2O + 2A$$

in which H_2A is a generalized hydrogen donor.

In the photosynthetic purple nonsulfur bacteria, other compounds, including alcohols, fatty acids, and keto acids, serve as electron (hydrogen) donors for the photosynthetic reaction.

Because of their requirement for hydrogen sulfide or a similar substrate, the photosynthetic sulfur bacteria are able to grow only in habitats that contain large amounts of decaying organic material, which is distinguishable to us by its sulfurous odor. In these bacteria, elemental sulfur may accumulate as deposits within the cell.

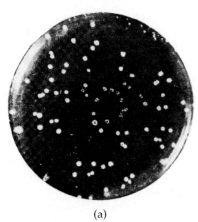 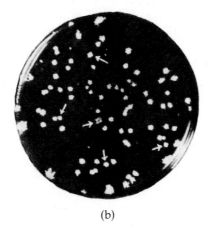 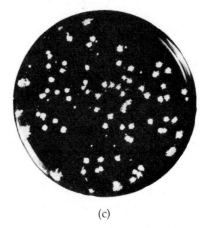

(a) (b) (c)

11–16
Replica plating as a means for detecting the presence of mutations in bacteria. The initial plate is shown in (a). Colonies were transferred from (a) to (b) and (c) by means of a sterile velveteen disk that was pressed on plate (a) and then pressed on (b) and (c), thus placing inoculum in the same relative positions. (a) and (b) show growth of the colonies on a complete medium, whereas a number of growth factors are lacking in (c). Under these conditions, the colonies indicated by arrows in (b) did not grow on plate (c); they are mutants that were already present but undetected in (a).

Halobacterium halobium, one of the purple nonsulfur bacteria, is an extraordinary organism. It is common in sunny places in concentrated brine, as around salt pools. These bacteria have patches of a purple pigment in their plasma membrane that is remarkably similar to rhodopsin, the visual pigment of the human eye. These patches are in effect light-driven pumps that enable the bacterium to pump out protons, much as in the oxidative phosphorylation that occurs in mitochondria. (Of the free-living bacteria, the purple nonsulfur bacteria are the most similar to mitochondria, organelles that may ultimately have been derived from this group of organisms.) This process seems to lead directly to ATP production, utilizing the enzyme ATPase, which is located in the nonpurple areas of the plasma membrane. *Halobacterium* is thus the only living unit other than chlorophyll-containing systems that is able to convert sunlight into chemical energy through the process of photosynthesis. It is motile and responds phototactically to violet light, suggesting that its pigment detects light (as does rhodopsin).

The fifth group of photosynthetic bacteria consists of the single genus *Prochloron,* which was discovered in the early 1970s by Ralph A. Lewin of the Scripps Institution of Oceanography. *Prochloron* contains chlorophylls *a* and *b* and carotenoids, the photosynthetic pigments of the green algae and plants (Figures 11–2 and 11–17). As far as is known, it lives only in association with colonial marine animals of the groups known as ascidians (sea squirts) along seashores in the tropics and subtropics. Biochemically, *Prochloron* has the set of pigments that would be expected in the group of bacteria that gave rise to the chloroplasts of the green algae by symbiosis (this relationship will be discussed further in Chapter 15). On the other hand, *Prochloron* resembles the cyanobacteria more closely than the chloroplasts of green algae and plants in its ribosomal RNA (rRNA) sequences. The relationships and possible evolutionary role of *Prochloron* are under active investigation.

Chemoautotrophic Bacteria

Chemoautotrophic bacteria obtain the energy they use to drive their synthetic reactions from the oxidation of inorganic molecules such as nitrogen, sulfur (see Figure 11–7d), and iron compounds, or from the oxidation of gaseous hydrogen. The reduced inorganic compounds that they oxidize provide a source of energy for them, just as light does in photosynthetic organisms. Their carbon source, however, is the same: carbon dioxide. The bacteria that live around deep-sea vents at temperatures as high as 360°C are chemosynthetic. They obtain their energy by converting hydrogen sulfide to elemental sulfur and by doing so provide the energy that supports the whole community of organisms that exists in complete darkness around the vents.

Archaebacteria

One group of chemoautotrophic bacteria that has attracted a great deal of attention in recent years is the methanogenic (methane-producing) bacteria. These bacteria are strictly anaerobic; they are common in the digestive tracts of cattle and other ruminants, in sewage treatment plants, bogs, and deep in the sea. Most of our reserves of natural gas were produced by the metabolic activities of methanogenic bacteria in the past. Some of these bacteria produce methane from carbon dioxide and hydrogen, obtaining their energy from the process; others are able to reduce elemental sulfur to form hydrogen sulfide.

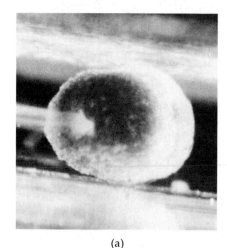

(a)

(b)

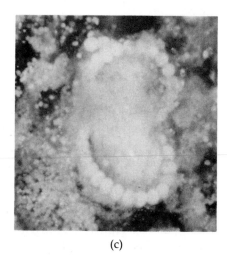

(c)

11–17
Prochloron is a very unusual photosynthetic bacterium that occurs only symbiotically in members of the group of marine animals known as sea squirts. (a)

Prochloron *in the tadpolelike larval stage of the sea squirt* Didemnum, *illustrating the means by which the bacterium is spread.* (b) *An adult individual*

of Didemnum. (c) Didemnum *broken open, showing a solid mass of* Prochloron *within.*

Methanogenic bacteria are diverse morphologically, but recently it has been shown—principally by Carl Woese of the University of Illinois and his colleagues—that the sequences of bases in a portion of their rRNA resemble one another closely. This is accepted as evidence that these bacteria are closely related by descent. What is surprising, however, is that these sequences differ sharply from those in the corresponding rRNA molecule in the other bacteria and also from those found in eukaryotes. In these sequences, for example, the cyanobacteria resemble *Escherichia coli* or the actinomycetes much more closely than any of them resemble the methanogenic bacteria. The cell walls of methanogenic bacteria do not contain muramic acid, as do those of all other bacteria, and their metabolism appears to be fundamentally different. In 1984, it was reported that one of the archaebacteria, *Methanosarcina*, was capable of fixing atmospheric nitrogen, an interesting discovery in view of the limited number of genera that possess this ability.

In view of this evidence, it has been suggested that the methanogenic bacteria might have originated more than 3 billion years ago, when there was an anaerobic atmosphere rich in CO_2 and H_2, and that they have persisted in certain favorable habitats. Their distinctiveness from other bacteria is so great that it has been proposed that the methanogenic bacteria should be considered as a separate kingdom—the Archaebacteria.

Halobacterium, just discussed, together with a series of other bacteria that occur at extremely high temperatures in acidic habitats, share a number of features with the methane-producing bacteria. For these reasons, all of these organisms are grouped together in the proposed kingdom Archaebacteria.

Bacterial Ecology

Soil Bacteria

Different groups of microorganisms are involved in specific stages of the processes of decomposition and recycling that occur in the soil, and such natural communities tend to be organized in a complex fashion (Figure 11–18). In one case, a strain of *Streptococcus* that was not itself resistant to penicillin was found to be protected by a penicillin-resistant strain of the bacterial genus *Bacteroides* with which it was growing: the *Bacteroides* individuals broke down the penicillin molecules so efficiently that they were unable to control the *Streptococcus* infection. Many bacteria and fungi break down carbon-containing compounds, releasing CO_2 into the atmosphere. The most important organic compounds originating from plants in nature are cellulose and lignin, and secondary ones are pectic substances, starch, and sugars. It has been estimated that more than 90 percent of the CO_2 production in the biosphere results from the activity of bacteria and fungi.

Some microorganisms break down proteins into peptides, which are subsequently broken into their constituent amino acids. Many microorganisms use a process called ammonification to break down amino acids, with the consequent release of ammonium ions (NH^+; see Chapter 26). Ammonia can be oxidized to nitrite ions (NO_2^{4-}) by the chemoautotrophic bacterium *Nitrosomonas*, and the nitrites are oxidized to nitrates (NO^{3-}) by the bacterium *Nitrobacter*. The conversion of ammonia to nitrites and nitrates constitutes the process of <u>nitrification</u>. This process releases energy, which is used by these chemoautotrophs to reduce carbon dioxide to carbohydrate. Several other kinds of bacteria are capable of reversing the process and changing the nitrates back into nitrites and, ultimately, ammonia.

Denitrification, the conversion of nitrates into nitrogen gas or nitrous oxide, results in the loss of nitrogen from the soil. The reverse of this process, which is extremely important biologically, is <u>nitrogen fixation</u>. Of all living organisms, only a few genera of bacteria are capable of nitrogen fixation. Outstanding among them is the symbiotic bacterium *Rhizobium* (see Chapter 26),

11–18
The complexity of interactions that occur among soil organisms is suggested by this photograph, which shows bacteria growing on an agar plate to which penicillin has been added. Down the center of the plate is a dense colony of Staphylococcus epidermides, *a strain of bacteria that is resistant to penicillin because it produces an enzyme called penicillinase, which breaks down the antibiotic. On either side of the* Staphylococcus *colony can be seen tiny colonies of* Neisseria gonorrhoeae, *the causative agent of gonorrhea.* Neisseria *is susceptible to penicillin but is able to grow in the part of the medium where the resistant* Staphylococcus *has broken down the antibiotic.*

BACTERIA AND FOSSIL FUEL

It has conventionally been thought that deposits of petroleum, coal, and natural gas originated primarily from marine plankton and masses of terrestrial plants growing in swamps and other similar places in warm, mild conditions. Now it appears that the bacteria that decomposed the remains of these organisms were a dominant force in the creation of the deposits. Workers in France have found biochemical evidence of the activities of these bacteria in petroleum, and more recently the bacteria themselves have been found with the aid of electron microscopy.

At the Australian National University of Canberra, geologist Miryam Glikson concentrated the organic material present in petroleum-rich shales about 120 million years old, dried it, embedded it in resin, and sectioned it for electron microscopy. By doing so, she found several kinds of bacteria that resemble living forms found in the sea, where they degrade algae and other organic material; other remains resemble cyanobacteria. Dr. Glikson's preliminary results suggest that the activities of bacteria may have been less important in forming petroleum deposits in freshwater sites than in marine sites.

 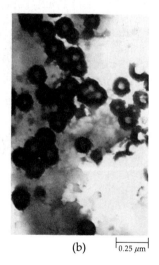

(a) $\overline{\quad 0.25\ \mu m \quad}$ (b) $\overline{\quad 0.25\ \mu m \quad}$

These bacteria are from the Toolebuc oil shales, Queensland, Australia, and illustrate Miryam Glikson's findings. These oil shales are from the Cretaceous period and are about 120 million years old. (a) Filamentous bacteria, probably cyanobacteria, in which partitions can sometimes be observed. Such bacterial remains, together with the abundance of calcareous remains in whole rock fragments in the oil shales have led Dr. Glikson to conclude that cyanobacteria were the major contributors to the organic matter that is present in them. (b) Doughnut-shaped bacteria resembling Flectobacillus marinus. *Such bacteria degrade planktonic protists and are occasionally preserved in the oil deposits that result, in part, from their activities.*

which forms nodules on the roots of legumes and a few other plants. The actinomycetes are involved in nodule formation in many woody plants, such as alders (*Alnus*), wax myrtles (*Myrica*), and mountain lilacs (*Ceanothus*). Actinomycetes, which are efficient nitrogen fixers, have a great influence on the accumulation of nitrogen in soils (Figure 11–19). In addition, certain nitrogen-fixing bacteria are regularly associated with the roots and even the leaves of plants, where they utilize the carbohydrate-rich exudates from the plants and in turn provide nitrogen in a usable form. Many free-living bacteria, including some cyanobacteria, also play an important role in nitrogen fixation.

Sulfur is made available to plants (which cannot utilize elemental sulfur) by chemoautotrophic bacteria such as *Thiobacillus*, which oxidize elemental sulfur to sulfates:

$$2S + 2H_2O + 3O_2 \longrightarrow 4H^+\ 2SO_4^{2-}$$

Sulfates are accumulated by plants, and the sulfur contained in them is incorporated into the structure of proteins. The degradation of proteins (discussed as one aspect of the nitrogen cycle in Chapter 26), liberates amino acids, some of which contain sulfur (Figure 11–20). A number of bacteria are capable of breaking down these amino acids, thus releasing hydrogen sulfide (H_2S). Sulfates are also reduced to H_2S by certain soil microorganisms, such as *Desulfovibrio*.

Parasitic and Symbiotic Bacteria

Some heterotrophic bacteria break down organic material while it is still incorporated in the bodies of living organisms. The disease-causing (pathogenic) bacteria belong to this group, as do a number of other nonpathogenic forms. Some of these bacteria have little effect on their hosts, and others are beneficial. For example, if all the bacterial inhabitants of the human intestinal tract are eliminated—as can happen through prolonged antibiotic therapy—the tissues become much more vulnerable to disease-causing bacteria and fungi. In cattle and other ruminants, bacteria digest the cellulose that the animals ingest and make the sugars they derive from it available to their hosts, which would otherwise be unable to utilize grass and leaves as a source of food. Another critically important bacterial interaction is the mutually beneficial one that exists between the symbiotic bacteria of the genus *Rhizobium* and the legumes (discussed in Chapter 26).

In general, microbes and other parasites cause disease when they interfere with one or more essential physiological processes of the organisms they invade. When these effects are severe, the disease is considered serious. If death of the host organism results prematurely, the relationship will not be as favorable as possible for the parasite, which will then have to find another host individual. If it fails to do so, it will die itself.

(a)

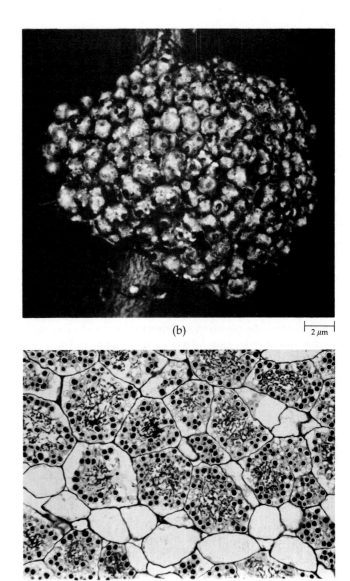

(b)

(c)

11-19

Alders form symbiotic relationships with actinomycetes of the genus Frankia, *which fix nitrogen. Three aspects of the relationship are shown here, all involving red alder,* Alnus rubra, *as studied by John Torrey and his colleagues at the Harvard Forest in Massachusetts. (a) Alder seedlings grown in different soil mixtures after inoculation with* Frankia. *The seedlings at the left were not inoculated and had no nodules; those at the right were inoculated and formed abundant nodules. (b) Nodule on alder root. (c) Micrograph of a section of a root nodule lobe, stained with toluidine blue, showing cortical cells filled with* Frankia. *The hyphae fill the center of the infected cells and the swollen terminal ends of the filaments, called vesicles, fill the periphery. The vesicles are the probable site of the N_2-fixing enzyme, nitrogenase.*

11-20

The sulfur cycle. Without the activities of bacteria and fungi, carbon, nitrogen, sulfur, and other elements would remain locked in the molecules into which they are incorporated. Soil-dwelling heterotrophs break down complex organic molecules into substances that can enter other biological cycles. Without the recycling of organic substances, all organisms would soon be overwhelmed by the products of their own metabolism. The release of large quantities of sulfur into the atmosphere by burning fossil fuels is placing a severe strain on the natural mechanisms of the sulfur cycle.

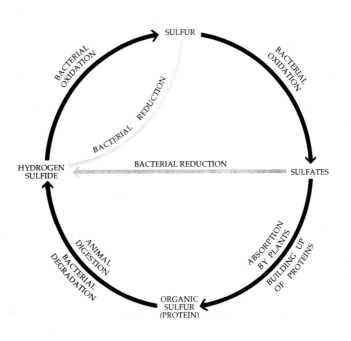

Human Diseases

Some human diseases are caused by airborne bacteria. Among the better known are bacterial pneumonia, caused by *Diplococcus pneumoniae,* whooping cough, caused by *Bordetella pertussis,* and diphtheria, caused by *Corynebacterium diphtheriae.* The latter organism produces a powerful toxic substance that circulates rapidly throughout the body and causes serious damage to the heart muscles, nervous tissue, and kidneys. Diphtheria is now rare because most children are immunized against it in infancy. Other serious diseases are caused by airborne bacteria of the genus *Streptococcus,* which cause scarlet fever, rheumatic fever, and other infections. Tuberculosis, caused by *Mycobacterium tuberculosis,* is still a leading cause of death in humans, despite improved methods of detection. The number of reported cases in the United States decreased, however, to about 30,000 per year by the 1970s and to just over 20,000 per year by the mid-1980s.

A number of other diseases of bacterial origin are spread in food or water. Examples are typhoid fever and paratyphoid (caused by bacteria of the genus *Salmonella*) and bacillary dysentery (caused by *Shigella dysenteriae*). Others, like typhus, which is caused by *Rickettsia prowazekii,* are spread by insect vectors. Undulant fever, caused by bacteria of the genus *Brucella,* affects both cattle and humans and is usually contracted by humans through eating or drinking milk or milk products from an infected cow. Pasteurization of milk destroys *Brucella,* and the disease has become rare in those portions of the world where this process is employed. The most common form of travelers' diarrhea is caused by certain strains of the common colon bacterium *Escherichia coli.*

Legionnaire's disease, spread in water, was first detected in 1976 when a mysterious lung ailment killed 34 members of the American Legion attending a conference in Philadelphia. It is now known to be caused by small, flagellated, rod-shaped bacteria that have been assigned to the genus *Legionella.* These bacteria, common in warm water, enter and multiply rapidly in the monocytes, a kind of white blood cell that normally plays a major defensive role against most microorganisms. It is now estimated that Legionnaire's disease affects some 50,000 people in the United States annually, causing a severe form of pneumonia that proves fatal to some 15 to 20 percent of its victims if it is not treated. The species of *Legionella* are gram-negative; the disease they cause can be cured with erythromycin.

Bacteria play an important role in the spoilage of food and other stored organic products, and some of these organisms are very pathogenic to humans. Food poisoning by *Clostridium botulinum* is rare but extremely dangerous (Figure 11–21). The endospores of this species can survive a short period of boiling and may then give rise to flourishing colonies within sealed cans or jars. *Staphylococcus* food poisoning is fairly common but, fortunately, much less serious. In recent years, a number of virulent strains of *Staphylococcus* have caused serious infections. Many of these strains are resistant to penicillin, and some produce the enzyme penicillinase, which breaks down the antibiotic. These resistant bacteria frequently come into contact with penicillin-producing fungi and often grow with them in nature, so this resistance confers a natural advantage (see Figure 11–18).

Plant Diseases

Many economically important diseases of plants are associated with bacteria, and these diseases contribute substantially to the roughly one-eighth of crops worldwide that are lost to disease. Almost all kinds of plants can be affected by bacterial diseases, which can be extremely destructive (Figures 11–3 and 11–22).

Virtually all plant-pathogenic bacteria are bacilli, and almost all occur primarily in their host plants as parasites. The symptoms caused by plant-pathogenic bacteria are quite varied, with the most common appearing as spots of various sizes on stems, leaves, flowers, and

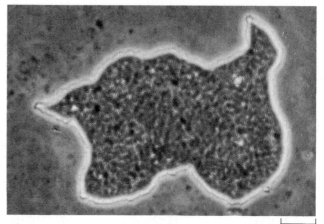

11–21
A colony of Clostridium botulinum. *The type of food poisoning (botulism) suffered by consumers of contaminated canned foods is associated with the survival of the spores of this bacterium. They are extremely resistant to heat and occasionally persist in food that has not been treated properly.* Clostridium botulinum, *which grows only in the absence of oxygen, produces its powerful toxin inside the can or jar. The toxin can be destroyed by boiling for about 15 minutes. It is the most powerful toxin known; 1 gram would be enough to kill 14 million people. Food botulism is rare in the United States; between 1980 and 1983, there were 114 reported cases.*

fruits (Figure 11–23). Almost all such bacterial spots are caused by two closely related genera, *Pseudomonas* and *Xanthomonas*.

Some of the most destructive diseases of plants—such as blights, soft rots, and wilts—are also caused by bacteria. Blights are characterized by rapidly developing necroses (dead, discolored areas) on stems, leaves, and flowers. Fire blight in apples and pears may result in the death of young trees within a single season (see (Figure 11–3). The pathogen of fire blight is the bacillus *Erwinia amylovora*. Bacterial soft rots occur most commonly in the fleshy storage organs of vegetables, such as potatoes and onions, or in fleshy fruits, such as tomatoes and eggplants. The most destructive soft rots are caused by bacteria of the genus *Erwinia*.

Bacterial vascular wilts affect only herbaceous plants. The bacteria invade the vessels of the xylem, where they move about and multiply. As a consequence, the bacteria interfere with the movement of water and inorganic nutrients, resulting in the wilting and death of the plant. The bacteria commonly destroy portions of the vessel walls and can even cause the vessels to rup-

11–22
This population of giant saguaro cactus (Carnegiea gigantea) *at an elevation of 1000 meters on the south slope of the Santa Catalina Mountains near Tucson, Arizona, was attacked by the bacterium* Erwinia carnegieana *after being weakened by severe freezing conditions in January 1962.*

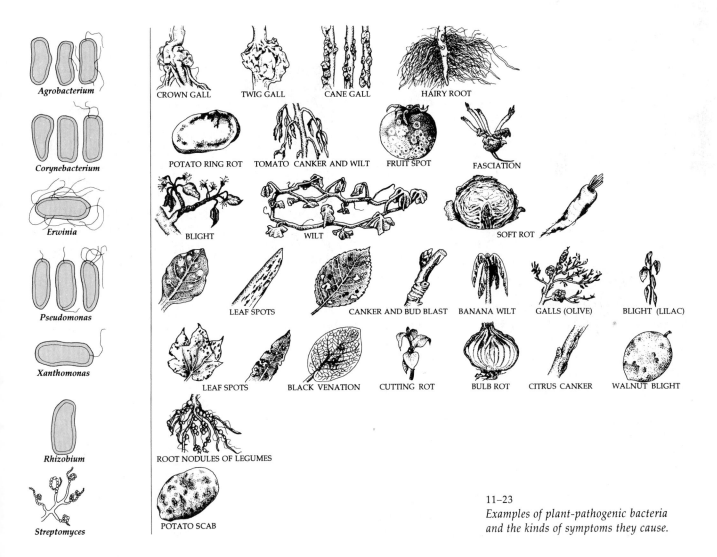

11–23
Examples of plant-pathogenic bacteria and the kinds of symptoms they cause.

CITRUS CANKER

One of the most dramatic outbreaks of a bacterial disease of plants in recent years began in August 1984, when diseased, mottled yellow and brown citrus seedlings were discovered at a nursery in Avon Park, Florida. The seedlings were diagnosed as having citrus canker, a disease caused by one of the more than 100 disease-causing varieties of the bacterium *Xanthomonas campestris*. Other varieties of this species are responsible for diseases in peaches, beans, cabbages, and other plants, but the citrus-infecting strain presents special problems to growers. Within four months of the initial discovery, over 7 million seedlings of citrus were destroyed in Florida in an effort to control the disease, and major funds were committed to research on the disease for the first time in the United States.

The symptoms of citrus canker, lesions on the stems, leaves, and fruits of citrus trees, spread rapidly through an infected grove, and the bacteria are easily spread on crates, clothing, transported plants, and in many other ways. There is considerable variation even in the citrus-infecting strains of the bacterium, which differ considerably in their symptoms, the particular kinds of citrus they infect, and the places they occur.

At the time of the 1984 outbreak, citrus canker had not been reported for some 40 years in Florida. Why did the new outbreak occur? Unfortunately, none of the bacteria that caused the disease through the 1930s had been preserved; it would have been important to compare their characteristics with those of the new strain. The new strain may have come from abroad: almost 5800 lots of canker-infested fruits were intercepted at U.S. ports of entry between 1971 and 1983, thus illustrating the great economic importance of careful agricultural inspection. Only the development of resistant citrus varieties, like those grown in Japan, will restore the Florida citrus industry, which amounted to $2.5 billion in 1983. About 58 percent of the orange juice that Americans drank in the early 1980s came from Florida, with another 35 percent from Brazil.

(a)

(b)

Citrus canker spread rapidly in Florida, starting in late 1984. (a) Cankers on a diseased orange and orange leaves. (b) Lesions on the stem of an orange tree.

ture. They then spread to the adjacent parenchyma tissues, where they continue to multiply. In some bacterial wilts, the bacteria ooze to the surface of the stems or leaves through cracks formed over cavities filled with cellular debris, gums, and bacteria. More commonly, however, the bacteria do not reach the surface of the plant until the plant has been killed by the disease. Among the most important examples are the bacterial wilt of alfalfa and bean plants (each caused by different species of *Corynebacterium*), the bacterial wilt of cucurbits, such as squashes and watermelons (caused by *Erwinia tracheiphila*), and the black rot of crucifers, such as cabbage (caused by *Xanthomonas campestris*).

The members of another important group of bacteria, the genus *Agrobacterium*, cause crown gall in plants. They are of great importance in genetic engineering and will be discussed in Chapter 30.

Mycoplasmas

Mycoplasmas—a group of bacteria that are the smallest known cells—can be as small as 0.1 micrometer in diameter. Each mycoplasma cell consists of a plasma membrane, DNA, RNA, ribosomes, soluble proteins, sugars, and lipids. Mycoplasmas lack nuclei and membrane-bound cellular organelles and do not have cell walls. An individual mycoplasma probably contains fewer than 650 genes, which is about one-fifth of the number found in a single common bacterium.

The mycoplasmas isolated thus far have been classified into six genera, with about 50 of the 60 recognized species assigned to the genus *Mycoplasma*. By sequencing a portion of the ribosomal RNAs in these organisms, C. R. Woese and his colleagues determined that all but one of the genera are related to one another and that they are probably derived from a single bacterial line that includes *Bacillus* and *Lactobacillus*. The remaining genus, *Thermoplasma*, apparently acquired the characteristics of mycoplasmas independently. All the mycoplasmas, however, seem to be bacteria that have been simplified in structure during their evolution.

In addition to the mycoplasmas identified in animal tissue, mycoplasmalike organisms (MLOs) have also been found in more than 200 plant species and have been implicated in, or associated with, more than 50 plant diseases, many with symptoms of yellowing and stunting. Most attempts to culture MLOs obtained from plant tissue have been unsuccessful.

Some spiroplasmas—long, thin, helical mycoplasmas that are motile in liquid and may be up to 10 micrometers in length and less than 0.2 micrometers in diameter—have been cultured on artificial media. These include *Spiroplasma citri*, the agent of stubborn and little-leaf diseases of citrus, and other plant pathogens as well (Figure 11–24). Spiroplasmas exhibit vigorous whirling and flexing movements in culture. Proof of pathogenicity—that the organisms isolated from diseased tissues actually caused the disease—has been obtained only with a few spiroplasma isolates.

MLOs are generally confined to the sieve tubes of the phloem. The organisms are believed to move passively from one sieve-tube member to another through the sieve-plate pores along with the solution of sugar being transported by the phloem (Figure 11–25). Motile spiroplasmas, however, may also be able to spread actively throughout the tissue.

Cyanobacteria

The cyanobacteria ("blue-green algae") deserve special mention because of their varied ecological roles, and in view of the fact that they have been regarded loosely as "plants" more often and more recently than other groups of bacteria. Although over 7500 names have been proposed for species of cyanobacteria, experimental studies suggest that there may be as few as 200 distinct nonsymbiotic species. These species often change remarkably in appearance in the different kinds of circumstances under which they can live. For example, *Microcoleus vaginatus* occurs in wet soil or fresh or brackish water from northern Greenland to Antarctica, and from the floor of Death Valley to the top of Pike's Peak. Studies of this organism have revealed that under changing environmental conditions, a single colony may undergo such extreme changes that its members come to resemble dozens of different species.

Like other bacteria, cyanobacteria sometimes grow in extremely inhospitable environments, from the water of hot springs to the frigid lakes of Antarctica, where they have recently been reported to form luxuriant mats, 2 to 4 centimeters thick, in water beneath more than 5 meters of permanent ice. The greenish color of

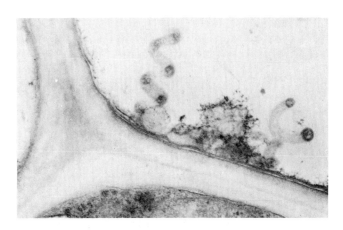

11–24

Spiroplasmas isolated from stunted corn (Zea mays) plants.

some polar bears in zoos has been shown to be due to colonies of cyanobacteria that develop within the hollow hairs of their fur. Cyanobacteria were the first colonists on the new island of Surtsey, near Iceland, following its eruption and appearance above the sea. On the other hand, they are absent in acidic waters, where eukaryotic algae are often abundant.

Layered chalk deposits called stromatolites (Figure 11–26), which have a continuous geologic record of 2.7 billion years, are produced when colonies of cyanobacteria bind calcium-rich sediments. Today, stromatolites are formed only in a few places, such as shallow pools in hot, dry climates. Their abundance in the fossil record is evidence of the frequency of such conditions in the past, when cyanobacteria played the decisive role in elevating the level of free oxygen in the atmosphere of the early earth.

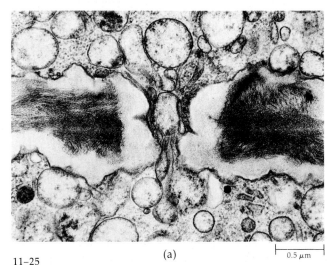

11–25

(a) *Mycoplasmalike organisms apparently traversing a sieve-plate pore in a young inflorescence of a coconut palm (Cocos nucifera) affected by lethal yellowing*

(a)

0.5 μm

disease. (b) *A devastated grove of coconut palms—now looking like telephone poles—in Jamaica. Lethal yellowing has*

(b)

been responsible for the death of palms belonging to many genera in southern Florida and elsewhere.

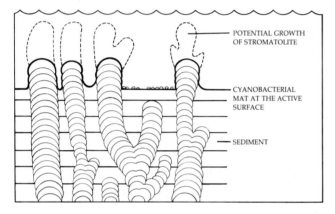

11–26
*Stromatolites are produced when
flourishing colonies of cyanobacteria bind
calcium carbonate into domed structures,
like the ones shown in the diagram and
photograph, or others with a more
intricate form. They are abundant in the
fossil record but today are being formed
only in a few, highly suitable environments such as that shown here on the
tidal flats of Shark Bay in Western
Australia.*

Many cyanobacteria have a mucilaginous sheath or coating, which is often deeply pigmented, particularly in species that sometimes occur in terrestrial habitats. The colors of the sheaths in different species include light gold, yellow, brown, red, emerald green, blue, violet, and blue-black. In addition, the carotenoids and phycobilins that are present in the cells of cyanobacteria modify their color. Despite their name, only about half of the species of the "blue-green algae" are actually blue-green in color. Indeed, the Red Sea apparently was given its name because of the dense concentrations, or "blooms," of the marine planktonic species of *Trichodesmium*—a cyanobacterium that contains vacuoles—that frequently occur in that body of water.

Many marine cyanobacteria occur in limestone or lime-rich substrates such as coralline algae (see page 247) and the shells of mollusks. Some freshwater species, particularly those that occur in hot springs, often deposit thick layers of lime in their colonies. In Yellowstone National Park, the filamentous *Mastigocladus* occurs in hot water at temperatures up to 55°C, and the unicellular *Synechococcus* occurs in temperatures up to 73° to 75°C. Cyanobacteria are abundant in soils, including desert soils, with 20,000 to 50,000 individuals per gram being a representative figure.

Many genera of cyanobacteria can fix nitrogen. In the warmer parts of Asia, rice can often be grown continuously on the same land without the addition of fertilizers because of the presence of nitrogen-fixing cyanobacteria in the rice paddies. Here the cyanobacteria, especially members of the genus *Anabaena*, often occur in association with the floating water fern *Azolla*, which has a nearly obligate relationship with them (see page 535). Because of their nitrogen-fixing capacities, cyanobacteria—as well as lichens, liverworts, and other organisms with which they may be symbiotic—are able to colonize bare areas of rock and soil. In the sea, species of *Trichodesmium* fix about a quarter of the total nitrogen fixed in the sea, an enormous amount.

Nitrogen fixation in cyanobacteria often occurs within *heterocysts*, enlarged cells that are formed in the filamentous species (see Figures 11–13 and 11–15). The walls of a heterocyst are identical to that of any other cyanobacterial cell, but outside the wall, the heterocysts form a bilayered envelope. The outer portion of this envelope consists of a polysaccharide, and the inner layer, of glycolipids. Within the heterocysts, the membranes of the cyanobacterium are reorganized into a concentric or reticulate pattern. Since the heterocysts lack photosystem II (Chapter 7), the cyclic phosphorylation that occurs in them does not result in the evolution of oxygen; the oxygen that is present is rapidly reduced by hydrogen, a by-product of nitrogen fixation, or expelled through the wall of the heterocyst. Nitrogenase is sensitive to the presence of oxygen, and nitrogen fixation is an anaerobic process. The formation of heterocysts in *Nostoc* and other genera is inhibited by the presence of ammonia or nitrates, but when these nitrogen-containing substances fall below a threshold, heterocysts begin to appear. As mentioned earlier, some cyanobacteria form resistant spores called akinetes, which are very different in structure and function from heterocysts.

Cyanobacteria may occur as symbionts within the bodies of some sponges, amoebas, flagellated protozoa, certain diatoms, green algae that lack chlorophyll, other cyanobacteria, bryophytes, vascular plants, and oomycetes. When they occur as symbionts, some cyanobacteria may lack a cell wall. Under such circumstances, they function as chloroplasts. The symbiotic cyanobacterium divides at the same time as the host cell by a process similar to chloroplast division. The relationship of cyanobacteria to the evolution of particular groups of eukaryotes will be examined in Chapter 15.

SUMMARY

Bacteria, the prokaryotes, lack an organized nucleus and membrane-bound cellular organelles and do not reproduce sexually. Their genetic material is incorporated into a single circular molecule of double-stranded DNA. Except for mycoplasmas and spiroplasmas, bacteria have rigid cell walls. In all groups except the Archaebacteria, this wall is made up of an enormous molecule of peptidoglycan; some also have a lipopolysaccharide coating over the main portion of their cell walls. Bacteria are fundamentally unicellular although their cells may adhere to one another after they divide, forming filaments or other seemingly multicellular structures.

Most species of bacteria are heterotrophic; they share with the fungi the role of decomposers in the world ecosystem, and they are enormously abundant. As a group, the bacteria are metabolically extremely versatile: most bacteria are heterotrophic, but some are photosynthetic (photoautotrophic), and others are chemoautotrophic. Some bacteria are aerobic, others are obligate anaerobes, and still others are facultative anaerobes. A number of genera play important roles in the cycling of nitrogen, sulfur, and carbon. Many bacteria are important pathogens, both in animals and in plants.

Photosynthetic bacteria comprise at least five distinct groups, two of which, the cyanobacteria and *Prochloron*, contain the same chlorophyll *a* that occurs in all photosynthetic eukaryotes. Chemosynthetic bacteria derive their energy from the oxidation of inorganic molecules. Some of those that oxidize proteins and amino acids convert ammonium into nitrites, while other genera are able to convert the nitrites into nitrates; thus bacteria play an indispensible role in the nitrogen cycle. In addition, many cyanobacteria are able to fix atmospheric nitrogen. Cyanobacteria were responsible for the deposition of the layered chalk formations called stromatolites.

Bacterial cells may be rod-shaped (bacilli), spherical (cocci), or coiled (spirilli). If the cell wall does not divide completely, the daughter cells may adhere in groups or in filaments or solid masses. Bacteria may have flagella and be motile; the flagella are located either all over their cells or at one end. The rotation of the flagella moves the bacteria through the water and allows them to move toward food and away from toxic substances efficiently. Bacteria may also have shorter rodlike structures known as pili. Some bacteria move by gliding.

The Archaebacteria are a group of bacteria, many of them methane-producing, that have a unique sequence of bases in their ribosomal RNA. They survive in unusual habitats that resemble those that existed early in the history of the world.

Mutation, combined with a high reproductive rate, is the most important source of variability in bacteria. Genetic recombination also occurs; it involves the transfer of DNA from one cell to another. This may come about by conjugation, by transformation (the passive incorporation of fragments of DNA in the medium), or by transduction (injection of DNA into a bacterial cell by a bacteriophage).

SUGGESTIONS FOR FURTHER READING

BLAKEMORE, R. P., and R. B. FRANKEL: "Magnetic Navigation in Bacteria," *Scientific American* 245 (6): 58–65; December 1981.
A fascinating account of the way that bacteria orient to magnetic fields.

BROCK, THOMAS D.: *Biology of Microorganisms*, 3rd ed., Prentice-Hall, Inc., Englewood Cliffs, N.J., 1979.
An interesting and often entertaining presentation of microbiology, including algae and protozoa, with emphasis on the whole cell and its ecology.

CARR, N. G., and B. B. WHITTON (Eds.): *The Biology of Cyanobacteria*, University of California Press, Berkeley, 1982.
A comprehensive treatment of cyanobacteria, with contributions by many experts.

DICKINSON, C. H., and J. A. LUCAS: *Plant Pathology and Plant Pathogens*, 2nd ed., Halsted Press, John Wiley & Sons, Inc., 1982.
A well-written introduction to plant diseases, a subject of considerable general interest and economic importance.

KROGMANN, DAVID W.: "Cyanobacteria (Blue-Green Algae)—Their Evolution and Relation to Other Photosynthetic Organisms," *BioScience* 31: 121–124; 1981.
An interesting and concise review of the group.

MARAMOROSCH, K.: "Spiroplasmas: Agents of Animal and Plant Diseases," *BioScience* 31: 374–380; 1981.
A good review of this fascinating group of pathogenic, wall-less bacteria.

MARGULIS, LYNN: *Early Life*, Science Books International, Publishers, Portola Valley, Calif., 1982.
An excellent account of the evolutionary history of bacteria.

ROBERTS, DANIEL A., and CARL W. BOTHROYD: *Fundamentals of Plant Pathology*, 2nd ed., W. H. Freeman and Company, New York, 1984.
An up-to-date account of the diseases of plants and the organisms that cause them.

SONEA, S., and M. PANISET: *A New Bacteriology*, Jones and Bartlett Publishers, Inc., Boston, Mass., 1983.*
A marvelous, short book that presents an informative view of bacteria as living organisms.

STARR, M. P., et al. (Eds.): *The Prokaryotes. A Handbook on Habitats, Isolation, and Identification of Bacteria*, Springer-Verlag, New York, 1981.
The definitive treatise on bacteria, this 169-chapter, two-volume work is a rich source of up-to-date, accurate information.

TORTORA, G. J., B. R. FUNKE, and CHRISTINE L. CASE: *Microbiology: An Introduction*, Benjamin/Cummings, Menlo Park, Calif., 1982.
A well-organized, readable text dealing with all aspects of the basic biology of bacteria, viruses, and microscopic eukaryotes.

WOESE, C. R.: "Archaebacteria," *Scientific American* 244 (6): 98–122; June 1981.
An interesting account of Carl Woese's theories concerning the distinctiveness of the group.

* Available in paperback.

C H A P T E R   1 2

Viruses

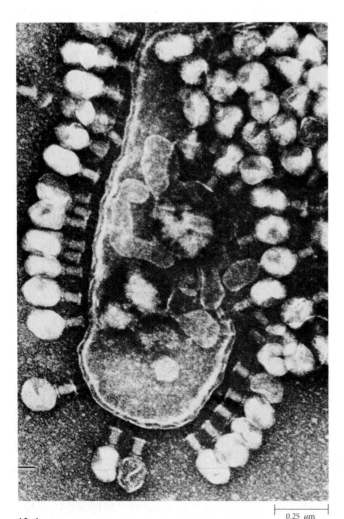

├─ 0.25 μm ─┤

12–1

Bacterial viruses (T2) infecting a cell of
Escherichia coli. *Some of the viruses
have discharged their DNA into the bacterial cell, which has begun to lyse, or
break down.*

A virus consists primarily of a genome that replicates itself within a host cell by directing the machinery of that cell to synthesize viral nucleic acids and proteins (Figure 12–1). Since no virus can grow or replicate on its own, viruses are not considered to be alive, in the usual definition of life. They are made up of highly organized sequences of nucleic acids, either DNA or RNA, double-stranded or single-stranded, depending on the virus. All viruses have a protein covering, or capsid, which encloses the nucleic acid, and some viruses also have an envelope, a lipid-rich outer layer surrounding the capsid. The protein or lipoprotein coat determines the surfaces to which a particular kind of virus will adhere and protects the nucleic acid as the virus passes from one host to another.

Viruses are important to us because they are so widespread in the cells of plants and other organisms, and because of the fact that they frequently cause disease. In humans, viruses are responsible for many diseases, including smallpox, chicken pox, measles, mumps, influenza, colds (often complicated by secondary bacterial infections), infectious hepatitis, yellow fever, polio, rabies, herpes, and AIDS (acquired immune deficiency syndrome). They also cause many diseases of other animals and plants; some of the viruses that infect plants are closely related to some of those that infect animals. Viruses cause diseases because they often seriously disrupt the normal functioning of the cells in which they occur. Many viral diseases of animals can be prevented by immunization, a process whose efficacy was first established, for smallpox, by the English country doctor Edward Jenner in the 1790s. Once established, however, viral diseases are relatively difficult to control because they do not respond to antibiotics. Eventually, the animal's immune system responds and limits further viral invasion.

The cells of every kind of organism may at times contain one or more distinctive kinds of virus; therefore, there may be literally millions of different kinds of viruses. Sometimes, however, viruses isolated from one kind of host will affect another, often very different kind of host. Therefore, it is difficult to draw specific conclusions about the number of distinct kinds of viruses that exist, but over 400 plant viruses, for example, have been described; these fall into about 25 distinct groups. Not only do viruses infect every species of eukaryotic organism that has been investigated for their presence, but they are also common in bacteria. Viruses that infect bacteria are called *bacteriophages* ("bacteria eaters"), or simply *phages.* Phages are one of the most important tools of modern molecular biology; we have learned a great deal about the details of genetics and protein synthesis by studying their interactions with bacteria. Viruses are widely used in studies of DNA replication, gene regulation, transcription, RNA processing, translation, and cancer biology, both in bacteria and in eukaryotes. They are natural gene transfer and expression systems; as such, they have become indispensable tools of modern biotechnology, as powerful gene vectors for bacterial, animal, and, more recently, plant cells. The use of viruses as plant gene vectors will undoubtedly expand considerably in the near future and is likely to have considerable impact on both basic botany and agriculture. In addition, scientists are beginning to consider the use of viruses, in addition to bacteria, to direct cells to manufacture useful molecules.

A number of different viruses that cause diseases in insects and other pests are under investigation as biological control agents. Already, there have been some notable successes: for example, in the United States, the corn earworm (*Heliothis zeae*) has been effectively controlled by a virus. Such methods of control do not have the harmful side effects that may accompany the repeated use of certain kinds of chemical pesticides, and the viruses appear to be highly specific for their target pests.

THE NATURE OF VIRUSES

The existence of viruses was first recognized when it was found that the causative agents of the tobacco mosaic disease could pass through the porcelain filters that were commonly used to trap bacteria. Viruses range in size from about 17 nanometers to more than 300 nanometers in diameter. Thus they are comparable in size to molecules, a hydrogen atom being about 0.1 nanometer in diameter and a protein molecule being at most a few hundred nanometers across.

As mentioned before, viruses can multiply only within a living host cell; many are highly specific with regard to the type of cell in which they can multiply. They essentially "take over" biosynthesis in the host cell, using their own nucleic acids to "command" the host to produce more virus particles, thereby competing

with the genetic material of the host cell in regulating cell functions (see Figure 8–3, page 119). For example, cold viruses multiply in the mucous membranes of the respiratory tract, producing the familiar symptoms of a cold. The ability of viruses to infect specific kinds of cells is limited; in fact, one of the techniques for rapid identification of unknown bacteria is to expose them to a spectrum of known bacterial viruses (bacteriophages) and observe which type destroys them (see Figure 12–1). Conversely, the responses of a range of plant species exposed to an unknown virus can be used, together with other tests, to identify the virus.

Until the 1930s, viruses were considered to be extremely small bacteria. Evidence against this point of view began to accumulate in 1933, when Wendell Stanley, a scientist working at the Rockefeller Institute, prepared an extract of a common virus—the tobacco mosaic virus—from infected plants and purified it. The purified virus precipitated in the form of crystals. Crystallization is one of the chief tests for the presence of a single, uncontaminated chemical compound, and so, clearly, virus particles were chemically much simpler than living organisms. When these needlelike crystals were put back into solution and reapplied to a tobacco leaf, however, the characteristic symptoms of tobacco mosaic disease reappeared. This indicated that the virus had retained its ability to infect the tobacco plants even after it had been crystallized and then resuspended.

Most plant viruses, like the tobacco mosaic virus, are RNA viruses, whereas many other viruses are DNA viruses. Unlike viruses, all independent cellular organisms contain both RNA and DNA. Viruses also lack ribosomes, as well as all of the enzymes necessary for protein synthesis and energy production. In these respects, viruses differ from independent cellular organisms.

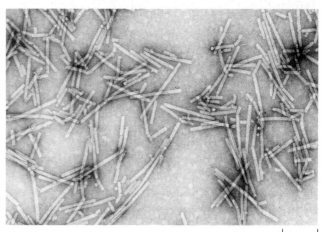

0.15 μm

12–2

Tobacco mosaic virus (TMV) particles as seen with the electron microscope.

12–3

A diagram of a portion of a tobacco mosaic virus (TMV) particle. This virus has a central core of RNA and a protein coat, or capsid, composed of 2200 identical protein molecules, each containing 158 amino acid residues folded into irregularly shaped subunits. The RNA fits into a groove at the narrower end of the subunit and is represented here by the dark chain exposed at the top.

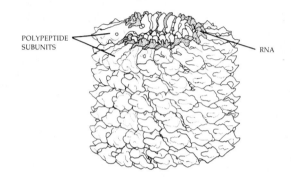

POLYPEPTIDE SUBUNITS

RNA

THE STRUCTURE OF VIRAL PARTICLES

By the use of electron microscopy, the structure of a large number of viruses has now been elucidated. The tobacco mosaic virus, for example, is a particle about 300 nanometers long and 15 nanometers thick, with a single molecule of RNA over 6000 bases long (Figure 12–2). The capsid consists of over 2000 identical protein molecules arranged with helical symmetry (Figure 12–3). The most common shape of a virus particle is that of an icosahedron, a 20-sided figure. An icosahedron is the most efficient arrangement with cubic symmetry that large numbers of identical subunits can take to form an outer shell with maximum internal capacity (geodesic domes are constructed along the same principles). Among the icosahedral viruses are those that cause many colds (Figure 12–4), polio, chicken pox, fever blisters, human warts, many types of cancer in mice and other animals, and many plant diseases, including tulare apple mosaic (Figure 12–5a), tobacco ringspot, cucumber mosaic, and bean pod mottle. The rhabdoviruses of plants and animals are bacilliform (bullet-shaped) with a lipoprotein outer envelope (Figure 12–5b). The elongated viruses with helical symmetry like that of tobacco mosaic virus (see Figure 12–2) are common in plant viruses, often assuming a loose helix and appearing as long, flexible threads (Figure 12–5c).

One common type of bacteriophage, called T4 phage (Figures 12–1 and 12–6), is much more complex in structure than the general kinds of viruses that have just been discussed. Each phage is about 100 nanometers long and is made of at least five separate proteins, the repeating protein subunits that make up each of the following: the hexagonal head, the tail core, the submolecules of the contractile sheath, the base plate of the tail, and the tail fibers. The long DNA molecule is coiled within the hexagonal head of the bacteriophage. T4, like other viruses, produces many proteins that are not part of the structure of the viral particles themselves. These proteins are utilized for replication of viral DNA and other purposes. One such protein is the enzyme lysozyme, which causes the breakdown (lysis) of the bacterial cell at the end of the infectious cycle. Clear spots called plaques appear in bacterial colonies in the areas where active lysis by bacteriophages is occurring.

Other large, complex viruses may consist of several different kinds of molecules of either DNA or RNA, together with many different kinds of proteins. Such viruses may contain as many as 200 different genes. Most viruses, however, have a relatively simple icosahedral or helical structure and contain only a few genes—for example, tobacco mosaic virus has three genes.

The Infective Properties of Viruses

The protein capsids of individual virus particles are often surrounded with or even replaced by a lipid-rich envelope, as mentioned earlier. Protruding through the envelopes of many viruses are spikes, which contain glycoproteins and lipids. The properties of the molecules that make up the external structures of a virus—capsid, envelope, or spikes—determine its infective

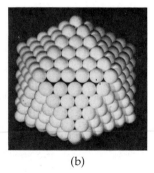

(a)   25 nm   (b)

12–4

(a) *An adenovirus, one of the many kinds of viruses that cause colds in humans. This virus is an icosahedron. Each of its 20 sides is an equilateral triangle made up of protein subunits. There is a total of 252 subunits. (b) A model of an adenovirus, made up of 252 tennis balls.*

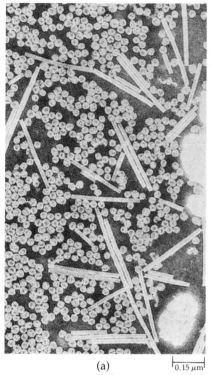

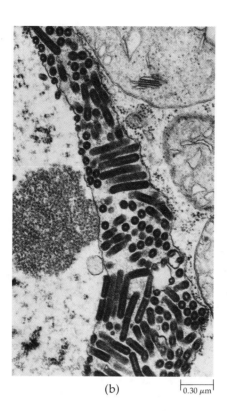

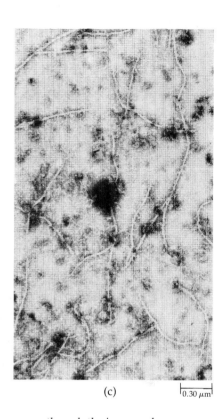

(a) ⊢0.15 μm⊣

(b) ⊢0.30 μm⊣

(c) ⊢0.30 μm⊣

12–5

Plant viruses; all of these are RNA viruses. (a) A mixture of Tulare apple mosaic virus and tobacco mosaic virus. Tulare apple mosaic virus is icosahedral, while tobacco mosaic virus is rigid and elongated. (b) Particles of a rhabdovirus *in a cell from a pepper plant (Capsicum frutescens); this is a typical bacilliform virus. The rhabdovirus particles are seen here in the space between the inner and outer membranes of the nuclear envelope. As the virus migrates to this space and* *passes through the inner nuclear membrane, it acquires a portion of that membrane, which then surrounds the particle. (c) Particles of a flexible, elongated virus in a cell from a Christmas cactus (Zygocactus truncatus).*

12–6

Electron micrograph (a) and model (b) of a T4 bacteriophage. The type T bacteriophages are a group of viruses that attack the bacterium Escherichia coli. *The tail is a hollow core encased in a contractile sheath containing molecules of ATP that provide the energy for the contraction. At the tip of the tail is a base plate, from which radiate six long fibers. The fibers attach to the bacterial cell wall, drawing the base plate to it. The tail protein then contracts, forcing the hollow tail core into the cell like a syringe. The DNA molecule, which is some 650 times longer than the protein head, then passes into the bacterial cell, leaving the protein coat outside.*

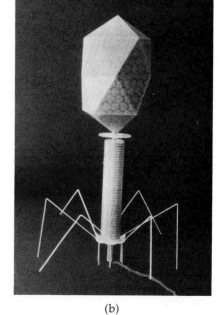

(a) ⊢0.05 μm⊣

(b)

properties. For example, a virus that enters an animal cell must first bind to a specific receptor molecule in the plasma membrane of that cell. Plant viruses, however, are introduced through wounds and are not known to have specific receptors.

The way in which viral infection occurs can be illustrated by reference to the influenza virus (Figure 12–7). The characteristics of the glycoproteins in the spikes of influenza virus particles determine their infectivity, and these properties are altered both by recombination and by mutation. Because immunity to a virus is, in effect, immunity to the specific proteins of its protective coat, new viral strains, produced as a result of recombination or mutation, are able to infect previously immune populations. The properties of the spikes of the influenza virus are changing constantly as a result of these processes, and the infectivity of the particular strains changes with them.

The length of time between flu epidemics presumably reflects the time required for a new strain of the virus to become established. Influenza is the only infectious disease that appears periodically in life-threatening global epidemics; in the winter of 1968–69, more than 50 million cases of Hong Kong flu were reported in the United States alone, and as many as 70,000 people died. The great influenza epidemic of 1918–19, which moved quickly around the world in three waves, caused more than 20 million deaths. Other viral diseases, including those of plants, presumably also spread when the characteristics of their capsids or envelopes change, but are less well known than influenza.

REPLICATION OF VIRUSES

Many viruses shed their capsids and, if they are also present, their envelopes, before they begin to replicate themselves. In some, such as the T4 bacteriophages, the capsid is left outside the host cell; in others, the capsid and, if present, the envelope, is shed within; and in still others, they are digested by the host cell's enzymes. In some viruses at least some of the viral proteins must enter the cell and remain associated with the viral nucleic acid, because they are enzymes such as polymerases, which are indispensable for viral replication. When the genome of the virus has been freed of its capsid and its envelope (if present) and is within the host cell, one of two things happens, provided the host-virus combination is permissive:

1. The virus may multiply by taking over the genetic machinery of the cell. Virus multiplication occurs in three steps. First, the viral nucleic acids direct the host cells to produce new viral enzymes. Second, the required viral nucleic acids and structural proteins are synthesized, each in the appropriate amount. Finally, these materials are assembled into virus particles. These steps generally overlap

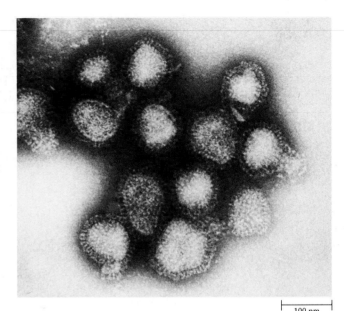

12–7
Influenza virus. The spikes in the envelope may be seen around many of the individual virus particles.

in time, often involve extensive genetic regulation, and lead to the production of many (sometimes thousands) new virus particles in each cell. When the process of virus multiplication is complete in a bacteriophage, the new virus particles escape from the host cell, which by then is generally dead. Such a cell is said to have been *lysed* (Greek *lysis*, a loosening), and viruses that have this kind of effect on a cell are said to be *lytic* viruses. Productive infections do not always end in lysis. Many animal viruses are released by budding through the cell membrane, whereas plant viruses move to adjacent cells via plasmodesmata and eventually through vascular tissues to the rest of the plant.

2. The virus may become established in the host cell as a stable part of its genome. Some bacteriophages, for example, consist of DNA fragments that carry one or more portions of the bacterial genome. By crossing over with the bacterial genome, these viruses may become incorporated into it. Such a bacteriophage is called a *temperate* bacteriophage, because it does not produce cell lysis while it is incorporated in the bacterial chromosome. In such a state, the virus is called a *prophage*. The viruses involved in transduction (page 174) are temperate phages. Only a few animal viruses and no plant viruses are known to behave in this way.

In DNA viruses—such as the vaccinia virus, which causes a poxlike disease in cattle—the viral DNA directs the synthesis of a series of different mRNA molecules, which in turn direct the production of different proteins. The DNA packaged by such viruses may be either double-stranded or single-stranded.

In most RNA viruses, such as the tobacco mosaic virus, the RNA is single-stranded. This viral RNA replicates itself by directing the formation of a complementary strand, which then serves as the template for new viral RNA molecules. The viral RNA utilizes the ribosomes of the host cell and acts as mRNA. In this role, it is responsible for the synthesis of enzymes and virus coat proteins.

Viral activity can profoundly affect the metabolism of the host cell. In one species of the bacterial genus *Clostridium,* the production of the lethal toxins associated with botulism in some strains occurs only with the active and continued participation of specific bacteriophages. Noninfected bacterial cells do not produce the toxin. Even more surprisingly, infection by other specific bacteriophages causes the same bacterial strain to produce the toxins that are associated with gas gangrene and many other diseases in animals. The causative organisms associated with botulism and gas gangrene had hitherto been considered to be different species of *Clostridium;* they are now thought to be the same species of bacterium, infected by different bacteriophages. Similarly, new species of plants and fungi have been described and named in which the characteristics have later been shown to be caused by a virus infection.

THE DIVERSITY OF VIRUSES

No one knows how many kinds of viruses exist, and new ones are found almost whenever additional kinds of organisms are investigated for their presence. Viruses are usually given distinctive names that apply to the different groups, not Latin binomials. Bacteriophages are named by formulas, such as T7 (the T standing for "type"). Plant viruses are usually simply designated by English names, such as "tobacco mosaic virus," or TMV; animal viruses are sometimes given Latin names, like bacteria. A few of the major kinds of viruses will now be discussed.

RNA Viruses

Single-stranded RNA viruses are called plus-strand or minus-strand viruses, depending, respectively, on whether the RNA acts directly as a messenger upon infection or a complementary strand is synthesized that acts as a template for producing viral mRNA. In turn, plus-strand RNA viruses are often divided into two groups, based on the presence or absence of an envelope. For example, tobacco mosaic virus is an unenveloped, plus-strand RNA virus. Poliomyelitis is also caused by an unenveloped plus-strand RNA virus, as is hoof-and-mouth disease, an important cattle disease.

In addition, unenveloped plus-strand viruses are responsible for about a third of the cases of colds in humans.

The enveloped, plus-strand RNA viruses include many of the *arboviruses,* or arthropod-borne viruses, a group that causes many important diseases, especially in the tropics. Yellow fever, which is spread from infected monkeys to humans by mosquitoes, is an arbovirus that belongs to this group, as are the retroviruses (discussed on page 194).

Minus-strand RNA viruses include those that cause rabies, measles and mumps, Newcastle disease in poultry, and distemper in various animals. All but rabies are caused by the paramyxoviruses, which are complex, enveloped viruses up to 300 nanometers in diameter, that are internally similar to tobacco mosaic virus in having a helical arrangement of protein around the RNA. The influenza viruses (see page 190) are likewise minus-strand RNA viruses. Like all of the RNA viruses already discussed, the minus-strand RNA viruses have single-stranded RNA. Double-stranded RNA viruses also exist, including wound tumor virus (Figure 12–8), which is unusual in replicating both in plants and in its leafhopper vector.

DNA Viruses

Double-stranded DNA viruses are the infectious agents responsible for warts (papilloma viruses) and herpes (herpesviruses). Different herpesviruses cause cold sores and fever blisters, genital herpes, chickenpox and shingles, mononucleosis and certain human cancers, and fatal systemic infections of newborn babies. Another group of double-stranded DNA viruses, the poxviruses, includes the largest and most complex viruses known, such as the vaccinia virus. Hepatitis B is caused by a *partly* double-stranded DNA virus (hepatitis A is caused by an RNA virus).

Two groups of DNA viruses, the caulimoviruses and geminiviruses, are known to infect plants. The molecular properties of the geminiviruses are unusual; the viruses are small particles that almost always exist as pairs formed from two incomplete icosahedra. Within each pair there are two single-stranded DNA molecules that are only infectious together, so that the geminiviruses appear to be divided-genome DNA viruses. Bean golden mosaic is an example of a plant disease caused by a geminivirus; it is spread by whiteflies. Another such disease is maize streak virus, which is spread by leafhoppers.

Cauliflower mosaic virus, which is spread by aphids, is an example of a plant disease caused by a caulimovirus, a double-stranded DNA virus. The virus particles are icosahedral. There are gaps in the DNA molecule, which prevents its major coiling. The caulimoviruses are now being studied intensively because, genetically altered, they may be a useful means of introducing desirable genes into plants.

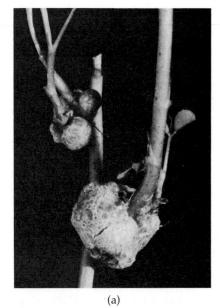

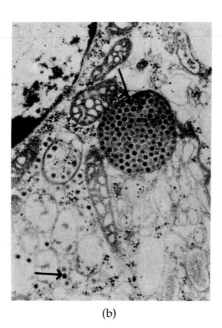

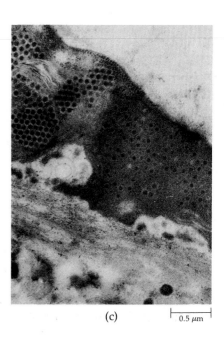

(a)   (b)   (c)

0.5 μm

12–8

(a) *Tumors produced by the wound tumor virus in sweet clover* (Melilotus alba). *(b) Wound tumor virus particles (indicated by arrows) in an electron micrograph of a cell of the host plant. (c)*

The clover leafhopper (Agallia constricta) *is the transmission vector of the wound tumor virus. Electron micrograph of an epidermal cell of the clover leafhopper. Viruses are being produced in*

the honeycomblike area at the upper left. Individual viruses can be seen in the dark area below. Like most plant viruses, wound tumor virus is a double-stranded RNA virus.

VIRAL DISEASES OF PLANTS

Well over a thousand kinds of diseases in plants are known to be caused by viruses, with several hundred different kinds of viruses involved (Figure 12–8). The viral diseases of plants are almost always transmitted from plant to plant by invertebrate animals, such as insects or nematodes. Sucking insects such as aphids and leafhoppers, for example, regularly transmit viruses from one plant to another as they move about sucking fluids from the phloem or merely from the epidermal cells (see Chapter 27). Some of these viruses multiply in the cells of the animal vector, as well as in those of the plant in which they cause diseases.

Necroses, plant diseases characterized by areas of dead tissue, and mosaic diseases, in which the leaves or other plant parts appear mottled, are among the common diseases of plants caused by viruses. In mosaic diseases, light green or yellow areas appear on the leaves or other green areas of the plants, ranging from small flecks to large stripes or bands. Sometimes the whole infected plant may be lighter green than normal. Almost all mosaic diseases are caused by viruses. The yellow blotches or borders that occur on the leaves of some prized horticultural varieties may be caused by viruses; the variegated appearance of some flowers is the result of viral infections that are passed on from generation to generation (Figure 12–9).

12–9

Streaked flowers of Rembrandt tulips. The streaking is caused by a virus infection.

Mosaic viruses affect primarily the parenchyma tissue of plants, reducing or sometimes eliminating the chloroplasts in them. Other viruses may occur in the sucrose-rich fluid circulating in the phloem, which they may ultimately kill. Viral diseases greatly reduce the productivity of many different kinds of crops throughout the world; they are present in a very wide array of plants. Virus-free plants can be produced by means of meristem culture; in this method a very small growing tip, which may be free of viruses, is dissected away from the parent plant and induced to form a whole new individual. Such methods have greatly increased yields in some crops, such as potatoes and rhubarb.

An interesting recent finding concerns a kind of virus called rice necrosis mosaic virus, which stunts infected rice plants. In 1982, S. K. Ghosh, from the Central Research Institute of India, found that other plants (such as jute, an important crop used to provide the raw material for burlap sacks and rope) grew better when they were infected with this virus than when they were not. No satisfactory explanation for this phenomenon has yet been found, but the example illustrates the complex nature of virus-plant interactions and their promise for future economic development.

With the exception of the geminiviruses and caulimoviruses discussed on page 191, plant viruses are RNA viruses. As you might suppose from the many kinds of diseases they cause, they are both abundant and diverse. With a few exceptions, among them the comoviruses (in which the capsids contain two kinds of protein), the capsids of plant viruses consist of a single protein. The most striking characteristic of plant RNA viruses is the frequent occurrence of divided genomes. Several plant viruses have more than one molecule of RNA comprising their genome, and each molecule is packaged within a separate but identical protein capsid. Many more viruses have two to four different-sized RNA molecules, each of which is contained within an individual capsid, so that the virus particles may be heterogeneous in shape or density. The ways in which these individual RNA molecules interact in promoting infectivity of individual plant viruses differ from group to group. As in other plus-strand RNA viruses, the replication of plant viruses involves two steps: the formation of a minus strand, using the original RNA as template, and then the formation of plus strands (mRNA) on that template.

VIROIDS AND OTHER INFECTIOUS PARTICLES

There are some molecular pathogens other than viruses that seem, like viruses, to have originated from the genomes of bacteria and eukaryotes. Particularly important among these are the viroids, which, despite their name, are quite distinct from viruses.

Viroids are the smallest known agents of infectious disease; they are much smaller than the smallest viral genomes, and they lack protein coats. Viroids are thus far known only from plants; they consist of small, single-stranded molecules of RNA that replicate autonomously in susceptible cells. They have been identified as the agents responsible for some very important plant diseases. One viroid has been responsible for the death of millions of coconut trees in the Philippines over the last half century, and another nearly destroyed the chrysanthemum industry in the United States in the early 1950s.

The first viroid to be characterized—potato spindle

12–10

Electron micrograph of potato spindle tuber viroid (arrows) mixed with DNA from a double-stranded DNA molecule from a bacteriophage. This micrograph illustrates the tremendous difference in size between a viroid and a virus.

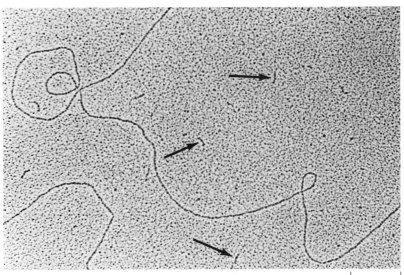

0.25 μm

tuber viroid, or PSTV—was identified by Theodor O. Diener of the U.S. Department of Agriculture in 1971 (Figure 12–10). Potatoes that have PSTV are elongated and gnarled; they sometimes have deep crevices in their surfaces. PSTV is one of the largest viroids of the dozen or so known; its RNA, which may exist either as a closed circle or a hairpin-shaped structure, open at one end, is 359 bases long. In either case, complementary base pairs are joined by hydrogen bonds, resulting in a double-stranded RNA structure similar to that of DNA. Under the electron microscope, both forms of PSTV appear as rods about 50 nanometers long. Despite the fact that it is a very large viroid, the genome of PSTV is only about a tenth the size of that of the simplest virus. Viroids are found almost exclusively in the nuclei of infected cells; they replicate themselves by forming a complementary strand that acts as a template, much like virus replication. Presumably host enzymes catalyze this process.

Because of their location in the nucleus and their apparent inability to act as messenger RNAs, it has been suggested that viroids may cause their symptoms by interference with gene regulation in the infected host cells. Certain plant proteins, found in healthy cells, are present in significantly larger quantities in infected cells. Although nucleotide sequences complementary to PSTV have not been found in uninfected plants of host species, it has been postulated that PSTV may have originated from alterations of genes normally present in certain species of the potato family, which are its primary hosts.

Molecular pathogens other than viroids are also active in biological systems. The existence of particles similar to viroids, but consisting of DNA, has been suspected in animal cells; such particles have been called "subviral particles." Even more surprising is the possibility that certain protein fragments may be able to direct their own replication in animal cells, without the presence of nucleic acids; such particles have been named "prions."

V I R U S E S   A N D   C A N C E R

Viruses cause cancer in many different kinds of animals; the retroviruses, a group of enveloped, plus-strand RNA viruses, have been the object of intensive investigations recently because of their role in causing human cancer. These viruses have the enzyme reverse transcriptase, which catalyzes the synthesis of a double strand of DNA, one strand of which is complementary to the RNA strand of the virus. This double-stranded fragment of DNA is incorporated into the chromosomes of the host cell, where it is often active in causing malignant tumors or leukemia in mammals, birds, and reptiles.

By careful studies of retroviruses such as Rous sarcoma virus—a retrovirus that causes tumors in chickens—scientists identified *oncogenes,* or cancer-causing genes, which interfere with the normal growth-regulating signals of the cell and thus cause cancer. Oncogenes, or *onc* genes, seem to occur normally in humans and other animals and cause cancer when they have been changed to a new form by mutagenic substances such as those that are abundant in cigarette smoke and other carcinogenic substances. Evidence is accumulating that more than one mutant *onc* gene may be needed to cause cancer.

In addition to the retroviruses, a group of DNA herpesviruses—the Epstein-Barr viruses—also has been linked with two types of human cancer; the same viruses also cause infectious mononucleosis. Epstein-Barr viruses can persist in cells without inducing the manufacture of progeny viruses and the disease symptoms with which they are associated.

T H E   O R I G I N   O F   V I R U S E S

Because of the simplicity of viral structure, some earlier students mistakenly thought that viruses might represent the direct descendants of the first self-replicating units from which the first cells evolved. This is clearly not so, for viruses exist only by virtue of their ability to utilize the machinery of their host cells. They compete with the nucleic acids of the host cells and take over the genetic and metabolic activities of the cells in directing the formation of new virus particles. They must have come into being after the evolution of cells in which the genetic code was already established.

Viruses are essentially similar to bacterial chromosomes or plasmids (double-stranded DNA viruses) or molecules of messenger RNA (single-stranded and double-stranded RNA viruses) packaged in a protein coat. The phenomenon of transformation, discussed on page 174, suggests the likely mode of the origin of viruses. Fragments of DNA or RNA occasionally find their way into cells in one way or another. When they do, they may exert an effect upon the genetic processes of the host cell. If they can replicate themselves and spread from cell to cell, they may persist. If they can produce a protective protein coat, they become, in effect, viruses. It is conceivable that the largest and most complex viruses, the poxviruses, evolved from degenerate bacteria.

Viruses have probably originated independently on many different occasions over the past 3 billion years. Moreover, current evidence shows clearly that established viruses can evolve with astonishing rapidity in response to strong selection pressures. It also seems certain that additional kinds of viruses are still evolving both from bacteria and from eukaryotes today.

SUMMARY

Viruses have genomes of either DNA or RNA, surrounded by an outer protein coat, or capsid, and sometimes also by a lipid-rich envelope, through which spikes, composed of glycoprotein and lipid molecules, may protrude. Viruses cannot reproduce themselves outside of living cells and so are not independent cellular organisms.

Viruses are responsible for many diseases of humans and other animals, including measles, mumps, influenza, and herpes; they also cause many diseases of plants. Important among these are necroses, diseases characterized by the presence of areas of dead tissue; and mosaic diseases, in which the plants develop blotches or streaks of light-colored tissue. Bacteriophages, or simply phages, are viruses that infect bacteria. Phages and other viruses are among the most important tools of molecular biology.

Viruses range from 17 to more than 300 nanometers in diameter and are thus comparable to macromolecules in size. Many are spherical with icosahedral symmetry, and a number of others are rod-shaped, with helical symmetry. When viruses replicate, they take over the genetic machinery of the cell and eventually lyse (destroy) it; or they may become incorporated in the chromosomes of the host cell and simply remain in that position.

RNA viruses sometimes act as mRNA in the host cells or direct the synthesis of a complementary strand that acts as a template in producing viral mRNA. DNA viruses simply compete with the DNA of the host cell. One group of RNA viruses, the retroviruses, enzymatically catalyzes the synthesis of a double strand of DNA, one strand of which is complementary to themselves, which is incorporated into the chromosomes of the host cell. Once this occurs, some of the retroviruses cause cancer; studies of the retroviruses have yielded valuable information about the action of oncogenes, or cancer-causing genes, in general.

Viruses seem to have originated as portions of the genetic apparatus of both bacteria and eukaryotes that broke loose, acquired the ability to synthesize a protein-rich covering, and began to function within the cells of organisms. They have originated many times in the past; new kinds of viruses are evidently still evolving from both bacteria and eukaryotes.

Viroids, which cause certain plant diseases, consist of small molecules of RNA. Unlike viruses, they lack protein coats. They are thought to interfere with gene regulation in infected host cells.

SUGGESTIONS FOR FURTHER READING

BROCK, THOMAS D., DAVID W. SMITH, and MICHAEL T. MADIGAN: *Biology of Microorganisms*, 4th ed., Prentice-Hall, Inc., Englewood Cliffs, N.J., 1984.

An excellent textbook that deals with the morphology, biochemistry, and ecological roles of bacteria, viruses, and selected eukaryotes.

DIENER, T. O.: "The Viroid—A Subviral Pathogen," *American Scientist* 71 (1983): 481–489.

An excellent review of the properties of these fascinating RNA particles.

DICKINSON, C. H., and J. A. LUCAS: *Plant Pathology and Plant Pathogens*, Halsted Press, John Wiley & Sons, Inc., New York, 1977.

A well-written introduction to plant diseases, a subject of considerable economic importance.

FRAENKEL-CONRAT, H., and P. C. KIMBALL: *Virology*, Prentice-Hall, Inc., Englewood Cliffs, N.J., 1982.

An outstanding basic reference on all aspects of virology.

GIBBS, A. J., and B. D. HARRISON: *Plant Virology: The Principles*, Halsted Press, John Wiley & Sons, Inc., New York, 1979.

A useful overall review of the basic nature of plant-virus interactions; it is well illustrated.

MATTHEWS, R.E.F.: *Plant Virology*, 2nd ed., Academic Press, New York, 1981.

The most comprehensive general text currently available.

MAYO, M. A., and K. A. HARRAP: *Vectors in Virus Biology*, Academic Press, London, 1984.

A report of a symposium to explore some of the diversity of virus-vector relationships, using examples from plant and animal virology.

TORTORA, G. J., B. R. FUNKE, and CHRISTINE L. CASE: *Microbiology: An Introduction*, Benjamin/Cummings, Menlo Park, Calif., 1982.

A well-organized, readable text dealing with all aspects of the basic biology of bacteria, viruses, and microscopic eukaryotes.

C H A P T E R 1 3

Fungi

13–1
Mycelium of a basidiomycete on a fallen tree trunk. The activities of fungi and bacteria make it possible for the material that is incorporated into the bodies of organisms to be recycled in the ecosystem.

Fungi are as distinct from algae, bryophytes, and vascular plants as they are from animals. They are discussed in this text because they have traditionally been grouped with plants. However, the only characteristic that fungi share with plants, other than those common to all eukaryotes, is a somewhat filamentous or elongate, multicellular growth form. (A few fungi—the yeasts—are unicellular.) Fungi have no motile cells at any stage of their life cycles and no direct evolutionary connection with the plants. Clearly the two groups were derived independently from different groups of single-celled eukaryotes. Therefore, we treat fungi as a distinct kingdom—the kingdom Fungi.

The fungi, together with the heterotrophic bacteria and a few other groups of heterotrophic organisms, are the decomposers of the biosphere (Figure 13–1). Their activities are as necessary to the continued existence of the world, as we know it, as are those of the food producers. Decomposition releases carbon dioxide into the atmosphere and returns nitrogenous compounds and other materials to the soil where they can be used again —recycled—by plants and eventually by animals. It is estimated that, on the average, the top 20 centimeters of fertile soil may contain nearly 5 metric tons of fungi and bacteria per hectare (2.47 acres).

As decomposers, fungi often come into direct conflict with human interests. A fungus makes no distinction between a rotten tree that has fallen in the forest and a fence post; the fungus is just as likely to attack one as the other. Equipped with a powerful arsenal of enzymes that break down organic products, fungi are often nuisances and are sometimes highly destructive (Figure 13–2). This is especially true in the tropics, because warmth and dampness promote fungal growth; it is estimated that during World War II less than 50 percent of the military supplies sent to tropical areas arrived in usable condition. Fungi attack cloth, paint, cartons, leather, waxes, jet fuel, insulation on cables and wires, photographic film, and even the coating of

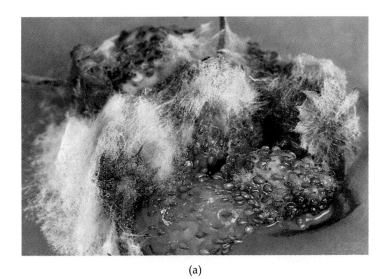

(a)

(b)

13–2
(a) *The common mold* Aspergillus *growing on strawberries.* (b) Aspergillus, *along with other fungi and bacteria, will eventually reduce this stump to soil.*

the lenses of optical equipment—in fact, almost any conceivable substance. Even in temperate regions, they are the scourge of food producers and sellers alike, for they grow on bread, fresh fruits, vegetables, meats, and other products. Fungi reduce the nutritional value, as well as the palatability, of such foodstuffs. They also produce toxins, some of which—the aflatoxins—are highly carcinogenic, showing their effects at concentrations as low as a few parts per billion.

The importance of fungi as commercial pests is enhanced by their ability to grow under a wide range of conditions. Thus some strains of *Cladosporium herbarum,* which attack meat in cold storage, can grow at −6°C. In contrast, one species of *Chaetomium* grows optimally at 50°C and can even be grown at 60°C.

The qualities of fungi that make them such important pests also make them commercially valuable. Many fungi, especially the yeasts, are useful because of their ability to produce substances such as ethanol and carbon dioxide, which plays a central role in baking. The commercial use of certain fungi in large-scale industrial processes is growing (Figure 13–3). Other fungi are of value as sources of antibiotics, including penicillin, the first antibiotic to be used widely.

A striking example of the potential of unknown compounds derived from fungi is cyclosporine, a new "wonder drug," which was isolated from a soil-inhabiting fungus. Developed by Sandoz Corporation in Switzerland, cyclosporine became available in 1979. This cyclic molecule is composed of 13 different amino acids, one of them not yet found anywhere else in nature. Cyclosporine suppresses immune reactions—reactions that arise in connection with organ transplants. It does not, however, have the undesirable side effects of other drugs used for this purpose, which kill bone marrow cells (the source of all blood cells) and thus tend to cause leukemia. Because of the discovery of this remarkable drug, it became possible to resume heart transplants and similar operations in the early 1980s.

13–3
This huge fermentor, about 30 meters tall, is used by Imperial Chemical Industries in England to produce yeast, which is sold as a protein-rich animal feed in place of soybean extract under the trade name Pruteen.

The kinds of relationships between fungi and other organisms are extremely diverse. For example, about four-fifths of all vascular plants form associations (called mycorrhizae) between their roots and fungi; these associations play a critical role in plant nutrition and distribution (see page 224). As another example, lichens (see page 211) are symbiotic associations of fungi and algal or cyanobacterial cells; lichens occupy some of the least hospitable habitats. Many fungi attack living organisms, rather than dead ones, and sometimes do so in highly unusual ways (see "Predaceous Fungi," page 210). They are the most important single cause of plant diseases; well over 5000 species of fungi attack economically valuable crop and garden plants, as well as many wild plants. Certain fungi attack living trees, causing enormous losses to the world's timber crop. Other fungi are the cause of serious diseases in humans and domestic animals.

Some 100,000 distinct species of fungi have been described, and it is estimated that as many as 200,000 more may await discovery. There may be as many species of fungi as there are species of plants, although far fewer fungi have been given names thus far.

Traditionally, the fungi have been considered to include not only the related divisions included in this chapter but also the heterotrophic protists discussed in Chapter 14. Because there is little evidence for a direct relationship between the fungi and the several groups of protists that have traditionally been included with them, they are separated in this book for clarity of discussion. No group of protists seems to be a likely ancestor of the fungi, even though that ancestor, were it known, would certainly be considered a member of the kingdom Protista.

BIOLOGY OF THE FUNGI

Fungi are primarily terrestrial. Although some fungi are unicellular, most are filamentous, and structures such as mushrooms consist of a great many such filaments, packed tightly together (Figure 13-4). Fungal filaments are known as _hyphae_, and a mass of hyphae is called a _mycelium_ (see Figure 13–1). Growth of hyphae occurs at their tips, but proteins are synthesized throughout the mycelium and are carried to the tips of the hyphae by cytoplasmic streaming, a phenomenon that is particularly well developed among the fungi (Figure 13–4). Within 24 hours, an individual fungus may produce more than a kilometer of new mycelium. (The words "mycelium" and _mycologist_—a scientist who studies fungi—are derived from the Greek word _myketos_, meaning "fungus.")

The cell walls of plants and many protists are built on a framework of cellulose microfibrils, which is interpenetrated by a matrix of noncellulosic molecules, such as hemicelluloses and pectic substances. In fungi, the

13–4

A mushroom, such as this one growing in a redwood forest in central California, is made up of densely packed hyphae, collectively known as mycelium. Mushrooms reproduce by spores, which are formed on structures that line the gills under the cap.

cell wall is composed primarily of another polysaccharide—_chitin_—which is the same material found in the hard shells, or exoskeletons, of arthropods, such as insects, arachnids, and crustaceans (Figure 13–5). Chitin is far more resistant to microbial degradation than is the cellulose found in plant cell walls.

With their rapid growth and filamentous form, fungi have a relationship to their environment that is very different from that of any other group of organisms.

CHITIN

13–5

The structure of chitin, which consists of β-1,4-linked N-acetylglucosamine units. A similar linkage is found in cellulose and in the cell walls of bacteria, suggesting that such a linkage provides a particularly strong polysaccharide. Chitin is characteristic of the cell walls of many fungi, as well as the exoskeletons of arthropods.

The surface-to-volume ratio of fungi is very high, which means that they are in as intimate a contact with the environment as are, for example, the bacteria. With a few exceptions, no part of a fungus is more than a few micrometers from its external environment, being separated from it only by a thin cell wall and the plasma membrane. With its extensive mycelium, a fungus can have a profound effect on certain aspects of its surroundings, for example, in binding soil particles together. Hyphae often fuse, even when they have grown from different spores, thus increasing the intricacy of the network.

The maintenance of this intimate relationship between fungus and environment requires that all parts of the fungus are metabolically active; the sorts of quiescent layers of tissue, such as wood, that are found in plants are absent in fungi. The enzymes and other substances that are secreted by fungi have an immediate effect on their surroundings and are of great importance for the maintenance of the fungus itself.

All fungi are heterotrophic. They obtain their food either as _saprobes_ (organisms that live on dead organic material) or as _parasites_ (organisms that feed on living matter). In either case, the food is ingested by absorption after it has been partially digested by enzymes that are secreted outside the fungal cell. Some fungi, mainly yeasts, can release energy by fermentation, such as in the production of ethyl alcohol from glucose. Glycogen is the primary storage polysaccharide in some fungi, as it is in animals and bacteria; lipids serve an important storage function in other fungi.

Saprobic fungi sometimes have a form of somewhat specialized hyphae, known as _rhizoids_, which anchor them to the substrate. Parasitic fungi often have specialized hyphae, called _haustoria_ (singular: haustorium), which absorb nourishment directly from the cells of other organisms (Figure 13–6).

All fungi have cell walls and most produce spores of some kind. Fungi are nonmotile throughout their life cycle, lacking cells with cilia or flagella. Many fungal spores are dispersed by the wind.

THE EVOLUTION OF FUNGI

The first fungi were probably unicellular eukaryotic organisms that apparently no longer have any living counterparts. From these organisms were derived the _coenocytic_ fungi, in which many nuclei are found in a common cytoplasm (_coenocytic_ means "contained in a common vessel," or multinucleate, the nuclei not separated by walls). The living coenocytic fungi are classified as members of the division Zygomycota, commonly known as the zygomycetes.

In members of the division Ascomycota, called ascomycetes, and the division Basidiomycota, or basidiomycetes, the mycelia are septate—divided by cell

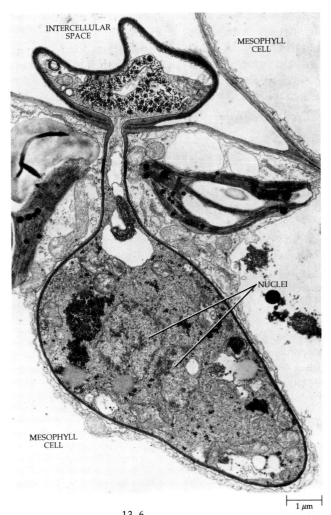

13–6

A haustorium of Melampsora lini, _a rust fungus, growing in a flax (_Linum usitatissimum_) leaf cell._

walls—but the septa, or cross walls, are perforated (Figure 13–7). In some fungi, the cytoplasm and its contents stream quite freely along the hyphae. The exact connections between the three classes of fungi are difficult to establish; there is no evidence of a close connection between the zygomycetes and the other two groups. Once it was widely held that basidiomycetes were derived from ascomycetes, but the opposite may be just as likely; these two groups are certainly closely related.

The oldest fossils that resemble fungi occur in strata about 900 million years old, but the oldest that have been identified with certainty as fungi are from the Ordovician period, 450 to 500 million years ago (see Appendix D). Coenocytic fungi that are presumed to have been zygomycetes were associated with the underground portions of the earliest vascular plants in the Silurian period, some 400 million years ago. Fungi may be among the oldest eukaryotes, but their early history is poorly understood. The three divisions of fungi were in existence by the close of the Carboniferous period, some 300 million years ago. Basidiomycetes seem to precede ascomycetes in the fossil record by a wide margin, a relationship that adds credibility to the hypothesis that they may be ancestral to the ascomycetes. The great majority of fungi, as discussed, are terrestrial, and the invasion of the land by plants and animals, which began about 410 million years ago, was the event that led to their major evolutionary diversification. Earlier fungi must have been aquatic, living in fresh water or the sea, and their fossils may be difficult to interpret in relation to modern forms.

FUNGAL REPRODUCTION

The reproductive structures of fungi are separated from the hyphae by complete septa. These reproductive structures are called *gametangia* if they are directly involved in the production of gametes, and *sporangia* (zygomycetes) or *conidiogenous cells* (ascomycetes and basidiomycetes) if they are involved in the production of asexual spores. The gametes are equal in size, and the fungi are therefore *isogamous*. Meiosis immediately follows the formation of the zygote in all fungi; in other words, meiosis in fungi is zygotic (see Figure 10–11, page 163).

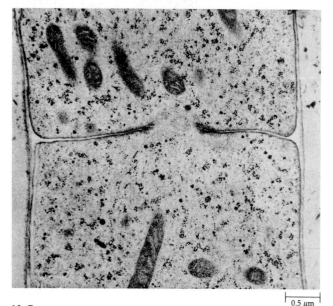

13–7

Electron micrograph of a septum in
Fusarium acuminatum, *an ascomycete.*
The large objects are mitochondria, and
the small black ones are ribosomes. This
specimen was thin-sectioned through the
central pore region of a septum in the
final stages of formation.

Nonmotile spores are the characteristic means of reproduction in fungi. Some of the spores are very small; they can remain suspended in the air for long periods, thus being carried to great heights and for great distances. This presumably is an important factor in the very wide distribution of many fungi. The spores of some fungi are spread by adhering to the bodies of insects and other animals. The bright colors and powdery textures often seen in association with many types of molds are produced by the spores (see Figure 13–17, page 209). The spores of a few fungi are propelled ballistically through the air (see "Phototaxis in a Fungus," page 203).

Mitosis and Meiosis

Fungi have many biological peculiarities that are just beginning to be understood. One of the most intriguing characteristics involves the process of nuclear division. In fungi, the processes of mitosis and meiosis are different from those that occur in plants, animals, and most protists. The nuclear envelope does not dissociate and re-form but is constricted near its midpoint between the two daughter nuclei, and the spindle apparatus is formed within the nuclear envelope. All fungi lack centrioles. This unique combination of features indicates that fungi are not directly related to other living eukaryotes and therefore deserve their present status as a separate kingdom.

Heterokaryosis and Parasexuality

Heterokaryosis

Among the genetic characteristics that distinguish fungi from other groups of organisms, none is more significant than the phenomenon of *heterokaryosis*, which was discovered by the mycologist H. Burgeff in 1912. A strain of fungus is heterokaryotic if the nuclei found in a common cytoplasm are genetically different, either because of mutation or the fusion of genetically distinct hyphae; the latter appears to be common in nature. If the nuclei are genetically similar, the fungal strain is *homokaryotic*.

Heterokaryosis is important in the genetics and evolution of fungi. If the genetically different nuclei in heterokaryotic fungi are incorporated in the cytoplasm of different derivative hyphae, these hyphae can be phenotypically distinct. Thus, even if only two different sorts of nuclei are present, three distinct phenotypes (one the same as the original phenotype, containing both types of nuclei) can be derived from that particular mycelium, depending on the type or types of nuclei that the derivative hyphae receive.

The effects of heterokaryosis are somewhat similar to those of the diploid condition that characterizes other organisms; that is, the appearance and physiological

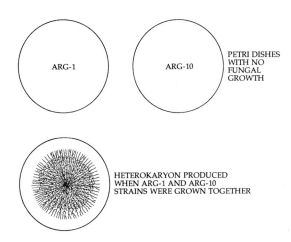

13–8
Arg-1 and Arg-10 are mutant strains of
Neurospora, *each of which lacks a
different enzyme involved in the
production of arginine. Neither will grow
on a minimally nutritious medium.
However, a heterokaryon formed between
the two strains grows readily on the
minimal medium. By combining the genetic information contained in the
original mutant nuclei, the heterokaryon
can synthesize the missing amino acid.*

characteristics of a heterokaryotic organism are determined by the interaction of the genetically different
nuclei. Recessive mutations may accumulate in some of
the nuclear lines, but their effects may be concealed by
the effects of the alleles or different genes present in
other nuclei. Finally, because many of the nuclear lines
are unable to exist or, at least, to compete successfully
in a homokaryotic state, heterokaryotic strains are favored by selection (Figure 13–8).

Parasexuality

The *parasexual cycle* in fungi was discovered in 1952 by
two investigators at the University of Glasgow. Working with *Aspergillus nidulans*, G. Pontecorvo and J. A.
Roper found that haploid nuclei may fuse in a heterokaryotic mycelium to produce diploid nuclei, some of
which are heterozygous (produced by the fusion of genetically different nuclei). It is estimated that *A. nidulans* contains 1 diploid heterozygous nucleus for every
1000 haploid nuclei.

Within the diploid nucleus, the chromosomes may
become associated. If this occurs, some crossing-over
can take place. Haploid nuclei may or may not re-form.
If they do re-form, they may be genetically different
from either of the parent haploid nuclei that fused initially. These new haploid nuclei can then take part in
other new heterokaryotic combinations.

It has recently been demonstrated that parasexual
cycles—cycles that have some of the genetic effects of
true sexual cycles but do not operate in the same way
—exist in several other groups of fungi. The significance of such cycles in nature has yet to be assessed
completely. However, the parasexual cycle appears to
be a flexible and commonly occurring system, especially among those fungi that either do not reproduce
sexually or in which sexual reproduction is an infre-
quent phenomenon.

THE MAJOR GROUPS OF FUNGI

Division Zygomycota

Most zygomycetes live on decaying plant and animal
matter in the soil; some are parasites of plants, insects,
or small soil animals; and a few occasionally cause severe infections in humans and domestic animals. There
are approximately 600 described species of zygomycetes. The terms "Zygomycota" and "zygomycetes"
refer to the chief characteristic of the division—the production of sexual resting structures called *zygosporangia*
(Figure 13–9). A zygosporangium is produced follow-

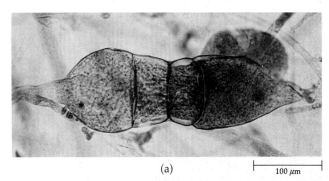

(a)   100 µm

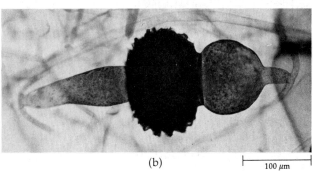

(b)   100 µm

13–9
*Rhizopus stolonifer, black bread mold,
colonizes the surface of moist substrates
exposed to air. (a) Gametangia, the
gamete-producing structures. (b) A zygosporangium, or sexual resting structure
(the dark mass in the center). Such a
zygosporangium contains one to several
diploid zygotes.*

ing the fusion of two multinucleate gametangia. Inside the zygosporangium, the gametes—which are simply nuclei—fuse, forming one or more diploid nuclei, or zygotes. In zygomycetes, the zygosporangium has often been called a zygospore, but since it contains several to many diploid nuclei, themselves zygotes, we prefer the terminology presented here. Asexual reproduction by means of spores produced in more or less specialized sporangia borne on the hyphae is almost universal in zygomycetes. Most members of the division have coenocytic (multinucleate and not divided by cross walls) hyphae, within which the cytoplasm can often be seen streaming rapidly.

One of the most common members of this division is *Rhizopus stolonifer*, a black bread mold that forms cot-

tony masses on the surface of moist, carbohydrate-rich foods and similar substances that are exposed to air. This organism is also a serious pest of stored fruits and vegetables, which it is capable of destroying quickly. The life cycle of *R. stolonifer* is illustrated in Figure 13–10. The mycelium of *Rhizopus* is composed of three different types of haploid hyphae. The bulk of the mycelium consists of rapidly growing hyphae that are coenocytic. These hyphae mostly grow within the substrate and function largely in the absorption of nutrients. From them, arching hyphae called *stolons* are formed. The stolons form rhizoids wherever their tips come into contact with the substrate. Sporangia form on the tips of the sporangiophores ("sporangia bearers"), which are erect branches formed directly

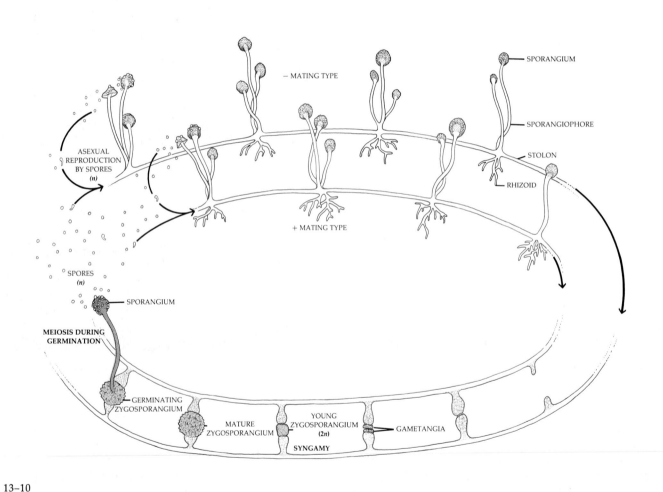

13–10

In Rhizopus stolonifer, *sexual reproduction occurs only between different mating strains, which have been traditionally labeled + and − types. (Although the mating strains are morphologically indistinguishable, they are shown here in two colors.) When the two strains are in close proximity, hormones are produced that cause outgrowths near their hyphal*

tips to come together and develop into gametangia, which become separated from the rest of the fungal body by the formation of septa (see Figure 13–9a). The walls between the two touching gametangia dissolve, and the two multinucleate protoplasts come together. The + and − nuclei fuse in pairs to form a young zygosporangium with several di-

ploid nuclei. The zygosporangium then develops a thick, rough, black coat and becomes dormant, often for several months. Meiosis occurs at the time of germination. The zygosporangium cracks open and produces a sporangium that is similar to the asexually produced sporangium, and the cycle begins again.

PHOTOTAXIS IN A FUNGUS

In *Pilobolus*, a zygomycete 5 to 10 millimeters tall that grows on dung, the sporangia are shot toward the light. The sporangium-bearing stalk, or sporangiophore, of this fungus grows toward the light so that all light rays entering the subsporangial swelling converge on a basal photoreceptive area. Light focused elsewhere promotes maximum growth of the sporangiophore on the side away from the light. The high turgor pressure of the sap in the vacuole of the subsporangial swelling eventually causes it to split, which blasts the sporangium off to a distance of 2 meters or more

at an initial velocity that may approach 50 kilometers per hour! Considering that the sporangia are only about 80 micrometers in diameter, this is an enormous distance. After the sporangium has been fired off, the sporangiophore collapses. The sporangium adheres where it lands, and if this is on a blade of grass, it may be eaten by an herbivore. It then passes through the digestive tract of the herbivore unharmed and is deposited in the dung to begin the cycle anew.

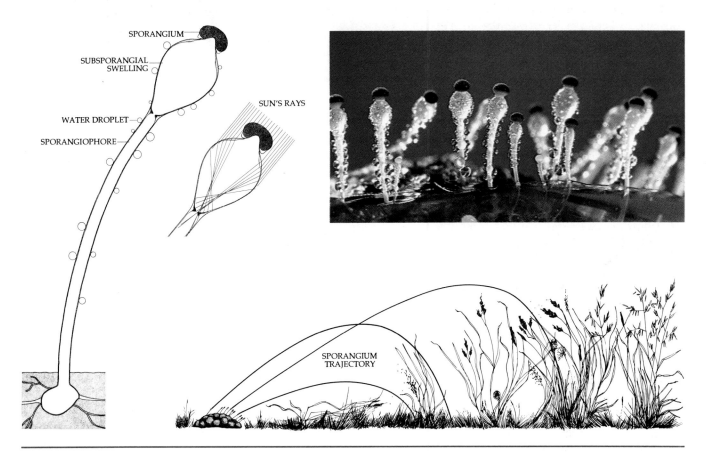

above the rhizoids. Each sporangium begins as a swelling, into which a number of nuclei flow; the sporangium is eventually cut off from the sporangiophores by the formation of a septum. The protoplasm within is cleaved, and a cell wall is formed around each spore. The sporangium becomes black as it matures, giving the mold its characteristic color. Each spore, when liberated, can germinate to produce a new mycelium.

One of the most important groups of zygomycetes includes *Glomus* and related genera, which grow in association with the roots of plants. These fungi occur in mycorrhizae, which will be discussed on page 224.

Division Ascomycota

The ascomycetes comprise about 30,000 described species, including a number of familiar and economically important fungi. Most of the blue-green, red, and brown molds that cause food spoilage are ascomycetes, including the salmon-colored bread mold *Neurospora*, which has played an important role in the development of modern genetics. Ascomycetes are the cause of a number of serious plant diseases, including the powdery mildews, which primarily attack leaves; chestnut blight (caused by the fungus *Endothia parasitica*, acci-

(a)

(b)

(c)

13–11

Ascomycetes. (a) *Common morel,* Morchella esculenta. *The true morels are among the choicest edible fungi. Mushroom gatherers look for them when the oak leaves are "the size of a mouse's ear." Morels were first grown successfully in culture in 1983 but have not yet been developed into a commercial crop.* (b)

Scarlet cup, Sarcoscypha coccinea, *a beautiful fungus with an open ascocarp (apothecium).* (c) *The highly prized, edible ascocarp of a black truffle,* Tuber melanosporum. *In the truffles, this spore-bearing structure is produced below ground and remains closed, liberating its ascospores only when the ascocarp*

decays or is broken open by digging animals. Truffles are mycorrhizal, mainly on oaks, and are searched for by specially trained dogs and pigs. Recently, they have been cultivated commercially on a small scale by inoculating the roots of seedling host plants with their spores.

dentally introduced from northern China); and Dutch elm disease (caused by *Ceratocystis ulmi,* a fungus native to certain European countries). Many yeasts are also ascomycetes, as are the edible morels and truffles (see Figure 13–11). As a whole, this group of fungi is relatively poorly known, and thousands of additional species—some undoubtedly of great economic importance—await scientific description.

Characteristics of the Ascomycetes

Ascomycetes, with the exception of the unicellular yeasts, are filamentous when they are growing. In general, their hyphae are septate, or divided by cross walls. The septa are perforated, however (Figure 13–7), and the cytoplasm and its nuclei can move through the septal pores. The hyphal cells of the vegetative mycelium may be either uninucleate or multinucleate. Some species of ascomycetes are *homothallic* (self-fertile) and can produce sexual structures from a single genetic strain; others are *heterothallic* and require a combination of + and − strains.

In the majority of ascomycetes, asexual reproduction is by formation of specialized spores, known as *conidia* (a Greek word meaning "fine dust"). Conidia, which are usually multinucleate, are formed from conidio-

genous cells (Figure 13–12), as we mentioned above; such cells are cut off from the tips of modified hyphae called *conidiophores* ("conidia bearers").

Sexual reproduction in ascomycetes always involves the formation of an *ascus* (plural: asci), a saclike structure containing haploid *ascospores.* Both asci and ascospores are unique structures; they distinguish the ascomycetes from all other fungi (Figure 13–13). Ascus formation usually occurs within a complex structure composed of tightly interwoven hyphae—the *ascocarp.* Many ascocarps are macroscopic. An ascocarp may be open and more or less cup-shaped (an apothecium; see Figure 13–11b), closed and spherical (a cleistothecium; see Figure 13–13b), or flask-shaped, with a small pore through which the ascospores escape (a perithecium; see Figure 13–13c). The asci are usually borne on the inner surface of the ascocarp. The layer of asci is called the *hymenium,* or hymenial layer (see Figure 13–14).

Figure 13–15 illustrates the characteristic life cycle of an ascomycete. (Frequent reference to this illustration should prove helpful with the following material.) The mycelium is initiated with germination of an ascospore, and soon after, the mycelium begins to form conidiophores. Many crops of conidia are produced during the growing season, and it is the conidia that are primarily responsible for propagation of the fungus.

13–12
Conidia are the characteristic asexual spores of ascomycetes; they are multinucleate. In these electron micrographs, stages in the formation of conidia in No-muraea rileyi, which infects the velvetbean caterpillar, are shown. (a) Scanning electron micrograph of conidia at various stages of development. (b) Transmission electron micrograph of conidia. In this species, the formation of conidia proceeds toward the base.

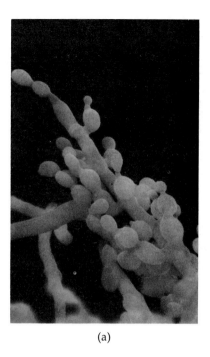

(a)

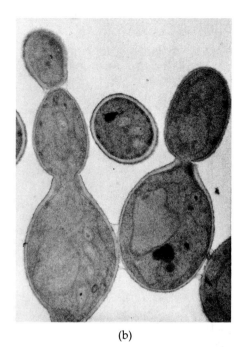

(b)

13–13
(a) An electron micrograph showing two asci of Ascodesmis nigricans *in which ascospores are maturing. (b) Ascocarp of* Erysiphe aggregata, *showing the enclosed asci and ascospores. This completely enclosed type of ascocarp is called a cleistothecium. (c) An ascocarp of* Chaetomium erraticum, *showing the enclosed asci and ascospores. Note the small pore at the top. This sort of ascocarp, with a small opening, is known as a perithecium.*

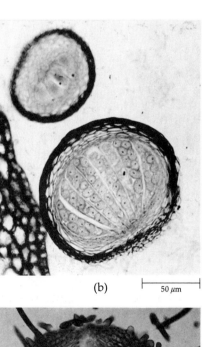

(b) 50 μm

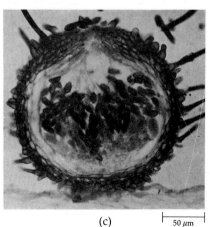

(c) 50 μm

13–14
A section through the hymenial layer of a morel (Morchella) *showing asci with ascospores.*

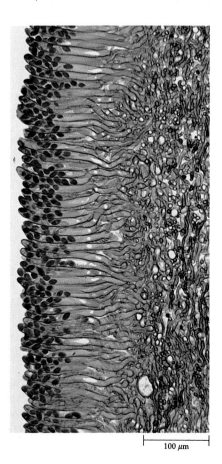

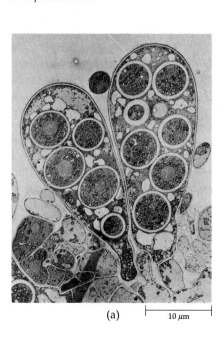

(a) 10 μm

100 μm

13–15

The typical life cycle of an ascomycete. Asexual reproduction occurs by way of specialized spores, known as conidia, *which are usually multinucleate. Sexual reproduction involves the formation of asci and ascospores. Karyogamy is* *followed immediately by meiosis in the ascus, resulting in the production of ascospores.*

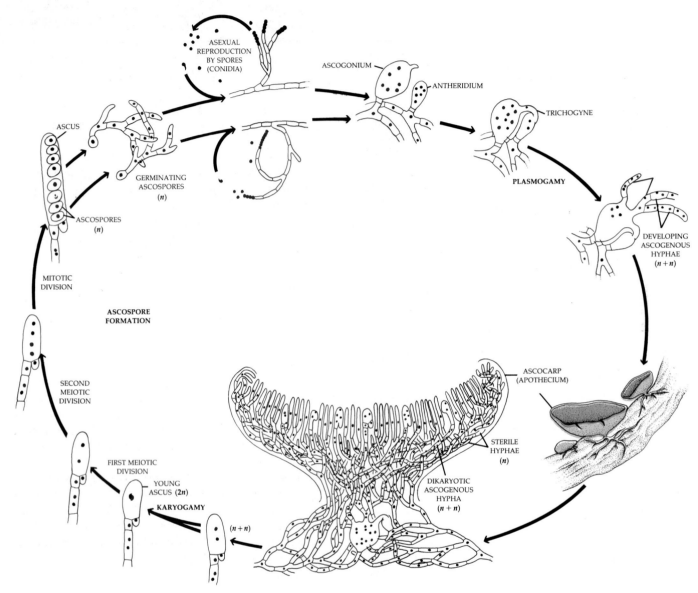

Ascus formation occurs on the same mycelium that produces conidia; it is preceded by the formation of generally multinucleate gametangia called antheridia and ascogonia. The male nuclei of the antheridium pass into the *ascogonium* via the *trichogyne*, which is an outgrowth of the ascogonium. *Plasmogamy*, the fusion of the two protoplasts, has now taken place. The male nuclei may then pair with the genetically different female nuclei within the common cytoplasm, but they do not fuse with them. Hyphal filaments now begin to grow out of the ascogonium and elongate into *ascogenous hy-*

phae. As the ascogenous hyphae develop, pairs of nuclei migrate into them and simultaneous mitotic divisions occur in the hyphae and ascogonium. Cell division in the developing ascogenous hyphae occurs in such a way that the resultant cells are *dikaryotic* (contain two haploid nuclei).

The ascus first forms at the tip of a developing dikaryotic, ascogenous hypha. In the formation of an ascus, one of the binucleate cells of the dikaryotic hypha grows over to form a hook, or *crozier*. In this hooked cell, the two nuclei divide in such a way that their spindle fibers are parallel and more or less vertical in orientation. Two of the daughter nuclei are close to one another at the top of the hook; one of the remaining two daughter nuclei is near the tip, and the other is near the basal septum of the hook. Two septa are then formed; these divide the hook into three cells, of which the middle one becomes the ascus. It is in this middle cell that *karyogamy* occurs: the two nuclei fuse to form a diploid nucleus (zygote), the only diploid nucleus in the life cycle of the ascomycetes (except occasionally in parasexual cycles). Soon after karyogamy, the young ascus begins to elongate. The diploid nucleus then undergoes meiosis, which is generally followed by one mitotic division. If this mitotic division takes place, the mature ascus is eight-spored; if not, it is four-spored. These haploid nuclei are then cut off in segments of the cytoplasm to form *ascospores*. In most ascomycetes, the ascus becomes turgid at maturity and finally bursts, sending its ascospores explosively into the air. The ascospores are generally propelled only about 2 centimeters from the ascus, but some species propel them as far as 30 centimeters.

Unicellular Ascomycetes: The Yeasts

Yeasts are predominantly unicellular; they reproduce asexually by fission or by the pinching off of small buds (Figure 13–16a) rather than by spore formation. Sexual reproduction in yeasts occurs when either two cells or two ascospores unite and form a zygote. The zygote may produce diploid buds or may function as an ascus, undergoing meiosis and producing four haploid nuclei. There may be a subsequent mitotic division. Within the zygote wall, which is now an ascus, walls are laid down around these nuclei so that four or eight ascospores are formed; the ascospores are liberated when the ascus wall breaks down (Figure 13–16b). The ascospores either bud asexually or fuse with another cell to repeat the sexual process.

Yeasts are important to humans because of their ability to ferment carbohydrates, breaking down glucose to produce ethyl alcohol and carbon dioxide in the process. Thus yeasts are utilized by vintners as a source of ethanol, by bakers as a source of carbon dioxide, and by brewers for both substances. Many domestically useful strains of yeast have been developed by selection and breeding, and the techniques of genetic engineering (see Chapter 30) are now being used to improve them further through the addition of useful genes from other organisms. In modern wine-making, pure strains of yeasts are added to relatively sterile grape juice; historically, wild yeasts that were present on the grapes were used to produce wine (see Figure 6–17, page 94). Some of the flavors of wine come directly from the grape, but most arise from the direct action of the yeast. In brewing beer, the medium is sterilized by heating prior to fermentation, and pure yeast cultures are added to it. Most of the yeasts important in the production of wine, cider, sake, and beer are strains of a single species—*Saccharomyces cerevisiae*—although other species also play a role. This yeast is now virtually the only one used in baking bread (Figure 13–16a). Some species of yeast are important as human pathogens, causing such diseases as thrush and cryptococcosis.

A number of yeasts, especially *Saccharomyces cerevisiae*, have become important laboratory organisms for genetic research. Like *Escherichia coli* among the bacteria, this yeast has become the organism of choice for studies of the metabolism, genetics, and development of eukaryotic cells. Recently, scientists have even succeeded in synthesizing a functional yeast microchromosome that is transmitted through successive mitotic cycles; this has not been achieved in any other eukaryote. Such detailed knowledge, coupled with the ease of manipulating the genetics of yeasts, will doubtless increase their importance in industry greatly in the future (see Figure 13–3).

13–16
Yeasts. (a) *Budding cells of bread yeast,* Saccharomyces cerevisiae. (b) *Asci with ascospores of* Schizosaccharomyces octosporus.

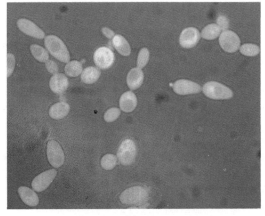

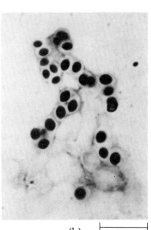

(a) (b) 10 μm

ERGOTISM

A number of ascomycetes are parasitic on higher plants. The disease of plants called ergot is caused by *Claviceps purpurea,* a parasite of rye *(Secale cereale)* and other grasses. Although ergot seldom causes serious damage to the crop of rye, it is dangerous because a small amount mixed with rye grains is enough to cause severe illness among domestic animals or among the people who eat bread containing flour from such grain. Ergotism, the toxic condition caused by eating grain infected with ergot, is often accompanied by gangrene, nervous spasms, psychotic delusions, and convulsions. It occurred frequently during the Middle Ages, when it was known as St. Anthony's fire. In one epidemic in 994, more than 40,000 people died. In 1722, ergotism struck down the cavalry of Czar Peter the Great on the eve of battle for the conquest of Turkey and thus changed the course of history.

Recently, it has been suggested that the widespread accusations of witchcraft in Salem Village (now Danvers) and other communities in Massachusetts and Connecticut in 1692, which resulted in numerous executions of teenage girls, may have resulted from an outbreak of convulsive ergotism. Ergot contains lysergic acid amide (LDA), a precursor of lysergic acid diethylamide (LSD). LSD was first discovered in studies of the alkaloids—a class of nitrogen-containing

compounds that are physiologically active in vertebrates—of *Claviceps purpurea,* and their derivatives. The convulsions and gangrene associated with ergot are, however, caused by other, naturally occurring alkaloids.

As recently as 1951, there was an outbreak of ergotism in a small French village in which 30 people became temporarily insane, imagining that they were pursued by demons and snakes; 5 of the villagers died. Ergot, which causes muscles to contract and blood vessels to constrict, is used in medicine. Its properties, both harmful and beneficial, stem from its alkaloids. As early as 1552, ergot was used in obstetrics to hasten uterine contraction during childbirth; in general, its medical properties are based on its ability to enhance muscle contraction. By inhibiting the hormones adrenalin, noradrenalin, and serotonin, ergot alkaloids also dilate the veins and therefore decrease blood pressure. For this reason, these alkaloids are used to treat diseases such as angina pectoris (chest pain) and glaucoma, in which high blood pressure is an important factor. Efforts are now under way to produce improved strains of *Claviceps purpurea* and to grow them commercially.

The tough, dark resting structures of *Claviceps*—sclerotia —can be seen among the spikelets of rye in the accompanying photograph. After the grain shatters, they overwinter in the soil. Activated by frost, they produce a number of multicellular spore-bearing bodies, which contain abundant perithecia. Each perithecium contains up to 100 asci, from which the ascospores are shed at the time when rye and other grasses are flowering. The ascospores germinate among these flowers, and the resulting mycelia give rise to abundant conidiospores, which are imbedded in a sticky, sweet liquid and are spread further among the grass flowers by the insects they attract. Proliferating in individual immature grass fruits, the fungus converts them into the dark resting structures seen in the photograph; these mature at the same time that the grain is ripe and are dispersed with it. The characteristic alkaloids of *Claviceps* are produced only during the time in which the sclerotia are being formed.

Most yeasts are simple ascomycetes. However, a few genera are basidiomycetes—such as *Cryptococcus,* the causative agent of cryptococcosis, and some nonpathogenic species of *Candida.* The simplified and usually unicellular structure of yeasts evidently has been derived from that of the more complex mycelium-forming fungi. The resulting organisms are so reduced that it is difficult to determine their affinities to the different groups of fungi from which they ultimately derived. As far as we know, however, all yeasts were derived from multicellular ancestors, and the evolution of the single-celled, yeast form of growth has occurred more than

once, even among the ascomycetes. The yeasts include some 39 genera, with approximately 350 species. They are found in a wide range of terrestrial and aquatic habitats in which a suitable carbon source is available.

Fungi Imperfecti

Fungi Imperfecti, or Deuteromycetes, comprise about 25,000 described species of fungi in which the sexual phase is unknown (Figures 13–17 and 13–18). The sexual phase may be unknown either because these fungi have not been sufficiently studied or because it has

13–17
Penicillium *and* Aspergillus—*two of the common genera of Fungi Imperfecti.* (a) *A culture of* Penicillium notatum, *the penicillin-producing fungus, showing the distinctive colors produced during growth and spore development.* (b) *A culture of* Aspergillus fumigatus, *a fungus that causes respiratory disease in humans. Notice the concentric growth pattern produced by successive "pulses" of spore production.*

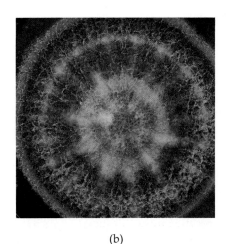

(a) (b)

been lost in the course of evolution. Most fungi lacking a sexual phase (the "imperfect fungi") are ascomycetes; they reproduce only by means of conidia. A few are basidiomycetes, as shown by the presence of the septa and clamp connections characteristic of that group (see Figure 13–29, page 216). Parasexual cycles are common among the Fungi Imperfecti and seem to compensate partially for the lack of sexual reproduction in maintaining variability. In general, the classification of the members of this class is based upon the mode of formation of their conidia (Figures 13–12, 13–18). Fungi Imperfecti is an artificial group made up of diverse organisms; it is maintained primarily as a convenience in identification. The group can be considered a class, but it is not equivalent to the other classes of fungi.

Among the Fungi Imperfecti are many organisms of great economic importance. For example, certain members of the genus *Penicillium* give some types of cheese the flavor, odor, and character so highly prized by gourmets. One such mold, *Penicillium roquefortii*, was first found in caves near the French village of Roquefort. Legend has it that a peasant boy left his lunch, a fresh piece of mild cheese, in one of these caves and on returning several weeks later found it marbled, tart, and fragrant. Only cheeses from the area around these particular caves are permitted to bear the name of Roquefort. Another species of the same genus, *Penicillium camembertii*, gives Camembert cheese its special qualities. In the Orient, soy paste (miso) is produced by fermenting soybeans with *Aspergillus oryzae*; soy sauce (shoyu) is produced by fermenting soybeans with a mixture of *A. oryzae* and *A. soyae*. Lactic acid bacteria and yeasts are also actively involved in producing the final product. (Tofu, or bean curd, and tempeh, a similar protein-rich food of tropical Asia, are produced following fermentation of soybeans with species of *Mucor* and *Rhizopus*, respectively; these fungi are zygomycetes.) *Aspergillus oryzae* is also important for the initial steps in brewing sake, the traditional alcoholic beverage of Japan; the yeast *Saccharomyces cerevisiae* is

13–18
The conidiophores of imperfect fungi are used in their classification. (a) *Penicillium (branching) and* (b) *Aspergillus (tightly clumped).*

(a)

(b)

important later in the process. Citric acid is produced commercially in large amounts from colonies of *Aspergillus* grown under very acid conditions. Currently, the enrichment of livestock feed by fermentation with *A. oryzae* (to increase the protein content) is being investigated in Europe and the United States.

Antibiotics are substances produced by one living organism that inhibit the growth of other organisms (such as bacteria) and so may be therapeutically useful to humans. Several important antibiotics are produced by Fungi Imperfecti. The first antibiotic was discovered by Sir Alexander Fleming, who noted in 1928 that a strain of *Penicillium* that had contaminated a culture of *Staphylococcus* growing on a nutrient agar plate had completely halted the growth of the bacteria. Ten years later, Howard Florey and his associates at Oxford University purified penicillin and later came to the United States to promote the large-scale production of the drug. The demand during World War II was so great that the production of penicillin was increased from a few million units per month in 1942 to more than 700 billion units in 1945. Penicillin is effective in curing a wide variety of bacterial diseases caused by gram-positive bacteria, including pneumonia, scarlet fever, syphilis, gonorrhea, diphtheria, rheumatic fever, and many others. Many of the substances that we use as antibiotics undoubtedly also play significant ecological roles in nature, allowing the organisms that produce them to outcompete others.

Not all substances produced by imperfect fungi are useful to humans. For example, the aflatoxins, a group of closely related secondary metabolites produced by certain strains of *Aspergillus flavus* and *Aspergillus parasiticus* are highly toxic and also carcinogenic (cancer-producing). These fungi sometimes contaminate stored food, and they constitute a significant public-health problem. Another group of fungal toxins, the trichothecenes, which are produced by a number of genera of imperfect fungi, are potent inhibitors of protein synthesis of eukaryotes.

PREDACEOUS FUNGI

Among the most highly specialized of the fungi are the predaceous fungi; they have developed a number of mechanisms for capturing the small animals they use as food. Although microscopic fungi with such habits have been known for many years, recently it has been learned that a number of species of gilled fungi also attack and consume nematodes (small roundworms). The oyster mushroom, *Pleurotus ostreatus*, for example, grows on decaying wood. Its hyphae excrete a substance that anesthetizes nematodes, after which its hyphae envelop and penetrate these tiny worms; the fungus apparently uses them primarily as a source of nitrogen, thus supplementing the low levels of nitrogen that are present in wood.

Some of the microscopic imperfect fungi secrete a sticky substance on the surface of their hyphae in which passing protozoa, rotifers, small insects, or other animals become stuck. More than 50 species of this class trap or snare nematodes. In their presence, the fungal hyphae produce loops that swell rapidly, closing the opening like a noose, when a nematode rubs against its inner surface. Presumably the stimulation of the cell wall increases the amount of osmotically active material in the cell, causing water to enter the cells and increase their turgor.

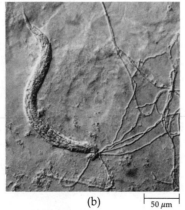
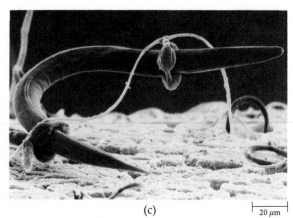

(a) | (b) $50 \mu m$ | (c) $20 \mu m$

(a) *The oyster mushroom,* Pleurotus ostreatus. (b) *The hyphae of the oyster mushroom converging on the mouth of an immobilized nematode.* (c) *The predaceous imperfect fungus* Arthrobotrys dactyloides *has* trapped a nematode. The traps consist of rings, each comprising three cells, which swell rapidly to about three times their original size and garrot the nematode. Once the worm has been trapped, fungal hyphae grow into its body and digest it. When triggered, the ring cells can expand completely in a period of less than a tenth of a second.

One group of Fungi Imperfecti, the dermatophytes (from the Greek words *dermatos*, skin + *phyton*, plant), are the cause of ringworm, athlete's foot, and other fungal skin diseases. Such diseases are especially prevalent in the tropics. The pathogenic stages of these fungi are asexual, but most of them have now been correlated with species of ascomycetes; nevertheless, they continue to be classified on the basis of their disease-causing forms. Athlete's foot flourishes under warm, wet conditions and usually disappears rapidly if a person switches from boots or shoes to sandals and keeps his or her feet dry. Although there are typically about 2 million bacteria per square centimeter on the sole of the foot, competing with the dermatophytes for the available nutrients, the fungi can outcompete them because they grow directly into the epidermal cells and also poison the bacteria with the antibiotics they secrete. They are much more active than bacteria in metabolizing keratin, the tough, fibrous protein to which the contents of epidermal cells are converted as they are sloughed off. Eventually, however, the bacteria may develop resistant strains and then proliferate, causing severe infection.

During World War II, more soldiers had to be sent back from the South Pacific because of skin infections than because of wounds received in battle. Many fungi are associated with particular diseases; one of these, *Candida albicans,* which is yeastlike under certain conditions, causes thrush and other infections of the mucous membranes. Fungal spores are continually inhaled, and some cause internal diseases, which may be serious or fatal, especially if they attack the lungs.

In recent years, fungal diseases have become much more prevalent in humans than formerly. Chemicals such as cyclosporine (see page 197), which was first isolated from an imperfect fungus, *Tolypocladium inflatum*, are now routinely administered to transplant patients to suppress their normal immune reactions so that they will be able to accept organ and tissue transplants. However, these chemicals tend to make the patients more susceptible to fungal and other diseases. Some chemical treatments, such as those for acute leukemia, also lower the defenses of the body against fungal infection; thus attention is increasingly directed toward the prevention and treatment of fungal diseases.

Lichens

Lichens are symbiotic associations between ascomycetes and certain genera of green algae or cyanobacteria.* The photosynthetic members of these associations provide the nutrients, and they are protected from environmental extremes by their fungal partners. Because of this symbiotic relationship, lichens are able to live in the harshest habitats on earth (Figure 13–19). Lichen-forming fungi constitute about 20,000 species of varied appearance, different from the roughly 30,000 named species of ascomycetes. Some 26 genera of photosynthetic organisms are found in combination with these ascomycetes. The most frequent are the green algae *Trebouxia, Pseudotrebouxia,* and *Trentepohlia* and the cyanobacterium *Nostoc;* about 90 percent of all lichens have one of these four genera in their photosynthetic component.

---

* There are also about a dozen species of basidiomycetes that form associations with algae, but they are closely related to free-living basidiomycetes and do not resemble the other lichen-forming fungi.

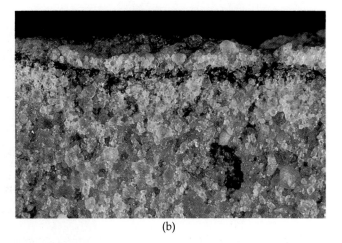

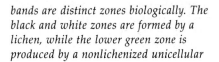

(a)   (b)

13–19
In this seemingly lifeless, dry region of Antarctica (a), lichens (b) live just beneath the exposed surface of the sandstone. In the fractured rock, the colored *bands are distinct zones biologically. The black and white zones are formed by a lichen, while the lower green zone is produced by a nonlichenized unicellular* *green alga. The temperatures in this part of Antarctica rise almost to freezing in the summer and probably fall to −60°C in winter.*

Lichens are extremely widespread in nature; they occur from arid desert regions to the arctic and grow on bare soil, tree trunks, sunbaked rocks, fence posts, and windswept alpine peaks all over the world (see Figures 13–20, 13–21, and 13–22). Some lichens are so tiny that they are almost invisible to the unaided eye; others, like the reindeer "mosses," may cover kilometers of land with ankle-deep growth. One species, *Verrucaria serpuloides*, is a permanently submerged marine lichen. Lichens are often the first colonists of newly exposed rocky areas. In Antarctica, there are more than 350 species of lichens (Figure 13–19) but only two species of vascular plants; seven species of lichens actually occur within 4° of the South Pole! The activities of lichens in forming the first soil on rock surfaces initiate biological succession (see page 666) in these areas. Those lichens in which the photosynthetic partner is a cyanobacterium are of special importance because they also contribute fixed nitrogen to the soils; such lichens are the primary factor in supplying nitrogen in many regions.

The colors of lichens range from white to black, through shades of red, orange, brown, yellow, and green, and they contain many unusual chemical compounds. Many lichens are used as sources of dyes around the world; for example, in early days the characteristic color of Harris tweed resulted from the treatment of the wool with a lichen dye. Many lichens have also been used as medicines, perfumes, or as minor sources of food by different groups of people.

Lichens have long been the subject of biological investigations because of the intriguing nature of the association between a fungus and its included algae or cyanobacteria. The fungal partner apparently plays the major role in determining the form of the lichen; recently, however, it has been demonstrated that a single fungus, associated with different kinds of algae can

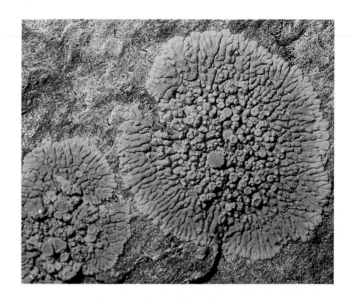

13–20
A crustose ("encrusting") lichen growing on a bare rock surface in central California.

produce morphologically very different individuals that have traditionally been placed in different genera. The algae or cyanobacteria found in lichens also occur as free-living species, whereas the lichen fungi are generally found in nature only in lichens.

Some lichens produce special fragments, known as *soredia*, which are composed of fungal hyphae and algae or cyanobacteria (Figure 13–23). The dispersal of these soredia serves to establish the lichens in new localities. Lichen fungi also often form ascocarps, which are similar to those of nonlichen fungi except that in lichens the ascocarps may endure and produce spores continuously over a number of years.

13–21
(a) Parmelia perforata, *a foliose ("leafy") lichen growing around a hummingbird's nest on a dead tree branch in Mississippi.* (b) Old-man's beard (Usnea), *a hanging lichen that often occurs in masses on the limbs of trees. Although it is superficially similar and also occupies a similar ecological niche, "Spanish moss," which is common throughout the southern United States, is actually a flowering plant—a member of the pineapple family (Bromeliaceae).*

(a)

(b)

(a)

(b)

(c)

13–22

Several fruticose ("shrubby") lichens. (a) Goldeye lichen, Teloschistes chrysophthalmus. *(b) British soldiers,* Cladonia cristatella, *are 1 to 2 centimeters tall. (c)* Cladonia subtenuis, *commonly called "reindeer moss," is actually a lichen. Lichens of this group, which are abundant in the Arctic, concentrated*

radioactive substances following atmospheric nuclear tests. Reindeer feeding on the lichens concentrated these radioactive substances still further and passed them on to the humans and other animals that consumed them or their products, especially milk and cheese.

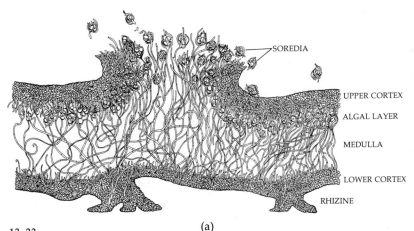

SOREDIA

UPPER CORTEX

ALGAL LAYER

MEDULLA

LOWER CORTEX

RHIZINE

(a)

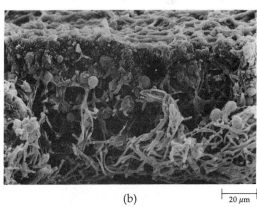

(b)

20 μm

13–23

(a) A cross section of the lichen Lobaria verrucosa. *The simplest lichens consist of a crust of fungal hyphae entwining colonies of algae. In the more complex lichens, however, the hyphae and the algal cells are organized in a thallus with a definite growth form and characteristic internal structure. The lichen shown here has four distinct layers: (1) the upper cortex, a protective surface of heavily gelatinized fungal hyphae; (2) the algal*

layer, which consists of algal cells and loosely interwoven, thin-walled hyphae; (3) the medulla, which is a thick layer of loosely packed, colorless, weakly gelatinized hyphae. This layer, which makes up about two-thirds of the thickness of the thallus, appears to serve as a storage area, with enlarged cells in the fungal hyphae; and (4) the lower cortex, which is thinner than the upper one and is covered with fine projections

that attach it to the substrate. Soredia are fragments of a lichen that include algal cells and fungal hyphae; they serve to disperse the lichen to new areas. (b) A scanning electron micrograph of a section of Cladonia cristatella, *the same lichen shown in Figures 13–22b and 13–25, showing upper cortex, algal layer, and medulla. This lichen has no lower cortex.*

Biology of the Lichens

How can the lichens survive under environmental conditions so adverse to any other form of life? At one time, it was thought that the secret of the lichen's success was that the fungal tissue protected the alga or cyanobacterium from drying out. Actually, one of the chief factors in lichen survival seems to be the fact that they dry out very rapidly. Lichens are frequently very desiccated in nature, with a water content ranging from only 2 to 10 percent of their dry weight. When the lichen dries out, photosynthesis ceases; in this state of "suspended animation," even blazing sunlight or great extremes of heat or cold can be endured by some species of lichens. Cessation of photosynthesis depends, in large part, on the fact that the upper cortex of the lichen becomes thicker and more opaque when dry, cutting off the passage of light energy. A wet lichen is destroyed by light intensities or temperatures that do not harm a dry lichen.

When a lichen is wetted by rain, it absorbs 3 to 35 times its own weight in water in a very short time. If a dry, brittle lichen is submerged in water, it becomes soft and pliable within a few minutes. This is the simple physical process of imbibition—the lichen takes up water just as blotting paper does.

The lichen reaches its maximum vitality, as judged by the rate of photosynthesis, after it has been soaked with water and has begun to dry. Its rate of photosynthesis reaches a peak when the water content is 65 to 90 percent of the maximum it can hold; below this level, if the lichen continues to lose water, the rate of photosynthesis decreases. In many environments, the water content of the lichen varies markedly in the course of a day, with most photosynthesis taking place only during a few hours, usually in the early morning after wetting by fog or dew. As a consequence, lichens have an extremely slow rate of growth, their radius increasing at a rate of about 0.1 to 10 millimeters a year. Calculated on this basis, some mature lichens have been estimated to be as much as 4500 years or more in age. They achieve their most luxuriant growth and development along seacoasts and on fog-shrouded mountains.

Lichens apparently absorb some mineral nutrients from their substrate (this is suggested by the fact that particular species are characteristically found on particular kinds of rocks or soil or tree trunks), but most of the elements absorbed by lichens enter them from the surrounding air and in rainfall. Lichens absorb elements from rainwater rapidly and concentrate them. In their activities, lichens clearly play an important role in the function of ecosystems. Because they have no means of excreting the elements that they absorb, however, lichens are particularly sensitive to toxic compounds; the toxins cause the chlorophyll in their algal or cyanobacterial cells to deteriorate. Lichen growth is a very sensitive indication of the toxic components of polluted air, and they are being used increasingly to monitor atmospheric pollutants, especially around cities. Lichens are particularly sensitive to sulfur dioxide, probably because it quickly destroys their limited supplies of chlorophyll. Both the state of health of the lichens and their chemical composition are used as indicators of the health of their environment. For example, the analysis of lichens can be used to trace the amounts of heavy metal and other pollution around industrial sites. Many lichens have the ability to bind heavy metals outside their cells and thus escape damage themselves.

When nuclear tests were being conducted in the atmosphere, lichens were used to monitor the fallout. Now lichens are considered to present a useful way to monitor the contamination by radioactive substances that may occur following the crash of satellites, especially when this occurs in remote areas that are difficult to monitor by other means.

The Nature of the Relationship between Fungi and Photosynthetic Organisms

What is the relationship between the two components of a lichen? It is clear that the fungus obtains organic carbon from the alga or cyanobacterium, since the lichen behaves like a photosynthetic organism, dependent only on light, air, and mineral nutrients. In fact, the movement of organic carbon from the included algae or cyanobacteria to the fungus has been established by tracing experiments using ^{14}C-labeled carbon dioxide. In those lichens that include cyanobacteria such as *Nostoc*, nitrogen fixation by the bacteria and transfer to the fungus are also important. In the lichen, the fungal hyphae form a close network around the included cells. Haustoria (the specialized hyphae of parasitic fungi), which are very common in lichens, penetrate the photosynthetic cells (Figure 13–24); other specialized organs, called appressoria, lie along the surface of the photosynthetic cells, penetrating them by means of specialized pegs. The association with a fungus profoundly affects the nature of the metabolic output from the included photosynthetic cells. For example, green algae excrete large quantities of two particular alcohols, D-sorbitol and D-ribitol, only when they are associated with fungi to form lichens.

It is possible to separate the algal or cyanobacterial and fungal components of a particular lichen and grow them alone in pure culture. Under such conditions, the fungus grows in compact colonies that look quite unlike the actual lichen. It requires a large number of complex carbohydrates for growth and does not ordinarily form spore-producing bodies. In contrast, isolated lichen algae or cyanobacteria grow more rapidly when they are free-living than when they are in the lichen association. In some ways it is appropriate to think of the li-

13–24

A haustorium of the fungal component of the lichen Strigula elegans, *penetrating the green algal component,* Cephaleuros virescens. *The densely packed chloroplasts are prominent in the algal cell.*

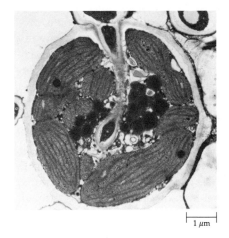

13–25

Scanning electron micrographs of early stages in the interaction between fungal and algal components of British soldiers (Cladonia cristatella) in laboratory culture. The photosynthetic component in this lichen (see Figure 13–23b) is

Trebouxia, *a green alga. (a) An algal cell surrounded by fungal hyphae. (b) Penetration of algal cells by fungal haustoria. (c) Mixed groups of fungal and algal components developing into the mature lichen.*

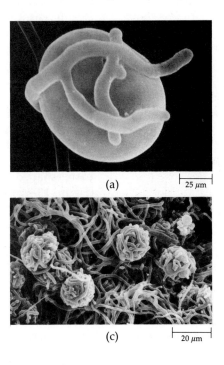

(a) 25 μm

(c) 20 μm

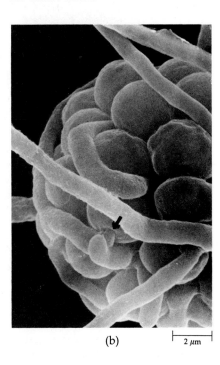

(b) 2 μm

chen partnership as controlled parasitism of the alga or cyanobacterium by the fungus, rather than symbiosis; nevertheless, a lichen does amount to more than the sum of its components, and in that sense it is a distinct "organism." When grown together in culture, the fungus seems first to bring its photosynthetic partner under control and then to establish the characteristic appearance of the mature lichen (Figure 13–25).

Division Basidiomycota

The most familiar fungi are members of this large division, which includes some 25,000 described species—not only mushrooms, toadstools, stinkhorns, puffballs, and shelf fungi but also two important groups of plant pathogens: the rusts and smuts (see Figure 13–26). Basidiomycetes are distinguished from other fungi by their production of *basidiospores*, which are borne outside a club-shaped spore-producing structure called the *basidium* (plural: basidia) (Figure 13–27). In nature, most basidiomycetes reproduce primarily through the formation of basidiospores. Even though the larger basidiomycetes are the best known fungi, many more species undoubtedly await discovery.

13–26

Corn smut is a familiar plant disease in which the fungus Ustilago maydis *produces black, dusty-looking masses of spores in ears of corn.* Ustilago *is a smut, a member of the class Teliomycetes of the division Basidiomycota.*

13-27

*Scanning electron micrograph of the top
of a basidium of* Aleurodiscus amor-
phus, *with an elongate, warty basidiospore attached to each of its four
sterigmata.*

13-28

*Characteristic septum of the secondary
mycelium of a basidiomycete, shown in an
electron micrograph of* Laetisaria arvalis,
*a common wood-rotting fungus. As usual
in the basidiomycetes, the septum is
perforated by a pore.*

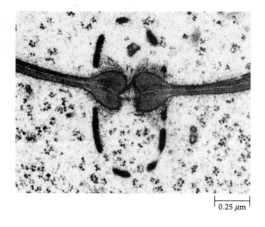

The mycelium of the basidiomycetes is always septate and in most species, passes through two distinct
phases—primary and secondary—during the life cycle
of the fungus. When it germinates, a basidiospore produces the *primary mycelium.* Initially, this mycelium
may be multinucleate, but septa are soon formed and
the mycelium is divided into monokaryotic (uninucleate) cells. Commonly, the *secondary mycelium* is produced by fusion of primary hyphae from different
mating types (in which case it is heterokaryotic) or it
may arise when septa are not formed after nuclear formation (in which case it is homokaryotic). In either
case, the result is formation of a dikaryotic (binucleate)
mycelium, since karyogamy does not immediately follow plasmogamy.

The septa in the secondary mycelia of basidiomycetes
are perforated; however, they are lined with a thick,
barrel-shaped structure (absent in the rusts and smuts)
(Figure 13–28). Nuclei may not be able to pass through
such septa. In those heterokaryotic secondary mycelia
in which nuclear migration appears to be taking place,
the septa have only simple, enlarged pores, similar to
those of ascomycetes (Figure 13–7).

The apical cells (those at the apex or tip) of the secondary mycelium usually divide by the formation of
clamp connections (Figure 13–29). These clamp connections, which ensure the allocation of one nucleus of
each type to the daughter cells, are highly characteristic
of the basidiomycetes.

The mycelium that forms the *basidiocarps*—fleshy,
spore-producing bodies, such as mushrooms and puffballs—is secondary mycelium. The formation of the
basidiocarps may require light. As it forms the basidiocarps, the secondary mycelium becomes differentiated
into specialized hyphae that play different functions
within the basidiocarp; such mycelium is sometimes
called *tertiary mycelium.*

The basidiomycetes are divided into three classes:
Hymenomycetes, Gasteromycetes, and Teliomycetes.

13-29

(a) *In the basidiomycetes, dikaryotic
hyphae are distinguished by the presence
of clamp connections over the septa.
Clamp connections are formed during cell
division; they presumably ensure the
proper distribution of the two genetically
distinct types of nuclei in the basidiocarp.* (b) *Clamp connections and
characteristic septa in a basidiomycete
hypha.*

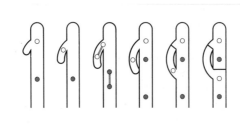

(a)

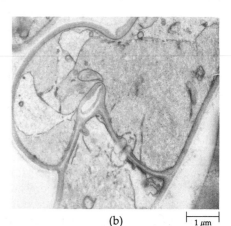

(b)

(a)

(b)

(c)

(d)

13-30

Basidiomycetes—members of the class Hymenomycetes. (a) The fly agaric, Amanita muscaria. *The mushrooms are at various stages of growth; one has been picked to show the gills. Among the characteristics of this genus of mushrooms, many members of which are poisonous, are the ring on the stalk and the cup around its base. (b)* Suillus bovinus, *a bolete. In the boletes, the gills found in most kinds of mushrooms are replaced by pores. (c) A shelf fungus,* Ganoderma tsugae. *Shelf fungi are responsible for most wood rot. (d) An edible coral fungus,* Hercium coralloides. *The hymenium, an outer spore-bearing layer of basidia, is borne on all sides of the basidiocarp.*

Class Hymenomycetes

The class Hymenomycetes contains the edible and poisonous mushrooms, coral fungi, and shelf or bracket fungi (see Figures 13–4 and 13–30). Included in this class are all fungi that produce basidiospores exposed *on* a basidiocarp; the Gasteromycetes include those fungi that form spores enclosed *in* a basidiocarp. The basidia of Hymenomycetes are always formed on a hymenium—the source of the group's name—and not from single dispersed cells, as in the Teliomycetes. The basidiocarps that are characteristic of the Hymenomycetes and Gasteromycetes are analogous to the ascocarps of the ascomycetes.

Some Hymenomycetes have club-shaped, aseptate (internally undivided) basidia, each of which usually bears four basidiospores, each on a minute projection called a *sterigma* (plural: sterigmata) (Figures 13–27 and 13–31). Others—the jelly fungi (Figure 13–32)—have septate (internally divided) basidia like those of the Teliomycetes.

The structure that one recognizes as a mushroom or toadstool is a basidiocarp. ("Mushroom" is sometimes popularly used to designate the edible forms of basidiocarps and "toadstool" is used to designate the inedible ones, but mycologists do not recognize such a distinction and use only the term "mushroom." In this book, all such forms are referred to as mushrooms; this does not mean that they are all edible.) The mushroom usually consists of a *cap*, or *pileus*, that sits atop a *stalk*, or *stipe*. The masses of hyphae in the basidiocarps usually form distinct tissues and layers. Early in its development—the "button" stage—the mushroom may be covered by a membranous tissue that ruptures as the mushroom enlarges. In some genera, remnants of this tissue are visible on the upper surface of the cap and, at the base of the stipe, as a cup, or *volva*. In many Hymenomycetes, the lower surface of the cap consists of radiating strips of tissue called *gills* (Figure 13–31). In other members of the class, the hymenium is located elsewhere; for example, in the coral fungi (Figure 13–30d) the hymenium covers the teeth and fingerlike projections. In the shelf fungi and boletes (see Figure 13–30b), the hymenium lines the pores.

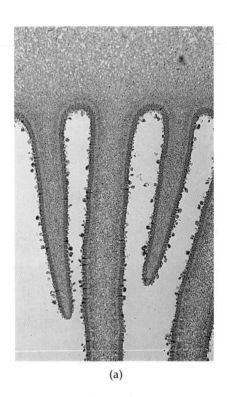

(a)

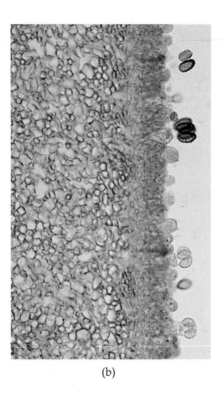

(b)

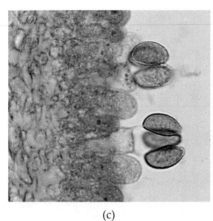

(c)

13–31
Coprinus, *a common mushroom.* (a) *Section through the gills. The relatively dark margins of the gills constitute the hymenial layer.* (b) *Section through the hymenial layer, showing developing basidia and basidiospores.* (c) *Nearly mature basidiospores attached to the basidium by sterigmata.*

In relatively uniform habitats such as lawns and fields, the mycelium from which mushrooms are produced spreads underground, growing downward and outward and forming a ring on the surface that may grow as large as 30 meters in diameter. In an open area, the mycelium expands evenly in all directions, dying at the center and sending up spore-producing bodies at the outer edges, where it grows most actively, because this is the area in which the nutritive material in the soil is most abundant. As a consequence, the mushrooms appear in rings, and as the mycelium grows, the rings become larger. Such circles of mushrooms are known in European folk legend as "fairy rings" (see Figure 13–36).

The best known basidiomycetes are the gill fungi. *Agaricus campestris,* the common field mushroom, is particularly abundant. The closely related *Agaricus bisporus* is one of the few mushrooms that can be cultivated commercially. It is now grown in more than 70 countries, and the value of the world crop exceeds $14 billion. Together with the Oriental shiitake mushroom, *Lentinus edodes, Agaricus bisporus* makes up about 86 percent of the world crop. Other mushrooms are also being cultivated, and some that are gathered in large quantities in nature are sold fairly regularly. Not all mushrooms are edible, however; the gill fungi also include many that are poisonous. For example, the genus *Amanita* includes the most highly poisonous of all mushrooms, as well as some that are edible. Even one bite of the "destroying angel," *Amanita phalloides,* can be fatal. Some basidiomycetes contain chemicals that cause humans who consume them to hallucinate.

A representative mushroom life cycle is shown in Figure 13–33. The basidia form on the hymenia that line the gills, each as a terminal cell of a dikaryotic hypha. Soon after the young basidium enlarges, karyogamy occurs. This is followed almost immediately by meiosis of each diploid nucleus, resulting in the formation of four haploid nuclei. Each of the four nuclei then migrates into a sterigma, which enlarges at its tip to form a uninucleate basidiospore. The reproductive capacity of a single mushroom is tremendous, with billions of spores being produced by a single basidiocarp.

13–32
A jelly fungus growing on a dead limb in the Amazon forest of Brazil. Jelly fungi are the only Hymenomycetes that form septate basidia.

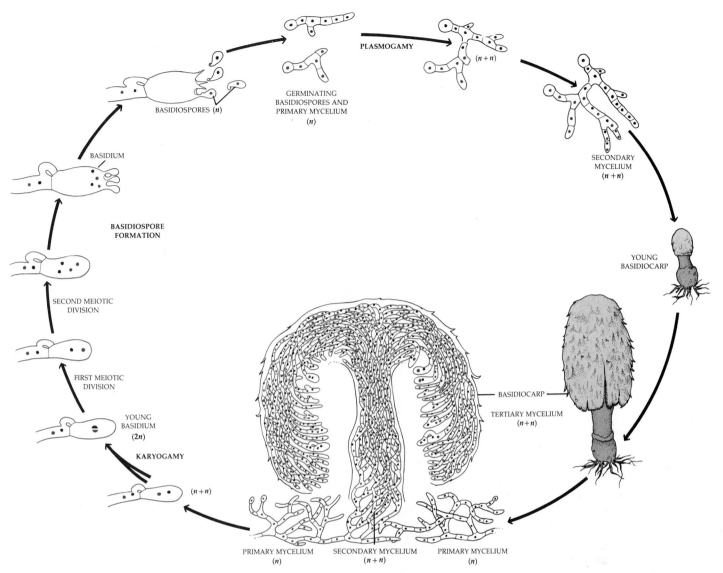

13-33
Life cycle of a mushroom (division Basidiomycota, class Hymenomycetes). The primary mycelia are produced from basidiospores. Secondary dikaryotic mycelia are formed from the primary mycelia. Some secondary mycelia are formed by the fusion of hyphae from different mating types, in which case the mycelia are heterokaryotic. The secondary mycelia divide and differentiate to form the tertiary mycelia that make up the basidiocarp.

13-34
Mushrooms figure prominently in the religious ceremonies of several groups of Indians in southern Mexico and Central America; the Indians eat certain basidiomycetes for their hallucinogenic qualities. One of the most important of the mushrooms is Psilocybe mexicana, *shown here growing in a pasture near Huautla de Jiménez, Oaxaca, Mexico. The shaman María Sabina is eating* Psilocybe *in the course of a midnight religious ceremony. Psilocybin is the chemical responsible for the colorful visions experienced by those who eat these "sacred" mushrooms; it is a structural analogue of LSD and mescaline (see Figure 29-50, page 618).*

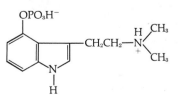

PSILOCYBIN

13–35

Basidiomycetes, members of the class Gasteromycetes. (a) *Puffballs,* Lycoperdon ericetorum. *The spores are being discharged from the pore at the top of each puffball and dispersed by the wind.* (b) *Stinkhorn,* Phallus impudicus. *The basidiospores are released in a foul-smelling, sticky mass at the top of the fungus. Flies, such as the individuals of* Mydaea urbana *shown here, visit it for food and spread the spores, which adhere to their legs and bodies in great numbers.* (c) *Bird's-nest fungus,* Cyathus striatus. *The round structures of the "nests" of these puffballs contain the basidiospores, which are splashed out and dispersed by raindrops.* (d) *Earthstar,* Geastrum triplex. *The outer layer of a typical puffball is folded back in this genus.*

(a)

(c)

(b)

(d)

13–36

A partial "fairy ring" formed by the mushroom Agaricus arvensis. *Some fairy rings are estimated to be up to 500 years old. Because of the exhaustion of key nutrients, the grass immediately inside such a ring is often stunted and lighter green than that outside of it.*

Class Gasteromycetes

The Gasteromycetes produce basidiospores inside basidiocarps, which are completely enclosed for at least part of their development. This class probably evolved from the Hymenomycetes. Among the more familiar members of this class are the stinkhorns, earthstars, false truffles, bird's-nest fungi, and puffballs (Figure 13–35). Stinkhorns (Figure 13–35b) have a remarkable morphology. They develop underground in tough sacs. At maturity, they differentiate into a lower receptacle, derived from the sac in which they develop; a stalk; and a gleba, which is a convoluted, external hymenium. On the gleba is borne an evil-smelling, sticky mass of spores that is regularly visited by flies and beetles, which disperse the spores in their feces.

The puffballs are familiar Gasteromycetes. At maturity, the interior of a puffball dries up, and it releases a cloud of spores when touched (Figure 13–35a). A single puffball may produce several trillion basidiospores. The bird's-nest fungi (Figure 13–35c) begin their development like puffballs, but the disintegration of much of their internal structure leaves them looking like miniature bird's nests.

Class Teliomycetes

The class Teliomycetes consists of two large orders of fungi: the rusts (order Uredinales) and the smuts (order Ustilaginales). The members of this class differ from other basidiomycetes in that they do not form basidiocarps. They do, however, form dikaryotic hyphae and basidia, which are septate like those of the jelly fungi (members of the class Hymenomycetes). The rusts and smuts (see Figure 13–26) produce their spores in masses, which are smaller in rusts and are called *sori*. As plant pathogens, the rusts and smuts are of tremendous economic importance, causing billions of dollars of damage to crops throughout the world each year.

The life cycles of many rusts are complex, and these pathogens are a continual challenge to the plant pathologist whose task it is to keep them under control. Until recently, the rusts were thought to be obligate parasites on vascular plants, but now several species have been maintained in artificial culture. Some smuts are also capable of completing their development under laboratory conditions.

An example of a rust life cycle is provided by *Puccinia graminis*, the cause of black stem rust of wheat. It is one of some 7000 species of rusts. Numerous strains of *P. graminis* exist; in addition to wheat, they parasitize other cereal grains such as barley, oats, and rye, and various species of wild grasses. *Puccinia graminis* is a continual source of economic loss for the wheat grower. In a single year, the losses in Minnesota, North Dakota, South Dakota, and the prairie provinces of Canada amounted to nearly 8 million metric tons. As early as 100 A.D., Pliny described wheat rust as "the greatest pest of the crops." Today plant pathologists combat black stem rust largely by breeding resistant wheat varieties, but mutation and recombination in the rust make any advantage short-lived. Parasexual processes resulting in somatic recombination is one factor in the production of new infective strains of wheat rust.

Puccinia graminis is *heteroecious;* that is, it requires two different hosts to complete its life cycle (Figure 13–37). *Autoecious* parasites, in contrast, require only one host. *Puccinia graminis* can grow indefinitely on its grass host, but there it reproduces only asexually. In order for sexual reproduction to take place, the rust must spend part of its life cycle on barberry (*Berberis*) and part on a grass. One method of attempting to eliminate this rust has been to eradicate barberries. For example, the crown colony of Massachusetts passed a law ordering "whoever . . . hath any barberry bushes growing in his or their land . . . shall cause the same to be extirpated or destroyed on or before the thirteenth day of June, A.D. 1760."

Infection of the barberry occurs in the spring, when uninucleate basidiospores infect the plant by forming flask-shaped *spermogonia,* primarily on the upper surfaces of the leaves. The form of *Puccinia graminis* that grows on barberry consists of separate + and − strains, so that the rust's basidiospores and the spermogonia derived from them are either + or −. Each spermogonium contains two types of hyphae: one that produces chains of small cells called *spermatia* and a second, the so-called *receptive hyphae*. The spermatia are discharged from an opening in the spermogonium. If the + spermatium of one spermogonium comes in contact with the − receptive hypha of another spermogonium, or vice versa, plasmogamy occurs and dikaryotic hyphae are produced. Aecial initials are produced from the dikaryotic hyphae that extend downward from the spermogonium. *Aecia* are then formed primarily on the lower surface of the leaf, where they produce chains of *aeciospores*. The dikaryotic aeciospores then infect the wheat.

The first external manifestation of infection on the wheat is the appearance of rust-colored, linear streaks on the leaves and stems (the red stage). These streaks are *uredinia*, containing unicellular, dikaryotic *urediniospores*. Urediniospores are produced throughout the summer and reinfect the wheat; they are also the primary means by which wheat rust has spread throughout the wheat-growing regions of the world. In late summer and early fall, the red-colored sori gradually darken and become *telia* with two-celled, dikaryotic *teliospores* (the black stage). The teliospores are overwintering spores, which infect neither wheat nor barberries. In early spring, prior to germination, the two haploid nuclei fuse with one another to form a diploid nucleus. With the onset of germination, meiosis takes place, apparently in the short, cylindrical basidia that emerge from the two cells of the teliospore. Septa are formed between the resultant nuclei, which then migrate into the sterigmata and develop into basidiospores. Thus, the year-long cycle is completed.

In certain regions, the life cycle of wheat rust can be short-circuited through the persistence of the uredinal state when actively growing plant tissues are continuously available. In the North American plains, urediniospores from winter wheat in the southwestern states and Mexico drift northward until they reach southern Manitoba. Later generations scatter westward to Alberta, and finally there is a southern drift at the end of summer, apparently moving along the east flank of the Rocky Mountains, and so back to the wintering grounds. Under such circumstances, wheat rust does not depend on barberries to persist. In contrast, the spread of urediniospores from the south is prevented in Eurasia, where there are extensive east-west mountain ranges, and barberries *are* necessary for survival.

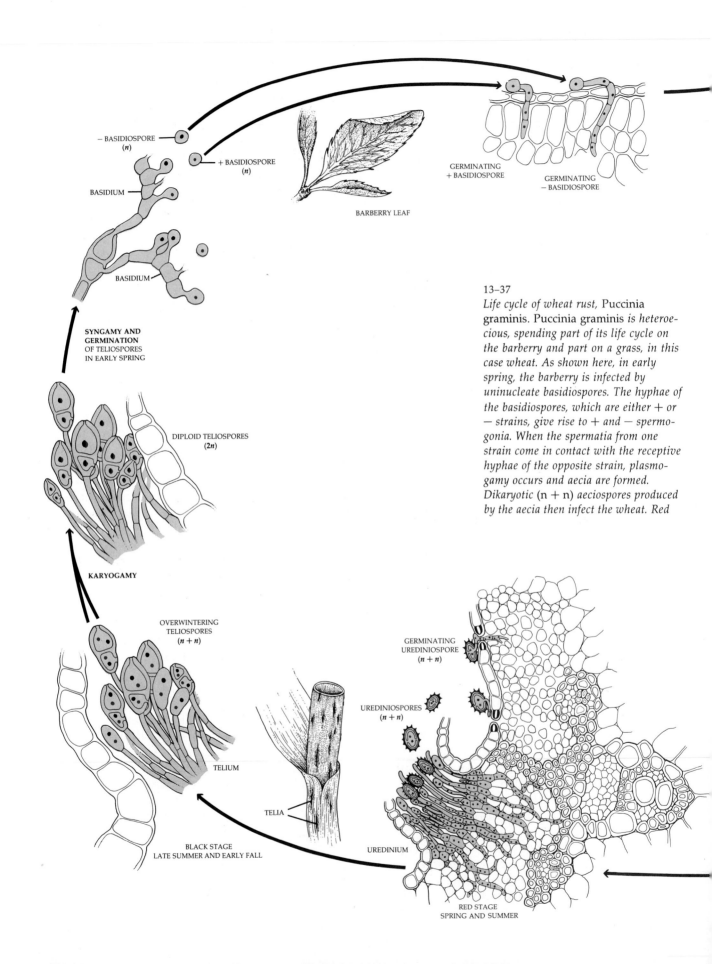

– BASIDIOSPORE
(*n*)

+ BASIDIOSPORE
(*n*)

BASIDIUM

BASIDIUM

BARBERRY LEAF

GERMINATING
+ BASIDIOSPORE

GERMINATING
– BASIDIOSPORE

**SYNGAMY AND
GERMINATION**
OF TELIOSPORES
IN EARLY SPRING

DIPLOID TELIOSPORES
(2*n*)

KARYOGAMY

OVERWINTERING
TELIOSPORES
(*n* + *n*)

GERMINATING
UREDINIOSPORE
(*n* + *n*)

UREDINIOSPORES
(*n* + *n*)

TELIUM

TELIA

BLACK STAGE
LATE SUMMER AND EARLY FALL

UREDINIUM

RED STAGE
SPRING AND SUMMER

13–37

Life cycle of wheat rust, Puccinia
graminis. Puccinia graminis *is heteroecious, spending part of its life cycle on
the barberry and part on a grass, in this
case wheat. As shown here, in early
spring, the barberry is infected by
uninucleate basidiospores. The hyphae of
the basidiospores, which are either + or
– strains, give rise to + and – spermogonia. When the spermatia from one
strain come in contact with the receptive
hyphae of the opposite strain, plasmogamy occurs and aecia are formed.
Dikaryotic* (n + n) *aeciospores produced
by the aecia then infect the wheat. Red*

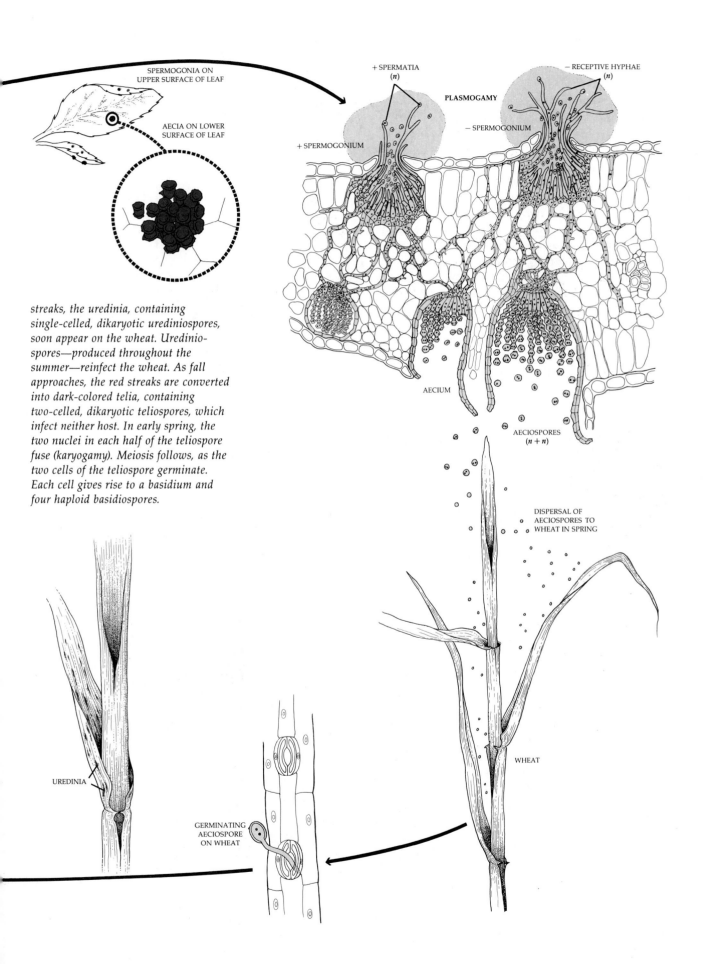

SPERMOGONIA ON
UPPER SURFACE OF LEAF

AECIA ON LOWER
SURFACE OF LEAF

+ SPERMATIA
(n)

− RECEPTIVE HYPHAE
(n)

PLASMOGAMY

+ SPERMOGONIUM

− SPERMOGONIUM

AECIUM

AECIOSPORES
(n + n)

streaks, the uredinia, containing
single-celled, dikaryotic urediniospores,
soon appear on the wheat. Urediniospores—produced throughout the
summer—reinfect the wheat. As fall
approaches, the red streaks are converted
into dark-colored telia, containing
two-celled, dikaryotic teliospores, which
infect neither host. In early spring, the
two nuclei in each half of the teliospore
fuse (karyogamy). Meiosis follows, as the
two cells of the teliospore germinate.
Each cell gives rise to a basidium and
four haploid basidiospores.

DISPERSAL OF
AECIOSPORES TO
WHEAT IN SPRING

WHEAT

UREDINIA

GERMINATING
AECIOSPORE
ON WHEAT

MYCORRHIZAE

Certain fungi play a crucial role in the mineral nutrition of vascular plants. If seedlings of many forest trees are grown in a sterile nutrient solution and then are transplanted to grassland soils, they grow poorly and may eventually die from malnutrition, even if abundant nutrients are present in the soil (Figure 13–38). However, if a small amount (0.1 percent by volume) of forest soil containing the appropriate fungi is added to the soil around the roots of the seedlings, they will grow normally. The restoration of normal growth is caused by the functioning of _mycorrhizae_ ("fungus-roots"), which are intimate and mutually beneficial symbiotic associations between roots and fungi.

Mycorrhizae occur in most groups of vascular plants. Only a few families of flowering plants characteristically lack mycorrhizae or form them very rarely; these include the mustard family (Brassicaceae) and the sedge family (Cyperaceae). The clumped, very fine roots of Proteaceae, the protea family, appear to play a function similar to that of mycorrhizae in other plants.

Many plants seem to grow normally when they are well supplied with essential elements, especially phosphorus, even if mycorrhizae are lacking; however, if the essential elements are present in limited supply, plants that lack mycorrhizae grow poorly or not at all. The role of mycorrhizae in the direct transfer of phosphorus from the soil to the roots of plants has been verified experimentally. In turn, the plant supplies carbohydrates to the symbiotic fungus. One of the most fascinating observations about mycorrhizae is that they may function, under certain circumstances, as a bridge through which photosynthate, phosphorus, and probably other substances can pass from one mycorrhizal host plant to another. This phenomenon, which may be of great importance in nature, needs further study.

Endomycorrhizae

There are two major types of mycorrhizae: _endomycorrhizae_ and _ectomycorrhizae_. Of these two, endomycorrhizae are by far the more common, occurring in about 80 percent of all vascular plants. The fungal component is a zygomycete, with fewer than 100 species of the fungi involved in such associations worldwide. Thus, endomycorrhizal relationships are not highly specific. The fungal hyphae penetrate the cortical cells of the plant root, where they form coils, swellings, or minute branches (Figure 13–39). The hyphae also extend out into the surrounding soil. Two of the characteristic intracellular swellings are called vesicles and arbuscles, and thus endomycorrhizae are often called vesicular-arbuscular mycorrhizae (abbreviated as V/A mycorrhizae).

Endomycorrhizae are of particular importance in the tropics, where soils tend to be positively charged and

13–38
Mycorrhizae and tree nutrition. Nine-month-old seedlings of white pine (Pinus strobus) were raised for two months in a sterile nutrient solution and then transplanted to prairie soil. The seedlings on the left were transplanted directly. The seedlings on the right were grown for two weeks in forest soil containing fungi before being transplanted to the grassland soil.

13–40
Network of roots and fungi lifted off a decomposing leaf in the Amazon forest. Such a network allows less than a thousandth of the nutrients reaching the forest floor to penetrate more than 5 centimeters into the soil. Mycorrhizal fungi transfer most of the nutrients back to the roots of the plants from which the leaves fell. When a tropical forest is cleared, this relationship is destroyed, and the soil generally becomes infertile quite rapidly.

13–39
Endomycorrhizae. Glomus versiforme, a zygomycete, is shown here growing in association with the roots of leeks,

Allium porrum. (a) *General view of a squashed leek root, showing vesicles.* (b) *Vesicles in leek root.* (c) *Arbuscles in leek*

root. Arbuscles predominate in young infections, with vesicles becoming common subsequently.

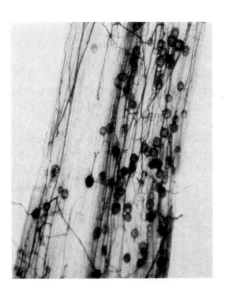

(a)

(b)

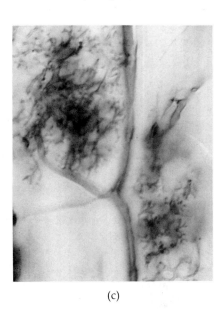

(c)

thus to retain phosphates so tightly that they are available only in very limited supplies for plant growth (see boxed essay, page 526). In contrast, most temperate soils are negatively charged. Since the poor farmers of the tropics are often unable to afford fertilizers, endomycorrhizae may play an especially critical role there in making phosphates available to crops (Figure 13–40). A better understanding of mycorrhizal relationships is certainly one key to lessening the amount of fertilizer (especially phosphates) that must be applied to agricultural lands to ensure satisfactory crops, and to managing rangelands and other natural plant communities effectively. The commercial application of selected strains of endomycorrhizae to crops to increase their yield appears to be an increasingly attractive possibility.

Ectomycorrhizae

Ectomycorrhizae (Figure 13–41) are characteristic of certain groups of trees and shrubs that are primarily found in temperate regions. These groups include the beech family (Fagaceae), the willow family (Salicaceae), and the pine family (Pinaceae), as well as certain groups of tropical trees that form dense stands of only one or a few species. The trees that grow at timberline in different parts of the world—such as pines in the northern mountains, *Eucalyptus* in Australia, and southern beech (*Nothofagus*; see Figure 29–8, page 590) —often are ectomycorrhizal. The ectomycorrhizal association apparently makes the trees more resistant to the harsh, cold, dry conditions that occur at the limits of tree growth.

13–41
Ectomycorrhizal rootlets from a western hemlock (Tsuga heterophylla). *In such ectomycorrhizae, the fungus commonly forms a sheath of hyphae, called a fungal mantle, around the root. Hormones secreted by the fungus cause the root to branch. This growth pattern and the hyphal sheath impart a characteristic branched and swollen appearance to the ectomycorrhizae. The narrow strands extending from the mycorrhizae are rhizomorphs—bundles of hyphae that function as extensions of the root system.*

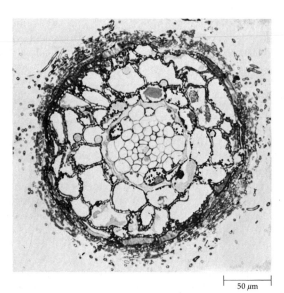

50 μm

13–42
Transverse section of an ectomycorrhizal
rootlet of Pinus. *The hyphae of the*
fungus are confined mostly to the spaces
between the cortical cells.

In ectomycorrhizae, the fungus surrounds but does not penetrate living cells in the roots (Figure 13–42). The extensive mycelium extends far out into the soil and plays an important role in transferring organic carbon to the plant. Root hairs are often absent, their function evidently replaced by the fungal hyphae. Ectomycorrhizae are mostly formed with basidiomycetes, but some involve associations with ascomycetes, including truffles. At least 5000 species of fungi are involved in ectomycorrhizal associations, often with a high degree of specificity.

Other Kinds of Mycorrhizae

Two other kinds of mycorrhizae are those characteristic of the heather family (Ericaceae) and a few closely related groups, and those associated with the orchids (Orchidaceae). In Ericaceae, the fungus may constitute up to 80 percent of the mycorrhizae by weight. It forms a prominent web of hyphae over the root surface, from which fine lateral hyphae penetrate the cortical cells of the plant. Although it is energetically expensive for the vascular plant to support such a large fungal biomass, the relationship seems to play a key role in allowing Ericaceae to colonize acidic soils that are poor in nutrients. Members of the family Ericaceae are common in such habitats. Both ascomycetes and basidiomycetes are involved in mycorrhizal associations with Ericaceae. In some subtypes of mycorrhizae that occur in Ericaceae, the fungi form a sheath around the root tissue and also occur within it; in others, the fungus is entirely internal.

The mycorrhizae that occur on the roots of Ericaceae seem to function primarily in enhancing nitrogen, rather than phosphate, intake by the plants, a factor of great importance in the acid soils where Ericaceae often grow. They also seem to increase the tolerance of some Ericaceae to heavy-metal poisoning, always a risk in acid conditions, whether natural or caused by acid precipitation.

The seeds of orchids germinate in nature only in the presence of suitable fungi. This mycorrhizal association is unique in that the fungus, which is internal, also supplies the host plant with carbon, at least when it is a seedling. The fungi involved in such associations are mostly basidiomycetes, of which more than 100 species may be involved.

Mycorrhizae and the History of Vascular Plants

A study of the fossils of early vascular plants (discussed in more detail in Chapters 17 and 18) has revealed that endomycorrhizal associations were as frequent then as they are in modern vascular plants. This finding led K. A. Pirozynski and D. W. Malloch to suggest that the evolution of mycorrhizal associations may have been a critical step in allowing colonization of the land by plants. Given the poorly developed soils that would have been available at the time of the first colonization, the role of mycorrhizal fungi (probably zygomycetes) may have been of crucial significance, particularly in facilitating the uptake of phosphorus and other nutrients. A similar relationship has been demonstrated among contemporary plants colonizing extremely nutrient-poor soils, such as slag heaps: those individuals that have endomycorrhizae have a much better chance of surviving. Thus it may not have been a single organism but rather a symbiotic association of organisms, comparable to a lichen, that initially invaded the land.

SUMMARY

Fungi, together with the heterotrophic bacteria, are the principal decomposers of the biosphere, breaking down organic products and recycling carbon, nitrogen, and other components to the soil and air. Fungi are rapidly growing nonphotosynthetic organisms that characteristically form filaments called hyphae, which may be septate or aseptate. In most fungi, the hyphae are highly branched, forming a mycelium.

Fungi, which are almost all terrestrial, reproduce by means of spores, which are usually dispersed by the wind. No motile cells are formed at any stage of their life cycle. Among the genetic peculiarities of fungi are the phenomena of heterokaryosis and parasexuality. A heterokaryotic fungus is one in which the nuclei found in a common cytoplasm are genetically different. Parasexuality involves the fusion of haploid nuclei to form diploid nuclei, crossing-over between associated chromosomes, and their ultimate separation to re-form the haploid nuclei. It provides an alternative means for genetic recombination, which is especially important in those fungi that do not reproduce sexually.

Glycogen is the primary storage polysaccharide of the fungi. The primary component of fungal cell walls is chitin. Most fungi are saprobes; that is, they live on organic materials from dead plants and animals. Rhizoids are specialized hyphae that serve to anchor some saprobic fungi to the substrate. Parasitic fungi often have specialized hyphae (haustoria) by means of which they extract organic carbon from the living cells of other organisms.

In addition to their role as decomposers, the fungi are economically important to humans as destroyers of foodstuffs and other organic materials. The group also includes the yeasts, *Penicillium* and other producers of antibiotics, cheese molds, and edible mushrooms.

The zygomycetes (division Zygomycota) have coenocytic mycelia. Their asexual spores are generally formed in sporangia—saclike structures in which all of the protoplasm becomes converted into spores. The zygomycetes are so named because of their formation of zygosporangia during the course of their sexual reproduction.

Division Ascomycota, the members of which are commonly called ascomycetes, has some 30,000 described species, more than any other group of fungi. Their distinguishing characteristic is the ascus—a saclike structure in which the meiotic spores, known as ascospores, are formed. In the characteristic life cycle, the protoplasts of male and female gametangia fuse, and the female gametangia produce filaments that are dikaryotic (containing two paired haploid nuclei). The ascus forms at the tip of a dikaryotic hypha. Asexual reproduction may also take place by spore formation; typically the ascomycetes form a type of asexual spore known as a conidium. Yeasts are single-celled ascomycetes in which asexual reproduction occurs by fission or budding, and sexual reproduction occurs by the formation of asci not enclosed in spore-producing bodies.

An artificial group of fungi—Fungi Imperfecti—includes many thousands of species with no known sexual cycle. Most Fungi Imperfecti are ascomycetes, and a few are basidiomycetes.

Lichens are symbiotic partnerships between ascomycetes (and a few basidiomycetes) and green algae or cyanobacteria. The fungi obtain their food from their photosynthetic partners, surrounding them with their hyphae and penetrating their cells. A lichen is morphologically and physiologically different from either organism as it exists separately. The ability of the lichen to survive under adverse environmental conditions is related to its ability to withstand desiccation and remain dormant when dry.

The division Basidiomycota, the members of which are called basidiomycetes, includes many of the largest and most conspicuous fungi. They are also the most familiar and include mushrooms, puffballs, and several important plant pathogens. Their distinguishing characteristic is the production of basidia. Like the ascus, the basidium is produced at the tip of a dikaryotic hypha and is the structure in which meiosis occurs. The resulting basidiospores are typically four in number, and they are the principal means of reproduction in the basidiomycetes. They differ from ascospores in that they are formed outside the basidium.

In the classes Hymenomycetes and Gasteromycetes, the basidia are incorporated into complex spore-producing bodies called basidiocarps. The basidia are borne externally, often lining gills and sometimes lining pores, in Hymenomycetes, and internally for at least part of their development in Gasteromycetes. The members of the class Teliomycetes, the rusts and smuts, do not form basidiocarps; they have septate basidia, as do the jelly fungi (class Hymenomycetes). Hymenomycetes other than the jelly fungi and Gasteromycetes form basidia that are not divided by septa.

Mycorrhizae—symbiotic associations between the roots of plants and fungi—characterize all but a few families of vascular plants. Of the two major types of mycorrhizae, endomycorrhizae, in which the fungal partner is a zygomycete, occur in about 80 percent of all vascular plants. In such associations, the fungus actually penetrates the cortical cells of the host. In the second major type, ectomycorrhizae, the fungus does not penetrate the cells of the host but forms a feltlike layer over its roots. Both basidiomycetes and ascomycetes are involved in ectomycorrhizal associations. Mycorrhizal associations are important in obtaining phosphorus, and probably other nutrients, for the plant and in providing organic carbon for the fungus.

SUGGESTIONS FOR FURTHER READING

BATRA, LEKH R. (Ed.): *Insect-Fungus Symbiosis: Nutrition, Mutualism and Commensalism,* John Wiley & Sons, Inc., New York, 1979.

A fascinating collection of articles about the many and varied interactions between these abundant and successful organisms.

CHRISTENSEN, CLYDE M.: *The Molds and Man: An Introduction to the Fungi,* 3rd ed., University of Minnesota Press, Minneapolis, 1965.*

A highly informative account of the fungi and especially of their complex and important interrelationships with humans.

CHRISTENSEN, CLYDE M.: *Molds, Mushrooms and Mycotoxins,* University of Minnesota Press, Minneapolis, 1975.

An engagingly written account of some fungi and their importance to humans; supplements, rather than duplicates, his earlier book.

COURTENAY, BOOTH, and HAROLD H. BURDSALL, JR.: *A Field Guide to Mushrooms and their Relatives,* Van Nostrand Reinhold, New York, 1984.

A beautifully illustrated guide to some 350 common species of mushrooms of temperate North America.

DEMAIN, ARNOLD L.: "Industrial Microbiology," *Science* 214: 987–995, 1981.

Valuable review of the prospects for greatly expanded use of fungi and bacteria in industrial processes.

HALE, MASON E.: *The Biology of Lichens,* 3rd ed., University Park Press, Baltimore, Md., 1983.

An outstanding, concise summary of all aspects of the morphology, physiology, systematics, ecology, and economic uses and applications of the lichens.

* Available in paperback.

HARLEY, J. L., and S. E. SMITH: *Mycorrhizal Symbiosis,* Academic Press, New York, 1983.

The standard work in the field; an invaluable reference.

LARGE, E. C.: *The Advance of the Fungi,* Dover Publications, Inc., New York, 1962.*

A fascinating popular account of the closely interwoven histories of fungi and humans, first published in 1940.

LAWREY, J. D.: *Biology of Lichenized Fungi,* Praeger Scientific, New York, 1984.

An excellent, contemporary account of the biology of lichens.

"Microbes for Hire": *Science 85,* July/August 1985.

An outstanding series of articles on the actual and potential industrial uses of microorganisms, including yeasts and other fungi.

MILLER, ORSON K., JR.: *Mushrooms of North America,* E. P. Dutton, New York, 1977.*

A pocket-sized field guide, beautifully illustrated, to the fleshy fungi of temperate North America.

MOORE-LANDECKER, E.: *Fundamentals of the Fungi,* Prentice-Hall, Inc., Englewood Cliffs, N.J., 1972.

An excellent outline of the fungi.

PHAFF, H. J., et al.: The Life of Yeasts, 2nd ed., Harvard University Press, Cambridge, Mass., 1978.

An authoritative account of the properties and uses of yeasts.

ROSS, IAN K.: *Biology of the Fungi: Their Development, Regulation and Associations,* McGraw-Hill Book Company, New York, 1979.

A concise, biologically oriented overview of the fungi.

SCAGEL, R. F., et al.: *Nonvascular Plants: An Evolutionary Survey,* Wadsworth Publishing Company, Belmont, Calif., 1982.

This overall survey of all groups that are included in our book, except for the vascular plants, provides a well-illustrated overview of fungi.

STROBEL, GARY A., and GERALD N. LANIER: "Dutch Elm Disease," *Scientific American* 245(2): 56–66, August 1981.

Integrated methods of control are proving helpful in controlling this deadly fungal infection, which has devastated the elm trees of Europe and North America.

STRUHL, KEVIN: "The New Yeast Genetics," *Nature* 305: 391–397, 1983.

A concise review of the many ways in which yeasts are becoming model systems for investigations of the metabolism, genetics, and development of eukaryotic cells.

WEBB, A. DINSMOOR: "The Science of Making Wine," *American Scientist* 72: 360–367, 1984.

Fascinating account of the ways in which an ancient practical art is being turned into a modern science.

* Available in paperback.

C H A P T E R   1 4

Unicellular Protista: Water Molds, Slime Molds, Chytrids, and Unicellular Algae

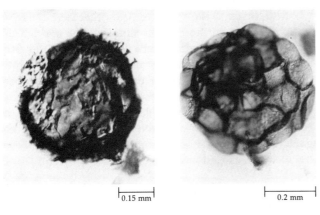

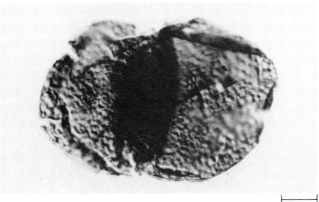

0.15 mm

0.2 mm

0.1 mm

14–1

These three very different kinds of acritarchs occurred together about 700 million years ago in what is now the Grand Canyon region of Arizona. Acritarchs, which appeared in the fossil record about 1.5 billion years ago, were the first eukaryotes; for hundreds of millions of years, they were the only eukaryotes. The acritarchs, which were very diverse, apparently were planktonic, autotrophic organisms; some of the later ones appear to be directly related to the green algae.

Eukaryotes appeared in the fossil record about 1.5 billion years ago (Figure 14–1); by that time, bacteria had been in existence for more than 2 billion years. For hundreds of millions of years, all eukaryotes were unicellular. The first known fossils of multicellular eukaryotes are about 630 million years old, although the earlier existence of multicellular organisms is indicated by what appear to be worm burrows that are dated at about 700 million years old. For at least 800 million years of the time that eukaryotes have existed on earth, all were organisms that we would place in the kingdom Protista.

A number of evolutionary lines of multicellular organisms have originated from unicellular ancestral eukaryotes, and some multicellular members occur in almost every division of protists. Four of these wholly or partly multicellular groups are photosynthetic. Among them, the plants, which arose from multicellular green algae, are so distinct that they are treated as a separate kingdom (kingdom Plantae) and are discussed in Chapters 16, 17, and 18. The remaining three groups —the green, brown, and red algae—which are not directly related to one another, are placed in separate divisions in the kingdom Protista and are covered in Chapter 15. In this chapter, we shall describe the seven predominantly unicellular divisions of protists that have been treated as plants or fungi in the past—the water molds, slime molds, chytrids, and unicellular algae. Those groups of protists that have traditionally been regarded as Protozoa—such as ciliates and amoebas—are not treated in this book, since they are not normally studied by botanists.

Of the seven divisions of protists discussed in this chapter, four consist of heterotrophic organisms. They have traditionally been the concern of the *mycologist*, or student of fungi. Most scientists now interpret the available evidence as indicating that there is no direct relationship between these heterotrophic protists and the fungi. Nevertheless, they continue to be described by terminology used for the fungi. For example, these organisms are called "molds," a traditional word for some fungi, and their filaments are called *hyphae* (singular, hypha) and a mass of hyphae is called a *mycelium*, as in the fungi.

The other three divisions covered in this chapter, along with the three divisions discussed in the following chapter, consist principally of photosynthetic organisms. They are studied by *phycologists*, or students of algae.

It is important to note that there is little evidence to indicate a direct relationship between any of the ten divisions of the kingdom Protista discussed in this book. Each is best considered a separate evolutionary line among the eukaryotes. They are linked with one another by nothing more than their basic eukaryotic structure and lack of the distinguishing features of plants, animals, or fungi.

ECOLOGY OF UNICELLULAR PROTISTA

In all bodies of water, minute photosynthetic cells and tiny animals occur as suspended *plankton* (from the Greek word for "wanderer"). The planktonic algae and cyanobacteria, or *phytoplankton*, are the beginning of the food chain for most of the heterotrophic organisms that live in the deep waters of the ocean and large bodies of fresh water. Phytoplankton are usually composed of single cells—some are quite simple in appearance, and others have intricate and delicately detailed forms sometimes joined into colonies or filaments (Figure 14–2). The smallest members of the phytoplankton —*nannoplankton*—are so small that they normally pass

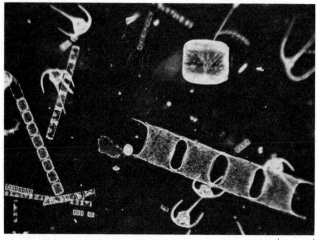

14–2

Phytoplankton. The organisms shown here are dinoflagellates and filamentous and unicellular diatoms.

through the openings of a plankton net, which has mesh openings of 40 to 76 micrometers. The nannoplankton include algae, as well as cyanobacteria, which are also important contributors to marine and freshwater productivity.

The heterotrophic plankton—*zooplankton*—consists mainly of tiny crustaceans, the larvae of many different phyla of animals, and many heterotrophic protists and bacteria. In the sea, small fish and some larger ones, as well as most of the great whales, feed on the phytoplankton and zooplankton; and still larger fish feed on the smaller ones. In this way, the "great meadow of the sea," as the phytoplankton is sometimes called, can be likened to the meadows of the land, serving as the source of nourishment for heterotrophic organisms.

Of the four heterotrophic divisions discussed in this chapter, two consist primarily of aquatic organisms: Oomycota, the water molds and related organisms, and Chytridiomycota, the chytrids. The other two divisions —Acrasiomycota, the cellular slime molds, and Myxomycota, the plasmodial slime molds, are primarily terrestrial. The water molds and chytrids have swimming spores; they occur as parasites or as saprobes, consumers of organic material derived from other organisms. The cellular and plasmodial slime molds, in contrast, feed actively on bacteria and small bits of organic debris. They are the most "animal-like" of the organisms included in this book.

SYMBIOSIS AND THE ORIGIN OF THE CHLOROPLAST

In all organisms that produce oxygen during photosynthesis—the cyanobacteria, eukaryotic algae, and plants —chlorophyll *a* is involved in the process. The chlorophyll *a* is found in chloroplasts in all these organisms except cyanobacteria, which lack chloroplasts. The similarity of these chloroplasts suggests a common origin, yet the various groups of autotrophs differ so greatly in their other characteristics that they could not actually have had such an origin. The similarities of their chloroplast biochemistry and structure are related to the mode of origin of these organelles.

Based on the close structural and biochemical similarities between certain groups of bacteria and chloroplasts (reviewed in Chapter 2), the great majority of scientists accept the view that chloroplasts originated as a result of a series of independent events involving different groups of photosynthetic bacteria.

The ease and frequency with which such symbiotic relationships occur can be seen in the wide variety of such relationships that exist today. These modern examples include symbioses of algae with invertebrates, fungi (mostly in lichens), other algae, vertebrates, bryophytes, and a few kinds of vascular plants (in which

14–3

Symbiosis in algae. The organism shown here is a unicellular green alga, Chlorella, *discussed in more detail in Chapter 15. (a) Each bell-shaped cell of the protozoan* Vorticella *contains numerous cells of the symbiotic alga. (b) Electron micrograph of a* Vorticella *that contains cells of* Chlorella. *Each algal cell is found in a separate vacuole bounded by a single membrane. The protozoan provides the algae with protection and shelter, while the algae probably produce a carbohydrate that serves as nourishment for their host cell.*

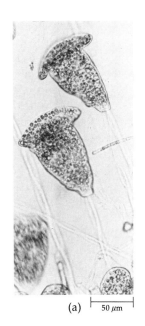

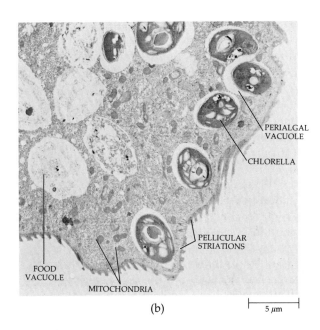

PERIALGAL VACUOLE

CHLORELLA

PELLICULAR STRIATIONS

FOOD VACUOLE

MITOCHONDRIA

(a) ⊢ 50 μm ⊣

(b) ⊢ 5 μm ⊣

the algae grow in cavities in stems, petioles, and leaves). Among invertebrates, for example, some 150 genera belonging to eight different phyla are known to have algal symbionts growing within their cells (Figure 14–3). Symbiotic dinoflagellates (photosynthetic protists) are responsible for much of the productivity of coral reefs, and other symbiotic associations are of equal importance ecologically.

Photosynthetic protists have traditionally been regarded as related to one another; it was realized only recently that they had acquired the ability to manufacture their food quite independently. Thus autotrophic eukaryotes other than plants have traditionally been grouped as "algae"; filamentous, heterotrophic ones as "fungi"; and nonfilamentous, heterotrophic ones as "protozoa." However, the members of these informal groups are not necessarily related; for example, some algae are more closely related to some protozoa than they are to other algae, and vice versa.

For example, the euglenoids (organisms that will be discussed in this chapter), consist of about 40 genera of aquatic, unicellular organisms characterized by, among other features, a flexible, protein-rich covering called a pellicle. About a third of the genera of euglenoids have chloroplasts, and for this reason they have sometimes been considered "algae," or even plants. However, except for their possession of chloroplasts, the euglenoids are virtually identical to some of the zoomastigotes (phylum Zoomastigina), a very large and diverse group of protozoans. The chloroplasts of euglenoids biochemically resemble those of the green algae, which otherwise differ from the euglenoids in almost every way. This similarity may indicate that the ancestral green algae and euglenoids acquired a symbiotic bacterium (possibly one similar to *Prochloron*) (see Figures 11–2, page 166, and 11–17, page 176), as a result of independ-

ent symbiotic events. Thus, even though green algae and euglenoids do have similar chloroplasts, they must otherwise be considered as distantly related and distinct divisions of the kingdom Protista.

An alternative suggestion (which if confirmed would not change our evaluation of the relationships of the euglenoids) is that the euglenoids may have acquired the chloroplasts of green algae by ingesting whole green algal cells. Following the incorporation of such cells, there would have been a progressive loss of the other elements of the symbiotic green alga, ultimately leaving only the chloroplast and the plasma membrane. Such a relationship might explain why the chloroplasts of euglenoids are surrounded by three membranes, and those of green algae by two, when they are otherwise remarkably similar.

CHARACTERISTICS OF THE DIVISIONS

Four divisions of heterotrophs and three divisions of autotrophs are considered in this chapter. Considering the heterotrophs first, they are a highly diverse group. Among them, the Chytridiomycota (chytrids) and Oomycota (oomycetes) have coenocytic (multinucleate and undivided) hyphae, which can be extensive and highly branched in some oomycetes. Oomycetes produce motile asexual spores—zoospores—with two flagella, one tinsel and one whiplash (see Figure 2–26, page 31), whereas chytrid zoospores have a single posterior whiplash flagellum. The Myxomycota spend most of their nonreproductive phases as streaming masses of cytoplasm. The Acrasiomycota are amoeba-like organisms that do not form flagellated cells at any stage of their life cycle. The principal features of these divisions, along with those of their autotrophic relatives, are summarized in Table 14–1.

The autotrophic divisions of protists also differ widely from one another in the nature of their flagella (when present) and in their biochemical characteristics, such as pigmentation, food reserves, and cell wall components. The characteristics of all divisions of protists, including those treated in Chapter 15, are presented in Table 14–1. The names of some of the algal divisions are derived from the colors of the predominant accessory pigments, which mask the grass-green color of the chlorophylls. Accessory pigments contribute energy absorbed from light primarily to photosystem II in photosynthesis. A rich diversity of storage products is found among these organisms, with most having distinctive carbohydrate food reserves, often in addition to lipids. Reflecting the heterotrophic ancestry of these divisions is the fact that some genera of chrysophytes, dinoflagellates, and euglenoids regularly ingest solid particles of food, just like the members of many divisions that have not acquired chloroplasts symbiotically during the course of their evolution.

Different groups of protists accumulate a rich variety of carbohydrates and lipids through photosynthesis. In green algae, starch accumulates within the chloroplasts (as it does in plants, which are descendants of the green algae); in the members of all other divisions, the food reserves are accumulated in the cytoplasm outside the chloroplasts. In red algae and dinoflagellates, these reserves are starch; in brown algae, the food reserve is another glucose-containing polysaccharide, *laminarin*, which has different linkages between the glucose units. In the other divisions with chlorophyll *c*, the main storage product is *chrysolaminarin*, a more highly polymerized form of laminarin. Many brown algae also store mannitol, an alcohol derived from the sugar mannose. The euglenoids synthesize paramylon, a polysaccharide in a helical configuration, and differ sharply from all other divisions in this respect.

DIVISION OOMYCOTA

The division Oomycota, with about 475 species, is a very distinctive group whose members are commonly called oomycetes. The cell walls of these organisms are composed largely of cellulose or celluloselike polymers, thus differing markedly from the cell walls of the fungi. The chromosomes of oomycetes resemble those of most eukaryotic organisms and differ greatly from the highly condensed chromosomes found in fungi. Meiosis and mitosis resemble the same processes as found in other eukaryotes; centrioles are present. The organisms in this division range from unicellular forms to highly branched, coenocytic filamentous ones.

Most species of oomycetes can reproduce both sexually and asexually. In oomycetes, sexual reproduction is oogamous and involves an *oogonium*, which contains many eggs, and an *antheridium*, containing numerous male nuclei (Figure 14–4). Syngamy results in the formation of a thick-walled zygote, the *oospore*, which serves as a resting spore; the oospore is the structure for which this division is named. As mentioned previously, asexual reproduction among the oomycetes is by means of motile zoospores, which, like the motile sperm, have two flagella—one tinsel and one whiplash.

One large group of the division Oomycota is aquatic. The members of this group—the so-called water molds —are abundant in fresh water and are easy to isolate from it. Most of them are saprobic, but a few are parasitic, including species that cause diseases of fish and fish eggs.

In some water molds, such as *Saprolegnia* (Figure 14–5), sexual reproduction can occur with male and female sex organs borne on the same individual; in other words these water molds are *homothallic*. Other water molds, such as some species of *Achlya* (Figure 14–4), are *heterothallic*—male and female sex organs are borne by different individuals, or, if they are borne on one individual, that individual is genetically incapable of fertilizing itself. Both *Saprolegnia* and *Achlya* can reproduce sexually and asexually.

 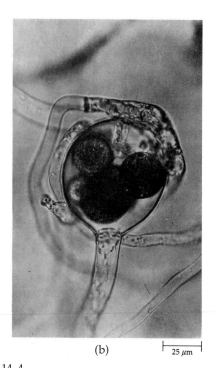

(a) 50 μm   (b) 25 μm

14-4
Achlya ambisexualis, *a water mold that reproduces both asexually and sexually.* (a) *Empty sporangium with zoospores encysted about its opening, a distinctive feature of* Achlya. (b) *Sex organs, showing fertilization tubes extending from the antheridium through the wall of the oogonium to the eggs.*

Table 14–1. *Comparative Summary of Characteristics in Ten Divisions of Protista.*

DIVISION	NUMBER OF SPECIES	PHOTOSYNTHETIC PIGMENTS	CARBOHYDRATE FOOD RESERVE	FLAGELLA	CELL WALL COMPONENT	HABITAT
Oomycota (water molds)	475	None	Glycogen	2, 1 tinsel, 1 whiplash; in reproductive cells only	Cellulose	Aquatic or need water
Chytridiomycota (chytrids)	750	None	Glycogen	1, posterior, whiplash; in reproductive cells only	Chitin, other polymers	Aquatic
Acrasiomycota (cellular slime molds)	70	None	Glycogen	None (amoeboid)	None	Terrestrial
Myxomycota (plasmodial slime molds)	450	None	Glycogen	2, whiplash; in reproductive cells only	None	Terrestrial
Chrysophyta (chrysophytes and diatoms)	6650	Chlorophyll *a*, chlorophyll *c*, carotenoids, including fucoxanthin	Chrysolaminarin	None or 1 or 2, apical, whiplash or tinsel, equal or unequal	None, or cellulose, with silica scales in some; silica in diatoms	Marine and freshwater
Pyrrhophyta (dinoflagellates)	1100	Chlorophyll *a*, chlorophyll *c*, carotenoids, including peridinin	Starch	None or 2, lateral, tinsel	Cellulose, other materials	Marine and freshwater
*Phaeophyta (brown algae)	1500	Chlorophyll *a*, chlorophyll *c*, carotenoids, including fucoxanthin	Laminarin, mannitol	2, lateral, tinsel forward whiplash behind; in reproductive cells only	Cellulose matrix with alginic acids (polysaccharides)	Almost all marine, flourish in cold ocean waters
*Rhodophyta (red algae)	4000	Chlorophyll *a*, carotenoids, phycobilins	Starch	None	Cellulose, pectic materials, calcium carbonate in many	Marine, some freshwater, many tropical species
*Chlorophyta (green algae)	7000	Chlorophyll *a*, chlorophyll *b*, carotenoids	Starch	None or 2 (or more); apical or lateral; equal; whiplash	Polysaccharides, sometimes cellulose	Mostly fresh-water, but some marine
Euglenophyta (euglenoids)	800	Chlorophyll *a*, chlorophyll *c*, carotenoids	Paramylon	1 (to 3), apical tinsel (one row of hairs)	No cell wall; have a proteinaceous pellicle	Mostly fresh-water

*Divisions consisting largely or entirely of multicellular organisms; treated in Chapter 15.

Another group of oomycetes is primarily terrestrial, although the organisms still form motile zoospores when open water is available. Among this group—the Peronosporales—are several forms that are economically important. As C. J. Alexopoulos has said, "At least two of them have had a hand—or should we say hypha!—in shaping the economic history of an important portion of mankind."

14–5

Life cycle of **Saprolegnia,** *an oomycete. The mycelium of this water mold is diploid. Reproduction is mainly asexual. Biflagellated zoospores released from a zoosporangium swim for a while and then encyst. Each eventually gives rise to a secondary zoospore, which also encysts and then germinates to produce a new mycelium.*

During sexual reproduction, oogonia and antheridia are formed on the same hypha. Meiosis occurs within these structures. The oogonia are enlarged cells in which a number of spherical eggs are produced. The antheridia develop from the tips of other filaments of the same individual and produce numerous male nuclei. In mating, the antheridia grow toward the oogonia and develop fertilization tubes, which penetrate the oogonia.

Male nuclei travel down the fertilization tubes to the female nuclei and fuse with them. Following each nuclear fusion, a thick-walled zygote—the oospore—is produced. On germination, the oospore develops into a hypha, which then produces a zoosporangium, beginning the cycle anew.

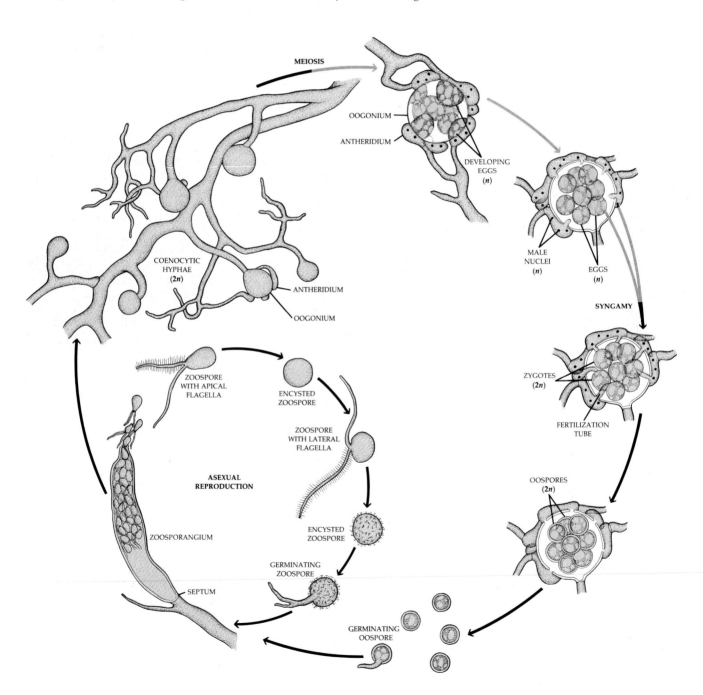

MEIOSIS

OOGONIUM

ANTHERIDIUM

DEVELOPING EGGS (*n*)

MALE NUCLEI (*n*)

EGGS (*n*)

SYNGAMY

COENOCYTIC HYPHAE (*2n*)

ANTHERIDIUM

OOGONIUM

ZYGOTES (*2n*)

FERTILIZATION TUBE

ZOOSPORE WITH APICAL FLAGELLA

ENCYSTED ZOOSPORE

ZOOSPORE WITH LATERAL FLAGELLA

OOSPORES (*2n*)

ASEXUAL REPRODUCTION

ENCYSTED ZOOSPORE

ZOOSPORANGIUM

GERMINATING ZOOSPORE

SEPTUM

GERMINATING OOSPORE

HORMONAL CONTROL OF SEXUALITY IN A WATER MOLD

Achyla ambisexualis is a heterothallic water mold; it produces male and female sex organs on different individuals. The late John R. Raper of Harvard University studied the hormonal control of sexuality in these organisms. The female hyphae secrete a substance that induces the initial development of antheridia on the hyphae of the male organism. (a) Appearance of undifferentiated hyphae before the addition of this substance—characterized and named antheridiol by Alma Barksdale of The New York Botanical Garden. (b) Antheridial branches produced within two hours after the addition of crystalline antheridiol. (c) Antheridial branches growing toward a plastic particle containing antheridiol.

After the developing antheridia appear, the male organism secretes its own hormone, called oogoniol, which induces the formation of oogonia on the hyphae of the female organism. With the appearance of young oogonia, the antheridial hyphae are attracted to them, and distinct, mature antheridia develop. Raper attributed this latter response to a substance that he called "hormone C," presumably secreted by the female, but subsequent research suggested that antheridiol elicits this developmental reaction. After the antheridia are formed, there is a differentiating response in the female oogonia, presumably hormonally induced, which leads to the maturation of the oogonium and its enclosed female gametes (eggs). Thus, even in protists, sexual reproduction can involve a highly coordinated sequence of events, which are hormonally controlled.

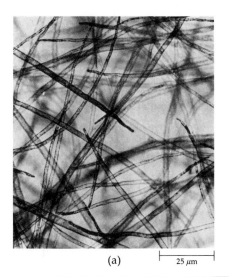
(a)   25 μm

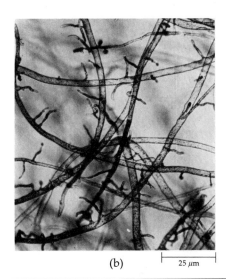
(b)   25 μm

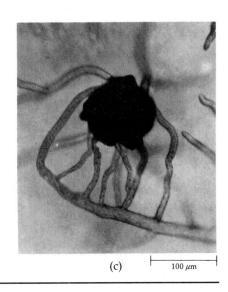
(c)   100 μm

One of these is *Plasmopara viticola*, which causes downy mildew in grapes. Downy mildew was accidentally introduced into France in the late 1870s on American grape stock that had been imported because of its resistance to other diseases. The mildew soon threatened the entire French wine industry. It was eventually brought under control by a combination of good fortune and skillful observation. French vineyard owners in the vicinity of Medoc customarily put a disagreeable mixture of copper sulfate and lime on vines growing along the roadside to discourage passersby from picking them. A professor from the University of Bordeaux, who was studying the problem of downy mildew, noticed that these plants were free from symptoms of the disease. After conferring with the vineyard owners, the professor prepared his own mixture of chemicals—the Bordeaux mixture—which was made generally available in 1882. The Bordeaux mixture was the first chemical used in the control of a plant disease.

Another economically important member of this group is the genus *Phytophthora* (which means "plant destroyer"). *Phytophthora*, with about 35 species, is a particularly important plant pathogen that causes widespread destruction of many crops, including cacao, pineapples, tomatoes, rubber, papayas, onions, strawberries, apples, soybeans, tobacco, and citrus. A widespread member of this genus, *Phytophthora cinnamomi*, which occurs in soil, has killed or rendered unproductive millions of avocado trees in southern California and elsewhere in recent years. It has also destroyed tens of thousands of hectares of valuable eucalyptus timberland in Australia. The zoospores of *P. cinnamomi* are attracted to the plants they infect by chemicals exuded by the roots. This oomycete also produces resistant spores that can survive for up to six years in moist soil. Extensive breeding efforts are now under way with avocados and other susceptible crops to produce strains resistant to this oomycete.

The best-known species of *Phytophthora*, however, is *Phytophthora infestans* (Figure 14–6), the cause of the late blight of potatoes, which produced the great potato famine of 1846–47 in Ireland. The population of Ireland, which had risen from 4.5 million people in 1800 to about 8.5 million in 1845, fell as a result of this famine to 6.5 million people in 1851. About 800,000 people starved to death, while the remainder emigrated, many of them to the United States. Virtually the entire Irish potato crop was wiped out in a single week in the summer of 1846. This was a disaster for the Irish peasants, who depended almost entirely on potatoes for their food, each adult consuming between four and six kilograms of potatoes every day. They had to eat such a large amount to obtain sufficient protein to maintain their bodies.

Another member of this division, blue mold *(Peronospora hyoscyami)*, devastated tobacco crops in the United States and Canada in 1979, causing an estimated loss of almost a quarter of a billion dollars. Blue mold spreads by means of airborne, multinucleate spores. Epidemics such as this will become increasingly abundant in the future as the genetic variability of major crops is narrowed, a problem that we shall consider in Chapter 30.

DIVISION CHYTRIDIOMYCOTA

The Chytridiomycota, or chytrids, are a predominantly aquatic group of protists, consisting of about 750 species. These organisms are extremely varied in form, in the nature of their sexual interactions, and in their life histories. They have cell walls that seem to consist mainly of chitin, although other polymers are also found. The mitotic and meiotic cycles, as far as is known, are like those in Oomycota, and all chytrids are coenocytic with few septa at maturity.

The principal distinguishing characteristic of chytrids is that their motile cells (zoospores and gametes) possess a single posterior whiplash flagellum. This characteristic alone separates the chytrids from all other protists and indicates a relationship between the diverse organisms that are grouped together in this division.

Some chytrids are simple, unicellular organisms that do not develop a mycelium. In these chytrids, the whole organism is transformed into a reproductive structure at the appropriate time. Other chytrids have slender rhizoids that extend into the substrate and serve as an anchor (Figure 14–7). Different species of chytrids are parasites of algae, aquatic oomycetes, spores, pollen grains, or other parts of vascular plants; other species are saprobic on such substrates as dead insects.

One of the chytrids, the genus *Coelomomyces*, is an obligate parasite of the larvae of mosquitoes and other

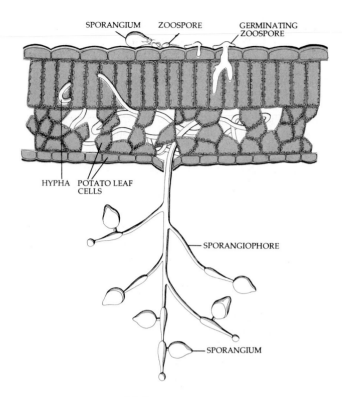

14–6
Phytophthora infestans, *the cause of the late blight of potatoes. The cells of the potato leaf are shown in gray. In the presence of water and low temperatures, zoospores are released from the sporangia and swim to the germination site (as shown here), or the sporangia germinate directly through a germ tube.*

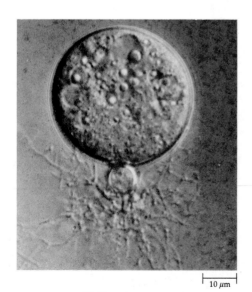

14–7
Chytridium confervae, *a common chytrid, as seen with the aid of Nomarski system optics. Note the slender rhizoids extending downward.*

flies. It exhibits a life cycle comparable to that of the rust fungi, in that it has an alternation of hosts between the small aquatic crustaceans known as copepods and the mosquito larvae. *Coelomomyces* is being investigated as a possible biological control for mosquitoes.

Other chytrids are much more complex in their structure and reproduction. Consider, for example, the life cycle of the genus *Allomyces*. Some species have an alternation of isomorphic generations comparable to that shown in Figure 14–8, whereas in other species the al-

ternating generations are heteromorphic—the haploid and diploid individuals do not resemble one another closely. Alternation of generations is characteristic of plants and of many algae but is otherwise found only in *Allomyces*, in one other closely related genus of chytrids (see Figure 10–10, page 162), and in a very few other heterotrophic protists not considered in this book. In terms of its life cycle, morphology, and physiology, *Allomyces* is one of the best-known protists; its sexual hormones have been investigated extensively.

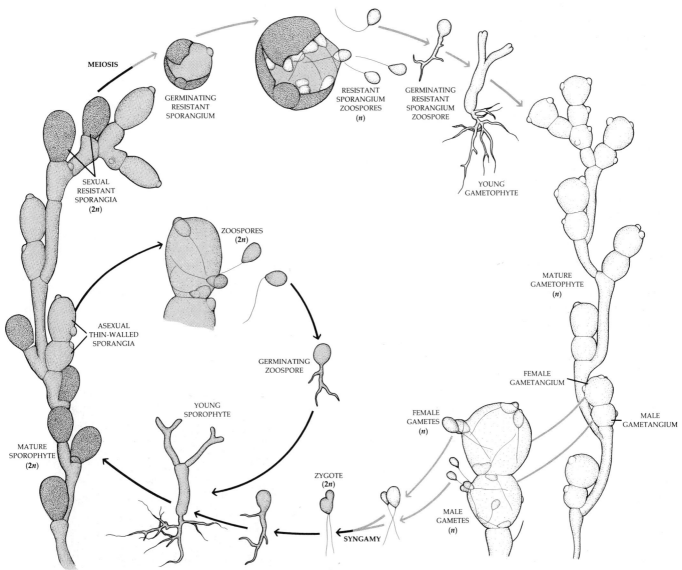

14–8

In the chytrid Allomyces arbusculus, *whose life cycle is shown here, there is an alternation of isomorphic generations. The haploid and diploid individuals are indistinguishable until they begin to form reproductive organs. The haploid individuals produce approximately equal numbers of colorless female gametangia and orange male gametangia. The male gametes, which are about half the size of* the female gametes, *are attracted by sirenin, a hormone produced by the female gametes. The zygote loses its flagella and germinates to produce a diploid individual. This sporophyte forms two kinds of sporangia: (1) asexual sporangia—colorless, thin-walled structures that release diploid zoospores—which in turn germinate and repeat the diploid generation; and (2) sexual* sporangia—thick-walled, reddish-brown structures that are able to withstand severe environmental conditions. After a period of dormancy, meiosis occurs in these sexual resistant sporangia, resulting in the formation of haploid zoospores. These zoospores develop into haploid individuals that produce gametangia at maturity.

DIVISION ACRASIOMYCOTA

The cellular slime molds—a group of about 65 species in several genera—are probably more closely related to the amoebas (phylum Rhizopoda) than to any other group. Unlike fungi, they have cellulose-rich cell walls and normal mitosis, in which the nuclear envelope breaks down; centrioles are present. One member of this division, *Dictyostelium discoideum*, was discussed as a model of multicellular differentiation in Chapter 8 (see page 129). In this species and in other members of the division, the amoebas (myxamoebas) retain their individuality even in the aggregating stage of their life cycle, when they come together to form pseudoplasmodia, or "slugs" (Figure 14–9). Cellular slime molds are also able to encyst individually (microcysts) and thus to survive during short periods of drought or starvation.

Asexual reproduction involving asexual spores is common in the cellular slime molds. Sexual reproduction also occurs frequently and involves structures known as macrocysts. In *Dictyostelium mucoroides*, for example, the macrocysts are somewhat flattened, irregularly circular to ellipsoidal, multicellular structures from 25 to 50 micrometers in diameter (Figure 14–10). During the formation of macrocysts, pairs of haploid amoebas first fuse, forming zygotes; subsequently, these zygotes attract and then engulf amoebas that are nearby. The macrocysts are formed by smaller aggregations of amoebas than are involved in the formation of "slugs," and the aggregations are rounded in outline, rather than being elongate. A thin membrane laid down by the amoebas encircles each macrocyst. Later, a heavy cell wall, rich in cellulose, is laid down around the whole group within the membrane. Within the macrocyst, the zygote—the only diploid cell of the life cycle—undergoes meiosis and several mitotic divisions before germination and the release of new, haploid amoebas.

The cellular slime molds were once considered rare curiosities. Then, in 1933, Kenneth B. Raper discovered *Dictyostelium discoideum;* his subsequent research on this organism brought a great deal of attention to the cellular slime molds. It is now known that they are normal inhabitants of most soils and are especially common in forest leaf litter, where they feed on bacteria. Like most kinds of organisms, they are better represented in the tropics than in temperate regions. Cellular slime molds are easily isolated and brought into culture and have proved to be useful experimental tools in studies of differentiation and molecular biology (see pages 128–129.)

In 1982, David Waddell of Princeton University reported the unusual habits of a predatory slime mold, *Dictyostelium caveatum*, which he brought into culture from bat guano in Blanchard Springs Cavern, Arkansas. Amoebas of this species formed common aggregates with all other cellular slime molds with which

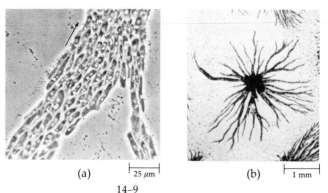

(a)   ⊢ 25 µm   (b)   ⊢ 1 mm

14–9
Dictyostelium discoideum. (a) *Aggregating myxamoebas. Note that each amoeba has retained its individuality. The arrow indicates the direction in which the stream of amoebas is moving.* (b) *Low-magnification view of a large number of myxamoebas, showing the overall pattern of aggregation.*

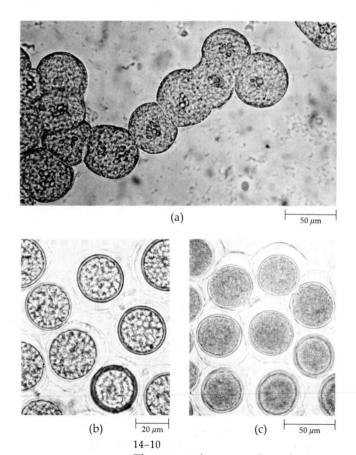

(a)   ⊢ 50 µm

(b)   ⊢ 20 µm   (c)   ⊢ 50 µm

14–10
The process of macrocyst formation in Dictyostelium mucoroides. (a) *Each zygote, or giant cell, is beginning to engulf the amoebas around it.* (b) *Giant cells have engulfed all of the amoebas, and each has laid down a cellulose wall.* (c) *Mature macrocysts; at this stage they appear homogeneous.*

they were tested, delayed their development, and consumed all of the amoebas of the other species, eventually forming fruiting bodies consisting only of cells of their own species.

DIVISION MYXOMYCOTA

This division comprises the plasmodial slime molds, or myxomycetes, a group of about 450 species that seems to have no direct relationship to the cellular slime molds or any other group. When conditions are appropriate, the plasmodial slime molds exist as thin, streaming masses of protoplasm that creep along in amoeboid fashion. Lacking a cell wall, this "naked" mass of protoplasm is called a _plasmodium._ As these plasmodia travel, they engulf and digest bacteria, yeast cells, fungal spores, and small particles of decayed plant and animal matter. The successful culture of plasmodia in media that do not contain particulate matter suggests that they can also obtain food by direct absorption.

The plasmodium may grow to weigh as much as 20 to 30 grams, and because slime molds are spread out thinly, this amount can cover an area of up to several square meters (see Figure 10–5a, page 154). The plasmodium contains many nuclei but is not divided by cell walls. As it grows, the nuclei divide repeatedly and synchronously; that is, all the nuclei in a plasmodium divide at the same time. Centrioles are present and mitosis is normal, although the chromosomes are very small.

Typically, the moving plasmodium is fan-shaped, with flowing protoplasmic tubules that are thicker at the base of the fan and that spread out, branch, and become thinner toward their outer ends. The tubules are composed of slightly solidified protoplasm through which the more liquefied protoplasm flows rapidly. The foremost edge of the plasmodium consists of a thin film of gel separated from the substrate only by a plasma membrane and a slime sheath of unknown chemical composition.

Plasmodial growth continues as long as an adequate food supply and moisture are available. Generally, when either of these is in short supply, the plasmodium migrates away from the feeding area. At such times, plasmodia may be found crossing roads or lawns, climbing trees, or in other unlikely places. In many species, when the plasmodium stops moving, it divides into a large number of small mounds. The mounds are similar in size and volume, so their formation is probably controlled by chemical effects within the plasmodium. Each mound produces a mature sporangium, usually borne at the tip of a stalk; this sporangium is often extremely ornate (Figures 14–11c, d). Meiosis occurs in the young diploid spores after cleavage and wall formation, and four nuclei are produced. After meiosis, three of the four nuclei disintegrate, leaving the spore with a single haploid nucleus. In some

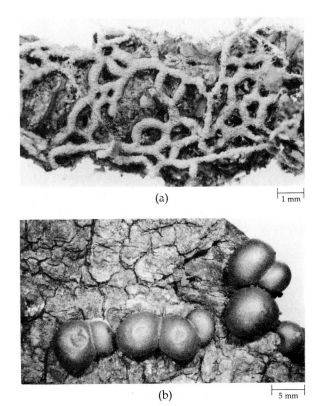

(a)

|‾ 1 mm ‾|

(b)

|‾ 5 mm ‾|

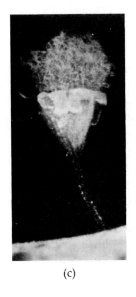

(c)

(d)

14–11
Spore-producing structures in the division Myxomycota. (a) _Plasmodiocarp of_ Hemitrichia serpula. (b) _Aethalia of_ Lycogala _growing on bark._ (c) _Sporangium of_ Arcyria cineria. (d) _Sporangia of_ Stemonitis splendens.

members of this group, discrete sporangia are not produced, and the entire plasmodium may develop either into a *plasmodiocarp* (Figure 14–11a), which retains the former shape of the plasmodium, or into an *aethalium* (Figure 14–11b), in which the plasmodium forms a large mound that is essentially a single large sporangium.

The spores of myxomycetes are resistant to environmental extremes; some have germinated after being kept in the laboratory for more than 60 years. Thus spore formation in this group seems to make possible not only genetic recombination but also survival under adverse conditions.

Under favorable conditions, the spores split open and the protoplast slips out. The protoplast may remain amoeboid, or it may develop one or two whiplash flagella; the amoeboid and flagellated stages are readily interconvertible. The amoebas feed by the ingestion of bacteria and organic material and multiply by mitosis and cell cleavage. If the food supply is used up, or conditions are otherwise unfavorable, an amoeba may

cease moving about, become round, and secrete a thin wall to form a microcyst. These microcysts can remain viable for a year or more, resuming activity when favorable conditions return.

After a period of growth, plasmodia appear in the amoeba population. Their appearance is governed by a number of factors, including cell age, environment, density of amoebas, and chemical inducers that play a role similar to those discussed for the cellular slime mold *Dictyostelium discoideum* (see pages 128–129). One method of plasmodium formation is by the fusion of gametes, which are usually genetically different from one another and are ultimately derived from different haploid spores. These gametes are simply some of the amoebas or flagellated cells that are now playing a new role. In some species and strains, however, the plasmodium is known to form directly from a single amoeba; such plasmodia are usually haploid, like the amoebas that originally gave rise to them.

The life cycle of a typical plasmodial slime mold is summarized in Figure 14–12.

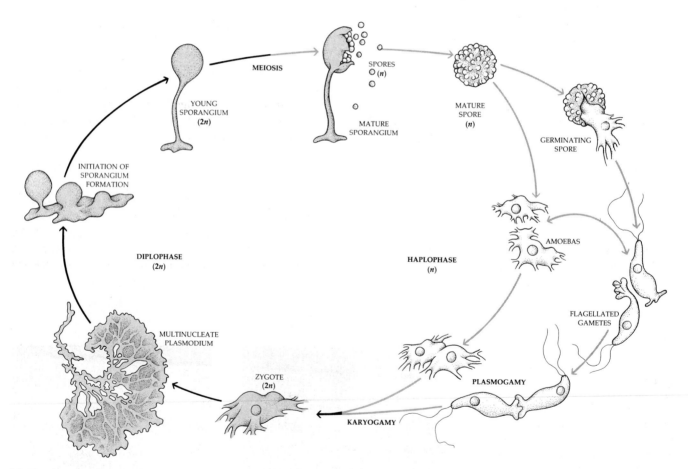

14–12

Life cycle of a typical myxomycete. Sexual reproduction in the plasmodial slime molds consists of three distinct phases: plasmogamy, karyogamy, and meiosis. Plasmogamy consists of the union of two protoplasts, bringing two *haploid nuclei together in the same cell. Karyogamy is the fusion of these two nuclei, resulting in the formation of a diploid zygote and the initiation of the so-called diplophase of the life cycle. The plasmodium is a multinucleate, free-* *flowing mass of protoplasm that can pass through a silk cloth or a piece of filter paper and remain unchanged. Meiosis restores the haploid condition and initiates the haplophase of the cycle.*

CHRYSOPHYTES: DIVISION CHRYSOPHYTA

Chrysophytes are mostly autotrophic, unicellular organisms, which are abundant in fresh and salt water throughout the world. They have chlorophylls *a* and *c*, the color of which is largely masked (in the members of the classes Chrysophyceae and Bacillariophyceae) by the abundance of the accessory pigment fucoxanthin, a carotenoid. The carbohydrate food reserve of Chrysophyta is a substance called chrysolaminarin, which often accumulates as a large granule near the posterior end of the cell. The cells of chrysophytes may be naked and lack a cell wall, or they may have cell walls, made primarily of cellulose, with prominent silica scales or shells in many members of the division. The members of one of the three classes of Chrysophyta, Xanthophyceae, contain the pigment vaucherixanthin, which is related to fucoxanthin; members of the class Xanthophyceae are a bright yellow-green.

The chloroplasts of chrysophytes are biochemically and structurally similar to those of the brown algae (see Chapter 15) and dinoflagellates. This implies that the chloroplasts of these three groups may be derived from the same kind of bacterium. No bacterium with precisely these characteristics has yet been identified, although one recently discovered possibility is *Heliobacterium chlorum*, a strictly anaerobic, brownish, nitrogen-fixing bacterium described by Indiana University scientists in 1983. Another biochemical similarity between the chrysophytes and brown algae is that they all store food as the polysaccharide laminarin (or its more polymerized form chrysolaminarin) outside of the chloroplast; in addition, the brown algae store a different substance, mannitol. Both chrysophytes and brown algae have unequal flagella of similar structure (see Figure 2–26, page 31). The evidence is mounting that perhaps the chrysophytes and brown algae ought to be grouped in a single division.

The chloroplasts of the dinoflagellates (division Pyrrhophyta) are also biochemically and structurally similar to those of the chrysophytes and brown algae. However, the dinoflagellates differ from these groups in so many other respects, as we shall see, that it seems highly unlikely that they share an immediate common ancestor with them. The similarities between the dinoflagellates and the chrysophytes and brown algae probably stem from the fact that the same group of bacteria gave rise to the chloroplasts of both evolutionary lines, as mentioned above, and not to any other direct relationship between them.

The Golden Algae: Class Chrysophyceae

The largest class of Chrysophyta—the golden algae—consists of about 500 species (Figure 14–13). Until recently, they were thought to be primarily a freshwater group, but Chrysophyceae contribute greatly to the productivity of the marine plankton, particularly the nannoplankton.

Many Chrysophyceae lack a clearly defined cell wall but have silica scales or skeletal structures, which may be superficial or internal and often are exceedingly elaborate. Most members of this class are flagellated, unicellular organisms (see Figure 2–26, page 31), although a few lack flagella and some are amoeboid. Except for the presence of their chloroplasts, these amoeboid cells are indistinguishable from amoebas (phylum Rhizopoda), and the two groups may be closely related. A few Chrysophyceae actually ingest bacteria and other organic particles. In the members of this group, an individual cell usually contains one or two large chloroplasts; a large granule of chrysolaminarin often accumulates near the posterior end of the cell. Reproduction in most golden algae is asexual and involves zoospore formation. Some Chrysophyceae are colonial.

14–13

Scanning electron micrographs of two genera of marine golden algae. (a) Distephanus speculum, *a cold-water species that has a siliceous skeleton, is encased in an amoeboid protoplast containing many small chloroplasts.* (b) *A species of the genus* Gephyrocapsa, *extremely minute members of the group known as coccolithophorids. Although these nannoplankton are quite abundant, they are so small that they are not caught by conventional plankton nets; they also dissolve in acid fixatives, so they are difficult to find and to study.*

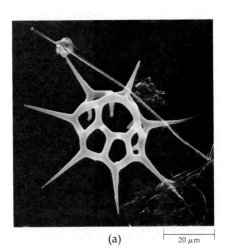

(a)   20 µm

(b)   1 µm

The Yellow-Green Algae: Class Xanthophyceae

The organisms of this class, which consists of about 550 species, are yellow-green. Yellow-green algae resemble golden algae in that chlorophyll *c* occurs in many species of the class; they differ from them, however, in lacking the accessory pigment fucoxanthin. Yellow-green algae are mostly nonmotile organisms, although some are amoeboid or flagellated, as are the gametes.

One of the better-known members of the class is *Vaucheria*, the "water felt," a coenocytic, little-branched, filamentous alga. *Vaucheria* reproduces both asexually, by the formation of large, compound zoospores that are multiflagellated, and sexually, by oogamy (Figure 14–14). *Vaucheria* is widespread in freshwater, brackish, and marine habitats, and often is found on mud that is alternately immersed in water and exposed to air.

The Diatoms: Class Bacillariophyceae

The diatoms are mostly unicellular organisms that are exceedingly important components of phytoplankton. As such, they are a primary source of food for water-dwelling animals both in marine and freshwater habitats (Figure 14–15). It is estimated that there may be at least 5600 living species in this class. Counting the great number of extinct species, the diatoms account for at least 40,000 species. There are often tremendous numbers of species in very small areas. In two small samples of mud from the ocean near Beaufort, North Carolina, for example, 369 species were identified. Most species of diatoms occur in plankton, but some are bottom dwellers or grow on other algae or plants. They occur both in fresh and salt water.

Diatoms differ from other Chrysophyta in lacking flagella, except for some male gametes, and in their unique shells, or cell walls. These thin double shells, or frustules, are made of polymerized, opaline silica ($SiO_2 \times nH_2O$); the two halves (valves) fit together, one on top of the other. The delicate markings of these shells, by which the species are identified, have been traditionally used by microscopists to test the quality of their lenses. Electron microscopy has shown that fine tracings on the diatom shells are actually composed of a large number of minute, intricately shaped depressions, pores, or passageways that connect the living protoplasm within the shell to the outside environment (Figure 14–16). The most conspicuous features within the protoplast of diatoms are the brownish plastids that contain chlorophylls *a* and *c*, as well as fucoxanthin. Diatoms reproduce primarily asexually by cell division (Figure 14–17).

On the basis of symmetry, two types of diatoms are recognized: the _pennate_ diatoms, which are bilaterally symmetrical (Figure 14–16d), and the _centric_ diatoms, which are radially symmetrical (Figure 14–16c). Centric diatoms are most abundant in marine waters. When

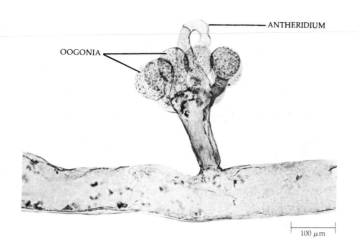

ANTHERIDIUM

OOGONIA

100 μm

14–14
Vaucheria, *the "water felt," is a coenocytic, filamentous member of the division Chrysophyta. Vaucheria is oogamous, producing oogonia and antheridia. The antheridium shown here is empty.*

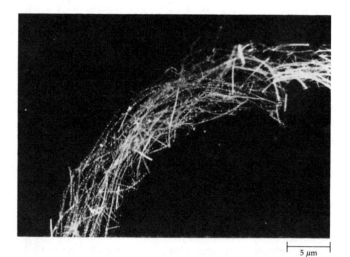

5 μm

14–15
Portion of a fluffy mat formed of intertwining chains of diatom cells, comprising two species of the genus Rhizosolenia. Mats of this kind are often abundant in nutrient-poor ocean waters; this mat, for example, came from the central North Pacific, just such an area. Within the vacuoles of the diatom cells live high concentrations of symbiotic, nitrogen-fixing bacteria. Such associations make a very important contribution to the productivity of nutrient-poor waters.

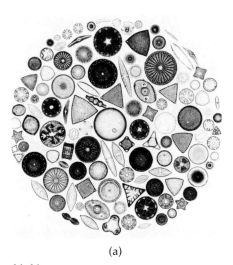

(a)

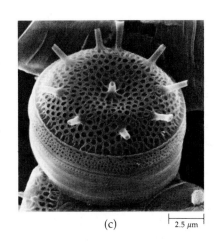

(b) ├─ 30 μm ─┤

(c) ├─ 2.5 μm ─┤

14–16

(a) *A selected array of marine diatoms, as seen with a light microscope. Scanning electron micrographs of (b) one valve of an* Entogonia *shell and (c)* Thalassioria nordenskioeldii, *a centric diatom. (d)* Pinnularia, *a pennate diatom, as seen with a light microscope.*

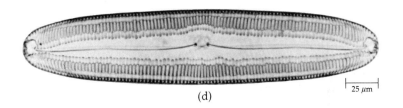

├─ 25 μm ─┤

(d)

sexual reproduction occurs in the centric diatoms, it is oogamous. The male gametes may have a single tinsel flagellum; they are the only flagellated cells found in the diatoms at any stage of the life cycle. In the pennate diatoms, sexual reproduction is isogamous, and both gametes are nonflagellated.

Although most species of diatoms are autotrophic, many can become heterotrophic and then exist by absorbing organic carbon; these are primarily pennate diatoms that occur on the bottom of the sea in relatively shallow habitats. A few diatoms are obligate heterotrophs; they lack chlorophyll and are no longer capable of providing their own food through photosynthesis. On the other hand, some diatoms, lacking their characteristic shells, live symbiotically in large marine protozoa (Foraminifera) and provide organic carbon for their hosts.

Despite their lack of flagella or other locomotor organelles, many species of pennate diatoms are motile. Their locomotion results from a rigorously controlled secretion that occurs in response to a wide variety of physical and chemical stimuli. All motile diatoms seem to possess along each shell a fine groove called the raphe, which is basically a pair of pores connected by a complex slit in the siliceous wall of the diatom. Many nonmotile diatoms are attached to one another, with their shells arranged in long filaments (Figure 14–15). The raphe apparently evolved as a locomotor device by modification of the apical pores, which secrete the substances that unite the nonmotile diatoms into filaments.

A diatom moves in response to an external stimulus —such as mechanical disturbance, light, heat, or a toxic

chemical—by initiating contractions in contractile bundles that lie adjacent to the raphe system. This contraction moves dehydrated crystalline bodies to reservoir areas adjacent to pores in the raphe, and the crystalline bodies are then discharged into the pores. Here they take up water and expand into twisting fibrils. The fibrils move along the raphe until they touch a surface. They immediately adhere to anything they touch and then contract. If the object to which they adhere is large enough, the diatom moves toward it, depositing a trail of secreted material analogous to the slime trail of a snail. If the object is small, it is transported along the raphe, and the diatom remains stationary. Even motile diatoms are usually at rest. Each can move only a limited distance, for they have available at any one time only a limited supply of the required crystalline bodies.

The silica shells of diatoms have accumulated over millions of years and form the fine, crumbly substance known as diatomaceous earth, which is used as an abrasive in silver polish and for filtering and insulating materials. It is estimated that 1 cubic centimeter of diatomaceous earth contains some 4.6 million diatom shells. In the Santa Maria, California, oil fields there is a subterranean deposit of diatomaceous earth that is 900 meters thick, and near Lompoc, California, more than 270,000 metric tons of diatomaceous earth are quarried annually for industrial use.

Diatoms first became abundant in the fossil record some 100 million years ago, during the Cretaceous period. Many of the fossil species are identical to those still living today, which indicates an unusual persistence through geological time.

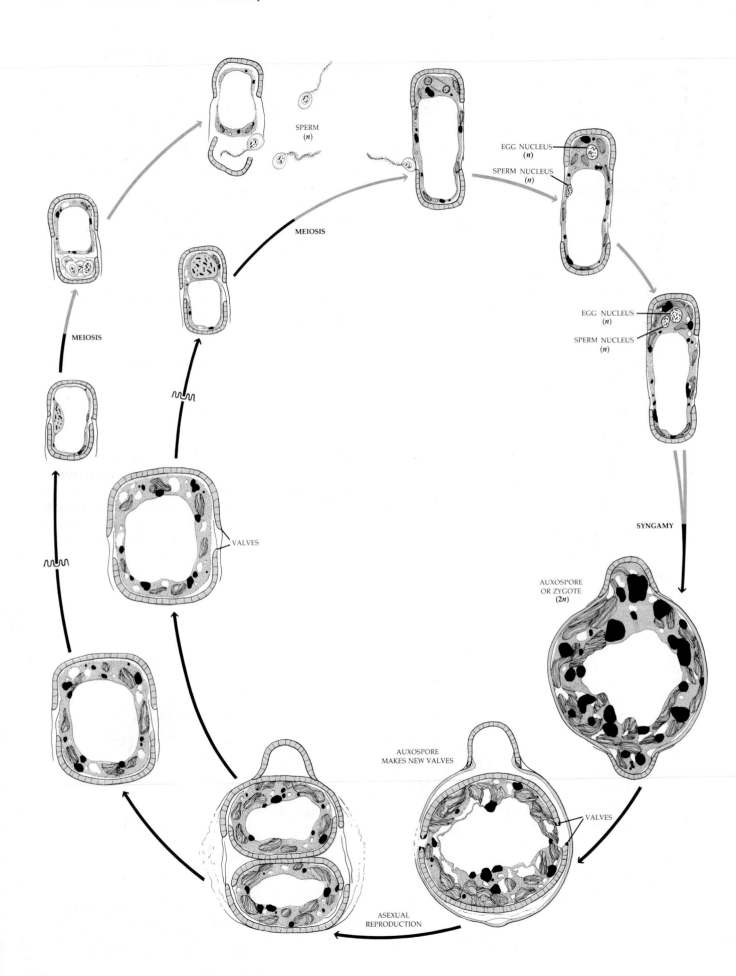

SPERM
(*n*)

EGG NUCLEUS
(*n*)

SPERM NUCLEUS
(*n*)

MEIOSIS

MEIOSIS

EGG NUCLEUS
(*n*)

SPERM NUCLEUS
(*n*)

SYNGAMY

VALVES

AUXOSPORE
OR ZYGOTE
(2*n*)

AUXOSPORE
MAKES NEW VALVES

VALVES

ASEXUAL
REPRODUCTION

14–17

Reproduction in the diatoms is mainly asexual and occurs by cell division. Each daughter cell (bottom left) receives one of the valves of the previous shell (bottom right) and constructs a second valve. The old valve always forms the large part (the lid) of the silica "box," with the new valve fitting inside it. Thus one cell of each new pair tends to be smaller than the one before it. In some species, the shells are expandable and are enlarged by the growing protoplasm within them. In other species, however, the shells are more rigid. When the individuals of these species have decreased in size to about 30 percent of the maximum diameter, sexual reproduction may occur. Certain cells function as male gametangia; they each produce four sperm through meiosis. Other cells function as female gametangia; in these, three of the four products of meiosis are nonfunctional, so that one egg is produced per cell. After syngamy (fertilization), the resulting auxospore, or zygote, expands to the full size characteristic of the species. The walls formed by the auxospores are often different from those of the asexually reproducing cells of the same species. Once the auxospore is mature, it divides and produces new shells identical in all their intricate markings to the previous ones. This diagram illustrates asexual reproduction in a centric diatom.

THE DINOFLAGELLATES: DIVISION PYRRHOPHYTA

Most dinoflagellates are unicellular biflagellates (see Table 14–1, page 233). More than 1000 species are known, many of them abundant and highly productive members of the marine plankton; others occur in fresh water. In the dinoflagellates, the flagella beat within two grooves; one groove circles the body like a belt, and the second groove is perpendicular to the first. The beating of the flagella in their respective grooves causes the dinoflagellate to spin like a top as it moves. The encircling flagellum is ribbonlike. There are also numerous nonmotile dinoflagellates, some of which are nonflagellated. Some genera of dinoflagellates ingest solid food particles, obtaining part or all of their nutrition in this way.

Many of the dinoflagellates are bizarre in appearance, with stiff cellulose plates forming a wall (theca), which often looks like a strange helmet or part of an ancient coat of armor (Figures 14–18 and 14–19). The plates of the wall are in vesicles inside the plasma membrane, not outside it like the cell wall of most algae.

Most dinoflagellates contain chlorophylls *a* and *c*, which are generally masked by carotenoid pigments, including peridinin, which is similar to fucoxanthin. Their chloroplasts probably were derived from the same bacterial group that gave rise to the chloroplasts of the chrysophytes and brown algae, as mentioned above. In other respects, the dinoflagellates are very different from the members of these other divisions.

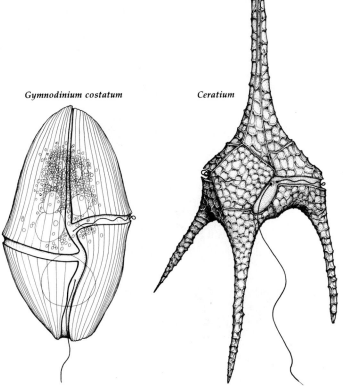

Gymnodinium costatum

Ceratium

Exuviaella

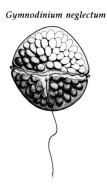

Gymnodinium neglectum

14–18

The "armor" of some dinoflagellates consists of cellulose plates in vesicles inside the plasma membrane. The vesicles of genera that appear to be naked may or may not contain cellulose plates .

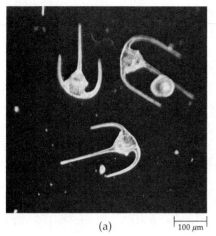

(a) | 100 μm

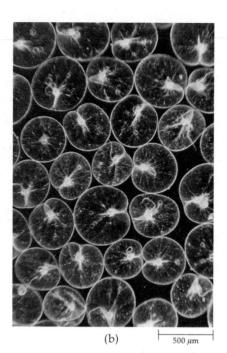

(b) | 500 μm

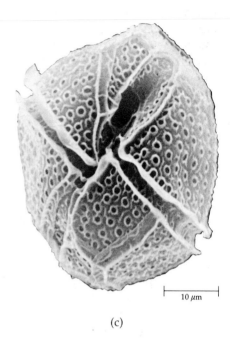

| 10 μm

(c)

14–19

(a) Ceratium triops, *an armored dinoflagellate.* (b) Noctiluca scintillans, *a bioluminescent marine dinoflagellate.* (c) Gonyaulax polyedra, *the dinoflagellate responsible for the spectacular red tides along the coast of southern California.*

MITOSIS IN DINOFLAGELLATES

Dinoflagellates have a unique type of mitosis in which some of the features of cell division in bacteria may have been retained. In the dinoflagellate cell, as shown in (a), the chromosomes are always visible and do not condense prior to mitosis. There are extremely large amounts of DNA in each cell. The chromosomes are attached to the nuclear envelope, which persists during mitosis. Dinoflagellate chromosomes have a much lower ratio of protein to DNA than other eukaryotes, and they may have evolved from bacterial chromosomes independently from the chromosomes of other eukaryotes.

Cytoplasmic channels, shown extending through the center of (b), invade the dividing nucleus at the time of mitosis. The chromosomes remain attached to the nuclear envelope and are carried on the sides of these channels, which contain bundles of microtubules similar to those of a eukaryotic spindle apparatus. The microtubules are all oriented in one direction and presumably regulate the separation of the portions of the nuclear envelope with its attached chromosomes. The organism illustrated in both (a) and (b) is *Cryptothecodinium cohnii.*

At least two species of dinoflagellates have binucleate cells, in which one nucleus is called *dinokaryotic* and the second is eukaryotic. In this second nucleus—which was presumably obtained by the host as a result of the invasion of a chrysophytelike symbiotic organism—the chromosomes do not become condensed at any stage during the cell cycle, an observation that indicates they are not identical with "typical" eukaryotic chromosomes.

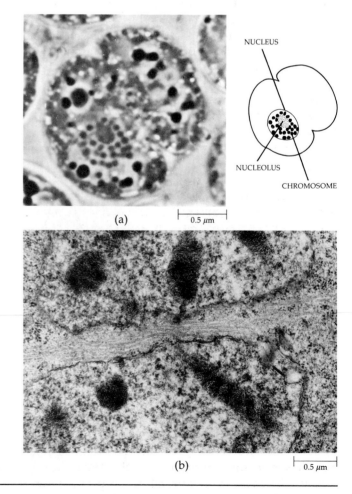

NUCLEUS

NUCLEOLUS

CHROMOSOME

(a) | 0.5 μm

(b) | 0.5 μm

The carbohydrate food reserve of dinoflagellates is starch. Some of the species do not contain chlorophyll and are heterotrophic, but their structure clearly allies them to the other members of the division. Even the autotrophs usually have strong requirements for vitamin B_{12}, like many diatoms, and cannot be regarded as autotrophs without this qualification. Some dinoflagellates are capable of ingesting other cells. It is assumed that the colorless forms are grazers that obtain their nutrition by ingesting other cells or small particulate organic material.

Cyanobacteria are common symbionts in tropical dinoflagellates. Dinoflagellates in turn occur as symbionts in many other kinds of organisms, including sponges, jellyfish, sea anemones, tunicates, corals, octopuses and squids, gastropods, turbellarians, and certain protists. In the giant clams of the family Tridachnidae, the dorsal surface of the inner lobes of the mantle may appear chocolate-brown as a result of the presence of symbiotic dinoflagellates. When they are symbionts, the dinoflagellates lack thecae and appear as golden spherical cells called _zooxanthellae_ (Figure 14–20).

Zooxanthellae are primarily responsible for the photosynthetic productivity that makes possible the growth of coral reefs in tropical waters, which are notoriously poor in nutrients. Coral tissues may contain as many as 30,000 symbiotic dinoflagellates per cubic millimeter, primarily within cells in the lining of the gut of the coral polyps. Since the algae require light for photosynthesis, the corals that contain them grow mainly in ocean waters less than 60 meters deep. Many of the variations in the shapes of coral are related to the light-gathering properties of different geometrical arrangements, somewhat similar to the ways that various branching patterns of trees act to expose their leaves to sunlight.

Dinoflagellates also play an important role in human affairs. During the winter and spring of 1974, the west coast of Florida was ravaged by its twenty-fifth major red tide since 1844. Hundreds of thousands of dead fish littered the beaches, and millions of tourist dollars were lost. Such red tides are caused by unusual outbreaks, or "blooms," of dinoflagellates, which color the sea red or brown (Figure 14–21). These dinoflagellates are ingested not only by fish, which may be poisoned directly, but also by shellfish, such as mussels and clams, which are generally not harmed. Shellfish accumulate and concentrate the toxins produced by these organisms and, depending on the species of dinoflagellate, may become dangerous for human consumption. In the fall of 1972, for example, the New England coast from Maine to Cape Cod experienced its first red tide. Twenty-six people were poisoned by shellfish contaminated by _Gonyaulax excavata,_ and public confidence was shaken to such an extent that the Massachusetts shellfish industry had returned to only about two-

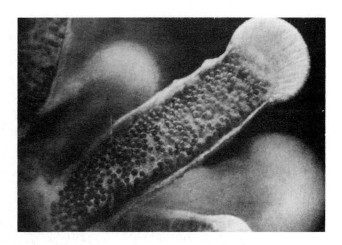

14–20
Zooxanthellae, the symbiotic form of dinoflagellates, shown here in a tentacle of a coral animal, are responsible for much of the productivity of coral reefs.

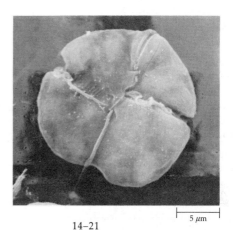

14–21
Ptychodiscus brevis, _the unarmored dinoflagellate responsible for the outbreaks of red tide along the Florida coast. The curved transverse flagellum lies in a groove encircling the organism. Its longitudinal flagellum, only a portion of which is visible, extends from the middle of the organism toward the lower left. The apical groove at the top is an identifying characteristic of_ Ptychodiscus.

thirds of its former level four years later. Red tides are a recurrent problem. In the worst outbreak since 1972, the entire coast of Maine was legally closed to shellfishing from mid-August to mid-October 1980. Commercial fishermen were unable to harvest an estimated $7 million worth of clams, oysters, and mussels.

The poisons produced by some dinoflagellates, such as _Gonyaulax catenella,_ are extraordinarily powerful nerve toxins. The chemical nature and biological activity of most of these toxins are relatively well known.

On the other hand, the factors that cause the red tides themselves are poorly understood. Levels of nutrients and certain trace metals, sewage runoff, ocean salinity and temperature, winds, light, and many other factors all seem to play some role. The periodic recurrence of red tides is compounded by the observation that once a "bloom" has occurred, if conditions shift for the worse, the dinoflagellates may lose their flagella and form resting cysts that sink to the bottom and lie dormant until conditions are more favorable for their renewed activity.

The chief method of reproduction in dinoflagellates is by longitudinal cell division, with each daughter cell receiving one of the flagella and a portion of the theca and then constructing the missing parts in a very intricate sequence (see "Mitosis in Dinoflagellates," page 246). Some nonmotile species form zoospores. In some of these species only the zoospore has the typical dinoflagellate structure, whereas the mature forms have cells with no flagella or may even be joined together in filaments. Sexual reproduction has been found in a number of dinoflagellates; it is generally isogamous, but anisogamy is also present.

THE EUGLENOIDS: DIVISION EUGLENOPHYTA

There are more than 800 species of euglenoids, most of which are found in fresh water, especially water rich in organic material (see Table 14–1, page 233). The euglenoids range from less than 10 micrometers to more than 500 micrometers (0.5 millimeter) in length and are quite variable in form. All are unicellular except for the colonial genus *Colacium.*

As discussed earlier, the similarities between the chloroplasts of the euglenoids and those of the green algae suggest that they were derived independently from the same sort of bacterium, possibly one similar to *Prochloron* (see Figures 11–2, page 166, and 11–17, page 176). In other respects, the members of the two divisions are utterly different. About a third of the approximately 40 genera of euglenoids have chloroplasts, and these have chlorophylls *a* and *b* together with several carotenoids.

Euglenoids store their carbohydrate food reserves in the form of paramylon, a polysaccharide that is not found in any other group of organisms. As in all organisms except green algae and plants, this storage product is formed outside the chloroplast. Among the euglenoids that lack chloroplasts, some absorb organic matter and others ingest it. As mentioned above, euglenoids are essentially zoomastigote protozoa that have acquired chloroplasts in the course of their evolution; the two divisions (phyla) probably ought to be merged.

Euglenoids reproduce by cell division, with the individual cells remaining motile throughout the process. The nuclear envelope remains intact during mitosis, as it does in most of the green algae, the dinoflagellates, many fungi, and some ciliated protozoa. The centrioles function as basal bodies and organize a normal spindle apparatus within the nuclear envelope. The chromosomes, like those of dinoflagellates, remain condensed during interphase and throughout the mitotic cycle. No sexual reproduction is known to occur among the euglenoids.

The division Euglenophyta takes its name from *Euglena,* a common genus. Many species of *Euglena* are elongated, as shown in Figure 14–22. The cell is complex and contains numerous small chloroplasts. A long,

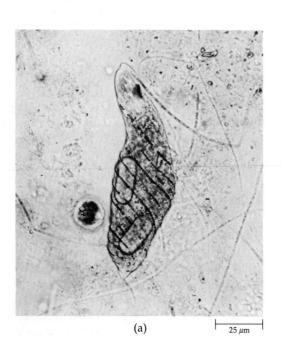

(a)

25 μm

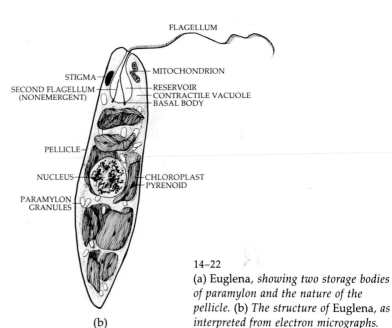

(b)

14–22

(a) Euglena, *showing two storage bodies of paramylon and the nature of the pellicle.* (b) *The structure of* Euglena, *as interpreted from electron micrographs.*

emergent flagellum with very fine hairs is present along one side, as well as a short, nonemergent flagellum. The emergent flagellum is usually held in front of the cell like a spinning lasso.

In *Euglena*, the flagella are attached at the base of the flask-shaped opening—the *reservoir*—at the anterior end of the cell. The contractile vacuole collects excess water from all parts of the cell and discharges it into the reservoir. The cell is delimited by a plasma membrane inside of which is a series of flexible, interlocking proteinaceous strips that are arranged helically. These strips, together with the plasma membrane, form a structure called the *pellicle*. Unlike the stiff walls of plant cells, the flexible pellicle permits *Euglena* to change its shape, providing an alternative means of locomotion for mud-dwelling forms.

If one leaves a culture of *Euglena* near a sunny window, a clearly visible green cloud will form in the water, and this will move as the light changes, always toward a spot that is bright (but not too bright). If the light is too bright, the individuals of *Euglena* will swim away from it. *Euglena* is probably able to orient with respect to light because of the presence of a pair of special structures—the stigma and an associated photoreceptor, which is a swelling at the base of the flagellum. When the individual euglenoid is in certain orientations with respect to light, the "shading" of the photoreceptor by the pigmented stigma probably aids in its directional movement. Only a few nongreen euglenoids have a stigma.

Some species of *Euglena* can survive if they are kept in the dark, where they cannot photosynthesize, as long as they are provided with a carbon source in addition to their other vitamin and mineral requirements. If some strains of *Euglena* are kept in the light, at an appropriate temperature and in a rich medium, the cells may replicate faster than the chloroplasts, producing nonphotosynthetic cells, which can nonetheless survive indefinitely in a suitable medium. The euglenoids are basically a group of protozoa with unstable chloroplasts.

SUMMARY

The four divisions of heterotrophic protists that have often been treated as fungi are Oomycota, Chytridiomycota, Acrasiomycota, and Myxomycota. They have a more or less typical mitotic cycle, and centrioles are present. The first two groups are primarily aquatic; the latter two are terrestrial.

Members of the division Oomycota range from unicellular forms to highly branched coenocytic filamentous forms. The sperm and zoospores of this group have two flagella—one tinsel and one whiplash. The cell walls are composed largely of cellulose or celluloselike polymers. Sexual reproduction involves a large, immobile egg and small, motile sperm. Two of the partly terrestrial members of this division are *Phytophthora*—a very important causative agent of plant diseases, in-

cluding the blight that caused the potato famine of 1846–47 in Ireland—and *Plasmopara viticola*, which causes downy mildew of grapes.

The Chytridiomycota, or chytrids, are unicellular or coenocytic organisms that are aquatic; in most species that have been examined, the cell walls are rich in chitin. The motile spores and gametes have a single posterior whiplash flagellum. *Allomyces* and one closely related genus are the only nonphotosynthetic organisms known to have an alternation of generations similar to that of plants and many algae.

The cellular slime molds, division Acrasiomycota, are a small group of amoebalike organisms that aggregate together at one stage of their life cycle to form pseudoplasmodia, or "slugs." Flagellated cells are not known. Sexual reproduction occurs in at least some members of this group; it takes place by means of macrocysts.

The plasmodial slime molds, division Myxomycota, may exist as streaming, multinucleate masses of protoplasm called plasmodia, which are usually diploid. These plasmodia ultimately form sporangia in which diploid spores are formed. Meiosis occurs within each of the spores, and three of the resulting nuclei disintegrate, leaving one haploid nucleus in each spore. Under favorable conditions, the spores split open, producing amoebas, which may become flagellated. These amoebas or flagellated cells may function as gametes. Plasmodium formation often, but not always, follows fusion of the gametes.

Among the photosynthetic, primarily unicellular protists, we have discussed three divisions here. Two of these—Chrysophyta and Pyrrhophyta—have chloroplasts that are biochemically and, to some extent, structurally similar. They are characterized by the presence of chlorophylls *a* and *c* and the accessory pigment fucoxanthin. These relationships suggest that the ancestors of these divisions, and of the brown algae (division Phaeophyta) as well, may have acquired their chloroplasts through a series of independent symbiotic events involving a bacterium with characteristics somewhat like those of the newly discovered *Heliobacterium chlorum*. Alternatively, they may have ingested a eukaryote that had chloroplasts with these characteristics. In an analogous example, the chloroplasts of the division Euglenophyta resemble those of the green algae (division Chlorophyta) and were almost certainly acquired independently from those of the other photosynthetic, unicellular protists discussed in this chapter.

Chrysophyta are important components of freshwater and marine phytoplankton. A number of the members of one class, Chrysophyceae, are small algae (1 to 3 micrometers), which are found in large quantities in nannoplankton. These organisms are very important contributors to photosynthetic productivity in the sea. A second class, the Xanthophyceae, consists of organisms that lack the pigment fucoxanthin. The third class, the diatoms, class Bacillariophyceae, consist of at least 5600 species of unicellular organisms with unique double silica cell walls. Diatoms have contributed in a major way to the marine and freshwater plankton for the past 200 million years. The only flagellated cells in the life cycle of diatoms are the male gametes, which have been observed only in a few species.

The dinoflagellates (division Pyrrhophyta) are unicellular biflagellates, which are mostly marine. Dinoflagellates are characterized by two flagella, which beat in different planes, causing the organism to spin. Stiff cellulose plates, frequently bizarre in appearance, are often present in vesicles beneath the plasma membrane. In their symbiotic form, in which they are called zooxanthellae, dinoflagellates are major contributors to the productivity of coral reefs; they also occur widely as symbiotes in many other marine animals.

Euglenophyta, the euglenoids, are a small group of organisms, most of which occur in fresh water and are unicellular. They contain chlorophylls *a* and *b* and store carbohydrates as an unusual polysaccharide called paramylon. Euglenoids lack a cell wall but have a flexible series of protein strips, which make up the pellicle, inside the plasma membrane. The cells are highly differentiated, containing chloroplasts, a contractile vacuole, and flagella. No sexual reproduction is known. Euglenophyta are fundamentally similar to protozoa of the phylum Zoomastigina and perhaps should be included in that group.

SUGGESTIONS FOR FURTHER READING

BOLD, HAROLD C., and MICHAEL J. WYNNE: *Introduction to the Algae: Structure and Reproduction,* 2nd ed., Prentice-Hall, Inc., Englewood Cliffs, N.J., 1985.

A detailed reference work on the algae that contains a wealth of information on all groups; taxonomically oriented.

BONNER, JOHN T.: *The Cellular Slime Molds,* 2nd ed., Princeton University Press, Princeton, N.J., 1968.

A record of experimental work with a small but fascinating group of organisms.

CHRISTENSEN, CLYDE M.: *The Molds and Man: An Introduction to the Fungi,* 3rd ed., University of Minnesota Press, Minneapolis, 1965.*

A highly informative account of the fungi and especially of their complex and important interrelationships with humans.

DARLEY, W. MARSHALL: *Algal Biology: A Physiological Approach,* Blackwell Scientific Publications, Boston, Mass., 1982.

A comprehensive account of the biology of the photosynthetic protists, emphasizing the way in which the individual organism functions in its environment.

DAWES, CLINTON J.: *Marine Botany,* Wiley-Interscience, New York, 1981.

A survey of the marine algae in an ecological context.

GRAY, W. D., and CONSTANTINE J. ALEXOPOULOS: *Biology of the Myxomycetes,* Ronald Press Co., John Wiley & Sons, Inc., New York, 1968.

A well-written book about the biology of the plasmodial slime molds, emphasizing ultrastructural, biochemical, and physiological aspects.

LARGE, E. C.: *The Advance of the Fungi,* Dover Publications, Inc., New York, 1962.*

A fascinating popular account of the closely interwoven histories of fungi and humans, first published in 1940.

LOOMIS, WILLIAM F. (Ed.): *The Development of* Dictyostelium discoideum, Academic Press, New York, 1982.

Comprehensive reviews by a number of authors of all aspects of the biology of this species and of the division to which it belongs.

MARGULIS, LYNN, and KARLENE V. SCHWARTZ: *Five Kingdoms. An Illustrated Guide to the Phyla of Life on Earth,* W. H. Freeman and Company, New York, 1982.

The outstanding short guide to biological diversity.

RAPER, KENNETH B.: *The Dictyostelids,* Princeton University Press, Princeton, N.J., 1984.

An excellent discussion of the growth, morphogenesis, and systematics of these fascinating organisms.

ROSS, IAN K.: *Biology of the Fungi: Their Development, Regulation, and Associations,* McGraw-Hill Book Company, New York, 1979.

An outstanding treatment of the fungi, biologically oriented, emphasizing the divisions treated in this chapter.

ROUND, F. E.: *The Ecology of Algae,* Cambridge University Press, New York, 1981.

A comprehensive account of freshwater and marine algal ecology, organized according to principles and very well written.

SCAGEL, R. F., R. J. BANDONI, J. R. MAZE, G. E. ROUSE, W. B. SCHOFIELD, and J. R. STEIN: *Nonvascular Plants. An Evolutionary Survey,* Wadsworth Publishing Company, Belmont, Calif., 1982.

A thorough and outstanding treatment of the diversity of those protists that are included in this book and of fungi.

SPECTOR, DAVID L. (Ed.): *Dinoflagellates,* Academic Press, Inc., New York, 1984.

An up-to-date review of contemporary studies of this fascinating group of protists, with each article by an expert in the field.

STEIDINGER, K. A., and K. HADDAD: "Biologic and Hydrographic Aspects of Red Tides," *BioScience* 31: 814–19, 1981.

Through the use of satellite imagery, the blooms of dinoflagellates that lead to the production of red-tide conditions can be traced from the time of their initiation.

VIDAL, GONZALO: "The Oldest Eukaryotic Cells," *Scientific American,* February 1984, pages 48–57.

An excellent review of the nature and occurrence of acritarchs, the oldest known eukaryotic fossils.

* Available in paperback.

C H A P T E R   1 5

Multicellular Protista: Red, Brown, and Green Algae

15–1
Algae on the rocks at low tide along the coast of North Carolina.

The open sea, the shore, and the land are the three life zones that make up our biosphere. Algae play a role in these first two ancient dwelling places, the sea and the shore, comparable to the role played by plants in the far younger terrestrial world (Figure 15–1). Often, algae are also dominant in freshwater habitats—ponds, streams, and lakes—where they may be the most important contributors to the productivity of these ecosystems. Everywhere they grow, algae play an ecological role comparable to that of the plants in land habitats.

Along the rocky shore can be found the larger, more complex algae, or seaweeds, typically arranged in fairly distinct visible bands or layers in relation to intertidal levels (Figure 15–2). Their structural complexity reflects their ability to survive in this challenging life zone, where twice each day they are subject to great fluctuations of humidity, temperature, salinity, and light, in addition to the pounding action of the surf and the abrasive action of suspended sand particles churned up by the waves.

Anchored offshore beyond the zone of waves, algae provide shelter for a rich diversity of microscopic organisms, as well as for larger fish and invertebrate animals that feed on the microorganisms and on each other. These extensive algal growths may be so dense as to be called forests. For example, off the coast of California, there are huge beds of giant kelps—brown algae whose broad, 15-meter-long fronds are buoyed upward on sinuous stalks 30 meters or more from their holdfasts on the bottom. Many large carnivores, including sea otters and tuna, find food and refuge in these kelp beds, which are also harvested by humans for food and fertilizer.

The three divisions of algae that are wholly or partly multicellular—the larger algae—form the subject of this chapter. These are the algae that form masses along the coasts and at times float freely in the open ocean, as in the so-called Sargasso Sea. One of these divisions, the green algae (division Chlorophyta), also includes many unicellular algae, some of which are important contributors to the productivity of the plankton, both in the sea and in fresh water. This exceedingly diverse division is of special interest because it is almost certainly the group that gave rise to the plants, an event that occurred more than 430 million years ago. Although there is probably no existing green alga that resembles the ancestor of plants exactly, many retain some of its probable features.

In addition to the green algae, which are better represented in fresh water than in the sea, and are common even in terrestrial habitats, there are two other divisions of primarily multicellular algae: red algae (division Rhodophyta) and brown algae (division Phaeophyta). Both of these divisions are almost exclusively marine. There are also a few freshwater genera of red algae and brown algae. All of the brown algae are multicellular, whereas a few red algae are unicellular.

CHARACTERISTICS OF RED, BROWN, AND GREEN ALGAE

As discussed in the preceding chapter, different bacteria seem to have been involved in the origins of chloroplasts in the red, brown, and green algae. In red algae, the chloroplasts contain chlorophyll *a* and phycobilins; these chloroplasts are similar to cyanobacteria, an ancient group. Most scientists have concluded that the evidence for an origin of the chloroplasts of red algae from symbiotic cyanobacteria is very strong. Green algae, in contrast, have chloroplasts with chlorophylls *a* and *b;* they share these features with the euglenoids and with the plants, which are descended from green algae. The biochemical characteristics of these chloroplasts are similar to those of *Prochloron* (see Figures 11–2, page 166, and 11–17, page 176), another bacterium, and they may have been derived from an organism of this kind. Finally, the golden-brown chloroplasts of the closely related brown algae, the chrysophytes, and the dinoflagellates, which have chlorophylls *a* and *c*, may have been derived from a bacterium with characteristics similar to those of the newly discovered *Heliobacterium chlorum*. These chloroplasts may have originated from eukaryotic cells ingested by the algae, but these organelles would still ultimately have had a bacterial origin.

The members of the three divisions considered in this chapter also differ greatly in many other characteristics, some of which are summarized in Table 14–1 (page 233). Although cellulose is found in the cell walls of some members of each division, these walls are of extremely varied constitution, both within and between these groups. The red algae lack flagella completely at all stages of their life cycles, and they also lack centrioles. Red algae have structures called *polar rings*,

15–2
Rocks along the coast of Cornwall, in England, showing the different layers of algal species from lower levels to higher levels.

which are associated with the nucleus and move during mitotic prophase like the centrioles of other algae; whether polar rings and centrioles are homologous is unknown. The similarities between red algae and ascomycetes have been noted for many years and deserve further consideration in the light of additional characteristics. The flagellated cells of the brown algae, like those of the closely related chrysophytes and diatoms, have unequal flagella—a tinsel flagellum directed forward and a trailing whiplash flagellum (see Figure 2–26, page 31). In sharp contrast, the flagella of green algae (when present) are almost always equal whiplash flagella.

Green algae, like plants, store their food reserves as starch in their plastids (Figure 15–3a). All other groups of photosynthetic eukaryotes, including those discussed in Chapter 14, as well as the red and brown algae described in this chapter, store their food reserves outside of the chloroplasts in the cytoplasm. The food reserves of red algae are stored as starch, while those of brown algae are stored as laminarin, another polymer of glucose but with a linkage between the glucose units different from that in starch. The alcohol mannitol is another storage product in the brown algae.

When algal cells divide, the plasma membranes

generally pinch inward from the margin of the cell (furrowing), just as in animals, fungi, and protists generally. However, cell plates like those of plants have been found in one brown alga and a few genera of filamentous green algae. Most algal cells—except those of the red algae—have centrioles, which become the basal bodies for flagella of any motile cells that are produced (see page 32).

In general, the multicellular members of the divisions treated here do not have a complex array of tissues like that found in vascular plants. In certain kelps (brown algae), however, a central conducting strand consisting of cells that resemble sieve elements is found in the stalk and blade. The reproductive structures of algae are generally single cells, not multicellular structures with sterile protective jackets such as are found in bryophytes and vascular plants (see Chapter 16).

The three divisions of algae discussed in this chapter are ancient groups, and they differ sharply in their characteristics. Fossils similar to unicellular green algae occur in the Bitter Springs Formation of central Australia, which is about 900 million years old. By the Cambrian period, at least 550 million years ago, large, siphonous (coenocytic) green algae as well as red algae had evolved. Other multicellular lines of green algae appeared throughout the Paleozoic era; for example, the oldest stoneworts, representatives of the most distinct branch of green algae, date from the late Silurian period, about 420 million years ago. Those green algae and red algae that are best represented in the fossil record have calcareous cell walls; only one genus of brown algae is calcified. Consequently, the fossil record of brown algae is poor; they are not preserved as easily as some of the green and red algae.

RED ALGAE: DIVISION RHODOPHYTA

Red algae have no flagellated cells, are structurally complex (Figures 15–4, 15–5), and have complex life cycles. Red algae lack centrioles, in contrast to green and brown algae, which usually have centrioles even in their vegetative cells; the structures called polar rings play the role of centrioles in red algae. There are some 4000 species of red algae, and they are particularly abundant in tropical and warm waters, although many are found in the cooler regions of the world. Fewer than 100 species of red algae are found in fresh water, but in the sea the number of species is greater than that of all other groups of seaweeds combined. The members of this division usually grow attached to rocks or other algae; there are a few floating forms and a few that are unicellular or colonial.

The water-soluble phycobilins, which mask the color of chlorophyll *a* and give the red algae their distinctive color, are accessory pigments. They are particularly well suited to the absorption of the green, violet, and blue light that penetrates into deep water. Their chloroplasts seem to have originated as symbiotic cyanobacteria, which they resemble both biochemically and structurally.

The cell walls of most red algae include a rigid inner component composed of microfibrils, which may be cellulose or another polysaccharide, and a mucilaginous matrix, usually a sulfated polymer of galactose, such as agar or carrageenan (see "The Economic Uses of Seaweeds," page 262). It is these latter components that give red algae their characteristic flexible, slippery texture.

15–3

In green algae and plants, food reserves are stored as starch in plastids, whereas other photosynthetic prokaryotes store their food reserves in the cytoplasm. (a) Chlorella, *a green alga, with large, light-colored starch grains in its plastids. (b) A mass of starch grains surrounded by plastids are seen in the cytoplasm of* Batrachospermum moniliforme, *a red alga.*

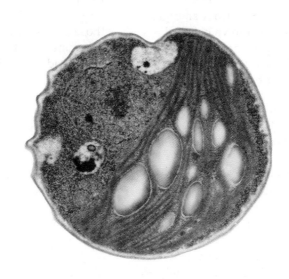
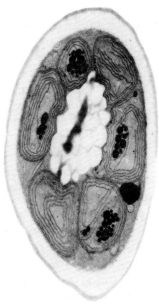

(a)   (b)

In addition, many red algae deposit calcium carbonate in their cell walls. Many of these algae are especially tough and stony, and constitute the family Corallinaceae, the coralline algae. Coralline algae are common throughout the oceans of the world on stable surfaces that receive enough light (see Figure 15–5b, c). Recently, an encrusting coralline alga was found at 268 meters, a greater depth than had previously been recorded for any photosynthetic organism (Figure 15–6). This is more than 100 meters below the depth to which sunlight normally penetrates.

Coralline algae play an important role in building coral reefs. Indeed, the productivity of such reefs, and their ability to grow in relatively infertile tropical waters, depends directly on the coralline algae and the symbiotic dinoflagellates, or zooxanthellae, found within the coral animals themselves (see Figure 14–20, page 247). The reef-building coralline algae are usually unjointed (Figure 15–5c), whereas many of the other coralline algae are conspicuously jointed (Figure 15–5b). Some coralline algae form crusts on rocks (Figure 15–6). Coralline algae are an ancient group, with records going back to the late Cambrian period, more than 500 million years ago.

Most red algae are composed of filaments, which are often densely interwoven and are held together by the mucilage of the intercellular matrix. Simpler forms, such as the freshwater red alga *Batrachospermum,* are filamentous (see Figure 15–4). Growth in filamentous

red algae is initiated by a single, dome-shaped apical cell, which cuts off segments sequentially to form an axis. This axis in turn forms whorls of lateral branches (Figure 15–4b). Most red algae are multiaxial or apparently parenchymatous, with a three-dimensional body. In such forms, the filaments are interconnected secondarily, forming a network that is held together by the mucilaginous matrix. In many red algae, pit connections are a distinctive feature of the cell walls. These are lens-shaped plugs held in the walls by their equatorial grooves (see Figure 15–8). True parenchyma, in which the cells are densely packed together like those of the plants, is found in a few genera, such as *Porphyra* (see page 262).

The life history of most red algae is unusual in that it consists of three phases: (1) a haploid gametophyte; (2) a diploid phase, called a *carposporophyte;* and (3) another diploid phase, called a *tetrasporophyte* (Figure 15–7). Male algae (gametophyte generation) bear spermatangia, which release nonmotile spermatia, male gametes that float passively in the water. The female sex organ, the *carpogonium,* develops a long, hairlike outgrowth, the trichogyne, which is similar in structure and function to the trichogynes found in the ascomycetes (see Chapter 13). The spermatia reach the trichogyne by chance, by means of currents in the water. After attachment, the sperm nucleus enters the trichogyne through a pore, migrates to the egg nucleus, and fuses with it.

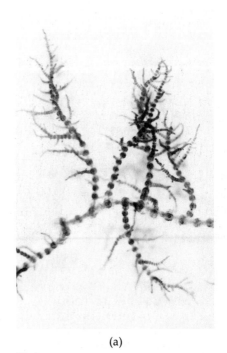

(a)

(b)

(c)

15–4

(a) *The simple, filamentous red alga* Batrachospermum. *The soft, gelatinous, branched axes of this freshwater red alga*

are most frequent in cold streams, ponds, and lakes, where they occur throughout the world. (b) *The apex of an individual*

of Batrachospermum, *showing the whorls of lateral branches.* (c) *Two cystocarps of* Batrachospermum.

(a)

(b)

15–5

(a) *In* Bonnemaisonia hamifera, *the basically filamentous structure of the red algae is clearly evident. It appears that multicellularity in red algae evolved separately from that in green and brown algae, where no similar patterns of organization are found. The branched filaments of this red alga are hooked, enabling it to cling to other seaweeds.* (b) *Jointed coralline algae in a tidal pool in California.* (c) *The reef-building, unjointed coralline red alga* Porolithon craspedium. (d) *Irish moss (*Chondrus crispus*), an important source of carrageenan and other colloids.*

(c)

(d)

(a)

(b)

(c)

15–6

(a) *Scanning electron micrograph of an unidentified species of crustose red alga collected at a depth of 268 meters by Mark Littler of the Smithsonian Institution and his colleagues in October 1984. This alga formed patches about a meter across on a seamount in the Bahamas, at a depth where the light intensity was estimated as 0.0005 percent of its value* at the ocean surface; here the alga covered about 10 percent of the rock surfaces. When tested in the laboratory, it was found to be about 100 times more efficient than its shallow-water relatives in capturing and using light energy. This alga, which grew continuously up to the top of the seamount at about 70 meters depth, occurred nearly 100 meters below *the lowest limits that had previously been established for any anchored photosynthesizing organism.* (b) *Purple crustose coralline algae and other deep-water algae from the seamount.* (c) *The research submersible Johnson Sea Link I; the deepest records of algae were made with the aid of this vessel.*

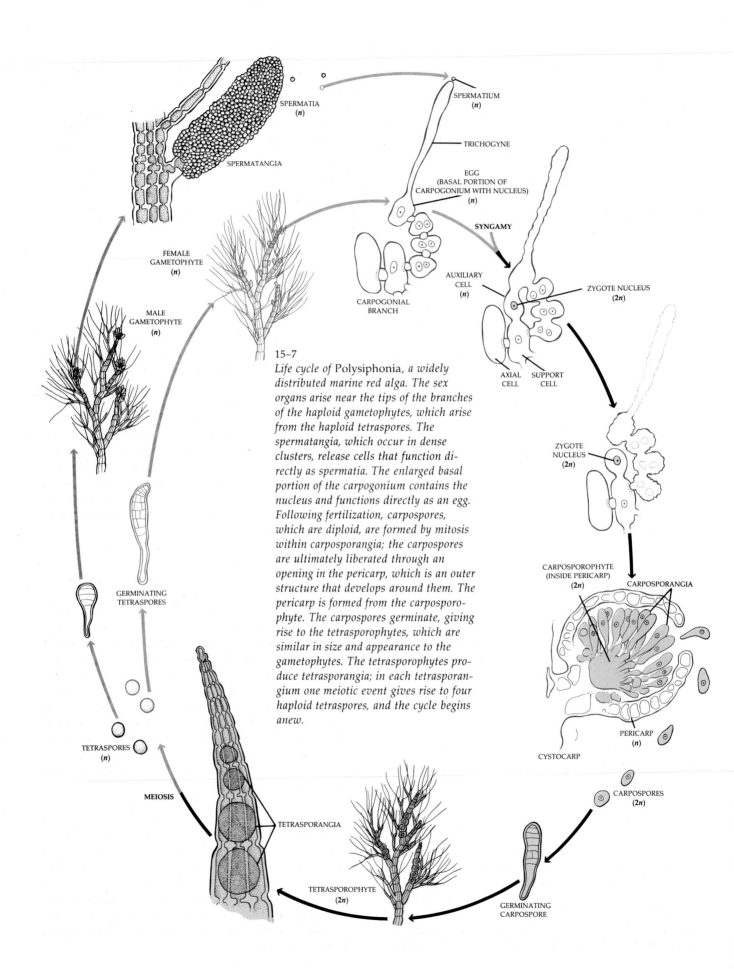

SPERMATIA
(*n*)

SPERMATIUM
(*n*)

SPERMATANGIA

TRICHOGYNE

EGG
(BASAL PORTION OF
CARPOGONIUM WITH NUCLEUS)
(*n*)

SYNGAMY

FEMALE
GAMETOPHYTE
(*n*)

AUXILIARY
CELL
(*n*)

ZYGOTE NUCLEUS
(2*n*)

MALE
GAMETOPHYTE
(*n*)

CARPOGONIAL
BRANCH

AXIAL
CELL

SUPPORT
CELL

ZYGOTE
NUCLEUS
(2*n*)

15–7

Life cycle of Polysiphonia, *a widely
distributed marine red alga. The sex
organs arise near the tips of the branches
of the haploid gametophytes, which arise
from the haploid tetraspores. The
spermatangia, which occur in dense
clusters, release cells that function directly as spermatia. The enlarged basal
portion of the carpogonium contains the
nucleus and functions directly as an egg.
Following fertilization, carpospores,
which are diploid, are formed by mitosis
within carposporangia; the carpospores
are ultimately liberated through an
opening in the pericarp, which is an outer
structure that develops around them. The
pericarp is formed from the carposporophyte. The carpospores germinate, giving
rise to the tetrasporophytes, which are
similar in size and appearance to the
gametophytes. The tetrasporophytes produce tetrasporangia; in each tetrasporangium one meiotic event gives rise to four
haploid tetraspores, and the cycle begins
anew.*

GERMINATING
TETRASPORES

CARPOSPOROPHYTE
(INSIDE PERICARP)
(2*n*)

CARPOSPORANGIA

PERICARP
(*n*)

TETRASPORES
(*n*)

CYSTOCARP

CARPOSPORES
(2*n*)

MEIOSIS

TETRASPORANGIA

GERMINATING
CARPOSPORE

TETRASPOROPHYTE
(2*n*)

15-8

A pit connection in the red alga Pal-
maria. Pit connections are distinct,
lens-shaped plugs that form between the
cells of red algae. Their cores are protein
and their outer cap layers are, at least in
part, polysaccharide. Pit connections
form between both the progeny cells and
the cells of neighboring filaments, linking
together the bodies of red algae. They
also form between the many parasitic red
algae and their hosts, probably facilitat-
ing the transfer of photosynthate between
them.

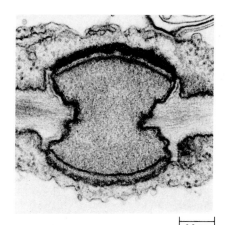

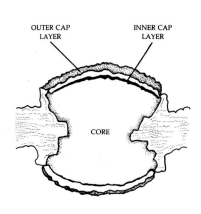

In the simpler red algae, like *Batrachospermum*, the diploid carposporophyte generation (Figure 15–4c) develops directly from the fertilized carpogonium. Terminal, spore-producing structures called carposporangia form on the carpogonium. In the more advanced groups of red algae, the diploid nucleus that is formed by syngamy within the carpogonium is transferred to a second cell, the auxiliary cell, from which the carposporophyte develops. *Polysiphonia* provides an example of this kind of life cycle (Figure 15–7).

The diploid carpospores, formed within the carposporangia are released and develop into free-living tetrasporophytes. The tetrasporophyte produces still another kind of spore-producing structure, the tetrasporangium; each tetrasporangium undergoes meiosis, giving rise to four haploid tetraspores. On germination, each tetraspore produces a gametophyte.

In most red algae, the gametophyte, carposporophyte, and tetrasporophyte generations resemble one another closely and are therefore said to be isomorphic, as in *Polysiphonia*. Coralline algae, for example, have isomorphic life cycles. An increasing number of heteromorphic life cycles are also being discovered; in them the tetrasporophytes are microscopic and filamentous, or they consist of a thin crust that is tightly attached to the rock substrate.

BROWN ALGAE: DIVISION PHAEOPHYTA

The brown algae, an almost entirely marine group, comprise most of the conspicuous seaweeds of temperate regions. Although there are only about 1500 species, the brown algae are of considerable interest because they dominate rocky shores throughout the cooler regions of the world (Figure 15–9). The larger brown algae of the order Laminariales are called kelps; a number of them form extensive beds offshore. In clear water, brown algae flourish from low-tide level to a depth of 20 to 30 meters, and on gently sloping shores, they may extend 5 to 10 kilometers from the coastline. Even in the tropics, where the brown algae are less common, there are immense floating masses of *Sargassum* (Figure 15–10) in such areas as the Sargasso Sea in the Atlantic Ocean northeast of the Caribbean Islands.

Brown algae range in size from microscopic forms to the largest of all seaweeds. The simplest members of the division are branched filaments like *Ectocarpus* (Figure 15–11), in which growth takes place by means of intercalary meristems, meristems within the filaments. In more specialized filamentous brown algae, the filaments are connected into a solid body, which in the most complex forms becomes a three-dimensional cellular structure like that of the plants.

Large kelps, such as *Laminaria*, are differentiated into regions known as the holdfast, stipe, and blade, with the meristematic region located between the blade and stipe (see Figure 15–9b). This pattern of growth is particularly important in the utilization of *Macrocystis*, which is harvested along the California coast; whenever the older blades are harvested at the surface by kelp-cutting boats, *Macrocystis* is able to regenerate from its deeper portions. Giant kelps, such as *Macrocystis* and *Nereocystis*, may be more than 60 meters long, and they grow very rapidly so that a considerable amount of material is available for harvest. One of the most important products derived from the kelps is a mucilaginous intercellular material called algin, which is important as a stabilizer and emulsifier for some foods and for paint, and as a coating for paper.

Rockweed, *Fucus* (see Figure 15–9c), is a dichotomously branching brown alga that has air bladders near the ends of its blades. The differentiation of *Fucus* otherwise resembles that of the kelps. Related to *Fucus* is *Sargassum* (see Figure 15–10). Individuals of *Sargassum* often break loose and form floating masses, in which the holdfasts have been lost. *Fucus* and *Sargassum* grow by means of repeated divisions from a single apical cell and not from a meristem located within the body, like that of the kelps.

(a)

(b)

(c)

15–9

*Brown algae. (a) Bull kelp (Durvillea
antarctica) exposed at low tide off a
rocky coast in New Zealand. (b) Detail of
the kelp* Laminaria, *showing holdfasts,
stipes, and the bases of several fronds. (c)
Rockweed (Fucus vesiculosus) densely
covers many rocky shores that are
exposed at low tide. When submerged,
the air-filled bladders on the blades carry
them up toward the light. Photosynthetic
rates of frequently exposed marine algae
are one to seven times as great in air as in
water, whereas they are higher in water
for those rarely exposed, which accounts
in part for the vertical distribution of
seaweeds in intertidal areas.*

The internal structure of kelps is complex. Some of them even have elongated cells in the center of their stipe that are modified for food conduction; these cells resemble those of phloem in the vascular plants (Figure 15–12). They have sieve plates and are able to conduct food material rapidly, at rates as high as 60 centimeters per hour, from the blades at the water surface to the poorly illuminated stipe and holdfast regions far below. Lateral translocation from the outer photosynthetic layers to the inner cells also takes place in many of the kelps that are relatively thick. The alcohol mannitol is the primary material that is translocated, along with amino acids.

As we discussed in Chapter 14, the brown algae are very closely related to the chrysophytes (division Chrysophyta), so much so that some scientists are now beginning to speak of combining these groups in a single division. The features that link these groups are summarized in Table 14–1 (page 233). Some chrysophytes certainly resemble the unicellular ancestors that must have given rise to the brown algae.

The life cycles of most brown algae involve an alternation of generations and, therefore, sporic meiosis (see Figure 10–11c, page 163). The gametophytes of brown algae produce multicellular reproductive structures called *plurilocular gametangia*, which may function as male or female gametangia or produce flagellated haploid spores that give rise to new gametophytes. The sporophytes are diploid and produce both *plurilocular* and *unilocular sporangia* (see Figure 15–11). The plurilocular sporangia form diploid zoospores that produce new sporophytes. Meiosis takes place within the unilocular sporangia, producing haploid zoospores that germinate to produce gametophytes.

15–10

The brown alga Sargassum *has a complex pattern of organization.* Sargassum, *like* Fucus, *is a member of the order Fucales and has a life cycle like that shown in Figure 15–14. Two species of this genus, which lack sexual reproduction, form the great free-floating masses of the Sargasso Sea; the others are anchored on rocks along shores.*

BLADE

FLOAT
(AIR-FILLED
BLADDER)

STIPE

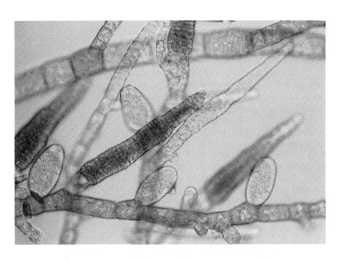

15–11

Ectocarpus, *a brown alga that has simple branched filaments. This micrograph of* Ectocarpus siliculosus *shows unilocular sporangia (the short, rounded, light-colored structures) and plurilocular sporangia (the longer, dark-colored structures), which are borne on sporophytes. Meiosis takes place within the unilocular sporangia; in this micrograph they have already discharged their zoospores.* Ectocarpus *occurs in shallow water and estuaries throughout the world, from cold arctic and antarctic waters to the tropics.*

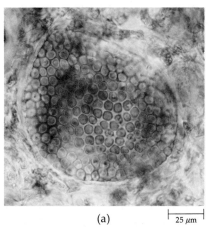

(a)   25 μm

(b)   0.1 mm

15–12

Some brown algae, such as the giant kelp Macrocystis pyrifera, *have evolved sieve tubes comparable to those found in vascular plants. (a) Cross section showing a sieve plate. (b) A longitudinal section of part of a stipe, with sieve tubes. In this view, the sieve plates appear thickened because they are covered with the wall substance callose.*

In *Ectocarpus,* the gametophyte and sporophyte are similar in size and appearance (isomorphic). Many of the larger brown algae, including the kelps, undergo an alternation of heteromorphic generations—a large sporophyte and a microscopic gametophyte, as in the common kelp *Laminaria* (Figure 15–13). In *Laminaria,* the unilocular sporangia are produced on the surface of the mature blades. Half of the zoospores that they produce grow into male gametophytes and half into female gametophytes. The plurilocular gametangia borne on these gametophytes have become modified during the course of their evolution into one-celled antheridia, each of which releases a single sperm, and one-celled oogonia, each with a single egg. The fertilized egg in *Laminaria* remains attached to the female gametophyte and develops into a new sporophyte. The molecules secreted by the female gametes that serve to attract the male gametes in several genera of brown algae are olefinic hydrocarbons (open-chain hydrocarbons with one or more double bonds).

Fucus has a unique kind of life cycle (Figure 15–14), in which meiosis is gametic and leads directly to the production of the gametes (see Figure 10–11b, page 163). Zygotic meiosis, which characterizes two large classes of green algae, is not known in brown algae.

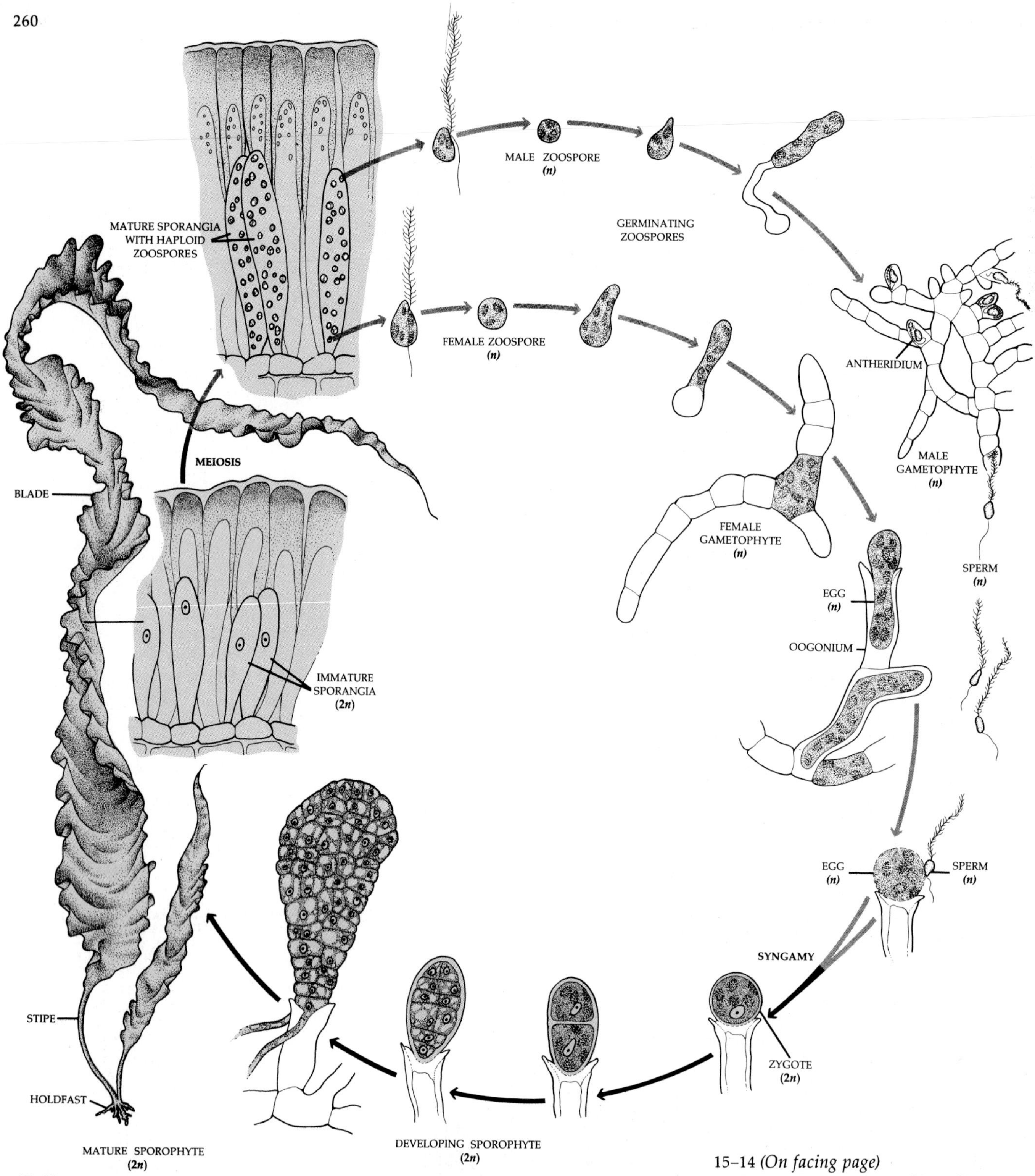

MATURE SPORANGIA
WITH HAPLOID
ZOOSPORES

MALE ZOOSPORE
(n)

GERMINATING
ZOOSPORES

FEMALE ZOOSPORE
(n)

ANTHERIDIUM

MALE
GAMETOPHYTE
(n)

MEIOSIS

BLADE

FEMALE
GAMETOPHYTE
(n)

SPERM
(n)

EGG
(n)

OOGONIUM

IMMATURE
SPORANGIA
(2n)

EGG
(n)

SPERM
(n)

SYNGAMY

STIPE

ZYGOTE
(2n)

HOLDFAST

MATURE SPOROPHYTE
(2n)

DEVELOPING SPOROPHYTE
(2n)

15–13
Life cycle of the kelp Laminaria. *Like
most of the brown algae,* Laminaria *has
an alternation of heteromorphic generations in which the sporophyte is
conspicuous. Motile haploid zoospores
are produced in the sporangia following
meiosis. From these zoospores grow the
microscopic, filamentous gametophytes,
which in turn produce the motile sperm
and nonmotile eggs. In the simpler
brown algae, the sporophyte and
gametophyte are often similar; they have
an alternation of isomorphic generations.*

15–14 *(On facing page)*
In Fucus, *gametangia are formed in
specialized hollow chambers (concepta-
cles), which are found in fertile areas
(receptacles), at the tips of the branches
of diploid individuals. There are two
types of gametangia—oogonia and
antheridia. Meiosis is followed immediately by mitosis to give rise to 8 eggs per
oogonium and 64 sperm per antheridium.
Eventually the eggs and sperm are set
free in the water, where fertilization
takes place. Meiosis is gametic, and the
zygote grows directly into the new
diploid individual.*

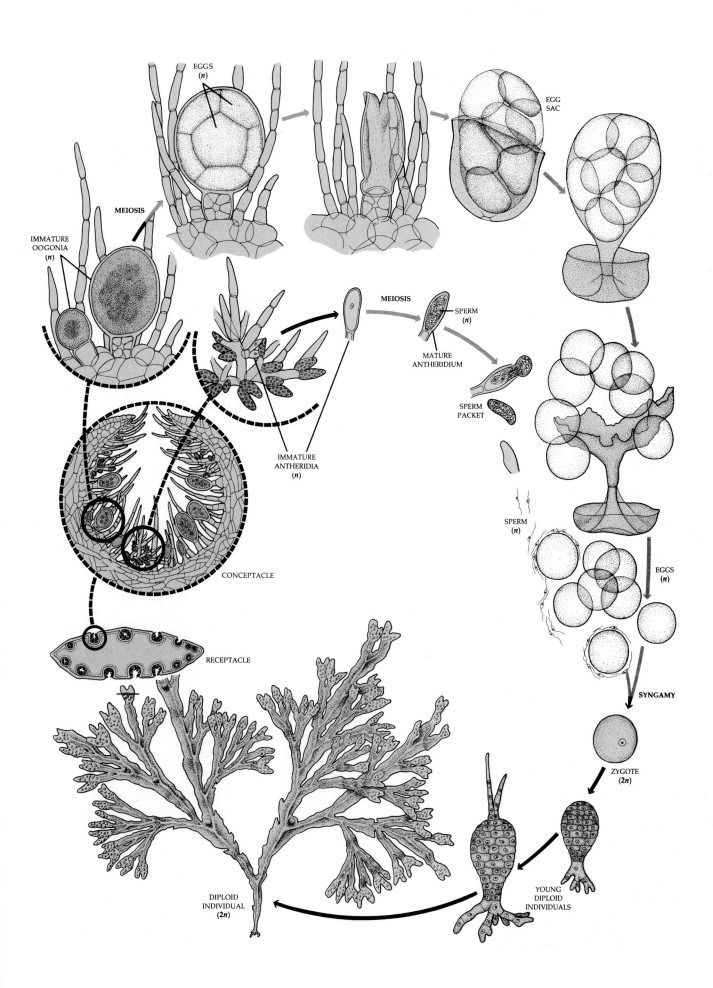

EGGS (*n*)

MEIOSIS

EGG SAC

IMMATURE OOGONIA (*n*)

MEIOSIS

SPERM (*n*)

MATURE ANTHERIDIUM

SPERM PACKET

IMMATURE ANTHERIDIA (*n*)

SPERM (*n*)

EGGS (*n*)

CONCEPTACLE

SYNGAMY

RECEPTACLE

ZYGOTE (2*n*)

YOUNG DIPLOID INDIVIDUALS

DIPLOID INDIVIDUAL (2*n*)

THE ECONOMIC USES OF SEAWEEDS

People of various parts of the world, especially in the Far East, eat both red and brown algae. Kelps ("kombu") are eaten regularly as vegetables in China and Japan; they are sometimes cultivated but are mainly harvested from natural populations. *Porphyra* ("nori"), one of the red algae, is eaten by many of the inhabitants of the north Pacific Basin and has been cultivated in Japan and China for centuries. The nori industry presently employs more than 30,000 persons in Japan alone and is valued at over $20 million annually. Various other red algae are eaten on the islands of the Pacific and also on the shores of the North Atlantic. Seaweeds are generally not of high nutritive value as a source of carbohydrates because humans, like most other animals, lack the necessary enzymes to break down most of the materials in cell walls, such as cellulose. Seaweeds do, however, provide necessary salts, as well as a number of important vitamins and trace elements, and so are valuable supplementary foods. Some green algae, such as *Ulva*, or sea lettuce, are also eaten as greens.

In many north temperate regions, kelp is harvested for its ash, which is rich in sodium and potassium salts and is therefore valuable for industrial processes. Iodine is also produced commercially from kelp. Algae are often harvested and used directly for fertilizer.

Alginates, which are a group of substances derived from kelp such as *Macrocystis*, are widely used as thickening agents and colloid stabilizers in the food, textile, cosmetic, pharmaceutical, paper, and welding industries. Off the west coast of the United States, *Macrocystis* kelp beds can be harvested several times a year by cropping them just below the surface. Attempts are currently under way to cultivate this giant kelp on a commercial scale.

One of the most useful direct commercial applications of any alga is the preparation of agar, which is made from a mucilaginous material extracted from the cell walls of a number of genera of red algae. Agar is used to make the capsules that contain vitamins and drugs, as a dental-impression material, as a base for cosmetics, and as a culture medium for bacteria and other microorganisms. It is also employed as an antidrying agent in bakery goods, in the preparation of rapid-setting jellies and desserts, and as a temporary preservative for meat and fish in tropical regions. Agar is produced in many parts of the world, but Japan is the principal source. A similar algal colloid called carrageenan is used in preference to agar for the stabilization of emulsions such as paints, cosmetics, and dairy products. In the Philippines, the red alga *Eucheuma* is cultivated on a commercial scale as a source of carrageenan.

(a)

(b)

(c)

(a) *A forest of giant kelp* (Macrocystis pyrifera) *growing off the coast of California.* (b) *Harvesting the seaweed* Nudaria *by hand from submerged ropes in Japan.* (c) *A kelp harvester operating in the nearshore waters of California. Cutting racks at the rear of the ship are lowered 3 meters below the water's surface, and the ship moves backward through the kelp canopy. The cut kelp is moved via conveyor belts to a collecting bin on board the ship.*

GREEN ALGAE: DIVISION CHLOROPHYTA

The green algae are the most diverse of all the algae, both in form and in life history. The group comprises at least 7000 species. Although most green algae are aquatic, they are found in a wide variety of habitats, including the surface of snow, on tree trunks, in the soil, and in symbiotic relationship with lichens, protozoa, and hydras. Some green algae—such as species of the unicellular genus *Chlamydomonas* found growing on the surface of snow, or *Trentepohlia*, a filamentous alga that grows on tree branches—produce large amounts of carotenoids as a shield against intense light. Because of the presence of these accessory pigments, they often appear red or rust-colored. Most aquatic green algae are found in fresh water, but a few groups are entirely marine. Many of the members of this division are microscopic, but some of the marine species are large; *Codium magnum* of Mexico, for example, sometimes attains a breadth of 25 centimeters and a length of more than 8 meters.

Chlorophyta resemble plants in several important characteristics. They have chlorophylls *a* and *b*; store starch, their food reserve, inside their plastids (green algae and plants are the only two groups to do so); and have firm cell walls composed, in some genera, of polysaccharides such as cellulose, with a matrix of hemicelluloses and pectic substances incorporated in it. In addition, the details of the way in which the basal bodies are associated with the flagella in plant sperm cells resembles that in certain green algae, supporting the hypothesis that the two groups are directly related. For these and other reasons, green algae are believed to be the group from which plants originated.

Detailed investigations with the electron microscope have added greatly to the overall impression of diversity among the green algae. The members of this division comprise several evolutionary lines that have been derived independently from unicellular, flagellated green ancestors. Although the characteristics of their chloroplasts are uniform, the green algae are otherwise highly diverse. If the division had a common ancestor, it was probably a scaly flagellate in which, in chlorophycean flagellates, the scales eventually became fused into a coherent cell wall during the course of evolution. Unicellular, flagellated green algae with such primitive characteristics are included in the class Micromonadophyceae.

Cell Division in Green Algae

The members of the largest class of green algae, Chlorophyceae, have a unique mode of cell division involving a phycoplast (Figure 15–15). In them, the daughter nuclei move toward one another as the nonpersistent

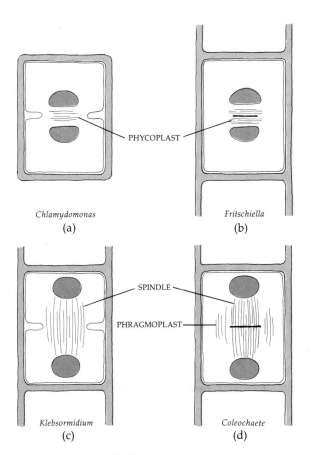

15–15
Cell division in two classes of the division Chlorophyta. In the class Chlorophyceae, (a) and (b), the spindle is absent (nonpersistent), and the daughter nuclei, which are relatively near one another, are separated by phycoplasts. (a) Cell division by furrowing; (b) cell division by cell-plate formation. In the class Charophyceae, (c) and (d), the spindle is persistent and the daughter nuclei are relatively far apart. (c) Cell division by furrowing; (d) phragmoplast present and cell division by cell-plate formation. Ulvophyceae have a persistent spindle like Charophyceae, but the spindle does not proliferate to form a phragmoplast.

spindle collapses, and a new system of microtubules, the *phycoplast*, develops parallel to the plane of cell division. Presumably, the role of the phycoplast is to ensure that the cleavage furrow will pass between the two daughter nuclei. The nuclear envelope persists throughout mitosis. If motile cells are present, the flagella are inserted at their anterior end. Internally, the motile cells have a system of flagellar roots arranged in a cross-shaped pattern of four narrow bands of microtubules (Figure 15–16). These bands of microtubules originate at or near the anterior basal bodies.

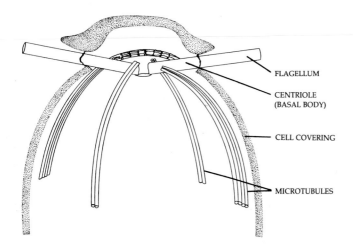

15–16
Diagram of the cross-shaped arrangement of microtubules, associated with the flagellar centrioles, that is characteristic of green algae of the class Chlorophyceae.

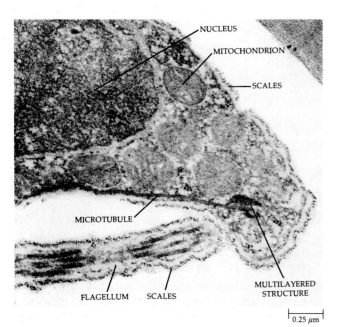

15–17
Electron micrograph of the anterior portion of a spermatozoid of the green alga Coleochaete, *showing the multilayered structure near the mitochondria. A layer of microtubules extends from the multilayered structure down into the posterior end of the cell and serves as a cytoskeleton for these wall-less cells. The flagellar and cellular membranes are covered by a layer of small diamond-shaped scales.*

In other green algae, the spindle apparatus persists during cell division, and in some of these, a *phragmoplast* is formed. The microtubules in a phragmoplast are oriented perpendicular to the plane of cell division (Figure 15–15). The spindle remains until it is "broken," either by the growing cell plate—which originates in the central region of the cell and grows outward to the margins of the cell—or by furrowing. This mode of cell division is less specialized than that involving a phycoplast, a structure that evolved within the green algae. The mode of cell division is of great importance in the classification of the green algae.

Classification of Green Algae

A small class of green algae, the unicellular Micromonadophyceae, which are scaly or naked flagellates, retain many of what seem to have been the ancestral features of the division. Aside from them, contemporary phycologists divide the green algae into four classes, each of which is believed to represent a separate evolutionary line. One of these, Pleurastrophyceae (see essay, page 274), is a small group that we shall not consider in detail, whereas the other three classes are large and include many familiar organisms.

The three major classes of green algae—Charophyceae, Ulvophyceae, and Chlorophyceae—differ from one another in a number of fundamental features. In Chlorophyceae, which have a phycoplast, the spindle fibers always shorten during mitotic anaphase, whereas in Charophyceae and Ulvophyceae (in which the nuclear envelope persists during mitosis) they generally do not. In the last two classes, the nuclear envelope disintegrates at the start of mitosis, as it does in plants; in Chlorophyceae, the nuclear envelope persists throughout mitosis. Each class includes unicellular flagellated cells, which may be whole unicellular organisms in Chlorophyceae and are reproductive cells in the other two classes. The original members of each class were certainly unicellular flagellates, with multicellular organisms having originated independently in each of them.

Charophyceae and Chlorophyceae occur mainly in fresh water, whereas Ulvophyceae are predominantly marine. The motile cells formed by members of these classes differ greatly; those of Charophyceae are asymmetrical, those of the other two classes radially symmetrical or nearly so. The flagella of Charophyceae are lateral or subapical (inserted near, but not exactly at, the apex), and they extend laterally at right angles from the cell. On the other hand, the flagella of Chlorophyceae and Ulvophyceae are apical and directed forward (Figure 15–17).

Internally, the cells of Charophyceae possess a system of microtubules in the form of a flat, broad band. This band originates in a multilayered structure near the laterally located basal bodies and radiates downward into the cell. The same kind of structure is also found in the motile sperm cells of plants. In addition, Charophyceae manufacture the photorespiratory enzyme glycolate oxidase in peroxisomes; this enzyme is otherwise known to occur in peroxisomes only in plants, although it occurs outside of peroxisomes in a few nonchlorophycean green algae. On the basis of these and a number of other similarities, Charophyceae are more closely related to the plants than are the other classes of green algae, so that the living form most similar to the ancestor of plants should be sought among the members of the class Charophyceae.

Sexual reproduction in Charophyceae and Chlorophyceae always involves the formation of a dormant zygote (zygospore) and zygotic meiosis (see Figure 10–11a, page 163). In contrast, sexual reproduction in Ulvophyceae often involves alternation of generations and sporic meiosis, and dormant zygotes are rare. In terms of their life cycles, Ulvophyceae seem to be the most advanced of these classes, and Charophyceae are the most primitive.

Class Charophyceae

Charophyceae consist of unicellular, few-celled, filamentous, and parenchymatous genera. Their relationship with one another is revealed by many fundamental structural and biochemical similarities. These include their asymmetrical flagellated cells, with lateral or subapical flagella that extend at right angles from their cells; their zygotic meiosis and production of dormant zygospores; the presence of a multilayered structure in motile cells; and the details of their cell division. All Charophyceae have a persistent mitotic spindle, and some of them develop a phragmoplast that aids in the formation of the new cell plate.

Spirogyra (Figure 15–18) is a well-known genus of unbranched, filamentous Charophyceae, which often forms frothy or slimy floating masses in bodies of fresh water. Each filament is surrounded by a watery sheath that is slimy to the touch. The name *Spirogyra* refers to the helical arrangement of the one or more ribbonlike chloroplasts found within each uninucleate cell. The chloroplasts contain numerous *pyrenoids*, which are differentiated regions of the chloroplast that are centers of starch formation in the green algae. Pyrenoids are commonly found among the algae but are lacking in nearly all plant chloroplasts. Recent research has shown that in two other green algae the pyrenoid is a region in the chloroplast where the enzyme RuBP carboxylase (see page 105) is concentrated. Pyrenoids may also be associated with the conversion of sugars to starch, for deposits of starch are usually found surrounding them.

Asexual reproduction occurs in *Spirogyra* by cell division and fragmentation; there are no flagellated cells at any stage of the life cycle. Morphologically, sexual re-

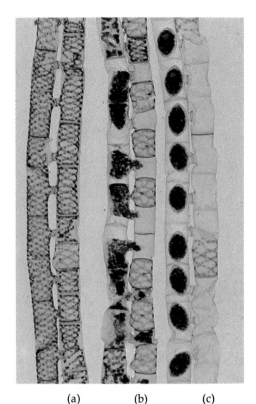

(a) (b) (c)

15–18

(a) *Sexual reproduction in* Spirogyra *follows the formation of conjugation tubes between the cells of adjacent filaments.* (b) *The contents of the cells of the — strain pass through these tubes into the cells of the + strain.* (c) *Syngamy occurs within these cells; the resulting zygote develops a thick, resistant cell wall and is termed a zygospore. The vegetative filaments of* Spirogyra *are haploid, and meiosis occurs during germination of the zygospores, as it does in all Charophyceae.*

production in *Spirogyra* must be termed isogamous. However, during sexual reproduction one of the isogametes behaves like a male gamete by migrating across a conjugation tube to fuse with the other isogamete. Meiosis is zygotic and results in the formation of resting zygospores, as it does in all Charophyceae.

The desmids are a large group of freshwater green algae related to *Spirogyra*; like *Spirogyra*, they lack flagellated cells. Some desmids are filamentous, but most are unicellular. The cell construction of desmids is unusual in that the cell wall is in two sections with a narrow constriction—the isthmus—between them (Figure 15–19). Cell division in desmids, as in *Spirogyra*, involves the formation of a persistent spindle and furrowing. Some estimates place the number of species of desmids as high as 10,000.

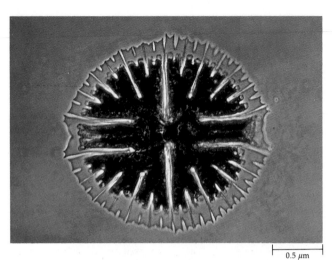

15-19
In this desmid, a species of Micrasterias, *each cell is deeply constricted, as is the case in most members of this large group of unicellular, freshwater Charophyceae.*

0.5 μm

Plantlike Charophyceae

The two groups of green algae that will be considered here are those that resemble plants most completely in the details of their cell division; that is, they have a phragmoplast and a nuclear envelope that disintegrates at the start of mitosis. In addition, they are oogamous, like plants. Neither group, however, is the direct ancestor of the plants, because each is too specialized in certain respects. Plants were probably derived from an extinct member of the Charophyceae, judged from the close resemblance between the two groups.

The first group of Charophyceae that we shall consider in connection with the ancestry of the plants is the order Coleochaetales, which includes branched filamentous genera, discoid (disk-shaped) ones, and some unicellular ones. *Coleochaete*, which grows on the surface of submerged, freshwater plants, is particularly complex (Figure 15–20; see also Figure 15–17). These algae reproduce asexually by zoospores; their zygotes are protected by a layer of sterile cells, which forms after fertilization. Individuals of *Coleochaete* may consist of a mass of dichotomously branched filaments, or they may be discoid in form. The vegetative cells are uninucleate, with one large chloroplast and one or more pyrenoids. Cell division may occur at the apices of the filaments or the margins of the disk, depending on the form of the alga.

Historically *Coleochaete* has been singled out as a possible ancestor of the plants. Some species are parenchymatous and grow by a meristem around the margin of the disk, reproduction is oogamous, and the zygotes are protected by a layer of cells. Like a few other Charophyceae, *Coleochaete* also resembles plants in the presence of a phragmoplast, the photorespiratory enzyme glycolate oxidase localized in peroxisomes, and a multilayered structure associated with the flagella of its reproductive cells (Figure 15–17). In addition, there are certain fossils from about the time when plants first evolved that seem to resemble *Coleochaete* in their characteristics (Figure 15–21). Although *Coleochaete* itself does not seem to have been the ancestor of plants, its features, which include a number of structural and biochemical adaptations that are useful in the land environment, suggest the way in which the evolution of plants probably occurred.

The stoneworts, order Charales, are a second group of plantlike Charophyceae (Figure 15–22). These distinctive green algae occur in fresh or brackish water. Some of them have heavily calcified cell walls and thus are well represented as fossils. There are about 250 living species. In their growth pattern, the stoneworts are complex: they have apical growth, like plants, and differentiation into nodal and internodal regions. Whorls of short branches arise in the nodal regions, and the in-

15-20
(a) Coleochaete, *collected from the stem of an aquatic flowering plant growing in shallow water in a lake.* (b) *Individuals of this species of* Coleochaete *consist of a parenchymatous disk, which is generally one cell thick. The large cells are zygotes, which are protected by a cellular covering. The hair cells extending from the disk are ensheathed at the base; "Coleochaete" means "sheathed hair." These hairs are thought to discourage aquatic animals from feeding on the alga.*

(a)

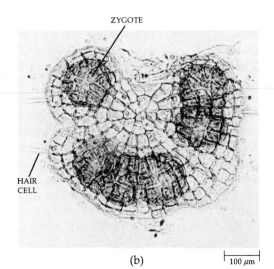

ZYGOTE

HAIR CELL

(b)

100 μm

15–21
Parka decipiens, *a fossil from the late*
Devonian period, about 380 million years
ago, closely resembles Coleochaete *in its*
shape, tissue structure, and chemistry.
This organism shows many features that
suggest it may be related to the group of
aquatic organisms that eventually gave
rise to the bryophytes and vascular
plants.

ternodal regions are coenocytic. The sperm of stoneworts are produced in multicellular antheridia that are
more complex than those found in any other group of
protists. Their eggs are borne in oogonia of simpler construction than the antheridia. Sperm are the only flagellated cells in the stonewort life cycle.

Class Ulvophyceae

Alone among the classes of green algae, the Ulvophyceae are primarily marine organisms. Their flagellated
cells are scaly or naked like those of Charophyceae, but
they are nearly radially symmetrical and have apical,
forward-directed flagella like those of Chlorophyceae.
The flagellated cells of Ulvophyceae may have two,
four, or many flagella, like Chlorophyceae; those of
Charophyceae are always biflagellated. In the details of
their cell division, Ulvophyceae have a closed mitosis,
in which the nuclear envelope persists; the spindle is
persistent through cytokinesis. They are the only group
of green algae in which an alternation of generations,
with sporic meiosis, occurs; unlike the other two classes
discussed in detail in this book, Ulvophyceae rarely
form dormant zygospores.

Ulvophyceae may be few-celled, filamentous, flat
sheets of cells, parenchymatous, or coenocytic. All filamentous and complex marine green algae that have
been examined in detail share the characteristics of this
class, which probably originated in the sea, unlike other
classes of green algae.

The first evolutionary line of Ulvophyceae consists of
filamentous algae that have large, multinucleate cells

15–22
(a) Chara, *a stonewort (class Charophy-*
ceae) that grows in shallow waters of
temperate lakes. Its determinate growth
pattern is clearly evident. (b) *A segment*
of Chara *showing gametangia. The top*
structure is an oogonium, and that below
is an antheridium.

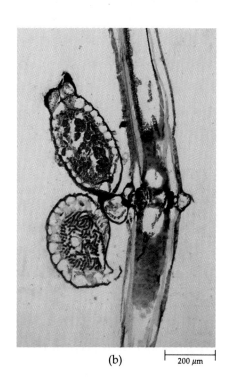

(a)   (b)   200 μm

separated by septa. Of these, *Cladophora* (Figure 15–23) is widespread both in fresh and salt water. Its filaments commonly grow in dense mats, which are either free-floating or attached to rocks or vegetation; they elongate and branch near the ends. Each cell contains many nuclei and a single peripheral, netlike chloroplast with many pyrenoids. Marine species of *Cladophora* have an alternation of isomorphic generations; the freshwater species do not have an alternation of generations, apparently having lost this characteristic during the course of their evolution.

Another member of this evolutionary line is *Ulothrix*, an alga common in cold-water streams, ponds, and lakes, whose filaments are attached to stones or other objects by means of a holdfast (Figure 15–24). All cells of the filament are essentially similar in appearance and contain a single, ringlike chloroplast, pyrenoid, and nucleus. Asexual reproduction occurs through formation of zoospores with four flagella; sexual reproduction occurs by means of biflagellated isogametes. Meiosis occurs prior to the germination of the zygote; the

Ulothrix filament is haploid. Although several groups of green algae that were once classified as species of *Ulothrix* are actually members of the Chlorophyceae (see Figure 15–37), *Ulothrix* itself belongs to the Ulvophyceae.

A second kind of growth habit among the Ulvophyceae is that of *Ulva*, commonly known as sea lettuce. This familiar alga is common along temperate seashores throughout the world (Figure 15–25). Individuals of *Ulva* consist of a glistening, flat *thallus* (a simple, relatively undifferentiated vegetative body), which is two cells thick and up to a meter or more long in exceptionally large individuals. The thallus is anchored to the substrate by a holdfast produced by protuberances of the basal cells. Each cell of the thallus contains a single nucleus and chloroplast. *Ulva* is anisogamous and has an alternation of isomorphic generations like that of many Ulvophyceae (Figure 15–26).

Algae of the marine order Siphonales—characterized by very large, branched, coenocytic cells that are rarely septate—develop as a result of repeated nuclear divi-

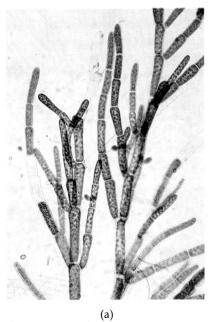

(a)

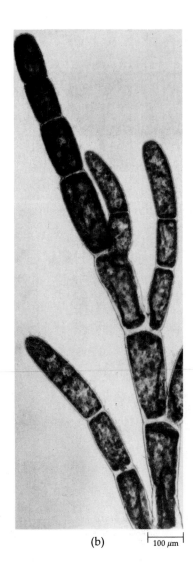

(b) ⊢ 100 μm ⊣

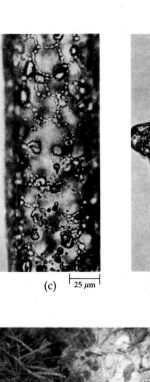

(c) ⊢ 25 μm ⊣   (d) ⊢ 25 μm ⊣

(e)

15–23

Cladophora, *a member of the class Ulvophyceae, is widespread in marine and freshwater habitats. The marine species have an alternation of generations like most Ulvophyceae, while the freshwater species do not. (a) Branched filaments of* Cladophora. *(b) Portion of a branched filament. (c) Part of an individual cell, showing the netlike chloroplast. (d) Commencement of branching at the apical end of a cell. (e) An individual of* Cladophora *growing in a sluggish stream in California.*

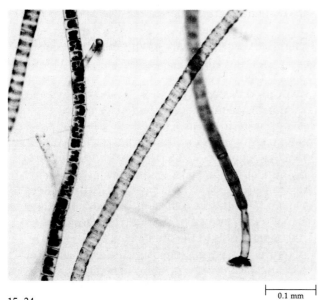

15–24
Ulothrix, *an unbranched, filamentous member of the class Ulvophyceae. The filament on the left with dense contents consists of sporangia in which zoospores are forming. The other filaments are vegetative. A holdfast can be seen on the filament on the right.*

0.1 mm

15–25
Sea lettuce, Ulva, *a common member of the class Ulvophyceae that grows on rocks, pilings, and similar places in shallow seas worldwide.*

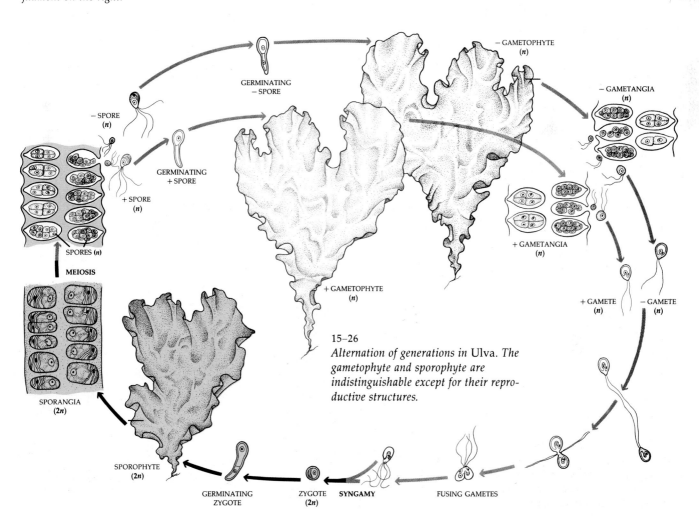

15–26
Alternation of generations in Ulva. *The gametophyte and sporophyte are indistinguishable except for their reproductive structures.*

(a)

(b)

10 mm

(c)

15–27
*Three genera of siphonous green algae
(class Ulvophyceae). (a) A species of
Codium, abundant along the Atlantic
coast. (b) Valonia, common in tropical
waters; individuals are often about the
size of a hen's egg. (c) Acetabularia, the
"mermaid's wine glass," a mushroom-
shaped siphonous green alga. The
siphonous green alga in the background
is Dasycladus. This photograph was
taken in the Bahamas.*

sion without the formation of cell walls (Figure 15–27). *Codium magnum,* mentioned above, is a member of this order. Cells with walls are produced only in the reproductive phase of the Siphonales. One of these algae, *Valonia,* which is common in tropical waters, has been widely used in studies of cell walls and in physiological experiments requiring large amounts of cell sap. *Valonia* appears to be unicellular but is actually a large multinucleate vesicle, with separate rhizoids and young branches (Figure 15–27b); it grows to the size of a hen's egg. One of the best-known Siphonales is *Acetabularia* (Figure 15–27c), which has been widely used in experiments on the genetic control of differentiation. The Siphonales are primarily diploid; the gametes are the only haploid cells in the life cycle.

The chloroplasts of some siphonaceous green algae, including *Codium,* are symbiotic in the bodies of sea slugs (nudibranchs), marine mollusks that lack shells. The sea slugs eat the algae, and their chloroplasts persist and divide in the cells that line the respiratory chamber of the nudibranch. In the presence of light, these chloroplasts carry on photosynthesis so efficiently that individuals of the nudibranch *Placobranchus ocellatus* are reported to evolve oxygen more rapidly than it is consumed by the mollusk.

Recently it has been found that at least one genus of siphonaceous green algae, *Halimeda* (Figure 15–28), contains a secondary metabolite that significantly reduces feeding in herbivorous fishes. This metabolite apparently represents a chemical defense in this widespread tropical alga, which is often the most abundant photosynthetic organism in the reef communities where it occurs. Such toxic compounds are apparently widely distributed among the algae, with more being discovered each year.

15–28
Halimeda, *a siphonous green alga that is often dominant in reefs in warmer waters throughout the world. This alga produces distasteful compounds that retard grazing by fishes and other marine herbivores.*

Class Chlorophyceae

Most of the green algae belong to this diverse group, which employs a mode of cell division involving a phycoplast. This unique characteristic indicates that no other groups of organisms have arisen from members of this class of green algae. Chlorophyceae include flagellated and nonflagellated unicellular algae, few-celled algae, motile and nonmotile colonial algae, filamentous algae, and parenchymatous algae. The members of this class occur mainly in fresh water, although a few unicellular, planktonic species occur in coastal marine waters. A few Chlorophyceae are essentially terrestrial, occurring in habitats such as soil or on wood.

Motile Unicellular Chlorophyceae

Among the least complex of the Chlorophyceae are its unicellular, biflagellated members, which belong to the order Chlamydomonadales. The best known of these organisms is the genus *Chlamydomonas*, which is one of the most common freshwater green algae. Individuals of this genus are small (usually less than 25 micrometers long), green, and rounded or pear-shaped (Figure 15–29). They move rapidly, with a characteristic darting motion that results from the beating of the two equal whiplash flagella protruding from its smaller, anterior pole. The flagella propel the organism through the water by beating in opposite directions.

Each cell of *Chlamydomonas* has a single massive chloroplast, containing a red pigment body—the *stigma*, or eyespot—which may be a shading device associated with a site of light perception. Recent studies have revealed that the photoreceptor in *Chlamydomonas* is homologous with rhodopsin, a visual pigment that is ubiquitous in multicellular animals. It is suspected, but has not yet been proved, that the visual pigment in the eyespots of dinoflagellates is also rhodopsin. These findings suggest a very ancient origin for the pigment. Individuals of *Chlamydomonas* are able to move toward light of appropriate intensities. Recently, it has also been claimed that they can orient themselves in a magnetic field, but such orientation needs to be confirmed and its basis clarified.

The chloroplast of *Chlamydomonas* also contains a roughly spherical pyrenoid. The uninucleate protoplast is surrounded by a thin glycoproteinaceous cell wall rich in hydroxyproline, inside of which is the plasma membrane. There is no cellulose in the cell wall of *Chlamydomonas*. At the anterior end of the cell, there are two contractile vacuoles, which collect excess water from the cell and ultimately discharge it.

Under some environmental conditions, cells of *Chlamydomonas* become nonmotile. Cells in this state are usually nonflagellated, and their walls become gelatinous. When conditions change, the flagella may reappear and the cells again become free-swimming.

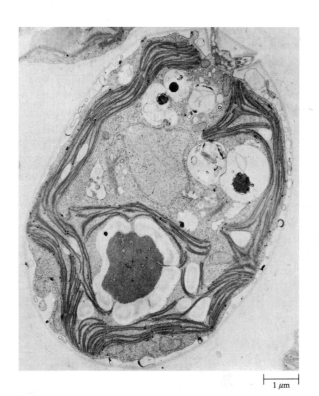

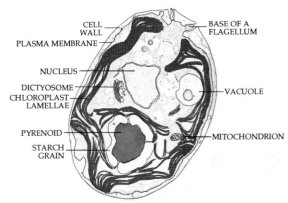

15–29
Chlamydomonas, *a unicellular green alga. Only the bases of the flagella can be seen in this electron micrograph.*

Chlamydomonas reproduces both sexually and asexually. During asexual reproduction, the haploid nucleus usually divides by mitosis to produce four daughter cells within the parent cell wall. Each cell then secretes a wall around itself and develops flagella. The cells secrete an enzyme that breaks down the original mother wall, and the daughter cells can then escape, although fully formed daughter cells are often retained for some time within the parent cell wall. In ancient flagellates, such aggregations of daughter cells may have been the forerunners of colonial organisms.

Sexual reproduction, which occurs in some species of *Chlamydomonas*, involves the fusion of individuals belonging to different mating types (Figure 15–30). The vegetative cells are induced to form gametes by nitrogen starvation. The gametes, which resemble the vegetative cells, first become aggregated in clumps. Within these clumps, pairs are formed that stick together first by their flagellar membranes and later by a slender protoplasmic thread that connects them at the base of their flagella. As soon as this protoplasmic connection is formed, the flagella become free, and one or both pairs of flagella beat, thus propelling the partially fused gametes through the water. The two gametes fuse completely, forming the zygote. Soon the four flagella shorten and eventually disappear, and a thick cell wall forms around the diploid zygote. This thick-walled, resistant zygote (zygospore) then undergoes a period of dormancy. Meiosis occurs at the end of the dormant period, resulting in the production of four haploid cells, each of which develops two flagella and a cell wall.

These cells can either divide asexually or mate with a cell of another mating strain to produce a new zygote. Thus, *Chlamydomonas* exhibits zygotic meiosis (see Figure 10–11a, page 163), and the haploid phase is the dominant phase in its life cycle.

In most species of *Chlamydomonas*, the cells of the two different mating types (conventionally designated + and −) are isogamous—identical in size and structure. In addition to these isogamous species, there are anisogamous species of *Chlamydomonas* (in which the female gametes are larger than the male gametes but are still motile) and oogamous ones (in which the female gametes are nonmotile) (Figure 15–31). Thus, the entire range of differences between gametes that occurs among algae is exhibited in the various species of the single genus *Chlamydomonas*.

Historically, *Chlamydomonas* has been considered primitive among the green algae, but it shares many advanced features with other Chlorophyceae and is not similar in detail to the ancestor of all green algae. *Chla-*

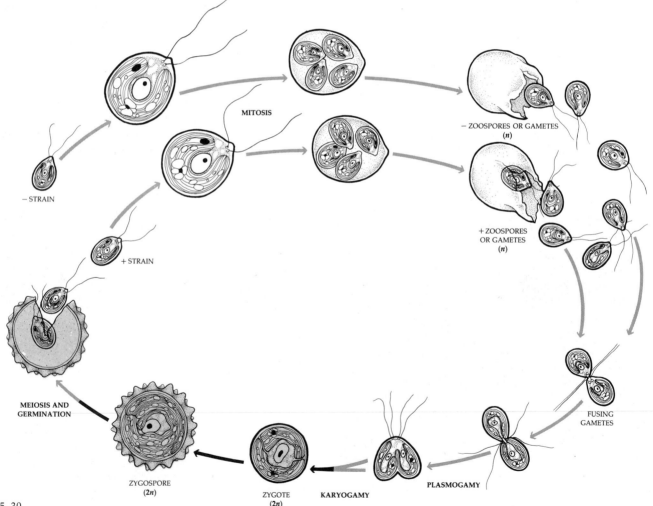

MITOSIS

− STRAIN

+ STRAIN

− ZOOSPORES OR GAMETES
(n)

+ ZOOSPORES OR GAMETES
(n)

FUSING GAMETES

MEIOSIS AND GERMINATION

ZYGOSPORE
(2n)

ZYGOTE
(2n)

KARYOGAMY

PLASMOGAMY

15–30

Life cycle of Chlamydomonas. *Sexual reproduction occurs when gametes of different mating types come together, cohering at first by their flagellar membranes and then by a slender protoplasmic*

thread—the conjugation tube. The protoplasts of the two cells fuse completely (plasmogamy), followed by the union of their nuclei (karyogamy). A thick wall is then formed around the diploid

zygote. After a period of dormancy, meiosis occurs and four haploid cells emerge. Asexual reproduction of the haploid individuals by cell division is the most frequent mode of reproduction.

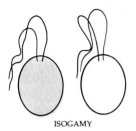
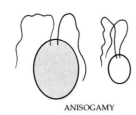
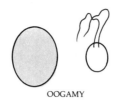

ISOGAMY   ANISOGAMY   OOGAMY

15–31
Types of sexual reproduction, based on gamete form; each is found in at least one species of Chlamydomonas. *Isogamy— the gametes are equal in size and shape. Anisogamy—one gamete, conventionally termed male, is smaller than the other. Oogamy—the female gamete is nonmotile.*

mydomonas-like cells do, however, resemble what must have been some of the specialized evolutionary lines within its class, some of which will be discussed under the following four headings: (1) nonmotile unicellular genera, (2) nonmotile colonial genera, (3) motile colonial genera, and (4) filamentous and parenchymatous genera.

Nonmotile Unicellular Chlorophyceae

Chlorella is a unicellular green alga that lacks flagella, eyespots, and contractile vacuoles; individuals are rounder and smaller than those of *Chlamydomonas* (Figure 15–32). In nature, *Chlorella* is widespread in both fresh and salt water and in soil. Each *Chlorella* cell contains a single cup-shaped chloroplast, with or without a pyrenoid, and a single minute nucleus. The only known method of reproduction in *Chlorella* is asexual, each haploid cell dividing mitotically two or three times to give rise to either four or eight nonmotile cells.

Chlorella, the first alga to be grown in culture, was used extensively in the studies that revealed some of the basic steps involved in photosynthesis. The fact that it is so easy to maintain in culture makes it an ideal experimental organism. Currently, it is under investigation as a potential food source for humans (Figure 15–33). Pilot farms have been established in the United States, Germany, Japan, and Israel. The Japanese have been able to process *Chlorella* into a tasteless white powder, which is rich in vitamins and protein and can be mixed with flour for the preparation of baked goods. Recently, *Chlorella* has also been investigated as a possible means of producing energy; in these experimental systems, it is grown together with a bacterium that converts the starch it produces to lipids. Such systems can be used on barges or platforms in the open ocean, or even in space, which is one reason for the current interest in their development. Other algae that produce large amounts of hydrocarbons are also under investigation for possible commercial production.

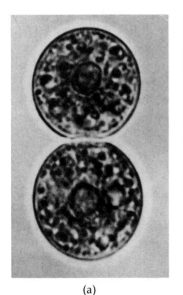
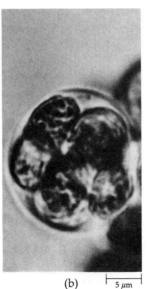

(a)   (b) |—— 5 μm ——|

15–32
Chlorococcum echinozygotum, *a close relative of* Chlorella. Chlorococcum *is the commonest soil alga.* (a) *Two individuals.* (b) *Formation of asexual spores following mitosis within the cell.*

15–33
These outdoor ponds at Sede Boqer in Israel are being used for experiments regarding the commercial feasibility of producing protein from Chlorella. *Although the cost is higher than producing an equivalent amount of protein from soybeans, owing in part to problems of light distribution throughout the ponds, continued experimentation may well lead to the commercial production of protein by this green alga,* Chlorella.

SYMBIOTIC GREEN ALGAE

Many algae, including green algae are symbiotic in other organisms. Most of the symbiotic green algae resemble the genus *Chlorella* (see Figure 15–32); they can be found living within many freshwater protozoa, sponges, hydra, and some flatworms. Most of these algae reproduce by simple cell division and divide with the cells of their hosts, with which they remain in intimate contact.

Another green alga, *Tetraselmis convolutae,* a member of the class Pleurastrophyceae, is found mostly in the subepidermal cells of the marine flatworm *Convoluta roscoffensis.* Within the cells of the flatworm, *Tetraselmis* has no cell wall and is irregularly shaped; its plasma membrane, greatly increased in surface area by fingerlike projections, is in more or less direct contact with the vacuolar membrane of the host cell. When removed from the flatworm and cultured, *Tetraselmis* has a cell wall, four flagella, and an eyespot, all of which are missing when it is symbiotic. Another example of symbiosis involving green algal chloroplasts is discussed on page 270.

Nonmotile Colonial Chlorophyceae

Hydrodictyon, the "water net," is a nonmotile, colonial member of the Chlorophyceae (Figure 15–34). Under favorable conditions it accumulates in massive aggregates in ponds, lakes, and gentle streams. Each colony consists of many cylindrical cells arranged in the form of a large hollow cylinder. Initially uninucleate, each cell eventually becomes multinucleate. At maturity, each cell contains a large central vacuole and peripheral cytoplasm in which the nuclei and a large reticulate chloroplast with numerous pyrenoids are located. *Hy-drodictyon* reproduces asexually through the formation of uninucleate, biflagellated zoospores. Eventually, the zoospores form into groups of four to nine (most typically six) within the cylindrical parent cell, lose their flagella, and form daughter colonies. Sexual reproduction in *Hydrodictyon* is isogamous, and meiosis is zygotic, as in all sexually reproducing Chlorophyceae.

Motile Colonial Chlorophyceae

In the motile, colonial genera of Chlorophyceae, *Chlamydomonas*-like cells adhere in colonies that are propelled by the beating of the flagella of the individual cells (Figure 15–35). The members of this group have often been called the "Volvocine line" after the genus *Volvox,* the largest, most complex organism exhibiting this colonial existence (Figure 15–36). In the colonies of some algae of this type, the cells are connected by cytoplasmic strands that provide for the integration of the whole organism. The simplest member of this group is *Gonium* (Figure 15–35a, b). The *Gonium* colony consists of separate cells held together within a gelatinous matrix. Each colony is made up of 4, 8, 16, or 32 cells (depending on the species) arranged in a slightly curved, shield-shaped disk. The flagella of each cell beat separately, pulling the entire colony forward. Each cell in *Gonium* can divide to produce an entire new colony.

A closely related colonial organism is *Pandorina,* which forms a tightly packed ovoid or ellipsoid colony, usually consisting of 16 or 32 cells held together within a matrix (Figure 15–35c). The colony is polar, the eyespots being larger in the cells at one end of the colony. Each cell has two flagella, and because all the flagella point outward, *Pandorina* rolls through the water like a ball. When the cells attain their maximum size, the colony sinks to the bottom of the pond and each of the

15–34
(a) *The "water net," Hydrodictyon, a colonial member of the Chlorophyceae.*
(b) *Scanning electron micrograph of a flattened young colony of* Hydrodictyon reticulatum.

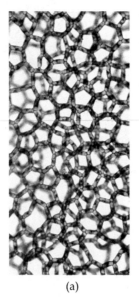

(a)

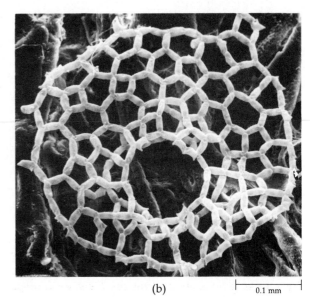

(b)

0.1 mm

15–35
Several colonial Chlorophyceae. (a) and (b) Two views of Gonium, *(c) Pandorina, and (d) Eudorina. In these algae, cells similar to those of* Chlamydomonas *adhere in a gelatinous matrix to form multicellular colonies propelled by the beating of the flagella of the individual cells. Varying degrees of cellular specialization are found in different genera.*

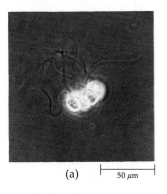

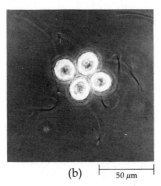

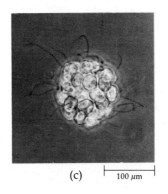

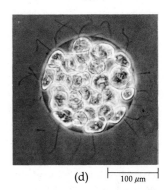

(a)   50 µm   (b)   50 µm   (c)   100 µm   (d)   100 µm

cells divides to form a daughter colony. The latter remain together until all have developed flagella. The parent matrix then breaks open like Pandora's box (which suggested its name), releasing new daughter colonies.

Eudorina is another spherical colony made up of green flagellates, the number of which is 32, 64, or 128 in the most common species (Figure 15–35d). It differs from *Gonium* and *Pandorina* in that some of the cells in the colony, which are smaller than the others and are located in the anterior part of the colony (relative to the direction of movement), are incapable of reproducing to form new colonies. Here we see a beginning of specialization of function.

The most spectacular of these colonial green algae, however, is *Volvox* (Figure 15–36). *Volvox* consists of a hollow sphere made up of a single layer of 500 to 60,000 vegetative, biflagellated cells and a small number of reproductive cells. The number of cells in the sphere varies from species to species. When *Volvox* whirls through the water, it appears like a spinning universe of individual stars fixed in an invisible firmament.

Volvox exhibits polarity; that is, it has anterior and posterior poles. The flagella of each cell beat in such a way as to spin the entire colony around its axis in a clockwise direction as it moves forward (usually toward the light at most stages of the life cycle).

Most of the cells of a spherical *Volvox* colony are strictly vegetative, and the few that do engage in asexual reproduction are usually confined to the posterior hemisphere in a characteristic pattern. In some species of *Volvox*, the cells that will fulfill the reproductive function are not clearly distinguishable in juvenile spheroids, but they later become apparent as they undergo successive cycles of growth and division. Ultimately, these dividing cells give rise to new daughter spheroids, which "hatch" from the parental spheroid by releasing an enzyme that dissolves its transparent

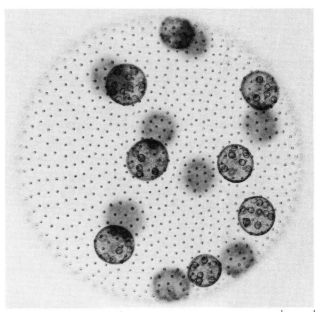

0.1 mm

15–36
Asexual reproduction in Volvox carteri. *The approximately 2000 small cells at the periphery of the transparent sphere are the* Chlamydomonas-*like somatic cells; in this species, somatic cells are not interconnected in mature colonies as they are in some members of the genus. In the specimen shown here, the 16 reproductive cells in the interior of the spheroid have already undergone division to produce juvenile spheroids, each containing about 2000 tiny somatic cells and 16 reproductive cells. Eventually, each juvenile colony digests a passageway out of the parent colony, swims away, and the cycle then repeats.*

matrix. In more advanced members of the genus, however, asexual reproductive cells are set aside early in each reproductive cycle and are at all times structurally and functionally distinct from the vegetative, or somatic, cells. In these species, such as *Volvox carteri* (Figure 15–36), there is a true "division of labor" between two mutually dependent cell types. The nonmotile reproductive cells depend on the motile nonreproductive somatic cells to carry them to the surface, where they can obtain adequate light and CO_2 for photosynthesis. Thus, the more advanced species of *Volvox* are not simply colonial like the other species of the genus. They are instead multicellular and differentiated, as are plants and animals.

Sexual reproduction in *Volvox* is always oogamous, but substantial variation in details of reproductive pattern are seen among the various species. In some species, both eggs and sperm may be produced within a single spheroid. In other species, members of a single genetically uniform clone (a culture derived by asexual means from a single individual) may become sexually reproducing spheroids—either male or female—but spheroids that include both sexes are unknown. In the more advanced species, however, sex is genetically determined, and each clone derived by asexual reproduction is either male or female. In all species that have been studied, sexual reproduction is synchronized within the population of colonies by a sexual inducer molecule. This molecule is produced by a spheroid that has itself become sexual by some other, as yet poorly understood, mechanism. One male colony of *V. carteri* may produce enough inducer to induce over half a billion other colonies to reproduce sexually.

Volvox is one of the simplest multicellular organisms to show clear division of labor, and the sexual inducers of *Volvox* are among the most potent biologically active substances known. Thus the genus (particularly *V. carteri*) has become the focus of a great deal of research in recent years. Like all of its relatives, *Volvox* is haploid; therefore, mutant genes are not masked by dominant alleles, and mutations affecting development can be readily detected. Hundreds of strains with specific, heritable developmental defects have been isolated and are being studied with the hope of learning how specific genes regulate cellular differentiation.

To summarize, increasing specialization in the members of the colonial Chlorophyceae is evident in several ways. First, there is an increase in the cell number and in the size of the colonies. Second, there is increasing specialization in cell morphology and function. Finally, there is increasing sexual specialization paralleling that within the genus *Chlamydomonas*. Thus, *Gonium* and *Pandorina* have isogamous reproduction, whereas species of *Eudorina* and *Volvox* are oogamous. Nevertheless, this line clearly represents an evolutionary "dead end," in that it has not given rise to a more complex group of organisms.

Filamentous and Parenchymatous Chlorophyceae

The filamentous and parenchymatous Chlorophyceae, members of the orders Chaetophorales and Oedogoniales, include algae that have the most complex structures found in the class. Their cells, which are often specialized with respect to particular functions or positions in the algal body, are connected by plasmodesmata, like plant cells. Although many genera of these groups have a cell plate (Figure 15–37), like plants, they also have a phycoplast, like all Chlorophyceae, and thus cannot be ancestral to plants. In addition, the nuclear envelope persists through the mitotic cycle, as it does in all Chlorophyceae.

15–37

Stigeoclonium, a member of the Chlorophyceae that was formerly considered to belong to the genus Ulothrix *(Ulvophyceae). Cell division in* Stigeoclonium *occurs by cell-plate formation, which is nearly complete at the stage of cytokinesis shown in this electron micrograph. The nuclei are close together during cytokinesis because the spindle collapses at telophase of the preceding mitosis. The phycoplast is not visible at this relatively low magnification.*

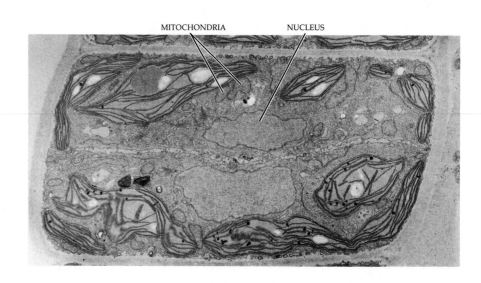

MITOCHONDRIA   NUCLEUS

Among the filamentous Chlorophyceae, *Stigeoclonium* (Figure 15–37) superficially resembles *Ulothrix* (Figure 15–24) but has all of the characteristics of Chlorophyceae. The two genera, together with other Chlorophyceae now recognized as distinct, were once considered to be the same. They are now placed in different classes, however, based on detailed studies with the electron microscope. *Stigeoclonium*, unlike *Ulothrix*, has branched filaments.

Fritschiella is a terrestrial alga that resembles *Stigeoclonium* in most of its microscopic features but is more complex. The body of *Fritschiella* consists of subterranean rhizoids, a parenchymatous prostrate system near the soil surface, and two kinds of erect branches (Figure 15–38). *Fritschiella* occurs on damp surfaces such as tree trunks, moist walls, and leaf surfaces. It has, like the plants, become adapted to a terrestrial existence, and it exhibits some of the features that were probably also found in ancestral plants.

Oedogonium is an unbranched filamentous alga; individuals are attached to the substrate by a holdfast. *Oedogonium* has very unusual features, especially in regard to cell division. The cells of this genus are uninucleate, with a netlike chloroplast around its periphery. As a cell of *Oedogonium* divides, a doughnut-shaped ring of wall material forms near its upper (apical) end. Following division of the nucleus, one of the daughter nuclei migrates into the upper end of the cell, and the wall ruptures violently and precisely at this ring. The ring material is then drawn out into a cylinder, forming the wall of the upper daughter cell, which is nearer the apex of the filament. A new cross wall forms in the plane of the phycoplast, just past the edge of the ruptured parental wall. As a consequence of ring expansion, the rims of the original mother cell bend outward, giving rise to the characteristic "caps," or annular scars (Figure 15–39). These scars reflect the number of divisions that have occurred earlier.

Asexual reproduction in *Oedogonium* takes place by means of zoospore formation, with a single zoospore produced per cell. Each zoospore has a crown of about 120 flagella. Sexual reproduction is oogamous (Figure 15–40). Each antheridium produces two multiflagellated sperm, and each oogonium produces a single egg. Meiosis in *Oedogonium* is zygotic, as in all Chlorophyceae.

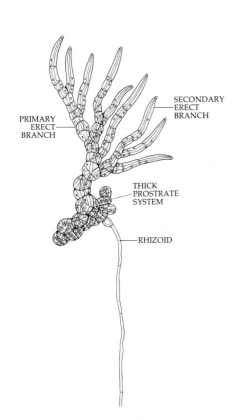

PRIMARY ERECT BRANCH

SECONDARY ERECT BRANCH

THICK PROSTRATE SYSTEM

RHIZOID

15–38
Fritschiella, *a terrestrial member of the Chlorophyceae. In its adaptation to a terrestrial habitat, Fritschiella has independently evolved some of the features that are characteristic of plants.*

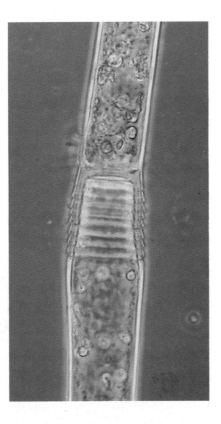

15–39
Oedogonium, *an unbranched, filamentous member of the Chlorophyceae. A section of vegetative filament showing annular scars.*

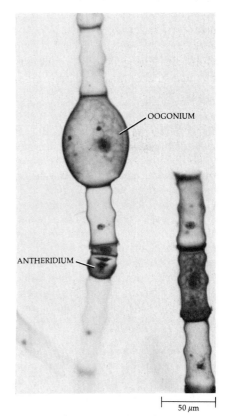

OOGONIUM

ANTHERIDIUM

50 μm

15–40
Sexual reproduction in Oedogonium is *oogamous. Each oogonium produces a single egg, whereas each antheridium produces two multiflagellated sperm.*

SUMMARY

The three wholly or partly multicellular divisions of algae treated here are the red algae (division Rhodophyta), brown algae (division Phaeophyta), and green algae (division Chlorophyta). The members of these divisions are largely aquatic; the brown and red algae are of great importance in marine habitats, and the green algae are often abundant in freshwater habitats. A number of genera of green algae also occur in the sea, and a few genera of red and brown algae live in fresh water. In all of these habitats, the members of these divisions of algae play an ecological role comparable to that of the plants in land habitats.

The various divisions of algae appear to have had their origins in symbiotic relationships between nonphotosynthetic, protozoalike eukaryotic cells and photosynthetic bacteria. Specifically, it appears likely that the chloroplasts of red algae originated as symbiotic cyanobacteria, while those of green algae may have been derived from bacteria with characteristics somewhat like those of *Prochloron*. The golden-brown chloroplasts of brown algae—in which the xanthophyll fucoxanthin and similar accessory pigments are abundant— have characteristics somewhat like that of the newly discovered bacterium *Heliobacterium chlorum*. They are also very similar to the chloroplasts of chrysophytes (which resemble brown algae in many respects), as well as those of the dinoflagellates, a very different group of protists.

Rhodophyta (red algae) are a large group particularly common in warmer marine waters. They nearly always grow attached to a substrate, and some grow at great depths (down to 268 meters). They contain phycobilins, which give them their characteristic colors, and chlorophyll *a*.

Phaeophyta (brown algae) include the largest and most complex of the marine algae. In many types the vegetative body is well differentiated into holdfast, stipe (stalk), and blade. Some have food-conducting tissues that approach those of the vascular plants in their complexity. The chloroplasts of brown algae contain chlorophylls *a* and *c*, together with abundant quantities of fucoxanthin, which gives them their olive-green to dark-brown color. The sporophyte is usually larger than the gametophyte.

Chlorophyta (green algae) are the largest and most diverse of the three divisions considered here, and the one from which the plants probably evolved. Their chloroplasts contain chlorophylls *a* and *b*, together with carotenoids, and store starch, their reserve food; in these respects, they are identical to those of plants. The ancestral green algae appear to have been scaly, flagellated unicellular organisms. The division comprises at least five classes, of which we have considered three in detail.

Charophyceae and Chlorophyceae occur primarily in fresh water, whereas Ulvophyceae occur primarily in marine habitats. All three of these classes include unicellular, few-celled, filamentous, and parenchymatous genera, and Chlorophyceae include motile and nonmotile colonial genera as well. The motile spores of Charophyceae are asymmetrical and have lateral or subapical flagella that project at right angles, whereas those of the other two classes are radially symmetrical and have apical flagella.

In Charophyceae and Ulvophyceae, the mitotic spindle persists during telophase; in Chlorophyceae, the mitotic spindle collapses at telophase and a phycoplast develops. A phycoplast is a system of cytokinetic microtubules oriented parallel to the plane of cell division. In the other two classes, the spindle is persistent; and in Charophyceae, there is a phragmoplast, a system of microtubules oriented at right angles to the plane of cell division, as in plants.

Sexual reproduction in Charophyceae and Chlorophyceae always involves the production of a dormant zygospore and is characterized by zygotic meiosis; sexual reproduction in Ulvophyceae often involves alternation of generations (sporic meiosis), and dormant zygotes are rare.

SUGGESTIONS FOR FURTHER READING

BOLD, HAROLD C., and MICHAEL J. WYNNE: *Introduction to the Algae: Structure and Reproduction*, 2nd ed., Prentice-Hall, Inc., Englewood Cliffs, N.J., 1985.
A detailed reference work on algae, which contains a wealth of information on all groups; taxonomically oriented.

BONEY, A. D.: *A Biology of Marine Algae*, Hutchinson and Co., Ltd., London, 1966.*
An excellent treatise on marine algae that is ecologically and physiologically oriented.

CHAPMAN, A.R.O.: *Biology of Seaweeds: Levels of Organization*, University Park Press, Baltimore, Md., 1979.*
A biologically oriented view of the green, brown, and red marine algae; the discussion ranges from cellular ultrastructure and function to community ecology.

CHAPMAN, V. J., and D. J. CHAPMAN: *Seaweeds and Their Uses*, 3rd ed., Methuen, New York, 1980.
A full account of the traditional and modern uses of seaweeds.

DARLEY, W. MARSHALL: *Algal Biology: A Physiological Approach*, Blackwell Scientific Publications, Boston, Mass., 1982.
An introduction to algal physiological ecology, stressing the way in which individual algae function in their environment.

DAWES, CLINTON J.: *Marine Botany*, Wiley-Interscience, New York, 1981.
A survey of the marine algae, presented in an ecological context.

ESSER, KARL: *Cryptogams: Cyanobacteria, Algae, Fungi, Lichens*, Cambridge University Press, New York, 1982.
A good account of the groups enumerated in the title of the book, including a comprehensive manual of laboratory techniques for these organisms.

GRAHAM, LINDA E.: "*Coleochaete* and the Origin of Land Plants," *American Journal of Botany* 71:603–8, 1984.
This brief review paper presents the latest information on a green alga that bears an intriguing resemblance to the plants.

GRAHAM, LINDA E.: "The Origin of the Life Cycle of Land Plants," *American Scientist* 73:178–86, 1985.

* Available in paperback.

This paper presents an outstanding discussion of the way in which a simple modification in the life cycle of an extinct green alga could have been responsible for the origin of plants.

LOBBAN, CHRISTOPHER S., and MICHAEL J. WYNNE (Eds.): *The Biology of Seaweeds,* Blackwell Scientific/University of California Press, Berkeley, Calif., 1982.

An excellent advanced treatise on the structure, reproduction, ecology, physiology, biochemistry, and uses of seaweeds.

PICKETT-HEAPS, JEREMY D.: *Green Algae: Structure, Reproduction and Evolution in Selected Genera,* Sinauer Associates, Inc., Sunderland, Mass., 1975.

A magnificently illustrated treatise on the green algae, which contains a wealth of interesting information.

ROUND, F. E.: *The Ecology of Algae,* Cambridge University Press, New York, 1981.

A comprehensive account of freshwater and marine algal ecology.

SCAGEL, R. F., et al.: *Nonvascular Plants: An Evolutionary Survey,* Wadsworth Publishing Company, Belmont, Calif., 1982.

An excellent synoptical treatment of all groups of photosynthetic organisms except for vascular plants.

TRAINOR, F. R.: *Introductory Phycology,* John Wiley & Sons, Inc., New York, 1978.

An outline of the systematics of algae; clearly written and well illustrated.

VAN DER MEER, JOHN P.: "The Domestication of Seaweeds," *BioScience* 33:172–76, 1983.

An interesting analysis of the way in which the intelligent selection of new strains and improved management of crops could lead to much better yields from seaweeds.

* Available in paperback.

CHAPTER 16

Bryophytes

16-1
Dense growth of the moss Fissidens *on limestone rocks in a waterfall. This photograph was taken in a nature reserve just west of Yalta on the Crimean Peninsula in the Soviet Union.*

A number of lines of evidence reviewed in the last chapter indicate that plants—bryophytes and vascular plants—evolved from some ancient group of green algae. Like green algae, plants have chlorophyll *a* as their primary photosynthetic pigment and chlorophyll *b* and carotenoids as accessory pigments. In all three groups, starch is the primary carbohydrate food reserve, and the starch is deposited within the chloroplasts, not in the cytoplasm as in other photosynthetic eukaryotes. Cellulose is the principal component of plant cell walls and is an important cell wall component in some green algae also. Finally, plants form a phragmoplast and a cell plate during the process of cell division. The only other living organisms that share this characteristic are one genus of brown algae and a few genera of green algae. Since bryophytes (Figure 16–1) and vascular plants share all of these characteristics, it seems likely that they are descended from a remote common ancestor that successfully invaded the land.

If this hypothesis is correct, the evolutionary lines giving rise to the bryophytes and the vascular plants must have diverged long ago. The oldest known fossils that resemble bryophytes are from the Devonian period, some as much as 400 million years old (Figure 16–2), and definite fossil bryophytes go back at least 370 million years. Moreover, remains that resemble vascular plants have been found in the early part of the Silurian period, about 430 million years ago (see Figure 1–5, page 4). In view of these findings, the hypothetical common ancestor of bryophytes and vascular plants—a relatively complex green alga (see Figure 15–21, page 267)—presumably invaded the land more than 430 million years ago. Spore tetrads and cuticlelike fragments from the Ordovician period, about 450 million years ago, suggest an even earlier origin for plants.

Apparently the common ancestor of the bryophytes and vascular plants had a well-developed alternation of generations, for this is the type of life cycle exhibited by all bryophytes and vascular plants. Its gametophytes almost certainly bore multicellular gametangia like those of all living plants. All plants have embryos: their zygotes begin to divide while they are held within, and are nutritionally dependent on, the gametophyte generation. Algae lack embryos, and it is not known whether the first plants formed embryos or not. We also do not know what kind of conducting tissue, if any, may have been present in the ancestor of the plants. Some fossils from the early Silurian period, however, have elongate cells that resemble tracheids, with annular or helical thickenings, and there is also evidence of ligninlike substances in the walls of these cells. The first land plants were probably endomycorrhizal, with a probable zygomycete as the fungal partner (see page 224).

The organisms that were successful in making the transition from water to land had developed features that prevented them from drying out. One of these was the development of a sterile jacket layer around the sperm-producing and egg-producing cells of the male and female gametangia, call *antheridia* and *archegonia*, respectively. Similarly, a sterile jacket layer was formed around the spore-producing cells of the *sporangia*.

The invasion of the land by the ancestors of plants was also correlated with retention of the zygote within the female gametangium and its development there into an embryo. Thus, during its critical early stages of development, the young embryo, or sporophyte, is protected by the female gametophyte. In the green algae, by contrast, the development of the zygote is essentially independent of the female gametophyte.

The aerial parts of most vascular plants are covered with a waxy protective layer, the cuticle, which helps prevent drying out. This cuticle is closely correlated with the presence of stomata, the specialized pores that function primarily in the regulation of gas exchange. A cuticle is apparently absent in most bryophytes, but stomata are present on the sporophytes of hornworts and mosses. Some of the stomata in mosses consist only of a single, doughnut-shaped guard cell, which is a very different structure from that found in vascular plants. In hornworts the stomata apparently remain open until late in their development, and those of the mosses close only after the moss sporophytes are completely dried up. These stomata, therefore, do not function like those found in contemporary vascular plants.

All plants are oogamous, and all possess alternation of heteromorphic generations.

CHARACTERISTICS OF THE BRYOPHYTES

The bryophytes—the liverworts, hornworts, and mosses—are relatively small plants, many less than 2 centimeters long and most less than 20 centimeters long (Figure 16–3). Bryophytes are often abundant in relatively moist areas, where a variety of species and a luxuriance of individuals often can be found. Mosses sometimes dominate the terrain to the exclusion of other plants over large areas of the far north and far south, as well as on rocky slopes above timberline in the mountains. A significant number of mosses are even able to withstand the long periods of severe temperatures on the Antarctic continent (Figure 16–4). Like the lichens, the bryophytes are remarkably sensitive to air pollution, especially sulfur dioxide, and they are often absent or represented by only a few species in highly polluted areas. A number of species of mosses are found in deserts, and some form extensive masses on dry exposed rocks where very high temperatures

16–2

(a) *Longitudinal section through a gametophyte from the Rhynie chert of Scotland. This plant lived in the early Devonian period about 400 million years ago. The arrows indicate the position of antheridia on the upper surface of the gametophyte.* (b) *Mature sperm are shown in a section through an antheridium. In their size and complexity, such gametophytes resemble those of the bryophytes. It is not known whether any of the early Devonian sporophytes known from the same rock formations are related to these gametophytes or not.*

(a) |— 1 mm —|

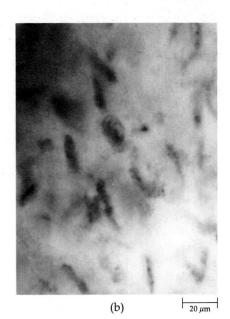

(b) |— 20 μm —|

16–3

Hairy-cap moss, Polytrichum, *showing sporophytes attached to the female gametophytes. An individual sporophyte consists of a stalk, capsule, and foot, which connects the sporophyte to the gametophyte. The leaves at the bottom of the photograph are part of the gametophytes on which these sporophytes were produced, following a fertilization event that occurred at least six months earlier.*

can be reached. Many mosses can remain alive for years in a dry condition, recovering quickly if moistened; other mosses and liverworts are aquatic and die within a day or so if kept dry. A few bryophytes are even found on seaside, wave-splashed rocks, although none is truly marine. There are about 16,000 species, more than in any other group of plants except the flowering plants.

Two important characteristics distinguish bryophytes from vascular plants. First, bryophytes lack the kinds of specialized vascular tissues—xylem and phloem—that characterize vascular plants. Accordingly, all bryophytes, strictly speaking, lack true leaves, true stems, and true roots, since these structures are defined, in part, by the presence of vascular tissue in them. Nevertheless, the terms "leaf" and "stem" are commonly used when referring to the leaflike and stemlike structures of the gametophytes of leafy liverworts and mosses, a practice that will be followed in this book.

The stalks, or *setae* (singular, *seta*), of the sporophytes of most mosses contain a central strand of water-conducting cells known as *hydroids*; similar conducting cells are also present in the gametophytes of many of these species. Hydroids are elongated cells with inclined end walls that are thin and highly permeable to water, making the hydroids preferential pathways for water and solutes. Like the tracheids and vessel elements of vascular plants, hydroids lack a living protoplast at maturity and so appear empty. Unlike the water-conducting cells of vascular plants, however, they lack specialized wall thickenings (pages 389, 390). In some genera of mosses, food-conducting cells

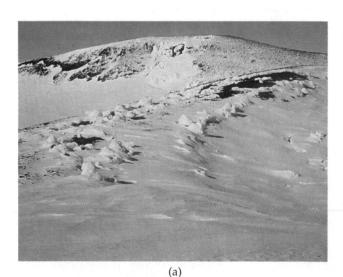

(a)

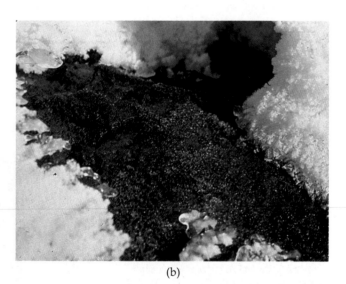

(b)

16–4

(a) *Mount Melbourne, Antarctica, at about 75°S latitude. At about 3000 meters elevation on Mount Melbourne, the daily temperatures in summer mainly range from −10° to −30°C. In this incredibly harsh environment, in the bare*

patches visible in the photograph, botanists from New Zealand in late 1984 discovered patches of a moss of the genus Campylopus (b), *where volcanic activity produces temperatures that may reach*

30°C. The dispersal of Campylopus *to this locality demonstrates the remarkable dispersal powers of mosses more than their ability to survive in harsh habitats, which is considerable.*

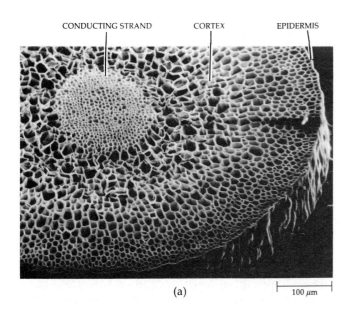

CONDUCTING STRAND   CORTEX   EPIDERMIS

(a)

100 μm

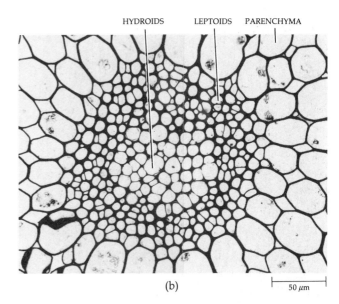

HYDROIDS   LEPTOIDS   PARENCHYMA

(b)

50 μm

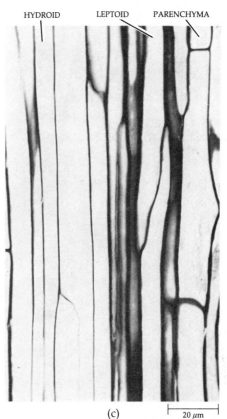

HYDROID   LEPTOID   PARENCHYMA

(c)

20 μm

16–5

Conducting strands in the seta of a sporophyte of the moss Dawsonia superba. *(a) General organization of the seta as seen in transection with the scanning electron microscope. (b) Transection showing central column of hydroids surrounded by a sheath of leptoids, and parenchyma of the cortex. (c) Longitudinal section of a portion of central strand, showing (from left to right) hydroids, leptoids, and parenchyma.*

known as *leptoids* surround the strand of hydroids (Figure 16–5), much as phloem surrounds a strand of xylem in some vascular plants. Leptoids are elongate cells that have degenerate nuclei but living protoplasts at maturity. Some leptoids closely resemble the food-conducting cells in the phloem of primitive vascular plants. On balance, it seems likely that the water- and food-conducting cells of mosses and vascular plants are derived from similar cells that were present in the common ancestor of these two groups.

In most bryophytes, the gametophyte is attached to the substrate by means of elongate single cells or filaments of cells called *rhizoids*. These rhizoids generally serve only to anchor the plants, since absorption of water and inorganic ions commonly occurs directly and rapidly throughout the gametophyte. Rootlike organs are lacking, although the underground stems of a few mosses are structurally quite complex.

The second important distinguishing characteristic of the bryophytes is the nature of their alternation of generations: the gametophytes are always nutritionally independent, whereas the sporophytes are permanently attached to the gametophytes and vary in their dependence on them. In other words, the gametophyte is the conspicuous and dominant generation in the bryophytes. Among vascular plants, the sporophyte is the conspicuous and dominant generation (see Figure 10–11c, page 163).

Bryophytes occur in habitats that are moist, at least seasonally. In order for fertilization to occur, the biflagellated sperm must swim through water to reach the egg, which is located inside an archegonium. The archegonium, which may be stalked, is flask-shaped, with a long neck and swollen basal portion, the *venter*, which encloses a single egg (Figure 16–6a). The central cells of the neck, the *neck canal cells*, disintegrate when

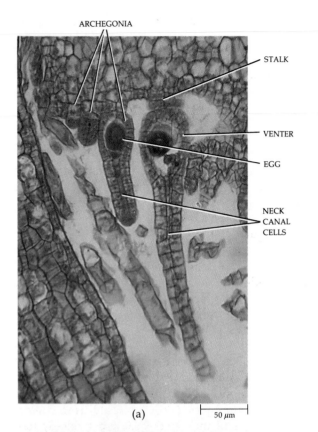

ARCHEGONIA

STALK

VENTER

EGG

NECK CANAL CELLS

(a)   50 μm

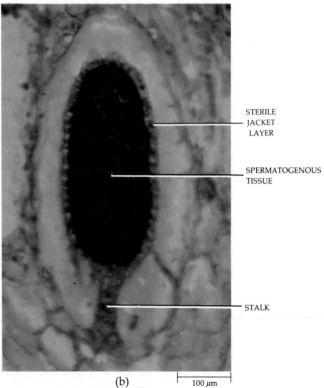

STERILE JACKET LAYER

SPERMATOGENOUS TISSUE

STALK

(b)   100 μm

16–6
Gametangia of Marchantia. *(a) Details of several archegonia at different stages of development. (b) One developing antheridium. In* Marchantia, *the archegonia and antheridia are borne on different gametophytes.*

the archegonium is mature, leaving a fluid-filled tube through which the sperm swim to the egg. The elongate or spherical antheridium is commonly stalked and consists of a sterile jacket layer, one cell thick, surrounding numerous *spermatogenous cells* (Figure 16–6b). Each spermatogenous cell forms a single biflagellated sperm.

In liverworts and hornworts, the sperm often actually swim to the vicinity of the archegonia through a continuous film of water. Once in the region of an archegonium, sperm are attracted into its neck by chemicals that it secretes. In mosses, the antheridia are often clustered within leafy structures called splash cups (Figure 16–7). The sperm from several to many antheridia are discharged into a drop of water within such a cup and then are dispersed widely as a result of the action of raindrops falling into the cup. Insects may also carry drops of water, rich in sperm, from plant to plant. Since sperm can often reach them even over relatively great distances by these means, isolated female moss gametophytes often form sporophytes abundantly.

The fertilized egg, or zygote, is retained within the venter of the archegonium, where it develops into an embryo. For a period, the venter undergoes cell division, keeping pace with the growth of the young sporophyte within the archegonium. The enlarged archegonium is called a *calyptra*. At maturity the sporophyte of many bryophytes consists of a foot—which remains embedded in the archegonium—a seta, or stalk, and a *capsule* or *sporangium* (Figure 16–8).

16–7
Leafy male gametophytes of the moss Polytrichum piliferum, *showing the mature antheridia clustered together in heads. The sperm are discharged into drops of water held within these leafy heads and are then splashed out of them by raindrops, sometimes reaching the vicinity of archegonia on other gametophytic individuals.*

Generally, the cells of the young and maturing sporophyte contain chloroplasts and carry out photosynthesis, but by the time meiosis occurs in the capsule and the spores are produced, the chlorophyll has usually disappeared. In mosses, the calyptra is commonly lifted upward with the capsule as the seta elongates. Prior to spore dispersal, the protective calyptra falls off, and the spores are shed following the dehiscence (spontaneous bursting open of a structure and discharge of its contents) of the capsule.

The bryophytes are traditionally divided into three classes: Hepaticae (the liverworts, 6000 species), Anthocerotae (the hornworts, 100 species), and Musci (the mosses, 9500 species). The three groups differ sharply from one another, and the characteristics that they do have in common are in general those of simple plants. Most mosses have conducting tissue, and they could well be, in effect, reduced vascular plants; they seem to share a common ancestor with the vascular plants, whereas the liverworts and hornworts appear to have diverged earlier from this evolutionary line. There is no evidence that the liverworts are derived from ancestors with vascular tissue, and they may not be closely related to the mosses. The hornworts are very distinct from the other two groups and do not seem to be closely related to any known group of organisms. Faced with these differences, an increasing number of botanists is coming to regard the three groups of bryophytes as distinct divisions, whose similarities are the result of convergent evolution or the retention of primitive characteristics, rather than an indication of an immediate common ancestor. For the present, however, we shall follow tradition and retain them in a single division.

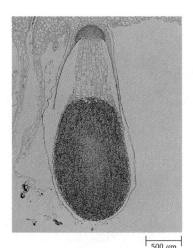

16–8

A nearly mature sporophyte of Marchantia. *Note the elaters—helical, filamentous structures that assist in the dispersal of spores—within the spore-filled capsule.*

THE LIVERWORTS: CLASS HEPATICAE

Liverworts, or hepatics, are small plants that are generally less conspicuous than mosses. Their name dates from the ninth century, when it was thought, because of the liver-shaped outline of the gametophyte in some genera, that these plants might be useful in treating diseases of the liver. According to the medieval "Doctrine of Signatures," the outward appearance of a body signaled the possession of special properties. The Anglo-Saxon ending "-wort" (originally *wyrt*) means "herb"; it appears as a part of many plant names in the English language.

The gametophytes of some liverworts are flattened, dorsiventral *thalli* (singular: *thallus*; a term traditionally used for undifferentiated plant bodies having distinct upper and lower surfaces), which grow from an apical meristem. The gametophytes of most species, however, are leafy. They grow from a single apical cell, which resembles an inverted pyramid with a base and three sides. Daughter cells are cut off from the sides of this single cell. The rhizoids of liverworts are single-celled, unlike those of mosses, which contain several cells each. The gametophytes develop directly from spores. The sporophytes of most liverworts are generally less complex than those of mosses, and their capsules have very different mechanisms for the release of spores than those of the mosses.

Thallose Liverworts

Thallose liverworts are a diverse group of nonleafy hepatics. They can be found on moist, shaded banks and in other suitable habitats, such as flowerpots in a cool greenhouse. The thallus is many cells thick—about 30 at the midrib and approximately 10 in the thinner portions—and is sharply differentiated into a thin, chlorophyll-rich upper (dorsal) portion and a thicker, colorless lower (ventral) region (Figure 16–9a). The lower surface bears two kinds of rhizoids, as well as rows of scales. The upper surface is divided into raised regions, each of which marks the limits of an underlying air chamber and has a large pore that leads to this chamber (Figure 16–9b).

One of the most familiar liverworts is *Marchantia*, a widespread, terrestrial genus, which grows on moist soil and rocks (Figure 16–10). Its dichotomously branched gametophytes are mostly one to a few centimeters long; its gametangia are restricted to specialized erect structures called gametophores. The gametophytes of *Marchantia* are unisexual, and the male and female gametophytes can be readily identified by the distinct structures that they bear. The antheridia are borne on disk-headed stalks called <u>antheridiophores</u>, whereas the archegonia are borne on umbrella-headed

16–9

(a) *Transverse section of the gametophyte of* Marchantia, *a thallose liverwort. Numerous chloroplast-bearing cells are evident in the upper layers, and there are several layers of colorless cells below them, as well as rhizoids that anchor the plant body to the substrate. Pores permit the exchange of gases in the air-filled chambers that honeycomb the photosynthetic dorsal layer. The specialized cells that surround each pore are usually arranged in four or five superimposed tiers of four, and the whole structure is barrel-shaped. Under dry conditions, the cells of the lowermost tier, which usually protrude into the cavity, become juxtaposed; while under moist conditions, they separate. Thus the pores serve a function similar to that of the stomata of vascular plants. (b) A scanning electron micrograph of a pore on the dorsal surface of a gametophyte of* Marchantia.

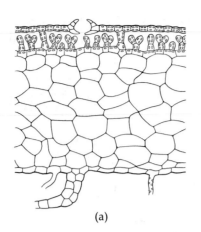

(a)

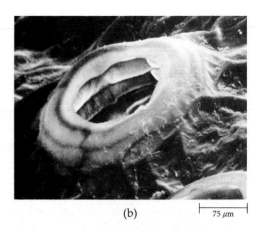

(b)   75 μm

stalks called *archegoniophores* (Figure 16–10). The life cycle of *Marchantia* is outlined in Figure 16–13. In this genus, the sporophyte generation consists of a foot, a short seta, and a capsule, or sporangium (see Figure 16–8). In addition to spores, the mature sporangium contains elongate cells called *elaters*, which have helically arranged hygroscopic (moisture-absorbing) wall thickenings (Figures 16–8 and 16–11; see also the boxed essay on page 287). The walls of these cells are sensitive to slight changes in humidity and produce a twisting action that aids in spore dispersal after the capsule first dehisces into a number of petal-like segments.

Fragmentation is the principal means of asexual reproduction in liverworts. Another widespread means of asexual reproduction in both liverworts and mosses is the production of *gemmae*—multicellular bodies that can give rise to new gametophytes. In *Marchantia*, the gemmae are produced in special cuplike structures—called gemma cups—located on the dorsal surface of the gametophyte (Figure 16–12). The gemmae are dispersed primarily by splashing out in the rain.

(a)

(b)

16–10

Gametophytes of Marchantia. *The antheridia (a) and archegonia (b) are elevated on stalks above the thallus.*

16–11

Mature spores and elaters from a capsule of Marchantia.

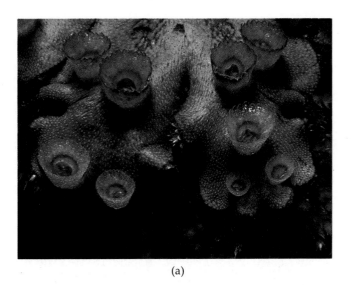

(a)

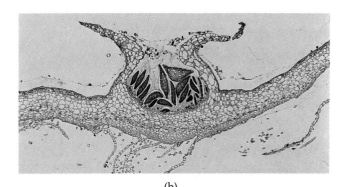

(b)

16–12
(a) *Gametophytes of* Marchantia, *with cup-shaped gemma cups containing gemmae, which are splashed out by the rain and may then grow into new gametophytes, each genetically identical to the parent plant from which they were derived by mitosis.* (b) *Longitudinal section of a gemma cup.*

SPORE DISCHARGE IN LIVERWORTS

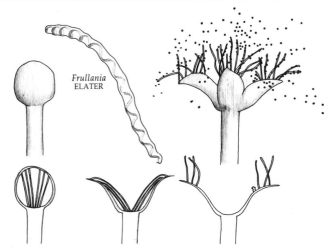

Frullania
ELATER

Frullania SPORANGIUM, UPON DRYING, PEELS OPEN IN FOUR SECTIONS. THE ELATERS, ONCE ATTACHED AT BOTH ENDS, SNAP LOOSE FROM THE CENTER OF THE SPORANGIUM, SPREADING THEIR SPORES.

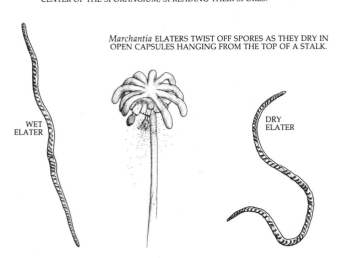

Marchantia ELATERS TWIST OFF SPORES AS THEY DRY IN OPEN CAPSULES HANGING FROM THE TOP OF A STALK.

WET
ELATER

DRY
ELATER

Three methods of spore discharge in liverworts. Most liverworts are similar to *Cephalozia* in their method of spore discharge, whereas some resemble *Marchantia*. The method employed by *Frullania* is less common.

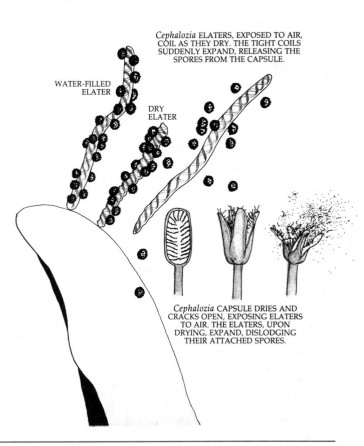

Cephalozia ELATERS, EXPOSED TO AIR, COIL AS THEY DRY. THE TIGHT COILS SUDDENLY EXPAND, RELEASING THE SPORES FROM THE CAPSULE.

WATER-FILLED
ELATER

DRY
ELATER

Cephalozia CAPSULE DRIES AND CRACKS OPEN, EXPOSING ELATERS TO AIR. THE ELATERS, UPON DRYING, EXPAND, DISLODGING THEIR ATTACHED SPORES.

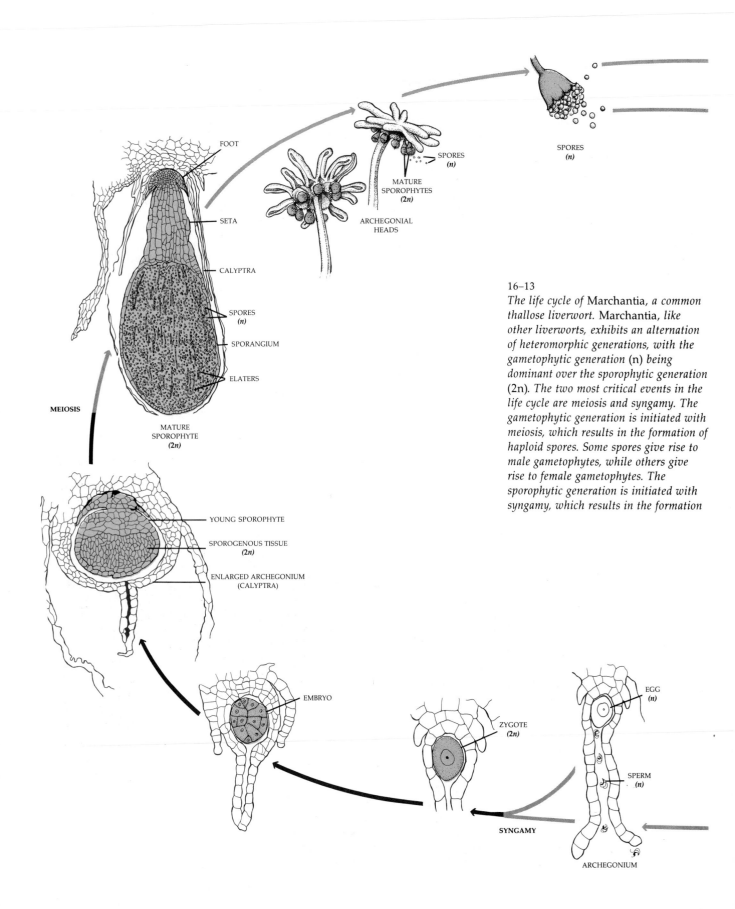

FOOT

SETA

CALYPTRA

SPORES
(n)

SPORANGIUM

ELATERS

MEIOSIS

MATURE
SPOROPHYTE
(2n)

YOUNG SPOROPHYTE

SPOROGENOUS TISSUE
(2n)

ENLARGED ARCHEGONIUM
(CALYPTRA)

SPORES
(n)

MATURE
SPOROPHYTES
(2n)

ARCHEGONIAL
HEADS

SPORES
(n)

EMBRYO

ZYGOTE
(2n)

EGG
(n)

SPERM
(n)

SYNGAMY

ARCHEGONIUM

16–13
The life cycle of Marchantia, *a common
thallose liverwort.* Marchantia, *like
other liverworts, exhibits an alternation
of heteromorphic generations, with the
gametophytic generation* (n) *being
dominant over the sporophytic generation
(2n). The two most critical events in the
life cycle are meiosis and syngamy. The
gametophytic generation is initiated with
meiosis, which results in the formation of
haploid spores. Some spores give rise to
male gametophytes, while others give
rise to female gametophytes. The
sporophytic generation is initiated with
syngamy, which results in the formation*

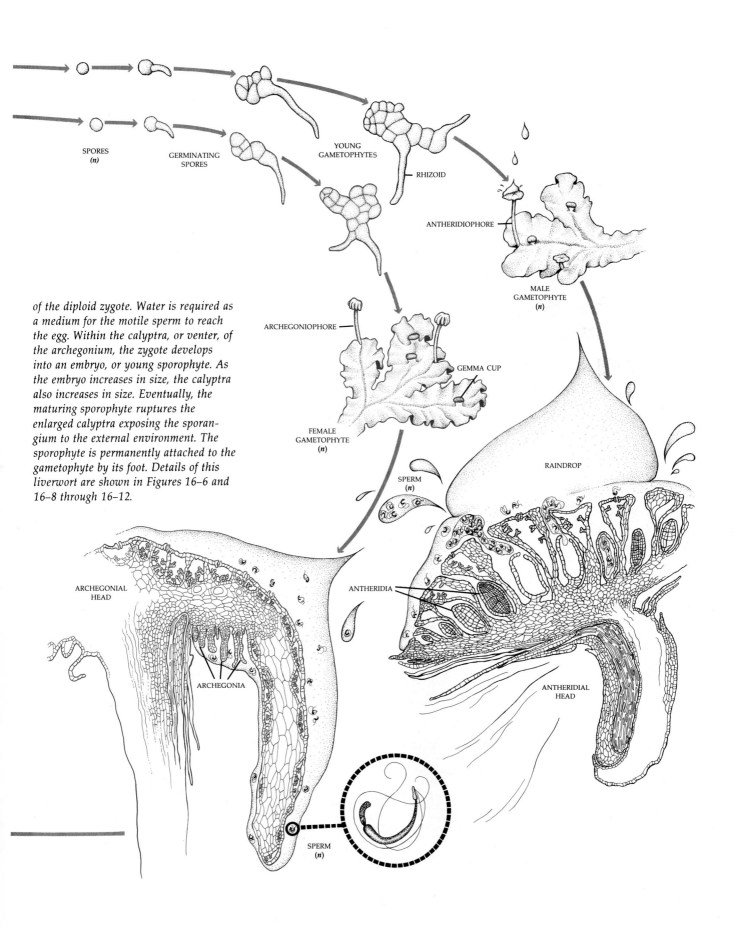

SPORES
(n)

GERMINATING
SPORES

YOUNG
GAMETOPHYTES

RHIZOID

ANTHERIDIOPHORE

MALE
GAMETOPHYTE
(n)

ARCHEGONIOPHORE

GEMMA CUP

FEMALE
GAMETOPHYTE
(n)

of the diploid zygote. Water is required as a medium for the motile sperm to reach the egg. Within the calyptra, or venter, of the archegonium, the zygote develops into an embryo, or young sporophyte. As the embryo increases in size, the calyptra also increases in size. Eventually, the maturing sporophyte ruptures the enlarged calyptra exposing the sporangium to the external environment. The sporophyte is permanently attached to the gametophyte by its foot. Details of this liverwort are shown in Figures 16–6 and 16–8 through 16–12.

SPERM
(n)

RAINDROP

ARCHEGONIAL
HEAD

ANTHERIDIA

ARCHEGONIA

ANTHERIDIAL
HEAD

SPERM
(n)

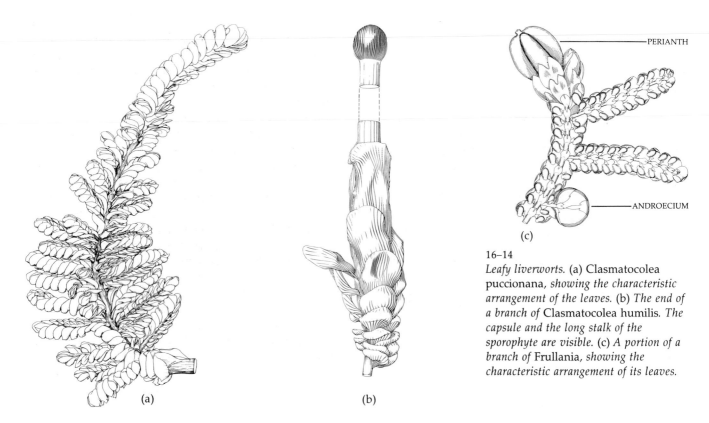

PERIANTH

ANDROECIUM

(a)   (b)   (c)

16–14
Leafy liverworts. (a) Clasmatocolea
puccionana, *showing the characteristic
arrangement of the leaves.* (b) *The end of
a branch of* Clasmatocolea humilis. *The
capsule and the long stalk of the
sporophyte are visible.* (c) *A portion of a
branch of* Frullania, *showing the
characteristic arrangement of its leaves.*

Leafy Liverworts

The leafy liverworts are a diverse group that includes more than 4000 of the 6000 known species of the class (Figure 16–14). The leafy liverworts are especially abundant in the tropics and subtropics, in regions of heavy rainfall or high humidity (Figure 16–15), but they are also present in large numbers in temperate regions. The plants are usually well branched and form small mats.

The leaves of liverworts, like those of mosses, generally consist of only a single layer of nondifferentiated cells. The leaves of many genera are arranged in two rows, with a third row of reduced leaves along the lower surface of the gametophyte. The leaves are often two-lobed, and each grows by means of two distinct apical growing points. In *Frullania*, a common liverwort that grows on bark, the leaves consist of a large, undivided dorsal lobe and a small helmet-shaped ventral lobe (Figure 16–14c).

In the leafy liverworts, the antheridia generally occur in a packetlike swelling, the _androecium_, which develops on the lower portion of a modified leaf. The developing sporophyte, as well as the archegonium from which it develops, are characteristically surrounded by a tubular sheath, the _perianth_ (Figure 16–14c).

16–15
*A leafy liverwort growing on the leaf of
an evergreen tree in the rainforest of the
Amazon Basin, near Manaus, Brazil.*

THE HORNWORTS: CLASS ANTHOCEROTAE

The class Anthocerotae consists of only about 100 species in six genera, the most familiar being *Anthoceros*, which is distributed worldwide and generally occurs in moist, shaded habitats. The gametophyte of *Anthoceros* superficially resembles that of a thallose liverwort (Figure 16–16a), but there are many features that indicate a relatively distant relationship. For example, each cell usually has a single large chloroplast, like those of many algae, rather than the many small, discoid chloroplasts found in the cells of all other plants. Each chloroplast contains a pyrenoid, making the resemblance to algae even more striking. The gametophytes have a strong dorsiventral orientation; they are often rosettelike and are usually about 1 to 2 centimeters across. Individuals of *Anthoceros* have extensive internal cavities filled with mucilage, rather than with air, as in the gametophytes of thallose liverworts. These mucilage-filled cavities are often inhabited by cyanobacteria of the genus *Nostoc*, which fix nitrogen and supply it to their host plants.

The gametophytes of some species of *Anthoceros* are unisexual; others are bisexual. The antheridia and archegonia are sunken on the dorsal surface of the gametophyte, with the antheridia clustered in chambers. Numerous sporophytes may develop on the same gametophyte.

The sporophyte of *Anthoceros*, which is an upright elongated structure, consists of a foot and a long cylindrical sporangium (Figures 16–16b and 16–17a). Very early in its development, a meristem, or zone of actively dividing cells, develops between the foot and sporangium; this meristem remains active as long as conditions are favorable for growth. As a result, the sporophyte continues to elongate for a prolonged period of time. It is green, having several layers of photosynthetic cells. It is also covered with a cuticle and has stomata (Figure 16–16c). Maturation of the spores and, ultimately, dehiscence of the sporangium begins near its tip and spreads toward its base as the spores mature (Figure 16–16d, e). Among the spores are sterile, elongate structures, often multicellular, that resemble the liverwort elaters (Figure 16–17b). The dehiscing sporangium splits longitudinally into ribbonlike halves.

(a)

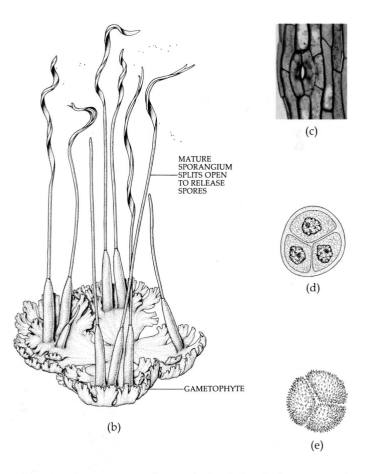

MATURE SPORANGIUM SPLITS OPEN TO RELEASE SPORES

(c)

(d)

GAMETOPHYTE

(b)

(e)

16–16

Anthoceros, a hornwort. (a) A gametophyte with attached sporophytes. (b) When mature, the sporangium splits and the spores are released. (c) A stoma; stomata are abundant on the sporophytes of the hornworts, which are green and photosynthetic. (d) Developing spores and (e) mature spores.

16–17
Anthoceros. (a) *Longitudinal section of a sporophyte, showing its foot embedded in the tissue of the gametophyte.* (b) *Longitudinal section of a sporangium, showing tetrads of spores with elaterlike structures among them. The column in the central lower part of the sporangium consists of tissue that may function in conduction.*

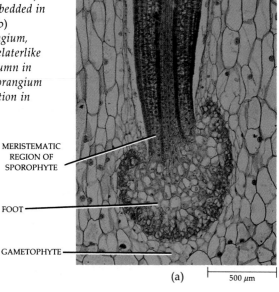

MERISTEMATIC
REGION OF
SPOROPHYTE

FOOT

GAMETOPHYTE

(a)   ⊢——⊣ 500 μm

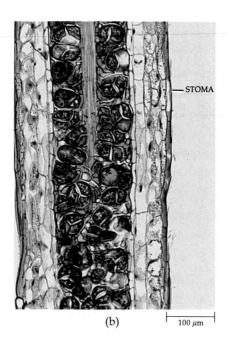

STOMA

(b)   ⊢——⊣ 100 μm

THE MOSSES: CLASS MUSCI

Many groups of plants contain members that are commonly called "mosses" (reindeer "mosses" are lichens, club "mosses" and Spanish "moss" are vascular plants, and sea "moss" and Irish "moss" are algae), but the genuine mosses are members of the class Musci, which consists of three subclasses: Bryidae (the "true" mosses), Sphagnidae (the peat mosses), and Andreaeidae (the granite mosses).

The True Mosses

The gametophytes of all mosses are represented by two distinct phases: the *protonema* (plural: *protonemata*, from the Greek words *protos,* "first," and *nema,* "thread"), which arises directly from a germinating spore, and the *leafy gametophyte.* In the true mosses, the cells of the protonema occur in a single layer, and the branching filament resembles a filamentous green alga (Figure 16–18). Leafy gametophytes develop from minute budlike structures on the branches of the protonemata. In a few genera of mosses the protonema is persistent and assumes the major photosynthetic role, while the leafy shoots of the gametophyte are minute. Protonemata, which are characteristic of all mosses, are also found in some liverworts.

In the true mosses, the gametophyte is leafy and usually upright, rather than dorsiventrally flattened as it is in the leafy liverworts. It grows from an apical initial cell, resembling an inverted, three-sided pyramid, that is similar to that of the leafy liverworts. Although three ranks of leaves are produced initially, subsequent twisting of the axis results in displacement of these ranks to give the appearance of a spiral arrangement. In some genera (*Fontinalis,* an aquatic moss, for example) the original, three-ranked condition of the leaves is still obvious in the mature gametophytes.

Moss gametophytes, which exhibit varying degrees of complexity, range from as little as 0.5 millimeter to 50 centimeters or more long. All have multicellular rhizoids, and the leaves are normally only one cell layer thick except at the midrib (which is lacking in some genera). As mentioned above, many mosses have a central strand of water-conducting hydroids in the stem, and some also have food-conducting leptoids.

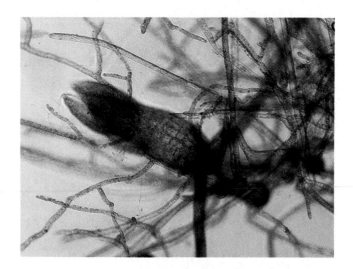

16–18
Protonema of a moss with a budlike structure. Protonemata are the first stage of the gametophyte generation of mosses and some liverworts. They often resemble filamentous green algae.

Two patterns of growth are common among moss gametophytes (Figure 16–19). In the first, the "cushiony" mosses, the gametophytes are erect and little-branched, usually bearing terminal sporophytes. In the second, the gametophytes are much-branched and "feathery," and the plants are creeping; the sporophytes are borne laterally. This second type of growth pattern is found in those mosses that hang in masses from the branches of trees in moist regions.

At maturity, gametangia are produced by most leafy gametophytes, either at the tip of the main axis or on a lateral branch. In some genera the gametophytes are unisexual (Figure 16–20), while in other genera both archegonia and antheridia are produced by the same plant.

Moss sporophytes, which usually take 6 to 18 months to reach maturity in temperate species, are borne on the gametophytes (see Figure 16–3). The capsules are generally elevated on a seta, which may reach 15 to 20 centimeters in length in a few species; some mosses lack a seta entirely. A short foot at the base of the seta is embedded in the tissue of the gametophyte, and the seta usually elongates early in the development of the sporophyte. Moss sporophytes are able to photosynthesize when they emerge from the gametophytic tissue. Typically, they are therefore less dependent on the gametophyte nutritionally than are liverwort sporophytes, which generally remain enclosed within the gametophyte. Stomata, which are lacking on liverwort sporophytes, are normally present on those of mosses. Moss sporophytes have a high degree of internal organization. As mentioned previously, the setae of most have a central strand of hydroids and many also have food-conducting leptoids (see Figure 16–5).

When a moss sporophyte is mature, it gradually loses its ability to photosynthesize and turns yellow, then orange, and finally brown. Eventually the lid, or _operculum_, of the capsule bursts off, revealing a ring of teeth—the _peristome_—surrounding the opening (Figure 16–21). The teeth of the peristome are formed by the splitting of a cellular layer near the apex of the capsule along a zone of weakness. The peristome plays a role in regulating spore discharge. The peristome is a characteristic of the subclass Bryidae and is lacking in the other two subclasses. Each capsule sheds up to 50 million haploid spores, each of them capable of giving rise to a new gametophyte! A representative life cycle of a true moss is shown in Figure 16–22.

Asexual reproduction in the true mosses usually takes place by fragmentation. Virtually any part of the gametophyte is capable of regeneration, including the sterile parts of the sex organs. Many species produce gemmae, which are able to give rise to new gametophytes.

(a)

(b)

16–19
The two common growth forms found in the gametophytes of different genera of mosses. (a) "Cushiony" form, in which the gametophytes are erect and have few branches, shown here in Polytrichum juniperinum. _Spore capsules atop long, slender seta—sporophytes—can be seen rising above the gametophytes. (b) "Feathery" form, with matted, creeping gametophytes, shown here in_ Thuidium delicatulum.

16–20

Gametangia of Mnium, *a moss. (a)*
Longitudinal section through archegonial
head showing the pink-stained arche-
gonia surrounded by sterile structures
called paraphyses. (b) Longitudinal
section through antheridial head showing
antheridia surrounded by paraphyses.

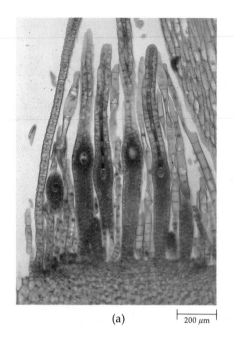

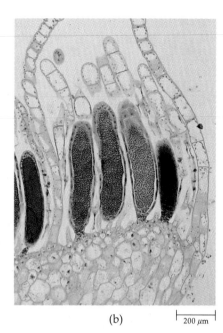

(a) 200 μm

(b) 200 μm

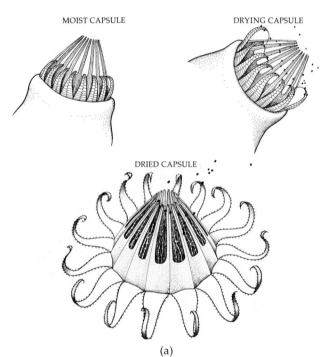

MOIST CAPSULE

DRYING CAPSULE

DRIED CAPSULE

(a)

(b)

16–21

Peristome teeth in true mosses. (a)
Brachythecium *has a peristome consisting*
of two rings of teeth, which open to
release the spores in response to changes
in moisture. The outer set of peristome
teeth interlock with the inner set under
damp conditions. As the capsule dries out,
the outer teeth pull away, allowing the
dispersal of spores by the wind.
(b) Scanning electron micrographs of the
peristome teeth of Orthotrichum,
showing the inner teeth curved inward
and the outer teeth curved outward in dry
conditions.

The Peat Mosses

The approximately 350 species of peat mosses, all belonging to the genus *Sphagnum*, constitute a very distinct group, which clearly diverged from the main line
of moss evolution a very long time ago (Figure 16–23).
The stems of the gametophyte of *Sphagnum* bear clusters of branches, often five at a node, which are more
densely tufted near the apex of the stem, forming a
moplike head. These gametophytes form large, bright
green or occasionally reddish clumps in boggy ground.
The gametophytes of *Sphagnum* arise from a protonema
that is platelike instead of filamentous and forms buds
along its margins, from which the leafy gametophytes

INSECT DISPERSAL OF MOSS SPORES

The angiosperms are distinctive among the plants in the nature and strength of their relationships with insects. In one group of mosses, however, a number of structural and physiological features have evolved that enable these plants to take advantage of insects to spread their haploid spores also. The parallelisms between these mosses and the flowers of the angiosperms are fascinating.

The unusual genus of mosses that is involved in these relationships, *Splachnum,* grows on dung or carrion, especially that which is dropped into moist situations such as bogs. The mosses form dense cushions up to 25 centimeters across, above which the capsules are borne on thick, translucent stalks. From the neck of the capsule develops a swollen ring of tissue, called the apophysis, which is purple, yellow, red, or otherwise colored, and may be up to 2 centimeters across. The spores of these mosses are sticky, unlike those of other mosses, and they are spread by flies, which are attracted by their color and the odors that are secreted by the apophysis in some species.

The flies that visit these mosses also visit and lay their eggs on fresh dung, the habitat where the mosses grow. By doing so, they spread the mosses to new habitats. It is not known whether individual species of *Splachnum*—there may be up to six species in a single locality—are attractive to different dung flies, but studies on this and other aspects of their unique adaptive systems are currently underway.

(a)

(b)

(c)

(a) Splachnum luteum. *With their prominent cream-colored apophyses, plants of this species resemble small flowering plants.* (b) *The colors of the plants themselves, as well as the chemicals they produce, attract flies.* (c) Splachnum rubrum, *which occurs on moose dung, shown here growing in Alberta.*

arise. The leaves of *Sphagnum* lack midribs, and the mature plants lack rhizoids. In the boggy places where they grow, *Sphagnum* plants are nearly always turgid and thus erect; they are also usually densely packed together. The distinctive leaves of this genus consist of large, dead cells surrounded by narrow green, or occasionally red-pigmented, living cells (Figure 16–23b). The dead cells contain pores and thickenings and readily become filled with water, so that the water-holding capacity of the peat mosses is up to 20 times its dry weight. By comparison, cotton absorbs only 4–6 times its dry weight.

As a result of their superior absorptive qualities, *Sphagnum* mosses were used in Europe from the 1880s onward as dressings for wounds and boils; however, cotton dressings have been used almost exclusively for such purposes since World War I, probably because of their cleaner appearance. Gardeners mix peat moss with soil to increase its water-holding capacity and make it more acid.

Sphagnum sporophytes (Figure 16–23a) are also distinctive. The red to blackish-brown capsules are nearly spherical and are raised on a stalk, the *pseudopodium,* which is part of the gametophyte and may be up to 3

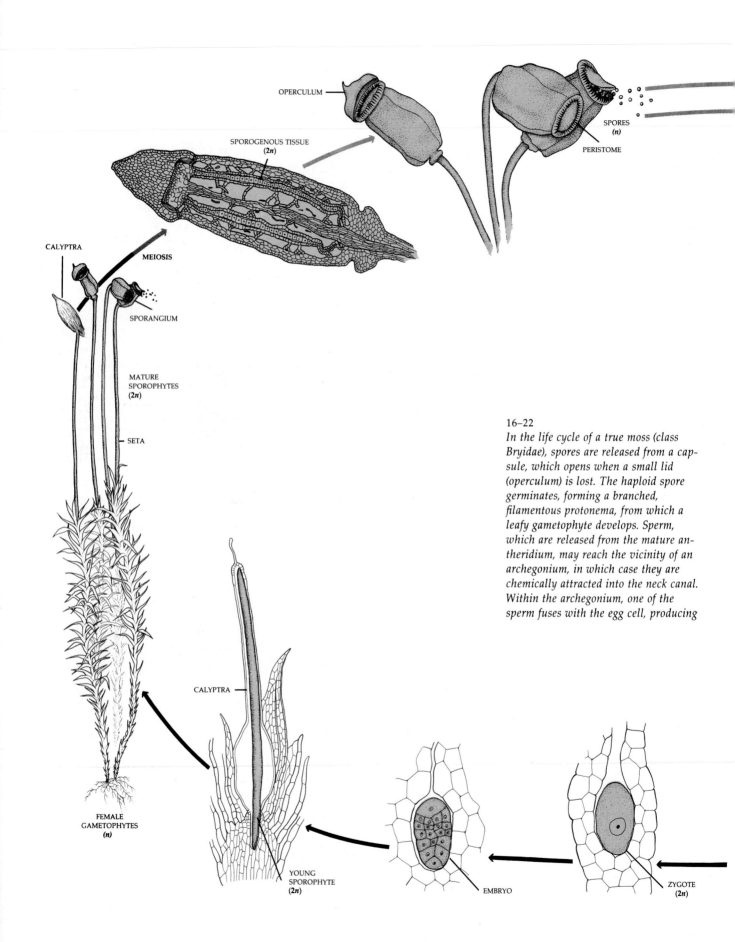

OPERCULUM

SPORES
(n)

PERISTOME

SPOROGENOUS TISSUE
(2n)

CALYPTRA

MEIOSIS

SPORANGIUM

MATURE
SPOROPHYTES
(2n)

SETA

CALYPTRA

FEMALE
GAMETOPHYTES
(n)

YOUNG
SPOROPHYTE
(2n)

EMBRYO

ZYGOTE
(2n)

16–22
In the life cycle of a true moss (class Bryidae), spores are released from a capsule, which opens when a small lid (operculum) is lost. The haploid spore germinates, forming a branched, filamentous protonema, from which a leafy gametophyte develops. Sperm, which are released from the mature antheridium, may reach the vicinity of an archegonium, in which case they are chemically attracted into the neck canal. Within the archegonium, one of the sperm fuses with the egg cell, producing

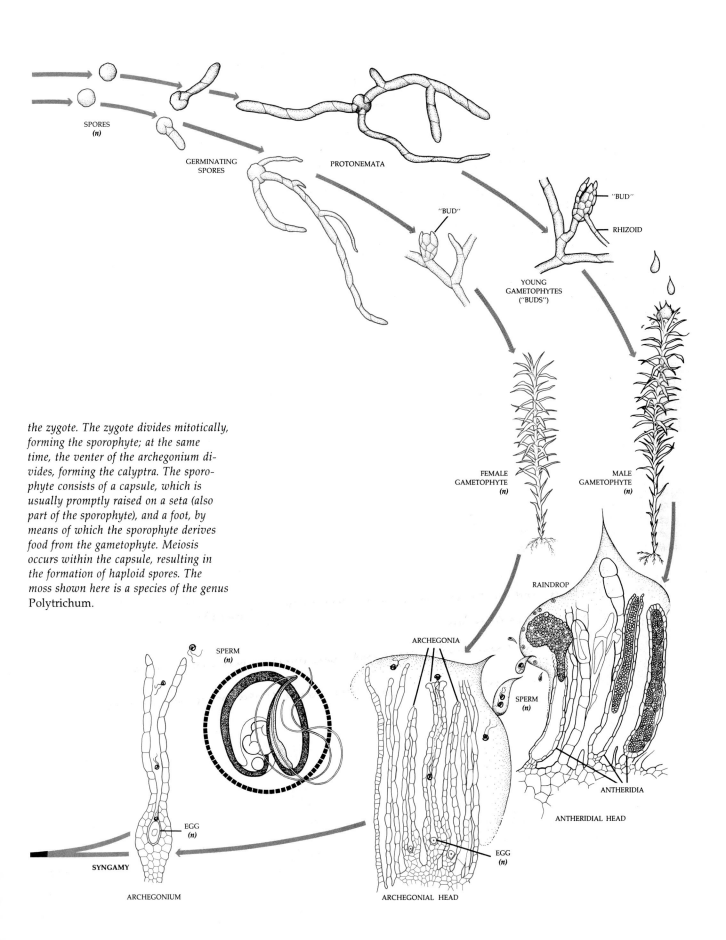

SPORES
(n)

GERMINATING
SPORES

PROTONEMATA

"BUD"

"BUD"

RHIZOID

YOUNG
GAMETOPHYTES
("BUDS")

FEMALE
GAMETOPHYTE
(n)

MALE
GAMETOPHYTE
(n)

RAINDROP

the zygote. The zygote divides mitotically,
forming the sporophyte; at the same
time, the venter of the archegonium divides, forming the calyptra. The sporophyte consists of a capsule, which is
usually promptly raised on a seta (also
part of the sporophyte), and a foot, by
means of which the sporophyte derives
food from the gametophyte. Meiosis
occurs within the capsule, resulting in
the formation of haploid spores. The
moss shown here is a species of the genus
Polytrichum.

SPERM
(n)

ARCHEGONIA

SPERM
(n)

ANTHERIDIA

ANTHERIDIAL HEAD

EGG
(n)

SYNGAMY

ARCHEGONIUM

EGG
(n)

ARCHEGONIAL HEAD

16–23

A peat moss, Sphagnum. (a) A gameto-phyte, with many attached sporophytes. Some of the capsules, such as the two on the left, have already discharged their spores. (b) Structure of a leaf. The interior, consisting of large, dead cells, is surrounded by smaller living cells, rich in chloroplasts. (c) Dehiscence of a capsule. As the capsule dries, it is filled with air. The stomata, through which the air enters, close in the process of drying; with further shrinking, the pressure blows off the capsule's operculum, re-leasing a gaseous cloud of spores.

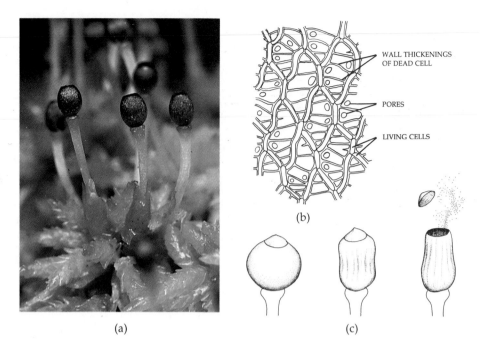

WALL THICKENINGS OF DEAD CELL

PORES

LIVING CELLS

(b)

(a)

(c)

millimeters long. The sporophyte itself has a very short seta. Spore discharge in *Sphagnum* is spectacular (Figure 16–23c). At the top of the capsule is a disk-shaped operculum, separated from the rest of the capsule by a circular groove. As the capsule matures, its internal tissues shrink and air is drawn in, presumably through the stomata, which cannot close. When the capsule wall dries, the air is trapped within. Contraction of the maturing capsule results in increased internal pressure, and the operculum is eventually blown off with an audible click. This happens in warm, sunny conditions. When the operculum is blown off, the escaping gas carries a cloud of spores out of the capsule. The most distinctive features of the subclass Sphagnidae are the lack of a peristome and the peculiar morphology of the gametophyte.

Ecology of Sphagnum

The peat mosses form extensive peat bogs, which are a conspicuous feature of cold and temperate regions throughout the world. The mosses contribute to the acidity of their own environment by releasing H^+ ions; the pH in the center of the bogs is often less than 4—very unusual for a natural situation. Peat is formed from the accumulation and compression of the mosses themselves, as well as the sedges, reeds, grasses, and other plants that grow among them. In Ireland and some other northern regions, dried peat is widely used as fuel, as well as for domestic heating. At a conservative estimate, peat bogs cover at least 1 percent of the world's land surface, an enormous area equivalent to that of half of the United States. Because the world supply of peat is so great, it is being considered increasingly as a possible major source of industrial energy, although the effects on the regions where the peat will be dug, as well as the effects of releasing so much carbon dioxide into the atmosphere by burning it, need careful consideration.

The Granite Mosses

The genus *Andreaea* consists of about 100 species of small, blackish-green or dark reddish-brown tufted rock mosses (Figure 16–24), which in their own way are as peculiar as *Sphagnum*. The members of this group occur in mountainous or arctic regions, often on granite rocks—which is the reason for their common name. Although the gametophyte closely resembles that of the true mosses, it arises from a thick structure with many lobes, rather than a filamentous one. The sporophyte lacks a true seta and is elevated above the leaves on a stalk of gametophytic tissue, or pseudopodium, as in the peat mosses. The minute capsules of *Andreaea* are marked by four vertical lines of weaker cells along which the capsule splits. The capsule remains intact above and below the dehiscence lines. The resulting four valves are very sensitive to the humidity of the surrounding air, opening widely when it is dry—the spores can be carried widely by the wind under such circumstances—and closing when it is moist. This mechanism of spore discharge, by means of slits in the capsule, is different from that of any other moss (see Figure 16–24).

Andreaeobryum, a second genus of granite mosses, was discovered in Alaska in 1976. It differs from *Andreaea* in that its sporophyte has a true seta and its capsule splits to the apex.

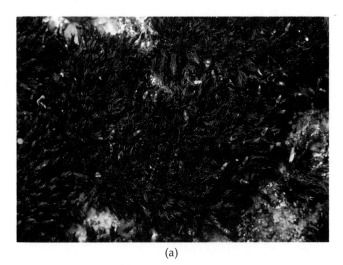

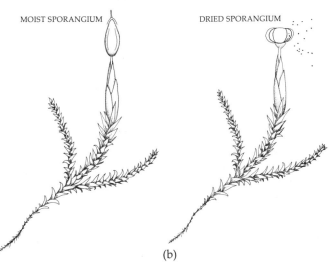

MOIST SPORANGIUM   DRIED SPORANGIUM

(b)

16–24
(a) Andreaea rothii *growing on granite
rock in Devon, England.* (b) *The*
Andreaea *sporangium contracts and
splits into four parts as it dries out,
allowing the spores to fall out under dry
conditions.*

SUMMARY

Bryophyta are a distinct division of plants, which is divided
into three classes: Hepaticae (liverworts), Anthocerotae
(hornworts), and Musci (mosses). In their pigments and food
reserves, they resemble both green algae and vascular plants.
Some hornworts and most mosses have stomata on their
sporophytes, as do vascular plants, and some mosses and liverworts have a cuticle on their leaves. Although many bryophytes lack specialized vascular strands, absorbing water
directly through the leaves and stems, many others have
strands of hydroids (water-conducting cells) and some also
have leptoids (food-conducting cells). Bryophytes and vascular plants probably share a remote common ancestor.

All bryophytes have a well-defined alternation of heteromorphic generations in which the gametophyte is the dominant generation. The male sex organs, antheridia, and female
sex organs, archegonia, are multicellular; both have sterile

jacket layers. Each archegonium contains a single egg. Numerous sperm are produced by each antheridium; the sperm
are free-swimming and biflagellated. The sporophyte develops within the archegonium but usually grows out of it in
mosses and hornworts, although remaining permanently attached to it. In mosses and some liverworts, the sporophyte is
differentiated into a foot, a seta, and a capsule, or sporangium. Hornwort sporophytes are unique in having a basal
meristem, which adds tissue to the sporangium over a prolonged period. Hornwort and moss sporophytes become nutritionally independent of the gametophyte. On the other
hand, liverwort sporophytes, most of which remain surrounded by gametophyte tissue at maturity, usually are completely dependent on the gametophyte for their food supply.
Upon germinating, the spores of mosses produce a filamentous or platelike gametophyte known as a protonema, from
which leafy gametophytes usually arise. Protonemata are
found in some liverworts but are lacking in hornworts.

SUGGESTIONS FOR FURTHER READING

CONARD, HENRY S., and PAUL L. REDFEARN, JR.: *How to Know
the Mosses and Liverworts,* 2nd ed., William C. Brown Co.,
Dubuque, Iowa, 1979.*

*A good beginner's guide for use in identifying many of the more
common bryophytes. It includes a profusely illustrated key and an
excellent glossary.*

CRANDALL-STOTLER, BARBARA: "Morphogenetic Designs and a
Theory of Bryophyte Origins and Divergence," *BioScience* 30:
580–85, 1980.

*A clear exposition of the reasons for regarding the three main
groups of bryophytes as only distantly related to one another.*

FLOWERS, SEVILLE: *Mosses: Utah and the West,* Brigham Young
University Press, Provo, Utah, 1973.

*An excellent local flora, with outstanding introduction, covering
the morphology and ecology of mosses, methods of collecting and
preserving them, and many other matters of interest.*

PAOLILLO, DOMINICK J.: "The Swimming Sperms of Land
Plants," *BioScience* 31:367–73, 1981.

*A fascinating account of the similarities and differences between
the sperm of different groups of plants.*

RICHARDSON, D.H.S.: *The Biology of Mosses,* John Wiley & Sons,
Inc., New York, 1981.*

*A well-written and well-illustrated short book on the mosses,
covering all aspects of the group.*

SCHUSTER, RUDOLF M.: *New Manual of Bryology,* Vols. 1 and 2,
The Hattori Botanical Laboratory, Nichinan, Miyazaki, Japan,
1983–84.

*An outstanding overview of all aspects of current bryological re-
search.*

SMITH, A.J.E. (Ed.): *Bryophyte Ecology,* Chapman and Hall, New
York, 1982.

*An excellent collection of articles describing current research on
bryophyte ecology.*

* Available in paperback.

C H A P T E R 1 7

Seedless Vascular Plants

Plants, like all living things, had aquatic ancestors, and the story of their evolution is inseparably linked with their progressive occupation of the land. One of the key events in their early invasion of the land was the development of spores with durable protective walls that allowed them to tolerate dry conditions. This permitted the dispersal of the spores over the surface of the land by wind. As plant bodies became larger, structures that allowed for more efficient release and dispersal of the spores evolved. An additional important evolutionary advance that accompanied this increase in size was the origin of cutin, a fatty substance that protects the plant body from excessive evaporation. Stomata must have evolved at the same time as the cuticle to allow the exchange of gases by these early plants.

In the vascular plants, the evolution of efficient conducting systems, consisting of xylem and phloem, solved the problem of water and food conduction on the land. Many scientists believe that the evolution of the ability to synthesize lignin, which was incorporated in the cells that made up systems for support and conduction, was a pivotal step in the evolution of land plants. Eventually, the belowground parts of the sporophyte evolved into roots, which function in absorption and anchorage, and portions of the aboveground parts evolved into leaves. This morphological differentiation stabilized the early plants in their environment and permitted efficient photosynthesis. Meanwhile, the gametophytic generation underwent a progressive reduction in size and became increasingly more protected by and nutritionally dependent on the sporophyte. Finally, a number of lines developed seeds—structures that protect the embryonic sporophyte from the rigors of a life on land, nourish it, and enable it to withstand unfavorable conditions.

As discussed in Chapter 16, the common ancestor of the different groups of bryophytes and vascular plants was probably a relatively complex, multicellular green alga (see Figure 15–21, page 267) that invaded the land about 450 million years ago, in the Ordovician period,

17–1

Simple primitive vascular plants had emerged from the water to colonize the barren landscape during the early Devonian period, some 408 to 387 million years ago. Shown here, reading from left to center, are Cooksonia, Rhynia, *and* Zosterophyllum, *all of which lacked leaves. By the beginning of the mid-Devonian period (387 to 374 million years ago), a more diverse plant community had evolved, with larger and more complex types of plants. Examples seen here on the right, reading from back to front, are* Psilophyton, *a robust trimerophyte;* Drepanophycus, *an early lycophyte; and* Protolepidodendron, *a lycophyte. The leaves of the lycophytes were microphylls. All of these early vascular plants produced spores.*

probably as part of an endomycorrhizal relationship (see page 226). An alga resembling *Coleochaete* (Figure 15–20, page 266) appears to be the best living model for this ancestor. Individuals of *Coleochaete* are haploid except for the zygote (they have zygotic meiosis). Many scientists now favor the hypothesis that the sporophytic generation, as it exists in plants, evolved independently from the diploid phase that is typical of many algal life cycles. Additional groups of vascular plants are found in the Silurian period (438 to 408 million years ago); for example, well-preserved strands that are evidently vascular tissue are known from North American rocks as much as 430 million years old. Vascular plants became numerous and diverse by the Devonian period (408 to 360 million years ago; Figure 17–1).

Because of their adaptations for existence on land, the vascular plants are dominant in terrestrial habitats. There are nine divisions with living representatives, including about 250,000 living species. In addition, there are several divisions that consist entirely of extinct vascular plants. In this chapter, we shall describe some of the evolutionary advances of the vascular plants and discuss seven divisions of seedless vascular plants, three of them extinct. In Chapter 18, we shall discuss the seed plants, which include five divisions with living representatives.

ORGANIZATION OF THE VASCULAR PLANT BODY

The sporophytes of early vascular plants were dichotomously branched axes that lacked leaves and roots. With evolutionary specialization, morphological and physiological differences arose between various parts of the plant body, bringing about the differentiation of roots, stems, and leaves—the organs of the plant (Figure 17–2). Collectively, the roots make up the *root system,* which anchors the plant and absorbs water and minerals from the soil. The stems and leaves together make up the *shoot system,* with the stems raising the specialized photosynthetic organs—the leaves—to-

17–2

A diagram of a young sporophyte of the club moss Lycopodium clavatum, *which is still attached to its subterranean gametophyte. The tissues of each organ are shown in transverse sections of (a) leaf, (b) stem, and (c) root. In all three organs, the dermal tissue system is represented by the epidermis, and the vascular tissue system, consisting of xylem and phloem, is embedded in the ground tissue system. The ground tissue in the leaf—in* Lycopodium, *a microphyll—is represented by the cortex, which surrounds a solid strand of vascular tissue, or protostele. The leaf is specialized for photosynthesis; the stem for support of the leaves and for conduction; and the root for absorption and anchorage.*

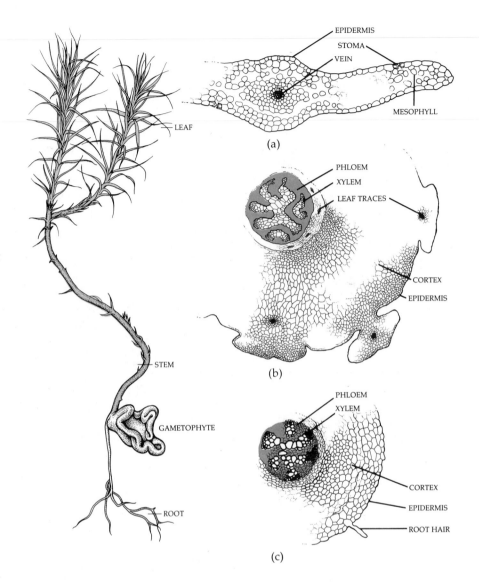

ward the sun and the vascular system conducting water and minerals to the leaves and the end products of photosynthesis away from them.

The different kinds of cells of the plant body are organized into tissues, and the tissues are organized into still larger units called tissue systems. Three tissue systems—*dermal*, *vascular*, and *ground tissues*—occur in all organs of the plant; they are continuous from organ to organ and reveal the basic unity of the plant body. The dermal tissue system makes up the outer protective covering of the plant. The vascular tissue system comprises the conductive tissues—xylem and phloem—and is embedded in the ground tissue system (Figure 17–2). The principal differences in the structures of root, stem, and leaf lie primarily in the relative distribution of the vascular and ground tissue systems, as will be discussed in Section 5.

Primary and Secondary Growth

Primary growth may be defined as the growth that occurs relatively close to the tips of roots and stems. It is initiated by their apical meristems and is primarily involved with extension of the plant body. The tissues arising during primary growth are known as *primary tissues*; the part of the plant body composed of these tissues is called the *primary plant body*. Primitive vascular plants, and many modern vascular plants as well, consist entirely of primary tissues.

In addition to primary growth, many plants undergo additional growth that thickens the stem and root. Such growth is termed *secondary growth*; it results from the activity of lateral meristems, the *vascular cambium*, which produces *secondary vascular tissues*, *secondary xylem*, and *secondary phloem* (see Figure 23–6, page 449). The production of secondary vascular tissues is commonly supplemented by the activity of the *cork cambium*, which forms a *periderm*, composed mostly of cork tissue. The periderm replaces the epidermis as the dermal tissue system of the plant. The secondary vascular tissues and periderm make up the *secondary plant body*. Secondary growth appeared in the mid-Devonian period about 380 million years ago in several unrelated groups of vascular plants.

Tracheary Elements

Sieve elements, the conducting cells of the phloem, have soft walls and often collapse after they die, so they are rarely well preserved as fossils. In contrast, *tracheary elements,* the conducting cells of the xylem, have rigid, persistent walls and are usually well preserved. Because of their various wall patterns, the tracheary elements provide valuable clues to the interrelationships of the different groups of vascular plants.

In fossil vascular plants from the Silurian and Devonian periods, the tracheary elements are elongated cells with long tapering ends. Such tracheary elements, called *tracheids,* were the first type of water-conducting cell to evolve; they are the only type of water-conducting cell in most living seedless vascular plants and gymnosperms. Tracheids not only provide channels for the passage of water and minerals, but they also provide support for stems. Water-conducting cells are rigid, mostly because of the lignin in their walls. It was this rigidity that allowed plants to evolve an upright habit and eventually for some of them to become trees.

Tracheids are more primitive (less specialized) than *vessel members*—the principal water-conducting cells in angiosperms. Vessel members apparently evolved independently from tracheids in several groups of vascular plants, including the angiosperms, in the gnetophytes (page 351), in several unrelated species of ferns, and in certain species of *Selaginella* (a lycophyte) and *Equisetum* (a sphenophyte). The evolution of vessel members in these different groups provides an excellent example of convergent evolution—the independent development of similar structures by unrelated or only distantly related organisms (see "Convergent Evolution," page 443). The evolution of conducting cells in the brown algae provides another excellent example of this phenomenon.

Steles

The primary vascular tissues—primary xylem and primary phloem—and the pith, if present, make up the central cylinder, or *stele,* of the stem and root in the primary plant body. Several types of stele are recognized, among them the protostele, the siphonostele, and the eustele.

The *protostele*—the most primitive type of stele—consists of a solid strand of vascular tissue, in which the phloem either surrounds the xylem or is interspersed within it (Figures 17–2, 17–3a). It is found in the extinct groups of seedless vascular plants discussed below, as well as in the Psilophyta, Lycophyta, and in the juvenile stems of some other living groups. In addition, it is the type of stele found in most roots.

The *siphonostele,* or tubular stele, is characterized by a central column of ground tissue, the *pith,* which is surrounded by the vascular tissue (Figure 17–3b). The phloem may form only outside of the cylinder of xylem or on both sides of it. Most ferns have a siphonostele. In these siphonosteles, the departure from the stem of the vascular strands leading to the leaves—the *leaf traces*—generally is marked by gaps—*leaf gaps*—in the siphonostele (as in Figure 17–3c).

If the vascular cylinder consists of a system of discrete strands around a pith, as in *Equisetum* and in many gymnosperms and angiosperms, the stele is called a *eustele.* The leaf gaps of eusteles are not as distinct as those of siphonosteles (Figure 17–3d). Comparative studies of living and fossil vascular plants have suggested that the eusteles of seed plants evolved directly from fairly primitive protosteles of the sort found in early fossil gymnosperms. Eusteles and siphonosteles, in other words, appear to have evolved independently from protosteles, although the exact relationships are not yet clear.

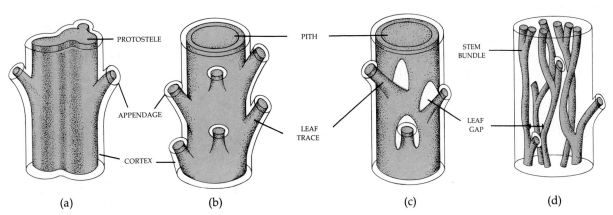

17–3
(a) *A protostele, with diverging appendages, the evolutionary precursors of leaves. (b) A siphonostele with no leaf gaps; the vascular traces leading to the* leaves simply diverge from the solid cylinder. This sort of siphonostele is found in Selaginella, *among other plants. (c) A siphonostele with leaf gaps, found in* many angiosperms and gymnosperms. (d) *A eustele. Siphonosteles and eusteles appear to have evolved independently from protosteles.*

Origins of Roots and Leaves

Although the fossil record reveals little information on the origins of roots as we know them today, they must have evolved from the lower, often subterranean, portions of the axis of primitive vascular plants. For the most part, roots are relatively simple structures that seem to have retained many of the primitive structural characteristics no longer present in the stems of modern plants.

Leaves are the principal lateral appendages of the stem. Regardless of their ultimate size or structure, they arise as protuberances (leaf primordia) from the apical meristem of the shoot. There are two fundamentally distinct kinds of leaves, microphylls and megaphylls.

Microphylls are usually relatively small leaves; they contain only a single strand of vascular tissue (Figure 17–4a). Microphylls are associated with stems possessing protosteles and are characteristic of the lycophytes. The leaf traces leading to microphylls are not associated with leaf gaps. Even though the name *microphyll* means "small leaf," some species of *Isoetes* have fairly long leaves. In fact, certain Carboniferous and Permian lycophytes had microphylls a meter or more in length.

Microphylls probably evolved as superficial lateral outgrowths of the stem (Figure 17–5). At first these structures were small scalelike or spinelike outgrowths devoid of vascular tissue. Gradually, rudimentary leaf traces developed, which initially extended only to the base of the outgrowth. Finally, the leaf traces extended into the appendage, resulting in formation of the primitive microphyll.

Most *megaphylls*, as the name implies, are larger than most microphylls. With few exceptions, megaphylls are associated with stems that have either siphonosteles or eusteles, and the leaf traces leading to them are associated with leaf gaps (Figure 17–4b). In contrast to the situation in microphylls, the blade, or *lamina*, of most megaphylls has a complex system of veins.

Megaphylls are believed to have evolved from entire branch systems by a series of steps similar to that shown in Figure 17–5. The earliest plants had a leafless, dichotomously branching axis; unequal branching resulted in more aggressive branches "overtopping" the weaker ones. This was followed with a flattening out, or "planation," of the subordinated lateral branches. The final step was fusion, or "webbing," of the separate lateral branches to form a primitive lamina.

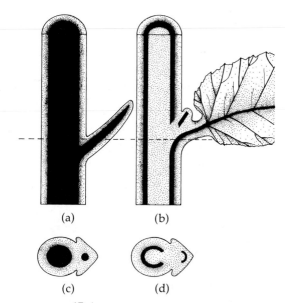

(a)   (b)

(c)   (d)

17–4

Longitudinal sections through a stem with a protostele and a microphyll (a), and one with a siphonostele and a megaphyll (b), emphasizing the nodes, or regions where the leaves are attached. (c) and (d) show transverse sections through these respective nodes, in the planes indicated by the dashed lines in (a) and (b). Note the presence of pith and leaf gap in (b) and (d) and their absence in (a) and (c). Microphylls are characteristic of lycophytes, while megaphylls are found in all other vascular plants.

17–5

According to one widely accepted theory, microphylls (above) evolved as outgrowths of the main axis of the plant. Megaphylls (below) evolved by fusion of branch systems.

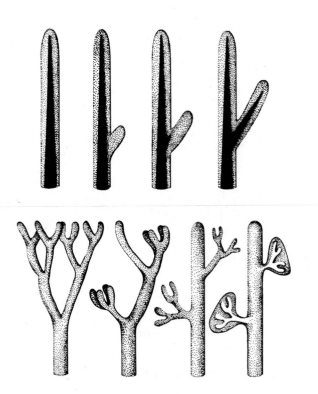

REPRODUCTIVE SYSTEMS

All vascular plants are oogamous and have an alternation of generations in which the gametophyte is nutritionally dependent on the sporophyte (Figure 17–6). Oogamy is clearly favored in land plants, since only one of the kinds of gametes must navigate across the hostile environment outside of the plant, rather than two. In vascular plants, the sporophyte is the dominant phase in the life cycle; it is larger and structurally much more complex than the gametophyte. In bryophytes, as we saw in Chapter 16, the gametophyte is almost always larger than the sporophyte, although, as in the vascular plants, it is usually not as complex as the sporophyte.

Homospory and Heterospory

Early vascular plants produced only one kind of spore; such vascular plants are said to be *homosporous*. Among living vascular plants, homospory is found in the psilophytes, sphenophytes, some of the lycophytes, and almost all ferns. Following meiosis, homosporous plants produce only one kind of spore. Upon germination such spores produce bisexual gametophytes—that is, gametophytes that bear both antheridia and archegonia.

Heterospory—the production of two types of spores in two different kinds of sporangia—is found in some of the lycophytes, in a few ferns, and in all seed plants. Heterospory arose many times in unrelated groups during the evolution of vascular plants. It was common as early as the Devonian period, more than 360 million years ago. The two types of spores are called *microspores* and *megaspores*, and are produced in *microsporangia* and *megasporangia*, respectively. Although

"micro" implies smallness and "mega" implies largeness, megaspores are not always larger than microspores, especially in the seed plants. The two types of spores are actually defined on the basis of function, not relative size. Microspores give rise to male gametophytes (microgametophytes), and megaspores give rise to female gametophytes (megagametophytes). Both of these types of unisexual gametophytes are much reduced in size as compared with the gametophytes of homosporous vascular plants. In heterosporous plants, the gametophytes develop within the spore wall; in homosporous plants, they develop outside the spore wall.

In *Equisetum*, although all of the spores are identical morphologically, they are physiologically differentiated, some giving rise to unisexual gametophytes. *Equisetum* might therefore be considered to exhibit physiological heterospory, a possible stage in the evolution of morphological heterospory of the kind we have just discussed.

Gametophytes and Gametes

The relatively large gametophytes of homosporous plants are independent of the sporophyte for their nutrition, although the subterranean gametophytes of some species—such as those of *Psilotum* and a number of species of club mosses (*Lycopodium*)—are heterotrophic, depending on endomycorrhizal fungi for their nutrients. Other species of *Lycopodium*, like most ferns and the horsetails, have free-living, photosynthetic gametophytes. In contrast, the gametophytes of heterosporous vascular plants are dependent on the sporophyte for their nutrition.

The evolution of the gametophyte of vascular plants is characterized by a progressive reduction in size and complexity, and the gametophytes of angiosperms are the most reduced of all (see page 361). Archegonia are found in all seedless vascular plants and most gymnosperms but are absent in all angiosperms; antheridia are also present in seedless vascular plants but are absent in all gymnosperms and angiosperms. In the seedless vascular plants, including the ferns, the motile sperm swim through water to the archegonium. These plants must therefore grow in habitats where water is at least occasionally plentiful.

The seed—a unique structure in which the embryo is shed from the parent plant enclosed within a resistant coat together with a supply of food that aids in its establishment—has evolved in a number of evolutionary lines of vascular plants, starting at least 360 million years ago. Among these groups were the lycophytes and sphenophytes, which are discussed in this chapter because all living members of these divisions lack seeds. The evolution of the seed—one of the principal factors responsible for the dominance of seed plants—and the major features of the groups in which they have evolved, will be discussed in Chapter 18.

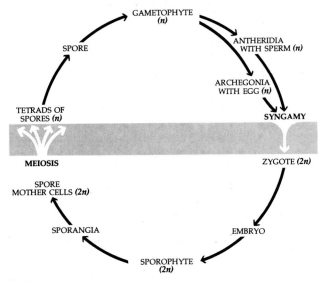

17–6

Generalized life cycle of a vascular plant.

THE DIVISIONS OF SEEDLESS VASCULAR PLANTS

Three divisions of seedless vascular plants—the Rhyniophyta, Zosterophyllophyta, and Trimerophyta—flourished in the Devonian period and became extinct by about 360 million years ago, the end of that period. Of these, the rhyniophytes appeared at least 420 million years ago, in the late Silurian period. All three divisions consist of seedless plants that are relatively simple in structure. A fourth division of seedless vascular plants, Progymnospermophyta, or progymnosperms, will be discussed in Chapter 18, because the members of this group may have been ancestral to the seed plants, both gymnosperms and angiosperms (Figure 17–7). In addition to these extinct divisions, we shall discuss in this chapter the Psilophyta, Lycophyta, Sphenophyta, and Pterophyta, the four divisions of seedless vascular plants that have living representatives.

The overall pattern of diversification of plants may be interpreted in terms of the successive rise to dominance of four major plant groups; these largely replaced the groups that were dominant earlier. In each instance, numerous species evolved in the groups that were rising to dominance. The major groups involved were:

(1) Early vascular plants—characterized by a simple and presumably primitive morphology. These included the rhyniophytes, zosterophyllophytes, and trimerophytes (Figure 17–8). Primitive vascular plants were dominant from the late Silurian period until the mid-Devonian period, from about 420 to 370 million years ago (see Figure 17–1).

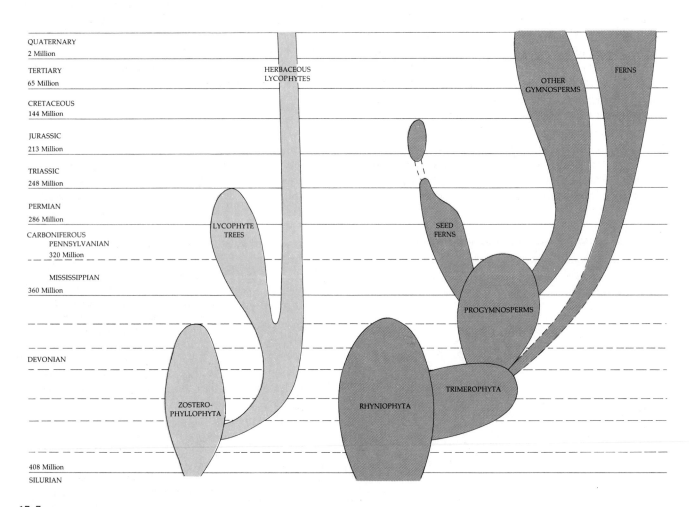

17–7

Possible course of evolution of vascular plants from the zosterophyllophytes and rhyniophytes. The length of Devonian time is exaggerated, and the possible relationships with the angiosperms, or flowering plants, are not indicated. What the ancestral groups of the Psilopsida and Sphenopsida may have been is not known. Only a few genera of fossil plants have been described that appear to be intermediate between the rhyniophytes, which appeared in the Silurian period (about 420 million years ago), and the zosterophyllophytes, which appeared in the fossil record about 10 million years later. The exact nature of the relationships between these two groups, and of their common ancestor, remains to be established.

(2) Ferns, lycophytes, sphenophytes, progymnosperms. These more complex groups were dominant from the late Devonian period through the Carboniferous period (see Figures 17–9, 18–1), from about 380 to about 290 million years ago. The essay on pages 324 and 325 illustrates plants of this period.

(3) Seed plants arose starting in the late Devonian period, at least 360 million years ago, and evolved many new lines by the Permian period. Gymnosperms dominated land floras throughout most of the Mesozoic era until about 100 million years ago.

(4) Flowering plants, which appeared in the fossil record about 127 million years ago, became dominant within 20 to 30 million years and have remained dominant ever since.

DIVISION RHYNIOPHYTA

The earliest known vascular plants that we understand in detail belong to the division Rhyniophyta, a group that dates back to the late Silurian period, at least 420 million years ago. The group became extinct in the mid-Devonian period (about 380 million years ago). Earlier vascular plants were probably similar; their remains go back at least another 15 million years. Rhyniophytes were seedless plants, consisting of simple, dichotomously branching axes with terminal sporangia. Their plant bodies were not differentiated into roots, stems, or leaves, and they were homosporous. The name of the division comes from the good representation of these primitive plants in chert near the village of Rhynie, Scotland.

Cooksonia, a rhyniophyte that is believed to have inhabited mud flats, has the distinction of being the oldest known vascular plant (see Figures 1–5, 17–1). Its slender, leafless aerial stems ranged up to about 6.5 centimeters long; their sporangia were globose. Although nothing is known about its basal parts, it is likely that *Cooksonia* possessed a rhizome, or underground stem, which gave rise to the aerial branches. Tracheids have been identified in macerated bits of the axis. *Cooksonia* had become extinct by the early Devonian period, about 390 million years ago.

The best-known rhyniophyte is *Rhynia* (Figure 17–8a). Probably a marsh plant, its leafless, dichotomously branching aerial stems were attached to a rhizome with tufts of water-absorbing rhizoids. The aerial stems of *Rhynia*, which were about 20 to 50 centimeters long and 3 to 6 millimeters thick, were covered with a cuticle, bore stomata, and served as the photosynthetic organs.

The internal structure of *Rhynia* was similar to that of many of today's vascular plants. A single layer of superficial cells—the epidermis—surrounded the photosynthetic tissue of the cortex, and the center of the axis consisted of a solid strand of xylem surrounded by one or two layers of cells, which may or may not have been phloem cells. Apparently, the first xylem cells to mature occupied a central portion of the strand, while the last to mature were peripheral.

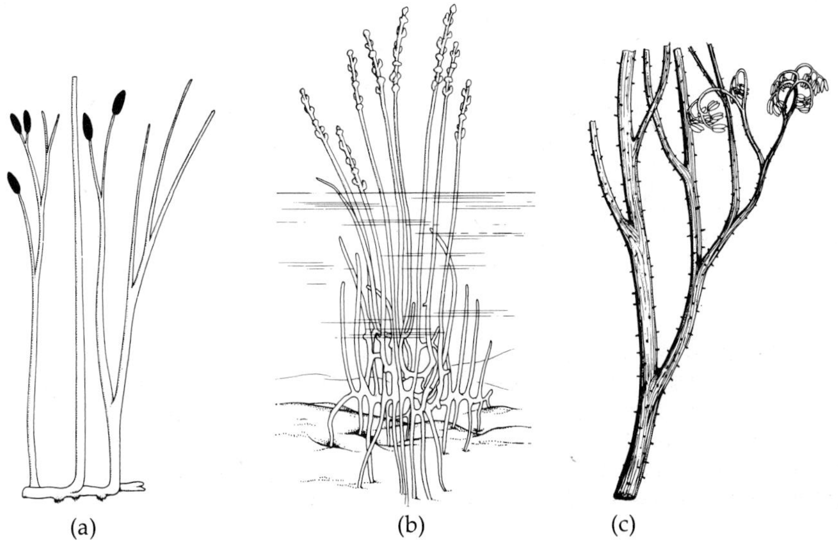

(a) (b) (c)

17–8

Early vascular plants. (a) Rhynia major is a rhyniophyte, one of the simplest of the known vascular plants. The stem was leafless and dichotomously branched. The sporangia, which were terminal, apparently released their spores by splitting longitudinally. (b) In Zosterophyllum and the other zosterophyllophytes, the sporangia, which were aggregated into a terminal spike, split along definite slits that formed around the outer margin. The zosterophyllophytes were larger than the rhyniophytes, but like the latter, they were mostly dichotomously branched plants that were either naked, spiny, or toothed. (c) The trimerophytes were larger and more complex plants with a strong central axis with smaller side branches. The side branches were dichotomously branched and often had terminal masses of paired sporangia that tapered at both ends. The best-known genera included in this group are Psilophyton and Trimerophyton. A reconstruction of Psilophyton princeps is shown here. Individuals of Rhynia major ranged up to about 0.5 meter tall, while some of the trimerophytes were up to 1 meter or more in height. See also Figure 17–1.

17–9
*Reconstruction of a Carboniferous period
swamp forest. See also Figure 18–1.*

DIVISION ZOSTEROPHYLLOPHYTA

The fossils of a second division of extinct seedless vascular plants—the Zosterophyllophyta—have been found in strata from the early to late Devonian period, from about 408 to 370 million years ago. Like the rhyniophytes, the zosterophyllophytes were leafless and dichotomously branched. It is possible that the group was aquatic. The aerial stems were covered with a cuticle, but only the upper ones contained stomata, indicating that the lower branches may have been embedded in mud. In *Zosterophyllum*, it has been suggested that the lower branches frequently produced lateral branches that forked into two axes, one that grew upward, the other downward (Figure 17–8b). The downward-growing branches may have permitted the plant to spread outwardly from the center by providing support. The zosterophyllophytes are named because of their general resemblance to the modern seagrasses, *Zostera*, marine angiosperms that superficially resemble grasses.

Unlike those of the rhyniophytes, the globose or kidney-bean-shaped sporangia of the zosterophyllophytes were borne laterally on short stalks. These plants were homosporous. Although the internal structure of the zosterophyllophytes was essentially similar to that of the rhyniophytes, the first xylem cells to mature in the zosterophyllophytes were located around the periphery of the xylem strand, and the last to mature were located in the center.

The zosterophyllophytes were almost certainly the ancestors of the lycophytes. The sporangia of early lycophytes were borne laterally, like those of the zosterophyllophytes, and the xylem of both divisions differentiated centripetally (from the outside of the strand to the center). The members of these two divisions are quite distinct from the rhyniophytes and trimerophytes.

DIVISION TRIMEROPHYTA

The division Trimerophyta probably evolved directly from the rhyniophytes; these plants seem to represent the ancestral stock of the ferns, the progymnosperms, and perhaps the horsetails as well. The trimerophytes, which were larger and more complex plants than the rhyniophytes or zosterophyllophytes (Figure 17–8c), first appeared in the early Devonian period about 395 million years ago and had become extinct by the end of the mid-Devonian, about 20 million years later—a relatively short period of existence.

Although trimerophytes were evolutionarily more advanced than the rhyniophytes, they still lacked leaves. The main axis formed lateral branch systems that dichotomized several times. The trimerophytes, like the rhyniophytes and zosterophyllophytes, were homosporous. Some of their smaller branches terminated in elongate sporangia, while others were entirely vegetative. Besides their more complex branching pattern, the trimerophytes had a more massive vascular strand than the rhyniophytes. Together with a wide cortex composed of thick-walled cells, the large vascular strand probably was capable of supporting a fairly large plant. As in the rhyniophytes, the xylem of the trimerophytes matured first in the center. The name of the division comes from the Greek words "tri," "meros," and "phyton," meaning three-parted plant, because of the three-parted branching of the secondary branches in the genus *Trimerophyton*.

DIVISION PSILOPHYTA

The Psilophyta include two genera, *Psilotum* and *Tmesipteris*, both living. *Psilotum* is tropical and subtropical in distribution; in the United States, it occurs in Florida, Louisiana, Arizona, Texas, Hawaii, and Puerto Rico, and as a common greenhouse weed. *Tmesipteris* is restricted in distribution to Australia, New Caledonia, New Zealand, and other islands of the South Pacific. Both genera are very simple plants that resemble the rhyniophytes in their basic structure.

Psilotum is unique among living vascular plants in that it lacks both roots and leaves. The sporophyte consists of a dichotomously branching aerial portion with small scalelike outgrowths and a branching underground portion, or a system of rhizomes with many rhizoids (Figure 17–11). An endomycorrhizal zygomycete is present in the outer cortical cells of the rhizomes. *Psilotum* has a protostele; consequently, it lacks leaf gaps (Figure 17–10).

Psilotum is homosporous, the spores being produced in sporangia borne on the ends of short, lateral

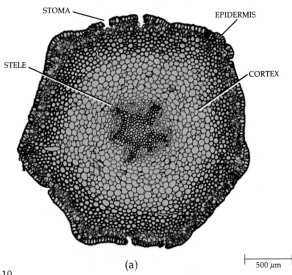

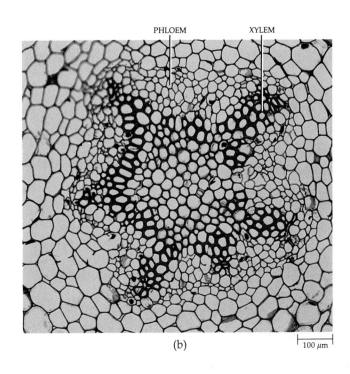

17–10
Psilotum nudum. (a) *Transverse section of stem, showing mature tissues.* (b) *Detail of protostele, showing xylem and phloem.*

branches. Upon germination, the spores give rise to bisexual gametophytes, which resemble portions of the rhizome (Figure 17–11). Like the rhizome, the subterranean gametophyte contains a symbiotic fungus. In addition, some gametophytes contain vascular tissue. The sperm of *Psilotum* are multiflagellated and require water to swim to the egg. Initially, the sporophyte is attached to the gametophyte by a foot, a structure that absorbs nutrients from the gametophyte. Eventually it

becomes detached from the foot, which remains embedded in the gametophyte.

Tmesipteris grows as an epiphyte on tree ferns and other plants (Figure 17–12). The leaflike appendages of *Tmesipteris* are larger than the scalelike outgrowths of *Psilotum,* but in other respects, *Tmesipteris* is essentially similar to *Psilotum.*

The life cycle of *Psilotum* is illustrated in Figure 17–13.

17–11
The subterranean gametophyte of Psilotum nudum. *The gametophytes are bisexual; that is, they bear both antheridia and archegonia.*

17–12
(a) Tmesipteris parva *growing on the trunk of the tree fern* Cyathea australis *in New South Wales, Australia.* (b) Tmesipteris lanceolata, *from New Caledonia.*

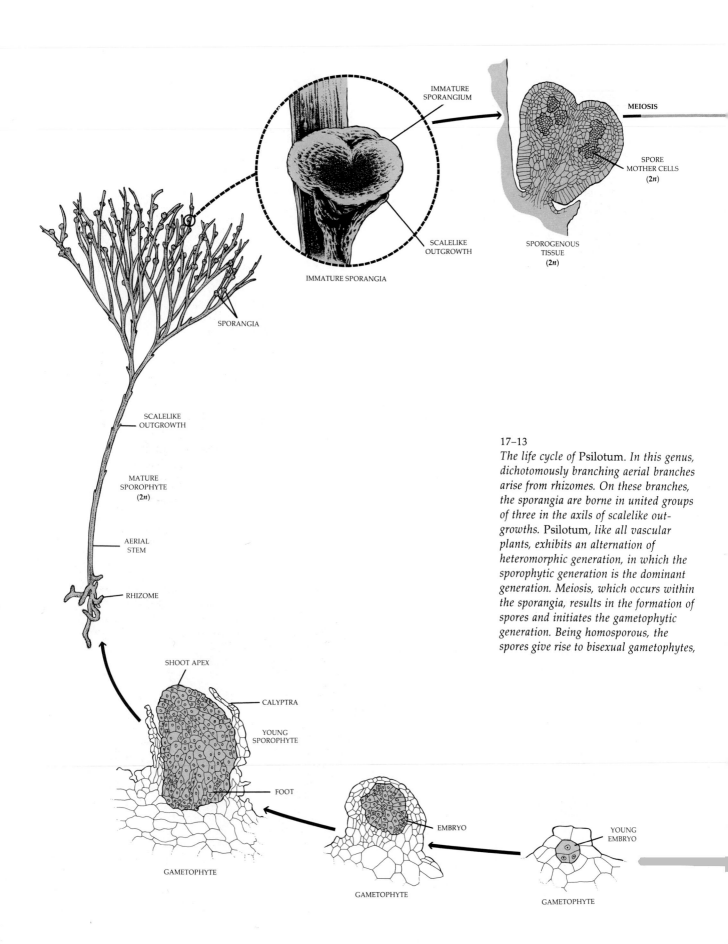

IMMATURE
SPORANGIUM

SCALELIKE
OUTGROWTH

IMMATURE SPORANGIA

MEIOSIS

SPORE
MOTHER CELLS
(2n)

SPOROGENOUS
TISSUE
(2n)

SPORANGIA

SCALELIKE
OUTGROWTH

MATURE
SPOROPHYTE
(2n)

AERIAL
STEM

RHIZOME

SHOOT APEX

CALYPTRA

YOUNG
SPOROPHYTE

FOOT

GAMETOPHYTE

EMBRYO

GAMETOPHYTE

YOUNG
EMBRYO

GAMETOPHYTE

17–13

The life cycle of Psilotum. *In this genus, dichotomously branching aerial branches arise from rhizomes. On these branches, the sporangia are borne in united groups of three in the axils of scalelike out-growths.* Psilotum, *like all vascular plants, exhibits an alternation of heteromorphic generation, in which the sporophytic generation is the dominant generation. Meiosis, which occurs within the sporangia, results in the formation of spores and initiates the gametophytic generation. Being homosporous, the spores give rise to bisexual gametophytes,*

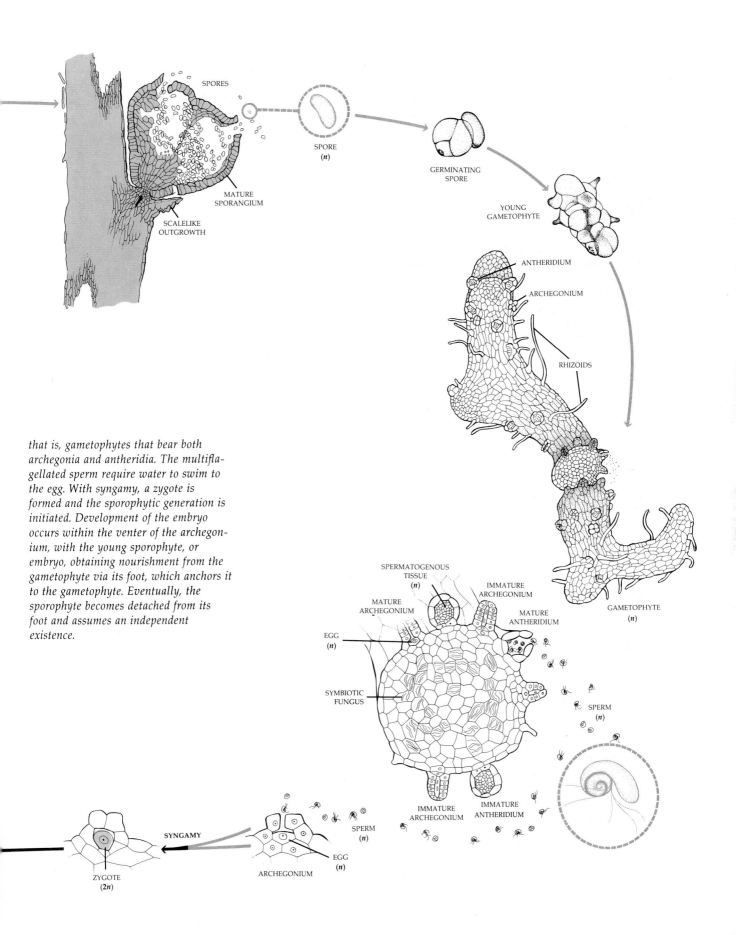

SPORES

SPORE
(n)

GERMINATING
SPORE

MATURE
SPORANGIUM

SCALELIKE
OUTGROWTH

YOUNG
GAMETOPHYTE

ANTHERIDIUM

ARCHEGONIUM

RHIZOIDS

GAMETOPHYTE
(n)

that is, gametophytes that bear both
archegonia and antheridia. The multiflagellated sperm require water to swim to
the egg. With syngamy, a zygote is
formed and the sporophytic generation is
initiated. Development of the embryo
occurs within the venter of the archegonium, with the young sporophyte, or
embryo, obtaining nourishment from the
gametophyte via its foot, which anchors it
to the gametophyte. Eventually, the
sporophyte becomes detached from its
foot and assumes an independent
existence.

SPERMATOGENOUS
TISSUE
(n)

IMMATURE
ARCHEGONIUM

MATURE
ARCHEGONIUM

MATURE
ANTHERIDIUM

EGG
(n)

SYMBIOTIC
FUNGUS

SPERM
(n)

IMMATURE
ARCHEGONIUM

IMMATURE
ANTHERIDIUM

ZYGOTE
(2n)

SYNGAMY

SPERM
(n)

EGG
(n)

ARCHEGONIUM

DIVISION LYCOPHYTA

The four living genera and approximately 1000 living species of Lycophyta are the representatives of an evolutionary line that extends back to the Devonian period. The progenitors of the lycophytes were almost certainly zosterophyllophytes (see Figure 17–8b). There are a number of orders of lycophytes, and at least three of the extinct ones included small to large trees. The three orders of living lycophytes, however, consist entirely of herbs. All lycophytes, living and fossil, possess microphylls, and this type of leaf is highly characteristic of the division. Tree lycophytes were among the dominant plants of the coal-forming forests of the Carboniferous period (see essay, pages 324 and 325, and Figure 18–1). Some of them bore seedlike structures analogous to those of modern seed plants. Most lines of woody lycophytes became extinct before the end of the Paleozoic era, 248 million years ago.

Lycopodium

Perhaps the most familiar living lycophytes are the club mosses, *Lycopodium* (see Figure 10–7c, page 156). The approximately 200 species of this genus extend from arctic regions into the tropics, but they rarely form conspicuous elements in any plant community. Most tropical species are epiphytes and are thus rarely seen, but several of the temperate species form mats that may be evident on forest floors. Because they are evergreen, they are most noticeable in winter.

The sporophyte of *Lycopodium* consists of a branching rhizome from which aerial branches and adventitious roots arise. Both stem and root are protostelic (Figure 17–14). The microphylls of *Lycopodium* are usually spirally arranged. *Lycopodium* is homosporous; the sporangia occur singly on the upper surface of fertile microphylls called *sporophylls*—modified leaf or leaflike organs that bear sporangia. In some species, the sporophylls are similar to ordinary microphylls and are interspersed among the sterile microphylls (Figure 17–15). In others, the nonphotosynthetic sporophylls are grouped into *strobili*, or cones, at the ends of the aerial branches (see Figure 10–7c, page 156).

Upon germination, the spores of *Lycopodium* give rise to bisexual gametophytes that, depending on the species, are either green, irregularly lobed masses or branching, subterranean, nonphotosynthetic structures. Like the gametophytes of *Psilotum* and *Tmesipteris*, those of the species of *Lycopodium* with subterranean gametophytes are associated with a symbiotic fungus. The development and maturation of archegonia and antheridia in a *Lycopodium* gametophyte may require from 6 to 15 years, and they may even produce a series of sporophytes in successive archegonia as they continue to grow.

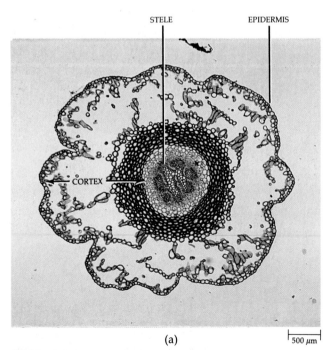

STELE   EPIDERMIS

CORTEX

(a)   500 μm

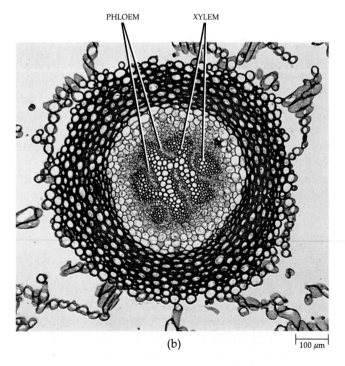

PHLOEM   XYLEM

(b)   100 μm

17–14

(a) *Transverse section of* Lycopodium complanatum *stem, showing mature tissues.* (b) *Detail of protostele of* L. complanatum, *showing xylem and phloem. Both the stem and root of* Lycopodium *show this protostelic organization. See also Figure 17–2.*

(a)

(b)

17–15
Lycopodium lucidulum is a representative of those species of the genus that lack strobili; the sporangia are interspersed among sterile microphylls.

Water is required for fertilization; the biflagellated sperm swim through water to the archegonium and then down its neck. Following fertilization, the zygote develops into an embryo, which grows within the venter of the archegonium. The young sporophyte may remain attached to the gametophyte for a long time, but eventually it becomes independent.

The life cycle of *Lycopodium* is illustrated in Figure 17–17 on pages 314 and 315.

Selaginella

Among the living genera of lycophytes, *Selaginella* has the most species, with about 700. These are mainly tropical in distribution. Many grow in moist situations; a few inhabit desert regions, becoming dormant during the driest part of the year. Among the latter is the so-called resurrection plant, *Selaginella lepidophylla*, which ranges from Mexico north to Texas and New Mexico.

Basically, the herbaceous sporophyte of *Selaginella* is similar to that of *Lycopodium*; it bears microphylls, and its sporophylls are arranged in strobili (Figure 17–16).

(c)

17–16
(a) Selaginella kraussiana, *a prostrate, creeping plant. Adventitious roots can be seen arising from stems.* (b) Selaginella rupestris *with strobili.* (c) Selaginella willdenovii, *a fern of the Old World tropics found in Vietnam, Malaysia, and the Himalayas. Shade-loving, it climbs to 7 meters and has peacock-blue leaves with a metallic sheen.*

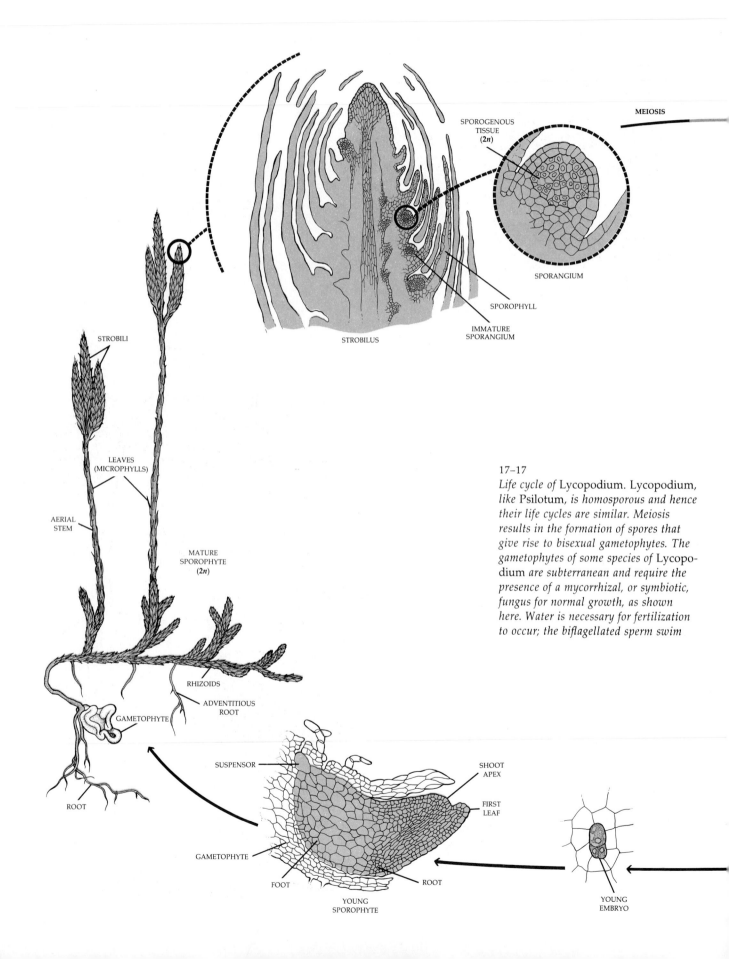

MEIOSIS

SPOROGENOUS
TISSUE
(2n)

SPORANGIUM

SPOROPHYLL

IMMATURE
SPORANGIUM

STROBILUS

STROBILI

LEAVES
(MICROPHYLLS)

AERIAL
STEM

MATURE
SPOROPHYTE
(2n)

RHIZOIDS

ADVENTITIOUS
ROOT

GAMETOPHYTE

ROOT

SUSPENSOR

SHOOT
APEX

FIRST
LEAF

GAMETOPHYTE

FOOT

ROOT

YOUNG
SPOROPHYTE

YOUNG
EMBRYO

17–17

Life cycle of Lycopodium. Lycopodium, *like* Psilotum, *is homosporous and hence their life cycles are similar. Meiosis results in the formation of spores that give rise to bisexual gametophytes. The gametophytes of some species of* Lycopodium *are subterranean and require the presence of a mycorrhizal, or symbiotic, fungus for normal growth, as shown here. Water is necessary for fertilization to occur; the biflagellated sperm swim*

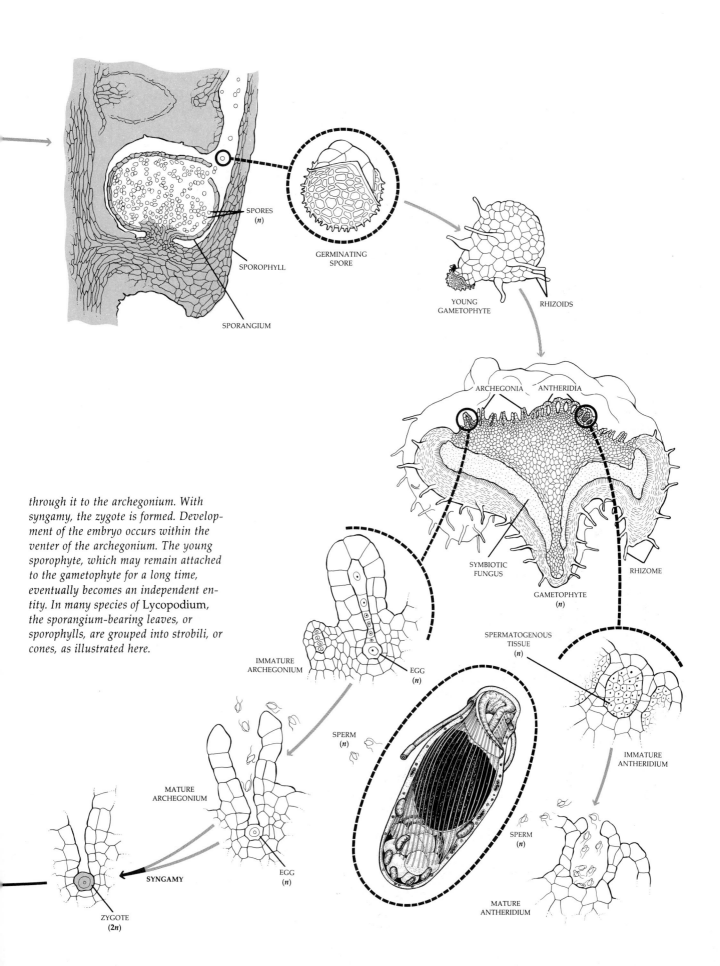

SPORES
(n)

SPOROPHYLL

SPORANGIUM

GERMINATING
SPORE

YOUNG
GAMETOPHYTE

RHIZOIDS

ARCHEGONIA   ANTHERIDIA

SYMBIOTIC
FUNGUS

RHIZOME

GAMETOPHYTE
(n)

through it to the archegonium. With syngamy, the zygote is formed. Development of the embryo occurs within the venter of the archegonium. The young sporophyte, which may remain attached to the gametophyte for a long time, eventually becomes an independent entity. In many species of Lycopodium, the sporangium-bearing leaves, or sporophylls, are grouped into strobili, or cones, as illustrated here.

SPERMATOGENOUS
TISSUE
(n)

IMMATURE
ARCHEGONIUM

EGG
(n)

IMMATURE
ANTHERIDIUM

SPERM
(n)

MATURE
ARCHEGONIUM

SPERM
(n)

EGG
(n)

SYNGAMY

MATURE
ANTHERIDIUM

ZYGOTE
(2n)

Unlike *Lycopodium,* in *Selaginella* a small scalelike outgrowth, called a ligule, develops near the base of the upper surface of each microphyll and sporophyll. The stem and root are protostelic (Figure 17–18).

Whereas *Lycopodium* is homosporous, *Selaginella* is heterosporous, with unisexual gametophytes. This constitutes the most significant difference between the two genera. Each sporophyll bears a single sporangium on its upper surface. Megasporangia are borne by *megasporophylls,* and microsporangia are borne by *microsporophylls.* Both kinds of sporangia occur in the same strobilus.

The male gametophytes (microgametophytes) in *Selaginella* develop from microspores. Four microspores are produced by meiosis from each microspore mother cell. The male gametophyte develops within the microspore and lacks chlorophyll. At maturity the male gametophyte consists of a single prothallial, or vegetative, cell and an antheridium, which gives rise to many biflagellated sperm. The microspore wall must rupture for the sperm to be liberated.

During development of the female gametophyte (megagametophyte), the megaspore wall ruptures, and the gametophyte protrudes through the rupture to the outside. This is the portion of the female gametophyte in which the archegonia develop. It has been reported that the female gametophytes sometimes develop chloroplasts, although it is more likely that the gametophytes derive their nutrition largely from food stored within the megaspores.

Water is required for the sperm to swim to the archegonia and fertilize the eggs. Commonly this occurs after the gametophytes have been shed from the strobilus. During development of the embryos of both *Lycopodium* and *Selaginella,* a structure called a suspensor is formed. Although inactive in *Lycopodium* and some species of *Selaginella,* in other *Selaginella* species the suspensor serves to thrust the developing embryo deep within the nutrient-rich tissue of the female gametophyte. Gradually, the developing sporophyte emerges from the gametophyte and becomes independent.

The life cycle of *Selaginella* is illustrated in Figure 17–22 on pages 318 and 319.

Isoetes

A very distinctive member of the Lycophyta is *Isoetes,* or quillwort. Plants of *Isoetes* may be aquatic, or they may grow in seasonally wet places that become dry at certain seasons. The sporophyte of *Isoetes* consists of a short, fleshy underground stem (corm) bearing quill-like microphylls on its upper surface and roots on its lower surface (Figure 17–19). In *Isoetes,* each leaf is a potential sporophyll.

Like *Selaginella, Isoetes* is heterosporous. The megasporangia are borne at the base of certain leaves (megasporophylls), and the microsporangia are borne at the base of other leaves (microsporophylls) nearer the center of the plant (Figure 17–20). A ligule is found just above the sporangium of each sporophyll.

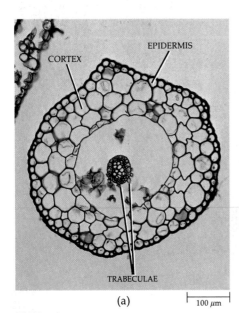

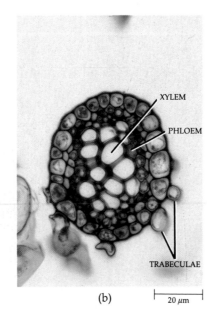

17–18
Selaginella. (a) *Transverse section of stem, showing mature tissues. The protostele is suspended in the middle of the hollow stem by elongate cortical cells* (endodermal cells), called trabeculae. *Only a portion of each trabecula can be seen here.* (b) *Detail of protostele.*

17–19
Isoetes muricata. *View of sporophyte showing quill-like leaves, stem, and roots.*

17–20
A diagram of a vertical section of an
Isoetes *plant. Leaves are borne on the*
upper surface, and roots on the lower
surface, of a short, fleshy underground
stem. Some leaves (megasporophylls) bear
megasporangia and others (microsporo-
phylls) bear microsporangia. The
microsporophylls are located nearer the
center of the plant.

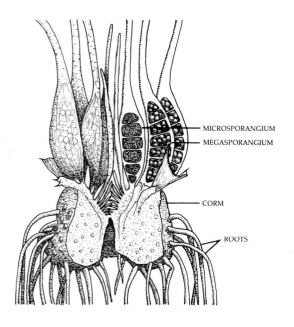

MICROSPORANGIUM
MEGASPORANGIUM
CORM
ROOTS

One of the distinctive features of *Isoetes* is the presence of a specialized cambium that adds secondary tissues to the corm. Externally the cambium produces only parenchyma tissue, while internally it produces a peculiar vascular tissue consisting of sieve elements, parenchyma cells, and tracheids in varying proportions.

In 1984 it was shown that some species of *Isoetes* from high elevations in the tropics have the unique characteristic of obtaining their carbon for photosynthesis from the sediment rather than from the atmosphere. The leaves of these plants lack stomata, have a thick cuticle, and carry on essentially no gas exchange with the atmosphere. Like at least some of the other species of *Isoetes* in which the plants dry out for part of the year, these species have CAM photosynthesis (see page 111).

DIVISION SPHENOPHYTA

Like the Lycophyta, the Sphenophyta extend back to the Devonian period. The sphenophytes reached their maximum abundance and diversity later in the Paleozoic era, about 300 million years ago. During the late Devonian and Carboniferous periods, they were represented by the calamites (see page 324), a group of trees that reached 15 meters in height, with a trunk that could be more than 20 centimeters thick. Today the Sphenophyta are represented by a single herbaceous genus, *Equisetum*, which consists of 15 species (Figure 17–21).

(a)

(b)

17–21
A species of Equisetum *in which the*
fertile shoots essentially lack chlorophyll
and are very different in appearance from
the vegetative shoots. (a) Fertile shoots,
each with a terminal strobilus. Notice
the whorls of scalelike leaves at each
node. (b) Vegetative shoots.

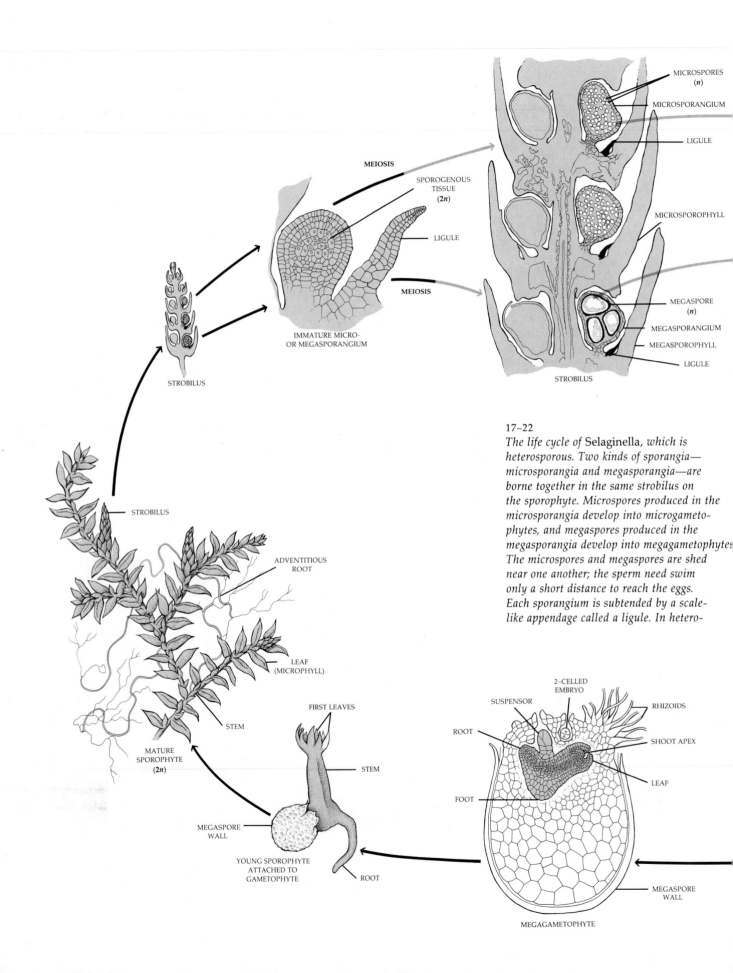

MICROSPORES (*n*)

MICROSPORANGIUM

LIGULE

MICROSPOROPHYLL

MEGASPORE (*n*)

MEGASPORANGIUM

MEGASPOROPHYLL

LIGULE

STROBILUS

MEIOSIS

SPOROGENOUS TISSUE (2*n*)

LIGULE

MEIOSIS

IMMATURE MICRO- OR MEGASPORANGIUM

STROBILUS

STROBILUS

LEAF (MICROPHYLL)

ADVENTITIOUS ROOT

STEM

MATURE SPOROPHYTE (2*n*)

FIRST LEAVES

STEM

MEGASPORE WALL

YOUNG SPOROPHYTE ATTACHED TO GAMETOPHYTE

ROOT

2-CELLED EMBRYO

SUSPENSOR

ROOT

FOOT

RHIZOIDS

SHOOT APEX

LEAF

MEGASPORE WALL

MEGAGAMETOPHYTE

17–22

The life cycle of **Selaginella,** *which is heterosporous. Two kinds of sporangia— microsporangia and megasporangia—are borne together in the same strobilus on the sporophyte. Microspores produced in the microsporangia develop into microgameto- phytes, and megaspores produced in the megasporangia develop into megagametophytes. The microspores and megaspores are shed near one another; the sperm need swim only a short distance to reach the eggs. Each sporangium is subtended by a scale- like appendage called a ligule. In hetero-*

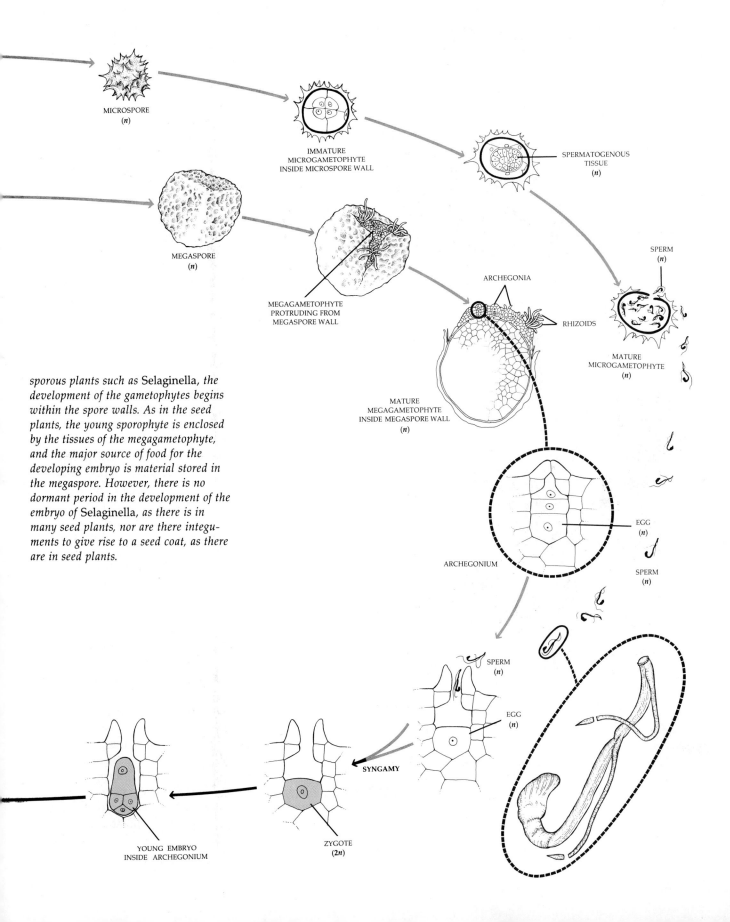

MICROSPORE
(*n*)

IMMATURE
MICROGAMETOPHYTE
INSIDE MICROSPORE WALL

SPERMATOGENOUS
TISSUE
(*n*)

MEGASPORE
(*n*)

MEGAGAMETOPHYTE
PROTRUDING FROM
MEGASPORE WALL

ARCHEGONIA

RHIZOIDS

SPERM
(*n*)

MATURE
MICROGAMETOPHYTE
(*n*)

MATURE
MEGAGAMETOPHYTE
INSIDE MEGASPORE WALL
(*n*)

EGG
(*n*)

ARCHEGONIUM

SPERM
(*n*)

SPERM
(*n*)

EGG
(*n*)

SYNGAMY

YOUNG EMBRYO
INSIDE ARCHEGONIUM

ZYGOTE
(2*n*)

sporous plants such as Selaginella, *the
development of the gametophytes begins
within the spore walls. As in the seed
plants, the young sporophyte is enclosed
by the tissues of the megagametophyte,
and the major source of food for the
developing embryo is material stored in
the megaspore. However, there is no
dormant period in the development of the
embryo of* Selaginella, *as there is in
many seed plants, nor are there integuments to give rise to a seed coat, as there
are in seed plants.*

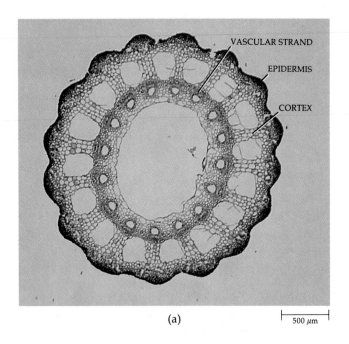

VASCULAR STRAND

EPIDERMIS

CORTEX

(a)
500 μm

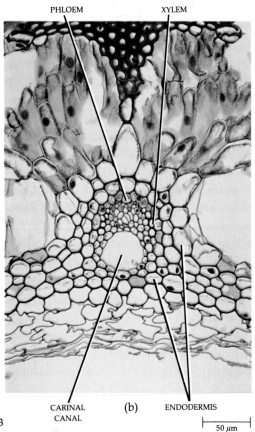

PHLOEM   XYLEM

CARINAL   (b)   ENDODERMIS
CANAL
50 μm

17-23

Stem anatomy of Equisetum. *(a)
Transverse section of stem, showing
mature tissues. (b) Detail of vascular
strand, showing xylem and phloem.*

The species of *Equisetum* are known as the horsetails; they are widespread in moist or damp places, by streams, or along the edge of woods (Figure 17–21). Horsetails are easily recognized because of their conspicuously jointed stems and rough texture. The small, scalelike leaves, which have a simple structure but are probably reduced megaphylls, are whorled at the nodes. When present, the branches arise laterally at the nodes and alternate with the leaves. The internodes (the portions of the stems between successive nodes) are ribbed, and the ribs are tough and strengthened with siliceous deposits in the epidermal cells. Horsetails have been used to scour pots and pans, particularly in colonial and frontier times, and have thus earned the name "scouring rushes." The roots are adventitious, arising at the nodes of the rhizomes.

The aerial stems of *Equisetum* arise from branching underground rhizomes and, although they may die back during unfavorable seasons, the rhizomes are perennial. The aerial stem is complex anatomically (Figure 17–23). At maturity, its internodes contain a hollow pith, surrounded by a ring of smaller canals called carinal canals. Each of these smaller canals is associated with a strand of primary xylem and primary phloem.

Equisetum is homosporous. Sporangia are borne in groups of five to ten along the margins of small umbrellalike structures known as *sporangiophores* (sporangia-bearing branches), which are clustered into strobili at the apex of the stem (Figures 17–21a and 17–26). The fertile stems of some species do not contain much chlorophyll, and in these species the fertile stems are sharply distinct from the vegetative stems, often appearing before them early in spring (see Figure 17–21). In other species of *Equisetum,* the strobili are borne at the tips of otherwise vegetative stems (Figure 10–7d, page 156). When the spores are mature, the sporangia contract and split along their inner surface, releasing numerous spores. Elaters, which arise from the outer layer of the spore wall, coil when moist and uncoil when dry, thus presumably playing a role in spore dispersal.

The gametophytes of *Equisetum* are green and free-living, most being about the size of a pinhead; they become established mainly on mud that has recently been flooded and is rich in nutrients. The gametophytes (Figure 17–24), which reach sexual maturity in 3 to 5 weeks, are either bisexual or male. In bisexual gametophytes, the archegonia develop before the antheridia; this developmental pattern increases the probability of cross-fertilization. The sperm are multiflagellated and require water to swim to the eggs. The eggs of several archegonia on a single gametophyte may be fertilized and develop into embryos, or young sporophytes.

The life cycle of *Equisetum* is illustrated in Figure 17–26 on pages 322 and 323.

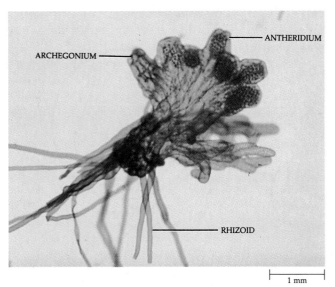

17-24
Equisetum. *Bisexual gametophyte showing male and female gametangia. Compare this gametophyte with that shown in Figure 16–2, a 400-million-year old one from Scotland.*

DIVISION PTEROPHYTA

Ferns have been relatively abundant in the fossil record from the Carboniferous period (see essay, pages 324 and 325, and Figure 18–1) to the present, and closely related groups occurred as far back as the Devonian period. About two-thirds of the approximately 12,000 living species are found in tropical regions, with the other third inhabiting temperate regions of the globe, including desert areas. Ferns are by far the largest group of living seedless vascular plants.

In both form and habitat, ferns exhibit great diversity (Figure 17–25). Some ferns are quite "unfernlike" in appearance; for example, *Salvinia*, an aquatic fern, has undivided leaves up to 2 centimeters long. More fernlike is another floating water fern, *Azolla* (see Figure 26–16, page 535), which is important in agricultural and natural ecosystems because of the symbiotic nitrogen-fixing cyanobacteria it harbors in cavities at the bases of its leaves. At the other extreme in size are the tree ferns (Figure 17–25b), such as those of the genus *Cyathea*, some of which have been recorded to reach heights of more than 24 meters and to have leaves of 5 meters or more in length. Although the trunks of such tree ferns may be 30 centimeters or more in diameter, their tissues are entirely primary in origin. Only *Botrychium*, the relatively small, herbaceous "grape fern," is known to have a vascular cambium.

(a)

(b)

(c)

(d)

17-25
The diversity of ferns. (a) *Cinnamon fern,* Osmunda cinnamomea. (b) *A tree fern,* Dicksonia squarros, *growing in New Zealand.* (c) *A cloak fern,* Northolaena neglecta, *which grows on limestone and is found in Texas, Arizona, and New Mexico.* (d) Marsilea, *a heterosporous water fern.*

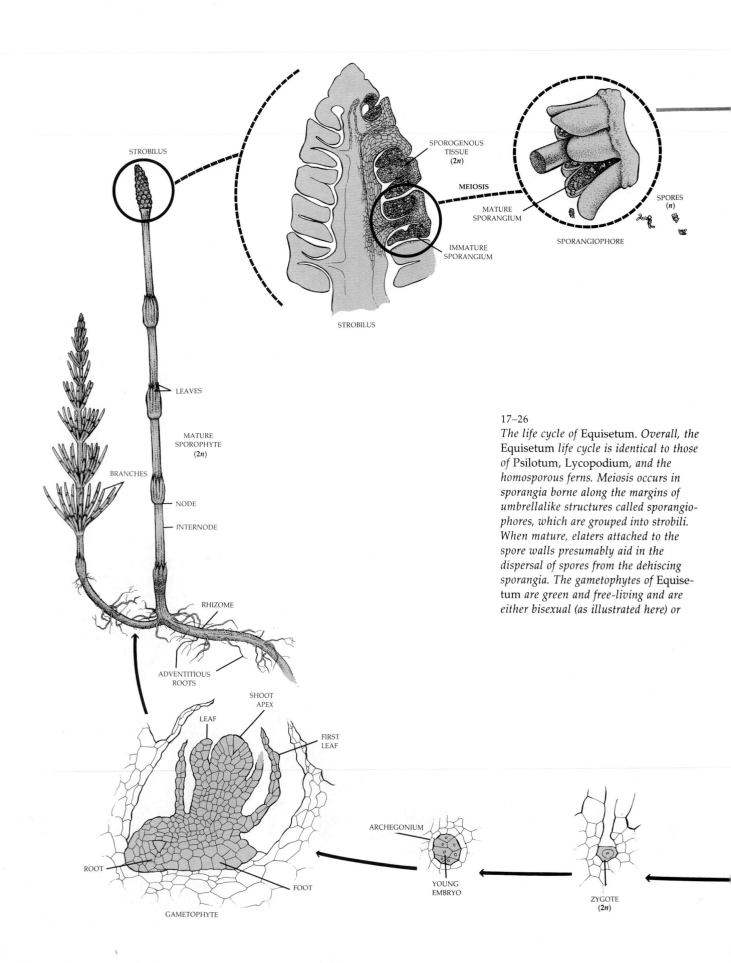

STROBILUS

SPOROGENOUS
TISSUE
(2*n*)

MEIOSIS

MATURE
SPORANGIUM

SPORANGIOPHORE

SPORES
(*n*)

IMMATURE
SPORANGIUM

STROBILUS

LEAVES

MATURE
SPOROPHYTE
(2*n*)

BRANCHES

NODE

INTERNODE

RHIZOME

ADVENTITIOUS
ROOTS

SHOOT
APEX

LEAF

FIRST
LEAF

ARCHEGONIUM

ROOT

FOOT

YOUNG
EMBRYO

ZYGOTE
(2*n*)

GAMETOPHYTE

17–26
The life cycle of Equisetum. *Overall, the* Equisetum *life cycle is identical to those of* Psilotum, Lycopodium, *and the homosporous ferns. Meiosis occurs in sporangia borne along the margins of umbrellalike structures called sporangiophores, which are grouped into strobili. When mature, elaters attached to the spore walls presumably aid in the dispersal of spores from the dehiscing sporangia. The gametophytes of* Equisetum *are green and free-living and are either bisexual (as illustrated here) or*

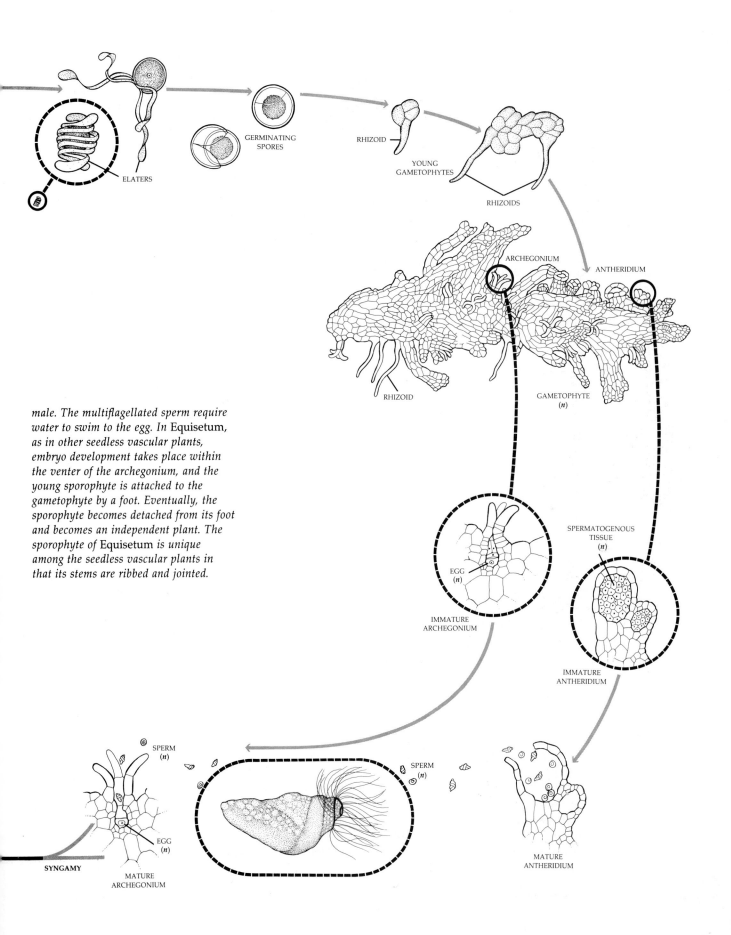

ELATERS

GERMINATING
SPORES

RHIZOID

YOUNG
GAMETOPHYTES

RHIZOIDS

ARCHEGONIUM

ANTHERIDIUM

RHIZOID

GAMETOPHYTE
(*n*)

SPERMATOGENOUS
TISSUE
(*n*)

EGG
(*n*)

IMMATURE
ARCHEGONIUM

IMMATURE
ANTHERIDIUM

male. The multiflagellated sperm require
water to swim to the egg. In Equisetum,
as in other seedless vascular plants,
embryo development takes place within
the venter of the archegonium, and the
young sporophyte is attached to the
gametophyte by a foot. Eventually, the
sporophyte becomes detached from its foot
and becomes an independent plant. The
sporophyte of Equisetum is unique
among the seedless vascular plants in
that its stems are ribbed and jointed.

SPERM
(*n*)

SPERM
(*n*)

EGG
(*n*)

SYNGAMY

MATURE
ARCHEGONIUM

MATURE
ANTHERIDIUM

COAL AGE PLANTS

The amount of carbon dioxide involved in photosynthesis on an annual basis is about 100 billion metric tons, about a tenth of the total carbon dioxide present in the atmosphere. The amount returned as carbon dioxide as a result of oxidation of these living materials is about the same, differing only by 1 part in 10,000. This very slight imbalance is caused by the burying of organisms in sediment or mud under conditions in which oxygen is excluded and decay is only partial. This accumulation of partially decayed plant material is known as peat (see page 298). The peat may eventually become covered with sedimentary rock and is thus placed under pressure. Depending on time, temperature, and other factors, peat may become compressed into soft or hard coal, one of the so-called fossil fuels.

During certain periods in the earth's history, the rate of fossil-fuel formation was greater than at other times. One such time was the Carboniferous period, which extended from 360 to 286 million years ago (see Figures 17–9 and 18–1). The lands were low, covered by shallow seas or swamps, and, in what are now temperate regions of Europe and North America, conditions were favorable for year-round growth. These regions were tropical to subtropical with the equator then arcing across the Appalachians, over northern Europe, and through the Ukraine. Five groups of plants dominated the swamplands, and three of them were seedless vascular plants: lycophytes, sphenophytes (calamites), and ferns. The other two were gymnospermous types of seed plants—seed ferns (Pteridospermales) and cordaites (Cordaitales).

Lycophyte Trees

Two-thirds of the "Age of Coal" in the late Carboniferous period (Pennsylvanian) was dominated by lycophyte trees, most of which grew to heights of 10 to 35 meters and were sparsely branched (a). After the plant attained over half of its total height, the trunk branched dichotomously. Successive branching produced progressively smaller branches until, finally, the tissues of the branch tips lost their ability to grow further. The branches bore long microphylls. The tree lycophytes were largely supported by a massive periderm surrounding a relatively small amount of xylem. Inasmuch as their root systems were shallow, these tall swamp plants could have been blown over easily by winds.

Like *Selaginella* and *Isoetes*, lycophyte trees were heterosporous, and their sporophylls were aggregated into cones. Some of these trees produced seedlike structures.

As the swamplands began to dry up and the climate in Euramerica began to change toward the end of the Carboniferous period, the lycophyte trees vanished almost overnight, geologically speaking. Herbaceous lycophytes essentially similar to *Lycopodium* and *Selaginella* existed in the Carboniferous period and representatives of some of them have survived to the present; there are four living genera.

(a)
One of the dominant trees of the late Carboniferous period was the lycophyte Lepidodendron, *some of which grew to heights of 40 meters or more.*

Calamites

Calamites, or giant horsetails, were plants of treelike proportions, reaching heights of 18 meters or more (see Figure 18–1). Like the plant body of *Equisetum*, that of the calamites consisted of a branched aerial portion and an underground rhizome system. In addition, the leaves and branches were whorled at the nodes. Even the stems were remarkably similar to those of *Equisetum*, except for the presence of secondary xylem, which accounted for most of the great diameter of the stems (trunks up to one-third meter in diameter). The calamites, of which there were a number of genera, are now regarded as belonging to the same order as the living genus *Equisetum*.

The fertile appendages, or sporangiophores, of the calamites were aggregated into cones. Although most were homosporous, a few giant horsetails were heterosporous. Unlike most of the lycophyte trees, the giant horsetails survived the Carboniferous period and were abundant in the Permian.

(b)

One of the most interesting of all gymnosperm groups are the seed ferns, a large group of primitive seed-bearing plants that appeared in the late Devonian period and flourished for about 125 million years. The remnants of these bizarre plants are common in rocks of Carboniferous age and have been well known to paleobotanists for a century or more. Their vegetative parts are so fernlike that for many years they were grouped with the ferns. This drawing is a reconstruction of the Carbon-iferous seed fern Medullosa noei. *The plant is about 5 meters tall.*

(c)

Tip of young branch of the primitive conifer Cordaites, *with long, straplike leaves and cones.*

Ferns

Many of the ferns represented in the fossil record are recognizable as members of today's primitive fern families. The "Age of Ferns" in the late Carboniferous period was dominated by tree ferns such as *Psaronius*. Up to 8 meters tall, *Psaronius* had a stele that expanded toward the apex; the stele was covered below with adventitious roots, which played the key role in supporting the plant. The stem of *Psaronius* ended in an aggregate of large, pinnately compound fronds (see Figure 18–1).

Seed Plants

The two remaining plant groups that dominated the tropical lowlands of Euramerica were the seed ferns and the cordaites. Remnants of seed ferns are common in rocks of Carboniferous age (b). Their large, pinnately compound fronds are so fernlike that they were long regarded as ferns. Then in 1905, F. W. Oliver and D. H. Scott demonstrated that these plants bore seeds and so were gymnosperms. Many species were small, shrubby or scrambling plants. Other probable seed ferns were tall woody trees. The fronds

of seed ferns were borne at the top of their stem, or trunk, with microsporangia and seeds borne on them. The seed ferns survived into the first part of the Mesozoic era. It has been suggested that the seed ferns and the ferns evolved from a common ancestor, but Figure 17–7 shows a more probable hypothesis.

The cordaites were widely distributed during the Carbon-iferous period both in swamps and in drier environments. Although some members of the order were shrubs, many were tall (15 to 30 meters), highly branched trees that formed extensive forests. Their long (up to 1 meter), straplike leaves were spirally arranged at the tips of the youngest branches (c). The center of the stem was occupied by a large pith, and a vascular cambium gave rise to a complete cylinder of secondary xylem. The root system, located at the base of the plant, also contained secondary xylem. The plants bore pollen-bearing cones and seed-bear-ing conelike structures on separate branches. The cordaites were also abundant during the Permian period (286 to 248 million years ago), the drier and colder period that followed the Carboniferous period.

In Conclusion

The dominant tropical coal-swamp plants of the Carbon-iferous period in Euramerica—the lycophyte trees—became extinct prior to the Permian, a time of increasing tropical drought and extensive temperate to polar glaciation. Only the herbaceous relatives of the tree lycophytes and horsetails of the Carboniferous period continued to flourish and exist today, as do several families of ferns that appeared in the Carboniferous period. Both the seed ferns and cordaites eventually disappeared. Only one group of Carboniferous gymnosperms, the conifers (not a dominant group at the time), survived and went on to produce new types during the Permian period. The living conifers are discussed in detail, along with the other seed plants, in Chapter 18.

17–27

The anatomy of fern rhizomes. (a) Adiantum, or maidenhair fern. Transverse section of a rhizome, showing the siphonostele. *Note the wide leaf gap. (b) Transverse section of part of the vascular region of a rhizome of the tree fern* Dicksonia. *The phloem is composed mainly of sieve elements; the xylem is composed entirely of tracheids.*

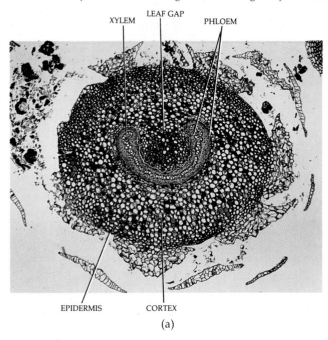

XYLEM   LEAF GAP   PHLOEM

EPIDERMIS   CORTEX

(a)

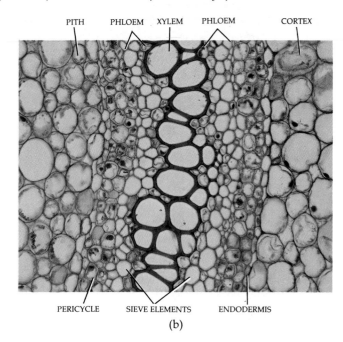

PITH   PHLOEM   XYLEM   PHLOEM   CORTEX

PERICYCLE   SIEVE ELEMENTS   ENDODERMIS

(b)

Most garden and woodland ferns of temperate regions have fleshy, siphonostelic rhizomes (Figure 17–27), which produce new sets of leaves each year. The roots are adventitious, arising from the rhizomes, near the bases of the leaves. The leaves, or *fronds,* are megaphylls and represent the most conspicuous part of the sporophyte. Their high surface to volume ratio marks them as much more efficient photosynthetic organs than the microphylls of the lycophytes. The ferns are the only seedless vascular plants to possess megaphylls. Commonly, the fronds are compound; that is, the lamina is divided into leaflets, or *pinnae,* which are attached to the *rachis,* an extension of the leaf stalk, or petiole. In nearly all ferns, the young leaves are coiled in the bud and are commonly referred to as "fiddleheads" (Figure 17–28). This type of leaf development is known as *circinate vernation.* It results from more rapid growth on the lower than the upper surface of the leaf early in development and is mediated by the hormone auxin produced by the young pinnae on the inner side of the fiddlehead.

All but a few genera of ferns are homosporous. The sporangia are variously placed on the lower surface of the leaves, on specially modified leaves, or on separate stalks (Figure 17–29). The sporangia commonly occur in clusters called *sori.* In many genera, the sori are covered by specialized outgrowths of the leaf, the *indusia* (singular: *indusium*), which may shrivel when the sporangia are ripe (Figure 17–30). At this time, the mature spores—the result of meiosis in the spore mother cells

17–28

"Fiddleheads" of the cinnamon fern (Osmunda cinnamonea). *Although they were formerly sometimes cooked and eaten, there is now considerable suspicion that fiddleheads may be somewhat toxic.*

17–29
Sori are clusters of sporangia found on the undersides of the leaves of ferns. (a) In Dennstaedtia punctilobula *and other ferns of this genus, the sori are bare. (b) In the bracken fern (*Pteridium aquilinum*), shown here, as well as the maidenhair ferns (*Adiantum*), the sori are located along the margins of the leaf blades, which are rolled back over them. (c) In the evergreen wood fern (*Dryopteris marginalis*), the sori, which are also located near the margins of the leaf blades, are completely covered by kidney-shaped indusia. (d) In* Onoclea sensibilis, *the globular sporangia are borne on stalks that differ greatly from the leaves. Many sporangia have specialized dehiscence mechanisms and walls that consist of only a single layer of cells. However, in some more primitive groups of ferns, the sporangia have walls made up of several layers of cells and are split along a simple line of thin-walled cells.*

(a)

(b)

(c)

(d)

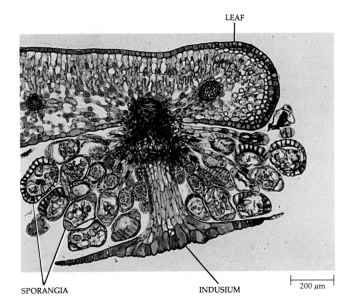

LEAF

SPORANGIA   INDUSIUM   200 μm

17–30
Cyrtomium falcatum, *a homosporous fern. Transverse section of a leaf, showing a sorus on the lower surface. The sporangia are in different stages of development and are protected by an umbrellalike indusium.*

—are ejected through a crack in the so-called *lip cells* of the sporangium. The sporangia are stalked, and each contains a special layer of unevenly thick-walled cells called an *annulus*. Contraction of the annulus causes tearing of the lip cells. Sudden expansion of the annulus then results in a catapultlike discharge of the spores.

Heterospory in living ferns is restricted to two specialized groups of water ferns (Figure 17–25d); a number of extinct ferns were heterosporous.

The spores of most homosporous ferns give rise to free-living bisexual gametophytes. The gametophyte begins development as a small, pale green, algalike chain of cells called the germ filament, or protonema. It then develops into a flat, heart-shaped, membranous structure, the *prothallus*, with numerous rhizoids on its central, lower surface. Both antheridia and archegonia develop on the ventral surface of the prothallus. The antheridia generally appear earlier than archegonia, primarily among the rhizoids. The archegonia then are formed near the notch, or indentation at the anterior end of the gametophyte. The difference in the timing of appearance of the two kinds of gametangia promotes outcrossing in ferns. Water is required for the multiflagellated sperm to swim to the eggs in both homosporous and heterosporous ferns.

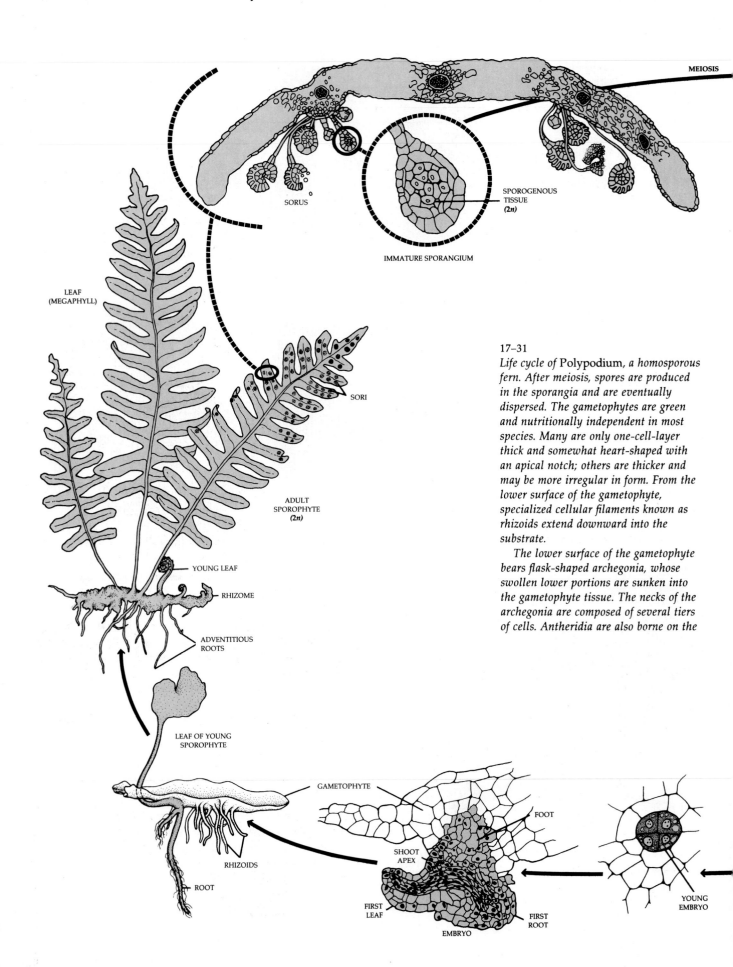

17-31
Life cycle of Polypodium, *a homosporous fern. After meiosis, spores are produced in the sporangia and are eventually dispersed. The gametophytes are green and nutritionally independent in most species. Many are only one-cell-layer thick and somewhat heart-shaped with an apical notch; others are thicker and may be more irregular in form. From the lower surface of the gametophyte, specialized cellular filaments known as rhizoids extend downward into the substrate.*

The lower surface of the gametophyte bears flask-shaped archegonia, whose swollen lower portions are sunken into the gametophyte tissue. The necks of the archegonia are composed of several tiers of cells. Antheridia are also borne on the

MEIOSIS

SORUS

SPOROGENOUS
TISSUE
(2n)

IMMATURE SPORANGIUM

LEAF
(MEGAPHYLL)

SORI

ADULT
SPOROPHYTE
(2n)

YOUNG LEAF

RHIZOME

ADVENTITIOUS
ROOTS

LEAF OF YOUNG
SPOROPHYTE

GAMETOPHYTE

FOOT

RHIZOIDS

ROOT

SHOOT
APEX

FIRST
LEAF

FIRST
ROOT

EMBRYO

YOUNG
EMBRYO

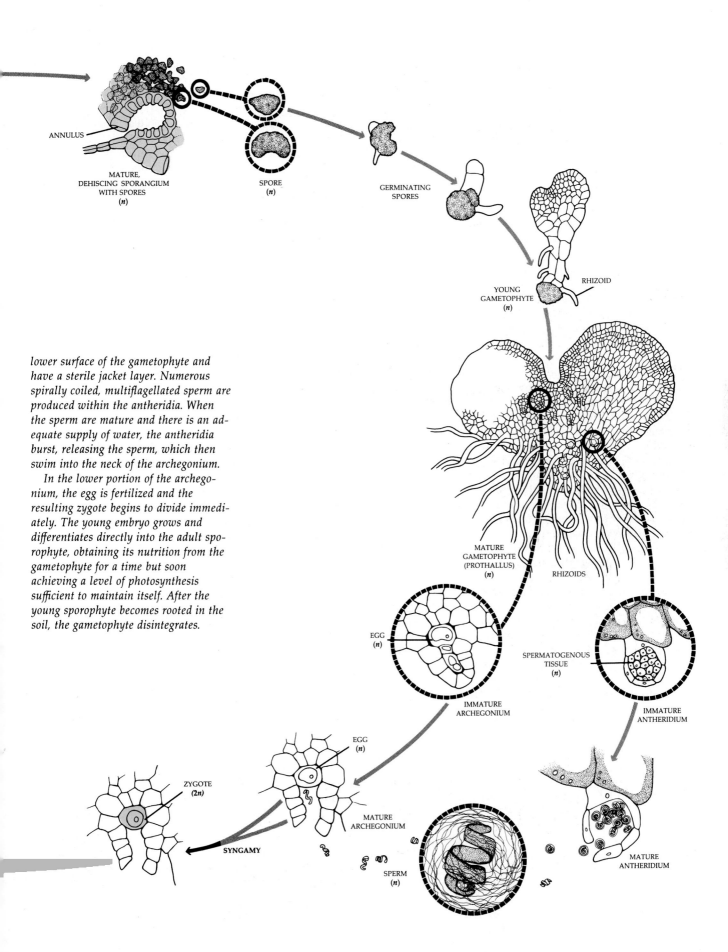

ANNULUS

MATURE,
DEHISCING SPORANGIUM
WITH SPORES
(*n*)

SPORE
(*n*)

GERMINATING
SPORES

YOUNG
GAMETOPHYTE
(*n*)

RHIZOID

lower surface of the gametophyte and have a sterile jacket layer. Numerous spirally coiled, multiflagellated sperm are produced within the antheridia. When the sperm are mature and there is an adequate supply of water, the antheridia burst, releasing the sperm, which then swim into the neck of the archegonium.

In the lower portion of the archegonium, the egg is fertilized and the resulting zygote begins to divide immediately. The young embryo grows and differentiates directly into the adult sporophyte, obtaining its nutrition from the gametophyte for a time but soon achieving a level of photosynthesis sufficient to maintain itself. After the young sporophyte becomes rooted in the soil, the gametophyte disintegrates.

MATURE
GAMETOPHYTE
(PROTHALLUS)
(*n*)

RHIZOIDS

EGG
(*n*)

IMMATURE
ARCHEGONIUM

SPERMATOGENOUS
TISSUE
(*n*)

IMMATURE
ANTHERIDIUM

EGG
(*n*)

ZYGOTE
(**2n**)

MATURE
ARCHEGONIUM

SYNGAMY

SPERM
(*n*)

MATURE
ANTHERIDIUM

Early in its development, the embryo, or young sporophyte, receives nutrients from the gametophyte via a foot. However, development is rapid, and the sporophyte soon becomes an independent plant, at which time the gametophyte disintegrates.

The life cycle of a homosporous fern is shown in Figure 17–31.

SUMMARY

Vascular plants contain xylem and phloem and exhibit an alternation of generations in which the sporophyte is the dominant and nutritionally independent phase.

The plant bodies of many vascular plants consist entirely of primary tissues. Today, secondary growth is confined largely to the seed plants, although it occurred in several unrelated fossil groups of seedless vascular plants. The primary vascular tissues and associated ground tissues exhibit three basic arrangements: (1) the protostele, which consists of a solid core of vascular tissue; (2) the siphonostele, which contains a pith, surrounded by vascular tissue; and (3) the eustele, which consists of a system of strands surrounding a pith and separated from one another by ground tissue.

Roots evolved from the underground portions of the primitive plant body. Leaves originated in more than one way. Microphylls, single-veined leaves whose leaf traces are not associated with leaf gaps, evolved as superficial lateral outgrowths of the stem. They are associated with protosteles and are characteristic of the lycophytes. Megaphylls, leaves with complex venation and leaf traces associated with leaf gaps, evolved from branch systems. Megaphylls are associated with siphonosteles and eusteles.

Vascular plants are either homosporous or heterosporous. Homosporous vascular plants produce only one type of spore, which gives rise to a bisexual gametophyte. Heterosporous plants produce microspores and megaspores, which germinate and give rise to male gametophytes and female gametophytes, respectively. The gametophytes of heterosporous plants are much reduced in size, compared with those of homosporous plants. In the history of the vascular plants, heterospory has evolved a number of times. There has been a long, continuous evolutionary trend toward a reduction in the size and the complexity of the gametophyte, which culminated in the angiosperms. Seedless vascular plants have archegonia and antheridia; archegonia have been lost in all but a few gymnosperms, and both archegonia and antheridia have been lost in all angiosperms.

Vascular plants go back at least 430 million years; the earliest ones about which we have many structural details belong to the division Rhyniophyta, the oldest fossils of which are from the late Silurian period, about 420 million years ago. The plant bodies of the rhyniophytes and other contemporary plants were simple, dichotomously branching axes lacking roots and leaves. With evolutionary specialization, morphological and physiological differences arose between various parts of the plant body, bringing about the differentiation of root, stem, and leaf.

The living seedless vascular plants are classified in four divisions: the Psilophyta (*Psilotum* and *Tmesipteris*), the Lyco-

phyta (which includes *Lycopodium, Selaginella,* and *Isoetes*), the Sphenophyta (*Equisetum*), and the Pterophyta (ferns). Most of the seedless vascular plants are homosporous. Heterospory is exhibited by *Selaginella, Isoetes,* and some water ferns.

The life cycles of the seedless vascular plants are essentially similar to one another: an alternation of heteromorphic generations in which the sporophyte is dominant and free-living. The gametophytes of the homosporous species are bisexual, producing both antheridia and archegonia, and are independent of the sporophyte for their nutrition. Those of heterosporous species, except for the few genera of heterosporous ferns, are unisexual, much reduced in size, and dependent on stored food derived from the sporophyte for their nutrition. All of the seedless vascular plants have motile sperm, and the presence of water is necessary for them to swim to the eggs.

Psilophytes differ from other living vascular plants in their lack of leaves (with the possible exception of *Tmesipteris*) and roots. Lycophytes are characterized by microphylls, associated with protosteles; the members of the other divisions have megaphylls, associated with siphonosteles or eusteles.

Two of the four divisions of seedless vascular plants that contain living representatives, the lycophytes and sphenophytes, extend back to the Devonian period. Among the seedless vascular plants, only the ferns, which first appear in the fossil record in the Carboniferous period, are represented by a large number of living species, about 12,000.

Five groups of vascular plants dominated the swamplands of the Carboniferous period ("Age of Coal"), and three of them were seedless vascular plants: lycophytes, sphenophytes, and ferns. The other two were gymnosperms—the seed ferns and the cordaites.

SUGGESTIONS FOR FURTHER READING

BANKS, HARLAN P.: *Evolution and Plants of the Past,* Wadsworth Publishing Co., Inc., Belmont, Calif., 1970.*
An introduction to plant evolution as it is revealed by the fossil record.

BIERHORST, DAVID W.: *Morphology of Vascular Plants,* The Macmillan Company, New York, 1971.
A detailed, profusely illustrated study of the morphology of the vascular plants.

DOYLE, WILLIAM T.: *The Biology of Higher Cryptogams,* The Macmillan Company, New York, 1970.*
A concise account of the biology of the bryophytes and seedless vascular plants, with an emphasis on development and evolutionary relationships.

ESAU, KATHERINE: *Plant Anatomy,* 2nd ed., John Wiley & Sons, Inc., New York, 1965.
The standard work in the field; a well-illustrated book that considers all aspects of plant anatomy.

* Available in paperback.

ESAU, KATHERINE: *Anatomy of Seed Plants*, 2nd ed., John Wiley & Sons, Inc., New York, 1977.
A shorter book than the preceding entry; an excellent textbook and reference.

FAHN, ABRAHAM: *Plant Anatomy*, 3rd ed., Pergamon Press, Inc., Elmsford, N.Y., 1982.*
A well-illustrated, up-to-date textbook considering all aspects of plant anatomy.

FOSTER, ADRIANCE S., and ERNEST M. GIFFORD: *Comparative Morphology of Vascular Plants*, 2nd ed., W. H. Freeman and Company, New York, 1974.
A well-organized, general account of the vascular plants, containing much interpretative material.

GENSEL, PATRICIA G., and HENRY N. ANDREWS: *Plant Life in the Devonian*, Praeger Publishers, New York; 1984.
This advanced book traces what is known about the earliest vascular plants at the time of their most active evolution.

MILNE, DAVID, et al. (Eds.): *The Evolution of Complex and Higher Organisms*, NASA Special Publication 478, U.S. Government Printing Office, 1985.
A concise summary of the history of multicellular life on earth, including an up-to-date review of the history of fossil plants.

RADFORD, ALBERT E., et al.: *Vascular Plant Systematics*, Harper & Row, Publishers, Inc., New York, 1974.*
A useful source book, including glossaries, techniques, bibliographies, indices, and useful discussions of all the various aspects of plant systematics—the scientific study of the kinds and diversity of plants, and of the relationships among them.

STEWART, WILSON N.: *Paleobotany and the Evolution of Plants*, Cambridge University Press, New York, 1983.
An excellent and well-illustrated account of all groups of fossil plants.

TAYLOR, THOMAS N.: *Paleobotany: An Introduction to Fossil Plant Biology*, McGraw-Hill Book Company, New York, 1981.
A readable and up-to-date survey that conveys both traditional knowledge and the excitement of new discoveries in paleobotany.

* Available in paperback.

18–1

A reconstruction of an Upper Carboniferous swamp forest dominated by the lycophyte tree Lepidodendron, forming the forest stand from left to center. Shown here are unbranched (resembling bottle brushes) and sparsely branched juvenile trees, as well as tall, many-branched canopy trees. Sigillaria, a lycophyte adapted to drier sites, is seen on the higher ground at the right (tree with tufts of long leaves). Other seedless trees include Calamites (the giant horsetails; far left and center foreground) and Psaronius (tree ferns with tapering trunks; far right). Seed plants also thrived, among them

CHAPTER 18

Seed Plants

Medullosa *(a seed fern with bifurcated fronds; center and far right) and* Cordaites, *both mangrovelike forms (with roots in the water; ground level on the right) and tree forms (tall trees; far right).*

One of the most dramatic innovations to arise during the evolution of the vascular plants was the seed. Seeds seem to be one of the factors responsible for the dominance of seed plants in today's flora—a dominance that has become progressively greater over a period of several million years (Figure 18–1). The reason is simple: the seed has survival value.

All seed plants are heterosporous, and the highly reduced megagametophyte is retained within the megaspore. The megaspore, in turn, is retained within the megasporangium, which is fleshy and is called the *nucellus* (plural: nucelli) in seed plants. The megasporangia of seed plants, unlike those of seedless heterosporous plants, are enveloped by one or two additional layers of tissue, the *integuments*. The integuments completely enclose the megasporangium except for an opening at the apex called the *micropyle*. This entire structure—the nucellus plus its integument(s)—is known as the *ovule* (Figure 18–2).

Following fertilization, the integument(s) develop into a *seed coat*, and a seed is formed. In other words, it is the ovule that develops into a seed; indeed, a seed is often called a mature ovule. In most modern seed plants, an embryo, or young sporophyte, develops within the seed before dispersal; in many of the ancient seed plants, such as those shown in Figures 18–2 to 18–4, the seeds may have been shed before the embryo developed. Perhaps the development of the embryo before dispersal gives the seed a better chance of survival in cold and harsh conditions, and the Permian period, when conifers, cycads, and *Ginkgo* arose, was certainly a period of climatic extremes. In addition to a megaspore or embryo and a seed coat, all seeds contain stored food, which further increases the young plant's chances to survive.

18–2

Sectional view of an ovule of Eurystoma angulare, *showing the spatial relationship of the integument, the megasporangium (nucellus), and the megaspore. The fertilization of ovules results in their maturation into seeds; in other words, seeds are mature, fertilized ovules.*

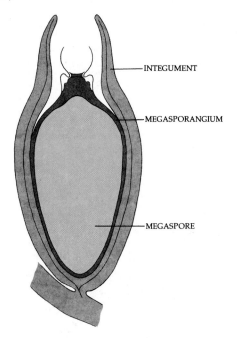

The oldest known seeds are from the late Devonian period, some 360 million years ago (Figure 18–3). During the next 50 million years, a wide array of seed-bearing plants evolved, including the seed ferns, cordaites, and conifers (Figure 18–4). The earliest vascular plants, including the three extinct divisions described in Chapter 17, had very wide distributions compared with those of their seed-bearing descendants. The reason for this is that seeds are less widely dispersed than spores. With the advent of plants that bore seeds, the floras—assemblages of plants—found in different regions of the world became more distinct from one another.

The seed plants, all of which possess megaphylls, include five divisions with living representatives: the Cycadophyta, the Ginkgophyta, the Coniferophyta, the Gnetophyta, and the Anthophyta. Before beginning our discussion of seed plants, we shall briefly examine one more group of seedless vascular plants—the entirely fossil progymnosperms. They are discussed at this time, rather than in Chapter 17, because they are the likely progenitors of the gymnosperms.

The angiosperms, which are overwhelmingly the most successful vascular plants at the present time, have not been identified with certainty in the fossil record until about 127 million years ago, in the early part of the Cretaceous period. Their evolution in relation to that of the gymnosperms will be considered in Chapter 29. Although the angiosperms actually must be somewhat older than our current understanding of the fossil record indicates, they are still relative newcomers in the broad picture of vascular plant evolution.

18–3

(a) *Reconstruction of a fertile branch of the late Devonian plant* Archeosperma arnoldii, *showing its seedlike structures. The cupules—cup-shaped structures that partly enclose the megasporangia—are arranged in pairs, and each cupule contains two flask-shaped seeds about 4 millimeters long. The apex of each seed was dissected into lobes. (b) Diagram showing the position of the megaspore within the seed. (c) A megaspore released from a seed by maceration. This fossil, from Pennsylvania, is the oldest known seed—about 360 million years in age.*

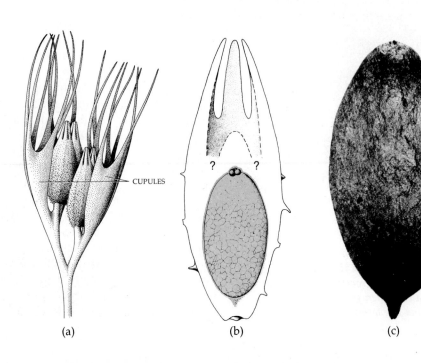

(a)   (b)   (c)

18–4

Seedlike structures in several Paleozoic plants, showing some stages in the evolution of seeds. In Genomosperma kidstonii *(the Greek word* genomein *means "to become," and* sperma *means "seed"), the nucellus was surrounded by 8 to 11 fingerlike projections, which arose at the base of the megasporangium and are separated for their entire length. No integuments were present in this genus. In* Genomosperma latens *the integumentary lobes are fused from the base of the megasporangium for about a third of their length, forming a simple seed coat. In* Eurystoma angulare, *the fusion is almost complete, while in* Stamnostoma huttonense *it' is complete.*

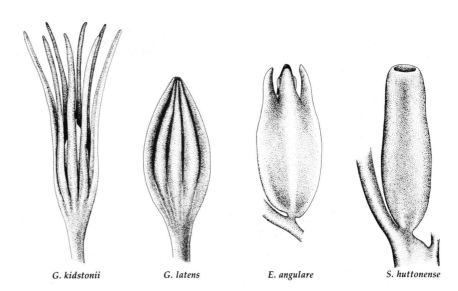

G. kidstonii   G. latens   E. angulare   S. huttonense

THE PROGYMNOSPERMS

In the late Paleozoic era, there existed a group of plants called the progymnosperms, which had characteristics intermediate between those of the trimerophytes and the gymnosperms. Although the progymnosperms reproduced by freely dispersed spores, they produced secondary xylem remarkably similar to that of gymnosperms (Figure 18–5); progymnosperms were unique among woody Devonian plants in producing secondary phloem also. The progymnosperms and Paleozoic ferns may have evolved from the more ancient trimerophytes (see Figure 17–8c, page 307), from which they differed primarily by having more elaborate branch systems and correspondingly more complex vascular systems. Another possibility is that the ferns evolved from the progymnosperms. Since the trimerophytes gave rise to the progymnosperms, and the ferns evolved from one of

these groups, the question concerns the timing of the origin of the ferns.

In the progymnosperms, the most important evolutionary advance over the trimerophytes is the presence of a bifacial vascular cambium—that is, one that produces both secondary xylem and secondary phloem. Vascular cambia of this type are characteristic of seed plants and apparently evolved first in the progymnosperms.

One kind of progymnosperm, the *Aneurophyton*-type, which occurred in the Devonian period approximately 360 to 380 million years ago, featured three-dimensional branching (Figure 18–6) and had protosteles. Their organization was similar to that of some early seed ferns, leading some paleobotanists to suggest that the branch systems of the *Aneurophyton*-type progymnosperm may have been the precursors of the fernlike leaves of early seed ferns.

18–5

Radial view of the secondary xylem, or wood, of the progymnosperm Callixylon newberryi. *This fossil wood, with its regular series of pitted tracheids, is remarkably similar to that of certain gymnosperms.*

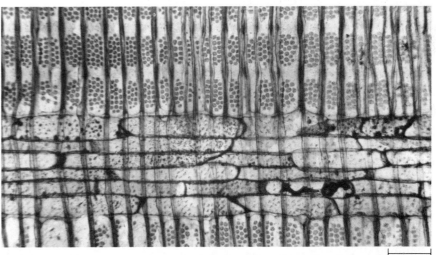

100 μm

18–6

*Reconstruction of a portion of the branch
system of* Triloboxylon ashlandicum, *an*
Aneurophyton-*type progymnosperm.
The main axis bears vegetative branches
at the top and bottom and fertile organs
with sporangia in between.*

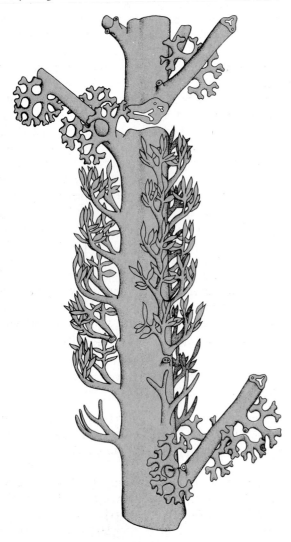

18–7

Reconstruction of the progymnosperm
Archaeopteris, *which is common in the
fossil record of eastern North America.
Specimens of* Archaeopteris *attained
heights of 20 meters or more, and some
of them seem to have formed forests.*

A second major kind of progymnosperm, the *Archaeopteris*-type, also appeared in the Devonian period, about 370 million years ago, and extended into the early Mississippian period, about 340 million years ago (see Figure 18–7). This group is considered advanced because their lateral branch systems were flattened in one plane and bore laminar structures considered to be leaves (Figure 18–8). These leafy branch systems resembled those of early conifers. The larger branches of *Archaeopteris*-type progymnosperms had a pith. Although most progymnosperms were homosporous, some species of *Archaeopteris* were heterosporous.

Some fossil logs of *Archaeopteris* are up to a meter or more in diameter and 10 meters long, indicating that at

least some species of this group were large trees. The way in which fossils of the group occur suggests that *Archaeopteris* dominated extensive forests in some regions. As the restoration in Figure 18–7 suggests, individuals of *Archaeopteris* may have resembled conifers in their branching patterns, but this is not certain.

Morphological evidence that has been amassed during the past several decades strongly supports the contention that the gymnosperms evolved from the progymnosperms upon the evolution of the seed. It is by no means certain, however, whether the seed evolved only once in this evolutionary line, or several times, nor is it certain from which group(s) of progymnosperms the gymnosperms evolved.

18–8

Reconstruction of a frondlike lateral branch system of the progymnosperm Archaeopteris macilenta. *Fertile leaves can be seen bearing maturing sporangia on centrally located primary branches.*

THE GYMNOSPERMS

There are four divisions of gymnosperms with living representatives: Cycadophyta (cycads), Ginkgophyta (maidenhair tree), Coniferophyta (conifers), and Gnetophyta (gnetophytes). The name *gymnosperm,* which literally means "naked seed," points to one of the principal characteristics of all seed plants except for the angiosperms: their ovules and seeds are exposed on the surface of sporophylls or analogous structures. The four divisions of gymnosperms probably represent the achievement of a particular stage in evolutionary progression by various descendants of the progymnosperms.

With few exceptions, the female gametophytes of gymnosperms produce several archegonia each. As a result, more than one egg may be fertilized, and several embryos may begin to develop within a single ovule—a phenomenon known as *polyembryony*. In most cases,

only one embryo survives, so that relatively few fully developed seeds contain more than one embryo.

In the seedless vascular plants, water is required for the motile, flagellated sperm to reach and fertilize the eggs. In the gymnosperms, water is not required as a medium of transport of the sperm to the eggs. Instead, the partly developed male gametophyte, the *pollen grain,* is transferred bodily (usually passively, by the wind) to the vicinity of a female gametophyte within an ovule. This process is called *pollination*. After pollination, the male gametophyte produces a tubular outgrowth, the *pollen tube*. In the conifers and gnetophytes, the sperm are nonmotile, and the pollen tubes convey them directly to the archegonia. In the cycads and *Ginkgo,* the sperm are multiflagellated, and the pollen tube apparently functions largely as a haustorial structure (analogous to the haustoria of parasitic fungi), which penetrates the ovule and absorbs nourishment directly from it. The pollen tube may grow for several months in the tissue of the nucellus, or megasporangium, before reaching a cavity above the female gametophyte. At that time, the pollen tube bursts and liberates its two sperm into the cavity. The sperm then swim to an archegonium, and one of them fertilizes the egg. With the development of sperm-conveying pollen tubes, the evolving vascular plants no longer were dependent on the presence of free water to assure fertilization, as are all seedless vascular plants.

The Conifers

By far the most numerous and widespread of the gymnosperm divisions living today, the Coniferophyta comprise some 50 genera with about 550 species. The tallest vascular plant, which is the redwood (*Sequoia sempervirens*) of coastal California and southwestern Oregon, is a member of this group (see Figure 31–1, page 653). Redwood trees attain heights of up to 117 meters and trunk diameters in excess of 11 meters. The conifers, which also include pines, firs, and spruces, are of great commercial value; their stately forests provide the wealth of vast regions of the North Temperate Zone. During the early Tertiary period, some genera were more widespread than they are now, and they dominated huge expanses on all of the northern continents.

The history of the conifers extends back at least to the late Carboniferous period, some 290 million years ago; the cordaites of that period (see boxed essay, pages 324 and 325) were primitive conifers. The leaves of modern conifers have many drought-resistant features, which may be related to the diversification of the division during the relatively dry and cold Permian period (286 to 248 million years ago). At that time, increasing worldwide aridity must have favored structural adaptations like those of conifer leaves.

The Pines

The pines (genus *Pinus*) include perhaps the most familiar of all gymnosperms (Figure 18–9); they dominate broad stretches of North America and Eurasia and are widely cultivated even in the Southern Hemisphere. There are about 90 species of pines, all of which are characterized by an arrangement of the leaves that is unique among living conifers. The foliage leaves of pines are needlelike. In the seedlings, they are spirally arranged and borne singly on the stems (Figure 18–10). After a year or two of growth, a pine begins to produce its leaves in bundles, or fascicles, each of which contains a specific number of long, needlelike leaves—from one to eight, depending on the species. These fascicles, wrapped at the base by a series of short, scale-like leaves, are actually short shoots in which the activity of the apical meristem is suspended (Figure 18–11). Thus, a fascicle of needles in a pine is morphologically a *determinate* (restricted in growth) branch. Under unusual circumstances, the apical meristem within a fascicle of needles in a pine may be reactivated and grow into a new shoot with *indeterminate* growth, or sometimes may even produce roots and grow into an entire pine tree (Figure 18–12).

The leaves of pines, like those of many other conifers, are impressively suited for growth under arid conditions (Figure 18–13). The epidermis is covered with a thick cuticle, beneath which are one or more layers of compactly arranged, thick-walled cells—the hypodermis. The stomata are sunken below the surface of the leaf. The mesophyll, or ground tissue of the leaf, consists of parenchyma cells with conspicuous wall ridges that project into the cells like the pieces of a puzzle. Commonly, the mesophyll is penetrated by two or more resin ducts. One vascular bundle, or two bundles side-by-side, are found in the center of the leaf. The veins are surrounded by transfusion tissue, composed of living parenchyma cells and short, nonliving tracheids. The transfusion tissue is believed to conduct materials between the mesophyll and vascular bundles.

18–9

Longleaf pines, Pinus palustris, *growing in North Carolina.*

18–10

(a) *Seedlings of longleaf pine,* Pinus palustris, *in Georgia, showing the juvenile leaves (long needles, borne singly) and the first mature leaves borne in this species in fascicles, or bundles, of three.* (b) *A seedling of pinyon pine,* Pinus edulis, *showing juvenile leaves and a young taproot system. The mature leaves of this species are borne in fascicles of two needles each.*

(a)

(b)

(a)

(b)

18–11

(a) *Bristlecone pine,* Pinus longaeva, *at Bryce Canyon, Utah. Branch showing clusters of five mature needles and a mature ovulate cone. The individual needles* of this species may remain functional for up to 45 years. It is also the longest-lived tree; see also Figure 23–27a, page 464. (b) *Branch of red pine,* Pinus resinosa, with young ovulate cones. Note the fascicles of needlelike leaves characteristic of the adult plant.

18–12

*One-year-old Monterey pines (*Pinus radiata*) grown from rooted fascicles of needles. This experiment emphasizes that a fascicle of pine needles is actually a short shoot in which the activity of the apical meristem has been suspended but can be regenerated.*

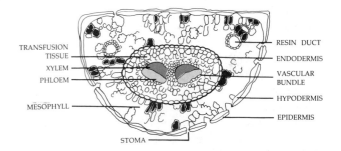

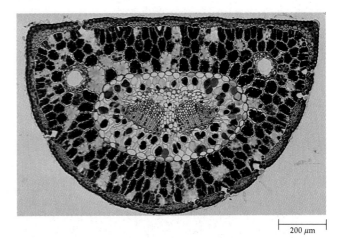

200 μm

18–13

Pinus. *Transverse section of needle, showing mature tissues.*

An endodermis surrounds the transfusion tissue, so that the transfusion tissue and mesophyll are not in direct contact with each other.

Most pine species retain their needles for two to four years, and the overall photosynthetic balance of a given plant depends on the health of several years' crop of needles. In bristlecone pine (*Pinus longaeva*), the longest-lived tree (see Figures 18–11a and 23–27a), the needles are retained up to 45 years and remain photosynthetically active the entire time. Because pines and other evergreens depend on their leaves functioning for more than one season, they are exposed to damage by drought, freezing, or air pollution for much longer than the leaves of deciduous plants and are more often injured.

In the stems of pines and other conifers, secondary growth begins early and leads to the internal formation of substantial amounts of secondary xylem (Figure 18–14). Secondary xylem is produced toward the inside of the vascular cambium, and secondary phloem is produced toward the outside. The xylem of conifers consists primarily of tracheids; the phloem consists of sieve cells, which are the typical food-conducting cells of gymnosperms and seedless vascular plants (see page 390). Both kinds of tissues are traversed radially by narrow rays. With the initiation of secondary growth, the epidermis is eventually replaced with a periderm, which has its origin in the outer layer of cortical cells.

As secondary growth continues, subsequent periderms are produced by active cell division deeper in the bark.

Reproduction in Pines.   The microsporangia and megasporangia in pines and most other conifers are borne in separate cones on the same tree. Ordinarily the microsporangiate cones are borne on the lower branches of the tree, and the megasporangiate, or ovulate, cones are borne on the upper branches; in some pines, they are borne on the same branch, with the ovulate cones closer to the ends. Since the pollen is not normally blown straight upward, the ovulate cones will normally be pollinated by pollen from another tree, thus enhancing outcrossing.

Microsporangiate cones in the pines are relatively small, usually 1 to 2 centimeters long (Figure 18–15). The microsporophylls (Figure 18–16) are spirally arranged and more or less membranous; each bears two microsporangia. A young microsporangium contains many microsporocytes, or microspore mother cells; in early spring the microsporocytes undergo meiosis, each giving rise to four haploid microspores. Each microspore develops into a winged pollen grain, consisting of two *prothallial cells*, a *generative cell*, and a *tube cell* (Figure 18–17). This four-celled pollen grain comprises the immature male gametophyte. It is at this stage that the pollen grains are shed in enormous quantities; some are carried by the wind to the ovulate cones.

18–14
Pinus. Cross section of stem, showing secondary xylem and secondary phloem separated from one another by vascular cambium. All of the tissues outside the vascular cambium, including the phloem, comprise the bark.

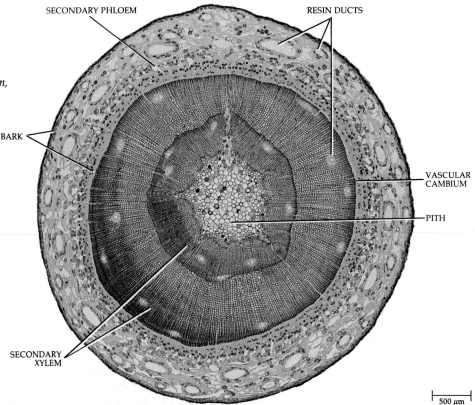

SECONDARY PHLOEM

RESIN DUCTS

BARK

VASCULAR CAMBIUM

PITH

SECONDARY XYLEM

500 μm

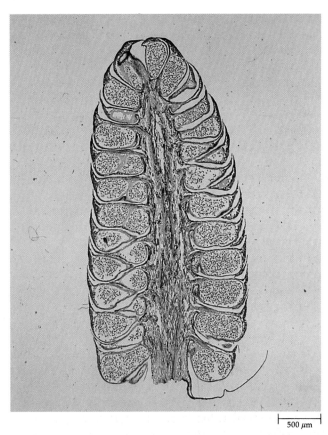

500 μm

18–15
Monterey pine, Pinus radiata. Micro-sporangiate cones shedding pollen, which is blown about by the wind. Some of the pollen reaches the vicinity of the ovules in the ovulate cones and then germinates, producing pollen tubes and eventually bringing about fertilization.

18–16
Pinus. *Longitudinal view of a pollen-producing cone, showing microsporophylls and microsporangia containing mature pollen grains.*

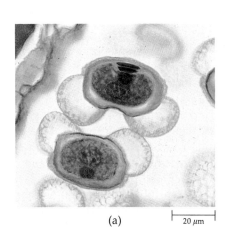

(a)   20 μm

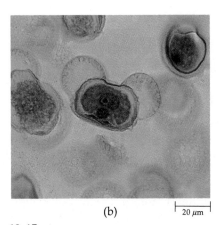

(b)   20 μm

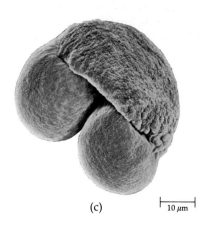

(c)   10 μm

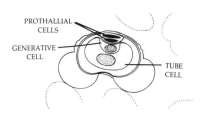

PROTHALLIAL CELLS

GENERATIVE CELL

TUBE CELL

18–17
Pinus. (a) *Pollen grains with enclosed immature male gametophytes. Each gametophyte consists of two prothallial cells, a relatively small generative cell, and a relatively large tube cell. (b) A somewhat older pollen grain than that shown in (a). Here the prothallial cells,* which have no apparent function, have degenerated. (c) *Scanning electron micrograph of a pine pollen grain, with its two bladder-shaped wings. When the pollen grain germinates, the pollen tube emerges from the lower end of the grain between the wings.*

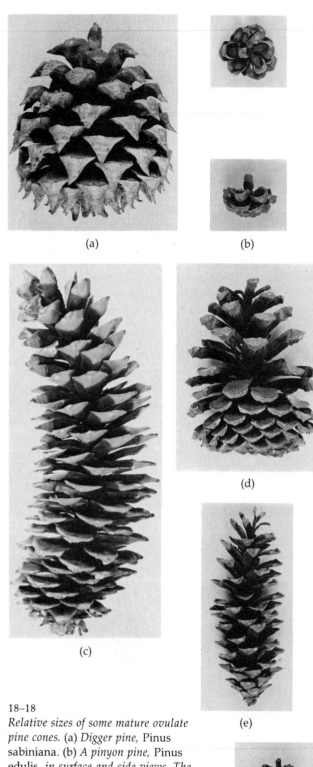

(a)

(b)

(c)

(d)

(e)

(f) ⌐ 25 μm

18–18

Relative sizes of some mature ovulate pine cones. (a) Digger pine, Pinus sabiniana. (b) A pinyon pine, Pinus edulis, in surface and side views. The edible seeds of this and certain other pines are called "pine nuts." They are wingless and are spread from place to place largely by nutcrackers, birds of the crow family. (c) Sugar pine, Pinus lambertiana. (d) Yellow pine, Pinus ponderosa. (e) Eastern white pine, Pinus strobus. (f) Red pine, Pinus resinosa.

The ovulate cones of pines are much larger and more complex in their structure than are the pollen-bearing cones (Figure 18–18). The ovuliferous scales (cone scales), which bear the ovules, are not megasporophylls; instead, they are entire modified determinate branch systems properly known as *seed-scale complexes*. Each seed-scale complex consists of the ovuliferous scale—which bears two ovules on its upper surface—and a subtending sterile bract (Figure 18–19). The scales are arranged spirally around the axis of the cone. (The ovulate cone is, therefore, a compound structure, whereas the microsporangiate cone is a simple one, in which the microsporangia are directly attached to the microsporophylls.) Each ovule contains a multicellular nucellus (the megasporangium) surrounded by a massive integument with an opening, the micropyle, facing the cone axis (Figure 18–19). Each megasporangium contains a single megasporocyte, or megaspore mother cell, which ultimately undergoes meiosis, giving rise to a linear series of four megaspores. However, only one of these megaspores is functional; the three nearest the micropyle soon degenerate.

Pollination in pines occurs in the spring; the pollen adheres to a drop of sticky fluid about the micropyle. At this stage, the scales of the ovulate cone are widely separated. As the micropylar fluid evaporates, the pollen grain is drawn into the micropyle and comes in contact with the nucellus. After pollination, the scales grow together and help protect the developing ovules. Shortly after the pollen grain comes in contact with the nucellus, it germinates, forming a pollen tube. At this time, meiosis has not yet occurred in the megasporangium. About a month after pollination, the four megaspores are produced, only one of which develops into a megagametophyte. The development of the megagametophyte is sluggish; it often does not begin until some 6 months after pollination, and it may then require another 6 months for completion. In the early stages of megagametophyte development, mitosis proceeds without immediate cell wall formation. About 13 months after pollination, when the female gametophyte contains some 2000 free nuclei, cell wall formation begins. Then, approximately 15 months after pollination occurs, archegonia, usually two or three in number, differentiate at the micropylar end of the megagametophyte. The stage is now set for fertilization.

About 12 months earlier, the pollen grain had germinated, producing a pollen tube that slowly digested its way through the tissues of the nucellus toward the developing megagametophyte. About a year after pollination, the generative cell of the four-celled male gametophyte undergoes division, giving rise to two kinds of cells—a *sterile cell* (stalk cell) and a *spermatogenous cell* (body cell). Subsequently, before the pollen tube reaches the female gametophyte, the spermatogenous cell divides, producing two sperm. The male gameto-

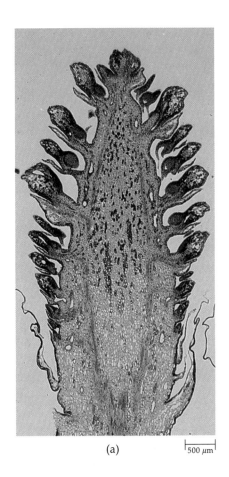

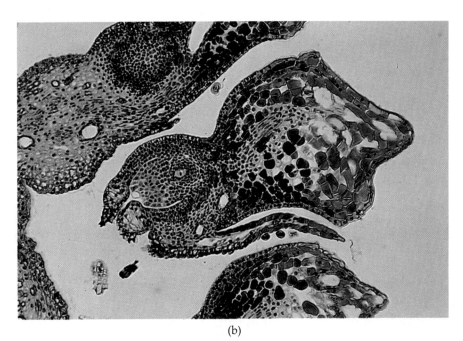

18–19
Pinus. (a) *Longitudinal view of young female cone, or strobilus, showing its complex structure.* (b) *Detail of a portion* of the strobilus. *Note the megasporocyte (megaspore mother cell) surrounded by the nucellus.*

(a)

|500 μm|

(b)

phyte, or germinating pollen grain, is now mature. Seed plants do not form antheridia.

Some 15 months after pollination, the pollen tube reaches the egg cell of an archegonium, where it discharges much of its cytoplasm and both of its sperm into the egg cytoplasm (Figure 18–20). One sperm nucleus unites with the egg nucleus, and the other degenerates. Commonly, the eggs of all archegonia are fertilized and begin to develop into embryos (the phenomenon of polyembryony). However, only one embryo usually develops fully.

During early embryogeny, four tiers of cells are produced near the lower end of the archegonium. Each of the four cells of the uppermost tier (that is, the tier farthest from the micropylar end of the ovule) begins to form an embryo. Simultaneously the four cells of the tier below the embryos, the suspensor cells, elongate greatly and force the four developing embryos into the female gametophyte. Thus, a second type of polyembryony is found in the pine life cycle. Once again, however, only one of the embryos develops fully. During embryogeny, the integument develops into a seed coat.

The conifer seed is a remarkable structure, for it consists of a combination of two sporophytic generations —the seed coat (and remnants of the nucellus) and the embryo—and one gametophytic generation. The game-

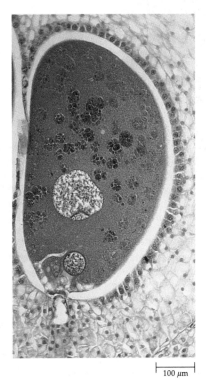

|100 μm|

18–20
Pinus. *Fertilization: union of a sperm nucleus with the egg nucleus. The second sperm nucleus (below) is nonfunctional; it will eventually disintegrate.*

18–21
Pinus. *Longitudinal section of seed.*

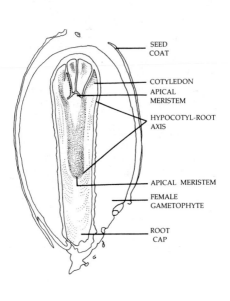

SEED
COAT

COTYLEDON
APICAL
MERISTEM

HYPOCOTYL-ROOT
AXIS

APICAL MERISTEM

FEMALE
GAMETOPHYTE

ROOT
CAP

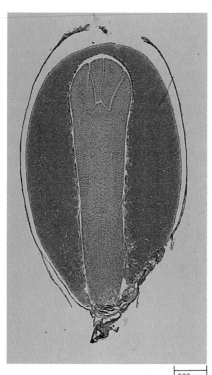

500 µm

tophyte serves as reserve food or nutritive tissue (Figure 18–21). The seed coat and embryo are diploid, and the female gametophyte is haploid. The embryo consists of a hypocotyl-root axis, with a root cap and apical meristem at one end and an apical meristem and several (generally eight) cotyledons, or seed leaves, at the other. The integument consists of three layers, of which the middle layer becomes hard and serves as the seed coat.

The seeds of pines are often shed from the cones during the autumn of the second year following the initial appearance of the cones and pollination. At maturity, the cone scales separate; the winged seeds of most species flutter through the air and are sometimes carried considerable distances by the wind. In some species of pines, such as jack pines (*Pinus banksiana*), the scales do not separate until the cones are subjected to extreme heat. When a forest fire sweeps rapidly through the pine grove and burns the parent trees, most of the fire-resistant cones are only scorched. These cones open, release the seed crop they have accumulated over many years, and reestablish the species in the recently burned soil. In other species of pines, including the limber pine (*Pinus flexilis*), whitebark pine (*Pinus albicaulis*), and pinyon pines of western North America, as well as a few similar species of Eurasia, the wingless, large seeds are harvested, transported, and stored for later eating by large, crowlike birds called nutcrackers.

The pine life cycle is summarized in Figure 18–25.

Other Conifers

Although other conifers (see Figures 18–22 through 18–29) lack the needle clusters of the pines and may also differ in a number of relatively minor details of their reproductive systems, the living conifers form a fairly homogeneous group. Among their important representatives are the firs (*Abies*; Figure 18–22a), spruces (*Picea*), hemlocks (*Tsuga*), Douglas fir (*Pseudotsuga*), cypresses (*Cupressus*; Figure 18–23), and junipers (*Juniperus*; Figure 18–24). In the yews (family Taxaceae), the ovules are not borne in cones but are solitary and surrounded by a fleshy, cuplike structure—the *aril* (Figure 18–26a). A number of other unique genera of conifers are found primarily in the Southern Hemisphere. Some of them, like the Norfolk Island pine (*Araucaria heterophylla*) and the monkey-puzzle tree (*A. araucana*), are frequently cultivated in areas where the climate is mild enough, and others cross the equator naturally.

One of the most interesting groups of conifers is the family Taxodiaceae, which appeared in the fossil record about 150 million years ago and is represented today by widely scattered species that are the remnants of populations that were much more widespread during the Tertiary period. One of the most remarkable of these is the coast redwood, *Sequoia sempervirens*, the tallest living plant (see Figure 31–1). The famous "big tree," *Sequoiadendron giganteum*, which forms spectacular, widely scattered groves along the west slope of the Sierra Nevada of California, also belongs to this family.

(a)

(b)

18–22
Two genera of the pine family (Pinaceae). (a) Ovulate cone of balsam fir (Abies balsamea). The upright cones, which are 5 to 10 centimeters long, do not fall to the ground whole, as they do in the pines; rather, they shatter and fall apart, *scattering the winged seeds. (b) European larch (Larix decidua). The leaves of larch are needlelike, as in the pines; they are borne singly on short branch shoots* *and are spirally arranged. Unlike most conifers, the larches are deciduous; that is, they shed their leaves at the end of each growing season.*

18–23
Gowen cypress (Cupressus goveniana). The small trees of this species—only about 6 meters tall at maturity—are extremely local, being found only near Monterey, California.

18–24
The common juniper (Juniperus communis) has spherical ovulate cones like those of the cypresses, but the scales are fleshy and fused together. Juniper "berries" give gin its distinctive taste and aroma.

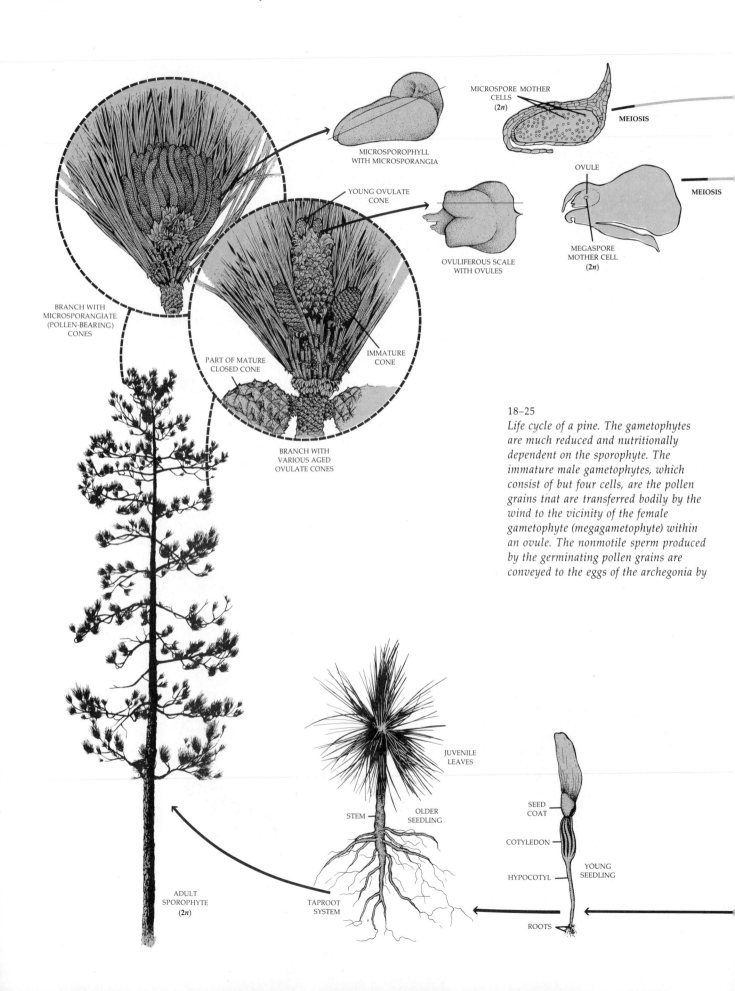

MICROSPORE MOTHER
CELLS
(2n)

MEIOSIS

MICROSPOROPHYLL
WITH MICROSPORANGIA

OVULE

MEIOSIS

YOUNG OVULATE
CONE

OVULIFEROUS SCALE
WITH OVULES

MEGASPORE
MOTHER CELL
(2n)

BRANCH WITH
MICROSPORANGIATE
(POLLEN-BEARING)
CONES

PART OF MATURE
CLOSED CONE

IMMATURE
CONE

BRANCH WITH
VARIOUS AGED
OVULATE CONES

18–25
*Life cycle of a pine. The gametophytes
are much reduced and nutritionally
dependent on the sporophyte. The
immature male gametophytes, which
consist of but four cells, are the pollen
grains tnat are transferred bodily by the
wind to the vicinity of the female
gametophyte (megagametophyte) within
an ovule. The nonmotile sperm produced
by the germinating pollen grains are
conveyed to the eggs of the archegonia by*

JUVENILE
LEAVES

SEED
COAT

COTYLEDON

STEM

OLDER
SEEDLING

HYPOCOTYL

YOUNG
SEEDLING

ADULT
SPOROPHYTE
(2n)

TAPROOT
SYSTEM

ROOTS

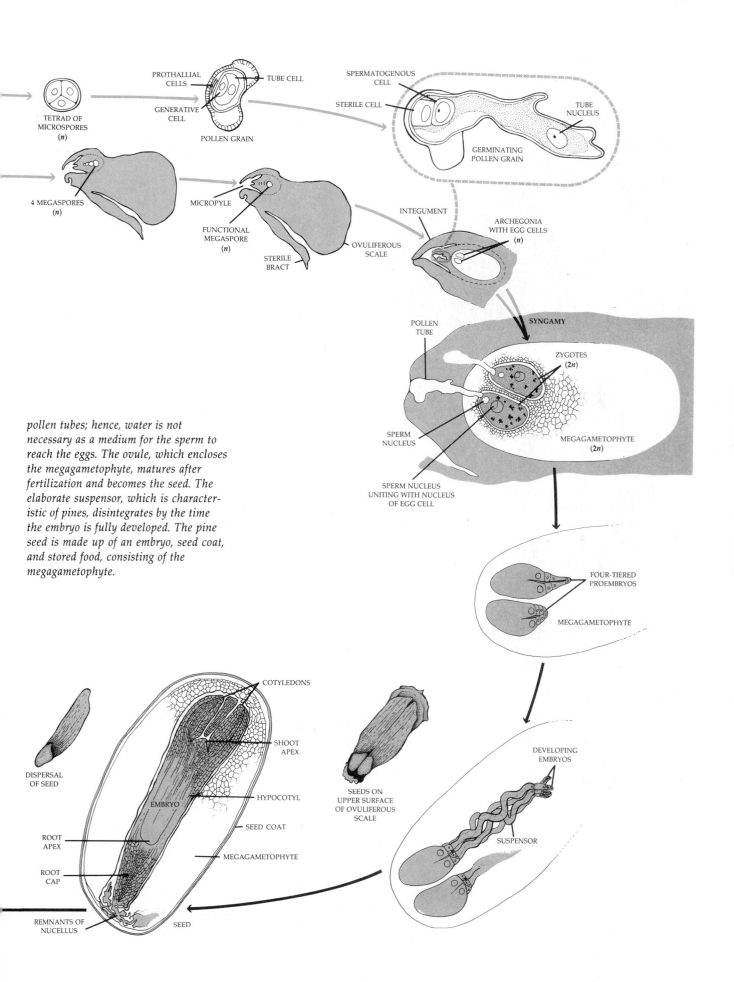

TETRAD OF
MICROSPORES
(*n*)

PROTHALLIAL
CELLS

TUBE CELL

GENERATIVE
CELL

POLLEN GRAIN

SPERMATOGENOUS
CELL

STERILE CELL

TUBE
NUCLEUS

GERMINATING
POLLEN GRAIN

4 MEGASPORES
(*n*)

MICROPYLE

FUNCTIONAL
MEGASPORE
(*n*)

STERILE
BRACT

OVULIFEROUS
SCALE

INTEGUMENT

ARCHEGONIA
WITH EGG CELLS
(*n*)

POLLEN
TUBE

SYNGAMY

ZYGOTES
(2*n*)

SPERM
NUCLEUS

SPERM NUCLEUS
UNITING WITH NUCLEUS
OF EGG CELL

MEGAGAMETOPHYTE
(2*n*)

*pollen tubes; hence, water is not
necessary as a medium for the sperm to
reach the eggs. The ovule, which encloses
the megagametophyte, matures after
fertilization and becomes the seed. The
elaborate suspensor, which is characteristic of pines, disintegrates by the time
the embryo is fully developed. The pine
seed is made up of an embryo, seed coat,
and stored food, consisting of the
megagametophyte.*

FOUR-TIERED
PROEMBRYOS

MEGAGAMETOPHYTE

COTYLEDONS

SHOOT
APEX

DISPERSAL
OF SEED

EMBRYO

HYPOCOTYL

SEED COAT

SEEDS ON
UPPER SURFACE
OF OVULIFEROUS
SCALE

DEVELOPING
EMBRYOS

ROOT
APEX

ROOT
CAP

MEGAGAMETOPHYTE

SUSPENSOR

REMNANTS OF
NUCELLUS

SEED

(a)

(b)

18–26

*The conifers of the yew family (Taxaceae)
have seeds that are surrounded by a
fleshy cup—the aril. The arils attract
birds and other animals that eat them and
thus spread the seeds. (a) Members of the
genus* Taxus, *the yews—which occur
around the Northern Hemisphere—produce red, fleshy ovulate structures. (b)
Sporophylls and microsporangia of the
pollen-bearing cones of a yew. Ovulate
cones and pollen-bearing cones are found
on separate individuals. The arils of
yews contain a toxic substance and are
the major cause of poisoning by plants in
children in the United States, although
fatalities are extremely rare.*

18–27

The bald cypress (Taxodium distichum)
*is a deciduous member of the redwood
family that grows in the swamps of the
southeastern United States. In this
autumn photograph, the leaves have
begun to change color. Spanish "moss"*
(Tillandsia usneoides), *actually a
flowering plant related to the pineapple,
is seen hanging from the branches of
these trees in masses.*

The bald cypresses (*Taxodium*) of the southeastern
United States and Mexico (Figure 18–27) are still another genus. All these genera were more widespread in
the Tertiary period than they are now (see Figure
18–29).

   Another genus that was abundant in the Tertiary in
both Eurasia and North America was *Metasequoia*, the
dawn redwood (Figure 18–28). Indeed, *Metasequoia*
was the most abundant conifer in western and arctic
North America from the late Cretaceous period to the
Miocene epoch (in other words, from about 90 to about
15 million years ago). It survived both in Japan and in
eastern Siberia until a few million years ago. The genus
Metasequoia was first described from fossil material by
the Japanese paleobotanist Shigeru Miki in 1941 (Figure 18–29). Three years later the Chinese forester
Tsang Wang, from the Central Bureau of Forest Research of China, visited the village of Mo-tao-chi in re-

18–28
The dawn redwood (Metasequoia glyptostroboides). *This tree, growing in Hubei Province in central China, is more than 400 years old.*

mote Sichuan Province in southwestern China. There he discovered a huge tree of a kind he had never seen before. The natives of the area had built a temple around the base of the tree. Tsang collected specimens of the tree's needles and cones, and studies of these samples revealed that the fossil *Metasequoia* had "come to life." In 1948, paleobotanist Ralph Chaney of the University of California led an expedition down the Yangtze River and across three mountain ranges to valleys where a thousand dawn redwoods were growing, the last remnant of the once great *Metasequoia* forest. Thousands of seeds were subsequently distributed widely, and this "living fossil" can now be seen growing in parks and gardens all over the world.

Other Living Gymnosperms

Cycads

The other groups of living gymnosperms are remarkably diverse and scarcely resemble one another at all. Among them are the cycads, division Cycadophyta, which are palmlike plants found mainly in tropical and subtropical regions. These bizarre plants, which appeared at least 285 million years ago near the start of the Permian period, were so numerous in Mesozoic times that this era is often called the "Age of Cycads and Dinosaurs." The members of this division are related to the seed ferns (see essay, pages 324, 325), and there are apparently transitional forms in the fossil record. Living cycads comprise 10 genera, with about 100 species. *Zamia pumila*, which occurs commonly in the sandy woods of Florida, is the only cycad native to the United States (Figure 18–30).

18–29
Fossil branchlet of Metasequoia, *about 50 million years old. The accompanying map shows the geographic distribution of some living and fossil members of the redwood family (Taxodiaceae).*

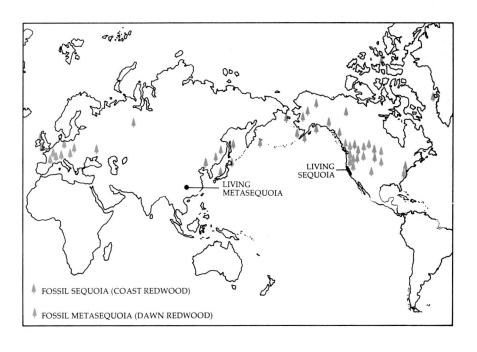

LIVING METASEQUOIA

LIVING SEQUOIA

⬧ FOSSIL SEQUOIA (COAST REDWOOD)

⬧ FOSSIL METASEQUOIA (DAWN REDWOOD)

18–30
Ovulate and microsporangiate plants of
Zamia pumila, *the only species of cycad
native to the United States. The stems
are mostly or entirely underground and,
along with the rootstalks, were used by
the Seminole Indians as food. The two
large grey cones in the foreground are
ovulate cones; the smaller brown cones
are microsporangiate cones.*

Most cycads are fairly large plants; some reach 18 meters or more in height. Many have a distinct trunk that is densely covered with the bases of shed leaves. The functional leaves characteristically occur in a cluster at the top of the stem; thus the cycads resemble palms (indeed, a common name for some cycads is "sago palms"). Unlike palms, however, cycads exhibit true, if sluggish, secondary growth from a vascular cambium; the central portion of their trunks consists of a great mass of pith. The reproductive units of cycads are more or less reduced leaves with attached sporangia that are loosely or tightly clustered into conelike structures near the apex of the plant. The pollen and seed cones of cycads are borne on different plants (Figure 18–31).

Ginkgo

The maidenhair tree, *Ginkgo biloba*, is easily recognized by its fan-shaped leaves with their openly branched, dichotomous (forking) pattern of veins (Figure 18–32). It is an attractive and stately, but slow-growing, tree that may reach a height of 30 meters or more. The leaves on the numerous spur shoots of *Ginkgo* are more or less entire, whereas those of the long shoots and seedlings are deeply lobed. Unlike most other gymnosperms, *Ginkgo* is deciduous; its leaves turn a beautiful golden color before falling in the autumn.

(a)

(b)

18–31
(a) Encephalartos altensteinii, *a cycad native to South Africa. Shown here is a microsporangiate plant with male cones.*
(b) *An ovulate plant of* Cycas siamensis. *The top of one ovulate cone has been removed to reveal seeds on upper surfaces of the megasporophylls.*

Ginkgo biloba is the sole living survivor of a genus that has changed little for more than 80 million years and is the only living member of the division Ginkgophyta. The living species shares features with other genera of gymnosperms that range back to the early Permian period, some 280 million years ago. There are probably no wild stands of *Ginkgo* anywhere in the world, but the tree was preserved in temple grounds in China and Japan. Introduced into cultivation from there, it has been an important feature of the gardens of the temperate regions of the world for more than 150 years. *Ginkgo* is especially resistant to air pollution and so is commonly cultivated in urban parks.

As do the cycads, *Ginkgo* bears its ovules and microsporangia on different individuals. The ovules of *Ginkgo* are borne in pairs on the end of short stalks and ripen to produce fleshy-coated seeds in autumn. In *Ginkgo*, fertilization within the ovules may not occur until after they have been shed from their parent tree. Embryos are formed during the later stages of seed maturation, which occur on the ground. The seeds have a rancid odor as a result of the butyric acid in their fleshy coats, and for this reason only male trees are usually cultivated in parks or gardens. The trees are propagated vegetatively, so that male trees can be obtained in this way. The microsporophylls are grouped in conelike structures; each microsporophyll bears two microsporangia.

Gnetophyta

The gnetophytes comprise three living genera and about 70 species of very unusual gymnosperms: *Gnetum, Ephedra,* and *Welwitschia.* Although these genera are clearly related to one another and are appropriately placed together, they differ greatly in their characteristics. These genera do, however, have many angiospermlike features, such as the similarity of their strobili to some angiosperm inflorescences, the presence of very similar vessels in their xylem, and the lack of archegonia in *Gnetum* and *Welwitschia.* For decades, scientists have debated whether the gnetophytes are related to the angiosperms. At present, the evidence appears to favor such a relationship, although none of the living gnetophytes could possibly be an ancestor of any angiosperm—each of the three living genera of gnetophytes has its own unique specializations.

Gnetum, a genus of about 30 species, consists of trees and climbing vines with large leathery leaves that closely resemble those of dicotyledons (Figure 18–33). It is found throughout the moist tropics.

Most of the approximately 35 species of *Ephedra* are profusely branched shrubs with inconspicuous, small, scalelike leaves (Figure 18–34). With its small leaves and apparently jointed stems, *Ephedra* superficially resembles *Equisetum.* Most species of *Ephedra* inhabit arid or desert regions of the world.

Welwitschia is probably the most bizarre vascular plant (Figure 18–35). Most of the plant is buried in sandy soil. The exposed part consists of a massive woody, concave disk that produces only two strap-shaped leaves, with the cone-bearing branches arising from meristematic tissue on the margin of the disk. *Welwitschia* grows in the desert areas of southwestern Africa, in Angola, Namibia, and South Africa.

(a)

(b)

18–32
(a) Ginkgo biloba, *the maidenhair tree. This tree is given its English name because of the resemblance between its leaves and the leaflets of maidenhair fern.* (b) Ginkgo *leaves and fleshy seeds attached to spur shoot.*

(a)

(b)

(c)

18–33
The large leathery leaves of the tropical gnetophyte Gnetum *resemble those of certain dicots. The species of* Gnetum *grow as shrubs or woody vines in tropical or subtropical forests. (a) A functionally microsporangiate inflorescence. (b) Microsporangiate inflorescence and leaves. (c) Fleshy seeds with leaves. (b) and (c) were photographed in the Amazon Basin of southern Venezuela.*

(a)

(b)

(c)

18–34
Ephedra *is the only one of the three living genera of Gnetophyta found in the United States. (a) A male shrub of* Ephedra viridis, *in California. It is a densely branched shrub that has scalelike leaves, like other members of the genus. (b) Microsporangiate strobili of E. viridis. Note the scalelike leaves on the stem. (c) Ephedra trifurca, in Arizona, with microsporangiate strobili. (d) Female plant of E. viridis with seeds.*

(b)

(a)

(c)

18–35
The gnetophyte Welwitschia mirabilis, *found only in the Namib Desert of Namibia and adjacent regions of southwestern Africa. Welwitschia produces only two leaves, which continue to grow for the life of the plant. As growth continues, the leaves break off at the tips and split lengthwise; thus older plants appear to have numerous leaves. (a) A large, seed-producing plant. (b) Microsporangiate strobili. (c) Ovulate strobili; the insect is a fire bug, sucking sap from the strobili. Welwitschia is dioecious.*

(d)

THE ANGIOSPERMS

The flowering plants, division Anthophyta, comprise about 235,000 species, by far the largest number of species of any plant group. In their vegetative structures, they are enormously diverse. In size, they range from species of *Eucalyptus,* trees well over 100 meters tall and trunks nearly 20 meters in girth (Figure 18–36), to some duckweeds, which are simple floating monocots often scarcely 1 millimeter long (Figure 18–37). Some angiosperms are vines that climb high into the canopy of the tropical rain forest, while others are epiphytes that grow in the canopy. Many angiosperms, such as cacti, are adapted for growth in extremely arid regions. For more than 100 million years, the flowering plants have dominated the land.

The division Anthophyta includes two classes: the Monocotyledones (monocots), with about 65,000 species (Figure 18–38), and the Dicotyledones (dicots), with about 170,000 species (Figure 18–39). The similarities between these two groups are far greater than the differences; nonetheless the two classes are clearly recognizable natural units. Monocots include such familiar plants as the grasses, lilies, irises, orchids, cattails, and palms. The dicots include almost all the familiar trees and shrubs (other than conifers), and many of the herbs. The major differences between the monocots and dicots are summarized in Table 18–1.

Table 18–1   *Main Differences between Monocots and Dicots*

CHARACTERISTIC	DICOTS	MONOCOTS
Flower parts	In fours or fives (usually)	In threes (usually)
Pollen	Basically tricolpate (having three furrows or pores)	Basically monocolpate (having one furrow or pore)
Cotyledons	Two	One
Leaf venation	Usually netlike	Usually parallel
Primary vascular bundles in stem	In a ring	Complex arrangement
True secondary growth, with vascular cambium	Commonly present	Absent

18–36
The giant eucalyptus, or red tingle (Eucalyptus jacksonii), *growing in the Valley of the Giants in southwestern Australia. The enormous stature of this angiosperm is evident by comparison to the man standing in its burnt-out base.*

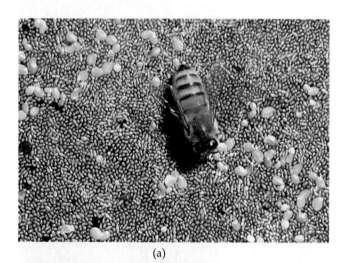

(a)

(c)

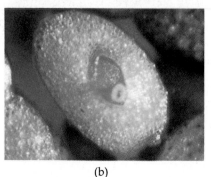

(b)

18–37
The duckweeds (family Lemnaceae) are the smallest flowering plants; their structural features mark them as extremely reduced derivatives of the aroid family (Araceae), the family that includes calla lilies (Zantedeschia) and Philodendron. (a) A honeybee resting on a dense floating mat of three species of duckweed. The larger plants are Lemna gibba, *about 2 to 3 millimeters long; the smaller ones are two species of* Wolffia, *up to 1 millimeter long. (b) A flowering plant of* Wolffia borealis, *with a circular concave stigma (looking like a tiny doughnut) and a minute anther just above it, both protruding from a central cavity. The whole plant is less than 1 millimeter long. (c) Flowering plant of* Lemna gibba; *two stamens and a style protrude from a pocket on the upper surface of the plant.*

(a)

(b)

(c)

18–38
Monocots. (a) A member of the palm family, the coconut palm (Cocos nucifera), growing in Tehuanatepec, Oaxaca, Mexico. The fruit of coconut is a drupe, not a nut (see Chapter 29). (b) *Flowers and fruits of the banana plant* (Musa × paradisiaca). *The banana flower has an inferior ovary, and the tip* of the fruit bears a large scar left by the fallen flower parts. (c) Rice (Oryza sativa) is a member of the grass family.

(a)

(b)

(c)

18–39
Dicots. (a) The water lily (Nymphaea odorata). The very fragrant flower of this species contains numerous petals and stamens and is regular, or radially symmetrical. The genus Nymphaea is widely distributed in tropical and temperate regions throughout the world. (b) Saguaro cactus (Carnegiea gigantea). *The cacti, of which there are about 2000 species, are almost exclusively a New World family. The thick, fleshy stems, which store water, contain chloroplasts and have taken over the function of the* leaves in photosynthesis. (c) Round-lobed hepatica (Hepatica americana), *which flowers in deciduous woodlands in the early spring. The flowers have no petals but have six to ten sepals and numerous spirally arranged stamens and carpels.*

18–40

*Parasitic and saprophytic angiosperms.
These plants have little or no chlorophyll;
they obtain their food as a result of the
photosynthesis of other plants. (a)
Dodder (Cuscuta salina), a parasite
that is bright orange or yellow. Dodder is
a member of the morning glory family
(Convolvulaceae). (b) Indian pipe
(Monotropa uniflora), a "saprophyte"
that obtains its food from the roots of
other plants via the fungal hyphae
associated with its roots. (c) The world's
largest flower,* Rafflesia arnoldii, *on
Mount Sago, Sumatra. Plants of this
genus are parasitic on the roots of a
member of the grape family (Vitaceae).
All of these plants are dicots.*

(a)

(b)

(c)

All but a few angiosperms are free-living, but both parasitic and saprophytic forms do exist (Figure 18–40). These plants mostly or completely lack chlorophyll. It has recently been shown that many, if not all, angiospermous "saprophytes" have obligate relationships with mycorrhizal fungi, which are associated with a second plant, in this case, a green, actively photosynthetic angiosperm. The fungus forms a bridge that actively transfers carbohydrates from the photosynthetic plant to the "saprophyte," which lacks chlorophyll. In addition to these "saprophytes," there are about 2800 species of parasitic dicots, like the mistletoes and *Cuscuta* (Figure 18–40a), and one parasitic gymnosperm (*Parasitaxus* of New Caledonia). Parasitic flowering plants form specialized absorptive organs called haustoria, by means of which they penetrate their hosts.

In Section 5, we shall consider in some detail the structure and development of the plant body, that is, the sporophyte of the flowering plant. The remainder of this chapter will be concerned with the most distinctive characteristic of angiosperms—the flower—and with the process of reproduction in flowering plants. In Chapter 29, the evolution of the flowering plants will be discussed in detail.

The Flower

The flower is a determinate shoot that bears sporophylls (Figure 18–41). The name "angiosperm" is derived from the Greek words *angeion*, meaning "vessel," and *sperma*, meaning "seed." Probably the most distinctive structure of the flower is the carpel—the "vessel." The carpel contains the ovules, which develop into seeds after fertilization.

Flowers may be clustered in various ways into aggregations called *inflorescences* (Figures 18–42 and 18–43). The stalk of an inflorescence or of a solitary flower is known as a *peduncle*. The stalk of an individual flower in an inflorescence is called a *pedicel*. The part of the flower stalk to which the floral parts are attached is termed the *receptacle*. Like any other shoot tip, the receptacle consists of nodes and internodes. In the flower, the internodes are very short, and consequently, the nodes are very close together.

Many flowers contain two sets of sterile appendages, the *sepals* and *petals*, which are attached to the receptacle below the fertile parts of the flower, the *stamens* and *carpels*. The sepals arise below the petals, and the stamens arise below the carpels. Collectively, the sepals form the *calyx*, and the petals form the *corolla*. Together, the calyx and corolla constitute the *perianth*. The sepals and petals are essentially leaflike in structure. Commonly the sepals are green and the petals are brightly colored, although in many flowers the members of both whorls are similar in color (Figure 18–41).

The stamens—collectively the *androecium* ("house of man")—are microsporophylls. In all but a few living angiosperms, the stamen consists of a slender stalk, or *filament*, upon which is borne a two-lobed *anther* containing four microsporangia, or pollen sacs.

The carpels—collectively the *gynoecium* ("house of woman")—are megasporophylls that are folded lengthwise, enclosing one or more ovules. A given flower may contain one or more carpels. If more than one carpel is present, they may be separate or fused together, in part or entirely. Sometimes the individual carpel or the group of fused carpels is called a pistil. The word "pistil" comes from the same root as "pestle," the instrument with a similar shape that pharmacists use for grinding substances into a powder in a mortar.

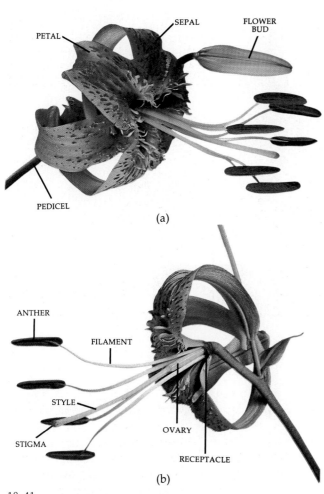

(a)

(b)

18–41

Parts of a lily (Lilium henryi) *flower. (a) An intact flower. In some flowers, such as lilies, the sepals and petals are similar to one another, and the perianth parts may then be referred to as tepals. Note that the sepals are attached to the receptacle below the petals. (b) A partly dissected flower with two tepals and two stamens removed to reveal the ovary. The gynoecium consists of the ovary, style, and stigma. The stamen consists of the filament and anther. Note that the sepals, petals, and stamens are attached to the receptacle below the ovary, which is said to be superior. Such a flower is said to be hypogynous.*

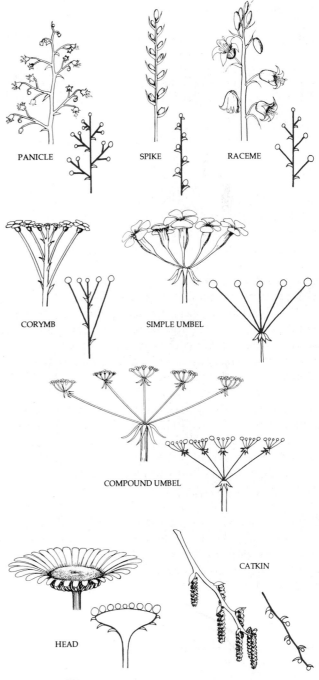

18–42

Illustrations of some of the common types of inflorescences found in the angiosperms, accompanied by simplified diagrams (in color).

(a)

(b)

(c)

(d)

(e)

18–43

Inflorescences of (a) *shooting star*
(Dodecatheon pauciflorum), (b) *butter-
and-eggs* (Linaria vulgaris), (c) *lupine*
(Lupinus diffusus), (d) *bluebells*
(Mertensia virginica), *and* (e) *water
hemlock* (Cicuta maculata). *Using
Figure 18–42 as a guide, can you identify
the types of inflorescences shown here?*

In most flowers, the individual carpels or group of fused carpels are differentiated into a lower part, the *ovary*, which encloses the ovules, and an upper part, the *stigma*, which receives the pollen. In many flowers a more or less elongated structure—the *style*—connects the stigma with the ovary. If the carpels are fused, there may be a common style or stigma, or each carpel may retain a separate style and stigma. The common ovary of such fused carpels is generally (but not always) partitioned into two or more locules—chambers of the ovary in which the ovules occur. The number of locules is usually related to the number of carpels in the gynoecium.

The portion of the ovary where the ovules originate and remain attached until maturity is called the *placenta*. The arrangement of the placentae (the placentation) and, consequently, the ovules, varies among different groups of flowering plants (Figure 18–44). In some flowers, the placentation is *parietal*; that is, the ovules are borne on the ovary wall or on extensions of it. In other flowers, the ovules are borne on a central column of tissue in a partitioned ovary with as many locules as there are carpels. This is *axile* placentation. In still others, the placentation is *free central*, the ovules being borne on a central column of tissue not connected by partitions with the ovary wall. And finally, in some flowers a single ovule occurs at the very base of a unilocular ovary. This is *basal* placentation. These differences are important in the classification of the flowering plants.

Given the basic structure of the flower, there are many differences between the flowers of different kinds of plants. The majority of flowers contain both stamens and carpels, and such flowers are said to be *perfect*. If either stamens or carpels are missing, the flower is *imperfect*, and depending on the part that is present, the flower is said to be either *staminate* or *carpellate* ("pistillate"; Figure 18–45). If both staminate and carpellate flowers occur on the same plant, as in corn (maize) and the oaks, the species is said to be *monoecious* (from the Greek words *monos*, "one," and *oecos*, "house"). If staminate and carpellate flowers are found on separate plants, the species is said to be *dioecious* ("two houses"), as in the willows and American holly. Seed plants other than angiosperms in which the pollen-bearing and seed-producing organs are produced on different individuals, such as *Ginkgo*, cycads, and junipers (conifers of the genus *Juniperus*), are also considered to be dioecious.

Any one of the floral whorls—sepals, petals, stamens, or carpels—may be lacking in the flowers of certain groups. Flowers with all four floral whorls are called *complete* flowers. If any whorl is lacking, the flower is said to be *incomplete*. Thus, an imperfect flower is also incomplete, but not all incomplete flowers are imperfect.

The particular arrangement of the floral parts may be either spiral on a more or less elongated receptacle, or similar parts—such as petals—may be attached in a whorl at a node. The parts may be united with other

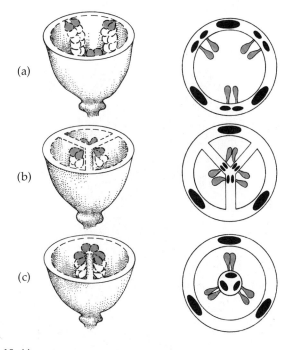

18–44
Types of placentation. (a) *Parietal.* (b) *Axile.* (c) *Free central. Basal placentation is not shown here.*

18–45
Staminate and carpellate flowers of a tan-bark oak (Lithocarpus densiflora). *Most members of the family Fagaceae, including the true oaks* (Quercus), *are monoecious, meaning that the staminate and carpellate flowers are separate but are borne on the same tree.*

members of the same whorl (*coalescence*) or with members of other whorls (*adnation*). An example of adnation is the union of stamens with the corolla, which is fairly common. When the floral parts of the same whorl are not joined, the prefixes *apo-* (meaning "separate") or *poly-* may be used to describe the condition. When the parts are coalesced, either *syn-* or *sym-* is used. For example, in an aposepalous or polysepalous calyx, the sepals are not joined; in a synsepalous calyx, they are joined.

In addition to the floral parts being either spirally arranged or whorled, the level of insertion of the sepals, petals, and stamens on the floral axis varies in relation to the level of the ovary or ovaries (Figure 18–46). If the sepals, petals, and stamens are attached to the receptacle below the ovary, the ovary is said to be *superior*, and the flower is said to be *hypogynous* (Figure 18–41). In some flowers with superior ovaries, the sepals, petals, and stamens are fused together to form a cup-shaped extension of the receptacle called the *hypanthium*. Such flowers are said to be *perigynous* (Figure 18–47). In them, the petals and stamens appear to arise from the margin of the cup. In other flowers the sepals, petals, and stamens apparently grow from the top of the ovary, which is *inferior*. Such flowers are said to be *epigynous* (Figure 18–48).

Finally, with regard to variation in floral structure, mention should be made of symmetry. In some flowers, the corolla is made up of petals of similar shape that radiate from the center of the flower and are equidistant from each other; they are radially symmetrical. Such flowers are said to be *regular*, or actinomorphic (from

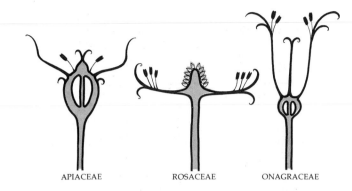

APIACEAE   ROSACEAE   ONAGRACEAE

18–46
Types of flowers in three common families of dicots, showing changes in the position of the ovary. Many Rosaceae have superior ovaries with the flower parts attached below the ovaries. Their flowers are said to be perigynous, because the upper whorls are fused together into a cup-shaped extension. The Apiaceae and Onagraceae have inferior ovaries; that is, the flower parts are attached above them. The flowers of these two families are said to be epigynous. If there is no fusion and the flower parts are attached below the ovary, the flower is said to be hypogynous.

(a)

18–47
(a) *and* (b) *Cherry* (Prunus) *flowers exhibit perigyny—their sepals, petals, and stamens are attached to a hypanthium (a cup-shaped extension of the receptacle). In* (b) *the stamens are crowded in the hypanthium because the flower has not yet opened.*

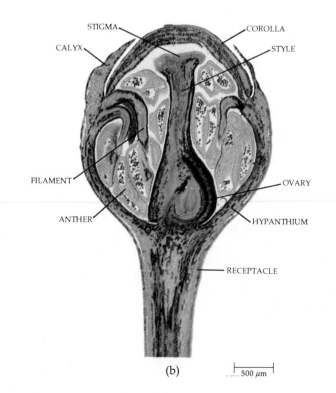

STIGMA   COROLLA
CALYX   STYLE
FILAMENT   OVARY
ANTHER   HYPANTHIUM
        RECEPTACLE

(b)   500 μm

(a)

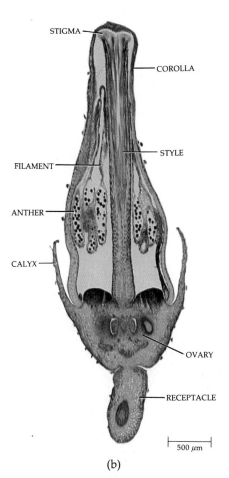

STIGMA

COROLLA

STYLE

FILAMENT

ANTHER

CALYX

OVARY

RECEPTACLE

500 μm

(b)

18–48
(a) *and* (b) *Apple* (Malus sylvestris)
*flowers exhibit epigyny—their sepals,
petals, and stamens apparently arise from
the top of the ovary. In* (b) *the flower is
very nearly open, but the stamens are not
yet erect.*

the Greek root *actinos,* ''star''). In other flowers, one or more members of at least one whorl are different from other members of the same whorl; they are bilaterally symmetrical. These flowers are said to be *irregular*, or zygomorphic (for an example, see Figure 18–43b, c).

The Angiosperm Life Cycle

Angiosperm gametophytes are very much reduced in size—more so than those of any other heterosporous plants, including the gymnosperms. The mature microgametophyte consists of only three cells; the mature megagametophyte, which is held for its entire existence within the tissues of the sporophyte, consists of only seven cells in most species. Both antheridia and archegonia are lacking. Pollination is indirect: pollen is deposited on the stigma, after which the pollen tube conveys two nonmotile sperm to the female gametophyte. After fertilization, the ovule develops into a seed, which is enclosed in the ovary. At the same time, the ovary (and sometimes additional structures associated with it) develops into a fruit.

Microsporogenesis and Microgametogenesis

Microsporogenesis is the formation of microspores within the microsporangia, or pollen sacs, of the anther. Microgametogenesis is the development of the microspore into the microgametophyte, or pollen grain.

After it has first differentiated, the anther consists of a uniform mass of cells, except for the partly differentiated epidermis. Eventually, four groups of fertile, or *sporogenous,* cells become discernible within the anther; each group is surrounded by several layers of sterile cells. The sterile cells develop into the wall of the pollen sac, including nutritive cells, which supply food to the developing microspores. The nutritive cells constitute the *tapetum,* the innermost layer of the pollen sac wall (Figure 18–49). The sporogenous cells become microsporocytes, which divide meiotically, each diploid microsporocyte giving rise to a tetrad of haploid microspores. Microsporogenesis is completed with formation of the single-celled microspores.

During meiosis, each nuclear division may be followed immediately by cell wall formation, or the four microspore protoplasts may be walled off simultaneously after the second meiotic division. The first condition is common in monocots, the second in dicots. Subsequently, the major features of the pollen grains are established (Figure 18–50). The pollen grains develop a resistant outer wall, called the *exine*, and a cellulosic inner wall, called the *intine*. The exine is composed of a very resistant substance known as *sporopollenin*, which apparently is derived partly from the tapetum and partly from the microspores. The intine, which is composed of cellulose and pectin, is laid down by the microspore protoplasts.

18–49

Two transverse sections of lily (Lilium) anthers. (a) Immature anther, showing the four pollen sacs containing micro- *sporocytes surrounded by the tapetum. (b) Mature anther containing pollen grains. The partitions between adjacent pollen* *sacs break down prior to dehiscence, as shown here.*

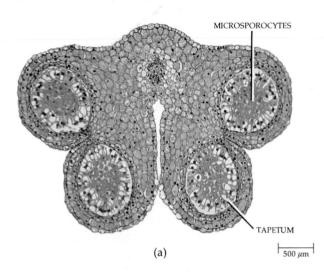

MICROSPOROCYTES

TAPETUM

(a)   500 µm

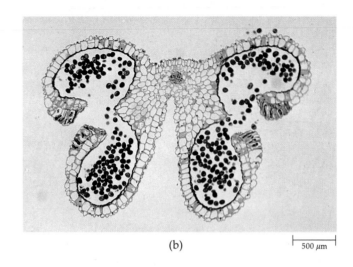

(b)   500 µm

18–50

The wall of the pollen grain serves to protect the male gametophyte on its often hazardous journey between the anther and the stigma. The outer layer, or exine, is composed chiefly of a substance known as sporopollenin, which appears to be a polymer composed chiefly of carotenoids. The exine, which is remarkably tough and resistant, is often elaborately sculptured.

The sculpturing of pollen grain walls is very precise and distinctly different from one species to another, as revealed in scanning electron micrographs of these pollen grains. (a) Pollen grains of the horse chestnut (Aesculus hippocastanum). (b) Pollen grains of a lily (Lilium longiflorum). (c) Detail of the surface of a lily (L. longiflorum) pollen

grain. (d) Pollen grain of the western ragweed (Ambrosia psilostachya). The pollen of ragweed is a primary cause of hay fever. Spiny pollen grains such as these are common among members of the sunflower family, Asteraceae, of which ragweed is a member.

(a)   10 µm

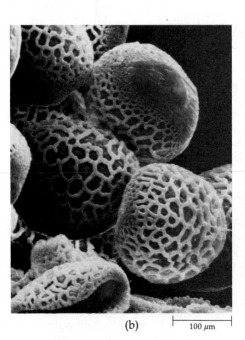

(b)   100 µm

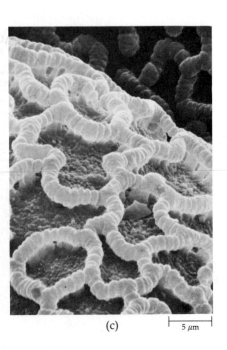

(c)   5 µm

Pollen grains, like spores, vary considerably in size and shape, ranging from less than 20 nanometers to over 250 nanometers in diameter. They also differ in the number and arrangement of the pores through which the pollen tubes ultimately grow. Nearly all families, many genera, and a fair number of species of flowering plants can be identified solely on the basis of their pollen grains, on the basis of characters such as size, number of pores, and sculpturing. In contrast with the larger portions of plants—such as leaves, flowers, and fruits—the pollen grains, because of the chemical nature of the exine, are extremely well represented in the fossil record. Thus pollen provides a valuable index to the kinds of plants and the nature of the climate that prevailed in the past.

Pollen grains and spores both have a sporopollenin wall and both are products of meiosis. Pollen grains, however, have two or three nuclei when shed, as a result of mitosis; spores have only one. Spores germinate from a suture centrally located in the body, while pollen grains germinate through their pores. For these reasons, pollen grains can often be distinguished from spores in the fossil record.

Microgametogenesis in angiosperms is uniform and begins when the uninucleate microspore divides mitotically, forming two cells within the original spore wall. One cell is called the *tube cell,* and the other is called the *generative cell* (Figure 18–51). In many species, the microgametophyte is in this two-celled stage at the time the pollen grains are liberated by dehiscence of the anther. In other species, the generative nucleus divides prior to the release of the pollen grains, giving rise to two male gametes, or sperm (Figure 18–52).

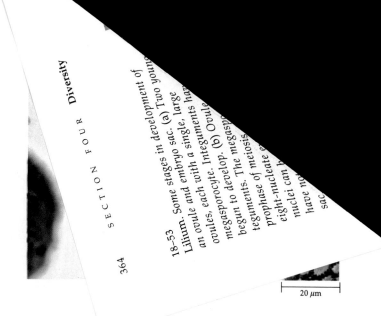

18–53. Some stages in development of Lilium. (a) Two young embryo sac. (a) Two young, large ovule and embryo sac with a single, large ovules, each with a single, large ovules, each with Integuments hav an ovule, each with a single, large ovules. Integuments hav ovules, each with Integuments hav Ovule megasporocyte. (b) Ovule megaspo begun to develop. The meiosis teguments. The meiosis prophase of meiosis eight-nucleate e nuclei can b have no sac

...ature pollen grain of Lilium, *containing a two-celled male gametophyte. The spindle-shaped generative cell will divide mitotically, giving rise to two sperm; the larger tube cell will form the pollen tube.*

20 μm

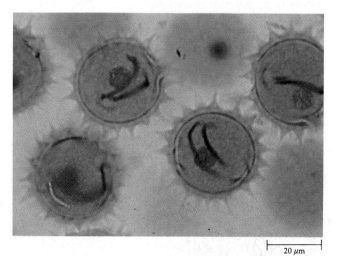

20 μm

18–52
Mature pollen grains with three-celled male gametophytes of telegraph plant, Silphium *(family Asteraceae). Prior to pollination, each pollen grain contains two filamentous sperm cells, which are suspended in the cytoplasm of the larger tube cell. In other words, the pollen of* Silphium *is shed at the three-celled stage, whereas that of* Lilium, *shown in the preceding figure, is shed at the two-celled stage.*

(d)

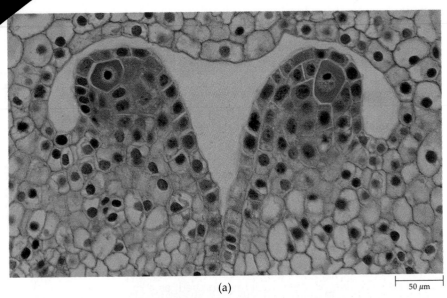

(a)

50 µm

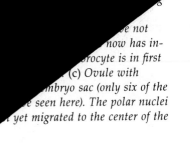

e not
now has in-
rocyte is in first
(c) Ovule with
mbryo sac (only six of the
e seen here). The polar nuclei
yet migrated to the center of the

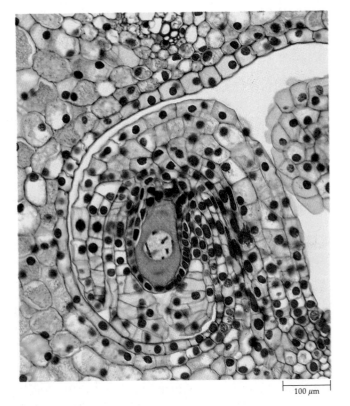

100 µm

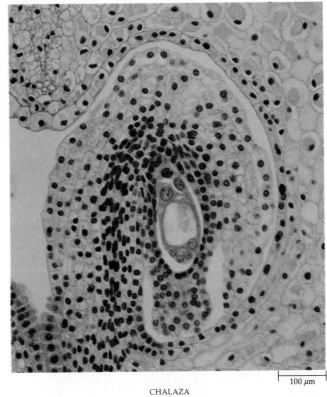

100 µm

CHALAZA

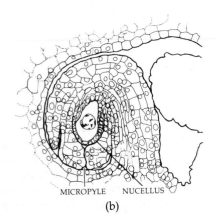

MICROPYLE   NUCELLUS

(b)

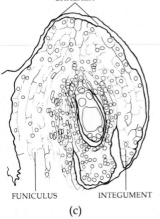

FUNICULUS   INTEGUMENT

(c)

Megasporogenesis and Megagametogenesis

Megasporogenesis is the process of megaspore formation within the nucellus (megasporangium). Megagametogenesis is the development of the megaspore into the megagametophyte.

The ovule is a relatively complex structure, consisting of a stalk—the *funiculus*—bearing a nucellus enclosed by one or two integuments. Depending on the species, one to many ovules may arise from the placentae, or ovule-bearing regions of the ovary wall. Initially, the developing ovule is entirely nucellus, but soon it develops one or two enveloping layers, the integuments, with a small opening, the micropyle, at one end (Figure 18–53).

Early in the development of the ovule, a single megasporocyte arises in the nucellus (Figure 18–53a). The diploid megasporocyte divides meiotically (Figure 18–53b) to form four haploid megaspores, which are generally arranged in a linear tetrad. With this, megasporogenesis is completed. Of the four megaspores, three usually disintegrate and the one farthest from the micropyle survives and develops into the megagametophyte.

The functional megaspore soon begins to enlarge at the expense of the nucellus, and its nucleus divides mitotically. Each of the resulting nuclei divides mitotically, followed by yet another mitotic division of the four resultant nuclei. At the end of the third mitotic division, the eight nuclei are arranged in two groups of four, one group near the micropylar end of the megagametophyte and the other at the opposite, or *chalazal*, end. One nucleus from each group migrates into the center of the eight-nucleate cell; these two nuclei are then known as the *polar nuclei*. The three remaining nuclei at the micropylar end become organized as the *egg apparatus*, consisting of an *egg cell* and two cellular *synergids*. Cell wall formation also occurs around the other three nuclei left at the chalazal end, forming the so-called *antipodals*. The *central cell*, containing the polar nuclei, remains binucleate. The eight-nucleate, seven-celled structure is the mature female gametophyte, or *embryo sac* (Figure 18–53c).

The pattern of embryo sac development just described is the most common one. Other patterns of embryo-sac development occur in about a third of the species of angiosperms that have been investigated.

Pollination and Fertilization

With *dehiscence* (shedding of contents) of the anthers, the pollen grains are transferred to the stigmas by a variety of vectors (see Chapter 29); the process whereby this transfer occurs is called pollination. Once in contact with the stigma, the pollen grains take up additional water; the water enters the pollen grains from the cells of the stigma surface along a water potential gradient.

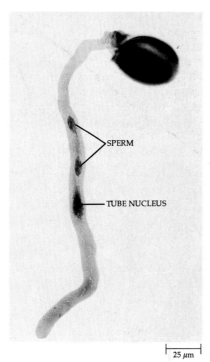

18–54

Mature male gametophyte of Solomon's seal (Polygonatum). The sperm and the tube nucleus can be seen in the pollen tube.

Following this rehydration, the pollen grain germinates, forming a pollen tube. If the generative cell has not already divided, it soon does so, forming the two sperm. The germinated pollen grain, with its tube nucleus and two sperm, constitutes the mature microgametophyte (Figure 18–54).

The stigma and style are modified both structurally and physiologically to facilitate the germination of the pollen grain and growth of the pollen tube. The surface of many stigmas is essentially glandular tissue (*stigmatic tissue*), which secretes a solution that becomes sugary. The stigmatic tissue is connected with the ovule by *transmitting tissue*, which serves as a path through the style for the growing pollen tubes. Some styles contain open canals, which are lined with transmitting tissue. In such styles, the pollen tubes grow either along or among the cells of the lining. In most angiosperms, however, the styles are solid, with one or more strands of transmitting tissue extending from the stigma to the ovules. The pollen tubes grow either between the cells of the transmitting tissue or within their thick walls, depending on the kind of plants involved.

Commonly, the pollen tube enters the ovule through the micropyle and penetrates one of the synergids, which begins to degenerate soon after pollination has taken place but before the pollen tube has reached the embryo sac. The two sperm and the tube nucleus are then released into the synergid through a subterminal

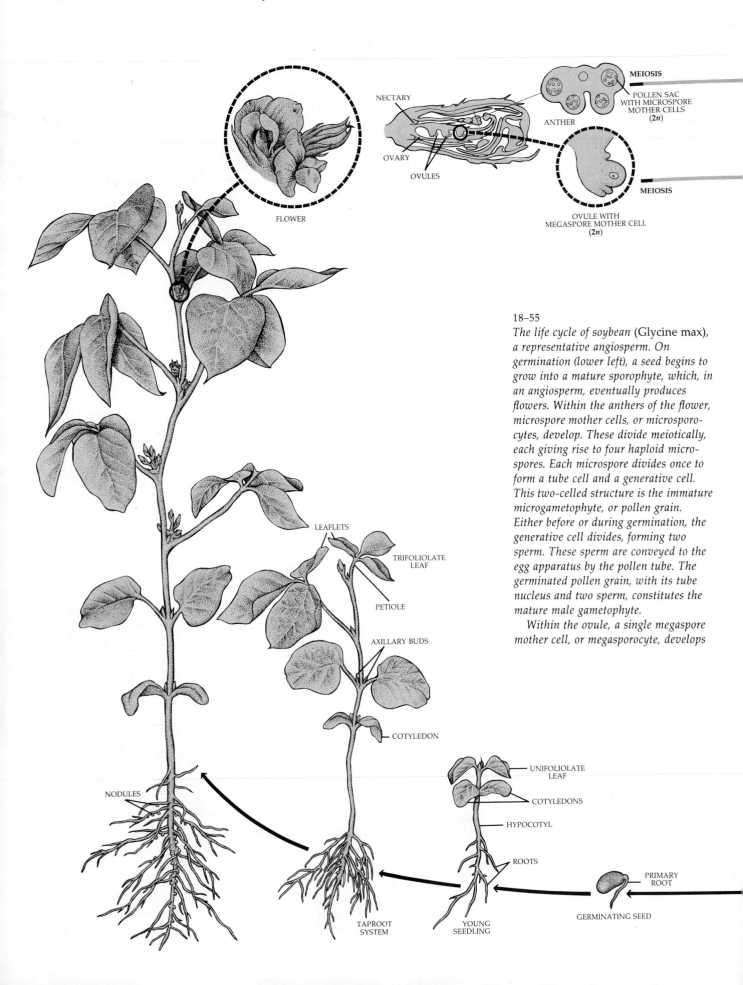

MEIOSIS

POLLEN SAC
WITH MICROSPORE
MOTHER CELLS
(**2n**)

ANTHER

NECTARY

OVARY

OVULES

MEIOSIS

OVULE WITH
MEGASPORE MOTHER CELL
(**2n**)

FLOWER

LEAFLETS

TRIFOLIOLATE
LEAF

PETIOLE

AXILLARY BUDS

COTYLEDON

UNIFOLIOLATE
LEAF

COTYLEDONS

HYPOCOTYL

NODULES

ROOTS

PRIMARY
ROOT

GERMINATING SEED

TAPROOT
SYSTEM

YOUNG
SEEDLING

18–55
*The life cycle of soybean (Glycine max),
a representative angiosperm. On
germination (lower left), a seed begins to
grow into a mature sporophyte, which, in
an angiosperm, eventually produces
flowers. Within the anthers of the flower,
microspore mother cells, or microsporocytes, develop. These divide meiotically,
each giving rise to four haploid microspores. Each microspore divides once to
form a tube cell and a generative cell.
This two-celled structure is the immature
microgametophyte, or pollen grain.
Either before or during germination, the
generative cell divides, forming two
sperm. These sperm are conveyed to the
egg apparatus by the pollen tube. The
germinated pollen grain, with its tube
nucleus and two sperm, constitutes the
mature male gametophyte.*

*Within the ovule, a single megaspore
mother cell, or megasporocyte, develops*

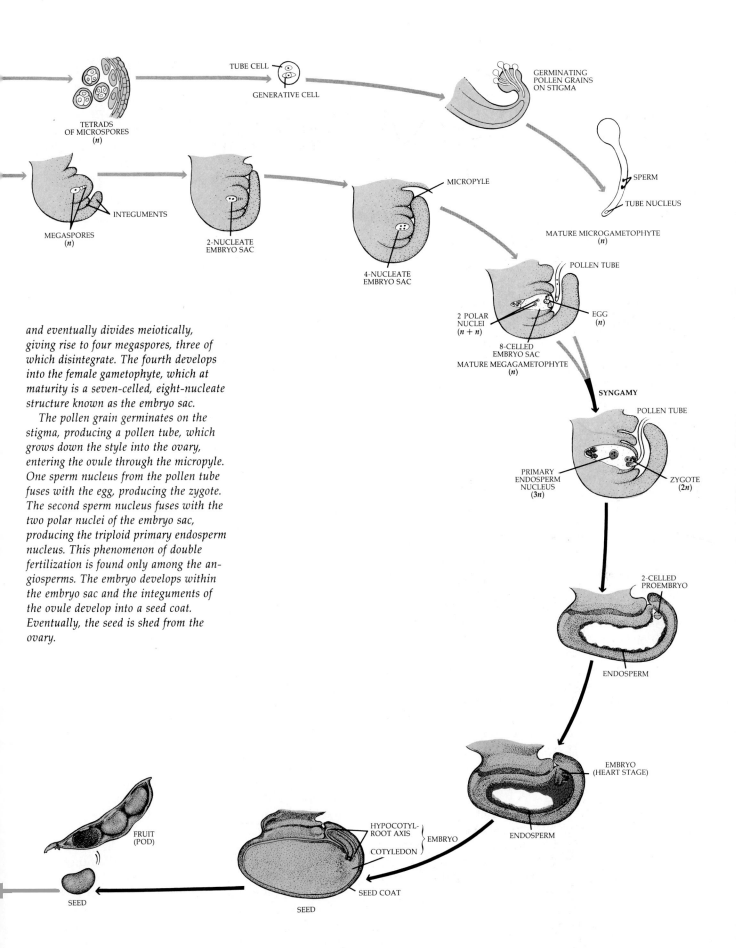

TUBE CELL

GENERATIVE CELL

TETRADS
OF MICROSPORES
(*n*)

GERMINATING
POLLEN GRAINS
ON STIGMA

SPERM

TUBE NUCLEUS

MATURE MICROGAMETOPHYTE
(*n*)

INTEGUMENTS

MEGASPORES
(*n*)

2-NUCLEATE
EMBRYO SAC

MICROPYLE

4-NUCLEATE
EMBRYO SAC

POLLEN TUBE

2 POLAR
NUCLEI
(*n* + *n*)

EGG
(*n*)

8-CELLED
EMBRYO SAC
MATURE MEGAGAMETOPHYTE
(*n*)

SYNGAMY

POLLEN TUBE

PRIMARY
ENDOSPERM
NUCLEUS
(3*n*)

ZYGOTE
(2*n*)

and eventually divides meiotically,
giving rise to four megaspores, three of
which disintegrate. The fourth develops
into the female gametophyte, which at
maturity is a seven-celled, eight-nucleate
structure known as the embryo sac.

The pollen grain germinates on the
stigma, producing a pollen tube, which
grows down the style into the ovary,
entering the ovule through the micropyle.
One sperm nucleus from the pollen tube
fuses with the egg, producing the zygote.
The second sperm nucleus fuses with the
two polar nuclei of the embryo sac,
producing the triploid primary endosperm
nucleus. This phenomenon of double
fertilization is found only among the angiosperms. The embryo develops within
the embryo sac and the integuments of
the ovule develop into a seed coat.
Eventually, the seed is shed from the
ovary.

2-CELLED
PROEMBRYO

ENDOSPERM

EMBRYO
(HEART STAGE)

ENDOSPERM

FRUIT
(POD)

HYPOCOTYL-
ROOT AXIS

EMBRYO

COTYLEDON

SEED COAT

SEED

SEED

SEED

18–56

Lilium. Double fertilization. Union of sperm and egg nuclei—"true" fertilization—can be seen in the lower half of the micrograph. Triple fusion of the other sperm nucleus and the two polar nuclei has taken place above.

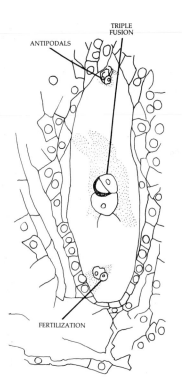

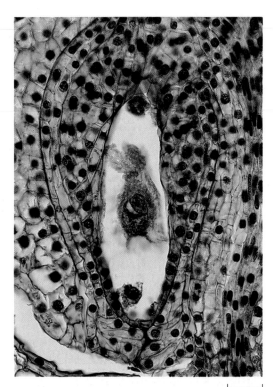

50 μm

pore that develops in the pollen tube. Ultimately, one sperm nucleus enters the egg cell and the other enters the central cell, where it unites with the two polar nuclei (Figure 18–56). Recall that in gymnosperms, only one of the two sperm is functional; one unites with the egg and the other degenerates. The involvement of both sperm in this process—the union of one with the egg and the other with the polar nuclei—is called *double fertilization*; it represents a unique characteristic of the angiosperms. As previously noted, "true" fertilization, or syngamy, involves only the union of gametes, in this case, sperm and egg. With the union of sperm and egg, a diploid zygote is formed. Union of the other sperm with the polar nuclei, called *triple fusion*, results in formation of a triploid *primary endosperm nucleus*. The tube nucleus disintegrates during the process of double fertilization, and the remaining synergid and antipodals disintegrate during the same process or early in the course of differentiation of the embryo sac.

Development of Seed and Fruit

With double fertilization, a number of processes are initiated: the primary endosperm nucleus divides, forming the *endosperm*; the zygote develops into an embryo; the integuments develop into a seed coat; and the ovary wall and related structures develop into a fruit.

In contrast to the embryogeny of the majority of gymnosperms, which begins with a free nuclear stage, embryogeny in angiosperms resembles that of the seedless vascular plants in that the first nuclear division of the zygote is accompanied by cell wall formation. In the early stages of development, the embryos of dicotyledons and monocotyledons undergo similar sequences of cell division, both becoming spherical bodies. It is with the formation of the cotyledons that a distinction first appears between dicot and monocot embryos: the dicot embryo develops two cotyledons, while the monocot embryo forms only one. The details of angiosperm embryogeny are presented in Section 5.

Endosperm formation begins with the mitotic division of the primary endosperm nucleus and usually is initiated prior to the first division of the zygote. In some angiosperms, a variable number of free nuclear divisions precede cell wall formation (nuclear-type endosperm formation); in other species the initial and subsequent mitoses are followed by cytokinesis (cellular-type endosperm formation). Although endosperm development may occur in a variety of ways, the function of the resulting tissue remains the same: to provide essential food materials for the developing embryo and, in many cases, the young seedling as well. In the seeds of some groups of angiosperms, the nucellus proliferates into a food-storage tissue known as *perisperm*. Some seeds may contain both endosperm and perisperm, as do those of the beet (*Beta*). In many dicots and some monocots, however, most or all of these storage tissues are absorbed by the developing embryo before the seed becomes dormant, as in peas or beans. The embryos of such seeds commonly develop fleshy food-storing cotyledons. The principal food materials stored in seeds are carbohydrates, proteins, and lipids.

Angiosperm seeds differ from those of gymnosperms in the origin of their stored food. In gymnosperms the stored food is provided by the female gametophyte; in angiosperms it is provided, at least initially, by endosperm, which is neither gametophytic nor sporophytic tissue.

Concomitantly with development of the ovule into a seed, the ovary (and sometimes other portions of the flower or inflorescence) develops into a fruit. As this occurs, the ovary wall, the *pericarp*, often thickens and becomes differentiated into distinct layers—the exocarp (outer layer), the mesocarp (middle layer), and the endocarp (inner layer), or into exocarp and endocarp only. These layers are generally more conspicuous in fleshy fruits than in dry ones. Fruits are discussed in greater detail in Chapter 29.

An angiosperm life cycle is summarized in Figure 18–55.

SUMMARY

The seed plants consist of gymnosperms and angiosperms. In addition to producing seeds, all seed plants bear megaphylls. The prerequisites of the seed habit include heterospory; retention of a single megaspore within the megasporangium; development of the embryo, or young sporophyte, within the megagametophyte; and integuments. All seeds consist of a seed coat derived from the integument, an embryo, and stored food. In gymnosperms the stored food is provided by the haploid female gametophyte; in angiosperms it is provided by generally triploid endosperm. The oldest known seeds occur in strata of the late Devonian period, about 360 million years ago. The likely progenitors of the gymnosperms are the progymnosperms, a Paleozoic group of seedless vascular plants, which may have given rise separately to the different groups of gymnosperms.

Living gymnosperms comprise four divisions: Cycadophyta, Ginkgophyta, Coniferophyta, and Gnetophyta. The angiosperms, or flowering plants, belong to the division Anthophyta, which is divided into two large classes—Monocotyledones and Dicotyledones.

The life cycles of gymnosperms and angiosperms are essentially similar: an alternation of heteromorphic generations with large, independent sporophytes and greatly reduced gametophytes. In gymnosperms the ovules (megasporangia plus integuments) are exposed on the surfaces of the megasporophylls or analogous structures, whereas in the angiosperms the ovules are held within the megasporophylls (the carpels). The distinctive reproductive structure of angiosperms—the flower—is characterized by the presence of carpels.

The gametophytes of angiosperms are more reduced than those of gymnosperms. At maturity the female gametophyte of most gymnosperms is a multicellular structure with several archegonia. In angiosperms the female gametophyte, commonly called an embryo sac, is most often a seven-celled,

eight-nucleate structure. Archegonia are lacking, and the egg cell is associated with two synergids; these three cells are called the egg apparatus.

The male gametophytes of both gymnosperms and angiosperms develop as pollen grains. Antheridia are lacking in both groups of seed plants. In gymnosperms, the male gametes, or sperm, arise directly from the spermatogenous cell; in angiosperms they arise directly from the generative cell. The germinated pollen grain, with its tube nucleus and two sperm, is the mature male gametophyte. Except for the cycads and *Ginkgo*, which have flagellated sperm, the sperm of seed plants are nonmotile and are conveyed to the egg by the pollen tube.

In seed plants, water is not necessary for the sperm to reach the eggs; instead, the sperm are conveyed to the eggs by a combination of pollination and pollen-tube formation. Pollination in gymnosperms is the transfer of pollen from microsporangium to megasporangium. In angiosperms, pollination is the transfer of pollen from anther to stigma.

In gymnosperms, one sperm of the male gametophyte (germinated pollen grain) unites with the egg of an archegonium. The second sperm has no apparent function, and it degenerates. In angiosperms, both sperm are functional: one unites with the egg (true fertilization, or syngamy), and the other unites with the two polar nuclei. The union of one sperm and the egg, as usual, results in the formation of a diploid ($2n$) zygote, whereas the union of the other sperm with the two polar nuclei gives rise to a triploid ($3n$) primary endosperm nucleus. This phenomenon, which is a unique characteristic of the angiosperms, is called double fertilization.

After fertilization, both gymnosperm and angiosperm ovules develop into seeds. In angiosperms, the ovaries (sometimes with some associated floral parts) develop into fruits, which enclose the seeds. The fruit is one of the principal characteristics of this great division of plants.

SUGGESTIONS FOR FURTHER READING

BIERHORST, DAVID W.: *Morphology of Vascular Plants*, The Macmillan Company, New York, 1971.

An encyclopedic and profusely illustrated treatise on the morphology of the vascular plants by one of their leading contemporary students.

FOSTER, ADRIANCE S., *and* ERNEST M. GIFFORD: *Comparative Morphology of Vascular Plants*, 2nd ed., W. H. Freeman and Company, New York, 1974.

A well-organized and general account of the vascular plants, containing much summary and interpretative material.

TAYLOR, THOMAS N.: "Reproductive Biology in Early Seed Plants," *BioScience* 32: 23–28, 1982.

A good review of the present state of knowledge about the reproductive features of early seed plants.

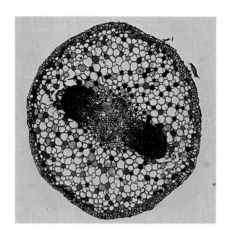

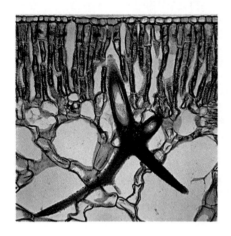

SECTION FIVE

The Angiosperm Plant Body:
Structure and Development

C H A P T E R 1 9

Early Development of the Plant Body

In the previous section, we traced the long evolutionary development of the angiosperms from their presumed ancestor—a relatively complex, multicellular green alga—through a series of early vascular plants whose forked axes were the forerunners of the leaves and roots of most modern vascular plants.

In this section, we shall be concerned primarily with the end result of this evolutionary history—the flowering plant. This chapter thus begins where the story of the angiosperm life cycle ended in Section 4—with the seed, consisting of a seed coat, stored food, and an embryo. We will follow the formation of the embryo because it is through this process, known as embryogeny, that the vegetative parts of the plant—root, stem, and leaf—have their origin and the organization of tissues is initiated.

THE MATURE EMBRYO AND SEED

The mature embryo of flowering plants consists of a stemlike axis bearing either one or two cotyledons (Figures 19–2 and 19–3). The cotyledons, sometimes referred to as the seed leaves, are the first leaves of the young sporophyte. As the names *dicotyledon* and *monocotyledon* imply, the embryos of dicotyledons have two cotyledons and those of monocotyledons have only one.

At opposite ends of the embryo axis are found the apical meristems of the shoot and root. As discussed previously (see page 6) the apical meristems are found at the tips of all shoots and roots. Meristems are composed of meristematic cells—cells that are physiologically young and capable of repeated division. In the embryo, the apical meristem of the shoot terminates the *epicotyl*, the stemlike axis above (*epi-*) the cotyledons. In some embryos, the epicotyl consists of little more than

19–1

Acorns—the fruits of an oak (Quercus)—*germinating on the forest floor. The first structure to emerge is the root, which will anchor the young plant to the soil and absorb water essential for growth of the developing seedling.*

19–2

*Seeds and stages in the germination of
some common dicotyledons. (a) Garden
bean (Phaseolus vulgaris); seed shown
open and from external edge view. (b)
Castor bean (Ricinus communis); seed
open, showing both flat and edge views of
embryo. (c) Pea (Pisum sativum);
external view of seed only.*

*Seed germination in both the garden
bean (a) and the castor bean (b) is
epigeous; that is, during germination the
cotyledons are carried above ground by
the elongating hypocotyl. Note that in
both of these seedlings, the elongating
hypocotyl forms a hook, which then
straightens out, pulling the cotyledons
and plumule above ground. By contrast,
seed germination in the pea (c) is
hypogeous—the cotyledons remain underground. In the pea seedling, it is the
epicotyl that elongates and forms a hook,
which then straightens out and pulls the
plumule above ground.*

the apical meristem (Figures 19–2b and 19–3b),
whereas in others it bears one or more young leaves
(Figures 19–2a and 19–3a). The epicotyl, together with
its young leaves, is called a *plumule*.

The stemlike axis below *(hypo-)* the cotyledons is referred to as the *hypocotyl*. At the lower end of the hypocotyl there may be an embryonic root, or *radicle*,
with distinct root characteristics (Figure 19–4). In many
plants, however, the lower end of the axis consists of
little more than an apical meristem covered by a root
cap. If a radicle cannot be distinguished in the embryo,
the embryo axis below the cotyledons may be called the
hypocotyl-root axis.

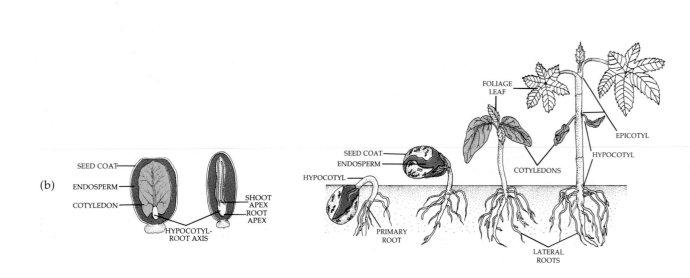

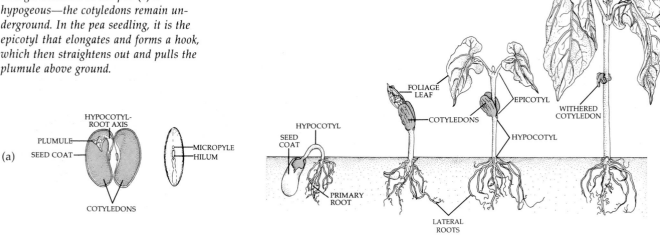

19–3
Seeds and stages in the germination of some common monocotyledons: (a) corn *(*Zea mays*) and (b)* onion *(*Allium cepa*). Both seeds are shown in longitudinal section.*

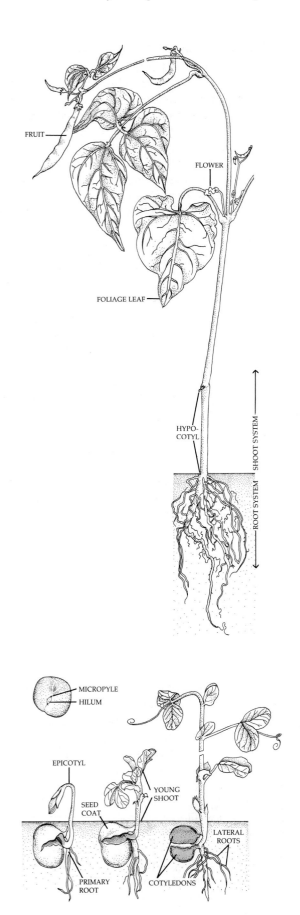

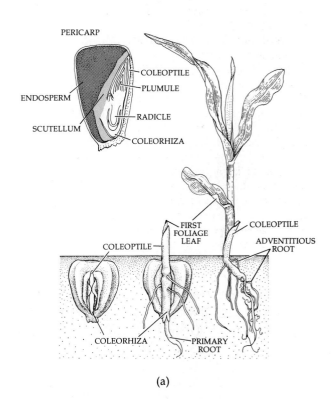

(a)

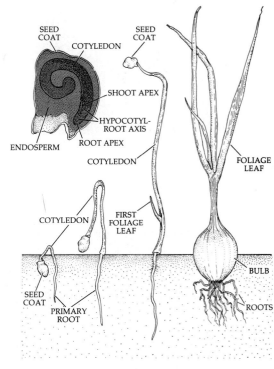

(b)

(c)

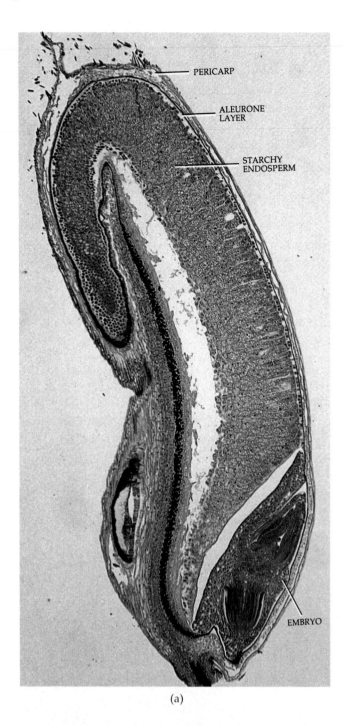

(a)

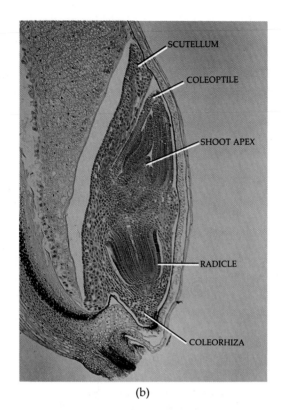

(b)

19–4

(a) *Longitudinal section of mature grain, or kernel, of wheat* (Triticum aestivum). *The covering layers of the wheat kernel consist largely of the pericarp. The seed coat, which becomes fused with the pericarp, disintegrates during development of the kernel.* (b) *Detail of the mature wheat embryo.*

In the discussion of the development of the angiosperm seed in Chapter 18, it was noted that in many dicotyledons most or all of the food-storing endosperm and the perisperm (if present) is absorbed by the developing embryo and that the embryos of such seeds develop fleshy, food-storing cotyledons. The cotyledons of most dicotyledonous embryos are fleshy and occupy the largest volume of the seed. Familiar examples of seeds lacking endosperm are sunflower, walnut, garden bean, and pea (Figure 19–2a, c). In dicots with large amounts of endosperm (for example, the castor bean), the cotyledons are thin and membranous (Figure 19–2b) and serve to absorb stored food from the endosperm.

In the monocotyledons, the single cotyledon, although somewhat fleshy, usually performs an absorbing rather than a food-storing function (Figure 19–3). Embedded in endosperm, the cotyledon absorbs food digested by enzymatic activity. The digested food is then moved by way of the cotyledon to the growing regions of the embryo. Among the most highly differentiated of monocot embryos are those of grasses (Figures 19–3a and 19–4). When fully formed, the grass embryo possesses a massive cotyledon, the *scutellum*, which is pressed close to the endosperm. The scutellum, like the cotyledons of most monocots, functions in the absorption of food stored in the endosperm. The scutellum is attached to one side of the axis of the embryo, which has a radicle at its lower end and a plumule at its upper end. Both the radicle and the plumule are enclosed by sheathlike, protective structures, called the *coleorhiza* and the *coleoptile*, respectively (Figures 19–3a and 19–4b).

All seeds are enclosed by a *seed coat*, which develops from the integuments of the ovule and provides protection for the enclosed embryo. The seed coat is usually much thinner than the integuments were originally.

The thin, dry seed coat may have a papery texture, but in many seeds it is very hard and highly impermeable to water. The micropyle (the opening in the integuments) is often visible on the seed coat as a small pore. Commonly, the micropyle is associated with a scar, called the _hilum_, which is left on the seed coat after the seed has separated from its stalk, or _funiculus_ (see the seeds on the left in Figure 19–2a, c).

FORMATION OF THE EMBRYO

The early stages of embryo development are essentially the same in dicotyledons and monocotyledons (Figures 19–5 and 19–6). Formation of the embryo begins with the division of the fertilized egg, or zygote, within the embryo sac of the ovule. In most flowering plants, the first division of the zygote is transverse, or nearly so, with regard to the long axis of the zygote (Figures 19–5a and 19–6a). With this division, the polarity of the embryo is established: the upper (chalazal) pole is the main seat for growth of the embryo; the lower (micropylar) pole produces a stalklike _suspensor_, which anchors the embryo at the micropyle.

Through an orderly progression of divisions, the embryo eventually differentiates into a nearly spherical structure—the embryo proper—and the suspensor (Figures 19–5b through d and 19–6b through d). Before this stage is reached, the developing embryo is often referred to as the _proembryo_.

Suspensors were mentioned on pages 316 and 343, in relation to the embryos of _Lycopodium_, _Selaginella_, and _Pinus_, as structures that merely push the developing embryos into nutritive tissues. Until recently, it was believed that the suspensors of angiosperm embryos played a similarly limited role. It now appears that the suspensors of angiosperms are also actively involved in absorption of nutrients from the endosperm. In addition, in some embryos, proteinaceous substances manufactured in the suspensor apparently are utilized by the embryo proper during periods of rapid growth.

When first formed, the embryo proper consists of a mass of relatively undifferentiated cells. Soon, however, changes in the internal structure of the embryo result in the initial development of the tissue systems of the plant. The future epidermis, the _protoderm_, is formed by periclinal divisions in the outermost cells of the embryo proper (Figures 19–5d and 19–6d). (Periclinal divisions are those in which the cell plates that form between the two new cells are parallel with the surface of the plant part in which they occur.) In addition, differences in the degree of vacuolation and den-

WHEAT

Like all grasses, bread wheat _(Triticum aestivum)_ is a monocotyledon, and its fruit—the grain or kernel—is one-seeded. The covering layers of the wheat kernel are composed of the pericarp and the remains of the seed coat. The endosperm and the embryo are located within the covering layers. The endosperm represents more than 80 percent of the bulk of the wheat kernel. The outermost layer of the endosperm is called the aleurone layer, which contains lipid and protein reserves. The aleurone layer surrounds the starchy endosperm, as well as the embryo.

White flour is made from the starchy endosperm. In the milling of wheat, the bran—consisting of the covering layers and the aleurone layer—is removed. The bran constitutes about 14 percent of the kernel. Actually, the bran somewhat decreases the nutritional value of the wheat kernel; because it is mostly cellulosic, it cannot be digested by humans and tends to speed the passage of food through the intestinal tract, resulting in lower absorption. The embryo ("wheat germ"), which represents about 3 percent of the kernel, is also removed, because its high oil content reduces the length of time that the flour can be stored. The bran and wheat germ, which contain most of the vitamins found in wheat, are being used more and more for human consumption, as well as for livestock feed.

19–5

Stages in the development of the embryo of shepherd's purse (Capsella bursa-pastoris), a dicot. (a) Two-celled stage, resulting from transverse division of the zygote. (b) Three-celled proembryo. (c) Six-celled proembryo. The suspensor is now distinct from the two terminal cells, which develop into the embryo proper. (d) The embryo proper is globular and has a protoderm. The large cell near the bottom is the basal cell of the suspensor. (e) Embryo at heart-shaped (emergence of cotyledons) stage. (f) Embryo at torpedo stage. In Capsella the embryos curve. (g) Mature embryo. The dark layer of cells lining the embryo sac in (a) through (f) is the endothelium, or innermost layer of the integument.

EMBRYO SAC   ENDOTHELIUM

EMBRYO SAC   ENDOTHELIUM

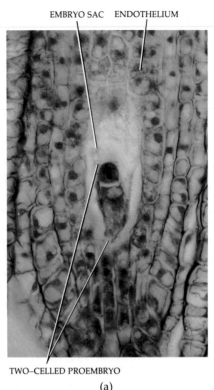

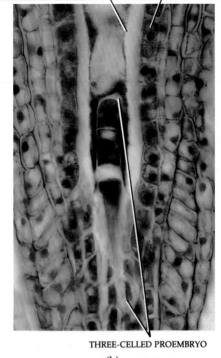

TWO–CELLED PROEMBRYO

THREE-CELLED PROEMBRYO

(a)

(b)

PROTODERM   ENDOTHELIUM

ENDOSPERM

NUCELLAR TISSUE

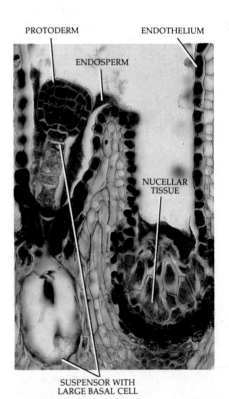

SUSPENSOR WITH LARGE BASAL CELL

(d)

ROOT TIP   EMERGING COTYLEDONS

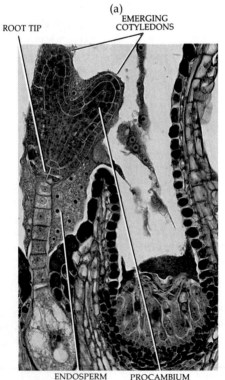

ENDOSPERM   PROCAMBIUM

(e)

ENDOTHELIUM   ENDOSPERM   BENDING COTYLEDONS

PROTODERM

PROCAMBIUM

GROUND MERISTEM

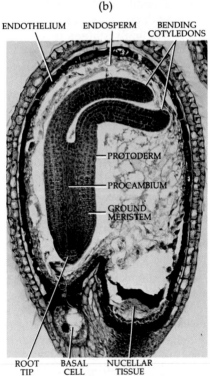

ROOT TIP   BASAL CELL   NUCELLAR TISSUE

(f)

ENDOSPERM   ENDOTHELIUM

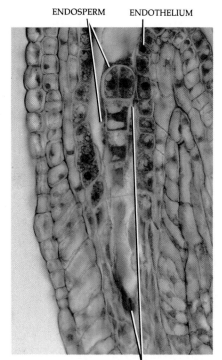

SUSPENSOR WITH LARGE BASAL CELL

(c)

APICAL MERISTEM

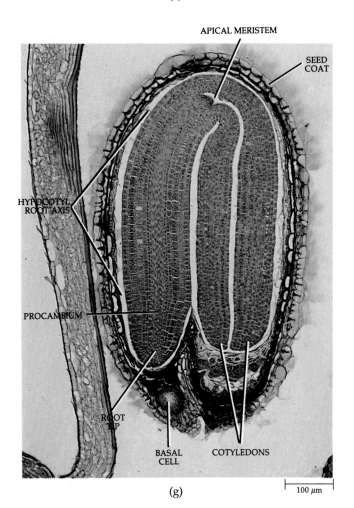

SEED COAT

HYPOCOTYL ROOT AXIS

PROCAMBIUM

ROOT TIP

BASAL CELL   COTYLEDONS

(g)   100 μm

sity of the cells within the embryo result in the initiation of the *procambium* and *ground meristem*. The highly vacuolated, less dense ground meristem gives rise to the *ground tissue*, which surrounds the less vacuolated and denser procambium, the precursor of the vascular tissues, xylem and phloem. The protoderm, ground meristem, and procambium—the so-called *primary meristems*—are continuous between the cotyledons and the axis of the embryo (Figure 19–5f, g and Figure 19–6e, f).

Development of the cotyledons may begin either during or after initiation of the primary meristems. (The stage of embryo development preceding cotyledon development is often referred to as the globular stage.) With the initiation of the cotyledons, the globular embryo in dicots gradually assumes a two-lobed form. This stage of the development of the embryo in dicots is often called the heart-shaped stage (Figure 19–5e).

Inasmuch as the embryos of monocots form only one cotyledon, they do not have a heart-shaped stage. Instead, their embryos become cylindrical in shape (Figure 19–6e). As embryo development continues, the cotyledons and axis elongate (the so-called torpedo stage of embryo development), and the primary meristems extend along with them (Figures 19–5f and 19–6f). During elongation, the embryo either remains straight or becomes curved. The single cotyledon of the monocot often becomes so large in comparison with the rest of the embryo that it is the dominating structure (see Figure 19–3b). With continued enlargement of the embryo, the cells of the suspensor are gradually crushed.

During the early stages of embryogeny, cell division takes place throughout the young sporophyte. However, as the embryo develops, the addition of new cells gradually becomes restricted to the apical meristems of the shoot and root. In dicotyledons, the apical meristem of the shoot arises between the two cotyledons (Figure 19–5g). In monocotyledons, it arises on one side of the cotyledon and is completely surrounded by a sheathlike extension from the base of the cotyledon (Figure 19–6f). The apical meristems of shoot and root are of great importance, since these tissues are the source of virtually all of the new cells responsible for the development of the seedling and the adult plant from the embryo.

Throughout the period of embryo formation, there is a continual flow of nutrients between the parent plant and tissues of the ovule, resulting in a massive buildup of food reserves within the endosperm, perisperm, or cotyledons of the developing seed. Eventually, the stalk, or funiculus, connecting the ovule to the ovary wall separates from the ovule, and the ovule becomes a nutritionally closed system. Finally, the seed becomes desiccated as it loses water to the surrounding environment, and the seed coat hardens, encasing the embryo and stored food in ''protective armor.''

19–6

*Some stages in the development of the
embryo of arrowroot (*Sagittaria*), a
monocot. Early stages are shown in (a)
through (d). (a) The two-celled stage,
resulting from transverse division of the
zygote. (b) The three-celled proembryo. (c)
Disregarding the large basal cell, the
proembryo is now at the four-celled stage.
All four of these cells, through a series of
divisions, contribute to formation of the
embryo proper. (d) The protoderm has
been initiated at the terminal end of the
embryo proper. At this stage, the
suspensor consists of only two cells, one
of which is the large basal cell. In (e)
and (f) late stages in embryo development
are shown. (e) A depression, or notch
(the site of the future shoot apex), has
formed at the base of the emerging
cotyledon. (f) Curving cotyledon; embryo
approaching maturity.*

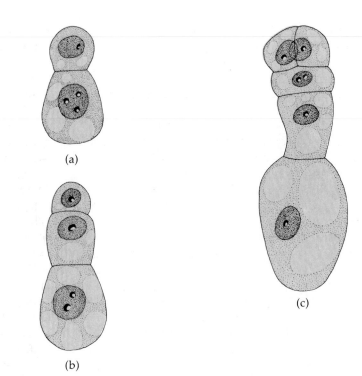

(a)

(b)

(c)

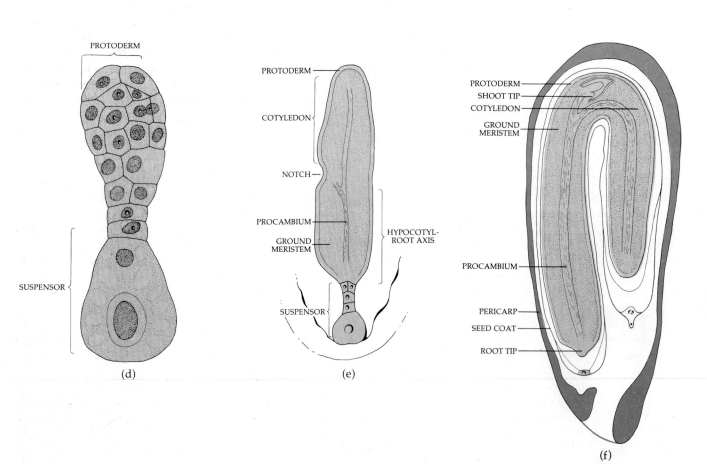

(d)

(e)

(f)

REQUIREMENTS FOR SEED GERMINATION

The growth of the embryo is usually delayed while the seed matures and is dispersed. Resumption of growth of the embryo, or _germination_ of the seed, is dependent upon many factors, both external and internal. Among the external, or environmental, factors, three are especially important: water, oxygen, and temperature. In addition, small seeds, such as those of lettuce (_Lactuca sativa_) and many weeds, commonly require exposure to light for germination (see page 501).

Most mature seeds are extremely dry, normally containing only 5 to 20 percent of their total weight as water. Thus, germination is not possible until the seed imbibes the water required for metabolic activities. Enzymes already present in the seed are activated, and new ones are synthesized for the digestion and utilization of the stored foods accumulated in the cells of the seed during the period of embryo formation. The same cells that earlier had been synthesizing enormous amounts of reserve materials now completely reverse their metabolic processes. Cell enlargement and cell division are initiated in the embryo and follow patterns characteristic of the species. Further growth requires a continuous supply of water and nutrients. As the seed imbibes water, it swells, and considerable pressure may develop within it (see boxed essay on "Imbibition," page 60).

During the early stages of germination, respiration may be entirely anaerobic, but as soon as the seed coat is ruptured, the seed switches to aerobic respiration, which requires oxygen. If the soil is waterlogged, the amount of oxygen available to the seed may be inadequate for aerobic respiration to occur, and the seed will fail to germinate.

Although many seeds germinate over a fairly wide range of temperatures, they usually will not germinate below or above a certain temperature range specific for the species. The minimum temperature for many species is 0°–5°C; the maximum is 45°–48°C; and the optimum range is 25°–30°C.

Even when external conditions are favorable, some seeds will fail to germinate. Such seeds are said to be _dormant_. The most common causes of dormancy in seeds are the physiological immaturity of the embryo and the impermeability of the seed coat to water and, sometimes, to oxygen. Some physiologically immature seeds must undergo a complex series of enzymatic and biochemical changes, collectively called _after-ripening_, before they will germinate. In temperate regions, after-ripening is triggered by the low temperatures of winter. Thus, the necessity for a period of after-ripening helps prevent germination of the seed during the inclement period of winter when it would be unlikely to survive.

Dormancy is of great survival value to the plant. As in the example of after-ripening, it is a method of ensuring that conditions will be favorable for growth of the seedling when germination does occur. Some seeds must pass through the intestines of birds or mammals before they will germinate, resulting in wider dispersal of the species. Some seeds of desert species will germinate only when inhibitors in their coats are leached away by rainfall; this adaptation ensures that the seed will germinate only during those rare intervals when desert rainfall provides sufficient water for the seedling to mature. Similarly, others must be cracked mechanically, as by tumbling along in the rushing water of a gravelly streambed. Still other seeds lie dormant in cones or fruits until the heat of a fire releases the seeds. For example, in certain species of _Pinus_, the heat melts the resin sealing the cone scales and the seeds are released, thus promoting persistence of the species in areas frequently swept by fires (Figure 19–7). Finally, the seeds of species that live in woodland clearings depend on the death of a canopy tree or some other disturbance to remove leaf cover before they will germinate. Hence, the germination strategies of plants are closely linked to the ecological problems existing in their particular habitats. (See Chapter 25, page 506, for further discussion of dormancy in seeds.)

19–7
Serotinous cones of pitch pine (Pinus rigida) _before_ (a) _and after_ (b) _exposure to fire. Serotinous cones remain closed on the tree after the seeds mature. The immediate cause of serotiny is the high melting point of the resin sealing the cone scales, and it appears to be under genetic control. Fire melts the resin, opening the cones and releasing the seeds onto sites cleared by the fire._ Pinus rigida _with cones that open normally are favored in habitats where most clearings are created by disturbances other than fire._

(a)   (b)

FROM EMBRYO TO ADULT PLANT

When germination occurs, the first structure to emerge from most seeds is the radicle, or embryonic root (Figure 19–1); this enables the developing seedling to become anchored in the soil and to absorb water. The continuation of this first root, called the *primary root*, develops branch roots, or *lateral roots*, and these roots in turn may give rise to additional lateral roots. In this manner, a much-branched root system develops. Commonly, the primary root in monocots is short-lived, and the root system of the adult plant develops from *adventitious roots*, which arise at the nodes (the parts of the stem at which the leaves are attached) and then produce lateral roots.

The way in which the shoot emerges from the seed during germination varies from species to species. For example, after the root emerges from the seed of the garden bean (*Phaseolus vulgaris*), the hypocotyl elongates and becomes bent in the process (Figure 19–2a). Thus, the delicate shoot tip is protected from injury by being pulled rather than pushed through the soil. When the bend, or hook, as it is called, reaches the soil surface, it straightens out and pulls the cotyledons and plumule up into the air. This type of seed germination, in which the cotyledons are carried above ground level, is called *epigeous*.

During germination and subsequent development of the seedling, the food stored in the cotyledons is digested and transported to the growing parts of the young plant. The cotyledons gradually decrease in size, wither, and eventually drop off. By this time, the seedling has become established; that is, it is no longer dependent upon the stored food of the seed for its nourishment. The seedling is now a photosynthesizing, autotrophic organism.

Germination of the castor bean (*Ricinus communis*) (Figure 19–2b) is essentially similar to that of the garden bean, except that in the castor bean the stored food is found in the endosperm. As the hook straightens out, the endosperm and often the seed coat are carried upward, along with the cotyledons and plumule. During this period, the digested foods of the endosperm are absorbed by the cotyledons and transported to the growing parts of the seedling. In both the garden bean and castor bean, the cotyledons become green upon exposure to light, but they do not play an important photosynthetic function.

In the pea (*Pisum sativum*), the epicotyl is the structure that elongates and forms the hook. As the epicotyl straightens out, the plumule is raised above the soil surface. The cotyledons remain in the soil (Figure 19–2c), where they eventually decompose. This type of seed germination, in which the cotyledons remain underground, is called *hypogeous*.

In the large majority of monocot seeds, the stored food is found in the endosperm. In relatively simple monocot seeds, such as that of the onion (*Allium cepa*), it is the single tubular cotyledon that emerges from the seed and forms the hook (Figure 19–3b). When the cotyledon straightens, it carries the seed coat and enclosed endosperm upward. Throughout this period, and for some time afterward, the embryo obtains much of its nourishment from the endosperm by way of the cotyledon. Furthermore, the green cotyledon in onion functions as a photosynthetic leaf, contributing significantly to the food supply of the developing seedling. Soon the plumule, enclosed within the protective, sheathlike base of the cotyledon, elongates and emerges from the cotyledon.

Our last example of seedling development is provided by corn (*Zea mays*), a monocot that has a highly differentiated embryo (see Figure 19–8). Both radicle and plumule are enclosed in sheathlike structures, the coleorhiza and the coleoptile, respectively. The coleorhiza is the first structure to grow through the pericarp (mature ovary wall) of the corn grain. (In corn, the integuments disintegrate during development of the seed and fruit, and hence the pericarp functions as a "seed coat.") The coleorhiza is then followed by the radicle, or primary root, which elongates very rapidly and quickly penetrates the coleorhiza. After the primary root emerges, the coleoptile is pushed upward by elongation of the first *internode*. (An internode is the part of the stem between two successive nodes.) When the base of the coleoptile reaches the soil surface, its edges spread apart at the tip, and the first leaves of the plumule begin to emerge. In addition to the primary root, two or more adventitious roots, which arise from the cotyledonary node, grow through the pericarp and then bend downward.

Regardless of the manner in which the shoot emerges from the seed, the activity of the apical meristem of the shoot results in the formation of an orderly sequence of leaves, nodes, and internodes. Apical meristems, which develop in the axils (the upper angles between leaves and stems) of the leaves, produce axillary shoots, and these, in turn, may form additional axillary shoots.

The period from germination to the time the seedling becomes established as an independent organism constitutes the most crucial phase in the life history of the plant. During this period, the plant is most susceptible to injury by a wide range of insect pests and parasitic fungi, and water stress can very rapidly prove fatal.

This type of growth of the root system and the shoot system is called vegetative growth. Eventually, one or more of the vegetative apical meristems of the shoot changes into a reproductive apical meristem, that is, a meristem that develops into a flower or an inflorescence (cluster of flowers). When the flowers are formed, the plant is prepared for sexual reproduction.

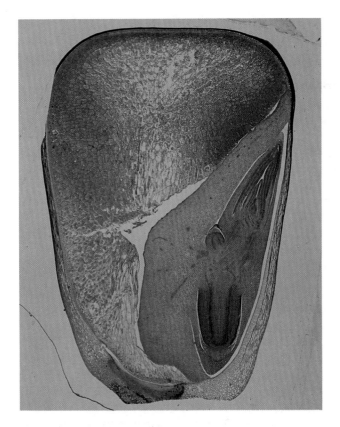

19–8

Longitudinal section of mature grain, or kernel, of corn (Zea mays). *By comparison with the corn grain diagram in Figure 19–3a, and with the photomicrographs of the wheat* (Triticum aestivum) *grain and embryo in Figure 19–4, you should be able to identify the pericarp, endosperm, and various parts of the embryo. Grain embryos commonly contain two or more seminal (seed) adventitious roots. Although at first pointed upward, these roots become inclined downward with further growth.*

SUMMARY

The seeds of flowering plants consist of an embryo, a seed coat, and stored food. When fully formed, the embryo consists basically of a hypocotyl-root axis bearing either one or two cotyledons and an apical meristem at the shoot apex and the root apex. The cotyledons of most dicots are fleshy and contain the stored food of the seed. In other dicots and in most monocots, the stored food resides in the endosperm, and the cotyledons function to absorb digested food from the endosperm. The digested food is then transported to growing regions of the embryo.

During embryo development, the shoot and root of the young plant are initiated as one continuous structure, beginning with the zygote. Through an orderly progression of divisions, the embryo differentiates into a suspensor and an embryo proper, within which the so-called primary meristems—the precursors of the epidermis, ground tissue, and vascular tissues—are formed. Development of the cotyledons may begin either during or after initiation of the primary meristems.

As the embryo develops, the addition of new cells is gradually restricted to the apical meristems. Germination of the seed—resumption of growth of the embryo—is dependent upon environmental factors, including water, oxygen, and temperature. Many seeds must pass through a period of dormancy before they are able to germinate.

Following a period of vegetative growth, one or more apical meristems of the shoot are changed into reproductive apical meristems, which develop into flowers or inflorescences.

C H A P T E R   2 0

Cells and Tissues of the Plant Body

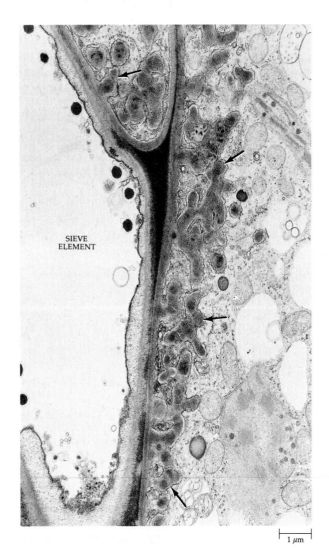

SIEVE
ELEMENT

├─ 1 μm ─┤

20–1

Transverse section from the phloem of the fern Platycerium bifurcatum, *showing portions of two transfer cells, with their numerous wall ingrowths (see arrows), bordering a relatively clear sieve element, or food-conducting cell.*

Following the development of the embryo, the formation of new cells, tissues, and organs becomes restricted almost entirely to the meristems—the perpetually young tissues that play such a central role in growth. As discussed in Chapter 1, there are two main types of meristems—apical meristems and lateral meristems. The apical meristems are involved primarily with extension of the plant body; they are found at the tips of shoots and roots. This type of growth is called *primary growth*, during which primary tissues are formed. The part of the plant body composed of these tissues is called the primary plant body.

The lateral meristems, the *vascular cambium* and the *cork cambium*, produce the secondary tissues, which make up the secondary plant body. The vascular cambium produces secondary xylem and secondary phloem; the cork cambium produces mostly cork.

In both the apical and the lateral meristems, certain cells are able to divide repeatedly; after each division one of the sister cells remains in the meristem while the other moves into the plant body. The self-perpetuating cells that remain in the meristem are called *initials*; their sister cells are called *derivatives*. It is important to note that the derivatives commonly divide one or more times before they begin to differentiate into specific types of cells. Consequently, the meristem includes the initials and their immediate derivatives.

Cell division is not limited to the apical and lateral meristems. For example, the protoderm, procambium, and ground meristem, which are partly differentiated tissues, are called primary meristems because (1) they give rise to the primary tissues and (2) many of their cells remain meristematic for some time before they begin to differentiate into specific cell types. With the exception of their initiation in the embryo proper during embryogeny, the primary meristems are derived from the apical meristems.

Growth of the plant body involves both cell division and cell enlargement. As one progresses from younger to older meristematic tissues, the overall size of the cells increases; eventually, the main factor involved in the increase in size of the particular region of the root, stem, or leaf is cell enlargement.

Differentiation—the process by which cells that have identical genetic constitution become different from one another and from the meristematic cells from which they originated—often begins while the cell is still enlarging. At maturity, when differentiation is complete, some cells are living and others are dead. Among these living and dead cells are many different cell types. How cells that have a common origin come to be so different is generally considered one of the crucial questions of modern biology. Some factors involved in the control of cellular differentiation are discussed in Chapter 8 (pages 128 through 131) and in Chapter 24 (page 478).

In the plant body, basic tissue patterns are established by early meristematic activity. The shape of the plant and organization of its tissues are influenced greatly by both cell division and cell enlargement. The acquisition of a particular shape or form is known as *morphogenesis* (from *morphè*, meaning "form," and *genere*, meaning "to create").

THE TISSUE SYSTEMS

Botanists have long recognized that the principal tissues of vascular plants are organized into larger units in all parts of the plant. These groups of tissues are known as *tissue systems*, and their presence in the root, stem, and leaf reveals both the basic similarity of the plant organs and the continuity of the plant body. The three tissue systems are: (1) the fundamental, or *ground tissue system*, (2) the *vascular tissue system*, and (3) the *dermal tissue system*. As discussed in Chapter 19, the tissue systems are initiated during development of the embryo, where they are represented by the primary meristems—ground meristem, procambium, and protoderm, respectively.

The ground tissue system consists of the three so-called ground tissues—parenchyma, collenchyma, and sclerenchyma. Parenchyma is by far the most common of the ground tissues. The vascular tissue system consists of the two conducting tissues—xylem and phloem. The dermal tissue system is represented by the epidermis, the outer protective covering of the primary plant body, and later by the periderm, in the secondary plant body.

TISSUES AND THEIR COMPONENT CELLS

Tissues may be defined as groups of cells that are structurally and/or functionally distinct. Tissues composed of only one type of cell are called *simple tissues*, whereas those composed of two or more types of cells are called *complex tissues*. Parenchyma, collenchyma, and sclerenchyma are simple tissues; xylem, phloem, and epidermis are complex tissues.

Parenchyma

Parenchyma tissue, the progenitor of all other tissues, is composed of *parenchyma cells*. In the primary plant body, parenchyma cells commonly occur as continuous masses in the cortex of stems (Figure 20–2) and roots, in the pith of stems, in leaf mesophyll (see Figure 20–23), and in the flesh of fruits. In addition, parenchyma cells occur as vertical strands of cells in the primary and secondary vascular tissues and as horizontal strands called *rays* in the secondary vascular tissues (see Chapter 23).

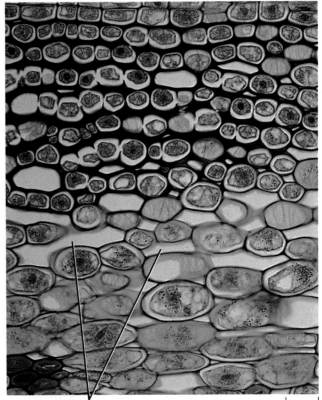

INTERCELLULAR SPACES   25 μm

20–2
Transverse section of the cortex of an elderberry (Sambucus canadensis) *stem, showing collenchyma cells with unevenly thickened walls (above) and parenchyma cells (below). Clear areas between the cells are intercellular spaces. In some of the parenchyma cells, a meshwork of lines is visible in the facing walls. The light areas within the meshwork are primary pit-fields, thin areas in the walls.*

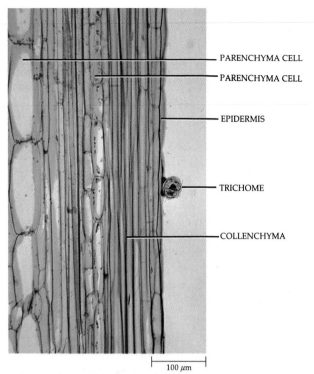

PARENCHYMA CELL

PARENCHYMA CELL

EPIDERMIS

TRICHOME

COLLENCHYMA

100 μm

20–3
Longitudinal section showing the elongated collenchyma cells in the stem of a squash (Cucurbita maxima). *A trichome is an epidermal appendage; several types of trichomes are shown in Figure 20–22 on page 396.*

Characteristically found living at maturity, parenchyma cells are capable of cell division, and although their walls are commonly primary, some parenchyma cells also have secondary walls. Because of their ability to divide, parenchyma cells with primary walls only play an important role in regeneration and wound healing. It is these cells that initiate adventitious structures, such as adventitious roots on stem cuttings. Parenchyma cells are involved in such activities as photosynthesis, storage, and secretion—activities dependent upon living protoplasm. In addition, parenchyma cells may play a role in the movement of water and the transport of food substances in plants.

Transfer Cells

During the past few years, considerable attention has been given to a special kind of parenchyma cell containing ingrowths of the cell wall, which often greatly increase the surface area of the plasma membrane (see Figure 20–1). These parenchyma cells, called *transfer cells*, are believed to play an important role in the transfer of solutes over short distances. Although such cells have long been known to exist in various plant parts, botanists have only recently realized that they

are exceedingly common and probably serve a similar function throughout the plant body. Transfer cells occur in association with the xylem and phloem of small, or minor, veins in cotyledons and in the leaves of many herbaceous dicotyledons. Transfer cells are also associated with the xylem and phloem of leaf traces at the nodes in both dicotyledons and monocotyledons. In addition, they are found in various tissues of reproductive structures (placentae, embryo sacs, endosperm) and in various glandular structures (nectaries, salt glands, the glands of carnivorous plants). Each of these locations is a potential site of intensive short-distance solute transfer.

Collenchyma

Collenchyma tissue is composed of *collenchyma cells* which, like parenchyma cells, are found living at maturity (Figures 20–2 through 20–5). Collenchyma tissue commonly occurs in discrete strands or as continuous cylinders beneath the epidermis in stems and petioles (leaf stalks). It is also found bordering the veins in dicot leaves. (The "strings" on the outer surface of celery stalks, or petioles, consist almost entirely of collenchyma.) The typically elongated collenchyma cells (Figure 20–3) contain unevenly thickened, nonlignified primary walls, making them especially well adapted for the support of young, growing organs (see the description of the primary wall on page 34). The name *collenchyma* derives from the Greek word *colla*, meaning "glue," which refers to their characteristic thick, glistening walls in fresh tissue (Figure 20–5). Being primary in nature, the walls of collenchyma cells are readily

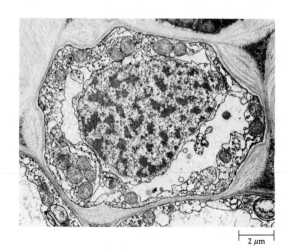

2 μm

20–4
Electron micrograph of a transverse section of a mature collenchyma cell from the filament of a wheat (Triticum aestivum) *stamen. Notice the protoplasmic contents and unevenly thickened wall of this cell.*

stretched and offer relatively little resistance to elongation of the plant part in which they are found. In addition, because collenchyma cells are living at maturity, they can continue to develop thick, flexible walls while the organ in which they are found is still elongating.

Sclerenchyma

Sclerenchyma tissue is composed of *sclerenchyma cells*, which may develop in any or all parts of the primary and secondary plant bodies; they often lack protoplasts at maturity. The term *sclerenchyma* is derived from the Greek *skleros,* meaning "hard," and the principal characteristic of sclerenchyma cells is their thick, often lignified secondary walls. Because of these walls, the sclerenchyma cells are important strengthening and supporting elements in plant parts that have ceased elongating (see the description of the secondary wall on page 34).

Two types of sclerenchyma cells are recognized: *fibers* and *sclereids*. Fibers are generally long, slender cells, which commonly occur in strands or bundles (Figure 20–6). The so-called bast fibers—such as hemp, jute, and flax—are derived from the stems of dicotyledons. Other economically important fibers, such as Manila hemp, are extracted from the leaves of monocots. Sclereids are variable in shape and are often branched (Figure 20–7), but compared with most fibers, sclereids are relatively short cells. Sclereids may occur singly or in aggregates throughout the ground tissue. They make up the seed coats of seeds, the shells of nuts, the stone, or endocarp, of stone fruits, and they give pears their characteristic gritty texture (Figure 20–8).

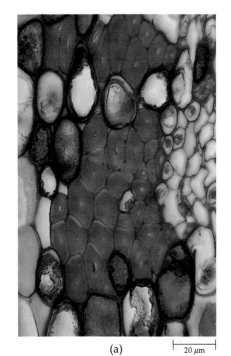

(a) 20 μm   (b) 25 μm

20–6
Primary phloem fibers from the stem of a linden, or basswood (Tilia americana), seen here in both (a) cross-sectional and (b) longitudinal views. The secondary walls of these long, thick-walled fibers contain relatively inconspicuous pits. Only a portion of the length of these fibers can be seen in (b).

CELL WALL   PROTOPLAST
25 μm

20–5
Transverse section of collenchyma tissue from a petiole in rhubarb (Rheum rhaponticum). In fresh tissue like this, the unevenly thickened collenchyma cell walls have a glistening appearance.

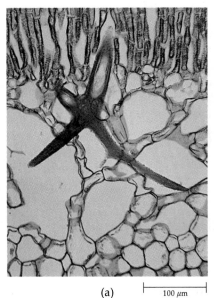

(a) 100 μm

20–7
Branched sclereid from a leaf of the water lily (Nymphaea odorata), as seen in (a) ordinary light and (b) polarized

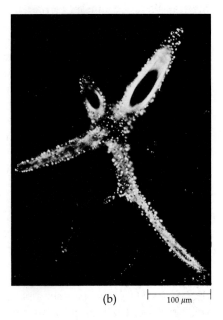

(b) 100 μm

light. Numerous small angular crystals are embedded in the wall of this sclereid.

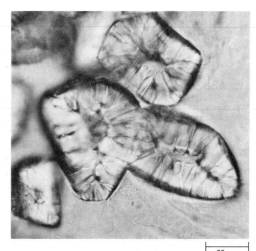

|———| 25 µm

20–8

Sclereids (stone cells) from fresh tissue of the pear (Pyrus communis) *fruit. The secondary walls contain conspicuous simple pits, with many branches. Branched pits such as these are called ramiform pits. During formation of the clusters of stone cells in the flesh of the pear fruit, cell divisions occur concentrically around some of the sclereids formed earlier. The newly formed cells differentiate as stone cells, adding to the cluster.*

EXTERNAL   VASCULAR
PHLOEM   CAMBIUM

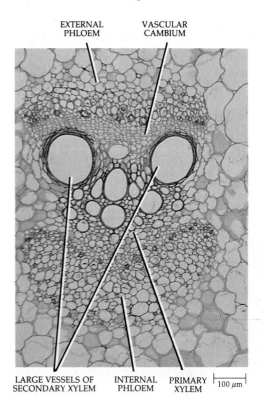

LARGE VESSELS OF   INTERNAL   PRIMARY |———| 100 µm
SECONDARY XYLEM   PHLOEM   XYLEM

20–9

Transverse section of a vascular bundle from the stem of a squash (Cucurbita maxima). *Phloem occurs on both sides of the xylem, and a vascular cambium has developed between the external phloem and the xylem.*

Xylem

Xylem is the principal water-conducting tissue found in vascular plants. It is also involved in the conduction of minerals, in food storage, and in support. Together with the phloem, the xylem forms a continuous system of vascular tissue extending throughout the plant body (Figure 20–9). Xylem may be primary or secondary in origin. The primary xylem is derived from the procambium, and the secondary xylem is derived from the vascular cambium (see Chapter 23).

The principal conducting cells of the xylem are the *tracheary elements*, of which there are two types, the *tracheids* and the *vessel members*, or vessel elements. Both are elongated cells that have secondary walls and lack protoplasts at maturity, and each type may have *pits* in their walls (Figure 20–10a through d). In addition to the presence of pits, vessel members contain *perforations*, which are areas lacking both primary and secondary walls; perforations are literally holes in the cell wall. Perforations generally occur on the end walls, but they may also be found on the lateral walls. The part of the wall bearing the perforation or perforations is called the *perforation plate* (Figure 20–11).

Vessel members are joined into long, continuous columns, or tubes, called *vessels* (Figure 20–12). The pits of the long, tapering tracheids are concentrated on the overlapping ends of the cells. The tracheid is the only type of water-conducting cell found in most seedless vascular plants and gymnosperms; the xylem of the great majority of angiosperms contains vessel members in addition to tracheids.

Extensive comparative studies of the tracheary elements in a wide range of vascular plants have clearly established that the tracheid is a more primitive (less specialized) type of cell than the vessel member—the principal water-conducting cell in the xylem of angiosperms. It is quite probable that vessel members evolved independently from tracheids in several groups of vascular plants, including the dicotyledons and monocotyledons (the two groups of angiosperms), the Gnetophyta (the vessel-containing gymnosperms), several unrelated species of ferns, and certain species of *Selaginella* (of the Lycophyta) and *Equisetum* (of the Sphenophyta). The evolution of the vessel member is an excellent example of convergent evolution, or the independent development of similar structures by unrelated or only distantly related organisms (see "Convergent Evolution," page 443).

The term *tracheary element* dates back to the seventeenth century and the Italian physician Marcello Malpighi, one of the founders of plant anatomy. Malpighi was especially interested in finding similarities between plants and animals. During his examination of xylem, he observed air bubbles escaping from a vessel with spiral wall thickenings. He immediately compared the vessel with the tracheae, or air ducts, of insects and later applied the same term to the vessels of the xylem.

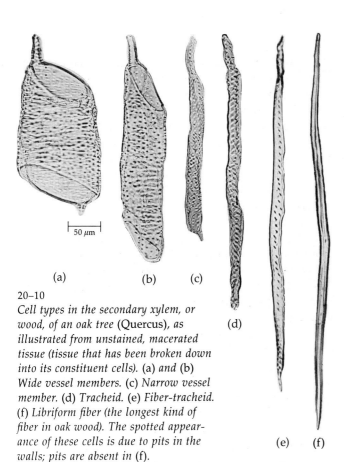

20–10
Cell types in the secondary xylem, or wood, of an oak tree (Quercus), as illustrated from unstained, macerated tissue (tissue that has been broken down into its constituent cells). (a) and (b) Wide vessel members. (c) Narrow vessel member. (d) Tracheid. (e) Fiber-tracheid. (f) Libriform fiber (the longest kind of fiber in oak wood). The spotted appearance of these cells is due to pits in the walls; pits are absent in (f).

The term has been used ever since for the water-conducting cells of the xylem.

Vessel members are generally thought to be more efficient conductors of water than the tracheids, because water can flow relatively unimpeded from vessel member to vessel member through the perforations. When flowing through tracheids, water must pass through the membranes of the pit-pairs (see pages 34 and 35). However, the pit membranes probably offer relatively little resistance to the flow of water, because during the final stages of differentiation they are partially hydrolyzed, leaving only a highly permeable meshwork of cellulose microfibrils.

In the secondary xylem and late-formed primary xylem (metaxylem), the pitted secondary walls of the tracheids and vessel members cover the entire primary wall, except at the pits and the perforations of the vessel members (Figure 20–10a through d). Consequently, these walls are rigid and cannot be stretched. During the period of elongation or expansion of the roots, stems, and leaves, many of the first-formed tracheary elements of the early-formed primary xylem (protoxylem) have their secondary walls deposited in the form of rings or spirals (Figure 20–13). These thickenings, which may be annular (ringlike) or helical (spiral), make it possible for such tracheary elements to be stretched or extended, although they are frequently destroyed during the overall elongation of the organ. In the primary xylem, the nature of the wall thickening is greatly influenced by the amount of elongation. If little elongation occurs, pitted elements rather than extensible types appear. On the other hand, if much elonga-

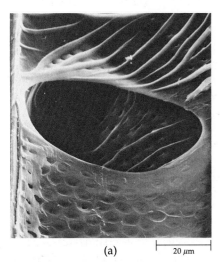

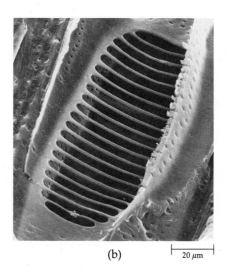

20–11
Scanning electron micrographs of the perforated end walls of vessel members from secondary xylem. (a) A simple perforation plate, with its single large opening, seen here between two vessel members in linden, or basswood (Tilia americana). (b) The ladderlike bars of a scalariform perforation plate between vessel members of red alder (Alnus rubra). Pits can be seen in the wall below the perforation plate in (a) and in portions of the wall in (b).*

20–12
Scanning electron micrographs of vessel members from secondary xylem. External view of parts of three vessel members of red oak (Quercus rubra). Notice the rims between vessel members.

tion takes place, many elements with annular and helical thickenings will appear. Figure 20–14 illustrates some stages of differentiation of a vessel member with helical thickenings.

In addition to tracheids and vessel members, xylem contains parenchyma cells that store various substances. The xylem parenchyma cells commonly occur in vertical strands. In the secondary xylem, they also are found in rays. Fibers (Figure 20–10e, f) and sclereids also occur in xylem. Many xylem fibers are living at maturity and serve a dual function of storage and support.

Phloem

Phloem is the principal food-conducting tissue found in vascular plants (Figure 20–9). The phloem may be primary or secondary in origin, and as with primary xylem, the first-formed primary phloem (protophloem) is frequently stretched and destroyed during elongation of the organ.

The principal conducting cells of the phloem are the *sieve elements*, of which there are two types, the *sieve cells* (Figure 20–15) and the *sieve-tube members* (Figures 20–16 and 20–17). The term "sieve" refers to the clusters of pores, the *sieve areas*, through which the protoplasts of adjacent sieve elements are interconnected.

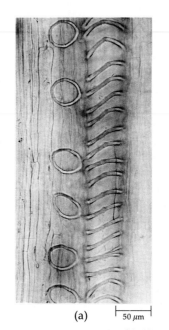 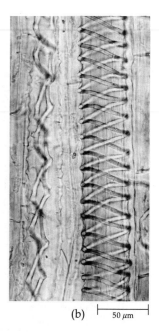

(a) 50 µm (b) 50 µm

20–13
Parts of tracheary elements from the first-formed primary xylem (protoxylem) of the castor bean (Ricinus communis). (a) Annular (the ringlike shapes at left) and helical wall thickenings in partly extended elements. (b) Double helical thickenings in elements that have been extended. The element on the left has been greatly extended, and the coils of the helices have been pulled far apart.

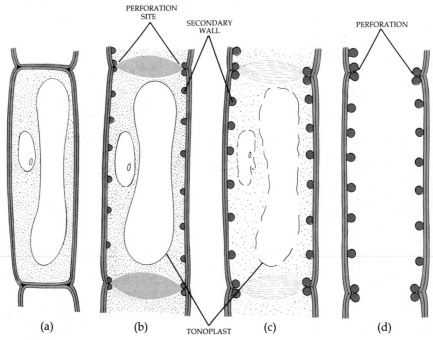

(a) (b) (c) (d)

20–14
Diagram illustrating development of a vessel member. (a) Young, highly vacuolated vessel member without secondary wall. (b) The cell has expanded laterally, secondary wall deposition has begun, and the primary wall at the perforation site has increased in thickness. (c) Secondary wall deposition has been completed, and the cell is at the stage of lysis. The nucleus is deformed, the tonoplast is ruptured, and the wall at the perforation site is partly disintegrated. (d) The cell is now mature; it lacks a protoplast and is open at both ends.

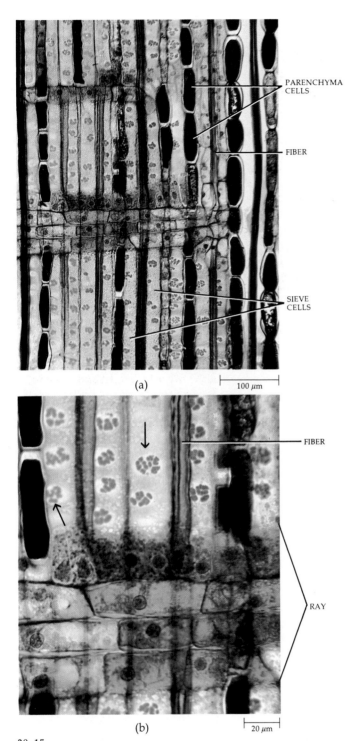

PARENCHYMA
CELLS

FIBER

SIEVE
CELLS

(a)

100 μm

FIBER

RAY

(b)

20 μm

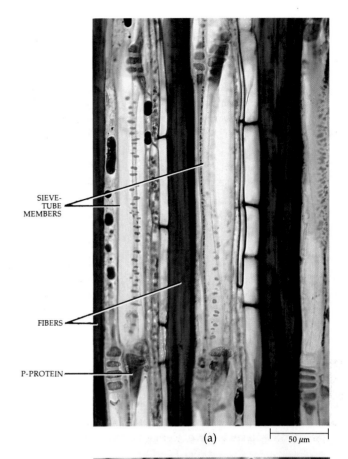

SIEVE-
TUBE
MEMBERS

FIBERS

P-PROTEIN

(a)

50 μm

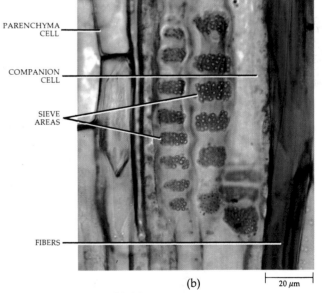

PARENCHYMA
CELL

COMPANION
CELL

SIEVE
AREAS

FIBERS

(b)

20 μm

20–15
(a) *Longitudinal (radial) view of secondary
phloem of yew (Taxus canadensis),
showing vertically oriented sieve cells,
strands of parenchyma cells, and fibers.
Parts of two horizontally oriented rays
can be seen traversing the vertical cells.*
(b) *Detail of portion of the secondary
phloem of yew, showing sieve areas, with
callose (stained blue) on the walls of the
sieve cells, and albuminous cells (see
page 394), which here constitute the top
row of cells in the ray.*

20–16
(a) *Longitudinal (radial) view of secondary
phloem of linden, or basswood (Tilia
americana), showing sieve-tube members
and conspicuous groups of thick-walled
fibers.* (b) *Compound sieve plates of
linden sieve-tube members. (Compound
sieve plates consist of two or more sieve
areas.) Each sieve area is composed of
pores bordered by cylinders of callose,
which are stained blue in this section.*

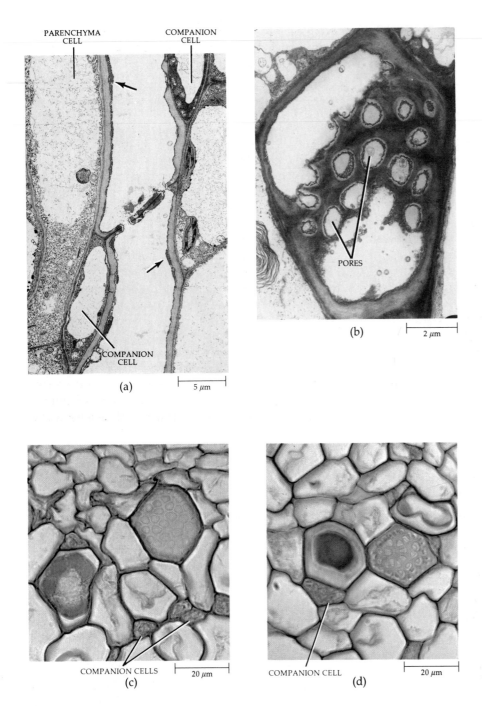

PARENCHYMA CELL   COMPANION CELL

COMPANION CELL

(a)   5 μm

PORES

(b)   2 μm

COMPANION CELLS   20 μm
(c)

COMPANION CELL   20 μm
(d)

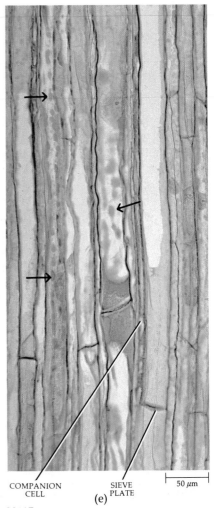

COMPANION CELL   SIEVE PLATE
(e)   50 μm

20–17
*The phloem in the stem of squash
(Cucurbita maxima), as seen in electron
micrographs, (a) and (b), as well as in
photomicrographs, (c) through (e). (a)
Longitudinal view of parts of two mature
sieve-tube members and a sieve plate,
showing distribution of P-protein (see
arrows) along the wall. (b) Face view of
simple sieve plate (one sieve area per
plate) between two mature sieve-tube
members. The sieve-plate pores are open.
(c) Transverse section, showing two
immature sieve-tube members. Slime
bodies, or P-protein bodies, can be seen
in the sieve-tube member on the left, an
immature sieve plate in the one to the
right, above. The small, dense cells are
companion cells. (d) Transverse section,
showing some mature sieve-tube
members. A slime plug can be seen in the
sieve-tube member on the left; a mature
sieve plate can be seen in the one on the
right. The small, dense cells are companion cells. (e) Longitudinal section,
showing mature and immature sieve-tube
members. The arrows point to P-protein
bodies in immature cells.*

In sieve cells, the pores are narrow and the sieve areas
are rather uniform in structure on all walls. Most of the
sieve areas are concentrated on the overlapping ends of
the long, slender cells (Figure 20–15a). In sieve-tube
members, the sieve areas on some walls have larger
pores than those on other walls of the same cell. The
part of the wall bearing the sieve areas with larger
pores is called a *sieve plate* (Figures 20–16b and 20–17a,
b). Although sieve plates may occur on any wall, they
generally are located on the end walls. Sieve-tube
members occur end-on-end in longitudinal series called

sieve tubes. Thus, the principal distinction between the two types of sieve elements is the presence of sieve plates in sieve-tube members and their absence in sieve cells.

The sieve cell is a more primitive type of cell than the sieve-tube member. Sieve cells are the only type of food-conducting cell in most seedless vascular plants and gymnosperms, whereas only sieve-tube members occur in angiosperms.

The walls of sieve elements are generally described as primary. In cut sections of phloem tissue, the pores of the sieve areas and sieve plates are generally blocked by a wall substance called *callose*, which is a polysaccharide composed of spirally wound chains of glucose residues. The presence of callose in the pores of conducting sieve elements long puzzled botanists; it seemed illogical for the pores to contain any substance that would impede the movement of other substances from cell to cell. It is now known that most, if not all, of the callose found in the pores of conducting sieve elements is deposited there in response to injury during preparation of the tissue for microscopy.

Unlike tracheary elements, sieve elements have living protoplasts at maturity (Figure 20–17). The protoplasts of mature sieve elements are unique among the living cells of the plant in that they either lack nuclei entirely or contain only remnants of them. In addition, most mature sieve elements lack a clear boundary between their cytoplasm and vacuoles. When young, the sieve element contains several vacuoles, each separated from the cytoplasm by a tonoplast, or vacuolar membrane. In the final stages of differentiation, the tonoplasts disappear and the distinction between cytoplasmic and vacuolar contents no longer exists. At maturity, all of the remaining components of the sieve-element protoplast are distributed along the wall; they consist of the plasma membrane, a network of smooth endoplasmic reticulum next to the plasma membrane, and some plastids and mitochondria. Ribosomes, dictyosomes, microtubules, and nuclei are lacking.

The protoplasts of the sieve-tube members of dicotyledons (and some monocotyledons) are characterized by the presence of a proteinaceous substance known as *slime*, or *P-protein*. (The "P" in P-protein stands for phloem.) P-protein has its origin in the young sieve-tube member in the form of discrete bodies, called slime bodies or P-protein bodies (Figure 20–17c, e). During the late stages of differentiation, the P-protein bodies elongate and disperse. In cut sections of phloem tissues, "slime plugs" of P-protein are usually found

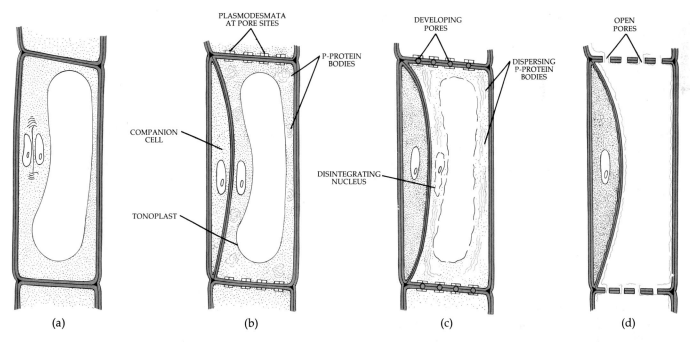

(a) (b) (c) (d)

20–18

Diagram illustrating differentiation of a sieve-tube member. (a) The mother cell of the sieve-tube member undergoing division. (b) Division has resulted in formation of a young sieve-tube member and a companion cell. After division, one or more P-protein bodies arise in the cytoplasm, which is separated from the vacuole by a tonoplast. The wall of the young sieve-tube member has thickened, and the sites of the future sieve-plate pores are represented by plasmodesmata. (c) The nucleus is degenerating, the tonoplast is breaking down, and the P-protein bodies are dispersing in the cytoplasm lining the wall; concomitantly, the plasmodesmata of the developing sieve plates are beginning to widen into pores. (d) At maturity, the sieve-tube member lacks a nucleus and a vacuole. All of the remaining protoplasmic components, including the P-protein, line the walls, and the sieve-plate pores are open.

near the sieve plates (Figure 20–17d). Like callose, slime plugs are not found in undisturbed cells and so are considered to be the result of the surging of the contents of the sieve-tube members that occurs when the tissues and sieve tubes are severed from the plant. In normal, mature sieve-tube members, P-protein is apparently distributed along the walls and is continuous from one sieve-tube member to the next through the pores of the sieve areas and sieve plates. The sieve-plate pores are lined with P-protein but are not plugged by it (Figure 20–17a, b). The function of P-protein has not been determined. However, some botanists believe that, together with wound callose, P-protein serves to seal the sieve-plate pores at the time of wounding, thus preventing the loss of contents from sieve tubes.

Sieve-tube members are characteristically associated with specialized parenchyma cells called *companion cells* (Figure 20–17c through e), which contain all of the components commonly found in living plant cells, including a nucleus. The sieve-tube member and its associated companion cells are closely related developmentally (they are derived from the same mother cell), and they have numerous connections with one another. Figure 20–18 illustrates some stages of differentiation of a sieve-tube member with P-protein. Functionally, the companion cells are very important, for they are largely responsible for the active secretion of substances into (and their removal from) the sieve-tube members. This subject will be discussed further in Chapter 27 when the mechanism of phloem transport in angiosperms is considered in detail.

The sieve cells of gymnosperms are characteristically associated with specialized parenchyma cells called *albuminous cells* (Figure 20–15b). Although generally not derived from the same mother cell as their associated sieve cells, the albuminous cells are believed to perform the same roles as companion cells. Like the companion cell, the albuminous cell contains a nucleus, in addition to other cytoplasmic components characteristic of living cells.

The sieve elements in most species apparently are short-lived and die in less than a year after their origin. This is not true, however, of all sieve elements. In the secondary phloem of the basswood, or linden tree *(Tilia americana)*, some sieve elements remain alive and presumably function as conducting elements for five to ten years. Sieve elements are known to remain alive for many years in perennial monocots. In certain palms, some sieve elements at the base of the main stem may be more than a century old. When the sieve elements die, their associated companion cells or albuminous cells also die, which is one more indication of the intimate relationship that exists between sieve elements and their companion cells or albuminous cells.

Other parenchyma cells occur in the primary and secondary phloem (Figures 20–15 through 20–17). They are largely concerned with the storage of various

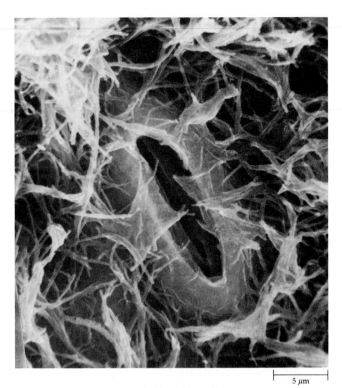

20–19
Surface view of lower epidermis of a Eucalyptus globulus *leaf, taken with the scanning electron microscope. A single stoma and numerous filaments of epicuticular wax deposits can be seen here.*

substances. Fibers (Figures 20–15 and 20–16) and sclereids may also be present.

Epidermis

The *epidermis* is the outermost layer of cells of the primary plant body, and it constitutes the dermal tissue system of leaves, floral parts, fruits, and seeds, and of stems and roots until they undergo considerable secondary growth. The epidermal cells are quite variable both functionally and structurally. In addition to the ordinary epidermal cells, which form the bulk of the epidermis, the epidermis may contain *stomata* (Figures 20–19 through 20–21), many types of appendages, or *trichomes* (Figure 20–22), and other kinds of cells specialized for specific functions.

In most plants, the epidermis is only one layer of cells in thickness. However, in some plants, divisions in the protoderm of the leaf are parallel with the surface (periclinal divisions), and an epidermis with several layers (a multiple epidermis) is formed (Figure 20–23). A multiple epidermis is found in the leaves of such familiar house plants as the rubber plant *(Ficus elastica)* and *Peperomia*. The multiple epidermis is believed to serve as a water-storage tissue.

The main mass of epidermal cells is closely knit and affords considerable mechanical protection to the plant part. The walls of the epidermal cells of the aerial parts are covered with a cuticle, which minimizes water loss. The cuticle consists mainly of cutin and wax (see page 50). In many plants the wax is exuded over the surface of the cuticle, either in smooth sheets or as rods or filaments extending upward from the surface, the so-called epicuticular wax (Figure 20–19; see also Figure 3–11). It is this wax that is responsible for the whitish or bluish "bloom" on the surface of some leaves and fruits.

Interspersed among the flat, tightly packed epidermal cells are specialized cells that contain chloroplasts. These are the *guard cells* (Figures 20–20 and 20–21), which regulate the small openings, or stomata (singular: stoma), in the aerial parts of the plant and, hence, control the movement of gases, including water vapor, into and out of those parts. (The mechanism of stomatal opening and closing is discussed in Chapter 27.) Although stomata occur on all aerial parts, they are most abundant on leaves. Stomata are often associated with epidermal cells that differ in shape from the ordinary epidermal cells. Such cells are termed *subsidiary*, or accessory, cells (Figure 20–20).

Trichomes have a variety of functions. Root hairs facilitate the absorption of water and minerals from the soil. Recent studies of arid-land plants indicate that in-

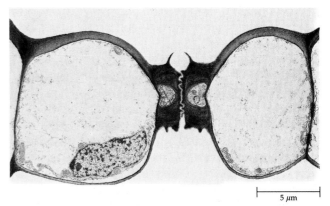

20–20
Electron micrograph of corn (Zea mays) stoma. Transverse section through mature, thick-walled guard cells, each of which is attached to a subsidiary cell.

20–21
(a) *Surface and* (b) *sectional views of developing stomata.* (c) *Diagram of a mature stoma, showing its relationship to the epidermis and underlying cells. The guard cells originate from unequal division of a protodermal cell. The smaller of the two cells is called a stoma mother cell, or guard-cell mother cell, and it is this cell that divides to give rise directly to the two guard cells (a) and (b). After the guard cells are formed, the intercellular substance in the median part of their common wall swells and then dissolves to form the pore. During this process the guard cells develop unevenly thickened walls. The walls next to the pores are generally thicker than those adjacent to other epidermal cells. While the stomata are being formed, they may be raised or lowered below the surface of the epidermis. Often there is a large air space, or substomatal chamber, just behind the stoma. Unlike ordinary epidermal cells, guard cells contain chloroplasts.*

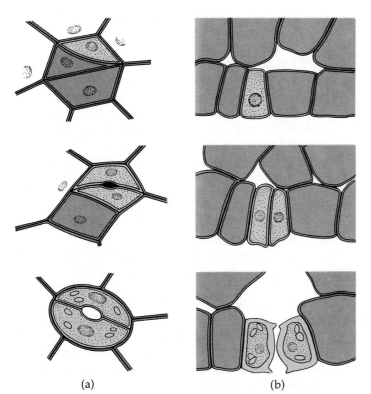

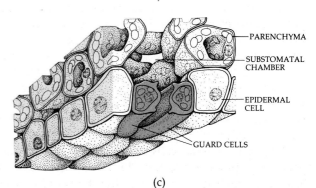

(a) (b) (c)

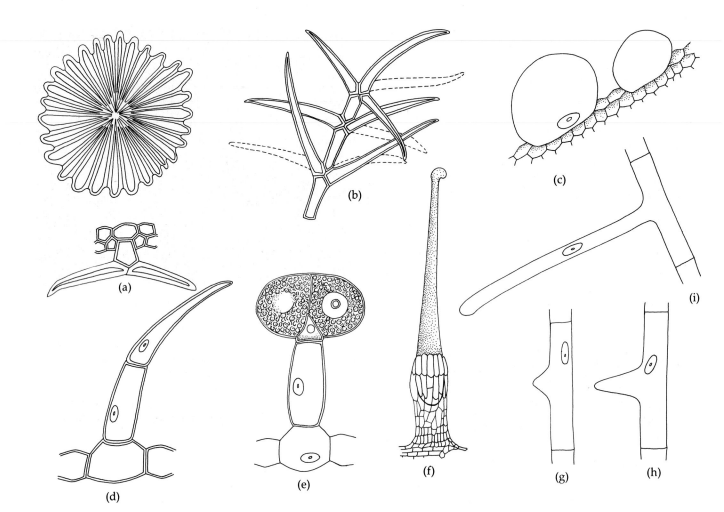

20–22

Trichomes. (a) Surface (above) and sectional (below) views of the peltate hair, or scale, of the olive (Olea europaea) *leaf. (b) Dendroid (treelike) hair from the leaf of the sycamore, or plane tree* (Platanus orientalis). *(c) Water vesicle of the "ice plant"* (Mesembryanthemum crystallinum). *(d) Short, unbranched hair from tomato* (Lycopersicon esculentum) *stem. (e) Glandular hair from tomato stem. (f) Stinging hair of the nettle,* Urtica. *The stinging hair consists of a long needlelike part and a broad base surrounded by other epidermal cells. When the hair is touched, the tip breaks off and poisonous cell contents (histamine and acetylcholine) are injected into the skin. (g)–(i) Stages in the development of a root hair, a simple tubular extension of an epidermal cell.*

crease in leaf pubescence (hairiness) results in increased reflectance of solar radiation, lower leaf temperatures, and lower rates of water loss. Many "air plants" such as atmospheric bromeliads utilize foliar trichomes for the absorption of water and minerals. In contrast, in the saltbush (*Atriplex*), salt-secreting trichomes remove salts from the leaf tissue, preventing an accumulation of toxic salts in the plant. Trichomes may provide a defense against insects. For example, in many species a positive correlation exists between hairiness and resistance to attack by insects. The hooked hairs of some species impale insects and larvae (Figure 20–24). Glandular (secretory) hairs may provide a chemical defense.

Periderm

A *periderm* commonly replaces the epidermis in stems and roots having secondary growth. The periderm consists largely of protective *cork tissue*, which is nonliving and has walls that are heavily suberized at maturity; of the *cork cambium*; and of *phelloderm*, a living parenchyma tissue (Figure 20–25). The cork cambium forms cork tissue on its outer surface and phelloderm on its inner surface. The origin of the cork cambium is variable, depending on the species and plant part. The periderm will be considered in detail in Chapter 23.

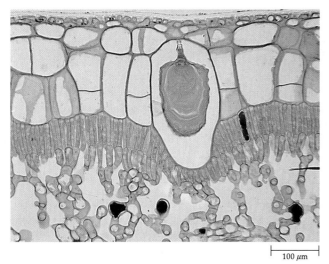

20–23
Transverse section showing upper portion of leaf blade of the rubber plant (Ficus elastica). *Notice the thick cuticle covering the multiple epidermis of mostly large cells. The club-shaped structure in the largest epidermal cell consists mostly of calcium carbonate deposited on a cellulose stalk. The cells below the large, clear epidermal cells are mesophyll cells.*

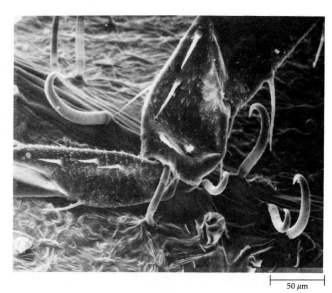

100 μm

50 μm

20–24
Both nymphs and adult leafhoppers (Empoasca sp.) *are frequently captured on leaves of field beans* (Phaseolus vulgaris) *bearing hooked trichomes. This scanning electron micrograph shows a trichome embedded in membranous tissue between leg segments of an adult.*

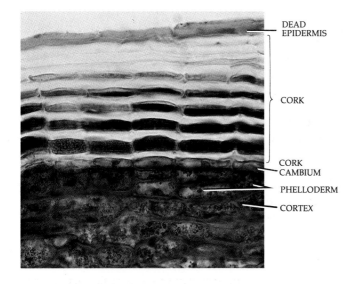

DEAD EPIDERMIS

CORK

CORK CAMBIUM

PHELLODERM

CORTEX

20–25
Transverse section of the periderm from the stem of apple (Mahis sylvestris). *The periderm shown here consists largely of cork cells, which are laid down to the outside (above) in radial rows by the cells of the cork cambium. A single layer of phelloderm cells occurs inside the cork cambium.*

SUMMARY

A summary of plant tissues and their cell types is found in Table 20–1, and the characteristics, location, and function of these types of plant cells are given in Table 20–2.

Table 20–1 *Summary of Plant Tissues and Their Cell Types*

TISSUE	CELL TYPES
Epidermis	Generally parenchyma cells; guard cells and trichomes; sclerenchyma cells
Periderm	Generally parenchyma cells; sclerenchyma cells
Xylem	Tracheids; vessel members; sclerenchyma cells; parenchyma cells
Phloem	Sieve cells or sieve-tube members; albuminous cells or companion cells; parenchyma cells; sclerenchyma cells
Parenchyma	Parenchyma cells
Collenchyma	Collenchyma cells
Sclerenchyma	Fibers or sclereids

Table 20–2   *Summary of Cell Types*

CELL TYPE	CHARACTERISTICS	LOCATION	FUNCTION
Parenchyma	Shape: commonly polyhedral (many-sided); variable Cell wall: primary, or primary and secondary; may be lignified, suberized, or cutinized Living at maturity	Throughout the plant, as parenchyma tissue in cortex, as pith and pith rays, or in xylem and phloem	Such metabolic processes as respiration, digestion, and photosynthesis; storage and conduction; wound healing and regeneration
Collenchyma	Shape: elongated Cell wall: unevenly thickened; primary only—nonlignified Living at maturity	On the periphery (beneath the epidermis) in young elongating stems; often as a cylinder of tissue or only in patches; in ribs along veins in some leaves	Support in primary plant body
Fibers	Shape: generally very long Cell wall: primary and thick secondary—often lignified Often (not always) dead at maturity	Sometimes in cortex of stems, most often associated with xylem and phloem; in leaves of monocotyledons	Support
Sclereids	Shape: variable; generally shorter than fibers Cell wall: primary and thick secondary—generally lignified May be living or dead at maturity	Throughout the plant	Mechanical; protective
Tracheid	Shape: elongated and tapering Cell wall: primary and secondary; lignified; contains pits, but not perforations Dead at maturity	Xylem	Chief water-conducting element in gymnosperms and seedless vascular plants; also found in angiosperms
Vessel member	Shape: elongated, generally not as long as tracheids Cell wall: primary and secondary; lignified; contains pits and perforations; several vessel members end-to-end constitute a vessel Dead at maturity	Xylem	Chief water-conducting element in angiosperms

Table 20–2   *(Continued)*

CELL TYPE	CHARACTERISTICS	LOCATION	FUNCTION
Sieve cell	Shape: elongated and tapering Cell wall: primary in most species; with sieve areas; callose often associated with wall and pores Living at maturity; either lacks or contains remnants of a nucleus at maturity; lacks distinction between vacuole and cytoplasm	Phloem	Chief food-conducting element in gymnosperms and seedless vascular plants
Albuminous cell	Shape: generally elongated Cell wall: primary Living at maturity; associated with sieve cell, but generally not derived from same mother cell as sieve cell; has numerous connections with sieve cell	Phloem	Believed to play a role in the movement of food into and out of the sieve cell
Sieve-tube member	Shape: elongated Cell wall: primary, with sieve areas; sieve areas on end wall with much larger pores than those on side walls—this wall part is termed a sieve plate; callose often associated with wall and pores Living at maturity; either lacks a nucleus at maturity, or contains only remnants of nucleus; contains a proteinaceous substance known as slime, or P-protein, in dicots and some monocots; several sieve-tube members in a vertical series constitute a sieve tube	Phloem	Chief food-conducting element in angiosperms
Companion cell	Shape: variable, generally elongated Cell wall: primary Living at maturity; closely associated with sieve-tube members; derived from same mother cell as sieve-tube member; has numerous connections with sieve-tube member	Phloem	Believed to play a role in the movement of food into and out of the sieve-tube member

C H A P T E R   2 1

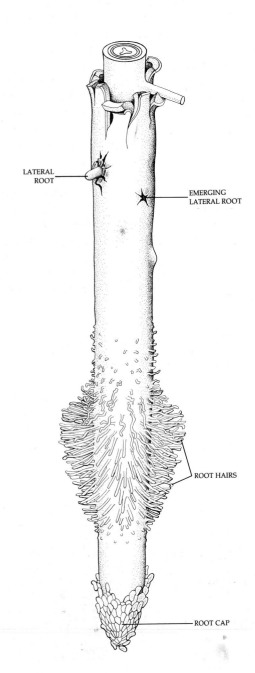

LATERAL
ROOT

EMERGING
LATERAL ROOT

ROOT HAIRS

ROOT CAP

21–1
A portion of a dicot root, showing the spatial relationship between the root cap and region of root hairs, and (near the top) the sites of emergence of lateral roots, which arise from deep within the parent root. New root hairs arise just beyond the region of elongation at about the same rate as the older hairs die off.

The Root: Primary Structure and Development

In most vascular plants, the roots constitute the underground portion of the sporophyte and are involved primarily in anchorage and absorption (Figure 21–1). Two other functions associated with roots are storage and conduction. Most roots are important storage organs and some, such as those of the carrot, sugar beet, and sweet potato, are specifically adapted for the storage of food. Foods manufactured above ground, in photosynthesizing portions of the plant body, move through the phloem to the storage tissues of the root. This food may eventually be used by the root itself, but more often the stored food is digested and transported back through the phloem to the aboveground parts. In biennials (plants that complete their life cycle over a two-year period) such as the sugar beet and carrot, large food reserves accumulate in the storage regions of the root during the first year and then are used during the second year to produce flowers, fruits, and seeds. Water and minerals, or inorganic ions, absorbed by the roots move through the xylem to the aerial parts of the plant. In addition, hormones (particularly cytokinins and gibberellins) synthesized in meristematic regions of the roots are transported upward in the xylem to the aerial parts, which depend on the presence of such hormones to stimulate growth and development (see Chapter 24).

R O O T   S Y S T E M S

The first root of the plant originates in the embryo and is usually called the *primary root*. In gymnosperms and dicotyledons, this root becomes a *taproot*; it grows directly downward, giving rise to branch roots, or *lateral roots*, along the way. The older lateral roots are found nearer the neck of the root (where the root and stem meet), and the younger ones are found nearer the root tip. This type of root system—that is, one that develops from a taproot and its branches—is called a *taproot system* (Figure 21–2a).

In monocotyledons, the primary root is usually short-lived, and the root system develops from adventitious roots that arise from the stem. These adventitious roots and their lateral roots give rise to a *fibrous root system*, in which no one root is more prominent than the others (Figure 21–2b). Taproot systems generally penetrate deeper into the soil than fibrous root systems. The shallowness of fibrous root systems and the tenacity with which they cling to soil particles make them especially well suited as ground cover for the prevention of soil erosion.

The extent of a root system—that is, the depth to which it penetrates the soil and spreads laterally—is dependent upon several factors, including the moisture, temperature, and composition of the soil. The bulk of most so-called "feeder roots" (roots actively engaged in the uptake of water and minerals) occurs in the upper meter of soil. The majority of the feeder roots of most trees occur in the upper 15 centimeters of soil, the part of the soil normally richest in organic matter. Some trees, such as spruces, beeches, and poplars, rarely produce deep taproots, whereas others, such as oaks and many pines, commonly produce relatively deep taproots, making such trees rather difficult to transplant. The record for depth of penetration by roots belongs to the desert shrub mesquite *(Prosopis juliflora)*. Mesquite roots were found growing at a depth of 53.3 meters at a new open-pit mine 30 kilometers southwest of Tucson, Arizona, in 1960. Roots of *Tamarix* and *Acacia* trees were found in Egypt at a depth of 30 meters during the digging of the Suez Canal. The lateral spread of tree roots is usually greater than the spread of the crown of the tree. The root systems of corn plants *(Zea mays)* often reach a depth of about 1.5 meters, with a lateral spread of about a meter on all sides of the plant. The roots of alfalfa *(Medicago sativa)* may extend to depths of up to 6 meters or more.

As a plant grows, it needs to maintain a balance between the total surface area available for the manufacture of food (the photosynthesizing surface) and the surface area available for the absorption of water and minerals. In a young plant, that is, one just becoming established, the total water- and mineral-absorbing surface usually far exceeds the photosynthesizing surface. However, this relationship tends to change in favor of the photosynthesizing surface as the plant ages, a fact that gardeners must consider. Even when plants are carefully transplanted, the balance between shoot and root is invariably disturbed. Most of the fine feeder roots are left behind when the plant is removed from the soil; cutting back the shoot helps to reestablish a balance between the root system and the shoot system.

One of the most detailed studies conducted on the extent of the root and shoot systems of any one plant was made on a 4-month-old rye plant *(Secale cereale)*. The total surface area of the root system, including root hairs, measured 639 square meters, or 130 times the surface area of the shoot system. What is even more amazing is the fact that the roots occupied only about 6 liters of soil.

ORIGIN AND GROWTH OF PRIMARY TISSUES

The growth of many roots is apparently a continuous process that stops only under such adverse conditions as drought and low temperatures. During their growth through the soil, roots follow the path of least resis-

21–2
Two types of root systems. (a) *Taproot system of dandelion* (Taraxacum officinale). (b) *Fibrous root system of a grass.*

(a)   (b)

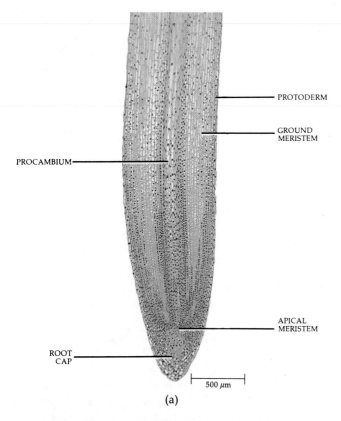

PROTODERM

GROUND MERISTEM

PROCAMBIUM

APICAL MERISTEM

ROOT CAP

500 μm

(a)

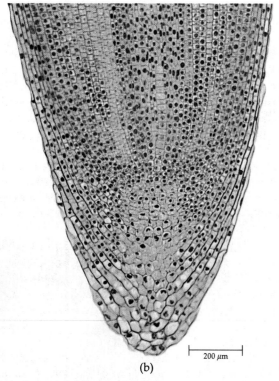

200 μm

(b)

21–3
Longitudinal sections of onion (Allium cepa) *root tip.* (a) *Primary meristems can be distinguished close to the apical meristem.* (b) *Detail of the apical meristem. Compare the organization of this apical meristem with that of the corn root tip shown in Figure 21–4.*

tance and frequently follow spaces left by earlier roots that have died and rotted.

The tip of the root is covered by a *root cap* (Figures 21–1, 21–3, and 21–4), a thimblelike mass of cells that protects the apical meristem behind it and aids the root in its penetration of the soil. As the root grows longer and the root cap is pushed forward, the cells on the periphery of the root cap are sloughed off. These sloughed-off cells form a slimy covering around the root and lubricate its passage through the soil. As quickly as root cap cells are sloughed off, new ones are added by the apical meristem. The longevity (from time of initiation to sloughing off) of root cap cells ranges from four to nine days, depending upon the length of the cap and the species.

The slimy substance of the root cap is a highly hydrated polysaccharide, probably a pectic substance, that is secreted by the outer root cap cells. The slimy substance accumulates in dictyosome vesicles, which fuse with the plasma membrane and then release the slime into the wall. Eventually the slime passes to the outside, where it forms droplets.

In addition to protecting the apical meristem and aiding the root in its penetration of the soil, the root cap plays an important role in controlling the response of the root to gravity (geotropism, or gravitropism; see Chapter 25).

Growth Regions of the Root

Apart from the root cap, the most striking structural feature of the root apex is the arrangement of longitudinal files of cells that emanate from the apical meristem. The apical meristem is composed of relatively small (10 to 20 micrometers in diameter), many-sided cells—the initials and their immediate derivatives (page 384)—with dense cytoplasm and large nuclei (Figures 21–3 and 21–4). The organization and number of initials in the apical meristems of roots are both quite variable.

Two main types of apical organization are found in the roots of seed plants. In one, the root cap, the vascular cylinder of xylem and phloem, and the cortex emerge from a common group of cells in the apical meristem (Figure 21–3). In the other, each region can be traced to an independent layer of cells (Figure 21–4). In the second type of organization, the epidermis has a common origin with either the root cap or the cortex.

Although the region of initials in the apical meristem of the root was once considered to be a region of active cell division, studies on the apical meristems of many roots indicate that this region is relatively inactive and that most cell division occurs a short distance beyond the relatively quiescent initials. The relatively inactive region of the apical meristem is known as the *quiescent center* (Figure 21–5).

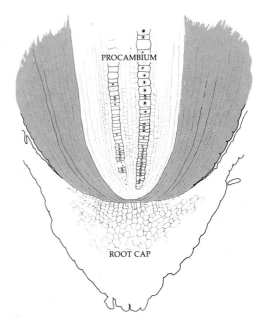

21-4
*Apical meristem of a corn (Zea mays)
root tip. Notice the three distinct layers
of initials. The lower layer gives rise to
the root cap, below; the middle layer to
the protoderm and to the ground
meristem or cortex; and the upper to the
procambium or vascular cylinder.*

| 100 μm

| 50 μm

21-5
*Apical meristem of a corn (Zea mays)
root tip, showing the quiescent center. To
prepare this autoradiograph, the root tip
was supplied for one day with thymine
labeled with radioactive tritium (^{3}H). In
the rapidly dividing cells around the
quiescent center (within the dashed oval),
the radioactive material was quickly
incorporated into the nuclear DNA and
left its marks (the dark grains) on this
autoradiograph.*

As indicated by the term "relatively," this quiescent center is not totally devoid of divisions under normal conditions. Moreover, the quiescent center is able to repopulate the bordering meristematic regions when they are injured. In a recent study, for example, the isolated quiescent centers of corn (*Zea mays*) grown in sterile culture were found to be able to form whole roots without first forming callus, or wound, tissue. In another study of corn roots, a striking correlation was found between the size of the quiescent center and the complexity of the primary vascular pattern of the root. These and other studies indicate that the quiescent center plays an essential role in organization and development of the root.

The distance beyond the apical meristem at which most cell division takes place varies from species to species and also within the same species, depending on the age of the root. The combination of the apical meristem and the nearby portion of root in which cell division does occur is called the *region of cell division* (Figure 21–6).

Behind the region of cell division, but not sharply delimited from it, is the *region of elongation*, which usually measures only a few millimeters in length (Figure 21–6). The elongation of cells in this region results in most of the increase in length of the root. Beyond this region the root does not increase in length. Thus, growth in length of the root occurs near the root tip and results in a very limited portion of the root constantly being pushed through the soil.

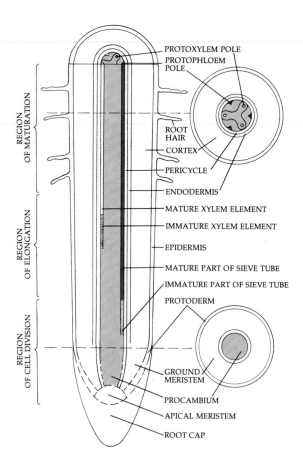

21–6
*A diagram illustrating early stages in the
primary development of a root tip.
(Compare this figure with Figure 21–1.)*

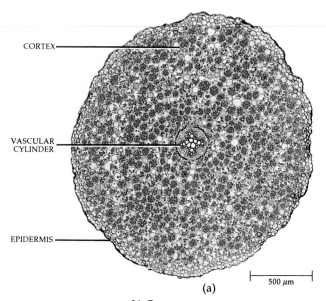

21–7
*Transverse sections of the root of a
buttercup (*Ranunculus*). (a) Overall
view of mature root. (b) Detail of
immature vascular cylinder. Notice the
intercellular spaces among the cortical
cells. (c) Detail of mature vascular
cylinder. Note the numerous starch grains
in the cortical cells.*

21–8
*Transverse sections of a corn (*Zea mays*)
root. (a) Overall view of mature root.
Part of a lateral root can be seen at the
lower right. The vascular cylinder, with
its pith, is quite distinct. (b) Detail of
immature vascular cylinder. (c) Detail
of mature vascular cylinder.*

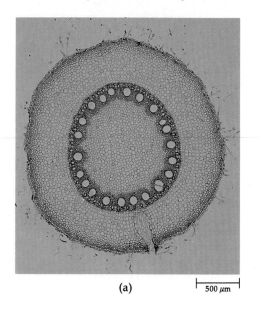

The region of elongation is followed by the *region of
maturation*, in which most of the cells of the primary
tissues mature (Figure 21–6). Root hairs are also produced in this region, and sometimes this part of the root
is called the root-hair zoné (see Figure 21–1).

It is important to note that there is a gradual transition from one region of the root to another. The regions
are not sharply delimited from one another. Some cells
begin to elongate and differentiate in the region of cell
division, whereas others reach maturity in the region of
elongation. For example, the first-formed elements of
the phloem and xylem mature in the region of elongation and are often stretched and destroyed during elongation of the root.

The protoderm, procambium, and ground meristem
can be distinguished in very close proximity to the apical meristem (Figures 21–3 and 21–6). These are the
primary meristems that differentiate into the epidermis,
the primary vascular tissues, and the cortex, respectively (Chapter 19).

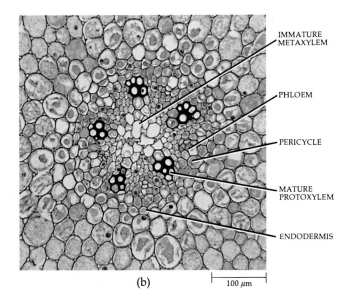

IMMATURE
METAXYLEM

PHLOEM

PERICYCLE

MATURE
PROTOXYLEM

ENDODERMIS

(b) 100 μm

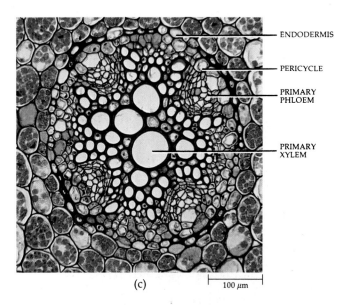

ENDODERMIS

PERICYCLE

PRIMARY
PHLOEM

PRIMARY
XYLEM

(c) 100 μm

PRIMARY STRUCTURE

Compared with that of the stem, the internal structure of the root is usually relatively simple. This is due in large part to the absence of leaves in the root and the corresponding absence of nodes and internodes. Thus, the arrangement of tissues shows very little difference from one level of the root to another.

The three tissue systems of the root in the primary stage of growth can be readily distinguished in both transverse (Figures 21–7 and 21–8) and longitudinal sections (Figure 21–3). The epidermis (dermal tissue system), the cortex (ground tissue system), and the vascular tissues (vascular tissue system) can be seen to be clearly distinguished from one another. In most roots, the vascular tissues form a solid cylinder (Figure 21–7),

but in some they form a hollow cylinder around a pith (Figure 21–8).

Epidermis

The epidermis in young roots absorbs water and minerals, and this function is facilitated by *root hairs*—tubular extensions of the epidermal cells—which greatly increase the absorbing surface of the root (Figure 21–9). (See pages 224–226 in Chapter 13 for a discussion of the role of mycorrhizae in absorption.) As part of the previously mentioned study of a four-month-old rye plant, it was estimated that the plant contained approximately 14 billion root hairs, with an absorbing surface of 401 square meters. Placed end to end, these root hairs would extend well over 10,000 kilometers.

ENDODERMIS   PROTOXYLEM   PHLOEM   PERICYCLE   MATURE
METAXYLEM

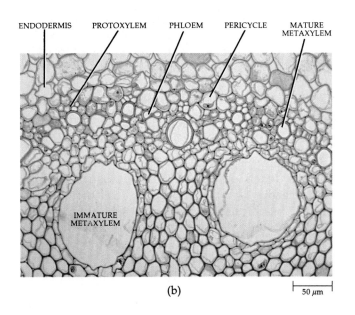

IMMATURE
METAXYLEM

(b) 50 μm

ENDODERMIS   PROTOXYLEM   PHLOEM   PERICYCLE

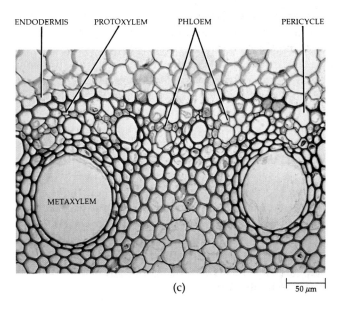

METAXYLEM

(c) 50 μm

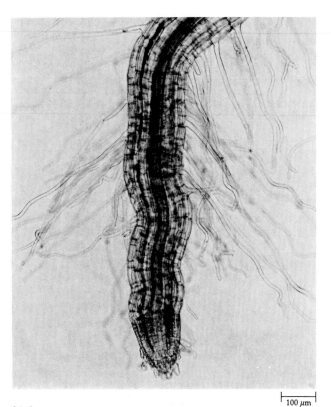

100 μm

21–9
*Root of a bent grass (Agrostis tenuis)
seedling, showing root hairs. Root hairs
may be as much as 1.3 centimeters long
and may attain their full size within
hours. Each hair is comparatively
short-lived, but the formation of new
root hairs and the death of old ones
continues as long as the root is growing.*

Root hairs are relatively short-lived and are confined largely to the region of maturation of the root. The production of new root hairs occurs just beyond the region of elongation (Figures 21–1 and 21–6) and at about the same rate as that at which the older root hairs are dying off at the upper end of the root hair zone. As the tip of the root penetrates the soil, new root hairs develop immediately behind it, providing the root with surfaces capable of absorbing new supplies of water and minerals, or inorganic ions. (See Chapter 27 for a discussion of absorption of water and inorganic ions by roots.) Obviously, it is the new and growing roots—the feeder roots—that are primarily involved in the absorption of water and minerals. For this reason, great care must be taken by gardeners to move as much soil as possible along with the root system when transplanting a plant. If the plant is simply "torn" from the soil, most of the feeder roots will be left behind and the plant probably will not survive.

The epidermal cells of the root, including those with root hairs, are parenchyma cells, which are closely packed together. In young portions of the root, the epidermis has at most a thin cuticle; hence, the cell wall offers little resistance to the passage of water and minerals into the root. In addition, the surface of many roots is covered by a slime sheath, or *mucigel*, which enables the roots to establish a much more intimate contact with the particles of soil. The exact origin of mucigel has not yet been determined, but it has been suggested that it is formed, at least in part, by the root cap. The mucigel has been shown to provide an environment favorable to beneficial bacteria. It may also influence the availability of ions to the root, as well as providing the root with short-term protection from drying out (desiccation).

Cortex

As seen in transverse section (Figure 21–7a), the cortex occupies by far the greatest area of the primary body of most roots. The cells of the cortex store starch and other substances but usually lack chloroplasts. Roots that undergo considerable amounts of secondary growth—those of gymnosperms and most dicots—shed their cortex early. In such roots, the cortical cells remain parenchymatous. In monocots, by contrast, the cortex is retained for the life of the root, and many of the cortical cells develop secondary walls that become lignified.

Regardless of the degree of differentiation, the cortical tissue contains numerous intercellular spaces—air spaces essential for aeration of the cells of the root (see Figures 21–7 and 21–8). The cortical cells have numerous contacts with one another, and their protoplasts are connected by plasmodesmata. Thus substances moving across the cortex may move from cell to cell by way of the protoplasts and plasmodesmata or by way of the cell walls.

Unlike the rest of the cortex, the innermost layer is compactly arranged and lacks air spaces. This layer, the *endodermis* (Figures 21–7 and 21–8), is characterized by the presence of *Casparian strips* in its anticlinal walls (the walls perpendicular to the surface of the root). The Casparian strip is a bandlike portion of the primary wall that is impregnated with a fatty substance called suberin and is sometimes lignified. The protoplasts of endodermal cells are quite firmly attached to the Casparian strips (Figures 21–10 and 21–11). Inasmuch as the endodermis is compact and the Casparian strips are impermeable to water, all substances entering and leaving the vascular cylinder must enter the protoplasts of the endodermal cells. This is accomplished either by crossing their plasma membranes or by way of the numerous plasmodesmata connecting the endodermal cells with the protoplasts of neighboring cells of the cortex and vascular cylinder.

The effectiveness of the Casparian strip as a barrier to the movement of substances across the walls of the endodermis has been demonstrated in corn (*Zea mays*) roots that had absorbed the element lanthanum, a positively charged ion that cannot penetrate cell membranes. When the roots were examined with the electron microscope, the lanthanum was found only in the cell walls of the cortex; its progress across the root was abruptly and completely stopped by the Casparian strip. (See Chapter 27 for further discussion of the role of the endodermis in the movement of water and solutes across the root.)

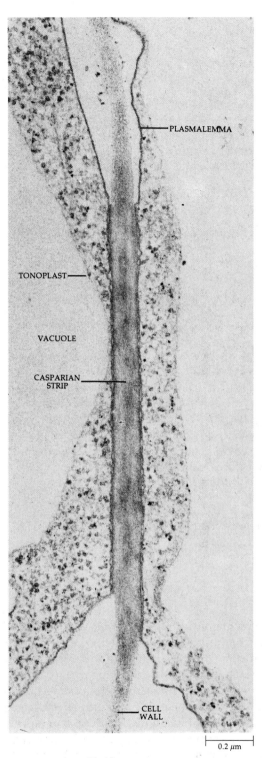

21–11
Transverse section of a radial wall between two endodermal cells. The region of the Casparian strip is more intensely stained than the rest of the wall. Notice how firmly the plasma membrane adheres to the wall in the region of the Casparian strip in each of these plasmolyzed cells.

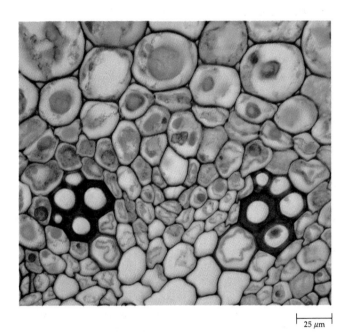

25 μm

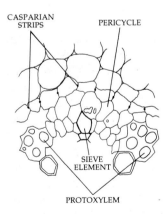

21–10
High-magnification view of a portion of an immature buttercup (Ranunculus) root, showing Casparian strips in the endodermal cells. Notice that the plasmolyzed protoplasts of the endodermal cells cling to the strips.

21–12
Three-dimensional diagrams showing three developmental stages of an endodermal cell from a root that remains in a primary state. (a) Initially, the endodermal cell is characterized by the presence of a Casparian strip in its anticlinal walls. (b) Then a suberin lamella is deposited internally over all wall surfaces. (c) Finally, the suberin lamella is covered by a thick, often lignified, layer of cellulose.

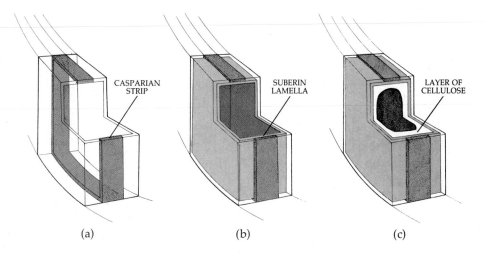

(a)   (b)   (c)

As mentioned previously, in roots that undergo secondary growth, the cortex and its endodermis are shed early. In roots in which the cortex is retained, a suberin lamella, consisting of alternating layers of suberin and wax, is eventually deposited internally over all wall surfaces of the endodermis. This is followed by the deposition of cellulose, which may become lignified (Figure 21–12). These changes in the endodermis begin opposite the phloem strands and spread toward the protoxylem (Figure 21–7a). Opposite the protoxylem, some of the endodermal cells may remain thin-walled and retain their Casparian strips for a prolonged period of time. Such cells are called *passage cells*, and in most species, they eventually become suberized. In conifers, the modification of the endodermal cell wall is terminated with the deposition of the suberin lamella (Figure 21–13).

Vascular Cylinder

The vascular cylinder of the root consists of vascular tissues and one or more layers of cells, the *pericycle*, which completely surrounds the vascular tissues (Figures 21–7 and 21–8). In the young root, the pericycle is composed of parenchyma cells with primary walls, but as the root ages, the cells of the pericycle may develop secondary walls (Figure 21–8).

The pericycle plays several important roles. In most seed plants, lateral roots arise in the pericycle. In plants undergoing secondary growth, the pericycle contributes to the vascular cambium and generally gives rise to the first cork cambium. Pericycle often proliferates—that is, gives rise to more pericycle.

The center of the vascular cylinder of most roots is occupied by a solid core of primary xylem from which ridgelike projections extend toward the pericycle (Figure 21–7). Nestled between the ridges of xylem are strands of primary phloem. Obviously, the vascular cylinder of the root is a protostele.

The number of ridges of primary xylem varies from species to species, sometimes even varying along the axis of a given root. If two ridges are present, the root is said to be *diarch*; if three are present, *triarch*; four, *tetrarch* (Figure 21–7a, c); and if many are present, *polyarch* (Figure 21–8). The first (*proto-*) xylem elements to mature in roots are located next to the pericycle, and the tips of the ridges are commonly referred to as *protoxylem poles* (Figures 21–7 and 21–8). The metaxylem (*meta-*, meaning "after") occupies the inner portions of the ridges and the center of the vascular cylinder and matures after the protoxylem. The roots of some monocotyledons (for example, corn) have a pith (Figure 21–8), which some botanists interpret as potential vascular tissue.

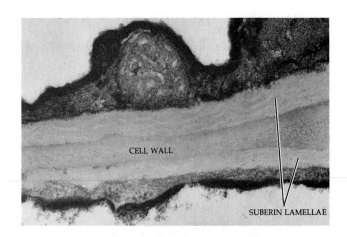

21–13
*Electron micrograph showing the suberin lamellae in the walls between two endodermal cells of a red pine (*Pinus resinosa*) root. Note the alternating light and dense bands, which are interpreted as consisting of wax and suberin, respectively.*

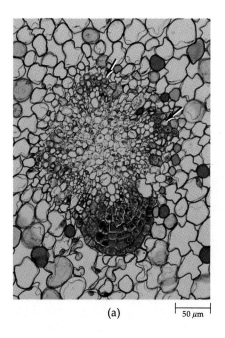

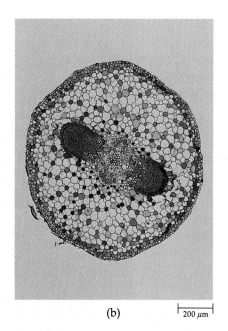

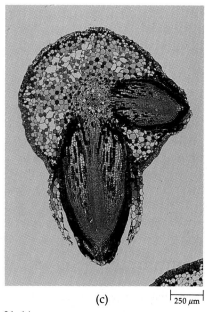

(a)   ⊢ 50 μm ⊣

(b)   ⊢ 200 μm ⊣

(c)   ⊢ 250 μm ⊣

21–14

Three stages in the origin of lateral roots in a willow (Salix). (a) One root primordium is present (below) and two others are being initiated in the region of the pericycle (see arrows). The vascular cylinder is still very young. (b) Two root primordia penetrating the cortex. (c) One lateral root has reached the outside, and the other nearly has.

ORIGIN OF LATERAL ROOTS

In most seed plants, lateral roots (branch roots) arise in the pericycle. Because lateral roots originate deep from within the parent root, they are said to be endogenous, meaning "originating within an organ" (Figures 21–1 and 21–14).

Divisions in the pericycle that initiate lateral roots occur some distance beyond the region of elongation in partially or fully differentiated root tissues. In angiosperm roots, derivatives of both the pericycle and endodermis commonly contribute to the new root primordium, although in many cases the derivatives of the endodermis are short-lived. As the young lateral root, or *root primordium*, increases in size, it pushes its way through the cortex (Figure 21–14c), possibly secreting enzymes that digest some of the cortical cells lying in its path. While still very young, the root primordium develops a root cap and apical meristem, and the primary meristems appear. Initially, the vascular cylinders of lateral root and parent root are not connected to one another. The two vascular cylinders are joined later, when derivatives of intervening parenchyma cells differentiate into xylem and phloem.

AERIAL ROOTS

Aerial roots are adventitious roots produced from aboveground structures. The aerial roots of some plants serve as *prop roots* for support, as in corn (Figure 21–15). When they come in contact with the soil, they branch and function also in the absorption of water and minerals. Prop roots are produced from the stems and branches of many tropical trees, such as the red man-

21–15

Prop roots of corn (Zea mays), a type of adventitious root.

grove *(Rhizophora mangle),* the banyan tree *(Ficus benghalensis),* and some of the palms. Other aerial roots, as in the ivy *(Hedera helix),* cling to the surface of objects such as walls and provide support for the climbing stem.

Roots require oxygen for respiration, which is why most plants cannot live in soil in which there is not adequate drainage and which consequently lacks air spaces. Some trees that grow in swampy habitats develop roots that grow out of the water and serve not only to anchor but to aerate the plant. For example, the root system of the black mangrove *(Avicennia germinans)* develops negatively geotropic extensions called *pneumatophores* (air roots), which grow upward out of the mud and so provide adequate aeration (Figure 21–16). The "knees" of the bald cypress *(Taxodium distichum;* Figure 18–27, page 348) have had a similar function attributed to them, but this idea is now in doubt.

Special Adaptations

Many special adaptations of roots are found among epiphytes—plants that grow on other plants but are not parasitic on them. The epidermis of the orchid root, for example, is several layers thick (Figure 21–17) and, in some species, is the only photosynthetic organ of the plant. This multiple epidermis, called velamen, provides mechanical protection for the cortex, as well as reducing water loss. The velamen also may function in the absorption of water.

Among epiphytes, *Dischidia rafflesiana,* the "flower pot plant," has a very unusual modification. Some of its leaves are flattened, succulent structures, but others

form hollow containers—the "flower pots"—that collect debris and rainwater (Figure 21–18). Ant colonies live in the pots and add to the nitrogen supply of the plant. Roots, formed at the node above the modified leaf, grow downward and into the pot, from which they absorb water and minerals.

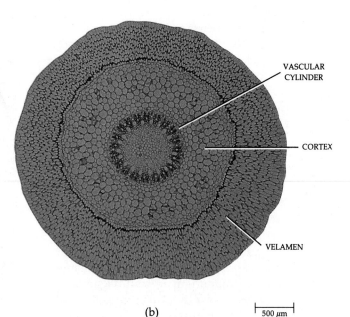

21–16
Pneumatophores (air roots) of the white mangrove (Laguncularia racemosa) *protruding from the mud near the base of a tree.*

(a)

VASCULAR CYLINDER

CORTEX

VELAMEN

(b)   500 μm

21–17
(a) *Aerial roots of an orchid* (Oncidium sphacelatum). (b) *Transverse section of* *an orchid root, showing the multiple epidermis, or velamen.*

21–18
The epiphyte Dischidia rafflesiana, *or "flower pot plant." (a) A modified leaf, or "pot," which collects debris and rainwater. (b) Modified leaf cut open to show roots that have grown down into it.*

(a)

(b)

ADAPTATIONS FOR FOOD STORAGE

Most roots are storage organs, and in some plants the roots are specialized for this function. Such roots are fleshy because of an abundance of storage parenchyma, which is permeated by vascular tissue. The development of some storage roots, such as that of the carrot *(Daucus carota),* is essentially similar to that of nonfleshy roots, except for a predominance of parenchyma cells in their secondary xylem and phloem. The root of the sweet potato *(Ipomoea batatas)* develops in a manner similar to the carrot; however, in the sweet potato, additional vascular cambium cells develop within the secondary xylem around individual vessels or groups of vessels (Figure 21–19). These additional cambia (plural of cambium), while producing a few tracheary elements toward the vessels and a few sieve tubes away from them, mainly produce storage paren-

21–19
Transverse sections of the root of a sweet potato (Ipomoea batatas). *(a) Overall view. (b) Detail of xylem, showing cambium around vessels.*

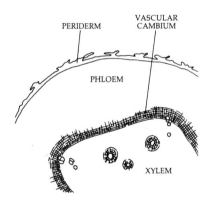

PERIDERM

VASCULAR CAMBIUM

PHLOEM

XYLEM

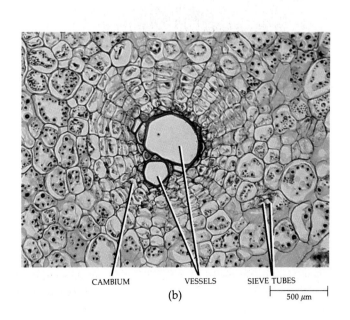

(a)   500 μm

CAMBIUM   VESSELS   SIEVE TUBES

(b)   500 μm

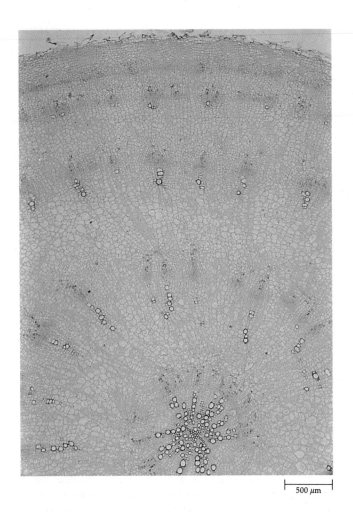

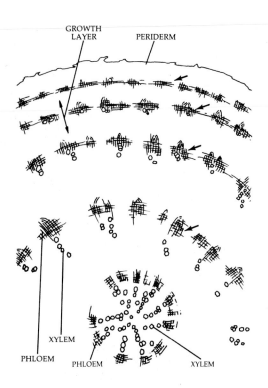

21–20
Transverse section of a sugar beet (Beta
vulgaris) *root, with supernumerary
cambia indicated by arrows. The original
cambium produces relatively little xylem
and phloem (in the center of the root).*

chyma cells in both directions. In the sugar beet (*Beta
vulgaris*), for example, most of the increase in thickness
of the root results from the development of extra cambia (supernumerary cambia) around the original vascular cambium (Figure 21–20). These concentric layers of
cambia, which superficially resemble growth rings in
woody roots and stems, produce parenchyma-domi-
nated xylem and phloem toward the inside and outside,
respectively. The upper portion of most fleshy roots actually develops from the hypocotyl.

SUMMARY

Roots are organs specialized for anchorage, absorption, storage, and conduction. Gymnosperms and dicots commonly
produce taproot systems, whereas monocots usually produce
fibrous root systems. The extent of the root system is dependent upon several factors, but the bulk of most feeder roots is
found in the upper meter of the soil.

The apical meristems of most roots contain a quiescent
center; most meristematic activity, or cell division, occurs a
short distance from the apical initials. During primary
growth, the apical meristem gives rise to the three primary
meristems—protoderm, ground meristem, and procambium
—which differentiate into epidermis, cortex, and vascular
cylinder, respectively. In addition, the apical meristem produces the root cap, which serves to protect the meristem and
aid the root in its penetration of the soil.

Many epidermal cells of the root develop root hairs, which
greatly increase the absorbing surface of the root. With the
exception of the endodermis, the cortex contains numerous
intercellular spaces. The compactly arranged endodermal
cells contain Casparian strips on their anticlinal walls. Consequently, all substances moving between the cortex and vascular cylinder must pass through the protoplasts of the
endodermal cells.

The vascular cylinder consists of pericycle and the primary
vascular tissues, which are completely surrounded by the pericycle. The primary xylem usually occupies the center of the
vascular cylinder and has radiating ridges that alternate with
strands of primary phloem. Branch roots originate in the pericycle and push their way to the outside through the cortex
and epidermis.

CHAPTER 22

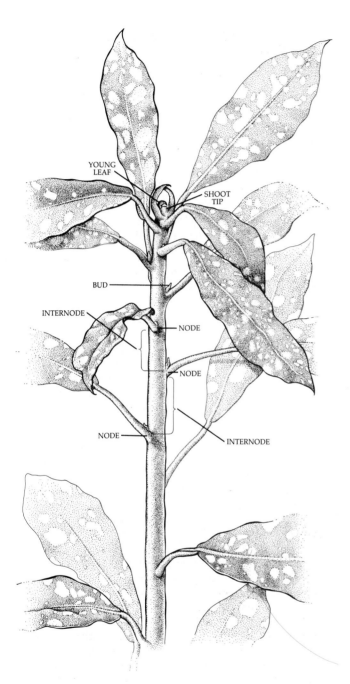

YOUNG
LEAF

SHOOT
TIP

BUD

INTERNODE

NODE

NODE

NODE

INTERNODE

22–1

A portion of a Croton *shoot. The leaves of* Croton, *a dicot, have a mottled appearance and are spirally arranged along the stem. At the apex the leaves are close together so that nodes and internodes do not exist as separate regions of the stem. Growth in length of the stem between successive leaves, which are attached to the stem at the nodes, results in formation of the internodes.*

The Shoot: Primary Structure and Development

The shoot, which consists of the stem and its leaves, is initiated during development of the embryo, where it may be represented by a plumule, consisting of a stem (the epicotyl), one or more *leaf primordia* (rudimentary leaves), and an apical meristem. (The plumule can be thought of as the first bud of the plant.) With resumption of growth of the embryo during germination of the seed, new leaves develop from the apical meristem, and the stem elongates and differentiates into nodes and internodes. Gradually *bud primordia* form in the axils of the leaves (Figures 22–1 and 22–2), and eventually they follow a sequence of growth and differentiation more or less similar to that of the first bud. This pattern is repeated many times as the shoot system of the plant is produced.

Commonly, the growing terminal bud of a shoot inhibits development of lateral buds, a phenomenon known as *apical dominance* (see Chapter 24). As the distance between shoot tip and lateral buds increases, the retarding influence of the terminal bud is lessened and the lateral buds can proceed with their development. (This is why pinching off the shoot tips, a common practice of home gardeners, results in fuller and bushier plants.)

The two principal functions associated with stems are conduction and support. Substances manufactured in the leaves are transported through the stems by way of the phloem to sites of utilization of those substances, including growing leaves, stems, and roots, as well as developing flowers , seeds, and fruits. Much of the food material is stored in parenchyma cells of roots, seeds, and fruits, but stems are also important storage organs, and some, such as the underground stems of the white potato, are specifically adapted for storage. The principal photosynthetic organs of the plant—the leaves—are supported by stems, which act to place the leaves in favorable positions for exposure to light. In addition, most of the plant's loss of water vapor occurs through

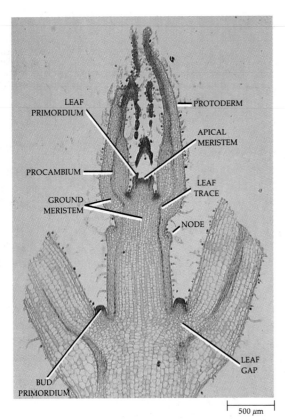

22–2
Longitudinal section of shoot tip of the common houseplant Coleus blumei, *a dicot. The leaves in* Coleus *are arranged opposite one another at the nodes, each successive pair at right angles to the previous pair; thus the leaves of the labeled node are at right angles to the plane of section.*

the leaves (Chapter 27). Water is conducted upward in the xylem from the roots and into the leaves via the stem.

ORIGIN AND GROWTH OF THE PRIMARY TISSUES OF THE STEM

The organization of the apical meristem of the shoot is more complex than that of the root. In addition to adding cells to the primary plant body, the apical meristem of the shoot is involved in the formation of leaf primordia, as well as the bud primordia (Figure 22–2) that develop into lateral branches. The apical meristem of the shoot also differs in its lack of a protective covering comparable to the root cap.

The vegetative shoot apex of most flowering plants has what is termed a *tunica-corpus* type of organization (Figure 22–3). The two regions—tunica and corpus—are usually distinguished by the planes of cell division present in them. The tunica consists of the outermost layer or layers of cells, which divide in planes perpendicular to the surface of the meristem (anticlinal divisions) and contribute primarily to surface growth. The corpus consists of a body of cells that lies beneath the tunica layers. In the corpus, the cells divide in various planes and add bulk to the developing shoot. The corpus and each layer of tunica have their own initials.

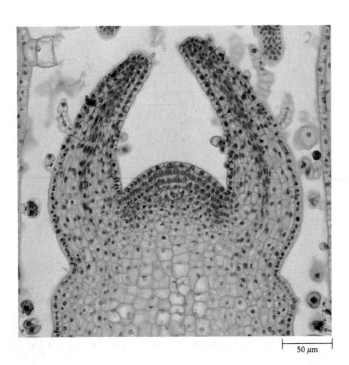

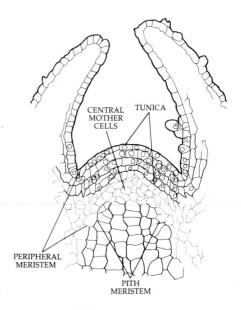

22–3
Detail of Coleus blumei *shoot apex, showing tunica-corpus organization. The zone of central mother cells roughly corresponds to the corpus.*

(a)   (b)   (c)

22-4
*Stages in growth of the terminal bud and
two lateral buds of the horse chestnut*
(Aesculus hippocastanum). (a) *The
young shoots are tightly packed in the
buds and are protected by bud scales. (b)
The buds open to reveal the oldest
rudimentary leaves. (c) Internodal elongation has separated the nodes from one
another. The terminal bud of the horse
chestnut is a mixed bud, containing both
leaves and flowers, although the flowers
are not visible here. The lateral buds
produce only leaves. (d) Detail of the
lower portions of the young shoots; the
bud scales are separated and folded back.*

(d)

In the shoot apices of many angiosperms, the bulk of
the corpus corresponds to an area of conspicuously vacuolated cells called the zone of *central mother cells*
(Figure 22–3). This zone of vacuolated cells is surrounded by the *peripheral meristem*, which originates
partly from the tunica and partly from the corpus, or
central mother cell zone. Beneath the central mother
cells is the *pith meristem*. Cell divisions are relatively in-
frequent in the central mother cell zone; by contrast,
the peripheral zone is very active mitotically. The protoderm always originates from the outermost tunica
layer, whereas the procambium and part of the ground
meristem (the cortex and sometimes part of the pith)
are derived from the peripheral meristem. The rest of
the ground meristem (all or most of the pith) is formed
by the pith meristem.

Although the primary tissues of the stem pass
through periods of growth similar to those of the root,
the stem cannot be divided along its axis into regions of
cell division, elongation, and maturation as in the case
of roots. When actively growing, the apical meristem of
the shoot gives rise to leaf primordia in such rapid succession that nodes and internodes cannot at first be distinguished. Gradually, growth begins to occur between
the levels of leaf attachment; the elongated parts of the
stem take on the appearance of internodes; and the
portions of the stem at which the leaves are attached
become recognizable as nodes (Figure 22–4). Thus, increase in length of the stem occurs largely by internodal
elongation.

Commonly, the meristematic activity causing the
elongation of the internode is more intense at the base
of the developing internodes than elsewhere. If elongation of the internode takes place over a prolonged period, the meristematic region at the base of the
internode may be called an *intercalary meristem* (a meristem region between two more highly differentiated regions). Certain elements of the primary xylem and
primary phloem—specifically the protoxylem and protophloem—differentiate within the intercalary meristem and connect the more highly differentiated
regions of the stem above and below the meristem.

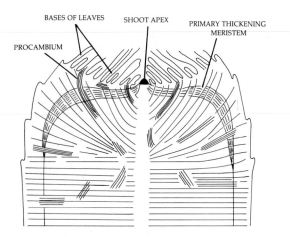

BASES OF LEAVES   SHOOT APEX   PRIMARY THICKENING MERISTEM

PROCAMBIUM

22–5
*Diagrammatic representation of the
anatomy of the top, or crown, of a
thick-stemmed monocot without
secondary growth. Increase in thickness
of such stems is due to the activity of a
primary thickening meristem. The apical
meristem and youngest leaf primordia
are sunken below surrounding stem
tissues. At the top of the stem, files of
cells are arranged like a stack of saucers.*

Growth in thickness of the primary body of the stem involves longitudinal divisions and cell enlargement in both cortex and pith. In plants with secondary growth this primary thickening is moderate. Many monocots, such as the palms, have massive primary growth. This growth usually occurs close to the apical meristem (which often appears to lie in a depression at the stem apex) and is localized in a relatively narrow region near the periphery of the stem called the *primary thickening meristem* (Figure 22–5).

As in the root, the apical meristem of the shoot gives rise to the primary meristems—protoderm, ground meristem, and procambium (Figure 22–2). These primary meristems in turn develop into the epidermis, ground tissue, and primary vascular tissues, respectively.

PRIMARY STRUCTURE OF THE STEM

Considerable variation exists in the primary structure of stems of seed plants, but three basic types of organization can be recognized: (1) In some conifers and dicots, the narrow, elongated procambial cells—and consequently the primary vascular tissues that develop from

them—appear as a more or less continuous hollow cylinder within the ground tissue (Figure 22–6). The outer region of ground tissue is called the *cortex*, and the inner region is called the *pith*. (2) In other conifers and dicots, the primary vascular tissues develop as a cylinder of discrete strands separated from one another by ground tissue (Figure 22–7). The ground tissue separating the procambial strands (and later the mature vascular bundles) is continuous with the cortex and pith, and is called the *interfascicular parenchyma*. (Interfascicular means "between the bundles.") The interfascicular regions are often called *pith rays*. Narrow interfascicular regions also interconnect the cortex and pith in the first type of organization, but there they are inconspicuous. (3) In the stems of most monocotyledons and of some herbaceous dicotyledons, the arrangement of the procambial strands and vascular bundles is more complex. The vascular tissues do not appear as a single ring of bundles between a cortex and a pith. Instead, the vascular tissues commonly develop in more than one ring of bundles or as a system of strands scattered throughout the ground tissue. In the latter instance, the ground tissue often cannot be distinguished as cortex and pith (Figure 22–8).

The *Tilia* Stem

The stem of the linden (*Tilia americana*) exemplifies the first type of organization (Figure 22–6). As in most stems, the epidermis is a single layer of cells covered by a cuticle. The stem epidermis generally contains far fewer stomata than the leaf epidermis.

The cortex of the *Tilia* stem consists of collenchyma and parenchyma cells. The several layers of collenchyma cells, which provide support to the young stem, form a continuous cylinder beneath the epidermis. The rest of the cortex consists of parenchyma cells that will contain chloroplasts when mature. The innermost layer of cortical cells, which are dark in color, sharply delimits the cortex from the cylinder of primary vascular tissues.

In the great majority of stems, including those of *Tilia*, the primary phloem develops from the outer cells of the procambium, and the primary xylem develops from the inner ones. However, not all of the procambial cells differentiate into primary xylem and primary phloem. A single layer of cells between the primary xylem and the primary phloem remains meristematic and becomes the vascular cambium. *Tilia* is also an example of a woody stem—a stem that produces much secondary xylem. (Secondary growth in stems is discussed in Chapter 23.) After internodal elongation is completed in the *Tilia* stem, fibers develop in the primary phloem. These fibers are called *primary phloem fibers* (see Figure 23–10, page 452).

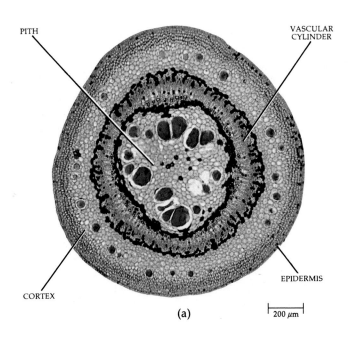

PITH

VASCULAR CYLINDER

CORTEX

EPIDERMIS

200 μm

(a)

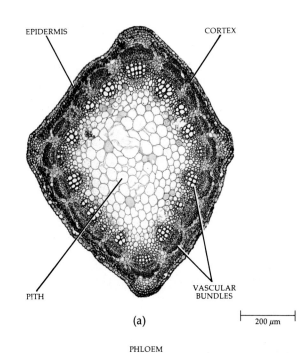

EPIDERMIS

CORTEX

PITH

VASCULAR BUNDLES

200 μm

(a)

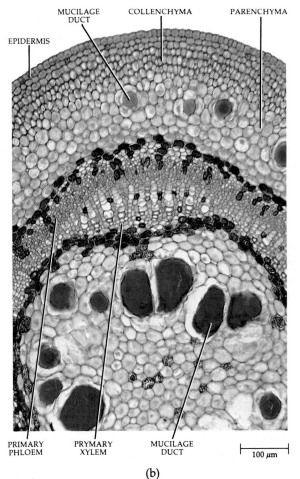

MUCILAGE DUCT

COLLENCHYMA

PARENCHYMA

EPIDERMIS

PRIMARY PHLOEM

PRYMARY XYLEM

MUCILAGE DUCT

100 μm

(b)

22–6

(a) *Transverse section of linden (Tilia americana) stem in a primary stage of growth. The vascular tissues appear as a continuous hollow cylinder that divides the ground tissue into an inner pith and an outer cortex. (b) Detail of a portion of the same linden stem.*

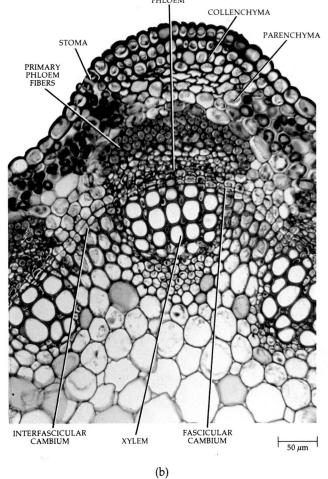

PHLOEM

COLLENCHYMA

PARENCHYMA

STOMA

PRIMARY PHLOEM FIBERS

INTERFASCICULAR CAMBIUM

XYLEM

FASCICULAR CAMBIUM

50 μm

(b)

22–7

(a) *Tranverse section of stem of alfalfa (Medicago sativa), a dicot with discrete vascular bundles. (b) Detail of a portion of the same alfalfa stem.*

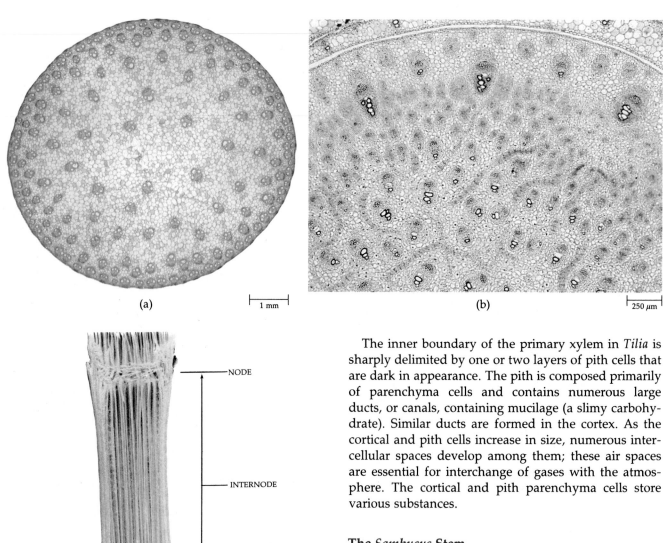

(a) ⊢—⊣ 1 mm

(b) ⊢—⊣ 250 μm

NODE

INTERNODE

NODE

ADVENTITIOUS
ROOTS

(c)

22–8
*Stem of corn (Zea mays). (a) Transverse
section of the internodal region, showing
numerous vascular bundles scattered
throughout the ground tissue. (b)
Transverse section of the nodal region of
a young corn stem, showing horizontal
procambial strands that interconnect
with vertical bundles. (c) A mature stem
split longitudinally; the ground tissue
has been removed to expose the vascular
system.*

The inner boundary of the primary xylem in *Tilia* is
sharply delimited by one or two layers of pith cells that
are dark in appearance. The pith is composed primarily
of parenchyma cells and contains numerous large
ducts, or canals, containing mucilage (a slimy carbohydrate). Similar ducts are formed in the cortex. As the
cortical and pith cells increase in size, numerous intercellular spaces develop among them; these air spaces
are essential for interchange of gases with the atmosphere. The cortical and pith parenchyma cells store
various substances.

The *Sambucus* Stem

In the stem of the elderberry (*Sambucus canadensis*), the
procambial strands and primary vascular bundles form
a system of discrete strands around the pith. The epidermis, cortex, and pith are essentially similar in organization to those of *Tilia*, so the following discussion of
the elderberry stem will be used to explain in more detail the development of the primary vascular tissues of
stems.

Figure 22–9a shows three procambial strands in
which the primary vascular tissues have just begun to
differentiate. The strand on the left is somewhat older
than the two on the right and contains at least one mature sieve element and one mature tracheary element.
Notice that the first mature sieve element appears in
the outer part of the procambial strand (next to the cortex), and that the first mature tracheary element appears in the inner part (next to the pith). Comparing
Figures 22–9a and 22–9c, we see that the more recently
formed sieve elements appear closer to the center of the
stem and that the xylem differentiates in the opposite
direction.

The first-formed primary xylem and primary phloem
elements (protoxylem and protophloem, respectively)

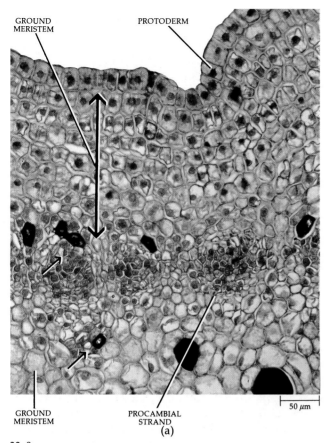

GROUND
MERISTEM   PROTODERM

GROUND   PROCAMBIAL
MERISTEM   STRAND

50 μm

(a)

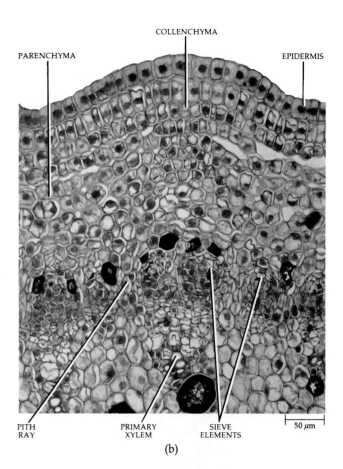

PARENCHYMA   COLLENCHYMA   EPIDERMIS

PITH   PRIMARY   SIEVE
RAY   XYLEM   ELEMENTS

50 μm

(b)

22–9

Transverse sections of the stem of the elderberry (Sambucus canadensis) *in a primary stage of growth. (a) A very young stem, showing protoderm, ground meristem, and discrete procambial strands. The procambial strand on the left contains one mature sieve element (arrow, above) and one mature tracheary element (arrow, below). (b) Primary tissues further along in development. (c) Stem near completion of primary growth. Fascicular and interfascicular cambia are not yet formed. (For further stages in the growth of the elderberry stem, see Figures 23–8, 23–9, and 23–11.)*

are stretched during elongation of the internode and are frequently destroyed. As in the *Tilia* stem, fibers develop in the primary phloem after internodal elongation is completed (see Figure 23–9).

Like the stems of *Tilia*, those of *Sambucus* become woody. In *Tilia*, almost all of the vascular cambium originates from procambial cells between the primary xylem and primary phloem, because the interfascicular regions are very narrow. In *Sambucus*, the interfascicular regions are relatively wide. Consequently, a substantial portion of the vascular cambium in *Sambucus* develops from the interfascicular parenchyma between the bundles.

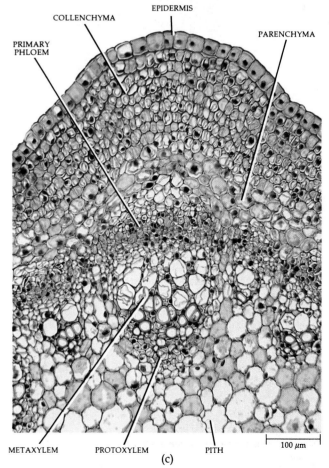

COLLENCHYMA   EPIDERMIS

PRIMARY   PARENCHYMA
PHLOEM

METAXYLEM   PROTOXYLEM   PITH

100 μm

(c)

The *Medicago* and *Ranunculus* Stems

The stems of many dicots undergo little or no secondary growth and therefore are _herbaceous,_ or nonwoody (see Chapter 23). Examples of herbaceous dicot stems can be found in alfalfa (*Medicago sativa*), as well as in the buttercups (*Ranunculus*).

Medicago is an example of an herbaceous dicot that exhibits some secondary growth (Figure 22–7). The structure and development of the primary tissues of the *Medicago* stem are essentially similar in organization to those of *Sambucus* and other woody dicots. The vascu-

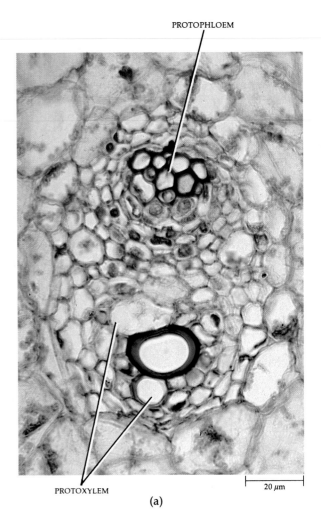

PROTOPHLOEM

PROTOXYLEM

20 µm

(a)

22–11
*Three stages in the differentiation of the vascular bundles of corn (Zea mays), as seen in transverse sections of the stem.
(a) The protophloem elements and two protoxylem elements are mature. (b) The*

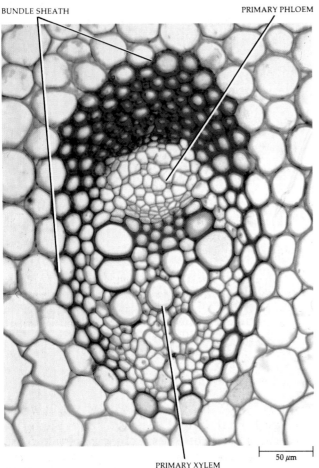

BUNDLE SHEATH

PRIMARY PHLOEM

PRIMARY XYLEM

50 µm

22–10
Transverse section of vascular bundle of the buttercup (Ranunculus), an herbaceous dicot. The vascular bundles of the buttercup are closed, in that all of the procambial cells mature; consequently, no secondary tissues develop. The primary phloem and primary xylem are surrounded by a bundle sheath of thick-walled sclerenchyma cells. Compare the vascular bundle shown here with the mature vascular bundle of corn shown in Figure 22–11c.

lar bundles are separated by wide interfascicular regions and surround a large pith. The vascular cambium is partly procambial and partly interfascicular in origin, but secondary vascular tissues are formed mainly in the vascular bundles. The interfascicular cambium generally produces only sclerenchyma cells on the xylem side.

The herbaceous stem of *Ranunculus* is an extreme example, and its vascular bundles resemble those of many monocots. The vascular bundles retain no procambium after the primary vascular tissues mature; consequently, the bundles never develop a vascular cambium and lose their potential for further growth. Vascular bundles such as those of *Ranunculus* (Figure 22–10) and the monocots, in which all the procambial cells mature and the potential for further growth within

METAPHLOEM   PROTOPHLOEM   METAXYLEM VESSEL

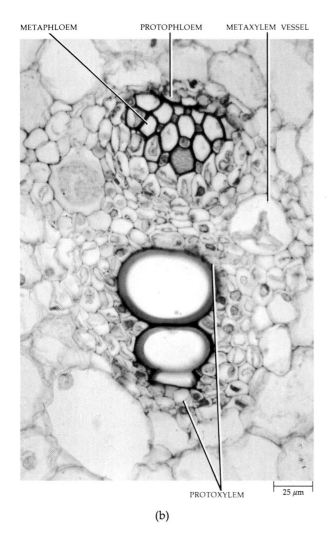

PROTOXYLEM   25 μm

(b)

SIEVE   COMPANION
ELEMENT   CELL

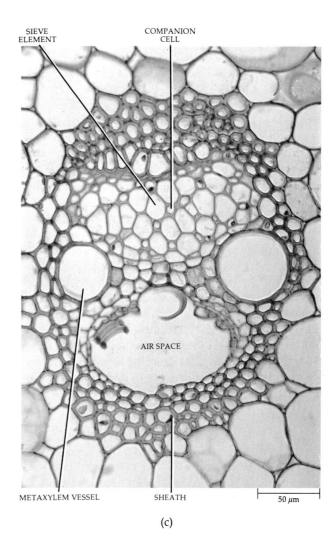

AIR SPACE

METAXYLEM VESSEL   SHEATH   50 μm

(c)

protophloem sieve elements are now crushed, and much of the metaphloem is mature. Three protoxylem members are now mature, and the two metaxylem vessel members are almost fully ex-

panded. (c) Mature vascular bundle surrounded by a sheath of thick-walled sclerenchyma cells. The metaphloem is composed entirely of sieve-tube members and companion cells. The portion of the

vascular bundle once occupied by the protoxylem elements is now a large air space. Note the wall thickenings of destroyed protoxylem elements bordering the air space.

the bundle is lost, are said to be "closed." Vascular bundles that do give rise to a cambium are said to be "open." In most dicots, the vascular bundles are of the open type; they produce some secondary vascular tissues.

The *Zea* Stem

The herbaceous stem of corn (*Zea mays*) exemplifies the stems of monocots in which the vascular bundles form a system of strands scattered throughout the ground tissue (Figure 22–8). As in other monocots, the vascular bundles of corn are closed.

Figure 22–11 shows three stages in the development of a corn vascular bundle. As in the bundles of dicot stems, the phloem develops from the outer cells of the procambial strand, and the xylem develops from the inner cells. Also, differentiation of the phloem and the xylem are in opposite directions; as seen in transverse sections, the phloem differentiates from outside to inside, and the xylem differentiates from inside to outside. The first-formed phloem and xylem elements (protophloem and protoxylem) are stretched and destroyed during elongation of the internode. This results in the formation of a very large air space on the xylem side of the bundle (Figure 22–11c). The mature vascular bundle contains two large vessel members (the metaxylem vessels), and the phloem (metaphloem) tissue is composed of sieve-tube members and companion cells. The entire bundle is enclosed in a sheath of sclerenchyma cells.

RELATION BETWEEN THE VASCULAR TISSUES OF THE STEM AND THE LEAF

The pattern formed by the vascular bundles in the stem reflects the close structural and developmental relationship that exists between the stem and its lateral appendages, the leaves. The term "shoot" serves not only as a collective term for these two vegetative organs but also as an expression of their intimate association.

The procambial strands of the stem arise behind the apical meristem just below the developing leaf primordia and sometimes are present below the sites of future leaf primordia even before they have begun to develop. As the leaf primordia increase upward in length, the procambial strands also differentiate upward within them. From its inception, the procambial system of the leaf is continuous with that of the stem.

At each node, one or more vascular bundles diverge from the cylinder of strands in the stem, cross the cortex, and enter the leaf or leaves attached at that node (Figures 22–12 and 22–13). A vascular bundle in the

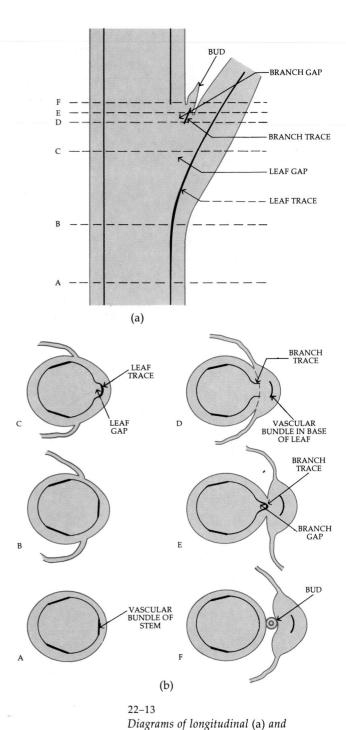

(a)

(b)

22–13
Diagrams of longitudinal (a) and transverse (b) sections of part of the stem of tobacco (Nicotiana tabacum), *illustrating the relationship of the vascular systems of the leaf and the stem. The stem of tobacco has a "continuous" vascular cylinder, portions of which are wider than others. At each node a single leaf trace diverges toward a leaf. Branch traces also occur at the nodes and are often closely associated with the leaf traces. In tobacco, two branch traces extend from the vascular system of the stem to the bud, and the branch gap is continuous with the leaf gap.*

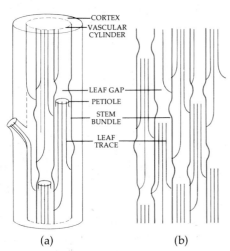

(a)   (b)

22–12
Diagrams of the primary vascular system of a stem of the oak-leaved goosefoot (Chenopodium glaucum), *in which discrete bundles form a cylinder of interconnected strands around a pith. (a) A three-dimensional diagram showing the arrangement of the bundles within the stem. (b) Vascular system spread out in one plane. Notice the leaf traces diverging outward from the vascular cylinder and the relation of the traces to the vascular bundles of the stem. In C. glaucum each leaf has three leaf traces.*

stem that extends from a longitudinal *stem bundle* to the base of a leaf, where it connects with the vascular system of the leaf, is called a *leaf trace*, and the wide gap or region of ground tissue found in the vascular cylinder, where the leaf trace diverges toward the leaf, is called a *leaf gap*. A single leaf may have one or more leaf traces connecting its vascular system with that in the stem. The number of internodes that leaf traces traverse before they join with stem bundles varies, so the traces vary in length.

If the stem bundles are followed either upward or downward in the stem, they will be found to be associated with several leaf traces. A stem bundle and its associated leaf traces are called a *sympodium* (plural: sympodia). In some stems, some or all of the sympodia are interconnected, whereas in others, all of the sympodia are independent units of the vascular system. Regardless, the pattern of the vascular system in the stem is a reflection of the arrangement of the leaves on the stem.

MORPHOLOGY OF THE LEAF

Leaves vary greatly in form and in internal structure. In dicots, the leaf commonly consists of an expanded portion, the *blade*, or lamina, and a stalklike portion, the *petiole* (Figure 22–14). Small scalelike or leaflike structures called *stipules* develop at the base of some leaves (Figure 22–15). Many leaves lack petioles and are said to be sessile (Figure 22–16). In most monocots and certain dicots, the base of the leaf is expanded into a *sheath,* which encircles the stem (Figure 22–16b). In some grasses, the sheath extends the length of an internode. The arrangement of leaves on the stem may be spiral (alternate), opposite (in pairs), or whorled (three or more at a node). For example, mulberry (*Morus alba*) and oak (*Quercus*) leaves are spiral, maple (*Acer*) leaves are opposite, and Culver's-root (*Veronicastrum virginicum*) leaves are whorled (Figure 22–14).

The leaves of dicotyledons are either simple or compound. In simple leaves, the blades are not divided into

22–14

Several examples of simple leaves. (a) *Mulberry* (Morus alba). (b) *Culver's-root* (Veronicastrum virginicum). (c) *Sugar maple* (Acer saccharum). (d) *Silver maple* (Acer saccharinum). (e) *Red oak* (Quercus rubra). *Note the spiral, or alternate, arrangement of the leaves in mulberry, and the whorled arrangement of those in Culver's-root. Leaf arrangement in the maples is opposite, and in oak it is spiral, although only single leaves of these trees are shown here.*

22–15

The pinnately compound leaf of the pea (Pisum sativum). Notice the stipules at the base of the leaf and the slender tendrils at the tip of the leaf. In the pea leaf, the stipules are often larger than the leaflets.

distinct parts, although they may be deeply lobed (Figure 22–14). The blades of compound leaves are divided into leaflets, each usually with its own small petiole (which is called a petiolule). Two types of compound leaves can be distinguished: pinnately compound leaves and palmately compound leaves (Figure 22–17). In pinnately compound leaves, the leaflets arise from either side of an axis, the <u>rachis</u>, like the pinnae of a feather. (The rachis is an extension of the petiole.) The leaflets of a palmately compound leaf diverge from the tip of the petiole, and a rachis is lacking.

Since leaflets are similar in appearance to simple leaves, it is sometimes difficult to determine whether the structure is a leaflet or a leaf. Two criteria may be used to distinguish leaflets from leaves: (1) buds are found in the axils of leaves—both simple and compound—but not in the axils of leaflets; and (2) leaves extend from the stem in various planes, whereas the leaflets of a given leaf all lie in the same plane.

STRUCTURE OF THE LEAF

Variations in the structure of angiosperm leaves are to a great extent related to the habitat, and the availability of water is an especially important factor affecting their form and structure. On the basis of their water requirements or adaptations, plants are commonly characterized as: *mesophytes* (plants that require abundant soil water and a relatively humid atmosphere), *hydrophytes* (plants that depend on an abundant supply of moisture

22–16

Sessile leaves (leaves without a petiole) are often found among dicots, such as Moricandia, a member of the mustard family (a), but are particularly characteristic of grasses and other monocots. (b) In corn (Zea mays), a monocot, the base of the leaf forms a sheath around the stem. The ligule, a small flap of tissue extending upward from the sheath, is visible.

(a)

(b)

22–17

Some examples of compound leaves. A palmately compound leaf is shown in (a); all the others are pinnately compound. (a) Red buckeye (Aesculus pavia). (b) Shagbark hickory (Carya ovata). (c) Green ash (Fraxinus pennsylvanica var. subintegerrima). (d) Black locust (Robinia pseudo-acacia). (e) Honey locust (Gleditsia triacanthos). In the honey locust, each leaflet is subdivided into smaller leaflets.

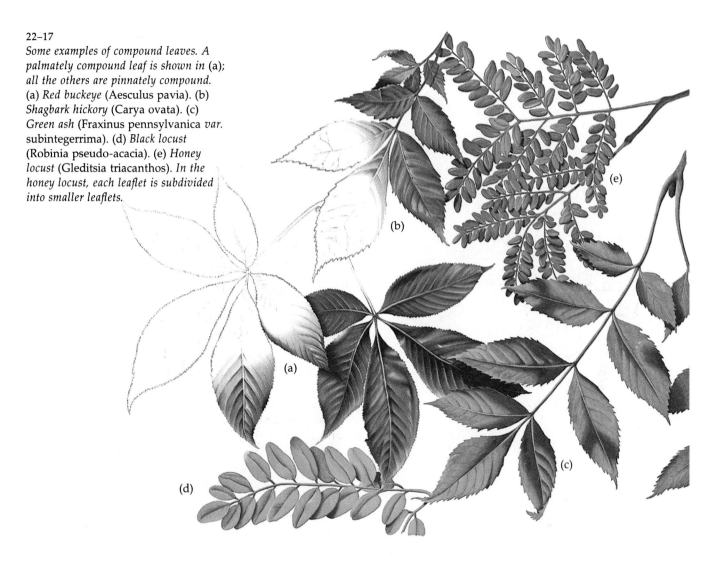

or grow wholly or partly submerged in water), and *xerophytes* (plants that are adapted to arid habitats). Such distinctions are not sharp, however, and leaves often exhibit a combination of features that are characteristic of different ecological types. Regardless of their varying forms, the foliage leaves of angiosperms are specialized as photosynthetic organs and, like roots and stems, consist of dermal, ground, and vascular tissue systems.

Epidermis

The ordinary epidermal cells of the leaf, like those of the stem, are compactly arranged and covered with a cuticle that reduces water loss (see pages 50 and 395). Stomata may occur on both sides of the leaf but are usually more numerous on the lower surface (Figure 22–18). In leaves of hydrophytes that float on the surface of the water, stomata may occur in the upper epidermis only (Figure 22–19); the immersed leaves usually lack stomata entirely. The leaves of xerophytes

generally contain greater numbers of stomata than those of other plants. Presumably these numerous stomata permit a higher rate of gas exchange during the relatively rare periods of favorable water supply. In many xerophytes, the stomata are sunken in depressions on the lower surface of the leaf (Figure 22–20). The depressions may also contain many epidermal hairs. Together these two features may serve to reduce water loss from the leaf. Epidermal hairs, or trichomes, may occur on either or both surfaces of a leaf. Thick coats of epidermal hairs may also retard water loss from leaves.

In the leaves of dicotyledons, the stomata are scattered and often appear to be randomly arranged (Figure 22–21); their development is mixed—that is, mature and immature stomata occur side by side in a partially developed leaf. In monocotyledons, the stomata are arranged in rows parallel with the long axis of the leaf (Figure 22–22). Their development begins at the tips of the leaves and progresses downward.

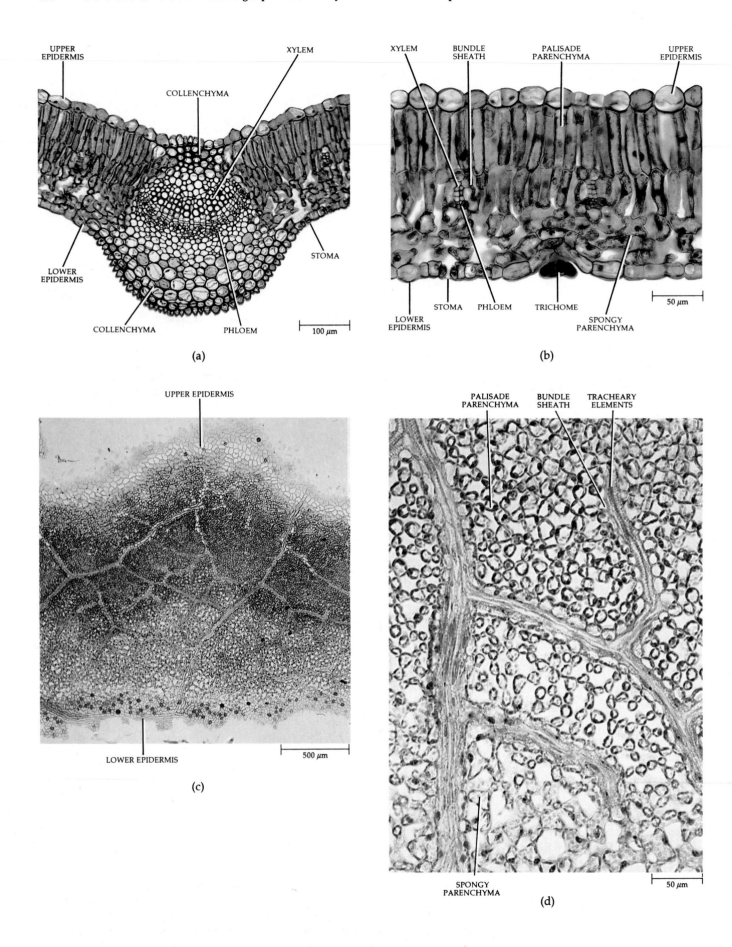

(a)

(b)

(c)

(d)

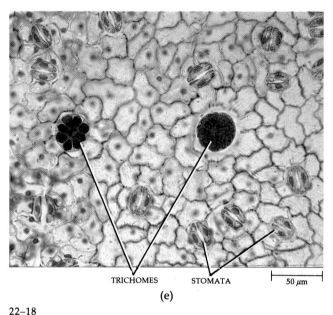

TRICHOMES   STOMATA   50 μm

(e)

22–18
Sections of lilac (Syringa vulgaris) leaf.
(a) A transverse section through a midrib
showing the midvein. (b) A transverse
section through a portion of the blade.
Two small veins (minor veins) can be seen
in this view. (c) This is a paradermal
section of the leaf—a section cut
approximately parallel with the leaf
surface. Progressing from the top to the
bottom of this micrograph, the section
cuts deeper into the leaf. Thus, part of the
upper epidermis can be seen in the light
area at the top of the micrograph, and
part of the lower epidermis can be seen
at the bottom. Notice the greater number
of stomata in the lower epidermis. The
venation in lilac is netted. (d) and (e) are
enlargements of portions of (c). Portions
of palisade parenchyma (above) and
spongy parenchyma (below) are shown in
(d). A vein ending, sectioned through
some tracheary elements and surrounded
by a bundle sheath, can be seen at the
upper right of this micrograph. A portion
of the lower epidermis, with two
trichomes (epidermal hairs) and several
stomata, in addition to many ordinary
epidermal cells, is shown in (e).

PALISADE   SCLEREID   UPPER   STOMA
PARENCHYMA   EPIDERMIS

SPONGY   TRICHOME   LOWER   VEIN   200 μm
PARENCHYMA   EPIDERMIS

22–19
Transverse section of the water lily
(Nymphaea odorata) leaf, which floats
on the surface of the water and has
stomata in the upper epidermis only. As
is typical of hydrophytes, the vascular
tissue in the Nymphaea leaf is much
reduced, especially the xylem. The
palisade parenchyma consists of several
layers of cells above the spongy paren-
chyma.

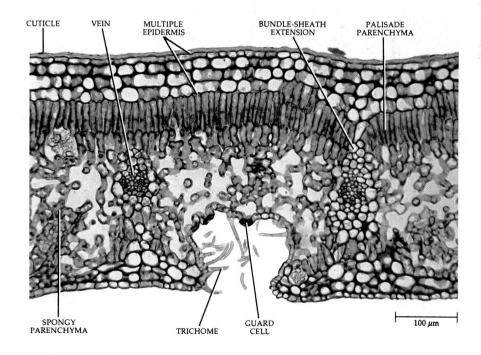

CUTICLE   VEIN   MULTIPLE   BUNDLE-SHEATH   PALISADE
                     EPIDERMIS   EXTENSION   PARENCHYMA

SPONGY   TRICHOME   GUARD   100 μm
PARENCHYMA   CELL

22–20
Transverse section of oleander (Nerium
oleander) leaf. Oleander is a xerophyte,
and this is reflected in the structure of
the leaf. Note the very thick cuticle
covering the multiple (several-layered)

epidermis on the upper and lower
surfaces of the leaf. The stomata and
trichomes are restricted to invaginated
portions of the lower epidermis called
stomatal crypts.

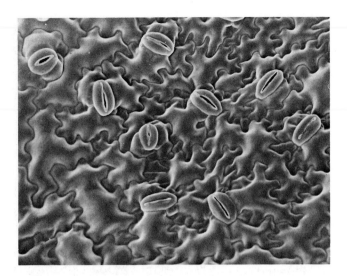

22–21
*Scanning electron micrograph of a potato
(Solanum tuberosum) leaf showing
random arrangement of stomata typical
of the leaves of dicots. The guard cells in
potato are kidney-shaped and are not
associated with subsidiary cells.*

22–22
*Scanning electron micrograph of a corn
(Zea mays) leaf showing parallel
arrangement of stomata typical of the
leaves of monocots. In corn each pair of
narrow guard cells is associated with two
subsidiary cells, one on each side of the
stoma.*

Mesophyll

It is the *mesophyll*—the ground tissue of the leaf—with
its large volume of intercellular spaces and numerous
chloroplasts, that is particularly specialized for photosynthesis. The intercellular spaces are connected with
the outer atmosphere through the stomata, which facilitates rapid gas exchange, an important factor in photosynthetic efficiency. In mesophytes, the mesophyll is
differentiated into *palisade parenchyma* and *spongy parenchyma*. The cells of the palisade tissue are columnar
in shape, with their long axes oriented at right angles to
the epidermis, and the spongy parenchyma cells are irregular in shape (Figure 22–18b, d). Although the palisade parenchyma appears more compact than the
spongy parenchyma, most of the vertical walls of the
palisade cells are exposed to intercellular space, and in
some leaves the palisade surface may be two to four
times greater than the spongy surface. Chloroplasts are
also more numerous in palisade cells than in spongy
cells. Most of the photosynthesis in the leaf, therefore,
apparently takes place within the palisade parenchyma.

The palisade parenchyma is usually located on the
upper side of the leaf, and the spongy parenchyma is
generally on the lower side (Figure 22–18). In leaves of
xerophytes, palisade parenchyma often occurs on both
sides of the leaf. In some plants—for instance, corn (see
Figure 7–19, page 108) and other grasses (Figures 22–25
through 22–27)—the mesophyll cells are of more or
less similar shape, and a distinction between spongy
and palisade parenchyma does not exist.

Vascular Bundles

The mesophyll of the leaf is thoroughly permeated by
numerous vascular bundles, or *veins*, which are continuous with the vascular system of the stem. In most
dicots, the veins are arranged in a branching pattern,
with successively smaller veins branching from somewhat larger ones. This type of vein arrangement is
known as *netted venation* (Figure 22–23). Often the
largest vein extends along the long axis of the leaf as a
midvein, which, with its associated ground tissue, comprises the so-called midrib of such leaves (Figure
22–18a). By contrast, most monocot leaves have many
veins of fairly similar size that are oriented parallel to
one another along the leaf. This vein arrangement is
called *parallel venation* (Figure 22–16b). In parallel-
veined leaves, the longitudinal veins are interconnected
by smaller veins, forming a complex network (Figure
22–24).

The veins contain xylem and phloem, which generally are entirely primary in origin. (The midvein and
sometimes the coarser veins undergo secondary growth
in some dicot leaves.) The vein-endings in dicot leaves
often contain only tracheary elements, although both
xylem and phloem elements may extend to the ends of
the vein. Commonly, the xylem occurs on the upper
side of the leaf, and the phloem occurs on the lower
side (Figure 22–18a, b).

The small veins of the leaf that are more or less completely embedded in mesophyll tissue are called *minor
veins*, while the large veins associated with ribs (protru-

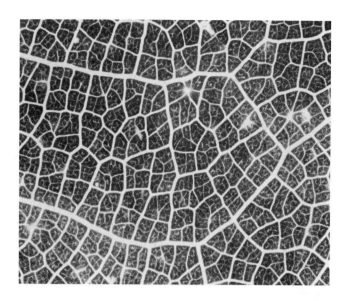

22–23
A cleared leaf (with the chlorophyll removed) of a cottonwood (Populus deltoides), *at two successive magnifications. The cottonwood leaf has the netted venation characteristic of a dicot. The small areas of mesophyll delimited by veins are called areoles. No mesophyll cell of the leaf is far from a vein. Water and dissolved minerals are carried to the leaf through the xylem; organic molecules produced in the leaf are carried out of the leaf through the phloem.*

sions on the underside of the leaf) are referred to as *major veins*. It is the minor veins that play the principal role in the collection of photosynthates from the mesophyll cells. With increasing vein size, the veins become less closely associated spatially with the mesophyll and increasingly embedded in nonphotosynthetic rib tissues. Hence, as the veins increase in size, their function changes from one of collection to one of primarily transport.

The vascular tissues of the veins are rarely exposed to the intercellular spaces of the mesophyll. The large veins are surrounded by parenchyma cells that contain few chloroplasts, whereas the small veins are enclosed by one or more layers of compactly arranged cells that form a *bundle sheath* (see Figures 22–18 through 22–20). The cells of the bundle sheath often resemble the mesophyll cells in which the small veins are located. The bundle sheaths extend to the ends of the veins, assuring that no part of the vascular tissue is exposed to air in the intercellular spaces and that all substances entering and leaving the vascular tissues must pass through the sheath (Figure 22–18d). Thus the bundle sheath performs a function analogous to that of the endodermis of the root.

In many leaves, the bundle sheaths are connected with either or both upper and lower epidermis by cells resembling the sheath cells (Figure 22–20). Such connections (actually, extensions of the sheaths) are called *bundle-sheath extensions*. Besides offering mechanical support to the leaf, in dicotyledons they apparently conduct water to the epidermis.

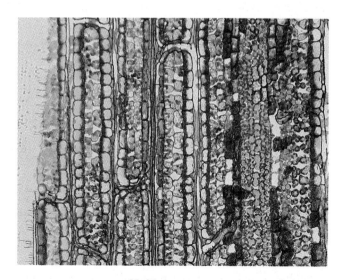

22–24
Paradermal section of a corn (Zea mays) *leaf showing the parallel venation characteristic of a monocot. Note the numerous transverse veins that interconnect the system of longitudinal (parallel) veins.*

PLANTS, AIR POLLUTION, AND ACID RAIN

The plant leaf, like the human lung, can function only when it is able to exchange gases with the surrounding air. As a consequence, the leaf, like the lung, is an organ that is exceedingly susceptible to air pollution.

Air pollution has many forms. Some pollutants are particulate. The particles may be organic, such as those present in the smoke produced by burning fossil fuels and garbage, or inorganic, such as cement kiln dusts, foundry dusts, and the lead compounds released in the combustion of leaded gasoline. As a major component of smog, these particles reduce the amount of sunlight reaching the earth's surface. Such particles also have direct ill effects on plants. They may clog stomata and prevent them from functioning, or they—particularly metallic particles—may act as plant poisons.

Fluorides, which enter the air as waste products from the manufacture of phosphates, steel, aluminum, and other industrial products, act as cumulative poisons, entering the leaf through the stomata and causing collapse of leaf tissue, apparently by inhibiting enzymes concerned with cellulose synthesis. Thousands of acres of Florida citrus groves have been damaged by fluorides discharged from phosphate fertilizer plants.

When ores containing sulfur are processed, sulfur dioxide is produced:

$$2CuS + 3O_2 \longrightarrow 2CuO + 2SO_2$$

Sulfur dioxide has a disagreeable pungent-sweet taste; it is unusual among air pollutants in that it can be tasted at lower concentrations than it can be smelled. Sulfur oxides are also produced by the burning of fossil fuels containing sulfur. In moist air, sulfur oxides react with water to form droplets of sulfuric acid, a strong corrosive acid and a component of acid rain. In several parts of the United States, virtual deserts have been created by the emission of sulfur dioxide combined with metals. As long ago as 1905, air pollution controls were instituted in areas surrounding copper smelters in Tennessee, but the surrounding area, once covered by luxuriant forest, remains barren to this day. Not only was all the vegetation killed but the acid leached the soil of nutrients. Similarly, in a copper-smelting area of the Sacramento Valley, California, all vegetation was killed over an area of 260 square kilometers, and growth was severely affected over an additional 320 square kilometers.

The most familiar type of air pollution to Californians is photochemical smog, which is produced by sunlight acting on automobile exhausts. The Los Angeles area offers an ideal setting for the formation of photochemical smog because the life of that area is heavily geared to the use of the automobile, and the mountains to the north and east form a "basin" that prevents the dispersal of reactants. Not only are many species of plants unable to survive in the city itself (as is true in many other major cities of the world), but smog moving out of the Los Angeles basin is damaging agricultural crops and killing pine forests in mountains as far away as 160 kilometers.

One of the principal ingredients of photochemical smog is nitrogen dioxide, NO_2, which is produced by any combustion process that occurs in air (dry air is 77 percent nitrogen) and so is present in automobile exhausts. Under the influence of light, NO_2 is split into NO and atomic oxygen. The latter is extremely reactive and forms ozone, O_3, by reaction with normal (molecular) oxygen.

Similar reactions, powered by ultraviolet light, occur at the outer layers of the atmosphere, producing the ozone shield described on page 4. Ozone can also be produced by an electric spark and is the source of the "clean" smell after a lightning storm. Ozone is a highly toxic substance. In plants, it damages the thin-walled palisade cells, apparently affecting the permeability of the membranes of both the cells and their chloroplasts. Another component of photochemical smog is PAN (peroxyacetyl nitrate, $C_2H_3O_5N$). It is several times more toxic than ozone but is normally present in much lower concentrations. Photosynthesis is reduced 66 percent by photochemical smog concentration of 0.25 part per million.

At the present time, there is much concern over the adverse effects of acid rain on the environment. Under normal conditions, rain arising in a nonpolluted area should have a pH of 5.6. With the burning of fossil fuels and the smelting of sulfide ores, the large quantities of sulfur oxides and nitrogen oxides emitted into the atmosphere react with water to form strong acids (sulfuric acid and nitric acid). Rain and snow formed in such areas have a pH of less than 5.6 and, by definition, are acidic. The phenomenon of acid precipitation is widespread today and occurs over large areas of western Europe, eastern United States, and southeastern Canada, where the yearly average pH of precipitation ranges from 4 to 4.5. Moreover, the rain of individual storms is often much more acid. Acid rains have been recorded in Scotland, Norway, and Iceland with pH values of 2.4, 2.7, and 3.5, respectively. The move to reduce local pollution problems through the building of tall stacks on smelters and industrial plants has created regional problems. Pollutants emanating from tall stacks are transported over long distances through the atmosphere. For example, more than 75 percent of the sulfur in rain that falls in the Scandinavian countries is believed to originate in the British Isles and central Europe. The effect of acid rain on vegetation is not well understood but is currently being investigated throughout the Northern Hemisphere. Acid rains have been shown to decrease growth of forest trees in Sweden. In addition, simulated acid rains have been shown to injure leaves and inhibit seed germination. In just a few years the observed

22–27

Transverse sec
(Poa annua, *a*
view of leaf. (l
of leaf includi
leaf, the meso
palisade and s

occurrences of damage in the forests of West Germany have increased from just a few percent to more than 50 percent.

The clearest effect of acid rain has been on fish populations, which have been virtually eliminated in acidified lakes in some parts of the world. It has been suggested that much of the toxicity to fish is actually due to increased aluminum concentrations and is not directly attributable to acidic water. Aluminum, which comprises about 5 percent of the

earth's crust, is almost completely insoluble in neutral or alkaline water and thus has not been biologically available. As a result of acid rain, however, concentrations of dissolved aluminum in some lakes may increase to levels toxic to fish and other aquatic organisms. The solubility of other toxic metals, such as lead, cadmium, and mercury, also increases sharply with decreasing pH.

(a) *Sulfur dioxide injury on a blackberry*
(Rubus) *leaf is characterized by areas of injured leaf tissue surrounded by healthy tissue.*
(b) *Ozone injury on a tobacco* (Nicotiana
tabacum) *leaf appears as stipples or flecks of dead tissues on the upper surface of the leaf. In cases of severe ozone injury, the flecks coalesce into larger lesions that are visible on both leaf surfaces.* (c) *Acid rain arises when sulfur oxides and nitrogen oxides react with water in the atmosphere to form sulfuric acid and nitric acid.*

(a)   (b)

unfolding or
thick-walled
with subsidia
20, page 395)

DEVEL

The first stru
angiosperms
beneath the
shoot apex. A
ther divisions
leaf buttress (
ity is establis
apex relative
fore or durin
appears bene

With contin
velops into an
dium (Figure
soon develops
on approxima

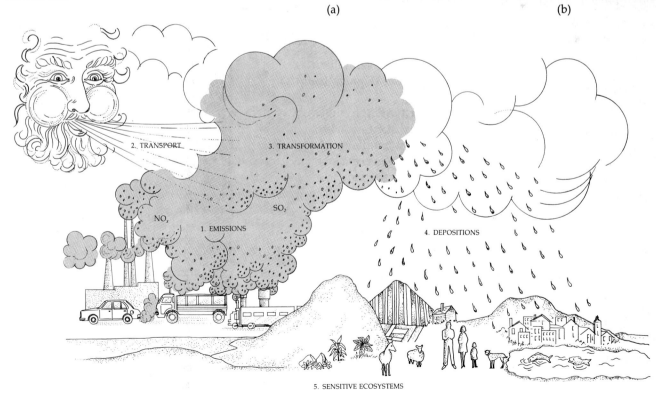

(c)

The epi
the leaf b
In additio
bordered
provide s
veins ma
and fibers
dicot and

GRAS

After the
sis in sug
were dev
leaves in
discovered
rather cor
in the lea
bundle-sh
around th
tions (Figu
sheath cell
cells that
This conc
bundle-sh
(German f

In the l
phyll cells
cally arrar
the paren
have rathe
tions the
inner, mo
mestome s
ure 22–26)

Another
leaves of C
that is, the
sheaths. Ir
intervene
C_3 grasses
species in
adjacent b

The leav
ates both
of C_3 plan
known, bu
physical di
phloem of
of loading

The epic
cell types.
gate cells.
cells, occu
participate

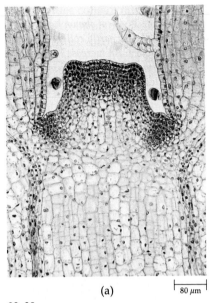

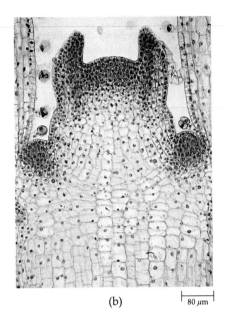

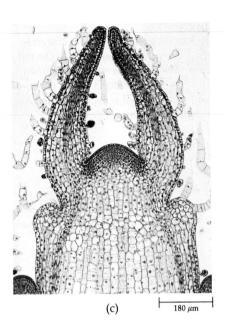

22–28
Some early stages of leaf development in Coleus blumei, *as seen in longitudinal sections of the shoot tip. The leaves in* Coleus *occur in pairs, opposite one another at the nodes (see Figure 22–2). (a) Two small bulges, or leaf buttresses, can be seen opposite one another on the flanks of the apical meristem. In*

addition, a bud primordium can be seen arising in the axil of each of the two young leaves, below. (b) Two erect, peglike leaf primordia have developed from the leaf buttresses. Notice the procambial strands extending upward into the leaf primordia. The bud primordia, below, are further along in development than those

in (a). As the leaf primordia elongate (c), the procambial strands, which are continuous with the vascular bundles in the stem, continue to develop into the leaves. Trichomes, or epidermal hairs, develop from certain protodermal cells very early, long before the protoderm matures to become the epidermis.

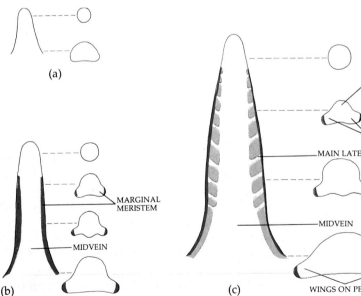

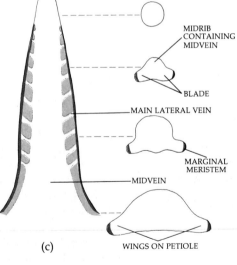

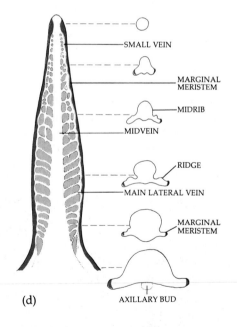

22–29
*Diagrams of longitudinal and transverse sections illustrating some early stages of leaf development in tobacco (*Nicotiana tabacum*). (a) A young, peglike leaf primordium without a blade, or lamina. (b) Marginal meristem activity has begun on opposite sides of the primordium to form the blade. The presence of a marginal meristem in the tobacco leaf*

has recently been questioned. Just as apical growth of the leaf primordium is of short duration, the activity of the marginal meristems may be comparatively short-lived. (c) The main lateral veins (narrow, clear areas) can be seen extending from the midvein as the blade grows. (d) The primordium has increased in height. With further growth of the

blade, some ridges appear in association with some of the larger veins. Small veins have begun to develop at the tip of the leaf. The bottom diagram in (a) through (d) is a transverse section through the "winged" petiole. Compare these diagrams with sections in Figure 22–30.

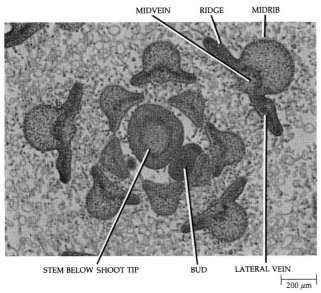

MIDVEIN   RIDGE   MIDRIB

STEM BELOW SHOOT TIP   BUD   LATERAL VEIN

200 µm

22–30
*Transverse section of developing leaves
of tobacco* (Nicotiana tabacum) *grouped
around the shoot tip, sectioned below the
apical meristem. The younger leaves are
nearer the axis. A leaf primordium at first
lacks differentiation into midrib and blade.
Some early stages in development of
blade and midrib can be seen here. Compare
these sections with the diagrams of
transverse sections in Figure 22–29.
(Portions of numerous trichomes surround
the developing leaves seen here.)*

22–31
Scanning electron micrographs of celery
(Apium graveolens), *showing early stages of
leaf development in the shoot apex.*
*(a) The nearer a developing leaf is to the
center of the apex, the younger it is.*
*(b) Two of the leaves shown here (right and
left) have developed further than any shown in (a).*

ABSCISSION
LAYER

PROTECTIVE
LAYER

50 µm

22–32
Abscission zone in maple (Acer) *leaf, as
seen in longitudinal section through the
base of the petiole.*

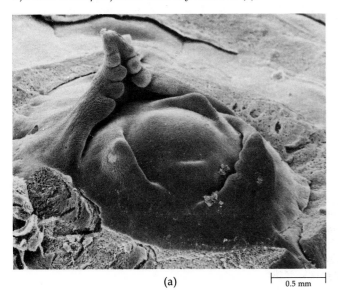

(a)   0.5 mm

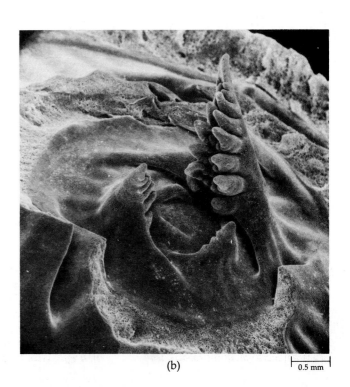

(b)   0.5 mm

cambial strand beneath develops upward into it (Figure 22–28c) and is continuous with the coarse veins emanating from the main vein, or midvein, of the blade (Figure 22–29). The smaller veins of the leaf are initiated at the tip of the leaf. They develop from the top to the bottom of the leaf in continuity with the coarser veins. Thus, the tip of the leaf is the first part to have a complete system of veins. This course of development reflects the overall maturation of the leaf, which is from the tip to the base of the leaf.

Figure 22–31 provides three-dimensional views of the shoot apex and some early stages of leaf development in celery *(Apium graveolens)*. The leaf primordia and young leaves are helically arranged on the shoot apex, with the younger leaves nearer the center of the apex. Two leaf buttresses can be seen, top right and left, near the center of the apex in Figure 22–31a. As a leaf primordium arises from a buttress, it broadens into a shell-like form while growing vigorously at its tip for a relatively short time to form the petiole-rachis part of what is to become a bipinnately compound leaf. Beneath the tip, protuberances arise on both sides of the rachis—first at the bottom and then in succession toward the tip—to form leaflets. Further protuberances develop from the earlier ones, becoming either lobes of the primary leaflets or distinctly separated secondary leaflets. These can be seen arising from the primary leaflets of the relatively large, rudimentary leaf in Figure 22–31b. Each leaflet resembles a simple leaf in its development.

Sun and Shade Leaves

Environmental factors, especially light, can have substantial developmental effects on the size and thickness of leaves. In many species, leaves grown under high light intensities—the so-called *sun leaves*—are smaller and thicker than the so-called *shade leaves* that develop under low light intensities. The increased thickness of the sun leaves is due mainly to a greater development of the palisade parenchyma. The vascular system of sun leaves is greater, and the walls of their epidermal cells are thicker than those of shade leaves. In addition, the ratio of the internal surface of the mesophyll to the area of the leaf blade is much higher in sun leaves. One effect of these differences is that, although both leaf types have similar photosynthetic rates at low light intensities, shade leaves are not adapted to high light intensities, and consequently they have considerably

22–33

The transition region—the connection between root and cotyledons—in the seedling of a dicotyledon with a diarch root. In the root, the primary vascular system is represented by a single cylinder of vascular tissue. In the hypocotyl-root axis, the vascular system branches and diverges into the cotyledons, and the xylem and phloem become reoriented along the hypocotyl-root axis.

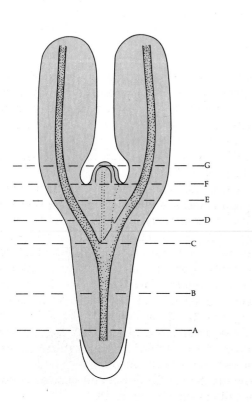

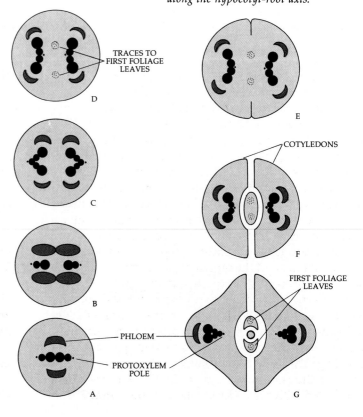

lower photosynthetic maxima under high light conditions.

Because light intensities vary greatly in different parts of the crowns of trees, extreme forms of sun and shade leaves can be found there. Sun and shade leaves also occur in shrubs, as well as in herbaceous plants. They can be induced to form by growing plants under high or low light intensities.

LEAF ABSCISSION

In many plants, the normal separation of the leaf from the stem—the process of _abscission_—is preceded by certain structural and chemical changes near the base of the petiole (see page 479). These changes result in the formation of an _abscission zone_ (Figure 22–32). In woody dicotyledons, two layers can be distinguished in the abscission zone: a separation (abscission) layer, and a protective layer. The separation layer consists of relatively short cells with poorly developed wall thickenings that make it structurally weak. A protective layer is formed by the deposition of suberin in the cell walls and intercellular spaces beneath the separation layer. After the leaf falls, the protective layer is recognized as a _leaf scar_ on the stem (see Figure 23–18, page 457). Hormonal factors associated with abscission are discussed in Chapter 24.

TRANSITION BETWEEN VASCULAR SYSTEMS OF THE ROOT AND THE SHOOT

As discussed in previous chapters, the distinction between plant organs is based primarily on the relative distribution of the vascular and ground tissues. For example, in dicot roots the vascular tissues generally form a solid cylinder surrounded by the cortex. In addition, the strands of primary phloem alternate with the radiating ridges of primary xylem. In contrast, in the stem the vascular tissues often form a cylinder of discrete strands around a pith, with the phloem on the outside of the vascular bundles and the xylem on the inside. Obviously, somewhere in the primary plant body, a change from the type of structure found in the root to that found in the shoot must take place. This change is a gradual one, and the region of the plant axis through which it occurs is called the _transition region_.

As described in Chapter 19, the shoot and root are initiated as a single continuous structure during the development of the embryo. Consequently, vascular transition occurs in the axis of the embryo or young seedling. This transition is initiated during the appearance of the procambial system in the embryo and is completed with the differentiation of the variously distributed procambial tissues in the seedling. Vascular continuity between the root and shoot systems is maintained throughout the life of the plant.

The structure of the transition region can be very complex, and much variation exists in the transition regions of different kinds of plants. In most gymnosperms and dicotyledons, vascular transition occurs between the root and cotyledons. Figure 22–33 depicts a type of transition region commonly found in dicotyledons. Notice the diarch (having two protoxylem poles) structure of the root; the branching and reorientation of the primary xylem and the primary phloem, which in the upper part of the axis results in the formation of a pith; and the traces of the first leaves of the epicotyl.

DEVELOPMENT OF THE FLOWER

The development of the flower or inflorescence terminates the meristematic activity of the vegetative shoot apex.

During the transition to flowering, the vegetative shoot apex undergoes a sequence of physiological and structural changes and is transformed into a reproductive apex. Consequently, flowering may be considered as a stage in the development of the shoot apex and of the plant as a whole. Inasmuch as the reproductive apex exhibits a determinate growth pattern, flowering in annuals indicates that the plant is approaching completion of its life cycle. By contrast, flowering in perennials may be repeated. Various environmental factors, including the length of day and the temperature, are known to be involved in the induction to flowering (see Chapter 25).

The transition from a vegetative to a floral apex is often preceded by an elongation of the internodes and the early development of lateral buds below the shoot apex. The apex itself undergoes marked increase in mitotic activity, accompanied by changes in dimensions and organization: from a relatively small apex with a tunica-corpus type of organization, the apex becomes broad and domelike.

The initiation and early stages of development of the sepals, petals, stamens, and carpels are quite similar to those of leaves. Commonly, the initiation of the floral parts begins with the sepals, followed by the petals, then the stamens, and finally the carpels (Figure 22–34). This usual order of appearance of the floral parts may be modified in certain flowers, but the floral parts always have the same relative spatial relation to one another (Figures 22–35 and 22–36). The floral parts may remain separate during their development, or they may become united within different whorls (coalescence) and between whorls (adnation).

The basic structure of the flower and some of its variations were discussed in Chapter 18.

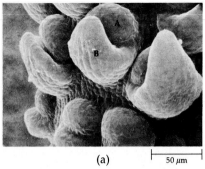

(a)   50 μm

(b)   50 μm

(c)   50 μm

22–34

*Scanning electron micrographs showing
some stages in the development of the
perfect flower of* Neptunia pubescens, *a
legume with radial symmetry and
whorled floral parts. Tips of the subtending bracts have been removed in (b)–(i).
In (f)–(i) most sepals and petals have
also been removed. (a) Floral apex (A) in
axil of a bract (B). (b) Five sepal primordia (S) have been initiated around the
floral apex. (c) Five petal primordia (P)
have been initiated around the floral
apex, alternating with the sepals (S).
During their development, the sepals will
form a calyx tube. (d) Five stamen
primordia (arrows) have been initiated
around the floral apex, alternating with
the petals (P). (e) A second whorl of
stamens (arrow) has been initiated, its
members alternating with members of the
first stamen whorl (ST$_1$). The carpel (C)
has been initiated at the center of the
floral apex. All floral parts are now
present. (f) The carpel has now developed
a cleft, which will form the locule of the
ovary. The outer, or first, whorl of
stamens (ST$_1$) are beginning to differentiate anthers and filaments. (ST$_2$
designates inner, or second, whorl of
stamens.) (g) The carpel is now beginning
to differentiate style and ovary. (h) Older
flower with both whorls of stamens in
view. (i) Older flower with some stamens
removed to reveal the carpel, which has
differentiated ovary (O), style, and stigma
(arrow).*

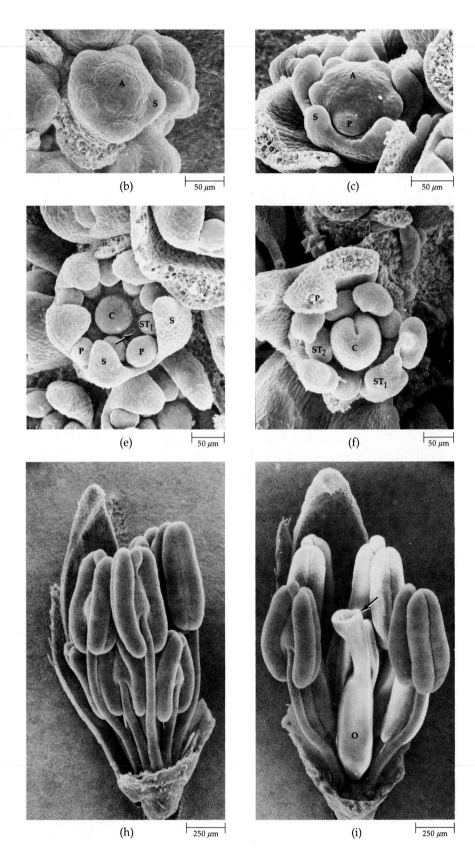

(e)   50 μm

(f)   50 μm

(h)   250 μm

(i)   250 μm

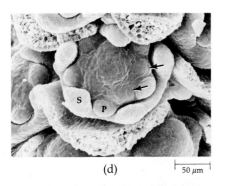

(d) 50 μm

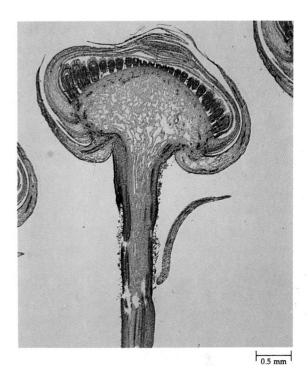

0.5 mm

22–35
Longitudinal section of young inflorescence of fleabane (Erigeron), a member of the Asteraceae. The inflorescence contains numerous flower primordia. The older flower primordia are on the periphery of the inflorescence, the younger in the center.

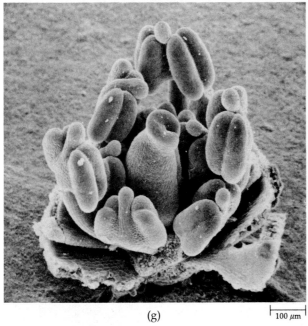

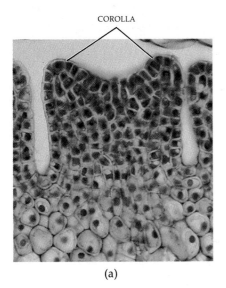

(g) 100 μm

COROLLA

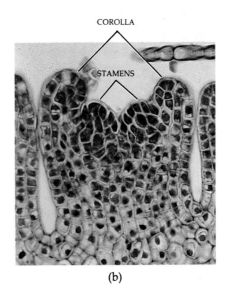

(a)

COROLLA
STAMENS

(b)

CARPELS COROLLA STAMEN

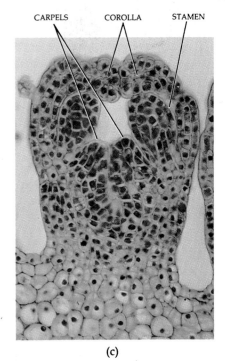

(c)

22–36
Stages in development of the fleabane (Erigeron) flower, as seen in longitudinal sections. (a) Flower primordium with developing corolla. (b) Initiation of stamens, which are adnate to the corolla. (c) Flower with developing corolla, stamens, and carpels. (Continued on next page.)

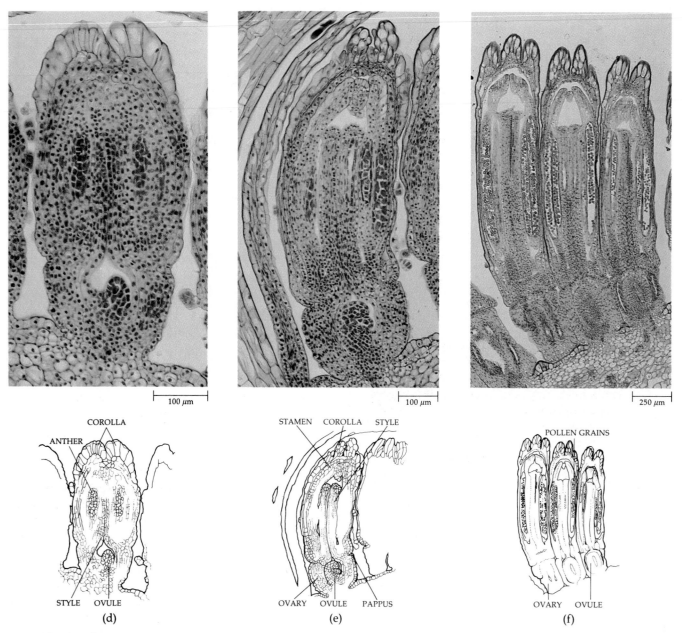

22–36 (Continued)
(d) and (e) Flowers with developing
ovules and styles. The two carpels
overarch the cavity containing the ovule
and then become prolonged into a solid
style with a two-part stigma. The calyx,
or pappus, is initiated at about the same
time as the anthers but develops slowly.
(f) Flowers approaching maturity. Pollen
grains can now be seen in the anthers.
Notice that the fleabane flower has an
inferior ovary.

STEM AND LEAF MODIFICATIONS

Stems and leaves may undergo modifications and perform functions quite different from those commonly
associated with these two components of the shoot.
One of the most common modifications is the formation of *tendrils*, which aid in support. Some tendrils are
modified stems. In ivy (*Hedera*), for example, the tendrils produce enlarged, cuplike structures—holdfasts—
at their tips. The tendrils of the grape (*Vitis*) (Figure
22–37), as well as those of Virginia creeper (*Partheno-
cissus quinquefolia*) are also modified stems that coil
around the supporting structure. In the grape, the tendrils sometimes produce small leaves or flowers.

LEAF DIMORPHISM IN AQUATIC PLANTS

In the natural environment, the leaves of aquatic flowering plants may develop into two distinct forms. Under water they develop into narrow and often highly dissected structures (water forms), and above the surface they develop into rather ordinary-looking leaves (land forms). It is possible to force immature leaves to develop into the form atypical for a given environment by applying any one of a wide range of treatments.

During a recent study on the aquatic plant *Callitriche heterophylla*, it was found that the plant hormone gibberellic acid induced shoots growing in the air (emergent shoots) to produce water-form leaves. Abscisic acid, another plant hormone (see Chapter 24), led to the formation of land-form leaves on submerged shoots. Higher temperatures or the addition of the sugar alcohol, mannitol, to the water also caused submerged shoots to produce land-form leaves.

In nature, the cellular turgor pressure of the submerged (water-form) leaves is relatively high, while that of the emergent (land-form) leaves is relatively low. The low values of the emergent leaves may be due in part to transpirational loss of water vapor via the numerous stomata on the surfaces of the leaves. Associated with the higher turgor pressures in developing water-form leaves are long epidermal cells in the mature leaves.

Gibberellic acid caused the cells of emergent leaves to elongate by increasing the water uptake and, hence, the turgor pressure. At maturity these leaves had all the characteristics of typical water forms, including long epidermal cells. The limited cell expansion in submerged shoots exposed to abscisic acid or high temperatures apparently was not the result of low turgor; instead, the treated cells became less yielding so that high turgor failed to promote cell expansion. Growing submerged shoots in a mannitol solution resulted in turgor pressures similar to those of the emergent controls

LAND-FORM LEAVES

WATER-FORM LEAVES

and the production of leaves with short epidermal cells.

The results of these experiments suggest that the relative magnitude of cellular turgor pressure determines the final leaf size and form in *Callitriche heterophylla*. Thus, merely the presence or absence of water surrounding the developing leaf ensures that the leaf will be suitably adapted to life above or below the water.

22–37
The tendrils of grape (Vitis) *are modified stems.*

Most tendrils are leaf modifications. In legumes, such as the garden pea (*Pisum sativum*), the tendrils constitute the terminal part of the pinnately compound leaf (Figure 22–15). Only a few legumes form tendrils. One legume, the peanut (*Arachis hypogaea*), has another interesting adaptation. After fertilization takes place, the stamens and corolla of the flower fall off, and the internode between ovary and receptacle begins to elongate. With continued elongation, the stalk bends downward and buries the developing fruit several centimeters into the ground, where it ripens. If the ovary is not buried, it withers and fails to mature.

Branches that assume the form of and closely resemble foliage leaves are called *cladophylls*. The filmy, leaf-like branches of asparagus (*Asparagus officinalis*) are

22–38
The filmy branches of the common edible asparagus (Asparagus officinalis) *resemble leaves. Such modified stems are called cladophylls.*

familiar examples of cladophylls (Figure 22–38). The thick and fleshy aerial shoots ("spears") of asparagus are the edible portion of the plant. The scales found on the spears are true leaves. If the asparagus plants are allowed to continue growing, cladophylls develop in the axils of the minute, inconspicuous scales and then act as photosynthetic organs. In some cacti, the branches resemble leaves (Figure 22–39).

In some plants, the leaves are modified as spines, which are hard and dry and nonphotosynthetic. The terms "spine" and "thorn" are frequently used interchangeably, but technically thorns are modified branches that arise in the axils of leaves (Figure 22–40).

(a)

(b)

22–40
(a) *Spines, as in this* Ferocactus melocactiformis, *are modified leaves.* (b) *Thorns are modified branches, which, as this photograph of a hawthorn (Cra-taegus sp.) shows, arise in the axils of the leaves.*

22–39
The branches of the spineless cactus Epiphyllum *resemble leaves but are actually modified stems called cladophylls.*

CONVERGENT EVOLUTION

Comparable selective forces, acting on plants growing in similar habitats but different parts of the world, often cause totally unrelated species to assume a similar appearance. The process by which this happens is known as convergent evolution.

Let us consider some of the adaptive characteristics of plants growing in desert environments—fleshy, columnar stems (which provide the capacity for water storage), protective spines, and reduced leaves. Three fundamentally different families of flowering plants—the spurge family (Euphorbiaceae), the cactus family (Cactaceae), and the milkweed family (Asclepiadaceae)—have members that have evolved in this direction. The cactuslike representatives of the spurge and milkweed families shown here evolved from

leafy plants that look quite different from one another.

Native cacti occur (with one exception) exclusively in the New World. The comparably fleshy members of the spurge and milkweed families occur mainly in desert regions in Asia and especially Africa, where they play an ecological role similar to that of the New World cacti.

Although the plants shown here—(a) *Euphorbia,* a member of the spurge family; (b) *Echinocereus,* a cactus; (c) *Hoodia,* a fleshy milkweed—have CAM photosynthesis, all three are related to and derived from plants that have C₃ photosynthesis, which indicates that the physiological adaptations involved in CAM photosynthesis also arose as a result of convergent evolution (see page 111).

(a)

(b)

(c)

Another term commonly used interchangeably with thorn and spine is "prickle." A prickle, however, is neither a stem nor a leaf but a small, more or less slender, sharp outgrowth from the cortex and epidermis. The so-called thorns on rose stems are prickles. All three structures—spines, thorns, and prickles—may serve as defensive structures, reducing predation by herbivores (plant-eating animals). In a special plant-herbivore interaction, the spines of the bull's-horn acacias provide shelter for ants that kill other insects attempting to feed on the acacias (see page 654).

Among the most spectacular of modified or specialized leaves are those of the carnivorous plants, such as the pitcher plant, the sundew, and the Venus flytrap, which capture insects and digest them with enzymes secreted by the plant. The available nutrients are absorbed by the plant (see Chapter 26).

Food Storage

Stems, like roots, serve food-storage functions. Probably the most familiar type of specialized storage stem is the *tuber,* as exemplified by the Irish, or white, potato *(Solanum tuberosum).* In the white potato, tubers arise at the tips of *stolons* (slender stems growing along the surface of the ground) of plants grown from seed. However, when cuttings of tubers are used for propagation, the tubers arise at the ends of long, thin *rhizomes,* or underground stems (Figure 22–41). Except for vascular tissue, almost the entire mass of the tuber inside the periderm ("skin") is storage parenchyma. The so-called "eyes" of the white potato are depressions containing groups of buds. The depression is the axil of a scalelike leaf.

A *bulb* is a large bud consisting of a small, conical

22–41
White potato (Solanum tuberosum), with tubers attached to a rhizome, or underground stem.

22–42
Examples of modified leaves or stems. (a) The fleshy storage stem of kohlrabi (Brassica oleracea *var.* caulorapa). *(b) An onion* (Allium cepa) *bulb, which consists of a conical stem with scalelike, food-containing leaves attached. The leaves are the part of the onion we eat. (c) A gladiolus* (Gladiolus grandiflorus) *corm, which is a fleshy stem with small, thin leaves.*

stem with numerous modified leaves attached to it. The leaves are scalelike and contain thickened bases where food is stored. Adventitious roots arise from the bottom of the stem. Familiar examples of plants with bulbs are the onion (Figure 22–42b) and the lily.

Although superficially similar to bulbs, _corms_ consist primarily of stem tissue. Their leaves commonly are thin and much smaller than those of bulbs; consequently, the stored food of the corm is found within the fleshy stem. Several well known plants, such as gladiolus (Figure 22–42c), crocus, and cyclamen, produce corms.

Kohlrabi (*Brassica oleracea* var. *caulorapa*) is one example of an edible plant with a fleshy storage stem. The short, thick stem stands above the ground and bears several leaves with very broad bases (Figure 22–42a). The common cabbage (*Brassica oleracea* var. *capitata*) is closely related to kohlrabi. The so-called "head" of cabbage consists of a short stem bearing numerous thick, overlapping leaves. In addition to a terminal bud, several well-developed axillary buds may be found within the head.

The leaf stalks, or petioles, of some plants become quite thick and fleshy. Celery (*Apium graveolens*) and rhubarb (*Rheum rhaponticum*) are familiar examples.

Water Storage: Succulency

Succulent plants are plants that have juicy tissues, that is, tissues specialized for the storage of water. Most of these plants, such as the cacti of the American deserts, the *Euphorbia* of similar appearance of the African deserts (see "Convergent Evolution," page 443), and the century plant (*Agave*), normally grow in arid regions, where the ability to store water is necessary for their

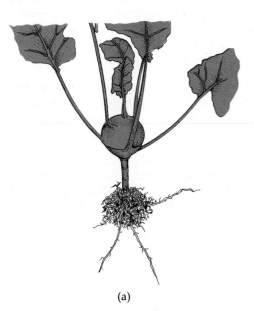

(a)

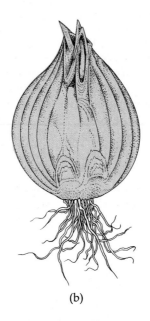

(b)

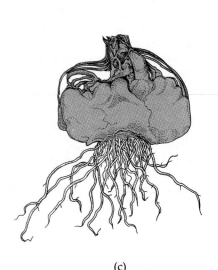

(c)

survival. The green, fleshy stems of the cacti serve both as photosynthetic and storage organs. The water-storing tissue consists of large, thin-walled parenchyma cells that lack chloroplasts.

In the century plant, the leaves are succulent. As in succulent stems, nonphotosynthetic parenchyma cells of the ground tissue constitute the water-storing tissue. Other examples of plants with succulent leaves are the ice plant *(Mesembryanthemum crystallinum)*, the stonecrops *(Sedum)*, and certain species of *Peperomia*. In the ice plant, large epidermal cells with appendages (trichomes), called water vesicles, which superficially resemble beads of ice, serve a water-storage function (see Figure 20–22c, page 396). The water-storing cells of the *Peperomia* leaf are part of a multiple (several-layered) epidermis derived by anticlinal divisions of the protoderm (Figure 22–43).

SUMMARY

The vegetative shoot apices of most flowering plants have a tunica-corpus type of organization consisting of one or more peripheral layers of cells (the tunica) and an interior (the corpus). Although the primary tissues of the stem pass through periods of growth similar to those of the root, the stem cannot be divided into regions of cell division, elongation, and maturation in the same manner as roots. The stem increases in length largely by internodal elongation.

As in the root, the apical meristem of the shoot gives rise to protoderm, ground meristem, and procambium, which develop into the primary tissues. Three basic variations exist in stems with regard to the relative distribution of ground and primary vascular tissues: the primary tissues may develop (1) as a more or less continuous hollow cylinder, (2) as a cylinder of discrete strands, or (3) as a system of strands scattered throughout the ground tissue. Regardless of the type of organization, the phloem is commonly located outside the xylem.

In dicotyledons, most leaves consist of a blade and petiole. The blades of some leaves are divided into leaflets. Stomata are commonly more numerous on the lower than the upper surface of the leaf. The ground tissue, or mesophyll, of the leaf is specialized as a photosynthetic tissue and, in mesophytes, is differentiated into palisade parenchyma and spongy parenchyma. The mesophyll is thoroughly permeated by air spaces and by veins, which are composed of xylem and phloem surrounded by a parenchymatous bundle sheath. The xylem commonly occurs on the upper side of the vein, whereas the phloem occurs on the lower side.

In most monocots, including the grasses, the leaf consists of a blade and a sheath, which encircles the stem. The leaves of C_3 and C_4 grasses have rather distinct anatomical differences. Most notable among the differences is the presence in C_4 grasses and absence in C_3 grasses of a Kranz anatomy, that

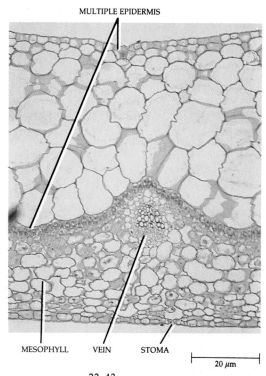

MULTIPLE EPIDERMIS

MESOPHYLL   VEIN   STOMA

20 μm

22–43
Transverse section of leaf blade of Peperomia. *The very thick multiple epidermis, visible on its upper surface, presumably functions as water-storage tissue.*

is, one in which the mesophyll cells and the bundle-sheath cells are arranged in two concentric layers around the vascular bundles.

Leaves have their origin in the peripheral region of the shoot apex, and their position on the stem is reflected in the pattern of the vascular system in the stem. Leaves are determinate in growth; that is, their development is of relatively short duration. However, vegetative shoot apices may exhibit unlimited or indeterminate growth. In many species, leaves grown under high light intensities are smaller and thicker than those grown under low light intensities. The former are called sun leaves, and the latter are called shade leaves.

In many plants, leaf abscission is preceded by formation of an abscission zone at the base of the petiole.

The change in the type of structure found in the root to that in the shoot occurs in a region of the plant axis of the embryo and young seedling called the transition region.

At flowering, the vegetative shoot apex is directly transformed into a reproductive apex.

Stems, like roots, may serve food-storage functions. Examples of fleshy stems are tubers, bulbs, and corms. Water-storing plants are known as succulents. Water-storage tissue of succulent plants is made up of large parenchyma cells. Stems or leaves or both may be succulent.

C H A P T E R 2 3

Secondary Growth

23-1

A solitary shagbark hickory (Carya ovata) in the winter condition. Plants have been able to achieve such great stature because of the ability of their roots and stems to increase in girth, that is, to undergo secondary growth. Most of the tissue produced in this manner is secondary xylem, or wood, which not only conducts water and minerals to the far reaches of the shoot but also provides great strength to the roots and stems.

In many plants (for instance, most monocotyledons and certain herbaceous dicotyledons such as *Ranunculus*), growth in a given part of the plant body ceases with maturation of the primary tissues. At the other extreme are the gymnosperms and woody dicotyledons, in which the roots and stems continue to increase in diameter in regions that are no longer elongating (Figure 23-1). This increase in thickness or girth of the plant body—termed secondary growth—results from activity of the two *lateral meristems*: the *vascular cambium* and the *cork cambium*.

Herbs, or herbaceous plants, are plants with shoots that undergo little or no secondary growth. In temperate regions either the shoot or the entire plant lives for only one season, depending on the species. Woody plants—trees and shrubs—can live for many years. At the start of each growing season, primary growth is resumed, and additional secondary tissues are added to the older plant parts through reactivation of the lateral meristems. Although most monocots lack secondary growth, some (such as the palms) may develop thick stems by primary growth alone (see page 416).

Plants are often classified according to their seasonal growth cycles as annuals, biennials, or perennials. In the *annuals*—which include many weeds, wildflowers, garden flowers, and vegetables—the entire cycle from seed to vegetative plant to flowering plant to seed again occurs within a single growing season, which may be only a few weeks in length. Only the dormant seed bridges the gap between one season of growth and the next.

In the *biennials*, two seasons are needed for the period from seed germination to seed formation. The first season of growth results in the formation of a root, a short stem, and a rosette of leaves near the soil surface. In the second growing season, flowering, fruiting, seed formation, and death occur, completing the life cycle.

In temperate regions, annuals and biennials seldom become woody, although both their stems and roots may undergo a limited amount of secondary growth.

Perennials are plants in which the vegetative structures live year after year. The herbaceous perennials pass unfavorable seasons as dormant underground roots, rhizomes, bulbs, or tubers. The woody perennials, which include vines, shrubs, and trees, survive above ground but usually stop growing during the unfavorable seasons. Woody perennials flower only when they become adult plants, which may take many years. For example, the horse chestnut, _Aesculus hippocastanum_, does not flower until it is about 25 years old. _Puya raimondii_, a large (up to 10 meters high) relative of the pineapple that is found in the Andes, takes about 150 years to flower. Many woody plants are deciduous, losing all their leaves at the same time and developing new leaves from buds when the season again becomes favorable for growth. In evergreen trees and shrubs, leaves are also lost and replaced but not simultaneously.

THE VASCULAR CAMBIUM

Unlike the many-sided initials of the apical meristems, which contain dense cytoplasm and large nuclei, the meristematic cells of the vascular cambium are highly vacuolated. They exist in two forms: as vertically elongated _fusiform initials_, and as horizontally elongated or squarish _ray initials_. The fusiform initials are much longer than they are wide and appear flattened or brick-shaped in transverse section. In the white pine, _Pinus strobus_, the fusiform initials average 3.2 millimeters in length; in the apple, _Malus sylvestris_, 0.53 millimeters (Figure 23–2); and in the black locust, _Robinia pseudo-acacia_, 0.17 millimeters (Figure 23–3).

Secondary xylem and secondary phloem are produced through periclinal divisions of the cambial initials and their derivatives. In other words, the cell plate that forms between the dividing cambial initials is parallel to the surface of the root or stem (Figure 23–4a). If the derivative of a cambial initial is divided off toward the outside of the root or stem, it eventually becomes a phloem cell; if it divides off toward the inside, it becomes a xylem cell. In this manner a long, continuous radial file, or row, of cells is formed, extending from the cambial initial outward to the phloem and inward to the xylem (Figure 23–5).

The xylem and phloem cells produced by the fusiform initials have their long axes oriented vertically and make up what is known as the _axial system_ of the secondary vascular tissues. The ray initials produce horizontally oriented _ray cells_, which form the _vascular rays_ or _radial system_ (Figure 23–5). The rays are composed largely of parenchyma cells and are variable in length.

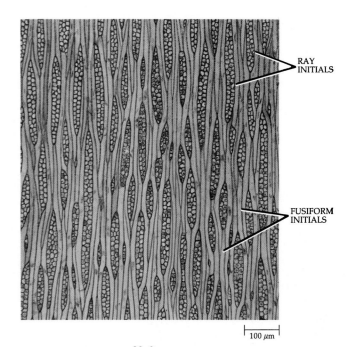

RAY INITIALS

FUSIFORM INITIALS

100 μm

23–2
Tangential view of a portion of vascular cambium of the apple (Malus sylvestris) _tree. Tangential sections are cut at right angles to the rays, so we see the rays here in transverse section. A cambium such as this, in which the fusiform initials are not arranged in horizontal tiers on tangential surfaces, is said to be nonstoried._

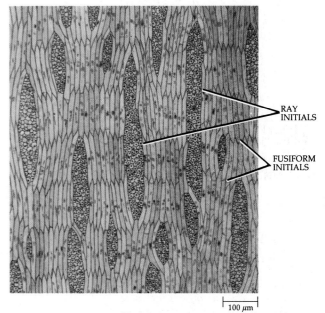

RAY INITIALS

FUSIFORM INITIALS

100 μm

23–3
Vascular cambium of the black locust (Robinia pseudo-acacia) _tree, seen here in tangential view. A cambium such as this, in which the fusiform initials are arranged in horizontal tiers on tangential surfaces, is said to be storied._

23–4

Periclinal and anticlinal divisions of fusiform initials. (a) Periclinal divisions are involved in the formation of secondary xylem and secondary phloem cells, and result in the formation of radial rows of cells (see Figure 23–5). When an initial divides periclinally, two cells appear, one behind (or in front of) the other. (b) Anticlinal divisions are involved in the multiplication of fusiform initials. When an initial divides anticlinally, two cells appear side by side where one was present previously.

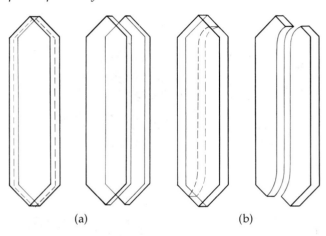

(a)   (b)

23–5

Diagram showing the relationship of the vascular cambium to its derivative tissues—secondary xylem and secondary phloem. The vascular cambium is made up of two types of cells—fusiform initials and ray initials—which give rise to the axial and radial systems, respectively, of the secondary vascular tissues. When the cambial initials produce secondary xylem and secondary phloem, they divide periclinally. Following division of an initial, one daughter cell (the initial) remains meristematic, and the other (the derivative of the initial) eventually develops into one or more cells of the vascular tissue. Cells produced toward the inner surface of the vascular cambium become xylem elements, and those produced toward the outer surface become phloem elements. The ray initials divide to form vascular rays, which lie at right angles to the derivatives of the fusiform initials. With the production of additional secondary xylem, the vascular cambium and secondary phloem are displaced in an outward direction. The diagrams (left to right) represent successively more mature stages.

Nutrients move from protoplast to protoplast via plasmodesmata *(symplastic movement)*, passing from the secondary phloem through the vascular cambium and to the living cells of the secondary xylem by way of the vascular rays. By contrast, water passes from the secondary xylem to the cambium and the secondary phloem largely by way of the walls *(apoplastic movement)* of both the rays and cells of the axial system. The rays also serve as storage centers for such substances as starch and lipids.

In a restricted sense, the term "vascular cambium" is used to refer only to the cambial initials, of which there is one per radial file. However, it is often difficult, if not impossible, to distinguish between the initials and their immediate derivatives, which may remain meristematic for a considerable period of time (Figure 23–5). Even in the winter condition, when the cambium is dormant, or inactive, several layers of undifferentiated cells that are similar in appearance can be seen between the xylem and phloem. Consequently, some botanists use the term "vascular cambium" in a broader sense to refer to the initials and their immediate derivatives, which are indistinguishable from the initials. Others refer to this region of initials and derivatives as the *cambial zone*.

As the vascular cambium adds cells to the secondary xylem and the core of xylem increases in width, the cambium is displaced outward. In order to accommodate to this change, the vascular cambium undergoes an increase in circumference, which is accomplished by anticlinal divisions of the initials (Figure 23–4b). Along with an increase in the number of fusiform initials, new ray initials and rays are added, so that a fairly constant ratio of rays to fusiform cells is maintained in the secondary vascular tissues. Obviously, the developmental changes that occur in the cambium are exceedingly complex.

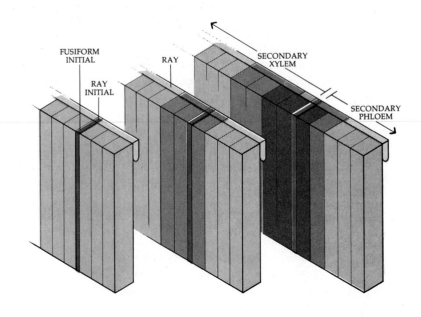

FUSIFORM INITIAL   RAY   SECONDARY XYLEM

RAY INITIAL   SECONDARY PHLOEM

In temperate regions, the vascular cambium is dormant during winter and becomes reactivated in the spring. New growth layers, or increments, of secondary xylem and secondary phloem are laid down during the growing season. Reactivation of the vascular cambium is triggered by the expansion of the buds and resumption of their growth. Apparently, the hormone auxin, produced by the developing shoots, moves downward in the stems and stimulates resumption of cambial activity. Other factors are also involved in cambial reactivation and in continued normal growth of the cambium (Chapter 24).

EFFECT OF SECONDARY GROWTH ON THE PRIMARY PLANT BODY

Root

In roots, the vascular cambium is initiated by the meristematic procambial cells that remain between the primary xylem and primary phloem. Thus, depending on the number of phloem strands present in the root, two or more independent regions of cambial activity are initiated more or less simultaneously (Figure 23–6). Soon

23–6

Comparison of primary and secondary structure in root and stem of a woody dicot. (a) Root and stem at completion of primary growth. In the triarch root represented here, cambial activity has been initiated in three independent regions from procambium between the three primary phloem strands and the primary xylem. (b) Origin of vascular cambium. The pericyclic cells opposite the three protoxylem poles will also contribute to the vascular cambium. Some secondary xylem has already been produced by the newly formed vascular cambium of procambial origin. (c) After formation of some secondary xylem and secondary phloem in root and stem, and periderm formation in root. (d) At end of first year's growth, showing the effect of secondary growth—including periderm formation—on the primary plant body. In (c) and (d), the radiating lines represent rays.

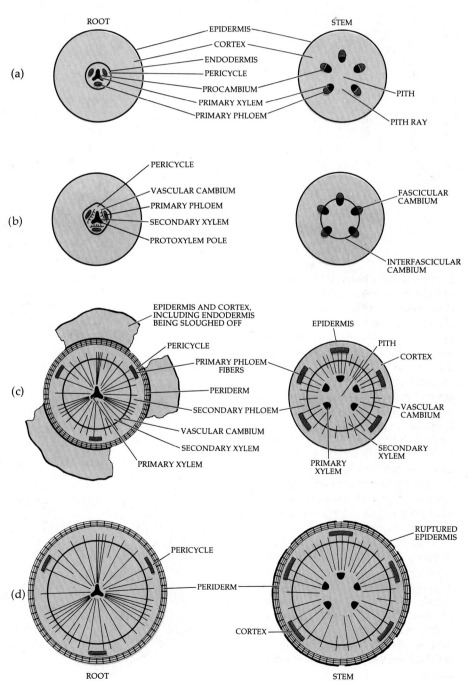

afterward, the pericyclic cells opposite the protoxylem poles divide periclinally, and the inner sister cells contribute to the vascular cambium. Now the cambium completely surrounds the core of xylem.

As soon as it is formed, the vascular cambium opposite the phloem strands begins to produce secondary xylem, and in the process the strands of primary phloem are displaced outwardly from their positions between the ridges of primary xylem. By the time the cambium opposite the protoxylem poles is actively dividing, the cambium is circular in outline and the primary phloem has been separated from the primary xylem (Figure 23–6).

By repeated divisions toward the inside and outside, secondary xylem and secondary phloem are added to the root (Figure 23–6 and 23–7). In some roots, the vascular cambium derived from the pericycle forms wide rays, whereas narrower rays are produced in other parts of the secondary vascular tissues.

With increase in width of the secondary xylem and phloem, most of the primary phloem is crushed or obliterated. Primary phloem fibers may be the only remaining distinguishable components of the primary phloem.

Stem

As mentioned previously, the vascular cambium of the stem arises from the procambium that remains undifferentiated between the primary xylem and primary phloem, as well as from parenchyma of the interfascicular regions. That portion of the cambium arising within the vascular bundles is known as *fascicular cambium*, and that arising in the interfascicular regions, or pith rays, is called *interfascicular cambium*. The vascular cambium of the stem, unlike that of the root, is essentially circular in outline from its inception (see Figure 23–6).

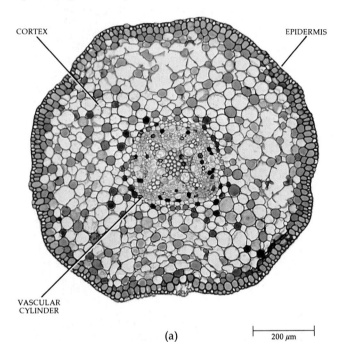

CORTEX

EPIDERMIS

VASCULAR
CYLINDER

(a)

200 μm

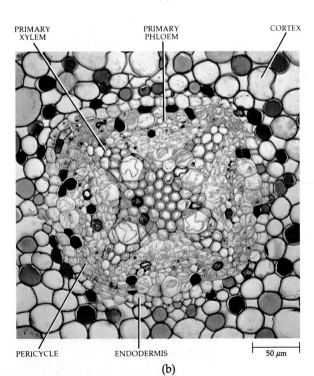

PRIMARY
XYLEM

PRIMARY
PHLOEM

CORTEX

PERICYCLE

ENDODERMIS

50 μm

(b)

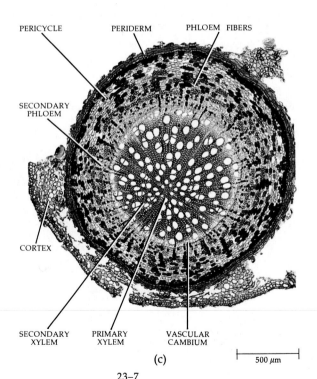

PERICYCLE

PERIDERM

PHLOEM FIBERS

SECONDARY
PHLOEM

CORTEX

SECONDARY
XYLEM

PRIMARY
XYLEM

VASCULAR
CAMBIUM

(c)

500 μm

23–7
Transverse sections of the willow (Salix) *root, which becomes woody. (a) Overall view of root near completion of primary growth. (b) Detail of primary vascular cylinder. (c) Overall view of root at end of first year's growth, showing effect of secondary growth on primary plant body.*

In woody stems, the production of secondary xylem and secondary phloem results in the formation of a cylinder of secondary vascular tissues, with the rays extending radially through the cylinder (Figure 23–6). Commonly, much more secondary xylem than secondary phloem is produced in the stem in any given year; this situation is also true in the root. As in the root, with secondary growth the primary phloem is pushed outward and its thin-walled cells are destroyed. Only the thick-walled primary phloem fibers remain intact (Figure 23–9).

Figures 23–8 and 23–9 show an elderberry (*Sambucus canadensis*) stem in two stages of secondary growth. (See pages 418 to 419 for a description of primary growth in the *Sambucus* stem.) Only a small amount of secondary xylem and secondary phloem has been produced in the stem of Figure 23–8. Figure 23–9 shows a stem at the end of the first year's growth. Note that considerably more secondary xylem than secondary phloem has been formed. The thick-walled cells outside the secondary phloem are primary phloem fibers.

Figure 23–10 shows one-year-old, two-year-old, and

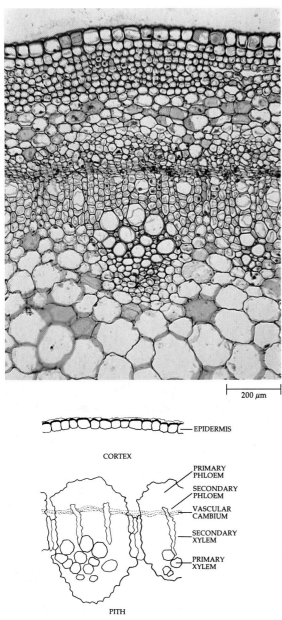

23–8
Transverse section of an elderberry (Sambucus canadensis) *stem in which a small amount of secondary growth has taken place. A cork cambium has not yet been formed.*

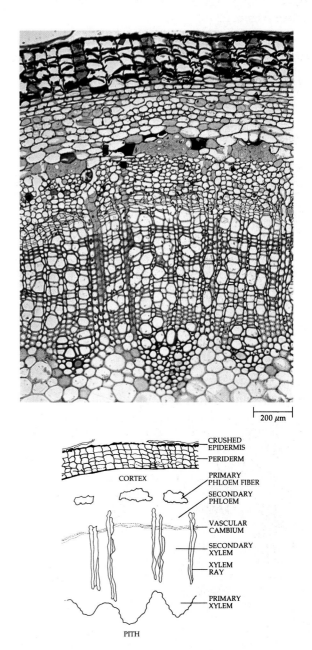

23–9
Transverse section of an elderberry (Sambucus canadensis) *stem at the end of first year's growth.*

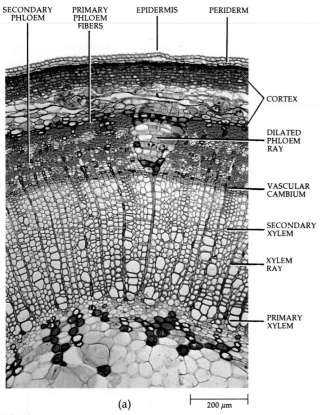

SECONDARY PHLOEM · PRIMARY PHLOEM FIBERS · EPIDERMIS · PERIDERM

CORTEX

DILATED PHLOEM RAY

VASCULAR CAMBIUM

SECONDARY XYLEM

XYLEM RAY

PRIMARY XYLEM

(a)   200 μm

23–10

Transverse sections of linden (Tilia americana) *tree stems. (a) One-year-old stem. (b) Two-year-old stem. (c) Three-year-old stem. Numbers indicate growth rings in the secondary xylem.*

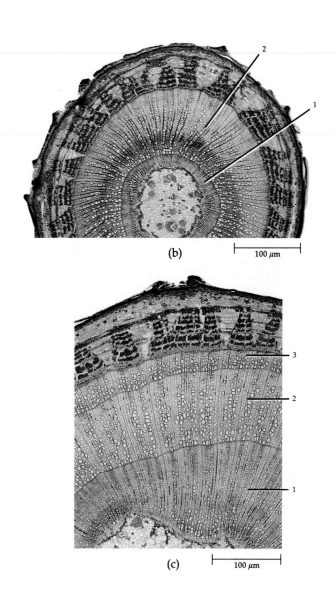

(b)   100 μm

(c)   100 μm

three-year-old stems of linden, or basswood *(Tilia americana).* In Chapter 22, the stem of *Tilia* was given as an example of one in which the primary tissues arise as an almost continuous hollow cylinder; thus most of the vascular cambium in the *Tilia* stem is fascicular in origin. Some of the rays in the secondary phloem of *Tilia* become very wide as the stem increases in girth. This is one way in which the tissues outside the vascular cambium keep up with the increase in girth of the core of xylem.

The vascular cambia and secondary tissues of root and stem are continuous with one another. There is no transition region in the secondary plant body as there is in the primary plant body (see page 437).

The Periderm

In most woody roots and stems, cork formation usually follows the initiation of secondary xylem and secondary phloem production, and the cork tissue replaces the epidermis as the protective covering on those portions of the plant. *Cork,* or phellem, is formed by a *cork cam-*

bium, or phellogen, which may also form *phelloderm* ("cork skin"). The cork is formed toward the outer surface of the cork cambium, and the phelloderm is formed toward the inner surface (see Figures 23–11 and 23–12). Together, these three tissues—cork, cork cambium, and phelloderm—make up the *periderm.*

In most dicots and gymnosperms, the first periderm commonly appears during the first year of growth in those portions of the root or stem that are no longer elongating. In stems, the first cork cambium most commonly originates in a layer of cortical cells immediately below the epidermis (Figures 23–6 and 23–11), although in many species it originates in the epidermis. In roots, the first cork cambium arises through periclinal division of pericycle cells, the outer sister cells combining to form a complete cylinder of cork cambium. Afterwards, the remaining cells of the pericycle may proliferate below the periderm and give rise to a tissue that resembles a cortex (Figures 23–6 and 23–7).

Repeated divisions of the cork cambium result in the formation of radial rows of compactly arranged cells, most of which are cork cells (Figures 23–11 and 23–12).

EPIDERMIS CORK CORK CAMBIUM PHELLODERM

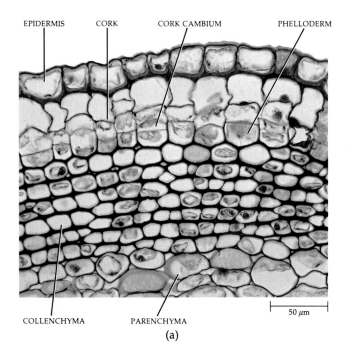

COLLENCHYMA PARENCHYMA

(a)

EPIDERMIS CORK PHELLODERM CORK CAMBIUM

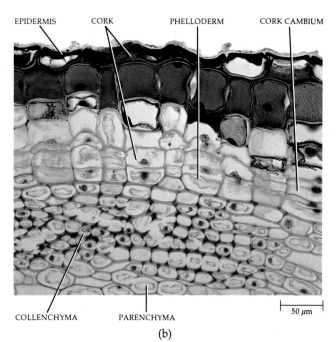

COLLENCHYMA PARENCHYMA

(b)

23–11
Some stages of periderm and lenticel development in elderberry (Sambucus canadensis), as seen in transverse sections. (a) Newly formed periderm beneath the epidermis; collenchyma and parenchyma of cortex. (b) Periderm in more advanced stage of development. (c) Initiation of lenticel; collenchyma of cortex beneath the developing lenticel. (d) Well-developed lenticel. The phelloderm in Sambucus generally consists of a single layer of cells.

COLLENCHYMA DEVELOPING LENTICEL EPIDERMIS

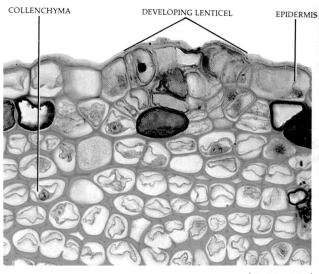

(c)

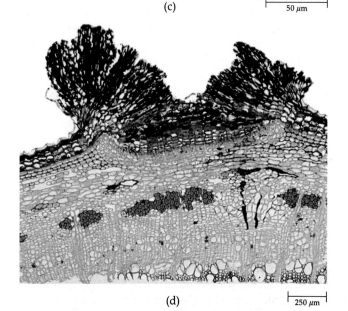

(d)

During differentiation of the cork cells, their inner walls are lined by a relatively thick layer of a fatty substance —*suberin*—which makes the tissue highly impermeable to water and gases. The walls of the cork cells may also become lignified. At maturity, the cork cells are dead.

The cells of the phelloderm are living at maturity, lack suberin, and resemble cortical parenchyma cells. The phelloderm cells may be distinguished from cortical cells by their inner position in the radial rows of other periderm cells (Figure 23–12).

With the formation of the first periderm in the root, the cortex (including the endodermis) and the epidermis are isolated from the rest of the root. Being separated from the supply of water and minerals by the impermeable barrier of cork, the cortex and epidermis eventually die and are sloughed off. Because the first periderm of the stem usually arises immediately below the epidermis, the cortex of the stem is not sloughed off during the first year (see Figures 23–6 and 23–9), although the epidermis does dry up and peel off.

23–12

Lenticel of the stem of Dutchman's pipe (Aristolochia), *as seen in transverse section. Unlike* Sambucus, *the phelloderm of* Aristolochia *consists of several layers of cells.*

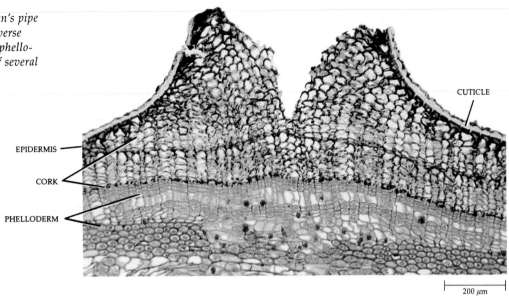

EPIDERMIS

CORK

PHELLODERM

CUTICLE

200 μm

At the end of the first year's growth, the following tissues are present in a woody root (from outside to inside): remnants of the epidermis and cortex, periderm, pericycle, primary phloem (fibers and crushed soft-walled cells), secondary phloem, vascular cambium, secondary xylem, and primary xylem. The following tissues are present in the stem: remnants of the epidermis, periderm, cortex, primary phloem (fibers and crushed soft-walled cells), secondary phloem, vascular cambium, secondary xylem, primary xylem, and pith (Figure 23–6).

The Lenticels

In the preceding discussion, it was noted that the suberin-containing cork cells are compactly arranged and, as a tissue, present an impermeable barrier to water and gases. However, the inner tissues of the stem, like all metabolically active tissues, need to exchange gases with the surrounding air; similarly, the inner tissues of the root need to exchange gases with the surrounding air spaces between soil particles. In stems and roots containing periderms, this necessary gas exchange is accomplished by means of *lenticels* (Figures 23–11 and 23–12)—portions of the periderm in which the phellogen (cork cambium) is more active than elsewhere, resulting in the formation of a tissue with numerous intercellular spaces. In addition, the phellogen itself contains intercellular spaces in the region of the lenticels.

Lenticels begin to form during development of the first periderm (see Figure 23–11) and, in the stem, generally appear below a stoma or group of stomata. On the surface of the stem or root, the lenticels appear as raised circular, oval, or elongated areas (see Figure 23–18, page 457). Lenticels are also formed on some fruits—for instance, the small dots on the surface of apples and pears are lenticels. As the roots and stems grow older, lenticels continue to develop at the bottom of cracks in the bark in newly formed periderms.

The Bark

The terms "periderm," "cork," and "bark" are often unnecessarily confused with one another. As previously discussed, cork is one of three parts of the periderm; it is a secondary tissue that replaces the epidermis in most woody roots and stems. The term *bark* refers to all the tissues outside the vascular cambium, including the periderm when present (Figures 23–13 and 23–14). When the vascular cambium first appears, and secondary phloem has not yet been formed, the bark consists entirely of primary tissues. At the end of the first year's growth, the bark includes any primary tissues still present, the secondary phloem, the periderm, and any dead tissues remaining outside the periderm.

Each growing season, the vascular cambium adds secondary phloem to the bark, as well as secondary xylem, or wood, to the core of the stem or root. Usually less secondary phloem is produced by the vascular cambium than secondary xylem. In addition, the soft-walled cells (sieve elements and various kinds of parenchymatic elements) of the old secondary phloem are commonly crushed (see Figures 23–15 through 23–17). Eventually, the old secondary phloem is separated from the rest of the phloem by newly formed periderms. As a result, considerably less secondary phloem accumulates in the stem or root than secondary xylem, which continues to accumulate year after year.

23–13
Diagram of part of a red oak (Quercus
rubra) *stem, showing the transverse,
tangential, and radial surfaces. The dark
area in the center is heartwood. The
lighter part of the wood is sapwood.*

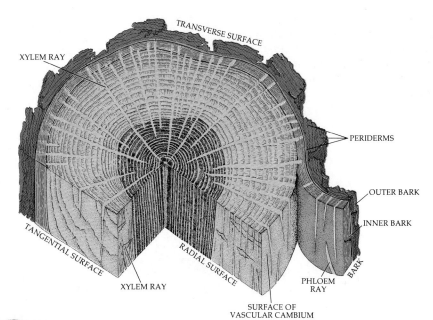

23–14
*Transverse section of the bark and some
secondary xylem from an old stem of
linden* (Tilia americana). *Several
periderms can be seen traversing the
mostly brownish outer bark in the upper
third of the section. Below the outer bark
is the inner bark, which is quite distinct
in appearance from the more lightly
stained xylem in the lower third of the
section.*

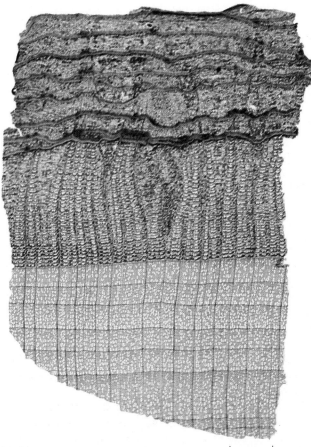

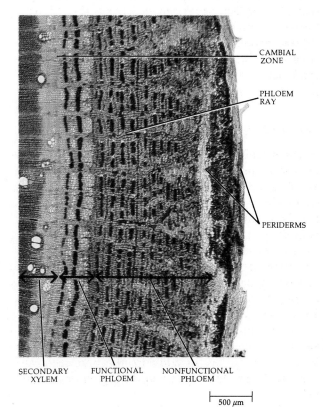

23–15
*Transverse section of the bark of the
black locust* (Robinia pseudo-acacia)
*stem, consisting mostly of nonfunctional
phloem.*

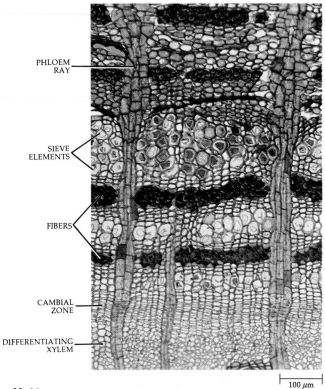

PHLOEM
RAY

SIEVE
ELEMENTS

FIBERS

CAMBIAL
ZONE

DIFFERENTIATING
XYLEM

100 μm

23–16
*Transverse section of secondary phloem
of the black locust, showing mostly
functional phloem. Sieve elements (indicated by arrows) of the nonfunctional
phloem have collapsed.*

As the stem or root increases in girth, considerable
stress is placed on the older tissues of the bark. In some
plants, tearing of these tissues results in the formation
of large air spaces. In many plants, the parenchyma
cells of the axial system and rays divide and enlarge; in
this manner, the old secondary phloem keeps up for a
while with the increase in circumference of the plant
part. It was noted earlier that certain rays in the *Tilia*
stem become very wide as the stem increases in girth;
these rays are called dilated rays.

The first-formed periderm may keep up with the increase in girth of the root or stem for several years, with
the cork cambium exhibiting periods of activity and in-
activity that may or may not correspond to the periods
of activity of the vascular cambium. In the stems of
apple *(Malus sylvestris)* and pear *(Pyrus communis)* trees,
the first cork cambium may remain active for up to 20
years. In most woody roots and stems, additional periderms are formed as the axis increases in circumference. After the first periderm, subsequently formed
periderms originate deeper and deeper in the bark (Figures 23–13 and 23–14) from parenchyma cells of the
phloem no longer actively engaged in the transport, or
translocation, of food substances. These parenchyma
cells become meristematic and form new cork cambia.

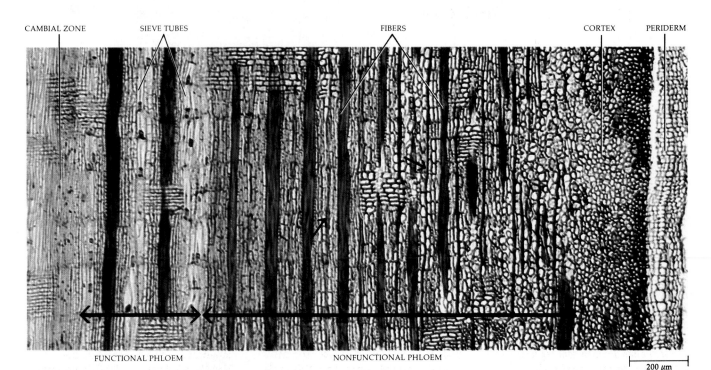

CAMBIAL ZONE   SIEVE TUBES   FIBERS   CORTEX   PERIDERM

FUNCTIONAL PHLOEM   NONFUNCTIONAL PHLOEM

200 μm

23–17
*Radial section of the bark of the black
locust. Most of the section consists of
nonfunctional phloem, in which the sieve*
*elements are collapsed (see arrows). In
black locust, the functional phloem
consists only of the current season's*
*growth increment, which becomes
nonfunctional in late autumn when its
sieve elements die and collapse.*

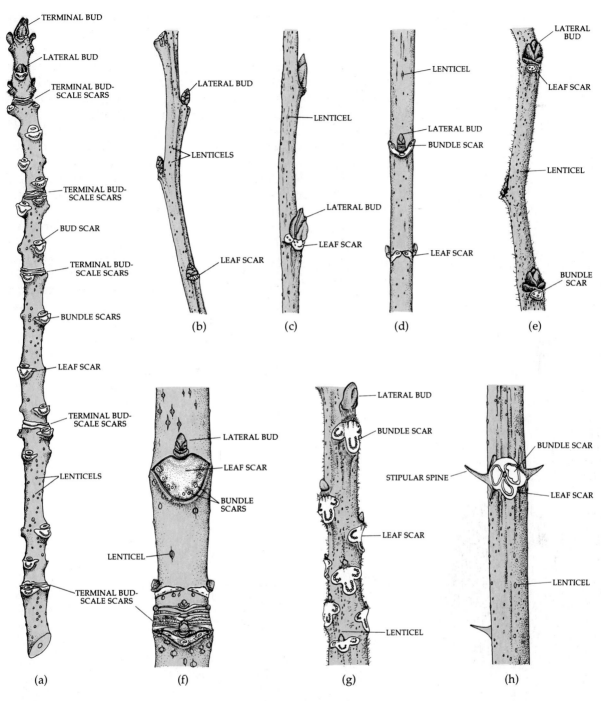

23–18

*External features of woody stems.
Examination of the twigs of deciduous
woody plants reveals many important
developmental and structural features
of the stem. The most conspicuous
structures on the twigs are the buds.
Buds occur at the tips—the terminal
buds—and in the axils of the leaves—
the lateral, or axillary, buds of the
twigs. In addition, accessory buds
occur in some species. Commonly
occurring in pairs, the accessory buds
are located one each on either side of an
axillary bud. In some species, the
accessory buds do not develop if their
associated axillary bud undergoes
normal development. In others, the
accessory buds give rise to flowers and
the axillary bud gives rise to a leafy
shoot.*

*After the leaves fall, leaf scars, with
their bundle scars, can be seen beneath
the axillary buds. The protective layer
of the abscission zone produces the leaf
scar. The bundle scars are the severed
ends of vascular bundles that extended
from the leaf traces into the petiole of
the leaf, prior to abscission.*

*Groups of terminal bud-scale scars
reveal the locations of previous terminal
buds, and until they are obscured by
secondary growth, these groups of scars
may be used to determine the age of
portions of the stem. The portion of
stem between two groups of such scars
represents one year's growth. The
lenticels appear as slightly raised areas
on the stem.*

*(a) Green ash (Fraxinus pennsylva-
nica var. subintegerrima). (b) White oak
(Quercus alba). (c) Linden (Tilia
americana). (d) Boxelder (Acer ne-
gundo). (e) American elm (Ulmus americana). (f) Horse chestnut (Aesculus
hippocastanum). (g) Butternut (Juglans
cinerea). (h) Black locust (Robinia
pseudo-acacia).*

(a)   (b)   (c)

(d)

23–19

Bark of four species of trees. (a) Thin, peeling bark of the paper birch (Betula papyrifera). Elongated areas on surface of bark are lenticels. (b) Shaggy bark of the shagbark hickory (Carya ovata). (c) Scaly bark of a sycamore, or plane tree (Platanus occidentalis). (d) Deeply furrowed bark of the black oak (Quercus velutina).

All of the tissues outside the innermost cork cambium—all of the periderms, together with any cortical and phloem tissues included among them—make up the *outer bark* (Figures 23–13 and 23–14). With maturation of the suberin-containing cork cells, the tissues outside them are separated from the supply of water and nutrients. Hence the outer bark consists entirely of dead tissues. The living part of the bark, which is inside the innermost cork cambium and extends inward to the vascular cambium, is called the *inner bark* (see Figures 23–13 and 23–14).

The manner in which new periderms are formed and the kinds of tissues isolated by them have a marked influence on the appearance of the outer surface of the bark (Figure 23–19). In some barks the newly formed periderms develop as discontinuous overlapping layers, resulting in formation of a scaly type of bark, called scale bark (see Figures 23–13 and 23–14). Scale barks, for example, are found on relatively young stems of pine *(Pinus)* and pear *(Pyrus communis)* trees. In other barks, the newly formed periderms arise as more or less continuous, concentric rings around the axis, resulting in formation of a ring bark. Grape *(Vitis)* and honeysuckle *(Lonicera)* are examples of plants with ring barks, which are less common than scale barks. The barks of many plants are intermediate between ring and scale barks.

Commercial cork is obtained from the bark of the cork oak, *Quercus suber,* which is native to the Mediterranean region. The first cork cambium of this tree has its origin in the epidermis, and the cork produced by it is of little commercial value. When the tree is about 20 years old, the original periderm is removed, and a new cork cambium is formed in the cortex, just a few millimeters below the site of the first one. The cork produced by the new cork cambium accumulates very rapidly and after about 10 years is thick enough to be stripped off the tree. Once again a new cork cambium arises beneath the previous one, and after about another 10 years the cork can be stripped again. This procedure may be repeated at about 10-year intervals until the tree is 150 or more years old. The spots and long dark streaks seen on the surfaces of commercial cork are lenticels.

In most woody roots and stems, very little secondary phloem is actually involved in the conduction of food. In most species, only the current year's growth increment, or growth ring, of secondary phloem is active in the long-distance transport of food through the stem. This is because the sieve elements are short-lived (Chapter 20); most of them die by the end of the same year in which they are derived from the vascular cambium. In some plants, such as black locust *(Robinia pseudo-acacia)*, the sieve elements collapse and are crushed relatively soon after they die (see Figures 23–15 through 23–17).

The part of the inner bark actively engaged in the transport of food substances is called *functional phloem.* Although the sieve elements outside the functional phloem are dead, the phloem parenchyma cells (axial parenchyma) and parenchyma cells of the rays may remain alive and continue to function as storage cells for many years. That part of the inner bark is known as *nonfunctional phloem.* Only the outer bark is composed entirely of dead tissue. (See Figures 23–15 and 23–17.)

THE WOOD: SECONDARY XYLEM

Apart from use of various plant tissues as food for humans, no single plant tissue has played a more indispensable role to human survival throughout recorded history than wood, or secondary xylem (see Table 23–1, page 468). Commonly, woods are classified as either *hardwoods* or *softwoods*. The so-called hardwoods are dicot woods, and the softwoods are conifer woods. The two kinds of woods have basic structural differences, but the terms "hardwood" and "softwood" do not accurately express the degree of density (weight per unit volume) or hardness of the wood. For example, one of the lightest and softest of woods is balsa *(Ochroma lagopus)*, a tropical dicot. By contrast, the woods of some conifers, such as slash pine *(Pinus elliottii)*, are harder than some hardwoods.

Conifer Wood

The structure of conifer wood is relatively simple compared with that of most dicots. The principal features of conifer wood are its lack of vessels (Chapter 20) and its relatively small amount of axial, or wood, parenchyma. Long, tapering tracheids constitute the dominant cell type in the axial system. In certain genera, such as *Pinus,* the only parenchyma cells of the axial system are those associated with *resin ducts*. Resin ducts are relatively large intercellular spaces lined with thin-walled parenchyma cells, which secrete resin into the duct. In *Pinus,* resin ducts occur in both the axial system and the rays (Figures 23–20 and 23–21). Wounding, pressure, and injuries by frost and wind can stimulate the formation of resin ducts in conifer wood, leading some investigators to suggest that all resin ducts are traumatic in origin. Resin apparently protects the plant from attack by decay-producing fungi and bark beetles.

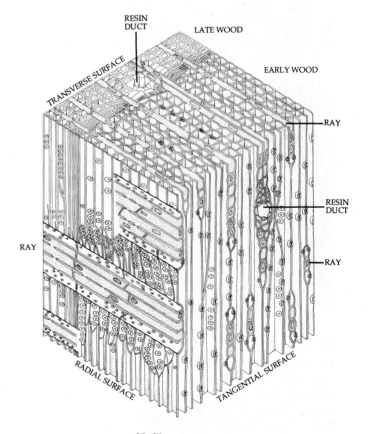

23–20
Block diagram of the secondary xylem of white pine (Pinus strobus). *With the exception of the parenchyma cells associated with the resin ducts, the axial system consists entirely of tracheids. The rays are only one cell wide, except for those containing resin ducts. Early and late wood are described on page 464.*

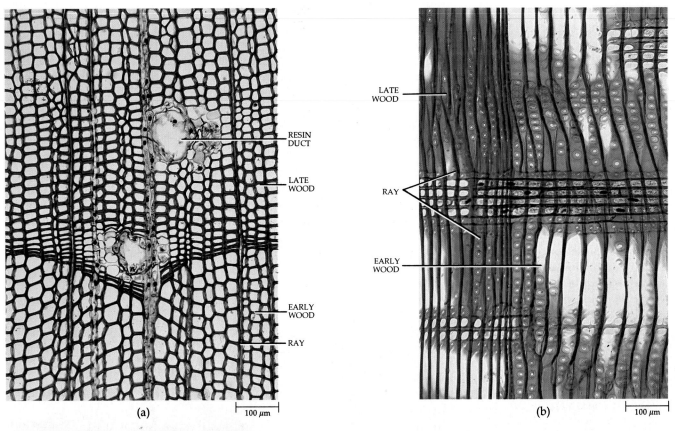

(a)   (b)

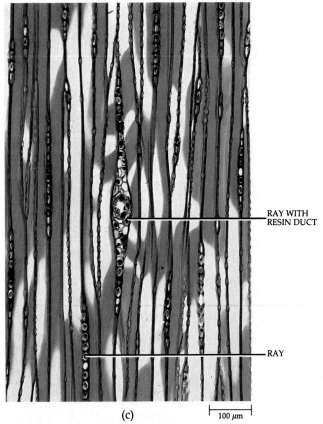

(c)

23–21

Wood of white pine (Pinus strobus), *a conifer, in (a) transverse, (b) radial, and (c) tangential sections.*

The tracheids of conifers are characterized by large, circular, bordered pits that are most abundant on the ends of the cells, where they overlap with other tracheids (Figures 23–20 through 23–22). The pairs of pits (pit-pairs; see Chapter 2, page 35) between conifer tracheids are each characterized by the presence of a *torus* (plural: tori). The torus is a thickened central portion of the pit-membrane (Figure 23–25) and is slightly larger than the openings or apertures in the pit-borders (Figure 23–22). The pit-membrane is flexible, and under certain conditions, the torus may block one of the apertures and prevent the movement of water or gases through the pit-pair (Figure 23–22).

Figure 23–20 is a three-dimensional diagram of the wood of white pine (*Pinus strobus*) based on the three wood sections shown in Figure 23–21. In sections cut at right angles to the long axis of the root or stem—transverse or cross sections—the tracheids appear angular or squarish, and the rays can be seen in their longitudinal extent traversing the wood (Figure 23–21a). There are two kinds of longitudinal sections—radial and tangential. Radial sections are cut parallel with the rays, and in such sections, the rays appear as sheets of cells oriented at right angles to the vertically elongated tracheids of the axial system (Figures 23–21b and 23–22d). Tangential sections are cut at right angles to the rays and reveal the width and height of the rays. In *Pinus* the rays are one cell wide, except for those containing resin ducts (Figure 23–21c). Details of white pine wood are shown in Figure 23–22.

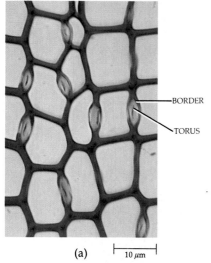

(a) 10 μm

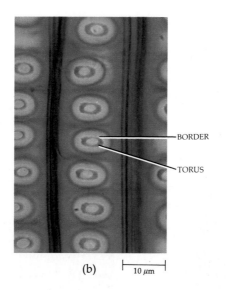

(b) 10 μm

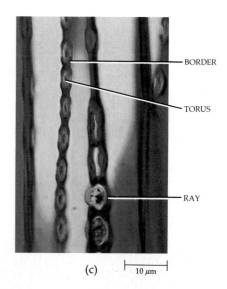

BORDER
TORUS
RAY

(c) 10 μm

23–22
Details of white pine (Pinus strobus) *wood. (a) Transverse section, showing bordered pit-pairs of tracheids. (b) Radial section, showing the face-view of bordered pit-pairs in walls of tracheids. (c) Tangential section, showing bordered pit-pairs of tracheids. (d) Radial section, showing ray. The rays of pine and other conifers are composed of ray tracheids and ray parenchyma cells. Notice the bordered pits of ray tracheids.*

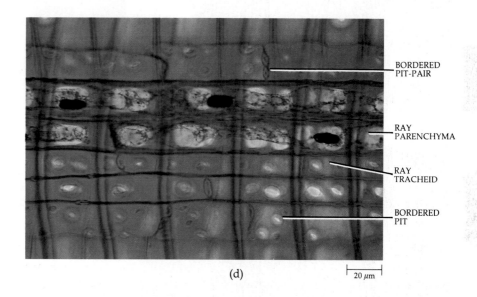

BORDERED PIT-PAIR
RAY PARENCHYMA
RAY TRACHEID
BORDERED PIT

(d) 20 μm

23–23
Transverse sections of wood, showing growth layers. (a) Red oak (Quercus rubra). *The large vessels of ring-porous wood such as red oak are found in the early wood. The dark vertical lines are rays. (b) Tulip tree* (Liriodendron tulipifera), *a diffuse-porous wood.*

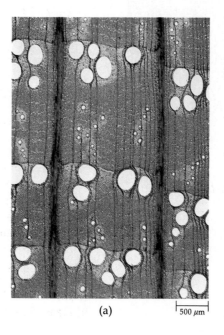

(a) 500 μm

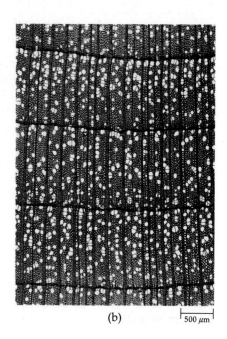

(b) 500 μm

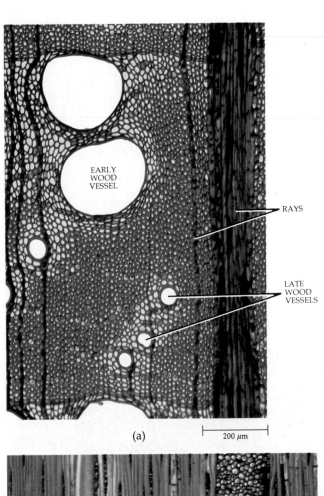

EARLY WOOD VESSEL

RAYS

LATE WOOD VESSELS

200 μm

(a)

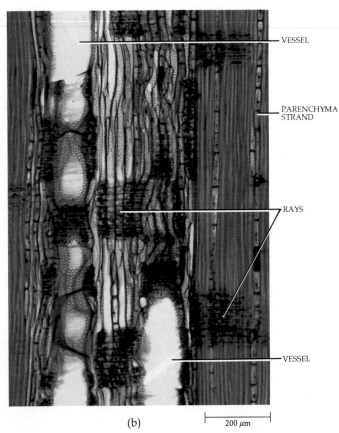

VESSEL

PARENCHYMA STRAND

RAYS

VESSEL

200 μm

(b)

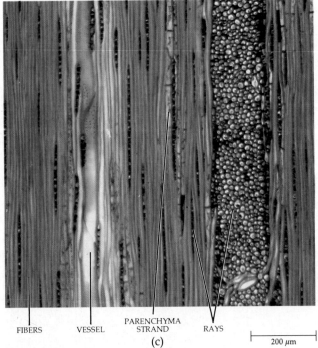

FIBERS   VESSEL   PARENCHYMA STRAND   RAYS

200 μm

(c)

23–24
Wood of red oak (Quercus rubra), *in (a) transverse, (b) radial, and (c) tangential sections.*

23–25
Scanning electron micrograph of pit membrane of bordered pit-pair in white pine (Pinus strobus) *tracheid. The thickened part of the membrane is the torus. The part of the membrane surrounding the torus is called the margo.*

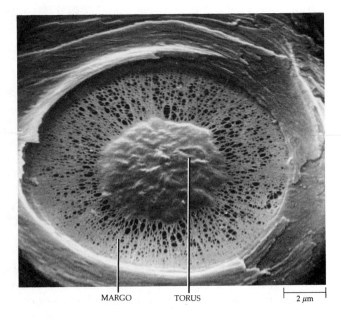

MARGO   TORUS   2 μm

Dicot Woods

Wood structure in dicotyledons is much more varied than in conifers, owing in part to a great number of cell types in the axial system, including vessel members, tracheids, several types of fibers, and parenchyma cells (Figures 23–23 and 23–24; see also Figures 20–10 and 23–26). The presence of vessel members, in particular, distinguishes dicot woods from conifer wood.

The rays of dicot woods are often considerably larger than those of conifer wood. In conifer wood, the rays are predominantly one cell wide, and most range from 1 to 20 cells high. The rays of dicot woods range from one to many cells wide and from one to several hundred cells high. In some dicot woods, such as oak, the large rays can be seen with the unaided eye (Figure 23–13). The large rays of the red oak wood illustrated in Figure 23–24c are 12 to 30 cells wide and hundreds of cells high. Besides the large rays, oak wood has numerous rays that are only one cell wide. In red oak wood, the rays make up, on average, about 21 percent of the volume of the wood. Overall, the rays of hardwoods average about 17 percent of the volume of the wood; the average for conifer wood is about 8 percent.

As in conifer wood, transverse sections of dicot woods reveal radial files of cells of both the axial and radial systems derived from the cambial initials (Figures 23–23 and 23–24). The files may not be as orderly as in conifer wood, however, for enlargement of the vessels and elongation of fibers tend to push many of the cells out of position. The displacement of rays by vessel members is particularly conspicuous in the transverse section of red oak (Quercus rubra) wood, as shown in Figure 23–23a.

Growth Rings

The periodic activity of the vascular cambium, which is a seasonally related phenomenon in temperate zones, produces growth increments, or _growth rings_, in both secondary xylem and secondary phloem (in the phloem the increments are not always readily discernible). If a growth layer represents one season's growth, it is called an _annual ring_. Abrupt changes in available water and other environmental factors may be responsible for the production of more than one growth ring in a given year; such rings are called false annual rings. Thus the age of a given portion of the old woody stem can be estimated by counting the growth rings, but the estimates may be inaccurate if false annual rings are included.

The width of individual growth rings may vary greatly from year to year as a function of such environmental factors as light, temperature, rainfall, available soil water, and length of the growing season. The width of a growth ring is a fairly accurate index of the rainfall of a particular year. Under favorable conditions—that is, during periods of adequate or abundant rainfall—the growth rings are wide; under unfavorable conditions, they are narrow.

In semiarid regions, where there is very little rain, the tree is a sensitive rain gauge. An excellent example of this is the bristlecone pine _(Pinus longaeva)_ of the western Great Basin (Figure 23–27). Each growth ring is different, and a study of the rings tells a story that dates back thousands of years. The oldest-known living specimen of bristlecone pine is 4900 years old. Dendrochronologists—scientists who conduct historical research through the growth rings of trees—have been able to match samples of wood from living and dead trees, and in this way they have built up a continuous series of rings dating back more than 8200 years. The

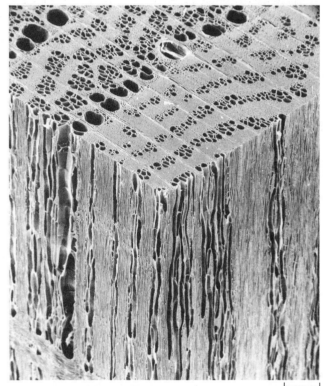

2.5 mm

23–26
Scanning electron micrograph of block of American elm (Ulmus americana) _wood, showing the three faces, or surfaces, of the wood. By comparison with Figures 23–20, 23–23, and 23–24, you should be able to identify each surface. This is a semi-ring-porous wood, with the late wood vessels arranged in wavy lines, a characteristic feature of the elms. Identify the early-wood and late-wood vessels and the rays in all three faces. The dense portion of the wood is composed largely of fibers. Axial parenchyma cells are also present but are not distinguishable at this magnification._

(a)

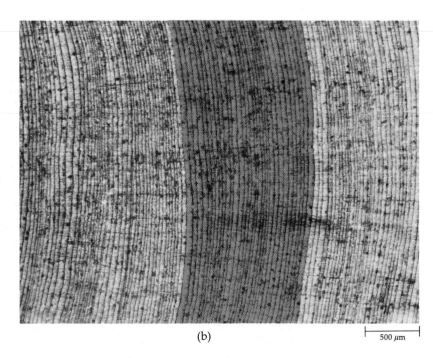

(b)   500 μm

23–27

(a) *Bristlecone pine* (Pinus longaeva) *in
the White Mountains of eastern California. These pines, which grow near the
timberline, are the oldest living trees; one
tree has reached an age of 4900 years.* (b)
*A transverse section of wood from a
bristlecone pine, showing the variation in
width of annual rings. This section
begins approximately 6260 years ago; the*

*band of rings in color represents the
thirty years from 4240* B.C. *to 4210* B.C.
*The overlapping patterns of rings in dead
trees have made it possible to determine
relative precipitation extending back
some 8200 years.*

*Despite its advanced age, the 4900-
year-old bristlecone pine may not be the
oldest living thing on earth. In fact, a*

ring of creosote bushes (Larrea divaricata), *all apparently derived from one
seed, has been estimated to be about
12,000 years old. Growing on the Mojave
Desert 150 miles northeast of Los
Angeles, California, the ring of bushes
has been dubbed the "King Clone." (See
also Figure 32–13.)*

widths of the growth rings of bristlecone pines at the
higher elevations (the upper tree line) have also been
found to be closely related to temperature changes, and
a record of average ring width in these trees provides a
valuable guide to past temperatures and climatic conditions. For example, in the White Mountains of California, the summers were relatively warm from 3500 B.C.
to 1300 B.C., and the tree line was about 150 meters
above its present level. Summers were cool from 1300
B.C. to 200 B.C.

The structural basis for the visibility of growth layers
in the wood is the difference in density of the wood
produced early in the growing season and that produced later (Figures 23–21, 23–23, and 23–24). The
early wood is less dense (with larger cells and proportionally thinner walls) than the *late wood* (with narrower cells and proportionally thicker walls). In a given
growth layer, the change from early to late wood may
be very gradual and almost imperceptible. However,
where the late wood of one growth layer abuts on the
early wood of a following growth layer, the change is
abrupt and thus clearly discernible.

In some dicot woods, size differences of the vessels,
or pores, in early and late woods are quite marked, the

pores of the early wood being distinctly larger than in
the late wood. (The term *pore* is used by the wood anat-
omist for a vessel seen in cross section.) Such woods are
called *ring-porous* woods (Figures 23–23a and 23–24a).
In other dicot woods, the pores are fairly uniform in
distribution and size throughout the growth layer.
These woods are called *diffuse-porous* woods (Figure
23–23b). In ring-porous woods, almost all the water is
conducted in the outermost growth layer, at speeds
about ten times greater than in diffuse-porous woods.

Sapwood and Heartwood

As the wood grows older and no longer serves as a
conducting tissue, its parenchyma cells eventually die.
Before this happens, however, the wood often undergoes visible changes, which involve the loss of reserve
food substances and the infiltration of the wood by
various substances (such as oils, gums, resins, and tannin), which color and sometimes make it aromatic. This
often darker, nonconducting wood is called *heartwood*,
while the generally lighter conducting wood is called
sapwood (see Figure 23–13, page 455). In many woods,
tyloses are formed in the vessels when they become

23–28
Tyloses, balloonlike outgrowths of parenchyma cells, which partially or completely block the lumen of the vessel. (a) Transverse and (b) longitudinal sections showing tyloses in vessels of white oak (Quercus alba), *as seen with a light microscope.*

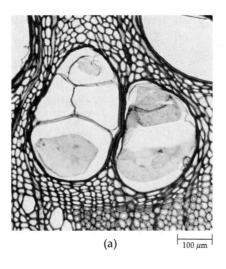

(a)   100 μm

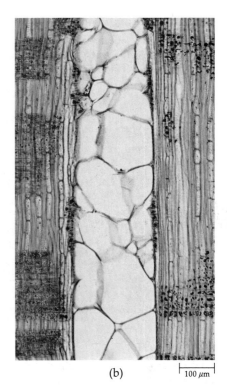

(b)   100 μm

nonfunctional (Figure 23–28). *Tyloses* are balloonlike outgrowths from ray or axial parenchyma cells through pit cavities in the vessel wall. They may completely occlude the lumen (space bounded by the cell wall) of the vessel. Tyloses often are induced to form prematurely or unnaturally by plant pathogens and serve as a defensive mechanism by inhibiting spread of the pathogen throughout the plant via the xylem.

The proportion of sapwood to heartwood and the degree of visible difference between them varies greatly from species to species. Some trees, such as maple (*Acer*), birch (*Betula*), and ash (*Fraxinus*), have thick sapwoods; whereas others, such as locust (*Robinia*), catalpa (*Catalpa*), and yew (*Taxus*), have thin sapwoods. Still other trees, such as the poplars (*Populus*), willows (*Salix*), and firs (*Abies*), have no clear distinction between sapwood and heartwood.

Reaction Wood

Reaction wood is a term applied to abnormal wood that develops in leaning trunks and limbs. Its formation is related to the process of straightening of those plant parts. In conifers, the reaction wood develops on the underside of the leaning part and is called *compression wood*. In dicotyledons, it develops on the upper side and is called *tension wood*.

Compression wood is produced by the increased activity of the vascular cambium on the lower side of the stem and results in the formation of growth rings that are eccentric. Portions of growth rings located on the lower surface are generally much wider than those located on the upper surface (Figure 23–29). Hence, compression wood causes straightening by expanding or pushing the trunk or limb upright. Compression wood has more lignin and less cellulose than normal wood, and its lengthwise shrinkage upon drying is often 10 or more times as great as normal wood. (Normal wood usually shrinks lengthwise not more than 0.1 to 0.3

23–29
Transverse section of the stem of a hemlock, showing compression wood with larger growth rings on the lower side. The cracks are due to drying.

percent.) The difference in relative lengthwise shrinkage of normal and compression wood in a drying board often causes the board to twist and cup. Such wood is virtually useless except as fuel.

Tension wood is produced by the increased activity of the vascular cambium on the upper side of the stem and, as in the case of compression wood, is recognized by the presence of eccentric growth rings (Figure 23–30). To straighten the stem, the tension wood must exert a pull; hence the name "tension wood." Positive identification of tension wood requires microscopic examination of wood sections. Anatomically, the principal distinguishing feature is the presence of gelatinous fibers—fibers with little or no lignification and in which part of the secondary wall has a gelatinous appearance. Lengthwise shrinkage of tension wood rarely exceeds 1 percent, but boards containing it twist out of shape in drying. When sawn green, tension wood tears loose in bundles of fibers, imparting a wooly appearance to the boards.

The Gross Features of Wood

The appearance of wood depends on several factors, among them color, grain, texture, and figure. Not only are some of these characteristics helpful in identification of various kinds of woods, but they are responsible for the decorative qualities of woods.

Color in wood is variable both between different kinds of wood and within a species. Color of heartwood can be important in identification of that wood; in addition, it can be responsible in part for the preferential uses of wood. The rich chocolate or purplish brown heartwood of black walnut (_Juglans nigra_) and the red-brown heartwood of black cherry (_Prunus serotina_) are examples of long-time favorites for high-quality furniture.

Grain in wood is a term used when referring to the direction of alignment of wood elements—fibers, tracheids, parenchyma cells, and vessel members—when considered _en masse_. When all the elements are oriented parallel to the longitudinal axis, the grain is said to be _straight_. When the alignment in a piece of wood does not coincide with the longitudinal axis of the piece, the wood is said to be _cross-grained. Spiral grain_ is applied to the spiral arrangement of elements in a log or tree trunk. The log or trunk with spiral grain has a twisted appearance after the bark has been removed (Figure 23–31). If the orientation of the spiral is reversed at more or less regular intervals along a single radius, the grain is said to be _interlocked_.

Texture of wood refers to the relative size and amount of variation in size of elements within the growth increments. _Coarse_ texture can result from the presence of wide bands of large vessels and broad rays as in some ring-porous woods, and _fine_ texture from the presence of small vessels and narrow rays. _Even_ texture is the result of uniformity in cell dimensions—there is no perceptible difference between the early and late woods. The texture is _uneven_ when distinct differences exist between early and late woods of a growth ring.

Figure refers to the patterns found on the longitudinal surfaces of wood. In a restricted sense, the term "figure" is used to refer to the more decorative woods prized in the furniture and cabinet-making industries. Figure depends on the grain and texture and the orientation of the surface that results from sawing.

23–30

Transverse section of the stem of red oak
(Quercus rubra), _showing tension wood_
with larger growth rings on the upper side.

23–31

Trunk of a dead white oak (Quercus
alba) _tree from which the bark has fallen,_
revealing the spiral grain of the wood.

DENSITY AND SPECIFIC GRAVITY OF WOOD

Density is the single most important indicator of the strength of wood and can be used to predict such characteristics as hardness, nailing resistance, and ease of machining. Dense woods generally shrink and swell more than light woods. In addition, the densest woods make the best fuel.

The specific gravity of a substance is the ratio of the weight of the substance to the weight of an equal volume of water. In the case of wood, the oven-dry weight of the wood is used in figuring its specific gravity:

$$\text{specific gravity} = \frac{\text{oven-dry weight of wood}}{\text{weight of the displaced volume of water}}$$

The specific gravity of dry solid wood substance (that is, oven-dry cell wall material) of all plants is about 1.5. The differences in specific gravity of woods depend, therefore, upon the proportion of the wood composed of wall substance and lumen (the space bounded by the cell wall).

Fibers are especially important in determination of the specific gravity. If the fibers are thick-walled and narrow-lumened, the specific gravity tends to be high. Conversely, if the fibers are thin-walled and wide-lumened, it tends to be low. The presence of numerous thin-walled vessels also tends to lower the specific gravity.

Density is expressed as weight per unit volume, either as pounds per cubic foot (English) or grams per cubic centimeter (metric). Water has a density of 62.4 lb./ft^3 or 1 g/cm^3. A wood weighing 31.2 lb./ft^3 or 0.5 g/cm^3 is, therefore, one-half as heavy as water and has a specific gravity of 0.5. *The Guinness Book of World Records* lists black ironwood (*Olea capensis*) of South Africa as the heaviest (93 lb./ft.^{3}) wood and *Aeschynomene hispida* of Cuba as the lightest (2.75 lb./ft.^{3}) wood. Their respective specific gravities are 1.49 and 0.044. The specific gravities of most commercially useful woods are between 0.35 and 0.65.

Specific Gravities of Some North American Hardwoods	
Ash, white (*Fraxinus americana*)	0.55
Aspen, quaking (*Populus tremuloides*)	0.35
Balsa (*Ochroma lagopus*)	0.12
Basswood, or linden (*Tilia americana*)	0.32
Beech, American (*Fagus americana*)	0.56
Birch, yellow (*Betula lutea*)	0.55
Buckeye, yellow (*Aesculus glabra*)	0.33
Butternut (*Juglans cinerea*)	0.36
Cherry, black (*Prunus serotina*)	0.47
Cottonwood (*Populus deltoides*)	0.37
Elm, American (*Ulmus americana*)	0.46
Hickory, shagbark (*Carya ovata*)	0.64
Honeylocust (*Gleditsia triacanthos*)	0.60
Ironwood, black (*Krugiodendron ferreum*)	1.30
Lignum vitae (*Guaiacum officinale*)	1.25
Locust, black (*Robinia pseudo-acacia*)	0.66
Magnolia, southern (*Magnolia grandiflora*)	0.46
Maple, red (*Acer rubrum*)	0.49
Maple, sugar (*Acer saccharum*)	0.56
Oak, live (*Quercus virginiana*)	0.81
Oak, red (*Quercus rubra*)	0.57
Oak, white (*Quercus alba*)	0.59
Osage-orange (*Maclura pomifera*)	0.76
Persimmon (*Diospyros virginiana*)	0.64
Sweetgum (*Liquidambar styracifula*)	0.46
Sycamore, American (*Platanus occidentalis*)	0.46
Walnut, black (*Juglans nigra*)	0.51
Willow, black (*Salix nigra*)	0.34

Specific Gravities of Some North American Softwoods	
Bald cypress (*Taxodium distichum*)	0.42
Douglas fir (*Pseudotsuga menziesii*)	0.45
Fir, balsam (*Abies balsamea*)	0.34
Hemlock, eastern (*Tsuga canadensis*)	0.38
Hemlock, western (*Tsuga heterophylla*)	0.42
Incense cedar (*Calocedrus decurrens*)	0.35
Larch, western (*Larix occidentalis*)	0.48
Pine, eastern white (*Pinus strobus*)	0.34
Pine, loblolly (*Pinus taeda*)	0.47
Pine, lodgepole (*Pinus contorta*)	0.38
Pine, ponderosa (*Pinus ponderosa*)	0.38
Pine, slash (*Pinus elliottii*)	0.64
Pine, sugar (*Pinus lambertiana*)	0.35
Redwood (*Sequoia sempervirens*)	0.38
Spruce, black (*Picea mariana*)	0.36
Spruce, Englemann (*Picea englemannii*)	0.38
Tamarack (*Larix laricina*)	0.49
Yew, Pacific (*Taxus brevifolia*)	0.60

Table 23–1 *Wood Uses of Some Common North American Trees*

Alder, red (*Alnus rubra*)—principal hardwood of Pacific Northwest. Used for furniture, especially chairs; sash, doors, and other millwork; plywood; charcoal; important source of pulpwood.

Ash, white (*Fraxinus americana*)—handles, especially long ones (shovel, spade, rake) because of its straightness of grain, strength, moderate weight, and other qualities; almost all baseball bats; oars, paddles, tennis-racket frames, hockey sticks; kitchen cabinets; toys and woodenware.

Basswood, or linden (*Tilia americana*)—veneer for plywood used as drawer panels and other concealed furniture parts; excelsior; sash, doors, and other millwork; piano keys; boxes and crates; caskets and coffins.

Beech, American (*Fagus americana*)—one of three important northern hardwoods, the others being yellow birch and sugar maple. Used for planing-mill products, especially flooring; veneer; fuel wood; hardwood distillation yielding acetic acid, methanol, and other chemicals; toys and woodenware.

Birch, yellow (*Betula alleghaniensis*)—veneer; hardwood distillation; railroad ties; furniture; toys and woodenware; musical instruments; toothpicks.

Cherry, black (*Prunus serotina*)—a superb cabinet wood; finished furniture; printers' blocks to which electrotypes are mounted; piano actions; interior trim; paneling; handles; toys and woodenware.

Cottonwood (*Populus deltoides*)—pulpwood for high-grade paper used in books and magazines; concealed parts of furniture; excelsior; tubs and pails for food products; boxes and crates.

Elm (*Ulmus* spp.)—has interlocked grain and, consequently, is difficult to split. Used for staves and hoops; boxes and crates; veneer for fruit and vegetable containers and round cheeseboxes; bent parts in furniture; interior trim.

Hickory, bitternut (*Carya cordiformis*)—tool handles, especially for axes, picks, and sledges; ladders; furniture; woodenware; famous for smoking meats; prime fuel wood.

Locust, black (*Robinia pseudo-acacia*)—mine timbers; railroad ties; fence posts; in construction where strength and durability are of great importance.

Maple, sugar (*Acer saccharum*)—veneer; hardwood distillation; railroad ties; fuel wood; furniture; flooring, especially bowling alleys and dance floors; toys and woodenware; musical instruments.

Oak, red and white (*Quercus rubra* and *Q. alba*)—railroad ties; veneer; flooring; sash, doors, and other millwork; firewood; ship and boat building; caskets and coffins.

Persimmon (*Diospyros virginiana*)—shuttles, spools, and bobbins in weaving industry; golf-club heads; boxes and crates; handles.

Sycamore, American (*Platanus occidentalis*)—has interlocked grain. Used for veneer; boxes and crates; interior trim and paneling; flour and sugar barrels; concealed parts of furniture.

Walnut, black (*Juglans nigra*)—the finest cabinet wood native to the continental United States; veneer for plywood faces used in manufacture of furniture; goes directly into high grade tables and chairs; leading wood for gunstocks; caskets and coffins.

Yellow-poplar (*Liriodendron tulipfera*)—veneer for plywood used for interior finish, furniture, and cabinetwork; pulpwood; boxes and crates; sash, doors, and other millwork.

Douglas fir (*Pseudotsuga menziesii*)—the western forests of the United States are about 50 percent Douglas fir, which furnishes more timber than any other single species grown in the United States. Used for building construction, including veneer converted largely into plywood; railroad ties; mine timbers; pulpwood; boxes and crates; ship and boat building.

Hemlock, eastern (*Tsuga canadensis*)—pulpwood; general construction; boxes and crates; sash and doors; kitchen cabinets.

Pine, western white (*Pinus monticola*)—matches; boxes and crates; building construction; sash, doors, and other millwork; core stock for plywood, especially tabletops.

Pine, ponderosa (*Pinus ponderosa*)—boxes and crates; sash, doors, and other millwork; building construction; turned work (posts, balusters, porch columns); poles; toys; caskets and coffins.

Pine, slash (*Pinus elliottii*)—pulpwood; heavy timbers; railroad ties; veneer; turpentine and rosin; boxes; baskets and crates.

Pine, sugar (*Pinus lambertiana*)—boxes and crates; sash, doors, and other millwork; signs; piano keys and organ pipes.

Redwood (*Sequoia sempervirens*)—building construction of all sorts; ship and boat building; cigar and candy boxes; garden furniture; shingles and shakes; caskets and coffins.

Spruce, red (*Picea rubens*)—most important use is for pulpwood; sounding boards for musical instruments; paddles and oars; ladder rails; ship and boat building; boxes and crates.

Boards are sawn from logs in two ways with reference to the radius of the log (Figure 23–32). In one, the wood is cut such that the broad faces of the board roughly parallel the tangential-longitudinal plane of the log. Tangentially cut boards are said to be *flat-cut* or *plain-sawn.* The growth rings in plain-sawn boards appear as wavy bands. In the other, a radial cut is made longitudinally through the center of the log. Radially cut boards are said to be *quarter-sawn.* In this type of stock the growth rings appear as parallel lines extending the length of the board with the rays crossing the growth increments at right angles. Quarter-sawn boards are preferred in many applications because the radial surfaces have more uniform wearing and finishing properties. Quartersawing is more time-consuming, however, and often more wasteful than plainsawing.

SUMMARY

Secondary growth (the increase in girth in regions that are no longer elongating) occurs in all gymnosperms and in most dicotyledons and involves the activity of the two lateral meristems—the vascular cambium and the cork cambium, or phellogen. Herbaceous plants may undergo little or no secondary growth, whereas woody plants—trees and shrubs—may continue to increase in thickness for many years. Figure 23–33 presents summaries of root and stem development of a woody plant, beginning with the apical meristem and ending with the secondary tissues produced during the first year's growth.

23–32
Diagram illustrating the direction in which logs are sawn to produce plain-sawn (A) and quarter-sawn (B) boards. In plain-sawn boards the growth rings are more or less parallel to the broad faces. In quarter-sawn boards, the rings are more or less at right angles to the broad faces.

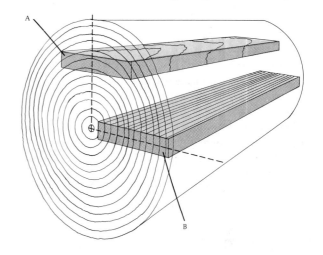

23–33
Summary of root and stem development in a woody dicot during the first year of growth.

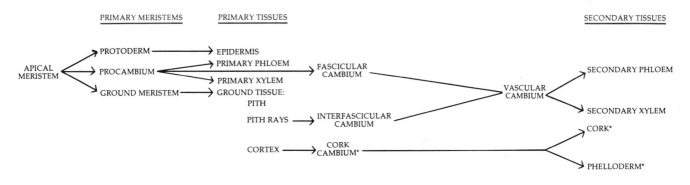

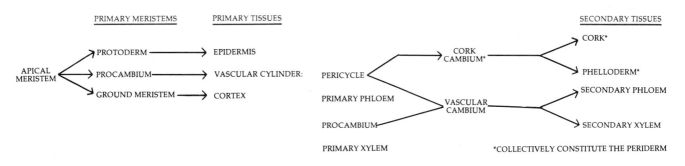

The vascular cambium contains two types of initials—fusiform initials and ray initials. Through periclinal divisions, the fusiform initials give rise to the components of the axial system, and the ray initials produce ray cells, which form the vascular rays, or radial system. Increase in circumference of the cambium is accomplished by anticlinal division of the initials.

The first cork cambium in most stems originates in a layer of cells immediately below the epidermis. In the root, the first cork cambium arises in the pericycle. The cork cambium produces cork toward the outside, and phelloderm toward the inside. Together, the cork cambium, cork, and phelloderm comprise the periderm. Although most of the periderm consists of compactly arranged cells, isolated areas called lenticels have numerous intercellular spaces.

The bark consists of all tissues outside the vascular cambium. In old roots and stems, most of the phloem comprising the bark is nonfunctional. Sieve elements are short-lived, and generally only the present year's growth increment contains conducting, or functional, sieve elements. After the first periderm, subsequently formed periderms originate deeper and deeper in the bark from parenchyma cells of nonfunctional phloem.

Woods are classified as either softwoods or hardwoods. All so-called softwoods are conifers and all so-called hardwoods are dicotyledons. Compared with dicot woods, those of conifers are simple, consisting of tracheids and parenchyma cells. Some contain resin ducts. Dicot woods may contain a combination of all of the following cell types: vessel members, tracheids, several types of fibers, and parenchyma cells.

Growth layers that correspond to yearly increments of growth are called annual rings. The difference in density between the late wood of one growth increment and the early wood of the following increment makes it possible to distinguish the growth layers. Density and specific gravity are good indicators of the strength of wood.

In many plants, the nonconducting heartwood is visibly distinct from the actively conducting sapwood.

Commonly, reaction wood develops on the underside of leaning trunks and limbs of conifers and on the upper side of similar parts in dicotyledons; its formation causes straightening of the trunk or limb. Reaction wood is called compression wood in conifers and tension wood in dicots.

SUGGESTIONS FOR FURTHER READING

CORE, HAROLD A., WILFRED A. CÔTÉ, and ARNOLD C. DAY: *Wood Structure and Identification*, 2nd ed., Syracuse University Press, Syracuse, N.Y., 1979.*

A beautifully illustrated manual of wood structure, with a key for wood identification and an illustrated glossary.

CUTLER, DAVID F.: *Applied Plant Anatomy*, Longman Inc., New York, 1978.*

An interestingly written textbook on the fundamentals of plant anatomy showing some of the ways in which plant anatomy can be applied to solve many important everyday problems.

CUTTER, ELIZABETH G.: *Plant Anatomy*, Part I: *Cells and Tissues*, 2nd ed., Addison-Wesley Publishing Co., Inc., Reading, Mass., 1978.

An introduction to plant cells and tissues, with emphasis on experimental work.

CUTTER, ELIZABETH G.: *Plant Anatomy*, Part II: *Organs*, Addison-Wesley Publishing Co., Inc., Reading, Mass., 1971.*

An introduction to the organs of the plant and possible causes underlying their development.

EPSTEIN, EMANUEL: "Roots," *Scientific American* 228(5): 48–58, 1973.

A discussion of root structure and function, including a discussion of the mechanism of ion uptake by the root.

ESAU, KATHERINE: *Plant Anatomy*, 2nd ed., John Wiley & Sons, Inc., New York, 1965.

The standard work in the field; a well-illustrated book that considers all aspects of plant anatomy.

ESAU, KATHERINE: *Anatomy of Seed Plants*, 2nd ed., John Wiley & Sons, Inc., New York, 1977.

A shorter book than the preceding entry; an excellent textbook and reference.

FAHN, ABRAHAM: *Plant Anatomy*, 3rd ed., Pergamon Press, Inc., Elmsford, N.Y., 1982.*

A well-illustrated, up-to-date textbook considering all aspects of plant anatomy.

HOADLEY, R. B.: *Understanding Wood*, The Taunton Press, Newtown, Conn., 1980.

Written by a wood technologist and woodworker, this well-illustrated book tells all about the properties of wood and how to work with it.

JOHRI, B. M. (Ed.): *Embryology of Angiosperms*, Springer-Verlag, Berlin, 1984.

A multiauthored text considering the significant advances made during the last 20 to 30 years in virtually all areas of reproduction in flowering plants.

O'BRIEN, TERENCE P., and MARGARET E. MCCULLY: *Plant Structure and Development*, The Macmillan Company, New York, 1969.

A pictorial and physiological approach to plant structure and development.

O'BRIEN, TERENCE P., and MARGARET E. MCCULLY: *The Study of Plant Structure: Principles and Selected Methods*, Termarcarphi and Pty. Ltd., Melbourne, Australia, 1981.

A very useful book dealing with methods for the study of plant structure, ranging from fresh to variously fixed tissues studied with simple lenses, at one extreme, and the electron microscope, at the other.

PANSHIN, A. J., and C. DE ZEEUW: *Textbook of Wood Technology*, vol. 1, 4th ed., McGraw-Hill Book Company, New York, 1980.

A technical reference on the identification, structure, and utilization of woods native to North America.

* Available in paperback.

RAY, PETER M.: *The Living Plant*, 2nd ed., Holt, Rinehart & Winston, Inc., New York, 1972.*

A short, readable account of plant growth and development by one of the leading workers in the field.

STEEVES, TAYLOR A., and IAN M. SUSSEX: *Patterns in Plant Development*, Prentice-Hall, Inc., Englewood Cliffs, N.J., 1972.

A structural approach to plant development, with emphasis on experimental and analytical data.

TORREY, JOHN G., and DAVID T. CLARKSON (Eds.): *The Development and Function of Roots*, Academic Press, Inc., New York, 1975.

A collection of articles by recognized specialists in the fields of root structure and function.

WARDLAW, CLAUDE W.: *Morphogenesis in Plants: A Contemporary Study*, 2nd ed., Methuen & Co. Ltd., London, 1968.

A fascinating book by an acknowledged master of the field.

WHITE, RICHARD A., and WILLIAM C. DICKINSON (Eds.): *Contemporary Problems in Plant Anatomy*, Academic Press, Inc., Orlando, Florida, 1984.

A collection of papers presented at a plant anatomy symposium held at Duke University and the University of North Carolina in early 1983, dealing with problems in contemporary anatomy. The central theme of all contributions is to view the plant as a developmental, structural, and functional whole.

ZIMMERMANN, MARTIN H.: *Xylem Structure and the Ascent of Sap*, Springer-Verlag, New York, 1983.

A delightfully written "idea" book on xylem structure and function by one who contributed a great deal to our understanding of functional xylem anatomy and sap movement in plants.

ZIMMERMANN, MARTIN H., and CLAUDE L. BROWN. *Trees: Structure and Function*, Springer-Verlag, New York, 1975.

An up-to-date discussion of how trees work, with emphasis on structure as it relates to function.

ZIMMERMANN, MARTIN H., and JOHN A. MILBURN (Eds.): *Transport in Plants I, Phloem Transport, Encyclopedia of Plant Physiology*, vol. I, Springer-Verlag, New York, 1976.

A collection of articles by leading researchers in the field of phloem structure and function.

* Available in paperback.

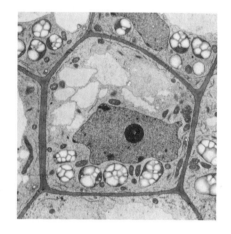

S E C T I O N   S I X

Growth Regulation and
Growth Responses

24–1

Photoperiodism is the phenomenon that brings the plants of many species into flower simultaneously, year after year. Flowering is induced by the interaction of external (environmental) and internal (hormonal) factors. This is a cherry tree (Prunus) blooming in the spring in Brooklyn, New York.

CHAPTER 24

Regulating Growth and Development: The Plant Hormones

A plant, in order to grow, needs light from the sun, carbon dioxide from the air, and water and minerals, including nitrogen, from the soil. From these things, it makes more of its own substance, turning simple materials into the complex organic molecules of which living things are composed. As discussed in the previous section, the plant does far more than simply increase its mass and volume as it grows. It differentiates, develops, and takes shape, forming a variety of cells, tissues, and organs. How can a single cell, the fertilized egg, be the source of the myriad tissues and organs that make up the extraordinary individual known as a "normal" plant? Many of the details of how these processes are regulated are not known, but it has become clear that normal development depends on the interplay of a number of internal and external factors. The principal internal factors that regulate plant growth and development are chemical, and they are the subject of this chapter. Some of the external factors—light, temperature, day length, gravity, and so on—that affect plant growth are discussed in Chapter 25.

Plant hormones play a major role in regulating growth. The term "hormone" was coined by investigators studying animal physiology; it refers to organic substances that are produced in one tissue and transported to another tissue, where their presence results in physiological responses. Hormones are active in very small quantities. In the shoot of a pineapple plant (*Ananas comosus*), for example, there are only 6 micrograms of indoleacetic acid, a common plant hormone, per kilogram of plant material. One enterprising plant physiologist calculated that the weight of the hormone in relation to that of the shoot is comparable to the weight of a needle in 20 metric tons of hay.

The word *hormone* comes from the Greek word *hormaein*, meaning "to excite." It is now clear, however, that many hormones have inhibitory influences. Therefore, rather than thinking of hormones as stimulators, it is perhaps more useful to consider them as chemical regulators. But this term also needs qualification, for the response to the particular "regulator" depends not only on its content (chemical structure) but on how it is "read" by the recipient (tissue specificity).

The following discussion reflects the historical development of our knowledge of plant hormones. Accordingly, it begins with auxin, which made the concept of hormones in plants a valid one.

AUXIN

Some of the first recorded experiments on growth-regulating substances were performed by Charles Darwin and his son Francis and were reported in *The Power of Movement in Plants*, published in 1881. The Darwins first made systematic observations of the bending or curving of plants toward light, which is known as phototropism (see page 493), using seedlings of canary grass (*Phalaris canariensis*) and oats (*Avena sativa*). They then showed that if they covered the upper portion of the *coleoptile* (the sheathlike, protective structure covering the shoot of grass seedlings) with a cylinder of metal foil or a hollow tube of glass blackened with India ink and exposed the plant to a lateral light (that is, light from the side), the characteristic bending did not occur (Figure 24–2). If, however, the tips were enclosed in transparent glass tubes, bending occurred normally. To the Darwins these experiments indicated the existence of a "communication" between the tip, that is, the tissue receiving the light stimulus, and the rest of the coleoptile, where the growth response occurs. They further stated, "We must therefore conclude that when seedlings are freely exposed to a lateral light some influence is transmitted from the upper to the lower part, causing the latter to bend."

In 1926, the Dutch plant physiologist Frits W. Went succeeded in isolating this "influence" from the coleoptile tips. Went cut off the coleoptile tips from a number of oat (*Avena*) seedlings and placed them for about an hour on a slice of agar with their cut surfaces in contact with the agar. (Agar is a gelatinous substance derived from certain red algae; it is commonly used as a neutral growth medium.) He then cut the agar into small blocks and placed the blocks on one side of the stumps of the decapitated plants, which were kept in the dark during the entire experiment. Within one hour, he observed a distinct curving *away* from the side on which the agar block was placed (Figure 24–3). Agar blocks that had not been exposed to a coleoptile tip produced either no curving or a slight curving toward the side on which the agar block had been placed. Agar blocks that had been exposed to a section of coleoptile from lower on the shoot also produced no physiological effect.

By these experiments, Went showed that the coleoptile tip exerted its effect by means of a chemical substance rather than a physical stimulus, such as an electrical one. Went named this chemical substance *auxin*, from the Greek word *auxein*, meaning "to increase."

24–2

The Darwins' experiment. (a) Seedlings normally bend toward the light. (b) When the tip of a seedling was covered by a light-proof collar, this bending did not occur. (Bending occurred normally when the tip of a seedling was covered with a transparent collar.) (c) When the collar was placed below the tip, the characteristic light response took place. From these experiments, the Darwins concluded that, in response to light, an "influence" that causes bending was transmitted from the tip of the seedling to the area below the tip, where bending normally occurs.

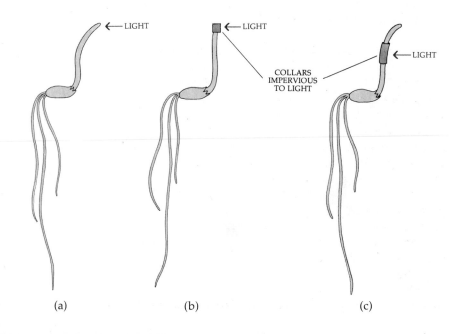

←LIGHT   ←LIGHT   ←LIGHT

COLLARS
IMPERVIOUS
TO LIGHT

(a)   (b)   (c)

24-3

Went's experiment. (a) Went removed the coleoptile tips from seedlings and placed the tips on agar for about an hour. (b) The agar was then cut into small blocks and placed on one side of the decapitated shoots of the seedlings. (c) The seedlings, which were kept in the dark during the entire experiment, subsequently were observed to bend away from the side on which the agar block was placed. From this result, Went concluded that the "influence" that caused the seedling to bend was chemical and that it accumulated on the side away from the light.

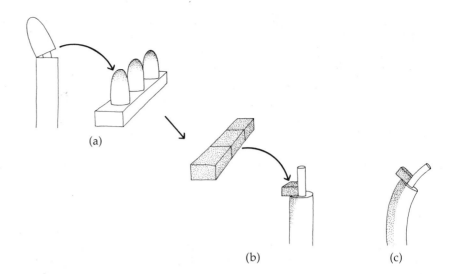

(a)

(b)   (c)

The curvature of an *Avena* coleoptile away from the side bearing an auxin-containing agar block is caused by asymmetric distribution of auxin, which in turn causes asymmetric increase in cell size in the coleoptile; the cells on the side bearing the auxin-containing block undergo a greater increase in length than those on the opposite side. The conditions for handling *Avena* seedlings and placing the agar block have been standardized to such an extent that the angle of curvature (measured with a protractor) can be used to determine the quantity of auxin in the agar block. Use of this technique, known as the *Avena* curvature bioassay, enabled investigators to isolate and identify the naturally occurring auxin, which is called indoleacetic acid (abbreviated IAA). (A *bioassay* is a method of quantitatively determining the concentration of a substance by its effect on the growth of a suitable organism under controlled conditions.)

As can be seen in Figure 24-4, IAA closely resembles the amino acid tryptophan (see Figure 3-14, page 52). Tryptophan is the precursor from which IAA is formed in the living plant, although four different pathways are known for IAA biosynthesis, each with different intermediates. Different plant groups use different metabolic routes to produce IAA from tryptophan. In addition, some plants, such as corn (*Zea mays*), use different pathways at different stages of development. Auxin is produced in the coleoptile tips of grasses and in shoot tips. Although IAA has been found in root tips, most evidence indicates that it is transported there via the vascular cylinder from the base of the root. It is probably abundant in embryos and is also found in young leaves and in flowers and fruits.

Shortly after the discovery of auxin and the recognition of its role in regulating cell elongation, an inhibitory effect of auxin was discovered in relation to the growth of lateral buds. For instance, if the apical meristem of a bean plant is removed, the lateral buds begin

CH₂—COOH

INDOLE RING   ACETIC ACID SIDE CHAIN

INDOLEACETIC ACID (IAA)

O—CH₂—COOH

Cl   Cl

2,4-DICHLOROPHENOXYACETIC ACID (2,4-D)

CH₂—COOH

α-NAPHTHALENACETIC ACID (NAA)

24-4

Indoleacetic acid (IAA) is the best-known naturally occurring auxin. Dichlorophenoxyacetic acid (2,4-D), a synthetic auxin, is widely used as an herbicide. Naphthalenacetic acid (NAA), another synthetic auxin, is commonly employed to induce the formation of adventitious roots in cuttings and to reduce fruit drop in commercial crops. The synthetic auxins, unlike IAA, are not readily broken down by natural plant enzymes and microbes and so are better suited for commercial purposes.

to grow. However, if auxin is immediately applied to the cut surface, the growth of the buds is inhibited. As with the phototropic response of *Avena* seedlings, this *apical dominance* demonstrates a form of "communication" between two plant tissues mediated by IAA. Furthermore, in both phenomena the "influence" moves from the growing tip to the base of the plant. This is because auxin is actively transported in plants from the tips of shoots to the base of the plant; that is, in a basipetal direction. The movement usually occurs in the tissue as a whole, rather than in the conduits (vessels and sieve tubes, respectively) of the xylem and phloem. The transport process is presumed to involve an interaction between IAA and the plasma membranes of the plant cells.

Application of auxin to plants causes a variety of effects, which differ from time to time, species to species, and most particularly, from tissue to tissue. Like many other physiologically active compounds, auxin is toxic at high concentrations. The weed killer 2,4-D is a synthetic auxin, one of many that have been manufactured commercially for a wide variety of applications (Figure 24–4).

Auxin and Cell Differentiation

Auxin influences the differentiation of the vascular tissue in the elongating shoot. If one cuts a wedge out of a stem of *Coleus* in such a way as to sever and remove portions of the vascular bundles, new vascular tissues will be formed from cells in the pith and will connect with the bundles in the uncut regions. If one takes away the leaves and buds above the excision, the formation of new cells is delayed. If one adds IAA to the petiole of the cut leaf just above the excision, formation of vascular tissue resumes. Auxin similarly plays an important role in the joining of vascular traces from developing leaves to the bundles in the stem.

Similar effects are seen in calluses. (A *callus* is a mass of undifferentiated cells that forms when a plant is wounded or when isolated cells are grown in tissue culture). If one takes a callus of lilac (*Syringa*) pith and grafts a bud onto it, vascular tissue is induced in the callus. Similarly, if the callus is grown in a medium containing auxin and sugar (the sugar is necessary because the callus does not contain photosynthetic cells), vascular tissues form. And, as discovered by R. H. Wetmore and his co-workers, by adjusting the amount of sugar in the medium, one can induce formation of xylem alone, xylem and phloem, or phloem alone. A low concentration of sucrose (1.5 to 2.5 percent) favors xylem, 4 percent favors phloem, and a percentage in between produces both. This illustrates the subtlety of growth regulator function and also draws attention to the critical fact that growth regulators *never* act alone; rather, they operate in concert with other internal factors such as sugar or other regulators of plant growth.

Auxin and the Vascular Cambium

In woody plants, auxin promotes the growth of the cambium. When the meristematic region of the shoot begins to grow in the spring, auxin moving down from the shoot tip stimulates cambial cells to divide, forming secondary phloem and secondary xylem. Again, these effects are modulated by other growth-regulating substances in the plant body.

Experiments with externally applied IAA and gibberellic acid indicate that, in the intact plant, interactions between auxins and gibberellins determine the relative rates of production of secondary phloem and secondary xylem. For example, IAA and gibberellic acid individually stimulate cambial activity when applied to a wide

24–5
The holly cuttings in the upper row were treated with auxin 21 days before the photograph was taken. The cuttings in the lower row were not. Note the growth of adventitious roots in the treated plants.

variety of woody plants. However, IAA in the absence of gibberellic acid stimulates only xylem development. Phloem development occurs with gibberellic acid alone, but both xylem and phloem development reach a maximum in the presence of both IAA and gibberellic acid.

Auxin and Root Growth

The first practical application of auxin involved its promoting effect on the formation of adventitious roots in cuttings (Figure 24–5). The practice of dipping cuttings in auxin solutions is commercially important, especially for the cultivation of woody plants that are produced on a large scale by vegetative propagation in nurseries. Application of large quantities of auxin to already growing roots, however, usually inhibits their growth.

Auxin and Fruit Growth

Auxin promotes the growth of fruit. Ordinarily, if the flower is not pollinated and fertilization does not take place, the fruit will not develop. In some plants, fertilization of one egg cell is sufficient for normal fruit development, but in others, such as apples or melons, which have many seeds, several must be fertilized for the ovary wall to mature and become fleshy. By treating the female flower parts of certain species with auxin, it is possible to produce parthenocarpic fruit (from *parthenos*, meaning "virgin"), which is fruit produced without fertilization, such as seedless tomatoes, cucumbers, and eggplants.

Apparently, developing seeds are a source of auxin. In the strawberry (*Fragaria ananassa*), if the seeds enclosed in the fruits, which are achenes, are removed during the fruit's development, the fruit stops growing altogether. If a narrow ring of seeds is left, the fruit (actually the fleshy receptacle) forms a bulging girdle of growth in the area of the seeds. If auxin is applied to the denuded receptacle, growth proceeds normally (Figure 24–6).

Auxin and Abscission

Auxin is produced in young leaves, but it does not appear to have direct effects on the rate of leaf growth. However, auxin does affect *abscission*—the dropping of leaves or other plant parts. As leaves grow older, certain reusable ions and molecules are returned to the stem, among them magnesium ions, amino acids (derived from proteins), and sugars (some derived from starch). Then, in some plants at least, enzymes break down the cell walls in the separation layer of the abscission zone, across the base of the petiole (see Figure 22–32, page 435). The cell wall changes may include weakening of the middle lamella and hydrolysis of the cellulosic walls themselves. Cell division may precede the actual separation. If cell division occurs, the newly formed cell walls are the ones largely affected by the degradation phenomena. Beneath the separation layer, a protective layer composed of heavily suberized cells is formed, further isolating the leaf from the main body of the plant before it drops. Eventually, the leaf is held to the plant by only a few strands of vascular tissue, which may be broken by the enlargement of parenchyma cells in the separation layer. Abscission has been correlated with a diminished production of auxin in the leaf, among other factors; under many circumstances, abscission can be prevented by the application of auxin.

(a) (b) (c) (d) (e)

24–6
(a) *Normal strawberry (Fragaria ananassa).* (b) *Strawberry from which all the seeds were removed.* (c) *Strawberry in which three horizontal rows of seeds were left.* (d) *Growth in strawberry induced by one developing seed.* (e) *Growth induced by three developing seeds. If a paste containing auxin is applied to the strawberry from which the seeds have been removed, the strawberry grows normally.*

The control of abscission of leaves, flowers, and fruits is extremely important in agriculture. Auxin and, more recently, ethylene have been used commercially for treatment of a number of plant species. For instance, auxin prevents leaf and berry drop from evergreen holly (*Ilex aquifolium*) and therefore minimizes losses during shipment. Auxin also prevents preharvest drop of citrus fruits, a commercially important effect in orange and grapefruit production. On the other hand, large amounts of auxin, by causing elevated production of ethylene, promote fruit drop. Based on this, auxin has been used for the thinning of fruit in the production of olives, apples, and other tree fruits.

Auxin and the Control of Weeds

Synthetic auxins have been used extensively for the control of weed pests in agricultural lands. In economic terms, this is the major practical use for plant growth regulators in the world today. Although a number of compounds can be employed, phenoxy auxins such as 2,4-D and its chemical derivatives are important weed control agents in that they represent approximately 20 percent of the total used. The continued use of these chemicals will depend on a number of factors, including cost-effectiveness and potential or real hazards to human health.

How Does Auxin Control Elongation?

Auxin increases the plasticity of the cell wall. When the cell wall softens, the cell enlarges because of its turgor pressure. As the turgor pressure is reduced by expansion of the cell, the plant cell takes up more water, and the cell thus continues to enlarge until it encounters sufficient resistance from the wall. (The factors involved in the uptake of water are discussed in Chapter 4.)

The softening of the cell wall is brought about by a complex series of interactions. One of the earliest effects of auxin treatment involves changes in cell metabolism that cause a rapid pumping of protons across the plasma membrane. The resulting acidification of the cell wall leads, by an unknown mechanism, to hydrolysis of restraining bonds within the wall and, consequently, to cell elongation driven by the cell's turgor pressure. Proof of this so-called *acid growth hypothesis* has been an important recent advance in the understanding of auxin action. It is supported by several lines of experimental evidence, including the demonstrated effect of auxin on proton excretion, the fact that acidic (pH 5.0–5.5) solutions can cause an elongation response mimicking auxin treatment, and the observation that neutral buffers, by preventing wall acidification, will inhibit auxin's effect on cell elongation.

Even though the acid growth hypothesis accounts for the initial elongation response to auxin, it cannot explain the continued effect of auxin treatments on plants. In fact, most investigators now agree that auxin has two different effects on cell elongation: a rapid, short-term effect caused by acid growth and a second continued effect due to the regulation of gene expression. In fact, it has been shown that auxin causes the expression of at least 10 specific genes, all presumably involved in the growth effect. It is further known that the step affected by auxin is transcription. Auxin's effect on gene expression in plants appears to be similar to that of several hormones in animals.

CYTOKININS

The discovery of auxin stimulated investigators to look for other types of chemicals regulating plant growth because, by analogy with animals, it seemed unlikely that growth and development in plants would be under the control of a single hormone. For instance, it was known that auxin often inhibited lateral bud growth when applied to decapitated plants. Were there natural hormones that counteracted the auxin effect?

Folke Skoog and his colleagues at the University of Wisconsin developed a method for studying hormonal regulation of bud growth in isolated plant tissues and organs in test tubes. When a stem segment of the tobacco plant (*Nicotiana tabacum*) was placed on a tissue culture medium containing sugar, vitamins, and various salts, cell division occurring at the cut surface resulted in the formation of an undifferentiated callus and, occasionally, an organized shoot that grew out as a leafy stem. Addition of auxin to the medium in sufficient amounts inhibited bud formation and growth, as expected. On the other hand, adenine promoted bud formation in the tissue culture and counteracted the inhibitory effects of auxin. The concentration of adenine required for bud formation was very high, however—too high for it to be considered a hormone—so Skoog and his colleagues set out to identify substances with greater biological activity in the test systems.

Initially, Skoog and his colleagues worked with the coconut (*Cocos nucifera*) because Johannes van Overbeek, and later F. C. Steward and his research group, had shown that coconut milk (which is a liquid endosperm) is a rich source of growth-promoting substances for tissue cultures. After many years of effort, Skoog and his co-workers succeeded in producing a thousandfold purification of a growth factor, but they could not isolate it. So, changing course, they tested a variety of purine-containing substances—largely nucleic acids—in the hope of finding a new source of the material.

Pursuing this new course, Carlos O. Miller, then a postdoctoral fellow in Skoog's laboratory, searched for bottles with a nucleic acid label. He tested one marked

"Herring Sperm DNA" and found that it caused tobacco cells to divide. More herring sperm DNA was ordered. To the consternation of the investigators, these fresh preparations did not work. As a last resort, they tried another old sample. This one, too, was active. Apparently, the factor was to be found only as a breakdown product of DNA. Subsequently, they found that a variety of preparations of DNA that had aged, or had been "aged" artificially by heating in acid solution, contained material that produced the cell-division response.

Miller, Skoog, and their co-workers eventually succeeded in isolating the growth factor from one of these DNA preparations and identifying its chemical nature. They called this substance kinetin and named the class of growth regulators to which it belongs the *cytokinins*, because of its involvement in cytokinesis, or cell division. As shown in Figure 24–7, kinetin resembles the purine adenine, which was the clue that led to its discovery. Kinetin, which probably does not exist in plants in nature, has a relatively simple structure, and biochemists were soon able to synthesize a number of other, related compounds that behaved like cytokinins. Eventually, a natural cytokinin was isolated from kernels of corn (*Zea mays*); called *zeatin*, it is the most active of the naturally occurring cytokinins, although a few still more active synthetic cytokinins have now been produced.

Cytokinins have now been isolated from many different species of seed plants, where they are found primarily in actively dividing tissues, including seeds, fruits, and roots. These hormones have also been found in bleeding sap—the sap that drips out of pruning cuts, cracks, and other wounds in many types of plants. Recently, cytokinins have been identified in two seedless vascular plants, a horsetail (*Equisetum arvense*) and the fern *Dryopteris crassirhizoma*.

Although practical applications for cytokinins are not as extensive as those for auxin, they have been important in plant development research. Inasmuch as cytokinins are central to tissue culture methods (see essay, pages 482 and 483), they undoubtedly will be extremely important for biotechnology in the future. Treatment of lateral buds with cytokinin often causes the buds to grow, even in the presence of auxin, thus modifying apical dominance.

Cytokinins and Cell Division

Studies of interactions involving auxin and cytokinins are helping physiologists understand how plant hormones work to produce the total growth pattern of the plant. Apparently, the undifferentiated plant cell has two courses open to it: either it can enlarge, divide, enlarge, and divide again, or it can elongate without undergoing cell division. The cell that divides repeatedly

24–7

Note the similarities between the purine adenine and these four cytokinins. Kinetin and 6-benzylamino purine (BAP) are synthetic cytokinins that are commonly used. Zeatin and i^6 Ade have been isolated from plants.

remains essentially undifferentiated, or embryonic, whereas the elongating cell tends to differentiate, or become specialized. In studies of tobacco stem tissues, the addition of IAA to the tissue culture produced rapid cell expansion, so that giant cells were formed. Kinetin alone had little or no effect. IAA plus kinetin resulted in rapid cell division, so that large numbers of relatively small, undifferentiated cells were formed. In other words, the addition of the kinetin, together with IAA (although not kinetin alone), switched the cells to a meristematic course.

In another tissue culture study, using tuber tissue of the Jerusalem artichoke (*Helianthus tuberosus*), it was shown that a third substance, the calcium ion, can modify the action of the auxin-cytokinin combination. In this study, IAA plus low concentrations of kinetin was shown to favor cell enlargement, but as Ca^{2+} was added to the culture, there was a steady shift in the growth pattern from cell enlargement to cell division. High concentrations of calcium prevent the cell wall from expanding, and at such concentrations the cell switches course and divides. Thus, not only do hormones modify the effects of hormones, but these combined effects are, in turn, modified by nonhormonal factors, such as calcium.

PLANT HORMONES, TISSUE CULTURE, AND BIOTECHNOLOGY

Advances in hormone research and DNA biochemistry have made it possible to manipulate the genetics of plants in specific ways. The term *biotechnology* is used to describe practical application of this capability. Among the most important procedures in biotechnology is tissue culture, which is feasible only because of our understanding of hormones that regulate development. In the ideal cases, tissue culture is used to obtain complete plants starting with single, genetically altered cells. Tissue culture can then be further used to produce as many identical copies (*clones*) of these plants as desired within a fairly short period of time.

At the present time, the greatest impact of tissue culture is in the area of plant multiplication, referred to as clonal propagation since the individuals produced are genetically identical. Basically, there are three different methods of tissue culture propagation, each taking advantage of the control of development that can be accomplished with plant hormones:

1. *Regeneration from callus and/or protoplasts.*   Plants that can be multiplied by this method represent the optimum organisms for biotechnology. Based on research conducted by Folke Skoog and his collaborators, this procedure uses cytokinin to cause the formation of growing shoots in unorganized masses of parenchymatous tissue known as callus. Once these form, the shoots can be treated with auxin to cause root formation, and the resulting plants can be transferred to soil. Regeneration from callus was demonstrated first with tobacco and has since been shown with related plants such as potato and petunia in the family Solanaceae. Even though a few other plants (e.g., sunflowers, snapdragons, and mustards) can also be regenerated from callus, the technique is still quite limited.

The most elegant use of tissue culture for biotechnology in recent years has involved plant regeneration from protoplasts that have had their walls removed by enzymatic digestion. Protoplasts can be used for plant genetics in two different ways. In one of these methods, known as protoplast fusion, protoplasts from two different plants are allowed to fuse together, producing a hybrid cell. The first plant created by this technique was a hybrid of tobacco. Interspecific hybrid plants have also been produced by protoplast fusion within the genera that include petunia, carrot, and potato. In addition, intergeneric hybrid plants have been obtained by protoplast fusion between the potato and tomato, both of which are members of the nightshade family (Solanaceae).

The second use of protoplasts in plant genetics employs the Ti plasmid from the crown gall bacterium *Agrobacterium tumefaciens* or some other DNA injection technique to introduce specific genes into protoplasts. Tissue culture is then utilized to regenerate plants from the individual protoplasts. This type of genetic manipulation has already been demonstrated in producing herbicide-resistant plants of tobacco and sunflower, and it is discussed further in Chapter 30.

2. *Somatic embryogenesis.*   In an important experiment performed by F. C. Steward of Cornell University, tissue culture was used to produce carrot embryos from single root cells. The procedure, known as *somatic embryogenesis,* because embryos arise from vegetative (i.e., somatic) cells, is a striking example of the *totipotency* of plant cells. Totipotency is the inherent capability of a single cell to provide the genetic program needed to direct the development of an entire individual. Somatic embryogenesis also represents a versatile tool for clonal propagation, inasmuch as it has been utilized successfully for a wide variety of plants, for example, corn, wheat, and sorghum.

3. *Shoot multiplication.*   This procedure takes advantage of the established role of cytokinin as a shoot growth factor and as an antagonist of auxin in the control of apical dominance. Growing shoot tips are exposed in tissue cultures to a high enough cytokinin concentration not only to sustain shoot growth but also to stimulate the development of lateral buds in the axils of leaves. By this technique, one can produce several shoots from one shoot during just a few weeks of growth in tissue culture. After subdivision, individual shoots can be used either for further shoot multiplication or they can be rooted using an auxin treatment. Toshio Murashige of the University of California at Riverside, in particular, has exploited this method of plant propagation. It is used commonly for multiplying ferns, orchids, and woody plants in the families Ericaceae (the heath family, including rhododendrons, azaleas, and mountain laurels) and Rosaceae (the rose family, including apples, roses, and strawberries).

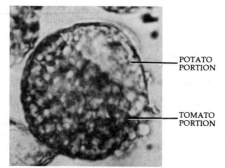 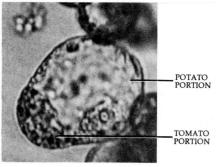

Protoplasts formed by the fusion of tomato and potato protoplasts. The tomato portion of the protoplasts can be recognized by the green chloroplasts, which are present in isolated tomato mesophyll protoplasts. The potato portion is colorless, because the isolated potato protoplasts contain only proplastids. (Research by Georg Melchers, Maria D. Sacristán, and Anthony A. Holder. Professor Melchers is at the Max-Planck-Institut für Biologie, Tübingen, Federal Republic of Germany.)

POTATO PORTION

TOMATO PORTION

POTATO PORTION

TOMATO PORTION

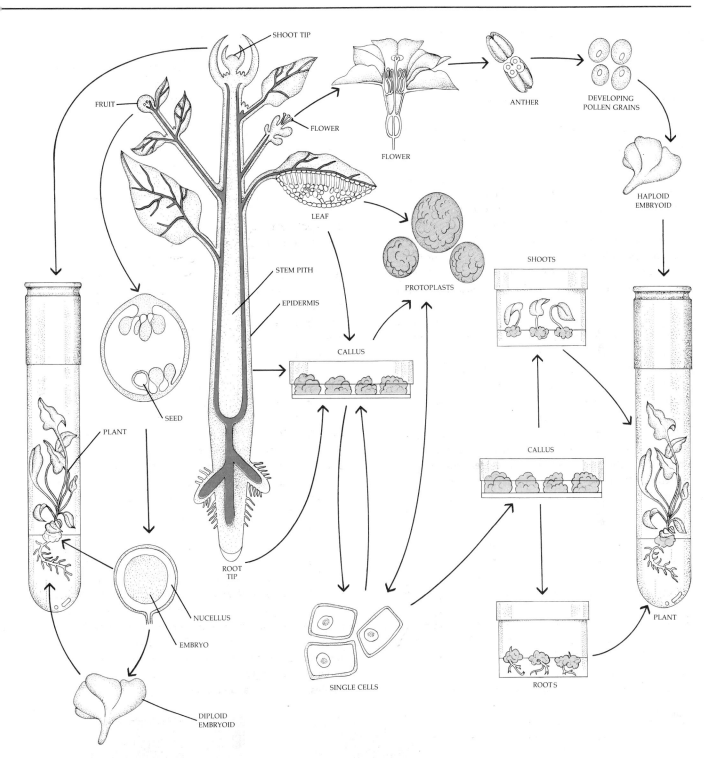

Theoretically, any plant cell, unless enucleate or enclosed by a rigid, lignified wall, is potentially capable of regenerating the organism from which it derived and is said to be totipotent. Collections of similar cells form tissues, tissues and tissue systems are organized into organs, and the spatial arrangement of organs constitute the organism. Plants can be regenerated in vitro from: organ explants (root and shoot tips, lateral and trace buds, leaf primordia, developing embryos, bud scales, etc.); tissue explants (pith, cortex, epidermis, phloem, nucellus); cells (parenchyma, collenchyma, uni- or binucleate pollen grains); and protoplasts. The scheme above illustrates some of the pathways by which whole plant regeneration can be achieved.

Cytokinins and Organ Formation in Tissue Cultures

In the presence of a high concentration of auxin, callus tissue frequently gives rise to organized roots. In tobacco pith callus, the relative concentrations of auxin and kinetin determine whether roots or buds form. With higher concentrations of auxin, roots are formed; with higher concentrations of kinetin, buds are formed; and when the two are present in roughly equal concentrations, the callus continues to produce undifferentiated cells (Figure 24–8).

Cytokinins and Leaf Senescence

In most species of plants, the leaves begin to turn yellow as soon as they are removed from the plant. This yellowing, which is due to a loss of chlorophyll, can be prevented by cytokinins. Leaves of the cocklebur (*Xanthium strumarium*), for example, when excised and floated on plain water, turn yellow in about ten days. If kinetin (10 mg per liter) is present in the water, much of the chlorophyll and the fresh appearance of the leaf are maintained. If excised leaves are spotted with kinetin-containing solutions, the spots remain green while the rest of the leaf yellows. Furthermore, if a cytokinin-spotted leaf contains radioactive amino acids, labeled with ^{14}C, it can be shown that the amino acids migrate from other parts of the leaf to the cytokinin-treated areas. Such studies, which have also been carried out on radishes and other plants, led to the hypothesis that senescence in leaves, and probably in other plant parts as well, results from the progressive "turning off" of segments of DNA, with a consequent loss of messenger RNA production and protein synthesis. These investigations have led to the proposal that cytokinins prevent the DNA from being turned off and so promote continued enzyme synthesis and continued production of such compounds as chlorophyll.

One interpretation of the prevention of senescence in detached leaves by cytokinins is that leaves normally do not synthesize sufficient cytokinin to meet their own requirements. Thus, an important and still unanswered question involves the site(s) of cytokinin production within plants. Based on several lines of evidence, one location is most certainly the root. A second site may be the region of rapid cell elongation behind the shoot tip. During reproduction growth the developing fruit also becomes a cytokinin-producing site, at least in the earliest stages of growth.

How Do Cytokinins Work?

Ever since the isolation of the first cytokinins from preparations of nucleic acids, plant physiologists have suspected that these hormones might in some way be involved with the nucleic acids. Molecules of transfer RNA (tRNA) contain a number of unusual bases (see Figure 8–8, page 124). In some types of transfer RNA, the natural cytokinin i[6]Ade ([6]N-isopentenyladenine), itself an unusual base, is incorporated into the molecule. For example, i[6]Ade is found in serine and tyrosine tRNA molecules, in which it is located immediately adjacent to the anticodon. It is still not known, however,

24–8

Effect of increasing IAA concentration at different kinetin levels on growth and organ formation of tobacco callus cultured on nutrient agar. Note that very little growth occurred without the addition of either IAA or kinetin. Higher levels of IAA alone promoted root formation, whereas they repressed bud formation when used alone or in combination with kinetin. The higher level of kinetin was more effective than the lower level in promoting bud development, but both kinetin levels were too high for promotion of root growth.

IAA CONCENTRATION (mg/liter)

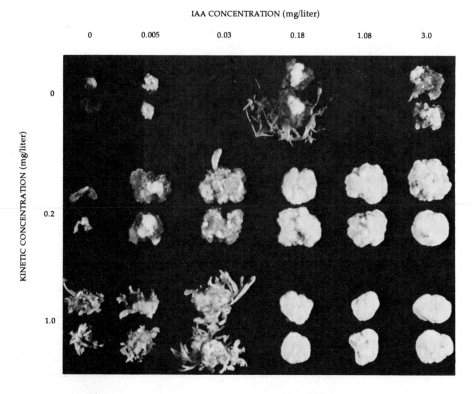

$$CH_3-S-CH_2-CH_2-CH-COO^-$$
$$|$$
$$NH_3^+$$

METHIONINE

ATP

PP$_i$ + P$_i$

$$CH_3-\overset{+}{S}-CH_2-CH_2-CH-COO^-$$
$$|\qquad\qquad\qquad\qquad|$$
ADENINE–RIBOSE $\qquad\qquad$ NH$_3^+$

S-ADENOSYLMETHIONINE

STIMULATED BY
HIGH AUXIN
CONCENTRATIONS,
AIR POLLUTION,
WOUNDING

$$H_2C\overset{\overset{+}{NH_3}}{\underset{COO^-}{\diagdown\,C\,\diagup}}$$
$$H_2C$$

1-AMINOCYCLOPROPANE-
1-CARBOXYLIC ACID

$$CH_2=CH_2$$
ETHYLENE

24–9
*Biosynthesis of ethylene, beginning with
methionine, which serves as the precursor
of ethylene in all higher plant tissues.
1-Aminocyclopropane-1-carboxylic acid
is the immediate precursor of ethylene.*

if its presence or its position in the tRNA molecule is related to its activity in promoting cell division. It is known that the ultimate effect of cytokinin involves changes in gene expression, probably at the transcriptional level of control.

ETHYLENE

Over a period of years, the discovery of auxin has led, more or less directly, to the isolation of kinetin and the recognition of cytokinin effects on plant growth and development. Ethylene, on the other hand, was known to have effects on plants long before its relationship to auxin was discovered; that is, long before it was considered to be a plant hormone.

The "botanical" history of *ethylene*, a simple hydrocarbon ($H_2C = CH_2$), goes back to the 1800s, when city streets were lighted with lamps that burned illuminating gas. In Germany, leaking illuminating gas from gas

mains was found to cause defoliation of shade trees along the streets. As gas became more extensively used for street illumination, this phenomenon was reported by many investigators.

In 1901, Dimitry Neljubov, a graduate student at the Botanical Institute of St. Petersburg University in Russia, demonstrated that ethylene was the active component of illuminating gas. Neljubov noticed that exposure of pea seedlings to illuminating gas caused stems to grow in a horizontal direction. When the gaseous components of illuminating gas were individually tested for effects, all were inactive except ethylene, which caused horizontal growth at concentrations as low as 0.06 parts per million (ppm) in air. Neljubov's findings have since been confirmed by many other investigators, and it is now known that ethylene exerts a major influence on many, if not all, aspects of growth, development, and senescence in plants.

Ethylene, although a gas under physiological conditions of temperature and pressure, does dissolve to some extent in cytoplasm. It is considered to be a plant hormone because it is a natural product of metabolism and because it interacts with other plant hormones in trace amounts. Ethylene effects can be observed particularly during critical periods—fruit maturation, fruit and leaf abscission, and senescence—in the life cycle of the plant.

Ethylene currently is the easiest hormone to assay. Since it is a gas that is evolved by tissues, it requires no extraction or purification prior to analysis by gas chromatography. The biosynthesis of ethylene begins with the amino acid methionine, which reacts with ATP to form a compound known as S-adenosylmethionine, abbreviated SAM (Figure 24–9). Next, SAM is split into two different molecules, one of which contains a ring composed of three carbon atoms. This compound, known as 1-aminocyclopropane-1-carboxylic acid (ACC), is then converted into ethylene, CO_2, and ammonium by enzymes on the tonoplast. Apparently, the reaction forming ACC is the step of the pathway that is affected by several treatments (e.g., high auxin concentrations, air pollution damage, wounding) that stimulate ethylene production by plant tissues.

Ethylene and Fruit Ripening

Ripening in fruit involves a number of changes. In fleshy fruits, the chlorophyll is degraded and other pigments may form, changing the fruit color. Simultaneously, the fleshy part of the fruit softens. This is a result of the enzymatic digestion of pectin, the principal component of the middle lamella of the cell wall. When the middle lamella is weakened, cells are able to slip past one another. During this same period, starches and organic acids or, as in the case of the avocado (*Persea americana*), oils are metabolized into sugars. As a consequence of these changes, fruits become conspicuous and palatable and thus attractive to animals that eat the fruit and so scatter the seed.

During the time of ripening of many fruits—including tomatoes, avocados, and pome fruits, such as apples and pears—there is a large increase in cellular respiration, evidenced by an increased uptake of oxygen. This phase is known as the *climacteric*, and such fruits are called climacteric fruits. (Fruits that show a steady decline, or gradual ripening, such as citrus fruits, grapes, and strawberries are called nonclimacteric fruits.) The relationship between the climacteric and the other events of fruit ripening is not known, but the ripening of fruits can be delayed by suppressing the intensity of the climacteric. For example, cold suppresses it, and in some fruits, exposure to low temperatures stops the climacteric permanently. Fruits can be stored for very long periods of time in a vacuum. Under such conditions, the amount of available oxygen is minimal, thus suppressing cellular respiration, and ethylene, which speeds up the onset of the climacteric, is held at low levels. After the climacteric, senescence sets in, and the fruit becomes susceptible to invasions by fungi and other microorganisms.

In the early 1900s, many fruit growers made a practice of improving the color and increasing the sweetness of citrus fruits by "curing" them in a room with a kerosene stove. (Long before, however, the Chinese ripened fruits in rooms where incense was being burned.) It was believed that the heat of the stove ripened the fruits. Ambitious fruit growers, who installed more modern equipment, found to their sorrow that this was not the case. As subsequent experiments demonstrated, it is actually the incomplete combustion products of the kerosene that are responsible. The most active combustion product was ethylene. As little as 1 ppm of ethylene in the air will speed up the onset of the climacteric.

As early as 1910, it was reported that gases emanating from oranges hastened the ripening of bananas, but it was not until almost 25 years later that ethylene was identified as a natural product of numerous fruits and plant tissues. The amounts produced by plants are very small, and new and extremely sensitive assay methods had to be developed before it could be proved that ethylene production began before the climacteric, even though the largest amounts coincide with the climacteric. When this was established, ethylene became generally accepted as a natural plant-growth regulator. It has now been found in fruits (in all the types tested), flowers, leaves, leafy stems, and roots of many different plant species and also in some species of fungi.

The effect of ethylene on fruit ripening has agricultural importance. A major use of ethylene is for promotion of ripening of tomatoes that are picked green and stored in the absence of ethylene and treated with ethylene just prior to marketing. It is also used to hasten ripening of walnuts and grapes.

Ethylene and Abscission

Ethylene promotes abscission of leaves, flowers, and fruits in a variety of plant species. In leaves, ethylene presumably triggers the enzymes that cause the cell wall dissolution associated with abscission. Ethylene is used commercially to promote fruit loosening in cherries, blackberries, grapes, and blueberries, thus making mechanical harvesting possible. It is also used as a fruit-thinning agent in commercial orchards of prunes and peaches.

Ethylene and Sex Expression

Ethylene appears to play a major role in determining the sex of flowers in monoecious plants (those plants having male and female flowers borne on the same individual). In cucurbits (family Cucurbitaceae), for example, high levels of gibberellins are associated with maleness, and treatment with ethylene changes the expression of sex to femaleness. In studies with cucumbers (*Cucumis sativus*), it was found that female buds evolved greater quantities of ethylene than male buds. In addition, cucumbers grown under light periods of short days, which promote femaleness, evolved more ethylene than those grown under long days. Hence, in cucurbits, ethylene apparently participates in the regulation of sex expression and is associated with the promotion of femaleness.

Ethylene and Auxin

In specific concentrations, auxin causes a burst of ethylene production in certain parts of some plants. It is believed that some of the effects on fruits and flowers once attributed to auxin are related to the effect of auxin on ethylene production.

Cell shape and size are controlled in part by the interaction of auxin and ethylene, which regulate cell expansion in opposing ways. In the garden pea (*Pisum sativum*), for example, concentrations of IAA that result in the deposition of transversely oriented cell-wall microfibrils enhance elongation of stem segments but restrict lateral growth. By contrast, higher concentrations of IAA, which induce much greater ethylene production and result in deposition of longitudinally oriented microfibrils, bring about considerable radial growth and swelling of the stem segments.

It is important to realize that the final shape and size of cells, as influenced by ethylene, are the result of its interaction not only with auxin but also with gibberellic acid and with cytokinin as well. Moreover, normal growth and development require complex hormonal interactions, which involve cytokinins, gibberellins, and abscisic acid, as well as auxin and ethylene.

ABSCISIC ACID

At certain times, the survival of the plant depends on its ability to restrain its growth or its reproductive activities. Following the early discovery of the growth-promoting hormones, plant physiologists began to speculate that regulatory hormones with inhibitory actions might also be found. Finally, in 1949, it was discovered that the dormant buds of ash and potatoes contained large amounts of growth inhibitors. These inhibitors blocked the effects induced by IAA in the *Avena* coleoptile. When dormancy in the buds was broken, the inhibitor contents declined. These inhibitors became known as *dormins*.

During the 1960s, several investigators reported the discovery in leaves and fruits of a substance capable of accelerating abscission. One of these, called abscisin, was identified chemically. In 1965, one of the dormins was also identified chemically, and abscisin and this dormin were found to be identical. The compound is now known as *abscisic acid*.

ABSCISIC ACID

Abscisic acid (abbreviated ABA) is collected largely from the ovary bases of fruits. The fruit of the cotton plant (*Gossypium*) has proved to be a particularly rich source. The largest amounts of abscisic acid are found at the time of fruit drop. Abscisic acid produced in core cells of the root cap has been implicated in the response of roots to gravity (see page 494).

Application of abscisic acid to vegetative buds changes them to winter buds by converting the leaf primordia into bud scales. Its inhibitory effects on buds can be overcome by gibberellin. The appearance of alpha-amylase, an enzyme induced by gibberellin in the seeds of barley (*Hordeum vulgare*), is inhibited by abscisic acid, which seems to depress protein production in general. Auxin, on the other hand, seems to act both by interacting with the plasma membrane and by accelerating the production of specific proteins. Hence, auxin is apparently antagonistic to abscisic acid in its action.

If a drop of abscisic acid is spotted on a leaf, the treated areas yellow rapidly, even though the rest of the leaf stays green, an effect that is opposite to that of the cytokinins. Whether this is a direct or indirect action is not known at present.

There are currently no practical uses for abscisic acid, perhaps because of the limited knowledge about its physiology and biochemistry. Abscisic acid may be-

come of tremendous importance in future agricultural practice, however, particularly in desert regions. There is reason to believe that the tolerance of some plants to stress conditions, such as drought, is directly related to their ability to produce abscisic acid. Furthermore, it is known that abscisic acid can cause the stomata of some plants to close, thus preventing water loss from leaves and lowering the plant's total water requirement. If plant geneticists can incorporate this characteristic related to abscisic acid into a wider range of plants, it may become possible to develop new crop plants that can be grown successfully in desert environments.

GIBBERELLINS

Unlike the other hormones, the discovery of the gibberellins is not closely related to the discovery of auxins. In fact, the early history of gibberellin research is quite independent of the work on auxin.

In 1926, the same year that Went first performed his experiments with blocks of agar, E. Kurosawa of Japan was studying a disease of rice (*Oryza sativa*) called "foolish seedling disease," in which the plants grew rapidly, were spindly, pale-colored, and sickly, and tended to fall over. The cause of these symptoms, Kurosawa discovered, was a chemical produced by a fungus, *Gibberella fujikuroi*, which was parasitic on seedlings. From the generic name of its source, this chemical substance was named *gibberellin*.

Gibberellin was isolated and identified chemically by biochemists in Japan in the 1930s, but for several decades it attracted little interest. Then, in 1956, gibberellin was successfully isolated from a plant (the seed of the bean *Phaseolus vulgaris*) rather than a fungus. Since that time, gibberellins have been isolated from many species of plants, and it is now generally believed that they probably occur in all plants. They are present in varying amounts in all parts of the plant, but the highest concentrations are found in immature seeds. More than 65 gibberellins now have been isolated and identified chemically. They vary slightly in structure (Figure 24–10), as well as in biological activity. The best studied of the group is GA$_3$ (known as gibberellic acid), which is also produced by the fungus *Gibberella fujikuroi*.

The gibberellins have dramatic effects on stem elongation in intact plants. A marked increase in the growth of the shoot is the most general response seen in higher plants; often the stems become long and thin and the leaves pale in color. The gibberellins stimulate both cell division and cell elongation and affect leaves as well as stems.

Gibberellins and Dwarf Mutants

The most remarkable results are seen when gibberellins are applied to certain plants that are single-gene dwarf mutants (Figure 24–11). Under gibberellin treatment, such plants become indistinguishable from normal tall, nonmutant plants. In corn, for example, four different types of dwarfs have been identified, each defective in a specific step of the gibberellin biosynthesis pathway. Actually, studies of the hormone biochemistry in these mutant plants has led to a very important conclusion. Even though corn plants contain nine different compounds that meet the operational definition of gibberellin (i.e., causation of a growth response), only the final compound in the pathway can cause effects directly. The other eight gibberellins need to be further metabolized before a growth response is obtained.

Gibberellins and Seeds

The seeds of most plants require a period of dormancy before they will germinate. In certain plants, dormancy usually cannot be broken except by exposure to cold or to light. In many species, including lettuce, tobacco, and wild oats, gibberellins will substitute for the dormancy-breaking cold or light requirement and promote the growth of the embryo and the emergence of the seedling. Specifically, the gibberellins enhance cell elongation, making it possible for the root to penetrate the growth-restricting seed coat or fruit wall. This effect of gibberellin has at least one practical application. Gibberellic acid hastens seed germination and thus insures uniformity in the production of the barley malt used in brewing.

Gibberellins and Juvenility

The juvenile stages of some plants are different from their adult stages, and this is often reflected in the juvenile and adult forms of leaves. Among annual dicots, the bean plant (*Phaseolus*), whose juvenile leaves are simple and adult leaves are compound (trifoliolate), provides an excellent example of such heterophylly (see Figure 19–2, page 374). Among the perennials, many species of *Eucalyptus* show striking differences between their juvenile and adult leaves (Figure 24–12).

24–11
The plant on the right was treated with gibberellin; the one on the left served as a control. The plants are Contender beans, a dwarf cultivar of the common bean (Phaseolus vulgaris).

Ivy (*Hedera helix*), a perennial vine, offers another familiar and extensively studied example. If mature plants of ivy are growing on a building or a wall near you, compare the upper branches with the lower ones. The form of the leaf is different, and the behavior is different. Juvenile branches root readily; adult branches do not. The adult branches flower; juvenile branches do not. If the apical meristem is cut off an adult branch, the axillary buds will develop and form new adult branches. If gibberellin is applied to such a bud, however, it will grow into a typical juvenile branch.

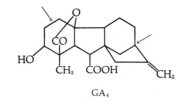

GIBBERELLIC ACID (GA₃)

GA₇

GA₄

24–10
Three of the more than 65 gibberellins that have been isolated from natural sources. Gibberellic acid (GA₃) is the most

abundant in fungi and the most biologically active in many tests. The arrows indicate minor structural differences that

distinguish the other two examples of gibberellins, GA₇ and GA₄.

(a)

(b)

24–12
(a) *Juvenile and* (b) *mature leaves of*
Eucalyptus globulus, *showing the great
differences that can occur within a given
species. The juvenile leaves are softer
and opposite to one another. Their only
layer of palisade parenchyma is just
below their upper epidermis. The mature
leaves are hard, spirally arranged, and
hang vertically. In the mature leaves,
both surfaces are equally exposed to the
light, and there are palisade parenchyma
on both sides.*

Gibberellins and Flowering

Some plants, such as cabbages (*Brassica oleracea* var. *capitata*), carrots (*Daucus carota*), and the biennial henbane (*Hyoscyamus niger*), form rosettes before flowering. (In a rosette, leaves develop but the internodes between them do not elongate.) In these plants, flowering can be induced by exposure to long days, to cold (as in the biennials), or both. Following the appropriate exposure, the stems elongate—a phenomenon known as bolting—and the plants flower. Application of gibberellin to such plants causes bolting and flowering without appropriate cold or long-day exposures (Figure 24–13). Elongation is brought about by an increase both in the number of cell divisions at certain locations and in the elongation of cells resulting from those divisions. Gibberellin can thus be used to produce early seed production of biennial plants. By treating a plant such as cabbage with gibberellic acid, seeds can be obtained after only one growing season.

24–13
*Bolting in cabbage (*Brassica oleracea
var. capitata*) produced by gibberellin
treatment. The plant on the right was
treated once a week for eight weeks.*

Gibberellins and Pollen and Fruit Development

Gibberellins have been shown to stimulate pollen germination and the growth of pollen tubes in a number of plants, including lilies, lobelias, petunias, and peas. Like auxin, gibberellins can cause the development of parthenocarpic fruits, including apples, currants, cucumbers, and eggplants. In some fruits, such as mandarin oranges, almonds, and peaches, the gibberellins have been effective in the promotion of fruit development where auxin was not. The major application of gibberellins, however, is in the production of table grapes. In the United States, large amounts of gibberellic acid are applied annually to such grape varieties as the Thompson Seedless, a cultivar, or cultivated variety, of *Vitis vinifera*. Treatment causes larger berries and a much looser appearance of the clusters of grapes (Figure 24–14).

How do Gibberellins Work?

The most important studies on the mechanism of gibberellin action were carried out simultaneously by investigators in Japan, Australia, and the United States. These studies, which trace the sequence of events in the germination of a barley seed and the early growth of its embryo, show the key role played by gibberellins in this sequence. In addition, they provide one of the best examples of how hormones integrate the biochemistry and physiology of the different tissues of a whole plant.

In barley (*Hordeum vulgare*) and other grass seeds, there is a specialized layer of endosperm cells, called the aleurone layer (see Figure 19–4, page 376), just inside the seed coat. The cells in the aleurone layer are rich in protein. When the seeds begin to germinate, triggered by the imbibition of water, the embryo releases gibberellins. In response to these hormones, the aleurone cells synthesize hydrolytic enzymes, the principal one being alpha-amylase, the enzyme that breaks down starch into sugar (Figure 24–15). The enzymes digest the stored food reserves of the starchy endosperm. These food reserves are released in the form of sugars and amino acids, which are absorbed by the scutellum and are then transported to the growing regions of the embryo (Figure 24–16). In this way, the embryo calls forth the substances needed for its growth at the moment it requires them.

Investigators believe that gibberellins activate certain genes, causing the synthesis of specific messenger RNA molecules, which, in turn, direct the synthesis of the enzymes. It has not been proved, however, that gibberellins act directly on the gene, although it has been shown that both RNA and protein synthesis take place and are necessary for the appearance of the enzymes. Whatever the details of the mechanism of gibberellin action in aleurone cells, it is clear that the cells of the aleurone layer constitute a highly differentiated tissue

24–14

The effect of gibberellic acid on growth of Thompson Seedless grapes, a cultivar of Vitis vinifera. *The slender bunch of grapes was untreated, whereas the fuller bunch was treated with gibberellic acid.*

24–15

The release of sugar from endosperm can be induced by gibberellin (GA_3) treatment. These data show that sugars are produced only when the aleurone layer is present. It is, in fact, the aleurone layer that is the source of the enzyme alpha-amylase, which digests the starches stored in the endosperm.

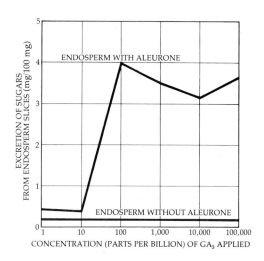

24–16

Action of gibberellin in barley seeds. (a) Gibberellic acid (GA) produced by the embryo migrates into the aleurone layer, stimulating the synthesis of hydrolytic enzymes. These enzymes are released into the starchy endosperm where they break down the endosperm reserves into sugars and amino acids, which are soluble and diffusible. The sugars and amino acids are then absorbed by the scutellum (cotyledon) and transported to the shoot and root for growth. (b) Each of these three seeds has been cut in half and the embryo removed. Forty-eight hours before the picture was taken, the seed at the top left was treated with plain water. The seed in the center was treated with a solution of 1 part per billion of gibberellin, and the seed at the bottom right was treated with 100 parts per billion of gibberellin. Digestion of the starchy storage tissue has begun to take place in the treated seeds.

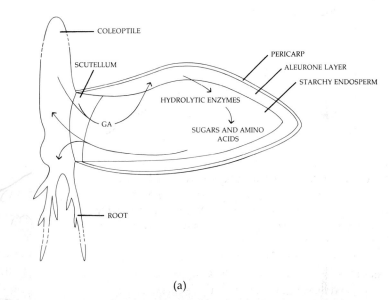

(a)

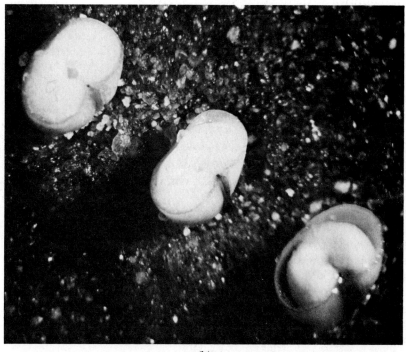

(b)

Table 24-1   *Plant Hormones (Phytohormones) and Their Effects*

		PHYSIOLOGICAL EFFECTS AND ROLES
Auxin	Indole-3-acetic acid, phenylacetic acid	Apical dominance, gravitropism and phototropism, vascular differentiation, inhibits abscission, stimulates ethylene synthesis, inhibits or promotes (in pineapples) flowering, stimulates fruit development (parthenocarpic fruit), induces roots on cuttings
Cytokinin	N⁶-adenine derivatives, phenyl urea compounds	Apical dominance, shoot growth, fruit development, delays leaf senescence
Ethylene	$CH_2 = CH_2$	Fruit ripening (especially in climacteric fruits such as apples, bananas, and avocados), leaf and flower senescence, abscission
Gibberellin	Gibberellic acid (GA_3), GA_1	Flowering stimulation in long-day plants and biennials, shoot elongation, regulates production of seed enzymes in cereals
Abscisic acid	ABA	Stomatal closure, may be necessary for abscission and dormancy in certain species

poised to respond to the demands—mediated by gibberellins—of the growing embryo. It is not known whether the way in which gibberellins work in seeds is related to their effects on other plant organs.

SUMMARY

Hormones are important chemical regulators of growth in both animals and plants. A hormone is a chemical produced in certain tissues of the organism and transported to other tissues, where it causes a physiological response. Hormones are biologically active in extremely small amounts.

Naturally occurring auxin is a hormone that is produced in the apical meristems of shoots and the tips of coleoptiles. Auxin travels unidirectionally toward the base of the plant, where it controls the lengthening of the shoot and the coleoptile, chiefly by promoting cell elongation. Studies indicate that its effect on cell elongation is achieved in some indirect way by a relaxation of the cellulose fibrils of the cell wall, permitting the cell to expand. Auxin also plays a role in differentiation of vascular tissue and initiates cell division in the vascular cambium. It often inhibits growth in lateral buds, thus maintaining apical dominance. The same quantity of auxin that promotes growth in the stem inhibits growth in the main root system. Auxin promotes the formation of adventitious roots in cuttings and retards abscission in leaves, flowers, and fruits. In fruits, auxin produced by seeds or the pollen tube stimulates growth of the ovary wall. Its capacity to produce such varied effects is believed to result from differential responses of the various target tissues.

The cytokinins, a second class of growth hormones, were first discovered as a consequence of their capacity to promote cell division and bud formation in cultures of plant tissues. They are chemically related to certain components of nucleic acids. Cytokinins can also act in concert with auxin to cause cell division in plant tissue culture. In tobacco pith cultures, a high concentration of auxin promotes root formation, while a high concentration of cytokinins promotes bud formation. In intact plants, cytokinins promote the growth of lateral buds, acting in opposition to the effects of auxin. Cytokinins prevent senescence in leaves by stimulating protein synthesis.

Ethylene is a gas produced by the incomplete combustion of hydrocarbons. It is also a natural growth regulator in plants, producing a number of distinct physiological responses, such as ripening in fruit and abscission.

Abscisic acid is a growth-inhibiting hormone that is found in dormant buds and in fruits. Abscisic acid has been shown to oppose the effects of all three types of growth-stimulating hormones in various laboratory tests.

The gibberellins were first isolated from a parasitic fungus that causes abnormal growth in rice seedlings. They were subsequently found to be natural growth hormones present in many plants. The most dramatic effects of gibberellins are seen in dwarf plants, in which the application of gibberellins restores normal growth, and in plants with a rosette form of growth, in which gibberellins cause bolting. Gibberellins cause seed germination in grasses. In the barley seed, the embryo releases gibberellins that cause the aleurone layer of the endosperm to produce several enzymes, including alpha-amylase, which breaks down the starch stored in the endosperm, releasing sugar. The sugar nourishes the embryo and promotes the germination of the seed.

The effects of plant hormones depend largely on the target tissues and the chemical environment in which these tissues are located.

C H A P T E R   2 5

External Factors and Plant Growth

Living things must regulate their activities in accordance with the world around them. Many animals, being mobile, can change their circumstances to some extent —foraging for food, courting a mate, and seeking or even making shelters in bad weather. A plant, by contrast, is immobilized once it sends down its first root. However, plants do have the ability both to respond and make adjustments to a wide range of changes in their external environment. This ability is manifested chiefly in changing patterns of growth.

THE TROPISMS

A growth response involving a bending or curving of a plant part toward or away from an external stimulus that determines the direction of movement is called a *tropism*. If the tropic response is toward the stimulus, it is said to be *positive*; if it is away, it is *negative*.

Perhaps the most familiar interaction between plants and the external world is the curving of growing shoot tips toward light (see Figure 24–2, page 476). This growth response, now known as *phototropism*, is caused by the elongation—under the influence of the plant hormone auxin—of the cells on the shaded side of the tip. What role does the light play in the phototropic response? Three possible answers to this question have been suggested: (1) light decreases the auxin sensitivity of the cells on the lighted side; (2) light destroys auxin; or (3) light drives auxin to the shaded side of the growing tip.

To choose among these hypotheses, Winslow Briggs and his co-workers carried out a series of experiments that were based on earlier work by Frits Went (see Figure 24–3, page 477). These investigators first showed that the same total amount of auxin is produced by the tip in the light as by the tip in the dark. Following the

(a)

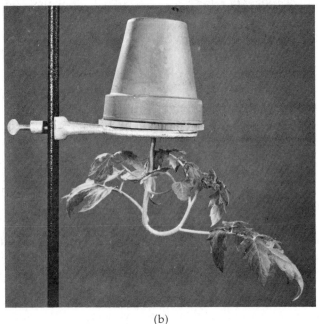

(b)

25–1

Gravitropic responses in the shoot of a young tomato (Lycopersicon esculentum) plant. The plant in (a) was placed on its side and kept in a stationary position, whereas that in (b) was placed upside down and held in position in a ring stand. Although originally straight, the stems bent and grew upward. If a horizontally placed potted plant is slowly rotated around the stem on its axis, no curvature (gravitropic response) will occur; that is, the plant will continue to grow horizontally. Can you explain the differences in the growth of a horizontally rotated plant and the plants shown here?

exposure to light, however, the amount diffused from the shaded side is greater than the amount diffused from the lighted side. If the tip is split and a barrier, such as a thin piece of glass, is placed between the two halves, this differential distribution of auxin diffusion no longer occurs. In other words, Briggs clearly demonstrated that auxin (or perhaps a precursor) migrates from the light side to the dark side and that the curving of the shoot is in response to this unequal distribution. Experiments using the auxin IAA (indoleacetic acid) labeled with ^{14}C have clearly shown that it is the auxin that migrates.

It has been determined that light of wavelengths from 400 to 500 nanometers is the most effective in inducing this hormone migration in a shoot tip. This observation indicates that a pigment absorbing blue light mediates the effect. The identity of the blue-light photoreceptor (which also mediates many other responses in plants and fungi) has not been definitively established, but present evidence indicates that it is a yellow-colored pigment called a flavin. According to a popular hypothesis, as yet unproven, blue light causes chemical reduction of a flavin which, in turn, causes electron transport on the plasma membrane, and this sets up a pH gradient.

Another familiar tropism is *gravitropism*, or geotropism, a response to gravity. Seedlings conspicuously manifest gravitropism. If a seedling is placed on its side, its root will grow downward (positive gravitropism) and its shoot will grow upward (negative gravitropism). The original explanation for this mechanism involved the asymmetric redistribution of auxin in favor of the lower side by means of a lateral polar transport of the molecule from the upper to the lower side of the organ. Under these conditions, the lower side of the shoot should grow upward (Figure 25–1). In the root, the upper side should elongate more than the lower side, and the root should grow downward. When either plant part becomes vertical, the lateral asymmetries in auxin concentration should disappear, and growth should continue in the vertical direction.

The validity of this hypothesis for the gravitropic response is now in doubt. Although it has been demonstrated that IAA does undergo lateral transport toward the lower side of corn (*Zea*) and oat (*Avena*) coleoptiles and thus becomes asymmetrically distributed, the situation is much less clear for the shoots of dicots. Moreover, it has not been clearly established that the asymmetrical distribution of IAA in the corn and oat coleoptiles alone could initiate the growth rate changes observed.

Asymmetries of growth-regulating agents are even less well understood in roots. It has now been demonstrated that IAA, cytokinins, gibberellins, and abscisic acid (ABA) are present in roots, though their physiological roles are unclear. Substances that inhibit growth are also present in roots, and at least one has its origin

25–2
(a) *Photomicrograph of a median longitudinal section of the root cap of the primary root of the common bean,* Phaseolus vulgaris. *Arrows point to amyloplasts (starch-containing plastids) sedimented near the transverse walls in the core (columella) cells of the root cap.* (b) *Electron micrograph of core cells from a root cap similar to that shown in* (a). *Here too the amyloplasts (arrows) are located at the bottom of each cell near the transverse walls.*

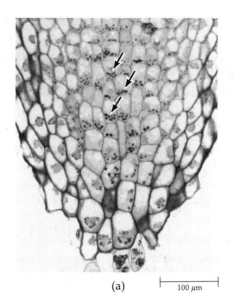

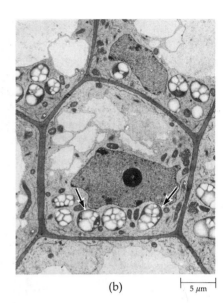

(a) |——————| 100 μm   (b) |——| 5 μm

in the root cap. Evidence indicates that an inhibitor from the root cap may be involved in bringing about the positive gravitropic response of the primary root.

Substantial evidence exists that the perception of gravity involves the sedimentation of amyloplasts (starch-containing plastids) within specialized cells of the shoot and root. Such cells are found all along the shoot, often in cells contiguous to or surrounding the vascular bundles (the bundle sheaths). In roots, however, they are localized in the root cap, particularly in the central core (columella) of the root cap (Figure 25–2). When a root is placed in a horizontal position, the plastids, which were sedimented near the transverse walls of the vertically growing roots, slide downward and come to rest near what were previously vertically oriented walls (Figure 25–3). After several hours, the root curves downward and the plastids return to their previous position along the transverse walls. How the movement of these gravity sensors (statoliths) is trans-

lated into hormonal gradients has yet to be convincingly explained. Recently, it has been suggested that calcium plays a key role in linking graviperception and gravitropism in roots through its control of hormone transport. Calcium ions are known to be present in amyloplasts of columella cells.

A less apparent, though common tropism, is *thigmotropism* (Greek *thigma*, touch), a response to contact with a solid object. One of the most common examples of thigmotropism is seen in tendrils, which are modified leaves in some species of plants and modified stems in others (see page 440). The tendrils wrap around any object with which they come in contact (Figure 25–4) and so enable the plant to cling and climb. The response can be rapid; a tendril may wrap around a support one or more times in less than an hour. Cells touching the support shorten slightly and those on the other side elongate. There is some evidence that auxin plays a role in this response.

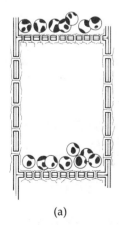

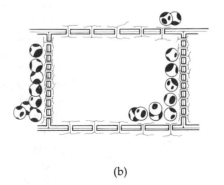

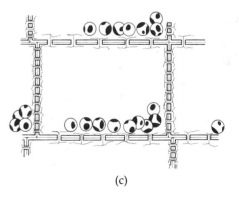

(a)   (b)   (c)

25–3
This diagram illustrates the response to gravity of amyloplasts (starch-containing plastids) in the core cells of a root cap. (a) *The amyloplasts are normally sedimented near the transverse walls in*

the root cap of a root growing vertically downward. When the same root is placed on its side, (b) *and* (c), *the amyloplasts slide toward the normally vertical walls that are now parallel with the soil*

surface. The evidence is convincing that this movement of particles plays an important role in the production of gradients of growth-controlling substances that lead to vertical root growth.

25–4
Tendrils of greenbrier (Smilax). Twisting is caused by different growth rates on the inside and outside of the tendril.

Studies by M. J. Jaffe of Wake Forest University, North Carolina, have revealed that the tendrils of young Alaska pea plants (*Pisum sativum*) can store sensory information and retrieve it and respond at a later time. For example, when tendrils were held in the dark for three days and then rubbed, they would not coil until they were subsequently illuminated. Moreover, the tendrils could be kept in the dark for as long as two hours after being rubbed and still coil immediately upon being illuminated. Although the sensory information could be stored during the dark period, the motor function was unable to proceed in the dark. Why the motor function could not proceed in the dark remains unexplained. It has been suggested, however, that ATP (which apparently is needed for coiling) may be used up during the dark period and is restored during illumination by photosynthesis; or that a coiling inhibitor accumulates in the dark and is quickly removed in the light.

CIRCADIAN RHYTHMS

It is a common observation that some plants open their flowers in the morning and close them at dusk or spread their leaves in the sunlight and fold them toward the stem at night (Figure 25–5). As long ago as 1729, the French scientist Jean-Jacques de Mairan noticed that these diurnal (daily) movements continue even when the plants are kept in dim light (Figure 25–6). More recent studies have shown that less evident activities, such as photosynthesis, auxin production, and the rate of cell division, also have regular daily rhythms, which continue even when all environmental conditions are kept constant. These regular, approximately 24-hour cycles are called *circadian rhythms*, from the Latin words *circa*, meaning "approximately," and *dies*, meaning "a day." Circadian rhythms, which

25–5
Leaves of the wood sorrel (Oxalis), during the day (a) and at night (b). One hypothesis concerning the function of such sleep movements is that they prevent the leaves from absorbing moonlight on bright nights, thus protecting the photoperiodic phenomena discussed later in this chapter. Another hypothesis, proposed by Darwin a century ago, is that the folding protects against heat loss from the leaves at night.

(a)   (b)

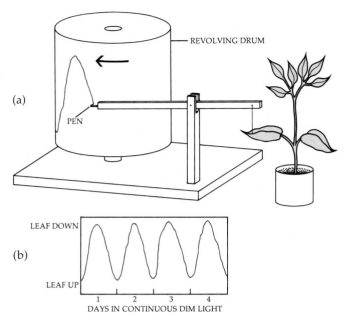

(a)

REVOLVING DRUM

PEN

(b)

LEAF DOWN

LEAF UP

1 2 3 4
DAYS IN CONTINUOUS DIM LIGHT

25–6
In many plants, the leaves move outward, to a position perpendicular to the stem and the sun's rays during the day and upward, toward the stem, at night. These sleep movements can be recorded on a revolving drum using a delicately balanced pen-and-lever system attached to a leaf by a fine thread (a). Many plants, such as the common bean (Phaseolus vulgaris) *shown here, will continue to exhibit these movements for several days even when kept in continuous dim light. (b) A recording of this circadian rhythm, showing its persistence under a condition of constant dim light.*

have been found to exist almost universally throughout the eukaryotes, appear to be absent in bacteria. Why this should be so is unknown.

Are the Rhythms Endogenous?

Are these rhythms actually internal—that is, caused by factors entirely within the organism—or is the organism keeping itself in tune with some external stimulus? For years now, biologists have debated whether it might not be some environmental force, such as cosmic rays, the magnetic field of the earth, or the earth's rotation, that was setting the rhythms. Two important observations bearing on this question were that the rhythms studied were circadian (that is, the period of a rhythm is *about* 24 hours) and that there were slight individual differences among organisms.

Attempts to settle this recurrent controversy have led to countless experiments under an extraordinary variety of conditions. Organisms have been taken down in salt mines, shipped to the South Pole, flown halfway

around the world in airplanes, and, most recently, orbited in satellites. Although there is still a vocal minority that believes that circadian rhythms are under the influence of a subtle geophysical factor, most workers now agree that the rhythms are endogenous—that is, they are controlled internally. The internal timing mechanism is often referred to as the organism's *biological clock*.

Setting the Clock

Under constant environmental conditions, the period of the circadian rhythm is *free-running*—that is, its natural period (usually between 21 and 27 hours) does not have to be reset at each cycle. It behaves as a self-sustained oscillator. Although circadian rhythms originate within the organisms themselves, the environment acts as a synchronizing agent (or "Zeitgeber," German for "time giver"). In fact, the environment is responsible for keeping a circadian rhythm in step with the daily 24-hour light-dark cycle. If a circadian rhythm of a plant were greater or less than 24 hours, the rhythm would soon get out of step with the 24-hour light-dark cycle. Then, if only the circadian rhythm were followed, a phenomenon such as flowering, which generally takes place in the light period, would occur at a different time each day, including the dark period. It is therefore necessary for the plant to become resynchronized—that is, to become *entrained* to the 24-hour day.

Entrainment is the process by which a periodic repetition of light and dark, or some external cycle, causes a circadian rhythm to remain synchronized with the same cycle as the entraining factor. Light-dark cycles and temperature cycles are the principal factors in entrainment (Figure 25–7).

Another feature of interest is that these rhythms do not automatically speed up as the temperature rises. The reason one might expect them to speed up is that biochemical activities—an internal biological clock certainly must have a biochemical basis—take place more rapidly at high temperatures than at low ones. Some clocks run slightly faster as the temperature rises, but others go more slowly, and many are almost unchanged. Thus, the biological clock must contain within its workings a compensatory mechanism—a feedback system that allows it to adjust to temperature changes. Such a temperature-compensating feature would be very important in plants.

Some recent evidence suggests that cell membranes may be the key to the biological clock. Although the mechanism is unknown, the established membrane functions that might conceivably be involved include the regulation of ion movements into and out of cells and subcellular compartments, and the regulation of energy metabolism via the energy-transducing membranes of chloroplasts and mitochondria.

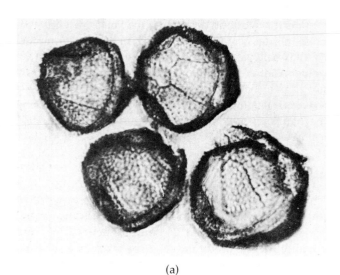

(a)

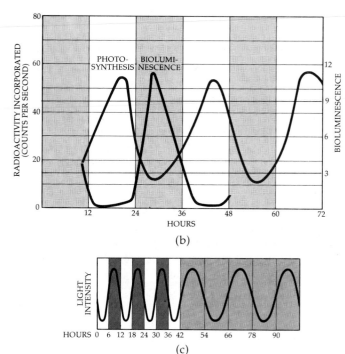

(b)

(c)

25–7

(a) *Photomicrograph of the dinoflagellate* Gonyaulax polyedra, *a single-celled marine alga.* (b) *In* G. polyedra, *three separate functions follow separate circadian rhythms: bioluminescence, which reaches a peak in the middle of the night (the curve in color); photosynthesis, which reaches a peak in the middle of the day (the black curve); and cell division (not shown), which is restricted to the hours just before dawn. If Gonyaulax is kept in continuous dim light, these three functions continue to occur with the same rhythm for days and even weeks, long after a number of cell divisions have taken place.* (c) *The rhythm of bioluminescence in* Gonyaulax, *like most circadian rhythms, can be altered by modifying the cycles of illumination. For example, if the investigators expose cultures of the algae to alternating light and dark periods, each 6 hours in duration, the rhythmic function will become entrained to this imposed cycle (left). If the cultures are then placed in continuous dim light, the organisms will return to their original rhythm of about 24 hours.*

Some biological rhythms regulate the interaction between organisms. For example, some plants secrete nectar at specific times of the day. As a result, bees—which have their own biological clocks—become accustomed to visiting these flowers at these times, thereby ensuring maximum rewards for themselves and cross-pollination for the flowers.

For most organisms, however, the use of the biological clock for such special purposes is probably a secondary one. The primary usefulness of the clock is that it enables the plant or animal to respond to the changing seasons of the year by accurately measuring changing day length. In this way, changes in the environment trigger responses that result in adjustments of growth, reproduction, and other activities of the organism.

PHOTOPERIODISM

Sixty-five years ago a mutant appeared in a field of tobacco plants (*Nicotiana tabacum*) growing near Washington, D.C. The new variety had unusually large leaves and eventually grew to a height of more than 3 meters. As the season progressed, the regular plants flowered, but Maryland Mammoth, as the variety came to be known, just grew bigger and bigger. Two investigators from the U.S. Department of Agriculture, W. W. Garner and H. A. Allard, took cuttings from the Maryland Mammoth and placed them in a greenhouse, where they would be protected from frost. These cuttings flowered in December, although by then they were only 1.5 meters tall, half the size of their parent. New Maryland Mammoths grew from their seed, and these, too, did not flower until almost winter.

Coincidentally, Garner and Allard were carrying out experiments with the Biloxi variety of soybean (*Glycine max*). Agriculturalists were interested in spacing out the soybean harvest by making successive sowings of seeds at two-week intervals from early May through June. But spacing out the planting had no effect; all the plants, no matter when the seeds were sown, came into flower at the same time—in September.

The investigators started growing these two kinds of

plants—Maryland Mammoth tobacco and Biloxi soybeans—under a wide variety of controlled conditions of temperature, moisture, nutrition, and light. They eventually found that the critical factor in both species was the length of day. Neither plant would flower unless the day length was shorter than a critical number of hours. Consequently, soybeans, no matter when they were planted, all flowered as soon as the days became short enough, which was in September; the Maryland Mammoth, no matter how tall it grew, would not flower until December, when the days had become even shorter.

Garner and Allard called this phenomenon *photoperiodism*. Photoperiodism is a biological response to a change in the proportions of light and dark in a 24-hour daily cycle. Although the concept of photoperiodism began with studies of plants, it has now been demonstrated in various fields of biology, including the mating behavior of such diverse animals as codling moths, spruce budworms, aphids, potato worms, fish, birds, and mammals.

Long-Day Plants and Short-Day Plants

Garner and Allard went on to test and confirm their discovery with many other species of plants. Following this single lead, they were able to answer a host of questions that had long troubled both professional botanists and amateur gardeners. Why, for example, is there no ragweed (*Ambrosia*) in northern Maine? *Answer:* Because ragweed starts producing flowers only when the duration of daylight is 14½ hours or less. The long summer days do not shorten to 14½ hours in northern Maine until August, and then there is not enough time for ragweed seed to develop before the frost. *Question:* Why does spinach (*Spinacia oleracea*) not grow in the tropics? *Answer:* Because spinach needs at least 14 hours of light per day for a period of at least two weeks in order to flower, and this combination of conditions never happens in the tropics.

The investigators found that plants are of three general types, which they called *short-day, long-day,* and *day-neutral.* Short-day plants flower in early spring or fall; they must have a light period shorter than a critical length. For instance, the common cocklebur (*Xanthium strumarium*) is induced to flower by 16 hours or less of light (Figures 25–8 and 25–9). Other short-day plants are some chrysanthemums (Figure 25–10a), poinsettias, strawberries, and primroses.

Long-day plants, which flower chiefly in the summer, will flower only if the light periods are longer than a critical length. Spinach, some potatoes, some wheat varieties, lettuce, and henbane (*Hyoscyamus niger*) are examples of long-day plants (Figures 25–9 and 25–10b).

Cocklebur and spinach will both bloom if exposed to 14 hours of daylight, yet only one is designated as a long-day plant. The important factor is not the absolute length of the photoperiod but whether it is longer or shorter than some critical interval. Day-neutral plants flower without respect to day length. Some examples of day-neutral plants are cucumber, sunflower, tobacco, rice, corn, and garden pea.

Within individual species of plants that cover a large north-south range, different photoperiodic ecotypes (locally adapted variants of an organism) have often been observed. Thus, in many prairie grasses, species that may occur from southern Canada to Texas, northern ecotypes flower before southern ones when they are grown together in a common environment. Different populations are precisely adjusted to the demands of the photoperiodic regime where they occur.

The photoperiodic response can be remarkably precise. At 22.5°C, the long-day plant *Hyoscyamus niger* (henbane) will flower when exposed to photoperiods of 10 hours and 20 minutes (Figure 25–9). But with a photoperiod of 10 hours, it will not flower at this temperature. Environmental conditions also affect photoperiodic behavior. For instance, at 28.5°C, henbane requires 11½ hours of light, whereas at 15.5°C, it requires only 8½ hours.

25–8

The relative length of day and night determines when plants flower. The four curves depict the annual change of day length in four North American cities at four different latitudes. The horizontal color lines indicate the effective photoperiod of three different short-day plants. The cocklebur, for instance, requires 16 hours or less of light. In Miami, it can flower as soon as it matures, but in Winnipeg, the buds do not appear until early August, so late that the frost will probably kill the plants before the seed is set.

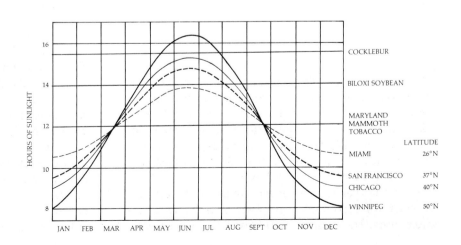

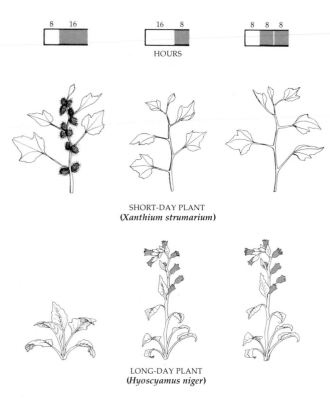

8 16   16 8   8 8 8

HOURS

SHORT-DAY PLANT
(Xanthium strumarium)

LONG-DAY PLANT
(Hyoscyamus niger)

25–9
*Short-day plants flower when the
photoperiod is less than some critical
value. The common cocklebur (Xanthium
strumarium) requires less than 16 hours
of light to flower. Henbane (Hyoscyamus niger) requires about 10 hours
(depending on temperature) or more to
flower. However, if the dark period is interrupted by a flash of light, Hyoscyamus
will flower even on a short-day
period. A "pulse" of light during the
dark period has the opposite effect in a
short-day plant—it prevents flowering.
The bars at the top indicate the duration
of light and dark periods in a 24-hour
day.*

The response varies with different species. Some
plants require only a single exposure to the critical day-
night cycle, whereas others, such as spinach, require
several weeks of exposure. In many plants, there is a
correlation between the number of induction cycles and
the rapidity of flowering or the number of flowers
formed. Some plants have to reach a certain degree of
maturity before they will flower, whereas others will
respond to the appropriate photoperiod when they are
seedlings. Some plants, when they get older, will eventually flower even if not exposed to the appropriate
photoperiod, although they will flower much earlier
with the proper exposure.

Measuring the Dark

In 1938, another pair of investigators, Karl C. Hamner
and James Bonner, began a study of photoperiodism,
using the cocklebur as their experimental tool. As mentioned previously, the cocklebur is a short-day plant,
requiring 16 hours or less of light per 24-hour cycle to
flower. It is particularly useful for experimental purposes because a single exposure under laboratory conditions to a short-day cycle will induce flowering two
weeks later, even if the plant is immediately returned to
long-day conditions. The cocklebur plant can withstand a great deal of rough treatment; for example, it
can survive even if all its leaves are removed. Hamner
and Bonner showed that it is the leaf blade of the cocklebur that perceives the photoperiod. A completely defoliated plant cannot be induced to flower. But if as
little as one-eighth of a fully expanded leaf is left on the
stem, the single short-day exposure induces flowering.

25–10
*Representative short-day and long-day
plants, each grown under short-day (on
left in a and b) and long-day (on right in
a and b) conditions. (a) Chrysanthe-*
*mum, a short-day plant. (b) Spinach
(Spinacia oleracea), a long-day plant.
Note that the plants exposed to long days
have longer stems than those exposed to
short days.*

(a)

(b)

In the course of these studies, in which they tested a variety of experimental conditions, Hamner and Bonner made a crucial and totally unexpected discovery. If the period of darkness is interrupted by as little as one-minute exposure to light from a 25-watt bulb, flowering does not occur. Interruption of the light period by darkness has absolutely no effect on flowering. Subsequent experiments with other short-day plants showed that they, too, required periods of uninterrupted darkness rather than uninterrupted light.

The most sensitive part of the dark with respect to light interruption is in the middle of the night period. If a short-day plant, such as cocklebur, is given an 8-hour light period and then an extended period of darkness, it can be shown that the plant proceeds from a stage of increasing sensitivity to light interruption lasting about 8 hours, followed by a period when light interruption has a diminishing effect. In fact, one-minute light exposure after 16 hours of darkness stimulates flowering. According to the German plant physiologist Erwin Bünning, the photoperiodic control of flowering is also regulated by the endogenous rhythm, the phases of the rhythm being distinguished according to whether light is stimulatory (the photophile phase) or inhibitory (the skotophile phase). The changes in sensitivity, therefore, correspond to different phases of the biological clock. So-called light interruption experiments, in fact, provide some of the best evidence for Bünning's hypothesis.

On the basis of the findings of Garner and Allard, commercial growers of chrysanthemums had found that they could hold back blooming in the short-day plants by extending the daylight with artificial light. On the basis of the new experiments by Hamner and Bonner, they were able to delay flowering simply by switching on the light for a short period in the middle of the night.

What about long-day plants? They also measure darkness. A long-day plant that will flower if it is kept in a laboratory in which there is light for 16 hours and dark for 8 hours will also flower on 8 hours of light and 16 hours of dark if the dark is interrupted by even a brief exposure of light.

CHEMICAL BASIS OF PHOTOPERIODISM

The next important clue to the response of plants to the relative proportions of light and darkness came from a team of research workers at the U.S. Department of Agriculture Research Station in Beltsville, Maryland. The clue was found in the report of an earlier study performed with lettuce (*Lactuca sativa*) seeds. Lettuce seeds germinate only if they are exposed to light. This requirement is true of many small seeds, which need to germinate in loose soil and near the surface in order for the seedlings to be sure of breaking through. The earlier workers, in studying the light requirement of germinating lettuce seeds, had shown that red light stimulated germination and that light of a slightly longer wavelength (far-red) inhibited germination even more effectively than did no illumination at all.

Hamner and Bonner had shown that when the dark period is interrupted by a single flash of light from an ordinary bulb, the cocklebur will not flower. The Beltsville group, following this lead, began to experiment with light of different wavelengths, varying the intensity and duration of the flash. They found that red light at about 660 nanometers (orange-red) was most effective in preventing flowering in the cocklebur and other short-day plants. It was also the most effective, they found, in promoting flowering in long-day plants.

The Beltsville group found that when a flash of red light was followed by a flash of far-red light, the seeds did not germinate. The red light most effective in inducing germination in seeds was light of the same wavelength as that involved in the flowering response —about 660 nanometers. Furthermore, they found that the light most effective in inhibiting the effect produced by red light was light of a wavelength of 730 nanometers. The sequence of red and far-red flashes could be repeated over and over; the number of flashes did not matter, but the nature of the final one did. If the sequence ended with a red flash, the majority of the seeds germinated. If it ended with a far-red flash, the majority did not germinate (Figure 25–11).

25–11

Light and the germination of lettuce seeds. (a) Seeds exposed briefly to red light; (b) seeds exposed to red light, followed by far-red light; (c) seeds exposed to a sequence of red, far-red, red; and (d) seeds exposed to a red, far-red, red, far-red sequence. Whether or not the seeds germinated depended on the final wavelength in the series of exposures— red light promoting and far-red light inhibiting germination.

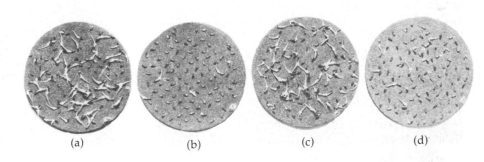

(a) (b) (c) (d)

Far-red light was tried on short-day and long-day plants, with the same on-off effect. Far-red light alone, when given during the dark period, had no effect. But a flash of far-red light immediately following a flash of red light canceled the effects of the red light.

Discovery of Phytochrome

Plants contain a pigment that exists in two different interconvertible forms; P_r (a form that absorbs red light) and P_{fr} (a form that absorbs far-red light). When a molecule of P_r absorbs a photon of red light of 660 nanometers in wavelength, it is converted to P_{fr} in a matter of milliseconds; when a molecule of P_{fr} absorbs a photon of far-red light of 730 nanometers, it is very quickly reconverted to the P_r form. These are called *photoconversion reactions*. The P_{fr} form is biologically active (that is, it will trigger a response such as seed germination), whereas P_r is inactive. The pigment molecule can thus function as a biological switch, turning responses on or off depending on which form it is in.

The lettuce-seed germination experiments are easily understood in these terms. Since P_r absorbs red light most efficiently (Figure 25–12), this wavelength will drive a high proportion of the molecules into the P_{fr} form, thereby inducing germination. Subsequent far-red light absorbed by the P_{fr} form will drive essentially all of the molecules back to P_r, thereby canceling the effect of the prior red light.

What about flowering under natural day-night cycles? Since white light contains both red and far-red wavelengths, both forms of the pigment are exposed simultaneously to photons that are efficient in driving their photoconversion to the opposite form. After a few minutes in the light, then, a photoequilibrium is established whereby the forward ($P_r \longrightarrow P_{fr}$) and backward ($P_{fr} \longrightarrow P_r$) reactions are balanced. A constant proportion of the phytochrome population is in each form at any instant under these conditions (about 60 percent P_{fr} in noontime sunlight), and this proportion is maintained while the light remains on.

When plants are switched to darkness at the end of the light period, the level of P_{fr} steadily declines over a period of several hours. If a high level of P_{fr} is regenerated by pulse irradiation with red light in the middle of

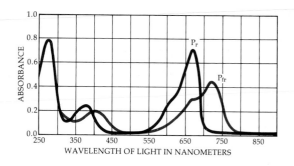

Absorption spectra of the two forms of phytochrome, P_r and P_{fr}. This difference in the absorption spectra made it possible to isolate the pigment.

the dark period (Figure 25–9), it will inhibit flowering in short-day (that is, "long-night") plants that otherwise would have flowered; and it will promote flowering in long-day (that is, "short-night") plants that otherwise would not have flowered. In either case, the effect of the red pulse in regenerating high levels of P_{fr} can be canceled by an immediate subsequent far-red pulse, which reconverts the P_{fr}.

In 1959, Harry A. Borthwick and his co-workers at Beltsville gave this pigment the name of *phytochrome* and presented conclusive physical evidence that it existed. The principal characteristics of the pigment, as they are presently understood, are summarized schematically in Figure 25–13. The molecule is synthesized continuously as P_r and accumulates in this form in dark-grown plants. Light causes photoconversion of P_r to P_{fr}, which *induces* a biological response. P_{fr} can be converted to P_r by photoconversion with far-red light or reversion to P_r in the dark, a process termed "dark reversion" that occurs over a period of time ranging from several minutes to hours. P_{fr} can also be lost through irreversible denaturation, by a process termed "destruction" that occurs over several hours and probably involves hydrolysis by a protease. All three alternate pathways for P_{fr} removal provide the potential for reversing induced responses. It should be noted, however, that dark reversion has only been noted in dicots, not in monocots.

25–13
Phytochrome is first synthesized in the P_r form from amino acids (designated P_p for precursor). P_r changes to P_{fr} when exposed to red light. P_{fr} is the active form that induces a biological response. P_{fr} is converted back to P_r when exposed to far-red light. In darkness, P_{fr} reverts to P_r or is destroyed (designated P_d for destruction product).

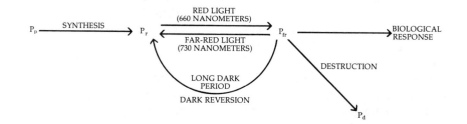

Isolation of Phytochrome

Phytochrome is present in plants in small amounts relative to such pigments as chlorophyll. To distinguish the phytochrome, one needs to use a spectrophotometer that is sensitive to extremely small changes in light absorbency. Such a spectrophotometer was not introduced until some seven years after the existence of phytochrome was proposed; in fact, the first use of the new instrument was to detect and isolate this pigment.

In order to avoid interference from chlorophyll which, like phytochrome, also absorbs light of about 660 nanometers, dark-grown seedlings (in which chlorophyll has not yet developed) were chosen as the source from which to isolate phytochrome. The pigment proved to be blue in color (why might you expect this color?), and it showed the characteristic red to farred conversion in the test tube by reversibly changing color slightly in response to red or far-red light.

The phytochrome molecule was found to contain two distinct parts: a light-absorbing portion (the chromophore) and a large protein portion (Figure 25–14). The chromophore is much like the phycobilins that serve as accessory pigments in cyanobacteria and red algae.

The specific mechanism by which phytochrome works has not yet been established. However, it is clear that phytochrome-regulated morphogenesis results from changes in gene transcription.

Other Phytochrome Responses

Phytochrome has now been shown to be involved in a number of other plant responses. The germination of many seeds, for instance, occurs in the dark. In the seedlings, the stem elongates rapidly, pushing the shoot (or, in most monocots, the cotyledon) up through the dark soil. During this early stage of growth, there is essentially no enlargement of the leaves, which would interfere with the passage of the shoot through the soil. Soil is not necessary for this growth pattern; any seedling grown in the dark will be elongated and spindly and will have small leaves. It will also be yellow to colorless, because plastids do not turn green until exposed to light. Such a seedling is said to be *etiolated* (Figure 25–15).

When the seedling tip emerges into the light, the etiolated growth gives way to normal plant growth. In dicots, the hook unbends, the stem growth rate may slow down somewhat, and leaf growth begins (see Figure 19–2, page 374). In grasses, growth of the mesocotyl (the part of the embryo axis between scutellum and coleoptile) stops, the stem elongates, and the leaves open.

A dark-grown bean seedling that receives five minutes of red light a day, for instance, will show these light effects beginning on the fourth day. If the expo-

25–14
The phytochrome chromophore (P$_r$ form), showing its attachment to the protein part of the molecule.

25–15
Dark-grown seedlings, such as the ones on the left, are thin and pale; they have longer internodes and smaller leaves than normal seedlings, such as those on the right. This group of physical characteristics, known as etiolation, has survival value for the seedling because it increases its chances of reaching light before its stored energy supplies are used up.

sure to red light is followed by a five-minute exposure to far-red, none of the changes usually produced by the red light will appear. (See also Figure 25–16.) Similarly, in the seedlings of grains, termination of mesocotyl growth is triggered by exposure to red light, and the effect of red light is canceled by far-red.

A recent report from England suggests that an important function of phytochrome in plants growing in a natural environment is the detection of shading by other plants. Radiation below 700 nanometers is almost completely reflected or absorbed by vegetation, whereas that between 700 and 800 nanometers (in the far-red range) is largely transmitted. This causes a dramatic upward shift in the equilibrium ratio of P_r to P_{fr} (that is, more P_{fr} driven to P_r) in shaded plants and results in a rapid increase in the rate of internodal elongation.

Reversible red/far-red reactions are also involved in anthocyanin formation in the apple, turnip, and cabbage; in the germination of seeds; in changes in the chloroplasts and other plastids; and in a tremendous variety of other plant responses in all phases of the plant life cycle.

25–16
All three bean plants received eight hours of light each day. The center plant was exposed to five minutes of elongation-promoting far-red light at the start of each dark period. The plant on the right received the same five-minute exposure to far-red light followed by a five-minute exposure to red light, which counteracted the effect of the far-red light. The plant on the left served as the control; its phytochrome is in the P_{fr} form as it enters the dark period because of its exposure to daylight.

Phytochrome and Photoperiodism

When the existence of phytochrome was first demonstrated, its discoverers hypothesized that its behavior might explain the phenomenon of photoperiodism—that is, that the reversible red/far-red reactions might be involved in the time-measuring mechanism, or biological clock. According to this hypothesis, P_{fr} inhibits flowering in short-day plants but promotes flowering in long-day plants. In short-day plants, P_{fr} would accumulate in the light and be removed in the subsequent dark period by destruction or dark reversion. When the nights were long enough, all (or a critical amount) of P_{fr} would be removed, and flowering would no longer be inhibited. Long-day plants, on the other hand, would require short nights, during which the P_{fr} would not be completely destroyed; if the night were short enough, sufficient P_{fr} would remain at the end of it to promote flowering.

Experiments have shown, however, that P_{fr} will disappear in many plants within three or four hours of darkness. On the basis of these experiments and other observations, it is now generally agreed that the time-measuring phenomenon of photoperiodism is not controlled by the interconversion of P_r and P_{fr} alone. A more complex explanation must be sought.

HORMONAL CONTROL OF FLOWERING

Hamner and Bonner, in their early cocklebur experiments, showed that the leaf "perceived" the light, which caused the bud to flower. Apparently, some substance that has profound effects on growth and development is transmitted from leaf to bud. This hypothetical substance has been termed the flowering hormone, or floral stimulus.

The early experiments on the floral stimulus were carried out independently in several laboratories in the 1930s. Some of the first, those of the Soviet plant physiologist M. Kh. Chailakhyan, carried out just a few years before the first cocklebur studies, are representative. Using the short-day *Chrysanthemum indicum*, Chailakhyan showed that if the upper portion of the plant was defoliated and the leaves on the lower part were exposed to a short-day induction period, the plant would flower. If, however, the upper, defoliated part was kept on short days and the lower, leafy part on long days, no flowering occurred. He interpreted these results as indicating that the leaves formed a hormone that moved to the stem apex and initiated flowering. Chailakhyan named this hypothetical hormone *florigen*, the "flower maker."

Further experiments showed that the flowering response will not take place if the leaf is removed immediately after photoinduction. But if the leaf is left on the plant for a few hours after the induction cycle is complete, it can then be removed without affecting flowering. The flowering hormone can pass through a graft from a photoinduced plant to a noninduced plant. Unlike auxin, however, which can pass through agar or other nonliving tissue, florigen can travel from one plant tissue to another only if there are anatomical connections of living tissue between them. If a branch is girdled, that is, if a circular strip of bark is removed, florigen movement ceases. On the basis of these data, it was concluded that florigen moves by means of the phloem system, the way most organic substances are transported in plants.

Subsequently, Anton Lang, then at the California Institute of Technology, showed that several long-day plants and biennials, such as celery and cabbage, could be made to flower by treatment with gibberellin, even when the plants were grown under the wrong photoperiod. This finding led Chailakhyan to modify his florigen hypothesis and to speculate that florigen actually consists of two classes of hormones, gibberellin and the as-yet-unidentified anthesin. According to this concept, long-day plants produce anthesin but not gibberellin during noninducing photoperiods. As such, gibberellin treatments at this time can cause flowering. On the other hand, short-day plants produce gibberellin but fail to make anthesin when they are grown under noninducing conditions. While the concept of florigen as a combination of gibberellin and anthesin has stimulated considerable research, it unfortunately cannot account for a critical observation: namely, that short-day plants grown under noninductive conditions will not cause flowering of grafted long-day plants that are also given a noninductive photoperiod.

In some plants—the Biloxi soybean (*Glycine max*) is an example—leaves must be removed from the grafted receptor plant or it will not flower. This observation suggests that the leaves of noninduced plants may produce an inhibitor. In fact, some investigators have concluded on the basis of such evidence that there is no substance that initiates flowering but rather a substance that inhibits flowering, until it is removed. Strong evidence now suggests that, at least in certain plants, both inhibitors and promoters are involved in the control of flowering.

The most convincing evidence for the existence of both flower-inducing and flower-inhibiting substances in the same plant is provided by the experimental studies of Anton Lang of Michigan State University and M. Kh. Chailakhyan and I. A. Frolova of the K. A. Timiryazev Institute of Plant Physiology in Moscow. These researchers chose three kinds of tobacco plants for their studies: the day-neutral *Nicotiana tabacum* cultivator Trabezond, the short-day tobacco cultivar Maryland Mammoth, and the long-day *Nicotiana silvestris*. They found that flower formation in the day-neutral tobacco was accelerated by a graft union with the long-day plant when the grafts were kept on long days, as well as by a graft union with the short-day tobacco when the grafts were kept on short days, as compared with grafted day-neutral controls. When long-day plants grafted to day-neutral plants were exposed to short days, flowering in the day-neutral receptor was greatly inhibited (Figure 25–17). By contrast, when short-day plants grafted to day-neutral plants were exposed to long days, there was little or no delay in flowering.

These results indicate that the leaves of the long-day plants are capable of producing flower-inducing substances on long days and flower-inhibiting substances on short days, and that both substances can be transferred through a graft union, an indication that both substances are transported through the plant. In the case of the short-day tobacco, the leaves apparently produce little or no flowering-inhibiting substance

25–17
Grafts of Nicotiana silvestris *(a long-day plant) onto the* Nicotiana tabacum *cultivar Trabezond (a day-neutral plant). Under long days, flower formation in the day-neutral tobacco was accelerated by the graft (as in the plant on the left). Under short days, the day-neutral tobacco remained vegetative (flowering inhibited) for the duration (90 to 94 days) of the experiment (as in the plant on the right).*

when exposed to long days. If such substances are produced by the short-day plant, they are much less effective in retarding flowering than those produced by the long-day plants.

Evidence for the existence of both flower-inducing and flower-inhibiting substances is very compelling, although attempts to isolate them have thus far been unsuccessful.

DORMANCY

Plants do not grow at the same rate all of the time. During unfavorable seasons, they limit their growth or cease to grow altogether. This ability enables plants to survive periods of water scarcity or low temperature.

Dormancy is a special condition of arrested growth that has evolved in both seeds and buds. Physiologically, a seed that fails to germinate, even though it has been provided with adequate water, sufficient oxygen, and a suitable temperature range for germination, is termed dormant. Dormancy in buds is defined as cessation of observable growth. Only certain, often quite precise, environmental cues can "activate" a dormant bud or embryo. This adaptation is of great survival importance to the plant. For example, the buds of plants expand, flowers are formed, and seeds germinate in the spring—but how do they recognize spring? If warm weather alone were enough, in many years all the plants would flower and all the seedlings would start to grow during Indian summer, only to be destroyed by the winter frost. The same could be said for any one of the warm spells that often punctuate the winter season. The dormant seed or bud does not respond to these apparently favorable conditions because of endogenous inhibitors, which must first be removed or neutralized before the period of dormancy can be terminated. In contrast to this "reluctance" to grow too rapidly, some seeds can germinate immediately. For example, this is true of willows (*Salix*) and sugar maple (*Acer saccharum*), whose seeds lose their viability within a week if kept in the air. In the mangrove (*Rhizophora mangle*), the seeds germinate on the parent plant. These seeds produce heavy radicles and fall like a dart, the radicle sometimes becoming embedded in the mud (Figure 25–18). Commercial seeds are artificially selected for their readiness to germinate promptly when they are exposed to favorable conditions, a trait that would be a great hazard for wild seeds.

Dormancy in Seeds

Almost all seeds growing in areas with marked seasonal temperature variations require a period of cold prior to germination. This requirement is normally satisfied by the temperatures of winter. The seeds of many ornamental plants have a similar cold requirement. If moist seed is exposed to low temperature for several days

(a)

(b)

25–18
(a) *Germinating seed and* (b) *young seedling of red mangrove* (Rhizophora mangle)*. The seeds germinate while still attached to the parent plant, producing a dartlike root, which may become embedded in the mud when the germinating seed falls from the plant.*

(average optimum temperature and time, 5°C for 100 days), the dormancy may be broken and the seed germinated. This horticultural procedure is termed *stratification*. Many seeds require drying before they germinate (although some may be nondormant before they dry out). This requirement prevents their germination within the moist fruit of the parent plant. Some seeds, as in the case of lettuce, require exposure to light, but others are inhibited by light.

Some seeds will not germinate in nature until they have become abraded, as by soil action. Such abrasion wears away the seed coat, permitting water or oxygen to enter the seed and, in some cases, removing the source of inhibitors. Hard seed coats that interfere with water absorption and embryo enlargement are common among legumes.

The seeds of some desert species germinate only when sufficient rain has fallen to leach away inhibitory chemicals present in the seed coat. The amount of rainfall necessary to wash off these germination inhibitors is directly related to the supply of water that the desert plant needs to become established as a seedling. Mechanical abrasion or breaking of the seed coat (*scarification*) with a knife, file, or sandpaper may allow the "hard seed" condition or inhibitor to be removed, or metabolic activity requisite to germination to be initiated. Germination may also be induced by soaking the seeds in alcohol or some other fat solvent (to dissolve waxy substances that impede entry of water) or concentrated acids. These procedures are widely used by horticulturists.

Some seeds may remain viable for a long time in the dormant condition, enabling them to exist for many years, decades, and even centuries under favorable conditions. In 1879, seeds of 20 species of common Michigan weeds were buried in moist, well aerated sand by W. J. Beal of Michigan State University as an experiment designed to determine their longevity. After 100 years, the seeds of three species were still viable. Seeds of 7 of 40 plant species recovered from the adobe walls of historic buildings constructed between 1769 and 1837 in California and northern Mexico were recently found to be viable. Although these demonstrations of endurance are impressive, they do not approach the record for seeds of the sacred lotus (*Nelumbo nucifera*) found by a Japanese botanist in a peat deposit in Manchuria. Radiocarbon dating showed the seeds to be some 2000 years old, but when the seed coats were filed to permit water to enter, every seed germinated.

In 1967, even this record reportedly was broken by seeds of the arctic tundra lupine (*Lupinus arcticus*). Some of these seeds, which were found in a frozen lemming burrow in the Yukon with animal remains estimated by carbon dating to be at least 10,000 years old, germinated within 48 hours. The age of the seeds themselves, however, was not determined.

During recent years, plant scientists have become increasingly interested in the factors involved in maintaining seed viability. Various enzyme systems may fail progressively in the stored seed, eventually leading to a complete loss of viability. Under what circumstances might viability best be prolonged? Such questions are relevant to the worldwide interest in developing seed banks with the goal of preserving the genetic characteristics of the wild and early cultivated varieties of various crop plants for use in future breeding programs. The need for such seed banks arises from the progressive replacement of older varieties with newer ones and the elimination of the replaced varieties, chiefly through the destruction of their habitats. In addition, many wild species of plants are in danger of becoming extinct for similar reasons, and the seeds of these plants should also be preserved, if possible (see "The Conservation of Plants," page 647).

Dormancy in Buds

Dormancy in buds is essential to the survival of temperate herbaceous and woody perennials that are exposed to low temperatures during winter. Although dormant buds do not elongate—that is, do not exhibit observable growth—they may undergo meristematic activity during various phases of dormancy.

Bud dormancy in many trees is initiated in midsummer, long before leaf fall in autumn. The dormant bud is an embryonic shoot consisting of an apical meristem, nodes and internodes (not yet extended), small rudimentary leaves or leaf primordia with buds or bud primordia in their axils, all enclosed by *bud scales* (Figure 25–19). The bud scales are very important because they help prevent desiccation, restrict the movement of oxygen into the bud, and insulate the bud from heat loss. In addition, growth inhibitors are known to accumulate in bud scales as well as in bud axes and leaves within buds. In many respects, therefore, the roles of the bud scales parallel those of the seed coat.

During and following the cessation of growth leading to the dormant condition, the tissues of the plant begin to undergo numerous physical and physiological changes to prepare the plant for winter, a process known as *acclimation*. Decreasing day length is the primary factor involved in the induction of dormancy in buds (Figure 25–20). Generally, more inhibitors are found in leaves and buds under short-day conditions than long-day conditions. Acclimation to cold leads to *cold hardiness*—the ability of the plant to survive the extreme cold and drying effects of winter weather.

As with seeds, the buds of many plant species require cold to break dormancy. If branches of flowering trees and shrubs are cut and brought inside in autumn, they do not flower; but if the same branches are left outside until late winter or early spring, they will bloom in the warmer indoor temperatures. Deciduous fruit trees,

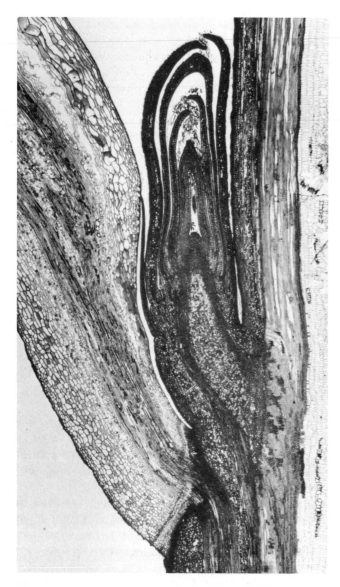

25–19
*Longitudinal section of a dormant,
axillary bud of maple (Acer). The bud
consists of an embryonic shoot enclosed
by bud scales.*

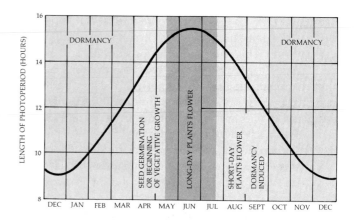

25–20
*Relationship between day length and the
developmental cycle of plants in the
North Temperate Zone.*

such as the apple, chestnut, and peach, cannot be grown in climates where the winters are not cold. Similarly, bulbs such as those of tulips, hyacinths, and narcissus can be "forced," that is, made to bloom inside in the winter, but only if they have previously been outside or in a cold place. As discussed in Chapter 22, such bulbs are actually large buds in which the leaves are modified for storage.

Investigations carried out under controlled conditions confirm these suggestions that cold is required for the breaking of dormancy in many species. Most varieties of peach, for example, must remain for 600 to 900

hours at temperatures below 4°C before they will respond to the activating influence of warmer temperatures and longer days. Some plants will respond to a brief exposure to freezing temperatures; if one bud of a greenhouse-cultivated lilac bush is briefly exposed to freezing temperatures that bud alone will break into bloom soon after. Cold is not required to break dormancy in all cases, however. In the potato, for instance, in which the "eyes" are dormant buds, at least two months of dry storage are the chief requirement; temperature is not a factor. In many plants, particularly trees, the photoperiodic response breaks winter dormancy, with the dormant buds being the receptor organs.

Application of gibberellins may sometimes break dormancy. For instance, gibberellin treatment of a peach bud may induce development after the bud has been kept for 164 hours below 8°C. Does this mean that under normal conditions, increase in gibberellin terminates the dormancy? Not necessarily. Dormancy may be a state of balance between growth inhibitors and growth stimulators. Addition of any growth stimulator (or removal of inhibitors, such as abscisic acid) may alter the balance so that growth begins.

There does not seem to be any common mechanism by which dormancy is induced or broken. This fact, although it considerably multiplies the problems of the plant physiologist, is certainly in keeping with our sense of evolutionary history. Dormancy became advantageous to plants only comparatively recently, when the seed plants began to spread into a variety of different ecological domains. Presumably, dormancy evolved independently in many groups of plants, each of which found its solution separately among the survivors, often in a different way.

COLD AND THE FLOWERING RESPONSE

Cold also may affect the flowering response. For example, if winter rye (*Secale cereale*) is planted in the autumn, it germinates during the winter and flowers the following summer, 7 weeks after growth begins. If it is planted in the spring, it does not flower for 14 weeks, and it remains vegetative throughout most of the growing season. In 1915, the German plant physiologist Gustav Gassner discovered that he could influence the flowering of winter rye and other cereal plants by controlling the temperature of the germinating seeds. He found that if the seeds of the winter strain are kept at near-freezing (1°C) temperatures during germination, the winter rye, even when planted in late spring, will flower the same summer it is planted. This procedure, which came to be known as *vernalization*, from the Latin *vernus*, meaning "spring," is now a common practice in agriculture.

Even after vernalization, the plant must be subjected to a suitable photoperiod, usually long days. The vernalized winter rye behaves like a typical long-day plant, flowering in response to the long days of summer. A similar example is seen in a biennial strain of henbane (*Hyoscyamus niger*). The vegetative rosette, which culminates the first year's growth, flowers only if it is exposed to cold. After cold exposure, it becomes a typical long-day plant, with the same photoperiodic response as that of the annual strain.

As the above examples indicate, in some plants the cold treatment affects the photoperiodic response. Spinach, ordinarily a long-day plant, does not usually flower until the days are 14 hours in length. If the spinach seeds are cold treated, however, they will flower when the days are only 8 hours long. Similarly, cold treatment of the clover *Trifolium subterraneum* can completely remove its dependency on day length for flowering.

In the biennial henbane and most other biennial long-day plants that form rosettes, gibberellin treatment can substitute for cold exposure. If gibberellin is applied, such plants will elongate rapidly and then flower. Application of gibberellin to short-day plants or to nonrosette long-day plants has little effect on (or inhibits) the flowering response. However, if gibberellin synthesis is inhibited when the plant is being exposed to the appropriate inductive cycle, the plant will not flower unless its gibberellin is replaced.

NASTIC MOVEMENTS

Nastic movements are plant movements that occur in response to a stimulus of some sort but whose direction is independent of the direction of the stimulus. Probably the most widely occurring nastic movements are the sleep movements mentioned previously, during our discussion of circadian rhythms. Known technically as *nyctinastic movements*, from the Greek meaning "night-closure," they constitute the up and down movements of leaves in response to the daily rhythms of light and darkness, with the leaves oriented vertically in darkness and horizontally in the light. These movements are especially common in leguminous plants.

Most nyctinastic leaf movements result from changes in the size of ground parenchyma cells in jointlike thickenings, the *pulvini*, at the base of each leaf (or, if the leaf is compound, at the base of each leaflet, also). The pulvinus is a flexible cylinder with the vascular system concentrated in the center. The largest volume of the pulvinus is occupied by thin-walled parenchyma cells that surround the vascular core. Pulvinar movement is associated with changes in turgor and the concomitant contractions and expansions of the ground parenchyma on opposite sides of the pulvinus. Turgor changes in the contracting and expanding cells are brought about by the shuttling of potassium ions between the two sides of the pulvinus, the site of the biological clock and the photoreceptor phytochrome.

Nastic movements resulting from touch are known as *thigmonastic movements*. Such movements are exemplified by the familiar sensitive plant (*Mimosa pudica*), whose leaflets and sometimes entire leaves droop suddenly when touched (Figure 25–21). As with sleep movements, this response is a result of a sudden change in turgor pressure in certain cells of pulvini at the base of leaflets and leaves. The loss of water from these cells follows the migration of potassium ions from them. Only a single leaflet needs to be stimulated; the stimulus then moves to other parts of the leaf and throughout the plant.

Two distinct mechanisms, one electrical and the other chemical, appear to be involved with the spread of the stimulus in the sensitive plant. There is some controversy about the survival value of this touch response to the plant. *Mimosa pudica* often grows in arid, exposed areas where it may be subjected to drying winds; strong winds may shake the leaves enough to make them fold up, thus conserving water. Another suggestion is that the wilting response makes the plant unattractive to large herbivores. Finally, it is conceivable that the folding response could startle herbivorous insects; there have been claims that other "nonsensitive" species of *Mimosa* growing near *Mimosa pudica* show more evidence of attack by insects.

Until recently, it was believed that the touch response in the carnivorous Venus flytrap (*Dionaea muscipula*), which enables it to capture prey, also involved changes in turgor pressure. It now appears, however, that the rapid closure of the leaves of the Venus flytrap

(a)

(b)

(c)

25–21

The sensitive plant (Mimosa pudica). *(a) The normal position of leaves and leaflets. Responses to touch with a needle*

are shown in (b) and (c). The responses result from changes in turgor pressure in certain cells of jointlike thickenings

(pulvini) at the base of the leaflets. Only a single leaflet need be stimulated for the response in (c) to occur.

involves irreversible cell enlargement, which can be initiated by acidifying the cell walls to pH 4.50 and below. During closure, the outer epidermal cells of the central portion of each leaf half expand, while the inner epidermal cells undergo no significant change in size. During reopening, the inner epidermal cells of the central portion expand, while the outer epidermal cells remain unchanged. These changes can be prevented by neutral buffers, which prevent the cell wall pH from reaching 4.5 to 4.75. Changes in ATP levels measured during closure indicate that about a third of the cellular ATP disappears during the period of time ranging from one to three seconds that is required for trap closure. It is likely that this ATP is used in the very rapid transport

of hydrogen ions from the cells.

Each half of the Venus flytrap leaf is equipped with three sensitive hairs. When an insect walks on one of these leaves, attracted by the nectar on the leaf surface, it brushes against the hairs, triggering the traplike closing of the leaf. The toothed edges mesh, the leaf halves gradually squeeze shut, and the insect is pressed against digestive glands on the inner surface of the trap (Figure 25–22).

The trapping mechanism is so specialized that it can distinguish between living prey and inanimate objects, such as pebbles and small sticks, that fall on the leaf by chance: the leaf will not close unless two of its hairs are touched in succession or one hair is touched twice.

25–22

Touch response in the Venus flytrap (Dionaea muscipula). *Here, an unwary fly, attracted by nectar secreted on the leaf surface, can be seen on a leaf, before and after its closure. Each leaf half has three sensitive hairs, which control the "trap." The fly triggered the traplike closing of the leaf halves when it brushed against one sensitive hair twice or two hairs in close succession.*

(a)

(b)

25–23
(a) *Leaves of a lupine (Lupinus arizonicus), orienting to track the course of the sun. This phenomenon is commonly known as solar tracking.* (b) *Solar tracking by a field of sunflowers (Helianthus annuus).*

SOLAR TRACKING

The leaves and flowers of many plants have the ability to move diurnally, orienting themselves either perpendicular or parallel to the sun's direct rays. Commonly known as *solar tracking,* this phenomenon is technically called *heliotropism* (from Greek *helios,* meaning "sun"). Unlike stem phototropism, leaf movement of heliotropic plants is not the result of asymmetric growth. In most cases the movements involve pulvini at the bases of leaves and/or leaflets. Some petioles appear to have pulvinal characteristics along most or all of their length. Some common plants exhibiting heliotropic leaf movements are cotton, soybeans, cowpeas, lupine, and sunflowers (Figure 25–23).

There are two types of heliotropism. In one (diaheliotropism), the movement of the leaves is such that the blades remain perpendicular to the sun's direct rays throughout the day. Such leaves have more quanta available for photosynthesis and appear to have higher photosynthetic rates throughout the day than nontracking leaves or leaves of the second type of heliotropism (Figure 25–24).

During periods of drought, some heliotropic plants actively avoid direct sunlight by orienting their leaf blades parallel to the sun's rays (paraheliotropism). Whereas this orientation minimizes absorption of solar radiation rather than maximizing it (Figure 25–24), it also decreases leaf temperature and transpirational water loss, enhancing survival during drought periods.

IN CONCLUSION

Each developmental event in the life of a plant is under the control not of any single factor but of a complex variety of factors. Internal and environmental factors in-

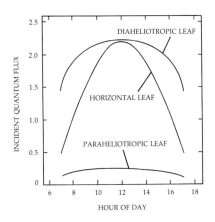

25–24
Comparison of the photosynthetically useful solar radiation between 400 and 700 nm incident on a diaheliotropic leaf, nontracking (horizontal) leaf, and paraheliotropic leaf over the course of a day.

teract. They may enhance, modify, or neutralize one another. In the words of William S. Hillman of the Brookhaven National Laboratory, "After all, if plants were as simple as the physiologist might wish, there would be nothing left to do."

SUMMARY

Plants possess a variety of adaptations that enable them to detect and respond to alterations in their environment. Phototropism, or the curving of the growing shoot toward light, is an example of such an adaptation. The differential growth of the seedling is caused by lateral migration of the growth hormone auxin under the influence of light. The photoreceptor for this response is a blue-light absorbing pigment. Gravitropism, or geotropism, is the response to the pull of the earth's gravity by a shoot or a root. The movement of auxin to the lower surface of a horizontally oriented shoot may play a role in the upward curvature of the shoot. The downward curving of a horizontally oriented root appears to result in part from a substance that originates in the root cap and inhibits growth. Thigmotropism is a response to contact with a solid object.

Circadian rhythms are cycles of activity in an organism that recur at intervals of about 24 hours under constant environmental conditions. They are probably endogenous, caused not by an external factor, such as alternating light and darkness or the earth's rotation, but by some internal timing mechanism within the organism. Such a timing mechanism, the chemical and physical nature of which is unknown, is called the biological clock. The possession of a biological clock makes it possible for an organism to perceive changes in external diurnal cycles, such as the lengthening and shortening of days as the seasons progress. In this way, activities such as dormancy, leaf abscission, and flowering can be brought into synchrony with the external environment.

Photoperiodism is the response of organisms to changing 24-hour cycles of light and darkness. Such responses control the onset of flowering in many plants. Some plants will flower only when the periods of light exceed a critical length. Such plants are known as long-day plants. Other plants, short-day plants, flower only when the periods of light are less than some critical period. Day-neutral plants flower regardless of photoperiods. Factors such as temperature and the age of the plant may affect the photoperiodic response. Interruption of the dark phase of the photoperiod, even by a brief flash of light, can serve to reverse the photoperiodic effects.

Phytochrome, a pigment commonly present in small amounts in the tissues of higher plants, is sensitive to the transitions between light and darkness. The pigment can exist in two forms, P_r and P_{fr}. The P_r form absorbs red light with a wavelength of 660 nanometers and is thereby converted to P_{fr}. The P_{fr} form absorbs far-red light (730 nanometers) and is converted to P_r. The P_{fr} form is also lost from the cell in the dark by reversion to P_r or by destruction. P_{fr} is the active form of the pigment; it promotes flowering in long-day plants and inhibits flowering in short-day plants. P_{fr} also is responsible for changes that take place in seedlings as they penetrate the soil to the light, for germination of seeds, and for development of anthocyanins.

In both long-day and short-day plants, the photoperiod is perceived in the leaves but the response takes place in the bud. Phytochrome is the photoreceptor that perceives the environmental stimulus and appears to interact with an endogenous circadian rhythm in an undetermined fashion to generate a floral stimulus. Although it has not yet been isolated or identified, this chemical substance—now termed florigen—moves from the leaves to the bud, where it induces flowering. Experiments have indicated that the chemical travels through the plant by way of the phloem system and that its structure and function are similar in long-day, short-day, and day-neutral plants. Strong evidence suggests that, at least in certain plants, both flower-inducing and flower-inhibiting substances are involved in flowering.

Alternation of periods of growth and cessation of growth permits the plant to survive water shortages and extremes of hot or cold. Dormancy is a special condition of arrested growth in which the entire plant, or such structures as seeds or buds, do not renew growth without special environmental cues. The requirement for such cues, which include cold exposure, dryness, and suitable photoperiod, prevents the tissue from breaking dormancy during superficially favorable conditions, such as those found within the succulent fruit of the parent plant or during the warmth of Indian summer. There is apparently no uniform mechanism, common to all plant groups, for the induction and breaking of dormancy. Decreasing day length is the primary factor involved in the induction of dormancy in buds. Acclimation to cold leads to cold hardiness, the ability of the plant to survive the extreme cold of winter weather. Vernalization refers to the promotion of flowering in winter strains by keeping seeds at low temperatures. Hormones, cold, and light interact to modify plant responses.

Plant movements that occur in response to a stimulus, but whose direction is independent of the direction of the stimulus, are called nastic movements. Among these are the widely occurring sleep movements—the up and down movements of leaves in response to the daily rhythm of light and darkness. Nastic movements resulting from touch include the triggered closing of the carnivorous Venus flytrap.

The leaves and flowers of certain plants follow or track the sun during the course of a day, either maximizing or minimizing absorption of solar radiation.

SUGGESTIONS FOR FURTHER READING

BRADY, JOHN: *Biological Clocks, Studies in Biology, No. 104*, University Park Press, Baltimore, Md., 1979.*

An interesting and well-written introduction to the subject of biological clocks and their experimental study.

ELLMORE, GEORGE S.: "The Organization and Plasticity of Plant Roots," *Scanning Electron Microscopy III*: 1083–1102, 1982.

An interesting account of the plant root as a dynamic organ—of the fluctuation of its structural and physiological components throughout its life history.

* Available in paperback.

GALSTON, ARTHUR W., PETER J. DAVIES, and RUTH L. SATTER: *The Life of the Green Plant*, 3rd ed., Prentice-Hall, Inc., Englewood Cliffs, N.J., 1980.*

A comprehensive, up-to-date, and basic description of the functioning of the green plant, especially suited to persons without an advanced background in biology or chemistry.

KENDRICH, RICHARD E., and BARRY FRANKLAND: *Phytochrome and Plant Growth, Studies in Biology, No. 68*, 2nd ed., Edward Arnold Publishers, London, 1983.*

A concise, well-illustrated introduction to the molecular properties and physiology of phytochromes.

LEOPOLD, A. CARL, and P. E. KRIEDEMAN: *Plant Growth and Development*, 2nd ed., McGraw-Hill Book Company, New York, 1975.

An important research work that provides much material of interest in connection with this and following chapters.

MOHN, HANS: *Lectures on Photomorphogenesis*, Springer-Verlag, New York, 1972.*

An in-depth treatment of the effects of light on plant development, primarily in seedlings of the wild mustard.

SALISBURY, FRANK B.: *The Biology of Flowering*, Natural History Press, Garden City, N.Y., 1971.

A semipopular account of research in the field, concentrating mainly on the short-day plant Xanthium.

SALISBURY, FRANK B., and CLEON W. ROSS: *Plant Physiology*, 3rd ed., Wadsworth Publishing Co., Inc., Belmont, Calif., 1985.

A detailed and useful review of the entire subject.

SMITH, HARRY: *Phytochrome and Photomorphogenesis*, McGraw-Hill Book Company, New York, 1975.

A particularly well-written account of the role of the phytochrome pigment in photomorphogenesis.

STEEVES, TAYLOR A., and IAN M. SUSSEX: *Patterns in Plant Development*, Prentice-Hall, Inc., Englewood Cliffs, N.J., 1972.

An exceptionally well-written and well-illustrated text that should be read by all students interested in plant morphogenesis.

THIMANN, KENNETH V.: *Hormone Action in the Whole Life of Plants*, University of Massachusetts Press, Amherst, Mass., 1977.

An outstanding review of the ways in which hormones initiate and control the growth and development of plants; written by a scholar who has participated in many of the fundamental discoveries in his field.

TORREY, JOHN G.: "The Development of Plant Biotechnology," *American Scientist* 73: 354–363, 1985.

An informative and interesting presentation of the history of plant biotechnology, beginning with the ideas of Justus von Liebig in the 1840s, and including consideration of present-day problems and prospects for the future.

VINCE-PRUE, DAPHNE: *Photoperiodism in Plants*, McGraw-Hill Book Company, New York, 1975.

A comprehensive treatment of the photoperiodic control of flowering, as well as other responses to light.

WILKINS, MALCOLM B. (Ed.): *Advanced Plant Physiology*, Pitman Press, Bath, Great Britain, 1984.

An up-to-date, advanced plant physiology textbook written by multiple authors.

* Available in paperback.

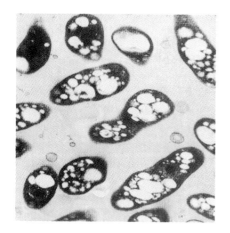

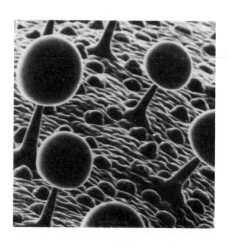

SECTION SEVEN

Uptake and Transport in Plants

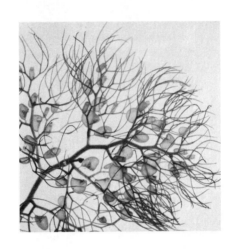

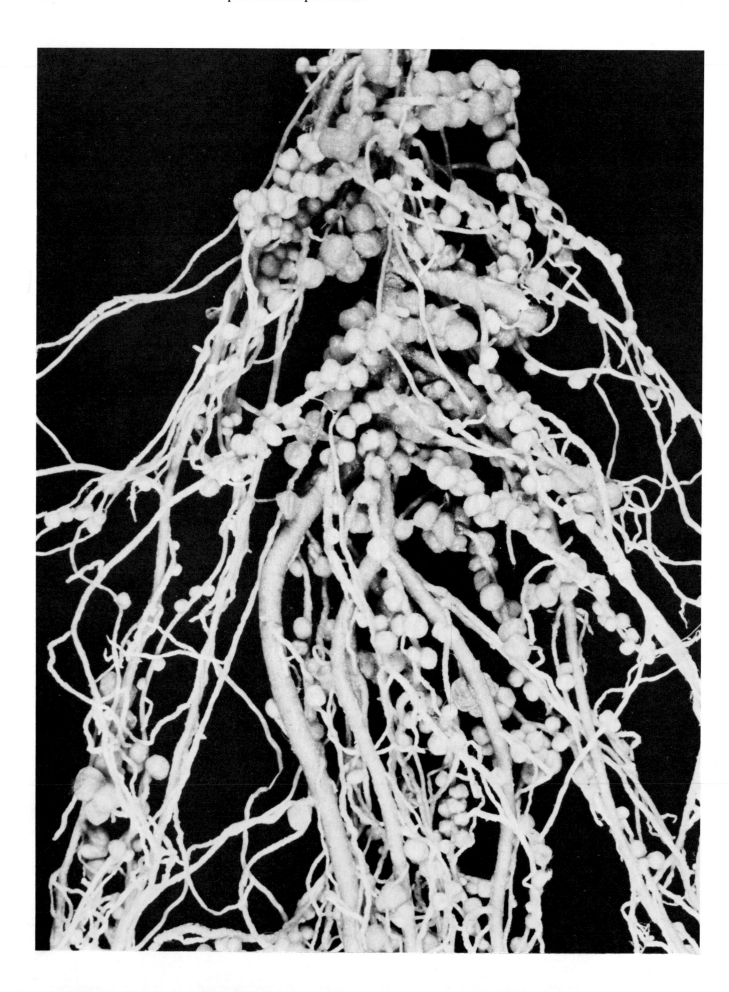

C H A P T E R   2 6

Plant Nutrition and Soils

GENERAL NUTRITIONAL REQUIREMENTS

Plants must obtain from the environment the specific raw materials required in the complex biochemical reactions necessary for the maintenance of their cells and for growth. In addition to light, higher plants require water and certain chemical elements for metabolism and growth. Much of the evolutionary development of plants has involved structural and functional specialization requisite to efficient acquisition of these raw materials and their distribution to living cells throughout the plant.

Like plants, animals need water and specific chemical elements for metabolism; animals, however, must also obtain external supplies of sugars or other compounds that can serve as an energy source, as well as certain amino acids and vitamins. By comparison, the nutritional demands of plants are relatively simple. Under favorable environmental conditions, most green plants can use light energy to transform CO_2 and H_2O into organic compounds for their energy source. They can also synthesize all of their required amino acids and vitamins, using inorganic nutrients drawn from the environment.

The self-sufficiency of nitrogen-fixing cyanobacteria is particularly impressive. Through the process of nitrogen fixation, they convert atmospheric nitrogen into forms of nitrogen that can be utilized in the synthesis of amino acids and proteins. In addition, of course, they carry out photosynthesis.

Plant nutrition involves the uptake from the environment of all the raw materials required for essential biochemical processes, the distribution of these materials within the plant, and their utilization in metabolism and growth.

26–1

The symbiotic relationship between soil bacteria of the genus Rhizobium *and the cortical root cells of the bird's-foot trefoil* Lotus corniculatus, *a legume, leads to the development of conspicuous root nodules. The bacteria living within the nodules use organic compounds supplied by the plant as an energy source for their metabolic activities, which include nitrogen fixation. The legume, in return, obtains a ready supply of nitrogen in a form it can use to manufacture plant proteins.*

Essential Inorganic Nutrients

As early as 1800, chemists and plant biologists had analyzed plants and demonstrated that certain chemical elements were absorbed from the environment. Opinions differed, however, on whether the absorbed elements were impurities or constituents required for essential functions. By the mid-1880s it had been established that at least ten of the chemical elements present in plants were necessary for normal growth. In the absence of any one of these elements, plants displayed characteristic abnormalities of growth or deficiency symptoms, and often such plants did not reproduce normally. These ten elements—carbon, hydrogen, oxygen, potassium, calcium, magnesium, nitrogen, phosphorus, sulfur, and iron—were designated as essential chemical elements for plant growth. They are also referred to as essential minerals, or *essential inorganic nutrients*.

In the early 1900s, it was established that manganese was also an essential element. During the next 50 years, with the aid of improved techniques for removing impurities from nutrient cultures, five additional elements—zinc, copper, chlorine, boron, and molybdenum—were determined to be essential, with chlorine being recognized only in 1954. At present, these 16 elements are generally considered to be essential for most plants (Figure 26–2).

Nutrient Concentrations in Plants

Chemical analyses for essential inorganic nutrients are a useful means of determining the relative amounts of various elements required for the normal growth of different plant species. Typical results of such chemical analyses are given in Table 26–1. Analyses of this sort are particularly valuable in agriculture as a guide to the nutritional well-being of plants and the need for applications of fertilizer. Potential nutritional deficiencies in livestock that consume specific plants can also be predicted by inorganic analyses.

Table 26–1   *Examples of Inorganic Analyses of Plants*

	CONCENTRATION OF ELEMENT		
ELEMENT	ALFALFA	CORN*	WHITE OAK**
Potassium	2.77%	1.86%	0.85%
Calcium	1.70	0.40	0.82
Magnesium	0.41	0.27	0.36
Nitrogen	3.12	2.81	2.19
Phosphorus	0.35	0.28	0.19
Sulfur	0.29	0.18	0.13
Iron	190 ppm***	110 ppm	126 ppm
Manganese	62	80	572
Zinc	57	27	22
Copper	9	6	8
Chlorine	8800	3100	43
Boron	35	14	38
Molybdenum	1.40	1.03	6.21
Sodium	4300 ppm	127 ppm	210 ppm
Cobalt	0.21	0.16	—

* Shoot only; grain not included in analysis; carbon, hydrogen, and oxygen not included.
** Leaf and twig growth of current year.
*** Abbreviation for parts per million; ppm equals units of an element by weight per million units of oven-dried plant material; 1% equals 10,000 ppm.

26–2
Deficiencies of one or more of the essential inorganic nutrients result in abnormal plant growth and development. (a) and (b), respectively, show the effects of zinc and boron deficiencies on a tomato plant. The plant to the left of the nutrient-deficient plant in (a) and (b) is a control plant.

(a)

(b)

The concentrations of specific elements in plants are known to vary over a wide range. On the basis of the usual concentrations in plants, the essential inorganic nutrients can be divided into two broad groups—macronutrients and micronutrients. In general, *macronutrients* are elements that are required in large amounts, and *micronutrients* ("trace elements") are those required in very small, or *trace*, amounts (Table 26–2).

Table 26–2   *A Summary of the Functions of Inorganic Nutrients in Plants*

ELEMENT	PRINCIPAL FORM IN WHICH ELEMENT IS ABSORBED	USUAL CONCENTRATION IN HEALTHY PLANTS (% OF DRY WEIGHT)	IMPORTANT FUNCTIONS
Macronutrients			
Carbon	CO_2	~44%	Component of organic compounds.
Oxygen	H_2O or O_2	~44%	Component of organic compounds.
Hydrogen	H_2O	~6%	Component of organic compounds.
Nitrogen	NO_3^- or NH_4^+	1–4%	Amino acids, proteins, nucleotides, nucleic acids, chlorophyll, and coenzymes.
Potassium	K^+	0.5–6%	Enzymes, amino acids, and protein synthesis. Activator of many enzymes. Opening and closing of stomata.
Calcium	Ca^{2+}	0.2–3.5%	Calcium of cell walls. Enzyme cofactor. Cell permeability. Component of calmodulin, a regulator of membrane and enzyme activities.
Phosphorus	$H_2PO_4^-$ or HPO_4^{2-}	0.1–0.8%	Formation of "high-energy" phosphate compounds (ATP and ADP). Nucleic acids. Phosphorylation of sugars. Several essential coenzymes. Phospholipids.
Magnesium	Mg^{2+}	0.1–0.8%	Part of the chlorophyll molecule. Activator of many enzymes.
Sulfur	SO_4^{2-}	0.05–1%	Some amino acids and proteins. Coenzyme A.
Micronutrients			
Iron	Fe^{2+} or Fe^{3+}	25–300 ppm	Chlorophyll synthesis, cytochromes, and nitrogenase.
Chlorine	Cl^-	100–10,000 ppm	Osmosis and ionic balance; probably essential in photosynthetic reactions that produce oxygen.
Copper	Cu^{2+}	4–30 ppm	Activator of certain enzymes.
Manganese	Mn^{2+}	15–800 ppm	Activator of certain enzymes.
Zinc	Zn^{2+}	15–100 ppm	Activator of many enzymes.
Molybdenum	MoO_4^{2-}	0.1–5.0 ppm	Nitrogen fixation. Nitrate reduction.
Boron	BO_3^- or $B_4O_7^{2-}$	5–75 ppm	Influences Ca^{2+} utilization. Functions unknown.
Elements Essential to Some Plants or Organisms			
Cobalt	Co^{2+}	Trace	Required by nitrogen-fixing microorganisms.
Sodium	Na^+	Trace	Osmotic and ionic balance, probably not essential for many plants. Required by some desert and salt-marsh species. May be required by all plants that utilize C_4 photosynthesis.

26–3

(a) *Plants of the mustard family, such as wintercress* (Barbarea vulgaris), *use sulfur in the synthesis of the mustard oils that give these plants both their name and their characteristic sharp taste.* (b) *Horsetails* (Equisetum) *incorporate silicon into their cell walls, making them indigestible to most herbivores but useful, as was the case in colonial North America, for scouring pots and pans.*

(a)

(b)

Potassium probably reaches higher concentrations in plants than any element other than carbon and oxygen. When the soil concentration is very high, as much as 10 percent of the oven-dried weight of plants may be potassium.

Certain plant species and taxonomic groups are characterized by unusually high or low amounts of specific elements (Figure 26–3). As a result, plants growing in the same nutrient medium may differ markedly in nutrient content. Compare the analyses in Table 26–1 for corn (a monocot) and alfalfa (a dicot). Dicots generally require greater amounts of calcium and boron than do monocots.

Nutritional studies have established that some elements are essential for only limited groups of plants or for plants grown under specific environmental conditions. Legumes, such as alfalfa (*Medicago sativa*), benefit from the addition of cobalt to the culture medium. It is not the alfalfa, however, that requires cobalt but the symbiotic nitrogen-fixing bacteria growing in association with the roots of the alfalfa. As indicated in Table 26–1, sodium is present in many plants in relatively high concentrations. It has been recognized for many years that sodium can partially replace the requirement for potassium in some species. Relatively recent studies have shown that sodium is an essential element for still other species. For example, sodium appears to be required by plants and by certain halophytes (plants that grow in salty soils). In addition, it has recently been demonstrated that soybean plants deprived of nickel accumulated toxic concentrations of urea in necrotic lesions (dead tissue) on the tips of their leaflets and exhibited reduced growth. Addition of nickel to the nutrient media prevented urea accumulation, necrosis, and reduction in growth. This evidence indicates that nickel is essential for soybeans.

Table 26–1 includes two elements—sodium and co-
balt—that are not essential for all crop plants but are essential for herbivorous animals that consume these plants. For example, if the cobalt concentration in a forage crop is less than 0.1 ppm, animals eating these plants are likely to display symptoms of cobalt deficiency.

FUNCTIONS OF INORGANIC NUTRIENTS IN PLANTS

Inorganic nutrients are essential in many aspects of growth and metabolism. Table 26–2 lists the inorganic nutrients required by most plants, the forms in which they most commonly are absorbed from the environment, their usual concentration range, and some of their functions.

Specific versus Nonspecific Functions

Inorganic ions affect osmosis (Chapter 4) and thus help to regulate water balance. Because several inorganic ions can serve interchangeably in this role, in many plants this particular requirement is described as *nonspecific.* On the other hand, an inorganic nutrient may function as part of an essential biological molecule; in this case the requirement is highly *specific.* An example of a specific function is the presence of magnesium in the chlorophyll molecule (see Figure 7–8, page 99). Some inorganic nutrients are essential constituents of cellular membranes, whereas others control the permeability of such membranes. Other inorganic nutrients are indispensable components of a variety of enzyme systems that catalyze biological reactions in the cell. Still others provide a proper ionic environment in which biological reactions can occur.

Because inorganic nutrients fill such basic needs and are involved in such fundamental processes, the deficiencies affect a wide variety of structures and functions in the plant body.

Catalysts

As key role of the inorganic nutrients is their participation as catalysts in some of the enzymatic reactions of the plant cell. In some cases, they are an essential structural part (a "prosthetic group") of the enzyme. In other cases, they serve as activators or regulators of certain enzymes. Potassium, for instance, which probably affects 50 to 60 enzymes, is believed to regulate the conformation of some proteins. Changing the shape of an enzyme could, for example, expose or obstruct reaction sites (see Figure 3–18, page 55).

Electron Transport

Many of the biochemical activities of cells, including photosynthesis and respiration, are oxidation-reduction reactions (see page 75). In such reactions, electrons are often transferred to or from a molecule that functions as an electron acceptor. The cytochromes, which contain iron (see Figure 6–9, page 89), are involved in electron transfer.

Structural Components

Some mineral elements serve as structural components of cells, either as part of a physical structure (see Figure 26–3b) or as part of the chemical structures involved in cellular metabolism. Calcium combines with pectic acid in the middle lamella of the plant cell wall. Phosphorus occurs in the sugar phosphate backbone of the DNA helix, in RNA, and in the phospholipids of the cellular membranes. Nitrogen is an essential component of amino acids, chlorophylls, and nucleotides. Sulfur is found in two amino acids, thus forming an important structural element in proteins.

Osmosis

The movement of water into and out of plant cells, as discussed in Chapter 4, is largely dependent on the concentration of solute in the cells and in the surrounding medium. The uptake of ions by the plant cells thus may result in the entry of water into the cell. The resultant turgor pressure from within the cell results in expansion of the immature cell, which is the chief cause of cellular growth, and is responsible for turgor in the mature cell (see page 62). This is an example of conversion of energy from one form to another by a living system; the chemical (ATP) energy expended in the active uptake of ions by the plant cell is translated into the physical energy of water movement.

Effects on Cell Permeability

Calcium has a direct effect on the physical properties of the cellular membranes. When there is a calcium deficiency, membranes seem to lose their integrity, and solutes within the membranes or cells leak out.

THE SOIL

Soil is the primary nutrient medium for plants. Soils must provide plants not only with physical support but also adequate inorganic nutrients at all times, as well as with adequate water and a suitable gaseous environment for the root systems. Understanding the origins of soils and their chemical and physical properties in relation to plant growth requirements is critical in planning for the nutrition of field crops.

Weathering of the Earth's Crust

The inorganic nutrients utilized by plants are derived from the atmosphere and from the weathering of rocks in the crust of the earth. The earth is composed of about 92 naturally occurring elements, which are often found in the form of minerals. *Minerals* are naturally occurring inorganic compounds that are usually composed of two or more elements in definite proportions by weight (see Appendix A). Quartz (SiO_2) and calcite ($CaCO_3$) are examples of minerals.

Most rocks consist of several different minerals and are divided into three groups based on origin and formation. *Igneous* rocks such as granite are derived directly from molten material and, for the most part, were originally formed when the earth cooled and solidified. As a result of weathering, igneous rocks and other kinds of rocks may be broken down into soluble and insoluble component parts. After being transported by water, wind, or glaciers, these components form new deposits—usually in water—that in time become cemented and solidified into *sedimentary* rocks, such as shale, sandstone, and limestone. Although sedimentary rocks form only about 5 percent of the earth's crust, they are of great importance because they occur extensively at or near the surface. Under the extreme heat and pressure deep within the earth, sedimentary and igneous rock may be transformed into a third type of rock—*metamorphic*. In this way, quartzite is formed from sandstone, slate from shale, and marble from limestone.

Weathering processes, involving the physical disintegration and chemical decomposition of minerals and rocks at or near the earth's surface, produce the inorganic materials from which soils are formed. Weathering involves freezing and thawing or heating and cooling, which cause substances in the rocks to expand and contract, splitting rocks apart. Water and wind often carry the rock fragments great distances, exerting a scouring action that breaks and wears the fragmented rock into even smaller particles. Water enters between the particles, and soluble materials dissolve in the water. Water, in combination with carbon dioxide and air impurities such as sulfur dioxide and oxides of nitrogen, forms dilute acids that help to dissolve materials that are less soluble in pure water. Soil formation may occur at the site of weathering, or the parent materials may be transported elsewhere by gravity, wind, water, or glaciers.

Soils also contain organic materials. If light and temperature conditions permit, bacteria, fungi, algae, lichens, and bryophytes and small vascular plants gain a foothold on or among the weathered rocks and minerals. Growing roots also split rocks, and their disintegrating bodies and those of the animals associated with them add to the accumulating organic material. Finally, larger plants move in, anchoring the soil in place with their root systems, and a new community begins (Figure 26–4).

Examining a vertical section of soil (Figure 26–5), one can see variations in the color, the amount of living and dead organic matter, the porosity, the structure, and the extent of weathering. These variations generally result in a succession of rather distinct layers that soil scientists refer to as *horizons*. A minimum of three horizons is recognized.

The A horizon (sometimes called the "topsoil") is the upper region, that of the greatest physical, chemical, and biological activity. The A horizon contains the greatest portion of the soil's organic material, both living and dead, such as large amounts of dead and decaying leaves and other plant parts, insects and other small arthropods, earthworms, protists, nematodes, and decomposer organisms (Figure 26–6).

The B horizon is a region of deposition. Iron oxide, clay particles, and small amounts of organic matter are among the materials carried from the A horizon to the B horizon by water percolating, or moving down, through the soil. The B horizon contains much less organic material and is less weathered than the horizon above it.

The C horizon is composed of the broken-down and weathered rocks and minerals from which the true soil in the upper horizons is formed.

Soil Composition

Soils consist of solid matter and pore space (the space around the soil particles). Different proportions of air and water occupy the pore space, depending on prevailing moisture conditions. The soil water is present primarily as a film on the surfaces of soil particles.

26–4

The fibrous root systems of grasses bind and anchor prairie soil in place.

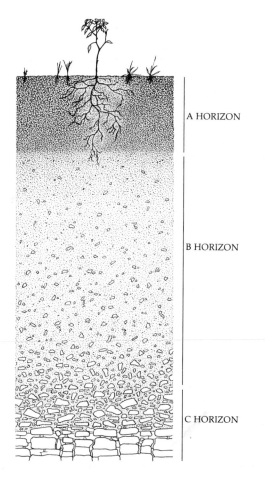

A HORIZON

B HORIZON

C HORIZON

26–5

The three primary horizons, or soil layers, recognized in a typical soil.

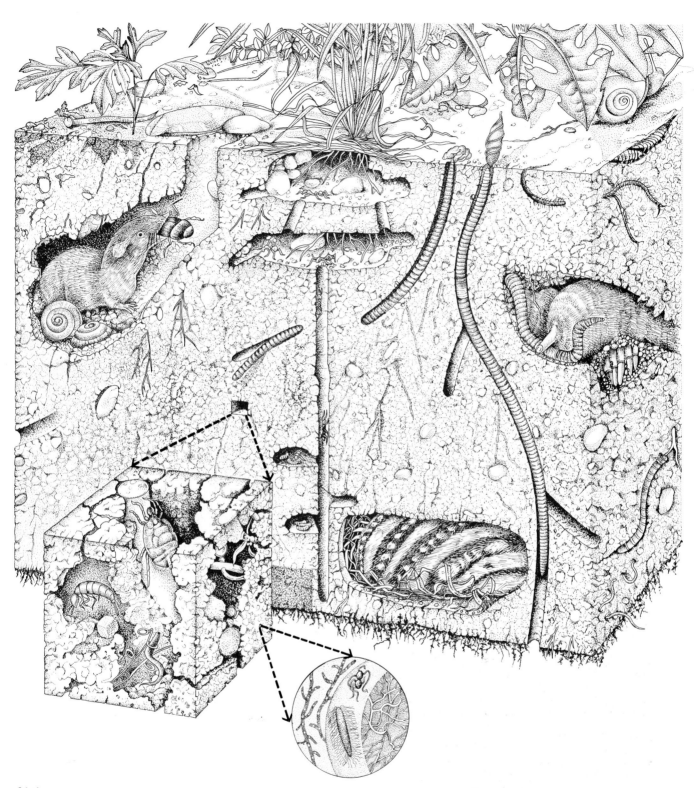

26–6

Plants share the soil with a vast number of living organisms, ranging from microbes to small mammals such as moles, shrews, and ground squirrels. Multitudes of burrowing creatures—most notably ants and earthworms—aerate the soil and improve its ability to absorb water. Called by Aristotle "the intestines of the earth," earthworms refine the soil by processing it through their gut. The refined soil is then deposited on the soil surface in the form of castings. In a single year, the combined activities of earthworms may produce as much as 500 metric tons of castings per hectare. The castings are very fertile, containing 5 times the nitrogen content of the surrounding soil, 7 times the phosphorus, 11 times the potassium, 3 times the magnesium, and 2 times the calcium. Bacteria and fungi are the principal decomposers of the organic matter in soils.

THE WATER CYCLE

The earth's supply of water is stable and is used over and over again. Most of the water (98 percent) is present in oceans, lakes, and streams. Of the remaining 2 percent, some is frozen in polar ice and glaciers, some is found in the soil, some is in the atmosphere as water vapor, and some is in the bodies of living organisms.

Sunshine evaporates water from the oceans, lakes, and streams, from the moist soil surfaces, and from the bodies of living organisms, drawing the water back up into the atmosphere, from which it falls again as rain. This constant movement of water from the earth into the atmosphere and back again is known as the water cycle. The water cycle is driven by solar energy.

Some of the water that falls on the land percolates down through the soil until it reaches a zone of saturation. In the zone of saturation, all holes and cracks in the rock are filled with water. Below the zone of saturation is solid rock through which the water cannot penetrate. The upper surface of this zone of saturation is known as the water table.

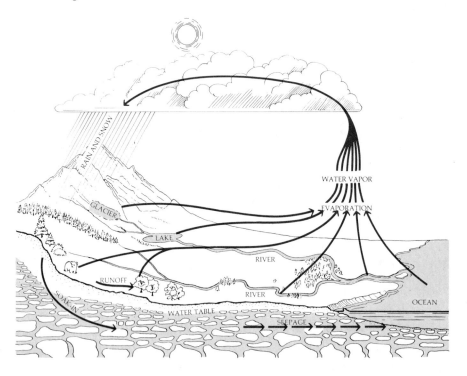

The fragments of rocks and minerals in the soil vary in size from sand grains, which can be seen easily with the naked eye, to clay particles too small to be seen even under the low power of a light microscope. The following classification is one scheme for categorizing soil particles (also known as soil separates) according to size:

Separate	Diameter (in micrometers)
Coarse sand	200–2000
Fine sand	20–200
Silt	2–20
Clay	Less than 2

Soils contain a mixture of particles of different sizes and are divided into textural classes according to the proportions of different particles present in the mixture. For example, soils that contain 35 percent or more clay and 45 percent or more sand are sandy clays; those containing 40 percent or more clay and 40 percent or more silt are silty clays. Loam soils contain sand, silt, and clay in proportions that result in ideal agricultural soils.

The solid matter of soils consists of both inorganic and organic materials, with the proportions varying greatly in different soils. The organic component includes the remains of organisms in various stages of decomposition, as well as a wide range of living plants and animals. Structures as large as tree roots may be included, but the living phase is dominated by fungi, bacteria, and other microorganisms.

Cation Exchange

The inorganic nutrients taken in through the roots of plants are present in the soil solution as ions. Most metals form positively charged ions, that is, cations, such as Ca^{2+}, K^+, and Na^+. Clay particles provide a reservoir of such cations for the plant—at various points on their crystalline lattice there is an excess of negative charge, where cations can be bound and thus held against the leaching action of percolating soil water.

The cations bound in this way to the clay particles can be replaced by other cations (in a process called _cation exchange_) and then released into the soil solution, where they become available for plant growth. This is one reason clay particles are an essential component of productive soils.

The principal negatively charged ions, or anions, found in soil are NO_3^-, SO_4^{2-}, HCO_3^-, and OH^-. Anions are leached out of the soil more rapidly than cations because they do not attach to clay particles. An exception is phosphate, which is retained against leaching because it forms insoluble precipitates and is specifically absorbed, or held on the surface, of compounds containing iron, aluminum, and calcium.

The acidity or alkalinity of soil is related to the availability of inorganic nutrients for plant growth. Soils vary widely in pH, and many plants have a narrow range of tolerance on this scale. In alkaline soils, some cations are precipitated, and such elements as iron, manganese, copper, and zinc may thereby become unavailable to plants.

Soils and Water

Approximately 50 percent of the total soil volume is represented by pore space, which is occupied by varying proportions of air and water, depending on moisture conditions. When no more than half the pore space is occupied by water, adequate oxygen is available for root growth and other biological activity.

Following a heavy rain or irrigation, soils retain a certain amount of water and remain moist even after gravity has removed loosely bound water. If the fragments that make up a soil are large, the pores and spaces between them will be large; water will drain through the soil rapidly, and relatively little will be available for plant growth in the A and B horizons. Because of their finer pores, clay soils are able to hold a much greater amount of water against the action of gravity. Thus, clay soils may retain three to six times more water than a comparable volume of sand. The percentage of water that a soil can hold against the action of gravity is called its _field moisture capacity_.

If a plant is allowed to grow indefinitely in a sample of soil and no water is added, the plant will eventually not be able to absorb water rapidly enough to meet its

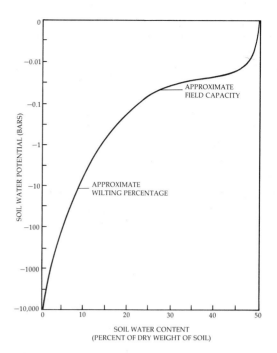

26–7
Relation between the soil water potential and the soil water content in a sandy loam soil.

needs, and it will droop and wilt. When wilting is severe, plants fail to recover even when placed in a humid chamber. The percentage of water remaining in a soil when such irreversible wilting occurs is called the _permanent wilting percentage_ of that soil.

Figure 26–7 shows the relationship between the soil water content and the potential with which that water is held by a sandy loam soil. The forces that retain water in the soil can be expressed in the same terms (in this case, water potential) as the forces for water uptake that develop in cells and tissues (see Chapter 4, pages 58–59). The potential of the soil water decreases gradually with a decrease in the soil moisture below field moisture capacity. The soil water potential then decreases sharply as the moisture content approaches the permanent wilting percentage of approximately −15 bars.

NUTRIENT CYCLES

The bulk of the organic matter of soil is made up of dead leaves and other plant material, together with the decomposing bodies of animals. This organic debris is mixed with the inorganic particles of the soil, and in this mixture live astonishing numbers of small organisms that spend all or part of their lives beneath the soil surface. A single teaspoon of soil may contain 5 billion bacteria, 20 million small filamentous fungi, and 1 million protists. The soil animals and microorganisms (Figure 26–6) break down the organic matter, releasing its inorganic nutrients, which then can be reutilized by the plants. Thus, except for the nutrients that leach out of the soil, are carried away by streams and rivers, and eventually precipitate in the ocean, substances that are taken from the soil are constantly returned to it. Both macronutrients and micronutrients are recycled constantly through plant and animal bodies, returned to the soil, broken down, and taken up into plants again. Each element has a different cycle, involving many different organisms and different enzyme systems. The end results are the same, however—a significant amount of the element is constantly returned to the soil and is available for plant use.

MYCORRHIZAE AND PLANT NUTRITION

Nutrient uptake from soil by most higher plants is greatly enhanced by naturally occurring mycorrhizal fungi associated with their root systems (see Chapter 13). Mycorrhizae are especially important in the absorption and transfer of phosphorus, but the increased absorption of zinc, manganese, and copper has also been demonstrated. These nutrients are relatively immobile in soil, and depletion zones for these nutrients quickly develop around roots and root hairs. The hyphal network of mycorrhizae extends several centimeters out from colonized roots, thus exploiting a larger volume of soil more efficiently. The ability of mycorrhizae to absorb and transport soil phosphorus was shown in experiments using radioactive ^{32}P (a). In addition to this physical increase in absorptive surface area, mycorrhizal fungi may also be able to extract phosphorus from more dilute soil solutions and utilize sources of phosphorus normally unavailable to plants.

The increase in phosphorus uptake attributable to mycorrhizae can be demonstrated by growing mycorrhizal and nonmycorrhizal plants over a range of nutrient conditions (b). In this example, mycorrhizal citrus plants receiving no phosphorus fertilizer were the same size as nonmycorrhizal plants fertilized with 560 kilograms of phosphorus per hectare. In this case, mycorrhizae could be substituted for the large addition of phosphorus fertilizer, substantially reducing costs and energy requirements for plant production.

Plant species vary in their dependence on mycorrhizal fungi. Some plants, such as citrus, are so dependent on mycorrhizae for phosphorus uptake that without them their growth is stunted at all but extremely high phosphorus levels (b). The percentage growth increase obtained for mycorrhizal compared to nonmycorrhizal plants has been reported for many species; representative values include wheat (220%), corn (122%), onions (3155%), strawberries

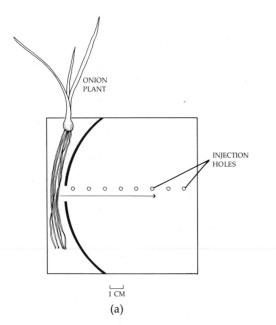

1 CM

(a)

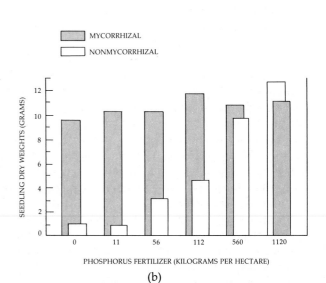

(b)

(a) *Translocation of ^{32}P by mycorrhizae. Mycorrhizal onion (Allium) plants were grown in specially constructed chambers with a small opening that allowed the fungal hyphae (arrow) but not the roots to grow into a separate volume of soil. ^{32}P was then injected at different distances from the root. ^{32}P was detected in the roots and shoots of mycorrhizal onion plants but not in nonmycorrhizal plants nor in treatments where the hyphal connections of mycorrhizal plants were severed. The great majority of mycorrhizal hyphae were found to occur within 2 to 5 millimeters of the supporting root. (b) Growth response of mycorrhizal and nonmycorrhizal citrus seedlings fertilized with different amounts of phosphorus.*

NITROGEN AND THE NITROGEN CYCLE

The nitrogen in soil is derived from the earth's atmosphere. Although the atmosphere is 78 percent nitrogen, most living things cannot use atmospheric nitrogen to make proteins and other organic substances. Unlike carbon and oxygen, nitrogen is chemically unreactive.

The highly specialized capacity for converting atmospheric nitrogen into a form that can be used by cells is limited to a few bacteria. This unique process, called nitrogen fixation, will be discussed shortly.

On a world-wide basis, usable nitrogen is the major limiting nutrient in crop plant growth. The processes by which nitrogen is circulated from the atmosphere through plants and the soil by the action of living organisms are illustrated in the nitrogen cycle shown in Figure 26–8.

(250%), and western red cedar (962%). At high phosphorus levels, where large amounts of soluble phosphorus fertilizer are added, mycorrhizae do not generally increase plant growth. In fact, mycorrhizal colonization is often inhibited in heavily fertilized soils.

Mycorrhizal fungi are present in most plant communities, but their populations can be reduced or eradicated by severe land disturbances such as strip mining. The indiscriminate use of fungicides and soil fumigants in agriculture and forest tree nurseries to eliminate plant pathogenic fungi can also reduce populations of mycorrhizal fungi, resulting in severe stunting and plant death. It is desirable to reinoculate these soils with mycorrhizal fungi. Ectomycorrhizal fungi can be cultured, and vegetative mycelium can be mass-produced and added to soil mixes or nursery beds. Conifer seedlings in the southeastern United States and the Pacific Northwest are often inoculated with particular species of ectomycorrhizal fungi to improve plant growth on harsh and infertile sites. Vesicular-arbuscular (V/A) mycorrhizal fungi—a type of endomycorrhizal fungus—cannot yet be grown axenically, so V/A inoculum production necessitates the use of host plant "pot cultures." Fungus spores or V/A-colonized plant roots obtained from these pot cultures are then used to inoculate greenhouse nursery soils. Because of the difficulty in mass-producing V/A fungi, the commercial use of these fungi has been limited to small-scale applications. Growers of native plant species for the revegetation of strip-mined areas preinoculate nursery plants before outplanting to obtain better growth and survival. Fumigated citrus nurseries in southern California and hardwood seedling nurseries in the southeastern United States are also inoculated with V/A mycorrhizal fungi. For most agricultural crops produced on large acreages, it is not practical to inoculate with mycorrhizal fungi. For high-value crops and especially for plant propagating nurseries, however, mycorrhizal fungi can be extremely beneficial to plant growth without the expensive addition of phosphorus fertilizer.

Although enhanced phosphorus uptake in nutrient-deficient soils has been well documented, other plant growth benefits occur as a result of the mycorrhizal condition. These include increased resistance to soil-borne plant pathogens, increased tolerance to highly acid soils, and improved tolerance to drought stress.

Ammonification

Much of the soil nitrogen is derived from dead organic materials in the form of complex organic compounds such as proteins, amino acids, nucleic acids, and nucleotides. These nitrogenous compounds are usually rapidly decomposed into simple compounds by soil-dwelling saprobic bacteria and various fungi, which incorporate the nitrogen into amino acids and proteins and release excess nitrogen in the form of ammonium ions (NH_4^+) by a process known as *ammonification*. The nitrogen may be given off as ammonia gas (NH_3), but this usually occurs only during the decomposition of large amounts of nitrogen-rich material, as in a manure pile or a compost heap. Usually the ammonia produced by ammonification is dissolved in the soil water, where it combines with protons to form the ammonium ion.

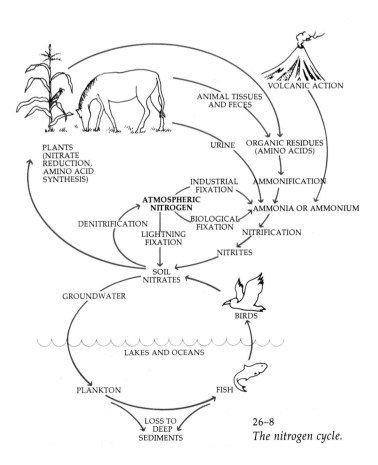

26–8
The nitrogen cycle.

Nitrification

Several species of bacteria common in soils are able to oxidize ammonia or ammonium ions. The oxidation of ammonia, or _nitrification_, is an energy-yielding process, and the energy released in the process is used by these bacteria to reduce carbon dioxide in much the same way the photosynthetic autotrophs use light energy in the reduction of carbon dioxide. Such organisms are known as chemosynthetic autotrophs (as distinct from photosynthetic autotrophs). The chemosynthetic nitrifying bacterium *Nitrosomonas* is primarily responsible for oxidation of ammonia to nitrite ions (NO_2^-).

$$2NH_3 + 3O_2 \longrightarrow 2NO_2^- + 2H^+ + 2H_2O$$

Nitrite is toxic to higher plants, but it rarely accumulates in the soil. *Nitrobacter*, another genus of bacteria, oxidizes the nitrite to form nitrate ions (NO_3^-), again with a release of energy:

$$2NO_2^- + O_2 \longrightarrow 2NO_3^-$$

Because of nitrification, nitrate is the form in which almost all nitrogen is absorbed by plants.

A few species of plants are able to use animal proteins directly as a nitrogen source. These carnivorous plants (Figure 26–9) have special adaptations that are used to lure and trap insects and other very small animals. The plants digest the trapped organisms, absorbing the nitrogenous compounds they contain, as well as other organic compounds and minerals, such as potassium and phosphate. Most of the carnivores of the plant world are found in bogs, a habitat that is usually quite acidic and thus not favorable for the growth of nitrifying bacteria.

26–9

Carnivorous plants entrap their prey in a variety of ways. (a) The common bladderwort (Utricularia vulgaris) is a free-floating aquatic plant. The traps are tiny, flattened, pear-shaped bladders. Each bladder has a mouth guarded by a hanging door. The tripping mechanism consists of four stiff bristles near the lower free edge of the door. When a small animal brushes against these bristles, the hairs distort the lower edge of the door, causing it to spring open. Water then rushes into the bladder, carrying the animal inside, and the door snaps shut behind it. A range of enzymes secreted by the internal wall of the bladder and by the residential bacterial population digest the animal, and released minerals and organic compounds are taken up by the cellular walls of the trap. The undigested exoskeletons remain within the bladders. (b) The sundew (Drosera intermedia) is a tiny plant, often only a few centimeters across, with club-shaped hairs on the upper surface of its leaf. The tips of these glandular hairs secrete a clear, sticky liquid, or mucilage, that attracts insects. When an insect is caught in the mucilage, the hairs bend inward until the leaf finally curves around the insect. The hairs are known to secrete at least six enzymes, which, together with bacteria-produced enzymes, especially chitinase, digest the insect. Nutrients released from the prey into the mucilage are resorbed by the same glands that secreted the digestive enzymes. (See also "Carnivorous Plants," on the facing page.)

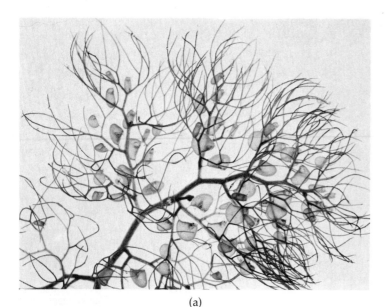

(a)

(b)

CARNIVOROUS PLANTS

Carnivorous plants obtain minerals, including fixed nitrogen, from animal prey. In some of these plants, flies and other insects are caught on the leaf surface, as in (a) the butterwort (*Pinguicula grandiflora*). The capture of small insects is accomplished by numerous stalked glands scattered over the leaf. Each of these glands, as seen in the scanning electron micrograph (b), bears a globular droplet of mucilaginous secretion, so that the leaf is sticky, like flypaper. Insects coming in contact with the secretion pull it out into strands that set to form strong cables, some of which can be seen holding the ant down in (c). The more the insect struggles, the more glands it touches, and the more tightly it is held onto the leaf. Scattered among the stalked glands are sessile glands (b), which bear no surface material until they are stimulated by the struggling of the captured prey. The sessile glands then pour out an enzyme-containing secretion that quickly forms a pool around the insect. These enzymes digest the prey, and the products accumulate in the secretion pool. After digestion is complete, the pool is absorbed into the leaf, and the digestion products are distributed to the growing parts of the plant.

The digestive enzymes are synthesized in the sessile glands. Until stimulation these enzymes are stored in enlarged vacuoles and in the wall, where they can be detected by suitable cytochemical techniques. As can be seen in (d), for example, the localization of the enzyme acid phosphatase in the walls of the gland head is revealed by the distribution of the dark reaction product. The discharge of the enzymes is brought about by the rapid passage of water through the gland head. This process is driven by the pumping of chloride ions from an underlying reservoir cell. The "ion pumps" are present in the membranes of an intervening endodermislike cell; they are quiescent in the inactive gland but are quickly activated by the stimulus of the prey.

Similar secretion mechanisms are found in other carnivorous plants, including the Venus flytrap (*Dionaea muscipula*) (see Figure 25–22) and the sundews (*Drosera*) (see Figure 26–9).

(a)   1 cm

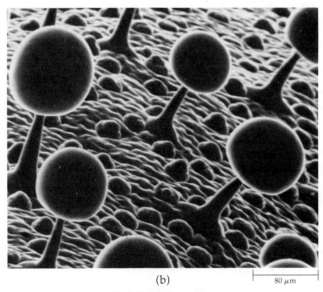

(b)   80 μm

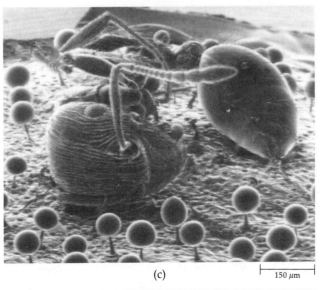

(c)   150 μm

(d)   10 μm

Assimilation of Nitrogen

Once the nitrate ions enter the plant cell, they are reduced to ammonium ions. This reduction process requires energy, in contrast to nitrification, which involves the oxidation of NH_4^+ and releases energy. The ammonium ions formed by reduction are transferred to carbon-containing compounds to produce amino acids and other nitrogen-containing organic compounds. This process is known as *amination*. The incorporation of nitrogen into organic compounds takes place largely in young, growing root cells. The initial stages in the metabolism of nitrogen also appear to occur in the root; almost all the nitrogen ascending the stem in the xylem is already in the form of organic substances, largely molecules of various amino acids.

Formation of Amino Acids

Amino acids are formed from ammonium ions and keto acids. The latter are usually products of the metabolic breakdown of sugars. The major amino acid that is formed in this fashion is glutamic acid—the chief carrier of nitrogen through the plant body. From the amino acid produced by the amination of a keto acid, other amino acids are formed by the process of *transamination*—the transfer of the amino group ($—NH_2$) on one amino acid to a keto acid, producing a second amino acid.

Plants, either by amination or transamination, can make all the amino acids they require, starting from inorganic nitrogen. Animals can make only about 8 of the 20 amino acids they require and must obtain the others in their diet. Thus, the animal world is completely dependent on the plant world for its proteins, just as it is for its carbohydrates.

Other Nitrogen-Containing Compounds

Other important organic nitrogen-containing compounds include the nucleotides, such as ATP, ADP, NAD, and NADP; chlorophyll and other similar organic molecules with porphyrin ring structures; and the nucleic acids, DNA and RNA. Many of the vitamins, such as the vitamin B group, contain nitrogen. These, like the amino acids, can be synthesized by plants, but animals must obtain them from plants.

Nitrogen Loss

As discussed, the nitrogen-containing compounds of green plants are returned to the soil with the death of plants (or animals that have eaten the plants) and are reprocessed by soil organisms. Nitrate dissolved in the soil water is then taken up by plant roots and reconverted to organic compounds. In the course of this cycle, a certain amount of nitrogen is always "lost," in the sense that it becomes unavailable to the plants in specific ecosystems.

A main source of nitrogen loss from specific ecosystems is the removal of plants from the soil. Soils under cultivation often show a steady decline of nitrogen content. Nitrogen may also be lost when topsoil is carried off by soil erosion or when ground cover is destroyed by fire. Nitrogen is also removed by leaching; nitrates and nitrites, both of which are anions, are particularly susceptible to being washed from the root zone by water percolating through the soil.

Under anaerobic conditions, nitrate is often reduced to volatile forms of nitrogen, such as nitrogen gas (N_2) and nitrous oxide (N_2O), which then return to the atmosphere. This process of reduction, called *denitrification*, is carried out by numerous microorganisms. The low oxygen conditions necessary for denitrification are characteristic of waterlogged soils and such habitats as swamps and marshes. A fresh supply of readily decomposable organic matter provides the energy source required by denitrifying bacteria and, if other conditions are appropriate, promotes denitrification.

Nitrogen Fixation

If the nitrogen that is removed from the soil were not steadily replaced, virtually all life on this planet would slowly disappear. Nitrogen is replenished in the soil by nitrogen fixation.

Nitrogen fixation is the process by which N_2 is reduced to NH_4 and made available for amination reactions. Nitrogen fixation, which can be carried out to a significant extent only by certain bacteria, is a process on which all living organisms are now dependent, just as most organisms are ultimately dependent on photosynthesis as the source of their energy.

Of the various classes of nitrogen-fixing organisms, the symbiotic bacteria are by far the most important in terms of total amounts of nitrogen fixed. The most common of the nitrogen-fixing bacteria is *Rhizobium*, a bacterium that invades the roots of legumes, such as alfalfa (*Medicago sativa*), clovers (*Trifolium*), peas (*Pisum sativum*), soybean (*Glycine max*), and beans (*Phaseolus*) (see Figures 26–1 and 26–10).

The beneficial effects on the soil that are derived from growing leguminous plants have been recognized for centuries. Theophrastus, who lived in the third century B.C., wrote that the Greeks used crops of broad beans (*Vicia faba*) to enrich the soils. Where leguminous plants are grown, some of the "extra" nitrogen may be released into the soil, becoming available for other plants. In modern agriculture, it is common practice to rotate a nonleguminous crop, such as corn (*Zea mays*), with a leguminous crop, such as alfalfa. The leguminous plants are then either harvested for hay, leaving behind the nitrogen-rich roots, or, better still, they

26–10
(a) *Nitrogen-fixing nodules on the roots of a soybean* (Glycine max) *plant. These nodules are the result of a symbiotic relationship between the root cells of this leguminous plant and the bacterium* Rhizobium. (b) *and* (c) *Scanning electron micrographs of nitrogen-fixing nodules on white clover* (Trifolium repens).

(a)

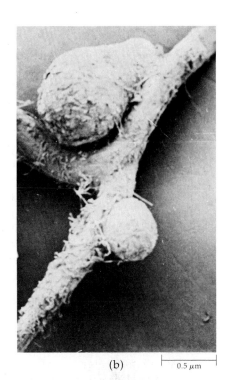

(b)   0.5 µm

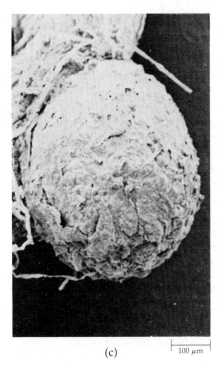

(c)   100 µm

are simply plowed under. A crop of alfalfa that is plowed back into the soil may add as much as 300 to 350 kilograms of nitrogen per hectare of soil. As a conservative estimate, 150 to 200 million metric tons of fixed nitrogen are added to the earth's surface each year by such biological systems.

An industrial process for the chemical fixation of atmospheric nitrogen was developed in 1914. Since that time, the commercial production of fixed nitrogen has increased steadily to the current level of approximately 50 million metric tons per year. Most of this nitrogen is used in agricultural fertilizers. Industrial fixation, unfortunately, is accomplished at a high energy cost in terms of fossil fuels.

Small amounts of nitrogen can also be fixed by lightning and brought to the earth in rainfall. Rainwater sometimes also brings down ammonia and oxides of nitrogen that have escaped into the atmosphere. Measurements at an experimental station in England over a five-year period showed that the rainwater brought down 7.1 kilograms of nitrogen per hectare each year.

Nitrogen-Fixing Symbioses

In the symbiotic associations between *Rhizobium* and legumes, the leguminous plant supplies the bacteria with carbon compounds as an energy source for nitrogen fixation and other metabolic activities and also provides a protective environment. The legume, in return, obtains nitrogen in a form usable for the production of plant proteins.

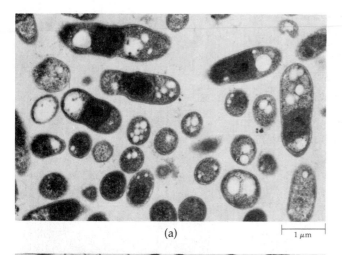

(a)   ⊢___⊣ 1 μm

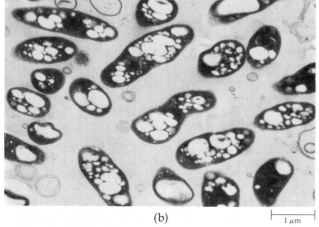

(b)   ⊢___⊣ 1 μm

26–11

Rhizobium. (a) *Bacteria in their
free-living form.* (b) *The bacteroid form
that the rhizobia assume when they enter
the root cells.*

26–12

*Early events in the infection of soybean
by* Rhizobium japonicum. (a) *Scanning
electron micrograph showing rhizobia
(arrows) attached to recently emerged root
hair.* (b) *Differential-interference
contrast photomicrograph showing short,
curled root hair containing multiple
infection threads (arrows).* (c) *Transverse
section of an infected root showing early
development of the nodule meristem. At
this stage, most of the cell divisions are
anticlinal (at right angles to the surface)
across the cortex. In the nodule meristem
adjacent to the infected root hair, cells
have divided in other planes as well.* (d)
*Higher magnification of a transverse
section showing an infected root hair and
adjacent nodule meristem. Infection
threads (arrows) can be seen in the root
hair and adjacent meristem.*

The *Rhizobium* bacteria, or rhizobia (Figure 26–11a), enter the root hairs of leguminous plants when the plants are still seedlings. Establishment of the nitrogen-fixing symbiosis between *Rhizobium japonicum* and soybean *(Glycine max)* begins with the attachment of rhizobia to the emerging root hairs (Figure 26–12a). Many of these root hairs develop into deformed, often tightly curled structures. Following attachment, the rhizobia penetrate the cell wall of the root hair. Degradation of the hair cell wall at the penetration site is probably due to rhizobia-produced enzymes. During penetration by a rhizobial cell, growth of the root hair is arrested at the tip and redirected to the penetration site, where a localized dome of new cell wall material is deposited. Progressive inward growth of the cell wall from this dome into the root hair results in formation of a tubular structure called an *infection thread* (Figure 26–12b). A single root hair may be penetrated by several rhizobia. Once inside the root hair, the rhizobia are carried through the cell by the infection threads, first to the base of the infected hair and then through the cell

walls into cells of the cortical region of the root. The bacterial symbiont also induces cell division in localized regions of the cortex, which it then enters through growth and branching of the infection thread (Figures 26–12c, d and 26–13). Release of the rhizobia from the infection thread into envelopes derived from the plasma membrane of the host root hair and continued proliferation of both *bacteroids* (the name given the now-enlarged rhizobia; Figure 26–11b) and cortical cells of the root result in the formation of tumorlike growths known as *nodules* (Figure 26–10). The manner of infection and nodule formation in the roots of other legumes is apparently similar to that for soybean.

The root nodules of legumes consist in part of cells that become greatly enlarged and infected with rhizobia and partly of many smaller uninfected cells interspersed among the infected ones (Figure 26–14). Until recently, the uninfected cells were largely ignored because it was thought they played an insignificant role in the metabolism of recently fixed nitrogen and provided only supportive functions in the nodule. In the case of soybean, however, it is now clear that the uninfected cells play a major role in the production of ureides (derivatives of urea) from N_2 recently fixed by the infected cells.

Legumes may be divided into several groups based on the major nitrogenous compounds transported from the nodules to the aboveground parts of the plant. For example, vetches *(Vicia)*, lupines *(Lupinus)*, and peas belong to a group that transports fixed nitrogen mainly as the amino acid asparagine. Soybeans, cowpeas *(Vigna)*, and beans, on the other hand, belong to a group that produces ureides for nitrogen export. The early steps in ureide production occur in the infected cells and the later ones in the uninfected cells.

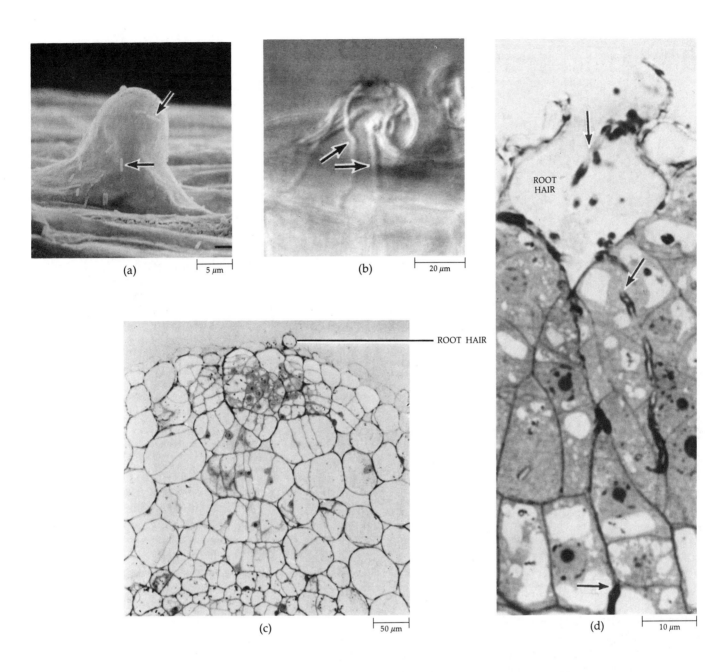

(a)   5 μm

(b)   20 μm

ROOT HAIR

(c)   50 μm

ROOT HAIR

(d)   10 μm

26–13
*Electron micrograph of a branched
infection thread containing bacteroids in
an infected nodule cell of soybean.*

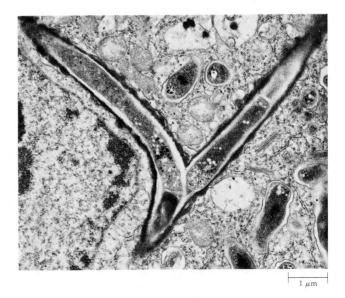

1 μm

26–14

Electron micrograph of a mature soybean root nodule. Vacuolate uninfected cells can be seen among the darkly stained infected cells.

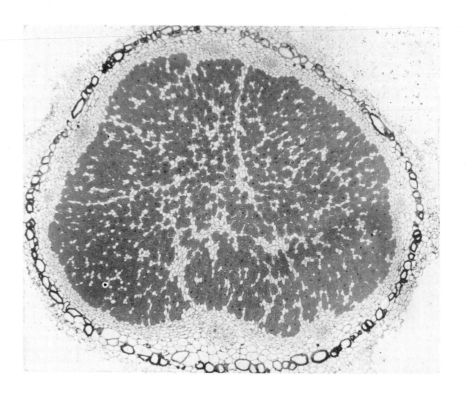

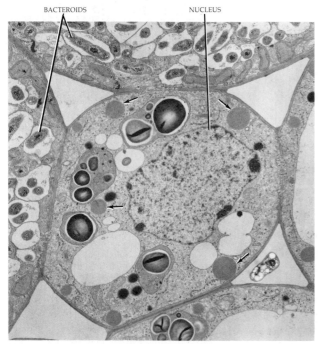

26–15

Electron micrograph of portion of nodule of soybean showing uninfected cell (middle) adjacent to infected cells (top and left). The infected cells contain many bacteroids. Note the numerous microbodies (arrows) in the uninfected cell. Compared to the uninfected cells, the volume of mitochondria in infected cells is four times greater—apparently a reflection of the enormous demand of the bacteroids for ATP.

Nodule cells uninfected by rhizobia develop large microbodies (peroxisomes) and abundant tubular endoplasmic reticulum (Figure 26–15). The enzymes involved in the final steps of ureide formation apparently are located in these cellular components.

The symbiosis between a species of *Rhizobium* and a legume is quite specific; for example, bacteria that invade and induce nodule formation in clover *(Trifolium)* roots will not induce nodules on the roots of soybeans *(Glycine)*. Specific laboratory-grown strains (inocula) of *Rhizobium* are now commercially available. By mixing particular bacteria with the legume seed at the time of planting, farmers can ensure that the appropriate bacteria are available in the soil to form effective associations for nitrogen fixation in the legume crop. The mechanism by which the *Rhizobium* bacteria interact with only the specific legume root is currently being investigated. The bacteria and legume may recognize each other via a plant protein (lectin) found on the root surface. Lectin binds to a polysaccharide on the surface of an effective *Rhizobium* cell but not to the surface polysaccharides of ineffective bacteria.

The legumes are by far the largest group of plants that enters into a nitrogen-fixing partnership with symbiotic bacteria. There are, however, a few nitrogen-fixing symbioses that involve plants other than legumes. Alder trees *(Alnus)*, for example, form nodules that are induced by and contain nitrogen-fixing actinomycetes, rather than *Rhizobium*. Sweet gale *(Myrica gale)*, sweet fern *(Comptonia)*, and mountain lilacs *(Ceanothus)* also form symbiotic associations with actinomycetes (see Chapter 11).

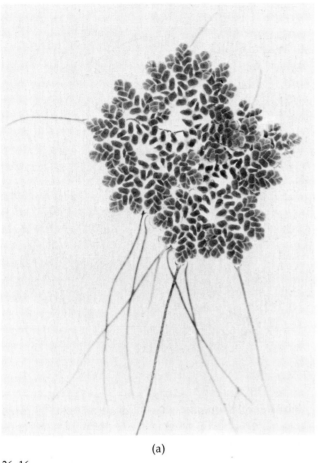

(a)

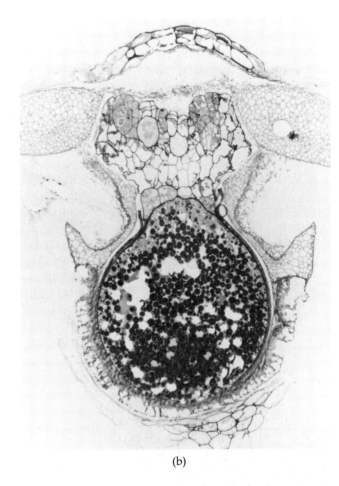

(b)

26–16

(a) Azolla caroliniana, *a water fern that grows in symbiotic association with the cyanobacterium* Anabaena. (b) *The* Azolla-Anabaena *symbiosis is unique* *among nitrogen-fixing symbioses in that the relationship is sustained throughout the life cycle of the host. Here filaments of* Anabaena *can be seen associated with* *a female gametophyte (megagametophyte) that has developed from a germinated megaspore.*

Another symbiotic relationship is of considerable practical interest in certain parts of the world. *Azolla* is a small floating water fern, and *Anabaena* is a nitrogen-fixing cyanobacterium that lives in the cavities of the *Azolla* fronds (Figure 26–16). *Azolla* infected with *Anabaena* may contribute as much as 50 kilograms of nitrogen per hectare. In the Far East, for example, heavy growths of *Azolla-Anabaena* are permitted to develop on rice paddies (see also Figure 11–12b). The rice plants eventually shade out the *Azolla*, and as the fern dies, nitrogen is released for use by the rice plants.

Free-Living Nitrogen-Fixing Microorganisms

Nonsymbiotic bacteria of the genera *Azotobacter* and *Clostridium* are both able to fix nitrogen. *Azotobacter* is aerobic, and *Clostridium* is anaerobic; both are common saprophytic soil bacteria. It is estimated that they probably add about 7 kilograms of nitrogen to a hectare of soil per year. Another important group includes many photosynthetic bacteria, such as cyanobacteria.

The differences between nitrogen fixation by free-living bacteria and nitrogen fixation by bacteria living in symbiotic associations are not always distinct. For example, the aerobic nitrogen-fixing bacteria *Azotobacter* regularly grow around the roots of certain grasses, such as sugarcane (*Saccharum officinarum*), where they play a significant role in providing nitrogen to the plants. The grasses (and other plants) presumably exude organic compounds that serve as an energy source for these "free-living" bacteria. Thus an association that approximates symbiosis results.

The enzyme that carries out nitrogen fixation is called nitrogenase. Nitrogenase contains molybdenum, iron, and sulfide prosthetic groups, and therefore these elements are required for biological systems to perform nitrogen fixation. Nitrogenase also uses large amounts of ATP as an energy source for nitrogen fixation, making it an expensive metabolic process.

In addition to reducing N_2 to NH_4^+, nitrogenase also converts acetylene, a triple-bonded molecule, to ethylene, a double-bonded molecule. The conversion of acetylene to ethylene—an example of an alternate substrate for an enzyme—is a widely used assay for nitrogenase activity.

As mentioned previously, although plants do not require cobalt as a micronutrient, the microorganisms that carry out nitrogen fixation do require cobalt. Therefore cobalt, although not a part of the nitrogenase enzyme, is required for symbiotic nitrogen fixation.

THE PHOSPHORUS CYCLE

In contrast to the nitrogen cycle, the phosphorus cycle (Figure 26–17) seems simpler because there are fewer steps, and particular steps are not dependent on specific groups of microorganisms. The phosphorus cycle also differs from the nitrogen cycle in that the earth's crust, rather than the atmosphere, is the primary reservoir of phosphorus. As previously discussed, the weathering of rocks and minerals over long periods of time is the source of most of the phosphorus in the soil solution.

Compared to nitrogen, the amount of phosphorus required by plants is relatively small (see Tables 26–1 and 26–2). Nevertheless, of all the elements for which the earth's crust is the primary reservoir, phosphorus is the most likely to limit plant growth. In Australia, for example, where the soils are extremely weathered and deficient in phosphorus, the distribution and limits of native plant communities are often determined by the available soil phosphate.

Phosphorus circulates from plants to animals and is returned to the soil in organic forms in residues and wastes; these organic forms of phosphorus are converted to inorganic phosphate and thus again become available to plants (Figure 26–17).

Through erosion, pollution, and loss in drainage water, large amounts of phosphorus are discharged into rivers and streams. This phosphorus eventually reaches the oceans, where it is deposited in sediments as precipitates and in the remains of organisms. In the past, the use of guano (deposits of seabird feces) as agricultural fertilizer returned some of the ocean phosphorus to terrestrial ecosystems. However, most of the phosphorus in deep-sea sediments will become available only as a result of major geological uplifts. To counter this loss, deposits of phosphate rock are being mined on a large scale for use as agricultural fertilizer.

HUMAN IMPACT ON NUTRIENT CYCLES

Normal functioning of the phosphorus, nitrogen, and other nutrient cycles requires the orderly transfer of elements between steps in a cycle in order to prevent buildup or depletion of nutrients at any one stage. Over millions of years, organisms have been provided with

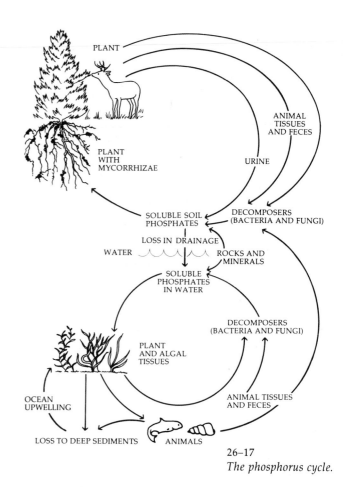

26–17
The phosphorus cycle.

needed quantities of essential inorganic nutrients as a result of the normal functioning of such cycles. However, in recent years, the activities of humans have had drastic effects on some cycles, and in time this may lead to harmful accumulations and depletions of nutrients at specific stages of one or more nutrient cycles. For example, increased soil erosion has accelerated phosphorus loss from soils. Also, rather than recycling the phosphorus present in sewage effluents, the effluents have often been discharged into streams, and the phosphorus has thus been transported (and lost) to the oceans.

Normal functioning of the nitrogen cycle involves a balance between fixation processes that remove nitrogen from the atmospheric reservoir and denitrification reactions that return nitrogen to the atmosphere. Recently there has been massive introduction of fixed nitrogen (nitrates) into the environment through the extensive use of commercial fertilizers. Because nitrogen in the form of nitrates is readily leached from soil, increased nitrogen pollution of groundwater and lakes and streams has resulted. To add to this problem, the marshes and wetlands that are primary sites of denitrification are being destroyed at an alarming rate through their conversion to building sites, agricultural land, and dump sites.

To reduce the loss of soil nitrates by leaching, another manipulation of the nitrogen cycle has been suggested and could become common agricultural practice.

COMPOST

Composting, a practice as old as agriculture itself, has attracted increased interest as a means of utilizing organic wastes by converting them to fertilizer. The starting product is any collection of organic matter—leaves, kitchen garbage, animal manure, straw, lawn clippings, sewage sludge, sawdust—and the population of bacteria and other microorganisms normally present. The only other requirements are oxygen and moisture. Grinding of the organic matter is not essential, but it provides greater surface area for microbial attack and thus speeds the process.

In a compost heap, microbial growth accelerates quite rapidly, generating heat, much of which is retained because the outer layers of organic matter act as insulation. In a large heap (2 meters $\times$ 2 meters $\times$ 1.5 meters, for instance), the interior temperature rises to about 70°C; in small heaps, it usually reaches 40°C. As the temperature rises, the population of decomposers changes, with thermophilic and thermotolerant forms replacing the organisms previously present. As the original forms die, their organic matter also becomes part of the product. A useful side effect of the temperature increase is that most of the common pathogenic bacteria that may have been present, for example, in sewage sludge, are destroyed, as are cysts, eggs, and other immature forms of plant and animal parasites.

With the passage of time, changes in pH also occur in a compost heap. The initial pH value is usually slightly acidic (about 6), which is comparable to the liquid portion of most plant material. During the early stages of decomposition, the production of organic acids causes a further acidification, with the pH decreasing to a value of about 4.5 to 5.0. However, as the temperature rises, the pH also increases; the composted material eventually levels off at slightly alkaline values (7.5 to 8.5).

An important factor in composting (as in any biological growth process) is the ratio of carbon to nitrogen. A ratio of about 30 to 1 (by weight) is optimal. If the carbon ratio is higher, microbial growth slows. If the nitrogen ratio is higher, some escapes as ammonia. If the compost materials are quite acidic, limestone (calcium carbonate) may be added to balance the pH; however, if too much of this buffer is added, it will increase the nitrogen loss.

Studies involving municipal compost piles in Berkeley, California, demonstrated that if large piles were kept moist and aerated, composting could be completed in as little as two weeks. Generally, though, three months or more during the winter is needed to complete the process. If compost is added to the soil before the composting process is complete, it may temporarily deplete the soil of soluble nitrogen.

Because it greatly reduces the bulk of plant wastes, composting can be a very useful means of waste disposal. In Scarsdale, New York, for example, leaves composted in a municipal site were reduced to one-fifth their original volume. At the same time, they formed a useful soil conditioner, improving both the aeration and water-holding capacity. Chemical analyses indicate, however, that a rich compost commonly contains, in dry weight, only about 1.5 to 3.5 percent nitrogen, 0.5 to 1.0 percent phosphorus, and 1.0 to 2.0 percent potassium, far less than commercial fertilizers. However, unlike such fertilizers, compost can be a source of nearly all the elements known to be needed by plants. Compost provides a continuous balance of nutrients, releasing them gradually as it continues to decompose in the soil.

As commercial fertilizers are becoming more expensive and less available and our waters are becoming increasingly polluted with fertilizer runoff and organic wastes, composting is an increasingly attractive alternative.

This involves the application of an organic chemical that selectively inhibits the nitrifying bacteria *Nitrosomonas* for a limited period of time and then is broken down in the soil. This practice would help retain most of the fertilizer nitrogen in the ammonium form until the nitrogen can be absorbed by plants. As is often true of such an environmental manipulation, this possibility involves trade-offs. For example, high levels of ammonia can be quite toxic to some crop plants.

SOILS AND AGRICULTURE

In natural situations, the elements present in the soil recirculate and so become available again for plant growth. As discussed previously, negatively charged clay particles are able to bind such positively charged ions as Ca^{2+}, Mg^{2+}, and K^+. The ions are removed from the particles by the roots of the plant, either directly or after they pass into the soil solution. In general, the cations that are required by plants are present in large amounts in fertile soils, and the amounts removed by a single crop are small. However, when a series of crops is grown on a particular field and the nutrients are continuously removed from the cycle as the crops are harvested, some of these cations (most commonly potassium) may become depleted to such an extent that fertilizers containing the missing element must be added.

Programs for supplementing nutrient supplies for agricultural and horticultural crops should be based on a bookkeeping approach that equates amounts of nutrients required to produce a specific crop and the nutrients available to the plants from all sources. Often the soil and plant residues cannot supply the required nutrients, and supplemental amounts must be provided in applications of commercial fertilizers, organic residues such as compost, or a combination of the two.

Nitrogen, phosphorus, and potassium are the three elements that are commonly included in commercial fertilizers. Fertilizers are usually labeled with a formula that indicates the percentage of each of these elements. A 10-5-5 fertilizer, for example, is one that contains 10 percent nitrogen (N), 5 percent phosphorus pentoxide (P_2O_5), and 5 percent potassium oxide (K_2O).

Other essential inorganic nutrients, although required in very small amounts, can sometimes become limiting factors in soils on which crops are grown.

PLANT NUTRITION RESEARCH

Research on the inorganic nutrients essential for crop plants—particularly on the quantities of those nutrients required for optimal crop yields and on the capacities of various soils to provide the nutrients—has been of great practical value in agriculture and horticulture. Because of the steady increase in worldwide food needs, this type of research undoubtedly will continue to be essential.

Soil Deficiencies and Toxicities

Modification and manipulation of soils by adding nutrients in fertilizers, by raising the pH with lime, or by removing excess salts by leaching with water may not be the only means of improving and maintaining crop production in below-optimum soils. By using the knowledge and techniques of plant breeding and plant nutrition, it may be possible to select and develop cultivars of crop species that are better adapted for growth in nutrient-deficient environments than are present commercial cultivars. The validity of this research approach is confirmed by the occurrence of wild plants in nutritional environments that are very different from the average soil environments in which crop plants are grown; examples are acidic sphagnum bogs, in which the pH may be less than 4.0, and mine tailings, which often contain high concentrations of potentially toxic metals, such as zinc and nickel.

Research to develop bean (*Phaseolus vulgaris*) strains that are tolerant of potassium-deficiency stress has recently been carried out. Strains of beans were collected from around the world and grown in a nutrient solution containing less potassium than is required for optimal bean growth. All other nutrients were at optimal concentrations. The extremes in bean growth under the imposed potassium-deficiency stress are shown in Figure 26–18. Strain 58 (on the left) is nearly normal; strain 63 (on the right) shows the symptoms of severe potassium-deficiency. In another aspect of this relatively new area of research, Emanuel Epstein and his associates at the University of California at Davis have had extraordinary success in isolating barley (*Hordeum vulgare*) strains tolerant of high salt concentrations. The most tolerant strains grew reasonably well even when irrigated with sea water rather than fresh water.

26–18

Comparison of potassium-deficiency symptoms in the bean (Phaseolus vulgaris). In the bean strain on the left, the plant is tolerant of a deficiency level of potassium, whereas the plant on the right is clearly a nontolerant strain.

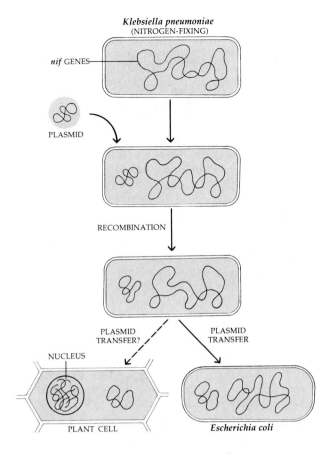

Klebsiella pneumoniae
(NITROGEN-FIXING)

nif GENES

PLASMID

RECOMBINATION

PLASMID
TRANSFER?

PLASMID
TRANSFER

NUCLEUS

PLANT CELL

Escherichia coli

26–19

Transfer of nitrogen-fixing (nif) genes from a nitrogen-fixing bacterium to an organism incapable of nitrogen fixation. The nif genes from Klebsiella pneumoniae *were incorporated into a segment of extrachromosomal DNA (a plasmid) and then were transferred to* Escherichia coli, *a bacterium that cannot fix nitrogen. Although an essential initial step, gene transfer does not ensure that the recipient organism will fix nitrogen. The genetically modified E. coli, for example, produced the enzyme nitrogenase, which is required for nitrogen fixation, but was able to fix nitrogen only when the enzyme was experimentally protected from oxygen. That is, the "new" nitrogen-fixing abilities of the bacteria did not include the ability to shield the critical enzyme from oxidation. In theory, nitrogen-fixing capabilities could be transferred via plasmids to cells of nonfixing plant species.*

Efficiency of Nitrogen Fixation

Manipulation of biological nitrogen fixation also offers tremendous potential for improved efficiency in nitrogen utilization. One aspect of research in this area is concerned with improving the efficiency of the *Rhizobium*-legume association, for example, through the genetic screening of both legumes and bacteria to identify combinations that would result in increased fixation in specific environments. This could result from a greater photosynthetic efficiency in legumes, so that more carbohydrate is available for bacterial nitrogen fixation and growth. In the pursuit of this possibility, however, it must be recognized that nitrogen fixation requires considerable energy and that any increase in fixation might be at the expense of shoot productivity.

A second research approach is to develop additional and more effective associations of free-living nitrogen-fixing bacteria and higher plants. In Brazil in the early 1970s, several types of nitrogen-fixing bacteria were found growing in association with the roots of certain tropical grasses; for example, the grass *Digitaria* was found to support populations of the bacterium *Azospirillum*. Similar associations with some of the world's major food crops, such as corn and sugarcane (*Saccharum*), have since been reported. While the practical benefits to agriculture could be tremendous, the effectiveness of associations of nitrogen-fixing bacteria and grasses such as corn is still uncertain.

Using the most sophisticated techniques of molecular biology, probably the most exciting research approach is that of genetic modifications and transfer of the genes necessary for nitrogen fixation from one organism to another. Transfer of the appropriate genes between organisms has already been accomplished. The cluster of genes that are responsible for nitrogen fixation was transferred from the bacterium *Klebsiella pneumoniae* to *Escherichia coli*. The transfer was accomplished by incorporating the genes in a plasmid (see page 643) and introducing the plasmid into *E. coli* cells (Figure 26–19). The genetically modified *E. coli* fixed nitrogen under some conditions.

Conceivably nitrogen-fixing genes could be transferred to nonfixing species, such as corn, but this involves a number of problems beyond transfer of the genes, such as the protection of the nitrogen-fixing enzyme from oxygen, which inhibits fixation. Consequently, research workers in this field are not optimistic that an effective transfer of this type will be easily accomplished.

Effects of Pollution

The toxic effects of various inorganic agents discharged into the environment as pollutants also are of current research interest. Crop plants, for example, may be adversely affected by heavy metals such as copper and cadmium. Aquatic ecosystems, in particular, have been damaged because they have become common disposal sites for industrial and municipal wastes. The introduction of nitrogen and phosphorus—the primary eutrophication nutrients—into freshwater ecosystems has resulted in massive growths of algae and aquatic flowering plants, thus seriously reducing the recreational value of affected lakes and streams.

Both terrestrial and, particularly, aquatic ecosystems are subject to widespread environmental damage because of acid rain. Acid rain results from the interaction of sulfur dioxide and oxides of nitrogen—derived principally from the combustion of fossil fuels—with atmospheric moisture to form sulfuric and nitric acids. These acids impart a high degree of acidity to the rainfall. In parts of the Scandinavian countries, in the northeastern and upper midwestern United States, and in southeastern Canada, the pH of rainwater is commonly in the range of 4.0 to 4.5 and on occasion is below 4.0. In contrast, the pH of rainwater in equilibrium with carbon dioxide in an unpolluted atmosphere is approximately 5.6. Acid rain can have adverse effects on plants, the weathering of rocks and minerals, the solubilities of potentially toxic metals in the environment, and even human health. Hundreds of soft-water lakes in Scandinavia, Canada, and the United States are subject to damage. These lakes lack the buffering provided by bicarbonates and carbonates that neutralize acid rain in hard-water lakes. Soft-water lakes are common at high altitudes, such as in the Adirondack Mountains of the northeastern United States and in other watersheds in which shallow soils overlie igneous rocks. The increased acidities now developing in such lakes can severely affect the reproduction of game fish. (See also pages 430 and 431.)

SUMMARY

A total of 16 inorganic nutrients are required by most plants for normal growth. Of these, carbon, hydrogen, and oxygen are derived from air and water. The rest are absorbed by roots in the form of ions. These 16 elements are categorized as either macronutrients or as micronutrients depending on the amounts in which they are required. The macronutrients are carbon, oxygen, hydrogen, nitrogen, potassium, calcium, phosphorus, magnesium, and sulfur. The micronutrients are iron, chlorine, copper, manganese, zinc, molybdenum, and boron. Some inorganic nutrients, such as sodium and cobalt, are essential only for specific organisms (Table 26–1).

Inorganic nutrients perform a number of important roles in cells. They regulate osmosis and affect cell permeability. Some also serve as structural components of cells, as components of critical metabolic compounds, and as activators and components of enzymes (Table 26–2).

The chemical and physical properties of soils are critical in determining their capabilities to provide the inorganic nutrients, water, and the conditions necessary for maximum

crop plant production. The weathering of rocks and minerals provides fragments that represent the inorganic component of soils. All of the inorganic nutrients except nitrogen are derived from weathering processes. In addition, soils contain organic matter and pore space occupied by varying proportions of water and gases. Under agricultural conditions, nitrogen, phosphorus, and potassium are the nutrients most often limiting to plant growth and most frequently added to soils in fertilizers.

Each essential inorganic nutrient is circulated in a complex cycle among organisms within ecosystems and between those organisms and environmental reservoirs of the nutrient. The circulation of nitrogen through the soil, through the bodies of plants and animals, and back to the soil again is known as the nitrogen cycle. Nitrogen reaches the soil in the form of organic material of plant and animal origin. These substances are decomposed by soil organisms. Ammonification—the release of ammonium (NH_4^+) ions from nitrogen-containing compounds—is carried out by soil bacteria and fungi. Nitrification is the oxidation of ammonia or ammonium ions to form nitrites and nitrates; one type of bacterium is responsible for the oxidation of the ammonia to nitrite, and another for the oxidation of the nitrite to nitrate. Nitrogen enters plants almost entirely in the form of nitrates. Within the plants, nitrates are reduced to ammonium ions. Amino acids are formed from reactions that produce glutamic acid (amination), or by the transfer of an amino group from one amino acid to a keto acid producing another amino acid (transamination). These organic compounds are eventually returned to the soil, completing the nitrogen cycle.

Nitrogen is lost from the soil by crop removal, erosion, fire, leaching, and the action of denitrifying bacteria. Nitrogen is added to the soil by nitrogen fixation, which is the incorporation of elemental nitrogen into organic components. Biological nitrogen fixation is carried out only by bacteria. These include bacteria (*Rhizobium*) that are symbionts of leguminous plants, free-living bacteria, and actinomycetes in symbiotic relationship with a few genera of nonleguminous plants. In agriculture, plants are removed from the soil. As a consequence, nitrogen and other elements are not recycled, as they are in nature; therefore, they must be replenished in either an organic form or an inorganic form.

CHAPTER 27

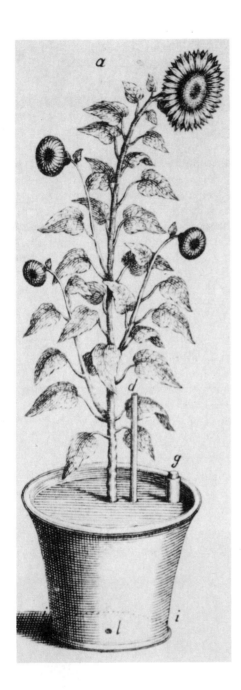

27–1

Diagram of the sunflower plant used by Stephen Hales in his experiments on the movement of water through the plant body. Hales found that most of the water "imbibed" by the plant was lost by "perspiration."

The Movement of Water and Solutes in Plants

The ability of the plant to transport both organic and inorganic nutrients, including water, throughout the plant body is critical in determining the ultimate structure and function of its component parts, as well as the development and form of the plant as a whole. In the first part of this chapter, we examine the movement of water and solutes through the plant body from the soil to the aerial plant parts. Then, toward the end of the chapter, we consider the movement of solutes and water from the sites of photosynthesis to the nonphotosynthetic parts of the plant body. We begin with a description of transpiration because this process is a major determining factor in the movement of water through the plant body.

MOVEMENT OF WATER THROUGH THE PLANT BODY

Transpiration

In the early eighteenth century, Stephen Hales, an English physician, noted that plants "imbibe" a much greater amount of water than animals. He calculated that one sunflower plant, bulk for bulk, "imbibes" and "perspires" 17 times more water than a human every 24 hours (Figure 27–1). Indeed, the total quantity of water absorbed by any plant is enormous—far greater than that used by any animal of comparable weight. An animal uses less water because much of its water is recirculated through its body over and over again, in the form (in vertebrates) of blood plasma and other fluids. In plants, more than 90 percent of the water taken in by the roots is released by the plant into the air as water vapor (Table 27–1). This process is known as *transpiration*, which is defined as the loss of water vapor by any part of the plant body, although leaves are by far the principal organs of transpiration.

Table 27–1   *Water Loss by Transpiration in One Plant in a Single Growing Season*

PLANT	WATER LOSS (LITERS)
Cowpea (*Vigna sinensis*)	49
Potato (*Solanum tuberosum*)	95
Wheat (*Triticum aestivum*)	95
Tomato (*Lycopersicon esculentum*)	125
Corn (*Zea mays*)	206

After J. F. Ferry, *Fundamentals of Plant Physiology,* The Macmillan Company, New York, 1959.

Why do plants lose such large quantities of water to transpiration? This question can be answered by considering the requirements for the chief function of the leaf, photosynthesis—the source of all the food for the entire plant body. The necessary energy for photosynthesis comes from sunlight. Therefore, for maximum photosynthesis, a plant must spread a maximum surface to the sunlight. But sunlight is only one of the requirements for photosynthesis; the chloroplast also needs carbon dioxide. Under most circumstances, carbon dioxide is readily available in the air surrounding the plant, but in order for carbon dioxide to enter the plant cell, which it does by diffusion, it must go into solution, because the plasma membrane is nearly impervious to the gaseous form of carbon dioxide. Hence, the gas must come into contact with a moist cell surface. However, wherever water is exposed to air, evaporation occurs. Plants have developed a number of special adaptations that limit evaporation, but all of these cut down the supply of carbon dioxide. In other words, the

uptake of carbon dioxide for photosynthesis and the loss of water by transpiration are inextricably bound together in the life of the green plant.

Absorption of Water by Roots

The root system serves to anchor the plant in the soil and, above all, to meet the tremendous water requirements of the leaves. Almost all the water that a plant takes from the soil enters through the younger parts of the root. Absorption takes place directly through the epidermis of the root. The root hairs, located several millimeters above the root tip, provide an enormous area for absorption (Figure 27–2; Table 27–2). From the root hairs, the water moves through the cortex, the endodermis (the innermost layer of cortical cells), and the pericycle, and into the primary xylem. Once in the conducting elements of the xylem, the water moves upward through the root and stem and into the leaves.

The pathway followed by water and solutes in the plant may be *apoplastic* (that is, via the cell walls) or *symplastic* (from protoplast to protoplast via plasmo-

Table 27–2   *Number of Root Hairs per Square Centimeter of Root Surface in Three Plant Species*

PLANT	ROOT HAIR DENSITY
Loblolly pine (*Pinus taeda*)	217
Black locust (*Robinia pseudo-acacia*)	520
Rye (*Secale cereale*)	2500

After J. F. Ferry, *Fundamentals of Plant Physiology,* The Macmillan Company, New York, 1959.

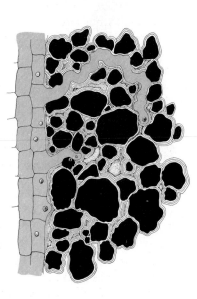

27–2

(a) *A primary root of radish* (Raphanus sativus) *seedling, showing root hairs.* (b) *Root hairs surrounded by soil particles with water adhering to them.*

(a)   (b)

27–3

Principal pathways for the movement of water (black line) and inorganic ions (colored line) from the soil, across the epidermis and cortex, and into the tracheary elements, or water-conducting elements, of the root. The water follows a mostly apoplastic pathway until it reaches the endodermis, where apoplastic movement is blocked by the Casparian strips. The Casparian strips force the water to cross the plasma membranes and protoplasts of the endodermal cells on its way to the xylem. Having crossed the plasma membrane on the inner surface of the endodermis, the water may once again enter the apoplastic pathway and make its way into the lumina of the tracheary elements. The inorganic ions are actively absorbed by the epidermal cells and then follow a symplastic pathway across the cortex and into parenchyma cells from which they are secreted into the tracheary elements.

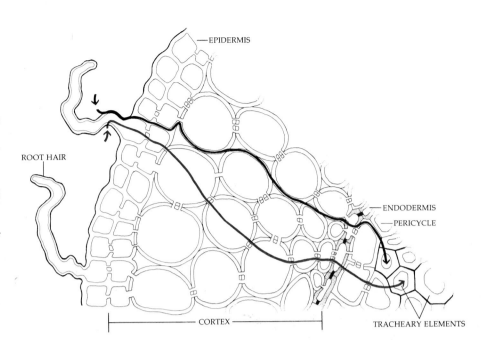

desmata) or a combination of the two. The main pathway for water across the epidermis and cortex of the root is apoplastic (Figure 27–3). At the endodermis, however, the water is forced to traverse the plasma membranes and protoplasts of the tightly packed endodermal cells because of the presence of the water-impermeable Casparian strip in their radial and transverse walls (see page 406). The endodermis, therefore, forms an osmotic barrier between the cortex and the vascular cylinder of the root.

During periods of rapid transpiration, water may be removed from around the root hairs so quickly that the soil becomes depleted of water; water will then move from some distance away toward the root hairs through fine pores in the soil. In general, however, the roots come into contact with additional water by growing, although roots will not grow in dry soil. For example, under normal conditions, roots of apple trees grow an average of about 3 to 9 millimeters a day; roots of prairie grasses may grow more than 13 millimeters a day; and the main roots of corn plants average 52 to 63 millimeters a day. The results of such rapid growth can be remarkable: a four-month-old rye *(Secale cereale)* plant has over 10,000 kilometers of roots and many billions of root hairs.

The *transpiration stream*, in addition to keeping the shoot provided with water, distributes inorganic ions to the shoot as well (Figure 27–4). After ions are absorbed by the outer cells of the root, they are transferred across the cortex and finally are secreted into the xylem. When transpiration is occurring, the ions are carried rapidly throughout the plant.

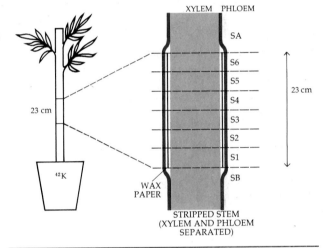

		ppm ^{42}K IN PHLOEM	ppm ^{42}K IN XYLEM
ABOVE STRIP	SA	53	47
STRIPPED SECTION	S6	11.6	119
	S5	0.9	122
	S4	0.7	112
	S3	0.3	98
	S2	0.3	108
	S1	20	113
BELOW STRIP	SB	84	58

27–4

Radioactive potassium (^{42}K) added to the soil water shows that the xylem is the channel for the upward movement of both water and inorganic ions. Wax paper was inserted between the xylem and the phloem to prevent lateral transport of the isotope. The relative amounts of radioactive potassium detected in each segment of the stem is given in the table.

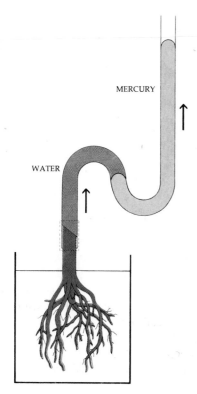

MERCURY

WATER

27–5
Demonstration of root pressure in the cut stump of a plant. Uptake of water by the plant roots causes the mercury to rise in the column. Pressures of 3 to 5 bars have been demonstrated by this method.

27–6
Guttation droplets at the leaf tips of barley (Hordeum vulgare) also demonstrate the presence of root pressure. These droplets are not condensation from water vapor in the surrounding air; rather, they are forced out of the leaf through special openings, called hydathodes, at the leaf tips.

Root Pressure and Guttation

When transpiration is very slow or absent, as it is at night, the root cells may still secrete ions into the xylem. Because the vascular tissue of the root is surrounded by the endodermis, ions do not tend to leak back out of the xylem. Therefore, the water potential (page 58) of the xylem becomes more negative, and water moves into the xylem by osmosis through the surrounding cells. In this manner a positive pressure, called *root pressure*, is created, and it forces both water and dissolved ions up the xylem (Figure 27–5).

Dewlike droplets of water at the tips of grass leaves in the early morning demonstrate the effects of root pressure (Figure 27–6). These droplets are not dew—which is water that has condensed from the air—but come from within the leaf by a process known as *guttation* (from Latin *gutta*, "drop"). They exude not through the stomata but through special openings called hydathodes, which occur at the tips and margins of leaves (Figure 27–7). The water of guttation is literally forced out of the leaves by root pressure.

Root pressure is least effective during the day, when the movement of water through the plant is the fastest, and the pressure never becomes high enough to force water to the top of a tall tree. Moreover, many plants, including conifers such as pine, develop no root pressure at all. Thus, root pressure can perhaps be regarded as a by-product of the mechanism of pumping ions into the xylem and as a subsidiary means of moving water into the shoot under special conditions.

Passive Water Absorption

During periods of high transpiration rates, ions accumulated in the xylem of the root are swept away in the transpiration stream, and the amount of osmotic movement across the endodermis decreases. At such times, the roots become passive absorbing surfaces through which water is pulled by bulk flow generated in the transpiring shoots. Some investigators believe that practically all of the absorption of water by the roots of transpiring plants occurs in this passive manner.

Water Transport

Water enters the plant by the roots and is given off, in large quantities, by the leaf. How does the water get from one place to another, often over large vertical distances? This question has intrigued many generations of botanists.

The general pathway that the water follows in its ascent has been clearly identified. You can trace this pathway in a simple experiment. Place a cut stem in water that is colored with any harmless dye (preferably, cut the stem under the water to prevent air from entering the conducting elements of the xylem) and then

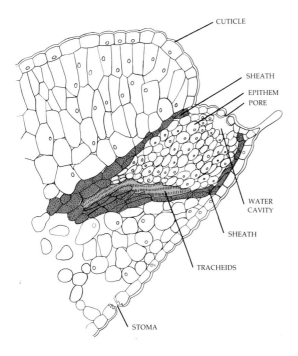

27–7
Longitudinal view of a hydathode of the leaf of Saxifraga lingulata. *The hydathode consists of the terminal tracheids of a bundle end, thin-walled parenchyma (the epithem) with numerous intercellular spaces, and epidermal pores. The tracheids are in direct contact with epithem. The epidermal pores commonly are stomata that lack the ability to open and close.*

trace the path of the liquid into the leaves. The stain quite clearly delineates the conducting elements of the xylem. Experiments using radioactive isotopes confirm that the isotope and, presumably, the water do indeed travel by way of vessel elements (or tracheids) in the xylem. In the experiment shown in Figure 27–4, care had to be taken to separate the xylem from the phloem. Earlier experiments in which this separation was not made produced ambiguous results, because there is a great deal of lateral movement from the xylem into the phloem. This lateral movement, however, as the experiment shows, is not necessary for the overall movement of water and minerals from soil to leaf.

Hence, this is the path that water takes, but how does the water move? Logic suggests two possibilities: it can be pushed from the bottom or pulled from the top. (A third possibility, involving active pumps, or "hearts," along the way has been proposed from time

to time but is no longer seriously considered by botanists.) The first of these possibilities has already been eliminated, however. Root pressure, as noted previously, does not exist in all plants, and in those plants in which it is present, it is not sufficient to push water to the top of a tall tree. Moreover, the simple experiment just described (that involving the cut stem) rules out root pressure as a crucial factor. So we are left with the hypothesis that water is pulled up through the plant body, and this hypothesis is correct according to all present evidence.

The Cohesion-Adhesion-Tension Mechanism

When water evaporates from the wall surfaces bordering the intercellular spaces in the interior of a leaf during transpiration, it is replaced by water from within the cell. This water diffuses across the plasma membrane, which is freely permeable to water but not to the solutes of the cell. As a result, the concentration of solutes within the cell increases, and the water potential of the cell decreases. A gradient of water potential then becomes established between this cell and adjacent, more saturated cells. These cells, in turn, gain water from other cells until eventually this chain of events reaches a vein and exerts a "pull" or tension on the water of the xylem. Because of the extraordinary cohesiveness of water, this tension is transmitted all the way down the stem to the roots, so that water is withdrawn from the roots, pulled up the xylem, and distributed to the cells that are losing water to the atmosphere (Figure 27–8). However, this loss makes the water potential of the roots more negative, thus increasing their ability to extract water from the soil.

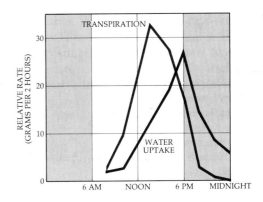

27–8
Measurements of water movement in ash (Fraxinus) *trees shows that a rise in water uptake follows a rise in transpiration. These data suggest that the loss of water generates the force needed for its uptake.*

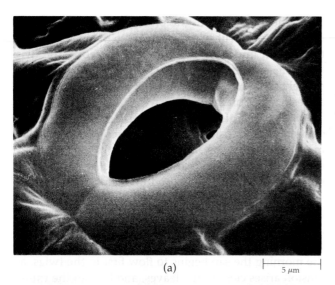

(a)

5 μm

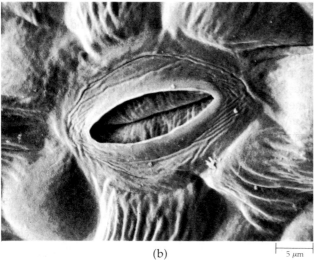

(b)

5 μm

27–13
*Scanning electron micrographs showing
(a)* open stoma in epidermis of cucumber
(Cucumis sativus) *leaf, and (b)* closed
stoma in epidermis of parsley (Apium
petroselinum) *leaf. The stomata lead into
a honeycomb of air spaces that surround
the thin-walled, photosynthetic mesophyll
cells within the leaf. The air spaces are
saturated with water vapor that has
evaporated from the surfaces of the mesophyll cells.*

Cuticle and Stomata

Leaves are covered by a cuticle that makes the surface of the leaf largely impervious both to water and to carbon dioxide. Only a small fraction of the water transpired by plants is lost through this protective outer coating, and another small fraction is lost through the lenticels in the bark. By far the largest amount of water transpired by a vascular plant is lost through the stomata (Figure 27–13). Stomatal transpiration involves two steps: (1) evaporation of water from cell wall surfaces bordering the intercellular spaces, or air spaces, of the leaf, and (2) diffusion of the resultant water vapor from the intercellular spaces into the atmosphere by way of the stomata (see Figure 27–18).

Stomata are small openings in the epidermis surrounded by two guard cells, which can change their shape to bring about the opening and closing of the pores. Stomata are also found on young stems, but they are far more abundant on leaves. The number of stomata may be quite large; for example, there are approximately 12,000 stomata per square centimeter of leaf surface in tobacco leaves. The stomata lead into a honeycomb of air spaces within the leaf that surround the thin-walled mesophyll cells. The air in these spaces—which make up 15 to 40 percent of the total volume of the leaf—is saturated with water vapor that has evaporated from the damp surfaces of the mesophyll cells. Although the stomatal openings take up only about 1 percent of the total leaf surface, more than 90 percent of the water transpired by the plant is lost through the stomata. The rest is lost through the cuticle.

Closing of stomata not only prevents the loss of water vapor from the leaf but, as mentioned, also prevents the entry of carbon dioxide into the leaf. A certain amount of carbon dioxide, however, is produced by the plant during respiration, and as long as light is available, this carbon dioxide can be used to sustain a very low level of photosynthesis even when the stomata are closed.

The Mechanism of Stomatal Movements

Stomatal movements result from changes in turgor pressure within the guard cells. Opening occurs when solutes are actively accumulated in the guard cells. The accumulation of solutes results in a movement of water into the guard cells and a buildup of turgor pressure in excess of that in the surrounding epidermal cells. Stomatal closing is brought about by the reverse process: with a decline in guard-cell solutes, water moves out of the guard cells and the turgor pressure decreases. Thus, turgor is maintained or lost due to the passive osmotic movement of water into or out of the cells along a gradient of water potential.

27–14

Quantitative changes in potassium concentrations across the stomatal complex (guard cells and subsidiary cells) and neighboring ordinary epidermal cells in the dayflower (Commelina communis) *leaf. The guard cells of the dayflower leaf are associated with six subsidiary cells, four lateral and two terminal. (a) Vacuolar potassium concentrations in the various cells when the stomata were closed. (b) Vacuolar potassium concentrations in the various cells when the stomata were open. Potassium-sensitive microelectrodes were used to determine the potassium content of the individual cells.*

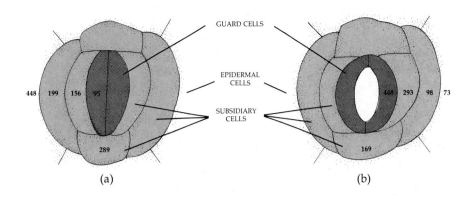

The major solute responsible for these gradients in water potential is the potassium ion (K^+). This ion has been found in the guard cells of open stomata of more than 50 species, including CAM plants whose stomata open at night. Techniques for estimating potassium levels within a single guard cell show that potassium levels rise when the stomata open and drop when the stomata close. The surrounding cells provide the required reservoir of potassium ions. The gradient of potassium between the guard cells and surrounding cells changes significantly and is accompanied by the osmotic flow of water and resultant turgor changes (Figure 27–14).

With the positively charged K^+ transported in such large amounts, negatively charged ions (anions) are needed to counter the charge. Two anions in particular have been implicated in this regard—chloride and malate.

The structure of the guard cell walls plays a crucial role in stomatal movements. During expansion of the paired guard cells, two physical constraints cause the guard cells to bend and thus to open the pore. One of these is the radial orientation of the cellulose microfibrils in the guard cell walls (Figure 27–15a). This *radial micellation* allows the guard cells to lengthen while preventing them from expanding laterally. The second constraint is found at the ends of the guard cells, where they are attached to one another. This common wall remains almost constant in length during opening and closing of the stomata. Consequently, increase in turgor pressure causes the outer (dorsal) walls of the guard cells to move outward relative to their common walls. As this happens, the radial micellation transmits this movement to the wall bordering the pore (the ventral wall) and the pore opens. Figure 27–15b through d depicts the results of some experiments with balloons that have been used in support of the role of radial micellation in stomatal movement.

27–15

(a) Diagram of a pair of guard cells. The radial lines indicate the radial arrangement of cellulose microfibrils in the guard cell walls. (b)–(d) Diagrams of models (balloons) used to study the effect of radial micellation on the opening of stomata. (b) Two partially inflated balloons that had been glued together near their ends. (c) The same balloons at higher pressure. A narrow slit is now visible. (d) A pair of inflated balloons after bands of tape had been added to simulate radial micellation. The opening is much greater than in (c).

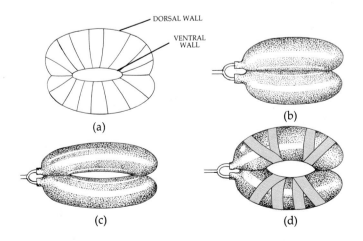

Factors Affecting Stomatal Movements

A number of environmental factors affect stomatal opening and closing, *water loss* being the major influence. When the turgor of a leaf drops below a certain critical point, which varies with different species, the stomatal opening becomes smaller. The effect of water loss overrides other factors affecting the stomata, but stomatal changes can occur independently of overall water gain or loss by the plant. The most conspicuous example is found in the many species in which the stomata open regularly in the morning and close in the evening, even though there may be no changes in water available to the plant.

During periods of water stress in many plants there is a marked increase in the level of abscisic acid (ABA). When "fed" or applied to leaves, ABA causes stomatal closure within a few minutes; moreover, the effect of ABA on stomatal movement is readily reversible. Experimental evidence suggests that solute (K^+) loss from guard cells begins when ABA of mesophyll origin arrives at the stomata, signaling the stomata that the mesophyll cells are experiencing water stress. How ABA acts on guard cells remains to be determined.

Other environmental factors that affect stomatal movement include *carbon dioxide concentration, light,* and *temperature.* In most species, an increase in CO_2 concentration causes the stomata to close. The magnitude of this response to CO_2, varies greatly from species to species and with the degree of water stress a given plant has suffered or is suffering. In corn (*Zea mays*), the stomata may respond to changes in CO_2 in a matter of seconds. The site for sensing the level of CO_2 has been demonstrated to be located within the guard cells.

In most species, the stomata open in the light and close in the dark. This can be explained in part by the photosynthetic utilization of CO_2, which brings about a reduction in the CO_2 level within the leaf. Light may, however, have a more direct effect on stomata. Blue light has long been known to stimulate stomatal opening independently of CO_2. For example, guard-cell protoplasts of onion (*Allium cepa*) swell in the presence of K^+ when illuminated with blue light. The blue-absorbing pigment (a flavin or a flavoprotein located in the tonoplast and possibly the plasma membrane) promotes K^+ uptake by the guard cells.

Within normal ranges of temperature (10 to 25°C), changes in temperature have little effect on stomatal behavior, but temperatures higher than 30 to 35°C can lead to stomatal closure. The closing can be prevented, however, by holding the plant in air that contains no carbon dioxide, which suggests that temperature changes work primarily by affecting the concentration of carbon dioxide in the leaf. An increase in temperature results in an increase in respiration and a concomitant increase in the concentration of intercellular carbon dioxide, which may actually be the cause of stomatal closure in response to heat. Many plants in hot climates close their stomata regularly at midday, apparently because of the effect of temperature on carbon dioxide accumulation and because of dehydration of leaves as the water loss of transpiration exceeds the water uptake by absorption.

Stomata not only respond to environmental factors but they also exhibit daily rhythms of opening and closing that appear to be controlled from within the plant—that is, they also exhibit circadian rhythms (see page 496).

Although the stomata of most plants are open during the day and closed at night, this is not true of all plants. A wide variety of succulents—including cacti, the pineapple (*Ananas comosus*), and members of the stonecrop family (Crassulaceae) among others—open their stomata at night, when conditions are least favorable to transpiration. The Crassulacean acid metabolism (CAM) characteristic of such plants has a pathway for carbon flow not substantially different from that of C_4 plants, as discussed in Chapter 7. At night, when their stomata are open, the CAM plants take in carbon dioxide and convert it into organic acids. During the day, when their stomata are closed, the carbon dioxide is released from these organic acids for use in photosynthesis.

Factors Affecting the Rate of Transpiration

Although stomatal opening and closing are the major factors affecting the rate of transpiration, there are a number of other factors both in the environment and in the plant itself. One of the most important of these is *temperature.* The rate of water evaporation doubles for every temperature rise of about 10°C. However, because evaporation cools the leaf surface, its temperature does not rise as rapidly as that of the surrounding air. As noted previously, stomata close when temperatures exceed 30 to 35°C.

Humidity is also important. Water is lost much more slowly into air already laden with water vapor. Leaves of plants growing in shady forests, where the humidity is generally high, typically spread large luxuriant leaf surfaces, because their main problem is getting enough light, not losing water. In contrast, plants of grasslands or other exposed areas often have narrow leaves, with relatively little leaf surface. They get all the light they can use but are constantly in danger of excess water loss.

Air currents also affect the rate of transpiration. A breeze cools your skin on a hot day because it blows away the water vapor that has accumulated near the skin surface and so accelerates the rate of evaporation of water from your body. Similarly, wind blows away the water vapor from leaf surfaces. Sometimes, if the air is very humid, wind may actually decrease the transpiration rate by cooling the leaf, but a dry breeze will greatly increase evaporation. Leaves of plants that grow in exposed, windy areas are often hairy; these hairs are believed to protect the leaf surface from wind action and so slow the rate of transpiration by stabilizing the boundary layer of air over the leaf surface.

It recently has been suggested that the main function of the sensitivity of stomata to CO_2 is to induce partial closure of the stomata as wind speed increases. Whereas wind transports water vapor away from the leaf surface, it transports CO_2 toward it; thus, if the stomata close in response to an increase in CO_2 concentration, they will lower the transpiration rate.

MOVEMENT OF INORGANIC NUTRIENTS THROUGHOUT THE PLANT BODY

Uptake of Inorganic Nutrients

The uptake, or absorption, of inorganic ions takes place through the epidermis of the root. Current evidence suggests that the major pathway followed by ions from the epidermis to the endodermis of the root is symplastic. Ion uptake by the symplastic route begins at the plasma membrane of the epidermal cells. The ions then move from the epidermal cell protoplasts to the first layer of cortical cells through the plasmodesmata in the epidermal-cortical cell walls (see Figure 27–3). Radial movement of the ions continues in the cortical symplast—from protoplast to protoplast via plasmodesmata—through the endodermis and into the parenchyma cells of the vascular cylinder by diffusion, possibly aided by cytoplasmic streaming within the individual cells.

How the ions enter the vessels (or tracheids) of the xylem from the parenchyma cells of the vascular cylinder has been the subject of considerable debate. It was suggested that the ions leak passively from the parenchyma cells into the vessels, but there is now substantial evidence that the ions are secreted into the vessels from the parenchyma cells by an active, carrier-mediated membrane transport (see page 64).

Active Solute Absorption

The mineral composition of root cells is far different from that of the medium in which the plant grows. For example, in one study, cells of pea (*Pisum sativum*) roots were found to have a concentration of potassium ions 75 times greater than that of the nutrient solution. Similarly, in another study, the vacuoles of rutabaga (*Brassica napus* var. *napobrassica*) cells were shown to contain 10,000 times more potassium than the external solution.

Since substances do not diffuse against a concentration gradient, it is clear that minerals are absorbed by *active transport*. Support for this hypothesis comes from observations indicating that the uptake of minerals is an energy-requiring process. For instance, if roots are deprived of oxygen, or poisoned so that respiration is curtailed, mineral uptake is drastically decreased. Also, if a plant is deprived of light, it will cease to absorb salts after carbohydrate reserves are exhausted and will finally release them back into the soil solution (Figure 27–16). Hence, ion transport from the soil to the vessels of the xylem requires two active, carrier-mediated membrane events: (1) uptake at the plasma membrane of the epidermal cells and (2) secretion into the vessels at the plasma membrane of the vascular parenchyma cells.

Transmembrane Potential

The active transport of ions across the plasma membrane may result in a difference of electrical charge on the two sides of the membrane. When this occurs, a voltage difference, called the *transmembrane potential*, develops across the membrane. The hydrogen ion, H^+, is one of the principal cations involved with the generation of the transmembrane potential; with the pumping of H^+ to the outside, a negative potential develops on the inside of the cell. Once generated, the transmembrane potential can have a marked effect on further ion movements. For example, a negative interior electrical potential will attract positively charged ions such as K^+ but will repel negatively charged ions such as Cl^-.

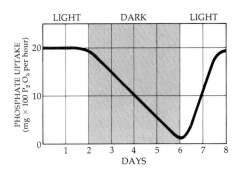

27–16
The rate of phosphate absorption by corn (Zea mays) plants fell to near zero after four days of continuous darkness. It began to rise again when the plants were reilluminated. These and other data indicate that salt uptake in plants is a process that requires energy.

Transport of Inorganic Nutrients

Once the inorganic ions have been secreted into the vessels of the xylem, they are rapidly distributed upward and throughout the plant in the transpiration stream. Some ions move laterally from the xylem into surrounding tissues of the roots and stems, while others are transported into the leaves (Figure 27–17).

Much less is known about the pathways followed by the ions in leaves than in roots. Within the leaf, the ions are transported along with the water in the leaf apoplast, that is, in the cell walls. Some ions may remain in the transpiring water and reach the main regions of water loss—the stomata and other epidermal cells. Most eventually enter the protoplasts of the leaf cells, probably by carrier-mediated transport mechanisms similar to those demonstrated in roots. The ions may then move symplastically to other parts of the leaf, including the phloem. Inorganic ions can also be absorbed in small amounts through the leaves; consequently, fertilization of some crop plants that involves the direct application of micronutrients to the foliage has become a standard agricultural practice.

Substantial amounts of the inorganic ions imported into the leaves through the xylem are exchanged with the phloem of the leaf veins and exported from the leaf with sucrose in the assimilate stream (Figure 27–17; see also the discussion of translocation that follows). For instance, in the annual white lupine (*Lupinus albus*), transport in the phloem accounted for more than 80 percent of the fruit's vascular intake of nitrogen and sulfur and 70 to 80 percent of its phosphorus, potassium, magnesium, and zinc. The uptake of such inorganic ions by developing fruits is undoubtedly coupled to the flow of sucrose in the phloem.

Recycling may occur in the plant as nutrients reaching the roots in the descending assimilate stream are transferred to the ascending transpiration stream of the xylem (Figure 27–17). Only those ions that can move in the phloem, and which are said to be *phloem-mobile*, can be exported from the leaves to any great extent. For example, K^+, Cl^-, and $H_2PO_4^{2-}$ are readily exported from leaves, whereas Ca^{2+} is not. Solutes such as calcium are said to be *phloem-immobile*.

27–17

Diagram showing basic elements in the circulation of water, inorganic ions, and assimilates in the plant. Water and inorganic ions taken up by the root move upward in the xylem in the transpiration stream. Some move laterally into tissues of the root and stem, while others are transported to growing plant parts and mature leaves. In the leaves, substantial amounts of water and inorganic ions are transferred to the phloem and are exported with sucrose in the assimilate stream. The growing plant parts, which are relatively ineffective in capturing water through transpiration, receive much of their nutrients and water via the phloem. Both water and solutes entering the roots in the phloem may be transferred to the xylem and recirculated in the transpiration stream. The symbol A designates sites specialized for the absorption and assimilation of raw materials from the environment. L and U designate sites of loading and unloading, respectively, and I, principal points of interchange between the xylem and phloem.

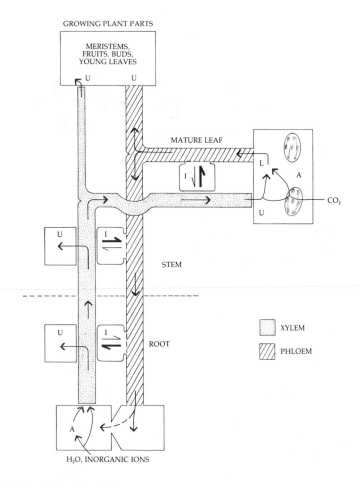

TRANSLOCATION: THE MOVEMENT OF SUBSTANCES IN THE PHLOEM

As discussed in Chapters 20 through 22, the xylem and phloem together form a continuous vascular system that penetrates practically every part of the plant. Whereas the water and inorganic solutes ascend the plant in the xylem, or transpiration stream, the sugars manufactured during photosynthesis move out of the leaf in the phloem, or *assimilate stream* (Figure 27–18), to sites where they are utilized, such as growing shoot and root tips, and to sites of storage, such as fruits, seeds, and the storage parenchyma of stems and roots (see Figure 27–17).

Assimilate movement is said to follow a source-to-sink pattern. The principal *sources* of assimilate solutes are the photosynthesizing leaves, but storage tissues may also serve as important sources. All plant parts unable to meet their own nutritional needs may act as *sinks*, that is, as importers of assimilates. Thus, storage tissues act as sinks when they are importing assimilates and as sources when they are exporting assimilates.

Source-sink relations may be relatively simple and direct, as in young seedlings, where cotyledons containing reserve food often represent the major source and growing roots represent the major sink. In older plants, the upper, most recently formed mature leaves commonly export assimilates primarily toward the shoot tip; the lower leaves export assimilates primarily to the roots; and those in between export assimilates in both directions (Figure 27–19a). This pattern of assimilate distribution is markedly altered during the change from vegetative to reproductive growth. Developing fruits are highly competitive sinks that monopolize assimilates from the nearest leaves, and frequently from distant ones as well, often causing a decline or virtual cessation of vegetative growth (Figure 27–19b).

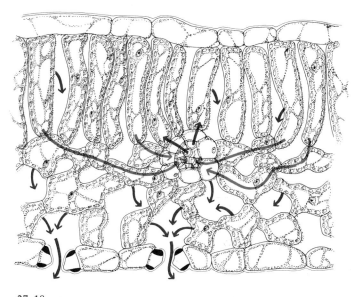

27–18

Diagram of leaf showing the pathways followed by water molecules of the transpiration stream as they move from the xylem of a minor vein to the mesophyll cells, evaporate from the surface of the mesophyll cell walls, and then diffuse out of the leaf through an open stoma (colored lines).

Also shown are the pathways followed by sugar molecules manufactured during photosynthesis as they move from mesophyll cells to the phloem of the same vein and enter the assimilate stream. The sugar molecules manufactured in the palisade parenchyma cells are believed to move to the spongy parenchyma cells and then laterally to the phloem via the spongy cells (gray lines).

(a)   (b)

27–19

Diagrams of a plant in (a) the vegetative stage and (b) the fruiting stage. The arrows indicate the direction of assimilate transport in each stage.

Evidence of Sugar Transport in Phloem

Early evidence supporting the role of the phloem in assimilate transport came from observations of trees from which a complete ring of bark had been removed. As noted in Chapter 23, the bark in older stems is composed largely of phloem and contains no xylem. When a photosynthesizing tree is ringed, or "girdled," in this manner, the bark above the ring becomes swollen, indicating an accumulation of assimilates moving downward in the phloem from the photosynthesizing leaves (Figure 27–20).

Much more convincing evidence of the role of the phloem in assimilate transport was obtained with radioactive tracers. Before such tracers were available, it was necessary to cut into the intact plant to introduce dyes and other substances in an attempt to study certain transport phenomena. However, when the high turgor (hydrostatic) pressures of the sieve tubes are released at the time the sieve tubes are severed, the contents of the sieve elements surge toward the cut surfaces, greatly disturbing the system. As discussed in Chapter 20, this phenomenon is responsible for the formation of slime (P-protein) plugs in injured sieve elements. With the use of radioactive tracers, it is now possible to experiment with entire plants and thus to obtain a fairly clear understanding of normal transport phenomena. The results of experiments with radioactive assimilates (such as ^{14}C-labeled sucrose) confirmed the movement of such substances in the phloem. More recently, such studies have shown conclusively that sugars are transported in the sieve tubes of the phloem (see "Radioactive Tracers and Autoradiography in Plant Research" on the facing page).

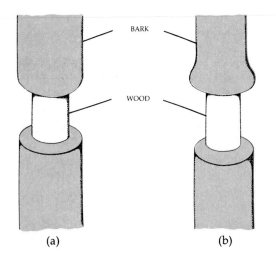

BARK

WOOD

(a)   (b)

27–20
As early as the seventeenth century, Marcello Malpighi of Italy noticed that, when a ringlike piece of bark was removed from a stem (a), the tissues above the ring became swollen (b). He correctly interpreted this phenomenon as new growth of wood and bark tissues stimulated by an accumulation of food moving down from the leaves and intercepted at the ring. Malpighi studied the effect of ringing at different times of the year and found that no swelling occurred during the winter months.

The Aphid in Phloem Research

Much valuable information on movement of substances in the phloem has come from studies utilizing aphids—small insects that suck the juices of plants. Most species of aphids are phloem feeders. When these aphids insert their modified mouth parts, or stylets, into a stem or leaf, they extend them until the tips of the stylets puncture a conducting sieve tube (Figure 27–21). The turgor pressure of the sieve tube then forces the sieve-tube sap through the aphid's digestive tract and out its posterior end as droplets of "honeydew." If feeding aphids are anesthetized and severed from their stylets, exudation often continues from the cut stylets for many hours. The sieve-tube exudate can then be collected from the cut ends of the stylets with a micropipette. Analyses of exudates obtained in this manner reveal that sieve-tube sap contains 10 to 25 percent dry matter, 90 percent or more of which is sugar—*mainly sucrose*—in most plants. Low concentrations (less than 1 percent) of amino acids and other nitrogen-containing substances are also present.

Data obtained from studies utilizing aphids and radioactive tracers indicate that the rates of longitudinal movement of assimilates in the phloem are remarkably fast. For example, in one series of experiments utilizing severed aphid stylets, it was estimated that the sieve-tube sap was moving at a rate of about 100 centimeters per hour at the sites of the stylet tips.

The Mechanism of Phloem Transport: Pressure Flow

Several mechanisms have been proposed over the years to explain assimilate transport in the sieve tubes of the phloem. Probably the earliest was that of diffusion, followed by that of cytoplasmic streaming. Normal diffusion and cytoplasmic streaming of the kind found in higher plant cells were largely abandoned as possible translocation mechanisms when it became known that the velocities of assimilate transport (typically 50 to 100 centimeters per hour) were far too great for either of these two phenomena to account for long-distance transport via sieve tubes.

RADIOACTIVE TRACERS AND AUTORADIOGRAPHY IN PLANT RESEARCH

Radioactive tracers can be used in a number of ways to study the synthesis, transport, and use of materials within plants. Initially, in any tracer study, the radioactive isotope must be incorporated into the plant. Radioactive carbon, for example, will be taken up by a plant if its leaves are exposed to carbon dioxide that contains ^{14}C, or radioactive phosphorus will be taken up by a plant if the roots are exposed to a solution containing ions of ^{32}P.

The length of time the plant must be exposed to the radioactive material is determined by the information that the investigators hope to obtain. For example, in studies designed to determine the time required for carbon dioxide to be incorporated into the various products of photosynthesis, a sequence of exposure times would be used. In studies focusing on the location of a particular product formed by the metabolic processes that involve the radioactive substance, the length of exposure would depend on the time required for the chemical reactions under study to occur.

In whole-plant autoradiography, the plant is quick-frozen and freeze-dried after exposure to the radioactive substance. The plant is then flattened and pressed against a sheet of X-ray film. Radiation given off by the isotope exposes the film adjacent to the portions of the plant in which the tracer is located. By comparing the flattened plant with the developed film, investigators can determine the location of the radioactive substance within the plant.

In tissue autoradiography (microautoradiography), the freeze-dried plant tissues are embedded in paraffin, resin, or a similar material. Next, they are sliced into very thin sections, which are mounted on microscope slides. The tissue sections are then placed in contact with a photographic emulsion or film. The radiation from the isotope exposes the film in contact with the portions of the tissue section containing the radioactive material. After an appropriate interval of time, the film is developed. Comparison, under a microscope, of the developed film and the underlying tissue section reveals the exact location of the radioactive substance in the plant tissues.

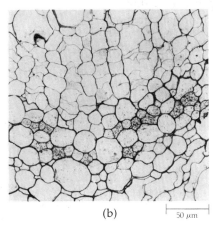

(b)
50 μm

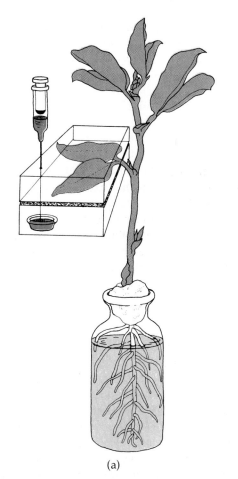

(a)

(c)
50 μm

(a) *Two leaflets of a broad bean plant (Vicia faba) were enclosed in a Plexiglas container in which $^{14}CO_2$ was generated and were exposed to radioactive carbon dioxide and light for 35 minutes. During that time, the $^{14}CO_2$ was incorporated into sugars, which were then being transported to other parts of the plant. A cross section (b) and a longitudinal section (c) from the stem were placed in contact with autoradiographic film for 32 days. When the film was developed and compared with the underlying tissue sections, it was apparent that the radioactivity (visible as dark grains on the film) was confined almost entirely to the sieve tubes.*

(a)

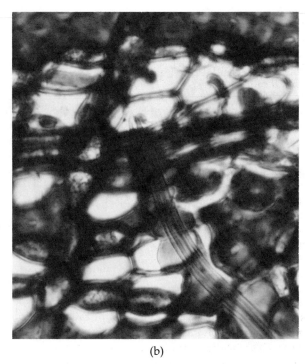

(b)

27–21

(a) Aphid (Longistigma caryae) *feeding on a linden* (Tilia americana) *stem. A droplet of "honeydew" can be seen emerging from the aphid. (b) A photomicrograph showing part of the modified mouth parts (stylets) of the aphid in a sieve tube of the secondary phloem of the linden stem.*

27–22

The Münch model for the basic mechanism of pressure flow. A and B are osmotic cells. A contains a higher concentration of sucrose than B. V is a glass tube that connects the two cells, and W is a trough containing water. Water enters A by osmosis, thus increasing the turgor pressure and pushing the sugar solution to bulb B.

Alternate hypotheses have been advanced to explain the mechanism of phloem transport, but only one, the pressure-flow hypothesis, satisfactorily accounts for practically all of the data obtained in experimental and structural studies on phloem. All of the other hypotheses have serious deficiencies.

Originally proposed in 1927 by the German plant physiologist Ernst Münch, and since modified, the pressure-flow hypothesis is clearly the simplest and, at present, the most widely accepted explanation for long-distance assimilate transport in sieve tubes. It is the simplest explanation because it depends only on osmosis as the driving force for assimilate transport.

Briefly stated, the *pressure-flow hypothesis* asserts that assimilates are transported from source to sink along a gradient of turgor pressure developed osmotically. The principle underlying this hypothesis can be illustrated by a simple physical model consisting of bulbs, or osmotic cells, permeable only to water and connected by glass tubes (Figure 27–22). Initially, the first bulb (A) contains a more concentrated sugar solution than the second bulb (B). When these interconnected bulbs are placed in water, water will enter the first bulb by osmosis, thereby increasing the turgor pressure within that bulb. This pressure will be transmitted through the tube to the second bulb, causing the sugar solution to move by bulk, or mass, flow to the second bulb and forcing water out of it. If the second bulb is connected with a third one containing a sucrose concentration lower than that in the second one, the solution will flow from the second to the third bulb by the same process, and so on indefinitely down the turgor pressure gradient.

27–23

Diagram of osmotically generated pressure-flow mechanism. Circles represent sugar molecules. Sugar is actively loaded into the sieve tube at the source. With the increased concentration of sugar, the water potential is decreased, and water from the xylem enters the sieve tube by osmosis. Sugar is removed (unloaded) at the sink, and the sugar concentration falls; as a result, the water potential is increased, and water leaves the sieve tube. With the movement of water into the sieve tube at the source and out of it at the sink, the sugar molecules are carried passively by the water along the concentration gradient and gradient of hydrostatic pressure between source and sink. Note that, unlike the Münch model depicted in Figure 27–22, the sieve tube between source and sink is bounded by a differentially permeable membrane, the plasma membrane. Consequently, water does not enter and leave the sieve tube only at the source and sink, but all along the pathway. Evidence indicates that few if any of the original water molecules entering the sieve tube at the source end up in the sink, because they are exchanged with other water molecules that enter the sieve tube from the phloem apoplast along the pathway.

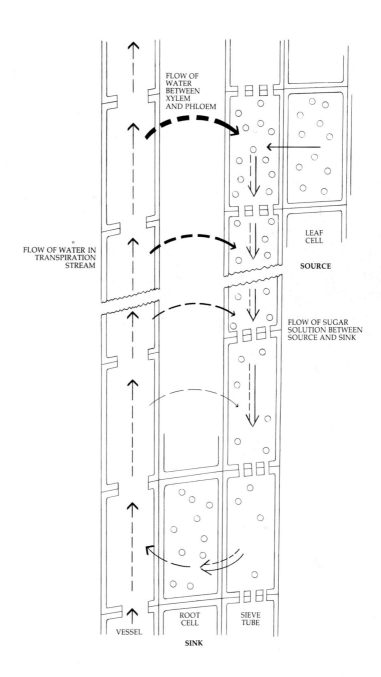

FLOW OF WATER BETWEEN XYLEM AND PHLOEM

FLOW OF WATER IN TRANSPIRATION STREAM

LEAF CELL

SOURCE

FLOW OF SUGAR SOLUTION BETWEEN SOURCE AND SINK

VESSEL

ROOT CELL

SIEVE TUBE

SINK

In the plant, sucrose produced by photosynthesis in a leaf is actively secreted into the sieve tubes of the minor veins (Figure 27–23). This active process, called *phloem loading*, decreases the water potential in the sieve tube and causes water entering the leaf in the transpiration stream to move into the sieve tube by osmosis. With the movement of water into the sieve tube at this source, the sucrose is carried passively by the water to a sink, such as a storage root, where the sucrose is removed (unloaded) from the sieve tube. The removal of sucrose results in an increased water potential in the sieve tube at the sink and the subsequent movement of water out of the sieve tube there. The sucrose may be either utilized or stored at the sink, but most of the water returns to the xylem and is recirculated in the transpiration stream.

Note that the pressure-flow hypothesis casts the sieve tubes in a *passive* role in the movement of the sugar solution through them. Active transport is also involved in the pressure-flow mechanism; however, it is not directly involved with the long-distance transport through the sieve tubes, but rather with the loading and possibly unloading of sugars and other substances into and out of the sieve tubes at the sources and sinks. Considerable evidence indicates that the driving force for sucrose accumulation (phloem loading) at the source is provided by a proton pump energized by ATP and mediated by ATPase at the plasma membrane, involving a sucrose-proton co-transport (symport) system (see page 64). The metabolic energy required for the loading and unloading is expended by companion cells or parenchyma cells bordering the sieve tubes, rather than by the sieve tubes. Until recently, it was assumed that loading occurred across the plasma membrane of the companion cell, which then transferred the sugar to its associated sieve tube via the many plasmodesmatal connections in their common wall. It now appears, however, that some sieve tubes are capable of loading themselves, the site of active transport being their own plasma membranes. Whatever the case, the mature sieve tube is dependent upon its companion cell or neighboring parenchyma cells for most or all of its energy needs.

Phloem loading is a selective process. As mentioned previously, sucrose is by far the most common sugar transported; in addition, all of the sugars found in sieve-tube sap are nonreducing sugars. Certain amino acids and ions also are selectively loaded into the phloem.

SUMMARY

In plants, more than 90 percent of the water taken in by the roots is given off into the air as water vapor. This loss of water vapor by plant parts is called transpiration, and most of the water transpired by vascular plants is lost through the stomata in the leaves.

Absorption of water takes place largely through the root hairs, which provide an enormous surface area for water uptake. In some plants, when the roots absorb water from the soil and transport it into the xylem, the water within the xylem builds up positive pressure, or root pressure. This osmotic uptake depends on the transport of inorganic ions from the soil into the xylem by the living cells of the root and can result in a phenomenon called guttation, in which liquid water is forced out through special openings in the tips or margins of the leaves. The water follows a largely apoplastic pathway across the epidermis and cortex until it reaches the endodermis, where further movement is blocked by the Casparian strips. The water must pass through the plasma membrane and protoplast of the endodermal cells on its way to the xylem.

From the roots, water moves to the leaves by means of the xylem. The current and widely accepted theory of water movement to the top of tall plants through the xylem is the cohesion-adhesion-tension theory. According to this theory, water within the vessels is under tension because the water molecules cling together in continuous columns pulled by evaporation from above. It has been demonstrated that water has sufficient tensile strength to withstand such tension when contained in small-diameter conduits. Other supporting evidence includes observations that water in the xylem is under tension, that water movement in trees begins in the topmost branches, and that the trunk of a tree shrinks slightly at the beginning of water movement.

The rate of transpiration is affected by such factors as carbon dioxide concentration in the intercellular spaces (and, conversely, the exposure of the leaf to carbon dioxide from the air), light, temperature, atmospheric humidity, air currents, and availability of soil water. Most of these factors have

an effect on the behavior of stomata. Stomatal opening and closing are controlled by changes in turgor of the guard cells, which are closely correlated with changes in the potassium ion level within the guard cells. Abscisic acid and blue light also play roles in stomatal movement. The stomata open when the guard cells become turgid and close when they become flaccid.

Inorganic nutrients become available to plants in soil solution in the form of ions. Plants employ metabolic energy to concentrate the ions they require. Most ions are taken up by active transport processes, whereas others flow in passively due to the water potential across the plasma membrane that is created by the actively moving ions and their pumps. Inorganic ions follow a mostly symplastic pathway from the epidermis to the xylem.

Research on the movement of substances in the phloem has been greatly aided by the use of aphids and radioactive tracers. Analyses of sieve-tube sap reveal that it contains sugar—mainly sucrose—and small quantities of nitrogenous substances. Rates of longitudinal movement of substances in the phloem are far in excess of the normal rate of diffusion of sucrose in water—the rates typically range from 50 to 100 centimeters per hour.

According to the pressure-flow hypothesis, assimilates move from source to sink along a turgor pressure gradient that is developed osmotically. Sugars are actively secreted into (loaded) and absorbed from (unloaded) the sieve tube by companion cells or parenchyma cells at source and sink, respectively, resulting in mass flow of solution in the sieve tube. The role of the sieve tube is a passive one.

SUGGESTIONS FOR FURTHER READING

BEM, ROBYN: *Everyone's Guide to Home Composting,* Van Nostrand Reinhold Company, New York, 1978.*

A comprehensive, easy-to-read manual that discusses the construction and monitoring of a compost heap, as well as the uses for compost. Included are such topics as the nutrient content of various compost materials, green manure crops, and essential plant nutrients.

BRADY, NYLE C.: *The Nature and Properties of Soils,* 8th ed., Macmillan Publishing Company, Inc., New York, 1974.

A comprehensive elementary text on soil science.

BRILL, WINSTON J.: "Biological Nitrogen Fixation," *Scientific American* 236:68–81, 1977.

————: "Nitrogen Fixation: Basic to Applied," *American Scientist* 67:458–66, 1979.

The above two articles are excellent reviews of a complex subject presented in an easy-to-understand and readable style.

BUNTING, BRIAN T.: *The Geography of Soils,* Aldine Publishing Co., Chicago, 1965*

A brief description of the main soil groups of the world and their influence on plant distribution.

* Available in paperback.

CRAFTS, ALDEN S., and CARL E. CRISP: *Phloem Transport in Plants*, W. H. Freeman and Co., New York, 1971.

A definitive text on phloem structure and function. Experimental methods employed in phloem research and the influence of environmental factors on translocation are discussed. Five hypotheses for the phloem-transport mechanism are reviewed, and strong evidence in support of pressure flow is presented.

EPSTEIN, EMANUEL: *Mineral Nutrition of Plants: Principles and Perspectives*, John Wiley & Sons, Inc., New York, 1972.

A thorough coverage of the whole field of mineral nutrition, well illustrated and written by one of the leading students of the subject.

FARB, PETER: *Living Earth*, Pyramid Publications, Inc., New York, 1969.*

An eloquent description of the teeming life that exists below the surface of the ground. A book to read for pleasure.

International Symposium on Nitrogen Fixation, Washington State University Press, Pullman, Wash., 1975.

A collection of outstanding papers on all aspects of this very active field of research.

LÜTTGE, ULRICH, and NOE HIGINBOTHAM: *Transport in Plants*, Springer-Verlag, New York, 1979.

A comprehensive, up-to-date review of the various processes by which inorganic and organic substances are transported in plants, from the level of cellular organelles to long-distance movements in trees.

MARTIN, E. STEPHEN, MARIA E. DONKIN, and R. ANDREW STEVENS: *Stomata, Studies in Biology*, No. 155, Edward Arnold Publishers, London, 1983.*

A concise introduction to many of the new developments on the mechanism of stomatal movements.

MENGEL, K., and E. A. KIRKBY: *Principles of Plant Nutrition*, 3rd ed., International Potash Institute, Worblaufen-Bern, Switzerland, 1982.

A textbook for students interested in plant science and crop production; integrates soil science, plant physiology, and biochemistry.

NOGGLE, FRITZ: *Introductory Plant Physiology*, Prentice-Hall, Inc., Englewood Cliffs, N.J., 1976.

A text that can be useful to the student with training only in general biology and general chemistry.

PEEL, A. J.: *Transport of Nutrients in Plants*, John Wiley & Sons, Inc., New York, 1974.

A concise, up-to-date account of the physiology of long-distance transport in plants. Presents a balanced view of transport in

phloem and xylem. A good text for the student who is unfamiliar with transport processes.*

QUISPEL, A. (Ed.): *Biology of Nitrogen Fixation*, Frontiers of Biology Ser., vol. 33, Elsevier-North Holland Publishing Co., New York, 1974.

Specialists in various aspects of research on nitrogen fixation contributed chapters to this volume.

SALISBURY, FRANK B., and CLEON W. ROSS (Eds.): *Plant Physiology*, 3rd ed., Wadsworth Publishing Co., Inc., Belmont, Calif., 1985.

A detailed and useful review of the entire subject.

SUTCLIFFE, JAMES: *Plants and Water*, St. Martin's Press, Inc., New York, 1968.*

An excellent short account of water relationships, including a review of experimental work in this area.

VIORST, JUDITH: *The Changing Earth*, Bantam Books, Inc., New York, 1967.*

A well-written introduction for the layman to the science of modern geology.

WARDLAW, IAN F., and J. B. PASSIOURA (Eds.): *Transport and Transfer Processes in Plants*, Academic Press, Inc., New York, 1976.

The proceedings of a symposium designed to examine how the various forms of long-distance and short-distance transport operate and interact in the whole plant.

ZIMMERMANN, MARTIN H.: *Xylem Structure and the Ascent of Sap*, Springer-Verlag, New York, 1983.

A delightfully written "idea" book on xylem structure and function by one who contributed a great deal to our understanding of functional xylem anatomy and sap movement in plants.

ZIMMERMANN, MARTIN H., and CLAUD L. BROWN: *Trees: Structure and Function*, Springer-Verlag, New York, 1975.

This book is devoted to those aspects of tree physiology that are peculiar to tall woody plants. The emphasis is on function and includes material not found in general physiology texts.

ZIMMERMANN, MARTIN H., and JOHN A. MILBURN (Eds.): *Transport in Plants I, Phloem Transport*, Encyclopedia of Plant Physiology, vol. I, Springer-Verlag, New York, 1976.

A collection of articles by leading researchers in the field of phloem structure and function.

* Available in paperback.

S E C T I O N   E I G H T

Evolution

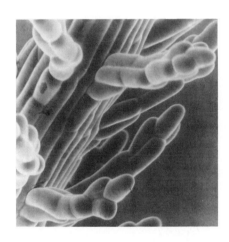

C H A P T E R   2 8

The Process of Evolution

In 1831, as a young man of 22, Charles Darwin (Figure 28–1) set forth on a five-year voyage as ship's naturalist on a British navy ship, HMS *Beagle*. The book he wrote about the journey, *The Voyage of the Beagle*, is not only a classic work of natural history but also provides us with insight into the experiences that led directly to Darwin's proposal of his theory of evolution by natural selection.

At the time of Darwin's historic voyage, most scientists—and nonscientists as well—still believed in the theory of "special creation." According to this idea, each of all the many different kinds of living organisms was created (or otherwise came into existence) in its present form. Some scientists, such as Jean Baptiste de Lamarck (1744–1829), had challenged the idea of special creation, but their arguments were not sufficiently convincing to most people to shake their belief. Special creation was not just a scientific hypothesis; it was firmly embedded in all of Western culture.

Darwin's theory was able to bring about a great intellectual revolution simply because the mass of evidence that he presented was so convincing that there was no longer room for reasonable scientific doubt. Particularly important in the genesis of Darwin's ideas were his experiences during a stay of some five weeks in the Galápagos Islands, an archipelago that lies in equatorial waters about 950 kilometers off the west coast of South America (Figure 28–2). There he made two particularly important observations: First, he noted that the plants and animals found on the islands, although distinctive, were similar to those on the nearby South American mainland. If each kind of plant and animal had been created separately and was unchangeable, as was then generally believed, why did the plants and animals of the Galápagos not resemble those of Africa, for example, rather than those of South America? Or, indeed, why were they not utterly unique, unlike organisms anywhere else on earth? Second, people familiar with the islands pointed out variations that occurred from island to island in such organisms as the giant tortoises.

28–1

"Afterwards, on becoming very intimate with FitzRoy (captain of the Beagle*), I heard that I had run a very narrow risk of being rejected on account of the shape of my nose! He . . . was convinced that he could judge of a man's character by the outline of his features; and he doubted whether anyone with my nose could possess sufficient energy and determination for the voyage. But I think he was afterwards well satisfied that my nose had spoken falsely"* (from Darwin's book The Voyage of the Beagle).

28–2

The Galápagos Islands are a small cluster of volcanic islands some 950 kilometers off the coast of Ecuador. Since they came into existence (starting a few million years ago), they have been colonized from time to time by plant and animal voyagers accidentally swept by wind or water from the mainland. Some of those organisms have managed to survive, reproduce, and adapt themselves to these bleak islands.

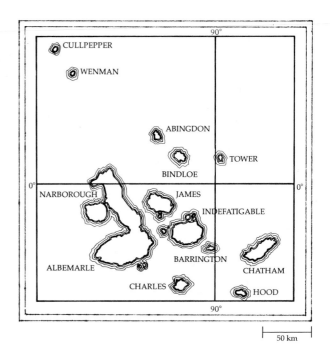

50 km

Sailors who took these tortoises on board and kept them as sources of fresh meat on their sea voyages were able to tell from which island a particular tortoise had come simply by noting its appearance. If the Galápagos tortoises had been specially created, why did they not all look alike?

Darwin began to wonder if all the tortoises and other strange animals and plants of the Galápagos might not have been derived at different times from organisms that existed on the mainland of South America. Once they reached this remote archipelago, they might have spread slowly from island to island, changing bit by bit in response to local conditions and eventually becoming distinct races that could be differentiated easily by a human observer.

In 1838, after returning from his historic voyage, Darwin read a book entitled *Essay on the Principles of Population*, by Thomas Malthus, an English clergyman. Published in 1798, eleven years before Darwin's birth, the book sounded an early warning about the explosive growth of the human population. Darwin saw that Malthus's reasoning was theoretically correct, not only for the human population but also for all other populations of organisms. For example, a single breeding pair of elephants—which have the slowest rate of reproduction of all animals—could build up to a population of some 19 million individuals in 750 years, if all of their progeny were to survive long enough to reproduce. Despite this, the number of elephants on earth remains relatively constant; where there were two individuals 750 years ago, there are, in general, only two at the present time. But what determines which two elephants, out of a possible 19 million, are the surviving progeny?

Darwin named the process by which such survivors are chosen *natural selection*. He used this expression in contrast with the process of *artificial selection*, by which breeders of domesticated plants and animals deliberately change the characteristics of the strains or races in which they are interested. They do this by allowing only those individuals with desirable characteristics to breed. Darwin recognized that wild organisms are also variable. Some individuals have characteristics that enable them to produce more progeny than others under natural conditions; if they do so, their characteristics will become more common in the population than the characteristics of individuals that produce fewer progeny. In the long run, this tendency can be counted on to produce slow but steady changes in the frequencies of different characteristics in populations.

In artificial selection, breeders can concentrate their efforts on one or a few characteristics of interest, such as fruit size. In natural selection, however, the entire organism must be "fit" in terms of the total environment in which it lives. Such a process might reasonably be expected to require a long period of time, and it is no coincidence that the writings of the geologist Charles Lyell, who postulated that the earth was much older than previously thought, had a profound influence on Darwin. Darwin needed such an earth as a stage on which to view the unfolding of the diversity of living things. The many and diverse kinds of fossils that were being discovered, which became more and more distinct from living organisms the older they were, also provided important evidence for the development of the theory of evolution. The process of natural selection soon was accepted by scientists as the basis for explaining the evolution of the living world.

THE BEHAVIOR OF GENES IN POPULATIONS

The Hardy-Weinberg Law

In the nineteenth century, when most biologists believed in some sort of blending inheritance (in which parental characteristics were presumed to blend together in the progeny), it was difficult to understand why rare characteristics did not simply become so diluted that they eventually disappeared. Darwin was unable to solve this problem because so little was known in his day about the mechanisms of heredity.

After the rediscovery of Mendel's work, which we discussed in Chapter 9, the question reappeared, framed in more modern terms: Why did the dominant alleles not eventually drive out the recessive ones, with a consequent loss of variability for the population as a whole? This question is extremely important, because it deals with the preservation of the genetic diversity that makes evolution possible. Its answer, although it was not immediately obvious, lay in a proper understanding of the particulate nature of the gene. The appropriate calculations were simultaneously presented in 1908 by

G. H. Hardy, an English mathematician, and G. Weinberg, a German physician.

The _Hardy-Weinberg law_, as it is now known, states that, in a large population in which random mating occurs, and in the absence of forces that change the proportions of alleles (discussed below), the original proportion of dominant alleles to recessive alleles will be retained from generation to generation.

For example, consider the alleles of a single gene—say, gene A. Let us make up an artificial population, so that half of the individuals are homozygous AA and the other half are homozygous aa. Figure 28–3 shows that the proportion of AA individuals (or aa, or Aa) in the third (or fourth or fifth) generation will, on the average, be the same as it was in the second generation.

For studies in population genetics, the elements of the Hardy–Weinberg law are usually stated in algebraic terms, with the fractions used in Figure 28–3 expressed as decimals. In the case of a gene for which there are two alleles in the gene pool, the frequency (p) of the dominant allele plus the frequency (q) of the recessive allele must together equal the frequency of the whole; $p + q = 1$. (The frequency of an allele is simply the pro-

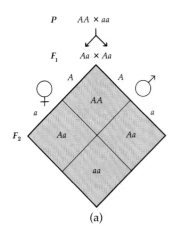

28–3

(a) If two populations, one homozygous for a dominant allele (AA) and one homozygous for a recessive one (aa), interbreed, all of the first generation (F₁) resemble the dominant parents, even though the progeny are heterozygous (Aa). When the F₁ generation interbreeds, however, the Mendelian (phenotypic) 3:1 ratio will appear in the F₂ generation, which, on the average, will be 1/4 AA, 1/2 Aa, and 1/4 aa.

Random breeding of the F₂ generation results in the following probabilities of producing AA individuals in the F₃ generation: (b) If AAs crossbreed, the chances of producing AAs are 4/4

(because all the offspring in the F₃ would be AA) multiplied by 1/16 (because the probability of AA breeding with AA is 1/4 × 1/4) for a total of 1/16. (c) If AAs breed with Aas, the chances of producing AAs are 1/2 × 1/8, or 1/16. (d) If Aas cross with AAs, the chances of producing AAs are, again, 1/16. (e) If Aas crossbreed, the chances of producing AAs are again 1/16. Thus, in the third generation, AA individuals, on the average, number 1/16 + 1/16 + 1/16 + 1/16, or 1/4, which is the same as in the second generation. In short, sexual recombination alone does not change the proportions of different alleles.

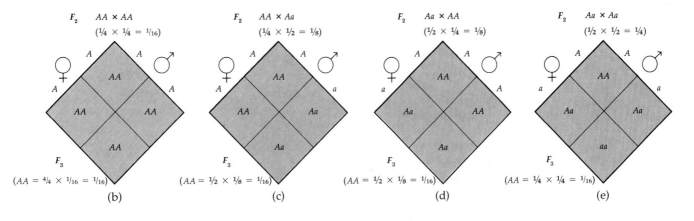

portion of that allele in a gene pool in relation to all alleles of the same gene.) This is equivalent to saying that, if there are only two alleles, A and a, of a particular gene, and if half (0.50) of the alleles in the gene pool are A, the other half must be a. Similarly, if 99 out of 100 (or 0.99) are A, then 1 out of 100 (or 0.01) is a.

How then do we find the relative proportions of individuals who are AA, Aa, and aa? These proportions can be calculated by multiplying the frequency of male A by that of female A [A^2]; male A by female a [Aa]; male a by female A [Aa]; and male a by female a [a^2]. These multiplications can be expressed in algebraic terms as $p^2 + 2pq + q^2$. This equals, as you may remember from the binomial theorem, $(p + q)^2$, and, if $p + q = 1$, then $(p + q)^2 = 1$. So, if half (0.50) of the gene pool is A and half is a, the proportion of AA will be 0.25, the proportion of Aa will be 0.50, and the proportion of aa will be 0.25, which is exactly what Mendel said—although in slightly different terms:

$$A^2 + 2Aa + a^2 = (0.50)^2 + 2(0.50)(0.50) + (0.50)^2$$
$$= 0.25 + 2(0.25) + 0.25$$
$$= 0.25 + 0.50 + 0.25$$
$$= 1$$

What happens in subsequent generations? As can be seen in Figure 28–4, the genotypic frequencies remain constant indefinitely. Similarly, the gene frequencies remain constant at $p(A) = 0.5$ and $q(a) = 0.5$, so that the Hardy–Weinberg law can be viewed as predicting a state of "genetic equilibrium."

We know that many populations do not exist in a state of equilibrium, however, for if they did no evolution could occur. Four factors operate to produce deviations from the Hardy–Weinberg equilibrium.

Population Size. The Hardy–Weinberg law operates only when there is a large population. In a small population, the random loss of one or more individual genotypes—by a failure to breed, for example—can lead to the elimination of one or more of the alleles from the population.

Migration. Further distortion is caused by migration into or out of a particular population. If individuals with particular genetic characteristics leave a population or enter it in a proportion different from that in which such individuals are already represented, the frequencies of particular alleles and genotypes will obviously change. Migration—by the dispersal of seeds, for example—is as characteristic of plants as it is of animals.

Mutation. If an allele of a particular gene is mutating to another allelic form at a rate higher than that at which the reverse mutation is occurring, the frequencies of those alleles in the population are obviously changing.

Selection. Selection is simply a term used to describe the nonrandom reproduction of genotypes. In any variable population, some individuals leave more progeny than others. As a result, certain alleles become more frequent in the population and others become less frequent. Selection is the major factor that causes deviations from Hardy–Weinberg equilibrium, and selection is also the primary reason for evolutionary change. Mutations are, of course, the basis for variability in individuals, but changes in populations occur as a result of selection acting on the variability provided by mutations.

Although selection is sometimes thought of as a creative force, it is important to remember that it is merely a description of events that have already taken place. When the proportion of a certain allele is higher in a given generation than it was in the preceding generation, and in the absence of an alternative explanation, it is said that selection has taken place. Selection does not actually *cause* the changes that occur. By continually eliminating certain alleles from a population under a given set of conditions, selection *channels* the pattern of variation in characteristics that is caused by mutation and recombination. As a result, in subsequent generations there is a proportionately higher representation of individuals that can survive better under given conditions than those that were eliminated. In unusual or marginal environments, genotypes that were poorly

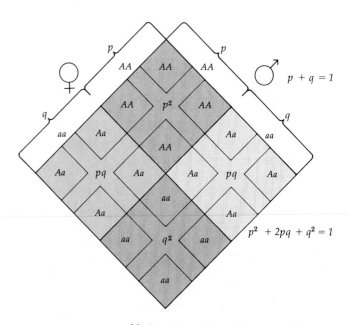

28–4

Possible combinations of gametes in a population consisting of AA, Aa, and aa individuals, illustrating the principle expressed in the Hardy–Weinberg law.

represented in the main part of the range of a population may be increased greatly by the process of selection.

In the case of a deleterious recessive allele, even one that causes the death of individuals homozygous for it, the lower the frequency of the allele in the population, the less it will be acted on by selection. Such a relationship occurs because the proportion of recessive alleles in homozygotes decreases precipitously as the frequency of the allele decreases in the population. In short, the lower the frequency of a recessive allele, the less it is exposed in a homozygous form to the action of selection. As the frequency of the recessive allele drops, the progress toward its removal from the population also slows down.

RESPONSES TO SELECTION

Genetic Factors

The response of a population to selection is affected by the principles of genetics (discussed in Chapter 9). In general, only the phenotype is being selected, and it is the relationship between the phenotype and the environment that determines the reproductive success of an organism. Because nearly all characteristics of natural populations are determined by the interactions of many genes, phenotypically similar individuals can have very different genotypes. When some feature, such as tallness, is strongly selected for, there is generally an accumulation of alleles that contribute to this feature and an elimination of those alleles that work in the opposite direction.

But selection for a polygenic characteristic is not simply the accumulation of one set of alleles and the elimination of others. Gene interactions, such as epistasis and pleiotropy (see pages 146–147), are of fundamental importance in determining the course of selection in a population. Some of the effects of one allele on another may be beneficial to a given individual or may contribute to the characteristic being selected for. Other effects of the allele may be harmful or may work against the characteristic being selected for. As a consequence of gene interactions, the phenotypic effects of a particular allele can be measured only within a particular genetic context. As selection for particular alleles changes the representation of the other alleles in the population, the interactive effects of the selected alleles gradually become changed. Thus, the value of the alleles in determining a particular characteristic, or in determining the fitness of the organism as a whole, also changes.

Another feature of importance in determining the nature of selective change is the necessity of producing an organism that "works." Strong selection for one feature may be prohibited because it would lead to the accumulation of so many undesirable side effects that the

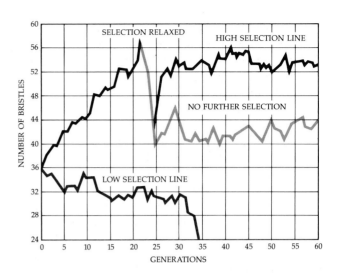

28–5
An experiment in which the number of abdominal bristles in the fly Drosophila melanogaster *was subjected to artificial selection. In the parental stock, the average number of bristles on the ventral surface of the abdomen was 36. One breeding group was selected for an increase in bristle number and one for a decrease in bristle number. The high selection line reached 56 bristles in 21 generations, but soon the stock began to become sterile. Selection was discontinued after generation 21 and was resumed at generation 25. This time the previously high bristle number was regained without loss of fertility. The low selection line, which was not permitted to achieve such a new internal genetic balance, died out because of sterility. Selection for this single "neutral" characteristic, with its associated genes, apparently had widespread effects on the genetic constitution of these organisms.*

organism would no longer be able to survive and reproduce. The results of the breeding experiment on fruit flies shown in Figure 28–5 suggest that, as a result of selection, the population reached a new internal genetic balance that allowed the production of fertile, otherwise normal individuals with 42, instead of 36, abdominal bristles. This balance, which can be considered a measure of the resistance of the population to genetic change, has been termed *genetic homeostasis* by I. Michael Lerner of the University of California, Berkeley. Homeostasis in physics is a sort of dynamic equilibrium—that is, a resistance to change. Genetic homeostasis tends to keep the population as a whole producing a high proportion of individuals that are well adapted to their environment.

Phenotypic Factors

Other kinds of limits to selection are imposed by the need for an individual organism to meet conflicting environmental demands. For example, the long tail of the male peacock, with its brilliant coloration, is highly correlated with the mating success of the bird. In this respect, therefore, it is clearly advantageous for the bird to have the largest and most showy feathers possible. On the other hand, the bird must also escape from its predators, and in this connection such tail feathers can be a disadvantage. Natural selection strikes some sort of balance between the two selective forces, and peacocks with tail feathers that are either too long and colorful or too short and drab probably do not make as great a reproductive contribution to subsequent generations as those with tails that are ''just right.''

Similarly, in harsh alpine environments, plants must grow and photosynthesize rapidly if they are to store enough carbohydrates to survive the long, severe winters. Therefore, they must respond with renewed metabolic activity to the first sign of favorable growing conditions in the spring. If their response is so finely tuned that they respond even to a temporary thaw in the middle of winter, however, they will be eliminated from the population. Desert plants are in a similar situation; their seeds must not germinate until conditions are appropriate for their survival.

As interesting evolutionary strategy among tropical woody members of the pea family (Fabaceae) has been detected by Daniel Janzen of the University of Pennsylvania. In Central America, Janzen found two sets of species: those whose seeds are numerous and small and those whose seeds are fewer but much larger. These plants can spend a certain amount of their food reserve each year on the production of seeds, and they can therefore produce either a large number of small seeds or a small number of large seeds. Why do some of the species follow the first path while others follow the second path?

Large seeds provide a more abundant food supply for the germinating seedlings. Because each parent tree or shrub will be replaced, on the average, by only one other tree or shrub, it would seem most efficient for it to produce a small number of large seeds with an abundant stored reserve of food. But this is not always possible in the tropics, owing to the very heavy predation on legume seeds by insects, especially the seed beetles (family Bruchidae). The adult bruchids lay their eggs on the fruits, and their larvae complete their development within the seeds. Bruchids are often so abundant that they virtually destroy the entire seed crop of a particular legume in a given year, except for the few seeds that are removed by dispersal agents, such as birds, before the bruchids have discovered and infested a particular fruit.

Janzen investigated bruchid infestation among large-seeded and small-seeded groups of legumes. To his surprise, he found that the small-seeded species were heavily attacked by bruchids, so that very few viable seeds remained, whereas nearly all the large-seeded species were untouched. Subsequently, he found that these large seeds, but not the small ones, contained chemical substances that apparently protected them from the beetles.

Summing up these results, it seems clear that, in an evolutionary sense, the legumes had two options. Either they could produce a very large number of small seeds, with a relatively limited food supply for the germinating seedlings but with a better chance that some of the seeds would escape attack by the beetles, or they could produce a smaller number of large seeds, with a good food supply for the germinating seedlings. This second option, however, would succeed only if the seeds were somehow protected from predation by beetles. This is an example of the sort of evolutionary compromise that is often found in nature.

In the light of Janzen's reasonable explanation of the situation in tropical legumes, it is interesting that the seeds and fruits of plants on oceanic islands are often larger than those of related plants on the mainland (Figure 28–6). The kinds of pests that attack the plants on the mainland are often absent on the islands. It seems likely that this provides the opportunity for the plants to produce larger fruits and seeds, which provide a better start for their seedlings because they contain greater amounts of stored food. This isolation from pests, like the biochemical defenses of the large-seeded legumes of the American tropics, is a factor that allows large seed size, which is selected for on other grounds, to prosper more than it would otherwise.

Changes in Natural Populations

In recent years, there have been many studies of changes in the characteristics of natural populations. Changes in climate and natural catastrophes have been a constant feature of the world since its beginning, and

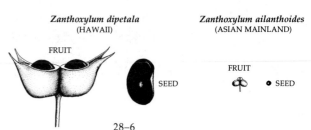

28–6
Island species, isolated from their natural enemies, are often able to produce larger fruits and seeds. Shown here are two related species of Zanthoxylum, *members of the citrus family (Rutaceae); one of them is found in Hawaii, the other on the Asian mainland.*

the populations of organisms present at earlier times responded to these changes in ways similar to those presently observed. The nature and rate of some of these changes have been such that we can speak about them as "evolution in action." It is not surprising that many of the relatively rapid changes in natural populations that have been observed over the last century or so have resulted from human activities, because in recent centuries human beings have become an ecologically dominant feature without parallel in the history of life on earth.

Some of the most spectacular evolutionary changes have been observed in populations of bacteria, in which only a single copy of the genetic material is present (that is, they are haploid). Their mutations are expressed immediately in the phenotype, and selection for or against such changes proceeds at once. All haploid organisms have a short generation time, which is probably essential for successful changes, since they do not have the store of variability for recombination that is found in diploid organisms.

In plants, strong selective forces have been observed to produce rapid changes in natural populations. For example, plants in the grazed part of a pasture that was studied in Maryland were much shorter than those in the ungrazed part, and it was thought that this might simply be a direct result of the grazing. This hypothesis was tested by digging up some of the plants from the grazed part and planting them elsewhere. It was assumed that, if they were short merely because they had been grazed, they would soon become tall in the ab-

sence of grazing, like the plants in the ungrazed part of the pasture. Some of them did, but plants of white clover *(Trifolium repens)*, Kentucky bluegrass *(Poa pratensis)*, and orchard grass *(Dactylis glomerata)* remained short, indicating that these populations had been modified genetically by the selective force of grazing. A similar example is illustrated in Figure 28–7.

In Wales, the tailings and dumps around a number of abandoned lead mines are rich in lead and almost bare of plants. One species of grass *(Agrostis tenuis)* colonizes this mine soil, which can contain up to 1 percent lead and 0.03 percent zinc. In an experiment, groups of *Agrostis tenuis* plants taken from mine tailings and others taken from a nearby pasture were grown together either in normal soil or in lead-mine soil. In normal soil, the lead-mine *Agrostis* plants were definitely slower growing and smaller than the pasture plants. On the lead-mine soil, however, the mine plants grew normally, but the pasture plants did not grow at all. Half the pasture plants were dead in three months and had misshapen roots that were rarely more than 2 millimeters long. But a few of the pasture plants (3 out of a sample of 60) showed some resistance to the effects of the lead-rich soil. These were doubtless similar to the plants originally selected in the development of the lead-resistant strain of *A. tenuis*. The mine was no more than 100 years old, so the lead-resistant race had developed in a relatively short period of time. The resistant plants were selected from the genetically variable plants found in adjacent habitats, and then constituted, through the agency of selection, as a distinct race.

(a)

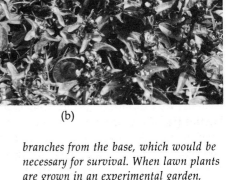

(b)

28–7
Prunella vulgaris *is a common herb of the mint family; it is widespread in woods, meadows, and lawns in temperate regions of the world. Most populations consist of erect plants, such as those shown in* (a), *which grow in open, often somewhat moist, grassy places throughout the cooler regions of the world. Popula-*

tions found in lawns, however, always consist of prostrate plants, such as those shown in (b), *growing in Berkeley, California. Erect plants of* Prunella vulgaris *cannot survive in lawns because they are damaged by mowing and do not have the capability for resprouting low*

branches from the base, which would be necessary for survival. When lawn plants are grown in an experimental garden, some remain prostrate whereas others grow erect. The prostrate habit is determined genetically in the first group and environmentally in the second group.

Asexual Reproduction and Evolution

Asexual reproduction results in progeny that are identical to their single parent. In plants, there is a wide variety of means of asexual reproduction, ranging from the development of an unfertilized egg cell to divisions of the parent organism into separate parts. In all such cases, however, the new plants are the product of mitosis and thus are genetically identical to the parent.

Plants often reproduce both sexually and asexually, thus hedging their evolutionary bets (Figure 28–8), but many species reproduce only asexually. Even among these, however, it is clear that their ancestors were capable of sexual reproduction. Therefore, vegetative reproduction represents an alternative—a "choice" made in response to evolutionary pressures calling for uni-

VEGETATIVE REPRODUCTION: SOME WAYS AND MEANS

The forms of vegetative reproduction in plants are many and varied. Some plants reproduce by means of runners, or stolons—long slender stems that grow along the surface of the soil. In the cultivated strawberry (*Fragaria ananassa*), for example, leaves, flowers, and roots are produced at every other node on the runner. Just beyond the second node, the tip of the runner turns up and becomes thickened. This thickened portion first produces adventitious roots and then a new shoot, which continues the runner.

Rhizomes are also important reproductive structures, particularly in grasses and sedges. Rhizomes invade areas near the parent plant, and each node can give rise to a new flowering shoot. The noxious character of many weeds results from this type of growth pattern, and many garden plants, such as irises, are propagated almost entirely from rhizomes. Corms, bulbs, and tubers are specialized for storage and reproduction. White potatoes are propagated artificially from tuber segments, each with one or more "eyes." It is the eyes of the "seed pieces" of potato that give rise to the new plant.

The roots of some plants—for example, cherry, apple, raspberry, and blackberry—produce "suckers," or sprouts, which give rise to new plants. Commercial varieties of banana do not produce seeds and are propagated by suckers that develop from buds on underground stems. When the root of a dandelion is broken, as it may be if one attempts to pull it from the ground, each root fragment may give rise to a new plant. (See also pages 440 through 445.)

In a few species, even the leaves are reproductive. One example is the house plant *Kalanchoë daigremontiana*, familiar to many people as the "maternity plant," or "mother of thousands." The name of this plant is based on the fact that numerous plantlets arise from meristematic tissue located in notches along the margins of the leaves. The maternity plant is ordinarily propagated by means of these small plants, which, when they are mature enough, drop to the soil and take root. Another example of vegetative propagation is provided by the walking fern (*Asplenium rhizophyllum*), in which young plants form where the leaf tips touch the ground.

A herbarium specimen of walking fern, Asplenium rhizophyllum, *showing the way the leaves root at their tips and produce new plants. In this way, the fern is capable of forming large colonies of genetically identical plants.*

In certain plants, including citrus, some grasses (such as Kentucky bluegrass, *Poa pratensis*, which is discussed on page 579), and dandelions, the embryos in the seeds may be produced asexually from the parent plant. This is a kind of vegetative reproduction, or apomixis. The seeds that are produced in this way give rise to individuals that are genetically identical to their parents and therefore provide another instance of asexual reproduction.

In general, asexual reproduction allows the exact replication of individuals that are particularly well suited to a certain environment or habitat. This suitability may include characteristics that are viewed as desirable in a cultivated plant or those that facilitate survival under a particular set of natural conditions.

form populations. This choice, if rigidly adhered to, severely restricts the ability of the population to adapt to varying conditions. Different asexually reproducing populations of a given species may occur in a wide variety of different habitats, however. Being set in their characteristics, each may differ sharply from the others and be better suited to growth in a particular kind of place.

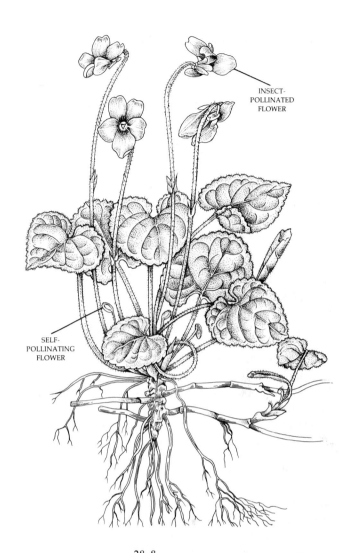

28–8
Violets reproduce both sexually and asexually. The large flowers are cross-pollinated by insects, and the seeds may be carried some distance from the parent plant by the wind or by water. The smaller flowers, closer to the ground, are self-pollinating and never open. Seeds from these flowers drop close to the parent plant and produce plants that are genetically similar to the parent. Presumably such plants are better able (on the average) to grow successfully near the parent. Both of these forms of reproduction are sexual and involve recombination. Asexual reproduction in violets occurs when horizontally creeping stems (either stolons or rhizomes) produce new, genetically identical plants close to the parent.

Kalanchoë daigremontiana, *showing small plants that have arisen in the notches along the margins of the leaf.*

*The strawberry (*Fragaria ananassa*) is propagated asexually by stolons. Strawberry plants also produce flowers and reproduce sexually.*

THE DIVERGENCE OF POPULATIONS

The world is physically and biologically complex, and individual species of plants almost always have discontinuous distributions. Their habitats are certainly discontinuous, whether they be lakes, streams, the tops of mountains, sunny spots in the woods, or a certain kind of soil. Given this observation, the movement of alleles between isolated populations is more or less limited, and the populations are correspondingly free to respond to the selective demands of their local situations in different ways. Even though normal *gene flow* (the movement of alleles from one population to another as a consequence of the migration of individuals) may provide a source of variability for a distant population, it cannot completely counteract the effects of local selection.

The distances that are necessary to isolate populations effectively from other populations of the same species vary with the dispersibility of the plants in question, but such distances are often quite small. For several kinds of insect-pollinated plants that grow in temperate regions, a gap of only 300 meters may effectively isolate two populations. Rarely will more than 1 percent of the pollen that reaches a given individual come from this far away. In plants whose pollen is spread by the wind, very little falls more than 50 meters from the parent plant under normal circumstances. The chance of pollen from such a plant reaching a receptive stigma on a plant that is farther away decreases very rapidly with increasing distance. This is not to say, of course, that two pine trees separated by 50 meters are out of genetic contact but, rather, that the sort of progeny that each produces will be affected much more by the demands of its local environment than by alleles introduced from a distance.

Any two separated populations diverge from one another because they are responding to different selective forces. If they regain contact, they may merge, or the differences that have accumulated between them may lead to some degree of genetic isolation. Genetic isolation, or reproductive isolation, can come about in a variety of ways, as is discussed later in this chapter.

Ecotypic and Clinal Variation

Developmental plasticity is the tendency of individuals to vary over time in response to different environmental conditions, or for genetically identical organisms to differ among themselves in response to different environmental stimuli. Such plasticity is much greater in plants than it is in animals, because the open system of growth that is characteristic of plants can more easily be modified in various ways that produce striking differences between the individuals.

Environmental factors, as every gardener knows, can cause profound changes in the phenotype of various species of plants. Leaves that develop in the shade are thinner and broader—and have a greater volume of airspace, thinner palisade tissue, and fewer stomata—than leaves of the same plant that develop in the sun (see page 436). Day length—that portion of each 24-hour period during which a plant is exposed to the light—can also affect leaf form. In *Kalanchoë,* for example, plants grown under short-day conditions (8 hours of light) have small succulent leaves with smooth edges, whereas plants grown under long-day conditions (16 hours of light) have large, thin leaves with notched edges. In view of such observations, it is not surprising that a number of scientists proposed, until about 1930, that much of the variability of plants observed in nature was brought about directly by the influence of the environment and did not have a genetic basis.

Are the differences between races of plants growing in different habitats genetically or environmentally controlled? The first definitive answer to this question was provided in the 1920s by the Swedish botanist Göte Turesson, who transplanted races of a number of the plant species that he found in southern Sweden into his experimental gardens. In the great majority of the 31 species that he investigated, the differences he observed in nature were genetically controlled; in a very few, they were the result of the direct action of the environment. Differences in plant stature, time of flowering, color of leaves, and so on, were usually genetically controlled. Turesson termed such genetic races, differentiated with respect to particular habitats, *ecotypes*.

Of particular interest in the study of ecotypes is the work of Jens Clausen, David Keck, and William Hiesey, which was carried out under the auspices of the Carnegie Institution of Washington at their Department of Plant Biology, at Stanford, California. These workers dealt experimentally with a number of species of plants from the western United States. They established transplant stations at three locations in California (Stanford, Mather, and Timberline; Figure 28–9). Working mainly with plants that could be propagated asexually, so that genetically identical individuals could be grown at all three locations, they expanded on Turesson's experiments.

The western United States is an area of great environmental contrasts, and it is not surprising that many plant species have developed sharply defined ecotypes in this region. One species studied by the Carnegie Institution group was the perennial herb *Potentilla glandulosa*. A close relative of the strawberry *(Fragaria)*, this herb ranges through a wide variety of climatic zones in California, and native populations occur near each of the three experimental stations mentioned above. When *P. glandulosa* plants from the different locations were grown side by side in gardens that the Carnegie

28–9

Transplant gardens of the Carnegie Institution, all in California. (a) Stanford, near sea level; (b) Mather, in the central Sierra Nevada at about 1400 meters elevation; and (c) Timberline, at 3050 meters elevation.

team set up at their three transplant stations, a number of ecotypic differences between the original strains became apparent. There appeared to be four distinct ecotypes of *Potentilla glandulosa,* each with morphological characteristics that are strongly correlated with particular physiological responses that are critical to the survival of the ecotype in its native environment.

For example, the Coast Range ecotype consists of plants that grow actively both in the winter and in the summer when cultivated at Stanford. Plants of this ecotype also survive at Mather, where they are subjected to about 5 months of cold winter weather. There they become winter-dormant, but they store enough food during their growing season to carry them through this long, unfavorable season. At Timberline, plants of this ecotype almost invariably die during the first winter; the short growing season at this high elevation does not permit them to store enough food to survive during the long winter. Other plants that occur in California's Coast Ranges produce ecotypes that have identical physiological responses. Indeed, it is often true that the strains of completely different and unrelated species of plants that are growing together naturally at a given place are more similar to one another physiologically than they are to other populations of their respective species.

The physiological and morphological characteristics of ecotypes usually have a very complex genetic basis, involving dozens (or in some cases perhaps even hundreds) of genes. Sharply defined ecotypes are characteristic of regions where the breaks between adjacent habitats are themselves sharply defined. On the other hand, if the environment changes gradually, with no clear breaks, then the characteristics of the populations of plants that occur in the area may do likewise. A gradual change of this sort is called a *cline.*

Clines are frequently encountered in organisms that live in the sea, where the temperature often rises or falls very gradually with changes in latitude. They are also characteristic of organisms that occur in such areas as the eastern United States, where rainfall gradients may extend over thousands of kilometers. When populations of plants are sampled along a cline, the differences are often proportional to the distance between them. On a small scale, populations of plants vary in similar ways—either gradually or abruptly, depending on the nature of the local environment.

Physiological Differentiation

In order to understand why ecotypes flourish where they do, we must understand the physiological basis for their ecotypic differentiation. For example, Scandinavian strains of goldenrod *(Solidago virgaurea)* from shaded and exposed habitats show experimental differences in the photosynthetic response to light intensity during growth. The plants from shaded environments grow rapidly under low light intensities, whereas their growth rate is markedly retarded under high light intensities. In contrast, plants from exposed habitats grow rapidly under conditions of high light intensity, but much less well at low light levels.

In another experiment, arctic and alpine populations of the widespread herb *Oxyria digyna* were studied, using strains from an enormous latitudinal gradient extending southward from Greenland and Alaska to the mountains of California and Colorado. Plants of northern populations had more chlorophyll in their leaves, as well as higher respiration rates at all temperatures, than plants from farther south. High-elevation plants from near the southern limits of the species' range carried out photosynthesis more efficiently at high light intensities than did low-elevation plants from farther north. Thus the respective races could function better in their own habitats, marked for example by high light intensities in high mountain habitats, lower ones in the far north, and so forth. The existence of *Oxyria digyna* over such a wide area and such a wide range of ecological conditions is made possible, in part, by differences in metabolic potential among its constituent populations.

Reproductive Isolation

The genetic system of a population of plants responds to selection as an integrated unit. As individual populations become more and more different from one another, the characteristics that originally allowed them to interbreed with one another may change along with characteristics of more immediate importance—those that allow them to grow successfully in their different habitats. For this reason, plants from widely different populations may be unable to produce hybrids (hybrids are the offspring of genetically dissimilar parents) or, if they do, the hybrids may not be fertile. In general, the greater the difference in appearance of two populations, the less likely it is that they will be able to form hybrids.

When two populations become reproductively isolated—unable to form fertile hybrids with one another —they have no effect on each other's subsequent evolution, at least in a genetic sense. Such isolation is therefore one of the more crucial steps in the evolutionary divergence of populations and the formation of species.

Reproductive isolation has often been used as a basis for defining the category *species*. This criterion is not generally applicable, however, because the very different-looking, ecologically distinct species in some groups of plants—particularly such long-lived plants as trees and shrubs—often can form fertile hybrids with one another. In contrast, hybrids between species of herbaceous plants are often sterile, or can be formed only with difficulty; similar patterns of differentiation often occur even between the individual populations of species of short-lived plants. Hybrids between such populations may be completely or partly sterile. Because of the differences between the genetic characteristics of the species that are recognized in different groups of plants, it is impossible to use a single criterion to define the category species, and a number of factors are important in reaching such a decision about the most appropriate classification for a particular set of organisms.

Populations of annual plants presumably change much more rapidly than populations of long-lived plants when they are subjected to similar selective pressures. Not only do annual plants have shorter life cycles, but each year they depend on re-establishment from seed, a factor that accelerates the effects of natural selection and, by doing so, causes the populations concerned to diverge from one another more rapidly. In areas where annual plants can grow successfully, such as deserts or regions that are dry in the summer (California, for example), they may constitute a third or more of the total species present. These species are, on the whole, much younger in terms of their time of origin than the species of trees, shrubs, and perennial herbs with which they are associated.

There are a number of factors, in addition to hybrid sterility or the inability to hybridize at all, that tend to keep different species of plants distinct when they grow together. Some of these factors prevent the formation of the hybrids in the first place. For example, two kinds of plants that are capable of forming fertile hybrids may occur in the same area but in different habitats. In the eastern United States, the scarlet oak (*Quercus coccinea*) can be found growing together with the black oak (*Q. velutina*) over a very wide area, and the two species, which are wind-pollinated, form fertile hybrids readily in cultivation. Despite these relationships, hybrids between them are rare in nature. In general, scarlet oaks are found in relatively moist, low areas with acidic soil, whereas black oaks are found in drier, well-drained habitats. Only where the habitat has been disturbed—as by burning or cutting the trees —do hybrids become more common, apparently because new kinds of habitats are created to which some hybrid individuals may be better suited than either of their parents.

Among the other mechanisms that can prevent the formation of hybrids between species that occur together are seasonal differences in the time of flowering. If two species do not flower together, they will not hybridize in nature even when they grow side by side. The many photoperiodic mechanisms discussed in Chapter 25 also allow differentiation of this kind. In addition, two species that occur together may differ in their pollination systems; various possibilities of this sort will be discussed in the next chapter. If two species are visited and pollinated by different kinds of insects, they will hybridize only when these insects mistakenly visit the "wrong" flower.

Clusters of Species

The evolutionary processes we have been discussing lead to the production of groups of related species in many different geographic areas. The clusters of species of animals on the Galápagos Islands, especially the tortoises we mentioned earlier and the Galápagos finches, are famous because of their role in the development of Darwin's theory of evolution. The development of such patterns is known as *adaptive radiation*. (The boxed essay on the next two pages illustrates this process.)

Differentiation on islands is particularly striking because, in the absence of competition, organisms seem more likely to produce forms that are highly unusual in comparison with those of related species that occur on the continents. Apparently, islands provide favorable situations for major evolutionary changes of the kind that occur when new genera and families arise. In such localities, the characteristics of plants and animals may change more rapidly than on the mainland, and features that are never encountered elsewhere may arise. Similar clusters of species may also arise in mainland areas, of course, and may involve spectacular differentiation. Species clusters of this sort are the inevitable result of the evolutionary processes that we have been discussing.

THE EVOLUTIONARY ROLE OF HYBRIDIZATION

Even if species hybridize only rarely in nature, the hybrids that they do produce may be important because of the way in which such hybrid individuals recombine the characteristics of their parents. The environment can change repeatedly, and individuals of hybrid origin often present new genetic combinations that are better suited to the new environment than those of either parent; they may even be able to colonize some habitat where neither parent can grow.

When the habitats of the parental species are found in the same area but are sharply distinct (as in the case of the scarlet oak and the black oak), there may be little opportunity for the establishment of hybrids; but where the habitats intergrade or are disturbed, the situation may be very different. Here the recombination of genetic material that was originally characteristic of the two differentiated species may have a greater potential for producing well adapted offspring than would changes within a single population (Figure 28–10). The characteristics of a population of hybrids may become stabilized if the hybrids are better adapted to some particular habitat conditions than either of their parents.

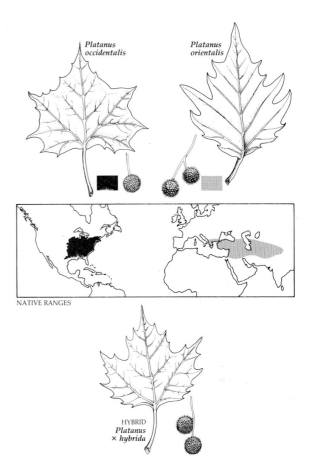

28–10

Sycamores (Platanus), which are called plane trees in Britain, offer examples of well-differentiated populations that have retained the ability to hybridize. Present-day species of this genus have been isolated from one another for at least 50 million years in widely scattered localities. One of these, the oriental plane (P. orientalis), is native from the eastern Mediterranean region to the Himalayas. This handsome tree has been widely cultivated in southern Europe since Roman times, but it cannot be grown in northern Europe away from the moderating influence of the sea. After the discovery of the New World, one of the North American species, P. occidentalis, was brought into cultivation in the colder portions of northern Europe, where it flourished. About 1670, these two very different trees hybridized spontaneously where they were cultivated together in England, producing the intermediate and fully fertile London plane, Platanus × hybrida. This hybrid, which is capable of growing in regions with cold winters, is now grown as a street tree throughout the temperate regions of the world.

ADAPTIVE RADIATION IN HAWAIIAN TARWEEDS

Some spectacular groups of native plants and animals occur in the Hawaiian Islands. On this archipelago, whose major islands arose from the sea in isolation over millions of years, the habitats are exceedingly diverse, and the distances to mainland areas from which plants and animals migrated to colonize the islands is great. Those organisms that did reach the Hawaiian Islands have often changed greatly in their characteristics as they have occupied the varied habitats that were available to them. The process by which such changes occur is called *adaptive radiation.*

The striking Hawaiian silverswords are the most remarkable example of a group of some 28 species of plants belonging to three closely related genera of the sunflower family, Asteraceae. Collectively, this group is known as the tarweeds, subtribe Madiinae; most of its members are found in California and adjacent regions. The silverswords themselves belong to the genus *Argyroxiphium,* which is closely related to two other genera also found only in Hawaii, *Dubautia* and *Wilkesia.* The species of these genera range in habit from small, mat-forming shrubs and rosette plants to large trees and climbing vines. They grow in habitats as diverse as exposed lava, dry scrub, dry woodland, moist forest, wet forest, and bogs. In these habitats, the annual precipitation ranges from less than 40 centimeters to more than 1230 centimeters. The rainiest of these habitats is among the wettest places on earth.

This enormous variation in habitats is paralleled by significant variation in leaf sizes and shapes among these plants. For example, species of *Dubautia* that grow in bright, dry habitats usually have very small leaves, while those that inhabit the shaded understories of wet forests have much larger ones. The silversword, *Argyroxiphium sandwicense,* which grows on the dry alpine slopes of Haleakala Crater on the island of Maui, has leaves that are covered with a dense, silvery mat of hairs. These hairs apparently provide protection from intense solar radiation and aid in the conservation of moisture. The leaves of the closely related greensword, *A. grayanum,* which also occurs on Maui, but in wet forest and bog habitats, lacks these hairs.

Important physiological differences likewise characterize the species of Hawaiian tarweeds, which must meet very different kinds of environmental changes. Robert Robichaux of the University of California, Berkeley, has shown, for example, that the species of *Dubautia* that occur in dry habitats have a much greater tolerance to water stress than those that occur in moist to wet habitats. The species that grow in dry habitats have more elastic cell walls in their leaves and under dry conditions are able to maintain higher turgor pressures (and consequently higher levels of metabolic activity) than their relatives.

Despite their great diversity in appearance and the very different habitats in which they grow, all 28 species of these three genera are very closely related to one another. Any

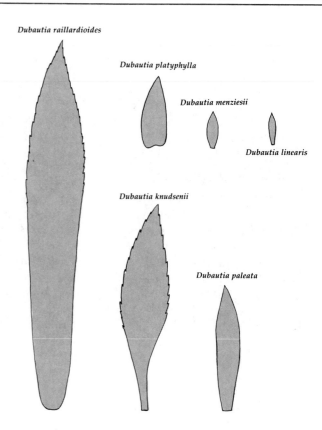

Dubautia raillardioides

Dubautia platyphylla

Dubautia menziesii

Dubautia linearis

Dubautia knudsenii

Dubautia paleata

Silhouettes of leaves of six species of Dubautia, *showing the great differences in leaf shape and size between species that grow in wet habitats and those that grow in dry ones.*

two of them can be hybridized, as far as we know, and all of the hybrids are at least partly fertile. Further studies of these relationships are being carried out by Gerald Carr of the University of Hawaii and Donald Kyhos of the University of California, Davis. The entire group of three genera appears certainly to have evolved in isolation on the Hawaiian Islands following the arrival of a single, original colonist, which presumably was from western North America.

The silversword, Argyroxiphium sandwicense, a remarkable plant that grows on the exposed, upper cinder slopes of Haleakala Crater on the island of Maui. Here the plants are exposed to high levels of solar radiation and low humidities.

Dubautia reticulata. *Species of the genus Dubautia (which includes 21 of the estimated 28 species of Hawaiian tarweeds) range from trees and shrubs to lianas and small, matted plants that are scarcely woody. Dubautia reticulata, shown here, grows in the wet forests of Maui, where it may reach a height of 8 meters or more and develop a trunk with a diameter of nearly 0.5 meter.*

Wilkesia gymnoxiphium. *This bizarre, yuccalike plant occurs only on the island of Kauai, where it is restricted to dry scrub vegetation along the fringes of Waimea Canyon. Kauai is the oldest of the main islands of the Hawaiian archipelago. Robert Robichaux, shown here, is studying the physiological ecology of this fascinating group of plants.*

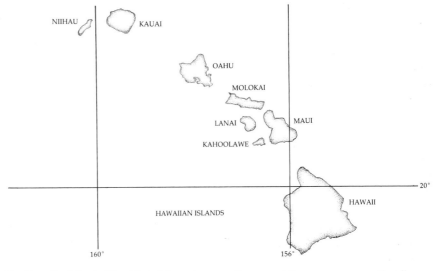

Dubautia scabra. *This low, matted, herbaceous member of the genus is found in moist to wet habitats on several of the islands in the Hawaiian archipelago. It is believed to be the progenitor of several additional species on the younger islands of the group.*

The Hawaiian Islands. *The oldest of the main islands, Kauai, includes some rocks that are as much as six million years old; the youngest, Hawaii, is still being formed. The Hawaiian archipelago is gradually moving northwest with the Pacific Plate, the older islands gradually being eroded below the surface of* the sea, and the younger ones continually being formed, apparently as they pass over a "hot spot": a thin spot in the earth's crust through which lava erupts. Thus we know that there were islands in approximately the present position of Hawaii much more than six million years ago.

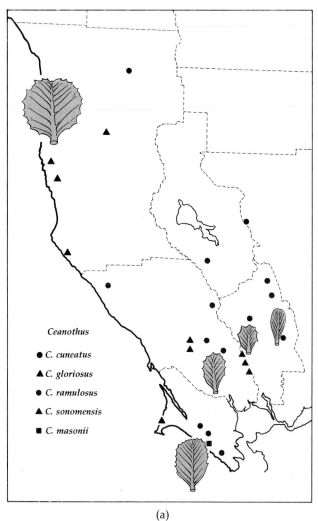

(a)

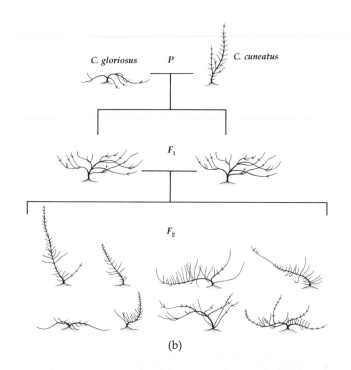

(b)

(c)

(d)

28–11

The establishment of hybrid populations is an important evolutionary mechanism in many groups of woody plants, including the mountain lilacs, Ceanothus, shown here. (a) Map showing a portion of central California (note San Francisco Bay at the bottom), a geologically complex region. Two relatively widespread and distinct species of mountain lilac, the coastal C. gloriosus *and the interior* C. cuneatus *(which ranges far to the east out of the area depicted), have produced three series of hybrid populations; each is variable but stabilized and is able to grow better than either parent in the areas in which they occur together. Leaves from representative populations are shown on the map. (b) Segregation of the extremes in appearance in the progeny of an artificial cross between the parental species. In habit, some of the intermediates resemble the named intermediate populations shown on the map to the left. (c) Flowering branch of* C. gloriosus. *(d) Flowering branch of* C. cuneatus.

Hybridization between species is an important evolutionary mechanism in many groups of plants. In some genera, the recombination of genetic material between species seems to be the chief means of developing new species that can grow in diverse habitats. These genera primarily comprise woody trees or shrubs, such as *Eucalyptus*, oaks (*Quercus*), manzanitas (*Arctostaphylos*), and mountain lilacs (*Ceanothus*; see Figure 28–11). Clusters of species that interact in this way tend to be most common on islands or in ecologically diverse areas, such as California and its neighboring states. Another example, in this case involving pollinating agents, is shown in Figure 28–12.

The establishment and stabilization of hybrid populations depends on the fertility of the hybrids. Even if the hybrids are sterile, however, they may still be able to propagate themselves, either asexually or by regaining their fertility through polyploidy.

28–12

Richard Straw, then of California State University, Los Angeles, hypothesized that hybridization accounts for the origin of Penstemon spectabilis, *a flowering plant found in the mountains of southern California. One of the parental species,* P. grinnellii, *has broad, two-lipped, pale blue flowers that are pollinated mainly by large bees, such as carpenter bees. Another species,* P. centranthifolius, *has long, slender red flowers that are visited mostly by hummingbirds. Their suspected hybrid derivative,* P. spectabilis, *shown here, is ecologically and morphologically intermediate between the two, having rose-purple flowers. It is pollinated by a specialized family of pollen-gathering wasps, which visit neither of the parental species.*

Asexual Reproduction in Hybrids

Sterile hybrids may reproduce by a variety of vegetative means. Systems in which apomixis or some other form of vegetative reproduction is prevalent, but in which occasional outcrossing (cross fertilization between different individuals) occurs, are the most flexible. As a consequence of outcrossing, hybridization between differentiated races or species may occur again, and may lead to the production of novel combinations of alleles, which may themselves be successful locally.

An outstanding example of such a system is the extremely variable Kentucky bluegrass (*Poa pratensis*), which in one form or another occurs all around the Northern Hemisphere. Occasional hybridization with a whole series of related species has produced hundreds of apomictic races, each well adapted to the ecological characteristics of the region in which it grows. In such a flexible system, new genotypes are produced constantly by hybridization, and the best of them are preserved by apomixis. In Arctic regions, apomictic plants, including bluegrasses of the genus *Poa*, are often particularly successful. Pollination by insects is difficult in the Arctic, because of the short growing season and harsh climatic conditions. In addition, certain narrowly defined genotypes may be more successful under the rigorous conditions of the Arctic than more variable populations of sexually reproducing organisms, and so populations produced as a result of vegetative reproduction or self-pollination may have a double advantage.

The hundreds of species of hawthorn (*Crataegus*) and blackberry (*Rubus*) found in the eastern United States are apomictic derivatives of groups of species in which occasional hybrids occur. In all these cases, the wholesale destruction of natural habitats by people has led to the establishment of many new genotypes that would have had no place in the primeval forests of the region.

Polyploidy

Cells or individuals that have more than two sets of chromosomes are called polyploids. Polyploid cells arise at a low frequency as the result of a "mistake" in mitosis in which the chromosomes divide but the cell does not. In this way, a cell with twice the usual number of chromosomes may be produced. If such cells then go through interphase and divide, they can give rise to a new individual, either sexually or asexually, that will have twice the number of chromosomes of its parents or parent. Polyploid individuals also can be produced by the use of colchicine, a drug that inhibits the formation of the spindle apparatus in mitosis through its disruptive effect on microtubule formation.

In a polyploid, the range of variation is often narrowed considerably from that in a related diploid because each gene is present twice as often. Individuals that are homozygous recessive for a given gene are represented by only 1/16 of the individuals, assuming random segregation, rather than 1/4 of the individuals, as in a diploid. (Both cases assume a frequency for the recessive allele of 0.50.) Polyploids are often self-pollinated, even when the related diploids are largely cross-pollinated, which further reinforces their decreased variability. Some polyploids are better suited to dry habitats or colder temperatures than their diploid relatives, whereas others do better in certain types of soils. In ways such as these, they may be adapted to colonize habitats near or beyond the tolerance limits of their diploid ancestors.

Polyploid plants occur at low frequencies in many natural populations. When such polyploids hybridize, which they often are able to do more easily than their diploid counterparts, a stable polyploid derivative, fertile from the start, may be formed. Less commonly, polyploids of hybrid origin have arisen by chromosome doubling in sterile diploid hybrids; such chromosome doubling sometimes functions to restore fertility.

This less common mode was involved in one of the earliest well-documented cases of polyploidy—the origin of a polyploid hybrid between the radish (*Raphanus sativus*) and the cabbage (*Brassica oleracea*). Both genera belong to the mustard family, Brassicaceae, and they are closely related. The two species both have 18 chromosomes in their vegetative cells and regularly form 9 pairs of chromosomes at meiotic metaphase I. The hybrid between them, which was obtained with some difficulty, had 18 unpaired chromosomes in meiosis and thus was completely sterile. In the polyploid that appeared spontaneously among these hybrid plants, there were 36 chromosomes in the vegetative cells, and 18 pairs of chromosomes were formed regularly in meiosis. In other words, the polyploid hybrid had all the chromosomes of the radish and the cabbage in its cells, and these functioned normally. Consequently, the polyploid had a relatively high fertility.

A number of polyploids originated as weeds in habitats associated with the activities of human beings, and sometimes these polyploids have been spectacularly successful. One of the best known is a salt-marsh grass of the genus *Spartina*. One native species, *S. maritima*, occurs in marshes along the coasts of Europe and Africa. A second species, *S. alterniflora*, was introduced into Great Britain from eastern North America in about 1800, spreading from the localities where it was first planted and forming large but local colonies.

In Britain, the native *Spartina maritima* is short in stature, whereas *S. alterniflora* is much taller, frequently growing to 0.5 meter and occasionally to 1 meter or even more in height. Near the harbor at Southampton, both the native species and the introduced species existed side by side throughout the nineteenth century. In 1870, botanists discovered a sterile hybrid between these two species that reproduced vigorously by rhizomes. Of the two parental species, *S. maritima* has a somatic chromosome number of $2n = 60$ and *S. alterniflora*, $2n = 62$; the hybrid, owing perhaps to some minor meiotic misdivision, also has $2n = 62$. This sterile hybrid, which was named *Spartina* × *townsendii*, still persists. About 1890, a vigorous seed-producing polyploid was derived naturally from it. This fertile polyploid, which had a diploid chromosome number of $2n = 122$ (one chromosome pair was evidently lost), spread rapidly along the coasts of Great Britain and northwestern France. It is often planted to bind mud flats, and such use has contributed to its further spread (Figure 28–13).

One of the most important polyploid groups of plants is the genus *Triticum*, the wheats. The most commonly cultivated crop in the world, bread wheat, *T. aestivum*, has $2n = 42$ chromosomes. Bread wheat originated at least 8000 years ago, probably in central Europe, following the natural hybridization of a cultivated wheat with $2n = 28$ chromosomes and a wild grass of the same genus with $2n = 14$ chromosomes. The wild grass probably occurred spontaneously as a weed in the fields where wheat was being cultivated. The hybridization that gave rise to bread wheat probably occurred between polyploids that arose from time to time within the populations of the two ancestral species.

It is likely that the desirable characteristics of the new, fertile, 42-chromosome wheat were easily recognized, and it was selected for cultivation by the early farmers of Europe when it appeared in their fields. One of its parents, the 28-chromosome cultivated wheat, had itself originated following hybridization between two wild 14-chromosome species in the Near East. Species of wheat with $2n = 28$ chromosomes are still cultivated at present, along with their 42-chromosome derivative. Such 28-chromosome wheats are the chief grains used in macaroni products because of the desirable agglutinating properties of their proteins.

Recent studies indicate that new strains produced by artificial hybridization can improve agricultural performance. Particularly promising is *Triticosecale*, a group of man-made hybrids between wheat *(Triticum)* and rye *(Secale)*. Some of these hybrids, combining the high yield of wheat with the ruggedness of rye, are more resistant to wheat rust—an economically important fungal disease (see pages 221–223). This characteristic could be particularly important in the subtropical and tropical highland regions of the world, where rust is the chief factor limiting wheat production. *Triticosecale* is now widely cultivated and gaining rapidly in popularity in France and other areas (see page 637). The most important strain of this new crop has $2n = 42$ chromosomes. It was derived by chromosome doubling following hybridization of 28-chromosome wheat with 14-chromosome rye.

In nature, polyploids are selected by the environment rather than directly by human beings. Polyploidy is a

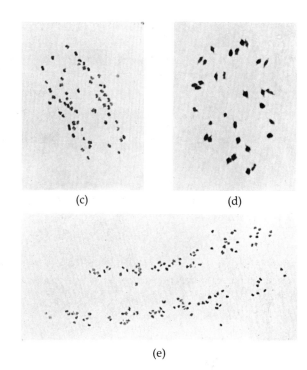

(a)

(b)

(c)

(d)

(e)

28–13

Polyploidy has been investigated extensively among grasses of the genus Spartina, which grow in salt-marsh habitats along the coasts of North America and Europe. (a) This salt marsh is on the coast of Great Britain. (b) A Spartina hybrid. (c) Spartina maritima, the native European species of salt marsh grass, has 2n = 60 chromosomes, shown here from a cell in meiotic anaphase I. (d) Spartina alterniflora is a North American species with 2n = 62 chromosomes (there are 30 bivalents and 2 unpaired chromosomes shown here from a cell in meiotic metaphase I). In Europe, it was first collected from the shores of Southampton Water in 1839, but the actual date of its introduction is unknown. Although the two species have never been crossed artificially, specimens of their sterile hybrid, Spartina × townsendii, were collected from along Southampton Water in 1870. (e) A vigorous polyploid, S. anglica, arose spontaneously from this sterile hybrid and was first collected in the early 1890s. This polyploid, which has 2n = 122 chromosomes, shown here from a cell in meiotic anaphase I, is now extending its range through the salt marshes of Great Britain and other temperate countries.

major evolutionary mechanism. Events similar to those we have reviewed in the history of wheat must have taken place well over 100,000 times just to account for the present representation of polyploids in the flora of the world, in which they amount to more than half of the total species of plants. Among these polyploids are many of our most important crops—not only wheat but also cotton, tobacco, sugarcane, bananas, potatoes, and safflower, just to mention a few. To this list can be added many of our most attractive garden flowers, such as chrysanthemums, pansies, and daylilies.

THE ORIGIN OF MAJOR GROUPS OF ORGANISMS

As knowledge about the ways in which species originate has become better developed, evolutionary scientists have turned to the question of the origin of genera and other higher taxonomic groups of organisms. As suggested by the boxed essay on Hawaiian tarweeds (pages 576–577), genera originate when the same kinds

of evolutionary processes that are responsible for the origin of species continue. If a particular species has a distinctive adaptation to a habitat that is very different from the habitat in which its progenitor occurs, it may come to be very different from that progenitor. It may then give rise in time to many new species, and come to constitute a novel evolutionary line. In time, it may become so distinct that this line may be classified as a new genus, family, or even class of organisms. No special mechanisms are necessary to account for the origin of taxonomic groups above the level of species—only discontinuity in habitat, range, or way of life.

Much has been written recently about whether evolution is uniformly a gradual process, as it has often been viewed in the past, or whether it takes place in spurts: long periods of gradual change or of no change at all, punctuated by periods of rapid change. The latter model is called the *punctuated equilibrium* model of evolution. Some of the proponents of this model have asserted that macroevolution—the process whereby higher taxonomic groups originate—takes place by principles different from those characteristic of microevolution, the sort of gradual change to which we have

(a)

• E. × ferrissii

E. laevigatum

E. hyemale

(b)

28–14

One of the most abundant and vigorous of the horsetails (see Figure 26–3b, page 520) found in North America is Equisetum × ferrissii, *a completely sterile hybrid of* E. hyemale *and* E. laevigatum. *Horsetails propagate readily from small fragments of underground stems, and the hybrid maintains itself over its wide range through such vegetative propagation.* (a) *Stems of* E. × ferrissii, *with strobili.* (b) *Range of* E. × ferrissii *and those of its parental species.*

devoted most of this chapter. All of the kinds of features that distinguish higher taxonomic categories, however, also distinguish some species, so the process of evolution seems to be of one kind only.

SUMMARY

Natural selection is the process by which organisms with more favorable characteristics with respect to their environment leave more surviving progeny. Natural selection occurs because of genetically determined variations in natural populations. It takes place as organisms progressively occupy different kinds of habitats. Reproductive success (fitness) can be determined only in relation to the habitat in which a particular population occurs.

The base line for the genetics of populations is the Hardy–Weinberg law. It states that, in a large population in which there is random mating, and in the absence of forces that change the proportions of alleles, the proportion of dominant to recessive alleles remains constant from generation to generation.

Four principal factors that can cause changes in the proportions of alleles and deviations from the Hardy–Weinberg equilibrium are population size, migration, mutation, and selection. Of these, selection is the most consequential. It is defined as the nonrandom reproduction of genotypes, some being favored over others in the particular environments in which they occur.

Recessive alleles are relatively inaccessible to selection in diploid organisms. The less frequent they become, the higher is the proportion of such alleles that are hidden (masked) in heterozygotes.

A population's response to selective pressures is complex for a number of reasons. Only the phenotype is accessible to selection, and similar phenotypes can result from very different combinations of alleles. Because of epistasis and pleiotropy, single alleles cannot be selected in isolation; selection affects the whole genotype.

Populations adjust to particular environments and form sharply defined units, or ecotypes, if the lines between the environments are sharply drawn. If environments gradually intergrade, populations of plants may form clines with respect to the characteristics under consideration.

While populations are changing, they also diverge in those factors that allow them to intercross successfully. After a period of isolation, two populations may be more or less incompatible or may produce sterile hybrids. Related species of relatively long-lived plants, such as trees and shrubs, are less apt to be reproductively isolated than are related species of annuals and other short-lived plants.

Hybrid populations derived from two species are common in plants, especially trees and shrubs. This is particularly true in environments with relatively few species, such as oceanic islands, where adjustment to environmental changes may be especially critical, and in regions with sharply defined environmental breaks and diverse climates, such as California.

Even if the hybrids between two species are sterile, they may be propagated by apomixis or other forms of vegetative reproduction or become fertile following doubling in chromosome number (polyploidy).

SUGGESTIONS FOR FURTHER READING

ANDERSON, EDGAR: *Plants, Man and Life,* University of California Press, Berkeley, 1967.*

A fascinating account of the evolution of certain groups of plants that reflects the personality of one of the greatest twentieth-century students of plants.

AYALA, FRANCISCO J., and JAMES W. VALENTINE: *Evolving: The Theory and Process of Organic Evolution,* Benjamin-Cummings Publishing Co., Menlo Park, Calif., 1979.

A very readable and modern account of the field of evolution.

COOK, ROBERT E.: "Clonal Plant Populations," *American Scientist* 71: 244–253, 1983.

Excellent discussion of the way in which vegetative reproduction in plants affects their ecology and evolution.

GOULD, STEPHEN JAY: *Ever Since Darwin: Reflections in Natural History,* W. W. Norton & Co., Inc., New York, 1979.*

A delightful collection of essays concerning the impact of Darwinism on modern science. Highly recommended.

GRANT, VERNE: *Plant Speciation,* Columbia University Press, New York, 1971.

This comprehensive treatment of plant evolution provides a thorough introduction to most aspects of the field.

MOORE, JOHN: "Science As a Way of Knowing—Evolutionary Biology," *American Zoologist* 23 (1):1–68, 1983.

A concise exposition of evolutionary theory, presented clearly as a series of hypotheses and deductions.

MOOREHEAD, ALAN: *Darwin and the Beagle,* Harper & Row Publishers, Inc., New York, 1969.

An engaging account of Darwin's work by a famous writer and historian.

STEBBINS, G. LEDYARD: *Processes of Organic Evolution,* 3rd ed., Prentice-Hall, Inc., Englewood Cliffs, N.J., 1977.*

A brief review of the entire field by one of its outstanding practitioners.

* Available in paperback.

C H A P T E R   2 9

Evolution of the Flowering Plants

29–1
(a) *A longhorn beetle (family Cerambycidae), laden with pollen, visiting the flower of a member of the parsley family (Apiaceae) in the mountains of northeastern Arizona. (b) Carapace of a beetle from the same strata as* Archaeanthus, *an extinct angiosperm from 95 to 98 million years ago (see ''Early Angiosperms and Their Flowers,'' page 592). The evolution of the flowering plants is, to a large extent, the story of increasingly specialized relationships between flowers and their insect pollinators, in which beetles played an important early role.*

The angiosperms make up much of the visible world of modern plants. Trees, shrubs, lawns, gardens, fields of wheat and corn, wildflowers, fruits and vegetables on the grocery shelves, the bright splashes of color in a florist's window, the geranium on a fire escape, duckweed and pond lilies, eel grass and turtle grass, saguaro cacti and prickly pears—wherever you are, flowering plants are there also.

In a letter to a friend, Charles Darwin referred to the apparent sudden appearance of this great group of plants in the fossil record as ''an abominable mystery.'' In the early fossil-bearing strata, one first finds primitive vascular plants with a very simple structure; then a Devonian and Carboniferous proliferation of ferns, lycophytes, sphenophytes, and progymnosperms. The early seed plants first appear in the late Devonian period and lead to the gymnosperm-dominated Mesozoic floras. Finally, during the first half of the Cretaceous period, toward the end of the Mesozoic era, the angiosperms or flowering plants appear, gradually assuming worldwide dominance in the vegetation. By about 75 million years ago, many modern families and some modern genera of this division already existed (Figure 29–2).

Why did the angiosperms rise to world dominance and then continue to diversify to such a spectacular extent? In this chapter, we shall attempt to answer this question, centering our discussion on three topics—the origin of the angiosperms, the evolution of the flower, and the role of certain chemical substances in angiosperm evolution. By doing so, we shall illustrate the process of evolution, which we began to describe in the preceding chapter.

ORIGIN OF THE ANGIOSPERMS

It is now generally agreed that the angiosperms evolved from some primitive gymnosperm, probably a shrub.

(a)

(b)

29–2

*Fossils of a gymnosperm (a) and two
angiosperms (b, c) from late Cretaceous
period deposits (about 70 million years ago)
in Wyoming. (a) Twig and individual cone
scales of* Araucarites longifolia, *an extinct
species of conifer belonging to a family
(Araucariaceae) now restricted to the
Southern Hemisphere. A familiar living*
*member of this family is the Norfolk Island
pine* (Araucaria heterophylla). *(b) Leaf of
a fan palm,* Sabalites montana, *an extinct
species distantly related to the palmettos
of the southeastern United States. (c) Leaf
of* Viburnum marginatum, *an extinct
species belonging to a widespread genus of
shrubs generally known as arrow-woods.*

(c)

No likely candidates are known from the Cretaceous
period, but there are a number of gymnosperms represented earlier in the Mesozoic and Paleozoic eras that
display certain combinations of angiospermlike traits.
This in itself suggests an earlier origin for the angiosperms than can be documented from the fossil record.

The first fossil remains that can definitely be assigned to the angiosperms date from the early part of
the Cretaceous period, some 125 million years ago.
These remains consist of pollen grains, each with a single pore. The pollen grains are similar to (but distinguishable from) the spores of ferns and the pollen of
gymnosperms. It is likely that most angiosperms that
existed more than 125 million years ago had pollen that
could not be distinguished from gymnosperm pollen.
All monocots, as well as the most primitive dicots, have
pollen that resembles the first known angiosperm pollen. By no less than 120 million years ago, the three-
pored angiosperm pollen characteristic of all but the
most primitive dicots had appeared in the fossil record.
By about 80 to 90 million years ago, angiosperms were
more abundant all over the world than the members of
any other group of plants. What do we know now
about the origin of this group?

Relationships of the Angiosperms

Since the time of Darwin in the nineteenth century, scientists have attempted to understand the ancestry of the angiosperms. One approach to this problem has been to search for their possible ancestors among fossil seed-bearing plants. In this effort, particular emphasis has been placed on the ease with which the ovule-bearing structures of various gymnosperms could be imagined as being transformed into a carpel. Recently, a different approach has been suggested, one that, instead of intensely searching for ancestral groups, simply attempts to define the major natural groups of seed plants and to understand their relationships with one another. The degree of relationship between these groups is then used to estimate how recently they may have been derived from a common ancestor. A leading scientist who has been pursuing this line of investigation recently is Peter Crane, of the Field Museum of Natural History in Chicago, who supplied much of the information on which this section is based.

The simplest situation in which to discuss phylogenetic (evolutionary) relationships is one in which only three different kinds (or groups) of plants are involved. Ferns, cycads, and pines, for example, all possess xylem composed of tracheids, and this suggests that at some point in the distant past they all shared a common ancestor that also possessed tracheids. The question can then be asked, "Which two of these three plants have a more recent common ancestor?" For three plants, there are only three possibilities from which to choose, and they can be conveniently summarized by three branching diagrams (Figure 29–3). Cycads and pines both have seeds and wood (secondary xylem), neither of which occurs in ferns, and this supports the idea that cycads and pines share a more recent common ancestor than either of these groups does with ferns (Figure 29–3a). Features that are thought to have been inherited from a common ancestor, such as the seeds, wood, and tracheids in this example, are said to be *homologous*. If an additional group of seed-bearing, woody plants is introduced into the picture—for example, the Douglas fir (*Pseudotsuga*)—more restricted homologies need to be found to clarify its relationship to cycads and pines. Pines and Douglas fir are similar in possessing resin ducts in their leaves, which suggests that they share a more recent common ancestor than either of these groups does with the cycads (Figure 29–4). As an analysis of this kind is extended by incorporating more similarities and more groups of plants, conflict frequently arises between conclusions about relationships that were derived by studying more limited groups, or by using fewer characteristics. In such cases, the simplest explanation (that is, the one favored by the largest number of homologies) is the one that should be adopted. The decision may then be tested and sometimes modified as additional features and additional groups of plants are considered.

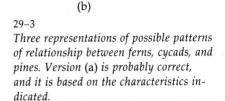

(a)

(b)   (c)

29–3

Three representations of possible patterns of relationship between ferns, cycads, and pines. Version (a) is probably correct, and it is based on the characteristics indicated.

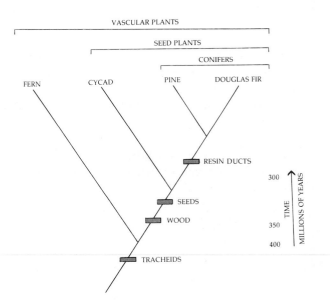

29–4

Phylogenetic relationships between ferns, cycads, pines, and Douglas fir, indicating the shared characteristics that support this particular pattern of relationship and the approximate times of divergence of these groups.

Estimating relationships in this way leads to a pattern of groups nested within groups, which is termed a *hierarchy*. Each group, or *clade*, is defined by one or more homologies and includes all the descendants of a single hypothetical common ancestor. The branching diagrams used to express the relationships between such plant groups are called *cladograms*. Vascular plants are a clade, within which seed plants form another clade, within which the conifers form yet another (Figure 29–4). The branching diagram also predicts the sequence in which homologies might be expected to appear in the fossil record. In this simple example, paleobotanical evidence conforms with these predictions: tracheids first appear at least 400 million years ago, seeds and wood about 350 million years ago, and resin ducts about 300 million years ago.

Certain kinds of similarities that may at first sight appear to indicate relationships do not play a part in estimating evolutionary relationships with a cladogram. For example, the ovulate cones of cycads, pines, and Douglas fir look similar, but the similarity is only superficial. In cycads, each cone scale is a simple, modified, seed-bearing leaf; in conifers, however, each cone scale is a complex structure, as we discussed in Chapter 18. The cones of cycads and conifers, therefore, do not appear to have evolved from a common ancestor and could not be used to define a clade; they are said to be *analogous*. Their superficial similarity is due to *convergent evolution* (see page 443). In contrast, the compound ovulate cones of pines and Douglas fir are homologous and could be used to define a conifer clade.

A further potential source of confusion in assessing similarities results from adopting too narrow a view of a particular homology. For example, *Lycopodium*, *Equisetum*, and ferns have free spores (that is, they all shed their spores, instead of retaining them on the parent sporophyte); but this feature could not be used to define a clade because the spores of these plants are also homologous with part of the seed in seed plants. A group defined by its possession of free spores would be incomplete and not include all the descendants of a single common ancestor. That same common ancestor would also have given rise to seed plants. In this example, there are two homologies, possession of spores and possession of megaspores modified into seeds; these define two clades, one of which is a subset of the other. Those vascular plants that form free spores do not form a clade by themselves.

Using the kind of reasoning just explained, which is known as *phylogenetic analysis* or *cladistics*, the problem of recognizing the ancestors of flowering plants (which may not yet have been discovered or may not have been preserved in the fossil record) is set aside. Instead, the objective is to define the major clades of seed plants and to assess their relationships in terms of relative recency of common ancestry.

Living seed plants provide the best place to begin the analysis because they are generally much better understood than their fossil relatives. One recently suggested cladogram of living seed plants is given in Figure 29–5. It is clear from this diagram that the gymnosperms are not a clade, and that the only features that could define the gymnosperms would also have to include the flowering plants. In effect, the gymnosperms are merely what is left of the clade of seed plants after the angiosperm clade has been removed from it.

After an analysis of living plants, the next stage is to incorporate what is known about fossil seed plants. While many fossil plants can be placed in groups that are known from their living members, others are representatives of major clades that are now extinct. These extinct clades can provide important clues to recognizing homologies and may therefore be extremely useful in clarifying ideas on evolutionary relationships. Two recently proposed cladograms expressing the relationships between the major kinds of living and fossil seed plants are given in Figure 29–6. These ideas are certain to be refined by future research, especially as more is discovered about some of the major extinct plant groups, but they do indicate how the flowering plants are currently thought to fit into the general pattern of evolution.

29–5

Phylogenetic relationships between the four divisions of living gymnosperms and the angiosperms.

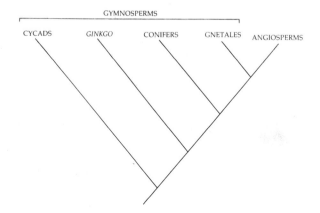

The Evolutionary Radiation of Angiosperms

Paleobotanist Daniel Axelrod, of the University of California, Davis, suggested a number of years ago that the early evolution of the angiosperms may have taken place away from the lowland basins where fossil deposits are normally found. In the hills and uplands of the tropics, the angiosperms could have evolved into a

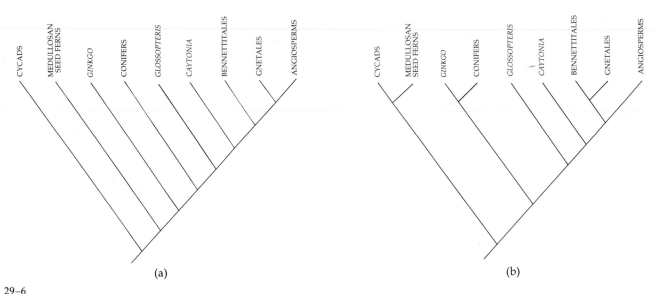

29–6
(a), (b) *Two possible schemes of phylo-genetic relationships between the living* *divisions of seed plants and some of the allied groups. These cladograms were* *prepared by Peter Crane of the Field Museum of Natural History, Chicago.*

variety of forms without much chance of their remains being preserved in the fossil record. In Axelrod's view, it was when angiosperms invaded the lowlands, about 120 million years ago, that they assumed the dominant role in the world's vegetation.

The early flowering plants possessed many adaptive traits that made them particularly resistant to drought and cold. Among these were tough leaves, commonly reduced in size; vessel members (efficient water-conducting cells); and a tough, resistant seed coat that protected the young embryo from drying out. The features that make many angiosperms particularly resistant to drought and cold are not found in all of the flowering plants, nor are they restricted to the angiosperms, but they have certainly played major roles in the diversification of the division.

There are a number of other factors that seem to have been important in the early and continuing success of the angiosperms, and we shall discuss some of them in more detail in this chapter. The evolution of sieve-tube members presumably made possible the more efficient conduction of sugars throughout the plant in the phloem, just as vessel members were more efficient than tracheids in the xylem. Perhaps of even greater importance, the precise systems of pollination and seed dispersal that are characteristic of the group allowed them to exist as widely scattered individuals in many different kinds of habitats. The enormous chemical diversity of the group, which includes many kinds of defenses against diseases and herbivores, has likewise been of great importance, as we shall see below.

These and other features bear directly on the fact that, compared with other plants, angiosperms reproduce rapidly and efficiently.

One important innovation in angiosperms was the evolution of the deciduous habit. There are some deciduous gymnosperms, including larches and bald cypresses, but very few in relation to the overall size of the group. In contrast, many groups of angiosperms are deciduous. The deciduous habit seems to have evolved first in tropical areas that experienced periodic drought. Descendants of these deciduous plants then spread to the north, where parts of the year were so cold that no water was available for growth. Other modifications of habit in the angiosperms, such as the evolution of herbaceous perennials and eventually of annuals, allowed their survival in conditions more extreme than those in which their woody progenitors could grow. All of these adaptations were of particular importance in a world in which the climate was becoming progressively more stressful during the past 50 million years or so—the most important period of angiosperm evolution.

About 125 million years ago—when angiosperm pollen first became part of the fossil record—Africa and South America were directly connected with each other and with Antarctica, India, and Australia in a great southern supercontinent called Gondwanaland (Figure 29–7). Africa and South America began to separate at about this time, forming the southern Atlantic Ocean, but they did not move completely apart in the tropical regions until about 90 million years ago. India began to move northward at about the same time, colliding with

Asia about 45 million years ago and causing the Himalayas to be thrust up in the process. Australia began to separate from Antarctica about 55 million years ago, but their separation did not become complete until more recently.

Within the central regions of West Gondwanaland, formed by what are now the continents of South America and Africa, the kinds of arid-to-subhumid habitats envisioned by Axelrod and others as an important early site of angiosperm evolution would have been well represented. With the final separation of these two continents, at about the time the angiosperms became abundant in the fossil record worldwide, the world climate changed greatly. This was especially the case in these equatorial regions, which became milder, with fewer extremes of temperature and humidity. These changes are thought to have been related to the evolutionary success of the early angiosperms, which increasingly dominated floras all over the earth.

When India and Australia were in the far south, they were covered with vegetation typical of a cool temperate climate. Certain unique groups of plants and animals are shared by southern South America and southeastern Australia–Tasmania at the present time, having achieved their ranges by migration across Antarctica long before the onset of full glaciation there, which occurred about 20 million years ago (Figure 29–8). As Australia moved northward during the past

55 million years, it reached the great zone of aridity flanking the tropics, and the sorts of arid plant communities that are now so widespread there expanded greatly. As the northern edge of Australia neared Asia, it reached truly tropical climates for the first time and was invaded by the plants and animals of tropical Asia. To a large extent, however, the plants and animals characteristic of Asia and those characteristic of Australia have remained geographically distinct, and the line separating the areas where each is predominant is known as "Wallace's line," after Alfred Russel Wallace, an early naturalist-explorer of these regions. (Wallace is also well remembered as the young man who proposed the theory of evolution by natural selection simultaneously with Darwin.) The original plants and animals of cool-temperate Australia have survived in the southeastern corner of the continent, in Tasmania, and in New Zealand, which separated from Australia–Antarctica about 80 million years ago and moved to the northeast.

The original flora and fauna of India did not fare so well, and only a few remnants of it survive today. India moved much farther than Australia, crossing the southern arid zone, the tropics, and the northern arid zone in the process. As a result, nearly all of India's original plants and animals became extinct; they were eventually replaced by desert, tropical, and mountain flora and fauna from Eurasia.

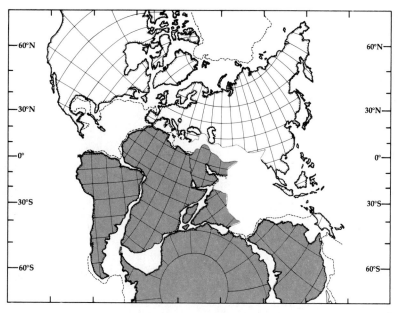

29–7

Relationship between the earth's southern land areas at the time of the first appearance of the angiosperms in the fossil record. In the middle of the

Cretaceous period, about 100 million years ago, South America was directly connected with Africa, Madagascar, and India, and, via Antarctica, with Australia.

These combined land masses, indicated here in color, formed the supercontinent of Gondwanaland.

(a)

(b)

29–8

(a) *This southern beech* (Nothofagus menziesii), *growing in Upper Caples Valley, Southland, New Zealand, is a relict from the cool temperate forest that extended from southern South America across Antarctica to Australia and New Zealand from about 80 million years ago* *until perhaps 30 million years ago, with increasingly large gaps developing throughout that entire period of time up to the present.* (b) *The palm* Nypa fruticans, *shown growing along the edge of a tidal swamp. Some 40 to 50 million years ago, when Africa and South* *America were in closer contact, Nypa was distributed widely throughout the warmer parts of the world. It has become progressively more restricted in distribution, and it is now confined to Southeast Asia, northern and northeastern Australia, and some Pacific islands.*

EVOLUTION OF THE FLOWER

The Parts of the Flower

The single characteristic that sets the angiosperms apart from all other groups is the flower, a structure that we have already discussed at some length in Chapter 18. As we noted previously, the flower is a determinate shoot bearing various leaflike appendages; how the various parts of the flower are related to leaves will be discussed in the following pages. Most of what we have concluded about the evolution of the flower is based on a comparative study of modern forms, although fossil flowers, as more and more of them have been found, have been of increasing importance as a record of angiosperm evolution.

The Carpel

The carpel in its most primitive form is a folded leaf blade. As shown in Figure 29–9, in some generalized carpels there is no localized area for the entrapment of pollen grains. Both margins of the folded blade are covered with stigmatic hairs. The folded carpel surrounds the ovules, which are located on its inner surface. The carpels are closed in all living angiosperms, although different groups illustrate different stages in the closure process. Primitive angiosperms have large stigmatic surfaces that run along the sides of their carpels. More specialized angiosperms, including nearly all of those that exist today, have a much smaller stigma that is held well above the ovary on a style.

The ovules, which were probably in rows along and near the edges of the inner surface of the carpel in the original angiosperms, became arranged in various ways in more specialized ones. Primitive angiosperms have many ovules, while advanced ones have relatively few. The gynoecium of the primitive angiosperm flower contained a number of separate carpels, which have been reduced in number, fused together, or both during the course of evolutionary specialization (Figure 29–10).

The Androecium

Although the stamens of modern plants (for an example, see Figure 18–39, page 355) rarely resemble leaves, it is possible to find leaflike stamens among the magnolias and their relatives. Like these, the primitive stamen may have been blade-shaped, with sporangia near the center of the blade (Figure 29–11). According to one theory, the blade became differentiated into a slender stalk (the filament), with the sporangia near its apex.

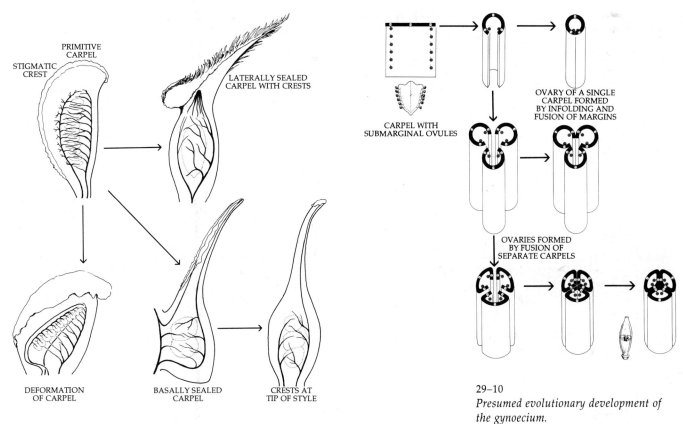

29–9
Some changes that have occurred during the evolution of the angiosperm carpel. The primitive carpel is conduplicate (folded together lengthwise). The folded blade encloses a number of ovules attached to the inner surface. The margins are not sealed, and there is no localized stigmatic surface, although stigmatic hairs are extensively developed on the crest, or margins, of the carpel. In the carpels derived from this primitive example, the margins of the blade are more completely sealed, and the pollen-trapping stigmatic surface is localized in a crest.

29–10
Presumed evolutionary development of the gynoecium.

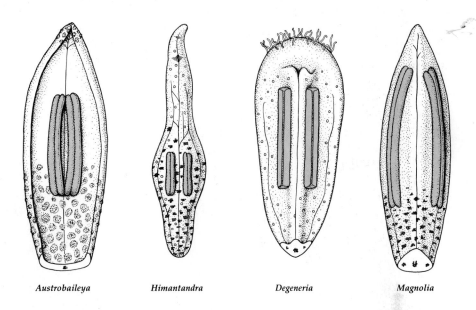

Austrobaileya   *Himantandra*   *Degeneria*   *Magnolia*

29–11
Stamens of primitive angiosperms. In these woody plants, the anthers are borne on the lower or upper surface of leaflike microsporophylls. Examples of the former, microsporophylls of Himantandra *and* Degeneria, *are here viewed from below; examples of the latter, those of* Austrobaileya *and* Magnolia, *are viewed from above. In most modern angiosperms, the stamens contain much*

less sterile tissue than those of primitive angiosperms, and the anther is borne at the end of a slender filament. Such stamens are not at all leaflike. These differences are difficult to explain and have led to the hypothesis that leaflike stamens such as the ones shown here may actually have been derived through the fusion of branch systems bearing terminal sporangia.

EARLY ANGIOSPERMS AND THEIR FLOWERS

Archaeanthus linnenbergeri, the only known member of an extinct family of angiosperms, lived in what is now central Kansas from about 98 to 95 million years ago, in the middle of the Cretaceous period. When *Archaeanthus* flourished on what were then the subtropical coastal plains of Kansas, dinosaurs were abundant in the same area. Its stout flowers had an elongate axis with 100–130 spirally arranged follicles, each producing 10–18 seeds. There were three outer perianth parts and six to nine inner ones; the stamens were numerous and spirally arranged. *Archaeanthus* was probably deciduous, and it seems to have been a small tree or shrub. In its general floral morphology, it resembles living members of the group of families related to Magnoliaceae (Figure 29–11). (a) Fossil reproductive axis. (b) Fossil leaf. (c) Reconstruction of a flowering branch. (d) Reconstruction of a fruiting branch. The reconstructions shown here are based on the studies of David Dilcher of Indiana University and were drawn by Megan Rohn. Careful studies by Dilcher and his students are revealing a great deal about the nature of early angiosperms and their flowers.

(a)

(c)

(b)

(d)

An alternative theory, which is also attractive, holds that the stamens were derived from slender branch systems bearing terminal sporangia. The branches gradually became fused and, in some instances, leaflike.

In some specialized flowers, the stamens, like the carpels, became fused. Stamens may be fused with one another into columnar structures, as they are in the members of the pea, melon, mallow (see Figure 29–13d), and sunflower families; or they may be fused with the corolla, as they are in the phlox, snapdragon, and mint families.

In some evolutionarily advanced flowers, the stamens have become secondarily sterile; that is, they have lost their sporangia and have become transformed into specialized structures, such as nectaries—glands that secrete nectar, a sugary fluid that attracts pollinators and provides food for them. (It is important to note, however, that most nectaries are not modified stamens but arise instead from other parts of the flower.) Stamens have also played a part in the evolution of petals, as we shall see.

The Perianth

The perianth consists of the sepals and the petals. The sepals of most flowers are green and photosynthetic. In this respect, and in being supplied by the same number of vascular strands as the leaves of the same plant (often, but not always, more than one), sepals resemble leaves. For this reason, sepals are considered to have been derived directly from leaves.

In a few plant families, including the water lilies, the petals seem to have been derived from sepals. In most angiosperms, however, the petals were probably derived originally from stamens that lost their sporangia and then became specially modified for their new role —that of attracting potential pollinators to the flower. Most petals, like stamens, are supplied by just one vascular strand; in contrast, sepals, like leaves, are normally supplied by the same number of vascular strands as the leaves of the same plant (often three or more), as we have mentioned. In sepals and petals alike, the vascular strands usually branch within the organ, so that the number of strands that enter it cannot be determined from the number of veins in its main body.

Petal fusion has occurred during the course of the evolution of many groups of angiosperms, resulting in the familiar tubular corolla that is characteristic of many families. When a tubular corolla is present, the stamens often fuse with it and appear to arise from the corolla. In a number of the evolutionarily advanced families, the sepals are similarly fused into a tube.

Evolutionary Trends among Flowers

A modern flower that may resemble the primitive flower in its general structure is that of the magnolia (Figure 29–12). A magnolia flower has numerous carpels and stamens and many perianth parts, all of which are well separated from one another. In addition, the spiral arrangement of these parts on the tip of the flower stalk, or receptacle, is still clearly evident. The

(a)

(b)

(c)

29–12
The flower structure of the southern magnolia (Magnolia grandiflora). *The cone-shaped receptacle bears numerous spirally arranged carpels from which curved styles emerge. Below the styles in*

(a) and (b) are the cream-white stamens. (a) An unopened bud cut open to reveal the pollen-receptive stigmas; the anthers have not yet shed their pollen. (b) The floral axis of a second-day flower,

showing stigmas that are no longer receptive and stamens that are shedding pollen. (c) Fruit, showing carpels and bright red seeds, each protruding on a slender stalk.

(a)

(b)

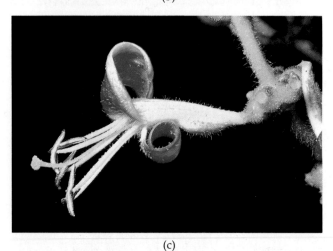

(c)

very elongated receptacle and the conelike nature of the carpel-bearing axis are, however, clearly secondary specializations within the magnolia family that are not shared by other primitive angiosperms.

By comparing primitive flowers of this sort with more specialized ones, four main trends in flower evolution can be traced (Figure 29–13; see also Figure 18–43 on page 358).

1. From flowers with many parts that are indefinite in number, flowers have evolved with few parts that are definite in number.
2. The number of kinds of parts has been reduced from four in the primitive flower to three, two, or sometimes one in more advanced flowers. The shoot has become shortened so that the original spiral arrangement of parts is no longer evident. The floral parts have become fused.
3. The ovary has become inferior rather than superior in position.
4. The radial symmetry (regularity), or actinomorphy, of the primitive flower has given way to bilateral symmetry (irregularity), or zygomorphy, in the more advanced flowers.

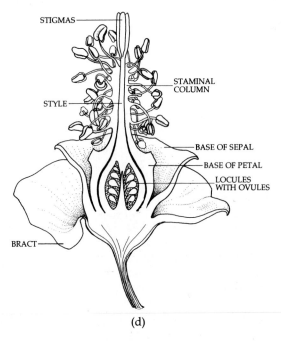

(d)

29–13
Examples of specialized flowers. (a) *Wintergreen,* Chimaphila umbellata. *The sepals (not visible) and petals are reduced to five each, the stamens to ten, and the five carpels are fused into a compound gynoecium with a single stigma.* (b) *Lotus,* Nelumbo lutea. *The undifferentiated sepals and the numerous petals and stamens are spirally arranged; the carpels are fused into a compound gynoecium.* (c) *Chaparral honeysuckle,* Lonicera hispidula. *The ovary is inferior and has two or three locules; the sepals are reduced to small teeth at its apex. The petals are fused into a corolla tube in the zygomorphic (bilaterally symmetrical) flower, and the five stamens, which protrude from the tube, are attached to its inner wall. The style is longer than the stamens, and the stigma is elevated above them; a pollinator visiting this flower would contact the stigma first, so that, if it were carrying pollen from another flower, it would deposit that pollen on the stigma before reaching the anthers of this flower. Fruits of this species are shown in Figure 29–46c.* (d) *A diagram of a cotton* (Gossypium) *flower, showing the column of stamens fused around the style.*

Examples of Specialized Families

Among the most evolutionarily specialized of the flowers are those of the family Asteraceae (Compositae), which are dicots, and those of the family Orchidaceae, which are monocots. In number of species, these are the two largest families of angiosperms.

Asteraceae

In the Asteraceae (the composites), the epigynous flowers are relatively small and closely bunched together into a head. Each of the tiny flowers has an inferior ovary composed of two fused carpels with a single ovule in one locule (Figure 29–14).

In composite flowers, the stamens are reduced to five in number and are usually fused to one another (coalescent) and to the corolla (adnate). The petals, also five in number, are fused to one another and also to the ovary, and the sepals are absent or reduced to a series of bristles or scales known as the *pappus*. The pappus often serves as an aid to dispersal by wind, as it does in the familiar dandelion, a member of the Asteraceae (Figure 29–14c; see also Figure 29–42). In other members of this family, such as beggar-ticks *(Bidens)*, the pappus may be barbed, serving to attach the fruit to a passing animal and thus to enhance its chances of being dispersed from place to place. In many members of the family Asteraceae, each head includes two types of

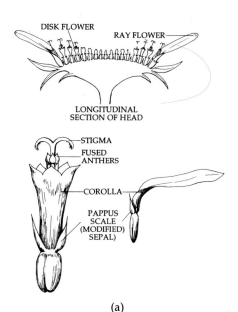

(b)

(c)

(a)

29–14

Composites (family Asteraceae). (a) A diagram of the organization of the head of a member of this family. The flowers are subordinated to the overall display of the head, which functions as a single large flower in attracting pollinators. (b) Thistle, Cirsium pastoris. *Members of the thistle tribe have only disk flowers. This particular species of thistle, with its bright red flowers, is regularly visited by hummingbirds, which are the primary agents of pollination for the plant. (c) Agoseris, a wild relative of the dandelion,* Taraxacum. *In the inflorescences of the chicory tribe (the group of composites to which dandelions and their relatives belong), there are no disk flowers; the marginal ray flowers, however, are often enlarged. (d) Sunflower,* Helianthus annuus.

(d)

flower; these are (1) disk flowers, which make up the central portion of the aggregate, and (2) ray flowers, which are arranged on the outer periphery. The ray flowers are often carpellate, but sometimes completely sterile. In some members of the Asteraceae, such as sunflowers, daisies, and black-eyed Susans, the fused corolla of each ray flower forms a long strap-shaped "petal."

In general, the composite head has the appearance of a single large flower. Unlike many single flowers, however, the head matures over a period of days, with the individual flowers opening serially in a centripetal spiral. As a consequence, the ovules in a given head may be fertilized by a number of different pollen donors. The success of this plan as an evolutionary strategy is attested to by the great abundance of the members of the Asteraceae, which, with about 22,000 species, is probably the largest family of flowering plants.

Orchidaceae

Another successful flower plan is that of the orchids (Orchidaceae), which, unlike the composites, are monocots. There are at least 17,000 species of orchids, but the classification of many of the genera is poorly established. In any case, the family is certainly one of the largest. Most of the species are tropical; only about 140 are native to the United States and Canada, for example. In the orchids, the three carpels are fused and, as in the composites, the ovary is inferior (Figure 29–15). Unlike the composites, however, each ovary contains many thousands of minute ovules; consequently, each pollination event may result in the production of a huge number of seeds. Usually only one stamen is present (in one subfamily, the lady-slipper orchids, there are two), and this stamen is characteristically fused with the style and stigma into a single complex structure—the _column_. The entire contents of an anther are held together and dispersed as a unit—the _pollinium_ (see Figure 29–26b). The three petals are modified so that the two lateral ones form wings and the third forms a cuplike lip that is often very large and showy. The sepals, also three in number, are often colored and similar to petals in appearance. The flower is always bilaterally symmetrical, and it is often bizarre in appearance.

Among the orchids are some species with flowers the size of a pinhead and others with flowers more than 20 centimeters in diameter. Several genera contain saprophytic species; two Australian species grow entirely underground, their flowers appearing in cracks in the ground, where they are pollinated by flies (Figure 29–16). In the commercial production of orchids, the plants are _cloned_ by making divisions of meristematic tissue, and thousands of identical plants can be produced rap-

(a)

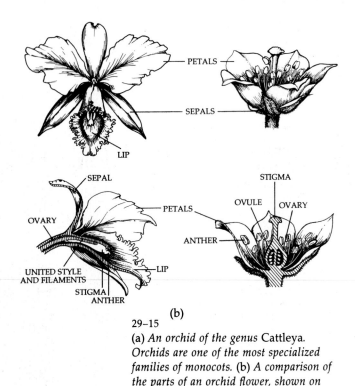

(b)

29–15

(a) _An orchid of the genus_ Cattleya. _Orchids are one of the most specialized families of monocots._ (b) _A comparison of the parts of an orchid flower, shown on the left, with those of a radially symmetrical flower, shown on the right. The "lip" is a modified petal that serves as a landing platform for insects._

(a)

(b)

29–16
Rhizanthella, *a Western Australian orchid, grows entirely underground. Cracks in the soil form during the dry season, revealing the flowers, which* never show above the soil surface. (a) A view from above, with leaves and debris cleared away, of the parted bracts of Rhizanthella, through which the plant's *pollinators (flies) enter. (b) A dozen flowers of* Rhizanthella *are seen here surrounded by the protective bracts.*

idly and efficiently. There are more than 60,000 registered hybrids of orchids, many of them involving two or more genera. The seed pods of orchids of the genus *Vanilla* are the natural source of the popular flavoring of the same name (Figure 29–17).

The Agents of Evolution

Plants, unlike most animals, cannot move from place to place to find food or shelter or to seek a mate. In general, plants must satisfy these needs by growth re-sponses and by the structures that they produce. The angiosperms, however, have evolved a set of features that, in effect, allows them some mobility in seeking a mate; this set of features is embodied in the flower. By attracting insects and other animals with their flowers, and by directing the activities of these animals so that cross-pollination (and therefore cross-fertilization) will occur at a high frequency, the angiosperms have tran-scended their rooted condition and, in a sense, become just as motile as animals in this one respect. How was this achieved?

29–17
Vanilla, *an orchid that is a commercial source of the flavoring of the same name. Originally used by the Aztecs in what is now Mexico, vanilla is now cultivated primarily on Madagascar and other islands in the western Indian Ocean and elsewhere in the old world. Chocolate is a blend of cacao and vanilla. Synthetic vanilla flavoring (vanillin) is now widely used in place of natural vanilla, which is extracted from the dried, fermented seed pods of this orchid. (a) Flowers of the vanilla orchid* (Vanilla planifolia). *(b) Hand pollination of vanilla plants in Mexico; this procedure is carried out, even in wild plants, to insure a good crop of the capsules from which vanilla is extracted.*

(a)

(b)

GENETIC SELF-INCOMPATIBILITY

By recombining genes from different individuals, sexual reproduction produces variability in natural populations, giving these populations the potential for adaptation to changes in their environment through progressive evolutionary modification. Plants that regularly self-pollinate lose some of this potential. Early in the history of the flowering plants, mechanisms evolved in many families that make cross-pollination almost mandatory, even when the flowers are bisexual or when both staminate and ovulate flowers are present on the same plant.

Two basic mechanisms that promote cross-pollination are found among modern families of flowering plants. In the more common one, present in such economically important groups as grasses and legumes, the behavior of the pollen grain is determined by its own (haploid) genotype. If it carries a gene at the incompatibility locus that is identical to one at either of the two corresponding loci in the diploid stigma and style, then the entry of the pollen tube is barred.

If it carries a gene at this locus that is different from either of those present in the stigma tissue, then it is accepted.

In the second kind of incompatibility system, found in the crucifer family (Brassicaceae) and composite family (Asteraceae), the behavior of the pollen is determined by the genetics of the pollen parent, not by the genes of the pollen grain itself; in other words, the control system is based on the correspondence between two kinds of diploid tissue. In both systems, the kind of "match" that occurs at the incompatibility locus determines acceptance or rejection.

Although much remains to be learned about the physiology of these mechanisms, it is known that they depend upon "recognition" reactions between specific incompatibility proteins carried by the pollen grains and matching proteins produced in the stigmas or styles. In grasses, the reaction often occurs on the stigma surface. The scanning electron micrograph (a) shows part of the stigma of a flower of orchard grass (*Dactylis glomerata*). The stigma has numerous

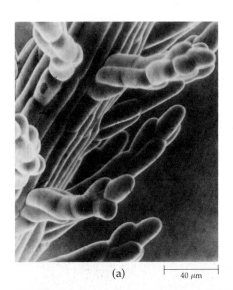

(a)   40 μm

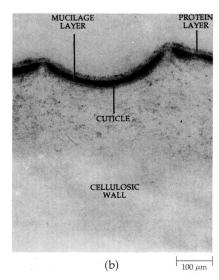

MUCILAGE LAYER   PROTEIN LAYER

CUTICLE

CELLULOSIC WALL

(b)   100 μm

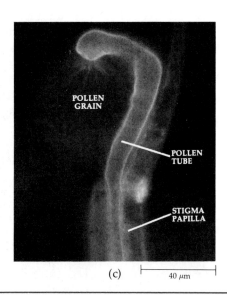

POLLEN GRAIN

POLLEN TUBE

STIGMA PAPILLA

(c)   40 μm

The earliest seed-bearing plants—various groups of gymnosperms—were pollinated passively. Their pollen was blown about by the wind, and it reached the vicinity of the ovules only by chance. The ovules, which were borne on the leaves or within the cones, exuded drops of sticky sap from their micropyles. These drops served to catch the pollen grains and to draw them to the micropyle, just as they do in modern gymnosperms. Insects, probably beetles (see Figure 29–1), feeding on the sap and resin of stems and on leaves, must have come across the protein-rich pollen grains and the sticky droplets that exuded from the ovules. Once these insects began returning regularly to these new-found sources of food, they began as a consequence of their

visits to carry pollen from plant to plant. Such a system would have been more efficient than passive pollination by the wind from the very beginning.

The more attractive the plants were to the beetles, the more frequently they would be visited, and the more seeds they would produce. Any changes in the phenotype that made such visits more frequent or more efficient offered an immediate selective advantage. Several important evolutionary developments followed. For example, plants that had flowers that provided special sources of food for their pollinators had a selective advantage. In addition to edible flower parts, pollen, and sticky fluid around the ovules, plants evolved floral *nectaries*, which secrete *nectar*, a nutri-

papillae, each capable of capturing several pollen grains. The transmission electron micrograph (b) shows the wall of a papilla of another grass species, rye *(Secale cereale)*, in section. Outside of the cuticle lie two other layers, an inner one composed of mucilaginous pectic materials and an outer proteinaceous one. The incompatibility response is shown when the tip of the pollen tube comes in contact with the outer layer, or very shortly thereafter. The other two micrographs (c and d), which were taken with the fluorescence microscope, show stigmas of foxtail grass *(Alopecurus pratensis)* stained with a fluorescent stain for the wall polysaccharide known as callose. In (c), the pollination has been a compatible one, and the tube is seen growing toward the ovary after penetrating the stigma. In (d), the pollination has been incompatible: after the tube tip touched the surface protein layer, its growth was halted and the tube filled with callose, which shows that it was rejected.

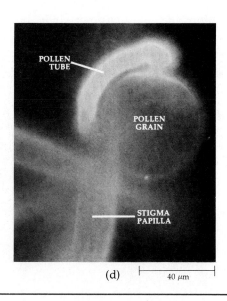

(d)   40 μm

tious, sugary fluid that provides a source of energy for insects and other animals.

Attraction of the insects to the flowers raised a new problem: protection of the ovule from predatory insects. The evolution of the closed carpel may have been a consequence of this problem. Further changes in the shape of the flower, such as the evolution of the inferior ovary, may also have arisen as additional means of protecting the ovules from being eaten by visiting insects.

Another important development was the bisexual flower. The presence of both carpels and stamens in a single flower (in contrast, for instance, to the separate microsporangiate and ovulate cones of conifers) serves

to make each visit by the pollinator more effective, because it can both pick up and deliver pollen at each stop.

In the early part of the Tertiary period, 40 to 60 million years ago, such specialized groups of flower-visiting insects as bees and butterflies, which had been evolving with the angiosperms for about 50 million years at that point, became even more abundant and diverse. The rise and diversification of these groups of insects were directly related to the increasing diversity of angiosperms. In turn, the insects profoundly influenced the evolutionary course of the angiosperms and contributed greatly to their diversification.

If a given plant species is pollinated by only one or a few kinds of visitors, it tends to become specialized relative to the characteristics of these visitors. Many of the modifications that have evolved in angiosperm flowers were special adaptations that promoted constancy to that particular kind of flower by their visitors.

Some of the special modifications of the flower that came about during the course of their evolution in response to specific pollinators will be described in the following pages.

Beetle-Pollinated Flowers

A number of modern species of angiosperms are pollinated solely or chiefly by beetles (Figure 29–18; see also Figure 29–1). The flowers of beetle-pollinated plants of one distinct type are large and borne singly, such as those of magnolias, some lilies, California poppies, and wild roses; the flowers of another type are small and aggregated in an inflorescence, such as those of dogwoods, elders, and spiraeas. Many species of the pars-

29–18
A pollen-eating beetle (Asclera ruficollis) *at the open, bowl-shaped flowers of round-leaved hepatica* (Hepatica americana). *All species of this family of beetles* (Oedemeridae) *are obligate pollen-feeders as adults.*

ley family (Apiaceae) also fall into this second group (see Figure 29–1a). Members of some 16 families of beetles are frequent visitors to flowers, although, as a rule, these beetles derive most of their nourishment from other sources, such as sap, fruit, dung, and carrion. In beetles, the sense of smell is much more highly developed than the visual sense, and beetle-pollinated flowers are often white or dull in color, but they typically have strong odors (Figure 29–19a). These odors are usually fruity, spicy, or similar to the foul odors of fermentation and are thus distinct from the sweeter odors of flowers that are pollinated by bees, moths, and butterflies. Some beetle-pollinated flowers secrete nectar; in others, the beetles chew directly on the petals or

on specialized food bodies (pads or clusters of cells on the surfaces of various floral parts), as well as eating the pollen. Most beetle-pollinated flowers have the ovules well buried beneath the floral chamber, out of reach of the chewing jaws of the beetles on which they depend for their pollination.

Bee-, Wasp-, and Fly-Pollinated Flowers

Bees are the most important group of flower-visiting animals, being responsible for the pollination of more species of plants than the members of any other animal group. Both male and female bees live on nectar, and the females also collect pollen to feed the larvae. Bees have mouthparts, body hairs, and other appendages that have special adaptations that make them suitable for collecting and carrying these food materials (Figure 29–20). As Karl von Frisch and other investigators of insect behavior have shown, bees can quickly learn to recognize colors, odors, and outlines. The color spectrum that most insects, including bees, can see is somewhat different from ours. Unlike human beings, bees perceive ultraviolet as a distinct color, but not red, which therefore fades into the background.

Many kinds of bees—especially solitary bees, which constitute a majority of species of the group (Figure 29–21)—are highly constant in their visits to flowers. Such constancy increases the efficiency of the particular species of bee—or of an individual bee, while it is visiting the flowers of one species of plant. In relation to this specialization, bee species with narrowly re-

(a)

(b)

(c)

29–19
(a) *Western skunk cabbage* (Lysichiton americanum) *is pollinated by small, actively flying beetles of the family Staphylinidae. The beetles are attracted by its very strong odor. Other species of this plant family (Araceae) have inflores-* *cences with odors resembling dead fish or carrion, and some of these are pollinated by carrion flies. One such plant is the eastern skunk cabbage,* Symplocarpus foetidus (b). *A number of plants in other families have similar odors and are* *pollinated by carrion flies; the foul-smelling milkweed* Stapelia schinzii (c) *and its relatives, most of which are African, are good examples.*

29–20

Bees have become as highly specialized as the flowers with which they have coevolved. Their mouthparts have become fused into a sucking tube containing a tongue. The first segment of each of the three pairs of legs has a patch of bristles on its inner surface. Those of the first and second pairs are pollen brushes that gather the pollen that sticks to the bee's hairy body. On the third pair of legs, the bristles form a pollen comb that collects pollen from these brushes and from the abdomen. From the comb, the pollen is forced up into pollen baskets, concave surfaces fringed with hairs on the upper segment of the third pair of legs. Shown here is a honeybee (Apis mellifera) foraging in a flower of rosemary (Rosmarinus officinalis). In the rosemary flower, the stamens and stigma arch upward out of the flower, and both come into contact with the hairy back of any visiting bee of the proper size; here the anthers can be seen depositing white pollen grains on the bee.

stricted foraging habits often feature conspicuous morphological and physiological adaptations, such as coarse bristles in their pollen-collecting apparatus (if they visit plants with large pollen grains) or elongated mouthparts (if they take nectar from plants with long-tubed flowers). When they are constant to this degree, bees exert a powerful evolutionary force for specialization on the plants that they visit. There are some 20,000 species of bees, the great majority of which visit flowers for food.

Bee flowers—that is, flowers that coevolved with bees—have showy, brightly colored petals, which are usually blue or yellow. They often have distinctive patterns by which bees can efficiently recognize them.

29–21

A sweat bee (family Halictidae) gathering pollen from the stamens of a flower of a cactus (Echinocereus) in Baja California, Mexico. The tall structures in the center of the flower are the stigmas.

Such patterns may include "honey guides," special markings that indicate the position of the nectar (Figure 29–22). Bee flowers are never pure red, and, as special photographic techniques have shown, they often have distinctive markings that are normally invisible to humans (Figure 29–23).

In bee flowers, the nectary is characteristically situated at the base of the corolla tube and is often set below the surface, where it is accessible only to such specialized organs as the mouthparts of bees and not, for example, to the chewing mouthparts of beetles. Bee flowers characteristically have a "landing platform" of some sort (see Figure 29–22).

Bumblebees are among the most familiar flower visitors in the North Temperate Zone (Figure 29–24). They are social bees that live in colonies. The queens (sexual females) overwinter and, when they emerge in the spring, lay the eggs to establish a new colony. Bumblebees cannot fly until their wing muscles reach a temperature of about 32°C; to maintain this temperature, they must forage constantly on flowers with a copious supply of nectar. Many plants of the cool parts of North America and Eurasia, including lupines, larkspurs, and fireweed, are regularly pollinated by bumblebees throughout their ranges.

Some of the evolutionarily more advanced flowers, in particular the orchids, have developed complex passageways and traps that force the bees that visit them to follow a particular route into and out of the flower. This ensures that both anther and stigma come into contact with the bee's body at a particular point and in the proper sequence (see Figure 29–20).

29–22
"Honey guides" on the flowers of the foxglove (Digitalis purpurea) *serve as distinctive signals to insect visitors. The lower lip of the fused corolla serves as a landing platform of the kind that is commonly found in bee flowers.*

(a)

(b)

29–23
The color perception of most insects is somewhat different from that of human beings. To a bee, for example, ultraviolet light (which is invisible to humans) is seen as a distinct color. These photographs show that a flower of marsh marigold (Caltha palustris), *which appears solid yellow to us* (a), *reflects ultraviolet light only in the portions of the flower that appear light in* (b). *The light portions of the flower reflect both yellow and ultraviolet, which combine to* form a color known as "bee's purple," *whereas the dark portions of the flower absorb ultraviolet and therefore appear pure yellow, when viewed by a bee. (See also page 611.)*

(a)

(b)

29–24
Bumblebees (Bombus). *These social bees are important pollinators of many genera of plants throughout the cooler parts of the Northern Hemisphere, and they have been introduced into regions where they are not native for the purpose of pollinating such plants as white clover* (Trifolium repens). *(a) A bumblebee gathering pollen from the flower of a California poppy* (Eschscholzia californica). *(b) Portion of the underground nest of a bumblebee, showing the cells in which the wormlike larvae complete their development. The bumblebees provision these cells with pollen and regurgitated nectar, which they obtain from flowers. Although a colony of bumblebees may visit a wide variety of flowers during the course of a season, an individual bee often visits only the flowers of one kind of plant on a single trip away from the nest.*

29–25
The beelike flowers of the orchid Ophrys speculum, *here illustrated from an individual in Sardinia, attract male bees, which are so deceived by the resemblance to female bees of their species that they attempt to copulate with the flowers. In doing so, they often pick up a pollen sac (pollinium) from a flower and may then carry it to another flower of the same species.*

An even more bizarre pollination strategy has been adopted by orchids of the genus *Ophrys*. The flower resembles a female bee, wasp, or fly rather closely (Figure 29–25). The males of these insect species emerge early in the spring, before the females. The orchids bloom early in the spring as well, and the male insects attempt to copulate with the orchid flower. During the course of their "sexual" visit, a pollinium may be deposited on the insect's body; when it visits another flower, the pollinium may be caught in the appropriate grooves on the stigma and thus bring about the pollination of that flower.

A whole additional spectrum of flowers with different characteristics is pollinated by flies of various kinds, including mosquitoes. These insects feed on nectar from the flower but do not gather pollen or store food for their larvae. Examples of flowers pollinated by mosquitoes and flies are shown in Figures 29–19b and c and 29–26.

(a)

(b)

29–26

Pollination by mosquitoes and other flies. (a) Some small-flowered orchids, such as Habenaria elegans, *in which the flowers are white or green and relatively inconspicuous, are visited and pollinated by mosquitoes in North Temperate and Arctic regions. The mosquitoes obtain nectar from the flowers. (b) A female mosquito of the genus* Aedes *with an orchid pollinium attached to its head. Other small-flowered orchids, such as those of the genus* Spiranthes, *are pollinated by bees. (c) A fly on a flower of a lily* (Zigadenus fremontii). *Notice the conspicuous yellow nectaries.*

(c)

29–27

Copper butterfly (Lycaena gorgon) *sucking nectar from the flowers of a daisy. The long sucking mouthparts of moths and butterflies are coiled up at rest and extended when feeding. They vary in length from species to species: only a few millimeters long in some of the smaller moths, they are 1–2 centimeters long in many butterflies, 2–8 centimeters in some hawkmoths of the North Temperate Zone, and as much as 25 centimeters in a few kinds of tropical hawkmoths.*

Flowers Pollinated by Moths and Butterflies

Flowers that have coevolved with butterflies and diurnal moths (those that are active during the day rather than at night) are similar in many respects to bee flowers, mainly because these insects are guided to flowers by a combination of sight and smell that is similar to the combination that guides bees to their flowers (Figure 29–27). Some species of butterflies, however, are able to perceive red as a distinct color, and some butterfly-pollinated flowers are red and orange.

Most moths are nocturnal, and the typical moth-pollinated flower—as seen, for example, in several species of tobacco (*Nicotiana*)—is white or pale in color and has a heavy fragrance, a sweet penetrating odor that often is emitted only after sunset. Well-known flowers that are pollinated by moths include the yellow-flowered species of evening primrose (*Oenothera*; see Figure 9–13b, page 145) and the pink-flowered amaryllis (*Amaryllis belladonna*).

29–28

Yucca moth (Tegeticula yucasella) *scraping pollen from a yucca flower. The female moth visits the creamy white flowers by night and gathers pollen, which it rolls into a tight little ball and carries in its specialized mouthparts to another flower. In the second flower, it pierces the ovary wall with its long ovipositor and lays a batch of eggs among the ovules. It then packs the sticky mass of pollen through the openings of the stigma. Moth larvae and seeds develop simultaneously, with the larvae feeding on the developing yucca seeds. When the larvae are fully developed, they gnaw their way through the ovary wall and lower themselves to the ground, where they pupate until the yuccas bloom again. It is estimated that only about 20 percent of the seeds are usually eaten.*

The nectary of a moth or butterfly flower is often located at the base of a long slender corolla tube or a spur, where it is usually accessible only to the long sucking mouthparts of moths and butterflies. Hawk-moths, for instance, do not usually enter flowers, as bees do, but hover above them, inserting their long mouthparts into the floral tube. Consequently, hawk-moth flowers do not have the landing platforms, traps, and elaborate internal structural modifications seen in some of the bee flowers. The more generalized moth–flower relationships, which typically involve smaller moths that do not require nearly as much energy as the hawkmoths, also involve shorter corolla tubes and often smaller flowers, over which the moths scramble.

One of the most specialized moth–flower relationships is shown in Figure 29–28.

Bird-Pollinated Flowers

Some birds regularly visit flowers to feed on nectar, floral parts, and flower-inhabiting insects; many of these birds also serve as pollinators. In North and South America, the chief pollinators among the birds are hummingbirds (Figure 29–29); in other parts of the world, flowers are visited regularly by representatives of other specialized bird families (Figure 29–30).

Although bird flowers have a copious thin nectar (some actually drip with nectar when the pollen is ma-

29–29

A male Anna's hummingbird (Calypte anna) *at a flower of the scarlet monkey-flower* (Mimulus cardinalis) *in southern California. Note the pollen on the bird's forehead, which is in contact with the stigma of the flower.*

29–30
A collared sunbird (Anthreptes collarii) *perching and feeding on a bird-of-paradise* (Strelitzia reginae) *flower in South Africa.*

(a)

ture), they usually have little odor, which is a corollary of the fact that the sense of smell is poorly developed in birds. However, birds do have a keen color sense that is much like our own; it is not surprising, therefore, that most bird flowers are colorful, with red and yellow ones being the most common (see Figure 29–14b). Bird-pollinated flowers include red columbine (Figure 29–31a), fuchsia, passion flower, eucalyptus, hibiscus, poinsettia (Figure 29–31b, c), and many members of the cactus, banana, and orchid families. Typically, these flowers are large or form parts of large inflorescences, features that can be correlated with their importance as visual stimuli and their ability to hold large amounts of nectar.

(b)

(c)

29–31
Examples of bird-pollinated flowers. (a) Columbine (Aquilegia canadensis). *Alternating perianth segments are modified into nectar-filled tubes. Hummingbirds visit these hanging flowers, taking nectar from them on the*

wing; the nectar of columbine flowers is inaccessible to most other kinds of animals. (b) and (c) Poinsettias (Euphorbia pulcherrima). *In this familiar plant, a native of Mexico, the flowers are small, greenish, and clustered, but each has a*

large, yellow nectary from which abundant nectar flows. Modified upper leaves, bright red in color, attract hummingbirds to the clusters of flowers.

Bird and other animal pollinators usually restrict their visits during a short period of time to the flowers of a particular plant species, but this is only one factor promoting *outcrossing* (cross-pollination between individuals of the same species). For outcrossing to result, it is also necessary that the pollinator not confine its visits to a single flower or to the flowers of a single plant. When flowers are visited regularly by large animals with a high rate of energy expenditure, such as birds, hawkmoths, or bats, they must produce large amounts of nectar to support the metabolic requirements of these animals and keep them coming back. On the other hand, if an abundant supply of this nectar is available to animals with a lower rate of energy expenditure, such as small bees or beetles, these will tend to remain at a single flower, and, being satisfied there, they will not move on to the flowers of other plants where they might bring about outcrossing. Consequently, flowers that are regularly pollinated by animals with a high rate of energy consumption, such as hummingbirds, have tended to evolve flowers with the nectar held in tubes or otherwise unavailable to smaller animals with lower rates of energy consumption. Similarly, the color red is a signal to birds but not to most insects, which do not perceive it as a distinct color. Birds, like ourselves, do not respond very strongly to odor clues. Thus, odorless, red flowers, being inconspicuous to insects, tend not to attract them, an adaptation that is advantageous in view of the copious production of nectar by such flowers.

Bat-Pollinated Flowers

Flower-visiting bats are found in tropical areas of both the Old World and the New World, and more than 250 species of bats—about a quarter of the total number of species in the group—include at least some nectar, fruit, or pollen in their diet. Those species of bats that derive all or most of their nourishment from flowers have slender and elongated muzzles and long extensible tongues, sometimes with a brushlike tip, and their front teeth are often reduced in size or are missing altogether.

Bat flowers are similar in many respects to bird flowers, being large, strong flowers that produce copious nectar (Figure 29–32). Because bats feed at night, bat flowers are typically dully colored, and many of them open only at night. Many bat-pollinated flowers are tubular or structurally modified to protect their nectar in other ways. Some bat-pollinated flowers and fruits in which the seeds are regularly dispersed by bats hang down on long stalks below the foliage, where the bats can fly more easily; others are borne on the trunks of trees. Bats are attracted to the flowers largely through their sense of smell, and bat-pollinated flowers characteristically have very strong fermenting or fruit-like odors. Bats fly from tree to tree, lapping up nectar,

eating pollen and other flower parts, and carrying pollen from flower to flower on their fur. At least 130 genera of angiosperms are pollinated by bats, or have their seeds dispersed by bats, or both.

It has recently been discovered that some bats derive a significant portion of their dietary protein from the pollen they consume. As yet another example of coevolution, the pollen of the flowers they visit has been found to contain significantly higher levels of protein than is found in insect-pollinated flowers.

Wind-Pollinated Flowers

At about the turn of the century, it was thought by botanists that wind-pollinated flowers are the most primitive of angiosperm flowers and that those of all other angiosperms had been derived from them. The conifers —thought by some scientists at the time to have been the direct ancestors of the angiosperms—have small, drab-colored, odorless, unisexual cones and are pollinated by the wind. Similarly, the many examples of

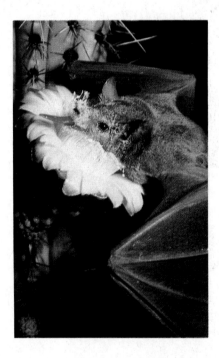

29–32

*By thrusting its face into the tubular corolla of a flower of an organ-pipe cactus (*Lemaireocereus*), this bat—a member of the genus* Leptonycteris—*is able to lap up nectar with its long, bristly tongue. Pollen grains clinging to the bat's face and neck are transferred to the next flower it visits. Bat-pollinated flowers have dingy colors and a musty scent (similar to the scent that bats produce to attract one another), and they open at night.*

wind-pollinated flowers have dull colors, are relatively odorless, and do not produce nectar; their petals are either small or absent; and the sexes are often separated. However, studies of other characteristics of these wind-pollinated angiosperms—in particular, their specialized wood—have convinced most botanists that all wind-pollinated angiosperms evolved not from the conifers but from insect-pollinated angiosperms.

According to current interpretations of the evidence, wind-pollinated angiosperms originated independently from several different ancestral stocks. They are best represented in temperate regions and are relatively rare in the tropics. In temperate climates, many individual trees of the same species are often found close together, and the dispersal of pollen by wind can occur readily in early spring, when the trees are leafless. In the tropics, on the other hand, many more kinds of trees are found in a given area, and the distance to the next nearest individual of a given species may be quite great. Furthermore, in many tropical communities, the trees are evergreen, and the dispersal of pollen by wind is therefore more difficult than it is in temperate deciduous forests when the trees are leafless. Under these circumstances, pollination by insects and other animals that have the ability to seek out other individuals of the same plant species, sometimes over relatively great dis-

tances (20 kilometers or more, in some cases), is much more efficient than wind pollination.

Because wind-pollinated angiosperms do not depend on insects to transport their pollen from place to place, they devote no energy to the production of nutritious rewards for insect visitors. Wind pollination is very inefficient, however, and it is successful only in situations where a large number of individuals grow fairly close to one another. Nearly all wind-borne pollen falls to the ground within a hundred meters of the parent plant. Thus, if the individual plants of a given species are widely scattered, the chance that a pollen grain will reach a receptive stigma is very slim. Many wind-pollinated plants are dioecious (having male and female flowers on separate plants), such as the willows; monoecious (having separate male and female flowers on the same plant), such as the oaks (see Figure 18–45, page 359); or genetically self-incompatible, such as many grasses. Even though their pollen moves about somewhat randomly, they have evolved devices that foster a high degree of outcrossing.

Wind-pollinated flowers usually have their stamens well exposed, where the pollen can easily be caught by the wind. In some, the anthers are suspended from long filaments hanging free from the flower (Figures 29–33 and 29–34). The abundant pollen grains, which

29–33
Unlike most angiosperms, the grasses have wind-pollinated flowers. Corn (Zea mays) has (a) staminate inflorescences (tassels) at the top of the stem and (b) ovulate inflorescences, with long protruding stigmas (the "silk" on the ears

of corn), lower on the stem. (c) Grasses characteristically have enlarged, feathery stigmas that efficiently catch the wind-blown pollen shed by the hanging anthers, as seen here in a grass of the genus Agropyron. (d) Scanning electron

micrograph of a pollen grain of corn (Zea mays), showing the smooth pollen wall found in most wind-pollinated plants and the single aperture characteristic of monocots.

(a)

(b)

(c)

29–34

Grass flowers (florets) usually develop in clusters. (a) As a cluster matures, a single pair of dry, chaffy bracts—the glumes—separate a little, exposing the elongating spikelet, with from one to many florets (depending on the species of grass) attached to a central axis, or rachilla. (b) Each floret is surrounded by two distinctive bracts of its own, the palea and the lemma. These are forced apart, exposing the inner parts of the flower (c), by the swelling of the lodicules—small, rounded bodies at the base of the carpel—and are spread wide when the grass is in flower. The stamens, usually three in number, have slender filaments and long anthers; and the stigmas are typically long and feathery and so are efficient at intercepting the wind-borne pollen.

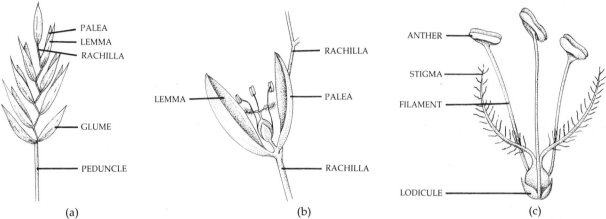

(a)   (b)   (c)

are generally smooth and small, do not adhere to one another as do the pollen grains of insect-pollinated species. The large stigmas are characteristically exposed, and they often have branches, or feathery outgrowths adapted for intercepting wind-borne pollen grains. Most wind-pollinated plants have ovaries with single ovules (and hence single-seeded fruits), because each pollination event consists of the meeting of one pollen grain with one stigma and leads to the fertilization of one ovule for each flower. Thus, each oak flower produces only a single acorn and each grass flower produces only one grain, but plants with very small flowers tend to have, in compensation, multiple inflorescences (Figures 29–33, 29–34, and 29–35).

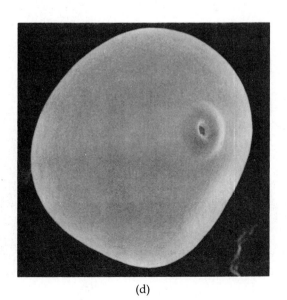

(d)

29–35

Most common species of trees in temperate regions are wind-pollinated. The staminate flowers of the paper birch (Betula papyrifera) hang down in catkins—flexible, thin tassels several centimeters long. These catkins are whipped by passing breezes, and the pollen, when mature, is scattered about by the wind.

29–36

Three anthocyanin pigments, the basic pigments on which flower colors in many angiosperms depend: pelargonidin (red), cyanidin (violet), and delphinidin (blue). Related compounds known as flavonols are yellow or ivory, and the carotenoids are red, orange, or yellow. Betacyanins (betalains) are red pigments that occur in one group of dicots. Mixtures of these different pigments, together with changes in cellular pH, produce the entire range of flower color in the angiosperms. Changes in flower color provide "signals" to pollinators, telling them which flowers have opened recently and are more likely to provide food.

PELARGONIDIN

CYANIDIN

DELPHINIDIN

29–37

Ipomopsis aggregata, scarlet gilia (Polemoniaceae), on Fern Mountain, near Flagstaff, in northern Arizona, flowers from mid-July to September. (a) In its early and middle period of flowering, individual plants and populations produce dark red flowers and are visited by hummingbirds. The population of hummingbirds in the area declines steadily through August, with none remaining by early September. (b) Late in the season, the plants produce much paler flowers, which are more attractive to hawkmoths, their principal pollinators at that time, than the dark red flowers produced earlier. These differences in flower color can almost certainly be attributed to differences in cellular pH rather than to the production of new kinds of anthocyanins.

Flower Colors

Surprisingly, all flower colors are produced by a very small number of pigments. Many red, orange, or yellow flowers owe their color to the presence of carotenoid pigments similar to those that occur in the leaves. The most important pigments in floral coloration, however, are *flavonoids*, which are compounds in which two six-carbon rings are linked by a three-carbon unit. Flavonoids probably occur in all angiosperms, but they are more sporadically distributed among the members of other groups of vascular plants, with a few reports even from algae and from animals. In the leaves, flavonoids block far-ultraviolet radiation (which is highly destructive to nucleic acids and proteins). They usually selectively admit light of blue-green and red wavelengths, which are important for photosynthesis.

One major class of flavonoids, the <u>anthocyanins</u> (Figure 29–36), consists of pigments that are important in determining flower color. Most red and blue plant pigments are anthocyanins, which are water-soluble and are found in vacuoles. By contrast, the carotenoids are fat-soluble and are found in plastids. The color of an anthocyanin pigment depends on the acidity of the cell sap of the vacuole; for example, cyanidin is red in acid solution, violet in neutral solution, and blue in alkaline solution. In some plants, the flower color changes after pollination, usually because the flowers are flushed with anthocyanin and become less conspicuous to insects. Recently, it has even been reported by Ken Paige and Thomas Whitman of Northern Arizona University that near Flagstaff, Arizona, individuals and populations of the perennial herb *Ipomopsis aggregata* produce darker-colored flowers early in the season, when they are pollinated primarily by hummingbirds, and lighter-colored ones later in the season, when hawkmoths are their primary pollinators (Figure 29–37). These extraordinary results presumably depend on differences in cellular pH, which the plant controls.

(a)

(b)

The *flavonols*, another group of flavonoids, are very commonly found in leaves and also in many flowers. Many of these compounds are colorless or nearly so, but they may contribute to the ivory or white hues of certain flowers.

For all flowering plants, different mixtures of flavonoids and carotenoids (as well as changes in cellular pH) and differences in the structural and thus the reflective properties of the flower parts produce the characteristic pigmentation. The bright fall colors of leaves come about when large quantities of colorless flavonols are converted into anthocyanins as the chlorophyll breaks down. In the all-yellow flower of the marsh marigold (*Caltha palustris*), the ultraviolet-reflective outer portion is colored by carotenoids, whereas the ultraviolet-absorbing inner portion is yellow to our eyes because of the presence of a yellow chalcone, one of the flavonoids. To a bee or other insect, the outer portion of the flower would appear to be a mixture of yellow and ultraviolet, a color called "bee's purple," whereas the nonreflective inner portion would appear pure yellow (see Figure 29–23). Most, but not all, ultraviolet reflectivity in flowers is related to the presence of carotenoids, and thus ultraviolet patterns are more common in yellow flowers than in those of any other group.

In the goosefoot, cactus, and portulaca families and other members of the order Chenopodiales (Centrospermae), the reddish pigments are not anthocyanins nor even flavonoids but a group of more complex aromatic compounds known as *betacyanins* (betalains). The red flowers of *Bougainvillea* and the red color of beets are due to the presence of betacyanins. No anthocyanins occur in these plants, and these three families have been shown by their biochemistry to be closely related.

THE DIVERSITY OF FRUITS

A fruit is a mature ovary, which may or may not incorporate some additional floral parts. A fruit in which such additional parts are retained is known as an *accessory fruit*. Although fruits usually have seeds within them, some—*parthenocarpic fruits*—may develop without seed formation. Bananas are a familiar example of this exceptional condition.

Fruits are generally classified as simple, multiple, or aggregate, depending on the arrangement of the carpels from which the fruit develops. *Simple fruits* develop from one carpel or from several united carpels. *Aggregate fruits*, such as those of magnolias, raspberries, and strawberries, consist of a number of separate carpels of one gynoecium. The individual parts of such aggregate fruits are known as *fruitlets*; they can be seen, for example, in the photograph of a magnolia fruit shown in Figure 29–12c. *Multiple fruits* consist of the gynoecia of more than one flower. The pineapple, for example, is a multiple fruit consisting of an inflorescence with many previously separate ovaries fused on the axis on which the flowers were borne (the other flower parts being squeezed between the expanding ovaries).

Simple fruits are by far the most diverse of the three groups. When ripe, they may be soft and fleshy, dry and woody, or papery. There are three main types of fleshy fruits—berries, drupes, and pomes. In *berries*—examples of which are tomatoes, dates, and grapes—there may be one to several carpels, each of which is typically many-seeded. The inner layer of the fruit wall is fleshy. In *drupes*, there may also be one to several carpels, but each carpel usually contains only a single seed. The inner layer of the fruit is stony and usually tightly adherent to the seed. Peaches, cherries, olives, and plums are familiar drupes. Coconuts are drupes whose outer layer is fibrous rather than fleshy, but in temperate regions we usually see only the coconut seed with the adherent stony inner layer of the fruit (Figure 29–38). *Pomes* are highly specialized fleshy fruits that are characteristic of one subfamily of the rose family. The pome is derived from a compound inferior ovary in which the fleshy portion comes largely from the enlarged base of the perianth. The endocarp of a pome resembles a tough membrane, as you know from eating apples and pears, the two most familiar examples of this kind of fruit.

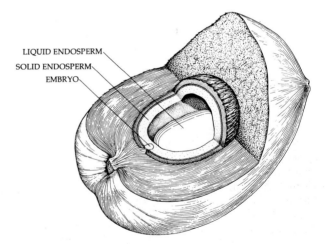

LIQUID ENDOSPERM
SOLID ENDOSPERM
EMBRYO

29–38
The fruit of the coconut (Cocos nucifera), a drupe. The coconut milk is liquid endosperm; it acquires cell walls by the time of germination. The entire fruit floats easily in the sea, and coconuts are dispersed widely by this means, reaching the most remote islands. When coconuts are transported commercially, their husks are usually removed first, so that in temperate regions people usually see only the stony inner shell of the fruit surrounding the seed.

(a)

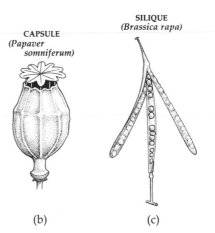

CAPSULE
(*Papaver somniferum*)

SILIQUE
(*Brassica rapa*)

(b)   (c)

29–39

Dehiscent fruits. (a) Bursting follicles of a milkweed (Asclepias). (b) In some members of the poppy family (Papaveraceae), such as poppies (genus Papaver), the capsule sheds its seeds through pores near the top of the fruit. (c) Plants of the mustard family (Brassicaceae) have a characteristic fruit known as a silique, in which the seeds arise from a central partition, and the two enclosing valves fall away at maturity.

Dry simple fruits are classified as either dehiscent (Figures 29–39 and 29–40) or indehiscent (Figure 29–41). In *dehiscent fruits*, the tissues of the mature ovary wall (the pericarp) break open, freeing the seeds. In *indehiscent fruits*, on the other hand, the seeds remain in the fruit after the fruit has been shed from the parent plant. Most fleshy fruits are indehiscent; they typically contain a single seed.

There are several kinds of dehiscent simple dry fruits. The *follicle* is derived from a single carpel that splits down one side at maturity, as in the columbines and milkweeds (Figure 29–39a). Follicles were also characteristic of the extinct, middle Cretaceous plant *Archaeanthus* (see boxed essay, page 592), and they are also found in magnolias (see Figure 29–12c). In the pea family (Fabaceae), the characteristic fruit is a *legume*.

29–40

The legume, a kind of fruit that is usually dehiscent, is the characteristic fruit of the pea family, Fabaceae (also called Leguminosae). With about 18,000 species, Fabaceae is one of the largest families of flowering plants. Many members of the family are capable of nitrogen fixation because of the presence of nodule-forming bacteria of the genus Rhizobium on their roots (see page 530). For this reason, these plants are often the first colonists on relatively infertile soils, as in the tropics, and they may grow rapidly there. The seeds of a number of plants of this family, such as peas, beans, and lentils, are important foods. (a) Legumes of the garden pea, Pisum sativum. (b) Legumes of Albizzia polyphylla, growing in Madagascar; each seed is in a separate compartment of the fruit. (c) Legume of Griffonia simplicifolia, a West African tree. The two valves of the legume are split apart, revealing the two seeds within.

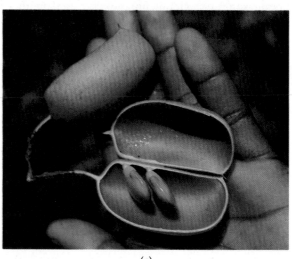

(a)   (b)   (c)

SAMARA
(Fraxinus)

(a)

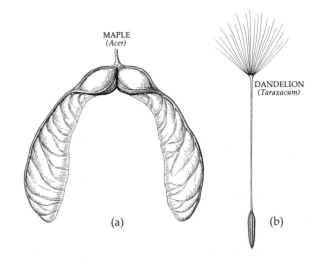

MAPLE
(Acer)

DANDELION
(Taraxacum)

(a)   (b)

29–42

Wind-dispersed fruits. (a) *In maples (Acer), each half of the schizocarp has a long wing.* (b) *The fruits of the dandelion (Taraxacum) and many other composites have a modified calyx, called the pappus, which is adherent to the mature cypsela and may form a plumelike structure that aids in wind dispersal.*

(b)

29–41

Indehiscent fruits. (a) *The samara, a winged fruit characteristic of ashes (Fraxinus) and elms (Ulmus), retains its single seed at maturity. Samaras are dispersed by wind.* (b) *Cypselas of burdock (Arctium), which attach themselves to passing animals and are dispersed in that way. Burdock is a member of the family Asteraceae.*

Legumes resemble follicles, but they split along both sides (Figure 29–40). In the mustard family (Brassicaceae), the fruit is called a *silique* and is formed of two fused carpels. At maturity, the two sides of the fruit split off, leaving the seeds attached to a persistent central portion (Figure 29–39c). The most common sort of dehiscent simple dry fruit is the *capsule*, which is formed from a compound ovary in plants with either a superior or an inferior ovary. Capsules shed their seeds in a variety of ways. In the poppy family (Papaveraceae), the seeds are often shed when the capsule splits longitudinally, but in some members of this family they are shed through holes near the top of the capsule (Figure 29–39b).

Indehiscent simple dry fruits are found in many different plant families (Figure 29–41). Most common is the *achene*, a small single-seeded fruit in which the seed

lies free in the cavity except for its attachment by the funiculus. Achenes are characteristic of the buttercup family (Ranunculaceae) and the buckwheat family (Polygonaceae). Winged achenes, such as those found in elms and ashes, are commonly known as *samaras* (Figure 29–41a). The achenelike fruit that occurs in grasses (members of the family Poaceae), is known as a *caryopsis*, or grain; in it, the seed coat is firmly united to the fruit wall. In the Asteraceae, the complex, achenelike fruit is derived from an inferior ovary; technically it is called a *cypsela* (Figure 29–41b; see also Figure 29–43). Acorns and hazelnuts are examples of *nuts*, which resemble achenes but have a stony fruit wall and are derived from a compound ovary. Finally, in the parsley family (Apiaceae) and the maples (Aceraceae), as well as a number of other unrelated groups, the fruit is a *schizocarp*, which splits at maturity into two or more one-seeded portions (Figure 29–42a).

Dispersal of Fruits and Seeds

Just as flowers may be grouped according to the pollinators that regularly visit them, fruits may be grouped according to their dispersal agents.

Wind-Borne Fruits and Seeds

Some plants have light fruits or seeds that are dispersed by the wind (Figures 29–39a, 29–41a, 29–42). The dustlike seeds of all members of the orchid family, for example, are wind-borne. Other fruits have wings, which are sometimes formed from perianth parts, that allow them to be blown from place to place. In maples, for example, which have a gynoecium composed of two fused carpels, each carpel of the resulting schizocarp develops a long wing (Figure 29–42a). The two carpels separate and fall when mature. Many members of the Asteraceae — dandelions, for example — develop a plumelike pappus, which aids in keeping the light fruits aloft (Figures 29–42b and 29–43). In some plants, the seed itself, rather than the fruit, carries the wing or plume; the familiar butter-and-eggs (*Linaria*) has a winged seed, and fireweed (*Epilobium*) and milkweed (*Asclepias*; see Figure 29–39a) have plumed seeds. In willows and the poplars (family Salicaceae), the seed coat is covered with woolly hairs. In tumbleweeds (*Salsola*), the whole plant (or a portion of it) is blown along by the wind, scattering seeds as it moves (Figure 29–44).

Other plants shoot their seeds aloft. In touch-me-nots (*Impatiens*), the valves of the capsules separate suddenly, throwing seeds for some distance. In the witch hazel (*Hammamelis*), the endocarp contracts as the fruits dries, discharging the seeds so forcefully that they sometimes travel as far as 15 meters from the plant. Another example of self-dispersal is shown in Figure 29–45. In contrast to these active methods of dispersal, the seeds or fruits of many plants simply drop to the ground and are dispersed more or less passively (or are dispersed by agents that operate only sporadically, such as floods).

Water-Borne Fruits and Seeds

The fruits and seeds of many plants, especially those growing in or near water sources, are adapted for floating, either because air is trapped in some part of the fruit or because the fruit contains tissue that includes large air spaces. Some fruits are especially adapted for dispersal by ocean currents; notable among these is the coconut (see Figure 29–38), which is why almost every newly formed Pacific atoll quickly acquires its own coconut tree. Rain is also a common means of fruit and seed dispersal; it is particularly important for plants that live on hillsides or mountain slopes.

Animal-Borne Fruits and Seeds

The evolution of sweet and often highly colored fleshy fruits was clearly involved in the coevolution of animals and flowering plants. The majority of fruits in which much of the pericarp is fleshy — cherries, raspberries, dogwoods, grapes — are eaten by vertebrates. When such fruits are eaten by birds or mammals, the seeds they contain are spread by being passed unharmed through the digestive tract, or by being regurgitated by the birds, at a distance from the place where they were ingested (Figure 29–46). Sometimes, partial digestion aids the germination of the seeds by weakening their seed coats.

When fleshy fruits ripen, they undergo a series of characteristic changes, mediated by the hormone ethyl-

29–43
The familiar small, indehiscent fruits of dandelions, which are technically known as cypselas (but often loosely called achenes), spread by their plumelike, modified calyx (the pappus). This photograph shows the fruiting heads of a plant of the genus Agoseris, which is closely related to the dandelions.

29–44
In tumbleweeds (Salsola), the plant breaks off and is blown across open country, scattering its seeds as it tumbles along. The tumbleweeds are natives of Eurasia, but they are widely naturalized as weeds in North America and elsewhere.

(a)

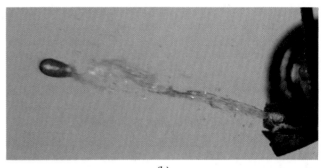

(b)

29–45

Dwarf mistletoe (Arceuthobium), a parasitic dicot that is the most serious cause for loss of forest productivity in the western United States. (a) A plant growing on a pine branch in California. (b) Seed discharge. Very high hydrostatic pressure builds up in the fruit and shoots the seeds as much as 15 meters laterally. The seeds have an initial velocity of about 100 kilometers per hour. This is one of the ways in which the seeds are spread from tree to tree, although they are also sticky and can be carried from one tree to another over much longer distances by adhering to the feet or feathers of birds.

ene, which was discussed in Chapter 24. Among these changes are a rise in sugar content, a softening of the fruit caused by the breakdown of pectic substances, and often a change in color from inconspicuous, leaf-like green to bright red (Figure 29–46a), yellow, blue, or black. The seeds of some plants, especially tropical ones, often have fleshy appendages, or arils, that have the bright colors characteristic of fleshy fruits and, like them, are aided in their dispersal by vertebrates. The arils of yews (*Taxus;* see Figure 18–26) are not homologous to the structures called arils that occur in angiosperms, because they are outgrowths of the seed, not of the fruit. Nevertheless, they play a similar role in seed dispersal.

(a)

(b)

(c)

29–46

The seeds of fleshy fruits are usually dispersed by vertebrates that eat the fruits and either regurgitate the seeds or pass them as part of their feces. Examples of vertebrate-dispersed fruits are shown here. (a) Strawberries (Fragaria), an example of an aggregate fruit. The achenes are borne on the surface of a fleshy

receptacle. Immature strawberries, like many bird- or mammal-dispersed fruits, are green, but they become red when the seeds are mature and thus ready for dispersal. (b) The berries of many cacti, such as this prickly pear (Opuntia), growing in southern Mexico, are conspicuous at maturity. (c) Berries of

chaparral honeysuckle, Lonicera hispidula. Such berries develop from the inferior ovaries of this species, and therefore have fused portions of the outer floral whorls incorporated in them. A flower of this species is shown in Figure 29–13c.

Unripe fruits are often green or colored in such a way that they are inconspicuous among the plant's green leaves, and thus they are somewhat concealed from birds, mammals, and insects. They may also be actually disagreeable to the taste — such as unripe cherries *(Prunus)*, which have a strong acidic taste — thereby discouraging animals from eating them before the seeds within them are ripe. The changes in color that accompany ripening are the plant's "signal" that the fruit is ready to be eaten — the seeds are ripe and ready for dispersal (Figure 29–46). It is no coincidence that red is such a prominent color among ripe fruits. Because it is red, the fruit is inconspicuous to insects, apparently blending with its background of green leaves. Insects are too small to disperse the large seeds of fleshy fruits effectively, and such concealment is therefore advantageous for the plants that bear red fruits. At the same time, the red fruits are very conspicuous to vertebrates, which then eat the fruits and may disperse their ripe seeds.

A number of angiosperms have fruits or seeds that are dispersed by adhering to fur or feathers (Figure 29–47; see also Figure 29–41b). These have hooks, barbs, spines, hairs, or sticky coverings that allow them to be transported, often for great distances, clinging to the bodies of animals.

29–47
The fruits of the African plant Harpago-phytum, *a member of the sesame family (Pedaliaceae) are equipped with "grappling hooks," by means of which they catch in the fur on the legs of large mammals and thus are spread from place to place.*

BIOCHEMICAL COEVOLUTION

Another important factor in the evolution of angiosperms has to do with the so-called secondary plant substances. Once thought of as waste products, these include an array of chemically unrelated compounds, such as alkaloids, quinones, essential oils (including terpenoids), glycosides (including cyanogenic substances and saponins), flavonoids, and even raphides (needlelike crystals of calcium oxalate). The presence of certain of these compounds can characterize whole families, or groups of families, of flowering plants (Figure 29–48).

In nature, these chemicals appear to play a major role in restricting the palatability of the plants in which they occur, or in causing animals to avoid them (Figure 29–49). When a given family of plants is characterized by a distinctive group of secondary plant substances, it is apt to be eaten only by insects belonging to certain families. The mustard family (Brassicaceae), for example, is characterized by the presence of mustard-oil glycosides and associated enzymes that break down these glycosides to release the pungent odors associated with cabbage, horseradish, and mustard. Plant-eating insects of most groups ignore plants of the mustard family and will not feed on them even if they are starving. However, certain groups of true bugs and beetles, and the larvae of some groups of moths, feed only on the leaves of plants of this family. The larvae of most of the members of the butterfly subfamily Pierinae (which includes the cabbage butterflies and orange-tips) also feed only on these plants. The same chemicals that act as deterrents to most groups of insect herbivores often act as feeding stimuli for these narrowly restricted feeders. For example, certain moth larvae that feed on cabbage will extrude their mouthparts and go through their characteristic feeding behavior when presented with agar or filter paper containing juices pressed from these plants.

It is clear that the ability to manufacture these chemicals and to retain them in their tissues is an important evolutionary step for the plants concerned and gives them biochemical protection from most herbivores. During the course of their evolution, the members of the family Brassicaceae were doubtless protected from most herbivores because they possessed mustard-oil glycosides. From the standpoint of herbivores in general, such protected plants, because they are not already heavily utilized by other plant feeders, represent an unexploited food source for any group of insects that can tolerate or break down the poisons manufactured by the plant. The main evolutionary development of the butterfly group Pierinae probably occurred after their ancestors had acquired the ability to feed on plants of the mustard family by breaking down these toxic molecules.

Herbivorous insects that are narrowly restricted in

29–48

Secondary plant substances: sinigrin, from black mustard, Brassica nigra; *calactin, a cardiac glycoside, from the milkweed* Asclepias curassavica; *nicotine, from tobacco,* Nicotiana tabacum, *a member of the nightshade family; caffeine, from* Coffea arabica *of the madder family; and theobromine, a prominent alkaloid in coffee, tea (*Thea sinensis), *and cocoa (*Theobroma cacao). *Nicotine, caffeine, and theobromine are alkaloids, members of a diverse class of nitrogen-containing ring compounds that are physiologically active in vertebrates.*

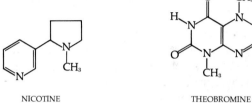

CALACTIN

SINIGRIN

CAFFEINE

NICOTINE

THEOBROMINE

their feeding habits to groups of plants with certain secondary plant substances are often brightly colored, which serves as a signal to their predators that they carry the noxious chemicals in their bodies and hence are protected. For example, the assemblage of insects found feeding on a milkweed plant on a summer day includes bright green chrysomelid beetles, bright red cerambycid beetles and true bugs, and orange and black monarch butterflies, among others. Milkweeds (Asclepiadaceae) are richly endowed with alkaloids and cardiac glycosides, heart poisons that have potent effects in vertebrates, the main potential predators of these insects. If a bird ingests a monarch butterfly, severe gastric distress and vomiting will follow, and orange-and-black patterns like that of the monarch will be avoided by the predator in the future. Other insects, such as the viceroy butterfly (which also has an orange-and-black pattern), have evolved similar coloration and markings. Thus, they escape predation by capitalizing on their resemblance to the poisonous monarch. This phenomenon, known as *mimicry*, is ultimately dependent upon the plant's chemical defenses. Various drugs and psychedelic chemicals, such as the active ingredients in marijuana (*Cannabis sativa*) and the opium poppy (*Papaver somniferum*), among others, are also secondary plant compounds that in nature play a role in discouraging the attacks of herbivores (Figure 29–50).

29–49

*Poison ivy (*Toxicodendron radicans) *produces a secondary plant substance, 3-pentadecanedienyl catechol, which causes an irritating rash on the skin of many people. The ability to produce this alcohol presumably evolved under the selective pressure exerted by herbivores. Fortunately, the plant is easily identifiable by its characteristic compound leaves with their three leaflets.*

(a)

(b)

(c)

(d)

29–50

Some plants that produce hallucinogenic and medicinal compounds. (a) Mescaline, from the peyote cactus (Lophophora williamsii), is used ceremonially by many Indian groups of northern Mexico and the southwestern United States. (b) Tetrahydrocannabinol (THC) is the most important active molecule in marijuana (Cannabis sativa). (c) *Quinine, a valuable drug used in the treatment and prevention of malaria, is derived from tropical trees and shrubs of the genus* Cinchona. (d) *Cocaine, a drug that has recently been abused to an unparalleled extent, is derived from coca (Erythroxylon coca), a cultivated plant of north-* *western South America. A Peruvian woman is shown here harvesting the leaves of cultivated coca. The secondary substances identified in these plants presumably protect them from the depredations of insects, but they are also physiologically active in vertebrates, including humans.*

Still more complex systems are known. When the leaves of potato or tomato plants are wounded, as by the Colorado potato beetle, the concentration of proteinase inhibitors, which interfere with the digestive enzymes in the gut of the beetle, rapidly increases in the tissues of the plant that have been exposed to air. Other plants manufacture molecules that resemble the hormones of insects or other predators and thus interfere with the predators' normal growth and development. One of the most useful of these is a complex molecule called diosgenin, obtained mainly from wild yams in Mexico, but obtainable in smaller amounts from yams native to India and China. Diosgenin is only two simple chemical steps away from the chemical known as 16-dehydropregnenolone (16D), the main active ingredient in many oral contraceptives, and wild yams have been a major source of 16D. Unfortunately, they grow very slowly, and the supply of wild yams may soon be exhausted. The Soviets and Ecuadorians, among others, have been studying the crop potential of certain species of *Solanum* (the potato and nightshade genus) that contain solasodine, a molecule that can be converted to 16D in two steps. Some of these plants are now being developed into commercial crops.

As mentioned previously, pollination systems have developed a particular coevolutionary pattern in which all the possible pollination types have evolved not once but usually several times in each group. The resulting array of forms gives the angiosperms an extremely wide variety. In the case of biochemical relationships, however, the evolutionary steps appear to have been large and definitive, and whole families of plants can be characterized biochemically and associated with major groups of plant-eating insects. These biochemical relationships may have played a key role in the early success of the angiosperms.

SUMMARY

The angiosperms, presently the dominant group of vascular plants, first appeared in the fossil record about 125 million years ago, in the early part of the Cretaceous period. The group became dominant worldwide some 80 to 90 million years ago, late in the Cretaceous period. Among the fossils formed at that time, many modern families of angiosperms and some modern genera can be recognized. The pollen of those primitive angiosperms that occurred more than 125 million years ago may have been indistinguishable from gymnosperm pollen or fern spores, so that it is difficult to be certain of the presence of angiosperms more than 125 million years ago, but the group is doubtless somewhat older.

Angiosperms may have evolved in the semiarid uplands and dry interior basins of West Gondwanaland, a supercontinent that included what are now South America and Africa. By the time these continents separated completely (about 90 million years ago), their climates had changed considerably and the angiosperms were approaching world dominance. Possible reasons for their success include various adaptations for drought resistance, as well as the evolution of efficient and often highly specialized pollination systems.

The flower is the most conspicuous feature of the angiosperms, and it has played a significant role in their evolution. The carpel is a leaflike structure that has undergone enfolding to enclose the ovules (which contain the megasporangia) and subsequent differentiation into a basal, swollen ovary, a stalklike style, and a stigma that is receptive to pollen. Similarly, the stamens have evolved from leaflike precursors or slender, branching systems with terminal sporangia to become specialized into the slender structures that are characteristic of most living angiosperms. Sepals are specialized leaves that protect the flower in bud, and petals in most angiosperms are sterilized stamens that have assumed a function in attracting insects. The petals of some angiosperms, however, were derived from sepals. The spiral arrangement of the flower parts of primitive angiosperms, coupled with their possession of numerous free parts, has given way to a whorled arrangement in most modern forms, with a definite number of parts that are often fused with one another or with the other whorls of the flower.

Examples of specialized families are Asteraceae (the composites), in which numerous highly specialized flowers are aggregated into a head, which functions as the attractive unit for insects; and Orchidaceae (the orchids), in which bizarre elaboration of the flower parts has resulted in a highly irregular flower with the most specialized pollination systems known.

Pollination by insects is basic in the angiosperms, and the first pollinating agents were probably beetles or similar insects. The closing of the carpel may have been a device to protect the ovules from being eaten by visiting insects. More specialized groups of insects evolved later in the history of the angiosperms, and wasps, flies, butterflies, and moths have each left their mark on the morphology of certain angiosperm flowers. The bees, however, are the most specialized and constant of flower-visiting insects and have probably had the greatest effect on the evolution of angiosperm flowers. Each group of flower-visiting animals is associated with a particular group of floral characteristics related to the visual and olfactory senses of the animals they attract. Some angiosperms have become wind-pollinated, shedding copious quantities of small, nonsticky pollen and having well-developed, often feathery stigmas that are efficient in collecting such pollen from the air.

Flowers that are regularly visited and pollinated by animals with high energy requirements, such as hummingbirds, must produce large amounts of nectar. They must then protect and conceal these sources of nectar from other potential visitors with lower energy requirements, which might satiate themselves with nectar from a single flower (or from the flowers of a single plant) and therefore fail to move on to another plant of the same species to effect cross-pollination. Wind pollination is inefficient, and the individual plants must grow close together in large groups for the system to work, whereas insects, birds, or bats can carry pollen great distances from flower to flower.

Fruits are as diverse as the flowers from which they are derived, and they can be classified either morphologically or anatomically or in terms of their methods of dispersal. They are basically mature ovaries, but, if additional floral parts are retained, they are said to be accessory fruits. Simple fruits are derived from one carpel or a group of united carpels, aggregate fruits from the free carpels of one flower, and multiple fruits from the fused carpels of several or many flowers. Dehiscent fruits split open to release the seeds, and indehiscent ones do not.

Wind-borne fruits or seeds are light and often have wings or tufts of trichomes that aid in their dispersal. The fruits of some plants expel their seeds explosively. Some seeds or fruits are borne away by water, in which case they must be buoyant and have water-resistant coats. Others are disseminated by animals, particularly by vertebrates, and have evolved fleshy coverings that are tasty and often conspicuous to attract feeding animals. Others adhere to the coats of mammals or the feathers of birds and are distributed in this manner.

A third aspect of the evolutionary success and diversification of the angiosperms has been biochemical coevolution. Certain groups of angiosperms have evolved various secondary plant substances, such as alkaloids, which protect them from most foraging herbivores. However, certain herbivores (normally those with narrow feeding habits) are able to feed on these plants and are regularly found associated with them. Potential competitors are excluded from the same plants by their toxicity. This pattern indicates that a stepwise pattern of coevolutionary interaction has occurred, and it appears likely that the early angiosperms also may have been protected by their ability to produce some chemicals that functioned as poisons for herbivores.

SUGGESTIONS FOR FURTHER READING

ALSTON, RALPH E., and BILLIE L. TURNER: *Biochemical Systematics,* Prentice-Hall, Inc., Englewood Cliffs, N.J., 1963.
Although increasingly out of date, this book still provides a useful outline of the classes of secondary plant substances.

BARTH, FRIEDRICH G.: *Insects and Flowers,* Princeton University Press, Princeton, N.J., 1985.
A well-written introduction to pollination biology, emphasizing many recent discoveries in the field.

BATRA, SUZANNE W. T.: "Solitary Bees," *Scientific American,* February 1984, pages 120–127.
The great majority of all bee species are solitary; as a group they are the most constant and numerous of pollinators.

BENTLEY, BARBARA, and THOMAS ELIAS (Eds.): *The Biology of Nectaries,* Columbia University Press, New York, 1983.
This volume presents a great deal of new and exciting information concerning the ways in which nectaries, both those located in flowers and those located elsewhere on the plants, function.

BLOCK, ERIC: "The Chemistry of Garlic and Onions," *Scientific American,* March 1985, pages 114–119.
Sulfur compounds account for both the odor and the medical properties attributed to these plants, and they also must play a role in deterring herbivores in nature.

CRAWFORD, DANIEL J., and DAVID E. GIANNASI: "Plant Chemosystematics," *BioScience* 32: 114–124, 1982.
This article provides an overview of the many applications of micromolecules and macromolecules to achieving an improved understanding of the relationships of plants.

DILCHER, DAVID, and WILLIAM CREPET (Eds.): "Historical Perspectives of Angiosperm Evolution," *Annals of the Missouri Botanical Garden* 71: 347–630, 1984.
A collection of fascinating articles about modern research into the early evolution of the flowering plants, including the description of Archaeanthus *and many other fossil flowers.*

FAEGRI, KNUT, and L. VAN DER PIJL: *The Principles of Pollination Ecology,* 3rd ed., Pergamon Press, Inc., Elmsford, N.Y., 1979.*
A rigorous and scholarly examination of the worldwide variety of pollination systems and their role in plant and animal ecology.

HEYWOOD, VERNON H. (Ed.): *Flowering Plants of the World,* Mayflower Books, Inc., New York, 1978.
The best available guide to the families of flowering plants for the student, a well-illustrated and accurate book that is a pleasure to use.

LEWIS, WALTER H., and P. F. ELVIN-LEWIS: *Medical Botany: Plants Affecting Man's Health,* John Wiley & Sons, Inc., New York, 1977.
A well-written review of plants that injure, heal, and nourish, or alter the conscious mind.

PIJL, L. VAN DER: *Principles of Dispersal in Higher Plants,* 2nd ed., Springer-Verlag, New York, 1972.
A brief and rather technical but informative book on all matters of seed and fruit dispersal.

PROCTOR, MICHAEL, and PETER YEO: *The Pollination of Flowers,* Taplinger Publishing Co., Inc., New York, 1973.
An outstanding and beautifully illustrated introduction to all aspects of pollination biology.

REAL, LESLIE (Ed.): *Pollination Biology,* Academic Press, New York, 1983.*
A comprehensive and fascinating account of modern developments in the rapidly expanding field of pollination biology.

TYRRELL, ESTHER QUESADA, and ROBERT A. TYRRELL: *Hummingbirds: Their Life and Behavior,* Crown Publishers, Inc., New York, 1985.
A perfectly magnificent book about these beautiful and highly specialized pollinators.

* Available in paperback.

C H A P T E R 3 0

Plants and People

30–1

Hunter-gatherers in the Congo Forest of Central Africa. The man has just caught the turtle, while the woman is carrying the family dog. People who must obtain their food in this way cannot build cities.

Our own species, *Homo sapiens*, has been in existence for at least 500,000 years, and we have been abundant for approximately 150,000 years. Like all other living organisms, we represent the product of at least 3.5 billion years of earlier evolution. Our immediate antecedents, members of the genus *Australopithecus*, first appeared no less than 5 million years ago. They apparently diverged within Africa from the evolutionary line that gave rise to the chimpanzees and gorillas, our closest living relatives, at about that time. The members of the genus *Australopithecus* were relatively small apes that often walked on the ground on two legs.

Larger, tool-using human beings—members of the genus *Homo*—first appeared about 2 million years ago. They almost certainly evolved from members of the genus *Australopithecus*, but they had much larger brains, apparently associated with their use of tools and reinforced by it. Early members of the genus *Homo* probably subsisted mainly by gathering food (picking fruits, seeds, and nuts, harvesting edible shoots and leaves, and digging roots), scavenging dead animals, and occasionally hunting. They learned how to use fire no less than 1.4 million years ago. Their style of life seems to have resembled that of some contemporary groups of humans (Figure 30–1). Our species, *Homo sapiens*, appeared in Africa by about 500,000 years ago and in Eurasia by about 250,000 years ago.

About 34,000 years ago, the powerfully built, short, stocky Neanderthal people, who were abundant earlier in Europe and western Asia, disappeared completely. They were replaced by human beings who were essentially just like us. From that time onward, our ancestors made increasingly complex tools of stone, and others of bone, ivory, and antler, materials that had not been used earlier. They were excellent hunters, preying on the herds of large animals with which they shared their environment. These people also began to create often magnificent ritual paintings on the walls of caves. The foundations of modern society had been laid.

THE AGRICULTURAL REVOLUTION

The Beginnings of Agriculture

The modern human beings who replaced the Neanderthals soon migrated over the entire surface of the globe. They colonized Siberia soon after they appeared in Europe and western Asia, and they reached North America by 12,000 to 13,000 years ago. Their migration eastward took place during one of the cold periods of the Pleistocene epoch, when savannas, with their large herds of grazing mammals, were widespread. As these people migrated, they seem to have been responsible for the extinction of many species of these animals. At any rate, extensive hunting by humans, together with major changes in climate, occurred at the same time that these animals were disappearing from many parts of the world.

About 18,000 years ago, the glaciers began to retreat, just as they had done 18 or 20 times before during the preceding two million years. Forests migrated northward across Eurasia and North America, while grasslands became less extensive and the large animals associated with them dwindled in number. Probably no more than 5 million humans existed throughout the world, and they gradually began to utilize new sources of food. Some of them lived along the seacoasts, where animals that could be used as sources of food were locally abundant; others, however, began to cultivate plants, thus gaining a new, relatively secure source of food.

The first deliberate planting of seeds was probably the logical consequence of a simple series of events. For example, the wild cereals (grain-producing members of the grass family, Poaceae) are weeds, ecologically speaking; that is, they grow readily on open or disturbed areas, patches of bare land where there are few other plants to compete with them. People who gathered these grains regularly might have spilled some of them accidentally near their campsites, or planted them deliberately, and thus created a more dependable source of food. When this sequence of events was initiated, cultivation had begun (Figure 30–2). In places where wild grains and legumes were abundant and readily gathered, humans would have remained for long periods of time, eventually learning how to increase their yields by saving and planting their seeds, by protecting their crops from mice, birds, and other pests, and by watering and fertilizing them.

Under the influence of cultivation, the characteristics of the domesticated crops would have changed gradually as people selected more seeds from plants with certain specific characteristics, because these characteristics made them easier to gather, store, or use. For example, the stalk (rachis) breaks readily in the wild wheats and their relatives, scattering the ripe seeds. In the cultivated species of wheat, the rachis is tough and

30–2
Harvesting (above) and winnowing (below) of wheat (Triticum) *in Tunisia, North Africa. Similar small-scale cultivation of wheat has been taking place around the Mediterranean basin for more than 10,000 years.*

holds the seeds until they can be harvested. Seeds held in this way would not be dispersed well in nature, but they can be gathered easily by humans for food and replanting. As a process of this sort is continued, a crop plant steadily becomes more and more dependent on the humans who cultivate it, just as the humans become more and more dependent on the plant.

Agriculture in the Old World

The domestication of plants and animals began about 11,000 years ago in the area known as the "Fertile Crescent" of the eastern Mediterranean, in lands that extend through parts of what are now Lebanon, Syria, Turkey, Iraq, and Iran. In this region, barley (*Hordeum vulgare*) and wheat (*Triticum*) were apparently the first plants to be brought into cultivation, with lentils (*Lens culinaris*) and peas (*Pisum sativum*) soon following (Figure 30–3). Additional plants that were domesticated early in this area were chickpeas, or garbanzos *(Cicer arietinum)*, vetch *(Vicia* spp.), olives *(Olea europaea)*, dates *(Phoenix dactylifera)*, pomegranates *(Punica granatum)*, and grapes *(Vitis vinifera)*. Wine made from the grapes and beer brewed from the grains were used from very early times. Flax *(Linum usitatissimum)* was cultivated very early, probably both as a source of food (the seeds are still eaten today in Ethiopia) and as a source of fiber for weaving cloth.

Among the early cultivated plants, the cereals provided a rich source of carbohydrates, while the legumes provided an abundant source of proteins. The seeds of legumes are among the richest in proteins of all plant parts, and their proteins, in turn, often are rich in particular amino acids that are poorly represented in the cereals. It is not surprising, therefore, that legumes have been cultivated, along with cereals, from the very beginnings of agriculture throughout the world. Of all the protein that is consumed by human beings worldwide, plants contribute about 70 percent, animals about 30 percent. Only 18 percent of the total plant protein comes from legumes; about 70 percent comes from cereals, even though they contain smaller amounts of protein. Despite this relationship, legume proteins are very important in human diets, and there seem to be great opportunities still for increasing the quality—in terms of their amino acid composition—of the legumes that we consume.

The cultivation of plants became more and more organized as time went by. For example, specialized implements associated with the harvesting and processing of grains, including flint sickle blades, grinding stones, and stone mortars and pestles, were in use more than 10,000 years ago. About 8000 years ago, humans began to make pottery vessels for storing grain. At the same time that plants were first being cultivated in the Near East, various animals—including dogs (which may have been the first animals kept regularly by humans), goats, sheep, cattle, and pigs—were also domesticated. Horses were domesticated later in southwestern Europe, cats in Egypt, and chickens in Southeast Asia, but all of these animals spread rapidly throughout the world.

Everywhere that they were kept, grazing animals ate the plants that were available to them, whether cultivated or not; produced such products as wool, hides, milk, cheese, and eggs; and could themselves be eaten by their owners. As human beings increased in number, they built up herds of grazing animals so large that they began to destroy their own pastures, causing widespread ecological destruction (Figure 30–4). Much of the Near East and other arid areas around the Mediterranean Sea are still badly overgrazed, and deserts continue to spread, as they have from the initial formation of large herds of domestic animals, up to the present.

30–3
Two of the first plants to be cultivated in the Near East were barley, Hordeum vulgare *(above), and peas,* Pisum sativum *(below).*

30–4
Herds of domesticated animals, like these karakul sheep in Afghanistan, devastated large areas of the eastern Mediterranean region as their numbers increased. In many areas, only spiny or poisonous plants survived, while previously fertile fields were converted to desert.

At any rate, humans now had steady sources of food, in the form of domesticated plants and animals, and they were able to organize villages, starting about 10,000 years ago, and eventually towns, about 4000 years later. Land that was productive, and on which humans could live permanently, could be owned, accumulated, and passed on to their descendants. In this way, the world became divided into semipermanent groups of haves and have-nots, just as it still is today.

The agriculture that had originated in the Near East spread northwestward, extending over much of Europe and reaching Britain by about 4000 B.C. At the same time, agriculture was developed independently in other parts of the world. There is evidence that agriculture may have been practiced in the subtropical Yellow River area of China nearly as long ago as it was in the Near East. Several genera of cereal grasses known as millets were cultivated for their grain, and eventually rice (*Oryza sativa*), now one of the most valuable cereals in the world, was added to them. Later, rice displaced the millets as a crop throughout much of their earlier range. Soybeans (*Glycine max*) have been cultivated in China for at least 3000 years (Figure 30–5).

In other parts of subtropical Asia, an agriculture based on rice and different legumes and root crops was developed. There is some archaeological evidence of the cultivation of rice in Thailand about 10,000 years ago, but much further research is needed before these dates can be established with certainty. The humid, rainy conditions of the tropics tend to destroy most of the evidence with which an archaeologist might trace such ancient events. Such animals as water buffaloes, camels, and chickens were domesticated in Asia long ago, and they became important elements in the systems of cultivation that were practiced there (Figure 30–6).

Eventually, such plants as the mango (*Mangifera indica*) and the various kinds of citrus (*Citrus* spp.) were brought into cultivation in tropical Asia, and others, such as rice and soybeans, began to be cultivated farther north. Taro (*Colocasia esculenta*) is a very important food plant in tropical Asia, where it is grown for its starchy tubers; *Xanthosoma* is a related food plant of the New World tropics. *Colocasia* and other similar genera, including *Xanthosoma*, are the source of poi, a staple starch food of the Pacific Islands, including Hawaii, where they were brought by Polynesian settlers about 1500 years ago.

Bananas (*Musa* × *paradisiaca*) are among the most important domesticated plants to come from tropical Asia; their fruits are a staple food throughout the tropical parts of the world. Starchy varieties of bananas, known as plantains, are much more important as a source of food in tropical countries, where two-thirds of the total banana crop is consumed, than are the sweet varieties, which are more familiar in temperate regions. Wild bananas have large, hard seeds, but the

30–5

Soybeans, Glycine max, *have been a major crop in the United States for only about 50 years. They are one of the richest sources of nutrients among food plants, the seeds consisting of 40 to 45 percent protein and 18 percent fats and oils. In the Orient, soybeans are used to make bean curd and soy sauce, among other products. Soybeans can be grown successfully only in temperate regions, and the United States produces more than half of the world's crop. Like many legumes, they harbor nitrogen-fixing bacteria in nodules on their roots, by means of which they are able to obtain nitrogen for their own growth and to enrich the soil. Soybeans are often rotated with corn in the United States, primarily to interrupt the life cycles of the important nematode and insect pests that attack these two major crops.*

(a)

(b)

30–6
Rice, Oryza sativa, *provides half of the food consumed by some 1.6 billion people and more than a quarter of the food consumed by another 400 million. It has been cultivated for at least 6000 years, and it is currently grown on 145 million hectares, about 11 percent of the arable* *land in the world. Rice is the source of several kinds of alcoholic beverages, including sake, a kind of wine that is a traditional drink in Japan. When it is grown wet, in paddy fields, fish are often grown in the flooded fields and harvested along with the rice. Since its establish-* *ment in 1962, the International Rice Research Institute, in the Philippines, has made important contributions to the improvement of this crop. Seen here are* (a) *rice terraces and* (b) *water buffalo being used to cultivate rice in Bali, Indonesia.*

cultivated varieties, like many kinds of cultivated citrus fruits, are seedless. The banana reached Africa about 2000 years ago, and it was brought to the New World soon after the voyages of Columbus.

There was also early domestication of food plants in Africa, but, again, we have little direct evidence of the timing of its origins. At any event, there is a gap of at least 5000 years between the origins of agriculture in the Fertile Crescent and its appearance at the southern tip of the African continent. Such grains as sorghum (*Sorghum* spp.) and various kinds of millet (*Pennisetum* spp. and *Panicum* spp.); a number of kinds of vegetables, including okra (*Hibiscus esculentus*); several kinds of root crops, but notably yams (*Dioscorea* spp.); and a species of cotton (*Gossypium*)—all were initially brought into cultivation in Africa. The various species of cotton are widespread as wild plants, mainly in seasonally arid but mild climates, and they are obviously useful; the long hairs on their seeds can easily be woven into cloth (Figure 30–7). Fragments of cotton cloth 4500 years old are known from India. Cotton seeds are also used as a source of oil, and the seed meal from which the oil has been expressed is used for feeding animals. Coffee (*Coffea arabica*) is another crop of African origin. It was brought into cultivation much later than the other plants mentioned here, but it is now a very important tropical crop.

30–7
Cotton, Gossypium, *one of the most useful of plants cultivated for fiber, seems to have been domesticated independently in Africa and/or India (the same species of cotton is cultivated in both areas), Mexico (another species), and western South America (yet a third species). Although cotton has been cultivated for thousands of years for cloth, it has also become an important source of edible oil during the past century. The cotton that is grown widely throughout the world today is a polyploid from the New World; the diploid species of the Old World are cultivated on a much more local basis.*

Agriculture in the New World

A parallel development of agriculture took place in North and South America. No domesticated plants appear to have been brought by human beings from the Old World to the New World prior to 1492. Dogs were certainly brought by people migrating to North America across the Bering Straits, but were the only kind of domestic animal that these humans brought with them. This attests to the great value of dogs for protection, hunting, herding, and, commonly, as a source of meat. It may also be related to the very early domestication of dogs, which were probably the first animals to be kept by humans on a more or less permanent basis.

The plants that were brought into cultivation in the New World were different from those that were first cultivated in the Old World. In place of wheat, barley, and rice, there was corn (*Zea mays*; Figure 30–8); and in place of lentils, peas, and chickpeas, the inhabitants of the New World cultivated kidney beans (*Phaseolus vulgaris*), lima beans (*Phaseolus lunatus*), and peanuts (*Arachis hypogaea*), among other legumes. Among the other important crop plants of Mexico were cotton (*Gossypium* spp.), red or chili peppers (*Capsicum* spp.), tomatoes (*Lycopersicon* spp.), tobacco (*Nicotiana tabacum*), cacao (*Theobroma cacao*, which yields cocoa, the major ingredient in chocolate), pineapple (*Ananas comosus*), pumpkins and squashes (*Cucurbita* spp.), and avocados (*Persea americana*). Cotton was domesticated independently in the New and in the Old World, with different species involved in the different centers of domestication. Its cultivation by humans in Mexico dates back at least 4000 years, and cotton has been cultivated even longer in Peru. The New World cottons are polyploid, and they are the source of nearly all the cotton that is cultivated throughout the world today; in contrast, the Old World cottons are diploid. After the diverse cultivated plants of the New World were discovered by Europeans following the voyages of Columbus, many of them were brought into cultivation in Europe and then spread from there all over the rest of the world. Not only were some of them totally new to Europeans, but others, such as the New World species of cotton, were better than the forms already in cultivation in Eurasia and soon replaced them.

The earliest evidence that we have of the existence of domesticated plants in Mexico dates from about 9000 years ago, but extensive agriculture appears not to have become well established until considerably later. The available evidence thus suggests that the domestication of plants began later in the New World than in Eurasia. Many of the crops that were originally domesticated in Mexico eventually became widespread from Central America and Mexico northward into Canada. Similar crops were also widely grown throughout the lowlands and middle elevations of South America. In fact, it is possible that agriculture was developed independently

30–8
Corn, or maize, Zea mays, *the most important crop plant in the United States. Originally used mainly to feed people, corn is now one of the world's chief foods for domestic animals. In the United States, some 80 percent of the crop is consumed by animals. At the time of Columbus, corn was cultivated from southern Canada to southern South America. Five main types are recognized: popcorn, flint corns, flour corns, dent corns, and sweet corn. Dent corns, in which there is a dent in each kernel, are primarily responsible for the productivity of the U.S. corn belt and are mainly used as animal food.*

in Mexico and in Peru, although the question cannot be settled with the evidence that is now available. The earliest evidence of agriculture in Peru is almost as ancient as that from Mexico, and some domestic plants, such as tomatoes and peanuts, may well have been brought from South America to Mexico by humans.

In the northern Andes of South America, a distinctive kind of agriculture was developed (Figure 30–9), based on such tuberous crops as potatoes (*Solanum tuberosum* and related species) and such seed crops as quinoa (*Chenopodium quinoa*) and lupines (*Lupinus* spp., of the family Fabaceae). Potatoes were cultivated throughout the highlands of South America at the time of Columbus, but they did not reach Central America or Mexico until they were brought there by the Spaniards. They have been one of the most important foods in Europe for two centuries, yielding more than twice as many calories per hectare as wheat.

(a)

(b)

(c)

(d)

30–9

A distinctive form of agriculture was developed at high elevations in the Andes of South America. (a) Cultivated fields in the mountains of northwestern Argentina. (b) A potato (Solanum tuberosum) field in Ecuador. Only a limited selection of the available genetic diversity of potatoes has been used for the improvement of the cultivated crop, one of the most important in the world. (c) Three of the four major root crops that are cultivated in the Andes, shown here for sale in the market at Tarma, Peru—potatoes (Solanum tuberosum), añu (Tropaeolum tuberosum), and ullucu (Ullucus tuberosus). Ullucu, which can be grown at higher elevations than potatoes, forms large, nutritious tubers; it might well become a useful crop elsewhere in the world. The fourth root crop that is common in the Andes is oca (Oxalis tuberosa), which is cultivated to a limited extent in New Zealand and elsewhere. The tubers of all four of these plants are naturally freeze-dried by Andean farmers, after which they can be easily stored for later consumption. (d) Quinoa (Chenopodium quinoa), a grain of the pigweed family (Chenopodiaceae), which is an important crop in the Andes—here shown in cultivation in northern Chile—and now being tested for more widespread cultivation.

THE ORIGIN OF CORN

Corn differs so greatly from its ancestors that for many years those ancestors could not be identified with certainty. We now know, however, that corn is the domesticated form of a wild Mexican grass, teosinte (*Zea mexicana*). Teosinte has narrow spikes with two rows of seeds enclosed in very hard seed coats. Its seeds are difficult to grind, but they are easy to use after they are heated and popped like popcorn. Populations of teosinte grow at scattered localities from northern Chihuahua, Mexico, to Honduras. Teosinte may be a weed of corn fields and field margins or a wild plant in winter-dry woodlands or steep slopes of middle elevations in Mexico. It is capable of forming fertile hybrids with corn, and it tends to do so wherever corn and teosinte occur together. Corn is known only as a cultivated plant.

The selection of the domesticated forms of corn began in Mexico more than 7000 years ago. The process primarily involved an increase in the number of rows of grains in the ear, much as the head of the modern sunflower differs from that of the wild sunflower in having many more flowers and, consequently, more seeds. Hugh Iltis of the University of Wisconsin has suggested that the actual "ear" of corn is homologous with the terminal portion of a lateral spike of teosinte, a structure that was originally entirely staminate (with male flowers only) but was converted to a pistillate spike (with female flowers only) by a mutation with drastic effects. The change was accompanied by the shortening and thickening of the whole inflorescence. It might have been related to the genetic-transposition phenomenon in corn, studies of which brought Barbara McClintock the Nobel Prize in Physiology or Medicine in 1983. The shallow cups in which the kernels are located in the ear of corn are very different from the deep ones in which the hard kernels of teosinte are located; and no wild teosinte has a pistillate spike in the central position in which it occurs in corn.

An exciting new development in the story of corn evolution has been the discovery of a new species of wild perennial corn, *Zea diploperennis*. It was found in 1978 in the mountains near Guadalajara, Mexico, by Rafael Guzmán, a student at the University of Guadalajara. Interfertile with annual corn, this newly discovered plant carries the genes for resistance to several of the major groups of viruses that infect corn in the United States, and for which no other source of resistance is known. By using it, plant breeders have also been able to develop a perennial corn, which might be useful in the often relatively infertile soils of subtropical regions, and it is already being tested in northern Argentina. *Zea diploperennis* occurs naturally only in a few small fields in the mountain range to which it is restricted, and it might easily have been destroyed by spreading agriculture there without ever having been known to science.

30–10
Sunflowers, Helianthus annuus, *an important crop because of the oil that is obtained from its seeds, were first domesticated at least 3000 years ago in what is now the central United States.*

Other crops were first brought into cultivation outside these main centers. For example, the sunflower (*Helianthus annuus*) was domesticated by American Indians living in what is now the United States (Figure 30–10). Another very important crop plant of the New World, manioc (also called cassava or yuca), was domesticated in the drier areas of South America but is now cultivated on a major scale throughout the tropics of the world (Figure 30–11).

Although the familiar white, or Irish, potato is a member of the nightshade family (Solanaceae), the sweet potato (*Ipomoea batatas*), as you might know if you had seen a field of it in flower, is a member of the morning-glory family (Convolvulaceae) and is therefore not closely related to the white potato. At the time of the voyages of Columbus, the sweet potato was widely cultivated in Central and South America, but it was also widespread on some Pacific islands, as far as New Zealand and Hawaii, to which it was apparently carried by human beings in the course of their early voyages. Subsequent to the time of Columbus, the sweet potato became a very important crop throughout most of Africa and tropical Asia.

Very few animals were domesticated in the New World. The so-called Muscovy duck (which, despite its name, does not come from Moscow), turkeys, guinea pigs, llamas, and alpacas are among the few to have originated there. Towns and eventually large cities were organized in the New World, just as they had

(a)

(b)

30–11

Manioc, Manihot esculenta, *one of the most important root crops throughout the tropics. Tapioca is produced by heating the starch extracted from the roots of this plant in warm water. Some of the* *cultivated strains, the bitter maniocs, contain poisonous cyanide compounds, which must be removed before the tuber is eaten. (a) Manioc cultivated on a* *forest clearing in southern Venezuela. (b) A Tirió woman in southern Surinam peeling a manioc root to prepare it for cooking.*

been earlier throughout Eurasia, wherever the agricultural systems were well enough developed to support them. Crops were extensively cultivated around such centers, but there were no large herds of domestic animals comparable to those that had become so prominent throughout Europe and Asia.

When Europeans colonized the Western Hemisphere, however, they brought their herds with them. In time, these herds brought about the same sort of widespread ecological devastation in parts of the New World that had taken place millennia earlier in the Near East and other parts of Eurasia. Many natural plant communities cannot readily be converted to pastures. For example, the widespread clearing of moist tropical forests to make pastures has been enormously destructive everywhere that it has been attempted. For the most part, pastures in such areas are productive only in the short term, until the nutrients in the soil have been exhausted, and they must often be abandoned after no more than ten or fifteen years.

Spices and Herbs

The molecules that are produced by plants primarily to defend themselves from insects and other herbivores were discussed in the preceding chapter. They contribute to the flavors, odors, and tastes of many plants and thus add features that have been utilized by human beings since prehistoric times. Some of these molecules make plants poisonous to humans and other animals, but others add characteristics that humans find desirable.

Spices—strongly flavored parts of plants, usually rich in essential oils—may be derived from the roots, bark, seeds, fruits, or buds of these plants. *Herbs,* on the other hand, are usually the leaves of nonwoody plants, although laurel or bay leaf and a few other condiments derived from trees or shrubs are considered herbs also. In practice, herbs and spices intergrade completely. Both have traditionally been used by humans to flavor food, especially if that food had become stale or somewhat spoiled.

Spices and herbs have been used widely in cooking for as long as we have records; the search for spices played a major role in the great Portuguese, Dutch, and English voyages that began in the thirteenth century and eventually led to the discovery of the entire world. The most important spices came from the tropics of Asia; they were responsible for the voyages and for a great deal of warfare as well. By the third century B.C., caravans of camels—often requiring two years for the trip—were carrying spices from tropical Asia back to the civilizations of the Mediterranean region. Included among these species were cinnamon (the bark of *Cinnamomum zeylanicum*), black pepper (the dried, ground fruits of *Piper nigrum;* Figure 30–12), cloves (the dried flower buds of *Eugenia aromatica*), cardamom (the seeds of *Elettaria cardamomum*), ginger (the rhizomes of *Zin-*

30–12
Black pepper (Piper nigrum) *has been known for thousands of years as an important spice.*

30–13
Nutmeg (Myristica fragrans) *is one of the most important traditional spices from tropical Asia. The spice nutmeg is derived from the ground seeds, whereas the spice mace comes from the fleshy seed coatings, seen here as bands of red tissue.*

giber officinale), and nutmeg and mace (the seeds and the dried outer seed coverings of *Myristica fragrans*; Figure 30–13). When the Romans learned that, by taking advantage of the seasonal shifts of the monsoon winds, they could reach India by sea from Aden, they shortened the trip to about a year—but it was still a highly dangerous and uncertain enterprise. A smaller number of additional spices, including vanilla (the dried, fermented seedpods of the orchid *Vanilla planifolia*; see Figure 29–17), red peppers *(Capsicum* spp.), and allspice (the berries of *Pimenta officinalis)*, came from the New World tropics after the voyages of Columbus.

In Europe and the Mediterranean region generally, there were many different kinds of native herbs, ones that were more familiar locally and, perhaps for that reason, not as highly prized as some of the spices that were available only from distant lands. Especially prominent among these herbs are members of the mint family (Menthaceae). These include thyme *(Thymus* spp.), mint *(Mentha* spp.), basil *(Ocimum vulgare)*, oregano *(Origanum vulgare)*, and sage *(Salvia* spp.). Also important were members of the parsley family (Apiaceae), including parsley *(Petroselinum crispum)*, dill *(Anethum graveolens)*, caraway *(Carum carvi)*, fennel *(Foeniculum vulgare)*, coriander *(Coriandrum sativum)*, and anise *(Anethum graveolens)*. Some members of this family (parsley, for example) are grown primarily for their leaves, and some (such as caraway) are grown for their seeds, but many (such as dill and coriander) are valued for both.

Tarragon *(Artemisia dracunculus)* is an herb that consists of the leaves of a plant belonging to the same

genus as wormwood and the sagebrush of the western United States and Canada. Mustard *(Brassica nigra)* seed, which can be ground into the spice we call mustard, likewise comes from a plant native to Eurasia. Bay leaves, traditionally taken from the tree *Laurus nobilis* of the Mediterranean region, but now also often taken from *Umbellularia californica* of California and Oregon, constitute another herb, one that is derived from temperate members of the largely tropical laurel family (Lauraceae). Saffron, popular in the Near East and adjacent regions, consists of the dried stigmas of *Crocus sativus*, a small, bulbous plant of the iris family (Iridaceae). The stigmas are gathered laboriously by hand, which accounts for the extremely high price of saffron and the fact that it is so prized—as much for its color as for its taste.

Coffee, *Coffea arabica* (Figure 30–14), and tea, *Camellia sinensis*, provide the two most important beverages in the world; both are consumed primarily because of the stimulating alkaloid, caffeine, that they contain. Coffee is made from seeds of the coffee plant that have been dried, roasted, and ground, whereas tea is prepared from the dried leafy shoots of the tea plant. Coffee, as mentioned earlier, was domesticated in the mountains of northeastern Africa, whereas tea was first cultivated in the mountains of subtropical Asia; but both are now widespread crops throughout the warm regions of the world. Coffee now provides the livelihood for some 25 million people around the world, and it provides a major source of income for the 50 tropical nations that export it. A third of the world's supply of coffee comes from Brazil.

30–14
Coffee, Coffea arabica, *is an important cash crop throughout the tropics of the world. It is a member of the madder family (Rubiaceae), as is cinchona (Cinchona),* which produces the medically important alkaloid quinine. The Rubiaceae is one of the larger families of flowering plants, with about 6000 species, mainly tropical.

Some tropical crops have also become widespread. For example, rubber (derived from several species of trees of the genus *Hevea,* of the family Euphorbiaceae) was brought into cultivation on a commercial scale about 150 years ago. The main area of rubber production is in tropical Asia. For rubber, as for many other crops, cultivation away from the areas where the plants are native seems to be favored; in such areas, they are often free of the pests and diseases that attack them in their native lands. Oil palms *(Elaeis guineensis)* are native to West Africa but are grown now in all tropical regions. Although they have been grown on a commercial scale for only about 75 years, oil palms are among the most important cash crops of the tropics today. Among the others are coffee and bananas, both widespread now. Cacao, which was first semidomesticated in tropical Mexico and Central America, is now most important as a crop in West Africa (Figure 30–15). Sugar cane *(Saccharum officinale)* was domesticated in New Guinea and adjacent regions, whereas sugar beets were developed from other cultivated members of their species in Europe. Yams *(Dioscorea* spp.) are an important tropical root crop. A number of species of yams are cultivated throughout the tropics, some of them from

Worldwide Agriculture

For the last 500 years, the important crops have been carried throughout the world and grown wherever they do best. The major grains—wheat, rice, and corn—are grown everywhere that the climate will permit. Plants unknown in Europe before the voyages of Columbus, including corn, tomatoes, and *Capsicum* peppers, are now cultivated throughout the world. Sunflowers, first domesticated in the area that is now the United States, now produce more than half of their total worldwide yield in the Soviet Union. Sunflowers are also displacing the traditional olives as a source of oil in many parts of Spain and elsewhere in the Mediterranean region. Throughout the world, the sunflower is being more and more extensively grown for its oil, and it is second only to the soybean among plants grown for this purpose.

30–15
Cacao, Theobroma cacao, *the source of chocolate and cocoa. The pods, shown here, contain several large seeds, or "beans." Cacao was first domesticated in Mexico, where chocolate was a prized drink among the Aztecs; at times, the beans were used as currency.*

West Africa, others from Southeast Asia, and a few—less important—are cultivated in Latin America. The better yams have now spread throughout the tropics, and they provide a staple food over wide areas. Manioc (*Manihot* spp.)—the source of tapioca—is another very important tropical root crop. Originally from seasonally dry tropical areas in South America, manioc is now grown abundantly in Africa and Asia (see Figure 30–11).

Another of the most important cultivated plants of the tropics is the coconut palm (*Cocos nucifera*), which seems to have originated in the western Pacific–tropical Asian region but was widespread in the western and central Pacific Ocean area before the European voyages of exploration; natural stands of coconuts are rare in the eastern Pacific, with a few known from Central America. The vast range of the coconut may be the result of natural dispersal of the fruits floating in the sea, rather than by human intervention. Each tree produces about 50 to 100 nuts each year, and they are a rich source of protein, oils, and carbohydrates. Coconut shells, leaves, husk fibers, and trunks are used to make many useful items, including clothing, buildings, and utensils.

Modern agriculture has become highly mechanized in temperate regions and also in some parts of the tropics. It has also become highly specialized, with just six kinds of plants—wheat, rice, corn, potatoes, sweet potatoes, and manioc—directly or indirectly (that is, after having been fed to animals) providing more than 80 percent of the total calories consumed by human beings. These plants are rich in carbohydrates, but they do not provide a balanced diet. They are usually eaten with legumes, such as common beans, peas, lentils, peanuts, or soybeans, which are rich in protein, and leafy vegetables, such as lettuce, cabbage, spinach (*Spinacia oleracea*), and chard, which are abundant sources of vitamins and minerals. Such plants as sunflowers and olives provide fats, which are also necessary in the human diet.

In addition to the six major food crops, there are eight others of considerable importance to human beings: sugar cane, sugar beets (Figure 30–16), common beans, soybeans, barley, sorghum, coconuts, and bananas. Taken together with the six mentioned earlier, these crops constitute the great majority of those widely cultivated as sources of food.

There are great regional differences in the human diet. For example, rice (see Figure 30–6) provides more than three-quarters of the diet in many parts of Asia, and wheat (Figure 30–17) is equally dominant in parts of North America and Europe. Corn can be grown successfully in many parts of the world only with supplementary irrigation, which is usually not necessary for wheat. Extending the productive ranges of these major grains and finding additional ones is a task of the greatest importance to the human species, as we shall see.

THE GROWTH OF HUMAN POPULATIONS

The roughly 5 million humans who lived 11,000 years ago were already the most widely distributed kind of large land mammal in the world. Subsequently, however, with the development of agriculture, these numbers have grown at an accelerating pace.

A striking characteristic of those groups of humans who make their living by hunting is that they strictly limit their own numbers. A woman on the move cannot carry more than one infant along with her household baggage, minimal though that baggage may be. When simple means of birth control—often simply abstention from sex—are not effective, such a woman may resort to abortion or, more commonly, infanticide. In addition, there is a high natural mortality, particularly among the very young, the old, the ill, the disabled,

30–16

Sugar beet, Beta vulgaris. *The sugar beet is simply a variety of the ordinary beet—selected from strains grown earlier for fodder, and not those cultivated as a root crop—in which the sucrose content has been increased by selection from about 2 percent to more than 20 percent. Beets were domesticated in Europe, where their leaves have long been used for food; Swiss chard is another variety of beet, one that is grown for its edible leaves. For about 300 years, beets have also been used as a source of sugar that has been competitive with sugar cane, which must be grown in the tropics and imported to the countries of the developed world. In 1984, U.S. production of raw sugar from beets amounted to more than 2.5 million metric tons, or about a third of the total sugar consumed.*

and women in childbirth. As a result of these factors, populations that are dependent on hunting tend to remain small. In addition, there is not much incentive for specialization of knowledge or skills; those basic skills on which individual survival depends are of the greatest importance.

Once most humans became sedentary, there was no longer the same urgent need to limit the number of births, and children may have become more of an asset to their families than previously, helping with agricultural and other chores. People could live much more densely on the land than ever before. For hunting and food-gathering economies, an area of 5 square kilometers, on the average, is required to provide enough food for one family to eat. In crop-based economies, only a small fraction of that amount of land is required. In the cities and towns that agricultural productivity made possible, human knowledge became increasingly spe-

cialized. Because the efforts of a few people could produce enough food for everyone, patterns of life became more and more diversified. People became merchants, artisans, bankers, scholars, poets, all the rich mixture of which a modern community is composed. The development of agriculture set human society on its modern course.

As a consequence of the development of agriculture, the numbers of human beings grew to about 130 million, distributed all over the earth, by the beginning of the Christian era. Over a period of about 8000 years, the human population had increased by about 25 times. By 1650, the world population had reached 500 million, with many people living in urban centers (Figure 30–18). The development of science and technology had begun, as had the process of industrialization, bringing about further profound changes in the lives of human beings and their relationships with the natural world. The human birth rate has remained essentially constant throughout the world from the seventeenth century onward, but the death rate has decreased dramatically on a regional basis, with the result that the population overall has grown at an unprecedented rate. In the twentieth century, the birth rate itself has dropped in developed countries.

30–17
Bread wheat, Triticum aestivum, *cultivated with modern techniques. First domesticated in the Near East, wheat has become the most widely grown crop in the world today. Along with barley, which is now largely used for animal food and as the source of malt for making beer, wheat was probably one of the first two plants to be cultivated. It is used more extensively than any other cereal for making bread because of the special properties of some of its proteins, which form a sticky substance called gluten that facilitates the handling of the dough and helps to hold the bread together.*

30–18
London was one of the thriving cities of seventeenth-century Europe, its influence reaching out to distant parts of the world as the global population increased, attaining a level of 1 billion people by 1850.

By 1986, there were about 5 billion humans on our planet; the most recent doubling of our population has occurred in just 36 years, since 1950. In 1986, about a quarter of us lived in the developed world (the United States, Canada, Europe, the Soviet Union, Japan, Australia, and New Zealand), some 22 percent in China, and the remainder in the less developed world, mostly consisting of countries that are subtropical or tropical, at least in part. The rate of growth of the world population is almost as incomprehensible as its enormous size. The birth rate in the United States has decreased drastically since 1972, and the U.S. population could stabilize during the next century, if present trends continue. Although similar patterns are found throughout much of the developed world, the picture elsewhere is unfortunately very different.

For the planet as a whole, the population is growing at about 1.7 percent per year. This means that about 160 humans are being added to the world population every minute—more than 230,000 each day, or 85 million every year. A high proportion—usually about 42 to 50 percent—of the humans living in the less developed countries are under 15 years of age; the comparable percentage for developed countries is about 22 percent. Such young people have not yet reached the age at which they usually bear children. Consequently, population growth in developing countries cannot soon be brought under control, even though government policy and individual choice often favor such a trend. Thus, it is predicted that there will be approximately 6.2 billion people on earth by the year 2000, and some 8 billion by the year 2020.

Even though the global population is projected to stabilize sometime in the twenty-first century at a level between 8 and 14 billion people, the next few decades are likely to be one of the most difficult periods ever faced by the human race (Figure 30–19). In 1983, the World Bank estimated that about a billion humans were living in absolute poverty, a condition that World Bank President Robert McNamara defined as ''so characterized by malnutrition, illiteracy, disease, squalid surroundings, high infant mortality, and low life expectancy as to be beneath any reasonable definition of human decency.'' At least half of these people are malnourished, and more than 10 million children in the tropics under the age of 5 years were starving to death each year in the mid-1980s, or dying of diseases directly linked with starvation.

As we shall see in more detail in the next chapter, we lack the agricultural technology to convert most of the lands in the tropics to sustainable productivity. Most of the world's lands that can be cultivated using available techniques are already in cultivation. Even though there is very little additional land that can be brought into cultivation, the rapidly growing majority of the world's population that lives in the tropics must somehow be fed. The job cannot be done by exporting surpluses from the more productive lands of the developed world. In 1983, just 8 percent of the food consumed in the developing world came from the developed world; yet that food constituted more than half of the total food that was exported from the developed countries that year. The solution to the problem of feeding the world's population must be found in the regions where most of the people live—the tropics and subtropics.

The Food and Agriculture Organization of the United Nations has estimated that an increase in world food production of 60 percent will be needed by the year 2000 if the world's population is to be fed adequately. Realistically, there appears to be little hope of attaining this goal, even though we must try to approach it as closely as possible. In some tropical areas, such as Africa south of the Sahara, food production per capita has actually been falling. Recently, even the *total* food production has declined in this vast area, whose population is well over 400 million and rapidly growing. For U.S. citizens, who spend on the average less than a fifth of their personal income on food, the increasing cost of food already occasions serious concern. For those in developing nations, who may spend 80 to 90 percent of their income on food, it can be a death sentence. Indeed, in such countries as Bangladesh and Haiti and in such regions as East Africa, humans are dying in increasing numbers because of the lack of food. How can the situation be improved?

30–19

These poor people, living on the outskirts of Tegucigalpa, Honduras, represent the condition of a majority of the world's population. Their prospects for the future depend directly on limiting population growth, incorporating the poor into the global economy, and finding new and improved methods for productive agriculture in the tropics and subtropics.

AGRICULTURE IN THE FUTURE

The Present Situation

The first significant advance that led to a massive increase in agricultural productivity was the development of irrigation (Figure 30–20). The necessity of providing water to crops has always been so evident that irrigation was practiced in the Near East as early as 7000 years ago, and it was developed independently in Mexico at least 3000 years ago. During the past two centuries, increasingly specialized and efficient machinery has been developed for agriculture, and great increases in productivity have been achieved. Fertilizers have been applied widely to the cultivation of crops, with their production drawing heavily on the use of fossil fuels. One of the great problems worldwide is how to utilize the gains in productivity that are made possible by increased mechanization, irrigation, and fertilization without simultaneously displacing millions of workers. Throughout much of the developing world, well over three-quarters of the people are directly engaged in food production, as compared with fewer than 3 percent in the United States in the late twentieth century.

Research has already done a great deal to make possible the improvement of agriculture. In the United States, the land-grant college system and the associated state agricultural experiment stations have made major contributions in this area. Nevertheless, many problems remain. The energy cost of producing crops in the United States and other developed countries is very high. Modern agriculture likewise depends on an elaborate distribution system, which is expensive in terms of energy and is easily disrupted. A significant proportion of each crop, the exact amount depending on the region and the year, is lost to insects and other pests. In many regions, an additional significant proportion is lost after harvest, either to spoilage or to insects, rats, mice, and other pests. Water is increasingly expensive in many areas, and the quality of local water supplies is often degraded by runoff from fields to which fertilizers and pesticides have been applied. Soil erosion is a problem everywhere, and it increases with the intensity of the agriculture (Figure 30–21). There are many significant efforts under way to achieve improvements in crop productivity, the protection of crops from pests, and the efficiency with which crops use water. Each of these will be discussed subsequently in this chapter.

30–21
The no-tillage cropping system, which combines ancient and modern agricultural practices, has been increasing rapidly in use. By the year 2000, as much as 65 percent of the acreage of crops grown in the United States may be grown under no-tillage conditions. Soil erosion is virtually eliminated in this system, as shown in the corn planted without tillage into a killed cover crop of clover (right), as compared with the evident erosion in corn planted conventionally (left). This photograph was taken soon after a spring rainstorm. When no-tillage methods are used, the energy input into corn and soybean production is reduced by 7 and 18 percent, respectively, and the crop yields are as high or higher than those obtained by conventional plowing and disking methods.

30–20
This irrigated cotton field in Texas is representative of modern, intensive agriculture. Irrigation may pose severe environmental problems over the long run, however, especially if it is coupled with the intensive use of pesticides and herbicides. More pesticides are used on cotton than on any other crop in the world.

Improving the Quality of Crops

The most promising approach to alleviating the world's food problem seems to lie in the further development of existing crops grown on land that is already in cultivation. Most land suitable for plant agriculture is already in cultivation, and increasing the supplies of water, fertilizers, and other chemicals for crops is not economically feasible in many parts of the world. Thus the improvement of existing crops is of exceptional importance. Such development entails not only improving the yield of these crops but also improving the quantity of proteins and other nutrients that they contain. The *quality* of the protein in food plants is also of the greatest importance to human nutrition: animals, including human beings, must be able to obtain from food the right amounts of all the essential amino acids—the ones that they cannot manufacture themselves. Eight of the 20 amino acids required by human beings must be obtained from food; the other 12 can be manufactured in the human body. Plants that have been selected for an improved protein content, however, inevitably have higher requirements for nitrogen and other nutrients than do their less modified ancestors. For this reason, improved crops cannot always be grown on the marginal lands where they would be especially useful.

The qualities of crops can also be improved in many respects other than yield and protein composition and quantity. Newly developed crop varieties may also be more resistant to disease, as when they contain secondary metabolites that their herbivores find distasteful; more interesting in form, shape, or color (such as apples of a brighter red color); more amenable to storage and to shipping (such as tomatoes that stack better in boxes); or improved in other features that are important for the crop in question.

For decades, the plant breeder, patiently sifting the available genetic diversity and producing hybrids of better promise, has produced thousands of improved strains of our major crops (Figure 30–22). Typically, thousands of hybrids are made and evaluated in order to find a few that represent genuine improvements on those already in wide cultivation. For example, the production of corn per hectare in the United States was increased about eightfold between the 1930s and the 1980s, even though very little of the genetic diversity of this remarkable plant was utilized in the process.

Hybrid Corn

The increase in the production of corn was made possible largely because of the introduction of hybrid corn seed. Inbred lines of corn (themselves originally of hybrid origin) are used as parents. When they are crossed, the result is seed that produces very vigorous hybrid plants. The strains to be crossed are grown in alternating rows, with the tassels (staminate inflorescences)

being removed from one of them by hand, so that all of the seeds set on those plants will be of hybrid origin. Through a careful selection of the best inbred lines, vigorous strains of hybrid corn that are suitable for cultivation at any given locality can be produced. Because of the uniform characteristics of the hybrid plants, they are easier to harvest, and they uniformly produce much higher yields than nonhybrid individuals. Less than 1 percent of the corn grown in the United States in 1935 was hybrid corn, but now virtually all of it is. Far less labor is now needed to produce much greater yields per hectare than was required earlier.

Progress at the International Plant-Breeding Centers

Extensive efforts have been made during the past few decades to increase the yields of wheat and other grains, particularly in warmer regions. These efforts have led to dramatic improvements in crop yields, largely at international crop-improvement centers located in subtropical areas. When the improved strains of wheat, corn, and rice produced at these centers were grown in such countries as Mexico, India, and Pakistan, they made possible the pattern of improved agricultural productivity that has been called the Green Revolution.

30–22

Norman Borlaug, who was awarded the Nobel Peace Prize in 1970. Borlaug was the leader of a research project sponsored by the Rockefeller Foundation under which new strains of wheat were developed at the International Center for the Improvement of Maize and Wheat (CIMMYT) in Mexico. Widely planted, these new strains changed the status of Mexico from that of a wheat importer when the program began in 1944 to that of an exporter by 1964.

The techniques of breeding, fertilizing, and irrigation that have been developed as a part of the Green Revolution have been applied in many countries of the developing world.

Any grain requires excellent conditions to produce high yields, and fertilization, mechanization, and irrigation have all been necessary components of the successes of the Green Revolution. Because of the distribution of credit, only relatively wealthy landowners have been able to cultivate the new crop strains, and in many areas the net effect has been to accelerate the consolidation of farmlands into a few large holdings by the very wealthy. Such consolidation has not necessarily provided either jobs or food for the majority of the people who live in the areas concerned.

Triticale

The traditional methods of the plant breeder can sometimes lead to surprising results. For example, hybrids of wheat (*Triticum*) with rye (*Secale*), which are called *triticale* (*Triticosecale* is the scientific name), have become crops of increasing importance in several areas, and they appear to hold substantial promise for the future (Figure 30–23). These triticale hybrids were produced by doubling the chromosome number in a sterile hybrid of wheat and rye (see page 580), a step that was taken in the mid-1950s by J. G. O'Mara of Iowa State University. O'Mara used the chemical colchicine, which arrests the formation of the cell plate and thus allows the chromosome number of treated plant cells to double.

Triticale combines the high yield of wheat with the rugged growth characteristics of rye. It is relatively resistant to wheat rust, an important fungus disease that is one of the major factors limiting productivity in wheat. Subsequent breeding and selection in triticale has produced improved strains for particular areas. In the mid-1980s, triticale was gaining rapidly in popularity in France, the largest grain producer in the European Economic Community, owing to its good yields, its resistance to climatic changes, and the excellent hay that is left behind when it is harvested. Most triticale is used for animal feed, but its use as human food is growing rapidly. In 1982, it was grown on more than 1 million hectares in the USSR, Europe, the United States, Canada, and South America.

Preserving and Using the Genetic Diversity of Crops

Intensive programs of breeding and selection have tended to narrow the genetic variability of crop plants with respect to all of their characteristics. Much artificial selection, for obvious reasons, has been concerned with yield, and sometimes disease resistance has been lost among the highly uniform members of a progeny that has been selected strongly for increased yield. Overall, individual crop plants have tended to become more and more uniform, as particular traits have been stressed more strongly than others, and these plants have therefore become more vulnerable to attack by diseases and pests. In 1970, for example, the fungus that causes southern leaf blight of corn, *Helminthosporium maydis*, destroyed approximately 15 percent of the U.S. corn crop—a loss of approximately 1 billion dollars (Figure 30–24). These losses were apparently re-

30–23
Triticale (Triticosecale), *a modern polyploid hybrid between wheat and rye, which combines the high-quality yield of wheat with the ruggedness of rye as a crop.*

30–24
Southern leaf blight of corn, a disease that is caused by the fungus Helminthosporium maydis.

lated to the appearance of a new race of the fungus that was highly destructive to some of the major strains of corn that were used extensively in the production of hybrid corn seed. The cytoplasm in many of the commercially important strains of corn was identical, having been introduced through the methods used to produce hybrid corn when identical pistillate parents were employed repeatedly.

In order to help guard against such losses, it is necessary to locate and to preserve distinct strains of our important crops, because these strains—even though their overall characteristics may not be attractive economically—may contain genes useful in the continuing fight against pests and diseases (Figure 30–25). Since the beginnings of agriculture, huge reserves of variability have accumulated in all crop plants by the processes of mutation, hybridization, artificial selection, and adaptation to a wide range of conditions. For such crops as wheat, potatoes, and corn, there are literally thousands of known strains. In addition, even more genetic variability exists among the wild relatives of the cultivated crops, often, however, in areas where it is being lost to advancing civilization. The problem lies in finding, preserving, and using the genetic variability of cultivated plants and their wild relatives, which is progressively being lost.

As an example of the role of genetic diversity in the history of, and the prospects for, a single crop plant, we shall consider the potato. There are more than 60 species of potatoes, most of them never cultivated, and thousands of different strains of cultivated potatoes are known (see Figure 30–9). Despite this, most of our cultivated potatoes are descended from a very few strains that were brought to Europe in the late sixteenth century. This genetic uniformity led directly to the Irish potato famine of 1846 and 1847, in which the potato crop was nearly wiped out by a blight caused by the mold *Phytophthora infestans* (Chapter 14). Within three years, the population of Ireland fell from 8.5 million to 6.5 million, with about one out of ten of the people who were there in early 1846 either starving to death or dying of diseases associated with starvation and an additional one out of five emigrating. The subsequent breeding of strains of potatoes resistant to the blight restored the status of the plant as a crop in Ireland and elsewhere. Looking to the future, the potential for developing potatoes as a crop still further through the use of additional cultivated and wild strains is enormous.

Striking examples of the potential of obtaining additional genetic material from the wild have been achieved by tomato breeders. The collection of strains of tomatoes, mostly carried out in recent years by Charles Rick and his associates at the University of California, Davis, has led to effective control of many of the important diseases of tomatoes, such as the rots caused by the imperfect fungi *Fusarium* and *Verticillium*, as well as several viral diseases. The nutritive

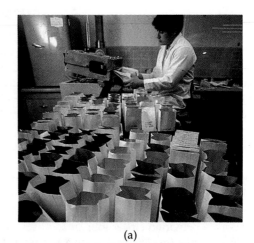

(a)

(b)

30–25
(a) *USDA seed bank at Fort Collins, Colorado, showing seeds sorted out to seal for long-term storage. Approximately 200,000 germ plasm lines are stored at this facility.* (b) *The seed potato farm at Three Lakes, Wisconsin, is the national repository for germ plasm of potatoes.*

value of tomatoes has been greatly enhanced, and their ability to tolerate saline or other basically unfavorable conditions has been increased greatly through the systematic collection, analysis, and use of strains of wild tomatoes in breeding programs.

Who Owns Crop Genetic Diversity?

A serious question has arisen concerning the ownership of the genetic resources of the world's food crops, nearly all of which are located in developing countries. Some of the nations of the developing world are concerned that their plant genes are being taken freely to industrialized countries, incorporated into food crops in breeding programs there, and then sold back to them at a profit. On the other hand, virtually everyone believes that access to the genetic diversity of plants ought to be relatively free. At a meeting of the Food and Agriculture Organization (FAO) of the United Nations, held in Rome in 1983, an intergovernmental committee was set up to oversee the conservation and use of plant genetic resources. It was decided at the same FAO meeting that individual countries should be allowed to withhold valuable strains of crops from free international exchange in order to benefit directly from their use and to offer them profit from their sale to other countries.

New Crops

In addition to the plant species that are already cultivated widely, there are many wild plants, or plants that are cultivated only locally, that could, if brought into cultivation extensively, make important contributions to the world economy. For example, we still derive more than 80 percent of our food, as mentioned earlier, from only six kinds of angiosperms out of the approximately 235,000 species that exist. Only about 3000 of them have ever been cultivated for food, and the great majority of those are not used at present, or are used only very locally. Only about 150 plant species have ever been cultivated widely.

Many plants other than this limited number, however, especially those that have been used before and have now been completely abandoned or are seen as having less importance, may prove to be very useful. Some of these even exist in cultivation in different areas of the world (Figure 30–26). Although we are accustomed to think of plants primarily as important sources of food, we tend to forget that they also produce oils, drugs, pesticides, perfumes, and many other products that are important to our modern industrial society. We have come to think that the production of such substances from plants is a rather archaic method that has now been fully supplanted by chemical synthesis in commercial laboratories. Their production by plants, however, requires no energy other than that of the sun: it occurs naturally. As our sources of nonrenewable energy near exhaustion, and as the price of such energy increases, it has become increasingly important to find ways to produce complex chemical molecules less expensively. Further, the vast majority of plants have never been examined or tested to determine their usefulness.

A few examples of crops that have recently been developed will demonstrate the great potential that exists in nature. Jojoba (pronounced "ho-*ho*-ba"), *Simmondsia chinensis*, despite its Latin name, is a shrub that is native to the deserts of northwestern Mexico and the adjacent United States (Figure 30–27). The large seeds of jojoba contain about 50 percent liquid wax, a substance that has impressive industrial potential. Wax of this sort is indispensable as a lubricant where extreme pressures are developed, as in the gears of heavy machinery and in automobile transmissions. It is difficult to produce

30–26
The winged bean, Psophocarpus tetragonolobus, *a legume, is a traditional crop in New Guinea, the Philippines, Indonesia, Thailand, Burma, Malaysia, and adjacent areas, but it is used primarily as a green vegetable in those regions. Recently introduced in many other areas throughout the world tropics, the winged bean shows great promise as a crop because of the high protein content of its seeds, which rivals that of soybeans. All parts of the plant, including the pods, seeds, leaves, and fleshy tubers, can be eaten.*

30–27
Jojoba, Simmondsia chinensis, *an important crop that is increasingly being planted throughout the arid areas of the world as a source of waxes with special lubricating characteristics and other useful properties.*

synthetic liquid wax commercially, and the endangered sperm whale is the only alternative natural source. Jojoba wax is likewise used in cosmetics and as a food additive, and other applications are still being discovered for this unusual substance. The plant thrives in hot deserts that are not suitable for the cultivation of most other crops, and some varieties are even highly salt-tolerant. Crops of jojoba are being produced in arid areas throughout the world, and they may make an important contribution to the economic development of these areas, especially by helping to provide employment for poor people in the cultivation of the crop.

Jojoba might also be able to help stop the spread of deserts, since it can grow in sandy soil when the rainfall is as little as 7.5 centimeters per year. In some areas, plantations of jojoba might make a useful contribution to stabilizing the soil. Such considerations are of prime importance, in view of the fact that the Sahara, for example, is spreading southward at the rate of about 5 kilometers per year and thus limiting the potential for food production of areas that are not now desert. At present, Saudi Arabia, Kuwait, Egypt, Morocco, Ecuador, and Nigeria are experimenting with jojoba for this purpose.

A second example of a possible interesting new crop is guayule, *Parthenium argentatum* (Figure 30–28). A member of the daisy, or sunflower, family (Asteraceae), guayule is closely related to the ragweed (*Ambrosia* spp.) that causes hay fever in millions of people during its period of flowering in August and September. Guayule is a low shrub native to northern Mexico and the southwestern United States. As much as 20 percent of the living plants may consist of rubber.

30–28
Guayule, Parthenium argentatum, *a desert shrub that produces natural rubber. First harvested from natural populations, guayule is now being cultivated widely.*

Synthetic rubber has replaced natural rubber for many purposes and now constitutes about two-thirds of the world supply. It is manufactured from petroleum and is therefore not renewable, as is rubber produced by plants. Almost all natural rubber at present comes from rubber trees, members of the genus *Hevea* of the spurge family (Euphorbiaceae). Although they are native to the Amazon Basin of South America, they are cultivated most successfully in tropical Asia. Guayule provides a promising alternative that can be cultivated in the desert and that has the potential of greatly increasing the world rubber supply. Wild plants of guayule have been used as a source of rubber for nearly a century, and cultivated fields in the United States produced more than 1.3 million kilograms of rubber in the United States during World War II. Although guayule received relatively little attention during the four decades following the end of the war, it is under active investigation again.

A third example of a crop that might well be grown more widely is provided by the grain amaranths (various species of *Amaranthus*), plants that have been grown for food for thousands of years in Latin America, but mainly on a relatively small scale. The grain amaranths have recently emerged as an excellent example of a potential new food crop. Weedy varieties of *Amaranthus* are called "pigweed" in English-speaking countries, but, in pre-Columbian times the seeds of these plants were one of the basic foods of the New World—almost as important as corn and beans. Some 20,000 metric tons of seeds were sent annually to Tenochtitlan (present-day Mexico City) in tribute to the chief ruler of the Aztecs from the different parts of his kingdom. The Spanish conquistadors forbade the Mexican people to use amaranth because of its association with pagan rituals and human sacrifice, and it survived as a crop only in very small, local areas. The protein content of amaranth seeds is as high as that of the cereals and, furthermore, amaranth protein contains large amounts of the amino acid lysine, making it nutritionally complementary to them. (Many cereals are low in lysine, which is an essential nutrient for human beings.) The leaves of some races of amaranth can also be used as nutritious vegetables. In view of all these virtues, it is not surprising that the cultivation of amaranth is spreading throughout the world.

One important area of investigation in the search for valuable new crops concerns those that are salt-tolerant. Increasingly, intensive agriculture is being practiced in the arid and semiarid areas of the world, primarily in response to the demands of the ever-increasing human population. Such practices have placed enormous demands on very limited local water supplies, which have become increasingly brackish as the water is used, reused, and polluted with fertilizer from adjacent cultivated areas. In addition, there are many parts of the world, especially near seashores, where the

(a)

(b)

30–29

The development of new crops, or new strains of presently cultivated crops, that can succeed at relatively high salt concentrations is important in many areas of the world, particularly those that are arid or semiarid. (a) Eggplants (Solanum melongena) cultivated in the Arava Valley of Israel using drip irrigation (their water reaches them through plastic tubes and is greatly con- *served in this way) with highly saline water (1800 parts per million). Ten years ago, this amount of salinity was thought to be incompatible with commercial crop production. (b) At an experimental site on the Mediterranean coast of Israel, a highly nutritious fodder shrub, Atriplex nummularia, is grown with 100 percent seawater. Yields are similar to those achieved with alfalfa, but the leaves and* *stems of the Atriplex are highly salty and thus not as valuable for fodder as alfalfa. Research being conducted by D. Pasternak and his co-workers at the Ben-Gurion University of the Negev aims at overcoming the problems of seawater irrigation, and it may one day make possible the cultivation of huge areas that are now coastal deserts.*

soil and the local water supplies are naturally saline. Unproductive when farmed using traditional agricultural methods, such areas might be brought under cultivation if the right kinds of plants could be found (Figure 30–29).

Finding the right kinds of plants for saline areas entails not only seeking out entirely new crop plants but also breeding salt-tolerance into traditional ones. In tomatoes, for example, a wild species that grows on the sea-cliffs of the Galápagos Islands and is therefore very salt-resistant, *Lycopersicon cheesmanii*, has been used as a source of salt-tolerance in new hybrids with the commonly cultivated tomato, *Lycopersicon esculentum*. The selection of these hybrids was carried out in a medium with half the salinity of seawater. Hybrid individuals have now been selected that will complete their development in water of that salinity. Salt-tolerant strains of barley have been developed using similar methods.

Drugs from Plants

In addition to their other uses, plants are also important sources of drugs. In fact, about a quarter of the prescriptions written in the United States contain at least one product that has been derived from some plant. For millennia, humans have been using plants for medici-

nal purposes. Botany, in fact, was traditionally regarded as a branch of medicine, and it has only been in the past 150 years or so that there have been professional botanists, as distinct from physicians. There has not, however, been any comprehensive effort to identify and bring into use previously unexploited secondary plant compounds of the sort that we discussed in the preceding chapter.

One of the reasons that plants will continue to be of importance as sources of drugs, despite the ease with which many drugs can be synthesized in the laboratory, is that plants manufacture the drugs inexpensively, without the provision of additional energy. Furthermore, the structure of some molecules—such as the steroids that make up the basic structure of cortisone and the hormones used in birth-control pills—is so complex that, although they can be synthesized, the products made from them are prohibitively expensive (Figure 30–30). For this reason, antifertility pills and cortisone were manufactured in the past principally from substances extracted from the roots of wild yams (*Dioscorea*), obtained largely from Mexico. When these sources were virtually exhausted, other plants, including *Solanum aviculare*, a member of the same genus as the nightshades and potatoes, were developed into crops for the same purpose.

CH₃
C=O

PROGESTERONE

30–30
The steroid progesterone is the precursor of both male and female sex hormones in humans. It is closely related in structure to cholestorol, another abundant human steroid, and to cortisol, which is marketed as cortisone and used to treat inflammation in humans. Other steroids are found in plants, and they are abundant in yams (Dioscorea), from which they are extracted and used as the starting material for the synthesis of the active ingredient in birth-control pills. Steroids are extremely active physiologically in vertebrates.

30–31
Periwinkle, Catharanthus roseus, *the natural source of the drugs vinblastine and vincristine. These drugs, which were developed in the 1960s, are highly effective against certain cancers. Vinblastine is typically used to treat Hodgkin's disease, a form of lymphoma, and vincristine is used in cases of acute leukemia. Before the development of vinblastine, a person with Hodgkin's disease had a one-in-five chance of survival; now, the odds have been increased to nine in ten. This periwinkle is widespread throughout the warmer regions of the earth, but it is native only to Madagascar, an island where only a small proportion of the natural vegetation remains undisturbed.*

In addition to considerations of cost, the marvelous variety of substances that plants produce is important: these provide a source of new products that is seemingly inexhaustible (Figure 30–31). One of the ways in which scientists have learned about potential new drugs is to study the medical uses to which plants are put by rural or indigenous people (Figure 30–32). For example, the contraceptive properties of Mexican yams were "discovered" in this way.

As we expand our search for useful plants, we should keep in mind the rapid loss of plant species that is related to (1) the rapid growth of the human population; (2) the poverty of many people, especially in the tropics, where about two-thirds of the world's species of plants occur; and (3) our incomplete understanding of how to construct productive agricultural systems in the tropics. With the complete destruction of undisturbed tropical forests, which seems almost certain to occur within a century, many kinds of plants, animals, and microorganisms will become extinct. Because our knowledge of plants, especially those of the tropics, is so rudimentary, we are faced with the prospect of losing many of them before we even learn of their existence, much less have the opportunity to examine them to see if they might be of some use to us. For this reason, the examination of wild plants for potential human use must be accelerated, and promising species must be preserved in seed banks, in cultivation, or, preferably, in natural reserves.

Genetic Engineering

One of the most important ways in which crops will be improved in the future is by the methods of genetic engineering. In recent years, molecular biologists have learned how to transfer foreign genes into plant cells. As we saw in Chapter 28, natural hybridization, which brings about the recombination of plant genetic material, plays an important role in the ongoing process of plant evolution. Similarly, the plant breeder has utilized the techniques of hybridization to recombine genetic material with the aim of developing crop plants with improved characteristics. What is new about the methods of genetic engineering is that they allow individual genes to be inserted into organisms in a way that is both precise and simple. Characteristics of interest can therefore be enhanced directly, with far less necessity for backcrossing and selection of the genetically engineered progeny than has been necessary in the past.

30–32
Mark Plotkin, a scientist from the World Wildlife Fund–U.S., collecting herbarium specimens of medicinal plants with the advice of a Wayana medicine man, southeastern Surinam. Although the study of the medicinal uses of forest plants has led to the discovery of important drugs such as D-tubocurarine chloride (used as a muscle relaxant in open-heart surgery) and ipecac (employed in the treatment of amoebic dysentery), the opportunities for gaining such knowledge are rapidly being lost as tribal cultures disappear and whole groups of people lose their traditional life-styles. Valuable knowledge that has accumulated through thousands of years of trial and error is being forgotten in a very short period of time. Much of this information has been transmitted orally, and written records of it do not exist.

Another important property of genetic engineering is that the species involved in the gene transfer do not have to be capable of hybridizing with one another. In normal plant breeding, they do; otherwise, the genes of one species cannot be combined with the genes of another. Genetic engineering, therefore, greatly increases the ability of the breeder to enhance the useful properties of plants or to create new ones. For example, attempts are being made to transfer nitrogen-fixing genes from bacteria into plants. This particular transfer has proved much more difficult than transferring the genes that are responsible for other growth characteristics, but developing nitrogen-fixing crops other than legumes would be one of the greatest conceivable accomplishments of agricultural research. In this connection, however, it should be noted that cereals, which do not fix nitrogen, have far higher yields than legumes; nitrogen-fixation has a definite cost in energy, which would

need careful evaluation if it proved possible to transfer nitrogen-fixing genes to plants in which they do not occur naturally. One would then need to make clear decisions about what sorts of productivity were most desirable.

Genetic engineering is based on the ability to cut DNA molecules precisely into specific pieces and to recombine those pieces to produce new combinations. The procedure depends on the existence of *restriction enzymes*, enzymes that break up DNA at sites where specific nucleotide sequences occur (Figure 30–33). These sequences are typically four to six nucleotides long, and they are always symmetrical. As a result, the DNA strands at the two broken ends of the fragment are complementary, and they can therefore pair with each other. Any two fragments produced by the same restriction enzyme can be joined with one another in this way, using a sealing enzyme called a *ligase*. This property makes possible virtually unlimited recombination of genetic material, since the original source of the DNA that is being recombined is almost completely irrelevant to whether the fragments will join with one another or not. Fragments of DNA derived from a giant redwood tree could be recombined with others from a bacterium that lives in the human intestine, and the resulting recombinant could then be introduced into the chromosomes of yet a third unrelated living organism, if such a movement of genes is desired. The expression of the recombined genes is not assured, however. Furthermore, changing some of the characteristics of a plant for the better may change other characteristics for the worse; for example, so much energy may be diverted to a new process, such as nitrogen fixation, that other important processes, such as sugar production, may be limited.

Since 1973, the technique of inserting fragments of foreign DNA into bacterial cells by using a virus or a plasmid to carry them into the new host cell has become routine. If human or other genes coding for proteins that are desired in quantity are inserted into bacteria, large quantities of those proteins (including many commercially important substances, such as insulin) can be produced relatively cheaply. The introns that are characteristic of eukaryotic cells need to be removed by molecular biological techniques before eukaryotic genes can function properly in bacteria, but that can be done readily in the laboratory.

Inserting foreign genes into plants has proved more difficult, however, since plants do not possess the many plasmids that are associated with viruses and that constitute a convenient vehicle for shifting blocks of genetic material. One way in which such transfers can be accomplished is by means of the bacterial plasmid *Ti*, which is associated with the crown-gall bacterium, *Agrobacterium tumefaciens*. When this bacterium infects plants, the *Ti* plasmid causes them to develop swollen portions called crown-gall tumors (Figure

PLANT GENETIC ENGINEERING

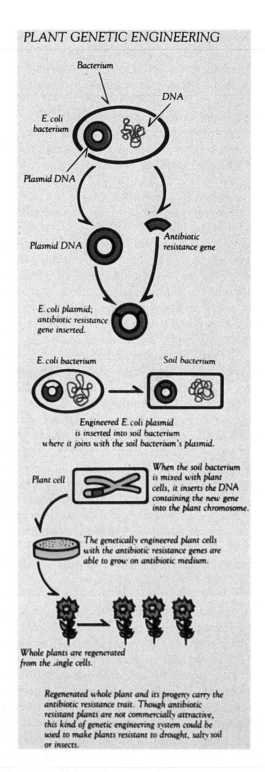

Bacterium

DNA

E. coli
bacterium

Plasmid DNA

Plasmid DNA

Antibiotic
resistance gene

E. coli plasmid;
antibiotic resistance
gene inserted.

E. coli bacterium   Soil bacterium

Engineered E. coli plasmid
is inserted into soil bacterium
where it joins with the soil bacterium's plasmid.

Plant cell

When the soil bacterium
is mixed with plant
cells, it inserts the DNA
containing the new gene
into the plant chromosome.

The genetically engineered plant cells
with the antibiotic resistance genes are
able to grow on antibiotic medium.

Whole plants are regenerated
from the single cells.

Regenerated whole plant and its progeny carry the
antibiotic resistance trait. Though antibiotic
resistant plants are not commercially attractive,
this kind of genetic engineering system could be
used to make plants resistant to drought, salty soil
or insects.

30–33
*This diagram illustrates how genetic
engineering is carried out in plants. In
the example illustrated, involving
plasmids from the bacterium* Escherichia
coli, *antibiotic-resistant plants are
produced. Although antibiotic-resistant
plants are not commercially attractive,
this kind of genetic-engineering system
could be used to make plants resistant to
drought, to salty soil, or even to insects.*

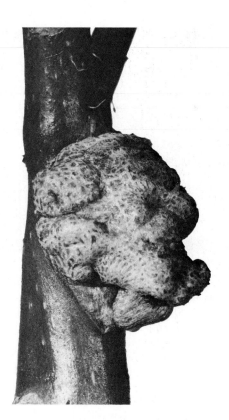

30–34
Crown gall on the stem of Begonia
semperforens. *This disease is caused by
an unusual form of parasitism of the
plant cells by the* Ti *plasmid, introduced
into the plants by the bacterium*
Agrobacterium tumefaciens. *The
uncontrolled cell divisions that lead to
the formation of the tumors continues
even after the bacteria themselves are
eliminated from the plant tissues.*

30–34). Each *Ti* plasmid is a closed circle of DNA that consists of about 100 genes. When these plasmids are incorporated into the plant's DNA, they first stimulate the plant to produce what appear to be greatly increased amounts of normal plant hormones, and these hormones are associated with the formation of the tumor. Other plasmid genes then cause the plant to synthesize unique amino acid derivatives called opines. The opines serve as nutrients for the bacteria, which carried the plasmids into the plant in the first place.

If genes associated with properties that we wish to introduce into a given plant are incorporated in the *Ti* plasmid, they can then be transferred to the plant following infection with the crown-gall bacterium. By using this method, yeast genes have been incorporated into plants, antibiotic resistance from petunias has been transferred into tobacco plants, and genes from beans have been transferred into sunflower tissues (Figure 30–35). It is not clear whether *Agrobacterium tumefa-*

30–35

This petunia plant (Petunia hybrida) *contains the gene for the human hormone known as human chorionic gonadotropin, and the plant produces the hormone in minute quantities. The gene was inserted by Monsanto scientists, using methods similar to those illustrated in Figure 30–33. It was linked with another gene, which controls resistance to the antibiotic kanamycin, so that the plants that had incorporated the foreign genetic material could be recognized easily in culture; only those cells containing both genes could survive in a medium containing kanamycin. Petunias are often used for such experiments because whole plants can readily be grown from single cells in the laboratory.*

ciens will infect monocots; evidence was presented in 1984 that it might do so without forming crown galls. Since many of our most important crops—the cereals, bananas, sugar cane, coconuts, and oil palms, for example—are monocots, the question is of great economic importance. Other ways of transferring genetic material dependably between plants, possibly including the use of other plasmids and viruses, must also be found, and that is the goal of numerous laboratory investigations now under way.

Attempts are currently being made to transfer genes to plants for the purpose of improving their resistance to disease, frost, and pests as well as improving their productivity. The plants that are produced by these methods may ultimately require smaller amounts of pesticides, fungicides, and fertilizers. It may prove possible to grow some of them productively in situations that are too salty, too wet, too dry, or otherwise unfavorable for existing crops (or for existing varieties of the same crop). A serious obstacle to those efforts lies in our imperfect understanding of the often complex genetic basis of many of the properties of plants that we might wish to enhance through genetic engineering, such as drought tolerance or salt tolerance. It is estimated that an average plant contains about 20,000 different genes, each of which may be present in as many as 5 million copies in a single cell: to determine which of these genes are responsible for agricultural productivity, and in what combinations, is an awesome task.

In addition to these problems, some genes, such as those responsible for nitrogen fixation in bacteria, do not seem to function properly when they are introduced into a new genetic environment. Experiments being carried out in a number of laboratories are starting to solve these problems, as we achieve a better understanding of the ways in which transferred genes work in new environments. The more we know about gene expression and gene regulation in plants, and the better we understand their biochemistry and physiology, the more able we will be to utilize the remarkable new techniques of genetic engineering in their improvement.

The development of new plant varieties by genetic engineering also entails some legal problems. Both established and newly formed companies are moving into the plant sciences in large numbers, stimulated by the potential of the methods of genetic engineering to increase crop yields. It is expensive to produce genetically engineered plant varieties, however, and it is not clear whether the companies that produce them will be able to patent them and thus realize any profits. There are also some who worry about the possible adverse effects of releasing genetically altered plants into the environment. Despite these obstacles, genetic engineering will obviously play a major role in the development of the improved crops of the future. It is expected that genetically engineered varieties of corn, tomato, and wheat, among other crops, will be cultivated fairly widely by the early 1990s.

A concrete example of the potential of genetic engineering concerns resistance to the herbicide atrazine, which is used widely to eradicate weeds in corn and other crops. Charles Arntzen, then of Michigan State University, demonstrated that atrazine kills most kinds of plants because it blocks electron transport in their chloroplasts. Corn, the crop with which atrazine is used most widely, is resistant to the herbicide because it contains enzymes that detoxify it. By 1970, however, farmers noticed that many common weeds had also become resistant to atrazine; such resistance is known to have evolved already in more than 25 species of weeds (Figure 30–36). The resistance of those weeds turns out to depend on a mutation in one of the chloroplast genes. As a result of this discovery, scientists at the University of Guelph, in Ontario, Canada, were able to produce atrazine-resistant oilseed rape *(Brassica rapus)*

30–36
Atrazine-resistant lamb's-quarters
(Chenopodium album) *growing luxuriantly in a cornfield after the application of 10 kilograms of atrazine per hectare. Corn is resistant to atrazine but not to many other herbicides. By producing crops that are resistant to specific herbicides, scientists are developing agricultural systems that promise high yields with minimal herbicide treatment. However, the acquisition of herbicide resistance by the weeds is a matter of concern in the development of such systems.*

through the introduction of chloroplasts from the related weed *Brassica campestris*. Since atrazine can be used to kill weeds in the resistant rape, this represents a valuable advance, and the new variety was grown on more than 400,000 hectares in 1985. Of course, if many of the weeds also become resistant to atrazine, then further selection will be needed to produce crops that can be treated effectively with specific mixtures of herbicides.

Ti plasmids cannot be used to transfer chloroplast genes, however, and thus far chloroplasts can be transferred easily only between plants that can be hybridized. Soybeans are not resistant to atrazine, and they are often grown alternately with corn in the same fields; their yield is depressed by any atrazine residues that remain in the soil. It would be valuable to be able to transfer atrazine resistance to soybeans, because the herbicide could be then used annually. Scientists are searching for spontaneous mutants for atrazine resistance among soybean progenies, and also for ways in which to increase the mutation rate. If they are successful in producing soybeans that are resistant to atrazine, farmers will need to alternate the atrazine with other herbicides, however, so as to lessen the chances for selection of additional resistant strains of weeds.

The Future: An Integrated Approach

The problem of how to ease the burdens of hunger and extreme poverty that are borne by at least a quarter of the world's people remains severe. The Green Revolution must of course go forward, but, at the same time, we must recognize that the broader solutions to these problems are social, political, and ethical. They involve not only the growth of food and its distribution but also the creation of jobs whereby the poor can earn the means to buy food. They involve not only limiting population growth but also raising the living standards of the poor to tolerable levels. No matter how striking the advances of agricultural science might be, they will never be adequate to eliminate hunger in a world in which the population is growing rapidly. In countries with large numbers of people living in poverty, there must be an organizational system that facilitates the introduction of new agricultural practices. There must be sources of fertilizers, pesticides, equipment, credit, and water that are available to all; the farmer must be able to sell his product; and he must be able to get it to market in order to do so. Genetic engineering, coupled with an increased knowledge of plant physiology, will certainly produce superior new crops, but will the billion poor farmers of the third world have access to them?

Plants also must be developed more fully as sources of nonfood products, such as medicines, other chemicals, and energy. Our major crop plants were mainly brought into widespread cultivation thousands of years ago, and many additional ones could be of great use if we were to identify them, work out their requirements for cultivation, and put them into production. However, many observers believe that, as vegetation is being destroyed rapidly throughout the world, 15 to 20 percent of the total number of plant species may become extinct during the next half century. Such a loss (of approximately 40,000 plant species) would tragically and unnecessarily limit our options, and it must be reduced as much as possible. Certainly, the prospect of a loss of this magnitude makes the search for new useful plants and their conservation seem especially urgent.

A broad knowledge of plant biology has become increasingly relevant to achieving solutions to some of society's most crucial problems. The stabilization of the human population must come first, but then our attention must turn both to the amelioration of poverty and malnutrition and to finding ways in which to provide food for the people of all countries. All the principles of plant growth and development must be brought into play in improving the practice of agriculture throughout the world, in making it possible for our planet to support unprecedented numbers of people in dignity and relative prosperity. Doing this will tax human imagination and ingenuity to the limit, but the task is of such great importance that we must make every effort to accomplish it.

THE CONSERVATION OF PLANTS

There are about 235,000 species of flowering plants in existence today. Approximately a third of these are native to temperate regions, and the balance are found in the tropics. As many as 40,000 species of tropical plants may be in danger of extinction in the wild within the next several decades because the human population of most tropical countries continues to double every 30 to 25 years and the forests are rapidly being cleared for agriculture. Approximately half of the world's tropical forests are already gone, and the remainder is disappearing rapidly.

So little is known of the plants of the tropics that many of them have not even been given scientific names. The samples of these plants that are preserved may well be all that are passed on to our descendants in the twenty-first century and beyond. The useful properties that these plants possess can certainly be determined better today, when most species are still in existence, than will be possible at any point in the future.

In temperate regions, about 5 percent of the native plant species are in current danger of extinction. Habitat destruction is only one of the causes. Overgrazing by domestic animals, the use of fertilizers and herbicides that drift into communities of native plants, the introduction of foreign plants without their natural controls, and the destruction of pollinators can all endanger plant species.

Of the approximately 20,000 kinds of plants native to the United States, at least 10 percent warrant concern, according to a study completed in 1978 by the Smithsonian Institution. About 90 species are thought to have become extinct during the past 200 years; about 850 are endangered (currently in danger of extinction throughout all or part of their range); and more than 1200 are threatened (likely to become endangered within the foreseeable future).

Only a few individuals of Hawaii's Clermontia pelleana, *with its spectacular dark wine-colored flowers, are known to survive, all on the island of Hawaii. It is pollinated by native birds, whose curved bills match the curve of the flowers. Approximately half of the roughly 950 species of native plants of the Hawaiian Islands are already extinct, endangered, or threatened. The future of plants that occur on islands all over the world is similarly doubtful; such plants have evolved in isolation, and they possess few natural defenses against alien plants or animals.*

*The Tiburon mariposa lily (*Calochortus tiburonensis*) is restricted to a single hilltop on the Tiburon Peninsula on the northern shore of San Francisco Bay. Discovered only in about 1970, this lily provides an excellent example of the extreme restriction of many plant species that occur in areas with a mediterranean, or summer-dry, climate.*

Extinct in the wild by 1806, Franklinia alatamaha, *a handsome small tree belonging to the camellia family (Theaceae), was discovered in 1765 near the Alatamaha River in Georgia. It was named in honor of Benjamin Franklin. Fortunately,* Franklinia *is widespread in cultivation and therefore is in no danger of extinction generally.*

SUMMARY

Human beings evolved in Africa. They have existed there for at least 5 million years, from which time we have our first records of the genus *Australopithecus*. Our genus, *Homo*, apparently evolved from *Australopithecus* about 2 million years ago, and our species, *Homo sapiens*, has existed for at least 500,000 years.

Starting about 11,000 years ago, in the Fertile Crescent—an area that extends from Lebanon and Syria through Iraq to Iran—humans began to cultivate such plants as barley, wheat, lentils, and peas. In cultivating and caring for these crops, the early farmers changed their characteristics, so that they were more nutritious, easier to harvest, and differed in other ways from their wild relatives. Agriculture spread from this center across Europe, reaching Britain by about 6000 years ago. It also seems to have spread southward through Africa, although it is possible that agriculture may have originated there independently in one or more centers. Many crops were brought into domestication in Africa for the first time, including yams, okra, coffee, and cotton, which was also domesticated independently in the New World and perhaps also in Asia. In Asia, agriculture based on such staples as rice and soybeans, and, farther south, citrus, mangos, taro, bananas, and other crops, was developed.

Domestic animals, starting with the dog, were an important feature of agriculture in the Old World from the earliest times. Herds of such animals as sheep, goats, cattle, horses, and hogs were ecologically destructive to many of the semi-arid areas of the Old World, especially as they increased in number, but they were also important sources of food. Other animals such as water buffaloes, camels, chickens, and elephants, were domesticated as agriculture spread. When the grazing animals that were so important in the Old World were introduced into Latin America following the voyages of Columbus, they proved to be enormously destructive in many habitats, including tropical forests.

Agriculture was developed independently in the New World. It began as much as 9000 years ago in Mexico and Peru. Dogs were brought to the New World by people migrating from Asia, but apparently no other domestic animals or plants were introduced in this way. Columbus and those who followed him found a virtual cornucopia of new crops to bring back to the Old World. Those crops included corn, kidney beans, lima beans, tomatoes, tobacco, chili peppers, potatoes, sweet potatoes, pumpkins and squashes, avocados, cacao, and the major cultivated cotton species.

Human populations have grown from an estimated 5 million people at the time when agriculture was first developed to about 5 billion people in the mid-1980s. The human population is growing very rapidly, and about 90 percent of the growth is taking place in the tropics, where the rural poor constitute about 40 percent of the population. As a result of this growth and of widespread poverty, and because relatively little has been done to develop agricultural practices suitable for tropical regions, the tropics are being devastated ecologically.

The world food supply can be improved by the traditional methods of plant breeding and selection; by the cultivation of new crops; and by the methods of genetic engineering. Among the more promising of recently developed new crops are such food plants as the winged bean (originally cultivated in New Guinea as a traditional crop and also known as a vegetable in the islands of tropical Asia but not cultivated there), and the grain amaranths. Also promising are such industrial crops as jojoba, valuable as a source of liquid wax for lubrication, and guayule, used as a source of rubber. Many drugs are also derived from plants, and there are doubtless numerous additional ones awaiting discovery. The destruction of habitats in the tropics, however, is threatening many potentially useful plants with extinction before they can be identified and exploited.

Genetic engineering utilizes restriction enzymes to break strands of DNA from one kind of organism into fragments that can be inserted (with the help of plasmids—circles of DNA) into the chromosomes of another organism. The *Ti* plasmid, which is associated with the bacterium of crown-gall disease in plants, is the only simple means presently available for carrying out such gene transfers in plants. It has been thought that the crown-gall bacterium attacks only dicots, although recent evidence suggests that it may infect monocots also, but without forming galls. Ways are being actively sought that would allow the transfer of genetic material to monocots, among which are many of our most important crops.

SUGGESTIONS FOR FURTHER READING

BALANDRIN, MANUEL F., et al.: "Natural Plant Chemicals: Sources of Industrial and Medicinal Materials," *Science* 228: 1154–1160, 1984.

An excellent review of the diversity of natural plant products and the promise they hold for industrial development.

BATIE, S. S., and R. H. HEALY: "The Future of American Agriculture," *Scientific American*, February 1983, pages 45–52.

Explores the factors that limit productivity in American agriculture as well as its ability to supply foreign needs.

BOARD OF AGRICULTURE, NATIONAL RESEARCH COUNCIL: *Genetic Engineering of Plants: Agricultural Research Opportunities and Policy Concerns*, National Academy Press, Washington, D.C., 1984.*

A survey of the field by a distinguished panel, indicating a number of important directions for the future.

BOARD ON SCIENCE AND TECHNOLOGY FOR INTERNATIONAL DEVELOPMENT, NATIONAL RESEARCH COUNCIL: *Underexploited Tropical Plants with Promising Economic Value*, National Academy Press, Washington, D.C., 1975.*

An interesting review of some of the kinds of useful plants that might be developed into important crops.

GRIERSON, DONALD, and SIMON COVEY: *Plant Molecular Biology*, Blackie/Chapman and Hall, London, 1985.

An excellent review of the state of knowledge about this rapidly expanding field in the first half of the 1980s.

HAWKES, JOHN G.: *The Diversity of Crop Plants*, Harvard University Press, Cambridge, Mass., 1983.

In this interesting short book, Professor Hawkes, one of the out-
* Available in paperback.

standing students of crop plants, outlines the evolution of crop plants and the ways in which their diversity can be maintained.

HEISER, CHARLES B., JR.: *The Sunflower,* University of Oklahoma Press, Norman, Okla., 1976.
A charming book about an important crop that was domesticated within the present boundaries of the United States.

HEISER, CHARLES B., JR.: *Seed to Civilization,* 2nd ed., W. H. Freeman & Co., New York, 1981.
An outstanding text that covers the origins of agriculture and the major crop plants in some depth.

HUXLEY, ANTHONY: *Green Inheritance,* Anchor Press (Doubleday), New York, 1985.
This book describes our heritage of wild and cultivated plants and landscapes, together with their potential for human beings; beautifully illustrated.

MYERS, NORMAN: *A Wealth of Wild Species: Storehouse for Human Welfare,* Westview Press, Boulder, Colo., 1983.
A thoughtful account of the ways in which wild plants, animals, and microorganisms have been, and continue to be, of use to humans.

OLDFIELD, MARGERY L.: *The Values of Conserving Genetic Resources,* U.S. Department of the Interior, National Park Service, Washington, D.C., 1984.*
An outstanding discussion of the value of genetic diversity and the ways in which we utilize it, can conserve it, and may benefit from it further.

PHILLIPS, RONALD E., et al.: "No-Tillage Agriculture," *Science* 208:1108–1113, 1980.
This article describes the advantages of a system of agriculture that is spreading rapidly in temperate regions.

RINDOS, DAVID: *The Origins of Agriculture: An Evolutionary Perspective,* Academic Press, Inc., New York, 1984.
Excellent, thoughtful essay on the first stages in the development of agriculture.

ROSENGARTEN, F., JR.: *The Book of Edible Nuts,* Walker and Company, New York, 1984.
The natural history, cultivation, and use of edible nuts. A book to be read for pleasure.

SCHERY, ROBERT W.: *Plants for Man,* 2nd ed., Prentice-Hall, Inc., Englewood Cliffs, New Jersey, 1972.
A balanced and comprehensive account of the many ways in which human beings use plants.

Science As a Way of Knowing—Human Ecology, American Society of Zoologists, Thousand Oaks, Calif., 1984.*
An outstanding series of review articles concerning population, food, health, and other important aspects of human ecology.

WETTSTEIN, DIETER VON: "Genetic Engineering in the Adaptation of Plants to Evolving Human Needs," *Experientia* 39: 687–713, 1983.
A useful review of genetic engineering that presents a balanced view of its potential for future contributions to plant improvement.

* Available in paperback.

S E C T I O N   N I N E

Ecology

CHAPTER 31

The Dynamics of Communities and Ecosystems

Ecology is the study of the interactions of organisms with one another and with their environment. As a science, it attempts to explain why particular plants and animals can be found living in one area and not in others; why there are so many organisms of one sort and so few of another; what changes one might expect the interactions among them to produce in a particular area; and how ecosystems function, with particular reference to the production and use of organic compounds and the cycling of elements.

In the following discussions, a few definitions will be helpful. A *community* consists of all the plants, animals, and other organisms that live in a particular area. An *ecosystem* includes not only the aggregation of living organisms (biotic factors) but also the nonliving (physical) elements of the environment with which they are interacting. *Biomes* are large terrestrial complexes of communities of living organisms that are characterized by distinctive vegetation and climate, such as deserts or grasslands. The principal biomes are described in Chapter 32.

INTERACTIONS BETWEEN ORGANISMS

No organism living in a community—whether it be a patch of woodland, a pasture, a pond, or a coral reef—exists in isolation. Each organism participates in a number of interactions, both with other organisms and with factors in the nonliving environment. In this chapter, three major kinds of interactions between organisms are discussed: mutualism, competition, and plant–herbivore interactions.

Mutualism

Mutualism is a biological interaction in which the growth and survival of both interacting species are en-

31–1
A forest of redwoods (Sequoia sempervirens) *along the coast of northern California.*

hanced. It is therefore one of the forms of symbiosis. In nature, neither species can survive without the other. Lichens, an outstanding example, were discussed in Chapter 13. Another is the relationship between legumes and the nitrogen-fixing bacteria that live in nodules on their roots (see Chapter 26, pages 530–534). Also, some of the closely linked pollination relationships discussed in Chapter 29, such as that between the yucca moth and the *Yucca* plant, represent mutualism.

One of the most interesting and ecologically significant examples of mutualism concerns the interaction between fungi and vascular plants. As discussed in Chapters 13 and 26, the roots of most vascular plants are associated with fungi, forming compound structures known as mycorrhizae. Without such mycorrhizae, the normal growth of the plants would be impossible. Mycorrhizal associations appear to have played a crucial role in the first invasion of the land by plants.

As more is learned about mycorrhizal relationships, their importance to the vascular plants becomes more evident. In many vascular plants, nonmycorrhizal individuals are rarely encountered under natural conditions, even though growth may be possible without fungi if other conditions are narrowly regulated. Most vascular plants are dual organisms in the same sense that lichens are dual organisms, although the relationship may not be obvious above ground. As University of Wisconsin soil scientist S. A. Wilde has stated, "A tree removed from the soil is only a part of the whole plant, a part surgically separated from its . . . absorptive and digestive organ." For most plants, the mycorrhizal fungi play a vital role in the absorption of phosphorus and other essential nutrients.

The fungi that form mycorrhizal associations in most plants are zygomycetes; as discussed in detail in Chapter 13, the associations formed are called endomycorrhizae, and they are characteristic of a majority of all herbs, shrubs, and trees. In some groups of conifers and dicotyledons—mainly trees—the associations are mostly with basidiomycetes, but also with certain ascomycetes; such associations are called ectomycorrhizae. Some of them are highly specific, with one species of fungus forming mycorrhizal associations only with vascular plants of a particular species or group of related species. The basidiomycete *Boletus elegans*, for example, is known to associate only with the larch (*Larix*), a conifer. Other fungi, such as *Cenococcum geophilum*, have been discovered living in ectomycorrhizal association with forest trees of more than a dozen genera. Ectomycorrhizae are particularly characteristic of relatively pure stands of trees growing at high latitudes in the Northern Hemisphere or at high elevations.

The most intricate examples of mutualism occur in the tropics, where many more kinds of organisms are found than in temperate regions. For example, trees and shrubs of the genus *Acacia* occur widely in the tropical and subtropical regions of the world. The interaction that occurs between certain species of *Acacia* in the lowlands of Mexico and Central America and the ants that inhabit their thorns provides a remarkable example of the complexity that plant–animal interactions can entail. The particular relationship between the bull's-horn acacias and the inhabitants of their thorns, ants of the genus *Pseudomyrmex*, is a good example (Figure 31–2).

The bull's-horn acacias have a pair of greatly swollen thorns more than 2 centimeters long at the base of each leaf. Nectaries occur on the petioles, and small nutritive organs known as Beltian bodies are located at the tip of each leaflet. The ants live inside the hollow thorns and obtain sugars from the nectaries and fats and proteins

31–2

Ants and acacias. (a) Acacia collinsii, showing the nectaries at the base of the petiole of one of the compound leaves. (b) Beltian bodies at the ends of the leaflets of Acacia collinsii. (c) Worker ants (Pseudomyrmex ferruginea) attacking the tendrils of a vine growing on bull's-horn acacia (Acacia cornigera). Obtaining all of their food from the plant, the ants in turn girdle all plants that come into contact with it and kill most other insects that attempt to feed on the plant. (d) A single individual of Acacia cornigera overtopping the dense second-growth vegetation in the tropical lowlands of Mexico. If this plant had not been occupied by ants, it would probably have died as a small seedling, overtopped by other vegetation and devoured by insects.

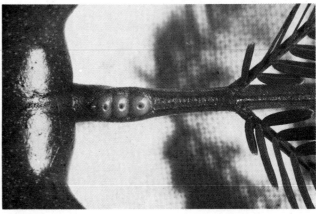

(a)

by eating the Beltian bodies. The acacias grow extremely rapidly and are particularly common in disturbed areas, where the competition between rapidly growing colonizing plants is often intense.

Thomas Belt first described the relationship between *Pseudomyrmex* and the swollen-thorn acacias in his book, *The Naturalist in Nicaragua* (1874). Following his observations, there was a prolonged controversy about whether the presence of the ants actually benefited the acacia plants. This question was finally and definitively solved in 1964 by Daniel Janzen, who was then a graduate student at the University of California, Berkeley, and is now a professor at the University of Pennsylvania. Janzen found that the worker ants, which swarm over the surface of the plant, bite and sting animals of all sizes that contact the plant, thus protecting it from the activities of herbivores and ensuring a home for themselves and their fellow ants. Moreover, whenever the branches of another plant touch an inhabited acacia tree, the ants girdle its bark. By doing so, they destroy the invading branches and produce a tunnel to the light through the rapidly growing tropical vegetation.

When Janzen removed the ants from a plant artificially—by poisoning them or clipping off the portions of the plant that contained ants—or when an acacia was naturally unoccupied by ants (a situation that occurs rarely), the plant's growth was extremely slow, and it usually died after a few months as a result of insect damage and shading by other species. Plants inhabited by ants grew very rapidly, soon reaching 6 meters or more in height and overtopping the other second-growth vegetation. These ants make their nests only in these particular acacias, and they are completely dependent on the acacias' nectaries and Beltian bodies for food. Therefore, it is clear that the ant–acacia system is as much a dual biological entity as, for example, a lichen. One element usually cannot survive without the other in the community in which it occurs.

There are many other kinds of mutualistic relationships that link organisms. For example, trees in a forest, and herbs as well, are often joined by their roots in a phenomenon known as root grafting. Nutrients present in one plant can thus be transferred to another in complex and unexpected patterns; the survival of one species in an area may literally depend on the presence of another species with which it forms root grafts. Stumps of trees may live indefinitely, even though they have no photosynthetic surface of their own, because their roots are linked with those of other individuals in the forest and they are able to obtain nutrients from them. Some diseases, such as oak wilt in the midwestern and eastern United States, can also be transmitted through root grafts.

Competition

Unlike animals, green plants are dependent upon a single process to obtain their energy—photosynthesis. Competition in plants is therefore manifested largely in terms of the "struggle for light," and plants that grow in the shade of others have evolved mechanisms for carrying on photosynthesis at low light intensities. Although competition for water and nutrients is also important for plants, competition for light is primary. Differences in plant height, in the arrangement of leaves, and in the shape of the crown seem to be significant factors in allowing individual kinds of plants to adapt to different habitats within a single community, whether it is low grassland or tall forest. In animals, competition seems to be expressed much more precisely than it is in plants, for which the kinds of competitive interactions we are discussing here may not be species-specific but may merely involve the pairs of plant species that happen to come into contact first. In most kinds of plants, seedlings are far more sensitive to competition than are established plants of the same species.

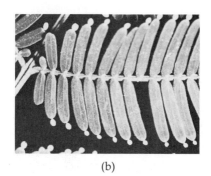
(b)

(c)

(d)

Under experimental conditions, when two kinds of organisms are grown together for a long enough time in a simple environment, one is always eliminated. From this observation and similar theoretical calculations, it has been deduced that, under natural conditions, two species cannot coexist indefinitely in exactly the same habitat utilizing exactly the same resources in the same way. This is a simplified statement of what Garrett Hardin has called the *principle of competitive exclusion*. If two species are growing together and utilizing the same essential resources, and if these resources are present in limiting quantities, the individuals of one or both of them will be smaller or fewer in number than they would be if they were growing alone. If the environment is complex, as in nature, various organisms may use it in different ways, in effect subdividing the habitat. They may then continue to coexist indefinitely.

In a bog, for example, mosses of the genus *Sphagnum* often appear to form a continuous cover, and several species are usually present. How can these several species continue to coexist? When the situation is examined in more detail, it is found that there are semiaquatic species growing along the bottoms of the wettest hollows; other species growing in drier places on the sides of hummocks, which they help to form; and still other species growing only under the driest conditions on the tops of hummocks, where they are eventually succeeded by one or more species of flowering plants. Therefore, although all the species of *Sphagnum* coexist, in the sense that they are all present in the same bog, they actually occupy different *microhabitats* and continually replace one another as the characteristics of each microhabitat change.

If the populations of coexisting species are kept at low levels, they may not eliminate one another. This effect was seen in England during this century, when a severe epidemic of myxomatosis, a disease caused by a virus, drastically reduced the population of rabbits. Formerly, the grasses growing on the chalky soils were kept closely cropped by the rabbits, and many different kinds of flowering plants were able to grow in this habitat. After the decline in the number of rabbits, the grass cover of the chalk soils became deeper and more dense, and many of the formerly abundant species of flowering plants became rare (Figure 31–3). Similar effects are often seen when comparing grazed and ungrazed pastures or grasslands, and they also arise in areas where natural disasters, such as hurricanes, are common. For these reasons, we might expect the diversity of species to be greater along a wave-lashed coast, where disturbance is continuous, than in a more stable environment.

Most competitive situations are quite complicated, and they affect relationships within, as well as between, species. There are a number of different ways of expressing the relative success of two species growing together. Figure 31–4, for example, shows the variations in performance (dry weight of yield per hectare) that occurred with changes in the density of corn (*Zea mays*) plants. With increasing density, the dry weight of the corn plants (shoots plus ears) increased within the limits of the experiment. When there were more than seven plants per square meter, the dry weight of the shoots less ears increased more rapidly than that of the shoots plus ears, but the dry weight of the ears decreased markedly.

Clonal reproduction, which is important in many plants, sometimes makes it difficult to define the limits of genetically identical individuals in nature. This, in turn, may impede the measurement of their competitive interactions. Genetically identical individuals may be widely dispersed but still occur in widely separated

(a)

(b)

31–3
Lullington Heath National Nature Reserve, East Sussex, England, an area of chalk grassland, before (a) and after (b) the elimination of rabbits by the viral disease myxomatosis. The first photograph was taken in 1954, the second in 1978. In an effort to restore the diversity of herbaceous plants, the authorities have introduced grazing programs for sheep and horses over much of the reserve, which has an area of 62 hectares.

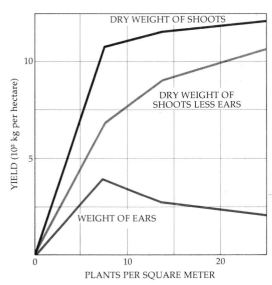

31–4

The influence of plant density on yield in corn (Zea mays) as measured in three different ways. If obtaining ears of corn of the maximum dry weight is the farmer's goal, an optimum density is between seven and eight plants per square meter.

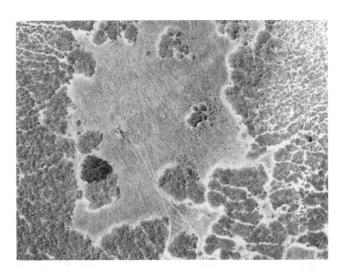

31–5

Purple sage (Salvia leucophylla) shrubs produce terpenes that evaporate and spread through the air, ultimately reaching the soil, where they inhibit the growth of other plants. In this aerial view, the bare zone around the individual Salvia shrubs and colonies can be seen plainly. Immediately around the shrubs is a completely bare zone, and then a zone of inhibited grassland with a few stunted annual herbs.

but similar environments within a single complex habitat. This is characteristic not only of rhizomatous perennials, such as many grasses and sedges, but also of dandelions and other plants in which the seeds are produced asexually and contain embryos that are genetically identical to their parents. It is characteristic also of those plants, such as white clover *(Trifolium repens)*, in which an original plant simply grows apart into distinct individuals.

In some competitive interactions, one or both of the competing organisms produce chemical substances that inhibit either the growth of members of their own species, resulting in increased spacing of like individuals, or the growth of other species. For example, the fungus *Penicillium chrysogenum*, which grows on such organic substrates as seeds, produces significant quantities of penicillin in nature. This penicillin inhibits the growth of bacteria, which compete directly with the fungus for the same nutrients. However, such bacteria as *Bacillus cereus*, which produce penicillinases (enzymes that break down penicillin), often replace the *Penicillium*.

Analogous relationships among plants are grouped under the general heading of *allelopathy*. In coastal California, for example, a bare zone normally occurs between shrub and grass communities (Figure 31–5). The feeding activity of rodents, rabbits, and birds, which find shelter in the shrubs, is concentrated in this zone. If the activity of these animals is prevented by means of wire-mesh exclosures, annual herbs grow vigorously within what normally would be a bare zone (Figure 31–6). In addition, such plants as purple sage

31–6

When mammals such as mice are excluded from the bare zone, annual plants grow luxuriantly right up to the edge of the shrubs in areas where they are normally absent. Such results, reported by Bruce Bartholomew of the California Academy of Sciences, indicate that the effect of these mammals would be sufficient to account for the bare zone, regardless of whether the plants are producing toxic chemicals.

(Salvia leucophylla) produce volatile terpenes, which inhibit the establishment of seedlings of many species of plants in their vicinity. Such seedlings may not grow in the bare zone even if exclosures are constructed to exclude herbivorous animals, but they often grow at the border between the bare zone and the fully developed grassland, where toxic substances produced by the *Salvia* are less concentrated and the grassland is still somewhat open.

Plant–Herbivore Interactions

Vast areas of Australia were at one time covered with spiny clumps of prickly-pear cactus *(Opuntia)*, a plant that was introduced to the continent from Latin America. Fertile lands became useless for grazing, and the economy of great stretches of the interior was severely threatened. Today, the cactus has been nearly eliminated by a cactus moth *(Cactoblastis cactorum)* discovered in South America and deliberately introduced into Australia. The larvae of this moth destroy the cactus plants by eating them. The moth, once abundant in Australia, can scarcely be found today, even by a careful inspection of the few remaining clumps of cactus; yet there is no doubt that it continues to exert a controlling influence over the populations of the plant in Australia (Figure 31–7).

In general, the effects of herbivores on plants are profound, both in the short term and in the long term. As discussed in Chapter 29, these interactions have led over the course of time to the evolution by plants of a wide variety of chemical defenses—in the form of biochemicals once commonly referred to as "secondary plant substances." The ability of plants to produce such toxic chemicals and to retain them in their tissues gives them a tremendous competitive advantage. Indeed, such chemicals are apparently the most important factor in controlling herbivorous insects in nature. This advantage is analogous to the advantage that is achieved by the production of thorns or tough, leathery leaves, and scientists working to improve the resistance of crops to herbivores are focusing much of their effort on such chemicals. In view of the importance of such protective compounds in terrestrial ecosystems, it is not surprising that compounds that play a similar function have also been found recently in many marine algae. Such compounds have been shown to protect the algae from herbivores in the sea.

Plant–herbivore and plant–parasite interactions may be quite complex. For example, pea plants *(Pisum sativum)* are largely protected from parasitic fungi by a substance that the plants produce called pisatin. Many strains of the important parasitic fungus *Fusarium*, however, have enzymes called monooxygenases, which convert the pisatin into a less toxic compound; such fungi then have the ability to infect peas. Human beings also utilize monooxygenases to detoxify certain chemicals that would otherwise be harmful to the body. In such ways, "chemical warfare" between plants and their herbivores is waged continuously.

The protective chemicals that plants produce are often distasteful, but some of these chemicals may display still other properties that deter herbivores. Chromenes, for example, can interfere with insect juvenile hormone (which is necessary for a normal life cycle) and thus can act as true insecticides. A Mexican sneeze-

31–7

(a) *Dense prickly-pear cactus* (Opuntia inermis), *growing in a mixed scrub forest in Queensland, Australia, in October 1926;* (b) *the same forest in October 1929, after the cacti were destroyed by the deliberately introduced South American moth* Cactoblastis cactorum. *First introduced in May 1925, the larvae of this moth destroyed the cacti on more than 120 million hectares of rangeland.*

(a)   (b)

weed (*Helenium* spp.) produces helenalin, which functions as a powerful insect repellent. Pyrethrum is an example of a natural insecticide that is produced commercially from a species of *Chrysanthemum.* Even the waxy surfaces of leaves, which are difficult to digest, may be important in retarding attacks by insects and fungi.

When plants are infected with fungi or bacteria, they often defend themselves by producing natural antibiotics called *phytoalexins*. These are lipidlike compounds whose synthesis can also be stimulated by leaf damage. They appear to be produced in response to the presence of specific carbohydrate molecules, called *elicitors*, that are present in fungal or bacterial cell walls. The elicitors, which are released from the fungal or bacterial cell walls by enzymes present in the plants being attacked, diffuse through the plant cells more or less like hormones. Ultimately the elicitors bind to specific receptors on the plasma membranes, there bringing about metabolic changes that result in the production of the phytoalexins. In principle, it should be possible to spray elicitors onto crops before they become infected and thereby protect them from fungal or bacterial pests. Such a process would be analogous to vaccination in human beings.

One possible problem, however, is that the energetic cost to the plant of producing the phytoalexins in large quantities might lower its ultimate yield more than would the fungal or bacterial infection. Nevertheless, understanding phytoalexin production in plants is of high importance for the protection of crops, and several synthetic elicitors have already been produced and tested. Manipulating the genetic basis of resistance, now possible through the methods of genetic engineering (described in Chapter 30), also offers new possibilities for enhancing crop resistance that do not carry a high energy cost.

Just as some plants produce phytoalexins, others produce tannins and other phenolic compounds, and these seem to play a similar role in nature. When gypsy moths (*Lymantria dispar*) attack and defoliate oak trees (*Quercus* spp.), the trees produce new leaves that are much higher in tannins and other phenolic compounds. The new leaves produced under such conditions are also tougher, and they contain less water than the ones that they replace. Indeed, the differences are great enough that feeding on the new leaves reduces larval growth and thus diminishes the intensity of further outbreaks of gypsy moths. The tannins apparently interfere with digestion in the insects by combining with other plant proteins and making them indigestible. Similar effects may be common in other plants as well. For example, when snowshoe hares heavily browse some trees and shrubs, such as the paper birch (*Betula papyrifera*), these plants produce new shoots that are much richer in resins and phenolic compounds than the shoots that were produced first.

Similar observations have been reported by David Rhoades of the University of Washington with willows and alders. Even more surprisingly, Rhoades found that these trees evidently produced a volatile substance when they were attacked by tent caterpillars or when a large enough portion of their total leaf surface was removed experimentally. This substance, which was apparently transmitted from tree to tree through the air, caused undamaged trees that were not in direct contact with the damaged ones to react as if they had been attacked. Both damaged and undamaged trees formed increased amounts of phenolic compounds and tannins, making them less palatable within 36 hours of the initial attack (Figure 31–8). The airborne substance has not yet been identified, although efforts are under way to do so.

That plants can protect themselves by producing toxic molecules has many other implications for agriculture. For example, wild plants of the squash family (Cucurbitaceae) produce bitter terpenes in their fruits and leaves, and these substances protect them from the

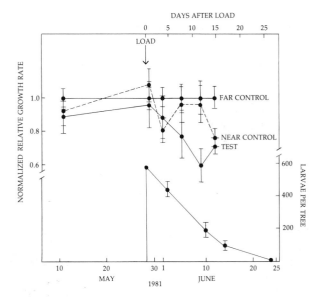

31–8

Normalized relative growth rates of tent caterpillars fed leaves from ten test trees of Sitka willow (Salix sitchensis) loaded with colonies of tent caterpillars on May 28, 1981; ten trees of the same species located nearby; and twenty trees located far away. Bottom: Number of tent caterpillars attacking the trees on various dates. Although the palatability of the nearby trees—as measured by their ability to support the growth of the tent caterpillars—fell markedly, that of the distant trees remained constant throughout the experiment.

PESTICIDES AND ECOSYSTEMS

Approximately half a billion metric tons of pesticides and herbicides are produced annually for application to crops in the United States alone. Of this enormous total, it has been estimated that only approximately 1 percent actually reaches the target organisms; most of the remainder either reaches soil, water, or nontarget organisms in the same ecosystem or spreads into neighboring ecosystems. When it does so, it may have important detrimental effects on the functioning of those ecosystems. It may result in the elimination of certain species, for example, which may affect the functioning of the system as a whole or lead indirectly to the elimination of other species. Important decomposers, such as earthworms and other soil organisms, may be greatly reduced in abundance by pesticides, so that the ecosystem as a whole ceases to function properly. The intensity of such effects depends on the toxicity of the chemicals and on their persistence in the environment.

One of the problems associated with some chemical pollutants is that they tend to be concentrated as they pass up through food chains, reaching their highest concentrations in the top predators. For example, chlorinated hydrocarbons, such as DDT (now outlawed in the United States and in most other developed countries), become concentrated in the tissues of predatory birds and cause the shells of the eggs they lay to be abnormally thin. Such an egg is likely to break before it is time for it to hatch, thereby causing the death of the young bird within. In addition, many strains of insects,

bacteria, and fungi have evolved that are resistant to the pesticides designed to control them. Of the estimated 2000 species of major insect pests, about a quarter have already evolved strains that are resistant to one or more insecticides. Similarly, a number of species of weeds have evolved resistance to herbicides, as discussed in Chapter 30.

Other effects are less direct. For example, important predator species, which naturally control populations of pest species, may be eliminated by pesticide poisoning, leading to outbreaks of the very pest organisms that the pesticides were employed to control. As with most human activities, the net effect of the application of pesticides and herbicides is to lessen the diversity of the ecosystems affected. Despite these difficulties, productive modern agriculture depends, to a large extent, on the application of these useful chemicals.

Because their effects are often so drastic, however, scientists are actively searching for less damaging substitutes. These include selective breeding to produce crops that are resistant to pests (the methods of genetic engineering described in the preceding chapter will be useful in achieving this goal); increased research to develop pesticides that are less toxic and less persistent than the ones currently used; and integrated pest-management systems, which involve combinations of control measures, including the encouragement of predators and diseases of pest species as well as the judicious application of pesticides.

---

attacks of most herbivores. Under cultivation these substances have been bred out in order to improve the flavor of the fruits; hence, the plants have again become palatable to herbivores. Special steps, such as spraying with insecticides, are now taken to protect them. The cultivated watermelon (*Citrullus vulgaris*) also can be attacked by a much wider range of insects than its wild relatives, and insects must be discouraged if the crop is to be grown successfully.

Pollination relationships are a specialized form of plant–herbivore interaction in which a particular portion of the plant (often nectar) is eaten by an animal that also effects pollination. In many ways, pollination relationships are certainly a form of mutualism, with both participants certainly benefiting from the relationship. Here, the emphasis is on attracting the herbivore, and, as described in Chapter 29, this has produced a great diversity of angiosperm flowers. In this sort of selective race, both plants and animals become better adapted to one another, and increasingly specialized pollination systems arise.

The secondary plant substances that animals ingest may in turn play a role in their ecological relationships with other animals. Some insects store these poisons

within their tissues and are thereby protected from their own predators (Figure 31–9). Some sex attractants in insects are derived from the plants on which they feed. The insects concentrate these substances and then use them to attract the opposite sex of their own species.

Of course, the ants that inhabit the swollen thorns of the bull's-horn acacias, which we discussed as an example of mutualism on pages 654 and 655, play a role for these plants similar to secondary chemical compounds—they protect the plants from herbivores. Significantly, species of *Acacia* that are not inhabited regularly by ants contain bitter-tasting chemicals, whereas those inhabited by ants do not. The chemicals and the ants appear to serve the same function.

Viewed as a whole, the relationships within a community are incredibly complex. Plants that occur together affect one another in an endless variety of ways, a few of which are just beginning to be understood. It may not be the individual of a species that is important in survival; rather, it may be that the linked group of individuals and the kinds of interactions that they have with one another determine the success of a particular species growing in a particular place.

(a)

(b)

31-9
(a) *The monarch butterfly* (Danaus
plexippus) *obtains cardiac glycosides
from the plants of the milkweed family*
(Asclepiadaceae) *upon which its larvae* (b)
*feed. As a result, the monarch is
unpalatable to birds and other vertebrates. It "advertises" this fact by bright
orange-and-black adult coloration and by
larvae that are conspicuously banded
with white, yellow, and black. Even the
eggs of the monarch, which are bright
yellow and conspicuous, contain enough
cardiac glycosides to be protected.*

CYCLING OF NUTRIENTS

In terms of its nutrient supply, an ecosystem is more or
less self-sustaining. One of the most important reasons
for this autonomy is the continuous cycling of chemical
elements between organisms and the environment. The
pathways of some of these essential elements, known
as nutrient cycles, were discussed in Chapter 26. Ide-
ally, nothing is lost, so that the pool of nutrients is continually renewed and continually available for the
growth of organisms. The rate of flow from the nonliving pool to the organisms and back again, the amount
of available material in the nonliving pool, and the
form of this pool differ from element to element.

Recycling in a Forest Ecosystem

Studies of a deciduous forest ecosystem have shown
that the plants of this community play a major role in
the retention of nutrient elements. The studies were
made in the Hubbard Brook Experimental Forest in the
White Mountain National Forest of New Hampshire.
The investigators first established a procedure for determining the mineral budget—input and output, or
"gain" and "loss"—of different areas in the forest. By
analyzing the nutrient content of rain and snow, they
were able to estimate input; and by constructing concrete weirs that channeled the water flowing out of selected areas, they were able to calculate output (Figure
31–10). A particular advantage of the site was that
bedrock is present just below the soil surface, so that
very little material leaches downward; that is, the soil
water percolates only a short distance.

The investigators discovered that the natural forest
was extremely efficient in conserving its mineral elements. For example, annual net loss of calcium from
the ecosystem was 9.2 kilograms per hectare. This represents only about 0.3 percent of the calcium in the system. In the case of nitrogen, the ecosystem was actually
accumulating this element at a rate of about 2 kilograms per hectare per year. There was a similar, though
somewhat smaller, net gain of potassium in the system.

At Hubbard Brook, the biological regulation of element cycling was tested by the following experiment.
In the winter of 1965–66, all trees, saplings, and shrubs
in one 15.6-hectare area in one small watershed of the
forest were cut down. No organic materials were removed, however, and the soil was undisturbed. During
the following spring, the area was sprayed with a herbicide to inhibit regrowth. During the four months
from June to September 1966, the runoff of water from
the area was four times greater than in previous years.
Net loss of calcium was 20 times higher than in the undisturbed forest, and potassium loss was 21 times
higher. The most severe disturbance was seen in the nitrogen cycle. The tissues of dead plants and animals

31–10
*Weir in the Hubbard Brook Experimental
Forest in New Hampshire. Water from
each of six experimental ecosystems was
channeled through a weir (such as this
one, built where the water leaves the
watershed) and was analyzed for chemical elements. The trees and shrubs in the
watershed behind this weir have been cut
down. The experiments showed that such
deforestation greatly increased the loss
of nutrient elements from the system.*

31–11
*A food chain. A three-toed box turtle
(Terrapene carolina triunguis) feeds on
a snail that has fed on the mushrooms
that are decomposing organic matter in
the soil.*

continued to be decomposed to ammonia or ammonium ions, which then were acted upon by nitrifying
bacteria to produce nitrates, the form in which nitrogen
is usually assimilated by plants. However, no plants
were present, and the nitrate ions were not held in the
soil. The net loss of nitrogen averaged 120 kilograms
per hectare per year from 1966 to 1968. As a side effect,
the stream that drained the area, polluted with nitrates,
supported an algal bloom. The nitrate concentration of
the stream increased to levels above those established
by the U.S. Public Health Service as safe for drinking
water.

Nitrogen is not always lost so readily from disturbed
forest ecosystems as it was at Hubbard Brook. The actual rates depend upon the details of the nitrogen cycle
in the particular forest before the disturbance takes
place. For example, if microorganisms have a high demand for the nitrogen released after forest clear-cutting, forest disturbance may not result in losses of the
sort measured at Hubbard Brook.

TROPHIC LEVELS

In addition to its *physical* (or nonliving) components,
each ecosystem includes two *biotic* (or living) components—autotrophs and heterotrophs. Autotrophs are
mainly photosynthesizing organisms, which are able to
use light energy to manufacture their own food. Be-

cause heterotrophs cannot manufacture their own food,
they must use the organic molecules made by the autotrophs. Several feeding levels, or *trophic levels*, occur
among the heterotrophs. First, there are the *primary
consumers*, or *herbivores*—animals that feed on plants.
Second, there are the *secondary consumers*, or carnivores and parasites—animals that feed on other
animals. Finally, there are the *decomposers*—fungi, bacteria, and various small animals. All these levels are
present in most ecosystems.

In a given ecosystem, organisms from each of the
trophic levels make up what is called a *food chain* (Figure 31–11). The relationships between the organisms in
a food chain regulate the flow of energy through the
ecosystem. The length and complexity of such food
chains vary a great deal. Usually an organism has more
than one source of food and is itself preyed upon by
more than one kind of organism. Under most circumstances, it is more nearly correct to speak of a *food web*
(Figure 31–12). The complexity of trophic relationships
has a number of important implications for the overall
properties of the ecosystem.

The Flow of Energy

In an ecosystem, the flow of energy typically begins
with solar energy that is captured in photosynthesis
and used to make carbohydrate molecules. Energy does
not flow in a cycle in ecosystems; rather, it flows

PLANT DEFENSE IN SOLANACEAE

Many of the plants of the family Solanaceae—which contains such economically important crops as potatoes (*Solanum*), tomatoes (*Lycopersicon*), and tobacco (*Nicotiana*)— have evolved glandular hairs as one major source of protection against insect pests. These hairs release a sticky and, in certain instances, toxic exudate. The cultivated potato (*Solanum tuberosum*) does not have such hairs, but a few related wild potato species, such as *Solanum berthaultii*, possess abundant glandular hairs on their leaves and stems. These hairs release a sticky substance when touched by insects, and the insects become trapped and eventually starve to death. This defense mechanism is effective against many kinds of insects, including such major potato pests as aphids, flea beetles, and leaf-hoppers. Attempts are currently under way in several countries to breed this characteristic into high-yielding varieties of the cultivated potato, which readily hybridizes with *S. berthaultii*.

These glandular hairs are only one element in the defensive arsenal of the Solanaceae. The leaves of plants of this family are rich in such alkaloids as nicotine, atropine, and hyoscyamine, which are poisonous to many kinds of herbivores and account for the hallucinogenic properties of Jimson weed (*Datura stramonium*) and other species of the family. Steroid glycosides found in their leaves also inhibit feeding by many herbivores, but the same compounds stimulate feeding in other insects that are specialized feeders on solanaceous plants. As mentioned in the preceding chapter, some species of *Solanum* are now being investigated as a possible source of steroid molecules to be used in the manufacture of birth-control drugs. In addition, tomatoes and potatoes have been shown by Clarence E. Nelson of Washington State University to produce two different proteinase inhibitors—enzymes that disrupt digestion in herbivores—at high concentrations within four hours after a single, severe wound. The concentrations of these inhibitors remain constant for another five hours before decreasing rapidly.

The first scanning electron micrograph (a) shows an insect-trapping hair on the foliage of *Solanum berthaultii*. A sticky exudate is released when the four-lobed head of the hair ruptures on contact with an insect. The other scanning electron micrographs (b, c, and d) show an aphid (*Myzus persicae*) glued to the stem of *S. berthaultii* by the exudate from the hairs, each at successively higher magnifications.

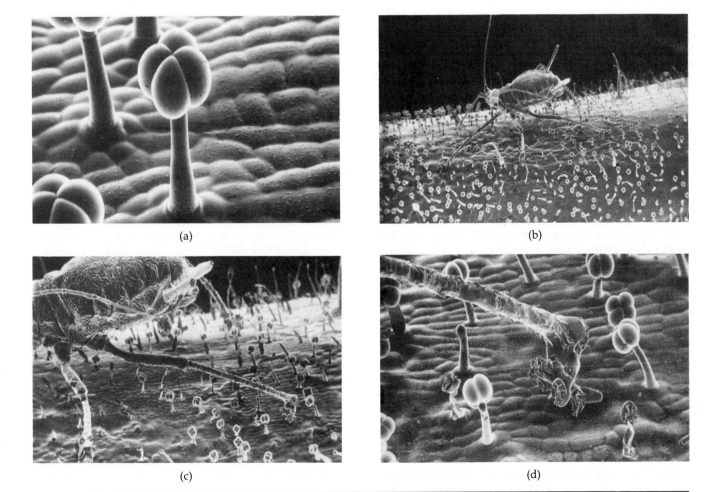

(a)

(b)

(c)

(d)

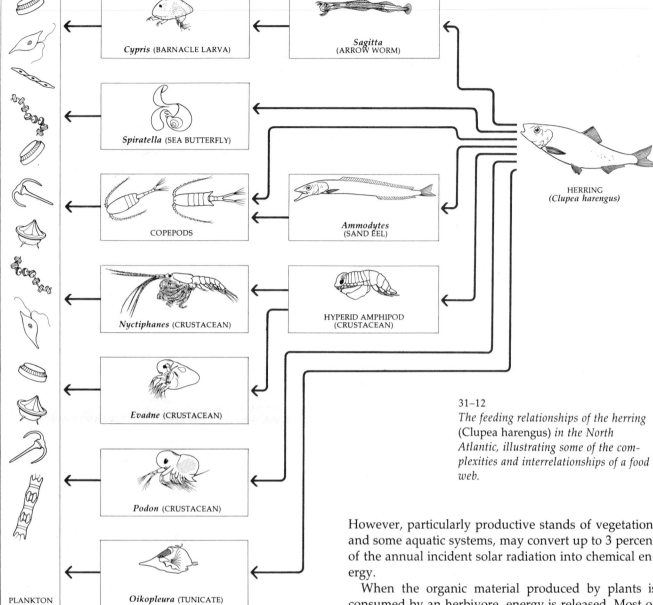

PLANKTON

Cypris (BARNACLE LARVA)

Sagitta (ARROW WORM)

Spiratella (SEA BUTTERFLY)

COPEPODS

Ammodytes (SAND EEL)

Nyctiphanes (CRUSTACEAN)

HYPERID AMPHIPOD (CRUSTACEAN)

Evadne (CRUSTACEAN)

Podon (CRUSTACEAN)

Oikopleura (TUNICATE)

HERRING (*Clupea harengus*)

31–12
The feeding relationships of the herring (Clupea harengus) *in the North Atlantic, illustrating some of the complexities and interrelationships of a food web.*

through autotrophs (usually photosynthesizing organisms, such as plants, algae, and some bacteria) to consumers (animals and heterotrophic protists) and then to decomposers. At each step, the great majority of the energy is dissipated in the form of heat energy, which eventually is returned to space as infrared radiation.

To begin with, a very large amount of *biomass* is produced on earth each year. (Biomass is a convenient shorthand term for organic matter; it includes woody parts of trees, stored food, bones, and so on.) It is currently estimated that the earth's annual production of biomass comes to about 200 billion metric tons. Despite this enormous figure, photosynthesizing organisms are not very efficient in converting the sun's energy into organic compounds. Generally, less than 1 percent of the light that falls on a plant is utilized (see Chapter 5).

However, particularly productive stands of vegetation, and some aquatic systems, may convert up to 3 percent of the annual incident solar radiation into chemical energy.

When the organic material produced by plants is consumed by an herbivore, energy is released. Most of this energy is lost as heat, and a fraction of the organic material consumed is converted to animal tissue. In general, only about 2 to 10 percent of the usable energy of the plant is added to the mass of the herbivore; the remainder is lost in respiration. A similar gain–loss relationship is found at each succeeding level. Thus, if an average of 1500 calories of light energy per square meter of land surface is utilized by plants per day, about 15 calories of energy are converted to plant material. Of this amount, about 1.5 calories are incorporated into the bodies of the herbivores that eat the plants, and about 0.15 calorie is incorporated into the bodies of the carnivores that prey on the herbivores.

To give a concrete example, Lamont Cole of Cornell University, in his studies of Cayuga Lake, calculated that, for every 1000 calories of light energy utilized by algae in the lake, about 150 calories are reconstituted as small aquatic animals and 30 calories as smelt (a kind of small fish). If we were to eat these smelt, we would gain about 6 calories from the original 1000 calories used by the algae. But if trout eat the smelt and we then eat the trout, we gain only about 1.2 calories from the original 1000 calories. From this example, it is clear that there is more energy available to us if we eat smelt rather than the trout that feed on the smelt; yet trout are considered a delicacy, smelt a much less desirable food fish for humans. Under conditions of starvation, people must turn to an all-plant diet, not being able to afford the tenfold loss in energy that occurs when plants are fed to animals. In order for us to make maximum use of the solar energy trapped by plants, we must become mainly herbivorous.

Food chains are generally limited to three or four links; the amount of food remaining at the end of a longer food chain is so small that few organisms can be supported by it. Body size also plays a role in the structure of food chains. For example, an animal constituting one link generally has to be large enough to capture prey from the previous, lower link on the food chain. However, most of the biomass in an ecosystem is used by decomposers, such as fungi and bacteria, and is recycled in that way.

Owing to the relationships just discussed, the total biomass at successive trophic levels in an ecosystem generally decreases sharply, setting up the sort of relationship described by the expression "pyramid of mass" (Figure 31–13). This relationship may not hold if there is rapid turnover among the primary producers, such as algae in a lake; the turnover rate then becomes the controlling factor, and the total biomass present at that level may be relatively small at any one time. If energy is measured, it follows the same rapid decrease characteristic of mass; there is a "pyramid of energy," with far less energy in the bodies of all the predators present in a given community, for example, than in all the plants. In general, there are also far more individuals at the lower levels than at the higher levels, which leads to a "pyramid of numbers." It also follows that, if all the organisms in an ecosystem are divided into size classes, the small animals will be far more numerous than the large ones.

A practical aspect of the flow of energy through ecosystems concerns human efforts to develop renewable sources of energy from plants, a process known as biological energy conversion. The waste materials that are available each year after the harvest of crop and forest products, it is estimated, could provide energy equivalent to 1 percent of the gasoline consumed annually in the United States, or 4 percent of the nation's annual consumption of electrical energy. This potential is limited, however, by the cost in energy of harvesting the material. Plantings of fast-growing trees and other plants may provide a major, renewable energy source in the future, and they may constitute one of the best ways of efficiently capturing solar energy.

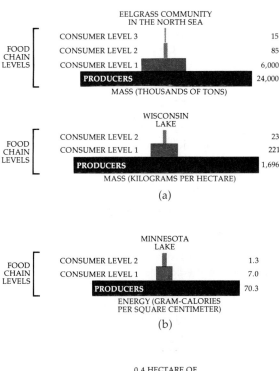

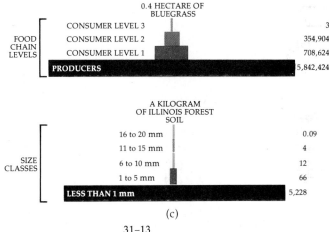

31–13

Pyramids of mass (a), energy (b), and numbers of organisms (c) in various communities. A relatively small amount of mass or energy is transferred to each successively higher level.

DEVELOPMENT OF COMMUNITIES AND ECOSYSTEMS

Succession

Some plant communities remain the same year after year, whereas others change rapidly. In the latter case, there is an orderly progression of changes in community composition that is called *succession.* A cleared woodlot is rapidly colonized by the remaining trees in the vicinity; in a similar fashion, a pasture eventually gives way to a forest. Analogous series of events occurs in naturally open areas, such as lakes, meadows, or rocky hillsides. The process of succession is both continous and worldwide in scope.

Succession occurs at a variable rate in all temporarily open areas. Some ponds, for example, fill with aquatic plant remains and debris; emergent vegetation builds soil; the site is taken over by meadow; moisture-loving shrubs may become established; and finally the forest characteristic of the region develops in the meadow that formed earlier where there had been a pond (Figure 31–14). In another example, rocks weather and break down because of freezing and thawing and other physical factors, and the process is sometimes expedited by the action of lichens, which secrete certain chemicals that erode the rocks directly; soil accumulates around the bases of the lichens; and they then give way to mosses and finally to flowering plants (Figure 31–15). The roots of the flowering plants probe cracks, breaking the rocks down further. Eventually, perhaps after many centuries, the rock is reduced to soil, and the soil is occupied by forest trees or other vegetation types characteristic of the region. In the early stages of succession, plants that have the ability to fix nitrogen are often prominent. Other examples of the process are shown in Figure 31–16.

An ecosystem undergoes various stages of succession until, finally, the sort of mature community that has been called a *climax community* is produced. Although the nature of the end point that is reached following succession varies according to the climate of the area, such a community is often more nearly self-sustaining than those that occur earlier in succession. The organisms in the community have developed an intricate set of interrelationships.

Some aspects of succession have great significance for people. For example, when European settlers first came to California in large numbers, they found a magnificent forest of sugar pine *(Pinus lambertiana)* along much of the length of the Sierra Nevada. Later, although conservationists tried to preserve some of this forest in national parks and national forests, many of

(a)

(b)

(c)

31–14
(a) *Emerging vegetation grows along the edge of a pond.* (b) *Aquatic plants with floating leaves, such as water lily* (Nymphaea odorata), *grow across the surface of a pond and eventually choke out bottom-dwelling plants.* (c) *Water hyacinths* (Eichhornia crassipes) *play a similar role in warmer climates.* (d) *Marsh grasses, sedges, and cattails* (Typha spp.) *growing on an old pond bed complete the process of succession.*

(d)

31–15

An early stage of succession on a rocky slope. Lichens have begun to break down the rocks, while ferns and bryophytes are accumulating soil in a small crevice.

31–16

(a) *Seedling trees of balsam fir* (Abies balsamea) *growing under and replacing quaking aspen* (Populus tremuloides) *in northern Minnesota—a stage in forest succession leading to a climax community of white spruce* (Picea glauca) *and balsam fir.* (b) *Seedling of a red maple* (Acer rubrum) *rising above needles of white pine* (Pinus strobus). *Mature white pines filter the light to such a degree that their own seedlings cannot survive beneath them and only those tolerant of shade, such as those of maples and oaks, can gain a foothold. On the other hand, the process of succession is slowed down by the suppression of nitrogen-fixing bacteria in the soil by the chemicals that leach from the fallen pine needles.*

the stands of pines were eventually replaced by other trees, such as white fir *(Abies concolor)* and incense cedar *(Calocedrus decurrens).* Why did this change take place?

The answer is that the sugar pine was a member of a certain stage in succession in the forests of this area, and this stage was maintained by periodic fires. These fires were greatly reduced in number and scope after the influx of Europeans to the area. Without periodic, lightning-set fires of low intensity racing through the groves, a thick growth of brush and smaller trees arose and created conditions so crowded that the sugar pines could not reproduce. Only a system of controlled burning can preserve the remaining groves of sugar pine in their original form (Figure 31–17b).

Recolonization

When people alter a landscape, changes are made in the community structure. In abandoned fields, on denuded sand dunes, and on the streets of the ghost towns of the American West, succession is taking place, and ecosystems that more and more closely resemble those of adjacent, less-disturbed areas are being produced (Figure 31–18). Given sufficient time, successional processes may gradually restore the original vegetation to the area. For example, in the northern hardwood forests of North America and Eurasia, it is estimated that 60 to 80 years may be required to replace the plant biomass and nutrients removed from the forest by harvesting the trees. In other communities, the process may be faster or slower. In any event, it requires a considerable period of time and a source of new seeds for recolonization to be successful. In many areas of the world, especially tropical ones, these conditions are not being met because of the enormous pressures caused by rapid human population growth and poverty, as discussed in Chapter 30.

(a)

(b)

(a)

(b)

31–17

(a) *When fire sweeps through a forest, secondary succession—with regeneration from nearby unburned stands of vegetation—is initiated. Some plants produce sprouts from the base, others seed abundantly on the burned area. In one* group of pines, the closed-cone (serotinous) *pines, the cones do not open to release their seeds until they have been exposed to fire. (See also Figure 19–7.)* (b) *Sugar pines* (Pinus lambertiana) *in the southern Sierra Nevada of California.*

With the prevention of forest fires by human beings, sugar pines are being replaced by other trees, such as incense cedar (Calocedrus decurrens), *the large tree on the right.*

(a)

(b)

31–18

(a) *View in a forest of loblolly pines* (Pinus taeda) *maintained on an experimental plot at the Tall Timbers* *Research Station, near Tallahassee, Florida. The last controlled burn of this area was carried out March 23, 1967.* (b)

In the absence of fire, succession proceeds rapidly. This photograph was taken in 1982 at the same spot.

Following natural disasters, recolonization produces similar changes. For example, in August 1883, a violent volcanic explosion destroyed half of the island of Krakatau, in the Java Straits about 40 kilometers from Java; the remaining half of the island was covered by a layer of pumice and ash more than 31 meters thick. Neighboring islands were also buried, and the entire assemblage of plants and animals on these islands was wiped out. Soon afterward, however, the recolonization of Krakatau began, and the expected number (based on the number originally occupying the area) of about 30 species of land and freshwater birds was reached within about 30 years. Recolonization by plants also proceeded rapidly, with a total of more than 270 species being recorded for the island of Krakatau by 1934.

In the state of Washington, the violent eruption of Mount St. Helens on May 18, 1980, sent a massive avalanche of volcanic debris from the top and north side of the mountain into the North Toutle River Valley. Within 15 minutes, more than 61,000 hectares of forest and recreation land were devastated by the lateral blast, which blew down forests on about 21,000 hectares and killed trees and other plants but left them standing on another 9700 hectares. In addition, the 9-hour eruption covered the whole area with up to 0.5 meter of ash, pumice, and rock pulverized by the blast (Figure 31–19).

Life began to reappear on the slopes affected by the eruption almost immediately, however, with plants sprouting up through the volcanic debris the following

(a)

(b)

(c)

(d)

31–19

The most recent eruption of Mount St. Helens in Washington State took place on May 18, 1980. (a) What had been productive forests of Douglas fir (Pseudotsuga menziesii), western hemlock (Tsuga heterophylla), and firs (Abies amabilis and A. procera) were blown
down by the lateral blast. (b) Over an area of more than 60 square kilometers, a deep layer of volcanic debris was deposited. (c) During the summer of 1980, many perennial plants, like fireweed (Epilobium angustifolium), sprouted up through the ash. This
photograph was taken a year later. (d) Douglas fir and many other plants with wind-dispersed seeds recolonized areas that had been so deeply covered by volcanic debris that the plants beneath were killed. This photograph was taken in 1984.

spring. Much of this debris soon eroded off, and wind-dispersed seeds and fruits soon blew back into the area. Such dispersal was especially important in areas that were buried by debris avalanches, which were so deep that they killed the plants buried beneath them, and on the volcanic flows. Many small animals also survived, both underground and in lakes and streams, and vertebrates also soon moved back into the area. The eruptive periods of Mount St. Helens have been separated by only 100 to 500 years in the last 35 centuries, and so the blast itself and the recovery from its effects that scientists were able to observe in the early 1980s were typical of natural phenomena that periodically affect living systems in volcanic areas.

Although Mount St. Helens has provided a recent and very dramatic example of the process, succession is a phenomenon that is typical of all communities, and it occurs throughout the world. It is one of the factors that accounts for the full extent of the diversity of life on earth.

SUMMARY

The ecosystem is the highest level of biological complexity and integration; it is a self-sustaining system, driven by energy from the sun, in which the regulated cycling of essential materials takes place. Within ecosystems, communities of organisms interact with the physical environment.

Some of the relationships that occur in communities can be grouped under three headings: mutualism, competition, and plant–herbivore relationships. In mutualism, both interacting populations benefit. Examples include the formation of lichens, the growth of nitrogen-fixing bacteria in nodules on the roots of legumes, mycorrhizal associations between fungi and the roots of higher plants, and close-knit relationships between certain flowering plants and their pollinators.

Competitive interactions are found between most plants that grow together. One of the most important kinds of interaction is competition for light. Plants have also evolved chemical defenses that limit the success of other plants in their vicinity, and such allelopathic relationships are of great importance in determining the future structure of communities.

Herbivores control the reproductive potential of plants by destroying their photosynthetic surfaces, their food-storage organs, or their reproductive structures. Plants counter these attacks through the evolution of spines, tough leaves, or, most importantly, chemical defenses. When an insect has overcome a plant's chemical defenses, it not only has a new and often largely untapped food resource at its disposal but may also utilize the toxic substances produced by the plant to gain a degree of protection from its own predators. Pollination relationships are a special case of plant–herbivore interactions in which the emphasis is on attraction rather than repulsion. Some plant–herbivore interactions involve a high degree of mutualism; the obligate interaction between the bull's-horn acacias and their associated ants in eastern Mexico is such a system.

An ecosystem consists of nonliving elements and two different kinds of living elements—autotrophs and heterotrophs. Among the heterotrophs are the primary consumers, or herbivores; the secondary consumers, or carnivores and parasites; and, finally, the decomposers. The organisms found at these levels are members of food chains or food webs.

Energy flows through ecosystems, with about 0.1 to 1 percent of the incident solar energy converted into chemical energy by green plants. When these plants are consumed, about 2 to 10 percent of their potential energy is stored at the next trophic level; a similar degree of efficiency characterizes transfers further up the food chain. The amounts of energy remaining after several transfers are so small that food chains are rarely more than three or four links long.

Succession occurs when an area is denuded by artificial or natural means, or when a new area is created, such as a lava bed, exposed rocks, or sand or gravel islands in a river. In the course of succession, the kinds of plants and animals in the area change continuously, some being characteristic only of early stages of succession and others only of later stages. The amount of organic material that is incorporated into organisms increases during succession. The diversity of species in the ecosystem increases greatly, and the relationships between species become more and more complex. Production reaches a peak early, and then respiration increases rapidly in the later stages of succession. Eventually, succession results in the production of a climax community, which reproduces itself indefinitely unless there are major environmental changes.

C H A P T E R   3 2

The Biomes

A *biome* is a terrestrial, climatically controlled set of ecosystems that are characterized by distinctive vegetation, and among which there is an exchange of water, nutrients, atmosphere, and biological components, including people. A single biome occupies large areas of land surface and usually occurs on more than one continent. Biomes can be classified in a number of different ways, but the categories that are discussed in this book provide a basis for an overall account of the vegetation of the world.

The distribution of biomes (Figure 32–1) results from three kinds of physical factors: (1) global patterns of air circulation, particularly the directions in which the prevailing moisture-bearing winds blow; (2) the distribution of heat from the sun and the relative seasonality of different portions of the earth; and (3) such geological factors as the distribution of mountains and their height and orientation. All of these factors interact to produce the varying patterns of vegetation over the face of the earth.

LIFE ON THE LAND

Land plants face a variety of problems. In most regions, they are subjected to periodic drought and to rapid diurnal and seasonal changes in temperature. They must survive during seasons unfavorable to their growth, they must often grow on substrates of unfavorable mineral composition, and they are subject to the action of gravity (which affects land plants much more strongly than it does water plants). Living on land has some advantages, however: except at high elevations, oxygen is more uniformly distributed in air than it is in water, and carbon dioxide is more readily available.

32–1
The distribution of biomes throughout the world. The biomes recognized are indicated in the legend. This map was specially prepared for this book by A. W. Küchler of the University of Kansas. Because of the global coverage of the map, its scale is relatively small and its content is generalized. For this reason, the various biomes shown are not always highly uniform; all of them include considerable variations in vegetation. The boundaries between the biomes may be sharp, but more often they are blurred, consisting frequently of broad zones of transition from one type to another.

1	RAINFORESTS
2	SAVANNAS
3	SUBTROPICAL MIXED FORESTS
4	MONSOON FORESTS
5	TROPICAL MIXED FORESTS
6	SOUTHERN WOODLAND AND SCRUB
7	DESERTS AND SEMIDESERTS
8	JUNIPER SAVANNA
9	GRASSLANDS
10	TEMPERATE DECIDUOUS FORESTS
11	MIXED WEST-COAST FORESTS
12	TEMPERATE MIXED FORESTS
13	ALPINE TUNDRA AND MOUNTAIN FORESTS
14	MEDITERRANEAN SCRUB
15	TAIGA
16	ARCTIC TUNDRA
17	ICE DESERT

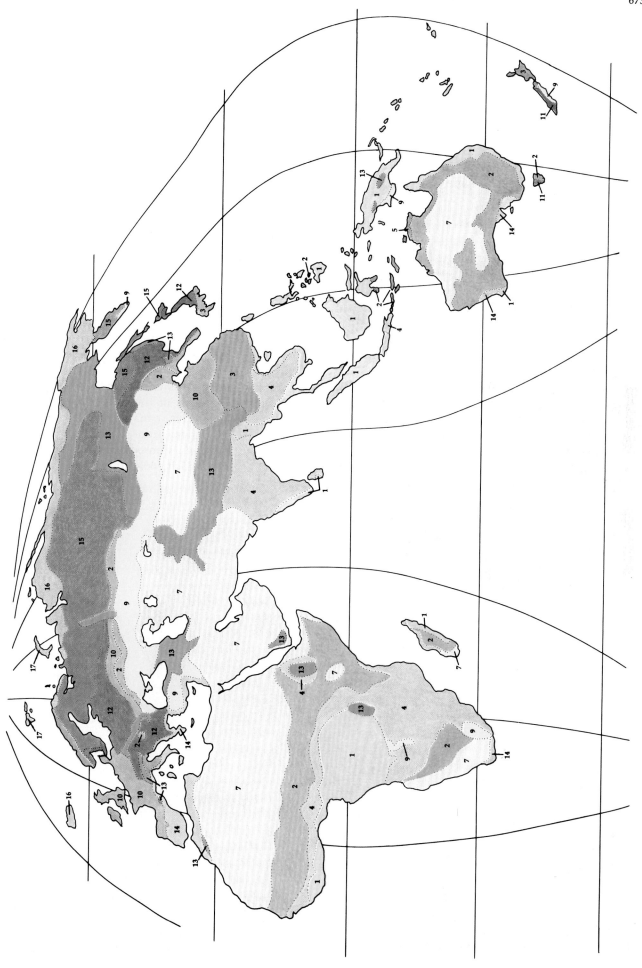

The earth's land areas are discontinuous, and this separation of land masses has an important effect on the distribution of organisms. In addition, there are often sharp differences in precipitation, substrate, climate, and other factors between different places on land. Such differences mean that the distribution of any particular kind of land organism is likely to be much more limited than that of an organism of similar size and motility in the sea (Figure 32–2).

Local Modifications in Climate

Plants and plant communities vary as the land varies. For instance, the mean atmospheric temperature decreases with increasing latitude. In terms of annual averages, the warmest atmospheric temperature is found not at the equator but at 10° north latitude; only during January is the equator the warmest circle of lati-

tude. In July, the warmest parallel of latitude is at 20° north. The Northern Hemisphere as a whole has a higher annual mean temperature than the Southern Hemisphere because of the much larger land surfaces in the north. The annual mean temperature at the equator is 26.3° C; that at 40° north latitude is 14°C; and that at 40° south latitude is 12.4°C. Just as one encounters lower temperatures in traveling north or south from the equator, one encounters them also when ascending to higher elevations. In general, a change in mean atmospheric temperature corresponding to an increase in latitude of one degree occurs with each increase in elevation of approximately 100 meters. This relationship has important consequences for the distribution of land organisms. For example, plants and animals characteristic of arctic regions may approach or even reach the equator at high elevations, particularly in mountain ranges trending north and south (Figure 32–3).

There are, however, important differences between high-latitude habitats and high-altitude habitats. In the mountains, the air is clearer and the solar radiation is more intense. Most of the water vapor in the atmosphere—which plays a major role in preventing heat from radiating away from the earth at night—occurs below 2000 meters. Consequently, nights in the mountains are often much cooler than those at lower eleva-

32–2

This fern, the spleenwort Asplenium viride, *is growing in rock crevices at Sierra Buttes, Sierra County, northern California. It occurs in similar habitats all around the Northern Hemisphere—across both North America and Eurasia—but in North America it occurs no closer to Sierra Buttes than central Washington and northeastern Nevada, both nearly 1000 kilometers away. Throughout its range,* Asplenium viride *occurs on cliffs at scattered localities, reaching new places by means of its wind-dispersed spores. Sierra Buttes, however, is far from the edge of the fern's more continuous range, and its occurrence there provides an excellent example of the disjunct and dispersed nature of the ranges of plants and animals in general.*

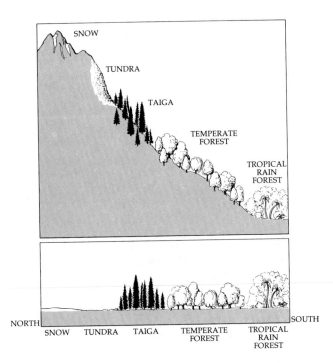

32–3

We can see a similar sequence of plant communities whether we travel north for hundreds of kilometers or merely ascend a tall mountain. This relationship between altitude and latitude was first pointed out by Alexander von Humboldt.

tions at the same latitude. Moreover, there is pronounced seasonal variation both in day length and in temperature near the poles and relatively little variation near the equator (Figure 32–4). Those "arctic" organisms that do range toward the equator in the mountains must make physiological adjustments for such differences between high-latitude (arctic) and high-altitude (alpine) environments.

There are often pronounced temperature variations from slope to slope on a particular mountain. In the middle latitudes of the Northern Hemisphere in winter, for example, the southern and western slopes of mountains are often drier than the northern and eastern ones, because the northern slopes receive no direct sunlight and the eastern slopes receive sunlight in the mornings, when it is cooler. Furthermore, the afternoon sun in the summer is much hotter than the morning sun because of a decrease in the amount of water vapor

32–4

In high mountains in the tropics, the temperature does not change appreciably with the season, but freezing may occur every night of the year. These unusual conditions have resulted in the evolution of plants with bizarre growth forms, such as these giant sunflowers (Espeletia grandiflora, family Asteraceae) in Colombia. In such plants, the apical meristem is protected from freezing by the tightly infolded leaves, the thick, fleshy construction of the entire plant, and a matted covering of reflective silvery hairs. Such adaptations are especially important in view of the very high ultraviolet radiation in tropical mountains, a form of radiation that is especially damaging to the proteins and nucleic acids in the leaves.

in the atmosphere. Because of the interaction of sunlight and clouds, the driest slope of a mountain in the Northern Hemisphere is usually that oriented toward the southwest. (Which would you expect to be the driest slope of a mountain in the Southern Hemisphere?) For the same reasons, mosses and lichens often grow, or are better developed, on the northeast side of trees in the Northern Hemisphere, leading to a well-known folk saying about how to identify the points of the compass when trying to find one's way in the woods. The relationship is valid, however, only in sufficiently well-lighted situations.

As a result of the complex interactions between varying local conditions, the distribution of biomes is not what we would expect if the earth were a perfect sphere. In addition, major shifts in vegetation types have occurred in the past, profoundly affecting the sort of vegetation that we see today. In the following pages, some examples are provided of the ways in which climate, topography, and soil type interact to control the distributions of the biomes. We discuss the 17 biomes shown on the map (see Figure 32–1) under the following categories: rainforests; savannas and deciduous tropical forests (including savannas, subtropical mixed forests, monsoon forests, tropical mixed forests, and southern woodland and scrub); deserts (including deserts and semideserts, and juniper savanna); grasslands; temperate deciduous forests; temperate mixed and coniferous forests (including mixed west-coast forests, temperate mixed forests, and alpine tundra and mountain forests); mediterranean scrub; taiga; and Arctic tundra (including Arctic tundra and ice desert).

RAINFORESTS

More species of plants and animals live in the tropical rainforest, which is dominated by broad-leaved evergreen trees, than in all the rest of the biomes of the world combined. Neither water nor temperature are limiting factors during any part of the year. Although there are many species, there are few individuals per species; a species of tree may be represented only once per hectare. Not only is there a large number of different organisms in the tropical rainforest but their interrelationships are more complex than those of plants and animals in any other biome (Figure 32–5).

Little light penetrates to the floor of the tropical rainforest, and the rainfall is generally between 200 and 400 centimeters per year. There is little accumulation of organic debris; decomposers rapidly break down leaves and stems and the bodies of animals that fall from the canopy. Nutrients released from this breakdown are quickly absorbed by mycorrhizal roots or are leached from the soil by the rain. Although there may be a notable variation in precipitation from month to month, there is no pronounced dry season.

ALEXANDER VON HUMBOLDT

Alexander von Humboldt (1769–1859) was perhaps the greatest scientific traveler who ever lived. A native of Germany, Humboldt ranged widely across the interior of Latin America from 1799 to 1804 and climbed some of its highest mountains. Exploring the region between Ecuador and central Mexico, Humboldt was the first to recognize the incredible diversity of tropical life and, consequently, the first to realize just how vast a number of species of plants and animals there must be in the world.

In his travels, Alexander von Humboldt was impressed with the fact that plants tend to occur in repeatable groups, or communities, and that wherever there are similar conditions—relating to climate, soil, or biological interactions—similar groupings of plants appear. He also discovered a second major ecological principle—the relationship between altitude and latitude. He found that climbing a mountain in the tropics is analogous to traveling farther north (or south) from the equator. Humboldt illustrated this point with his well-known diagram of the zones of vegetation on Mount Chimborazo in Ecuador, which he ascended. On this mountain, he reached the highest elevation that had been attained by any human being up to that point.

On his return from Latin America in 1804, Humboldt visited the United States for eight weeks. He spent three of these weeks as Thomas Jefferson's guest at Monticello, talking over many matters of mutual interest. It is thought that Humboldt's enthusiasm for learning about new lands encouraged Jefferson's own great scheme for the exploration of the western United States. Thus, it is fitting that Humboldt's name is commemorated in the names of several counties, mountain ranges, and rivers in the American West. After returning to Berlin, Humboldt lived for more than half a century, dying in his ninetieth year. He was one of the greatest writers and scientists of his era.

(a)

32–5
Tropical rainforest. (a) The diversity of the trees in the forest, which may reach several hundred species per hectare, is revealed when individual trees burst into bloom, like this guayacan tree (Tabebuia chrysantha) *in Panama. (b) A buttressed tree in a lowland forest in Thailand. (c) Stinkhorn fungus* (Dictyophora duplicata). *(d) Dense growth of epiphytes on a tree in Colombia. (e) An orchid growing on the trunk of a mango in Ecuador. (f) Lianas, like this hummingbird-pollinated, red-flowered passion flower* (Passiflora coccinea), *are abundant in the tropical rainforest (g).*

Generally, plants in the tropical rainforest have not evolved particular mechanisms that would permit them to survive unfavorable seasons of drought or cold. Nearly all the plants are woody, and woody vines are abundant. There is a large flora of *epiphytes*, which grow on the branches of other plants in the illuminated zone far above the forest floor. Epiphytes, which have no direct contact with the forest floor, obtain water from the humid air of the canopy, as well as directly from rain, and they obtain minerals from dust and the surfaces of the plants on which they grow. Along with epiphytes and climbing vines, many kinds of animals have moved into the treetops; this is the area of the tropical rainforest in which animal life is most abundant and diverse.

(b)

(c)

(d)

(e)

(f)

(g)

RODENT POLLINATION OF
TROPICAL PLANTS

New and complex interrelationships between plants and animals continue to be discovered in the tropics. In 1979, for example, Cecile Lumer of the New York Botanical Garden discovered the first New World example of pollination of flowers by rodents. In this nighttime photograph, a rice rat (*Oryzomus devius*) is shown drinking nectar from a flower of the epiphytic plant *Blakea chlorantha* in a Costa Rican cloud forest. The flowers of this plant exhibit several characteristics that can be related to the attraction of nonflying mammals—inconspicuous green blossoms that open at night and a well-hidden source of copious, sucrose-rich nectar. When the rodent grasps the flower and inserts its tongue to probe for nectar, the pollen is released explosively from the anthers, dusting the rodent's face. The pollen is then transferred to the stigma of the next flower visited. The pollination of flowers by mammals other than bats has also been discovered recently among marsupials in Australia and South America, lemurs in Madagascar, monkeys in South America, and rodents in Hawaii and South Africa. In all cases, the flowers produce copious amounts of nectar.

Because so little light reaches the forest floor, very few herbaceous plants grow there, and those that do so are found mostly in clearings. Many plants in the tropical rainforest are trees; they are usually larger than those found in temperate forests, often reaching 40 to 60 meters in height. Furthermore, the tree species are diverse; there are seldom fewer than 40 species per hectare. This is in sharp contrast with temperate zone forests, where there are rarely more than a few species per hectare. Nevertheless, the trees of the tropical rainforest are remarkably homogeneous in appearance. Generally, they branch only near the crown. Because their roots are usually shallow, they often have buttresses at the base of the trunk that provide a firm, broad anchorage. Their leaves are medium-sized, leathery, and dark green; their bark is thin and smooth; and their flowers are generally inconspicuous and greenish or whitish in color. Such forests often form several layers of foliage, the lower layers consisting of seedlings of the taller species, together with a relatively small number of lower-growing species.

There are three major areas of the world in which the tropical rainforest is well developed. The largest is in the Amazon Basin of South America, with extensions into coastal Brazil, Central America, and eastern Mexico, as well as some of the islands in the West Indies. In Africa, there is a large area of rainforest in the Zaïre Basin, with an extension along the west coast of Liberia. The third area of rainforest extends from Sri Lanka and eastern India to Thailand, the Philippines, the large islands of Malaysia, and a narrow strip along the northeastern coast of Australia (see Figure 32–1). In the North American region, tropical rainforest exists only in southern Mexico and on Puerto Rico and the northern side of Hispaniola in the West Indies (Figure 32–6). Temperate rainforests, dominated by broad-leaved evergreen trees like those in the tropics, occur along the eastern side of Australia, in Tasmania, and in parts of New Zealand, as shown in Figure 32–1.

32–6
Tropical rainforest in North America.
Such forest also occurs over part of
Hispaniola and Puerto Rico in the West
Indies.

The tropical rainforest now forms about half of the forested area of the earth, but it is in the process of being destroyed systematically by human activities. The rapidly expanding human population in the tropics, coupled with traditional patterns of land ownership, have made the usual practices of tropical agriculture—clear-cutting of the forest, followed by short-term cultivation—immensely destructive when carried out on such a wide scale. Many kinds of plants, animals, and microorganisms will become extinct in the course of this destruction, as we discussed in Chapter 30.

One reason for the rapidity with which tropical forests disintegrate under human pressure has to do with the nature of tropical soils. Many of these soils are conditioned by high and constant temperatures and abundant rainfall and are relatively infertile. More than half of the soils of the tropics are acidic and deficient in calcium, phosphorus, potassium, magnesium, and other nutrients. Furthermore, the phosphorus in such soils tends to combine with iron or aluminum to form insoluble compounds that are not available for plant growth, and they often contain toxic levels of aluminum ions. Plant roots tend to spread out in a thin layer, no more than a few centimeters in depth, and they rapidly transfer the nutrients released by the decomposition of fallen leaves and branches back into the living plants. Below this thin layer of topsoil, which is easily destroyed during the process of clearing away the trees, there is virtually no organic matter. For all of these reasons, the cultivation of tropical soils presents formidable problems, even though nutrients may be available in abundance soon after the soils have been cleared and crops may be good for two or three years. Despite these relationships, the tropical forests of the world are being cut and burned at an ever-increasing rate, mainly to produce fields that become completely useless to agriculture within a few years. It is estimated that, by early in the next century, most of the tropical rainforests will have disappeared, with the exception of those in the western Amazon Basin and central Africa.

SAVANNAS AND DECIDUOUS TROPICAL FORESTS

Savannas consist of grassland with scattered broad-leaved deciduous and evergreen trees, which may occur singly or in groves. Some savannas are dominated by trees, often thickly so, and others are dominated by shrubs; they occupy the region between the evergreen tropical rainforest biome and the desert biome (Figure 32–7). Savanna trees are generally deciduous and lose their leaves in the dry season. Savannas cover large areas of East Africa (see Figure 32–1), but they are also found on the margins of rainforests everywhere, where the rainfall is seasonal and limiting. Savannas also occur in the region between the prairies and temperate deciduous forests and between prairies and taiga in North America and over much of eastern Mexico, Cuba, and southernmost Florida (Figure 32–8).

32–7

A savanna in Kenya, with zebras, a giraffe, and an impala. The transitional nature of this biome, relative to the characteristic vegetation of the tropical rainforest biome and the desert biome, is evident in the grasses, shrubs, and short trees seen here. The trees in the background are acacias.

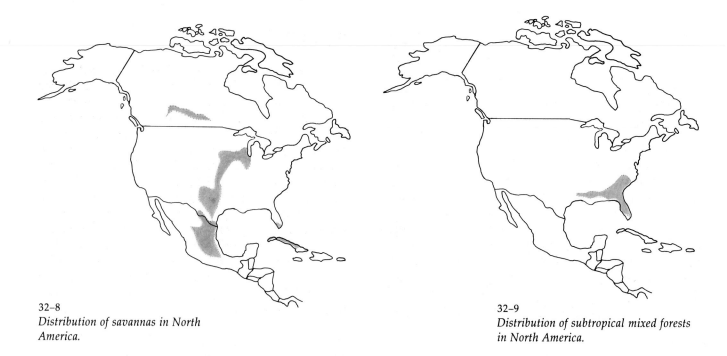

32–8
Distribution of savannas in North America.

32–9
Distribution of subtropical mixed forests in North America.

Savannas usually have much lower annual rainfall than tropical rainforests—frequently in the range of 90 to 150 centimeters a year. There is also a wider range in average monthly temperatures, owing to the seasonal drought and the sparse covering of vegetation. The thorn forest (cerrado) that covers wide areas in Brazil belongs to the savanna biome. Savannas and similar seasonally deciduous tropical and subtropical plant communities grade into tropical rainforests as the rainfall becomes higher and its seasonal distribution more even; and corridors of tropical rainforest (called gallery forest) extend out into the savannas along rivers. At their poleward margins, savannas and related plant communities grade into deserts.

In savannas, because of the scattered distribution of trees, the ground is generally well illuminated, and perennial herbs (mostly grasses) are common. Bulbous plants, which are able to withstand periodic burning, are abundant. Because of the dense cover of perennial herbs made possible by the abundant seasonal rainfall, there are few annual herbs. Epiphytes are also rare.

Savanna trees often have thick bark. They are well branched, but they are seldom more than 15 meters tall. Nearly all of them are deciduous; they lose their leaves at the start of the dry season and flower when they are leafless. Their leaves are generally smaller than those of the evergreen trees of the rainforest, and so they lose less water by transpiration.

In Southeast Asia, there are extensive areas of monsoon forest (see Figure 32–1), a kind of forest that resembles savanna in being seasonally dry. In the monsoon forest, however, the trees are much denser than they are in savannas, and the trees and shrubs are mainly broad-leaved and deciduous. Monsoon forests are one particularly widespread kind of seasonally deciduous forest, and a number of others are found in different parts of the tropics. Monsoon forests are also found in the northern Yucatán Peninsula in Mexico and in eastern Brazil (see Figure 32–1). In the areas where such forests occur, there is high precipitation during portions of the year, when the moisture-bearing monsoon is blowing off the ocean, but there is also a well-defined dry season during which the trees lose their leaves.

Other kinds of deciduous and partly deciduous forests occur in the tropics and subtropics, in areas where there is a pronounced dry season. Most of Florida and extensive areas throughout the southeastern United States are covered by subtropical mixed forests (Figure 32–9). In these forests, pines (Figure 32–10) and other evergreen trees are mixed with deciduous trees; the precipitation occurs mainly in the summer. Tropical mixed forests, in which evergreen trees and shrubs are better represented than in subtropical mixed forests, occur locally in eastern and southern Brazil and northern Australia (see Figure 32–1). Related communities—subtropical and tropical mixed forests and woodland and scrub—are found in Argentina (see Figure 32–1).

The existence of the savannas and monsoon forests of Africa (Figure 32–7) and other regions depends to a large extent on periodic burning. The winter is long and hot, and people frequently burn the plain to encourage the growth of young grass for the game that they hunt, as well as for their domestic herds of grazing animals.

32–10

Slash pine (Pinus elliottii) *is one of the widespread evergreen tree species that occur in the subtropical mixed forests of the southeastern United States.*

DESERTS

The great deserts of the world are all located in the zones of atmospheric high pressure that flank the tropics at about 30° north latitude and 30° south latitude, and they extend poleward in the interior of the large continents (see Figure 32–1). Many deserts are characterized by less than 20 centimeters of rainfall per year. In the Atacama Desert of coastal Peru and northern Chile, the average rainfall is less than 2 centimeters per year. Extensive deserts are located in North Africa and in the southern part of the African continent, where the Namib Desert is inhabited by some of the world's most extraordinary organisms, including *Welwitschia* (see Figure 18–35 on page 353). Other deserts occur in the Near East, in western North America and western South America, and in Australia. The Sahara, which extends all the way from the Atlantic coast of Africa to Arabia, is the largest desert in the world. Australia's desert, which covers some 44 percent of the continent, is the next largest. Less than 5 percent of North America is desert (Figure 32–11).

Many desert regions are characterized by very high temperatures; summer temperatures of more than 36°C are common in some deserts. Other deserts are cooler, on the average; in the Great Basin Desert of western North America, which lies between the Sierra Nevada–Cascade Mountain system and the Rocky Mountains, there are only a few weeks of warm temperatures each year. In general, however, the water vapor in desert air, even when it is relatively abundant, fails to condense because of the high temperatures. Because the cover of vegetation is generally sparse, heat normally radiates from deserts rapidly at night, creating large differences in temperature in the course of a single 24-hour period.

32–11

*North American deserts. The Sonoran Desert stretches from southern California to western Arizona and south into Mexico. Some of its characteristic plants are shown in Figure 32–12a through c. North of the Sonoran Desert is the Mojave Desert; one of its characteristic plants is the Joshua tree (*Yucca brevifolia; *Figure 32–12d). The Mojave Desert includes Death Valley, the lowest point on the continent (90 meters below sea level). Death Valley is only 130 kilometers from Mount Whitney, which, with an elevation of more than 4000 meters, is the highest point in the contiguous United States. The Mojave Desert blends into the Great Basin Desert, a cold desert that is bounded by the Sierra Nevada to the west and the Rocky Mountains to the east. Vast stretches of the Great Basin Desert are dominated by sagebrush (*Artemisia tridentata; *Figure 32–12e) and rabbitbush (*Chrysothamnus). *To the east of the Sonoran Desert is the Chihuahuan Desert.*

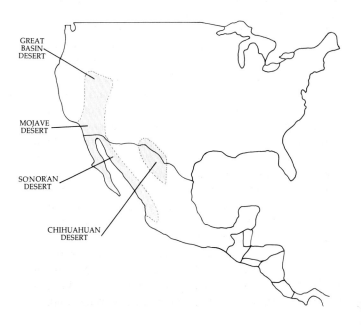

GREAT BASIN DESERT

MOJAVE DESERT

SONORAN DESERT

CHIHUAHUAN DESERT

The annual distribution of rainfall in desert areas generally reflects that of the bordering areas. On the equatorial side, it rains in the summer; on the poleward side, it rains in the winter. Between the two, as in the lowlands of Arizona, there may be two annual peaks of precipitation. As a result, in such an area there are generally two periods of active plant growth—one in the winter and one in the summer—and different plants are active in each period. In general, the patterns of activity of desert plants reflect their respective origins; plant species that have migrated into the desert from areas where they grew actively in the winter continue doing so in the desert, and plant species that have migrated into the deserts from areas where they grew during periods of summer rainfall continue to grow actively in the summer in the desert, provided that there is a sufficient amount of summer rainfall to make this possible.

Annual plants are better represented, both in number and in kind, in the deserts and semiarid regions of the world than anywhere else. Because of the erratic supply of water, perennial herbs do not succeed well in deserts and semideserts, and there is no dense covering of perennials, which would normally inhibit the growth of annuals. Because of their very rapid growth, annuals can spring up in the open areas during the limited periods when water is available. The seeds of these plants can survive in the soil during the long periods of drought, which sometimes extend over many years. Then, when sufficient water is available to induce their germination and support their growth to flowering, they can germinate rapidly.

The relatively few perennial herbs that occur in the desert are often bulbous and dormant for much of the year. Most of the taller plants are either succulents—such as the cacti, euphorbias, and other characteristic desert plants (Figure 32–12)—or have small leaves that either are leathery or are shed during unfavorable seasons (Figure 32–13). Usually the leaves have a thicker cuticle and fewer stomata than those of plants in less arid regions. Many desert plants have green stems, rich in chlorophyll, which may contribute large amounts of photosynthate to the plants; such plants often lack leaves, as do desert species of cacti. Many of the succulent plants have adopted CAM photosynthesis (see page 111) and absorb carbon dioxide only at night; their stomata remain closed during the day. C_4 photosynthesis is likewise more common among the plants of deserts and other seasonally dry, warm habitats than it is elsewhere (see page 110).

The temperatures at which the maximum photosynthetic rate occurs in desert plants are often much higher than the corresponding temperatures for plants that occur elsewhere. For example, in *Tidestromia oblongifolia*, a C_4 perennial herb that is common in the Death Valley region of California and Nevada, the maximum photosynthetic rate is achieved at temperatures be-

(a)

32–12

Desert plants. (a) Community of saguaro cactus (Carnegiea gigantea) and barrel cactus (Ferocactus) in the Sonoran Desert of southern Arizona. (b) Washington palms (Washingtonia filifera), growing along a creek in Palm Canyon, in the Sonoran Desert of southern California. (c) Coastal desert of Baja California, Mexico, which receives much of its moisture from fog, dominated by boojum trees (Fouquieria columnaris, formerly called Idria) and agaves (Agave shawii). (d) Joshua trees (Yucca brevifolia) at Joshua Tree National Monument in the Mojave Desert of southern California. (e) Great Basin Desert, dominated by sagebrush (Artemisia tridentata), in the John Day Basin of eastern Oregon.

(b)

(c)

(d)

(e)

32–13

One of the most characteristic plants of the Mojave, Sonoran, and Chihuahuan deserts of North America is the creosote bush (Larrea divaricata). Creosote bushes in the Mojave Desert may form circular or elliptical clones because of new branch production at the periphery of stem crowns and segmentation and death of older stem segments, resulting in a ring of satellite bushes around a central bare area. The latter usually accumulates a mound of sand, which may attain a depth of about a half meter and may function as a water-storage medium, thus contributing to the survival of creosote bushes in periods of drought. Some clones may attain extreme ages: The King Clone, shown here, is estimated to be nearly 12,000 years old. Such ancient clones, which are under investigation by Frank Vasek and his associates at the University of California, Riverside, started from seeds that germinated near the end of the last glacial expansion.

tween 45° and 50°C in full sunlight in midsummer. In addition, the leaves of many desert plants are oriented in such a way as to minimize heat absorption, or they may be covered with a dense coat of hairs for the same reason. The trees and shrubs that occur in deserts either have wide-ranging roots that effectively absorb the periodic rainfall or are restricted to washes and arroyos (watercourses that are usually dry) where moisture (when it is available) is concentrated.

The juniper savannas that are such a familiar sight throughout the interior of the western United States (Figures 32–14 and 32–15) occur in areas of cold desert,

(a)

(b)

32–14

Juniper savanna in North America.

32–15

(a) Juniper (Juniperus osteosperma) savanna in the Great Basin near Wellington, Utah. (b) Cold winters are characteristic of the pinyon-juniper savannas and woodlands of the Great Basin, as shown here south of Moab, Utah.

but generally at higher elevations. During the pluvial periods of the Pleistocene epoch, at the times of maximum expansion of the continental glaciers, such plant communities moved down into the lowland areas that are often occupied at present by severe deserts. The shifting patterns of vegetation present in a particular area can be traced in the dried stems and leaves in old pack-rat nests, because preservation of these plant materials is so good in the very dry desert air.

GRASSLANDS

The grasslands include a wide variety of plant communities; some intergrade with savannas, others with deserts, and still others with temperate deciduous forests. As the amount of precipitation decreases, the major grasslands of the world often grade into deserts toward the equator. The more fertile grasslands receive as much as 100 centimeters of precipitation per year; they grade into temperate deciduous forests, where the moisture supply is even more abundant. Unlike the savannas, the grasslands generally lack trees, except along streams, and are characterized by cold winters.

Grasslands form the zone that lies between deserts and temperate deciduous forests, occurring where the amount of rainfall is intermediate between the amounts characteristic of those two biomes. When disturbed, grasslands have often changed into either forests or deserts. The failure of people to recognize the existence of the delicate ecological balance in grasslands was the major cause of the ''dust bowl'' disasters of the central United States in the 1930s (Figure 32–16).

Grasslands have traditionally been heavily exploited for agriculture, both as pasture and, when cleared, for cultivation. Much effort has been devoted to learning how to convert other biomes into grassland and, once the conversion has been made, to learning how to maintain them in this condition. The most productive soils for temperate agriculture occur in areas formerly occupied by tall-grass prairies.

Grasslands generally occur over large areas in the interior portions of the continents (see Figure 32–1). In North America, there is a transition from the more desertlike, western, short-grass prairie (the Great Plains), through the moister, richer, tall-grass prairie (the Corn Belt), to the eastern temperate deciduous forest (Figures 32–17 and 32–18). Grasslands become drier and drier at increasing distances from the Atlantic Ocean and the Gulf of Mexico, which are the major sources of moisture-bearing winds in the eastern half of the North American continent. Farther north, they become moister again as evaporation becomes lower at relatively cool temperatures. "Typical" short-grass prairie occurs under near drought conditions (Figure 32–19); in moister years, taller grasses (but not the very tall ones of the tall-grass prairie) tend to predominate. In many for-

32–16

The prairie soils were once so bound together with the roots of grasses that they could not be cultivated until adequate plows were developed. But once the plants were removed by overgrazing or careless cultivation, prairie soils rapidly deteriorated and were carried away by the wind. This photograph of a badly eroded farmyard, taken in Oklahoma in 1937, vividly recalls the "dust bowl" conditions that led many thousands of people to migrate away from the central United States. John Steinbeck's novel, The Grapes of Wrath, *was based on the experiences of these migrants.*

32–17

Distribution of the grasslands of North America. Aside from the main areas of prairie in the center of the continent, there are disjunct areas of grassland in and around the Central Valley of California and along the Gulf Coast of southeastern Texas and Louisiana. The soils in all of these areas are fertile, of good structure, and excellent for sustained agriculture.

32–18

(a) *Tall-grass prairie in the Flint Hills of eastern Kansas. This beautiful area has been proposed as the site of a prairie national park.* (b) *Hill prairie near Alton, Illinois, dominated by big bluestem* (Andropogon gerardi). (c) *The rich hill prairies are almost entirely under cultivation, as in this Illinois scene. Soybeans* (Glycine max) *are being grown in the field.*

(a)

(b)

(c)

(a)

32–19

(a) *Short-grass prairie in northeastern Colorado, with a young pronghorn antelope* (Antilocapra americana). *The bluish clumps of vegetation are fringed* sagebrush (Artemisia frigida). *Healthy short-grass prairie of this sort is dominated by C_3 grasses of medium* stature. (b) *Overgrazed short-grass prairie in northeastern Colorado, dominated by drought-resistant C_4 grasses.*

(b)

32–20

Coreopsis *dominating a glade in temperate deciduous forest near St. Louis, Missouri. Such openings in the forest, which usually occur in areas of shallow soil, are often dominated by prairie plants that form continuous stands beyond the borders of the forest.*

ested areas, in the central and eastern United States, particularly where the soil is shallow, there are open areas, called glades, that are typically occupied largely by prairie plants (Figure 32–20).

Perennial bunchgrasses and sod-forming grasses are dominant, but other perennial herbs are common. Although the growth of grassland plants is seasonal, there is little room for the development of annual herbs, and these are essentially absent from this biome. Occasionally, annuals and weeds from other areas become established in disturbed areas, such as around the burrows of animals, near buildings, and along roadsides.

All of the great grasslands of the world were once inhabited by herds of grazing mammals associated with large predators. Such herds were widespread in the periods of maximum expansion of the glaciers during the Pleistocene epoch, when they were widely hunted by our ancestors. Many of these grazing mammals, such as the American bison, have since been hunted almost to the point of extinction. They survive today mainly in refuges, having given way to herds of domestic animals and cultivated fields.

(a)

(b)

(c)

32–21
*The deciduous forests of the southern
Appalachians are among the richest
temperate-zone forests in the world, containing many kinds of hardwoods. (a)
Mixed forest with hemlocks* (Tsuga
canadensis) *in western Tennessee. (b) A
deep cove, with rich deciduous forest. (c)
In fall, the leaves of the maples turn a
beautiful scarlet. (d) Several species of*
Trillium *are among the herbaceous
perennials that occur in these forests.*

(d)

TEMPERATE DECIDUOUS FORESTS

Temperate deciduous forest is almost absent in the
Southern Hemisphere, but it is represented on all the
major land masses of the north (see Figure 32–1). This
type of forest reaches its best development in areas
with warm summers and relatively cold winters (Figures 32–21 and 32–22). Annual precipitation generally
ranges from about 75 to 250 centimeters. The deciduous (leaf-dropping) habit may be related to the unavail-
ability of water during much of the winter, which is a
consequence of soil temperatures below the freezing
point. These conditions delay the movement of water
to the roots. In the colder regions occupied by temperate deciduous forest, the soils are frozen for many
months.

The ecology of the temperate deciduous forest differs
from that of evergreen forests in the nature of the annual cycle of growth (Figure 32–23). In winter, the trees
are leafless and their metabolic activity is greatly re-

32–22
*Distribution of the temperate deciduous
forests of North America.*

32–23

The four seasons in a temperate deciduous forest in Illinois. In such forests, the trees produce their leaves early in spring and begin to manufacture food; they lose them again in the autumn and enter an essentially dormant condition, in which they pass the unfavorable growing conditions of winter. Many herbs grow under the trees (see Figure 32–21d), and a number of these flower very early in the spring, before the leaves of the trees have reached full size and have begun to shade the forest floor. In the spring, most of the trees produce abundant pollen. It is carried by the wind, sometimes reaching the flowers of other trees of the same species.

duced. In spring, a variety of herbaceous plants bursts forth in profusion on the well-illuminated forest floor (Figure 32–21d). Before the appearance of tree leaves, which dramatically reduce the amount of light reaching the forest floor, most of the early spring herbs have set their seeds. Very few annual plants occur in the deciduous forest, probably because they lack the storage organs that enable the perennials to grow rapidly during the relatively short time that light and moisture are abundant.

The soils that occur in regions occupied by temperate deciduous forest are usually acid, and they tend to contain relatively low amounts of nutrients. For these reasons, they are easily depleted of nutrients when they are cleared, and they may become highly infertile following years of intensive cultivation. Prairie soils are far more fertile and have characteristics much more favorable to sustained cultivation.

One of the outstanding characteristics of the temperate deciduous forest is that the plants that are found in each of its three main regions in the Northern Hemisphere are similar to one another, often representing related species of particular genera, for example. In contrast, the deserts of various regions of the world are generally inhabited by different families and genera of plants that have converged in some of their morphological and physiological characteristics (see essay, page 443). The plants of the temperate deciduous forest are a much more uniform group. For example, the herbaceous plants of the forests of Japan closely resemble those of eastern North America—much more closely, in fact, than those of either area resemble those of western Northern America.

TEMPERATE MIXED AND CONIFEROUS FORESTS

Bordering the deciduous forests on the north are mixed forests, in which conifers form an important element along with the deciduous trees. Such temperate mixed forests (Figures 32–24 and 32–25) are characteristic of the Great Lakes–Saint Lawrence River region, much of the southeastern United States, eastern Europe, the northern and eastern border region of Manchuria and adjacent Siberia, eastern Korea, and northern Japan (see Figure 32–1). They occur in areas with colder winters than the temperate deciduous forests, and they represent a broad transitional zone between the temperate deciduous forests to the south and the taiga to the north.

Most of the deciduous trees and their associated herbs were eliminated in western North America during the latter half of the Cenozoic era as the amount of precipitation that fell during the summer was greatly reduced. In their place now are the mountain forests

32–24
Distribution of temperate mixed forests in North America.

32–25
Temperate mixed forest in southern Ontario, with the evergreen conifers appearing dark green and the deciduous trees showing their fall colors.

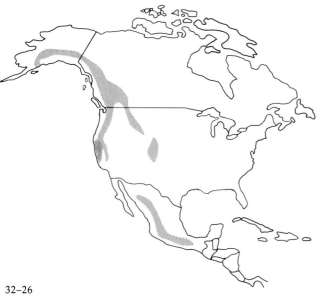

32–26
Distribution of mixed west-coast forests, alpine tundra, and mountain forests in North America.

and west-coast forests of western North America (Figure 32–26), which contain such trees as the coast redwood (*Sequoia sempervirens*; Figure 32–27), the big tree or giant sequoia (*Sequoiadendron giganteum*), the Douglas fir (*Pseudotsuga menziesii*; Figure 32–28), and the sugar pine (*Pinus lambertiana*; Figure 31–17), all of which were much more widely distributed in the past than they are at present. Similar kinds of vegetation are found in areas with similar climatic characteristics in Scandinavia, central Europe, the Pyrenees, the Caucasus, the Urals, southern Tibet, and the Himalayas northward to eastern Siberia. Vegetation types that resemble these are also found in disjunct areas in western South America, central New Guinea, southwestern New Zealand, the southern Arabian Peninsula, Ethiopia, and the mountains of Central Africa (see Figure 32–1). At higher elevations in these regions are found the open grassland communities called alpine tundra (Figure 32–29), which occurs intermixed with mountain forests from the Brooks Range in southern Alaska southward into the Rocky Mountains, the Cascades, and the Sierra Nevada (Figure 32–26).

32–27
Redwoods (Sequoia sempervirens) are a prominent feature of the mixed west-coast forests of California. Watered by the frequent fogs of the region during the dry summers and protected from freezing temperatures by their proximity to the ocean, redwoods often form spectacular groves, many of which, like the one shown here in Muir Woods near San Francisco, are now protected in parks and reserves. (See also Figure 31–1.)

32–28
Western coniferous forest in the Cascade Mountains of Washington, dominated by Douglas fir (Pseudotsuga menziesii).

32–29

Alpine tundra on the Olympic Peninsula in Washington. Such mountain tundra is comparable in many respects with the tundra found hundreds of miles to the north in the arctic. Here, however, forested slopes are found within a hundred meters or so of the alpine meadows.

MEDITERRANEAN SCRUB

Highly distinctive scrub communities have evolved from mixed deciduous–evergreen forests in areas with *mediterranean* climates—areas that are characterized by cool, moist winters and hot, dry summers (Figure 32–30). Such climates are found along the shores of the Mediterranean Sea, over a large part of California (and for a short distance to the north and south), in central Chile, in southwestern Africa, and along portions of the coast of southern and southwestern Australia (see Fig-

ure 32–1). The plants in these areas—often evergreen or summer-deciduous trees and shrubs—have relatively short growing seasons that are restricted to the cool part of the year, when moisture is relatively abundant. They may lock up nutrients efficiently in their evergreen leaves. In mediterranean climates, the luxuriant growth of spring is followed by drought and dormancy during the summer.

Fire is a prominent ecological factor in mediterra-

32–30

Chaparral in the mountains near Los Angeles, California. This sort of plant association, consisting of broad-leaved, drought-resistant, and, often, spiny evergreen shrubs, occurs only in limited areas, but it has evolved independently on five separate continents.

nean-type vegetation, and fires apparently occurred regularly in such plant communities long before they were invaded by human beings. Today, fire can be a serious problem in such areas as southern California, where dwellings extend far up into the chaparral. *Chaparral* is the word used in California and elsewhere in western North America for evergreen, often spiny shrubs that often form dense thickets. Similar vegetation types have evolved elsewhere, particularly in areas characterized by mediterranean climates, and have been given different names locally; the equivalent vegetation formation around the Mediterranean Sea is called *maquis*; in Chile, *matorral*; in South Africa, *fynbos*.

As is true of the deserts of the world, each of the areas of mediterranean climate is isolated from the others, and each has evolved its own distinctive assemblage of plants and animals. The degree of ecological convergence is high, however, and the chaparral of California is quite similar in appearance to the matorral of Chile or the maquis of the Mediterranean, even though the plants are mostly unrelated. Seasonal drought enhances the importance of edaphic (soil-related) and biotic variation, and small differences in precipitation often have profound effects on the vegetation and animal life present in the area. Hence, these areas often have high proportions of extremely local species of plants and animals, many of them now in great danger of extinction. In their modern form, these areas have already been profoundly changed by people; much of their vegetation occurs now in highly altered condition—for example, with more shrubs and fewer trees, or with more spiny and poisonous plants, than before people occupied those areas with their grazing animals.

TAIGA

This northern coniferous forest, or boreal forest, is characterized by a persistent cover of snow in the winter and a severe climate toward the north. In the southern reaches of the taiga, the trees are taller and more luxuriant, often reaching 75 meters or more in height (Figure 32–31), but in its main, northern area, they are shorter, comprising thousands of square kilometers of uniform forest with relatively few species of plants and animals (Figure 32–32). *Taiga* is the Russian word for vegetation of the sort found in this biome, which extends over much of the Soviet Union, Scandinavia, and northern North America (Figure 32–33; see also Figure 32–1). It is evergreen almost everywhere except in large areas of northeastern Siberia, where larches *(Larix)*, which are deciduous conifers, are dominant.

Taiga occurs in the interior of large continental masses at the appropriate latitudes. In such regions, extreme temperatures range from −50°C to 35°C. Taiga is flanked on the south by montane forests (as in western North America), deciduous forests, savannas, or grassland, depending on the amount of precipitation in the region. Because continental masses do not occur at the appropriate latitudes in the Southern Hemisphere, taiga is absent there (see Figure 32–1). Owing to the influence of the prevailing westerlies blowing over relatively warm currents between 40° and 50° north latitude, the western portions of North America and Eurasia are characterized by milder climates than their eastern portions. Consequently, taiga is found somewhat farther north toward the Pacific coast than it is along the Atlantic coast (Figure 32–33), and the same is true of the individual distributions of many kinds of

32–31
The dense, tall forest that grows on the rainy slopes and flats of the Olympic Peninsula in Washington is the southern expression of the taiga, or northern coniferous forest. Epiphytic mosses, liverworts, and lichens often grow luxuriantly on the trees, as seen here.

32–32
The northern taiga, which covers hundreds of thousands of square kilometers in the cooler part of the North Temperate Zone, is dominated by white spruce (Picea glauca) *and larch* (Larix), *which become smaller as they range northward. This photograph was taken in northern Manitoba, Canada.*

32–33
The distribution of taiga in North America.

plants and animals that inhabit this biome. The northern limits of taiga are ultimately determined by the severity of the arctic climate.

In the extensive northern reaches of the taiga, most of the precipitation falls in the summer; the cold winter air in these regions has a very low moisture content. The annual total precipitation usually amounts to less than 30 centimeters. The rate of evaporation is low, and lakes, bogs, and marshes are common (Figure 32–32). In addition, more than three-quarters of the northern reaches of the taiga is underlain by permanent ice, or *permafrost*, usually within less than a meter of the surface of the soil. For this reason, the ground is usually

moist, despite the low precipitation; the water that is present cannot percolate down through the soil. The only areas that are normally free of permafrost are south-facing slopes and floodplains adjacent to the major rivers, as well as the southern extensions of these boreal coniferous forests. Fires are common in the taiga, and they result in generally warmer, more productive sites for at least 10 to 20 years afterward, owing to local melting of the permafrost. In general, taiga soils are highly acidic, very low in nutrients, and poorly suited for agriculture.

Species of a few genera of trees are common in the northern taiga, including spruce *(Picea)*, larch *(Larix)*, fir *(Abies)*, and poplars *(Populus)*. Among the more common shrubs are dewberries *(Rubus)*, Labrador tea *(Ledum)*, willows *(Salix)*, birches *(Betula)*, and alders *(Alnus)*. Occasionally, in warm, dry areas, groves of pines *(Pinus)* are found. The members of all of these genera of trees and shrubs are ectomycorrhizal, and they occur in dense stands consisting of only one or a very few species. Perennial herbs are common; mosses and lichens are especially so, often forming luxuriant masses. Annual plants are essentially absent.

At its northern limits, taiga grades unevenly into tundra. In both of these biomes, the days are long during the relatively short growing season; far enough north, the sun does not set during at least part of the summer. Because of the abundant light and favorable temperatures that occur seasonally, cultivated plants, such as cabbages *(Brassica oleracea)*, may grow rapidly in cleared areas in the taiga, attaining large sizes in a remarkably short period of time.

(a)

(b)

32–34

(a) *Arctic wet coastal tundra near Prudhoe Bay, Alaska, in late summer. The reddish brown plants are the grass* Arctophila fulva, *the green ones are the sedge* Carex aquatilis. *Note that the water table is above the surface because* of the presence of permafrost; such conditions are characteristic of the arctic tundra. (b) *Arctic tundra at Barrow, Alaska. A clone of the sedge* Eriophrum angustifolium *growing out into a solid patch of another species of the same* genus, E. scheuchzeri. *Vegetative reproduction, like that illustrated by* E. angustifolium *here, is characteristic of many tundra plants.*

ARCTIC TUNDRA

Arctic tundra is a treeless biome that extends to the farthest northern limits of plant growth (Figure 32–34). It occupies an enormous area: fully one-fifth of the earth's land surface (see Figure 32–1). Arctic tundra is best developed in the Northern Hemisphere and is mostly found north of the Arctic Circle, although it extends farther south along the eastern sides of the continents (Figure 32–35). The Arctic tundra essentially constitutes one huge band across Eurasia and North America, with alpine tundra, more closely related to the adjacent mountain forests, extending southward in the mountains (Figure 32–29; see also Figure 32–3). Some species of plants that occur in the Arctic tundra have wide circumpolar ranges.

The entire region occupied by Arctic tundra is underlain by permafrost. The soils are acid to neutral, low in nutrients, and generally poor for agriculture. Even though precipitation is usually less than 25 centimeters per year, much of it is held near the surface, and the ground is usually wet. Fixed nitrogen is generally in very short supply, and there are only a few species of legumes and other plants with symbiotic bacteria that fix atmospheric nitrogen. Evaporation is low, because of the relatively high moisture content of the air at low temperatures. In contrast, some tundra areas are so dry as to constitute true polar deserts. Most of the land north of 75° north latitude is a desert or semidesert,

32–35

The distribution of Arctic tundra in North America.

with few plants taller than 5 centimeters. This is a result of the very low precipitation in both summer and winter and the very cold winter temperatures.

In order for plant growth to occur at all, the mean temperature must be above freezing for at least one month of the year. The growing season in many areas of Arctic tundra is less than two months. Several genera of low shrubs, including birch (*Betula*), willow (*Salix*), blueberry (*Vaccinium*), and Labrador tea (*Ledum*), are common in the tundra, and there are a number of genera of perennial herbs, but only one native annual species (*Koenigia islandica*). Many of the plants that grow in the Arctic tundra, including a number of the grasses and sedges, are evergreen, thus enabling them to initiate photosynthesis immediately when the conditions of light, moisture, and temperature allow it. Plant height is controlled primarily by snow depth in the winter; large woody plants are absent also because of their relatively high rates of respiration, which are especially unfavorable at low temperatures. A number of the plants have relatively large, showy flowers, which they must expend considerable quantities of energy to produce. Such flowers produce energy-rich rewards for their pollinators—necessary rewards, given the low temperatures that prevail at high latitudes. Vegetative propagation is characteristic of many of the perennials, and this may be correlated with the uncertainties of setting seed during the brief arctic summer. Much of the biomass of tundra plants—from half to as much as 98 percent—is underground, consisting not only of roots but also of rhizomes and other kinds of underground stems.

North of the Arctic tundra is ice desert, where physical conditions are even more extreme and vegetation is absent. Ice desert is characteristic of the interior of Greenland, Svalbard off Norway, and Novaya Zemlya off the north coast of Siberia. Much of Antarctica, not mapped in Figure 32–1, is also covered by ice.

SUMMARY

Tropical rainforest, in which neither water nor low temperature is a limiting factor, is so far the richest biome in terms of number of species. The trees are evergreen and characterized by large leathery leaves. There is a poorly developed layer of herbs on the forest floor, but there are many vines and epiphytes at higher levels. Tropical soils are often acidic and very poor in nutrients; such soils lose their fertility rapidly when the forest is cleared.

Mostly tropical and subtropical communities that are characterized by a seasonal drought are termed savannas, subtropical mixed forests, monsoon forests, tropical mixed forests, and southern woodland and scrub. The trees and shrubs of these communities are wholly or partly deciduous, shedding their leaves during times of drought. Herbaceous perennials are common. Savannas also occur between the prairies and temperate deciduous forests and between the prairies and the taiga in the United States and Canada. Subtropical mixed forests cover most of Florida and the Coastal Plain of the southeastern United States; in them, evergreen trees such as pines grow intermixed with deciduous ones. Away from the equator, communities of these sorts grade into deserts and semideserts, which are characterized by low precipitation and often high daytime temperatures during at least part of the year. Succulent plants and annual herbs are common in deserts.

Grasslands intergrade with savannas, deserts, and temperate forests; they are characterized by a general lack of trees, except along streams. The most productive soils for temperate agriculture are grassland soils.

In the temperate deciduous forests, most of the trees lose their leaves during the cold (usually snowy) winters, when moisture may be unavailable for growth. Many genera are common to the temperate deciduous forests of eastern North America and eastern Asia. Temperate deciduous forests are bordered by temperate mixed and coniferous forests to the north; in them, conifers play an important role.

Distinctive scrub communities, called chaparral in North America and maquis in the Mediterranean region, have evolved in the five areas of the world with a summer-dry climate and a cool, rainy growing season during the winter. In addition to the two areas mentioned, such vegetation types occur in the Cape Region of South Africa, in central Chile, and in southwestern and southern Australia.

The taiga is a vast northern coniferous forest that extends in unbroken bands across Eurasia and North America and down the Pacific Coast to northern California. In its southern reaches, the taiga is dominated by tall trees with a lush growth of epiphytic bryophytes and lichens; northward, it consists of vast monotonous stretches of forest with very few tree species. North of the taiga is the tundra, a treeless region that also extends around the Northern Hemisphere, mostly above the Arctic Circle, in a band that is broken only by bodies of water. Both the northern reaches of the taiga and all of the tundra are underlain by permafrost, which allows the soil to remain moist despite relatively low precipitation.

SUGGESTIONS FOR FURTHER READING

ATTENBOROUGH, DAVID: *The Living Planet: A Portrait of the Earth*, William Collins Sons & Co., Ltd., London, 1984.
Based on a BBC television series, this volume presents an accurate and exciting overview of life on earth.

BILLINGS, W. DWIGHT: *Plants, Man, and the Ecosystem*, 3rd ed., Wadsworth Publishing Co., Inc., Belmont, Calif., 1978*
An excellent summary of the field, with an emphasis on the physiology of individual organisms.

BORMANN, F. H.: "Air Pollution and Forests: An Ecosystem Perspective," *BioScience* 35: 434–441, 1985.
This article presents an excellent review of the decline being suffered by the forests of the Northern Hemisphere.

BORMANN, F. H., and GENE E. LIKENS: *Pattern and Process in a Forested Ecosystem*, Springer-Verlag, New York, 1979.
An integrated description of the structure, function, and development of the hardwood ecosystem in northern New England, based largely on the Hubbard Brook Ecosystem Study.

BROWN, DAVID E. (Ed.): "Biotic Communities of the American Southwest—United States and Mexico," *Desert Plants* 4 (1–4), 1982.
Well-illustrated account of the deserts, chaparral, and bordering communities of this biologically fascinating region.

EMLEN, J. MERRITT: *Ecology: An Evolutionary Approach*, Addison-Wesley Publishing Co., Menlo Park, Calif., 1973.
A theoretical consideration of ecology, including the construction of models, which are applied imaginatively in a variety of field situations.

* Available in paperback.

EMSLEY, MICHAEL, and KJELL SANDVED: *Rain Forests and Cloud Forests*, Harry N. Abrams, Inc., New York, 1979.
A magnificently illustrated account of two rapidly disappearing but extraordinarily rich ecosystems.

FINEGAN, BRYAN: "Forest Succession," *Nature* 312: 109–114, 1984.
A review of contemporary approaches to the problem of forest succession.

FORSYTH, ADRIAN, and KEN MIYATA: *Tropical Nature: Life and Death in the Rain Forests of Central and South America*, Charles Scribner's Sons, New York, 1984.
Conveys an excellent impression of the rainforests of Latin America.

FUTUYMA, DOUGLAS J., and MONTGOMERY SLATKIN (Eds.): *Coevolution*, Sinauer Associates Inc., Publishers, Sunderland, Mass., 1983.
Outstanding and up-to-date account of this rapidly expanding area of ecological research.

KREBS, CHARLES J.: *Ecology: The Experimental Analysis of Distribution and Abundance*, 3rd ed., Harper & Row, Publishers, Inc., New York, 1985.
Aptly described by its subtitle, this outstanding book provides a sound basis for understanding ecology.

LEIGH, EGBERT G., JR., A. STANLEY RAND, and DONALD M. WINDSOR (Eds.): *The Ecology of a Tropical Forest: Seasonal Rhythms and Long-Term Changes*, Smithsonian Institution Press, Washington, D.C., 1982.

This volume provides an excellent sampling of contemporary research in tropical forest ecology.

RICE, ELROY L.: *Allelopathy*, 2nd ed., Academic Press, Inc., New York, 1984.
An excellent review of chemical interactions between plants and microorganisms, including applications of the principles involved to agriculture, forestry, and other fields.

RICHARDS, PAUL W.: *The Tropical Rain Forest*, 2nd ed., Cambridge University Press, New York, 1974.
The standard reference on this complex, rapidly disappearing biome.

SOUTHWICK, CHARLES H. (Ed.): *Global Ecology*, Sinauer Associates, Inc., Sunderland, Mass., 1985.
An outstanding collection of readings on the topic, with critical comments by the editor.

VALE, THOMAS R.: *Plants and People: Vegetation Change in North America*, The Association of American Geographers, Washington, D.C., 1982.
A detailed and readable account of the changes in the vegetation of the continent associated with human activities.

WHITTAKER, ROBERT H.: *Communities and Ecosystems*, 2nd ed., Macmillan Publishing Co., Inc., New York, 1975.*
A concise introductory text that outlines the characteristics of interacting systems of organisms.

* Available in paperback.

Atoms also contain neutrons, which are uncharged particles of about the same mass as protons. The *atomic mass* of an element is essentially equal to the number of protons and neutrons in the nucleus of one atom. By comparison, electrons are so light that their weight is usually disregarded. For instance, when you weigh yourself, only about 30 grams, or about 1 ounce, of your total weight is made up of electrons.

Isotopes

Not all atoms of the same element have the same atomic mass. These different kinds of atoms are known as *isotopes*. They have the same number of protons—and hence the same atomic number—but different numbers of neutrons. For example, the common form of hydrogen, with its one proton, has an atomic mass of 1 and is symbolized as ^1H, or simply H. Deuterium, ^2H, is an isotope of hydrogen that contains one proton and one neutron, and so it has an atomic mass of 2. Tritium, ^3H, a third isotope of hydrogen, has one proton and two neutrons; it has an atomic mass of 3 (Figure A–1). Like many (but not all) of the less common isotopes, tritium is *radioactive*, which means that its nucleus is unstable and emits energy when it changes into a more stable form. Both deuterium and tritium have nearly the same chemical properties as the more common isotope of hydrogen (^1H), and either can substitute for it in chemical reactions. However, if an atom gains a proton along with two neutrons, it is no longer hydrogen but helium. It now has an atomic number of 2 and an atomic mass of 4 (Figure A–1). The fusion of hydrogen nuclei (protons) to form helium is the source of energy at the heart of the sun and also provides the terrible destructive force of the hydrogen bomb.

Many naturally occurring isotopes are radioactive. All the heavier elements—atoms that have 84 or more protons in their nucleus—are unstable and, therefore, radioactive. All radioactive isotopes emit nuclear particles at a rate proportional to the number of atoms present; they are said to undergo radioactive "decay" as they change to another element. The rate of decay is measured in terms of half-life: the half-life of a radioactive isotope is defined as the time in which half the atoms in a sample have changed into another isotope or

into a stable element. Because the half-life of an element is constant, it is possible to calculate the fraction of decay that will occur for a given isotope over a given period of time.

Half-lives vary widely, depending on the isotope. The radioactive nitrogen isotope ^{13}N has a half-life of only 10 minutes; tritium has a half-life of 12.25 years. The most common isotope of uranium (^{238}U) has a half-life of 4.5 billion years. The uranium atom decays through a series of isotopes and is eventually transformed to an isotope of lead (^{206}Pb).

Isotopes play a number of important roles in biological research. One use is in dating the age of fossils or the rocks in which fossils are found. For example, the proportion of ^{238}U to ^{206}Pb in a given rock sample is a good indication of how long ago that rock was formed. (The lead formed as a result of the decay of uranium is not the same as the lead commonly present in the original rock, ^{204}Pb.) Isotopes are also used as radioactive tracers. The use of radioactive carbon dioxide ($^{14}CO_2$) has played an important role in enabling plant physiologists to trace the path of carbon in photosynthesis, as described in Chapter 6 (see also "Radiocarbon Dating," page 47). A third use of isotopes is in autoradiography, a technique in which a sample of material containing a radioactive isotope is placed on a sheet of photographic film. Energy emitted from the isotope leaves traces on the film and so reveals the exact location of the isotope within the specimen; Figure 22–5, page 435, is an example of an autoradiograph (see also "Radioactive Tracers and Autoradiography in Plant Research," page 572).

Electrons and Orbitals

As early as 400 to 500 B.C., Greek philosophers suggested that matter cannot be forever divided into smaller and smaller parts. However, the modern concept of the atom as the fundamental unit of the chemical elements is less than 200 years old, and our ideas about its structure have undergone many changes in that time. These ideas, or hypotheses, have usually been presented in the form of models, as are many scientific hypotheses.

A–1

Diagrams of atoms of hydrogen, deuterium, tritium, and helium. Note that hydrogen, deuterium, and tritium each have only a single proton and a single electron, so they are chemically similar even though they differ in the number of neutrons in their nuclei. Helium, which has one more proton and one more electron, is chemically very different from hydrogen and its isotopes.

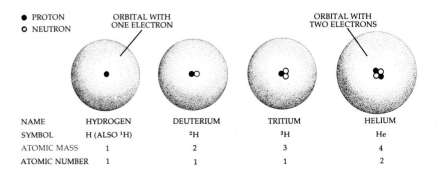

	PROTON	ORBITAL WITH ONE ELECTRON		ORBITAL WITH TWO ELECTRONS
NAME	HYDROGEN	DEUTERIUM	TRITIUM	HELIUM
SYMBOL	H (ALSO ^1H)	^2H	^3H	He
ATOMIC MASS	1	2	3	4
ATOMIC NUMBER	1	1	1	2

The earliest model was of an indivisible atom, resembling a billiard ball. When it was realized that electrons could be removed from the atom, the billiard-ball model gave way to the plum-pudding model, in which the atom was represented as a solid, positively charged mass with negatively charged particles—electrons—embedded in it. Subsequently, however, physicists found that an atom is in fact mostly open space. The distance from electron to nucleus, experiments indicated, is about 100,000 times the diameter of the nucleus; the electrons are so exceedingly small that the space is almost entirely empty. The physicist Niels Böhr proposed a type of planetary model in which the electrons in the atom were depicted as moving in definite orbits around the nucleus, with a specific energy associated with each orbit (Figure A–2).

The current model of electron configurations is quite different from all previous ones; it reflects our increased knowledge of the behavior of electrons. According to this model, the electron moves unpredictably around the nucleus, and its position at any given moment cannot be known with certainty. For convenience, its pattern of motion is defined as the volume of space in which the electron can be found 90 percent of the time. This volume is known as the electron's orbital. Each orbital can hold a maximum of two electrons. In this model, however, the electrons in the atom have definite energies, or energy levels.

The energy levels in the atom are roughly arranged into *shells*, each of which is composed of one or more *subshells*. Each subshell contains one or more variously shaped *orbitals*. The first two electrons occupy a single spherical orbital. Thus, 90 percent of the time, hydrogen's single electron moves about the nucleus within a single spherical orbital, as do the two electrons of helium. This single spherical orbital, with its maximum of two electrons, makes up the first shell.

The second shell is composed of two subshells and four orbitals, each of which can hold two electrons. The

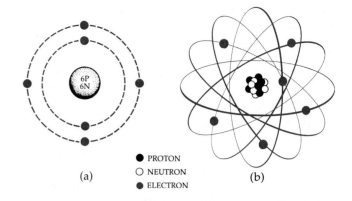

PROTON
NEUTRON
ELECTRON

(a)   (b)

A–2
The Böhr planetary model of a carbon atom, as seen in (a) *two dimensions and* (b) *three dimensions.*

first subshell has a single spherical orbital, and the second subshell has three dumbbell-shaped orbitals. The axes of these three orbitals are perpendicular to one another (Figure A–3). The spherical orbital is filled first, followed by the dumbbell-shaped ones. The second shell can hold a total of eight electrons. Atoms with electrons in as many as seven shells are known.

Atoms tend to stabilize, or complete their energy levels, and the chemical behavior of atoms is governed by this tendency. For instance, helium (atomic number 2), neon (atomic number 10), and argon (atomic number 18) have completely filled outer energy levels and so tend to be unreactive; they are called the "noble" gases because of this apparent "disdain" for reacting with other elements. Atoms of hydrogen (atomic number 1), lithium (atomic number 3), sodium (atomic number 11), and potassium (atomic number 19) have a single electron in their outermost energy level, and they tend to lose this electron. As a consequence of such a loss, each has one more proton than electron and therefore acquires a positive charge: H^+, Li^+, Na^+,

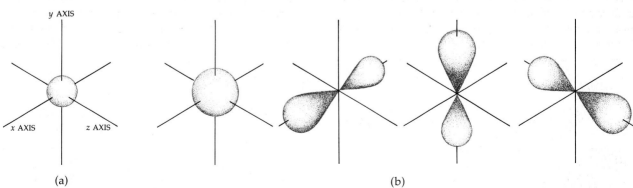

y AXIS

x AXIS   *z* AXIS

(a)   (b)

A–3
Orbital models. (a) *The two electrons in the first shell of an atom occupy a single spherical orbital.* (b) *The second shell has four orbitals, each containing two elec-* *trons. One of these orbitals is spherical and the other three are dumbbell-shaped. The axes of the dumbbell-shaped orbitals* *(indicated by the lines) are perpendicular to one another. The nucleus is at the intersection of the axes.*

and K^+. By contrast, fluorine and chlorine, with atomic numbers of 9 and 17, respectively, tend to gain an electron in order to complete an outer energy level and so become negatively charged: F^- and Cl^-. Similarly, an atom with two electrons in its outer energy level may lose both of them, thus acquiring a double positive charge. For example, magnesium (atomic number 12) and calcium (atomic number 20) become Mg^{2+} and Ca^{2+}. Such charged atoms are known as *ions*. Positively charged ions are called *cations*, and negatively charged ions are called *anions*.

Ions make up less than 1 percent of the weight of most living matter, but they play particularly crucial roles. For instance, K^+ is the principal positively charged ion in most cells, and many essential biological reactions cannot proceed in its absence. Both Na^+ and K^+ are involved in the active transport of sugar and amino acids across the plasma membrane in many plant and animal cells. The calcium cation, Ca^{2+}, has a direct effect on the physical properties of the membrane; Mg^{2+} forms a part of chlorophyll—the molecule that traps the radiant energy from the sun.

Electrons and Energy

As was noted previously, electrons, which are negatively charged, are attracted to the atomic nucleus because of the positive charge of the protons. The orbital occupied by an electron is determined by the amount of energy of the electron; this energy is in the form of potential energy. The following analogy may be useful: a boulder on flat ground may be said to have no energy. If you push it up a hill, you give it energy—potential energy. So long as it sits on the top of the hill, it neither gains nor loses energy. If it rolls down the hill, however, it loses its potential energy as it rolls back toward its original position on level ground.

The electron is like the boulder in that an input of energy can raise it to a higher energy level—to a position farther away from the nucleus. As long as it remains at this higher level, it possesses this added energy. Also, the electron tends to go to its lowest possible energy level, just as the boulder rolls downhill (Figure A–4).

In a given atom, the first spherical orbital is the lowest energy level. The four orbitals of the second level are occupied by electrons with more energy, and so on. Energy is needed to move a negatively charged electron farther away from the positively charged nucleus, just as energy is needed to push a boulder to the top of a hill. However, unlike the boulder on the hill, the electron cannot be pushed part way up. With an input of energy, electrons may move from a lower energy level to any one of several higher ones, but they cannot move to someplace in between. For an electron to move from one level to another, the atom must absorb a discrete packet of energy, known as a quantum, which contains precisely the amount of energy needed for the transition.

Electronegativity

The atomic nuclei of different elements have various degrees of attraction for electrons. The strength of the attraction depends on the number of protons in the nucleus, the number of electrons, and their proximity to the nucleus. The affinity of an atom for electrons is called *electronegativity*. Electronegativity is expressed on a scale of 0 to 4. Helium and the other unreactive noble gases have electronegativities of 0. At the other end of the scale is fluorine, which has an electronegativity of 4. The value for oxygen, the next most electronegative element, is 3.5. Electronegativity values for certain elements are given in Table A–2.

When an electron moves from an atom that is less electronegative to one that is more electronegative, it moves "downhill"—energetically speaking—and energy is released, as potential energy is released when a boulder rolls downhill.

Table A–2 *Electronegativity for Atoms of Some Common Elements*

Oxygen (O)	3.5
Nitrogen (N)	3.0
Chlorine (Cl)	3.0
Carbon (C)	2.5
Sulfur (S)	2.5
Hydrogen (H)	2.1
Phosphorus (P)	2.1
Sodium (Na)	0.9

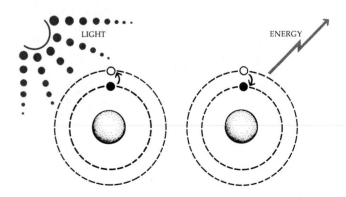

A–4
When an input of energy—such as light energy—boosts an electron to a high energy level, the electron possesses additional potential energy which is released when the electron returns to its previous energy level.

In the cells of photosynthesizing green plants and algae, the radiant energy of sunlight raises electrons to a higher energy level. In the course of a series of electron-transfer reactions (which are described in Section 2), the electrons are passed downhill, and the radiant energy of sunlight is changed to the chemical energy on which nearly all life on earth depends.

BONDS AND MOLECULES

Atoms may be held together by forces known as *chemical bonds*; an assembly of atoms held together by chemical bonds is called a molecule. When atoms interact to form a bond, only electrons in their outer shells (termed the valence shells) are involved. There are two general types of chemical bonds: ionic bonds and covalent bonds.

Ionic Bonds

Positive and negative ions attract one another. Bonds involving the mutual attraction of ions of opposite charge are known as *ionic bonds*. The ionic bond is a very common type of bond in inorganic molecules. Thus, the sodium ion (Na^+)—with its single positive charge—is attracted to the chloride ion (Cl^-)—with its single negative charge (Figure A–5). The calcium ion (Ca^{2+}), with its double positive charge, can attract and hold two Cl^- ions, forming $CaCl_2$—the subscript 2 indicates that two atoms of chlorine are present for each atom of calcium.

The combining capacity of an element is called its *valence*. A valence is the number of positive or negative charges on an ion that consists of a single atom (a monatomic ion) in an ionically bonded substance, or the number of pairs of electrons shared by an atom in a covalently bonded substance. The valences of Na^+ and Cl^- are 1, and the valence of Ca^{2+} is 2. Thus, Na^+ combines with Cl^- in a ratio of 1 to 1, and Ca^{2+} combines with Cl^- in a ratio of 1 to 2.

Covalent Bonds

Another way for an atom to complete its outer energy level is by *sharing* electrons with another atom. Bonds formed by shared pairs of electrons are known as *covalent bonds*. Covalent bonds figure prominently in organic chemistry, which is essentially the chemistry of carbon-containing compounds. The simplest covalent bond is that found between hydrogen atoms in the hydrogen molecule (Figure A–6). In a covalent bond, the shared pair of electrons forms a new orbital that encompasses both atoms. In such a bond, each electron spends part of its time around one nucleus and part of its time around the other. Thus, the electron sharing both completes the outer energy level and keeps the nuclear charge neutral.

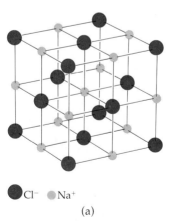

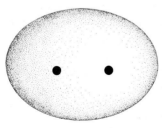

● Cl^- ● Na^+

(a)

(b)

A–5

Sodium (Na), which has only one electron in its outer energy level, becomes more stable if it loses the electron. Chlorine (Cl), which has seven electrons in its outer energy level, becomes more stable if it gains one. When sodium and chlorine interact, sodium loses its single electron and chlorine gains it; following this transaction, sodium has a positive charge (Na^+) and chlorine has a negative charge (Cl⁻). Such charged atoms are called ions. (a) Oppositely charged ions attract one another. Table salt is crystalline NaCl, a latticework of alternating Na^+ and Cl^- ions held together by their opposite charges. Such bonds between oppositely charged ions are known as ionic bonds. (b) The regularity of the latticework is reflected in the structure of salt crystals.

A–6

In a molecule of hydrogen, each atom shares its single first orbital electron with the other atom. As a result, both atoms effectively have a filled first orbital, containing two electrons—a highly stable configuration. The orbital shown here—a molecular orbital—indicates the volume in which the two electrons can be found 90 percent of the time.

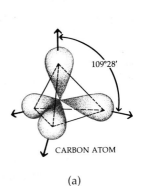

109°28'

CARBON ATOM

(a)

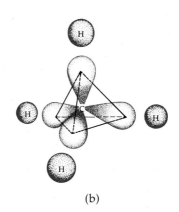

H

H   C   H

H

(b)

METHANE (CH₄)

(c)

A–7

(a) *When carbon forms covalent bonds with four other atoms, new orbitals are formed. These new orbitals, which are all the same shape, are oriented toward the four corners of a tetrahedron. Thus, the*

four orbitals are spatially separated as much as possible. (b) When a carbon atom reacts with hydrogen to form a methane molecule, the four unpaired electrons of the carbon atom form pairs

with each of the unpaired electrons in four hydrogen atoms. (c) Each pair of electrons moves in a new molecular orbital. The molecule has the shape of a regular tetrahedron.

Carbon, which has an atomic number of 6, needs to share four electrons to achieve a filled and therefore stable outer energy level; and so carbon is able to form covalent bonds with as many as four other atoms. When these bonds are formed, the electron pairs assume new orbitals. These orbitals are distributed symmetrically in space, forming a regular tetrahedron (Figure A–7). The capacity of carbon to form four covalent bonds is crucial to its central role in the chemistry of living systems.

Double and Triple Bonds

There are various ways in which atoms can participate in covalent bonds and satisfy their valence requirements. Oxygen, with two unpaired electrons in its outer electron orbitals, has a valence of 2. Carbon and oxygen can thus form a simple compound, carbon dioxide (CO_2), in which each oxygen atom shares two pairs of electrons with a central carbon atom. Such bonds in which two atoms are held together by two pairs of electrons (four electrons) are called double bonds. They are symbolized in a structural formula by two lines connecting the atomic symbols: $O=C=O$. Carbon atoms can form double or even triple bonds (in which three pairs of electrons are shared) with each other as well as with other atoms.

Electrons shared in double and triple bonds form orbitals that differ in shape from the orbitals filled by single electron pairs. For instance, when four bonds satisfy the electron requirements of carbon, they are directed toward the four corners of a tetrahedron that has the carbon atom at its center, as in Figure A–7. When two bonds are replaced by a double bond, the remaining single bonds form the arms of a Y with the double bond

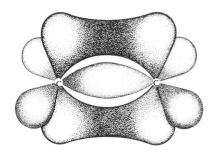

A–8

An orbital model showing carbon-carbon double bond of ethylene. One pair of electrons occupies the inner orbital between the two carbon atoms. The other pair of electrons occupies the outer orbital, which has two phases—one above the plane of the two nuclei and one below. This creates a rigid bond, about which the atoms cannot rotate. The two smaller orbitals extending from each carbon atom contain one electron each and can form covalent bonds with other atoms.

as its leg (Figure A–8). When two double bonds are made by a single carbon atom, as in carbon dioxide, the three bonded atoms lie in a straight line.

Single bonds are flexible, allowing atoms to rotate in relation to one another. Double and triple bonds hold the atoms relatively rigid in relation to one another. The presence of double bonds in a molecule can make a significant difference in its properties. For example, both fats and oils are composed of carbon and hydrogen atoms that are covalently bonded to one another, but in oils some of the bonds are double, and in fats the bonds are all single.

Polar Covalent Bonds

The electrons in covalent bonds are not always shared equally between the two atoms involved. Some atoms have a greater attractive force for the electrons than others—that is, they are more electronegative. Because of this differential attraction, the shared electrons tend to spend more time around the more electronegative atom. As a result, the more electronegative atom in the molecule will have a slightly negative charge, and the less electronegative atom will have a slightly positive one, since its nuclear charge is not entirely neutralized (Figure A–9). The polar properties of some covalent bonds have very important consequences for living things. For example, many of the special properties of water, upon which life depends, derive largely from its polar nature, as is explained below.

Ionic, covalent, and polar covalent bonds may be considered to be different versions of the same type of bond. The differences depend on the differences in electronegativity among the combining atoms. In a wholly nonpolar covalent bond, the electrons are shared equally; such bonds can exist only between identical atoms, as in H_2, Cl_2, O_2, and N_2. In polar covalent bonds, electrons are shared unequally. In ionic bonds, there is an electrostatic attraction between the negatively charged and positively charged ions as a result of their having gained or lost electrons.

Atomic and Molecular Weights

The *atomic weight* of any element is an average value for the naturally occurring mixture of isotopes of that element relative to the common isotope of carbon (^{12}C) which has an atomic mass of 12. Hypothetically, an element exactly twice as heavy as carbon would have an

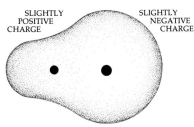

SLIGHTLY POSITIVE CHARGE   SLIGHTLY NEGATIVE CHARGE

HYDROGEN CHLORIDE (HCl)

A–9

In a polar molecule, such as hydrogen chloride (HCl), the shared electrons tend to spend more time around the more electronegative atom—in this case, the chlorine atom. As a result, chlorine has a slightly negative charge, and the less electronegative atom (hydrogen) has a slightly positive charge.

atomic weight of 24; an element one-half as heavy would have an atomic weight of 6. Because atoms are much too small to be weighed individually, they are measured in amounts called *gram atoms*. One gram atom of an element is that amount of the element in grams that equals its atomic weight. For example, 1 gram atom of carbon is that amount of carbon that weighs 12 grams, and 1 gram atom of hydrogen weighs about 1 gram (actually 1.008 grams because of the presence of isotopes in any naturally occurring sample). The *molecular weight* of a substance is the sum of the atomic weights of all the atoms in one molecule. A *gram molecule* is the amount of a substance, in grams, that equals its molecular weight. One gram molecule of oxygen gas (O_2) weighs about 32 grams; 1 gram molecule of hydrogen gas (H_2) weighs about 2 grams; and 1 gram molecule of CO_2 weighs about 44 grams.

The gram molecule is called a *mole*. The number of particles in 1 mole of any substance (whether atoms, ions, or molecules) is always the same: 6.022×10^{23}. For example, 1 mole of water contains 6.022×10^{23} water molecules, and 1 mole of glucose ($C_6H_{12}O_6$) contains 6.022×10^{23} glucose molecules. This numerical constant is known as *Avogadro's number*. The mole is useful for defining quantities of substances involved in chemical reactions.

In order to make water, for instance, one would combine 2 moles of hydrogen (4 grams) with 1 mole of oxygen (32 grams); in other words, four hydrogen atoms for every two oxygen atoms. Two moles of water would be produced, each weighing 18 grams. Similarly, to make table salt (NaCl), one would combine 1 gram atom of sodium (about 23 grams) and 1 gram atom of chlorine (about 35.5 grams).

Functional Groups

Sometimes clusters of atoms joined by covalent bonds tend to react together as a group. One such class of atomic clusters is represented by organic radicals, or *functional groups*. The specific chemical properties of organic molecules are determined primarily by their functional groups. The —OH group is an example (—OH, the functional group, is called hydroxyl; OH^-, the ion, is called hydroxide). When one hydrogen and one oxygen are bonded covalently, one electron on the oxygen is left over for sharing. A compound formed when a hydroxyl group replaces one or more of the hydrogens in a hydrocarbon (a compound consisting only of carbon and hydrogen) is known as an alcohol.

Thus, when one hydrogen atom in methane (CH_4) is replaced by a hydroxyl group, the compound becomes methanol (CH_3OH), a pleasant-smelling, poisonous alcohol noted for its ability to cause blindness and death. Ethane similarly becomes ethanol (C_2H_5OH), which is present in all alcoholic beverages. Glycerol, $C_3H_5(OH)_3$, is an alcohol that contains three hydroxyl groups.

Table A–3 *Some Functional Groups that Play Important Roles in Organic Compounds*

GROUP	NAME
—OH	Hydroxyl
—NH$_2$	Amino
—C—CH$_3$ $\overset{\displaystyle \|}{O}$	Acetyl
$\overset{\displaystyle O}{\overset{\displaystyle \|}{—C—C—C–}}$	Keto
—C—OH $\overset{\displaystyle \|}{O}$	Carboxyl
$\overset{\displaystyle OH}{—O—P—OH}$ $\underset{\displaystyle O}{}$	Phosphate

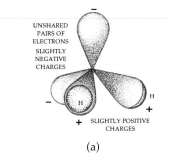

(a)

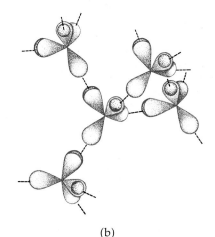

(b)

A–10
(a) *An orbital diagram of a single water molecule. Note the four-cornered distribution of positive and negative charges.* (b) *As a result of these slight positive and negative charges, each water molecule can form hydrogen bonds (indicated by dashed lines) with four other water molecules.*

The carboxyl group (—COOH) is a functional group that gives a compound the properties of an acid. Table A–3 shows some functional groups of considerable biological importance.

WATER AND THE HYDROGEN BOND

Water is composed of a number of very small molecules held together by the mutual attraction of positively and negatively charged atoms. Each water molecule is made up of two atoms of hydrogen and one atom of oxygen, held together by two covalent bonds.

The water molecule as a whole is neutral in charge, having an equal number of electrons and protons. However, the molecule is electrically asymmetric—that is, it is polar. This polarity occurs because oxygen is more electronegative than hydrogen. The paired electrons in the outer orbitals spend more time around the oxygen nucleus than they do around the hydrogen nuclei. Consequently, the region near the oxygen nucleus has two weakly negative zones, and each of the regions near the hydrogen nuclei has a weakly positive zone. Thus, the water molecule—in terms of its polarity—is four-cornered, with two positively charged corners and two negatively charged ones (Figure A–10a).

When any of these charged regions comes close to the oppositely charged region of another water molecule, the force of the attraction forms a bond between them—a *hydrogen bond*. Hydrogen bonds are found not only in water but can form between any hydrogen atom that is covalently bonded to an electronegative atom—usually oxygen or nitrogen—and the electronegative atom of another molecule. In liquid water, hydrogen bonds form between the negative "corners" of one water molecule and the positive "corners" of another. Thus, every water molecule can establish hydrogen bonds with four other water molecules. Liquid water is made up of water molecules bound together in this way, as shown in Figure A–10b.

Any single hydrogen bond is relatively weak and has an exceedingly short lifetime; on an average, such bonds last about 1/100,000,000,000th of a second. But, as one is broken, another is formed. As a group, however, hydrogen bonds have considerable strength, making water both liquid and stable under ordinary conditions of pressure and temperature.

WATER AS A SOLVENT

Many substances within living systems are found in solution. A *solution* is a uniform mixture of the molecules of two or more substances. The substance present in the greatest amount—usually a liquid—is called the *solvent*; the substances present in lesser amounts are

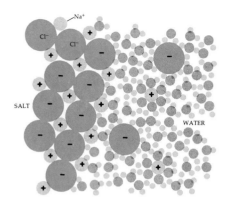

A–11

Because of the polarity of water molecules, water can serve as a solvent for polar atoms or molecules. This diagram shows sodium chloride (NaCl) dissolving in water as the water molecules cluster around the individual ions, separating them from each other. Substances that dissociate into ions in solution produce solutions that can conduct electricity. Such solutions are called electrolytes.

called *solutes*. The polarity of the water molecules is responsible for the capacity of water to act as a solvent. The polar water molecules tend to separate substances such as NaCl into their constituent ions. Then, as shown in Figure A–11, the water molecules cluster around and segregate the charged ions.

Many of the molecules important in living systems, such as glucose, also have polar areas. Such regions of partial positive or negative charges arise in the neighborhood of covalently bonded atoms of unequal electronegativity. Thus, these molecules attract water molecules and are dissolved in the water.

Molecules that readily dissolve in water are called *hydrophilic* ("water-loving"). Such molecules slip into aqueous solution easily because their partially charged regions attract water molecules and thus compete with the attraction between the water molecules themselves.

Molecules that lack polar regions, such as fats, tend to be insoluble in water. The hydrogen bonding between the water molecules acts as a force to exclude the nonpolar molecules. As a result of this exclusion, nonpolar molecules tend to cluster together in water, just as droplets of fats tend to coalesce on the surface of chicken soup, for example. Such molecules are said to be *hydrophobic* ("water-fearing").

ACIDS AND BASES

Acids taste sour, like sour milk, citrus fruits, and vinegar. Bases taste flat, like milk of magnesia, and feel slippery and soapy in solution.

To define "acid" and "base" in chemical terms, it is easiest to begin by looking at water. Water consists of two atoms of hydrogen and one of oxygen held together by covalent bonds. Water molecules also have a slight tendency to ionize—separate into H^+ and OH^- ions. (The H^+ ions tend to combine with other H_2O molecules to produce the hydronium ion, H_3O^+, but we shall omit the extra ion from our consideration in this discussion.) In any given volume of pure water, a very small but constant number of water molecules will be dissociated into ions. The number is constant because the tendency of water to dissociate is exactly offset by the tendency of the ions to reunite; thus, even as some are ionizing, an equal number of others are re-forming, a state known as dynamic equilibrium.

In pure water, the number of H^+ ions exactly equals the number of OH^- ions. This is necessarily the case, because neither ion can be formed without the other when only H_2O molecules are present. A solution acquires the properties we recognize as acidic when the number of H^+ ions exceeds the number of OH^- ions; conversely, a solution is basic when the concentration of OH^- ions exceeds that of the H^+ ions. There is always an inverse relationship between the concentration of H^+ and OH^- ions; when H^+ is high, OH^- is low, and vice versa. This is because the product of their concentrations is a constant. Thus, at $25\,°C$,

$$[H^+][OH^-] = 1 \times 10^{-14} \text{ mole}$$

We now can define our terms chemically:

1. An *acid* is a substance that donates H^+ ions to a solution. Because an H^+ ion is really a proton, acids can also be defined as *proton donors*.
2. A *base* is a substance that decreases the number of H^+ ions, or protons. More specifically, a base is a *proton acceptor*. The OH^- ion is a base because it can accept a proton and thus be neutralized:

$$H^+ + OH^- \longrightarrow H_2O$$

3. In any acid-base reaction, there is always a proton donor and a proton acceptor.

Strong and Weak Acids and Bases

Hydrochloric acid (HCl) is an example of a common acid. It is a strong acid, meaning that it tends to be almost completely ionized in an aqueous solution into H^+ and Cl^- ions. Sodium hydroxide (NaOH) is a common strong base; in an aqueous solution, it exists entirely as Na^+ and OH^- ions. Weak acids and weak bases are those that ionize only slightly. Compounds that contain the carboxyl group (—COOH) are often weak acids because the hydrogen atom may partially dissociate from the carboxyl group to yield a proton:

$$R-COOH \rightleftharpoons R-COO^- + H^+$$

In this reaction, R represents any chemical structure that may be attached to a carboxyl group.

Compounds that contain the amino group ($-NH_2$) act as weak bases, because the amino group has a weak tendency to accept hydrogen ions, thereby forming NH_3^+:

$$R-NH_2 + H^+ \rightleftharpoons R-NH_3^+$$

The pH Scale

Chemists define degrees of acidity by means of the pH scale. In the expression "pH," the "p" stands for "power" and the "H" stands for the concentration of hydrogen ions.

In a liter of pure water, 0.0000001 mole of hydrogen ions can be detected. For convenience, this is written in terms of a power of ten, 10^{-7}, and in terms of the pH scale, it is simply referred to as pH 7 (see Table A–4). At pH 7, the concentrations of free H^+ and OH^- ions are exactly the same, and pure water is thus "neutral." Any pH below 7 is acidic, and any pH above 7 is basic. The lower the pH number, the higher the concentration of hydrogen ions. Thus, pH 2 means 10^{-2} mole of hydrogen ions per liter of water, or 0.01 mole per liter—a much higher concentration than 0.0000001. A difference of one pH unit represents a tenfold difference in the concentration of hydrogen ions.

Lemon juice has a pH of about 2, as do the stomach contents of humans and other animals. Orange juice has a pH of about 3; human blood has a pH of 7.4. The best soil pH for most plants is about 6.4, but alkaline soils have a pH between 7 and 9, and peat bogs may have a pH as low as 3.

CHEMICAL REACTIONS

All chemical reactions involve the breaking of some bonds and the formation of new bonds. According to present concepts, atoms or molecules react with one another only when they collide with sufficient force to overcome the initial forces of repulsion. The force required varies with the nature of the atoms or molecules; the more stable their initial state, the more forceful the collision must be for a reaction to occur. In any given group of atoms, it is likely that some proportion of them is moving with sufficient energy for a reaction to occur, but often this proportion is so small that the reaction, for all practical purposes, does not take place.

Reaction rates can be increased by increasing the likelihood of forceful collisions. One way to do this is to raise the temperature, thereby increasing the average velocity at which the atoms or molecules move and so increasing their likelihood of colliding with sufficient force. Sometimes, as in the case of methane (natural gas), a spark is all that is needed. Once the reaction begins, it liberates heat that is transferred to the other CH_4 molecules until all are moving rapidly enough to react almost simultaneously with explosive force. Driving a reaction by heat is a method commonly used in chemical laboratories, as well as in industry.

A second way to increase the rate of a reaction is to increase the number of reacting molecules. Chemists working in research laboratories or in industry usually work with pure chemicals in high concentrations. By this means, the chemical reactions of interest are not only speeded up, but they are also easier to control.

A third way to increase a reaction rate is to use catalysts. Catalysts lower the _energy of activation_ of a reac-

Table A–4   *The pH Scale*

	CONCENTRATION OF H^+ IONS (MOLES PER LITER)		pH	CONCENTRATION OF OH^- IONS (MOLES PER LITER)	
Acidic	1.0	10^0	0	10^{-14}	
	0.1	10^{-1}	1	10^{-13}	
	0.01	10^{-2}	2	10^{-12}	
	0.001	10^{-3}	3	10^{-11}	
	0.0001	10^{-4}	4	10^{-10}	
	0.00001	10^{-5}	5	10^{-9}	
	0.000001	10^{-6}	6	10^{-8}	
Neutral	0.0000001	10^{-7}	7	10^{-7}	
Basic		10^{-8}	8	10^{-6}	0.000001
		10^{-9}	9	10^{-5}	0.00001
		10^{-10}	10	10^{-4}	0.0001
		10^{-11}	11	10^{-3}	0.001
		10^{-12}	12	10^{-2}	0.01
		10^{-13}	13	10^{-1}	0.1
		10^{-14}	14	10^0	1.0

A–12

Chemical reactions require a continuous input of energy—called the activation energy—in order to take place. An uncatalyzed reaction requires more "input," or activation energy, than does a catalyzed one, such as an enzymatic reaction. The lower activation energy in the presence of the catalyst is often within the range of energy possessed by the molecules, and so the reaction can occur with little or no added energy. Note, however, that the overall energy change from the initial state to the final state is the same with or without the catalyst.

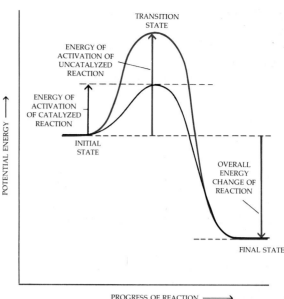

tion, that is, the minimum energy that must be available in a collision for a reaction to be initiated (Figure A–12). Metals, such as iron, nickel, and platinum, are commonly used as catalysts in industrial laboratories. Certain molecules apparently tend to cluster on the surfaces of such metals, thus increasing the likelihood of the necessary kinds of close encounters between the reactants. Although they participate in the reactions, catalysts are not used up, so they can be used over and over again (see Chapter 4, pages 77–81).

Types of Reactions

Chemical reactions can be classified into a few general types. One type can be represented by the expression:

$$A + B \longrightarrow AB$$

An example of this sort of reaction is the combination of hydrogen gas with oxygen gas to produce water:

$$2H_2 + O_2 \longrightarrow 2H_2O$$

A reaction may also take the form of a dissociation:

$$AB \longrightarrow A + B$$

For example, the equation for the formation of water can be reversed to show the breakdown of water into its component elements:

$$2H_2O \longrightarrow 2H_2 + O_2$$

This means that water molecules yield hydrogen and oxygen gases.

A reaction may also involve an exchange, taking the form:

$$AB + CD \longrightarrow AD + CB$$

An example of such a reaction is the combination of hydrochloric acid and sodium hydroxide to form table salt and water:

$$HCl + NaOH \longrightarrow NaCl + H_2O$$

Chemical Equilibrium

Some chemical reactions can go in either direction, as discussed previously. When net change ceases, the reaction is said to be at equilibrium. In the reaction

$$A + B \rightleftharpoons C + D$$

the point of equilibrium is reached when as many molecules of C and D are being converted to molecules of A and B as molecules of A and B are being converted to molecules of C and D.

The concentration of reactants does *not* have to equal the concentration of products in order for equilibrium to be established; only the *rates* of the forward and reverse reactions must be the same. Consider the reaction above. The different lengths of the arrows indicate that there is more C + D present at equilibrium than A + B. If only A and B molecules are present initially, the reaction occurs at first to the right, with A and B molecules converting into C and D molecules. Figure A–13 shows the relative changes in concentration as the reaction continues. As C and D accumulate, the rate of the reverse reaction increases, and at the same time, the rate

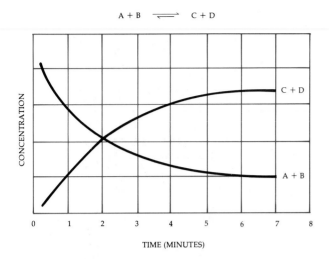

A–13

A graph of the changes in concentration of products and reactants in a reversible reaction. At first, only molecules of A and B are present. The reaction begins when A and B start to yield products C and D. At the end of two minutes, the concentrations of A + B and C + D are equal. As the reaction proceeds, the concentration of C + D will continue to increase to the point of chemical equilibrium (at about six minutes) and will thereafter remain greater than the concentration of A + B. This is the proportion at which the rates of forward and reverse reactions are the same.

of the forward reaction decreases because of the decreasing concentrations of A and B. After about six minutes, the rates of the forward and reverse reactions equalize, and no further changes in concentration take place. The proportions of A + B and C + D will remain the same, but there will always be more C and D molecules.

The Energy Factor

Although a chemical equation must be balanced in terms of its chemical components, one crucial factor is often missing—an indication of the energy changes that accompany the reaction. As discussed earlier, forces of electrical attraction between atoms produce the chemical bonds that hold a compound together. Depending on the strength of these forces, each compound has a particular energy content, or *bond energy* (Figure A–14). The total bond energy of a compound can be defined as the amount of energy required to break that compound into its constituent atoms. All chemical reactions involve the rearrangement of bonds; hence, they are accompanied by changes in energy.

Most chemical reactions result in the loss of energy to the surroundings.

Consider, for example, the burning of methane, represented by the following equation:

$$CH_4 + 2O_2 \longrightarrow CO_2 + 2H_2O$$

This reaction can be set in motion by a spark (which is what causes explosions in coal mines), and then energy is released in the form of heat. The amount of energy released (that is, 213 kilocalories per mole of methane) can be measured quite precisely (Figure A–15). This release of energy can be expressed by a simple equation: $\Delta H = -213$ kcal. The Greek letter delta (Δ) represents "change" and H represents "heat content." In general, the change in heat content is approximately equal to the change in potential energy. The minus sign indicates that energy has been released. (A calorie is defined as the amount of heat necessary to raise the temperature of 1 gram of water by 1°C; 1000 calories = 1 kilocalorie. In expressing the heat- or energy-producing value in a food that is oxidized, the calorie unit ordinarily used is actually a "large Calorie"—that is, a kilocalorie.)

Similarly, changes in energy occur in the chemical reactions that take place in living systems. However, liv-

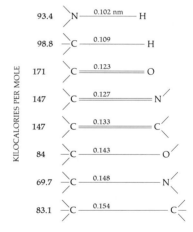

A–14

A chemical bond is a force holding atoms together. The strength of a bond is measured in terms of the energy required to break it. The figures at the left indicate the number of kilocalories per mole needed to break the bonds between the pairs of atoms shown. The lines connecting the atoms represent the bonds, and the figures above them represent the characteristic center-to-center distances between the atoms involved in the bond. Double lines indicate double bonds, which hold the atoms closer together and thus are stronger.

ing systems have evolved ways to minimize the activation energy required for molecules to react, as well as ways to convert a portion of the energy released in some reactions into more usable forms than heat (see Chapter 4).

Endergonic and Exergonic Reactions

A reaction that requires net input of energy is said to be *endergonic* (from the Greek *endon*, meaning "within," and *ergon*, meaning "work"), because energy must be put into it. A reaction that liberates energy is an *exergonic* reaction. An endergonic reaction can be thought of as an "uphill" reaction and an exergonic reaction as a "downhill" reaction. Only exergonic (downhill) reactions can proceed spontaneously.

Endergonic reactions do not occur by themselves; they must be coupled to some downhill process in such a way that the energy released by the downhill process can be used for the uphill process. Thus, the energy released in the downhill process must be at least slightly greater than that required for the uphill process. In living systems, the uphill and downhill movements are often accomplished in very small stages, so that large amounts of energy are not required or released all at once.

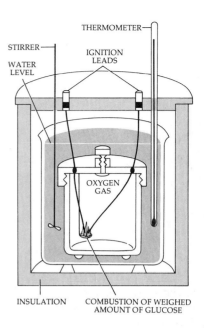

A–15
A calorimeter. A known quantity of glucose or some other material is ignited electrically. As it burns, the rise in the temperature of the water is measured. Based on the specific heat of water, one can then calculate the number of calories released by the burning of the sample.

A P P E N D I X   B

Metric Table

	FUNDAMENTAL UNIT	QUANTITY	NUMERICAL VALUE	SYMBOL	ENGLISH EQUIVALENT
Area		hectare	10,000 m²	ha	2.471 acres
Length	meter			m	39.37 inches
		kilometer	1000 (10³) m	km	0.62137 mile
		centimeter	0.01 (10⁻²) m	cm	0.3937 inch
		millimeter	0.001 (10⁻³) m	mm	
		micrometer	0.000001 (10⁻⁶) m	μm	
		nanometer (millimicron)	0.000000001 (10⁻⁹) m	nm (mμ)	
		angstrom	0.0000000001 (10⁻¹⁰) m	Å	
Mass	gram			g	0.03527 ounce
		kilogram	1000 g	kg	2.2 pounds
		milligram	0.001 g	mg	
		microgram	0.000001 g	μg	
Time	second			sec	
		millisecond	0.001 sec	msec	
		microsecond	0.000001 sec	μsec	
Volume (solids)	cubic meter			m³	35.314 cubic feet
		cubic centimeter	0.000001 m³	cm³	0.061 cubic inch
		cubic millimeter	0.000000001 m³	mm³	
Volume (liquids)	liter			l	1.06 quarts
		milliliter	0.001 liter	ml	
		microliter	0.000001 liter	μl	

Temperature Conversion Scale

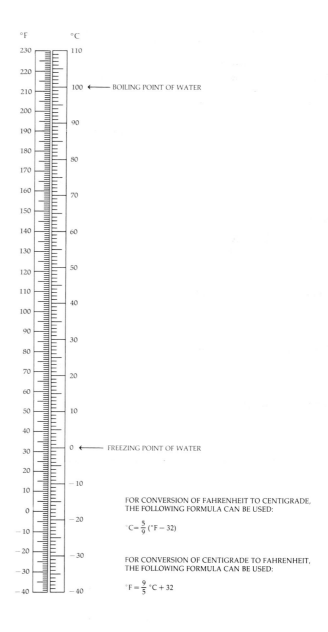

°F   °C

←— BOILING POINT OF WATER

←— FREEZING POINT OF WATER

FOR CONVERSION OF FAHRENHEIT TO CENTIGRADE, THE FOLLOWING FORMULA CAN BE USED:

$$°C = \frac{5}{9}(°F - 32)$$

FOR CONVERSION OF CENTIGRADE TO FAHRENHEIT, THE FOLLOWING FORMULA CAN BE USED:

$$°F = \frac{9}{5}°C + 32$$

APPENDIX C

Classification of Organisms

There are several alternative ways to classify organisms. The one presented here follows the overall scheme described at the end of Chapter 10, in which organisms are divided into five major groups, or kingdoms: Monera, Protista, Animalia, Fungi, and Plantae. The chief taxonomic categories are kingdom, division (phylum), class, order, family, genus, species.

The classification that follows includes the divisions of Protista, except those considered Protozoa, as well as the Fungi and Plantae. Certain classes given prominence in this book are also included, but the listings are far from complete. The number of species given for each group is the estimated number of living species that have been described and named. Only groups that include living species are described. Viruses are not included in this appendix, but are treated in Chapter 12.

KINGDOM MONERA

The bacteria; prokaryotic cells that lack a nuclear envelope, plastids and mitochondria and other membrane-bound organelles, and 9-plus-2 flagella. Bacteria are unicellular but sometimes aggregate into filaments or other superficially multicellular bodies. Their predominant mode of nutrition is absorption, but some groups are photosynthetic or chemosynthetic. Reproduction is primarily asexual, by fission or budding, but portions of DNA molecules may also be exchanged between cells under certain circumstances. They are motile by simple flagella, by gliding, or they may be nonmotile.

About 2500 species of bacteria and bacterialike organisms are recognized at present, but that is probably only a small fraction of the actual number. The recognition of species is not comparable with that in eukaryotes and is based largely upon metabolic features. One group, the class Rickettsiae—very small bacterialike organisms—occurs widely as parasites in arthropods and may contain tens of thousands of species, depending upon the classification criteria used; they have not been included in the estimate given here.

No satisfactory classification of the Monera has yet been proposed, but two groups recognized in the kingdom are the archaebacteria and the cyanobacteria. The methane-producing bacteria, or archaebacteria, are strictly anaerobic bacteria that produce methane from carbon dioxide and hydrogen; they are also able to reduce elemental sulfur to form hydrogen sulfide. Although they are diverse morphologically—they are either nonmotile or flagellated rods, cocci, or spirilla—and biochemically, the archaebacteria are linked by similar rRNA base sequences. Formerly and misleadingly called the "blue-green algae," cyanobacteria have a photosynthesis that is based on chlorophyll *a*. Cyanobacteria also have accessory pigments called phycobilins, like the red algae, the chloroplasts of which were probably derived from symbiotic cyanobacteria. Many cyanobacteria can fix atmospheric nitrogen, which often occurs in specialized cells called heterocysts. Some cyanobacteria form complex filaments or other colonies. Although some 7500 species of cyanobacteria have been described, a more reasonable estimate puts the total number of these specialized bacteria at about 200 distinct nonsymbiotic species.

KINGDOM FUNGI

Eukaryotic multicellular or rarely unicellular organisms in which the nuclei occur in a basically continuous mycelium; this mycelium becomes septate in certain groups and at certain stages of the life cycle. Fungi are heterotrophic; they obtain their nutrition by absorption. Members of all three divisions form important symbiotic relationships with the roots of plants, called mycorrhizae. Reproductive cycles typically include both sexual and asexual phases. There are about 100,000 valid species of fungi to which names have been given, and many more will eventually be found. Some have been named two or more times; this is particularly so for fungi that may be classified both as ascomycetes and as members of the Fungi Imperfecti.

DIVISION

DIVISION ZYGOMYCOTA: Terrestrial fungi with the hyphae septate only during the formation of reproductive bodies; chitin predominant in the cell walls. The class includes about 600 described species, of which about 100 occur as components of the endomycorrhizae that are found in about 80 percent of all vascular plants.

DIVISION ASCOMYCOTA: Terrestrial and aquatic fungi with the hyphae septate but the septa perforated; complete septa cut off the reproductive bodies, such as spores or gametangia. Chitin is predominant in the cell walls. Sexual reproduction involves the formation of a characteristic cell—the ascus—in which meiosis takes place and within which ascospores are formed. The hyphae in many ascomycetes are packed together into complex "bodies" known as ascocarps. Yeasts are unicellular ascomycetes that reproduce asexually by budding. There are about 30,000 species, in addition to some 25,000 described species of Fungi Imperfecti, in which sexual stages do not occur or are not known.

Lichens. The lichens are symbiotic associations between ascomycetes and certain genera of green algae or cyanobacteria that multiply within their densely packed hyphae. (There are also about a dozen species of basidiomycetes that form associations with algae, but they are closely related to free-living basidiomycetes and do not resemble other lichen-forming fungi.) There are about 20,000 described species of lichens.

DIVISION BASIDIOMYCOTA. Terrestrial fungi with the hyphae septate but the septa perforated; complete septa cut off reproductive bodies, such as spores. Chitin is predominant in the cell walls. Sexual reproduction involves formation of basidia, in which meiosis takes place and on which the basidiospores are borne. Basidiomycetes are dikaryotic during most of their life cycle, and there is often complex differentiation of "tissues" within their basidiocarps. They are components of most ectomycorrhizae. There are some 25,000 described species.

Class Hymenomycetes. Basidiomycetes that produce basidiospores in a hymenium exposed on a basidiocarp; the mushrooms, coral fungi, and shelf, or bracket, fungi. The basidia are either aseptate (internally undivided) or septate.

Class Gasteromycetes. Basidiomycetes that produce basidiospores inside basidiocarps, where they are completely enclosed for at least part of their development; puffballs, earthstars, stinkhorns, and their relatives.

Class Teliomycetes. Basidiomycetes that do not form basidiocarps and have septate basidia; the rusts and smuts.

KINGDOM PROTISTA

Eukaryotic unicellular or multicellular organisms. Their modes of nutrition include ingestion, photosynthesis, and absorption. True sexuality is present in most divisions. They move by means of 9-plus-2 flagella or are nonmotile. Fungi, plants, and animals are specialized multicellular groups derived from Protista. The divisions treated in this book are categorized as heterotrophic protists (water molds, slime molds, and chytrids; the first four divisions) and autotrophic protists (the algae). The characteristics of the divisions of Protisa are outlined in Table 14–1 (page 233).

DIVISION

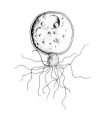

DIVISION OOMYCOTA: Water molds and related organisms. Aquatic or terrestrial organisms with motile cells characteristic at certain stages of their life cycle. The flagella are two in number—one tinsel and one whiplash. Their cell walls are composed of cellulose or celluloselike polymers, and they store their food as glycogen. There are about 475 species.

DIVISION CHYTRIDIOMYCOTA: Chytrids. Aquatic heterotrophic organisms with motile cells characteristic at certain stages in their life cycle. The motile cells have a single, posterior, whiplash flagellum. Their cell walls are composed of chitin, but other polymers may also be present, and they store their food as glycogen. There are about 750 species.

DIVISION ACRASIOMYCOTA: Cellular slime molds. Heterotrophic organisms that exist as separate amoebae (called myxamoebae). Eventually, the myxamoebae swarm together to form a pseudoplasmodium, within which they retain their individual identities. Ultimately, the pseudoplasmodium differentiates into a fruiting body. Sexual reproduction involves structures known as macrocysts. Pairs of amoebas first fuse, forming zygotes. Subsequently, these zygotes attract and then engulf nearby amoebas. The principal mode of nutrition is by ingestion. There are several genera, with about 65 known species.

DIVISION MYXOMYCOTA: Plasmodial slime molds. Heterotrophic amoeboid organisms that form a multinucleate plasmodium that creeps along as a mass and eventually differentiates into sporangia, each of which is multinucleate and eventually gives rise to many spores. Sexual reproduction is occasionally observed. The predominant mode of nutrition is by ingestion. There are about 450 species.

DIVISION CHRYSOPHYTA: Chrysophytes. Autotrophic organisms that possess chlorophyll *a*, chlorophyll *c*, and carotenoids, including fucoxanthin. Food is stored as the carbohydrate chrysolaminarin. The cell walls are absent, or consist of cellulose, with silica scales in some, and silica in diatoms. There are about 6650 living species.

Class Chrysophyceae. Golden algae. A diverse group of mostly unicellular organisms, most of which are flagellated; others lack flagella, and some are amoeboid. Many golden algae lack a clearly defined cell wall but have silica scales or skeletal structures. There are about 500 species.

Class Xanthophyceae. Yellow-green, mainly unicellular algae that have chlorophylls *a* and *c* but lack fucoxanthin. Most yellow-green algae are nonmotile, although some are amoeboid or flagellated. About 550 species.

Class Bacillariophyceae. Diatoms. Chrysophyta with double siliceous shells, the two halves of which fit together like a pillbox. Diatoms have chlorophylls *a* and *c*, as well as fucoxanthin. They are sometimes motile by the secretion of mucilage fibrils along a specialized groove, called the raphe. There are about 5600 living species, plus a large number of extinct ones.

DIVISION PYRRHOPHYTA: Dinoflagellates. Autotrophic organisms possessing chlorophylls *a* and *c* and carotenoids. Food is stored as starch. Cell walls contain cellulose. This division contains some 1100 species, mostly biflagellated organisms. These all have lateral flagella, one of which beats in a groove that encircles the organism. Sexual reproduction is generally isogamous, but anisogamy is also present. The mitosis of dinoflagellates is unique. Many—in a form called zooxanthellae—are symbiotic in marine animals, and they make important contributions to the productivity of coral reefs.

DIVISION EUGLENOPHYTA: The euglenoids. About a third of the approximately 40 genera of euglenoids have chloroplasts, with chlorophylls *a* and *b* and carotenoids; the others are heterotrophic and essentially resemble members of the phylum Zoomastigina, within which they would probably best be included. They store food as paramylon, an unusual carbohydrate. Euglenoids usually have a single apical flagellum of the tinsel variety and a contractile vacuole. The flexible pellicle is rich in proteins. Sexual reproduction is unknown. There are some 800 species, most of which occur in fresh water.

DIVISION RHODOPHYTA: Red algae. Primarily marine algae characterized by the presence of chlorophyll *a* and phycobilins. Their carbohydrate food reserve is starch, and the cell walls are composed of cellulose or pectic materials, with calcium carbonate in many. No motile cells are present at any stage in the complex life cycle. The vegetative body is built up of closely packed filaments in a gelatinous matrix and is not differentiated into roots, leaves, and stems. It lacks specialized conducting cells. There are some 4000 species.

DIVISION PHAEOPHYTA: Brown algae. Multicellular, nearly entirely marine algae characterized by the presence of chlorophyll *a*, chlorophyll *c,* and fucoxanthin. The carbohydrate food reserve is laminarin, and the cell walls have a cellulose matrix with alginic acids in it. Motile cells are biflagellated, with one forward flagellum of the tinsel type and one trailing flagellum of the whiplash type. A considerable amount of differentiation is found in some of the kelps (some of the large brown algae of the order Laminariales), with specialized conducting cells for transporting the products of photosynthesis to the dimly lighted regions of the body. There is, however, no differentiation into roots, leaves, and stems, as in the vascular plants. There are about 1500 species.

DIVISION CHLOROPHYTA: Green algae. Unicellular or multicellular photosynthetic organisms characterized by the presence of chlorophyll *a*, chlorophyll *b*, and various carotenoids. The carbohydrate food reserve is starch; only green algae and plants, which are clearly descended from the green algae, store their reserve food inside their plastids. The cell walls of green algae are formed of polysaccharides, sometimes cellulose. Motile cells have two lateral or apical whiplash flagella. True multicellular genera do not exhibit complex patterns of differentiation. Multicellularity has arisen at least three times, and quite possibly more often. There are about 7000 known species.

Class Charophyceae. Unicellular, few-celled, filamentous, or parenchymatous green algae in which cell division involves a phragmoplast—a system of microtubules perpendicular to the plane of cell division. The nuclear envelope breaks down during the course of mitosis. Motile cells, if present, are asymmetrical and possess two flagella that are lateral or subapical and extend laterally at right angles from the cell. Sexual reproduction always involves the formation of a dormant zygote and zygotic meiosis. Certain members of this class resemble plants more closely than do any other organisms. Predominantly occur in fresh water.

Class Ulvophyceae. Cell division like that of the Charophyceae; the nuclear envelope persists during mitosis. Motile cells, if present, are symmetrical and possess two, four, or many flagella that are apical and directed forward. Sexual reproduction often involves alternation of generations and sporic meiosis, and dormant zygotes are rare. Predominantly marine algae.

Class Chlorophyceae. Those green algae in which the unique mode of cell division involves a phycoplast—a system of microtubules parallel to the plane of cell division. The nuclear envelope persists throughout mitosis, and chromosome division occurs within it. Motile cells, if present, are symmetrical and possess two, four, or many flagella that are apical and directed forward. Sexual reproduction always involves the formation of a dormant zygote and zygotic meiosis. Predominantly occur in fresh water.

KINGDOM PLANTAE

The plants are autotrophic (some are derived heterotrophs), multicellular organisms possessing advanced tissue differentiation. All plants have an alternation of generations, in which the diploid phase (sporophyte) includes an embryo, and the haploid phase (gametophyte) produces gametes by mitosis. Their photosynthetic pigments and food reserves are similar to those of the green algae. Plants are primarily terrestrial.

DIVISION

DIVISION BRYOPHYTA: Liverworts, hornworts, and mosses. The bryophytes have multicellular gametangia with a sterile jacket layer; their sperm are biflagellated. The gametophytes and sporophytes of bryophytes both exhibit complex multicellular patterns of development; conducting tissues, which are not as specialized as the xylem and phloem of vascular plants, are present only in mosses. Most photosynthesis in these primarily terrestrial plants is carried out by the gametophyte, upon which the sporophyte is dependent, at least initially. There are about 16,000 species.

Class Hepaticae. The liverworts. The gametophytes are thallose or leafy, the rhizoids are single-celled, and the sporophytes, which lack stomata, are relatively simple structures. There are about 6000 species.

Class Anthocerotae. The hornworts. The gametophytes are thallose. The sporophyte grows from a basal intercalary meristem for as long as conditions are favorable. Stomata are present on the sporophyte. There are about 100 species.

Class Musci. The mosses. The gametophytes are leafy. Sporophytes have complex patterns of dehiscence. Rhizoids are multicellular. Stomata are present on the sporophyte. There are about 9500 species.

DIVISION PSILOPHYTA: Psilopsids. Homosporous vascular plants, one of which has leaflike appendages on the stem; both have extremely simple sporophytes, with no differentiation between root and shoot. Motile sperm. There are two genera and several species.

DIVISION LYCOPHYTA: The lycophytes. Homosporous and heterosporous vascular plants characterized by the presence of microphylls; the lycophytes are extremely diverse in appearance. All lycophytes have motile sperm. There are five genera, with about 1000 species.

DIVISION SPHENOPHYTA: The horsetails. A single genus of homosporous vascular plants, *Equisetum,* with jointed stems marked by conspicuous nodes and elevated siliceous ribs. Sporangia are borne in a strobilus at the apex of the stem. Leaves are scalelike. Sperm are motile. There are 15 living species of horsetails.

DIVISION PTEROPHYTA: The ferns. Mostly homosporous, although some are heterosporous. All possess a megaphyll. The gametophyte is more or less free-living and usually photosynthetic. Multicellular gametangia and free-swimming sperm are present. There are about 12,000 species.

DIVISION CONIFEROPHYTA: The conifers. Gymnosperms with active cambial growth and simple leaves; ovules and seeds exposed; sperm nonflagellated. The most familiar group of gymnosperms. There are some 50 genera, with about 550 species.

DIVISION CYCADOPHYTA: The cycads. Gymnosperms with sluggish cambial growth and pinnately compound, palmlike or fernlike leaves; ovules and seeds exposed. The sperm are flagellated and motile but are carried to the vicinity of the ovule in a pollen tube. There are 10 genera, with about 100 species.

DIVISION GINKGOPHYTA: Ginkgo. Gymnosperm with considerable cambial growth and fan-shaped leaves with open dichotomous venation; ovules and seeds exposed; seed coats fleshy. Sperm are carried to the vicinity of the ovule in a pollen tube but are flagellated and motile. There is only one species.

DIVISION GNETOPHYTA: Gnetophytes. Gymnosperms with many angiospermlike features, such as vessels; the gnetophytes are the only gymnosperms in which vessels occur. Motile sperm are absent. There are 3 very distinctive genera, with about 70 species.

DIVISION ANTHOPHYTA: The flowering plants. Seed plants in which ovules are enclosed in a carpel and seeds are borne within fruits. The angiosperms are extremely diverse vegetatively but are characterized by the flower, which is basically insect-pollinated. Other modes of pollination, such as wind pollination, have been derived in a number of different lines. The gametophytes are much reduced, with the female gametophyte often consisting of only seven cells at maturity. Double fertilization involving the two sperm of the mature microgametophyte gives rise to the zygote (sperm and egg) and to the primary endosperm nucleus (sperm and polar nuclei); the former becomes the embryo and the latter becomes a special nutritive tissue, called the endosperm. There are about 235,000 species.

Class Monocotyledons. The monocots. Flower parts are usually in threes; leaf venation is usually parallel; primary vascular bundles in the stem are scattered; true secondary growth is not present; there is one cotyledon. There are about 65,000 species.

Class Dicotyledons. The dicots. Flower parts are usually in fours or fives; leaf venation is usually netlike; primary vascular bundles in the stem are in a ring; many with a vascular cambium and true secondary growth; there are two cotyledons. There are about 170,000 species.

Some other common terms used to describe major groups of plants deserve mention here. In systems in which the algae and fungi are regarded as plants, they are often grouped as a subkingdom, Thallophyta, the thallophytes: organisms with no highly differentiated tissues, such as root, stem, or leaf, and no vascular tissues (xylem and phloem). The bryophytes and vascular plants are then grouped into a second subkingdom, Embryophyta, in which the zygote develops into a multicellular embryo still encased in an archegonium or an embryo sac. All embryophytes are marked by an alternation of heteromorphic generations.

Although they are no longer used in formal schemes of classification, terms such as "algae," "thallophytes," "vascular plants," and "gymnosperms" are still sometimes useful in an informal sense. An even earlier scheme divided all plants into "phanerogams," for those with flowers, and "cryptogams," for those lacking flowers; these terms are occasionally seen today as well.

A P P E N D I X   D

Geologic Eras

ERA*	PERIOD*	EPOCH*	LIFE FORMS	CLIMATES AND MAJOR PHYSICAL EVENTS
CENOZOIC (65)	Quaternary (2)	Recent (0.01) Pleistocene (2)	Age of human beings. Extinction of many large mammals and birds.	Fluctuating cold to mild. More than two dozen glacial advances and retreats; final uplift of many mountain ranges.
	Tertiary (65)	Pliocene (5.1)	Aridity, formation of deserts; climates diversify. First appearance of man-apes.	Cooler. Much uplift and mountain building; widespread glaciation in Northern Hemisphere. Uplift of Panama joins North and South America.
		Miocene (24.6)	Spread of grasslands as forests contract. Grazing animals, apes.	Moderate. Extensive glaciation begins again in Southern Hemisphere.
		Oligocene (38)	Browsing mammals, monkey-like primates; many modern genera of plants evolve.	Rise of Alps and Himalayas. South America separates from Antarctica. Volcanoes in Rocky Mountains.
		Eocene (54.9)	Extensive radiation of mammals and birds; initial formation of grasslands.	Mild to very tropical. Australia separates from Antarctica; India collides with Asia.
		Paleocene (65)	Early insectivorous mammals and primates.	Mild to cool. Wide, shallow continental seas largely disappeared.
MESOZOIC (248)	Cretaceous (144)		Angiosperms and many groups of insects appear and become dominant. Age of reptiles. Extinction of dinosaurs at end of period.	Climate tropical to subtropical throughout. Africa and South America separate.
	Jurassic (213)		Gymnosperms, especially cycads. Birds appear.	Mild. Continents low, with large areas covered by seas.
	Triassic (248)		Forests of gymnosperms and ferns. First dinosaurs and first mammals.	Continents mountainous, joined into a supercontinent. Large areas arid.
PALEOZOIC (590)	Permian (286)		Origin of conifers, cycads, and ginkgos; earlier forest types wane. Reptiles diversify.	Extensive glaciation in Southern Hemisphere. Appalachians uplifted. Worldwide aridity.
	Carboniferous (360) Pennsylvanian (320) Mississippian (360)		Amphibians appear on land; forests appear and become dominant. Origin of insects and reptiles. Age of amphibians.	Warm, with little seasonal variation; lands low, swampy, with formation of coal deposits.
	Devonian (408)		Age of fishes. Rise of land plants; extinction of primitive vascular plants.	Sea over most of the land, with mountains locally.
	Silurian (438)		Period starts with major extinction event. First fossil plants. First jawed fishes.	Mild. Continents generally flat.
	Ordovician (505)		Period starts with first major extinction event. Oldest fossil crustaceans. Diversification of mollusks. Possible invasion of land by plants. First fungi.	Mild. Shallow seas, continents generally flat; seas cover much of present United States.
	Cambrian (590)		Evolution of external skeletons in animals. Explosive evolution of phyla and divisions. Evolution of chordates.	Mild. Extensive seas spilling over present continents.
PRECAMBRIAN (4500)			Origin of life (at least 3.5 billion years ago). Origin of eukaryotes, at least 1.5 billion years ago. Multicellular animals by 700 million years ago.	Formation of earth's crust and beginning of continental movements.

* A number following the name of a geologic time division indicates the age (in millions of years) at which it began.

Glossary

Å: *See* ångstrom.

a- [Gk. *a-*, not, without]: Prefix, negates the succeeding part of the word; "an-" before a vowel.

abscisic acid [L. *abscissus*, to cut off]: A plant hormone that brings about dormancy in buds, maintains dormancy in seeds, is involved in the gravitropism of roots, and brings about stomatal closing, among other effects.

abscission (ăb·sizh'ŭn): The dropping off of leaves, flowers, fruits, or other plant parts, usually following the formation of an abscission zone.

abscission zone: The area at the base of a leaf, flower, or fruit, or other plant part containing tissues that play a role in the separation of a plant part from the plant body.

absorption spectrum: The spectrum of light waves absorbed by a particular pigment.

accessory bud: A bud generally located above or on either side of the main axillary bud.

accessory cell: *See* subsidiary cell.

accessory pigment: A pigment that captures light energy and transfers it to chlorophyll *a*.

acclimation: The process by which numerous physical and physiological processes prepare a plant for winter.

achene: A simple, dry, one-seeded indehiscent fruit in which the seed coat is not adherent to the pericarp.

acid: A substance that dissociates in water, releasing hydrogen ions (H^+) and thus causing a relative increase in the concentration of these ions; having a pH in solution of less than 7; a proton donor; the opposite of "base."

acid growth hypothesis: The hypothesis that acidification of the cell wall leads to hydrolysis of restraining bonds within the wall and, consequently, to cell elongation driven by the turgor pressure of the wall.

actinomorphic [Gk. *aktis*, ray of light, + *morphe*, form]: Pertaining to a type of flower that can be divided into two equal halves in more than one longitudinal plane; also called radially symmetrical or regular; *see* zygomorphic.

action spectrum: The spectrum of light waves that elicits a particular reaction.

active transport: Energy-requiring transport of a solute across a membrane in the direction of increasing concentration (against the concentration gradient).

ad- [L. *ad-*, toward, to]: Prefix meaning "toward" or "to."

adaptation [L. *adaptare*, to fit]: A peculiarity of structure, physiology, or behavior that aids in fitting an organism to its environment.

adaptive radiation: The evolution from one kind of organism to several divergent forms, each specialized to fit a distinct and diverse way of life.

adenine (ăd'e·nēn): A purine base present in DNA, RNA, and nucleotide derivatives, such as ADP and ATP.

adenosine triphosphate (ATP): A nucleotide consisting of adenine, ribose sugar, and three phosphate groups; the major source of usable chemical energy in metabolism. On hydrolysis, ATP loses one phosphate to become adenosine diphosphate (ADP), releasing usable energy.

adhesion [L. *adhaerere*, to stick to]: The sticking together of unlike objects or materials.

adnate [L. *adnatus*, grown together]: Said of fused unlike parts, as stamens and petals; *see also* connate.

ADP: *See* adenosine triphosphate.

adsorption [L. *ad-*, to, + *sorbere*, to suck in]: The adhesion of a liquid, gaseous, or dissolved substance to a solid, resulting in a higher concentration of the substance.

adventitious [L. *adventicius*, not properly belonging to]: Referring to a structure arising from an unusual place, such as buds at other places than leaf axils, or roots growing from stems or leaves.

aeciospore (ē'sĭ·o·spor) [Gk. *aikia*, injury, + *spora*, seed]: A binucleate spore of rust fungi; produced in an aecium.

aecium, *pl.* **aecia:** In rust fungi, a cuplike structure in which aeciospores are produced.

aerobic [Gk. *aer*, air, + *bios*, life]: Requiring free oxygen.

aerobic respiration: *See* respiration.

after-ripening: Term applied to the metabolic changes that must occur in some dormant seeds before germination can occur.

agar: A gelatinous substance derived from certain red algae; used as a solidifying agent in the preparation of nutrient media for growing microorganisms.

aggregate fruit: A fruit developing from the several separate carpels of a single flower.

akinete: A vegetative cell that is transformed into a thick-walled resistant spore in cyanobacteria.

albuminous cell: Certain ray and axial parenchyma cells in gymnosperm phloem that are closely associated with the sieve cells both morphologically and physiologically.

aleurone [Gk. *aleuron,* flour]: A proteinaceous material, usually in the form of small granules, occurring in the outermost cell layer of the endosperm of wheat and other grains.

alga, *pl.* **algae** (ăl'ga, ăl'je): Traditional term for a series of unrelated groups of photosynthetic eukaryotic organisms lacking multicellular sex organs (except for the charophytes); the "blue-green algae," or cyanobacteria, are one of the groups of photosynthetic bacteria.

alkali [Arabic *algili,* the ashes of the plant saltwort]: A substance with marked basic properties.

alkaline: Pertaining to substances that release hydroxyl ions (OH^-) in water, having a pH greater than 7.

alkaloids: Nitrogen-containing ring compounds produced by plants that are physiologically active in vertebrates. Many alkaloids have a bitter taste and some are poisonous; examples are nicotine, caffeine, and strychnine.

allele (ă·lēl') [Gk. *allelon,* of one another, + *morphe,* form]: One of the two or more alternative states of a gene.

allelopathy [Gk. *allelon,* of each other, + *pathos,* suffering]: The inhibition of one species of plant by chemicals produced by another plant.

allosteric interaction [Gk. *allos,* other, + *steros,* shape]: A change in the shape of a protein resulting from the binding to the protein of a nonsubstrate molecule; in its new shape, the protein typically has different properties.

alternate: Referring to bud or leaf arrangement in which there is one bud or one leaf at a node.

alternation of generations: A reproductive cycle in which a haploid (*n*) phase, the gametophyte, produces gametes, which, after fusion in pairs to form a zygote, germinate, producing a diploid (2*n*) phase, the sporophyte. Spores produced by meiotic division from the sporophyte give rise to new gametophytes, completing the cycle.

amino acids [Gk. *Ammon,* referring to the Egyptian sun god, near whose temple ammonium salts were first prepared from camel dung]: Nitrogen-containing organic acids, the units, or "building blocks," from which protein molecules are built.

ammonification: Decomposition of amino acids and other nitrogen-containing organic compounds, resulting in the production of ammonia (NH_3) and ammonium ions (NH_4^+).

amoeboid [Gk. *amoibe,* change]: Moving or eating by means of pseudopodia (temporary cytoplasmic protrusions from the cell body).

amphi- [Gk. *amphi-,* on both sides]: Prefix meaning "on both sides," "both," or "of both kinds."

amylase (ăm'ĭ·lās): An enzyme that breaks down starch into smaller units.

amyloplast: A leucoplast (colorless plastid) that forms starch grains.

an- [Gk. *an-,* not, without]: Prefix, equivalent to "a-," meaning "not" or "without"; used before vowels and "h."

anabolism [Gk. *ana-,* up,+ *-bolism* (as in metabolism)]: The constructive part of metabolism; the total chemical reactions involved in biosynthesis.

anaerobic [Gk. *an-,* without, + *aer,* air, + *bios,* life]: Any process that can occur without oxygen, or an organism that can live without oxygen; strict anaerobes cannot survive in the presence of oxygen.

analogous [Gk. *analogos,* proportionate]: Applied to structures similar in function but different in evolutionary origin, such as the phyllodes of an Australian *Acacia* and the leaves of an oak.

anaphase [Gk. *ana,* away, + *phasis,* form]: A stage in mitosis in which the chromatids of each chromosome separate and move to opposite poles; similar stages in meiosis in which chromatids or paired chromosomes move apart.

anatomy: The study of the internal structure of organisms; morphology is the study of their external structure.

andro- [Gk. *andros,* man]: Prefix meaning "male."

androecium [Gk. *andros,* man, + *oikos,* house]: (1) The floral whorl that comprises the stamens; (2) in leafy liverworts, a tubular sheath surround the antheridia.

angiosperm [Gk. *angion,* a vessel, + *sperma,* a seed]: Literally, a seed borne in a vessel (carpel); thus one of a group of plants whose seeds are borne with a mature ovary (fruit).

ångstrom [after A. J. Ångstrom, a Swedish physicist, 1814–74]: A unit of length equal to 10^{-10} meter; abbreviated Å.

anion [Gk. *anienae,* to go up]: A negatively charged ion.

anisogamy [Gk. *aniso,* unequal, + *gamos,* marriage]: The condition of having dissimilar motile gametes.

annual [L. *annulus,* year]: A plant in which the life cycle is completed in a single growing season.

annual ring: In wood, the growth layer formed during single year; *see also* growth layer.

annulus [L. *anus,* ring]: In ferns, a row of specialized cells in sporangium; in gill fungi, the remnant of the inner veil forming a ring on the stalk.

anterior: Situated before or toward the front.

anther [Gk. *anthos,* flower]: The pollen-bearing portion of a stamen.

antheridiophore [Gk. *anthos,* flower, + *phoros,* bearing]: In some liverworts, a stalk that bears antheridia.

antheridium: A sperm-producing organ, which may be multicellular or unicellular.

anthocyanin [Gk. *anthos,* flower, + *kyanos,* dark blue]: A water-soluble blue or red pigment found in the cell sap.

antibiotic [Gk. *anti,* against or opposite, + *biotikos,* pertaining to life]: Natural organic substances that retard or prevent the growth of organisms; generally used to designate substances formed by microorganisms that prevent growth of other microorganisms.

anticlinal: Perpendicular to the surface.

antipodals: Three (sometimes more) cells of the mature embryo sac located at the end opposite the micropyle.

apical dominance: The influence exerted by a terminal bud in suppressing the growth of lateral buds.

apical meristem: The meristem at the tip of the root or shoot in a vascular plant.

apomixis [Gk. *apo,* separate, away from, + *mixis,* mingling]: Reproduction without meiosis or syngamy; vegetative reproduction.

apoplast [Gk. *apo,* away from, + *plastos,* molded]: The cell wall continuum of a plant or organ; the movement of substances via the cell walls is called apoplastic movement or transport.

apothecium [Gk. *apotheke,* storehouse]: A cup-shaped or saucer-shaped open ascocarp.

arch-, archeo- [(Gk. *arche, archos,* beginning]: Prefix meaning "first," "main," or "earliest."

archegoniophore [Gk. *archegonos,* the first of a race, + *phoros,* bearing]: In some liverworts, a stalk that bears archegonia.

archegonium, *pl.* **archegonia:** A multicellular organ in which a single egg is produced; found in the bryophytes and some vascular plants.

aril (är'ĭl) [L. *arillus,* grape, seed]: An accessory seed covering, often formed by an outgrowth at the base of the ovule; often brightly colored, which may aid in dispersal by attracting animals that eat it and, in the process, carry the seed away from the parent plant.

artifact [L. *ars,* art, + *facere,* to make]: A product that exists because of an extraneous, especially human, agency and does not occur in nature.

artificial selection: The breeding of selected organisms to produce strains with desired characteristics.

ascocarp [Gk. *askos,* bladder, + *karpos,* fruit]: A multicellular structure in ascomycetes lined with specialized cells called asci, in which nuclear fusion and meiosis occur. Ascocarps may be open or closed.

ascogenous hyphae [Gk. *askos,* bladder, + *genous,* producing]: Hyphae containing paired haploid male and female nuclei; they develop from an ascogonium and eventually give rise to asci.

ascogonium: The oogonium or female gametangium of the ascomycetes.

ascospore: A spore produced within an ascus; found in ascomycetes.

ascus, *pl.* **asci:** A specialized cell, characteristic of the ascomycetes, in which two haploid nuclei fuse to produce a zygote that immediately divides by meiosis; at maturity, an ascus contains ascospores.

aseptate [Gk. *a-,* not, + L. *septum,* fence]: Nonseptate; lacking cross walls.

asexual reproduction: Any reproductive process, such as fission or budding, that does not involve the union of gametes.

atom [Gk. *atomos,* indivisible]: The smallest unit into which a chemical element can be divided and still retain its characteristic properties.

atomic nucleus: The central core of an atom, containing protons and neutrons, around which electrons orbit.

atomic number: The number of protons in the nucleus of an atom.

atomic weight: The weight of a representative atom of an element relative to the weight of an atom of carbon ^{12}C, which has been assigned the value 12.

ATP: *See* adenosine triphosphate.

ATP synthetase: An enzyme complex that forms ATP from ADP and phosphate during oxidative phosphorylation in the inner mitochondrial membrane.

auto- [Gk. *autos,* self, same]: Prefix, meaning "same" or "self-same."

autoecious [Gk. *auto,* self, + *oikia,* dwelling]: In some rust fungi, completing the life cycle on a single species of host plant.

autoradiograph: A photographic print made by a radioactive substance acting upon a sensitive photographic film.

autotroph [Gk. *autos,* self, + *trophos,* feeder]: An organism that is able to synthesize the nutritive substances it requires from inorganic substances in its environment; *see also* heterotroph.

auxin [Gk. *auxein,* to increase]: A plant growth-regulating substance; controls cell elongation, among other effects.

axial system: In secondary xylem and secondary phloem; the term applied collectively to cells derived from fusiform cambial initials. The long axes of these cells are oriented parallel with the main axis of the root or stem. Also called longitudinal system and vertical system.

axil [Gk. *axilla,* armpit]: The upper angle between a twig or leaf and the stem from which it grows.

axillary: Term applied to buds or branches occurring in the axil of a leaf.

bacillus, *pl.* **bacilli** (ba·sĭl′ŭs) [L. *baculum,* rod]: A rod-shaped bacterium.

backcross: The crossing of a hybrid with one of its parents or with a genetically equivalent organism; a cross between an individual whose genes are to be tested and one that is homozygous for all of the recessive genes involved in the experiment.

bacteriophage [Gk. *bakterion,* little rod, + *phagein,* to eat]: A virus that parasitizes bacterial cells.

bacterium, *pl.* **bacteria:** A prokaryotic organism.

bark: A nontechnical term applied to all tissues outside the vascular cambium in a woody stem; *see also* inner bark and outer bark.

basal body: A self-reproducing, cylinder-shaped cytoplasmic organelle from which cilia or flagella arise; identical in structure to the centriole, which is involved in mitosis and meiosis in most animals and protists.

base: A substance that dissociates in water, causing a decrease in the concentration of hydrogen ions (H^+), often by releasing hydroxyl ions (OH^-); bases have a pH in solution of more than 7; the opposite of "acid."

basidiocarp [L. *basidium,* a little pedestal, + *carpus,* a fruit]: A multicellular structure characteristic of the basidiomycetes within which basidia are formed.

basidiospore: A spore of the basidiomycetes, produced within and borne on a basidium following nuclear fusion and meiosis.

basidium, *pl.* **basidia:** A specialized reproductive cell of the basidiomycetes, often club-shaped, in which nuclear fusion and meiosis occur.

berry: A simple fleshy fruit that includes a fleshy ovary wall and one or more carpels and seeds; examples are the fruits of grapes, tomatoes, and bananas.

bi- [L. *bis,* double, two]: Prefix meaning "two," "twice," or "having two points."

biennial: A plant that normally requires two growing seasons to complete its life cycle, flowering and fruiting in its second year.

bilaterally symmetrical: *See* zygomorphic.

biological clock [Gk. *bios,* life, + *logos,* discourse]: The internal timing mechanism that governs the innate biological rhythms of organisms.

biomass: Total dry weight of all organisms in a particular population, sample, or area.

biome: A complex of terrestrial communities of very wide extent, characterized by its climate and soil; the largest ecological unit.

biosphere: The zone of air, land, and water at the surface of the earth that is occupied by organisms.

biotechnology: The practical application of advances in hormone research and DNA biochemistry to manipulate the genetics of plants.

biotic: Relating to life.

bisexual flower: A flower that has at least one functional stamen and one functional carpel.

bivalent [L. *bis,* twice, + *valere,* to be strong]: A pair of synapsed homologous chromosomes.

blade: The broad, expanded part of a leaf; the lamina.

body cell: The cell of the male gametophyte, or pollen grain, of gymnosperms, which divides mitotically to form two sperm.

bordered pit: A pit in which the secondary wall arches over the pit membrane.

bract: A modified, usually reduced leaflike structure.

branch root: *See* lateral root.

bryophytes (brī′o·fīts): The members of a division of nonvascular plants; the mosses, hornworts, and liverworts.

bud: (1) An embryonic shoot, often protected by young leaves; (2) a vegetative outgrowth of yeasts and some bacteria as a means of asexual reproduction.

bulb: A short underground stem covered by enlarged and fleshy leaf bases containing stored food.

bulk flow: The overall movement of water or some other liquid induced by gravity, pressure, or an interplay of both.

bulliform cell: A large epidermal cell present, with other such cells, in longitudinal rows in grass leaves; also called motor cell. Believed to be involved with the mechanism of rolling and unrolling of the leaves.

bundle scar: Scar or mark left on leaf scar by vascular bundles broken at the time of leaf fall, or abscission.

bundle sheath: Layer or layers of cells surrounding a vascular bundle; may consist of parenchyma or sclerenchyma cells, or both. Bundle sheaths are especially characteristic of C_4 plants.

bundle-sheath extension: A group of cells extending from a bundle sheath of a vein in the leaf mesophyll to either or both upper and lower epidermis; may consist of parenchyma, collenchyma, or sclerenchyma.

C_3 pathway: *See* Calvin cycle.

C_4 pathway: The set of reactions through which carbon dioxide is fixed to a compound known as phosphoenolpyruvate (PEP) to yield oxaloacetate, a four-carbon compound.

callose: A complex branched carbohydrate that is a common wall constituent associated with the sieve areas of sieve elements; may develop in reaction to injury in sieve elements and parenchyma cells.

callus [L. *callos*, hard skin]: Undifferentiated tissue; a term used in tissue culture, grafting, and wound healing.

calorie [L. *calor*, heat]: The amount of energy in the form of heat required to raise the temperature of one gram of water 1°C. In making metabolic measurements, the kilocalorie (kcal), or Calorie—the amount of heat required to raise the temperature of one kilogram of water 1°C—is generally used.

Calvin cycle: The series of enzymatically mediated photosynthetic reactions during which carbon dioxide is reduced to 3-phosphoglyceraldehyde and the carbon dioxide acceptor, ribulose 1,5-bisphosphate, is regenerated. For every six molecules of carbon dioxide entering the cycle, a net gain of two molecules of glyceraldehyde 3-phosphate results.

calyptra [Gk. *kalyptra*, covering for the head]: The hood or cap that partially or entirely covers the capsule of some species of mosses; it is formed from the expanded archegonial wall.

calyx (kā′lĭks) [Gk. *kalyx*, a husk, cup]: The sepals collectively; the outermost flower whorl.

CAM: *See* Crassulacean acid metabolism.

cambial zone: A region of thin-walled, undifferentiated meristematic cells between the secondary xylem and secondary phloem; consists of cambial initials and their recent derivatives.

cambium [L. *cambiare*, to exchange]: A meristem that gives rise to parallel rows of cells; commonly applied to the vascular cambium and the cork cambium, or phellogen.

capsid: The protein coat of a virion or virus particle.

capsule: (1) In angiosperms, a dehiscent, dry fruit that develops from two or more carpels; (2) a slimy layer around the cells of certain bacteria; (3) the sporangium of bryophytes.

carbohydrate [L. *carbo*, ember, + *hydro*, water]: An organic compound consisting of a chain of carbon atoms to which hydrogen and oxygen are attached in a 2:1 ratio; examples are sugars, starch, glycogen, and cellulose.

carbon cycle: Worldwide circulation and utilization of carbon atoms.

carbon fixation: The conversion of CO_2 into organic compounds during photosynthesis.

carnivorous: Feeding upon animals, as opposed to plants (herbivorous); also refers to plants that are able to utilize proteins obtained from trapped animals, chiefly insects.

carotene (kăr′o•tēn) [L. *carota*, carota, carrot]: A red, yellow, or orange pigment belonging to the carotenoid group.

carotenoids (kă•rŏt′e•noids): A class of fat-soluble pigments that includes the carotenes (yellow and orange pigments) and the xanthophylls (yellow pigments); found in chloroplasts and chromoplasts of plants. Carotenoids function as accessory pigments in photosynthesis.

carpel [Gk. *karpos*, fruit]: One of the members of the gynoecium, or inner floral whorl; each carpel encloses one or more ovules. One or more carpels forms a pistil.

carpellate: *See* pistillate.

carpogonium [Gk. *karpos*, fruit, + *gonos*, offspring]: In red algae, the female gametangium.

carposporangium [Gk. *karpos*, fruit, + *spora*, seed, + *angeion*, vessel]: In red algae, a carpospore-containing cell.

carpospore: In red algae, the single diploid protoplast found within a carposporangium.

carrier molecule: *See* permease.

caryopsis [Gk. *karyon*, a nut, + *opsis*, appearance]: Simple, dry, one-seeded indehiscent fruit, with the pericarp firmly united all around the seed coat; a grain characteristic of the grasses (family Poaceae).

Casparian strip [Robert Caspary, a German botanist]: A bandlike region of primary wall containing suberin and lignin; found in anticlinal—radial and transverse—walls of endodermal cells.

catabolism [Gk. *katabole*, throwing down]: Collectively, the chemical reactions resulting in the breakdown of complex materials and involving the release of energy.

catalyst [Gk. *katalysis*, dissolution]: A substance that accelerates the rate of a chemical reaction but is not used up in the reaction: enzymes are catalysts.

cation [Gk. *katienai*, to go down]: A positively charged ion.

catkin: A spikelike inflorescence of unisexual flowers; found only in woody plants.

cell [L. *cella*, small room]: The structural unit of organisms; in plants, cells consist of the cell wall and the protoplast.

cell division: The division of a cell and its contents into two roughly equal parts.

cell plate: The structure that forms at the equator of the spindle in the dividing cells of plants and a few green algae during early telophase. The cell plate, when mature, becomes the middle lamella.

cell sap: The fluid contents of the vacuole.

cellulase: An enzyme that hydrolyzes cellulose.

cellulose: A carbohydrate; the chief component of the cell wall in plants and some protists; an insoluble complex carbohydrate formed of microfibrils of glucose molecules attached end to end.

cell wall: The rigid outermost layer of the cells found in plants, some protists, and most bacteria.

central mother cells: Relatively large vacuolate cells in a subsurface position in apical meristems of shoots.

centriole [Gk. *kentron*, center, + L. *-olus*, little one]: A cytoplasmic organelle found outside the nuclear envelope, and identical in structure to a basal body; centrioles generally found in the cells of most eukaryotes other than fungi, red algae, and the nonflagellated cells of plants. Centrioles divide and organize spindle fibers during mitosis and meiosis.

centromere [Gk. *kentron*, center, + *meros*, a part]: That portion of the chromosome to which the spindle fibers become attached during cell division; also called the kinetochore.

chalaza [Gk. *chalaza*, small tubercle]: The region of an ovule or seed where the funiculus unites with the integuments and the nucellus.

chemical potential: The activity or free energy of a substance; it is dependent upon the rate of motion of the average molecule and the concentration of the molecules.

chemical reaction: The making or breaking of chemical bonds between atoms or molecules.

chemiosmotic coupling: Coupling of ATP synthesis to electron transport via an electrochemical H$^+$ gradient across a membrane.

chemoautotrophic: Refers to bacteria that are able to manufacture their own basic foods by using the energy released by specific inorganic reactions; *see* autotroph.

chiasma (kī·ăz′ma) [Gk. *chiasma*, a cross]: The X-shaped figure formed by the meeting of two nonsister chromatids of homologous chromosomes; the site of crossing-over.

chitin (kī′tĭn) [Gk. *chiton*, tunic]: A tough, resistant, nitrogen-containing polysaccharide forming the cell walls of certain fungi, the exoskeleton of arthropods, and the epidermal cuticle of other surface structures of certain other protists and animals.

chlor- [Gk. *chloros*, green]: Prefix meaning ''green.''

chlorenchyma: Parenchyma cells that contain chloroplasts.

chlorophyll [Gk. *chloros*, green, + *phyllon*, leaf]: The green pigment of plant cells, which is the receptor of light energy in photosynthesis; also found in algae and photosynthetic bacteria.

chloroplast: A plastid in which chlorophylls are contained; the site of photosynthesis. Chloroplasts occur in plants and algae.

chlorosis: Loss or reduced development of chlorophyll.

chroma- [Gk. *chroma*, color]: Prefix meaning ''color.''

chromatid [Gk. *chroma*, color, + L. *-id*, daughters of]: One of the two daughter strands of a duplicated chromosome, which are joined by a single centromere.

chromatin: The deeply staining complex of DNA and proteins that forms eukaryotic chromosomes.

chromatophore [Gk. *chroma*, color, + *phorus*, a bearer]: In some bacteria, a discrete vesicle delimited by a single membrane and containing photosynthetic pigments.

chromoplast: A plastid containing pigments other than chlorophyll, usually yellow and orange carotenoid pigments.

chromosome [Gk. *chroma*, color, + *soma*, body]: The organelle that carries the genes. Eukaryotic chromosomes are visualized as threads or rods of chromatin, which appear in contracted form during mitosis and meiosis, and otherwise are enclosed in a nucleus; bacterial chromosomes consist only of a closed circle of DNA.

chrysolaminarin: The storage product of the chrysophytes and diatoms.

cilium, *pl.* **cilia** (sĭl′ĭ·ŭm) [L. *cilium*, eyelash]: A short, hairlike flagellum, usually numerous and arranged in rows.

circadian rhythms [L. *circa*, about, + *dies*, a day]: Regular rhythms of growth and activity that occur on an approximately 24-hour basis.

circinate vernation [L. *circinare*, to make round, + *vernare*, to flourish]: As in ferns, the coiled arrangement of leaves and leaflets in the bud; such an arrangement uncoils gradually as the leaf develops further.

cisterna, *pl.* **cisternae** [L. *cistern*, a reservoir]: A flattened or saclike portion of the endoplasmic reticulum or a dictyosome (Golgi body).

cladistics: System of arranging organisms following an analysis of their primitive and advanced features so that their phylogenetic relationships will be reflected accurately.

cladophyll [Gk. *klados*, shoot, + *phyllon*, leaf]: A branch resembling a foliage leaf.

clamp connection: In the basidiomycetes, a lateral connection between adjacent cells of a dikaryotic hypha; ensures that each cell of the hypha will contain two dissimilar nuclei.

class: A taxonomic category between division and order in rank. A class contains one or more orders, and belongs to a particular division.

cleistothecium [Gk. *kleistos*, closed, + *thekion*, small receptacle]: A closed, spherical ascocarp.

climax community: The final stage in a successional series; its nature is determined largely by the climate and soil of the region.

cline: A graded series of changes in some characteristics within a species, often correlated with a gradual change in climate or another geographical factor.

clone [Gk. *klon*, twig]: A population of cells or individuals derived by asexual division from a single cell or individual; one of the members of such a population.

cloning: Producing a cell line or culture all of whose members are characterized by a specific DNA sequence; a key element in genetic engineering.

closed bundle: A vascular bundle in which a cambium does not develop.

coalescence [L. *coalescere*, to grow together]: The union of floral parts of the same whorl, as petals to petals.

coccus, *pl.* **cocci** (kŏk′ŭs) [Gk. *kokkos*, a berry]: A spherical bacterium.

codon (kō′dŏn): Sequence of three adjacent nucleotides in a molecule of DNA or mRNA that form the code for a single amino acid, or for the termination of a polypeptide chain.

coenocytic (se·nō·sī′tic) [Gk. *koinos*, shared in common, + *kytos*, a hollow vessel]: A term used to describe an organism or part of an organism that is multinucleate, the nuclei not separated by walls or membranes; also called siphonaceous, siphonous, or syncytial.

coenzyme: An organic molecule, or nonprotein organic cofactor, that plays an accessory role in enzyme-catalyzed processes, often by acting as a donor or acceptor of electrons; NAD$^+$ and FAD are common coenzymes.

cofactor: One or more nonprotein components required by enzymes in order to function; many cofactors are metal ions, while others are called coenzymes.

cohesion [L. *cohaerrere*, to stick together]: The mutual attraction of molecules of the same substance.

cold hardiness: The ability of a plant to survive the extreme cold and drying effects of winter weather.

coleoptile (kō′le·op′till) [Gk. *koleos*, sheath, + *ptilon*, feather]: The sheath enclosing the apical meristem and leaf primordia of the grass embryo; often interpreted as the first leaf.

coleorhiza (kō′le·o·rī′za) [Gk. *koleos*, sheath, + *rhiza*, root]: The sheath enclosing the radicle in the grass embryo.

collenchyma [Gk. *kolla*, glue]: A supporting tissue composed of collenchyma cells; often found in regions of primary growth in stems and in some leaves.

collenchyma cell: Elongated living cell with irregularly thickened primary cell wall.

colloid (kŏl′oid): A permanent suspension of fine particles.

community: All the organisms inhabiting a common environment and interacting with one another.

companion cell: A specialized parenchyma cell associated with a sieve-tube member in angiosperm phloem and arising from the same mother cell as the sieve-tube member.

competition: Interaction between members of the same population or of two or more populations to obtain a resource that both require and which is available in limited supply.

complete flower: A flower having four whorls of floral parts—sepals, petals, stamens, and carpels.

compound: A combination of atoms in a definite ratio, held together by chemical bonds.

compound leaf: A leaf whose blade is divided into several distinct leaflets.

compression wood: The reaction wood of conifers; develops on the lower sides of leaning trunks or limbs.

concentration gradient: The concentration difference of a substance per unit distance.

cone: *See* strobilus.

conidiophore: Hypha on which one or more conidia are produced.

conidium, *pl.* **conidia** [Gk. *konis,* dust]: An asexual fungal spore not contained within a sporangium; it may be produced singly or in chains; most conidia are multinucleate.

conifer: A cone-bearing tree.

conjugation: The temporary fusion of pairs of bacteria, protozoa, and certain algae and fungi during which genetic material is transferred between the two individuals.

conjugation tube: A tube formed during the process of conjugation to facilitate the transfer of genetic material.

connate (kŏn´āt): Said of similar parts that are united or fused, as petals fused in a corolla tube; *see also* adnate.

consumer: In ecology, an organism that derives its food from another organism.

continuous variation: Variation in traits to which a number of different genes contribute; the variation often exhibits a "normal" or bell-shaped distribution.

contractile vacuole: A clear, fluid-filled vacuole in some groups of protists that takes up water within the cell and then contracts, expelling its contents from the cell.

convergent evolution [L. *convergere,* to turn together]: The independent development of similar structures in organisms that are not directly related; often found in organisms living in similar environments.

cork, or **phellem:** A secondary tissue produced by a cork cambium; made up of polygonal cells, nonliving at maturity, with walls infiltrated with suberin, a waxy or fatty material resistant to the passage of gases and water vapor; the outer part of the periderm.

cork cambium, or **phellogen:** The lateral meristem that forms the periderm, producing cork (phellem) toward the surface (outside) of the plant and phelloderm toward the inside; common in stems and roots of gymnosperms and dicots.

corm: A thickened underground stem, upright in position, in which food is accumulated, usually in the form of starch.

corolla [L. *corona,* crown]: The petals collectively; usually the conspicuously colored flower whorl.

corolla tube: A tubelike structure resulting from the fusion of the petals along their edges.

cortex: Ground-tissue region of a stem or root bounded externally by the epidermis and internally by the vascular system; a primary-tissue region.

cotransport: Membrane transport in which the transfer of one solute depends on the simultaneous or sequential transfer of a second solute.

cotyledon (kŏt´ĭ·lē´dŭn) [Gk. *kotyledon,* cup-shaped hollow]: Seed leaf; generally stores food in dicotyledons and absorbs food in monocotyledons.

coupled reactions: reactions in which energy-requiring chemical reactions are linked to energy-releasing reactions.

covalent bond: A chemical bond formed between atoms as a result of the sharing of two electrons.

Crassulacean acid metabolism, or **CAM:** A variant of the C₄ pathway; phosphoenolpyruvate fixes CO_2 in C_4 compounds at night and then, during the day time, the fixed CO_2 is transferred to the ribulose bisphosphate of the Calvin cycle within the same cell. Characteristic of most succulent plants, such as cacti.

cristae, *sing.* **crista:** The enfoldings of the inner mitochondrial membrane, which form a series of crests or ridges containing the electron-transport chains involved in ATP formation.

crop rotation: The practice of growing different crops in regular succession to aid in the control of insects and diseases, to increase soil fertility, and to decrease erosion.

cross-fertilization: The fusion of gametes formed by different individuals; the opposite of self-fertilization.

crossing-over: The exchange of corresponding segments of genetic material between the chromatids of homologous chromosomes at meiosis.

cross-pollination: The transfer of pollen from the anther of one plant to the stigma of a flower of another plant.

cross section: *See* transverse section.

cryptogam: An archaic term for all organisms except the flowering plants (phanerogams), animals, and heterotrophic protists.

cultivar: A variety of plant found only under cultivation.

cuticle: Waxy or fatty layer on outer wall of epidermal cells, formed of cutin and wax.

cutin [L. *cutis,* skin]: Fatty substance deposited in many plant cell walls and on outer surface of epidermal cell walls, where it forms a layer known as the cuticle.

cyclic electron flow: In chloroplasts, the light-induced flow of electrons originating from and returning to photosystem I.

cyclosis (sī·klō´sis) [Gk. *kyklosis,* circulation]: The streaming of cytoplasm within a cell.

-cyte, cyto- [Gk. *kytos,* hollow vessel, container]: Suffix or prefix meaning "pertaining to the cell."

cytochrome [Gk. *kytos,* container, + *chroma,* color]: Heme proteins serving as electron carriers in respiration and photosynthesis.

cytokinesis: *See* cell division.

cytokinin [Gk. *kytos,* hollow vessel, + *kinesis,* motion]: A class of plant hormones that promote cell division, among other effects.

cytology: The study of cell structure and function.

cytoplasm: The living matter of a cell, exclusive of the nucleus; the protoplasm.

cytoplasmic ground substance: The least differentiated part of the cytoplasm, as seen with the electron microscope, and the part surrounding the nucleus and various organelles; also called hyaloplasm.

cytosine: One of the four pyrimidine bases found in the nucleic acids DNA and RNA.

cytoskeleton: The flexible network of cells composed of microtubules and microfilaments.

dark reactions: In photosynthetic cells, the light-independent enzymatic reactions concerned with the synthesis of glucose from CO_2, ATP, and NADPH.

day-neutral plants: Plants that flower without regard to day length.

de- [L. *de-,* away from, down, off]: Prefix meaning "away from," "down," or "off"; for example, dehydration means removal of water.

deciduous [L. *decidere,* to fall off]: Shedding leaves at a certain season.

decomposers: Organisms (bacteria, fungi, heterotrophic protists) in an ecosystem that break down organic material into smaller molecules that then are recirculated.

dehiscence [L. *de,* down, + *hiscere,* split open]: The opening of an anther, fruit, or other structure, which permits the escape of reproductive bodies contained within.

denitrification: The conversion of nitrate to gaseous nitrogen; carried out by a few genera of free-living soil bacteria.

deoxyribonucleic acid (DNA): Carrier of genetic information in cells; composed of chains of phosphate, sugar molecules (deoxyribose), and purines and pyrimidines; capable of self-replication as well as determining RNA synthesis.

deoxyribose [L. *deoxy,* loss of oxygen, + *ribose,* a kind of sugar]: A five-carbon sugar with one less atom of oxygen than ribose; a component of deoxyribonucleic acid.

dermal tissue system: The epidermis or the periderm.

desmotubule [Gk. *desmos*, to bind, + L. *tubulus*, small tube]: The tubule traversing a plasmodesmatal canal and uniting the endoplasmic reticulum of the two adjacent cells.

determinate growth: Growth of limited duration, characteristic of floral meristems and of leaves.

deuterium: Heavy hydrogen; a hydrogen atom, the nucleus of which contains one proton and one neutron. (The nucleus of most hydrogen atoms consists of only one proton.)

dichotomy: The division or forking of an axis into two branches.

dicotyledon: One of the two classes of angiosperms, Dicotyledones; often abbreviated as dicot. Dicotyledons are characterized by having two cotyledons, net-veined leaves, and flower parts usually in fours or fives.

dictyosome: In eukaryotes, a group of flat, disk-shaped sacs that are often branched into tubules at their margins; serve as collecting and packaging centers for the cell and concerned with secretory activities; also called Golgi bodies. The term "Golgi apparatus" is used to refer collectively to all the dictyosomes, or Golgi bodies, of a given cell.

differentially permeable membrane: A membrane through which different substances diffuse at different rates.

differentiation: A developmental process by which a relatively unspecialized cell undergoes a progressive change to a more specialized cell; the specialization of cells and tissues for particular functions during development.

diffuse-porous wood: A wood in which the pores, or vessels, are fairly uniformly distributed throughout the growth layers or in which the size of pores changes only slightly from early wood to late wood.

diffusion [L. *diffundere*, to pour out]: The net movement of suspended or dissolved particles from a more concentrated region to a less concentrated region as a result of the random movement of individual molecules; the process tends to distribute such particles uniformly throughout a medium.

digestion: The conversion of complex, usually insoluble foods into simple, usually soluble forms by means of enzymatic action.

dikaryon (Gk. *di*, two, + *karyon*, a nut]: In fungi, mycelium with paired nuclei, each usually derived from a different parent.

dikaryotic: In fungi, having pairs of nuclei within cells or compartments.

dimorphism [Gk. *di*, two + *morphe*, form]: The condition of having two distinct forms, such as sterile and fertile leaves in ferns, or sterile and fertile shoots in horsetails.

dioecious [Gk. *di*, two + *oikos*, house]: Unisexual; having the male and female (or staminate and ovulate) elements on different individuals of the same species.

diploid: Having two sets of chromosomes; the $2n$ (diploid) chromosome number is characteristic of the sporophyte generation.

disaccharide: A carbohydrate formed of two simple sugar molecules linked by a covalent bond; sucrose is an example.

disk flowers: The actinomorphic, tubular flowers in Asteraceae; contrasted with flattened, zygomorphic ray flowers. In many Asteraceae, the disk flowers occur in the center of the inflorescence, the ray flowers around the margins.

distal: Situated away from or far from point of reference (usually the main part of body); opposite of proximal.

division: One of the major kinds of groups used by botanists in classifying organisms, the equivalent of phyla in animals and hetrotrophic protists; kingdoms are divided into divisions (phyla).

DNA: *See* deoxyribonucleic acid.

dominant allele: One allele is said to be dominant with respect to an alternative allele if the homozygote for the dominant allele is indistinguishable phenotypically from the heterozygote; the other allele is said to be recessive.

dormancy [L. *dormire*, to sleep]: A special condition of arrested growth in which the plant and such plant parts as buds and seeds do not begin to grow without special environmental cues. The requirement for such cues, which include cold exposure and a suitable photoperiod, prevents the breaking of dormancy during superficially favorable growing conditions.

double fertilization: The fusion of the egg and sperm (resulting in a $2n$ fertilized egg, the zygote) and the simultaneous fusion of the second male gamete with the polar nuclei (resulting in a $3n$ primary endosperm nucleus); a unique characteristic of all angiosperms.

doubling rate: The length of time required for a population of a given size to double in number.

drupe [Gk. *dryppa*, overripe olive]: A simple, fleshy fruit, derived from a single carpel, usually one-seeded, in which the inner fruit coat is hard and may be adherent to the seed.

druse: A compound, more or less spherical crystal with many component crystals projecting from its surface; composed of calcium oxalate.

early wood: The first-formed wood of a growth increment; it contains larger cells and is less dense than the subsequently formed late wood: replaces the term "spring wood."

eco- [Gk. *oikos*, house]: Prefix meaning "house" or "home."

ecology: The study of the interactions of organisms with their physical environment and with one other.

ecosystem: A major interacting system that involves both living organisms and their physical environment.

ecotype [Gk. *oikos*, house, + L. *typus*, image]: A locally adapted variant of an organism, differing genetically from other ecotypes.

ectoplast: *See* plasma membrane.

edaphic [Gk. *edaphos*, ground, soil]: Pertaining to the soil.

egg: A nonmotile female gamete, usually larger than a male gamete of the same species.

egg apparatus: The egg cell and synergids located at the micropylar end of the female gametophyte, or embryo sac, of angiosperms.

elater [Gk. *elater*, driver]: (1) An elongated, spindle-shaped, sterile cell in the sporangium of a liverwort sporophyte (aids in spore dispersal); (2) clubbed, hygroscopic bands attached to the spores of the horsetails.

electrolyte: A substance that dissociates into ions in aqueous solution and so makes possible the conduction of an electric current through the solution.

electron: A subatomic particle with a negative electric charge equal in magnitude to the positive charge of the proton, but with a mass of $1/1837$ of that of the proton. Electrons orbit the atom's positively charged nucleus and determine the atom's chemical properties.

electron-dense: In electron microscopy, not permitting the passage of electrons and so appearing dark.

element: A substance composed of only one kind of atom; one of more than 100 distinct natural or synthetic types of matter that, singly or in combination, compose virtually all materials of the universe.

embryo [Gk. *en*, in, + *bryein*, to swell]: A young sporophytic plant, before the start of a period of rapid growth (germination in seed plants).

embryogeny: The formation of the embryo.

embryo sac: The female gametophyte of angiosperms, generally an eight-nucleate, seven-celled structure; the seven cells are the egg cell, two synergids and three antipodals (each with a single nucleus), and the central cell (with two nuclei).

endergonic: Describing a chemical reaction that requires energy to proceed; the opposite of "exergonic."

endo- [Gk. *endo*, within]: Prefix meaning "within."

endocarp [Gk. *endo*, within, + *karpos*, fruit]: The innermost layer of the mature ovary wall, or pericarp.

endocytosis [Gk. *endon*, within, + *kytos*, hollow vessel]: The uptake of material into cells by means of invagination of the plasma membrane; if solid material is involved, the process is called phagocytosis; if dissolved material is involved, it is called pinocytosis.

endodermis [Gk. *endon*, within, + *derma*, skin]: A single layer of cells forming a sheath around the vascular region in roots and some stems; the endodermal cells are characterized by a Casparian strip within radial and transverse walls. In roots and stems of seed plants, the endodermis is the innermost layer of the cortex.

endogenous [Gk. *endon*, within, + *genos*, race, kind]: Arising from deep-seated tissues, as in the case of lateral roots.

endomembrane system: Collectively, the cellular membranes that form a continuum (plasma membrane, tonoplast, endoplasmic reticulum, Golgi bodies, and nuclear envelope).

endoplasmic reticulum: A complex, three-dimensional membrane system of indefinite extent present in eukaryotic cells, dividing the cytoplasm into compartments and channels. Those portions that are densely coated with ribosomes are called rough endoplasmic reticulum, and other portions with fewer ribosomes are called smooth endoplasmic reticulum.

endosperm [Gk. *endon*, within, + *sperma*, seed]: A tissue, containing stored food, that develops from the union of a male nucleus and the polar nuclei of the central cell; it is digested by the growing sporophyte either before or after the maturation of the seed; found only in angiosperms.

entrainment: The process by which a periodic repetition of light and dark, or some other external cycle, causes a circadian rhythm to remain synchronized with the same cycle as the modifying, or entraining, factor.

entropy: A measure of the randomness or disorder of a system.

enzyme: A protein that is capable of speeding up specific chemical reactions by lowering the required activation energy, but is unaltered itself in the process; a biological catalyst.

epi- [Gk. *epi*, upon]: Prefix meaning "upon" or "above."

epicotyl: The upper portion of the axis of an embryo or seedling, above the cotyledons (seed leaves) and below the next leaf or leaves.

epidermis: The outermost layer of cells of the leaf and of young stems and roots; primary in origin.

epigyny [Gk. *epi*, upon, + *gyne*, woman]: A pattern of floral organization in which the sepals, petals, and stamens apparently grow from the top of the ovary; the opposite of hypogyny.

epiphyte (ĕp′ĭ·fīt): An organism that grows upon another organism but is not parasitic on it.

epistatic [Gk. *epistasis*, a stopping]: Term used to describe a gene the action of which modifies the phenotypic expression of a gene at another locus.

ergastic substance: A passive product of the protoplast; a storage or waste product.

ethylene: A simple hydrocarbon that is a plant hormone involved in the ripening of fruit; $H_2C = CH_2$.

etiolation (e′tĭ·o·lā′shŭn) [Fr. *etioler*, to blanch]: A condition involving increased stem elongation, poor leaf development, and lack of chlorophyll; found in plants growing in the dark or with a greatly reduced amount of light.

etioplast: Plastid of a plant grown in the dark and containing a prolamellar body.

eukaryote [Gk. *eu*, good, + *karyon*, kernel]: A cell that has a membrane-bound nucleus, membrane-bound organelles, and chromosomes in which the DNA is associated with proteins; an organism composed of such cells. Plants, animals, fungi, and protists are the four kingdoms of eukaryotes.

eustele [Gk. *eu-*, good, + *stele*, pillar]: A stele in which the primary vascular tissues are arranged in discrete strands around a pith: typical of gymnosperms and angiosperms.

evolution: The derivation of progressively more complex forms of life from simple ancestors; Darwin proposed that natural selection is the principal mechanism by which evolution takes place.

exergonic [L. *ex*, out, + Gk. *ergon*, work]: Energy-yielding, as in a chemical reaction; applied to a "downhill" process.

exine: The outer wall layer of a spore or pollen grain.

exocarp [Gk. *exo*, without, + *karpos*, fruit]: The outermost layer of the mature ovary wall, or pericarp.

exon [Gr. *exo*, outside]: A segment of DNA that is both transcribed into RNA and translated into protein; exons are characteristic of eukaryotes. Contrasts with intron.

eyespot: Also stigma; a small, pigmented structure in flagellate unicellular organisms that is sensitive to light.

F_1: First filial generation. The offspring resulting from a cross. F_2 and F_3 are the second and third generations resulting from such a cross.

facilitated diffusion: The carrier-assisted transport of solutes driven by a concentration gradient.

family: A taxonomic group between order and genus in rank; the ending of family names in animals and heterotrophic protists is *-idae*; in all other organisms it is *-aceae*. A family contains one or more genera, and each family belongs to an order.

fascicle (făs′ĭ·k′l) [L. *fasciculus*, a small bundle]: A bundle of pine leaves or other needlelike leaves of gymnosperms; an obsolete term for a vascular bundle.

fascicular cambium: The vascular cambium originating within a vascular bundle, or fascicle.

fats: A molecule composed of glycerol and three fatty acid molecules; the proportion of oxygen to carbon is much less in fats than it is in carbohydrates. Fats in the liquid state are called oils.

feedback inhibition: Control mechanism whereby an increase in the concentration of some molecule inhibits the further synthesis of that molecule.

fermentation: The extraction of energy from organic compounds without the involvement of oxygen.

ferredoxin: Electron-transferring proteins of high iron content; some are involved in photosynthesis.

fertilization: The fusion of two gamete nuclei to form a diploid zygote.

fiber: An elongated, tapering, generally thick-walled sclerenchyma cell of vascular plants; its walls may or may not be lignified; it may or may not have a living protoplast at maturity.

fibril: Submicroscopic threads composed of cellulose molecules, which constitute the form in which cellulose occurs in the cell wall.

field moisture capacity: The percentage of water a particular soil will hold against the action of gravity.

filament: (1) The stalk of a stamen; (2) a term used to describe the threadlike bodies of certain algae or fungi.

fission: Asexual reproduction involving the division of a single-celled individual into two new single-celled individuals of equal size.

fitness: The genetic contribution of an organism to future generations, relative to the contributions of organisms living in the same environment that have different genotypes.

flagellum, *pl.* **flagella** [L. *flagellum*, whip]: A long threadlike organelle that protrudes from the surface of a cell. The flagella of bacteria are capable of rotary motion and consist of a single protein fiber each; eukaryotic flagella, which are used in locomotion and feeding, consist of an array of microtubules with a characteristic internal 9 + 2 microtubule structure and are capable of a vibratory, but not rotary, motion. A cilium is a small eukaryotic flagellum.

flavoprotein: A dehydrogenase that contains a flavin and often a metal and plays a major role in oxidation; abbreviated FP.

floral tube: A cup or tube formed by the fusion of the basal parts of the sepals, petals, and stamens; floral tubes are often found in plants that have an inferior ovary.

floret: One of the small flowers that make up the composite inflorescence or the spike of the grasses.

florigen [L. *flor-*, flower, + Gk. *-genes*, producer]: A plant hormone that promotes flowering.

flower: The reproductive structure of angiosperms; a complete flower includes calyx, corolla, androecium (stamens), and gynoecium (carpels), but all flowers contain at least one stamen or one carpel.

fluid-mosaic model: Model of membrane structure, with the membrane composed of a lipid bilayer in which globular proteins are embedded.

follicle [L. *folliculus*, small ball]: A dry, dehiscent fruit derived from a single carpel and opening along one side.

food chain, food web: A chain of organisms existing in any natural community such that each link in the chain feeds on the one below and is eaten by the one above; there are seldom more than six links in a chain, with autotrophs on the bottom and the largest carnivores at the top.

fossil [L. *fossils*, dug up]: The remains, impressions, or traces of an organism that has been preserved in rocks found in the earth's crust.

fossil fuels: The altered remains of once-living organisms that are burned to release energy; oil, gas, and coal.

FP: *See* flavoprotein.

free energy: Energy available to do work.

frond: The leaf of a fern; any large, divided leaf.

fruit: In angiosperms, a mature, ripened ovary (or group of ovaries), containing the seeds, together with any adjacent parts that may be fused with it at maturity; sometimes applied informally, and misleadingly, as in "fruiting body," to the reproductive structures of other kinds of organisms.

fucoxanthin (fū′kŏ·zăn′thĭn) [Gk. *phykos*, seaweed, + *xanthos*, yellowish-brown]: A brownish carotenoid found in brown algae and chrysophytes.

fundamental tissue system: *See* ground tissue system.

funiculus [L. *funiculus*, small rope or cord]: The stalk of the ovule.

fusiform initials [L. *fusus*, spindle]: The vertically elongated cells in the vascular cambium that give rise to the cells of the axial system in the secondary xylem and secondary phloem.

gametangium, *pl.* **gametangia** [Gk. *gamein*, to marry, + L. *tangere*, to touch]: A cell or organ in which gametes are formed.

gamete [Gk. *gamete*, wife]: A haploid reproductive cell; gametes fuse in pairs, forming zygotes, which are diploid.

gametophore [Gk. *gamein*, to marry, + *phoros* bearing]: In the bryophytes, a fertile stalk that bears gametangia.

gametophyte: In plants, which have an alternation of generations, the haploid (*n*), gamete-producing phase.

gel: A mixture of substances having a semisolid or solid constitution.

gemma, *pl.* **gemmae** (jěm′ă) [L. *gemma*, bud]: A small mass of vegetative tissue; an outgrowth of the thallus, for example, in liverworts or certain fungi; it can develop into an entire new plant.

gene: A unit of heredity; a sequence of DNA nucleotides that codes for a protein, tRNA, or rRNA molecule, or regulates the transcription of such a sequence.

gene frequency: The relative occurrence of a particular allele in a population.

generative cell: (1) In many gymnosperms, the cell of the male gametophyte that divides to form the stalk and body cells: (2) in angiosperms, the cell of the male gametophyte that divides to form two sperm.

genetic code: The system of nucleotide triplets (codons) in DNA and RNA that dictates the amino acid sequence in proteins; except for three "stop" signals, each codon specifies one of 20 amino acids.

genetic recombination: The occurrence of gene combinations in the progeny that are different from the combinations present in the parents.

genotype: The genetic constitution, latent or expressed, of an organism, as contrasted with the phenotype; the sum total of all the genes present in an individual.

genus, *pl.* **genera:** The taxonomic group between family and species in rank; genera include one or more species.

geotropism: *See* gravitropism.

GERL: The Golgi associated–endoplasmic reticulum—lysosomal complex, which may be involved with lysosome or vacuole formation in some cells.

germination [L. *germinare*, to sprout]: The beginning or resumption of growth by a spore, seed, bud, or other structure.

gibberellins (jĭb·ĕ·rĕ′lĭns) [*Gibberella*, a genus of fungi]: A group of growth hormones, the best known effect of which is to increase the elongation of plant stems.

gill: The plates on the underside of the cap in basidiomycetes.

girdling: The removal from a woody stem of a ring of bark extending inward to the cambium; also called ringing.

glucose: A common six-carbon sugar ($C_6H_2O_6$); the most common monosaccharide in most organisms.

glycerol: A three-carbon molecule with three hydroxyl groups attached; glycerol molecules combine with fatty acids to form fats or oils.

glycogen [Gk. *glykys*, sweet, + *gen*, of a kind]: A carbohydrate similar to starch that serves as the reserve food in bacteria, fungi, and most organisms other than plants.

glycolysis: The anaerobic breakdown of glucose to form two molecules of pyruvic acid, resulting in the net liberation of two molecules of ATP.

glyoxylate cycle: A variant of the Krebs cycle; present in bacteria and some plant cells, for net conversion of acetate into succinate and, eventually, new carbohydrate.

glyoxysome: A microbody containing enzymes necessary for the conversion of fats into carbohydrates; glyoxysomes play an important role during the germination of seeds.

Golgi body (gôl′jē): *See* dictyosome.

grafting: A union of different individuals in which a portion, called the scion, of one individual is inserted into a root or stem, called the stock, of the other individual.

grain: *See* caryopsis; also a term used to refer to the alignment of wood elements.

grana, *sing.* **granum:** Structures within chloroplasts, seen as green granules with a light microscope and as a series of stacked thylaloids with an electron microscope; the grana contain the chlorophylls and carotenoids and are the sites of the light reactions of photosynthesis.

gravitropism [L. *gravis*, heavy, + Gk. *tropes*, turning]: The response of a shoot or root to the pull of the earth's gravity; also called geotropism.

ground meristem [Gk. *meristos*, divisible]: The primary meristem, or meristematic tissue, that gives rise to the ground tissues.

ground tissue system: All tissues other than the epidermis (or periderm) and the vascular tissues; also called fundamental tissue system.

growth layer: A layer of growth in the secondary xylem or secondary phloem; *see also* annual ring.

growth ring: A growth layer in the secondary xylem or secondary phloem, as seen in transverse section; may be called a growth increment, especially where seen in other than transverse section.

guanine [Sp. from Quechua, *huanu*, dung]: A purine base found in DNA and RNA. Its name is based on the fact that guanine is abundant in the form of a white crystalline base in guano and other animal excrements.

guard cells: Pairs of specialized epidermal cells surrounding a pore, or stoma; changes in the turgor of a pair of guard cells cause opening and closing of the pore.

guttation [L. *gutta*, a drop]: The exudation of liquid water from leaves caused by root pressure.

gymnosperm [Gk. *gymnos*, naked, + *sperma*, seed]: A seed plant with seeds not enclosed in an ovary; the conifers are the most familiar group.

gynoecium [Gk. *gyne*, woman, + *oikos*, house]: The aggregate of carpels in the flower of a seed plant.

habit [L. *habitus*, condition, character]: Characteristic form or appearance of an organism.

habitat [L. *habitare*, to inhabit]: The environment of an organism; the place where it is usually found.

haploid [Gk. *haploos*, single]: Having only one set of chromosomes (*n*), in contrast to diploid (2*n*).

hardwood: A name commonly applied to the wood of a dicot tree.

Hardy-Weinberg law: The mathematical expression of the relationship between the relative frequencies of two or more alleles in a population; it demonstrates that the frequencies of alleles and genotypes will remain constant in a random-mating population in the absence of inbreeding, selection, or other evolutionary forces.

haustorium, *pl.* **haustoria** [L. *haustus*, from *haurire*, to drink, draw]: A projection of fungal hypha that functions as a penetrating and absorbing organ.

heartwood: Nonliving and commonly dark-colored wood in which no water transport occurs; it is surrounded by sapwood.

heliotropism [Gk. *helios*, sun]: *See* solar tracking.

hemicellulose (hěm′ĭ·sěl′u·lōs): A polysaccharide resembling cellulose but more soluble and less ordered; found particularly in cell walls.

herb [L. *herba*, grass]: A nonwoody seed plant with a relatively short-lived aerial portion.

herbaceous: An adjective referring to nonwoody plants.

herbarium: A collection of dried and pressed plant specimens.

herbivorous: Feeding upon plants.

heredity [L. *heredis*, heir]: The transmission of characteristics from parent to offspring through the gametes.

hermaphrodite [Gk. for Hermes and Aphrodite]: An organism possessing both male and female reproductive organs.

hetero- [Gk. *heteros*, different]: Prefix meaning "other" or "different."

heterocyst [Gk. *heteros*, different, + *cystis*, a bag]: A kind of large, transparent, thick-walled, nitrogen-fixing cell that forms in the filaments of certain cyanobacteria.

heteroecious (hĕt′er·ē′shŭs) [Gk. *heteros*, different, + *oikos*, house]: As in some rust fungi, requiring two different host species to complete the life cycle.

heterogamy [Gk. *heteros*, other, + *gamos*, union or reproduction]: Reproduction involving two types of gametes.

heterokaryotic [Gk. *heteros*, other, + *karyon*, kernel]: In fungi, having two or more genetically distinct types of nuclei within the same mycelium.

heteromorphic [Gk. *heteros*, different, + *morphe*, form]: A term used to describe a life history in which the haploid and diploid generations are dissimilar in form.

heterosis [Gk. *heterosis*, alteration]: Hybrid vigor, the superiority of the hybrid over either parent in any measurable character.

heterosporous: Having two kinds of spores, designated as microspores and megaspores.

heterothallic [Gk. *heteros*, different, + *thallus*, sprout]: A term used to describe a species, the haploid individuals of which are self-sterile or self-incompatible; two compatible strains or individuals are required for sexual reproduction to take place.

heterotroph [Gk. *heteros*, other, + *trophos*, feeder]: An organism that cannot manufacture organic compounds and so must feed on organic materials that have originated in other plants and animals; *see also* autotroph.

heterozygous: Having two different alleles at the same locus on homologous chromosomes.

Hill reaction: The oxygen evolution and photoreduction of an artificial electron acceptor by a chloroplast preparation in the absence of carbon dioxide.

hilum [L. *hilum*, a trifle]: (1) Scar left on seed after separation of seed from funiculus; (2) the part of a starch grain around which the starch is laid down in more or less concentric layers.

histone: The group of five basic proteins associated with the chromosomes of all eukaryotic cells.

holdfast: (1) Basal part of a multicellular alga that attaches it to a solid object; may be unicellular or composed of a mass of tissue; (2) cuplike structures at the tips of some tendrils, by means of which they become attached.

homeo-, homo- [Gk. *homos*, same, similar]: Prefix meaning "similar" or "same."

homeostasis (hō′me·ō·stā′sĭs) [Gk. *homos*, similar, + *stasis*, standing]: The maintaining of a relatively stable internal physiological environment within an organism, or a steady-state equilibrium in a population or ecosystem. Homeostasis usually involves feedback mechanisms.

homokaryotic [Gk. *homos*, same, + *karyon*, kernel]: In fungi, having nuclei with the same genetic makeup within a mycelium.

homologous chromosomes: Chromosomes that associate in pairs in the first stage of meiosis; each member of the pair is derived from a different parent. Homologous chromosomes are also called homologues.

homology [Gk. *homologia*, agreement]: A condition indicative of the same phylogenetic, or evolutionary, origin, but not necessarily the same in present structure and or function.

homosporous: Having only one kind of spore.

homothallic [Gk. *homos*, same, + *thallus*, sprout]: A term used to describe a species in which the individuals are self-fertile.

homozygous: Having identical alleles at the same locus on homologous chromosomes.

hormogonium, *pl.* **hormogonia:** A portion of a filament of a cyanobacterium that becomes detached and grows into a new filament.

hormone [Gk. *hormaein*, to excite]: A chemical substance produced usually in minute amounts in one part of an organism, from which it is transported to another part of that organism on which it has a specific effect.

host: An organism on or in which a parasite lives.

humus: Decomposing organic matter in the soil.

hyaloplasm: *See* cytoplasmic ground substance.

hybrid: Offspring of two parents that differ in one or more heritable characteristics; offspring of two different varieties or of two different species.

hybridization: The formation of offspring between unlike parents.

hybrid vigor: *See* heterosis.

hydrocarbon [Gk. *hydro,* water, + L. *carbo,* charcoal]: An organic compound that consists only of hydrogen and carbon atoms.

hydrogen bond: A weak bond between a hydrogen atom attached to one oxygen or nitrogen atom and another oxygen or nitrogen atom.

hydrolysis [Gk. *hydro,* water, + *lysis,* loosening]: Splitting of one molecule into two by addition of the H^+ and OH^- ions of water.

hydrophyte [Gk. *hydro,* water, + *phyton,* a plant]: A plant that depends on an abundant supply of moisture or that grows wholly or partly submerged in water.

hydroxyl group: An OH^- group; a negatively charged ion formed by the dissassociation of a water molecule.

hymenium [Gk. *hymen,* a membrane]: The layer of asci on an ascocarp, or of basidia on a basidiocarp, together with any associated sterile hyphae.

hyper- [Gk. *hyper,* above, over]: Prefix meaning ''above'' or ''over.''

hypertonic: Refers to a solution that has a concentration of solute particles high enough to gain water across a permeable membrane from another solution.

hypha, *pl.* **hyphae** [Gk. *hyphe,* web]: A single tubular filament of a fungus, oomycete, or chytrid; the hyphae together comprise the mycelium.

hypo- [Gk. *hypo,* less than]: Prefix meaning ''under'' or ''less.''

hypocotyl: The portion of an embryo or seedling situated between the cotyledons and the radicle.

hypocotyl-root axis: The embryo axis below the cotyledon or cotyledons, consisting of the hypocotyl and the apical meristem of the root or the radicle.

hypodermis [Gk. *hypo,* under, + *derma,* skin]: One or more layers of cells beneath the epidermis, which are distinct from the underlying cortical or mesophyll cells.

hypogyny [Gk. *hypo,* under, + *gyne,* woman]: Floral organization in which the sepals, petals, and stamens are attached to the receptacle below the ovary; the opposite is epigyny.

hypothesis [Gk. *hypo,* under, + *tithenai,* to put]: A temporary working explanation or supposition based on accumulated facts and suggesting some general principle or relation of cause and effect; a postulated solution to a scientific problem that must be tested by experimentation and, if disproved or shown to be unlikely, is discarded.

hypotonic: Refers to a solution that has a concentration of solutes low enough to lose water across a permeable membrane to another solution.

IAA: *See* indoleacetic acid.

imbibition (ĭm·bĭ·bĭsh′ŭn): Adsorption of water and swelling of colloidal materials because of the adsorption of water molecules onto the internal surfaces of the materials.

imperfect flower: A flower lacking either stamens or carpels.

imperfect fungi: One of the Fungi Imperfecti, which reproduce only asexually, or in which the sexual cycle has not been observed; most Fungi Imperfecti are ascomycetes.

inbreeding: The breeding of closely related plants or animals: in plants, it is usually brought about by repeated self-pollination.

incomplete flower: A flower lacking one or more of the four kinds of floral parts, that is, lacking sepals, petals, stamens, or carpels.

indehiscent (ĭn′de·hĭs′ĕnt): Remaining closed at maturity, as are many fruits (samaras, for example).

independent assortment: *See* Mendel's second law.

indeterminate growth: Unrestricted or unlimited growth, as with a vegetative apical meristem that produces an unrestricted number of lateral organs indefinitely.

indoleacetic acid, or **IAA:** A naturally occurring auxin, a kind of plant hormone.

indusium, *pl.* **indusia** (ĭn·dū′zi·ŭm) [L. *indusium,* a woman's undergarment]: Membranous growth of the epidermis of a fern leaf that covers a sorus.

inferior ovary: An ovary that is completely or partially attached to the calyx; the other floral whorls appear to arise from its top.

inflorescence: A flower cluster, with a definite arrangement of flowers.

initial: (1) In the meristem, a cell that remains within the meristem indefinitely and at the same time, by division, adds cells to the plant body; (2) a meristematic cell that eventually differentiates into a mature, more specialized cell or element

inner bark: In older trees, the living part of the bark; the bark inside the innermost periderm.

integument: The outermost layer or layers of tissue enveloping the nucellus of an ovule; develops into the seed coat.

inter- [L. *inter,* between]: Prefix meaning ''between,'' or ''in the midst of.''

intercalary [L. *intercalare,* to insert]: Descriptive of meristematic tissue or growth not restricted to the apex of an organ; that is, growth in the regions of the nodes.

interfascicular cambium: The vascular cambium arising between the fascicles, or vascular bundles, from interfascicular parenchyma.

interfascicular region: Tissue region between vascular bundles in stem; also called pith ray.

internode: The region of a stem between two successive nodes.

interphase: The period between two mitotic or meiotic cycles; the cell grows and its DNA replicates during interphase.

intine: The inner wall layer of a spore or pollen grain.

intra- [L. *intra,* within]: Prefix meaning ''within.''

intron [L. *intra,* within]: A portion of mRNA as transcribed from eukaryotic DNA that is removed by enzymes before the mRNA is translated into protein. *See* exon.

ion: An atom or molecule that has lost or gained one or more electrons, and thus become positively or negatively charged.

irregular flower: A flower in which one or more members of at least one whorl differ in form from other members of the same whorl.

iso- [Gk. *isos,* equal]: Prefix meaning ''equal''; like ''homeo-'' or ''homo-.''

isogamy: A type of sexual reproduction in which the gametes (or gametangia) are alike in size; found in some algae and fungi.

isomer [Gk. *isos,* equal, + *meros,* part]: One of a group of compounds identical in atomic composition but differing in structural arrangement; for example, glucose and fructose.

isomorphic [Gk. *isos,* equal, + *morphe,* form]: Identical in form.

isotonic: Having the same osmotic concentration.

isotope: One of several possible forms of a chemical element that differ from other forms in the number of neutrons in the atomic nucleus, but not in chemical properties.

karyogamy [Gk. *karyon,* kernel, + *gamos,* marriage]: The union of two nuclei following plasmogamy.

kelp: A common name for any of the larger members of the order Laminariales of the brown algae.

kinetin [Gk. *kinetikos,* causing motion]: A purine that probably does not occur in nature but that acts as a cytokinin in plants.

kinetochore: *See* centromere.

kingdom: One of the five chief taxonomic categories; for example, Monera or Plantae.

Kranz anatomy [Ger. *Kranz,* wreath]: The wreathlike arrangement of mesophyll cells around a layer of large bundle-sheath cells, forming two concentric layers around the vascular bundle; typically found in the leaves of C_4 plants.

Krebs cycle: The series of reactions that results in the oxidation of pyruvic acid to hydrogen atoms, electrons, and carbon dioxide. The electrons, passed along electron-carrier molecules, then go through the oxidative phosphorylation and terminal oxidation processes. Also called the tricarboxylic acid cycle or TCA cycle.

lamella (la·měl′a) [L. *lamella,* thin metal plate]: Layer of cellular membranes, particularly photosynthetic, chlorophyll-containing membranes; *see also* middle lamella.

lamina: the blade of a leaf.

laminarin: One of the principal storage products of the brown algae; a polymer of glucose.

lateral meristems: Meristems that give rise to secondary tissue; the vascular cambium and cork cambium.

lateral root: A root that arises from another, older root; also called a branch root, or secondary root, if the older root is the primary root.

late wood: The last part of the growth increment formed in the growing season; it contains smaller cells and is denser than the early wood; replaces the term ''summer wood.''

leaching: The downward movement and drainage of minerals, or inorganic ions, from the soil by percolating water.

leaf buttress: A lateral protrusion below the apical meristem; represents the initial stage in the development of a leaf primordium.

leaf gap: Region of parenchyma tissue in the primary vascular cylinder above the point of departure of the leaf trace or traces.

leaflet: One of the parts of a compound leaf.

leaf primordium [L. *primordium,* beginning]: A lateral outgrowth from the apical meristem that will eventually become a leaf.

leaf scar: A scar left on a twig when a leaf falls.

leaf trace: That part of a vascular bundle extending from the base of the leaf to its connection with a vascular bundle in the stem.

legume [L. *legumem,* leguminous plant]: (1) A member of the Fabaceae, the pea or bean family: (2) a type of dry fruit that is derived from one carpel and opens along both sides.

lenticels (lĕn′ti·sĕls) [L. *lenticella,* a small window]: Spongy areas in the cork surfaces of stem, roots, and other plant parts that allow interchange of gases between internal tissues and the atmosphere through the periderm; occur in vascular plants.

leucoplast (lū′kō·plăst) [Gk. *leuko,* white, + *plasein,* to form]: A colorless plastid; leucoplasts are commonly centers of starch formation.

liana [F. *liane,* from *lier,* to bind]: A large, woody vine that climbs upon other plants.

life cycle: The entire sequence of phases in the growth and development of any organism from time of zygote formation to gamete formation.

light reactions: The reactions of photosynthesis that require light and cannot occur in the dark.

lignin: One of the most important constituents of the secondary wall of vascular plants, although not all secondary walls contain lignin; after cellulose, lignin is the most abundant plant polymer.

ligule [L. *ligula,* small tongue]: A minute outgrowth or appendage at the base of the leaves of grasses and those of certain lycophytes.

linkage: The tendency for certain genes to be inherited together owing to the fact that they are located on the same chromosome.

lipid [Gk. *lipos,* fat]: One of a large variety of nonpolar organic molecules that are insoluble in water (which is polar) but dissolve readily in nonpolar organic solvents; lipids include fats, oils, steroids, phospholipids, and carotenoids.

locule (lŏk′ūl) [L. *loculus,* small chamber]: A cavity within a sporangium or a cavity of the ovary in which ovules occur.

locus, *pl.* **loci:** The position on a chromosome occupied by a particular gene.

long-day plants: Plants that must be exposed to light periods longer than some critical length for flowering to occur; they flower in spring or summer.

lumen [L. *lumen,* light, an opening for light]: The space bounded by the plant cell wall.

lysis [Gk. *lysis,* a loosening]: A process of disintegration or cell destruction.

lysogenic bacteria: Bacteria-carrying viruses (phages) that eventually break loose from the bacterial chromosome and set up an active cycle of infection, producing lysis in their bacterial hosts.

lysosome [Gk. *lysis,* loosening, + *soma,* body]: An organelle, bounded by a single membrane, and containing hydrolytic enzymes that are released when the organelle ruptures and are capable of breaking down proteins and other complex macromolecules.

macrocyst: A flattened, irregular structure, encircled by a thin membrane, in which zygotes are formed during the life cycle of cellular slime molds.

macrofibril: An aggregation of microfibrils, visible with the light microscope.

macromolecule [Gk. *makros,* large]: A molecule of very high molecular weight; refers specifically to proteins, nucleic acids, polysaccharides, and complexes of these.

macronutrients [Gk. *makros,* large, + L. *nutrire,* to nourish]: Inorganic chemical elements required in large amounts for plant growth, such as nitrogen, potassium, calcium, phosphorus, magnesium, and sulfur.

maltase: An enzyme that hydrolyzes maltose to glucose.

mannitol: One of the storage molecules of the brown algae; an alcohol.

marginal meristem: The meristem located along the margin of a leaf primordium and forming the blade.

mating type: A particular genetically defined strain of an organism that is incapable of sexual reproduction with another member of the same strain but capable of such reproduction with members of other strains of the same organism.

mega- [Gk. *megas,* large]: Prefix meaning ''large.''

megagametophyte- [Gk. *megas,* large, + *gamos,* marriage, + *phyton,* plant]: In heterosporous plants, the female gametophyte; located within the ovule of seed plants.

megaphyll [Gk. *megas,* large, + *phyllon,* leaf]: A generally large leaf with several to many veins; its leaf trace is associated with a leaf gap; in contrast to microphyll.

megasporangium, *pl.* **megasporangia:** A sporangium in which megaspores are produced; *see* nucellus.

megaspore: In heterosporus plants, a haploid ($1n$) spore that develops into a female gametophyte; in most groups, megaspores are larger than microspores.

megaspore mother cell: A diploid cell in which meiosis will occur, resulting in the production of four megaspores; also called a megasporocyte.

megasporocyte: *See* megaspore mother cell.

megasporophyll: A leaf or leaflike structure bearing a megasporangium.

meiosis (mī·ō′sĭs) [Gk. *meioun,* to make smaller]: The two successive nuclear divisions in which the chromosome number is reduced from diploid ($2n$) to haploid ($1n$) and segregation of the genes occurs; gametes or spores (in organisms with an alternation of generations) may be produced as a result of meiosis.

Mendel's first law: The factors for a pair of alternate characteristics are separate and only one may be carried in a particular gamete (genetic segregation).

Mendel's second law: The inheritance of one pair of characteristics is independent of the simultaneous inheritance of other traits, such characteristics assorting independently as though there were no others present (later modified by the discovery of linkage).

meristem [Gk. *merizein,* to divide]: The undifferentiated plant tissue from which new cells arise.

meso- [Gk. *mesos,* middle]: Prefix meaning "middle."

mesocarp [Gk. *mesos,* middle, + *karpos,* fruit]: The middle layer of the mature ovary wall, or pericarp, between the exocarp and endocarp.

mesophyll: The ground tissue (parenchyma) of a leaf, located between the layers of epidermis; mesophyll cells generally contain chloroplasts.

mesophyte [Gk. *mesos,* middle, + *phyton,* a plant]: A plant that requires abundant soil water and a relatively humid atmosphere.

messenger RNA (mRNA): The class of RNA that carries genetic information from the gene to the ribosomes, where it is translated into protein.

metabolism [Gk. *metabole,* change]: The sum of all chemical processes occurring within a living cell or organism.

metaphase: The stage of mitosis or meiosis during which the chromosomes lie in the equatorial plane of the spindle.

metaxylem [Gk. *meta,* after]: The part of the primary xylem that differentiates after the protoxylem; the metaxylem reaches maturity after the portion of the plant part in which it is located has finished elongating.

micro- [Gk. *mikros,* small]: Prefix meaning "small."

microbody: An organelle bounded by a single membrane and containing a variety of enzymes; generally associated with the endoplasmic reticulum. Peroxisomes and glyoxysomes are kinds of microbodies.

microfibril: A threadlike component of the cell wall, composed of cellulose molecules, visible only with the electron microscope.

microfilament [Gk. *mikros,* small + L. *filum,* a thread]: A long filament, 5 to 7 nanometers thick, composed of actin and believed to play a causative role in cytoplasmic streaming.

microgametophyte [Gk. *mikros,* small, + *gamos,* marriage, + *phyton,* plant]: In heterosporous plants, the male gametophyte.

micrometer: A unit of microscopic measurement convenient for describing cellular dimensions; 1/1000 of a millimeter; its symbol is μm.

micronutrients [Gk. *mikros,* small, + L. *nutrire,* to nourish]: Inorganic chemical elements required only in very small, or trace, amounts for plant growth, such as iron, chlorine, copper, manganese, zinc, molybdenum, and boron.

microphyll [Gk. *mikros,* small, + *phyllon,* leaf]: A small leaf with one vein and one leaf trace not associated with a leaf gap; in contrast to megaphyll. Microphylls are characteristic of the lycophytes.

micropyle: In the ovules of seed plants, the opening in the integuments through which the pollen tube usually enters.

microsporangium: A sporangium within which microspores are formed.

microspore: In heterosporous plants, a spore that develops into a male gametophyte.

microspore mother cell: A cell in which meiosis will occur, resulting in four microspores; in seed plants, often called a pollen mother cell; also called a microsporocyte.

microsporocyte: *See* microspore mother cell.

microsporophyll: A leaflike organ bearing one or more microsporangia.

microtrabecular lattice: The three-dimensional lattice of slender strands comprising the ground substance of the cell.

microtubule [Gk. *mikros,* small, + L. *tubulus,* little pipe]: Narrow (about 25 nanometers in diameter), elongate, nonmembranous tubule of indefinite length; microtubules occur in the cells of eukaryotes. Microtubules move the chromosomes in cell division and provide the internal structure of cilia and flagella.

middle lamella: The layer of intercellular material, rich in pectic compounds, cementing together the primary walls of adjacent cells.

mimicry [Gk. *mimos,* mime]: The superficial resemblance in form, color, or behavior of certain organisms (mimics) to other more powerful or more protected ones (models), resulting in protection, concealment, or some other advantage for the mimic.

mineral: A naturally occurring chemical element or inorganic compound.

mitochondrion, *pl.* **mitochondria** [Gk. *mitos,* thread, + *chondrion,* small grain]: A double-membrane-bound organelle found in eukaryotic cells; contains the enzymes of the Krebs cycle and the electron-transport chain; the major source of ATP in nonphotosynthetic cells.

mitosis (mī·tō′sĭs) [Gk. *mitos,* thread]: A process during which the duplicated chromosomes divide longitudinally and the daughter chromosomes then separate to form two genetically identical daughter nuclei; usually accompanied by cytokinesis.

mole: The name for a gram molecule. The number of particles in 1 mole of any substance; always equal to Avogadro's number: 6.022×10^{23}.

molecular weight: The relative weight of a molecule when the weight of the most frequent kind of carbon atom is taken as 12; the sum of the relative weights of the atoms in a molecule.

molecule: Smallest possible unit of a compound, consisting of two or more atoms.

mono- [Gk. *monos,* single]: Prefix meaning "one" or "single."

monocotyledon: A plant whose embryo has one cotyledon; one of the two great classes of angiosperms, Monocotyledones; often abbreviated as monocot.

monoecious (mō·nē′shŭs) [Gk. *monos,* single, + *oikos,* house]: Having the anthers and carpels produced in separate flowers on the same individual.

monokaryotic [Gk. *monos,* one, + *karyon,* kernel]: In fungi, having a single haploid nucleus within one cell or compartment.

monosaccharide [Gk. *monos,* single, + *sakcaron,* sugar]: A simple sugar, such as five-carbon and six-carbon sugars, that cannot be disassociated into smaller sugar particles.

-morph, morph- [Gk. *morphe,* form]: Suffix or prefix meaning "form."

morphogenesis: The development of form.

morphology [Gk. *morphe,* form, + *logos,* discourse]: The study of form and its development.

mRNA: *See* messenger RNA.

mucigel: A slime-sheath covering the surface of many roots.

multigene family: A collection of related genes on a chromosome; most eukaryotic genes appear to be members of multigene families.

multiple epidermis: A tissue composed of several cell layers derived from the protoderm; only the outer layer assumes characteristics of a typical epidermis.

multiple fruit: A cluster of matured ovaries produced by a cluster of flowers, as in the pineapple.

mutagen [L. *mutare,* to change, + Gk. *genaio,* to produce]: An agent that increases the mutation rate.

mutant: A mutated gene or an organism carrying a gene that has undergone a mutation.

mutation: An inheritable change of a gene from one allelic form to another.

mutualism: The living together of two or more organisms in an association that is mutually advantageous.

myc-, myco- [Gk. *mykes,* fungus]: Prefix meaning ''pertaining to fungi.''

mycelium [Gk. *mykes,* fungus]: The mass of hyphae forming the body of a fungus, oomycete, or chytrid.

mycology: The study of fungi.

mycorrhiza, *pl.* **mycorrhizae:** A symbiotic association between certain fungi and plant roots; characteristic of most vascular plants.

NAD: *See* nicotinamide adenine dinucleotide.

NADP: *See* nicotinamide adenine dinucleotide phosphate.

nannoplankton: (năn'o·plăngk'tŏn) [Gk. *nanos,* dwarf, + *planktos,* wandering]: Plankton with dimensions of less than 70 to 75 micrometers.

nastic movement: A plant movement that occurs in response to a stimulus, but whose direction is independent of the direction of the stimulus.

natural selection: The differential reproduction of genotypes based on their genetic constitution.

nectary [Gk. *nektar,* the drink of the gods]: In angiosperms, a gland that secretes nectar, a sugary fluid that attracts animals to plants.

netted venation: The arrangement of veins in the leaf blade that resembles a net; characteristic of dicot leaves; also called reticulate venation.

neutron [L. *neuter,* neither]: An uncharged particle with a mass slightly greater than that of a proton, found in the atomic nucleus of all elements except hydrogen, in which the nucleus consists of a single proton.

niche: The role played by a particular species in its ecosystem.

nicotinamide adenine dinucleotide (NAD): A coenzyme that functions as an electron acceptor in many of the oxidation reactions of respiration.

nicotinamide adenine dinucleotide phosphate (NADP): A coenzyme that functions as an electron acceptor in many of the reduction reactions of biosynthesis; similar in structure to NAD except that it contains an extra phosphate group.

nitrification: The oxidation of ammonium ions or ammonia to nitrate; a process carried out by a specific free-living soil bacteria.

nitrogen fixation: The incorporation of atmospheric nitrogen into nitrogen compounds; carried out by certain free-living and symbiotic bacteria.

nitrogen-fixing bacteria: Soil bacteria that convert atmospheric nitrogen into nitrogen compounds.

nitrogenous base: A nitrogen-containing molecule having basic properties (tendency to acquire an H atom); a purine or pyrimidine; one of the building blocks of nucleic acids.

node [L. *nodus,* knot]: The part of a stem where one or more leaves are attached; *see* internode.

nodules: Enlargements or swellings on the roots of legumes and certain other plants inhabited by symbiotic nitrogen-fixing bacteria.

noncyclic electron flow: The light-induced flow of electrons from water to $NADP^+$ in oxygen-evolving photosynthesis; it involves both photosystems I and II.

nonseptate: *See* aseptate.

nucellus (nu·sěl'ŭs) [L. *nucella,* a small nut]: Tissue composing the chief part of the young ovule, in which the embryo sac develops; equivalent to a megasporangium.

nuclear envelope: The double membrane surrounding the nucleus of a cell.

nucleic acid: An organic acid consisting of joined nucleotide complexes; the two types are deoxyribonucleic acid (DNA) and ribonucleic acid (RNA).

nucleolar organizer: A special area on a certain chromosome associated with the formation of the nucleolus.

nucleolus (nu·klē'ō·lŭs) [L. *nucleolus,* a small nucleus]: A small, spherical body found in the nucleus of eukaryotic cells, which is composed chiefly of rRNA in the process of being transcribed from copies of rRNA genes; the site of production of ribosomes.

nucleoplasm: The ground substance of a nucleus.

nucleotide: A single unit of nucleic acid, composed of a phosphate, a five-carbon sugar (either ribose or deoxyribose), and a purine or a pyrimidine.

nucleus: (1) A specialized body within the eukaryotic cell bounded by a double membrane and contaning the chromosomes; (2) the central part of an atom of a chemical element.

nut: A dry, indehiscent, hard, one-seeded fruit, generally produced from a gynoecium of more than one fused carpel.

obligate anaerobe: An organism that is metabolically active only in the absence of oxygen.

-oid [Gk. *oid,* like, resembling]: Suffix meaning ''like'' or ''similar to.''

ontogeny [Gk. *on,* being, + *genesis,* origin]: The development, or life history, of all or part of an individual organism.

oo- [Gk. *oion,* egg]: Prefix meaning ''egg.''

oogamy: Sexual reproduction in which one of the gametes (the egg) is large and nonmotile, and the other gamete (the sperm) is smaller and motile.

oogonium (ō'o·gō'nĭ·ŭm): A unicellular female sex organ that contains one or several eggs.

oospore: The thick-walled zygote characteristic of the oomycetes.

open bundle: A vascular bundle in which a vascular cambium develops.

operator: A site of gene regulation; a sequence of nucleotides recognized by a repressor protein.

operculum (o·pûr'ku·lŭm) [L. *operculum,* lid]: In mosses, the lid of the sporangium.

operon [L. *operis,* work]: A group of adjacent genes transcribed as a unit by a single mRNA molecule. Operons are common in bacteria, but rare in eukaryotes.

opposite: Term applied to buds or leaves occurring in pairs at a node.

order: A category of classification between the rank of class and family; classes contain one or more orders, and orders in turn are composed of one or more families.

organ: A structure composed of different tissues, such as root, stem, leaf, or flower parts.

organelle (ôr'găn·ĕl) [Gk. *organella,* a small tool]: A specialized part of a cell.

organic: Pertaining to living organisms in general, to compounds formed by living organisms, and to the chemistry of compounds containing carbon.

organism: Any individual living creature, either unicellular or multicellular.

osmosis (ŏs·mō′sĭs) [Gk. *osmos*, impulse or thrust]: The diffusion of water, or any solvent, across a differentially permeable membrane; in the absence of other forces, the movement of water during osmosis will always be from a region of greater water potential to one of lesser water potential.

osmotic potential: *See* solute potential.

osmotic pressure: The potential pressure that can be developed by a solution separated from pure water by a differentially permeable or semipermeable membrane; it is an index of the solute concentration of the solution.

outcrossing: Cross-pollination between individuals of the same species.

outer bark: In older trees, the dead part of the bark; the innermost periderm and all tissues outside it; also called rhytidome.

ovary [L. *ovum*, an egg]: The enlarged basal portion of a carpel or of a gynoecium composed of fused carpels; a mature ovary, sometimes with other adherent parts, is a fruit.

ovule (ō·vūl) [L. *ovulum*, a little egg]: A structure in seed plants containing the female gametophyte with egg cell, all being surrounded by the nucellus and one or two integuments; when mature, an ovule becomes a seed.

ovuliferous scale: In certain conifers, the appendage or scalelike shoot to which the ovule is attached.

oxidation: The loss of an electron by an atom or molecule. Oxidation and reduction (gain of an electron) take place simultaneously, because an electron that is lost by one atom is accepted by another. Oxidation-reduction reactions are an important means of energy transfer in living systems.

oxidative phosphorylation: The formation of ATP from ADP and inorganic phosphate; oxidative phosphorylation takes place in the electron-transport chain of the mitochondrion.

pairing of chromosomes: Side-by-side association of homologous chromosomes.

paleobotany [Gk. *palaios*, old]: The study of fossil plants.

palisade parenchyma: A leaf tissue composed of columnar chloroplast-bearing parenchyma cells with their long axes at right angles to the leaf surface.

panicle (păn′ĭk′l) [L. *panicula*, tuft]: An inflorescence, the main axis of which is branched, and whose branches bear loose flower clusters.

para- [Gk. *para*, beside]: Prefix, meaning "beside."

paradermal section [Gk. *para*, beside, + *derma*, skin]: Section cut parallel with the surface of a flat structure, such as a leaf.

parallel evolution: The development of similar structures having similar functions in two or more evolutionary lines as a result of the same kinds of selective pressures.

parallel venation: The pattern of venation in which the principal veins of the leaf are parallel or nearly so; characteristic of monocots.

paramylon: The storage molecule of euglenoids.

paraphysis, *pl.* **paraphyses** [Gk. *para*, beside, + *physis*, growth]: As in certain fungi and brown algae, a sterile filament growing among the reproductive cells in the fruiting body.

parasexual cycle: The fusion and segregation of heterokaryotic haploid nuclei in certain fungi to produce recombinant nuclei.

parasite: An organism that lives on or in an organism of a different species and derives nutrients from it.

parenchyma (pa·rĕng′kĭ·ma) [Gk. *para*, beside, + *en*, in, + *chein*, to pour]: A tissue composed of parenchyma cells.

parenchyma cell: Living, generally thin-walled cell of variable size and form; the most abundant kind of cell in plants.

parthenocarpy [Gk. *parthenos*, virgin, + *karpos*, fruit]: The development of fruit without fertilization; parthenocarpic fruits are usually seedless.

passage cell: Endodermal cell of root that retains thin wall and Casparian strip when other associated endodermal cells develop thick secondary walls.

passive transport: Non-energy-requiring transport of a solute across a membrane down the concentration or electrochemical gradient by either simple diffusion or facilitated diffusion.

pathogen [Gk. *pathos*, suffering, + *genesis*, beginning]: An organism that causes a disease.

pathogenic: Disease-causing.

pathology: The study of plant or animal diseases, their effects on the organism, and their treatment.

pectin: A complex organic compound present in the intercellular layer and primary wall of plant cell walls; the basis of fruit jellies.

pedicel (pĕd′ĭ·sĕl): The stalk of an individual flower in an inflorescence.

peduncle (pe·dŭng′k′l): The stalk of an inflorescence or of a solitary flower.

pentose phosphate cycle: The pathway of oxidation of glucose 6-phosphate to yield pentose phosphates.

peptide: Two or more amino acids linked by peptide bonds.

peptide bond: The type of bond formed when two amino acid units are joined end to end by the removal of a molecule of water; the bonds always form between the carboxyl (—COOH) group of one amino acid and the amino (—NH$_2$) group of the next amino acid.

perennial [L. *per*, through, + *annuus*, a year]: A plant that persists and produces reproductive structures year after year.

perfect flower: A flower having both stamens and carpels; hermaphroditic flower.

perfect stage: That phase of the life history of a fungus that includes sexual fusion and the spores associated with such fusions.

perforation plate: Part of the wall of a vessel member that is perforated.

peri- [Gk. *peri*, around]: Prefix meaning "around" or "about."

perianth (pĕr′ĭ·ănth) [Gk. *peri*, around, + *anthos*, flower]: (1) The petals and sepals taken together; (2) in leafy liverworts, a tubular sheath surrounding an archegonium, and later, the developing sporophyte.

pericarp [Gk. *peri*, around, + *karpos*, fruit]: The fruit wall, which develops from the mature ovary wall.

periclinal: Parallel with the surface.

pericycle [Gk. *peri*, around, + *kykos*, circle]: A tissue, characteristic of roots that is bounded externally by the endodermis and internally by the phloem.

periderm [Gk. *peri*, around, + *derma*, skin]: Outer protective tissue that replaces epidermis when it is destroyed during secondary growth; includes cork, cork cambium, and phelloderm.

perigyny [Gk. *peri*, around, + *gyne*, female]: A form of floral organization in which the sepals, petals, and stamens are attached to the margin of a cup-shaped extension of the receptacle; superficially, the sepals, petals, and stamens appear to be attached to the ovary.

perisperm [Gk. *peri*, around, + *sperma*, seed]: Food-storing tissue derived from the nucellus that occurs in the seeds of some flowering plants.

peristome (pĕr′ĭ·stōm) [Gk. *peri*, around, + *stoma*, a mouth]: In mosses, a fringe of teeth around the opening of the sporangium.

perithecium: A spherical or flask-shaped ascocarp.

permanent wilting percentage: The percentage of water remaining in a soil when a plant fails to recover from wilting even if placed in a humid chamber.

permeable [L. *permeare,* to pass through]: Usually applied to membranes through which liquid substances may diffuse.

permease: A transport protein, or carrier molecule, that assists in the transport of substances across cellular membranes; not permanently altered in the process.

peroxisome: A microbody that plays an important role in glycolic acid metabolism associated with photosynthesis; the site of photorespiration.

petal: A flower part, usually conspicuously colored; one of the units of the corolla.

petiole: The stalk of a leaf.

pH: A symbol denoting the relative concentration of hydrogen ions in a solution; pH values run from 0 to 14, and the lower the value the more acidic a solution, that is, the more hydrogen ions it contains; pH 7 is neutral, less than 7 is acidic, and more than 7 is alkaline.

phage: *See* bacteriophage.

phagocytosis: *See* endocytosis.

phellem: *See* cork.

phelloderm (fĕl′o·dûrm) [Gk. *phellos,* cork, + *derma,* skin]: A tissue formed inwardly by the cork cambium, opposite the cork; inner part of the periderm.

phellogen (fĕl′lo·jĕn): *See* cork cambium.

phenotype: The physical appearance of an organism; the phenotype results from the interaction between the genetic constitution (genotype) of the organism and its environment.

phloem (flō′ĕm) [Gk. *phloos,* bark]: The food-conducting tissue of vascular plants, which is composed of sieve elements, various kinds of parenchyma cells, fibers, and sclereids.

phloem loading: The process by which substances (primarily sugars) are actively secreted into the sieve tubes.

phosphate: A compound formed from phosphoric acid by replacement of one or more hydrogen atoms.

phospholipids: A phosphorylated lipid; similar in structure to a fat, but with only two fatty acids attached to the glycerol backbone, with the third space occupied by a phosphorus-containing molecule; important components of cellular membranes.

phosphorylation (fŏs′fo·rĭl·ā′shŭn) [Gk. *phosphoros,* bringing light]: A reaction in which phosphate is added to a compound; e.g., the formation of ATP from ADP and inorganic phosphate.

photo-, -photic [Gk. *photos,* light]: Prefix or suffix meaning ''light.''

photolysis: The light-dependent oxidative splitting of water molecules that takes place in photosystem II of the light reactions of photosynthesis.

photon [Gk. *photos,* light]: The elementary particle of light.

photoperiodism: Response to duration and timing of day and night; a mechanism evolved by organisms for measuring seasonal time.

photophosphorylation [Gk. *photos,* light, + *phosphoros,* bringing light]: The formation of ATP in the chloroplast during photosynthesis.

photorespiration: The light-dependent production of glycolic acid in chloroplasts and its subsequent oxidation in peroxisomes.

photosynthesis [Gk. *photos,* light, + *syn,* together + *tithenai,* to place]: The conversion of light energy to chemical energy; the production of carbohydrates from carbon dioxide and water in the presence of chlorophyll by using light energy.

photosystem: A discrete unit of organization of chlorophyll and other pigment molecules embedded in the thylakoids of chloroplasts and involved with the light-requiring reactions of photosynthesis.

phototropism [Gk. *photos,* light, + *trope,* turning]: Growth in which the direction of the light is the determining factor, as the growth of a plant toward a light source; turning or bending in response to light.

phragmoplast: A spindle-shaped system of fibrils, which arises between two daughter nuclei at telophase and within which the cell plate is formed during cell division, or cytokinesis. The fibrils of the phragmoplast are composed of microtubules. Phragmoplasts are found in all green algae except the members of the class Chlorophyceae and in plants.

phycobilins: A group of water-soluble accessory pigments, including phycocyanins and phycoerythrins, which occur in the red algae and cyanobacteria.

phycology [Gk. *phykos,* seaweed]: The study of algae.

phycoplast: A system of microtubules that develops between the two daughter nuclei parallel to the plane of cell division. Phycoplasts occur only in green algae of the class Chlorophyceae.

phyllo-, phyll- [Gk. *phyllon,* leaf]: Prefix meaning ''leaf.''

phyllode (fĭl′ōd): A flat, expanded, photosynthetic petiole or stem; phyllodes occur in certain genera of vascular plants, where they replace leaf blades in photosynthetic function.

phylogeny [Gk. *phylon,* race, tribe]: Evolutionary relationships among organisms; the developmental history of a group of organisms.

phylum, *pl.* **phyla** [Gk. *phylon,* race, tribe]: A unit of classification equivalent to ''division,'' used in animals and heterotrophic protists.

physiology: The study of the activities and processes of living organisms.

phyto-, -phyte [Gk. *phyton,* plant]: Prefix or suffix meaning ''plant.''

phytochrome: A phycobilinlike pigment found in the cytoplasm of plants and a few green algae that is associated with the absorption of light; photoreceptor for red and far-red light; involved in a number of timing processes, such as flowering, dormancy, leaf formation, and seed germination.

phytoplankton: Autotrophic plankton.

pigment: A substance that absorbs light, often selectively.

pileus [L. *pileus,* a cap]: The caplike part of mushroom basidiocarps and of certain ascocarps.

pinna, *pl.* **pinnae** (pĭn′a) [L. *pinna,* feather]: A primary division, or leaflet, of a compound leaf or frond; may be divided into pinnules.

pinocytosis: *See* endocytosis.

pistil [L. *pistillum,* pestle]: Central organ of flowers, typically consisting of ovary, style, and stigma; a pistil may consist of one or more fused carpels.

pistillate: Pertaining to a flower with one or more carpels but no functional stamens; also called carpellate.

pit: A recessed cavity in a cell wall where the secondary wall does not form.

pit membrane: The middle lamella and two primary cell walls between two pits.

pit-pair: Two opposite pits plus the pit membrane.

pith: The ground tissue occupying the center of the stem or root within the vascular cylinder; usually consists of parenchyma.

pith ray: *See* interfascicular region.

placenta, *pl.* **placentae** [L. *placenta,* cake]: The part of the ovary wall to which the ovules or seeds are attached.

placentation: The manner of ovule attachment within the ovary.

plankton [Gk. *planktos,* wandering]: Free-floating, mostly microscopic, aquatic organisms.

plaque: Clear area in a sheet of cells resulting from the killing or lysis of contiguous cells by viruses.

-plasma, plasmo-, -plast [Gk. *plasma,* form, mold]: Prefix or suffix meaning ''formed,'' or ''molded''; examples are protoplasm, ''first-molded'' (living matter), and chloroplast, ''green-formed.''

plasma membrane or **plasmalemma:** Outer boundary of the protoplast, next to the cell wall; consists of a single membrane unit; also called cell membrane, and ectoplast.

plasmid: A relatively small fragment of DNA that can exist free in the cytoplasm of a bacterium and can be integrated into and then replicated with a chromosome. Plasmids make up about 5 percent of the DNA of many bacteria, but are rare in eukaryotes.

plasmodesma, *pl.* **plasmodesmata** [Gk. *plasma*, form, + *desma*, bond]: The minute cytoplasmic threads that extend through openings in cell walls and connect the protoplasts of adjacent living cells.

plasmodium (plăz·mo′dĭ·ŭm): Stage in life cycle of myxomycetes (plasmodial slime molds); a multinucleate mass of protoplasm surrounded by a membrane.

plasmogamy [Gk. *plasma*, form, + *gamos*, marriage]: Union of the protoplasts of gametes that is not accompanied by union of their nuclei.

plasmolysis (plăz·mŏl′ĭ·sĭs) [Gk. *plasma*, form, + *lysis*, a loosening]: The separation of the protoplast from the cell wall because of the removal of water from the protoplast by osmosis.

plastid (plăs′tĭd): Organelle in the cells of certain groups of eukaryotes that is the site of such activities as food manufacture and storage; plastids are bounded by a double membrane.

pleiotropy [Gk. *pleros*, more, + *trope*, a turning]: The capacity of a gene to affect more than one phenotypic characteristic.

plumule [L. *plumula*, a small feather]: The first bud of an embryo; the portion of the young shoot above the cotyledons.

pneumatophores [Gk. *pneuma*, breath, + *-phoros*, carrying]: Negatively geotropic extensions of the root systems of some trees growing in swampy habitats; they grow upward and out of the water and probably function to assure adequate aeration.

point mutation: An alternation in one of the nucleotides in a chromosomal DNA molecule.

polar molecule: A molecule with positively and negatively charged ends.

polar nuclei: Two nuclei (usually), one derived from each end (pole) of the embryo sac, which become centrally located; they fuse with a male nucleus to form the primary (3*n*) endosperm nucleus.

pollen [L. *pollen*, fine dust]: A collective term for pollen grains.

pollen grain: A microspore containing a mature or immature microgametophyte (male gametophyte); pollen grains occur in seed plants.

pollen mother cell: *See* microspore mother cell.

pollen sac: A cavity in the anther that contains the pollen grains.

pollen tube: A tube formed after germination of the pollen grain; carries the male gametes into the ovule.

pollination: The transfer of pollen from an anther to a stigma.

poly- [Gk. *polys*, many]: Prefix meaning "many."

polyembryony: Having more than one embryo within the developing seed.

polygenic inheritance: The inheritance of quantitative characteristics determined by the combined effects of multiple genes.

polymer (pŏl′ĭ·mer): A large molecule composed of many similar molecular subunits.

polymerization: The chemical union of monomers such as glucose or nucleotides to form polymers such as starch or nucleic acid.

polynucleotide: A single-stranded DNA or RNA molecule.

polypeptide: A molecule composed of amino acids linked together by peptide bonds, not as complex as a protein.

polyploid (pŏl′ĭ·ploid): Referring to an organism, tissue, or cell with more than two complete sets of chromosomes.

polysaccharide: A polymer composed of many monosaccharide units joined in a long chain, such as glycogen, starch, and cellulose.

polysome or **polyribosome:** An aggregation of ribosomes actively involved in the translation of the same RNA molecule, one after another.

pome (pōm) [Fr. *pomme*, apple]: A simple fleshy fruit, the outer portion of which is formed by the floral parts that surround the ovary and expand with the growing fruit; found only in one subfamily of the Rosaceae (apples, pears, quince, pyracantha, and so on).

population: Any group of individuals, usually of a single species, occupying a given area at the same time.

P-protein: Phloem-protein; a proteinaceous substance found in cells of angiosperm phloem, especially in sieve-tube members; also called slime.

preprophase band: A ringlike band of microtubules, found just beneath the plasma membrane, that delimits the equatorial plane of the future mitotic spindle of a cell preparing to divide.

primary endosperm nucleus: The result of the fusion of a sperm nucleus and the two polar nuclei.

primary growth: In plants, growth originating in the apical meristems of shoots and roots, as contrasted with secondary growth.

primary meristem or **primary meristematic tissue** (mĕr′ĭ·ste·măt′ĭk): A tissue derived from the apical meristem; of three kinds: protoderm, procambium, and ground meristem.

primary pit-field: Thin area in a primary cell wall through which plasmodesmata pass, although plasmodesmata may occur elsewhere in the wall as well.

primary plant body: The part of the plant body arising from the apical meristems and their derivative meristematic tissues; composed entirely of primary tissues.

primary root: The first root of the plant, developing in continuation of the root tip or radicle of the embryo; in gymnosperms and dicots it becomes the tap root.

primary thickening meristem: In many monocotyledons, the meristem responsible for the increase in thickness of the shoot axis.

primary tissues: Cells derived from the apical meristems and primary meristematic tissues of root and shoot; as opposed to secondary tissues derived from a cambium; primary growth results in an increase in length.

primary wall: The wall layer deposited during the period of cell expansion.

primordium, *pl.* **primordia** [L. *primus*, first, + *ordiri*, to begin to weave]: A cell or organ in its earliest stage of differentiation.

pro- [Gk. *pro*, before]: Prefix meaning "before" or "prior to."

procambium (pro·kăm′bĭ·ŭm) [L. *pro*, before, + *cambiare*, to exchange]: A primary meristematic tissue that gives rise to primary vascular tissues.

proembryo: Embryo in early stages of development, before the embryo proper and suspensor become distinct.

prokaryote [Gk. *pro*, before, + *karyon*, kernel]: A cell lacking a membrane-bound nucleus and membrane-bound organelles; a bacterium.

prolamellar body: Semicrystalline body found in plastids arrested in development by the absence of light.

promoter: A specific nucleotide sequence on a chromosome to which RNA polymerase is attached to start the transcription of mRNA from a gene.

prophage (prō′fāj′): Noninfectious phage units linked with the bacterial chromosome which multiply with the growing and dividing bacteria but do not bring about their lysis. Prophage is a stage in the life cycle of a temperate phage.

prophase [Gk. *pro*, before, + *phasis*, form]: An early stage in nuclear division, characterized by the shortening and thickening of the chromosomes and their movement to the metaphase plate.

proplastid: A minute self-reproducing body in the cytoplasm from which a plastid develops.

prop roots: Adventitious roots arising from the stem above soil level and helping to support the plant: common in many monocots, for example, corn (*Zea mays*).

prosthetic group: A heat-stable metal ion or an inorganic group (other than an amino acid) that is bound to a protein and serves as its active group.

protease (prō′te·ās): An enzyme that digests protein by the hydrolysis of peptide bonds; proteases are also called peptidases.

protein [Gk. *proteios*, primary]: A complex organic compound composed of many (100 or more) amino acids joined by peptide bonds.

prothallial cell [Gk. *pro*, before, + *thallos*, sprout]: The sterile cell or cells found in the male gametophytes, or microgametophytes, of vascular plants other than angiosperms; believed to be remnants of the vegetative tissue of the male gametophyte.

prothallus (pro·thăl′ŭs): In homosporous vascular plants, such as ferns, the more or less independent, photosynthetic gametophyte; also called the prothallium.

protist: A member of the kingdom Protista.

proto- [Gk. *protos*, first]: Prefix meaning "first," for example, Protozoa, "first animals."

protoderm [Gk. *protos*, first, + *derma*, skin]: Primary meristematic tissue that gives rise to epidermis.

proton: A subatomic, or elementary, particle, with a single positive charge equal in magnitude to the charge of an electron and a mass of 1; the basic component of every atomic nucleus.

protonema, *pl.* **protonemata** [Gk. *protos*, first, + *nema*, a thread]: The first stage in development of the gametophyte of mosses and certain liverworts; protonemata may be filamentous or platelike.

protoplasm: A general term for the living substance of all cells, but not their organelles.

protoplast: The protoplasm of an individual cell; in plants, the unit of protoplasm inside the cell wall.

protostele [Gk. *protos*, first, + *stele*, pillar]: The simplest type of stele, consisting of a solid column of vascular tissue.

protoxylem: The first part of the primary xylem, which matures during elongation of the plant part in which it is found.

proximal (prŏk′sĭ·măl) [L. *proximus*, near]: Situated near the point of reference, usually the main part of a body or the point of attachment; opposite of distal.

pseudo- [Gk. *pseudes*, false]: Prefix meaning "false."

pseudoplasmodium: A multicellular mass of individual amoeboid cells, representing the aggregate phase in the cellular slime molds.

pulvinus, *pl.* **pulvini:** Jointlike thickening at the base of the petiole of a leaf or petiolule of a leaflet, and having a role in the movements of the leaf or leaflet.

punctuated equilibrium: A model of evolutionary change that proposes that there are long periods of little or no change punctuated with brief intervals of rapid change.

purine (pū′rēn): The larger of the two kinds of nucleotide base found in DNA and RNA; a nitrogenous base with a double-ring structure, such as adenine or guanine.

pyramid of energy: Energy relationships among various feeding levels involved in a particular food chain; autotrophs (at the base of the pyramid) represent the greatest amount of available energy; herbivores are next; then primary carnivores; secondary carnivores; and so forth. Similar pyramids of mass, size, and number also occur in natural communities.

pyrenoid [Gk. *pyren*, the stone of a fruit, + *oides*, like]: Differentiated regions of the chloroplast that are centers of starch formation in green algae and hornworts.

pyrimidine: The smaller of the two kinds of nucleotide base found in DNA and RNA: a nitrogenous base with a single-ring structure, such as cytosine, thymine, or uracil.

quantasome [L. *quantus*, how much, + Gk. *soma*, body]: Granules located on the inner surfaces of chloroplast lamellae; believed to be involved in the light-requiring reactions of photosynthesis.

quantum: The ultimate unit of light energy.

quiescent center: The relatively inactive initial region in the apical meristem of a root.

raceme [L. *racemus*, bunch of grapes]: An indeterminate inflorescence in which the main axis is elongated but the flowers are borne on pedicels that are about equal in length.

rachis (rā′kĭs) [Gk. *rachis*, a backbone]: Main axis of a spike; the axis of a fern leaf (frond), from which the pinnae arise; in compound leaves, the extension of the petiole corresponding to the midrib of an entire leaf.

radially symmetrical: *See* actinomorphic.

radial section: A longitudinal section cut parallel to the radius of a cylindrical body, such as a root or stem; in the case of secondary xylem, or wood, and secondary phloem, parallel to the rays.

radial system: In secondary xylem and secondary phloem, the term applied to all the rays, the cells of which are derived from ray initials; also called the horizontal system, or ray system.

radicle [L. *radix*, root]: The embryonic root.

radioisotope: An unstable isotope of an element that decays or disintegrates spontaneously, emitting radiation; also called a radioactive isotope.

raphe [Gk. *raphe*, seam]: (1) Ridge on seeds, formed by the stalk of the ovule, in those seeds in which the stalk is sharply bent at the base of the ovule; (2) groove on the shell, or frustule, of a diatom.

raphides (răf′ĭ·dēz) [Gk. *rhaphis*, a needle]: Fine, sharp, needlelike crystals of calcium oxalate found in the vacuoles of many plant cells.

ray flowers: *See* disk flowers.

ray initial: An initial in the vascular cambium that gives rise to the ray cells of secondary xylem and secondary phloem.

reaction center: The chlorophyll molecule of a photosystem capable of using energy in the photochemical reaction.

reaction wood: Abnormal wood that develops in leaning trunks and limbs.

receptacle: That part of the axis of a flower stalk that bears the floral organs.

recessive: Describing a gene whose phenotypic expression is masked in the heterozygote by a dominant allele; heterozygotes are phenotypically indistinguishable from dominant homozygotes.

recombinant DNA: Fragments of DNA from two different species, such as a mammal and a bacterium, spliced together in the laboratory into a single molecule.

recombination: The formation of new gene combinations.

reduction [L. *reductio*, a bringing back; originally "bringing back" a metal from its oxide]: Gain of an electron by an atom; reduction takes place simultaneously with oxidation (the loss of an electron by an atom), because an electron that is lost by one atom is accepted by another.

regular: *See* actinomorphic.

regulator gene: A gene that prevents or represses the activity of the structural genes in an operon.

replicate: Produce a facsimile or a very close copy; used to indicate the production of a second molecule of DNA exactly like the first molecule or of a sister chromatid.

repressor: A protein that regulates DNA transcription; this occurs because RNA polymerase is prevented from attaching to the promoter and transcribing the gene. *See* operator.

resin duct: A tubelike intercellular space lined with resin-secreting cells (epithelial cells) and containing resin.

respiration: An intracellular process in which molecules, particularly pyruvate in the citric acid cycle, are oxidized with the release of energy. The complete breakdown of sugar or other organic compounds to carbon dioxide and water is termed aerobic respiration, although the first steps of this process are anaerobic.

restriction endonuclease: An enzyme that cleaves a DNA duplex molecule at a particular base sequence.

reticulate venation: *See* netted venation.

rhizobia [Gk. *rhiza,* root, + *bios,* life]: Bacteria of the genus *Rhizobium,* which may be involved with leguminous plants in a symbiotic relationship that results in nitrogen fixation.

rhizoids [Gk. *rhiza,* root]: (1) Branched rootlike extensions of fungi and algae that absorb water, food, and nutrients; (2) root-hairlike structures in liverworts, mosses, and some vascular plants, which occur on free-living gametophytes.

rhizome: A more or less horizontal underground stem.

ribonucleic acid (RNA): Type of nucleic acid formed on chromosomal DNA and involved in protein synthesis; composed of chains of phosphate, sugar molecules (ribose), and purines and pyrimidines; RNA is the genetic material of many kinds of viruses.

ribose: A five-carbon sugar; a component of RNA.

ribosome: A small particle composed of protein and RNA; the site of protein synthesis.

ring-porous wood: A wood in which the pores, or vessels, of the early wood are distinctly larger than those of the late wood, forming a well-defined ring in cross sections of the wood.

RNA *See* ribonucleic acid.

root: The usually descending axis of a plant, normally below ground, which serves to anchor the plant and to absorb and conduct water and minerals into it.

root cap: A thimblelike mass of cells that covers and protects the growing tip of a root.

root hairs: Tubular outgrowths of epidermal cells of the root in the zone of maturation.

root pressure: The pressure developed in roots as the result of osmosis, which causes guttation of water from leaves and exudation from cut stumps.

runner: *See* stolon.

samara: Simple, dry, one-seeded or two-seeded indehiscent fruit with pericarp-bearing, winglike outgrowths.

sap: (1) A name applied to the fluid contents of the xylem or the sieve elements of the phloem; (2) the fluid contents of the vacuole are called cell sap.

saprobe [Gk. *sapros,* rotten, + *bios,* life]: An organism that secures its food directly from nonliving organic matter.

sapwood: Outer part of the wood of stem or trunk, usually distinguised from the heartwood by its lighter color, in which active conduction of water takes place.

satellite DNA: A short nucleotide sequence repeated in tandem fashion many thousands of times; this region of the chromosome has a distinctive base composition and is not transcribed.

savanna: Grassland containing scattered trees.

scarification: The process of cutting or softening a seed coat to hasten germination.

schizo- [Gk. *schizein,* to split]: Prefix meaning "split."

schizocarp (skĭz'o·kärp): Dry fruit with two or more united carpels that split apart at maturity.

sclereid [Gk. *skleros,* hard]: A sclerenchyma cell with a thick, lignified secondary wall having many pits. Sclereids are variable in form but typically not very long; they may or may not be living at maturity.

sclerenchyma (skle·rĕng'kĭ·má) [Gk. *skleros,* hard, + L. *enchyma,* infusion]: A supporting tissue composed of sclerenchyma cells, including fibers and sclereids.

sclerenchyma cell: Cell of variable form and size with more or less thick, often lignified, secondary walls; may or may not be living at maturity; includes fibers and sclereids.

scutellum (sku·tĕl'ŭm) [L. *scutella,* a small shield]: The single cotyledon of a grass embryo, specialized for absorption of the endosperm.

secondary growth: In plants, growth derived from secondary or lateral meristems, the vascular and cork cambiums; secondary growth results in an increase in girth, and is contrasted with primary growth, which results in an increase in length.

secondary plant body: The part of the plant body produced by the vascular cambium and the cork cambium; consists of secondary xylem, secondary phloem, and periderm.

secondary root: *See* lateral root.

secondary tissues: Tissues produced by the vascular cambium and cork cambium.

secondary wall: Innermost layer of the cell wall, formed in certain cells after cell elongation has ceased; secondary walls have a highly organized microfibrillar structure.

seed: A structure formed by the maturation of the ovule of seed plants following fertilization.

seed coat: The outer layer of the seed, developed from the integuments of the ovule.

seedling: A young sporophyte, which develops from a germinating seed.

segregation: The separation of the chromosomes (and genes) from different parents at meiosis.

semipermeable membrane: A membrane that is permeable to water and some solutes, but not to other solutes. Also called a *selectively permeable membrane.*

sepal (se'păl) [L. *sepalum,* a covering]: One of the outermost flower structures, a unit of the calyx; sepals usually enclose the other flower parts in the bud.

septate [L. *septum,* fence]: Divided by cross walls into cells or compartments.

sessile (sĕs'ĭl) [L. *sessilis,* of or fit for sitting, low, dwarfed]: Attached directly by the base; referring to a leaf lacking a petiole or to a flower or fruit lacking a pedicel.

seta, *pl.* **setae** (sē'ta) [L. *seta,* bristle]: In bryophytes, the stalk that supports the capsule, if present; part of the sporophyte.

sexual reproduction: The fusion of gametes followed by meiosis and recombination of some point in the life cycle.

sheath: (1) The base of a leaf that wraps around the stem, as in grasses; (2) a tissue layer surrounding another tissue, such as a bundle sheath.

shoot: The above-ground portions, such as the stem and leaves, of a vascular plant.

short-day plants: Plants that must be exposed to light periods shorter than some critical length for flowering to occur; they usually flower in autumn.

shrub: A perennial woody plant of relatively low stature, typically with several stems arising from or near the ground.

sieve area: A portion of sieve-element wall containing clusters of pores through which the protoplasts of adjacent sieve elements are interconnected.

sieve cell: A long, slender sieve element with relatively unspecialized sieve areas and with tapering end walls that lack sieve plates; found in the phloem of gymnosperms and lower vascular plants.

sieve element: The cell of the phloem that is involved in the long-distance transport of food substances; further classified into sieve cells and sieve-tube members.

sieve plate: The part of the wall of sieve-tube members bearing one or more highly differentiated sieve areas.

sieve tube: A series of sieve-tube members arranged end-to-end and interconnected by sieve plates.

sieve-tube member: One of the component cells of a sieve tube; found primarily in flowering plants and typically associated with a companion cell; also called sieve-tube element.

silique [L. *siliqua*, pod]: The fruit characteristic of the mustard family; two-celled, the valves splitting from the bottom and leaving the placentae with the false partition stretched between. A small, compressed silique is called a silicle.

simple fruit: A fruit derived from one carpel or several united carpels.

simple leaf: An undivided leaf; as opposed to a compound leaf.

simple pit: A pit not surrounded by an overarching border of secondary wall; as opposed to a bordered pit.

siphonaceous, siphonous [Gk. *siphon*, a tube, pipe]: In algae, multinucleate cells without cross walls; coenocytic.

siphonostele [Gk. *siphon*, pipe, + *stele*, pillar]: A type of stele containing a hollow cylinder of vascular tissue surrounding a pith.

slime: *See* P-protein.

softwood: A name commonly applied to the wood of a conifer.

solar tracking: The ability of the leaves and flowers of many plants to move diurnally, orienting themselves either perpendicular or parallel to the sun's direct rays; also called heliotropism.

solute: A molecule dissolved in a solution.

solute potential: The change in free energy or chemical potential of water produced by solutes; carries a negative (minus) sign; also called osmotic potential.

solution: Usually liquid, in which the molecules of the dissolved substance, the solute (e.g., sugar), are dispersed between the molecules of the solvent (e.g., water).

somatic cells [Gk. *soma*, body]: All body cells except the gametes and the cells from which the gametes develop.

soredium, *pl.* **soredia** [Gk. *soros*, heap]: A specialized reproductive unit of lichens, consisting of a few cyanobacterial or green algal cells surrounded by fungal hyphae.

sorus, *pl.* **sori** (sō′rŭs) [Gk. *soros*, heap]: A group or cluster of sporangia or spores.

specialized: (1) Of organisms, having special adaptations to a particular habitat or mode of life; (2) of cells, having particular functions.

species, *pl.* **species** [L. kind, sort]: A kind of organism; species are designated by binomial names written in italics.

specificity: Uniqueness, as in proteins in given organisms or enzymes in given reactions.

sperm: A mature male gamete, usually motile and smaller than the female gamete.

spermagonium, *pl.* **spermagonia** (spûr′ma·gō′nĭ·ŭm) [Gk. *sperma*, sperm, + *gonos*, offspring]: In the rust fungi, the structure that produces spermatia.

spermatangium, *pl.* **spermatangia** (spûr′ma·tăn′jĭ·ŭm) [Gk. *sperma*, sperm, + L. *tangere*, to touch]: In the red algae, the cell that produces spermatia.

spermatium, *pl.* **spermatia** [Gk. *sperma*, sperm]: In the red algae and some fungi, a minute, nonmotile male gamete.

spermatophyte [Gk. *sperma*, seed, + *phyton*, plant]: A seed plant.

spherosome: Single, membrane-bound spherical sturctures in the cytoplasm of plant cells, many of which contain mostly lipids and apparently are centers of lipid synthesis and accumulation.

spike [L. *spica*, head of grain]: An indeterminate inflorescence in which the main axis is elongated and the flowers are sessile.

spikelet: The unit of inflorescence in grasses; a small group of grass flowers.

spindle fibers: A group of microtubules that extend from the centromeres of the chromosomes to the poles of the spindle or from pole to pole in a dividing cell.

spine: A hard, sharp-pointed structure; usually a modified leaf, or part of a leaf.

spirillum, *pl.* **spirilli** [L. *spira*, coil]: A long coiled or spiral bacterium.

spongy parenchyma: A leaf tissue composed of loosely arranged, chloroplast-bearing cells.

sporangiophore (spo·răn′jĭ·o·fōr′) [Gk. *spora*, seed, + *pherein*, to carry]: A branch bearing one or more sporangia.

sporangium, *pl.* **sporangia** (spo·răn′jĭ·ŭm) [Gk. *spora*, seed, + *angeion*, a vessel]: A hollow unicellular or multicellular structure in which spores are produced.

spore: A reproductive cell, usually unicellular, capable of developing into an adult without fusion with another cell.

spore mother cell: A diploid (2*n*) cell that undergoes meiosis and produces (usually) four haploid cells (spores) or four haploid nuclei.

sporophyll (spō′ro·fĭl): A modified leaf or leaflike organ that bears sporangia; applied to the stamens and carpels of angiosperms, fertile fronds of ferns, and other similar structures.

sporophyte (spōr′ro·fīt): The spore-producing, diploid (2*n*) phase in a life cycle characterized by alternation of generations.

sporopollenin: The tough substance of which the exine, or outer wall, of spores and pollen grains is composed.

stalk cell: One of two cells produced by division of the generative cell in developing pollen grains of gymnosperms; it is not a gamete, and eventually degenerates.

stamen (stā′měn) [L. *stamen*, thread]: The part of the flower producing the pollen, composed (usually) of anther and filament; collectively, the stamens make up the androecium.

staminate (stăm′ĭ·nat): Pertaining to a flower having stamens but no functional carpels.

starch [M.E. *sterchen*, to stiffen]: A complex insoluble carbohydrate, the chief food storage substance of plants; composed of a thousand or more glucose units.

statoliths [Gk. *statos*, stationary, + *lithos*, stone]: Gravity sensors; starch grains or other bodies in the cytoplasm.

stele (stē′le) [Gk. *stele*, a pillar]: The central cylinder, inside the cortex, of roots and stems of vascular plants.

stem: The part of the axis of vascular plants that is above ground, as well as anatomically similar portions below ground, such as rhizomes or corms.

stem bundle: Vascular bundle belonging to the stem.

sterigma, *pl.* **sterigmata** [Gk. *sterigma*, a prop]: A small, slender protuberance of a basidium, bearing a basidiospore.

stigma: (1) The region of a carpel that serves as a receptive surface for pollen grains and on which they germinate; (2) a light-sensitive eyespot, found in some kinds of algae.

stipe: A supporting stalk, such as the stalk of a gill fungus or the leaf stalk of a fern.

stipule (stĭp′ŭl): An appendage, often leaflike, that occurs on either side of the basal part of a leaf, or encircles the stem, in many kinds of flowering plants.

stolon (stō′lŏn) [L. *stolo*, shoot]: A stem that grows horizontally along the ground surface and may form adventitious roots, such as the runners of a strawberry plant.

stoma, *pl.* **stomata** (stō′ma) [Gk. *stoma*, mouth]: A minute opening bordered by guard cells in the epidermis of leaves and stems through which gases pass; also used to refer to the entire stomatal apparatus —the guard cells plus their included pore.

stratification: The process of exposing seeds to low temperatures for an extended period before attempting to germinate them at warm temperatures.

strobilus, *pl.* **strobili** (strōb′ĭ·lŭs) [Gk. *strobilos*, a cone]: A reproductive structure consisting of a number of modified leaves (sporophylls) or ovule-bearing scales grouped terminally on a stem; a cone. Strobili occur in many kinds of gymnosperms, lycophytes, and sphenophytes.

stroma [Gk. *stroma*, anything spread out]: The ground substance of plastids.

style [Gk. *stylos*, column]: A slender column of tissue that arises from the top of the ovary and through which the pollen tube grows.

sub- [L. *sub*, under, below]: Prefix meaning "under" or "below"; for example, subepidermal, "underneath the epidermis."

suberin (sū′ber·ĭn) [L. *suber*, the cork oak]: Fatty material found in the cell walls of cork tissue and in the Casparian strip of the endodermis.

subsidiary cell: An epidermal cell morphologically distinct from other epidermal cells and associated with a pair of guard cells; also called an accessory cell.

subspecies: The primary taxonomic subdivision of a species. Varieties are used as equivalent to subspecies by some botanists, or subspecies may be divided into varieties.

substrate [L. *substatus*, strewn under]: The foundation to which an organism is attached; the substance acted on by an enzyme.

substrate phosphorylation: Phosphorylation—the formation of ATP from ADP and inorganic phosphate—that takes place during glycolysis.

succession: In ecology, the orderly progression of changes in community composition that occurs during the development of vegetation in any area, from initial colonization to the attainment of the climax typical of a particular geographic area.

succulent: A plant with fleshy, water-storing stems or leaves.

sucker: A sprout produced by the roots of some plants and that gives rise to a new plant; erect sprouts that occur from the base of stems.

sucrase (sū′krās): An enzyme that hydrolyzes sucrose into glucose and fructose; also called invertase.

sucrose (sū′krōs): A disaccharide (glucose plus fructose) found in many plants; the primary form in which sugar produced by photosynthesis is translocated.

superior ovary: An ovary that is free and separate from the calyx.

suspension: A heterogeneous dispersion in which the dispersed phase consists of solid particles sufficiently large that they will settle out of the fluid dispersion medium under the influence of gravity.

suspensor: A structure at the base of the embryo of many vascular plants that pushes the terminal part of the embryo into the endosperm.

symbiosis (sĭm′bĭ·ō′sĭs) [Gk. *syn*, together with, + *bios*, life]: The living together in close association of two or more dissimilar organisms; includes parasitism (in which the association is harmful to one of the organisms) and mutualism (in which the association is advantageous to both).

symplast [Gk. *syn*, together with, + *plastos*, molded]: The interconnected protoplasts and their plasmodesmata; the movement of substances in the symplast is called symplastic movement, or symplastic transport.

sympodium, *pl.* **sympodia:** A stem bundle and its associated leaf traces.

syn-, sym- [Gk. *syn*, together with]: Prefix meaning "together."

synapsis [Gk. *synapsis*, a contract or union]: The pairing of homologous chromosomes that occurs prior to the first meiotic division; crossing-over occurs during synapsis.

synergids (sĭ·nûr′jĭds): Two short-lived cells lying close to the egg in the mature embryo sac of the ovule of flowering plants.

syngamy [Gk. *syn*, together with, + *gamos*, marriage]: The process by which two haploid cells fuse to form a diploid zygote; fertilization.

synthesis: The formation of a more complex substance from simpler ones.

systematics: Scientific study of the kinds and diversity of organisms and of the relationships between them.

taiga: The northern coniferous forest.

tandem clusters: Multiple copies of the same gene lying side by side in a series.

tangential section: A longitudinal section cut at right angles to the radius of a cylindrical structure, such as a root or stem; in the case of secondary xylem, or wood, and secondary phloem, at right angles to the rays.

tapetum (ta·pē′tŭm) [Gk. *tapes*, a carpet]: Nutritive tissue in the sporangium, particularly an anther.

taproot: The primary root of a plant formed in direct continuation with the root tip or radicle of the embryo: forms a stout, tapering main root from which arise smaller, lateral branches.

taxon: General term for any one of the taxonomic categories, such as species, class, order, or division.

taxonomy [Gk. *taxis*, arrangement, + *nomos*, law]: The science of the classification of organisms.

teliospore (tē′lĭ·ō·spōr′): In the rust fungi, a thick-walled spore in which karyogamy and meiosis occur and from which basidia develop.

telium, *pl.* **telia:** In the rust fungi, the structure that produces teliospores.

telophase: The last stage in mitosis and meiosis, during which the chromosomes become reorganized into two new nuclei.

temperate phage: A bacterial virus that may remain latent in its host bacterial cell; in this latent (prophage) state, it is associated with the bacterial chromosome and is replicated with it.

template: A pattern or mold guiding the formation of a negative or complement; a term applied especially to DNA duplication, which is explained in terms of a template hypothesis.

tendril [L. *tendere*, to extend]: A modified leaf or part of a leaf or stem modified into a slender coiling structure that aids in support of the stems; tendrils occur only in some angiosperms.

tepal: One of the units of a perianth that is not differentiated into sepals and petals.

test cross: A cross of a dominant with a homozygous recessive; used to determine whether the dominant is homozygous or heterozygous.

tetrad (tĕt′răd): A group of four spores formed from a spore mother cell by meiosis.

tetraploid (tĕt′ra·ploid) [Gk. *tetra*, four, + *ploos*, fold]: Twice the usual, or diploid (2*n*), number of chromosomes (that is, 4*n*).

tetrasporangium, *pl.* **tetrasporangia** [Gk. *tetra*, four, + *spora*, seed, + *angeion*, vessel]: In certain red algae, a sporangium in which meiosis occurs, resulting in the production of tetraspores.

tetraspore [Gk. *tetra*, four, + *spora*, seed]: In certain red algae, the four spores formed by meiotic division in the tetrasporangium of a spore mother cell.

tetrasporophyte [Gk. *tetra*, four, + *spora*, seed, + *phyton*, plant]: In certain red algae, a diploid individual that produces tetrasporangia.

texture: Of wood, refers to the relative size and amount of variation in size of elements within the growth rings.

thallophyte: A term previously used to designate fungi and algae collectively, now largely abandoned.

thallus (thăl′ŭs) [Gk. *thallos*, a sprout]: A type of body that is undifferentiated into root, stem, or leaf; the word *thallus* was used commonly when fungi and algae were considered to be plants, to distinguish their simple construction, and that of certain gametophytes, from the differentiated bodies of plant sporophytes and the elaborate gametophytes of the bryophytes.

theory [Gk. *theorein*, to look at]: A well-tested hypothesis; one unlikely to be rejected by further evidence.

thermodynamics [Gk. *therme*, heat, + *dynamis*, power]: The study of energy exchanges, using heat as the most convenient form of measurement of energy. The first law of thermodynamics states that in all processes, the total energy of the universe remains constant. The second law of thermodynamics states that the entropy, or degree of randomness, tends to decrease.

thorn: A hard, woody, pointed branch.

thigmotropism [Gk. *thigma*, touch]: A response to contact with a solid object.

thylakoid [Gk. *thylakos*, sac, + *oides*, like]: A saclike membranous structure in cyanobacteria and the chloroplasts of eukaryotic organisms; in chloroplasts, stacks of thylakoids form the grana; chlorophylls are found within the thylakoids.

thymine: A pyrimidine occurring in DNA but not in RNA; *see also* uracil.

tissue: A group of similar cells organized into a structural and functional unit.

tissue culture: A technique for maintaining fragments of plant or animal tissue alive in a medium after removal from the organism.

tissue system: A tissue or group of tissues organized into a structural and functional unit in a plant or plant organ. There are three tissue systems: dermal, vascular, and ground, or fundamental.

tonoplast [Gk. *tonos*, stretching, tension, + *plastos*, formed, molded]: The cytoplasmic membrane surrounding the vacuole in plant cells; also called vacuolar membrane.

torus, *pl.* **tori:** The central thickened part of the pit-membrane in the bordered pits of conifers and some other gymnosperms.

totipotent: Said of tissues or cells that are capable of developing into any structure of the mature plant.

tracheary element: The general term for a water-conducting cell in vascular plants; tracheids and vessel members.

tracheid (trā′ke·ĭd): An elongated, thick-walled conducting and supporting cell of xylem. It has tapering ends and pitted walls without perforations, as contrasted with a vessel member. Found in nearly all vascular plants.

transcription: The enzyme-catalyzed assembly of an RNA molecule complementary to a strand of DNA.

transduction: The transfer of genes from one organism to another by a virus.

transfer cell: Specialized parenchyma cell with wall in-growths that increase the surface area of the plasma membrane; apparently functions in the short-distance transfer of solutes.

transfer RNA (tRNA): Low-molecular-weight RNA that becomes attached to an amino acid and guides it to the correct position on the ribosome for protein synthesis; there is at least one tRNA molecule for each amino acid.

transformation: The transfer of naked DNA from one organism to another; also called "gene transfer." Transposons are often used as vectors in transformation when it is carried out in the laboratory.

transition region: The region in the primary plant body showing transitional characteristics between structures of root and shoot.

translation: The assembly of a protein on the ribosomes; mRNA is used to direct the order of the amino acids.

translocation: (1) In plants, the long-distance transport of water, minerals, or food; most often used to refer to food transport; (2) in genetics, the interchange of chromosome segments between nonhomologous chromosomes.

transpiration [Fr. *transpirer*, to perspire]: The loss of water vapor by plant parts; most transpiration occurs through stomata.

transport protein: A specific membrane protein responsible for transferring solutes across membranes; also called carrier proteins.

transposon [L. *transponere*, to change the position of something]: A DNA sequence that carries one or more genes and is flanked by sequences of bases that confer the ability to move from one DNA molecule to another; an element capable of transposition, which is the changing of a chromosomal location.

transverse section: A section cut perpendicular, or at right angles, to the longitudinal axis of a plant part.

tree: A perennial woody plant generally with a single stem (trunk).

tricarboxylic acid cycle or **TCA cycle:** *See* Krebs cycle.

trichogyne [Gk. *trichos*, a hair, + *gyne*, female]: In the red algae and certain ascomycetes and basidiomycetes, a receptive protuberance of the female gametangium for the conveyance of spermatia.

trichome [Gk. *trichos*, hair]: An outgrowth of the epidermis, such as a hair, scale, and water vesicle.

triose [Gk. *tries*, three, + *ose*, suffix indicating a carbohydrate]: Any three-carbon sugar.

triple fusion: In angiosperms, the fusion of the second male gamete, or sperm, with the polar nuclei, resulting in formation of a primary endosperm nucleus, which is triploid ($3n$) in most groups.

triploid [Gk. *triploos*, triple]: Having three complete chromosome sets per cell ($3n$).

tritium: A radioactive isotope of hydrogen, ^3H. The nucleus of a tritium atom contains one proton and two neutrons, whereas the more common hydrogen nucleus consists only of a proton.

tRNA: *See* transfer RNA.

-troph, tropho- [Gk. *trophos*, feeder]: Suffix or prefix meaning "feeder," "feeding," or "nourishing"; for example, autotrophic, "self-nourishing."

trophic level: A step in the movement of energy through an ecosystem, represented by a particular set of organisms.

tropism [Gk. *trope*, a turning]: A response to an external stimulus in which the direction of the movement is usually determined by the direction from which the most intense stimulus comes.

tube cell: In male gametophytes, or pollen grains, of seed plants, the cell that develops into the pollen tube.

tuber [L. *tuber*, swelling]: An enlarged, short, fleshy underground stem, such as that of the potato.

tundra: A treeless circumpolar region, best developed in the Northern Hemisphere and most found north of the Arctic Circle.

tunica-corpus: The organization of the shoot apex of most angiosperms and a few gymnosperms, consisting of one or more peripheral layers of cells (the tunica layers) and an interior (the corpus). The tunica layers undergo surface growth (by anticlinal divisions), and the corpus undergoes volume growth (by divisions in all planes).

turgid (tûr′jĭd) [L. *turgidus*, a swollen]: Swollen, distended, referring to a cell that is firm due to water uptake.

turgor pressure [L. *turgor*, a swelling]: The pressure within the cell resulting from the movement of water into the cell.

tylose [Gk. *tylos*, a lump]: A balloonlike outgrowth from a ray or axial parenchyma cell through the pit in a vessel wall and into the lumen of the vessel.

umbel (ŭm′bĕl) [L. *umbella*, sunshade]: An inflorescence, the individual pedicels of which all arise from the apex of the peduncle.

unicellular: Composed of a single cell.

unisexual: Usually applied to a flower lacking either stamens or carpels; a perianth may be present or absent.

unit membrane: A visually definable, three-layered membrane, consisting of two dark layers separated by a lighter layer.

uracil (ū′ra·sĭl): A pyrimidine found in RNA but not in DNA; *see also* thymine.

uredinium, *pl.* **uredinia** [L. *uredo,* a blight]: In rust fungi, the structure that produces urediniospores.

urediniospore [L. *uredo,* a blight, + *spora,* spore]: In rust fungi, a reddish, binucleate spore produced in summer.

vacuolar membrane: *See* tonoplast.

vacuole [L. *vaccus,* empty]: A space or cavity within the cytoplasm filled with a watery fluid, the cell sap; part of the lysosomal compartment of the cell.

variation: The differences that occur within the offspring of a particular species.

variety: A group of plants or animals of less than species rank; some botanists view varieties as equivalent to subspecies, and others consider them divisions of subspecies.

vascular [L. *vasculum,* a small vessel]: Pertains to any plant tissue or region consisting of or giving rise to conducting tissue; e.g., xylem, phloem, vascular cambium.

vascular bundle: A strand of tissue containing primary xylem and primary phloem (and procambium if still present) and frequently enclosed by a bundle sheath of parenchyma or fibers.

vascular cambium: A cylindrical sheath of meristematic cells, the division of which produces secondary phloem and secondary xylem.

vascular rays: Ribbonlike sheets of parenchyma that extend radially through the wood, across the cambium, and into the secondary phloem; they are always produced by the vascular cambium.

vascular tissue system: All the vascular tissues in a plant or plant organ.

vector [L., a bearer, carrier, from *vehere,* to carry]: (1) A pathogen that carries a disease from one organism to another; (2) in genetics, any virus or plasmid DNA into which a gene is integrated and subsequently transferred into a cell.

vegetative: Of, relating to, or involving propagation by asexual processes; also referring to nonreproductive plant parts.

vegetative reproduction: (1) In seed plants, reproduction by means other than by seeds; apomixis; (2) in other organisms, reproduction by vegetative spores, fragmentation, or division of the somatic body. Unless a mutation occurs, each daughter cell or individual is genetically identical with its parent.

vein: A vascular bundle forming a part of the framework of the conducting and supporting tissue of a leaf or other expanded organ.

velamen [L. *velumen,* fleece]: A multiple epidermis covering the aerial roots of some orchids and aroids; also occurs on some terrestrial roots.

venation: Arrangement of veins in leaf blade.

venter [L. *venter,* belly]: The enlarged basal portion of an archegonium containing the egg.

vernalization [L. *vernalis,* spring]: The induction of flowering by cold treatment.

vessel [L. *vasculum,* a small vessel]: A tubelike structure of the xylem composed of elongate cells (vessel members) placed end to end and connected by perforations. Its function is to conduct water and minerals through the plant body. Found in nearly all angiosperms and a few other vascular plants (e.g., gnetophytes).

vessel member: One of the cells composing a vessel; also called vessel element.

viable [L., *vita,* life]: Able to live.

volva [L. *volva,* a wrapper]: A cuplike structure at the base of the stalk of certain mushrooms.

wall pressure: The pressure of the cell wall exerted against the turgid protoplast; opposite and equal to the turgor pressure.

water potential: The algebraic sum of the solute potential and the pressure potential, or wall pressure; the potential energy of water.

water vesicle: An enlarged epidermal cell in which water is stored; a type of trichome.

weed [O.E. *weod,* used at least since the year 888 in its present meaning]: Generally an herbaceous plant not valued for use or beauty, growing wild, and regarded as using ground or hindering the growth of useful vegetation.

whorl: A circle of leaves or of flower parts.

wild type: In genetics, the phenotype or genotype that is characteristic of the majority of individuals of a species in a natural environment.

wood: Secondary xylem.

xanthophyll: (zăn′thō·fĭl) [Gk. *xanthos,* yellowish-brown, + *phyllon,* leaf]: A yellow chloroplast pigment; a member of the carotenoid group.

xerophyte [Gk. *xeros,* dry, + *phyton,* a plant]: A plant that has adapted to arid habitats.

xylem [Gk. *xylon,* wood]: A complex vascular tissue through which most of the water and minerals of a plant are conducted; characterized by the presence of tracheary elements.

zeatin (zē′ă·tĭn): Plant hormone; a natural cytokinin isolated from corn.

zoosporangium: A sporangium bearing zoospores.

zoospore (zō′o·spōr): A motile spore, found among algae, oomycetes, and chytrids.

zygomorphic [Gk. *zygo,* pair, + *morphe,* form]: A type of flower capable of being divided into two symmetrical halves only by a single longitudinal plane passing through the axis; also called bilaterally symmetrical.

zygospore: A thick-walled, resistant spore that develops from a zygote, resulting from the fusion of isogametes.

zygote (zī′gōt) [Gk. *zygotos,* paired together]: The diploid (2*n*) cell resulting from the fusion of male and female gametes.

Illustration Acknowledgments

All photographs not credited herein are by Ray F. Evert.

1-1 David Overcash/Bruce Coleman, Inc.

1-2 J. W. Schopf

1-3 After Helena Curtis, *Biology,* 4th ed., Worth Publishers, Inc., New York, 1983

1-4 (a) G. J. Breckon; (b) Larry West/Bruce Coleman, Inc.

1-5 Dianne Edwards

1-6 E. S. Ross

1-8 After W. Troll, *Vergleichende Morphologie der Höheren Pflanzen,* vol. 1, pt. 1, Verlag von Gebrüder Borntraeger, Berlin, 1937

1-9 E. S. Ross

Page 12 (bottom) R. D. Preston

Page 13 Mary Alice Webb

2-1 (photo) M. A. Walsh

2-2 A. Ryter

2-5 (a) D. Branton

2-6 After W. Braune, A. Leman, and H. Taubert, *Pflanzenanatomisches Prakticum,* VEB Gustav Fischer Verlag, Jena, 1967

2-10 L. M. Beidler

2-11 R. R. Dute

2-12 After W. W. Thomson and J. M. Whatley, *Annual Review of Plant Physiology,* vol. 31, pages 375–394, 1980

2-13 David Stetler

2-14 K. Esau

2-15 K. Esau

2-16 Mary Alice Webb

2-18 After F. Marty, *Proceedings of the National Academy of Sciences of the U.S.A.,* vol. 75, pages 852–856, 1978

2-19 S. E. Eichhorn

2-21 (photo) R. R. Dute

2-22 After D. J. Morré and H. H. Mollenhauer, in *Dynamic Aspects of Plant Ultrastructure,* A. W. Robards (Ed.), McGraw-Hill Book Company, New York, 1974

2-24 R. F. Evert and S. E. Eichhorn, *The American Journal of Botany,* vol. 63, pages 30–48, 1976

2-25 M. Kruatrachue and R. F. Evert, *The American Journal of Botany,* vol. 64, pages 310–325, 1977

2-26 R. R. Powers

2-27 H. J. Hoops and G. L. Floyd

2-28 R. D. Preston

2-30 After K. Esau, *Anatomy of Seed Plants,* 2nd ed., John Wiley and Sons, Inc., New York, 1977

2-31 After B. Alberts, D. Bray, J. Lewis, M. Raff, K. Roberts, and J. D. Watson, *Molecular Biology of the Cell,* Garland Publishing, Inc., New York, 1983

2-33 After R. D. Preston, in Robards, *op. cit.,* 1974

2-38 W. T. Jackson, *Physiologia Plantarum,* vol. 20, pages 20–29, 1967

2-41 R. R. Dute

2-42 R. R. Dute

2-43 J. Cronshaw

Page 41 (a), (b) S. M. Wick and J. Duniec, *Protoplasma,* vol. 122, pages 45–55, 1984; (c), (d) S. M. Wick, R. W. Seagull, M. Osborn, K. Weber, and B. E. S. Gunning, *The Journal of Cell Biology,* vol. 89, pages 685–690, 1981

2-44 P. K. Hepler, *Protoplasma,* vol. 111, pages 121–133, 1982

2-45 After M. C. Ledbetter, *Symposia of the International Society for Cell Biology,* vol. 6, pages 55–70, 1967

3-4 (c) After Albert L. Lehninger, *Biochemistry,* 2nd ed., Worth Publishers, Inc., New York, 1975

3-8 M. Kruatrachue and R. F. Evert, *The*

American Journal of Botany, vol. 64, pages 310–325, 1977

3-10 After S. H. Crowdy, in *Systemic Fungicides,* O. B. E. Marsh (Ed.), Longman, Inc., New York, 1977

3-11 B. E. Juniper

3-17 From Curtis, *op. cit.,* 1983

4-1 W. G. Whaley

4-2 Bruce Roberts/Photo Researchers, Inc.

4-3 From Curtis, *op. cit.,* 1983

4-4 After Lehninger, *op. cit.,* 1975

4-7 After S. J. Singer and G. L. Nicolson, *Science,* vol. 175, pages 720–731, 1972, © 1972, A.A.A.S.

4-8 After Alberts et al., *op. cit.,* 1983

4-9 After Alberts et al., *op. cit.,* 1983

4-11 K. B. Raper

4-13 E. B. Tucker, *Protoplasma,* vol. 113, pages 193–201, 1982

Page 68 (top) H. Towner; (bottom) S. G. Pallardy and T. T. Kozlowski, *New Phytologist,* vol. 85, pages 363–368, 1980

Page 69 John N. Telford

5-1 Yerkes Observatory, Harvard University

5-2 After Curtis, *op. cit.,* 1983

5-3 From Helena Curtis and N. Sue Barnes, *Invitation to Biology,* 4th ed., Worth Publishers, Inc., New York, 1985

Page 74 L. Jacobi

5-4 H. Wright/ Freelance Photographers Guild

5-5 From Curtis, *op. cit.,* 1983

5-6 After Lehninger, *op. cit.,* 1975

5-7 From Curtis, *op. cit.,* 1983

5-9 From Curtis, *op. cit.,* 1983

6-1 R. D. Warmbrodt

6–6 After A. J. Vander, J. N. Sherman, and D. S. Luciano, *Human Physiology*, McGraw-Hill Book Company, New York, 1969

6–7 After Lehninger, *op. cit.*, 1975

6–11 After Lehninger, *op. cit.*, 1975

6–13 (a) After Lehninger, *op. cit.*, 1975; (b) John N. Telford

6–14 After P. C. Hinkle and R. E. McCarty, *Scientific American*, vol. 238, pages 104–123, 1978

Page 92 Y. Haneda

6–16 After Lehninger, *op. cit.*, 1975

6–17 (b) Grant Heilman Photography

7–2 R. L. Gherna, American Type Culture Collection

7–3 H. Towner

7–4 From Curtis, *op. cit.*, 1983

7–7 Prepared by Govindjee

7–10 D. Branton

Page 103 (top) Cambridge University Library

7–14 S. G. Pallardy and T. T. Kozlowski, *New Phytologist*, vol. 85, pages 363–368, 1980

Page 109 From Curtis, *op. cit.*, 1983

7–21 After J. A. Teeri and L. G. Stowe, *Oecologia*, vol. 23, pages 1–12, 1976

Page 114 (top) H. R. Chen; (middle) P. B. Moens; (bottom) Don Kyhos

Page 115 G. Östergren

8–1 A. Sparrow

8–3 From Curtis, *op. cit.*, 1983

8–4 (a) J. D. Watson, *The Double Helix*, Atheneum, New York, 1968; (b) W. Etkin, *BioScience*, vol. 23, pages 652–653, 1973, as modified by R. A. Kellin and J. R. Gear, *BioScience*, vol. 30, pages 110–111, 1980

8–5 After Lehninger, *op. cit.*, 1975

8–6 From Curtis, *op. cit.*, 1983

Page 123 Alexander Rich et al., *Science*, vol. 211, pages 171–176, 1981. © by A.A.A.S.

8–8 (b) Sung Hou-Kim

8–9 O. L. Miller, Jr., B. A. Hamkalo, and C. A. Thomas, Jr., *Science*, vol. 169, pages 392–395, 1970

Page 129 (a) through (c) K. B. Raper; (d), (e) J. T. Bonner

8–12 H. R. Chen

9–1 V. Orel, The Moravian Museum, Brno, Czechoslovakia

9–2 Don Kyhos

9–3 P. B. Moens

9–4 B. John

9–5 G. Östergren

9–6 (photo) W. Tai

9–13 (a) C. G. G. J. van Steenis; (b) W. L. Wagner

9–14 AP/Wide World Photos

Page 148 (top) C. S. Webber; (middle) John Hodgin, from A. H. R. Buller, *Researches on Fungi*, vol. 6, Longman, Inc., New York

10–1 Larry West

10–2 The Bettmann Archive

10–3 (a) Larry West; (b) C. S. Webber; (c) Imagery

10–4 (a), (c) E. V. Gravé; (b) H. Forest

10–5 (a) D. S. Neuberger; (b) L. E. Graham; (c) Oxford Scientific Films/Animals, Animals; (d) Runk & Schoenberger/Grant Heilman Photography; (e) E. V. Gravé

10–6 (a), (b) E. S. Ross; (c) Larry West; (d) K. Sandved; (e) John A. Lynch/Photo NATS

10–7 (a), (d) R. Carr/Bruce Coleman, Inc.; (b), (i) J. W. Perry; (c) E. S. Ross/Bruce Coleman, Inc.; (e) J. Shaw/Bruce Coleman, Inc.; (f) through (h) J. Dermid; (j) E. Beals

Page 157 (top) G. R. Roberts; (bottom) Grant Heilman Photography

10–8 (a) L. V. Leak, *Journal of Ultrastructural Research*, vol. 21, pages 61–74, 1967; (b) K. Esau

10–10 After G. L. Stebbins, in *The Evolution of Life*, vol. 1, S. Tax (Ed.), The University of Chicago Press, Chicago, 1960

11–1 Dr. Huntington Potter and Dr. David Dressler, Harvard Medical School

11–2 T. D. Pugh and E. H. Newcomb

11–3 Purdue University

11–4 David O. Hall

11–5 (a) P. L. Grilione and J. Pangborn, *Journal of Bacteriology*, vol. 124, page 1558, 1975; (b) National Medical Audiovisual Center; (c) H. Lechevalier; (d) R. S. Wolfe

11–6 (a), (b) D. Greenwood; (c) J. L. Pate

11–7 (a) E. V. Gravé; (c) Turtox/Cambosco; (d) E. J. Ordal

11–8 C. Robinow

11–9 D. A. Cuppels and A. Kelman, *Phytopathology*, vol. 70, pages 1110–1115, 1980

11–10 H. Stolp and M. P. Starr, *Antoine van Leeuwenhoek*, vol. 29, pages 217–248; 1963

11–11 C. C. Brinton, Jr., and J. Carnahan

11–12 (a) N. J. Lang, *Journal of Phycology*, vol. 1, pages 127–134, 1965; (b) Bai Kezhi

11–13 R. D. Warmbrodt

11–14 A. Berkaloff, J. Bourquet, P. Favard, and M. Guinnebault, *Introduction à la Biologie: Biologie et Physiologie Cellulaire*, Hermann Collection, Paris, 1967

11–16 J. Lederberg and E. M. Lederberg

11–17 R. R. Olson

11–18 Walter Reed Army Institute of Research

Page 178 Miryam Glikson

11–19 (a) A. M. Berry and J. G. Torrey, *Plant and Soil*, (in press); (b) J. G. Torrey, in *New Root Formation in Plants and Cuttings*, M. B. Jackson (Ed.), Martinus Nijhoff/Junk Publishers, The Netherlands, 1985; (c) A. M. Berry

11–20 After M. C. Pelczar and R. D. Reid, *Microbiology*, McGraw-Hill Book Company, New York, 1965

11–21 National Medical Audiovisual Center

11–22 W. A. Niering

11–23 After G. N. Agrios, *Plant Pathology*, 2nd ed., Academic Press, Inc., New York, 1978

Page 182 (a) Florida Department of Agriculture, D. P. I.; (b) V. Jane Windsor, Florida Department of Agriculture, D. P. I.

11–24 J. F. Worley, U. S. Department of Agriculture

11–25 (a) M. V. Parthasarathy; (b) H. Donselman

11–26 (top) J. Wm. Schopf; (bottom) After M. R. Walter, *American Scientist*, vol. 65, pages 563–571, 1977

12–1 L. D. Simon

12–2 G. A. deZoeten and G. Gaard

12–3 After K. Namba, D. L. D. Caspar, and G. J. Stubbs, *Science*, vol. 227, pages 773–776, 1985

12–4 J. D. Almeida and A. F. Howatson, *Journal of Cell Biology*, vol. 16, pages 616–620, 1963

12–5 (a) G. Gaard and R. W. Fulton; (b), (c) G. Gaard and G. A. deZoeten

12–6 (a) L. D. Simon; (b) K. Allen

12–7 National Communicable Disease Center, Atlanta

12–8 (a) K. Maramorosch; (b) E. Shikata; (c) E. Shikata and K. Maramorosch

12–9 T. White

12–10 Th. Koller and J. M. Sogo, Swiss Federal Institute of Technology, Zürich

13–1 E. S. Ross

13–2 (a) Charles Marden Fitch/Taurus Photos; (b) E. S. Ross

13–3 Imperial Chemical Industries PLC Agricultural Division

13–4 E. S. Ross

13–6 M. D. Coffey, B. A. Palevitz, and P. J. Allen, *The Canadian Journal of Botany*, vol. 50, pages 231–240, 1972

13–7 R. J. Howard

Page 203 (photo) John Hodgin; from A. H. R. Buller, *Researches on Fungi*, vol. 6, Longman, Inc., New York

13–11 (a) A. E. Staffan; (b) J. Shaw; (c) G. J. Breckon

13–12 J. C. Pendland and D. G. Boucias

13–13 (a) C. Bracker; (c) J. Cooke, *The American Journal of Botany*, vol. 56, pages 335–340, 1969

13–15 After L. W. Sharp, *Fundamentals of Cytology*, McGraw-Hill Book Company, Inc., New York, 1943

13–16 (a) E. V. Gravé

Page 208 U. S. Department of Agriculture

13–17 (a) A. McClenaghan/Photo Researchers, Inc.; (b) John Durham/Photo Researchers, Inc.

13–18 G. L. Barron

Page 210 (a), (b) G. L. Barron, University of Guelph; (c) N. Allin and G. L. Barron, University of Guelph

13–19 E. Imre Friedmann

13–20 E. S. Ross

13–21 (a) Robert A. Ross; (b) E. S. Ross

13–22 (a), (b) E. S. Ross; (c) Larry West

13–23 (b) V. Ahmadjian and J. B. Jacobs

13–24 R. L. Chapman

13–25 (a), (b) V. Ahmadjian; (c) V. Ahmadjian and B. J. Jacobs

13–27 J. Keller

13–28 H. C. Hoch and R. J. Howard, *Experimental Mycology*, vol. 5, pages 167–172, 1981

13–29 (b) D. Niederpruem

13–30 (a), (b) J. Burton/Bruce Coleman, Inc.; (c), (d) J. W. Perry

13–32 E. S. Ross

13–33 After Sharp, *op. cit.*, 1943

13–34 (photos) R. Gordon Wasson, Botanical Museum of Harvard University

13–35 (a) J. Burton/Bruce Coleman, Inc.; (b) J. Markham/Bruce Coleman, Inc.; (c), (d) E. S. Ross

13–36 G. R. Roberts

13–38 S. A. Wilde

13–39 Design Network, Inc.

13–40 F. W. Went

13–41 B. Zak, U. S. Forest Service

13–42 R. D. Warmbrodt

14–1 G. Vidal and T. D. Ford, *Precambrian Research* (in press)

14–2 D. P. Wilson/Eric and David Hosking Photography

14–3 (a) L. E. Graham; (b) L. E. Graham and J. M. Graham, *Transactions of the American Microscopical Society*, vol. 99, pages 160–166, 1980

14–4 (a) A. W. Barksdale; (b) A. W. Barksdale, *Mycologia*, vol. 55, pages 493–501, 1963

Page 235 (a), (b) A. W. Barksdale; (c) A. W. Barksdale, *Mycologia*, vol. 55, pages 493–501, 1963

14–6 After J. H. Niederhauser and W. C. Cobb, *Scientific American*, vol. 200, pages 100–112, 1959

14–7 L. P. Gauriloff, in *Lower Fungi in the Laboratory*, M. S. Fuller (Ed.), © 1978 by Department of Botany, University of Georgia

14–8 After R. Emerson, *Lloydia*, vol. 4, pages 77–144, 1941

14–9 K. B. Raper

14–10 K. B. Raper

14–11 (c) Jonathan D. Eisenback/Phototake; (d) Runk & Schoenberger/Grant Heilman Photography

14–13 G. A. Fryxell

14–15 J. M. King

14–16 (a) G. D. Hanna, California Academy of Sciences, (b) F. Rossi; (c) G. A. Fryxell

14–19 (a), (b) D. P. Wilson/Eric and David Hosking Photography; (c) A. R. Loeblich, III

Page 246 D. F. Kubai and H. Ris

14–20 Design Network, Inc.

14–21 Florida Department of Natural Resources, Bureau of Marine Research

14–22 D. Longanecker

15–1 J. Dermid

15–2 D. P. Wilson/Eric and David Hosking Photography

15–3 (a) L. E. Graham; (b) R. G. Sheath, J. A. Hellebust, and T. Sawa, *Phycologia*, vol. 20, pages 22–31, 1981

15–4 (a), (c) H. C. Bold and M. J. Wynne, *Introduction to the Algae: Structure and Reproduction*, 2nd ed., Prentice-Hall, Inc., Englewood Cliffs, New Jersey, 1985; (b) R. Sheath

15–5 (a) D. P. Wilson/Eric and David Hosking Photography; (b) E. S. Ross; (c) R. C. Carpenter; (d) Jean Baxter/Photo NATS

15–6 (a), (b) M. Littler and D. Littler, Smithsonian Institution; (c) C. Chulamanis, M. Littler and D. Littler

15–8 (photo) C. M. Pueschel and K. M. Cole, *The American Journal of Botany*, vol. 69, pages 703–720, 1982

15–9 (a) G. R. Roberts; (b) D. P. Wilson/Eric and David Hosking Photography; (c) S. M. Carpenter

15–11 C. J. O'Kelly

Page 262 (a) Bob Evans/Peter Arnold, Inc.; (b) W. H. Hodge/Peter Arnold, Inc.; (c) Kelco Communications

15–14 After M. Knight and M. Parke, *Journal of the Marine Biological Association, U.K.*, vol. 29, pages 439–514, 1950

15–15 After G. L. Floyd

15–16 After K. R. Mattox and K. D. Stewart, in *Systematics of the Green Algae*, D. E. G. Irvine and D. M. John (Eds.), 1984

15–17 L. E. Graham

15–19 E. V. Gravé

15–20 L. E. Graham

15–21 K. J. Niklas

15–22 (a) W. H. Amos/Bruce Coleman, Inc.

15–23 (a) D. S. Neuberger; (b), (c), (d) K. Esser, *Cryptogams*, Cambridge University Press, Cambridge, 1982; (e) E. S. Ross

15–25 D. P. Wilson/Eric and David Hosking Photography

15–27 (a) Robert A. Ross; (b) Grant Heilman Photography; (c) L. R. Hoffman

15–28 V. Paul

15–29 (photo) G. E. Palade

15–32 K. Esser, *op. cit.*, 1982

15–34 (a) Carolina Biological Supply Company; (b) H. J. Marchant and J. D. Pickett-Heaps, *Australian Journal of Biological Science*, vol. 25, pages 1199–1213, 1972

15–35 L. E. Graham

15–36 D. K. Kirk

15–38 After R. F. Skagel, R. J. Bandoni, G. E. Rouse, W. B. Schofield, J. R. Stein, and T. M. C. Taylor, *An Evolutionary Survey of the Plant Kingdom*, Wadsworth Publishing Company, Inc., Belmont, California, 1966

15–39 L. E. Graham

Page 370 (bottom) E. S. Ross

Page 371 J. W. Perry

16–1 T. S. Elias

16–2 W. Remy, *Science*, vol. 219, pages 1625–1627, 1982

16–3 © 1973 Ken Brate/Photo Researchers, Inc.

16–4 D. R. Given

16–5 C. Hébant, *Journal of the Hattori Botanical Laboratory*, vol. 39, pages 235–254, 1975

16–6 (a) D. S. Neuberger

16–7 D. S. Neuberger

16–9 (a) After G. M. Smith, *Cryptogamic Botany*, vol. 2, *Bryophytes and Pteridophytes*, 2nd ed., McGraw-Hill Book Company, New York, 1955; (b) R. E. Magill, Botanical Research Institute, Pretoria

16–12 (a) E. S. Ross

Page 287 After C. T. Ingold, *Spore Discharge in Land Plants*, Clarendon Press, Oxford, 1939

16–14 (a), (b) J. J. Engel, *Fieldiana: Botany* (New Series), vol. 3, pages 1–229, 1980; (c) J. J. Engel

16–15 K. B. Sandved

16–16 (a) Robert A. Ross

16–18 D. S. Neuberger

16–19 (a) A. E. Staffan; (b) E. S. Ross

16–20 (a) D. S. Neuberger

16–21 (a) After Ingold, *op. cit.*, 1939; (b) R. Magill, Botanical Research Institute, Pretoria

Page 295 (a), (c) D. H. Vitt; (b) A. Klotz/ Design Network, Inc.

16–23 (a) Larry West; (b) after Skagel et al., *op. cit.*, 1966; (c) after Ingold, *op. cit.*, 1939

16–24 (a) M. C. F. Proctor; (b) after Ingold, *op. cit.*, 1939

17–1 Kristine Rasmussen and Stuart Naquin

17–2 (a) After A. S. Foster and E. M. Gifford, Jr., *Comparative Morphology of Vascular Plants*, 2nd ed., W. H. Freeman & Company, Publishers, New York, 1974

17–3 (a), (d) After K. K. Namboodiri and C. B. Beck, *The American Journal of Botany*, vol. 55, pages 464–472, 1968; (b), (c) after K. Esau, *Plant Anatomy*, 2nd ed., John Wiley & Sons, Inc., New York, 1965

17–5 After Smith, *op. cit.*, 1955

17–7 After H. P. Banks, *Evolution and Plants of the Past*, Wadsworth Publishing Company, Inc., Belmont, California, 1970

17–8 (a) After J. Walton, *Fossil Plants*, Macmillan Publishing Company, New York, 1940; (b) after J. Walton, *Phytomorphology*, vol. 14, pages 155–160, 1964; (c) after F. M. Heuber, *International Symposium on the Devonian System*, vol. 2, D. H. Oswald (Ed.), Alberta Society of Petroleum Geologists, Calgary, Alberta, Canada, 1968

17–9 Field Museum of Natural History

17–11 R. L. Peterson, M. J. Howarth, and D. P. Whittier, *The Canadian Journal of Botany*, vol. 59, pages 711–720, 1981

17–12 (a) D. Cameron; (b) R. Schmid

17–14 Specimen provided by Ripon Microslides, Ripon, Wisconsin

17–15 D. S. Neuberger

17–16 (a), (b) D. S. Neuberger; (c) K. B. Sandved

17–19 Milwaukee Public Museum

17–20 Kristine Rasmussen and Stuart Naquin

17–21 (a) R. Carr; (b) E. S. Ross

17–25 (a) Gene Ahrens/Bruce Coleman, Inc.; (b) K. B. Sandved; (c) C. Haufler; (d) E. S. Ross

Pages 324–325 (a) After M. Hirmer, *Handbuch der Paläobotanik*, vol. 1, Druck and Verlag von R. Oldenbourg, Munich and Berlin, 1927; (b) after W. N. Stewart and T. Delevoryas, *Botanical Review*, vol. 22, pages 45–80, 1956; (c) after Banks, *op. cit.*, 1970

17–28 R. B. Peacock/Photo Researchers, Inc.

17–29 C. Neidorf

18–1 Kristine Rasmussen and Stuart Naquin

18–2 After A. G. Long, *Transactions of the Royal Society of Edinburgh*, vol. 64, pages 29–44, 201–215, and 261–280, 1960

18–3 (a), (b) After J. M. Pettitt and C. B. Beck, *Contributions from the Museum of Paleontology*, University of Michigan, vol. 22, pages 139–154, 1968; (c) J. M. Pettitt and C. B. Beck, *Science*, vol. 156, pages 1727–1729, 1967

18–4 After H. N. Andrews, *Science*, vol. 142, pages 925–931, 1963; after A. G. Long, *op. cit.*, 1960

18–5 C. B. Beck, *Biological Reviews*, vol. 45, pages 379–400, 1970

18–6 After S. E. Scheckler, *The American Journal of Botany*, vol. 62, pages 923–934, 1975

18–7 Kristine Rasmussen and Stuart Naquin

18–8 Kristine Rasmussen and Stuart Naquin

18–9 J. Dermid

18–10 (a) J. Dermid

18–11 (a) R. Spurr and J. Spurr/Bruce Coleman, Inc.

18–12 J. Kummerow

18–15 B. Haley

18–17 (c) G. J. Breckon

18–22 (a) W. H. Hodge/Peter Arnold, Inc. (b) E. S. Ross

18–23 H. H. Iltis

18–24 Grant Heilman Photography

18–26 (a) Larry West; (b) J. Burton/Bruce Coleman, Inc.

18–27 Gene Ahrens/Bruce Coleman, Inc.

18–28 Sichuan Institute of Biology

18–29 (photo) Carolina Biological Supply Company

18–30 D. A. Steingraeber

18–31 (b) D. T. Hendricks and E. S. Ross

18–32 (a) J. W. Perry; (b) Runk & Schoenberger/ Grant Heilman Photography

18–33 (a) F. S. P. Ng; (b), (c) G. Davidse

18–34 (a) E. S. Ross; (b), (d) J. W. Perry; (c) K. J. Niklas

18–35 (a) C. H. Bornman; (b), (c) E. S. Ross

18–36 E. S. Ross

18–37 W. P. Armstrong

18–38 (a), (b) G. J. Breckon; (c) E. S. Ross

18–39 (a) D. S. Neuberger; (b) E. S. Ross; (c) E. R. Degginger/Earth Scenes

18–40 (a), (c) E. S. Ross; (b) R. Carr

18–43 (a), (c) Larry West; (b), (e) J. H. Gerard; (d) Grant Heilman Photography

18–45 E. S. Ross

18–47 (a) Runk & Schoenberger/Grant Heilman Photography

18–48 (a) Larry West

18–50 (a) P. Echlin; (b), (c) J. Heslop-Harrison and Y. Heslop-Harrison; (d) J. Mais

19–1 J. Dermid

Page 377 Karlene Schwartz

19–6 After Foster and Gifford, *op. cit.*, 1974

19–7 T. J. Givnish

20–1 R. D. Warmbrodt and R. F. Evert, *The American Journal of Botany*, vol. 66, pages 412–440, 1979

20–4 M. C. Ledbetter and K. B. Porter, *Introduction to the Fine Structure of Plant Cells*, Springer-Verlag, Inc., New York, 1970

20–11 H. A. Core, W. A. Coté, and A. C. Day, *Wood: Structure and Identification*, 2nd ed., Syracuse University Press, Syracuse, New York, 1979

20–12 I. B. Sachs, Forest Products Laboratory, U.S.D.A.

20–14 After Esau, *op. cit.*, 1977

20–17 (a), (b) R. F. Evert, W. Eschrich, and S. E. Eichhorn, *Planta*, vol. 109, pages 193–210, 1973

20–18 After Esau, *op. cit.*, 1977

20–19 J. S. Pereira

20–21 (c) After Esau, *op. cit.*, 1965

20–24 Ward M. Tingey, Cornell University, Agricultural Experiment Station

21–2 After Braune, Leman, and Taubert, *op. cit.*, 1967

21–2 (a) L. M. Chace/N.A.S., Photo Researchers, Inc.; (b) J. Dermid

21–5 F. A. L. Clowes

21–11 H. T. Bonnett, Jr., *Journal of Cell Biology*, vol. 37, pages 199–205, 1968

21–12 After Braune, Leman, and Taubert, *op. cit.*, 1967

21–15 E. R. Degginger/ Bruce Coleman, Inc.

21–16 Robert and Linda Mitchell

21–18 D. A. Steingraeber

22–4 M. E. Gerloff

22–5 After D. A. DeMason, *The American Journal of Botany*, vol. 70, pages 955–962, 1983

22–8 (c) W. Eschrich

22–12 After A. Fahn, *Plant Anatomy*, 2nd ed., Pergamon Press, Inc., Elmsford, New York, 1974

22–14 Rhonda Nass/Ampersand

22–16 (a) J. W. Perry

22–17 Rhonda Nass/Ampersand

22–21 Michele McCauley

22–23 W. A. Russin and R. F. Evert, *The American Journal of Botany*, vol. 71, pages 1398–1415, 1984

Page 431 (a) H. C. Jones, Tennessee Valley Authority; (b) J. S. Jacobson and A. C. Hill (Eds.), *Recognition of Air Pollution Injury to Vegetation: A Pictorial Atlas*, Air Pollution Control Association, Pittsburgh, 1970; (c) after T. H. Maugh, II, *Science*, vol. 226, pages 1408–1410, 1984

22–29 After Esau, *op. cit.*, 1965, and G. S. Avery, Jr., *The American Journal of Botany*, vol. 20, pages 565–592, 1933

22–31 R. D. Meicenheimer

22–34 Shirley C. Tucker

22–37 J. W. Perry

Page 441 after P. A. Deschamp and T. J. Cooke, *Science*, vol. 219, pages 505–507, 1983

22–39 D. A. Steingraeber

22–40 J. W. Perry

Page 443 E. S. Ross

22–41 G. R. Roberts

23–25 I. B. Sachs, Forest Products Laboratory, U.S.D.A.

23–26 Core, Coté, and Day, *op. cit.*, 1979

23–27 (a) U. S. Forest Service; (b) C. W. Ferguson, Laboratory of Tree-Ring Research, University of Arizona

23–29 R. B. Hoadley

23–30 R. B. Hoadley

23–32 After R. B. Hoadley, *Understanding Wood*, The Taunton Press, Newtown, Connecticut, 1980

Page 472 (top) J. Dermid; (middle) J. S. Ranson and R. Moore, *The American Journal of Botany*, vol. 70, pages 1048–1056, 1983; (bottom) Grant Heilman Photography

Page 473 D. A. Steingraeber

24–1 M. B. Winter

24–5 U. S. Department of Agriculture

24–6 J. P. Nitsch, *The American Journal of Botany*, vol. 37, pages 211–215, 1950

Page 482 G. Melchers, M. D. Sacristán, and A. A. Holder, *Carlsberg Research Communications*, vol. 43, page 203–218, 1978

Page 483 After art by Hilleshög, Laboratory for Cell and Tissue Culture, Research Division, Landskrona, Sweden

24–8 F. Skoog and C. O. Miller, *Symposia of the Society for Experimental Biology*, vol. 11, pages 118–131, 1957

24–11 S. W. Wittwer

24–12 H. Towner

24–13 S. W. Wittwer

24–14 Abbott Laboratories

24–15 J. van Overbeek, *Science*, vol. 152, pages 721–731, 1966

24–16 (a) After M. B. Wilkins (Ed.), *Advanced Plant Physiology*, Pitman Publishing Ltd., London, 1984; (b) J. E. Varner

25–2 J. S. Ranson and R. Moore, *The American Journal of Botany*, vol. 70, pages 1048–1056, 1983

25–3 After B. E. Juniper, *Annual Review of Plant Physiology*, vol. 27, pages 385–406, 1976

25–4 J. Dermid

25–5 R. F. Trump/Photo Researchers, Inc.

25–6 After A. W. Galston, *The Green Plant*, Prentice-Hall, Inc., Englewood Cliffs, New Jersey, 1968

25–7 (a) A. Loeblich, III; (b), (c) after B. Sweeney, *Rhythmic Phenomena in Plants*, Academic Press, Inc., New York, 1969

25–8 After A. W. Naylor, *Scientific American*, vol. 186, pages 49–56, 1952

25–9 After P. M. Ray, *The Living Plant*, Holt, Rinehart, & Winston, Inc., New York, 1963

25–11 U. S. Department of Agriculture

25–16 U. S. Department of Agriculture

25–17 A. Lang, M. Kh. Chailakhyan, and I. A. Frolova, *Proceedings of the National Academy of Sciences*, vol. 74, pages 2412–2416, 1977

25–18 D. A. Steingraeber

25–20 After Naylor, *op. cit.*, 1952

25–21 J. Dermid

25–22 Grant Heilman Photography

25–23 (a) J. Ehleringer and I. Forseth, *Science*, vol .210, pages 1094–1098, 1980; (b) Robert and Linda Mitchell

25–24 After Ehleringer and Forseth, *op. cit.*, 1980

Page 514 (top) R. E. Hutchins; (middle) R. R. Herbert, R. D. Holsten, and R. W. F. Hardy, E. I. duPont de Nemours Company; (bottom) Y. Heslop-Harrison

Page 515 Carolina Biological Supply Company

26–1 The Nitragin Company, Inc.

26–3 (a) J. H. Gerard/N.A.S., Photo Researchers, Inc.

26–4 R. H. Wright/N.A.S., Photo Researchers, Inc.

26–5 J. Bonner and A. W. Galston, *Principles of Plant Physiology*, W. H. Freeman & Company, Publishers, New York, 1952

26–6 After B. Gibbons, *National Geographic*, vol. 166, pages 350–388, 1984 (Ned M. Seidler, artist)

Page 524 From Curtis, *op. cit.*, 1983

26–7 After B. S. Meyer et al., *Introduction to Plant Physiology*, Van Nostrand Reinhold Company, Inc., New York, 1973

Page 526 (a) After L. R. Rhodes and J. W. Gerdemann, *New Phytologist*, vol. 75, pages 555–561, 1975; (b) J. A. Menge, *The Canadian Journal of Botany*, vol. 61, pages 1015–1024, 1983

26–9 (a) Carolina Biological Supply Company; (b) R. E. Hutchins

Page 529 Y. Heslop-Harrison

26–10 (a) The Nitragin Company, Inc.; (b), (c) Oxford University Botany School

26–11 R. R. Herbert, R. D. Holsten, and R. W. F. Hardy, E. I. duPont de Nemours Company

26–12 B. G. Turgeon and W. D. Bauer, *The Canadian Journal of Botany*, vol. 60, pages 152–161, 1982

26–13 R. R. Herbert, R. D. Holsten, and R. W. F. Hardy, E. I. duPont de Nemours Company

26–14 J. M. L. Selker and E. H. Newcomb, *Planta*, vol. 165, pages 446–454, 1985

26–15 E. H. Newcomb, Sh. R. Tandon, and R. R. Kowal, *Protoplasma*, vol. 125, pages 1–12, 1985

26–16 (b) H. E. Calvert

26–18 G. C. Gerloff and W. H. Gabelman, in *Encyclopedia of Plant Physiology* (New Series), vol. 15B, A. Lauchli and R. L. Bieleski (Eds.), Springer-Verlag New York, Inc., New York, 1983

26–19 After W. J. Brill, *Scientific American*, vol. 236, pages 68–81, 1977

27–1 S. Hales, *Vegetable Staticks*, London, 1727

27–2 (a) J. Dermid

27–4 After M. Richardson, *Translocation in Plants*, Edward Arnold Publishers, Ltd., London, 1968

27–5 After Richardson, *op. cit.*, 1968

27–7 After E. Haüsermann and A. Frey-Wyssling, *Protoplasma*, vol. 57, pages 37–80, 1963

27–8 After A. C. Leopold, *Plant Growth and Development*, McGraw-Hill Book Company, New York, 1964

27–9 After Richardson, *op. cit.*, 1968

27–10 After P. F. Scholander, H. T. Hammel, E. D. Bradstreet, and E. A. Hemmingsen, *Science*, vol. 148, pages 339–346, 1965

27–11 After M. H. Zimmermann, *Scientific American*, vol. 208, pages 132–142, 1963

27–12 *Ibid.*

27–13 (a) L. M. Beidler; (b) J. H. Troughton

27–14 After M. G. Penny and D. J. F. Bowling, *Planta*, vol. 119, pages 17–25, 1974

27–15 After D. E. Aylor, J. Y. Parlange, and A. D. Krikorian, *The American Journal of Botany*, vol. 60, pages 163–171, 1973

27–17 After J. S. Pate, *Transport in Plants I, Phloem Transport*, M. H. Zimmermann and J. A. Milburn (Eds.), Springer-Verlag Berlin, Inc., Berlin, 1975

27–20 After Malpighii, *Opera Posthuma*, London, 1675

Page 555 (b), (c) E. Fritz

27–21 (a) M. H. Zimmermann; (b) R. F. Evert, W. Eschrich, J. T. Medler, and F. J. Alfieri, *The American Journal of Botany*, vol. 55, pages 860–874, 1968

27–22 After E. Münch, *Die Stoffbewegungen in der Pflanze*, Gustav Fischer, Jena, 1930

Page 560 (top) Timothy Plowman; (middle) E. Dorf; (bottom) G. R. Carr

Page 561 J. Heslop-Harrison

28-1 The Bettmann Archive/B. B. C. Hulton

28-5 After K. Mather and B. J. Harrison, *Heredity,* vol. 3, pages 1–52, 1959

28-6 After S. Carlquist, *Island Life,* The Natural History Press, Garden City, New York, 1965

28-7 (a) G. M., British Museum (Natural History); (b) R. Ornduff

28-8 After S. Ross-Craig, *Drawings of British Plants,* Part IV, G. Bell and Sons, Ltd., London, 1950

Page 571 (bottom) From Curtis and Barnes, *Invitation to Biology,* 4th ed., Worth Publishers, Inc., New York, 1985; (left) A. M. Evans

28-9 B. Crandall, Carnegie Institution of Washington, Publication 540

Page 576 After R. H. Robichaux

Page 577 (photo) G. R. Carr

28-11 (a), (b) M. A. Nobs, Carnegie Institution of Washington, Publication 623; (c), (d) L. R. Heckard and C. S. Webber

28-12 D. Myrick, Jepson Herbarium, University of California, Berkeley

28-13 (a), (b) H. Angel/Biofotos; (c), (d), (e) C. J. Marchant

28-14 (a) L. Mellichamp

29-1 (a) E. S. Ross; (b) D. L. Dilcher

29-2 E. Dorf

29-3 After Peter Crane

29-4 After Peter Crane

29-5 After Peter Crane

29-6 After Peter Crane

29-8 (a) J. H. G. Johns, New Zealand Forest Service; (b) W. H. Hodge/Peter Arnold, Inc.

29-9 After I. W. Bailey and B. G. L. Swamy, in *Contributions to Plant Anatomy,* I. W. Bailey (Ed.), Ronald Press, New York, 1954

29-10 After G. H. M. Lawrence, *Taxonomy of Vascular Plants,* Macmillan Publishing Company, New York, 1951

Page 592 David L. Dilcher (reconstructions by Megan Rohn in consultation with D. Dilcher)

29-12 (a), (b) Populi and Paesi, Agenzia Fotografica, Luisa Ricciarini, Milan; (c) G. J. Breckon

29-13 (a), (c) E. S. Ross; (b) J. W. Perry

29-14 (b), (c) E. S. Ross; (d) A. Sabarese

29-15 (a) E. S. Ross

29-16 J. A. L. Cooke

29-17 W. H. Hodge

Pages 598–599 J. Heslop-Harrison

29-18 Larry West

29-19 (a) J. Dermid; (b) Larry West; (c) E. S. Ross

29-20 E. S. Ross

29-21 E. S. Ross

29-22 Larry West

29-23 T. Eisner

29-24 E. S. Ross

29-25 E. S. Ross

29-26 (a), (c) E. S. Ross; (b) L. B. Thein

29-27 E. S. Ross

29-28 M. P. L. Fogden/Bruce Coleman, Inc.

29-29 R. A. Tyrell

29-30 Oxford Scientific Films

29-31 E. S. Ross

29-32 D. J. Howell

29-33 (a), (b) T. Hovland/Grant Heilman Photography; (c) E. S. Ross; (d) J. F. Skvarla, University of Oklahoma

29-34 (b), (c) After D. B. Swingle, *A Textbook of Systematic Botany,* McGraw-Hill Book Company, New York, 1946

29-35 D. S. Neuberger

29-37 K. Paige, *Science 85,* page 6, 1985

29-38 After Skagel et al., *op. cit.,* 1966

29-39 (a) E. S. Ross; (b) after Skagel et al., *op. cit.,* 1966; (c) after L. Benson, *Plant Classification,* D. C. Heath and Company, Boston, 1957

29-40 (a) E. S. Ross; (b), (c) K. B. Sandved

29-41 (a) After Skagel et al., *op. cit.,* 1966; (b) E. S. Ross

29-43 E. S. Ross

29-44 E. S. Ross

29-45 (a) E. S. Ross; (b) U. S. Forest Service

29-46 E. S. Ross

29-47 E. S. Ross

29-49 Robert and Linda Mitchell

29-50 (a), (c), (d) Timothy Plowman; (b) E. S. Ross

30-1 James P. Blair, © 1983, National Geographic Society

30-2 E. S. Ross

30-3 E. S. Ross

30-4 E. S. Ross

30-5 W. H. Hodge

30-6 (a) James P. Blair, © 1983, National Geographic Society; (b) Susan Pierres/Peter Arnold, Inc.

30-7 Robert and Linda Mitchell

30-8 G. R. Roberts

30-9 (a) E. Zardini; (b) C. B. Heiser, Jr.; (c) A. Gentry; (d) M. K. Arroyo

30-10 E. S. Ross

30-11 (a) C. F. Jordan; (b) M. J. Plotkin

30-12 M. J. Plotkin

30-13 K. B. Sandved

30-14 W. H. Hodge/Peter Arnold, Inc.

30-15 E. S. Ross

30-16 W. H. Hodge/Peter Arnold, Inc.

30-17 Harvey Lloyd/Peter Arnold, Inc.

30-18 New York Public Library Picture Collection

30-20 R. Abernathy

30-21 C. A. Black

30-22 AP/Wide World Photos

30-23 W. H. Hodge

30-24 Agricultural Research Service, U.S.D.A.

30-25 (a) Agricultural Research Service, U.S.D.A.; (b) University of Wisconsin

30-26 N. Vietmeyer, National Academy of Sciences

30-27 Robert and Linda Mitchell

30-28 © 1982 Angelina Lax/Photo Researchers, Inc.

30-29 J. Aronson

30-30 After Lehninger, *op. cit.,* 1982

30-31 M. J. Plotkin

30-32 M. J. Plotkin

30-33 Monsanto Company, *Genetic Engineering,* St. Louis

30-34 Depatment of Plant Pathology, Cornell University

30-35 Monsanto Company, *Genetic Engineering,* St. Louis

30-36 C. J. Arntzen

Page 647 (top) W. Wagner; (bottom left) E. S. Ross; (bottom right) E. S. Ayensu, National Museum of Natural History, Smithsonian Institution

Page 650 (top) C. D. MacNeill; (bottom) E. S. Ross

Page 651 J. Reveal

31-1 Clyde H. Smith/Peter Arnold, Inc.

31-2 D. H. Janzen

31-3 D. H. Harvey

31-4 After J. L. Harper, *Symposia of the Society for Experimental Biology,* vol. 151, pages 1–39, 1961

31-5 C. H. Muller

31-6 B. Bartholomew

31-7 J. Mann, Australian Department of Lands

31-8 After D. F. Rhoades, in *Chemically Mediated Interactions Between Plants and Other Organisms,* G. A. Cooper-Driver, T. Swain, and E. E. Conn (Eds.), Plenum Publishing Company, New York, 1985

31-9 H. Harrison/Grant Heilman Photography

31-10 G. E. Likens

31-11 J. H. Gerard/N.A.S., Photo Researchers, Inc.

Page 663 R. W. Gibson, Rothamsted Experimental Station

31-12 After J. Phillipson, *Ecological Energetics*, Edward Arnold Publishers, Ltd., London, 1966

31-13 After G. G. Simpson and W. S. Beck, *Life: An Introduction to Biology*, 2nd ed., Harcourt Brace Jovanovich, Publishers, Inc., New York, 1965

31-14 (a), (c), (d) Larry West; (b) J. Dermid

31-15 Larry West

31-16 (a) U. S. Forest Service; (b) J. Dermid

31-17 (a) J. Dermid; (b) U. S. Forest Service

31-18 R. Komarek

31-19 (a), (c) P. Frenzen; (b), (d) D. K. Yamaguchi

32-1 After A. W. Kuchler

32-2 C. D. MacNeill

32-3 After Curtis, *op. cit.*, 1983

32-4 K. Sandved

Page 676 The Bettmann Archive

32-5 (a), (e) C. W. Rettenmeyer; (b), (c), (f) E. S. Ross; (d) P. Raven; (g) James P. Blair, © 1983, National Geographic Society

Page 678 R. Schoer

32-7 J. Van Wormer/Bruce Coleman, Inc.

32-10 J. Dermid

32-12 E. S. Ross

32-13 F. C. Vasck

32-15 J. Reveal

32-16 Soil Conservation Service

32-18 (a) H. E. Eversmeyer; (b), (c) J. H. Gerard

32-19 (a) M. Travis; (b) J. F. Dodd